The chemistry of

# sulphinic acids, esters and their derivatives

THE CHEMISTRY OF FUNCTIONAL GROUPS

*A series of advanced treatises under the general editorship of*
*Professor Saul Patai*

The chemistry of alkenes (2 volumes)
The chemistry of the carbonyl group (2 volumes)
The chemistry of the ether linkage
The chemistry of the amino group
The chemistry of the nitro and nitroso groups (2 parts)
The chemistry of carboxylic acids and esters
The chemistry of the carbon–nitrogen double bond
The chemistry of amides
The chemistry of the cyano group
The chemistry of the hydroxyl group (2 parts)
The chemistry of the azido group
The chemistry of acyl halides
The chemistry of the carbon–halogen bond (2 parts)
The chemistry of the quinonoid compounds (2 volumes, 4 parts)
The chemistry of the thiol group (2 parts)
The chemistry of the hydrazo, azo and azoxy groups (2 parts)
The chemistry of amidines and imidates
The chemistry of cyanates and their thio derivatives (2 parts)
The chemistry of diazonium and diazo groups (2 parts)
The chemistry of the carbon–carbon triple bond (2 parts)
The chemistry of ketenes, allenes and related compounds (2 parts)
The chemistry of the sulphonium group (2 parts)
Supplement A: The chemistry of double-bonded functional groups (2 volumes, 4 parts)
Supplement B: The chemistry of acid derivatives (2 parts)
Supplement C: The chemistry of triple-bonded functional groups (2 parts)
Supplement D: The chemistry of halides, pseudo-halides and azides (2 parts)
Supplement E: The chemistry of ethers, crown ethers, hydroxyl groups and their sulphur analogues (2 parts)
Supplement F: The chemistry of amino, nitroso and nitro compounds and their derivatives (2 parts)
The chemistry of the metal–carbon bond (5 volumes)
The chemistry of peroxides
The chemistry of organic selenium and tellurium compounds (2 volumes)
The chemistry of the cyclopropyl group
The chemistry of sulphones and sulphoxides
The chemistry of organic silicon compounds (2 parts)
The chemistry of enones (2 parts)

UPDATES
The chemistry of α-haloketones, α-haloaldehydes and α-haloimines
Nitrones, nitronates and nitroxides
Crown ethers and analogs

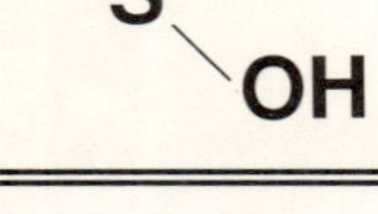

# The chemistry of
# sulphinic acids, esters and their derivatives

*Edited by*

**SAUL PATAI**

*The Hebrew University, Jerusalem*

1990

**JOHN WILEY & SONS**

**CHICHESTER – NEW YORK – BRISBANE – TORONTO – SINGAPORE**

*An Interscience® Publication*

*Other Wiley Editorial Offices*

John Wiley & Sons, Inc., 605 Third Avenue,
New York, NY 10158-0012, USA

Jacaranda Wiley Ltd, G.P.O. Box 859, Brisbane,
Queensland 4001, Australia

John Wiley & Sons (Canada) Ltd, 22 Worcester Road,
Rexdale, Ontario M9W 1L1, Canada

John Wiley & Sons (SEA) Pte Ltd, 37 Jalan Pemimpin 05–04,
Block B, Union Industrial Building, Singapore 2057

***Library of Congress Cataloging-in-Publication Data:***

The Chemistry of sulphinic acids, esters and their derivatives / edited by Saul Patai.
p. cm.—(The Chemistry of functional groups)
'An Interscience publication.'
Bibliography: p.
Includes index.
ISBN 0 471 91918 7
1. Sulphinic acids. 2. Esters. I. Patai, Saul. II. Series.
QD341.A2C4175 1990 89-31592
546'.72322—dc20 CIP

***British Library Cataloguing in Publication Data:***

Patai, Saul
The chemistry of sulphinic acids, esters and their derivatives.
1. Organic sulphur compounds
I. Title II. Series
547'.06

ISBN 0 471 91918 7

Typeset by Thomson Press (India) Ltd, New Delhi, India.
Printed in Great Britain by Courier International Ltd, Tiptree, Essex

# Contributing authors

| | |
|---|---|
| M. R. F. Ashworth | Organische und Instrumentelle Analytik, Universität des Saarlandes, D-6600 Saarbrücken, FRG |
| H. Basch | Department of Chemistry, Bar Ilan University, Ramat Gan 52100, Israel |
| A. Bassindale | POCRG, Department of Chemistry, The Open University, Walton Hall, Milton Keynes, MK7 6AA, UK |
| S. Braverman | Department of Chemistry, Bar-Ilan University, Ramat-Gan 59100, Israel |
| B. Bujnicki | Centre of Molecular and Macromolecular Studies, Polish Academy of Sciences, Department of Organic Sulphur Compounds, Sienkiewicza 112, 90-363 Łódź, Poland |
| G. Capozzi | Department of Organic Chemistry, University of Firenze, Via G. Carponi 9, 50121 Firenze, Italy |
| D. C. Dittmer | Department of Chemistry, Syracuse University, Syracuse, NY 13244-1200, USA |
| J. Drabowicz | Centre of Molecular and Macromolecular Studies, Polish Academy of Sciences, Department of Organic Sulphur Compounds, Sienkiewicza 112, 90-363 Łódź, Poland |
| T. Endo | Research Laboratory of Resources Utilization, Tokyo Institute of Technology, Nagatsuta-cho, Midori-ku, Yokohama 227, Japan |
| H. Fujihara | Department of Chemistry, The University of Tsukuba, Tsukuba, Ibaraki 305, Japan |
| N. Furukawa | Department of Chemistry, The University of Tsukuba, Tsukuba, Ibaraki 305, Japan |
| M. D. Hoey | Department of Chemistry, Syracuse University, Syracuse, NY 13244-1200, USA |

J. Hoyle — Chemistry–Soils Department, Nova Scotia Agricultural College, P.O. Box 550, Truro, Nova Scotia B2N 5E3, Canada

J. Iley — POCRG, Department of Chemistry, The Open University, Walton Hall, Milton Keynes, MK7 6AA, UK

A. Kalir — Department of Physiology and Pharmacology, Sackler Faculty of Medicine, Tel Aviv University, Tel Aviv 69978, Israel

H. H. Kalir — Department of Neurobiology, Mount Sinai School of Medicine, One Gustave L. Levy Place, New York, NY10029, USA

P. Kiełbasiński — Centre of Molecular and Macromolecular Studies, Polish Academy of Sciences, Department of Organic Sulphur Compounds, Sienkiewicza 112, 90-363 Łódź, Poland

M. M. Mikołajczyk — Centre of Molecular and Macromolecular Studies, Polish Academy of Sciences, Department of Organic Sulphur Compounds, Sienkiewicza 112,363 Łódź, Poland

A. Nudelman — Department of Chemistry, Bar Ilan University, Ramat Gan 51100, Israel

S. Oae — Department of Chemistry, Okayama University of Science, Ridai-cho 1-1, Okayama 700, Japan

T. Okuyama — Department of Chemistry, Faculty of Engineering Science, Osaka University, Toyonaka, Osaka 560, Japan

J. Omelańczuk — Centre of Molecular and Macromolecular Studies, Polish Academy of Sciences, Department of Organic Sulphur Compounds, Sienkiewicza 112, 90-363 Łódź, Poland

K. Pihlaja — Department of Chemistry, University of Turku, SF-20500 Turku, Finland

P. Sarti-Fantoni — Department of Organic Chemistry, University of Firenze, Via G. Carponi 9, 50121 Firenze, Italy

J. Shorter — Department of Chemistry, The University, Hull, HU6 7RX, UK

C. J. M. Stirling — Department of Chemistry, University College of North Wales, Bangor, Gwynedd LL7 2UW, UK

T. Takata — Research Laboratory of Resources Utilization, Tokyo Institute of Technology, 4259 Nagatsuta-cho, Midori-ku, Yokohama 227, Japan

J. G. Tillett
Department of Chemistry and Biological Chemistry, University of Essex, Wivenhoe Park, Colchester, CO4 3SQ, UK

H. Togo
Department of Chemistry, Faculty of Science, Chiba University, Yayoi-cho, Chiba City, 260 Japan

U. Zoller
Division of Chemical Studies, Haifa University, School of Education, Oranim, Kiryat Tivon 36910, Israel

# Foreword

This volume on sulphinic acids and their derivatives belongs to a subset on sulphur-containing functional groups within the framework of *The Chemistry of Functional Groups*. The first of this subset was *The Chemistry of the Thiol Group* (two parts, 1974), with much additional material on the subject published in *Supplement E: The Chemistry of Ethers, Crown Ethers, Hydroxyl Groups and their Sulphur Analogues* (two parts, 1980). A volume on *The Chemistry of the Sulphonium Group* appeared in two parts in 1981 and a volume on *The Chemistry of Sulphones and Sulphoxides* in 1988. The present volume deals with sulphinic acids and their esters, halides and amides. A volume on sulphenic acids is already in the proof stage and is scheduled to appear in the late spring of 1990, and manuscripts for a volume on sulphonic acids are reaching the editors now, and will be published, we hope, in early 1991.

Among the chapters originally planned for the present volume, three did not materialize. These are on structural chemistry, on electrochemistry and on free radical chemistry. We hope to include these subjects in a supplementary volume on the whole subset of sulphur-containing functional groups, to be published in a few years' time.

The references in almost all chapters cover the year 1987 and in many cases extend well into 1988.

I would like to thank my good friends, Professor C. J. M. Stirling FRS and Professor Zvi Rappoport, for their generous and unstinting advice and counsel during the preparation of the plan of the present volume.

I will be grateful to readers who would call my attention to omissions and mistakes in this volume.

Jerusalem
October 1989

Saul Patai

# The Chemistry of Functional Groups

# Preface to the series

The series 'The Chemistry of Functional Groups' was originally planned to cover in each volume all aspects of the chemistry of one of the important functional groups in organic chemistry. The emphasis is laid on the preparation, properties and reactions of the functional group treated and on the effects which it exerts both in the immediate vicinity of the group in question and in the whole molecule.

A voluntary restriction on the treatment of the various functional groups in these volumes is that material included in easily and generally available secondary or tertiary sources, such as Chemical Reviews, Quarterly Reviews, Organic Reactions, various 'Advances' and 'Progress' series and in textbooks (i.e. in books which are usually found in the chemical libraries of most universities and research institutes), should not, as a rule, be repeated in detail, unless it is necessary for the balanced treatment of the topic. Therefore each of the authors is asked not to give an encyclopaedic coverage of his subject, but to concentrate on the most important recent developments and mainly on material that has not been adequately covered by reviews or other secondary sources by the time of writing of the chapter, and to address himself to a reader who is assumed to be at a fairly advanced postgraduate level.

It is realized that no plan can be devised for a volume that would give a complete coverage of the field with no overlap between chapters, while at the same time preserving the readability of the text. The Editor set himself the goal of attaining reasonable coverage with moderate overlap, with a minimum of cross-references between the chapters. In this manner, sufficient freedom is given to the authors to produce readable quasi-monographic chapters.

The general plan of each volume includes the following main sections:

(a) An introductory chapter deals with the general and theoretical aspects of the group.

(b) Chapters discuss the characterization and characteristics of the functional groups, i.e. qualitative and quantitative methods of determination including chemical and physical methods, MS, UV, IR, NMR, ESR and PES—as well as activating and directive effects exerted by the group, and its basicity, acidity or complex-forming ability.

(c) One or more chapters deal with the formation of the functional group in question, either from other groups already present in the molecule or by introducing the new group directly or indirectly. This is usually followed by a description of the synthetic uses of the group, including its reactions, transformations and rearrangements.

(d) Additional chapters deal with special topics such as electrochemistry, photochemistry, radiation chemistry, thermochemistry, syntheses and uses of isotopically labelled compounds, as well as with biochemistry, pharmacology and toxicology. Whenever applicable, unique chapters relevant only to single functional groups are also included (e.g. 'Polyethers', 'Tetraaminoethylenes' or 'Siloxanes').

This plan entails that the breadth, depth and thought-provoking nature of each chapter will differ with the views and inclinations of the authors and the presentation will necessarily be somewhat uneven. Moreover, a serious problem is caused by authors who deliver their manuscript late or not at all. In order to overcome this problem at least to some extent, some volumes may be published without giving consideration to the originally planned logical order of the chapters.

Since the beginning of the Series in 1964, two main developments occurred. The first of these is the publication of supplementary volumes which contain material relating to several kindred functional groups (Supplements A, B, C, D, E and F). The second ramification is the publication of a series of 'Updates', which contain in each volume selected and related chapters, reprinted in the original form in which they were published, together with an extensive updating of the subjects, if possible, by the authors of the original chapters. A complete list of all above mentioned volumes published to date will be found on the page opposite the inner title page of this book.

Advice or criticism regarding the plan and execution of this series will be welcomed by the Editor.

The publication of this series would never have been started, let alone continued, without the support of many persons in Israel and overseas, including colleagues, friends and family. The efficient and pateint co-operation of staff members of the publisher also rendered me invaluable aid. My sincere thanks are due to all of them, especially to Professor Zvi Rappoport, who for many years, shares the work and responsibility of the editing of this Series.

The Hebrew University
Jerusalem Israel

SAUL PATAI

# Contents

# List of abbreviations used

| | |
|---|---|
| Ac | acetyl (MeCO) |
| acac | acetylacetone |
| Ad | adamantyl |
| All | allyl |
| An | anisyl |
| Ar | aryl |
| Bz | benzoyl ($C_6H_5CO$) |
| Bu | butyl (also *t*-Bu or $Bu^t$) |
| CD | circular dichroism |
| CI | chemical ionization |
| CIDNP | chemically induced dynamic nuclear polarization |
| CNDO | complete neglect of differential overlap |
| Cp | $\eta^5$-cyclopentadienyl |
| DBU | 1,8-diazabicyclo[5.4.0]undec-7-ene |
| DME | 1,2-dimethoxyethane |
| DMF | *N*,*N*-dimethylformamide |
| DMSO | dimethyl sulphoxide |
| ee | enantiomeric excess |
| EI | electron impact |
| ESCA | electron spectroscopy for chemical analysis |
| ESR | electron spin resonance |
| Et | ethyl |
| eV | electron volt |
| Fc | ferrocene |
| FD | field desorption |
| FI | field ionization |
| FT | Fourier transform |
| Fu | furyl($OC_4H_5$) |
| Hex | hexyl($C_6H_{11}$) |
| c-Hex | cyclohexyl($C_6H_{11}$) |
| HMPA | hexamethylphosphortriamide |
| HOMO | highest occupied molecular orbital |
| i- | iso |

| | |
|---|---|
| Ip | ionization potential |
| IR | infrared |
| ICR | ion cyclotron resonance |
| | |
| LCAO | linear combination of atomic orbitals |
| LDA | lithium diisopropylamide |
| LUMO | lowest unoccupied molecular orbital |
| | |
| M | metal |
| *M* | parent molecule |
| MCPBA | *m*-chloroperbenzoic acid |
| Me | methyl |
| MNDO | modified neglect of diatomic overlap |
| MS | mass spectrum |
| | |
| n | normal |
| Naph | naphthyl |
| NBS | *N*-bromosuccinimide |
| NMR | nuclear magnetic resonance |
| | |
| Pen | pentyl($C_5H_{11}$) |
| Pip | piperidyl($C_5H_{10}N$) |
| Ph | phenyl |
| ppm | parts per million |
| Pr | propyl (also *i*-Pr or $Pr^i$) |
| PTC | phase transfer catalysis |
| Pyr | pyridyl ($C_5H_4N$) |
| | |
| R | any radical |
| RT | room temperature |
| | |
| s- | secondary |
| SET | single electron transfer |
| SOMO | singly occupied monecular orbital |
| | |
| t- | tertiary |
| TCNE | tetracyanoethylene |
| THF | tetrahydrofuran |
| Thi | thienyl($SC_4H_3$) |
| TMEDA | tetramethylethylene diamine |
| Tol | tolyl($MeC_6H_4$) |
| Tos | Tosyl (*p*-toluenesulphonyl) |
| Trityl | Triphenylmethyl($Ph_3C$) |
| Xyl | xylyl($Me_2C_6H_3$' |

In addition, entries in the 'List of Radical Names' in *IUPAC Nomenclature of Organic Chemistry*, 1979 Edition, Pergamon Press, Oxford, 1979, pp. 305–322, will also be used in their unabbreviated forms, both in the text and in formulae instead of explicitly drawn structures.

We are sorry for any inconvenience to our readers. However, the rapidly rising costs of production make it absolutely necessary to use every means to reduce expenses—otherwise the whole existence of our Series would be in jeopardy.

The Chemistry of Sulphinic Acids, Esters and their Derivatives
Edited by S. Patai


CHAPTER **1**

# Sulphinic acids and carboxylic acids—a comparison

CHARLES J. M. STIRLING

*Department of Chemistry, University College of North Wales, Bangor, Gwynedd LL57 2UW, UK*



## I. INTRODUCTION

In looking for a guide to the reactivity and behaviour of the really rather unfamiliar sulphinic acids (**1**) the obvious, but as we shall see, superficial, analogy with carboxylic acids (**2**) may be drawn. The similarity of the structures as written is, of course, a delusion; the central atoms of carbon and sulphur respectively and the interaction of the respective carbonyl C=O and sulphinyl S=O groups with groups attached to carbon and sulphur

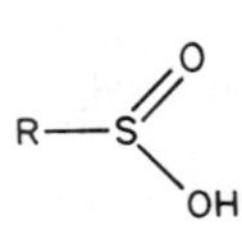


(**1**)

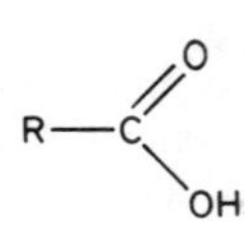


(**2**)

in place of the hydroxyl group in each case are crucially different in determining behaviour. While it will become apparent that the differences between these two classes of compound are greater than the similarities, nevertheless it is hoped that the comparison will serve to introduce some of the special features of an interesting class of compounds.

## II. DISCUSSION

### A. Structural Comparisons

The carboxyl group is planar, i.e. the four atoms of the

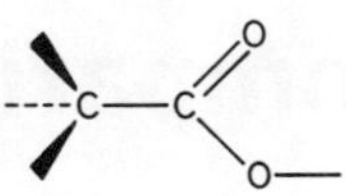

moiety lie in the same plane, a situation described by molecular orbital theory as $sp^2$-hybridization of the central carbon atom[1].

On ionization to give the carboxylate ion, delocalization of the charge over both oxygen atoms is indicated by the identical C—O bond distances and the loss of the 'normal' infrared carbonyl stretching frequency.

The situation for sulphinic acids is quite different. First, the $^{+}$S—O$^{-}$ vs S=O description for the sulphinyl group is to be preferred especially when considering stereochemistry. Sulphinic esters, amides and other derivatives are chiral; the sulphur atom is roughly tetrahedral (pyramidal discounting the filled orbital on S) in contrast to planar carbonyl carbon. Against the background of modern theory, these observations can be understood; orbital matching between oxygen and carbon is good but for oxygen and sulphur the latter's are much more diffuse. The dipole moment of benzenesulphinic acid (3.76 in dioxane)[2] is much larger than that of benzoic acid (1.0 in benzene)[3].

### B. Dissociation

The $pK_a$ values of simple carboxylic acids in water are around +4; those of the corresponding sulphinic acids are between +1 and +2[4]. Such dissociations are largely determined by two factors: (i) delocalization of negative charge in the anion, and (ii) solvation of the anion. For sulphinic acids, the first factor is probably unimportant; there is considerable negative charge on each oxygen atom in the sulphinate ion, a situation contributed to by the high polarizability of sulphur. This high charge density on oxygen increases the stabilization of the ion by hydrogen bonding.

### C. Oxidation–Reduction

The carboxylic acids lie at the end of the oxidation chain; the function is commonly derived by oxidation of alkyl, alkenyl, carbinol and aldehydic groups. Not so for the sulphinic acids; they lie on the oxidation sequence at oxidation state +4, being readily oxidized to the sulph*o*nic acids (+6) by expansion of the valence shell and reduced via sulph*e*nic acids (+2) to thiols (−2). The ability of sulphur to expand its valence shell confers upon the element its great versatility. The carboxylic acids are difficult to reduce whereas the +4 oxidation state of sulphur is a rather unstable one. Sulphinyl compounds are easily oxidized and reduced and disproportionation to a mixture of +6 and +2 oxidation states is common[5]. Free sulphinic acids, for example, decompose on standing, to mixtures of sulphonic acid and thiolsulphinate.

### D. Nucleophilicity

Carboxylate ions are notoriously poor nucleophiles[6]; the charge on the ion is heavily delocalized, the ion is usually heavily solvated and it is 'hard'. In bond formation to an electrophile, the resonance stabilization of the ion is substantially diminished, solvent has to be discarded and many electrophiles are 'soft'. The sulphinate ion stands in considerable contrast; the ion is not resonance stabilized in the same sense as the carboxylate ion, solvation is less and the ion is ambident. This last part is of great significance. Sulphur can expand the valence shell to produce a large, highly polarizable nucleophile with a 'soft' 'centre'. Notwithstanding the fact that the ion is around 3 orders of magnitude less reactive to the proton as shown by $pK_a$ data, nucleophilicity to carbon is much greater than that of the carboxylate ion. S-nucleophilicity is the predominant mode[7] particularly when the electrophile is also polarizable. Nucleophilic attack on halogens occurs extremely readily[8]. This pathway is seldom observed for carboxylates. Likewise, sulphinate ion is thiophilic in a way in which carboxylate is only poorly so. For example, disproportionation[9] of $AcNHCH_2CH_2SS(CH_2)_4X$ to symmetrical disulphides is 300 times faster with $X = SO_2^-$ than with $X = CO_2^-$. It is not silicophilic in the way in which carboxylate is[10].

The high thiophilicity of sulphinate can clearly be attributed to a polarizable–polarizable (soft–soft) interaction but a probable contributory factor is the weak S—O vs S—S bond strength. Interestingly, the situation is reversed for nucleophilic attack at silicon[10]. Oxy-anions are much more silicophilic than thiolate anions and the Si—O bond strength is very much greater than the Si—S bond strength.

Likewise, carboxylate ions are very feebly reactive towards electrophilic alkenes and can, under those conditions (dipolar aprotic solvent) in which addition *can* be effected[11], cause deprotonation and subsequent reactions because of their enhanced basicity. Sulphinate adds extremely readily[12], and is eliminated (reverse reaction) slowly by comparison with carboxylate (below) such that solutions of electrophilic alkenes and sulphinates in protic solvents (S—H) quickly become strongly basic because of generation of the lyate ion of the solvent, viz

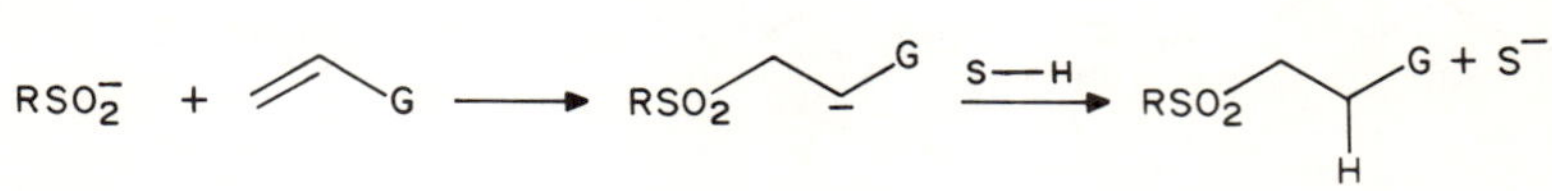


(G = carbanion stabilizing group)

Notice the important point that these reactions are *not* 'bond-strength' driven. C—S is substantially weaker than C—O. Nucleophilicity like nucleofugality (below) is a complex, solvent-dependent transition structure-dependent property to which several parameters contribute. Only rather seldom do bond strength differentials emerge as a controlling factor.

### E. Nucleofugality

Neither carboxylate nor sulphinate are very familiar participants as leaving groups in displacement reactions. Carboxyl esters, of course, react with nucleophiles primarily at the carbonyl group and so the alternative mode of attack at $sp^3$ carbon and alkyl oxygen fission is not generally observed. Displacement reactions can, however, be seen under appropriate conditions; for example, carboxylate ion is displaced from esters of carboxylic acids by iodide, thiolate and cyanide ions[13]. *Inter*molecular displacement of sulphinate from saturated carbon in sulphones or sulphinate esters is not a known reaction; *intra*molecular displacement of sulphinate from a sulphone under fairly brutal conditions has been observed[14] and the system, involving a sulphonyl stabilized nucleophile:

permitted quantitative comparison of true nucleofugalities. $PhSO_2^-$ is $10^5$ more nucleofugic in this reaction than $PhO^-$ and $10^7$ *less* nucleofugic than $TsO^-$. It is not possible to compare carboxylate under these conditions because of the supervening $sp^2$ rather than $sp^3$ carbon electrophilicity. A remarkably close correlation was, however, found between leaving-group nucleofugality and the $pK_a$ of the conjugate acid of the nucleophile determined in the experimental solvent. The value for the $pK_a$ of carboxylic acids in the solvent used (*t*-BuOH) is not known but can be guessed ($\approx 13$) from values in related solvents[15]. This comparison makes sulphinate a somewhat better nucleofuge than carboxylate in a reaction which all available evidence suggests has a very large degree of fission of the bond to the leaving group in the transition structure.

A totally different comparison emerges when these groups are compared in 1,2-alkene-forming eliminations. Again, an appropriate choice of system[16] has permitted comparison of nucleofugalities devoid of other effects on reactivity that have nothing to do with nucleofugality. The system is much more reactive than that for the displacement reactions. This time, acetoxy can be studied but the kinetics show that it is such a good nucleofuge that the rate-determining step is deprotonation and not leaving-group departure. For benzenesulphinate, ranking of the group derived either by C—O or C—S cleavage in the isomeric sulphinates and sulphones, respectively, shows it to be inferior to phenyldimethylammonium but comparable with phenoxy and thiophenoxy. In this system, all the available indicators as to the transition structure show that little fission of the bond to the leaving group is involved and that therefore, as would be expected, little correlation between nucleofugality and $pK_a$ of the conjugate acid of the leaving group is seen.

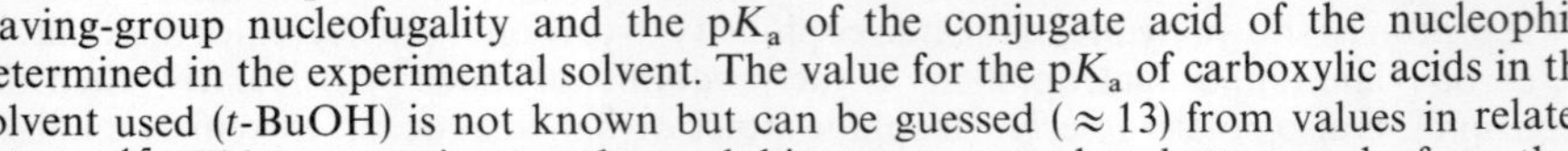


## F. Electrophilicity

Here the comparison has to be made between related derivatives because reactions of nucleophiles with the free acids are complicated by ionization. Superficially, the two classes of substrate behave similarly; alkaline hydrolysis of the esters yields the acid as its salt together with the alcohol by carbonyl- and sulphinyl-oxygen fission, respectively. Likewise, reaction with organometallics such as Grignard reagents yield ketones (initially) and sulphoxides[17] by attack at carbonyl carbon and sulphur, respectively. There the resemblance ends; for derivatives of carboxylic acids, the addition–elimination pathway via a tetrahedral intermediate is well established. The mechanistic details of substitution at sulphinyl sulphur have not been investigated to any extent. It is known that in the sulphinate ester–Grignard reactions, substitution occurs strictly with inversion of configuration but it is unclear whether or not a tetracoordinate intermediate is involved[17]. An important difference, which illustrates the significant contrast between the carbonyl group and the sulphinyl group, is seen in the behaviour of the amides. Carboxamides typically have $pK_a$ values (conjugate acid) close to 0 and hydrolyse rather slowly in acidic conditions. Sulphinamides are more basic. The insensitivity of the sulphinyl stretching frequency to substituents[18] suggests that interaction of the electron pair on nitrogen with the S—O bond is not involved, and the S—N bond is very much weaker.

They are, like phosphonamides, very labile in acidic conditions. Cleavage of sulphinamides probably occurs by dissociation of the conjugate acid:

$$RSO{-}\overset{+}{N}H_2R \rightarrow R\overset{+}{S}O + H_2NR$$

Sulphinamides of aromatic amines have long been known to rearrange[19] by a pathway presumably involving the sulphinium ion:

$$RSO_2NHPh \xrightarrow{H^+} p\text{-}RSO_2C_6H_4NH_2$$

This type of $A_{Ac}1$ mechanism is only seen for carboxamides in powerfully acidic conditions when other pathways are suppressed.

Reactions of carboxyl and of sulphinyl halides again appear superficially similar—attack of nucleophiles at carbon and at sulphur respectively produces the corresponding carbonyl and sulphinyl products. The products from sulphinyl halides, as has just been seen, are however labile and, for example, formation of sulphinamides from sulphinyl chlorides and amines often gives poor yields unless precautions are taken to safeguard the product from subsequent reactions.

### G. Disproportionation

This is a characteristic reaction of sulphinic acids and their derivatives which is not seen in the carboxylic acid series. It is a consequence of the greater basicity of sulphinyl derivatives rendering them prone to acid-catalysed processes and the much greater acidity of sulphinic acids. The combination of these properties with the low sulphinyl–heteroatom bond strength allows disproportionation to occur readily[20] (Scheme 1).

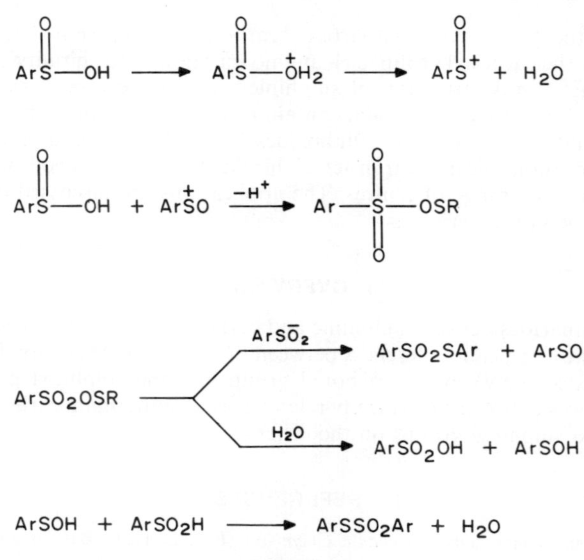

SCHEME 1

Thiol esters of carboxylic acids are stable compounds but their sulphinyl analogues disproportionate readily[21] (Scheme 2). Here the weak heteroatom–sulphinyl bond strength is responsible.

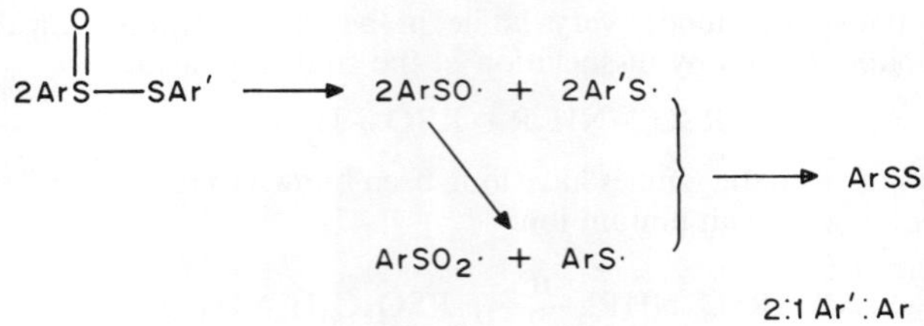


SCHEME 2

### H. Decarboxylation and Desulphination

Again these are formally analogous processes and here the resemblance is closer than in many other parallels that can be drawn between carboxylic and sulphinic acids. Very much more is known about the decarboxylation of carboxylic acids[22]. Simple carboxylic acids lose $CO_2$ only at high temperatures; in functionalized carboxylic acids, either some stable species (a resonance-stabilized carbanion or an enol) is produced, or loss of $CO_2$ is concerted with some other process, e.g. elimination of a *β*-group[23]. The carbon–carbon bond is stronger than the carbon–sulphur bond and sulphinic acids lose $SO_2$ readily notwithstanding the lower heat of formation of $SO_2$ ($-70.96\,\text{kcal mol}^{-1}$) than that of $CO_2$ ($-94.05\,\text{kcal mol}^{-1}$). Desulphinated products turn up frequently in reactions of free sulphinic acids[24].

### I. Chirality

The two substituents at the trigonal carbonyl carbon atom of carbonyl derivatives lie in the same plane as the carbonyl group itself and no question as to chirality of such species arises. By contrast, X-ray structures of sulphinic esters, for example, show[25] that the sulphur atom is tetrahedral (if the orbital containing the lone pair on sulphur is included). The situation is that which obtains for sulphoxides[26]. The chirality of sulphinyl derivatives has played an important role in the transfer of chirality from naturally occurring alcohols, e.g. menthol, to a wide range of centres. The applications are discussed elsewhere in a companion volume in this series[26].

## III. OVERVIEW

The apparent similarities between sulphinic and carboxylic acids are deceptive. The line structures belie the significant differences between these classes. These are dictated by the electronic differences between the carbonyl group and the sulphinyl group, and the enthalpic differences between superficially related species containing carbonyl groups on the one hand and sulphinyl groups on the other.

## IV. REFERENCES

1. E. W. Colvin, in *Comprehensive Organic Chemistry* (Eds. D. H. R. Barton and W. D. Ollis), Section 9.1, Pergamon, Oxford, 1979.
2. E. N. Gur'yanova and Ya. K. Syrkin, *Zh. Fiz. Khim.*, **23**, 105 (1949); *Chem. Abstr.*, **43**, 5245 (1949).
3. J. W. Williams and R. J. Allgeier, *J. Am. Chem. Soc.*, **49**, 2416 (1927).
4. E. P. Serjeant and B. Dempsey, *Ionisation Constants of Organic Acids in Aqueous Solution*, Pergamon, Oxford, 1979.
5. C. J. M. Stirling, *Int. J. Sulfur Chem.* (*B*), **6**, 277 (1971).

6. J. O. Edwards and R. G. Pearson, *J. Am. Chem. Soc.*, **84**, 16 (1962).
7. J. S. Meek and J. S. Fowler, *J. Org. Chem.*, **33**, 3422 (1958).
8. C. J. M. Stirling, *J. Chem. Soc.*, 3597 (1957); P. E. Pigou and C. J. M. Stirling, *J. Chem. Soc., Perkin Trans. 2*, 725 (1988).
9. Y. H. Khim and L. Field, *J. Org. Chem.*, **37**, 2714 (1972).
10. I. Fleming, in *Comprehensive Organic Chemistry* (Eds. D. H. R. Barton and W. D. Ollis), Section 13.1, Pergamon, Oxford, 1979.
11. G. D. Appleyard and C. J. M. Stirling, *J. Chem. Soc.* (*C*), 1904 (1969).
12. C. J. M. Stirling, *J. Chem. Soc.*, 5856 (1964).
13. J. McMurry, *Org. React.*, **24**, 187 (1976).
14. B. Issari and C. J. M. Stirling, *J. Chem. Soc., Perkin Trans. 2*, 1043 (1984).
15. L. Eberson, in *The Chemistry of Carboxylic Acids and Esters* (Ed. S. Patai), Chap. 6, Wiley, London, 1969.
16. C. J. M. Stirling, *Acc. Chem. Res.*, **12**, 198 (1979).
17. K. K. Andersen, in *Comprehensive Organic Chemistry* (Eds. D. H. R. Barton and W. D. Ollis), Section 11.18, Pergamon, Oxford, 1979.
18. H. H. Szmant, in *Sulfur in Organic and Inorganic Chemistry*, Volume 1 (Ed. A. Senning), Chap. 5, Dekker, New York, 1971.
19. J. Vonkennel and J. Kimmig, Deutsche Bundrepublik Patent 854, 802 (1900); *Chem. Zentralbl.*, 8143 (1953). O. Hinsberg, *Chem. Ber.*, **18**, 2493 (1855).
20. Work by J. L. Kice and coworkers summarized by S. Oae and N. Kuneida, in *Organic Chemistry of Sulfur* (Ed. S. Oae), Chap. 11, Plenum, New York, 1977.
21. D. Barnard, *J. Chem. Soc.*, 4675 (1957).
22. S. Patai (Ed.), *The Chemistry of Carboxylic Acids and Esters*, Chap. 12, Wiley London, 1969.
23. W. R. Vaughan, W. F. Cartright and B. Henzi, *J. Am. Chem. Soc.*, **94**, 4978 (1972) and previously cited papers.
24. F. Kurzer and J. R. Powell, *J. Chem. Soc.*, 3728 (1952).
25. K. Mislow, M. M. Green, P. Lauer, J. P. Melillo, T. Simmons and A. L. Ternay, *J. Am. Chem. Soc.*, **87**, 1958 (1965).
26. S. Patai, Z. Rappoport and C. J. M. Stirling, *The Chemistry of Sulphones and Sulphoxides*, Wiley, Chichester, 1988.

The Chemistry of Sulphinic Acids, Esters and their Derivatives
Edited by S. Patai


CHAPTER **2**

# General and theoretical

HAROLD BASCH

*Department of Chemistry, Bar-Ilan University, Ramat Gan 52100, Israel*



## I. INTRODUCTION

This chapter is concerned with a quantum mechanical description of the sulphinic acid group and its amide (sulphinamide) and halide (sulphinyl fluoride and chloride) derivatives. Within each of these groupings we will discuss the possible anions, radical and cation states, isomeric rearrangement compounds (tautomers) and, only in the case of the acid, also thio substitution. The 1:1 hydrogen—(H—) bonded complexes sulphinic acid–water, sulphinic acid–methanol and sulphinamide–water are also included in this study. In principle, such weakly bound complexes can have an existence of their own and can be probed spectroscopically either in the gas phase or in matrix isolation.

A computer data base search of these subjects, as well as direct perusal of the most recent review articles on the subject[1–3] reveal that almost all of the experimental literature deals with the aromatic sulphinic acids (and their derivatives). The simplest aliphatic sulphinyl compounds (RSOX) (X = OH, $NH_2$ or halide) tend to be unstable and disproportionate on standing. The small amount of experimental data found for these compounds is therefore either in matrix isolation or in adduct complexes.

Theoretically, the sulphinic acids and their derivatives seem to be virgin territory. With one exception, there has been no attempt to use conventional semi-empirical or *ab initio* molecular structure methods on this class of compounds, and even isolated studies are not to be found. The reason for this situation probably lies in the problematics of applying these theoretical methods to molecules containing second-row atoms in general, and to the large size of the aromatic systems for which almost all of the experimental data are known. Semi-empirical methods require careful paramaterization usually on a large number of well-known compounds for very specific structural properties. This information is clearly lacking in the case of the sulphinic acids and their derivatives. *Ab initio* methods, which can give reliable electronic and molecular structural information, cannot yet be routinely applied to aromatic compounds at a sufficiently high level.

We have therefore decided to explore the simplest parent (R = H) sulphinyl compounds using extended basis set *ab initio* molecular orbital theory methods with correlation effects (post-Hartree–Fock). The purpose of these calculations is to use a uniformly high-level theoretical treatment on a set of model compounds in the spirit of the very extensive work done by Pople and coworkers and summarized in their book[4]. We also hope that these calculations will stimulate theoretical interest in this very interesting class of chemical compounds. This review will just scratch the surface of the subject and poses more questions than answers. The properties examined here include geometric structures (bond lengths and angles), vibrational frequencies, isomerization energies, proton affinities, bond strengths, bond dissociation energies, dipole moments, atomic charges, excited states, and hydrogen-bond structures and strengths. Where possible, the calculated results are compared to similar experimental studies, although, as mentioned above, these are sparse.

## II. THEORETICAL MODEL AND RESULTS

*Ab initio* self-consistent-field (SCF) calculations were carried out on all the molecular systems reported here using the restricted Hartree–Fock (RHF) method for closed-shell systems and the unrestricted Hartree–Fock method (UHF) for the open-shell molecules. For each of the neutral or cation species the molecular (geometric) structure was gradient optimized at the SCF level using the standard 6-31G* basis set with the GAUSSIAN82 or GAUSSIAN86[5] set of computer programs. At each final optimized geometry the MP2 energy was calculated in both the 6-31G* and 6-31 + G* bases. For the anions, all calculations were done only in the 6-31 + G* basis while for the 1:1 water and methanol complexes the geometry optimizations and MP2 energies were obtained in both the 6-31G** and 6-31G* bases for sulphinic acid with water, and only in the latter basis for the other combination structures. A 6-311G* basis was used to probe excited state structures and energies.

Moller–Plesset perturbation theory carried to second order in the energy (MP2)[4] is the simplest post-Hartree–Fock method for eliminating defects of the SCF method, known as correlation effects. These defects are proportional to the degree that the single electronic configuration description is not valid. For example, $\pi$ bonds are usually less well described at the Hartree–Fock level than $\sigma$ bonds. Since the different species compared here can typically have different degrees of single and double bond character, the accuracy of the single configuration SCF methods (both RHF and UHF) differ accordingly. The MP2 method should go a long way towards mitigating these differences and make the energy comparisons more valid.

The 6-31G* basis is a standard valence electron double-zeta (or split valence) basis set of s- and p-type gaussian atomic orbitals augmented by a set of polarization (denoted by the $*$) d-type functions (5 components) on each of the first- and second-row atoms. This basis set is known to usually give accurate static property values such as geometries, charge densities and dipole moments. The 6-31 + G* basis adds diffuse s- and p-type basis functions (denoted by the $+$) for a better long-range description of lone-pair electrons, radicals and, especially, anions. The additional diffuse functions hardly affect the calculated geometries except for the anions. Recalculating the SCF and MP2 energies in the more extended basis set was done in order to obtain a more uniform description as a proper base for comparison of energies and properties. The 6-31G** basis adds a set of p-type polarization functions to each hydrogen atom, in addition to the d-type set on the heavier atoms. In contrast, a 6-311G* basis was used to study excited states. This basis set is of the 'triple-zeta' variety where the sp valence atomic orbitals are split into 3 basis functions (comprising 5 gaussians) instead of two functions (from 4 gaussians) in the 6-31G* basis, for added flexibility in the valence region. For comparison purposes the ground-state energies were also recalculated in this basis set.

At the 6-31G* SCF gradient optimized geometry, a full force-field calculation was carried out using the second derivative normal mode analysis to obtain the harmonic vibrational frequencies. For simplicity, we will present here only the stretch frequencies. At the basis set and SCF level model used here the calculated stretch frequencies are known to be overestimated by 10–12% and are usually reduced by this amount before comparison with experimental numbers. This is due to the known property of single determinant wave functions giving too steep a potential energy curve at the equilibrium geometry due to incorrect dissociation limits (usually to high-energy ionic fragments or atoms) for bond breaking.

The calculated results for the 35 structures examined here are presented in Tables 1–5 and, selectively, in Figures 1–26. Table 1 summarizes the calculated total SCF and MP2

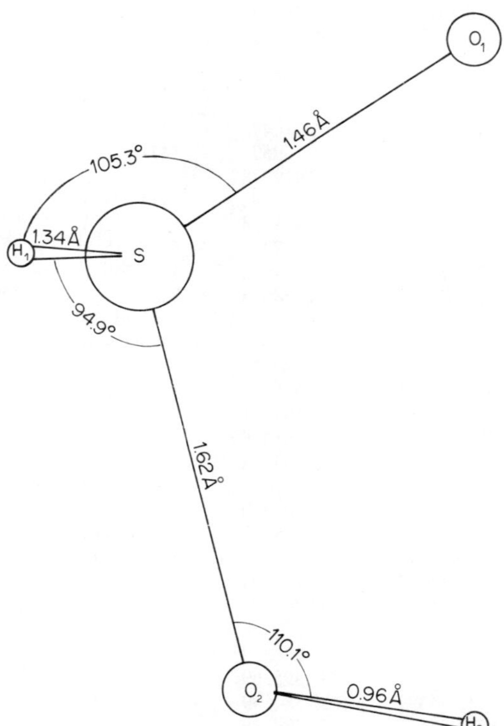


FIGURE 1. HSOOH, structure **1**, drawn in OSO plane, dihedral angles (deg): $O_2SH_1O_1 = 111.1$, $H_2O_2SO_1 = 32.0$

energies in the 6-31G* and 6-31 + G* basis sets as well as the dipole moments in the latter basis. Table 2 lists the gradient optimum bond lengths for each molecule, allowing a comparison of the same bond type in the different bonding situations and, alternatively, allowing the assignment of a bond order according to comparative bond length. Table 3 gives the standard Mulliken population analysis by atom which allows the tracking of charge shifts as a function of structural and composition changes in the molecules. Table 4 gives the calculated harmonic vibrational frequencies for the parent sulphinyl species (acid, amide and halide) for comparison with experiment. Table 5 summarizes the SCF and MP2 energies for the 1:1 water and methanol complexes of sulphinic acid and the water complexes of sulphinamide.

TABLE 1. Energies and dipole moments

| Molecule | Energy (a.u.) | | | | Dipole moment |
|---|---|---|---|---|---|
| | 6-31G* | | 6-31 + G* | | |
| | SCF | MP2 | SCF | MP2 | $D^b$ |
| **(1)** HSOOH | − 548.303809 | − 548.788598 | − 548.313765 | − 548.808984 | 2.818 |
| **(2)** $H_2SO_2$ | − 548.276711 | − 548.765030 | − 548.284668 | − 548.782472 | 4.115 |
| **(3)** $S(OH)_2$ | − 548.332607 | − 548.812164 | − 548.341007 | − 548.831155 | 0.395 |
| **(4)** $SOOH^-$ | | | − 547.763466 | − 548.272594 | 3.858 |
| **(5)** $HSO_2^-$ | | | − 547.767457 | − 548.276727 | 2.966 |
| **(6)** $\cdot SO_2H$ | − 547.746172 | − 548.216496 | − 547.754935 | − 548.235859 | 2.127 |
| **(7)** $HSO_2\cdot$ | − 547.686659 | − 548.169579 | − 547.695019 | − 548.186973 | 3.301 |
| **(8)** $HSOOH^+$ | − 547.985329 | − 548.419978 | − 547.988983 | − 548.428924 | 3.377 |
| | (− 547.970669 | − 548.410894 | − 547.974453 | − 548.420143 | 2.839)[a] |
| **(9)** $HSONH_2$ | − 528.478845 | − 528.948124 | − 528.488264 | − 528.967119 | 3.208 |
| **(10)** HSOHNH | − 528.442292 | − 528.917391 | − 528.451528 | − 528.936358 | 2.229 |
| **(11)** $SOHNH_2$ | − 528.515809 | − 528.979532 | − 528.523681 | − 528.996929 | 2.522 |
| **(12)** $H_2SONH$ | − 528.403458 | − 528.878635 | − 528.411405 | − 528.895579 | 3.031 |
| **(13)** $SONH_2$ | | | − 527.934859 | − 528.424877 | 3.144 |
| **(14)** $\cdot SONH_2$ | − 527.923966 | − 528.368959 | − 527.930979 | − 528.386265 | 2.136 |
| **(15)** $HSONH\cdot$ | − 527.852944 | − 528.289351 | − 527.861026 | − 528.305782 | 2.186 |
| **(16)** $HSONH_2^+$ | − 528.186354 | − 528.605398 | − 528.189260 | − 528.613250 | 3.421 |
| | (− 528.170846 | − 528.593631 | − 528.174254 | − 528.601775 | 2.882)[a] |
| **(17)** HSOF | − 572.300118 | − 572.781386 | − 572.312108 | − 572.805305 | 3.28 |
| **(18)** FSOH | − 572.322345 | − 572.794238 | − 572.334640 | − 572.818696 | 2.163 |
| **(19)** $SOF^-$ | | | − 571.776850 | − 572.281808 | 2.286 |
| **(20)** $\cdot SOF$ | − 571.743048 | − 572.209489 | − 571.754319 | − 572.232584 | 2.215 |
| **(21)** $HSOF^+$ | − 571.955613 | − 572.385797 | − 571.960416 | − 572.396872 | 1.805 |
| | (− 571.942130 | − 572.381017 | − 572.947790 | − 572.393362 | 1.943)[a] |
| **(22)** HSOCl | − 932.343052 | − 932.781104 | − 932.350718 | − 932.795646 | 3.270 |
| **(23)** ClSOH | − 932.381531 | − 932.807526 | − 932.387515 | − 932.820498 | 2.005 |
| **(24)** $SOCl^-$ | | | − 931.841093 | − 932.290959 | 2.120 |
| **(25)** $\cdot SOCl$ | − 931.791783 | − 932.208559 | − 931.797749 | − 932.222398 | 1.903 |
| **(26)** $HSOCl^+$ | − 932.011525 | − 932.395369 | − 932.015261 | − 932.403297 | 2.125 |
| | (− 931.994399 | − 932.384579 | − 931.998821 | − 932.392813 | 2.079)[a] |
| **(27)** HSOSH | − 870.952178 | − 871.377726 | | | 3.086 |
| **(28)** HSSOH | − 870.966485 | − 871.384621 | | | 3.305 |

[a]In the neutral species geometry.
[b]In the 6-31 + G* basis at the SCF level.

TABLE 2. Calculated optimized bond lengths

| Molecule | Bond lengths (Å) | | | | | | | |
|---|---|---|---|---|---|---|---|---|
| | S=O | S—O(H) | S—H | O—H | S—N | N—H | S—X[a] | S—S/ S=S |
| (**1**) HSOOH | 1.46 | 1.62 | 1.34 | 0.96 | | | | |
| (**2**) $H_2SO_2$ | 1.43 | | 1.33 | | | | | |
| (**3**) $S(OH)_2$ | | 1.63 | | 0.95 | | | | |
| (**4**) $SOOH^-$ | 1.54 | 1.72 | | 0.95 | | | | |
| (**5**) $HSO_2^-$ | 1.50 | | 1.37 | | | | | |
| (**6**) $\cdot SO_2H$ | 1.47 | 1.62 | | 0.95 | | | | |
| (**7**) $HSO_2\cdot$ | 1.44 | | 1.34 | | | | | |
| (**8**) $HSOOH^+$ | 1.55 | 1.56 | 1.33 | 0.96 | | | | |
| (**9**) $HSONH_2$ | 1.47 | | 1.34 | | 1.68 | 1.00 | | |
| (**10**) HSOHNH | | 1.66 | 1.34 | 0.95 | 1.53 | 1.00 | | |
| (**11**) $SOHNH_2$ | | 1.66 | | 0.95 | 1.66 | 1.00 | | |
| (**12**) $H_2SONH$ | 1.44 | | 1.32 | | 1.51 | 1.00 | | |
| | | | 1.34 | | | | | |
| (**13**) $SONH_2^+$ | 1.56 | | | | 1.74 | 1.00 | | |
| (**14**) $\cdot SONH_2$ | 1.51 | | | | 1.67 | 1.00 | | |
| (**15**) $HSONH\cdot$ | 1.47 | | 1.34 | | 1.68 | 1.01 | | |
| (**16**) $HSONH_2^+$ | 1.58 | | 1.33 | | 1.60 | 1.00 | | |
| (**17**) HSOF | 1.44 | | 1.34 | | | | 1.60 | |
| (**18**) FSOH | | 1.61 | | 0.95 | | | 1.61 | |
| (**19**) $SOF^-$ | 1.52 | | | | | | 1.72 | |
| (**20**) $\cdot SOF$ | 1.45 | | | | | | 1.59 | |
| (**21**) $HSOF^+$ | 1.54 | | 1.33 | | | | 1.53 | |
| (**22**) HSOCl | 1.45 | | 1.34 | | | | 2.08 | |
| (**23**) ClSOH | | 1.62 | | 0.95 | | | 2.03 | |
| (**24**) $SOCl^-$ | | | | | | | | |
| (**25**) $\cdot SOCl$ | 1.47 | | | | | | 2.04 | |
| (**26**) $HSOCl^+$ | 1.57 | | 1.33 | | | | 1.96 | |
| (**27**) HSOSH | 1.47 | | 1.34 | | | | | |
| | | | 1.33 | | | | | 2.09 |
| (**28**) HSSOH | | 1.62 | 1.33 | 0.96 | | | | 1.98 |

[a]X = F or Cl.

## III. SULPHINIC ACID

The 6-31G* calculated geometry for the simplest sulphinic acid, HSOOH (**1**), is presented in Figure 1. The numbers in parentheses refer to the list of structures in Tables 1–3. The only other known *ab initio* calculation of **1** is in the work of Boyd and coworkers[6] who used a STO-3G(*) basis set (d-type polarization functions on the sulphur atom only). The geometries compare reasonably well, with bond angles within 3–5 deg and bond lengths within 0.01–0.04 Å. The largest discrepancies are in the O—H bond length and S—O—H angle, which is to be expected considering the difference in basis sets. The geometry of **1** is, of course, non-planar due to the sulphur atom lone pair of electrons (which are absent in the HCOOH analogue) and, therefore, both the S—H and O—H bonds are pushed to the same side of the $SO_2$ plane. This non-planarity is one of the outstanding features of the sulphinic group and gives rise to the chiral properties around the sulphur atom in the sulphinyl derivatives, such as the esters.

One of the objectives of these calculations is to compare the relative stabilities of the simplest sulphinic acid **1**, the alternative sulphone **2** (see Figure 2) and the tautomeric

TABLE 3. Mulliken atomic population analysis

| | Mulliken atomic population | | | | | | | |
|---|---|---|---|---|---|---|---|---|
| Molecule | S | H(S) | H(O) | O | O(H) | N | H(N) | X[a] |
| (**1**) HSOOH | 14.94 | 0.96 | 0.47 | 8.81 | 8.82 | | | |
| (**2**) $H_2SO_2$ | 14.76 | 0.94 | | 8.68 | | | | |
| (**3**) $S(OH)_2$ | 15.50 | | 0.47 | 8.78 | | | | |
| (**4**) $SOOH^-$ | 15.78 | | 0.52 | 8.88 | 8.82 | | | |
| (**5**) $HSO_2^-$ | 15.12 | 1.08 | | 8.89 | | | | |
| (**6**) $\cdot SO_2H$ | 15.20 | | 0.47 | 8.55 | 8.78 | | | |
| (**7**) $HSO_2\cdot$ | 14.88 | 0.92 | | 8.60 | | | | |
| (**8**) $HSO_2H^+$ | 14.86 | 0.78 | 0.40 | 8.21 | 8.75 | | | |
| (**9**) $HSONH_2$ | 15.08 | 0.94 | | 8.79 | | 8.07 | 0.56 | |
| (**10**) HSOHNH | 15.19 | 0.93 | 0.47 | | 8.83 | 7.97 | 0.62 | |
| (**11**) $SOHNH_2$ | 15.59 | | 0.48 | | 8.77 | 8.02 | 0.57 | |
| (**12**) $H_2SONH$ | 14.95 | 0.90 | | 8.67 | | 7.93 | 0.60 | |
| | | 0.95 | | | | | | |
| (**13**) $SONH_2^-$ | 15.84 | | | 8.90 | | 8.04 | 0.61 | |
| (**14**) $\cdot SONH_2$ | 15.40 | | | 8.45 | | 8.03 | 0.56 | |
| (**15**) $HSONH\cdot$ | 15.09 | 0.92 | | 8.81 | | 7.58 | 0.60 | |
| (**16**) $HSONH_2^+$ | 15.03 | 0.77 | | 8.20 | | 8.02 | 0.49 | |
| (**17**) HSOF | 14.92 | 0.94 | | 8.73 | | | | 9.40 |
| (**18**) FSOH | 15.40 | | 0.46 | 8.77 | | | | 9.36 |
| (**19**) $SOF^-$ | 15.71 | | | 8.83 | | | | 9.47 |
| (**20**) $\cdot SOF$ | 15.15 | | | 8.53 | | | | 9.35 |
| (**21**) $HSOF^+$ | 14.78 | 0.75 | | 8.18 | | | | 9.29 |
| (**22**) HSOCl | 15.25 | 0.92 | | 8.66 | | | | 17.17 |
| (**23**) ClSOH | 15.68 | | 0.46 | 8.74 | | | | 17.12 |
| (**24**) $SOCl^-$ | 15.80 | | | 8.68 | | | | 17.52 |
| (**25**) $\cdot SOCl$ | 15.43 | | | 8.43 | | | | 17.14 |
| (**26**) $HSOCl^+$ | 15.28 | 0.75 | | 8.14 | | | | 16.84 |
| (**27**) HSOSH | 15.27 | 0.94 | | 8.74 | | | | |
| | 16.18 | 0.86 | | | | | | |
| (**28**) HSSOH | 15.37 | 0.88 | 0.50 | 8.80 | | | | |
| | 16.45 | | | | | | | |

[a] X = F or Cl.

sulphide $S(OH)_2$ (**3**) (see Figure 3). From Table 1 it is clear that the accepted isomer **1** is more stable than **2** by 0.7 eV for the best calculational level here (MP2/6-31 + G*). This latter number is to be compared with a 1.0 eV relative stability calculated using the RHF/STO-3G(*) level. The dipole moment of the sulphone is larger than that of the acid, so that polar solvents (in dilute solution where solute association is not a factor) will favour the former and reduce the 'gas phase' calculated energy difference. Experimental evidence has been interpreted as favouring the sulphinic over the sulphonic isomer for both aliphatic and aromatic compounds. However, R—SOOH with R = H has not been reported experimentally.

A possible explanation for HSOOH (**1**) not being observed is that, as can be seen in Table 1, $S(OH)_2$ (**3**) is calculated to be the most stable isomeric form of the sulphinic acid, 0.6 eV more stable than the classical form **1** at the MP2/6-31 + G* level. On the other hand, the large difference in molecular dipole moment between the two tautomeric structures favours the sulphinic acid form in dilute solution, although solvation may not be enough to overcome the intrinsic free molecule energy difference favouring **3**. In any

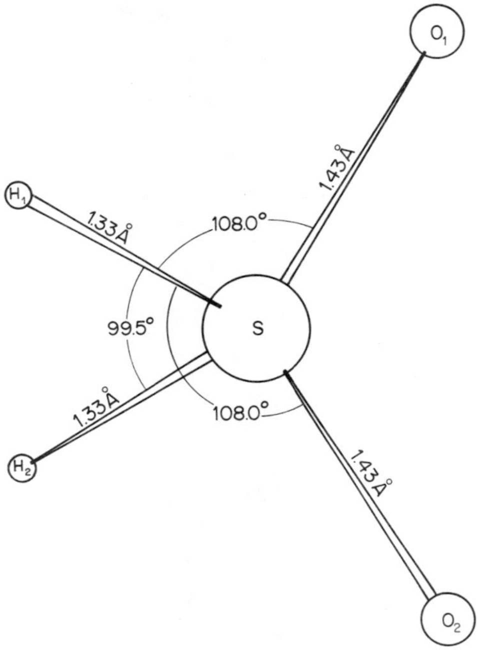


FIGURE 2. $H_2SO_2$, structure **2**, dihedral angles (deg): $O_2SH_1O_1 = 135.0$, $H_2SH_1O_1 = -112.5$

event, the alkyl sulphinic acids do exist as such and therefore the substitution of alkyl groups for the hydrogen atom bound to sulphur in **1** must preferentially stabilize the acid form over the sulfide structure. These questions bear further investigation.

The two possible anions resulting from the removal of a proton from either the sulphur atom, $SOOH^-$ (**4**), or from the oxygen atom, $HSO_2^-$ (**5**), were also examined in order to obtain their relative proton affinities. Both on the SCF and MP2 levels (Table 1) the hydroxyl protons are slightly more acidic. For $SOOH^-$ the calculated proton affinities are 15.0 eV (SCF) and 14.6 eV (MP2), while for $HSO_2^-$ the corresponding energies are 14.9 eV and 14.5 eV, in both cases a difference of only ~0.1 eV. This preferential stabilization is much smaller than the STO-3G(*) SCF difference calculated previously (0.6 eV)[6], which also favoured the hydroxyl proton. Another factor to be taken into account is the calculated dipole moment of each anion. Although dipole moments of anions are coordinate-origin dependent the central sulphur atom makes the centre-of-mass choice of origin a natural one and comparing the dipole moments of **4** and **5** should be valid. In our case it is actually $SOOH^-$ which has the larger dipole moment and, in solution, should be preferentially stabilized relative to $HSO_2^-$, in the opposite sense from the isolated molecule energy difference. Thus the higher acidity of the hydroxyl proton in sulphinic acid **1** is not clear-cut from these calculations.

Removal of one of the hydrogen atoms from either the sulphur or the oxygen atoms can lead to the $\cdot SO_2H$ (**6**) (see Figure 4) or $HSO_2\cdot$ (**7**) (see Figure 5) radicals, respectively. Here, from an energy perspective, the choice is unambiguous. The H—S homolytic dissociation energy is calculated at 2.0 eV (MP2/6-31 + G*), if we take the energy of the hydrogen atom

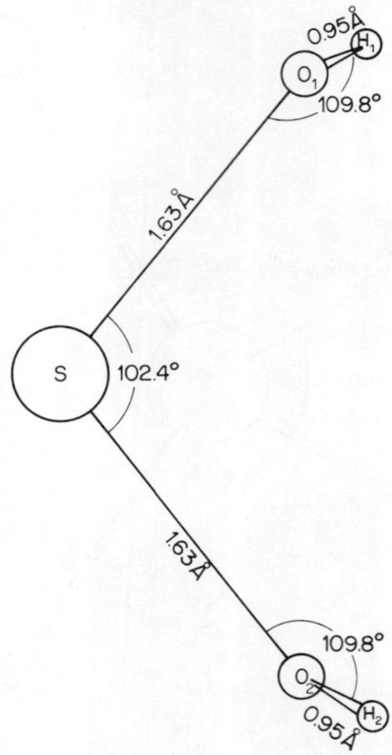


FIGURE 3. $S(OH)_2$, structure **3**, drawn in OSO plane, dihedral angles (deg): $H_1O_1SO_2 = 82.0°$, $H_2O_2SO_1 = 82.1°$

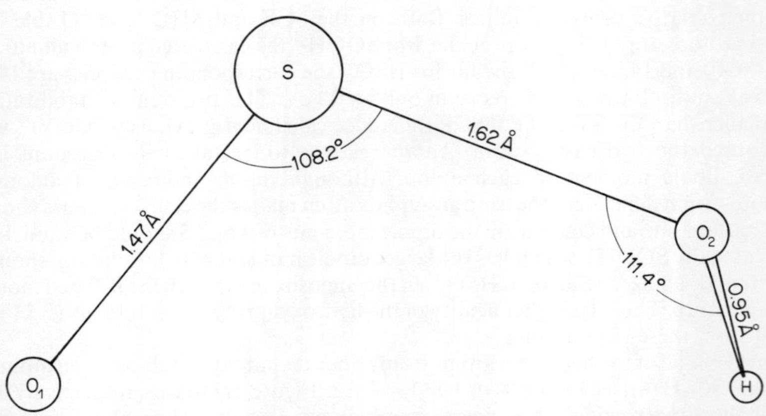


FIGURE 4. $\cdot SO_2H$, structure **6**, drawn in OSO plane, dihedral angle (deg): $HO_2SO_1 = 58.7$

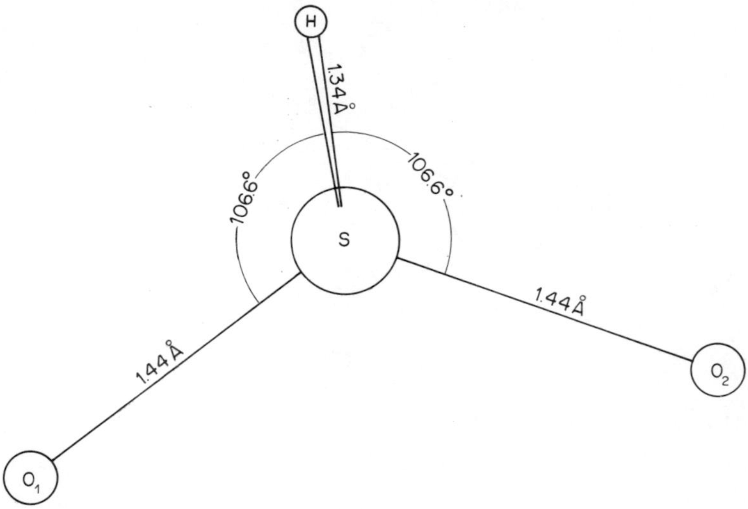


FIGURE 5. $HSO_2\cdot$, structure **7**, drawn in OSO plane, dihedral angle (deg): $O_2SHO_1 = 133.5$

at $-0.5$ a.u. exactly. In contrast, production of the $HSO_2\cdot$ radical absorbs 3.3 eV, a difference of 1.3 eV. On the SCF level (in the same basis set) this difference is 1.6 eV, similar to the published[6] STO-3G(*) difference of 1.8 eV. These large differences between isomeric species are in contrast to the almost equal heteronuclear dissociation energies for the two possible deprotonations of HSOOH (**1**). In radicals **6** and **7** the unpaired spin in both cases is distributed over the sulphur and oxygen (not OH) atoms. The radical $RSO_2\cdot$ is important in the mechanism of oxidation of $RSO_2H$ and the reaction of photochemically excited $SO_2$ with hydrocarbons in the gas phase[3]. This latter is relevant to environmental chemistry of the earth's atmosphere.

The adiabatic ionization potential from **1** to form $HSOOH^+$ (**8**) is calculated (Table 1) to be 8.8 eV (SCF) and 10.3 eV (MP2) at the 6-31 + G* basis set level. The MP2 value is, of course, expected to be the more accurate prediction since it includes a part of the correlation energy difference between the neutral and ion states. The electron comes out of a non-bonding orbital, as can also be determined by the verticality of the ionization process in Table 1; only 0.2 eV separates the MP2 calculated adiabatic and vertical ionization energies. The population analysis (Table 3) comparing HSOOH (**1**) and $HSOOH^+$ (**8**) shows that the ionized electron is actually coming mainly out of an oxygen atom (lone pair) and not from the sulphur atom (lone pair) as might be expected.

This is a good place to compare the calculated optimized bond lengths and angles in Table 2 with experiment, to the extent that this is possible. The most relevant sulphinyl compounds for which there are experimental structural data are the esters, RSOOR[7,8]. These esters have a S═O bond length range of 1.46–1.47 Å compared to the calculated 1.46 Å in HSOOH, a S—O bond length range of 1.62–1.63 Å compared to the calculated 1.62 Å in HSOOH, and a O═S—O bond angle of ~ 108 deg compared to 109 deg in Figure 1 for sulphinic acid. Thus the calculated structural values seem to be accurate.

## IV. SULPHINAMIDE

The simplest sulphinyl amide is formed by substituting the amino group for the hydroxyl group in sulphinic acid to form sulphinamide **9** whose geometry is displayed in Figure 6.

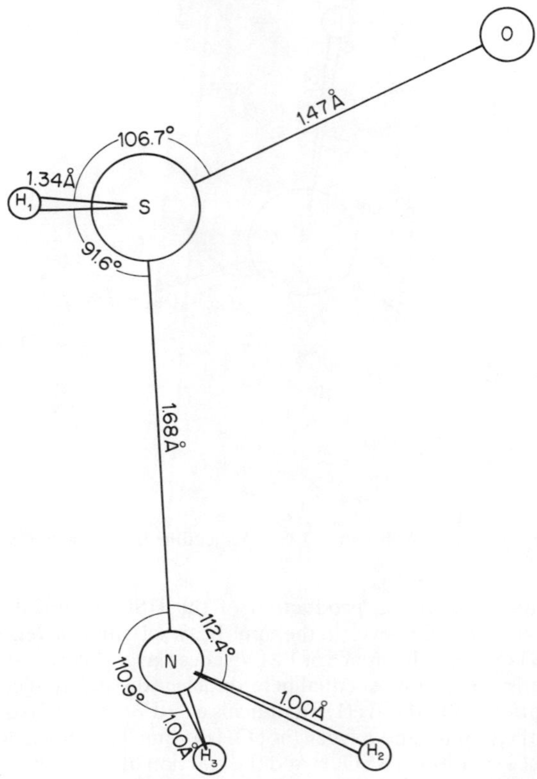


FIGURE 6. $HSONH_2$, structure **9**, drawn in NSO plane, dihedral angles (deg): $NSH_1O = 113.9$, $H_2NSO = 39.8$, $H_3NSO = 272.2$

Here, both the sulphur and nitrogen atoms each have two attached groups and a lone pair of electrons. The optimized geometry gives maximum staggering of the bonds and lone pair electrons, where the orientation of the $NH_2$ hydrogens as bracketing the oxygen atom on sulphur seems to determine the specific conformation about the S—N bond. The rotation profile, however, was not explored.

Three possible tautomers of sulphinamide **9** are possible. The first, HSOHNH (**10**) (see Figure 7), transfers a hydrogen atom from the amine to the oxygen atom to form a hydroxyl group. In classical bonding structures the sulphur–nitrogen bond thereby takes on double-bonding character. The consequent shortening of the S—N bond is evident in Table 2. The second tautomer transfers the sulphur-attached hydrogen atom to the oxygen atom to form the sulphide, $SOHNH_2$ (**11**) (see Figure 8). In both these cases the S—O bond length increases compared to **9**, from 1.47 to 1.66 Å, as shown in Table 2. The third isomeric alternative to **9** is the amide analogue to the sulphone form of sulphinic acid, $H_2SONH$ (**12**) (see Figure 9). Here, both S=O and S=NH have double-bond character, as indicated and as can be seen from a comparison of bond lengths in Table 2.

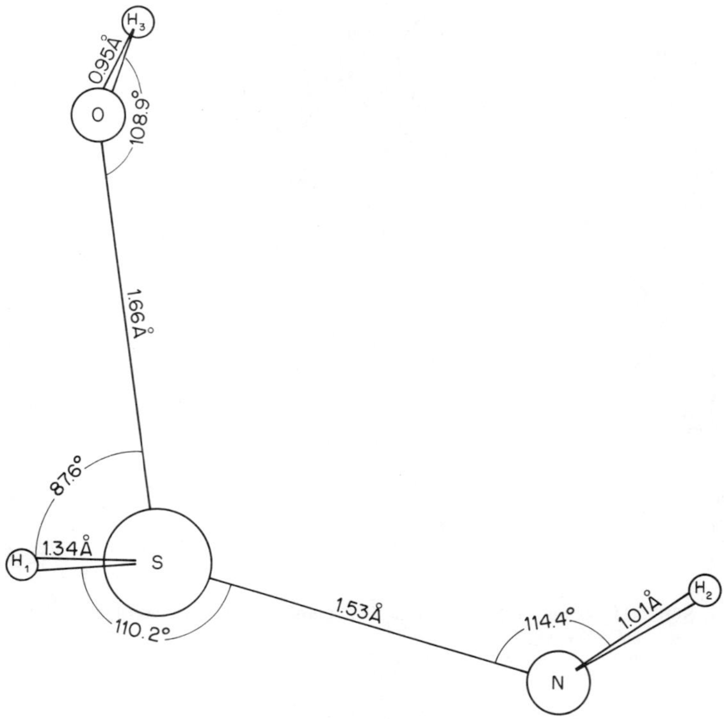


FIGURE 7. HSOHNH, structure **10**, drawn in NSO plane, dihedral angles (deg): $NSH_1O = 115.1$, $H_2NSO = 58.6$, $H_3OSH_1 = 189.4$

The relative stabilities of the four isomers are seen from Table 1 to be $SOHNH_2$ (**11**) > $HSONH_2$ (**9**) > HSOHNH (**10**) > $H_2SONH$ (**12**). Once again, the low-oxidation-state sulphur hydroxy compound (**11**) is MP2/6-31 + G* calculated to be most stable, by 0.8 eV over the classical $HSONH_2$ (**9**). Here, the difference in dipole moments is not large so that the effects of solvation are not clear-cut. Again, as with the acid, alkylation at the sulphur must preferentially stabilize the sulphinamide **9** form. These questions bear further investigation. The optimum S—N bond length for each of the four tautomers clearly distinguishes between the single and partial double-bond character structure for this bond. The two imide structures are highest in energy at 1.6 eV (**10**) and 2.8 eV (**12**) relative to the sulphide form.

Removal of a S—H proton to give the $SONH_2^-$ (**13**) anion is calculated to involve 15.1 eV (SCF) and 14.8 eV (MP2) energy, respectively. These numbers are close to the corresponding proton affinity values of $SOOH^-$. Thus, the substitution of the amine for the hydroxyl group does not greatly affect the (gas phase) acidity of the S—H proton.

Hydrogen atom dissociation from the sulphinamide can lead to two radicals, $\cdot SONH_2$ (**14**) (see Figure 10) and HSONH· (**15**) (see Figure 11). Energetically, the $\cdot SONH_2$ radical is calculated (MP2/6-31 + G*) to be more stable by an unequivocal 2.2 eV. As expected, the unpaired spin resides here on both the sulphur and oxygen atoms while for HSONH· the unpaired electron is localized on the nitrogen atom. The MP2/6-31 + G* calculated dissociation energy to form the $\cdot SONH_2$ radical is 2.2 eV compared to 4.4 eV for HSONH·.

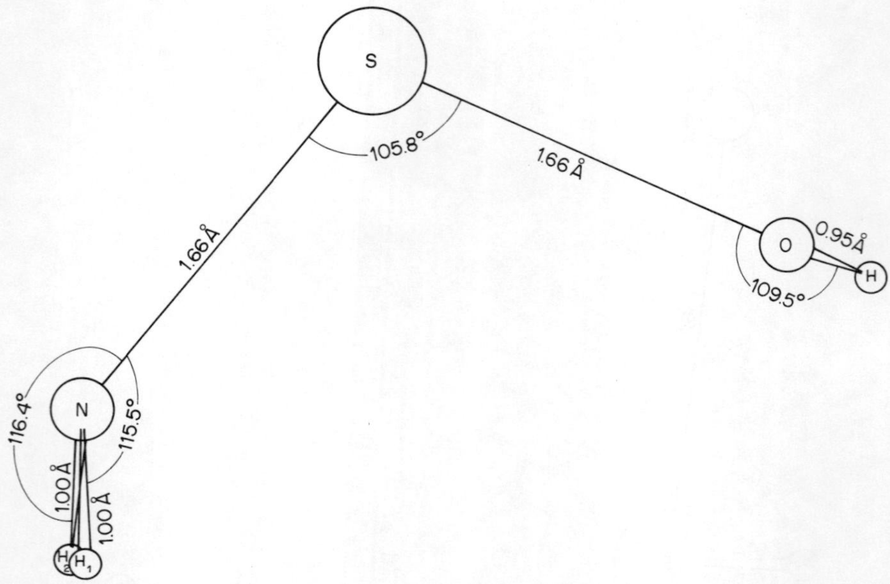


FIGURE 8. $SOHNH_2$, structure **11**, drawn in NSO plane, dihedral angles (deg): $H_1NSO = 66.0$, $H_2NSO = -69.1$, $HOSN = 90.9$

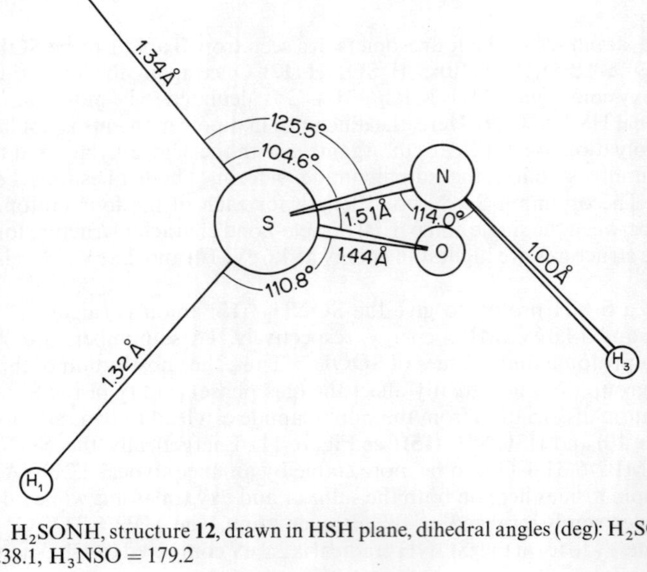


FIGURE 9. $H_2SONH$, structure **12**, drawn in HSH plane, dihedral angles (deg): $H_2SOH_1 - 105.5$, $NSOH_1 = 238.1$, $H_3NSO = 179.2$

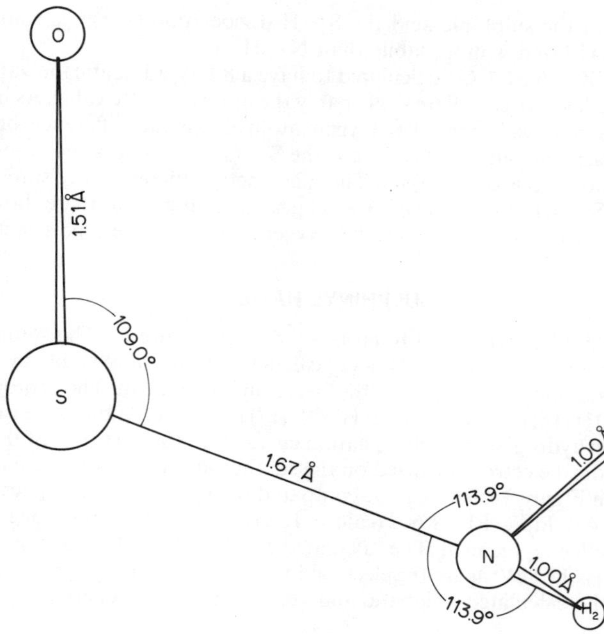


FIGURE 10. $\cdot SONH_2$, structure **14**, dihedral angles (deg): $H_1NSO = 64.2$, $H_2NSO = 296.1$

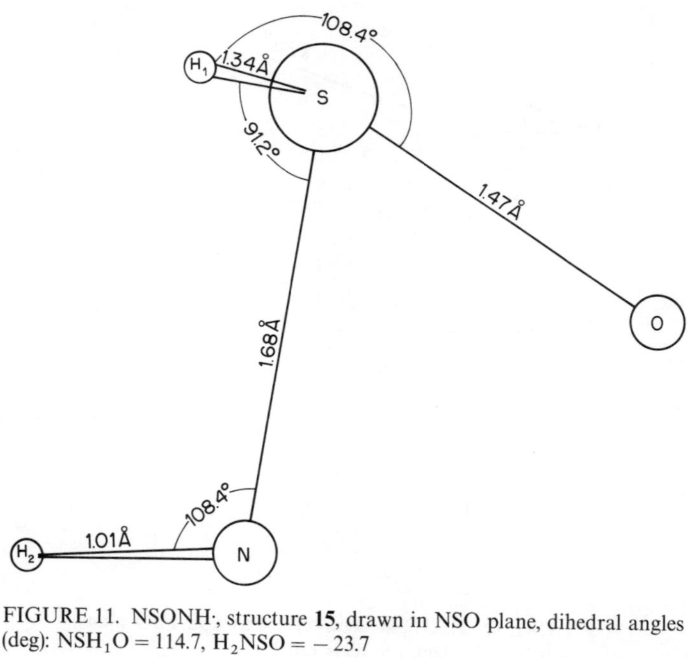


FIGURE 11. $NSONH\cdot$, structure **15**, drawn in NSO plane, dihedral angles (deg): $NSH_1O = 114.7$, $H_2NSO = -23.7$

In comparison with the sulphinic acid the S—H dissociation energy is similar but, as expected, the O—H bond is more labile than N—H.

$HSONH_2$ (**9**) is SCF/6-31 + G* calculated to have a 8.1 eV adiabatic ionization energy which, on the MP2 level, rises to 9.6 eV, probably the more accurate value. As in HSOOH (**1**) the electron is removed from the oxygen atom. Here the difference between the adiabatic and vertical energies is only 0.3 eV, the S=O bond length increases by 0.09 Å and the S—N bond decreases by 0.06 Å. These geometry differences are similar to those calculated for HSOOH and indicate the expected changes in these bond lengths accompanying electron ionization from the oxygen atom in these type systems.

## V. SULPHINYL HALIDES

The simplest sulphinyl fluoride, HSOF (**17**), is shown in Figure 12. The sulphide isomer FSOH (**18**) (see Figure 13) is MP2/6-31 + G* calculated to be more stable by only 0.4 eV where, again, the sulphinyl tautomer has the larger dipole moment. The proton affinity of $FSO^-$ (**19**) to FSOH (**18**) is 14.6 eV and to HSOF (**17**) is 14.2 eV. At the same calculational level the homolytic hydrogen atom dissociation energy for HSOF (**17**) → ·SOF (**20**) + H is 2.0 eV with the radical electron localized on the sulphur atom. The structure of the ·SOF radical is shown in Figure 14. This can be compared with experimental values[9], *R*(S—O) = 1.452 (calc. = 1.45), *R*(S—F) = 1.602 (calc. = 1.59) and < FSO = 108.3 deg (calc. = 107 deg), showing excellent agreement. The calculated MP2/6-31 + G* ionization potential of **17** to $HSOF^+$ (**21**) is 11.1 eV; again the electron ionized is from the oxygen atom, and only 0.1 eV separates the calculated adiabatic and vertical ionization energies.

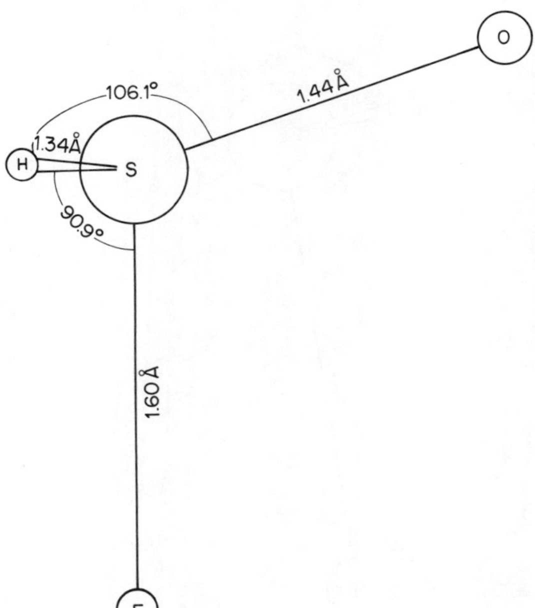


FIGURE 12. HSOF, structure **17**, drawn in FSO plane, dihedral angle (deg): FSHO = 110.2

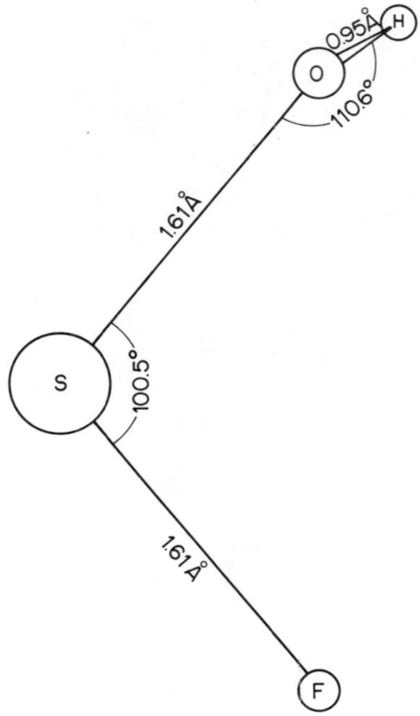


FIGURE 13. FSOH, structure **18**, drawn in FSO plane, dihedral angle (deg): FSOH = 83.1

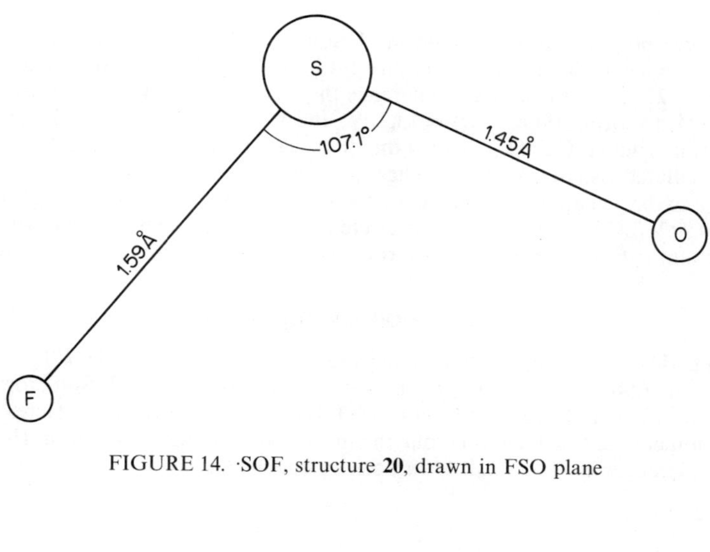


FIGURE 14. ·SOF, structure **20**, drawn in FSO plane

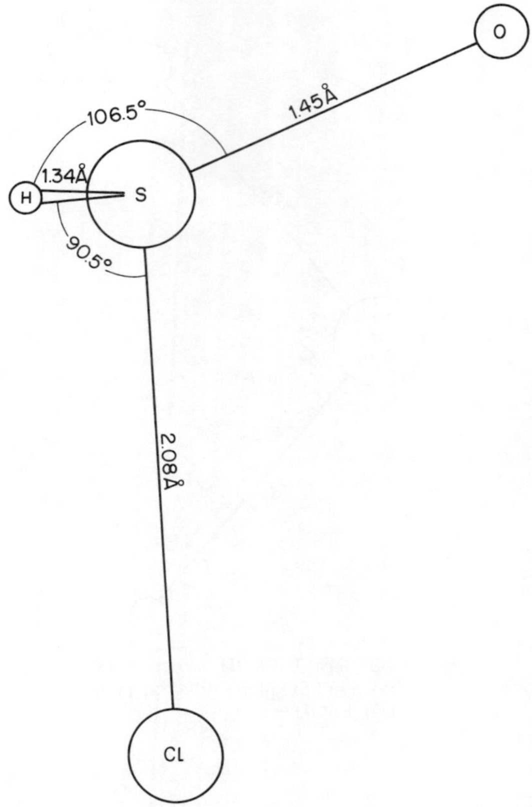


FIGURE 15. HSOCl, structure **22**, drawn in ClSO plane, dihedral angle (deg): ClSHO = 111.7

The corresponding sulphinyl chloride HSOCl (**22**) is shown in Figure 15. Here, the hydroxyl isomer ClSOH (**23**) (see Figure 16) is MP2/6-31 + G* calculated to be more stable than **22** by 0.7 eV, a larger difference than for the fluoride. The proton affinity of $SOCl^-$ (**24**) to form HSOCl (**22**) is 13.7 eV, and to ClSOH is 14.4 eV. Hydrogen atom dissociation from HSOCl (**22**) to form the ·SOCl (**25**) radical (see Figure 17) is MP2/6-31 + G* calculated also to take 2.0 eV, where the radical electron is distributed over both the sulphur and oxygen atoms. The calculated MP2/6-31 + G* ionization energy to $HSOCl^+$ (**26**) is 10.7 eV (SCF = 9.1 eV), with the ejected electron missing from the oxygen atom on the resultant cation. The adiabatic–vertical energy spread here is 0.3 eV.

## VI. THIOSULPHINIC ACID

Two isomeric forms of thiosulphinic acid were also examined, HSOSH (**27**) (see Figure 18) and HSSOH (**28**) (see Figure 19). From Table 1 we find that the HSSOH form is 0.4 eV (SCF) or 0.2 eV (MP2) more stable than HSOSH (6-31G* basis set level). Here, the relative dipole moments are such as to favour the more stable isomer in solution. However, the energy differences are too small for a decisive conclusion.

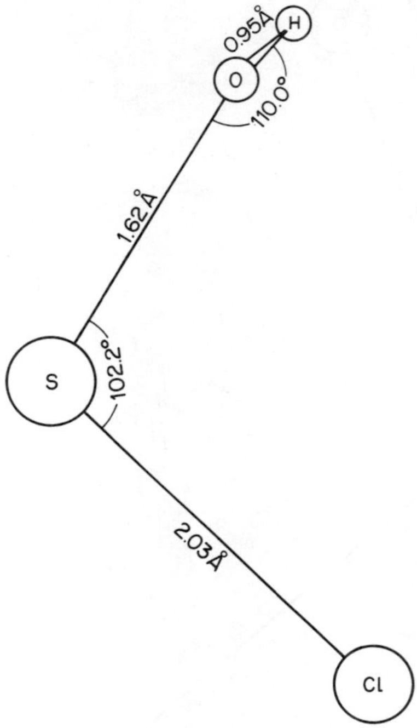


FIGURE 16. ClSHO, structure **23**, drawn in ClSO plane, dihedral angle (deg): ClSOH = 84.6

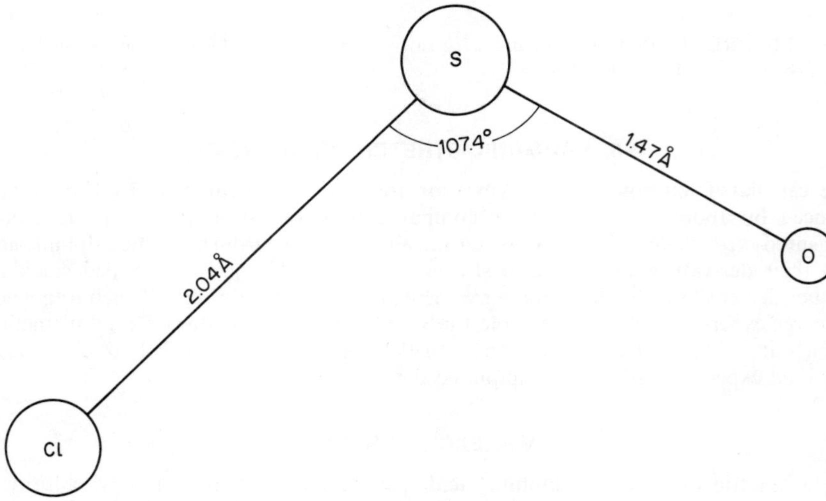


FIGURE 17. ·SOCl, structure **25**, drawn in ClSO plane

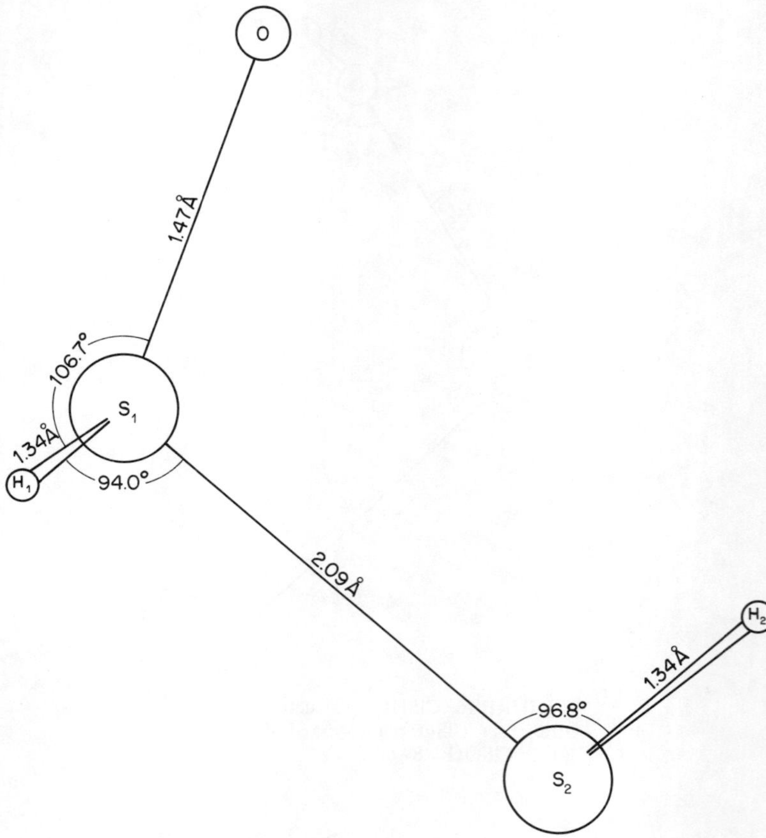


FIGURE 18. HSOSH, structure **27**, drawn in SSO plane, dihedral angles (deg): $S_2S_1H_1O = 112.8$, $H_2S_2S_1O = 45.2$

## VII. HARMONIC STRETCH FREQUENCIES

The calculated harmonic stretch vibration frequencies presented in Table 4, suitably reduced by about 10–12%, can be compared to infra-red spectroscopic frequencies measured experimentally[10–13]. These latter values are taken from aliphatic sulphinic acids and their derivatives and are also shown in Table 4. The properly scaled calculated frequencies are seen to be in good agreement with the general range of such frequencies observed experimentally. For example, the S=O stretch is calculated (after adjustment) to absorb at about 1,080 $cm^{-1}$ compared to the approximately 1,000–1,100 $cm^{-1}$ range observed experimentally for the sulphinyl derivatives[10].

## VIII. EXCITED STATES

The geometric structures of sulphinic acid **1**, sulphinamide **9** and sulphinyl chloride **22** were examined at the UHF/6-311G* level in their open-shell triplet states. Surprisingly, all three molecules were found to be dissociative. Sulphinic acid **1** dissociates smoothly in the

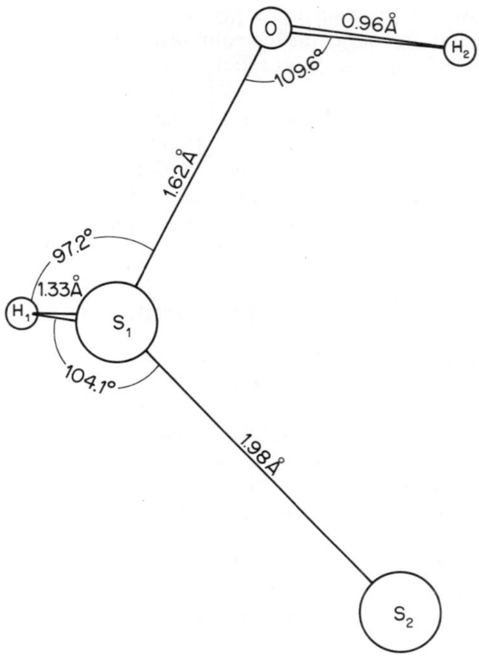


FIGURE 19. HSSOH, structure **28**, drawn in OSS plane, dihedral angles (deg): $OS_1H_1S_2 = 111.3$, $H_2OS_1S_2 = 22.9$

geometry optimization by homolytically breaking the S—OH bond. Even at a S—OH bond length of 2.97 Å the spin operator $\hat{S}^2$ value is 2.09 where the exact value of $S(S+1)$ with the spin quantum number $S = 1$ (triplet state) is 2.0. This small deviation from the theoretically correct spin-squared expectation value is a good indication that correlation effects are not playing a large role. ROHF calculations (which force the exact $\hat{S}^2$ value) give

TABLE 4. Calculated harmonic vibrational stretch frequencies[a]

| Group | HSOOH | $HSONH_2$ | HSOF | HSOCl | Experimental |
|---|---|---|---|---|---|
| S—O(H) | 892 | | | | 810–870[b] |
| S=O | 1,232 | 1,209 | 1,377 | 1,278 | 990–1,090[b] |
| S—H | 2,743 | 2,804 | 2,803 | 2,803 | 2,550[c] |
| O—H | 3,993 | | | | 3,700[b] |
| N—H | | 3,718/3,824 | | | 3,100–3,200[d] |
| S—N | | 797 | | | – |
| S—F | | | 877 | | 710–745[e] |
| S—Cl | | | | 506 | 438–489[e,f] |

[a]In $cm^{-1}$.
[b]From Reference 10.
[c]From Reference 1. Suggested to be possibly a mistaken assignment.
[d]From Reference 11.
[e]From Reference 12.
[f]From Reference 13.

the same S—OH dissociation result upon geometry optimization in the excited triplet state. The vertical excitation energy (at the equilibrium ground state geometry) is 4.9 eV (39,500 $cm^{-1}$) and corresponds to a S→O transition according to the charge shifts in the population analysis.

For the amide, UHF/6-311G* dissociation in the lowest energy triplet state is in the H—S bond and even at a H—S bond length of 3.74 Å, $\hat{S}^2 = 2.02$. Again, ROHF calculations show the same results as UHF with regard to which bond is dissociating. For the sulphinyl chloride the same level calculations predict S—Cl dissociation, analogous to the S—OH dissociation in HSOOH. For both the amide and the chloride the vertical excitation (energy = 4.9 eV and 3.1 eV, respectively) is in the O→S direction.

## IX. HYDROGEN-BONDED COMPLEXES

The characterization of the hydrogen bonding interaction between sulphinic acid or sulphinamide with water or methanol should give some primitive information on the solvation of these compounds, although the conformations that are important in the 1:1 complex may not be typical of actual solutions. Nonetheless, it is of interest to see the relative stabilities of each of the relevant groups in the sulphinic compounds to bind to water or methanol. Based on previous work on formamide with water or methanol[14] it is reasonable to expect that in the 1:1 complex the cyclic double hydrogen bond complexes will be most stable.

Two such gradient optimized complexes between sulphinic acid and water are shown in Figures 20 and 21 (structures **29** and **30**, respectively) and in Table 5. These structures involve simultaneously either the ···OS and ···OH or the ···OH and ···HS groups in H-

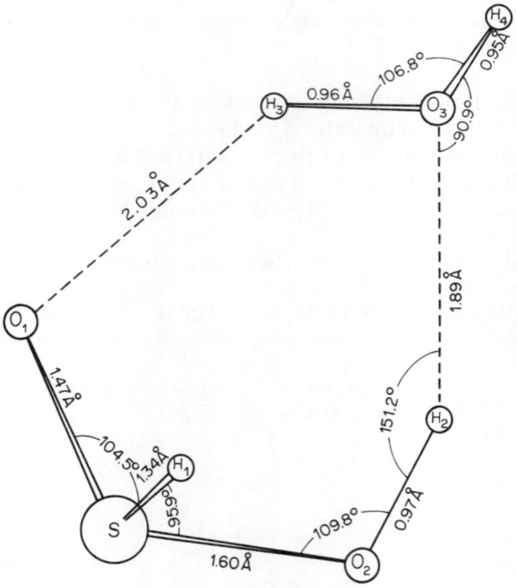


FIGURE 20. $H_2O + HSOOH$, structure **29**, drawn in OOO plane, dihedral angles (deg): $O_2SH_1O_1 = 111.8$, $H_2O_2SO_1 = 38.5$, $O_3H_2O_2S = -9.6$, $H_3O_3H_2O_2 = -17.4$, $H_4O_3H_3O_1 = 136.5$

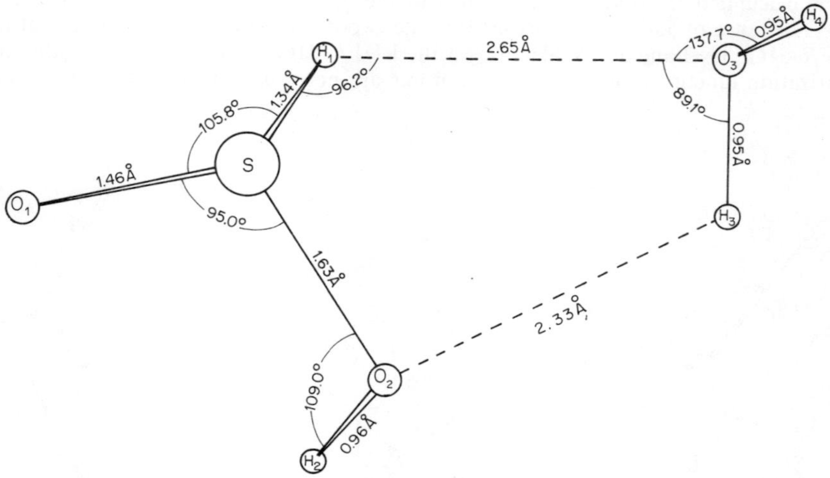


FIGURE 21. $H_2O + HSOOH$, structure **30**, drawn in OOO plane, dihedral angles (deg): $O_2SH_1O_1 = 110.0$, $H_2O_2SO_1 = 29.6$, $O_3H_1SO_2 = -62.2$, $H_3O_3H_1S = 59.3$, $H_4O_3H_1S = -53.6$

bonding with water. Attempts to find a simultaneous ···HS and ···OS or ···HS and ···O(H)S H-bonded structure with water eventually optimized to give one of the two structures shown in the Figures. However, their existence as stationary points on the multi-dimensional H-bonded surface cannot be ruled out.

TABLE 5. Hydrogen-bonded complexes

| | Energies (a.u.) | | | |
|---|---|---|---|---|
| | 6-31G* | | 6-31G** | |
| | SCF | MP2 | SCF | MP2 |
| $HSOOH + H_2O$ | | | | |
| 50 Å[a] | −624.313150 | −624.981708 | −624.337606 | −625.027722 |
| **(29)** | −624.334727 | −625.008114 | −624.358707 | −625.053183 |
| **(30)** | −624.321020 | −624.992493 | −624.345433 | −625.038247 |
| $HSOOH + CH_3OH$ | | | | |
| 50 Å[a] | −663.338045 | −664.129494 | | |
| **(31)** | −663.359429 | −664.156366 | | |
| **(32)** | −663.355205 | −664.151668 | | |
| **(33)** | −663.345752 | −664.140584 | | |
| $HSONH_2 + H_2O$ | | | | |
| 50 Å[a] | −604.488187 | −605.141220 | | |
| **(34)** | −604.505753 | −605.162942 | | |
| **(35)** | −604.495170 | −605.151525 | | |

[a]Optimization carried out at a fixed 50 Å distance between H and O.

The calculated H-bond energy (relative to the optimized dimer complex at a fixed H ··· O distance of 50 Å) for the stronger of the two complexes is 13.5 kcal mol$^{-1}$ at the RHF/6-31G* level and 13.2 kcal mol$^{-1}$ at the RHF/6-31G** level. Thus, the additional polarization functions on the hydrogen atoms are not crucial to the SCF result. The

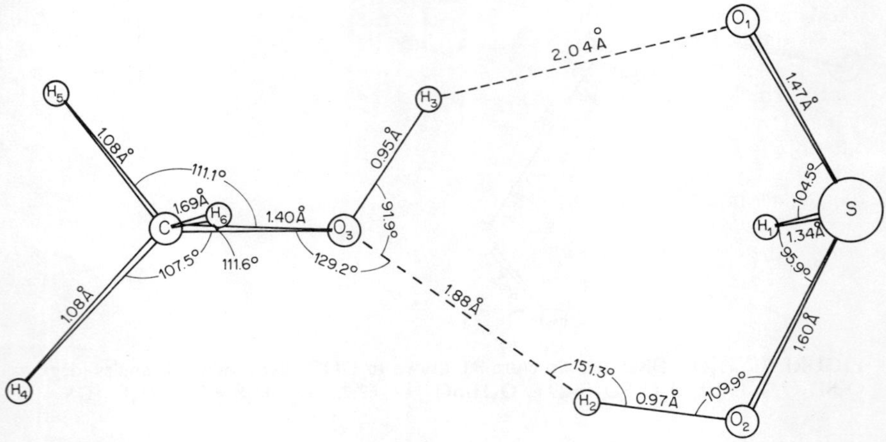


FIGURE 22. $CH_3OH + HSOOH$, structure **33**, drawn in OOO plane, dihedral angles (deg): $O_2SH_1O_1 = 111.9$, $H_2O_2S_1O_1 = 39.0$, $O_3H_2O_2S = -10.5$, $H_3O_3H_2O_2 = -17.4$, $CO_3H_2O_2 = 223.7$, $H_4CO_3H_3 = 180.8$, $H_5CO_3H_3 = 62.1$, $H_6CO_3H_3 = -59.9$

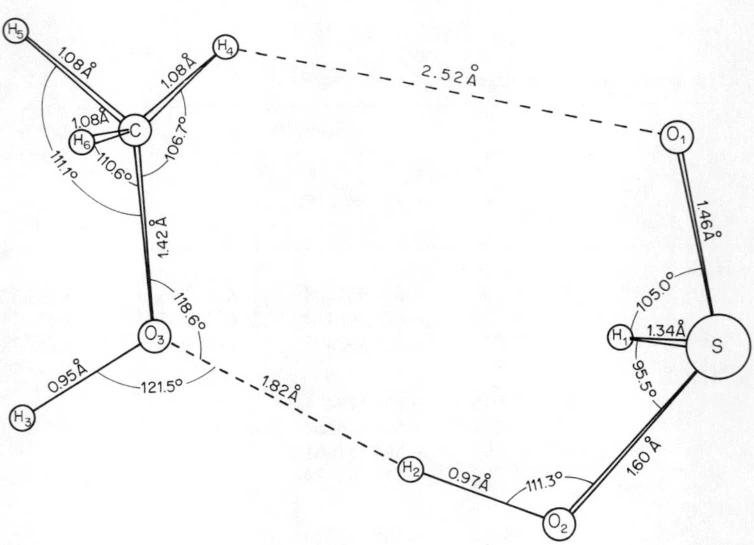


FIGURE 23. $CH_3OH + HSOOH$, structure **32**, drawn in OOO plane, dihedral angles (deg): $O_2SH_1O_1 = 112.7$, $H_2O_2S_1O_1 = 44.2$, $H_4CO_3H_3 = 174.6$, $H_5CO_3H_3 = 55.5$, $H_6CO_3H_3 = -62.2$

corresponding MP2 level calculations give H-binding energies of 16.6 and 16.0 kcal mol$^{-1}$, respectively, for the smaller and larger basis sets. The structure of the complex in Figure 20 shows two normal H-bonded distances (1.9–2.0 Å) where, in comparison with Figure 1, the sulphinic acid monomer geometry is only slightly perturbed. The weaker complex, on the other hand, shows unusually long hydrogen bonded distances (2.44–2.6 Å) and its H-bond energy (MP2/6-31G**) is only 6.6 kcal mol$^{-1}$, which is nearer a single hydrogen bond rather than a cyclic double bond.

Three stable 1:1 cyclic structures were also found for the methanol–sulphinic acid complex. The first, structure **31** (see Figure 22), corresponds to structure **29** in the water complex. The MP2/6-31G* H-bond energy for the methanol complex is 16.9 kcal mol$^{-1}$ compared to 16.6 kcal mol$^{-1}$ for the comparable level water complex. The second methanol complex **32** (see Figure 23) has a short single SOH $\cdots$ $OH_2$ bond length of 1.82 Å and a long oxygen (SO $\cdots$) methyl group hydrogen distance of 2.52 Å. The MP2/6-31G* calculated H-bond energy is 13.9 kcal mol$^{-1}$, which is stronger than a single H-bond energy of the short distance type. The nature of the interaction with the methyl group is not clear and the rotational barrier about the strong H-bond was not explored. The third

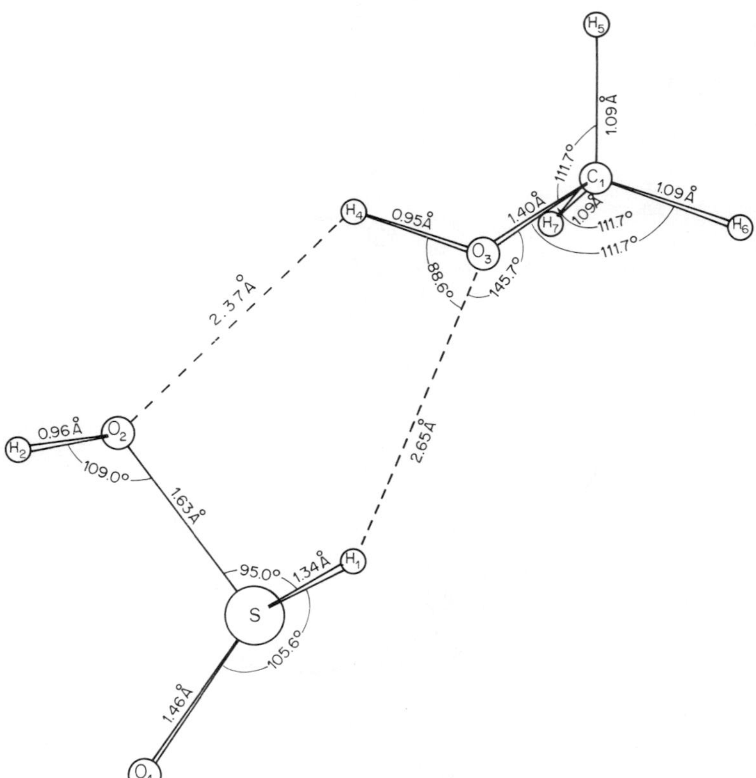


FIGURE 24. $CH_3OH + HSOOH$, structure **33**, drawn in SOO plane, dihedral angles (deg): $O_4S_1H_2O_3 = 109.9$, $H_5O_4S_1O_3 = 29.2$, $O_6H_2S_1O_4 = -61.8$, $H_7O_6H_2S_1 = 58.9$, $C_8O_6H_2S_1 = 65.6$, $H_9C_8O_6H_2 = 179.9$, $H_{10}C_8O_6H_2 = 57.6$, $H_{11}C_8O_6H_2 = -61.3$

methanol structure **33** (see Figure 24) couples SH···O and S(H)O···HO bonds with an H-bond energy of 7.0 kcal mol$^{-1}$. The H-bond lengths are longer than usual. The analogous complex with water instead of methanol shown in Figure 21 has very similar features.

Two stable 1:1 cyclic structures were also found for the sulphinamide–water complex. The more stable, structure **34** (see Figure 25), has two normal H-bonding distances with a MP2/6-31G* binding energy of 13.6 kcal mol$^{-1}$. This smaller H-bond energy relative to the corresponding acid complex is consistent with the observation[15] that the hydroxyl group generally makes a better hydrogen bond than the amino group. However, this H-bond energy for the sulphinamide is larger than for the corresponding formamide complex with water[14]. The second structure **35** (see Figure 26) involves the SH···O interaction simultaneously with SN···H, having an H-bond energy of 6.5 kcal mol$^{-1}$. Both H-bonds are weaker than in **34** and the result is essentially a single H-bond energy. The SH···O distance is very similar to that in structure **30** involving sulphinic acid.

In all these cyclic structures, constraints imposed by the ring conformation may force longer H-bond lengths. Some of the cyclic structures studied here have weak binding energies and there could be single H-bonded structures that are more stable. However, it is unlikely that any single H-bonded structures exist with binding energies as large as **29**, **31**

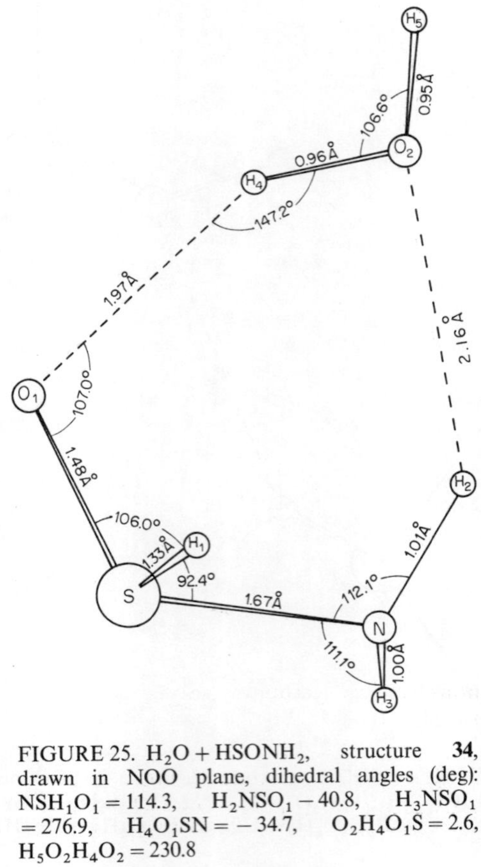


FIGURE 25. $H_2O + HSONH_2$, structure **34**, drawn in NOO plane, dihedral angles (deg): $NSH_1O_1 = 114.3$, $H_2NSO_1 - 40.8$, $H_3NSO_1 = 276.9$, $H_4O_1SN = -34.7$, $O_2H_4O_1S = 2.6$, $H_5O_2H_4O_2 = 230.8$

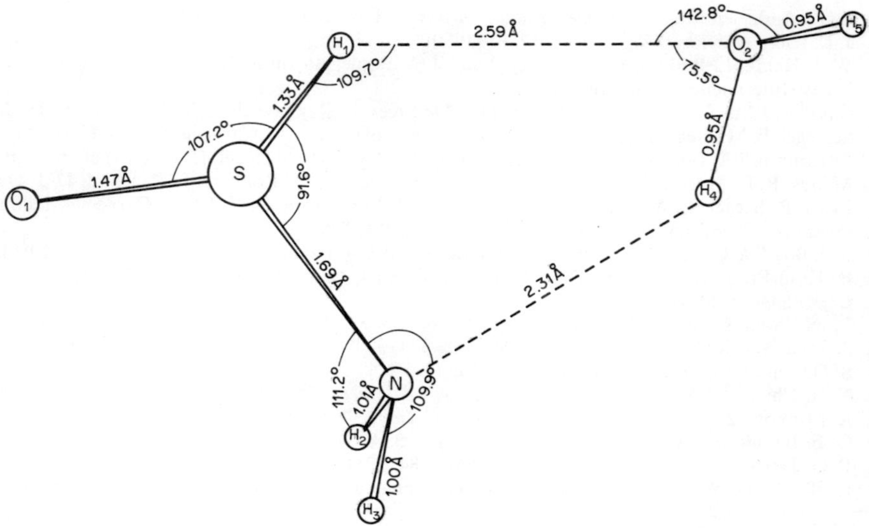


FIGURE 26. $H_2O + HSONH_2$, structure **35**, drawn in NOO plane, dihedral angles (deg): $NSH_1O_1 = 112.9$, $H_2NSO_1 = 40.9$, $H_3NSO_1 = 281.0$, $O_2H_1SN = -56.8$, $H_4O_2H_1S = 48.0$, $H_5O_2H_1S = -50.0$

and **34**. These relatively stable rigid cyclic structures may be observed experimentally. The analogous complexes for the (computed) more stable sulphide forms were not explored.

## X. EPILOGUE

A number of surprises were uncovered in this study of the simplest prototype sulphinic acid, sulphinamide and sulphinyl halides. Some of them are remarked upon in the text and clearly require further investigation. The work presented here just scratches the surface of this interesting class of compounds and even in what was presented here a great deal more analysis can be applied. For example, the different implications of the three-dimensional structurality (non-planarity) and conformational orientation of bonds and lone pairs, electronic structure (or frontier orbital) analysis of the relative stabilities of the various tautomers and the isomerization paths, electrostatic and charge density difference maps (for studying incipient nucleophilic or electrophilic attack), a better description of the S=O bond (double bond vs. $S^+O^-$), etc., are all fertile grounds for a deeper understanding of the systems studied here and a basis for looking at more complicated (and more realistic) sulphinyl systems. Further studies are currently being carried out here on the parent aromatic species, phenyl sulphinic acid and phenyl sulphinamide.

## XI. ACKNOWLEDGEMENTS

Many of the calculations were carried out by Ms. Marie Rose Hajnal. The Figures were generated by Dr. Tova Hoz and Ms. Marie Rose Hajnal.

## XII. REFERENCES

1. S. Oae and N. Kunieda, in *Organic Chemistry of Sulfur* (Ed. S. Oae), Plenum, New York, 1977, p. 603.

2. K. K. Andersen, *Comprehensive Organic Chemistry*, Barton and Ollis, 1979.
3. J. L. Kice, *Adv. Phys. Org. Chem.*, **17**, 65 (1980).
4. W. J. Hehre, L. Radom, P.v.R. Schleyer and J. A. Pople, *Ab Initio Molecular Orbital Theory*, Wiley–Interscience, New York, 1986.
5. *Gaussian 82*: J. S. Binkley, M. J. Frisch, D. J. De Frees, K. Raghavachari, R. A. Whiteside, H. B. Schlegel, E. M. Fleuder and J. A. Pople, Department of Chemistry, Carnegie-Melton University, Pittsburgh, PA. *Gaussian 86*: M. J. Frisch, J. S. Binkley, H. B. Schlegel, K. Raghavachari, C. F. Melius, R. L. Martin, J. J. P. Stewart, F. W. Bobrowicz, C. M. Rohlfing, L. R. Kahn, D. J. De Frees, R. Seeger, R. A. Whiteside, D. J. Fox, E. M. Fleuder and J. A. Pople, Carnegie-Mellon Quantum Chemistry Publishing Unit, Pittsburgh, PA, 1984.
6. R. J. Boyd, A. Gupta, R. E. Langler, S. P. Lownie and J. A. Pincock, *Can. J. Chem.*, **58**, 331 (1980).
7. B. Kojii-Prodii, A. L. Spek, D. Wijkens, G. Tadema, C. Y. Elsevier and D. Vermeer, *Acta Crystallogr.*, **C40**, 1841 (1984).
8. M. N. Ponnuswamy and J. Trotter, *Acta Crystallogr.*, **C41**, 915 (1985).
9. S. Sakai and K. Morokuma, *Chem. Phys.*, **52**, 33 (1980).
10. S. Detoni and D. Hadzi, *J. Chem. Soc.*, 3163 (1955).
11. Y. H. Chiang, Y. S. Luloff and E. Schipper, *J. Org. Chem.*, **34**, 2397 (1969).
12. R. Minkowitz and R. Lekies, *Z. Anorg. Allg. Chem.*, **537**, 169 (1968).
13. G. E. Binder and A. Schmidt, *Spectrochim. Acta*, **33A**, 815 (1977).
14. P. G. Jasien and W. J. Stevens, *J. Chem. Phys.*, **84**, 3271 (1986).
15. H. Basch and W. J. Stevens, *The structure of glycine–water H-bonded complexes*, submitted.

The Chemistry of Sulphinic Acids, Esters and their Derivatives
Edited by S. Patai


CHAPTER **3**

# Sulfinic acids and their derivatives. Stereochemistry and chiroptical properties

ABRAHAM NUDELMAN

*Department of Chemistry, Bar-Ilan University, Ramat Gan 52100, Israel*



## I. INTRODUCTION

The chemistry of chiral sulfinic acid derivatives has been reviewed up to 1979[1]. This chapter covers mainly the publications from 1979 to 1988. The reader is referred also to other relevant recent reports dealing with various aspects of chiral sulfinic derivatives, by Krauthausen[2], Drabowicz and Mikolajczyk and coworkers[3–6], Kice[7], Cinquini and Colonna[8,9], Solladie[10–12], Posner[13,14] and Hiroi[15].

The present review is restricted to aspects dealing with the chirality of the following types of sulfinic acid derivatives: (a) sulfinates, (b) sulfinamides, (c) sulfinimidamides and (d) sulfinyl halides. These sulfur derivatives, in addition to their interesting properties and chemistry, are frequently found as useful intermediates in stereoselective and stereospecific total syntheses of many natural products.

## II. SULFINATES

### A. Syntheses, Separation of Diastereomers, Resolution

Most frequently chiral sulfinates are isolated by separation of the diastereomeric mixtures obtained upon treatment of activated sulfinyl derivatives RS(O)X with optically active alcohols. Some of the sulfinates are oils, whereas others are crystalline materials. Moreover, acid-catalyzed equilibration of diastereomeric mixtures can sometimes afford a single enantiomeric sulfinate.

The pivotal member of the family of chiral sulfinates is menthyl *p*-toluenesulfinate. A large-scale synthesis and separation of the crystalline (*S*)-(−)-diastereomer **2** has been reported[16]. The (*R*)-(+)-enantiomer may be obtained analogously from (*R*)-menthol.

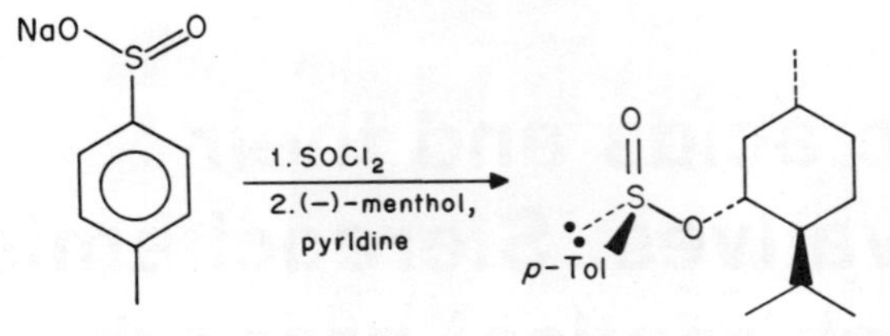


(1) (2)

Additional methods for the preparation of menthyl *p*-toluenesulfinate involve the conversion of *p*-toluene sulfinic acid to various activated derivatives when reacted with: (a) 2-chloro-1-methylpyridinium iodide, (b) diethyl azodicarboxylate/$Ph_3P$, (c) primary and secondary amines/DCC, (d) γ-saccharine chloride or (e) *p*-toluenesulfonic acid/DCC, followed by treatment with menthol[17–20] (Scheme 1).

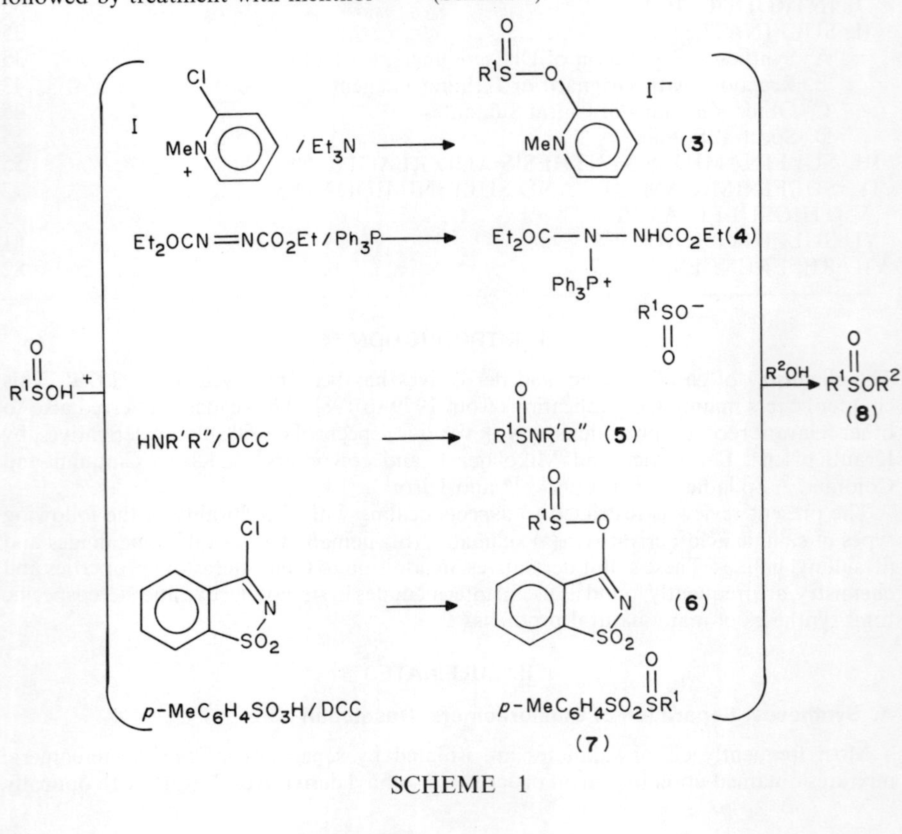


SCHEME 1

The $BF_3$ etherate-catalyzed reaction of alcohols with sulfinamide **9** proceeds with inversion of the configuration to give high yields (69–99%) of chiral sulfinates **10** in enantiomeric excesses ranging from 53–86%. The mildness of the conditions permit the use of alcohols bearing acid-labile acetal group[21] (Scheme 2).

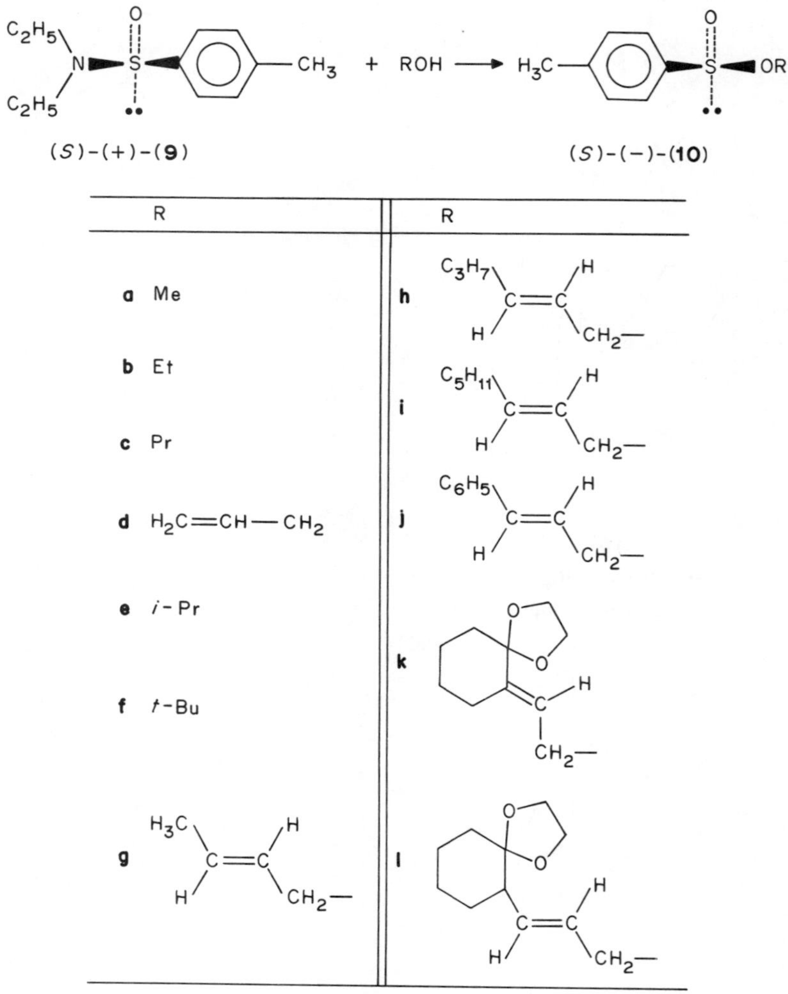

SCHEME 2

Diastereomeric mixtures of arene sulfinates **13** and **14** derived from D-glucose derivatives **11** and **12** were prepared analogously from the corresponding arenesulfinyl chlorides. Only the (*R*)-(+)-mesitylene ester **15** was obtained in a crystalline, optically pure state[22].

It has been established that nucleophilic substitution at chiral sulfinyl sulfur proceeds commonly with inversion of the configuration[1]. An unexpected high degree of retention of the configuration has been observed in the acid-catalyzed alcoholysis of sulfinamide **16**.

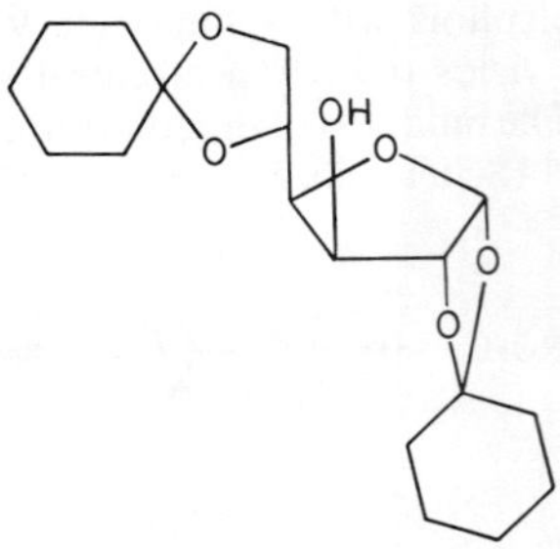


(11)

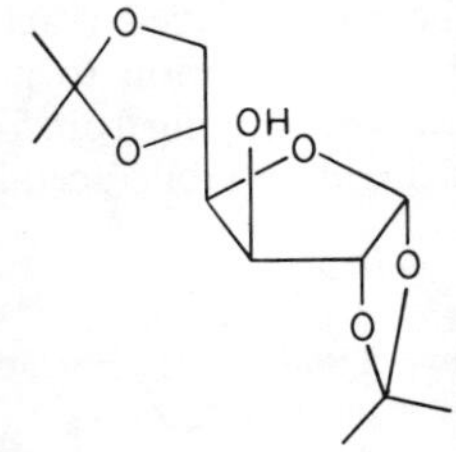


(12)

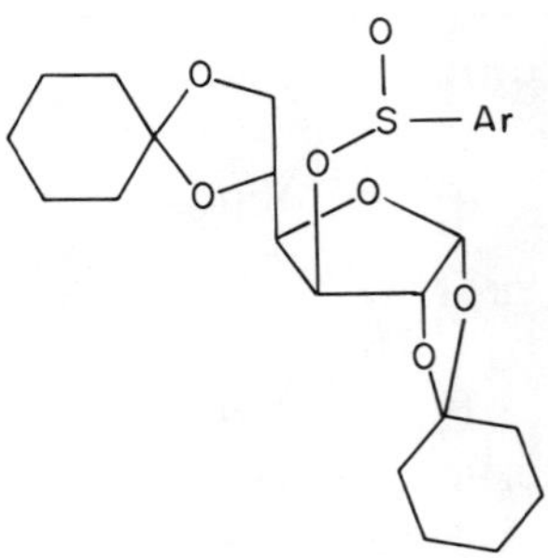

(13)

Ar = Ph; *p*-ClC$_6$H$_4$—;
*p*-An; 1-Naph; 2-Naph

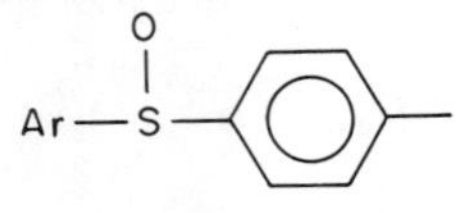

(14)

Ar = Ph; *p*-ClC$_6$H$_4$—; *p*-An;
1-Naph; 2-Naph

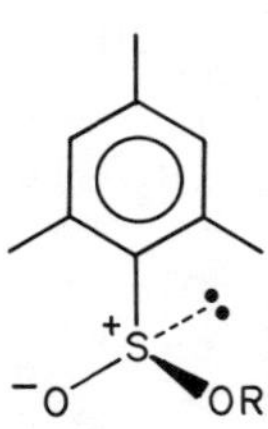

*R*-(15)

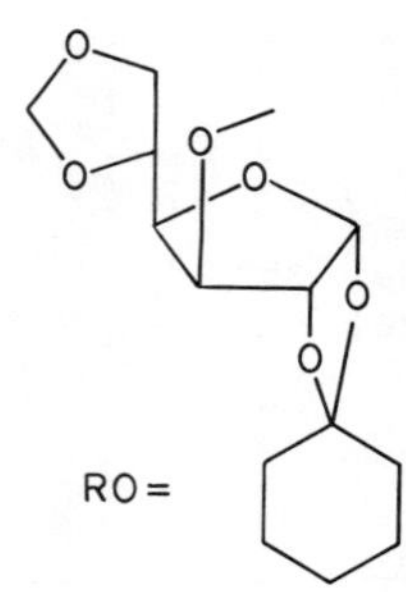


The stereochemical course of the reaction is influenced by the addition of silver perchlorate, whose presence favors the inversion product. Moreover, isopropyl alcohol and cyclohexanol, which in the absence of the salt gave predominant retention, in its presence gave mostly inversion. Addition of other inorganic salts had a dramatic effect on the stereochemical outcome of the reaction. Here the nature of the cation and the anion as well as the polarity of the solvent are of importance, where polar solvents favor retention. The retention of the configuration which is sometimes observed has been attributed to the formation of sulfurane intermediates that undergo rapid pseudorotation (Scheme 3)[23]. A review of these arguments has been published by Mikolajczyk[24]. (See Tables 1a–d.)

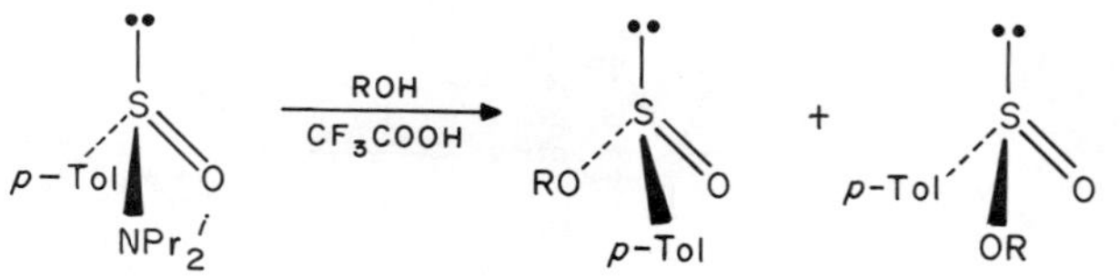


(+)-(S)-(**16**) (−)-(S)-(**17**) (+)-(R)-(**17**)

Other diastereomeric mixtures of sugar sulfinates **19** are formed from **18** upon displacement of triflate with the sulfinyl anion $CF_3SO_2^-$. However, the identity of the individual diastereomers was not established[25].

TABLE 1a. Reaction of (+)-(S)-**16** with alcohols catalyzed by trifluoroacetic acid

| Tol-S(O)NPr$^i_2$ [α]$_D$ (o.p.%) | Tol-S(O)OR R | [α]$_D$ (o.p.%) | Stereo selectivity | Inversion or retention |
|---|---|---|---|---|
| 94.4° (45.3) | Me | −35.0° (17.0) | 37.5% | 68.75% Inv |
| 94.4° (45.3) | Et | −7.1° (3.4) | 7.5% | 53.75% Inv |
| 94.4° (45.3) | Pr$^n$ | −13.9° (7.25) | 16.0% | 58.00% Inv |
| 95.0° (45.35) | Bu$^i$ | −2.7° (1.4) | 3.0% | 51.50% Inv |
| 94.4° (45.3) | Pr$^i$ | +15.8° (7.9) | 17.4% | 58.70% Ret |
| 86.9° (42.3) | Pr$^i$-D$_6$ | +10.1° (4.6) | 10.9% | 55.45% Ret |
| 86.9° (42.3) | Pr$^i$-F$_6$ | +3.4° (1.7) | 4.0% | 52.00% Ret |
| 95.0° (45.35) | Hex$^c$ | +41.0° (22.4) | 49.0% | 74.50% Ret |
| 95.0° (45.35) | Pen$^c$ | +3.3° (1.8) | 4.0% | 52.00% Ret |
| 95.0° (45.35) | Et$_2$CH | −4.4° (2.3) | 5.0% | 52.50% Inv |

TABLE 1b. Reaction in the presence of $AgClO_4$

| R | Inv/Ret ratio |
|---|---|
| Me | 100/0 |
| Et | 91/9 |
| Pr | 100/0 |
| *i*-Pr | 82/8 |
| *c*-Hex | 65.5/34.5 |

TABLE 1c. Reaction with *i*-PrOH in various solvents

| Solvent | Inv/Ret ratio |
|---|---|
| $CHCl_3$ | 55/45 |
| $C_6H_6$ | 56/44 |
| *n*-$C_6H_{12}$ | 58/42 |
| $CH_3CN$ | 49/51 |

TABLE 1d. Reaction with *i*-PrOH in the presence of various salts

| Salt | Prevailing stereochemistry | Salt | Prevailing stereochemistry |
|---|---|---|---|
| $CoCl_2$ | 55% Ret | $Co(NO_3)_3$ | 73% Inv |
| $NiC_2O_4$ | 71% Ret | $Ni(NO_3)_2$ | 66% Inv |
| $Ag_2CO_3$ | 67% Ret | $AgClO_4$ | 82% Inv |
| $Ag_2Cr_2O_7$ | 67% Inv | $AgNO_3$ | 53% Inv |
| $Ag_2SO_4$ | 63% Ret | $AgClO_4$ | 82% Inv |
| $HgBr_2$ | 69% Ret | $Ce(NO_3)_3$ | 71% Ontv |
| $Cd(OAc)_2$ | 68% Ret | $CrCl_3$ | 50.5% Inv |

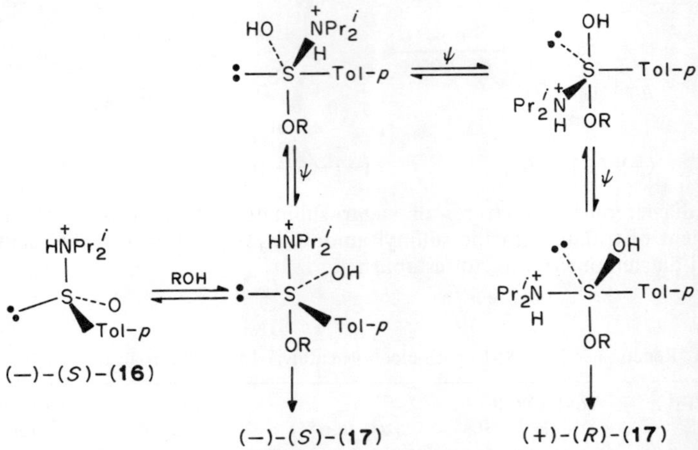

SCHEME 3. An A–E mechanism for acid-catalyzed alcoholysis of (−)-(*S*)-**16**

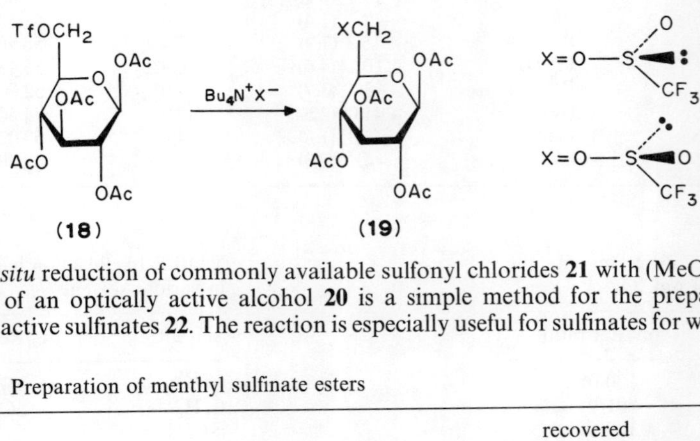

The *in situ* reduction of commonly available sulfonyl chlorides **21** with $(MeO)_3P$ in the presence of an optically active alcohol **20** is a simple method for the preparation of optically active sulfinates **22**. The reaction is especially useful for sulfinates for which there

TABLE 2. Preparation of menthyl sulfinate esters

| 1:R′ | Time (h) | Yield (%) | recovered menthol, (%) | diastereo-selectivity |
|---|---|---|---|---|
| *p*-Tol | 8 | 90 | 6 | 1.4:1 |
| 2-Naphthyl | 5 | 96 | 4 | 1.4:1 |
| *p*-An | 20 | 89 | | 1.3:1 |
| *p*-$ClC_6H_4$ | 1.5 | 92 | | 1.6:1 |
| 2,4,6-(*i*-Pr)$_3C_6H_2$ | 27 | 36 | 52 | |
| *p*-(*t*-Bu)$C_6H_4$ | 6.5 | 87 | | 1.5:1 |
| *o*-$MeO_2CC_6H_4$ | 6.5 | 48 | | 1.6:1 |
| 2,4,5-$Cl_3C_6H_2$ | 3 | 75 | 25 | 2.1:1 |
| 2,4,6-$Me_3C_6H_2$ | 15 | 70 | 21 | 1.5:1 |
| 8-Quinolyl | 4 | 52 | 41 | 1.9:1 |
| 2-Thienyl | 1.5 | 92 | | 1.8:1 |
| $CCl_3$ | 1 | 76 | 19 | 2.9:1 |
| $CH_3$ | 4 | 22 | | 1.7:1 |

are no readily available sulfinyl chloride precursors. A variety of menthyl sulfinates (Table 2) were thus prepared in up to 3:1 *S*/*R* diastereoselectivity[26].

$$\text{ROH} + \text{R'SO}_2\text{Cl} \xrightarrow{(\text{MeO})_3\text{P, Et}_3\text{N}} \text{R'S(O)OR}$$

**(20)** **(21)** **(22)**

A nonionic transition state **24** has been postulated for the coupling of sulfinyl halides with alkoxytrimethylsilanes[27].

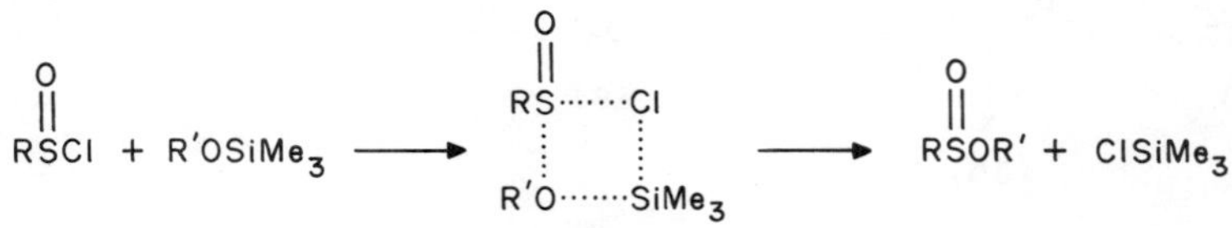


**(23)** **(24)**

In the presence of chiral amino alcohols **25**, *t*-butylmagnesium chloride has been found to be useful in two procedures for the stereoselective synthesis of optically active sulfinates[28,29]. The first method involves the reaction with symmetrical sulfites **26** to give chiral sulfinates **27** in up to 75% enantiomeric excess. The second procedure involves a kinetic resolution of racemic sulfinates **29** which under the reaction conditions gives chiral sulfoxides **30** (in up to 65% ee) leaving behind optically active unreacted sulfinate **29** (in up to 33 ee%). The optical purity of the sulfinates **29** was established by correlation with the optical rotation of the known chiral sulfoxide **31** (Scheme 4).

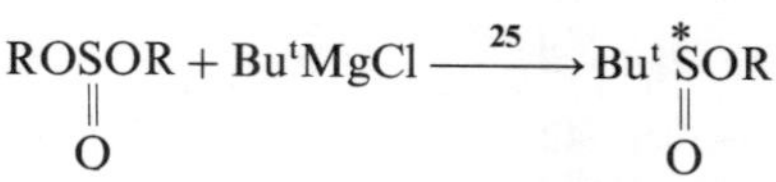


**(a)** (−)Quinine
**(b)** (−)Quinidine

**(26)** **(27)** **(25)**

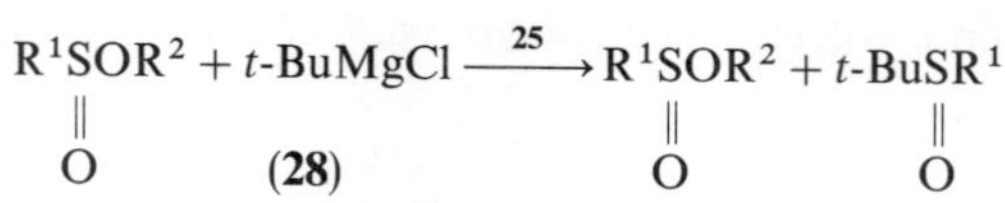


**(28)**

(±)-**(29)** (+) or (−)-**(29)** (+) or (−)-**(30)**

**(29)**

**(a)** $R^1 = p$-Tol, $R^2 = Me$
**(b)** $R^1 = p$-Tol, $R^2 = Et$
**(c)** $R^1 = p$-Tol, $R^2 = Pr^i$
**(d)** $R^1 = p$-Tol, $R^2 = Bu^t$
**(e)** $R^1 = p$-Tol, $R^2 = CH_2Bu^t$
**(f)** $R^1 = Ph$, $R^2 = Pr$
**(g)** $R^1 = Me$, $R^2 = Pr$
**(h)** $R^1 = Me$, $R^2 = CH_2Bu^t$

**(30)**

**(a)** $R^1 = Ph$
**(b)** $R^1 = p$-Tol
**(c)** $R^1 = Me$

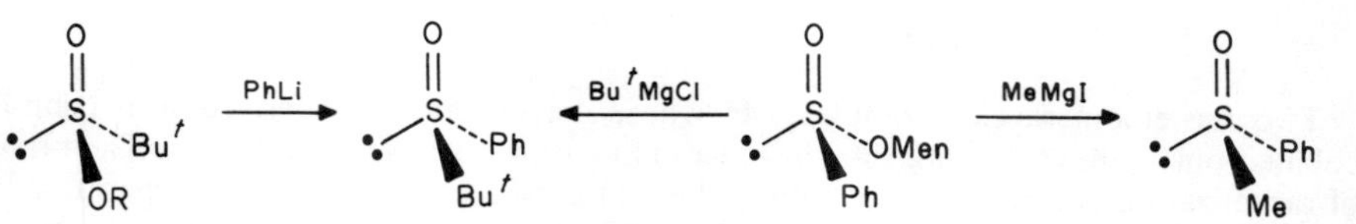


(+)-(*R*)-(**29**) (+)-(*R*)-(**31**) (−)-(*S*)-(**32**) (+)-(*R*)-(**33**)

SCHEME 4

Diastereomeric mixtures of sulfinate esters have been prepared by the reaction of α-halo-β-keto sulfones **35** with sulfinate anions[30]. The initial reaction, where the sulfinate anion attacks the halogen, produces sulfonyl halides **36** which react further with $ArSOO^-$ to give the reactive sulfinic–sulfonic mixed anhydrides **37**. Addition of an alcohol, such as menthol, provides the desired sulfinate ester **38**.

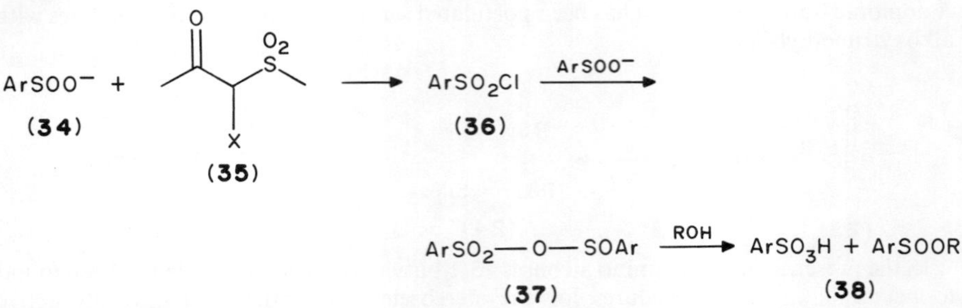


Arene sulfinic acids **39** react with alcohols in the presence of chiral carbodiimides to give optically active sulfinates **40**[31].

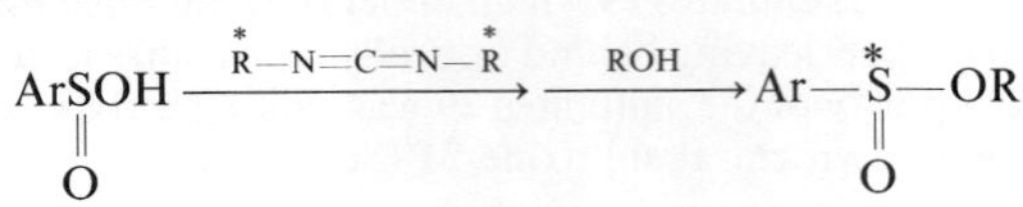


(39) (40)

Mixtures of menthyl sulfinate diastereomers **42** ($R^1$ = Me, Ph), with low asymmetric induction, were obtained[32] when disulfides **41** were oxidized with *m*-chloroperbenzoic acid and underwent subsequent oxidative alcoholysis promoted by NBS. The degree of asymmetric induction was determined by conversion of **43** to the corresponding sulfoxide **44**.

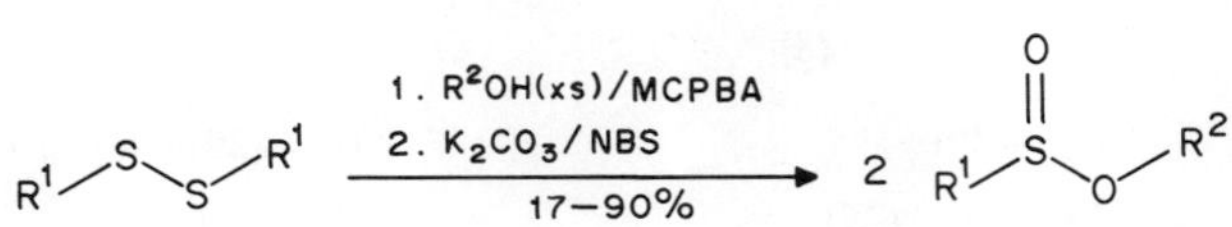


(41) (42)

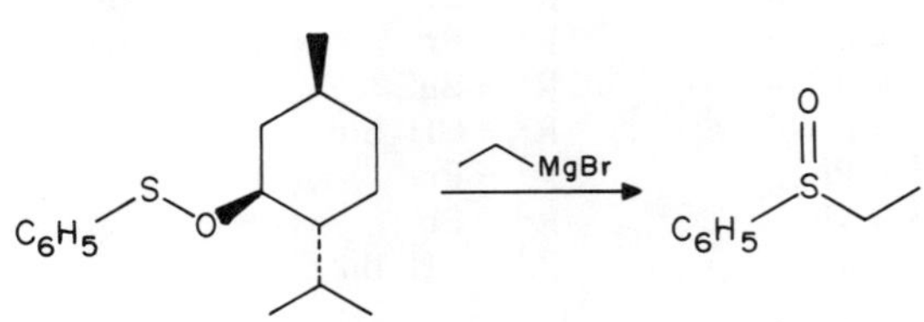


(43) (44)

Two diastereomeric $Co^{III}$ complexes **45**, whose stereochemistry is derived from a chiral cobalt atom (in the center, not shown) and a sulfinate group, have been isolated. Their rate of racemization is rather slow, enabling their successful chromatographic separation[33] (Scheme 5).

Direct partial resolution of various low molecular weight compounds, including sulfinate **47**, by selective formation of inclusion complexes with tri-O-thymotide (TOT) **46**

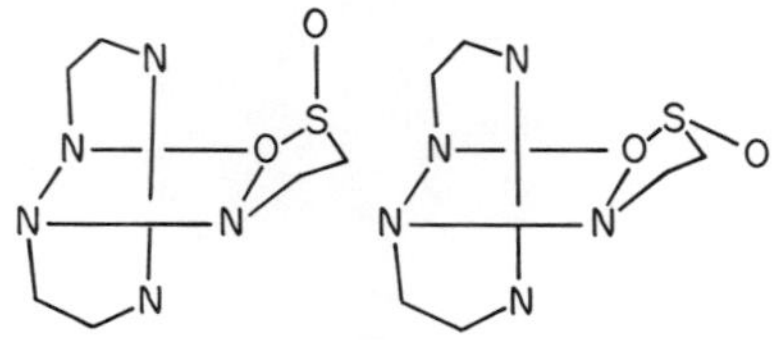

Λ-*R* (**45**) Λ-*S*

SCHEME 5. Two diastereomers of Λ-[Co(aesi-*N*, *O*)(en)$_2$]$^{2+}$

has been described. The clathrate crystals were stable up to 115 °C, but at 125 °C for 12 h complete racemization of the sulfinate took place. The enantiomerization was shown to proceed within the TOT cage cavity which provided appreciable stability to **47** toward racemization[34].

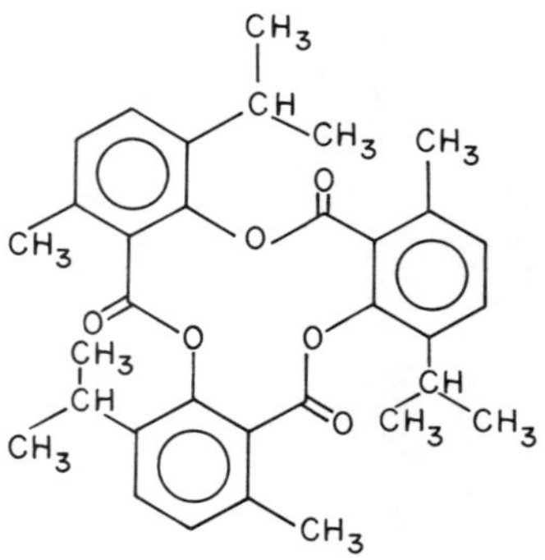

(**46**) TOT

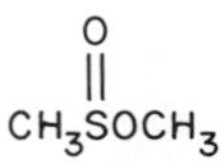

(**47**)

Thermolysis of penicillin sulfoxide **48** afforded desacetoxycephem **49** as well as a mixture of isomeric sulfinates **50** and **51**. Treatment of the sulfinates with methanesulfonic acid gave sulfoxide **52**, whereas reflux in DMF produced **49**[35] (Scheme 6).

An unexpected mixture of sulfinates **54a**, **b** in addition to sulfone **55** were obtained upon treatment of chloride **53** with sodium *p*-toluenesulfinate. The appearance of the sulfinates is attributed to the high reactivity of the alkylating reagent resulting in the formation of the products of kinetic control, i.e. sulfinates (58%), and lesser amounts of the thermodynamic controlled product, i.e. sulfone (28%). The absolute stereochemistry of **54a** was established by X-ray crystallography[36].

## B. Reactions with Grignard or Lithium Reagents

Optically active sulfoxides, commonly prepared by the reaction of chiral sulfinates with Grignard or lithium reagents, are the most important group of chiral sulfur derivatives. Many of them have been used as intermediates in synthetic sequences of natural products. The vast majority of the reported reactions have been carried out with (*S*)-(−)-menthyl *p*-toluensulfinate to give (*R*)-sulfoxides. A few examples of the use of the enantiomeric (*R*)-(+)-sulfinate to give the corresponding (*S*)-sulfoxides have been described. In Table 3 are listed the sulfoxides prepared from the (*S*)-(−)- and (*R*)-(+)-menthyl *p*-toluenesulfinates. Sulfoxides prepared analogously from other chiral sulfinates are presented in Table 4.

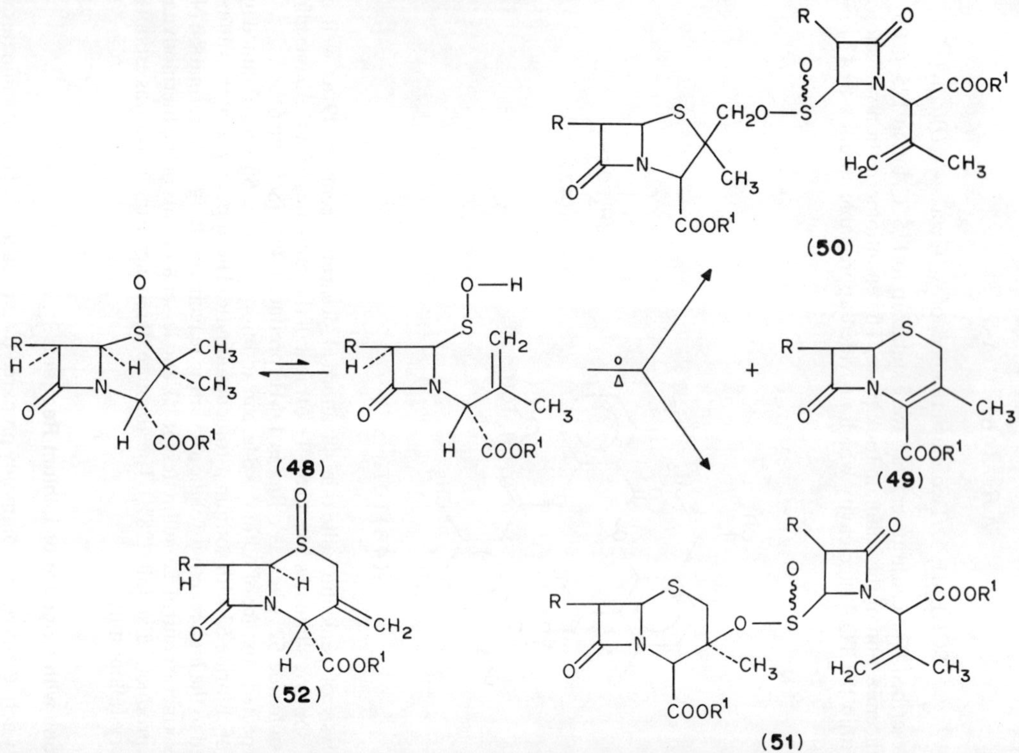


SCHEME 6. $R = C_6H_5OCH_2CONH$; $R^1 = CH_2C_6H_4$-$p$-$NO_2$; Q = *p*-benzoquinone or chloranil. Compounds **50a**, **b** and **51a**, **b** are epimeric at the sulfinyl sulfur atom.

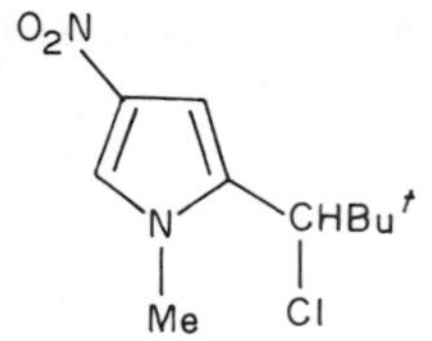


(53)

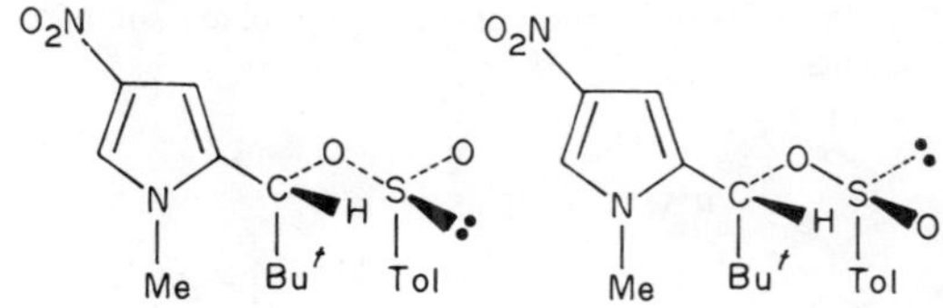


(a) (54) (b)

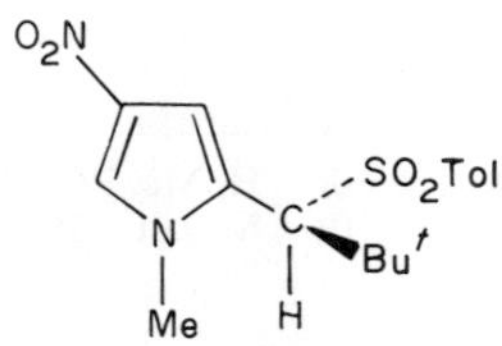


(55)

## C. Other Reactions of Chiral Sulfinates

An unusual double reaction of addition and substitution is instrumental in the synthesis of sulfoxides **56** when (*S*)-(−)-menthyl *p*-toluenesulfinate **2** is treated with acetylenic lithium compounds[102] (Scheme 7).

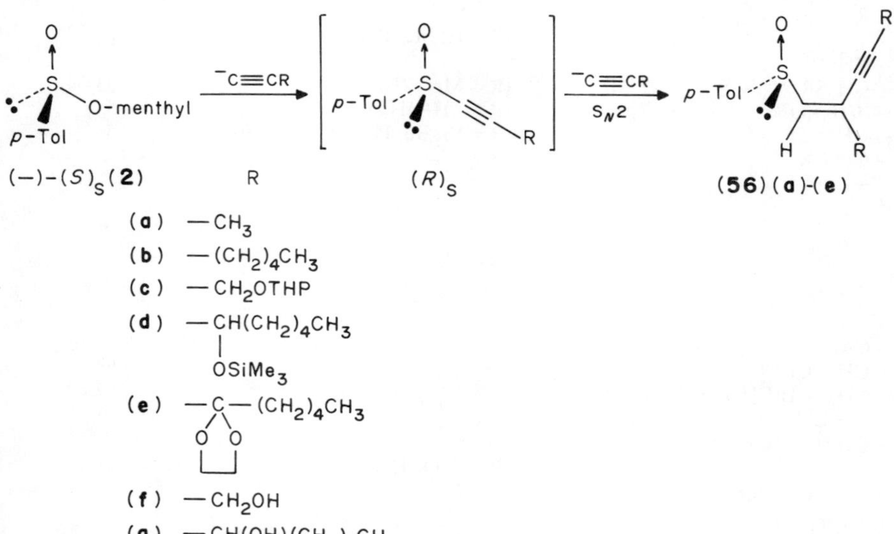


SCHEME 7

Chiral allylic sulfinates **59** obtained by $BF_3$ etherate catalyzed estrification of (*S*)-(+)-*N*,*N*-diethyl-*p*-tolylsulfinamide (*S*)-**57** undergo thermal rearrangement, preferably in DMF, to give sulfones **60** in good yield and with high stereospecificity[103]. The mechanism

TABLE 3. Products of reaction of menthyl *p*-toluenesulfinates with Grignard (R′MgX) or lithium (R′Li) reagents

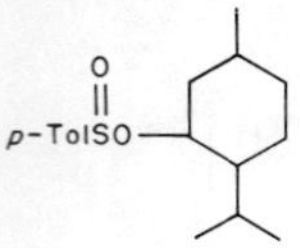

$+ R'M \longrightarrow p\text{-TolS(O)R}'$

(M = MgX or Li)

| Product *p*-TolS(O)R′ R′ | | References |
|---|---|---|
| $-CH_2R$ | (R = H, Me, Pr, *i*-Bu, Ph) | 37–41 |
| $-R$ | (R = Me, Et, *i*-Pr, Bu, Ph) | 42, 43 |
| $-CH(Me)Et$ | | 44 |
| 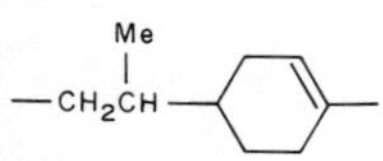 | | 45 |
| $-R$ | (R = Me, $MeOCH_2$, $ClCH_2$, $C_5H_{11}$, *t*-Bu) | 46, 47 |
| $-(CH_2)_5Me$ | | 48 |
| $-(CH_2)_5CHMe_2$ | | 49 |
| $-(CH_2)_{15}R$ | | 50 |
| $-CH(R)CN$ | (R = H, Me, Ph) | 51, 52 |
| $-CH_2SR$ | (R = Me, *p*-Tol) | 53, 54 |
| $-(CH_2)_3OR$ | (R = THP, H, TBDMS) | 55 |
| $-CH{=}CR^1R^2$ | (R = Pr, Bu, Pen, Hex, $Me_3Si-$) | 56 |
| $-C{\equiv}CR$ | | |
| 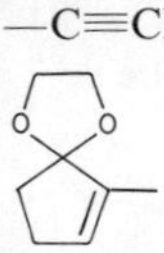 | | 16 |
| $-C(=CHMe)Ar'$ | | 57 |
| $-CH{=}CR^1R^2$ | | 58 |
| $-CH{=}CH(CH_2)_nOSi(t\text{-Bu})Me_2$ ($n = 1, 2$) | | 59, 60 |
| $-CH{=}C(CH_2OR)-C{\equiv}CCH_2OR$ | (R = $-OCH_2OMe$, 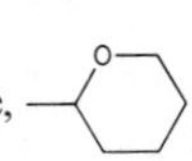) | 61 |
| $-CH_2P(O)R_2$ | (R = MeO, EtO, Ph, $Me_2N$) | 61–65 |
| $-CH(STol\text{-}p)P(O)(OEt)_2$ | | 66 |

TABLE 3. (*continued*)

| Product $p\text{-TolS(=O)R}'$ R′ | | References |
|---|---|---|
| $-CH_2C(=S)NMe_2$ | | 67 |
| $-CH_2C(=O)NR_2$ | [R = Me, $(CH_2)_4$, *i*-Pr, *t*-Bu] | 68, 69 |
| $-CH(R)C(=O)OBu\text{-}t$ | | 70–72 |
| 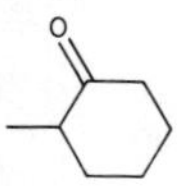 | | 73 |
| 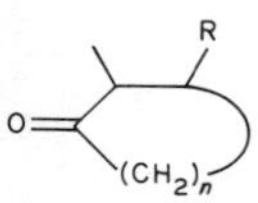 | R, *n*:<br>H 5, 6<br>Me 5 | 74 |
| $-CH_2C(=O)CH(R)COOMe$ | (R = H, Et) | 75 |
| $-CH_2S(=NMe)(=O)Ph$ | | 76 |
| $-CH_2C(OEt)=NOMe$ | | 77 |
| 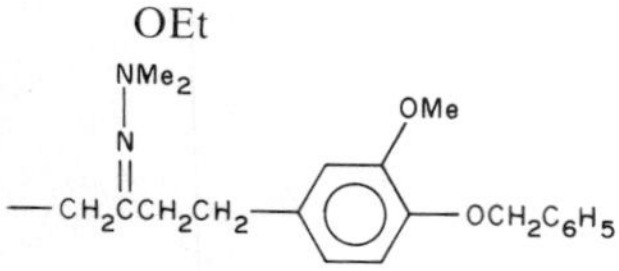 | | 78 |
| $-CH(R)C(R)=NNMe_2$ | | 79, 80 |
| $-C(R)=C(R)NHR$ | (R = H, Alk, Ar) | 81 |
| $-CH(R)C(R)=NR$ | | |
| 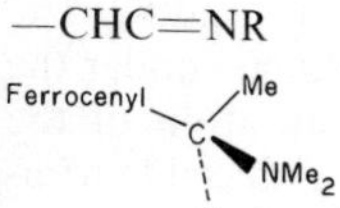 | | 82 |

TABLE 3. (*continued*)

| Product *p*-TolS(O)R′ R′ | | References |
|---|---|---|
| 4-Pyr; 3-Pyr | | 83 |
| 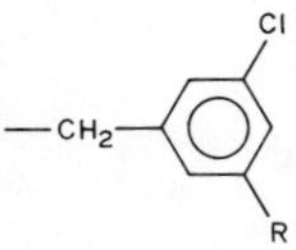 | (R = H, Cl) | 84 |
| 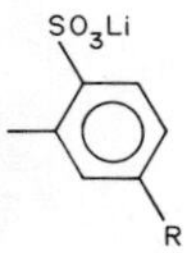 | (R = H, Me) | 85 |
| 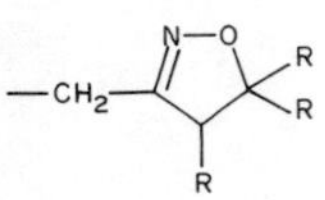 | (R = H, Alk, Ar) | 86–90 |
| 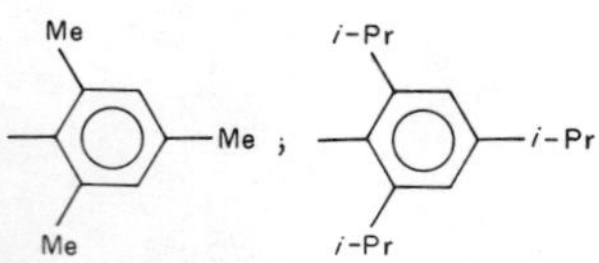 | | 91 |
| 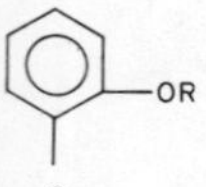 | (R = Me, $CH_2CH_2OMe$) | 92 |
| 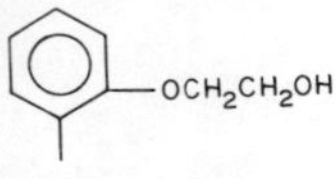 | | |
| 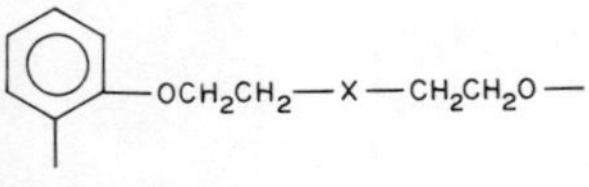 | (X = O, S, NMe, OP(O)Ph) | 93 |
| 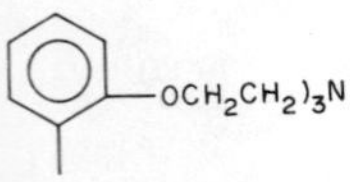 | | |

involves a concerted cyclic intramolecular [2, 3] rearrangement. Thus, (*S*)-**57a**, **c** and (*S*)-**57b**, **d** gave (*S*)-**60a**, **b** and (*R*)-**60a**, **b**, respectively. Subsequent reaction of the sulfones with sodium diethyl malonate in the presence of $Pd^0(PPh_3)_4$ gave mixtures of **61** and **62**. The direct $S_N2$ substitution reaction of the sulfinates with sodium diethyl malonate under the reaction conditions did not proceed at all[104,105]. The absolute configuration of the sulfones **60** was obtained by chemical correlations. Thus, sulfones **60** were reduced to **65a**–

**c**, which in turn had been prepared stereospecifically from chiral sulfinates **63a–c** via sulfides **64a–c** that underwent oxidation to **65a–c**.

Subsequently it was shown that the sulfinate–sulfone rearrangement is also catalyzed by Pd catalysts **66** and **67**, so that (*S*)-(−)-**59a, c, e** and (*S*)-(*m*)-**59b, d** gave, in high stereospecificity, good yields of the (*S*)-(+)- and (*R*)-(−)-sulfones **60**, respectively[106]. Surprisingly, sulfinate (*S*)-(−)-**59f** gave (*S*)-(+)-**60c**. The rearrangement took place even in THF, whereas in the absence of Pd catalyst the reaction proceeded only in hot DMF. In some cases, small amounts of sulfones **68** were also obtained. The mechanism of the

TABLE 4. Synthesis of sulfoxides by the reaction of chiral sulfinates with Grignard (RMgX) or lithium (RLi) reagents

$$R^a{-}\overset{O}{\overset{\|}{S}}{-}O{-}R^b + R^c{-}M \longrightarrow R^a{-}Tol{-}\overset{O}{\overset{\|}{S}}{-}R^c$$

(M = MgX or Li)

| $R^a$ | $R^b$ | $R^c$ | References |
|---|---|---|---|
| Ph<br>*p*-$ClC_6H_4$<br>*p*-$MeOC_6H_4$<br>1-Naph<br>2-Naph<br>mesityl | 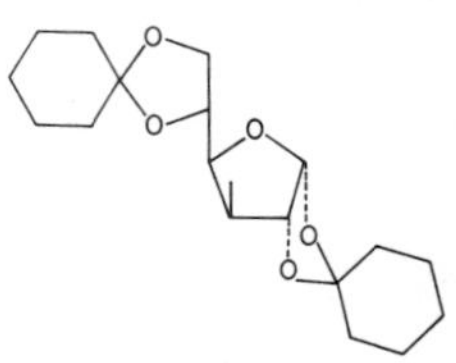 | *p*-Tol | 94 |
| *p*-$MeOC_6H_4$ | mesityl | 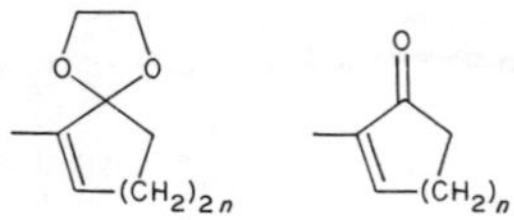 | 95 |
| 1-Naph | menthyl | 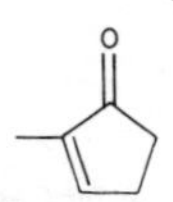 ; Me | 96, 97 |
| *p*-Tol | Me | 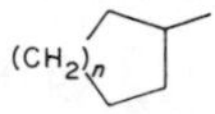 ($n$ = 1, 2, 3) | 98 |
| Me | cholesteryl | Pr, Bu, *i*-Bu, *p*-Tol, $PhCH_2$ | 99 |
| *p*-Tol<br>Ph | Me, Et, *i*-Pr | $MeS(O)CH_2$ | 100 |
| Me | $R^1C{\equiv}C{-}C(R^2)_2$<br>[$R^1$ = H, Ph;<br>$R^2$ = Me, $(CH_2)_5$] | Me, Et, *i*-Pr,<br>Bu, *t*-Bu, Ph | 101 |

| | (a) | (b) | (c) | (d) | (e) | (f) | (g) |
|---|---|---|---|---|---|---|---|
| $R^1$ = | Me | H | Pr | H | Pen | H | Ph |
| $R^2$ = | H | Me | H | Pr | H | Pen | H |

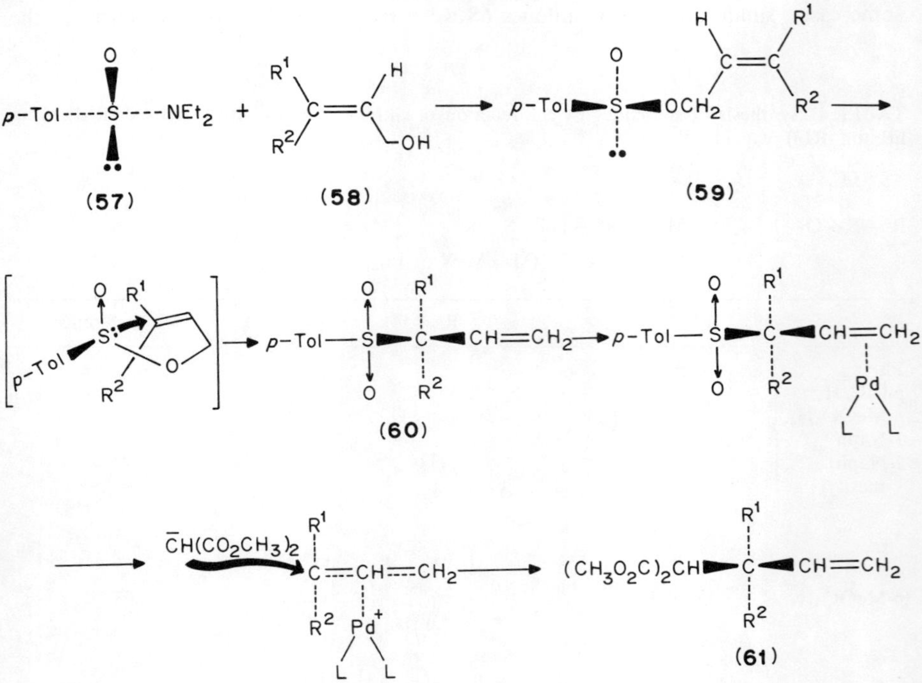


$RCH{=}CHCH_2CH(CO_2CH_3)_2$

(62) (a) $R = CH_3$
(b) $R = (CH_2)_2CH_3$

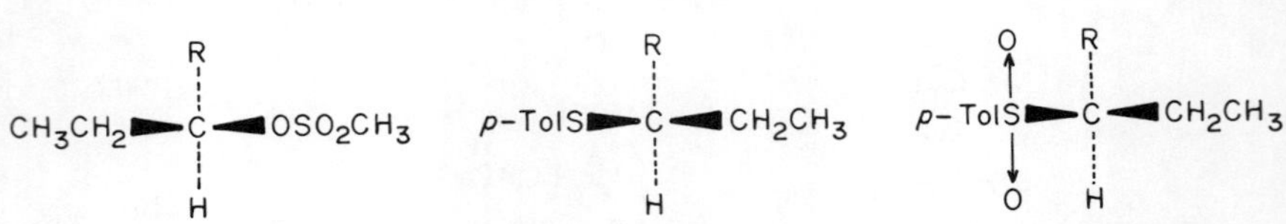


(63) (a)—(c) (64) (a)—(c) (65) (a)—(c)

(a) $R = CH_3$
(b) $R = (CH_2)_2CH_3$
(c) $R = (CH_2)_4CH_3$

reaction involves initial formation of a palladium chelate **69** followed by an ionic intermediate **70** (Scheme 8).

(*S*)-(−)-(**59**)

| | $R^1$ | $R^2$ |
|---|---|---|
| (**a**) | Me | H |
| (**b**) | H | Me |
| (**c**) | Pr | H |
| (**d**) | H | Pr |
| (**e**) | $[CH_2]_4Me$ | H |
| (**f**) | H | $[CH_2]_4Me$ |

(*S*)-(+)-(**60**)

| | R |
|---|---|
| (**a**) | Me |
| (**b**) | Pr |
| (**c**) | $[CH_2]_4Me$ |

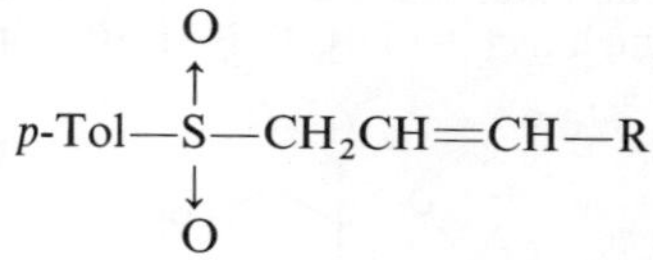

(**68**)

| | R |
|---|---|
| (**a**) | Me |
| (**b**) | Pr |
| (**c**) | $[CH_2]_4Me$ |

$[(Ph_3P)_4Pd)]$ (**66**)

$[(Ph_2PCH_2CH_2PPh_2)_2Pd]$

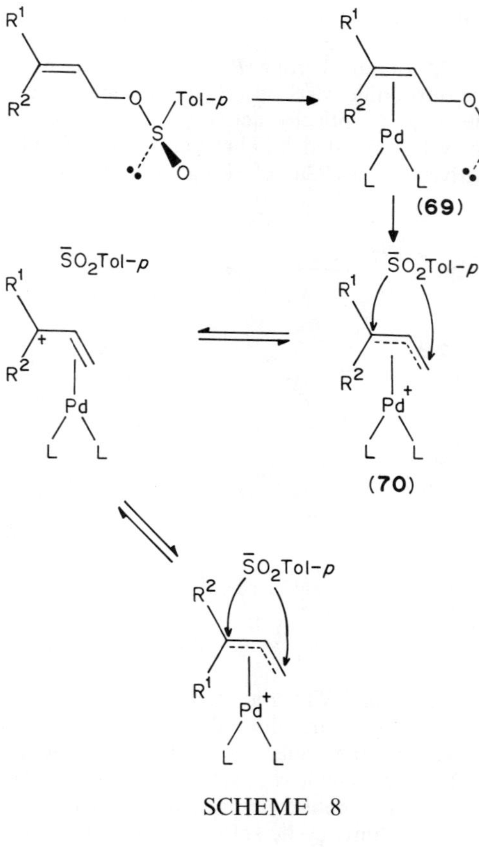

SCHEME 8

The high stereospecificity of the Pd-catalyzed sulfinate–sulfone rearrangement was further displayed with the synthesis of optically active sulfones **60a, b, c** from racemic sulfinates **59a, b, c, e**, in the presence of a mixture of $Pd^0(PPh_3)_4$ and a chiral Pd catalyst such as (−)-DIOP, NMDPP or (*S*)-MBDPP[107].

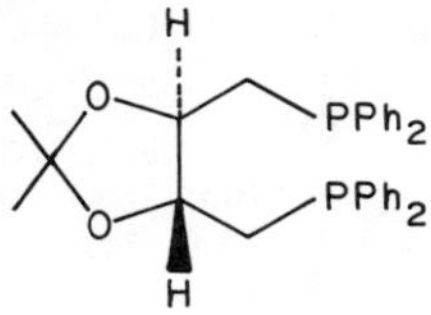


(−)-DIOP

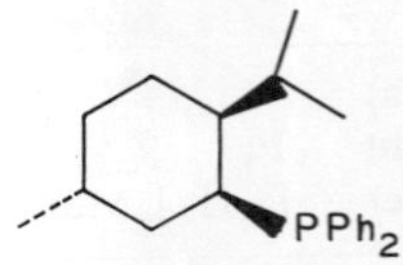


NMDPP

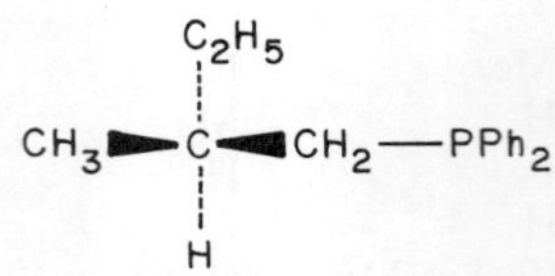


(*S*)-MBDPP

Irradiation of optically active methyl *p*-toluenesulfinate **71** resulted in a rapid decrease of optical activity but no photolysis. The photoracemization stems from rapid reversible formation of sulfinyl and menthyloxy radicals[108].

$$\underset{\substack{\downarrow\\ \text{O}}}{\text{R}\overset{*}{\text{S}}\text{OR}'} \xrightarrow{h\nu} [\underset{\substack{\downarrow\\ \text{O}}}{\text{RS}}\cdot\ \cdot\text{OR}'] \rightarrow \underset{\substack{\downarrow\\ \text{O}}}{\text{RSOR}'}$$

**(71)**
optically active racemic

Chiral allenic sufinates **73**, obtained from (*R*)- and (*S*)-1-butyn-3-ol (**72**) and sulfur dichloride, underwent electrophilic cyclization in the presence of bromine and methanesulfenyl chloride to give optically active γ-sultines **74a–d** as diastereomeric mixtures, some of which were separated by chromatography. Oxidation of the chiral sultines gave optically active sultones **75a–c** lacking a chiral sulfur[109].

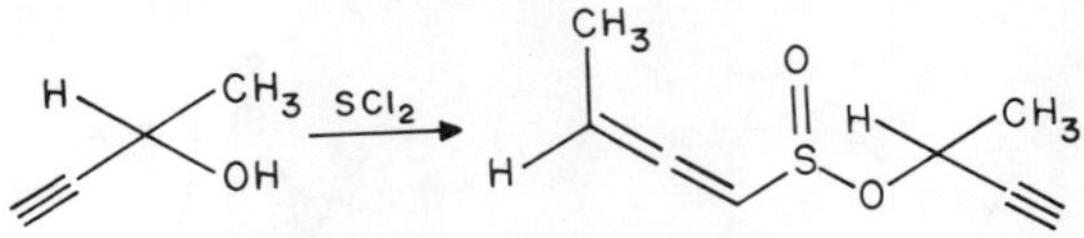


(**72**) (**73**)

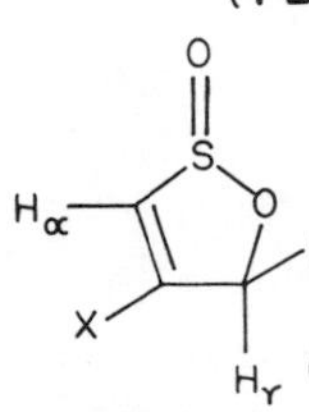


(**74**)

X = Br; R = Me
X = Br; R = *t*-Bu
X = MeS; R = Me
X = MeS; R = *t*-Bu

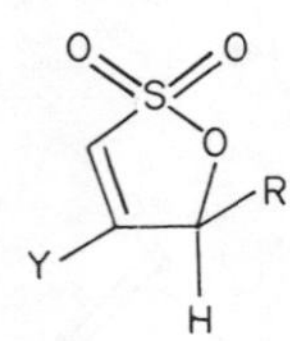


(**75**)

R = Me; Y = Br
R = *t*-Bu; Y = Br
R = Me; Y = $MeSO_2$

Under flash vacuum thermolysis (FVT) conditions at 700 °C the diastereomers **77a, b** underwent rapid epimerization at sulfur, followed by cleavage to *N*-allyl amide **78** and enamides **79** and **80**. The preferred mechanistic path for the ring cleavage involves initial formation of zwitterion **82** followed by loss of sulfur dioxide[110] (Scheme 9).

Chiral sulfinates such as **84**, under mild conditions and acid catalysis, reacted with enol silyl ethers **83** to give chiral α-sulfinyl cyclic ketones **85** with high stereospecificity. The

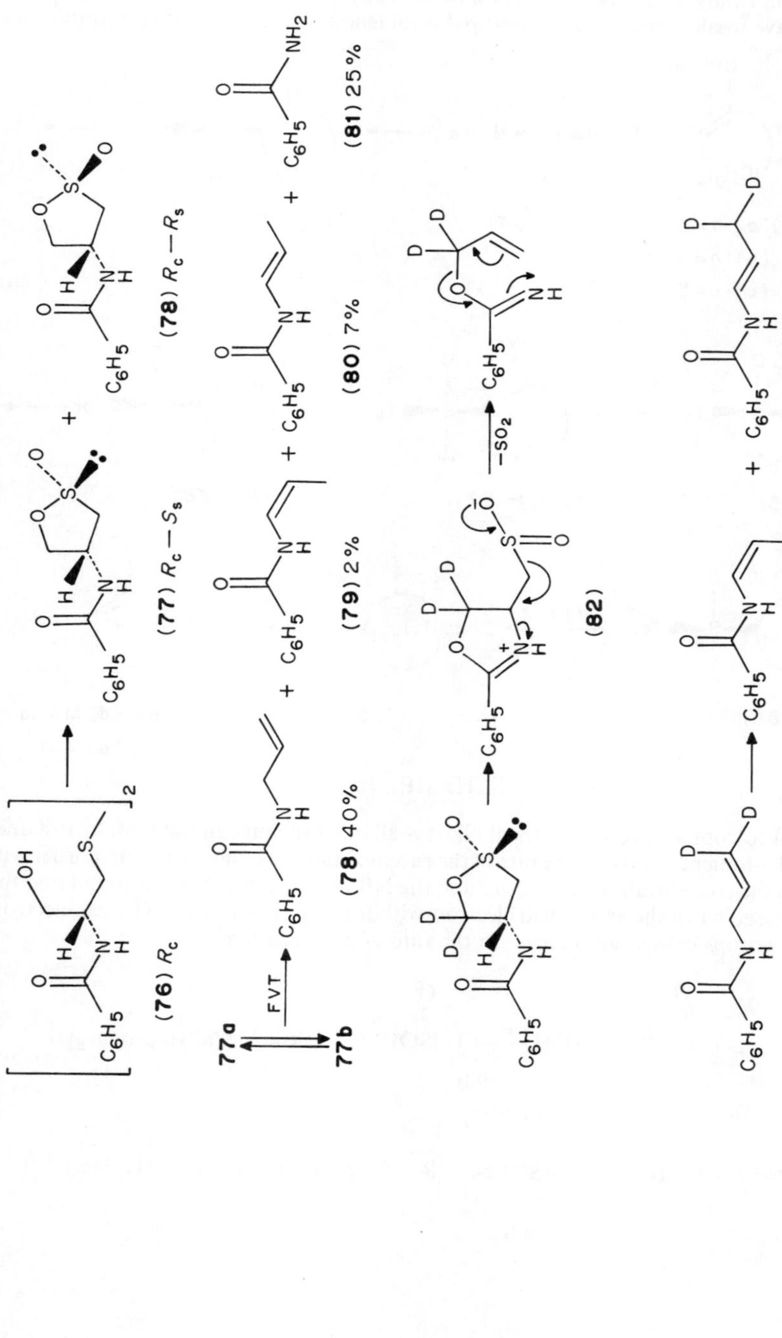


SCHEME 9

latter are valuable intermediates in asymmetric syntheses. The sulfur configuration in **85** was determined by correlation with **89**. Analogously, sodium or lithium enolates **91d, e** and chiral sulfinates also gave the α-sulfinyl ketones in high yield but with low stereospecificity[98] (Scheme 10).

The NBS-catalyzed transesterification of optically active sulfinates **92** with isopropyl alcohol gave in all cases recemic isopropyl *p*-toluenesulfinate **93**. The recemization was

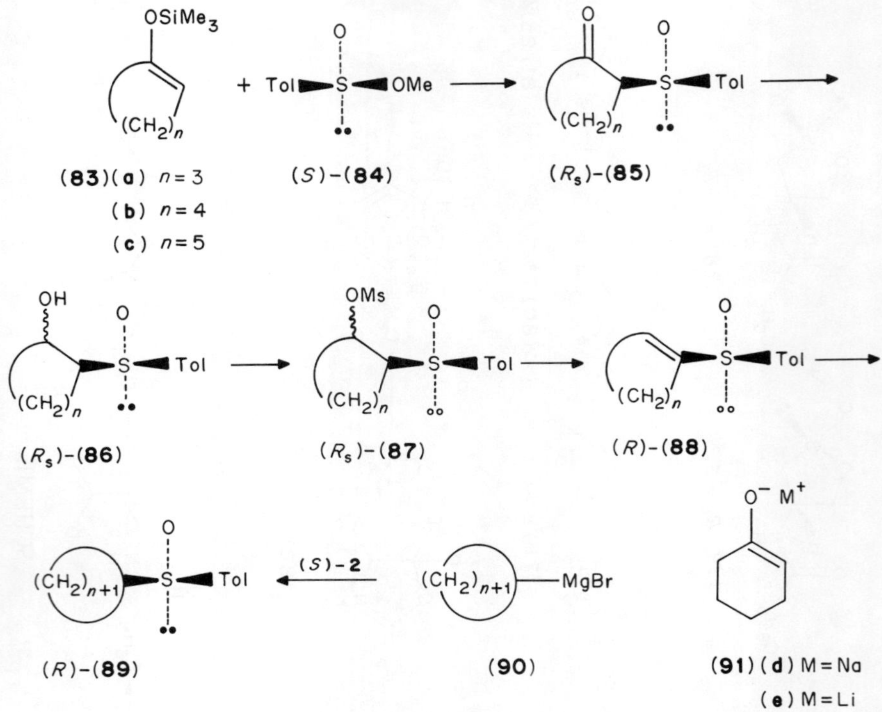


SCHEME 10

attributed to competitive symmetrical alkoxy–alkoxy exchange in the initially obtained isopropyl *p*-toluenesulfinate. The rate of the racemization was found to be first order with respect to the concentration of the ester and the NBS. Furthermore, it was found that this rate decreased when the arenes had electron-withdrawing substituents, whereas electron-donating groups caused an increase in the rate of racemization[111].

$$p\text{-TolS}(=\text{O})\text{OR} + i\text{-PrOH} \xrightarrow{\text{H}^+} \text{TolS}(=\text{O})\text{OPr-}i \qquad (\text{R} = \text{Me, allyl, propargyl})$$

(92) (93)

$$\text{R}\overset{*}{\text{S}}(=\text{O})\text{OPr-}i + i\text{-PrOH} \xrightarrow{\text{NBS}} \text{RS}(=\text{O})\text{OPr-}i \ (\text{racemic}) \qquad (\text{R} = \text{Ph}, p\text{-Tol}, p\text{-An}, p\text{-ClC}_6\text{H}_4, \text{Me}, i\text{-Pr})$$

(94) (95)

### D. Spectral Studies

The chiral nature of $\beta$- and $\gamma$-sultines has been determined by X-ray crystallography. Oxidative cyclization of 2-hydroxyalkyl *t*-butylsulfoxide **96** gave stable crystalline $\beta$-sultine **97** whose X-ray structure showed a nonplanar oxathietan ring, with sulfinyl oxygen assuming a pseudo-axial orientation[112].

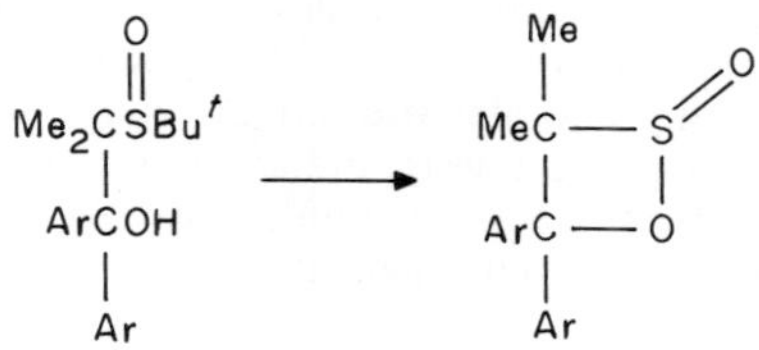

**(96)** **(97)**

X-ray analysis of $\beta$-amino-$\gamma$-sultine **98** indicated that the sultine possessed an envelope shape **99** and an *R*-configuration at sulfur[113].

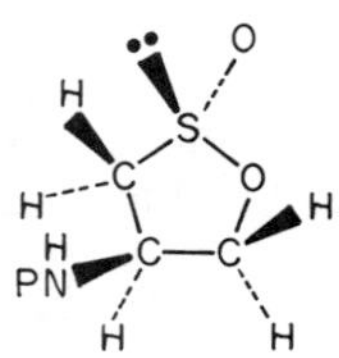

$R_c, R_s$

**(98)** P = BOC

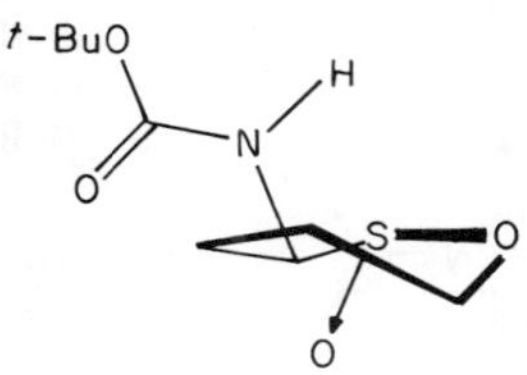

**(99)**

## III. SULFINAMIDES SYNTHESIS AND REACTIONS

Optically active sulfinamides are frequently prepared by the reaction of metal amine salts with chiral sulfinates[114]. Thus, treatment of ( − )-(*S*)-menthyl *p*-toluenesulfinate **2** with morpholinomagnesium bromide gave the ( + )-(*S*)-*N*-*p*-toluenesulfinylmorpholine **100**.

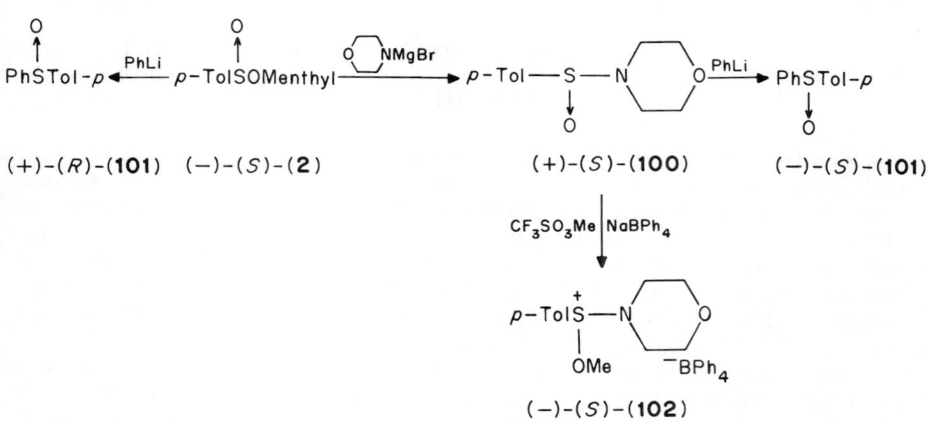


SCHEME 11

Subsequent reactions of **100** with PhLi gave the (–)-(*S*)-sulfoxide **101**. With inversion of configuration, and with methyl trifluoromethane sulfonate gave the (+)-(*S*)-methoxymorpholino-*p*-toluenesulfonium salt **102** (Scheme 11).

Chiral N-alkylidene sulfinamides were prepared similarly in optically pure form from imino-Grignard reagents, and were shown to undergo rapid *E–Z* interconversion at room temperature[115]. In the case of **103e**, obtained as a 3:2 mixture of diastereomers, the major component had a (–)-($S_sR_c$) configuration, which was determined by acid hydrolysis to (–)-(*R*)-**104**. Metal hydride reduction of the N-alkylidene sulfinamides to saturated sulfinamides proceeded readily and stereoselectively giving unequal amounts of diastereomers **105a–d**. The extent of asymmetric induction was established via conversion of **105** to the corresponding optically active sulfonamides **106** or amines **107**. The highest optical purity (60–80%) was observed when lithium aluminium hydride was used as reducing agent (Scheme 12).

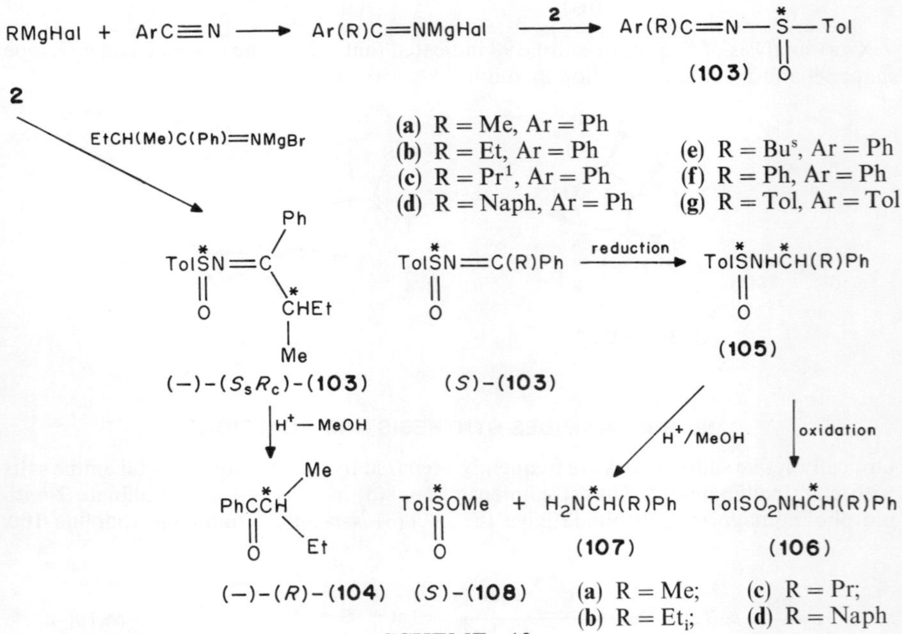


SCHEME 12

A high degree of stereospecificity has been observed in the synthesis of optically active sulfinamides **100** when chiral amidosulfites **109** were treated with Grignard reagents[116,117]. Thus, the reaction of **109a** with PhMgBr gave a 92:8 mixture of **110Ia** and **110Ib**. Conversely, the enantiomer **109b** gave the same mixture in the reverse ratio. Sulfinamides **110b, c** were analogously prepared with MeMgBr, MeLi or BuLi. The absolute configurations of **110Ia** and **110Ib** were established by conversion to the corresponding (*R*)-(+)- and (*S*)-(–)-butyl phenyl sulfoxides when reacted with BuLi, a reaction which is known to proceed with inversion of configuration (Scheme 13).

A moderate degree of transfer of chirality from carbon to sulfur was detected in the facile ene-reaction of alkene **111** with N-sulfinyl-*p*-toluenesulfonamide **112** to give sulfinamide **113**, where the configuration of the C=C double bond was always found to be *E*. Subsequent conversion of **113** to sulfinamide **114** and sulfoxide **115** indicated that the ene

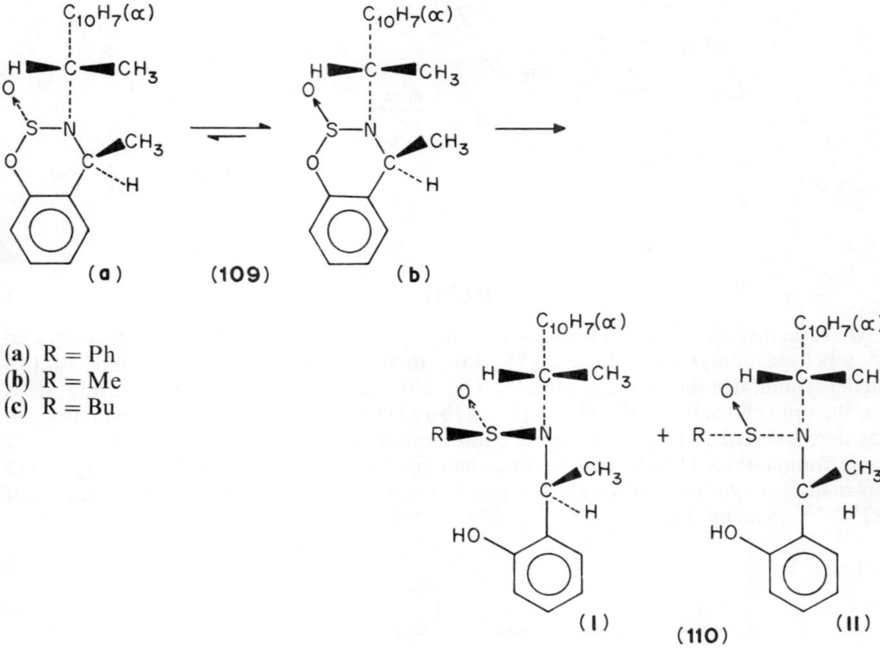

(a) R = Ph
(b) R = Me
(c) R = Bu

SCHEME 13

reaction proceeded with 40 ± 5% stereospecificity. The mechanism of the reaction is understood to involve the formation of a [2 + 2] complex between the reactants prior to the rate-determining step of allylic hydrogen abstraction by the lone electron pair of the nitrogen atom. The preferred cyclic *endo* transition state **116** is assumed, since it accounts for the selective formation of the (*R*)–(*E*)-sulfinamide **113**[118] (Scheme 14).

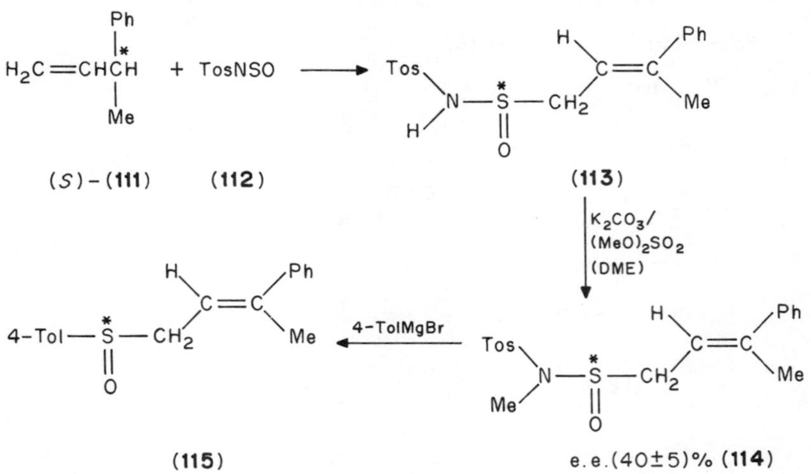

SCHEME 14 (*continued*)

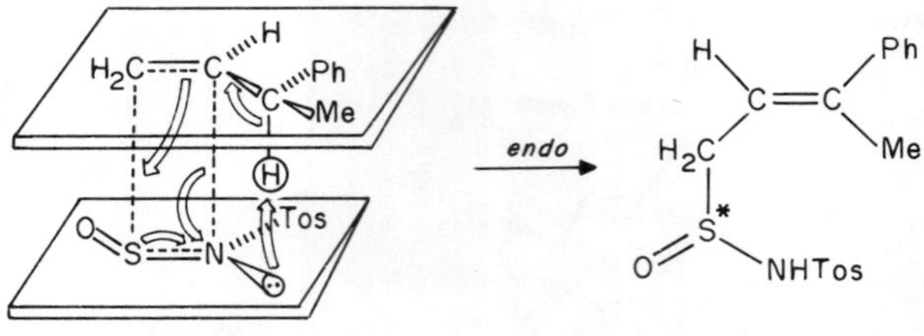


(116) (R)—(E)-(113)

SCHEME 14

Mixtures inseparable by TLC of diastereomeric sulfinamides **117** and **118** were prepared from benzensulfinyl chloride and (1*S*, 2*S*)-1-methoxy-1-phenyl-2-propylamine or (*S*)-phenethylamine in the presence of $Et_3N$. The sulfinamides were converted in several steps into the optically active sulfoximines (*SR*)-**119** and (*SS*)-**120**, whose absolute configuration was determined by their stereospecific, aluminium amalgam reduction to the optically active sulfinamides **117** and **118**. Independent confirmation of the stereochemistry of **117** was made via synthesis from (*S*)-( − )-menthyl benzenesulfinate **121** and the lithium salt **122**[119,120] (Scheme 15).

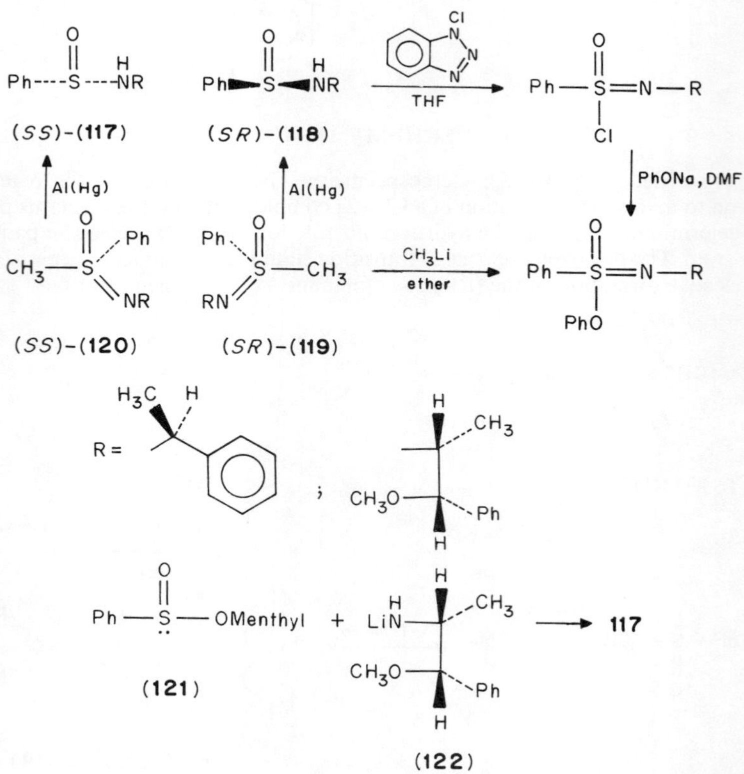


SCHEME 15

Deoxygenation of sulfoxide **123** by an acid chloride and ligand exchange on the sulfur atom with LDA gave ketone **124** and sulfinamide **125**[94].

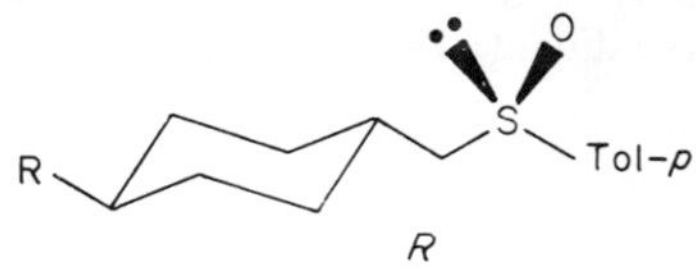


(**123**)

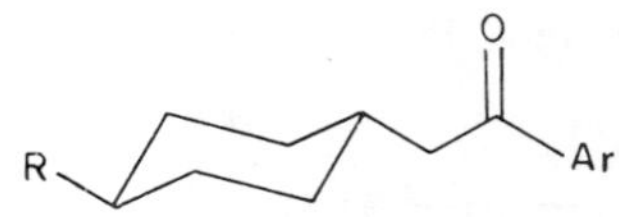


(**124**)

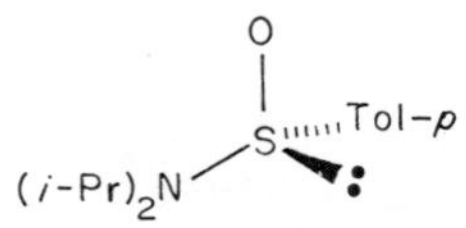


(**125**)

Chiral aminophenols **126** when treated with thionyl chloride produced benzoxathiazine 2-oxides **127**. Acid-catalyzed isomerization of **127-I** gave an equilibrium mixture comprised primarily of **127-II** (96%). Reaction of the latter with PhMgBr gave the sulfinamides **128**, which were converted *in situ* into sulfoxides **129** in high enantiomeric excess[121] (Scheme 16).

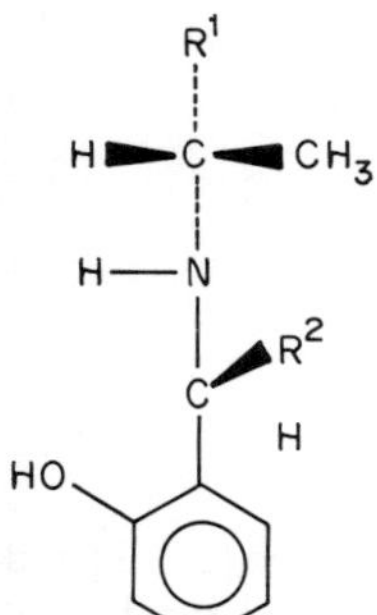

(**126**)

(**a**) $R^1$ = 1-Naph, $R^2$ = Me
(**b**) $R^1$ = Ph, $R^2$ = H
(**c**) $R^1$ = Ph, $R^2$ = Me
(**d**) $R^1$ = 1-Naph, $R^2$ = H
(**e**) $R^1$ = 1-Naph, $R^2$ = Et
(**f**) $R^1$ = 1-Naph, $R^2$ = Bu

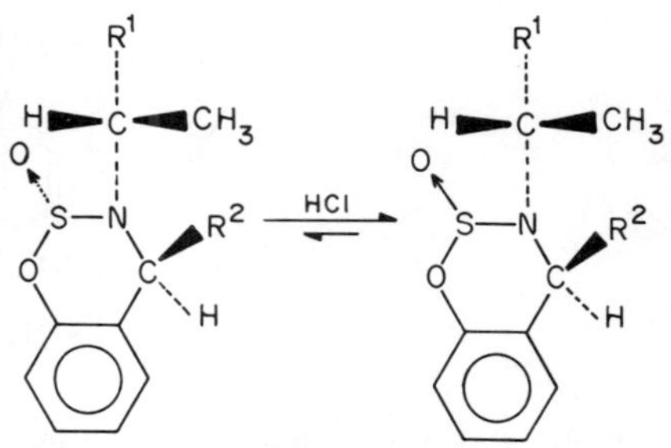


(**127-I**) (**127-II**)

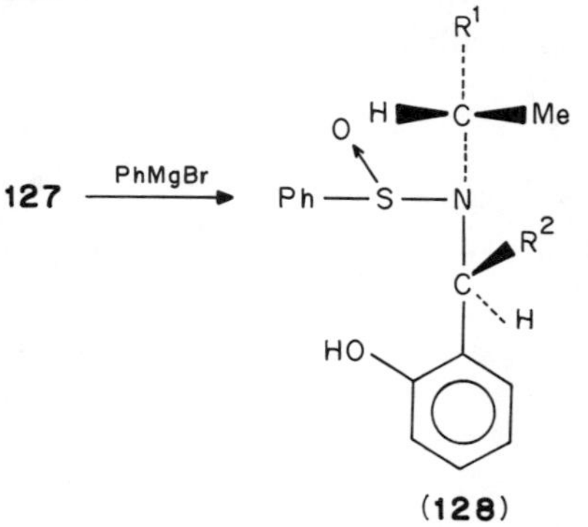


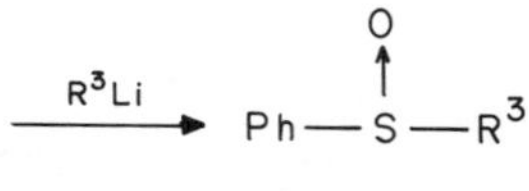


(**128**)

(**129**) (**a**) $R^3$ = Me
(**b**) $R^3$ = Bu

SCHEME 16

Diastereomeric mixtures of 1, 2, 3-thiadiazoline 1-oxides **131a, b** obtained upon treatment of **130** with $SOCl_2$ in DMF were separated by chromatography. They were shown to undergo cleavage to **132** when left in DMSO solution or treated with silica gel for prolonged time. Peracid oxidation afforded the dioxide **133**[122] (Scheme 17).

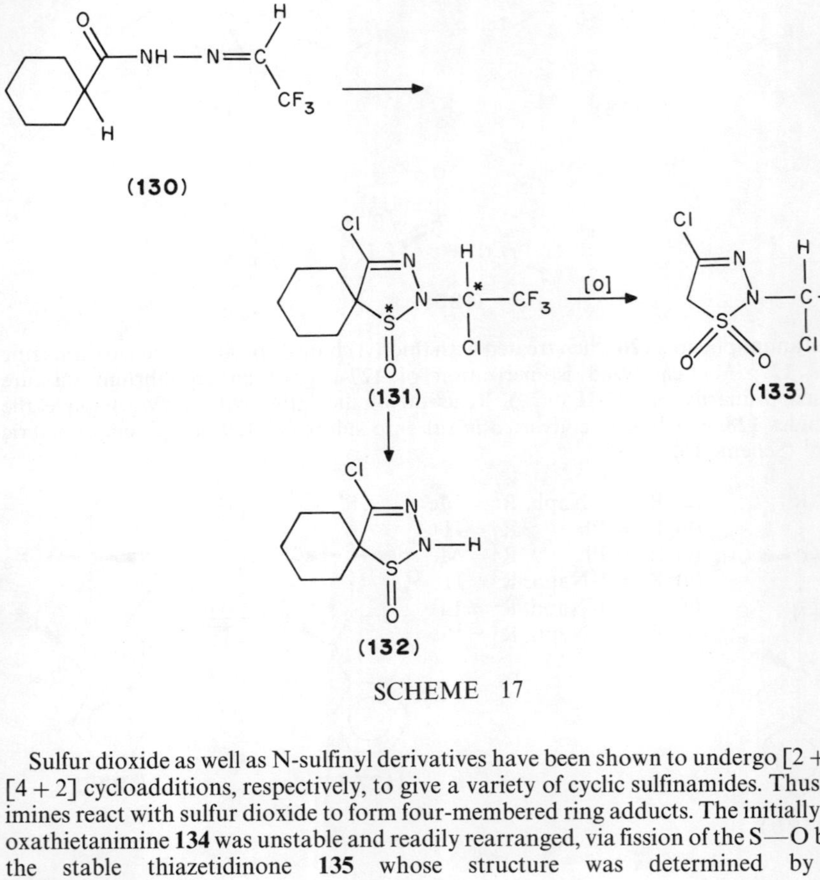


SCHEME 17

Sulfur dioxide as well as N-sulfinyl derivatives have been shown to undergo [2 + 2] and [4 + 2] cycloadditions, respectively, to give a variety of cyclic sulfinamides. Thus, ketene imines react with sulfur dioxide to form four-membered ring adducts. The initially formed oxathietanimine **134** was unstable and readily rearranged, via fission of the S—O bond, to the stable thiazetidinone **135** whose structure was determined by X-ray crystallography[123].

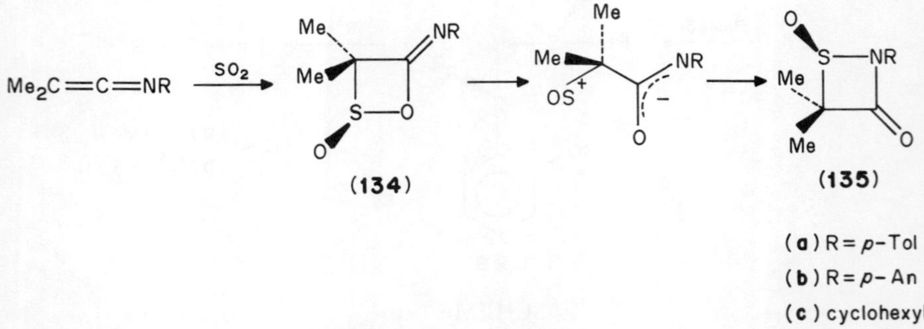

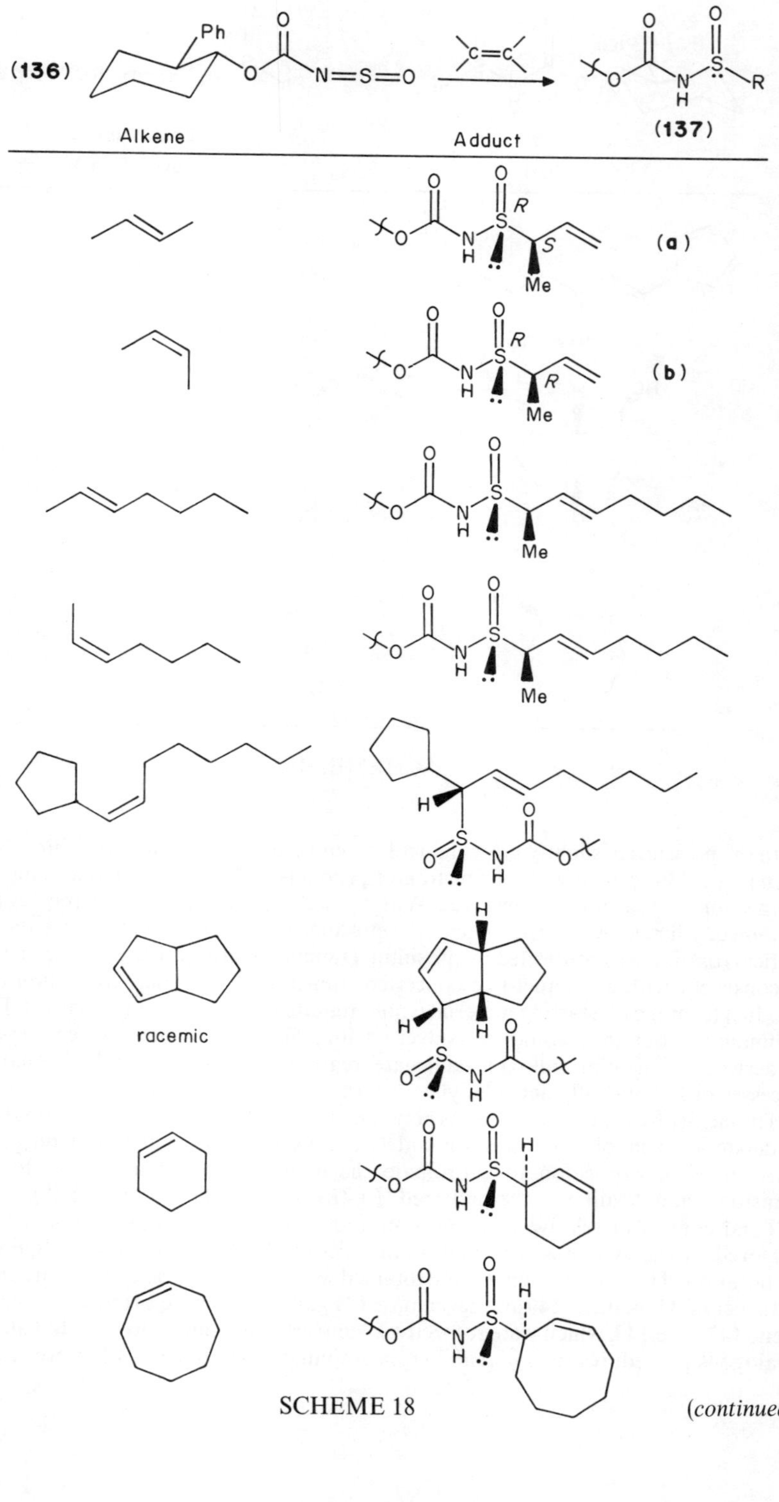


SCHEME 18 (*continued*)

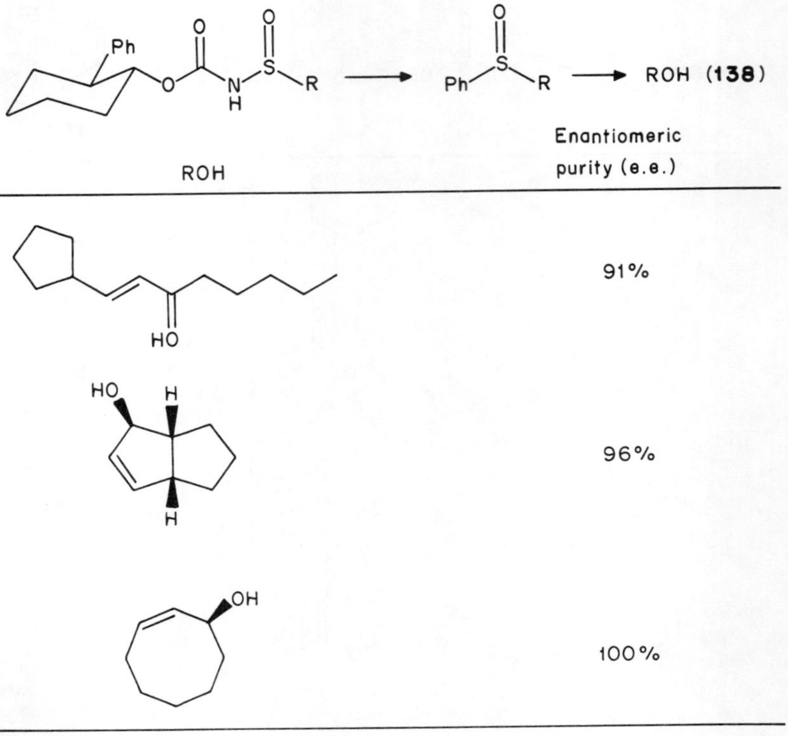


SCHEME 18

In the presence of slightly more than a 1 molar equivalent of $SnCl_4$ the chiral N-sulfinyl carbamate **136** reacted with olefins to give products **137** where asymmetric induction at both sulfur and carbon was observed. With *trans*-2-butene a single diastereomer **137a** was obtained, whereas with *cis*-2-butene the product was **137b**. In both cases the sense of optical rotation was controlled by the sulfinyl functional group. The proposed mechanism is consistent with a product-like concerted transition state, where formation of the *cis* product from the *cis* starting material would require serious steric interactions. The chiral sulfinamides obtained could be converted into allylic sulfoxides, which subsequently underwent sequential sulfoxide–sulfenate rearrangements to give high enantiomeric excesses of the optically active allylic alcohols **138**[124] (Scheme 18).

The facility by which ene reactions between olefins and N-sulfinyl enophiles take place is understood to involve a transition state whereby the nonbonded electron pair of the nitrogen is in a position to coordinate the allylic hydrogen in a 'pseudopericyclic' transition state leading to the preferred (*R*)–(*E*)-sulfinamide[125] (Scheme 19).

Total control of relative configuration and double-bond geometry could be readily achieved in the synthesis a vicinal amino alcohols and amino sugars obtained upon elaboration of the chiral sulfinamides obtained when N-sulfinyl carbamates are condensed with dienes. Thus, diene **140** and carbamate **139** gave sulfinamide **141** whereas the isomeric diene **142** gave **143**, which were respectively converted to amino alcohols **144** and **145**. By analogous procedures the *threo* and *erythro* sphingosines **148** and **149** were respectively

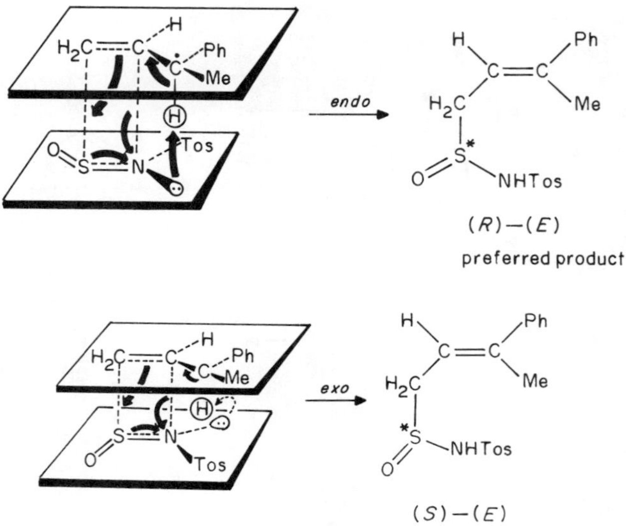


SCHEME 19

prepared from **146** and **147**. In all cases the products obtained were racemic since the starting materials were achiral[126,127] (Scheme 20).

Similar methodology has been used in the synthesis of amino sugars from noncarbohydrate precursors[129]. A quasi-boat conformation **151** is involved in the transition state leading to sulfinamide **152**, which was converted to desosamine **153**[128]. An inseparable mixture (15:1) of sulfinamides **157** epimeric at sulfur, obtained from **154** via N-sulfinylcarbamate **155**, which underwent cyclization by a preferred conformation **156**, were used in the synthesis of **158** (Scheme 21).

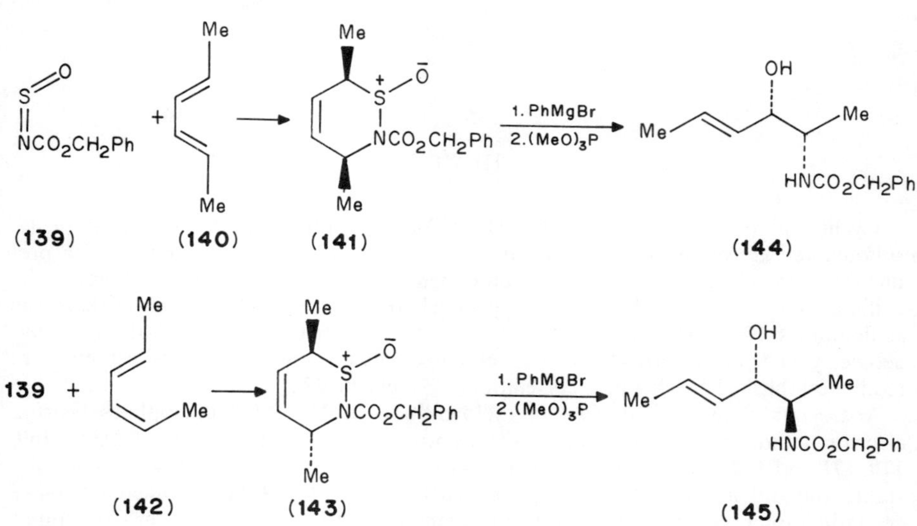


SCHEME 20 *(continued)*

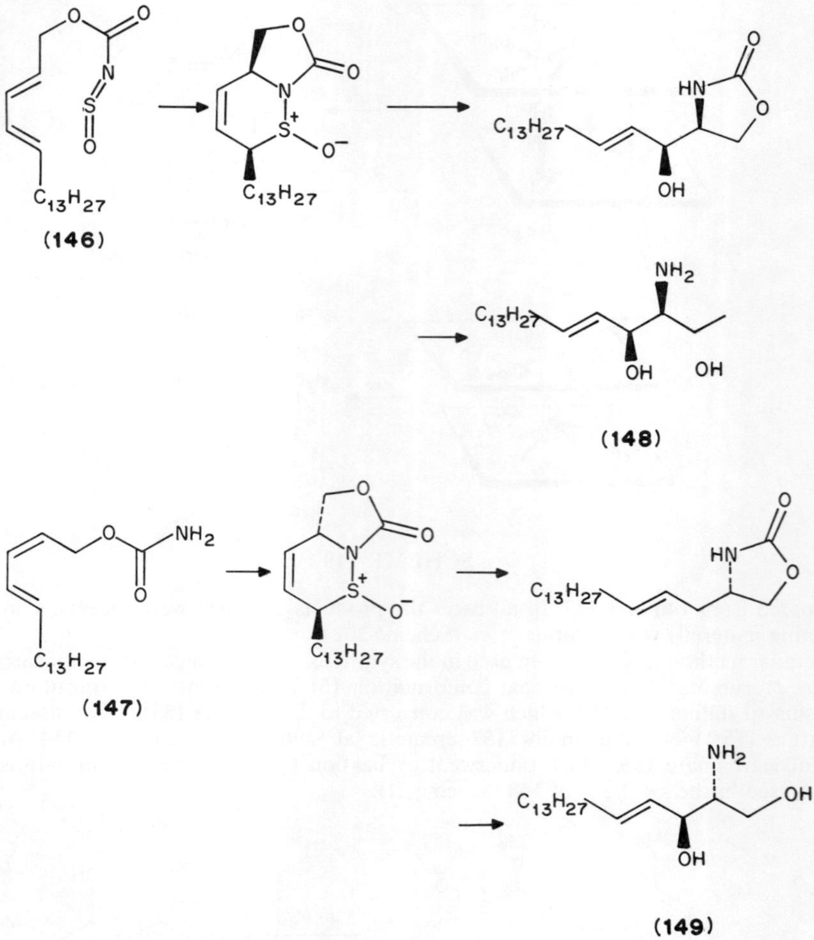


SCHEME 20

Cyclic sulfinamides obtained by Diels–Alder addition to *N*-sulfinyl-*p*-toluenesulfonamide are useful intermediates in the synthesis of homoallylic amines with predictable stereochemistry and double bond geometry. Single diastereomeric homoallylic sulfonamides **161**, **165** and **168** were respectively obtained upon hydrolysis of the cyclic sulfinamides **160**, **164** and **167**. In all cases the products possessed *E*-double-bond geometry as determined by NOE. The relative configurations of the chiral centers were readily established by chemical correlations[130] (Scheme 22).

Analogous stereoselective [4 + 2] cycloaddition of N-sulfinyl dienophiles bearing electron-withdrawing groups and 1, 3-dienes provided 3, 6-dihydrothiazine 1-oxides **169**, **170**, **171** and **172**[131]. The sulfinamide functional groups in **169** and **170** were thermally stable and did not interconvert. The stereochemistry of **169a**, determined by X-ray crystallography, was shown to have an approximate twist-boat conformation. Compound **172** was obtained as a single diastereomer. Fission of the S—N bond of the adducts

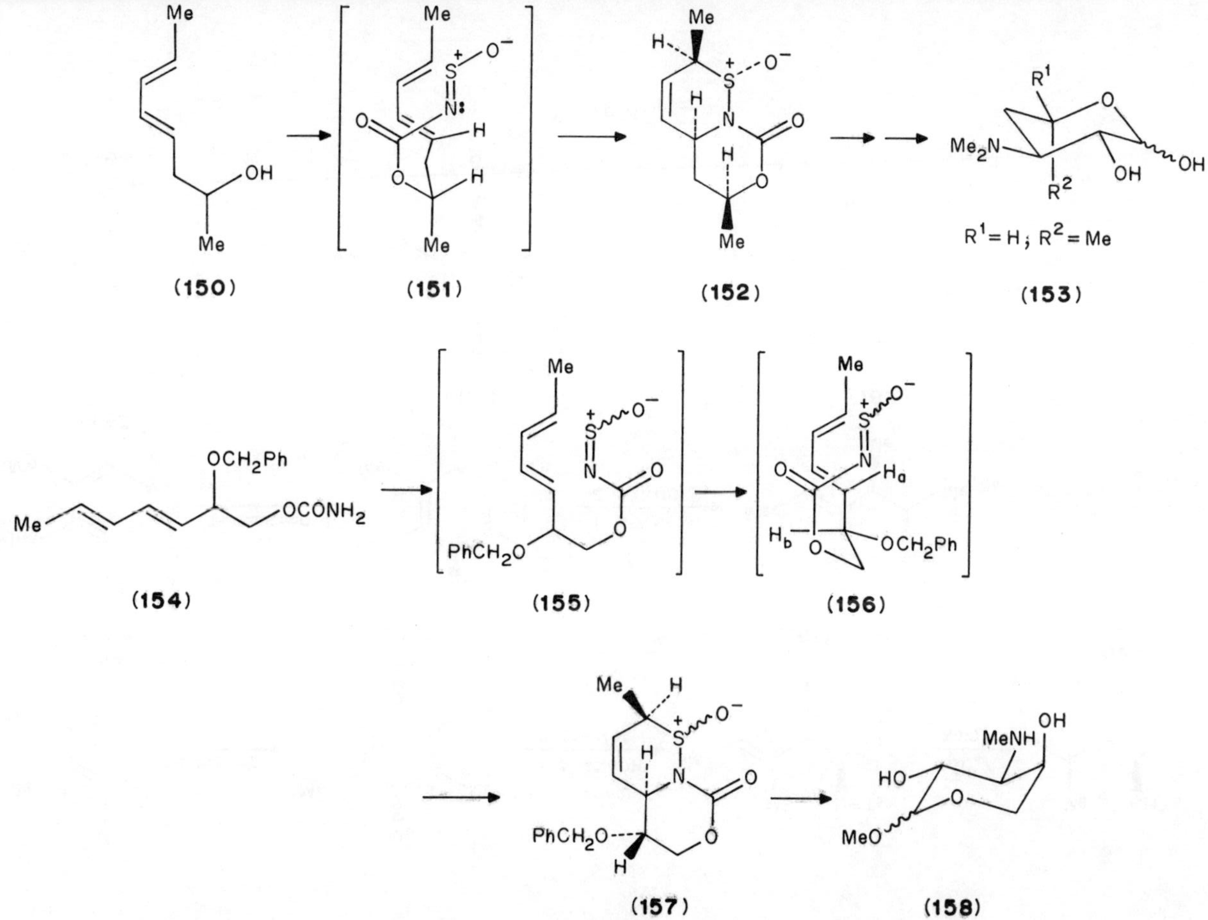


SCHEME 21

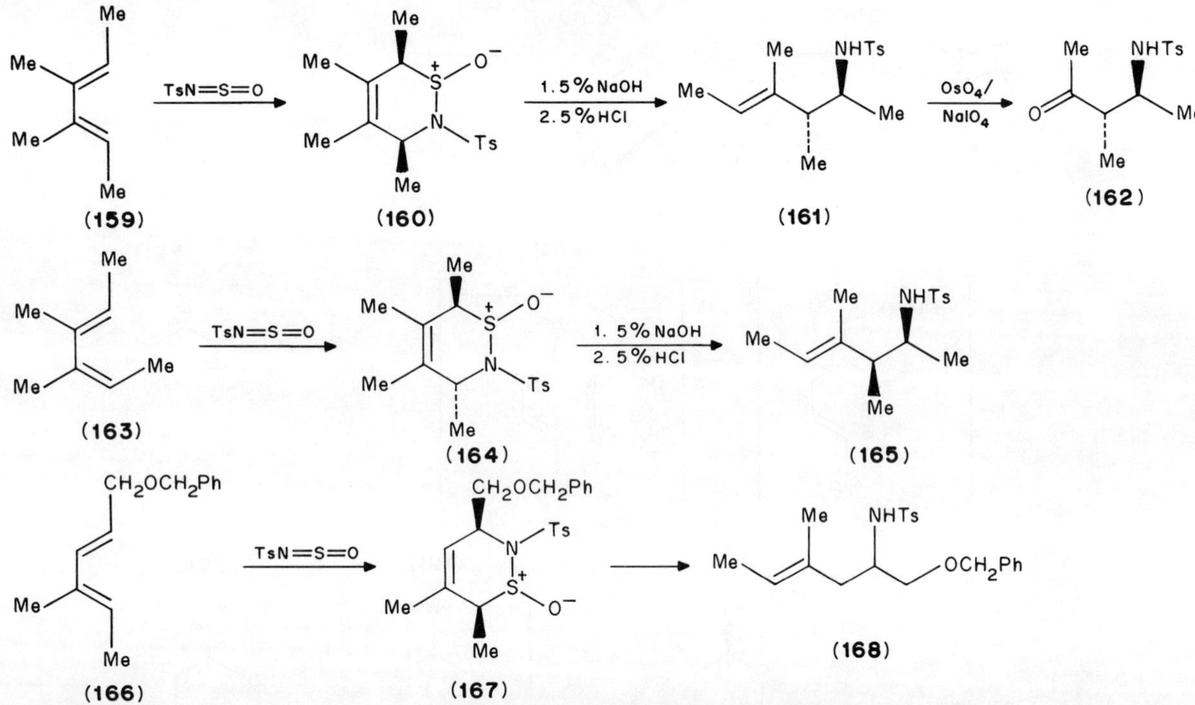
TsN=S=O
1. 5% NaOH
2. 5% HCl
OsO4/
NaIO4
(159)
(160)
(161)
(162)
(163)
(164)
(165)
(166)
(167)
(168)

SCHEME 22

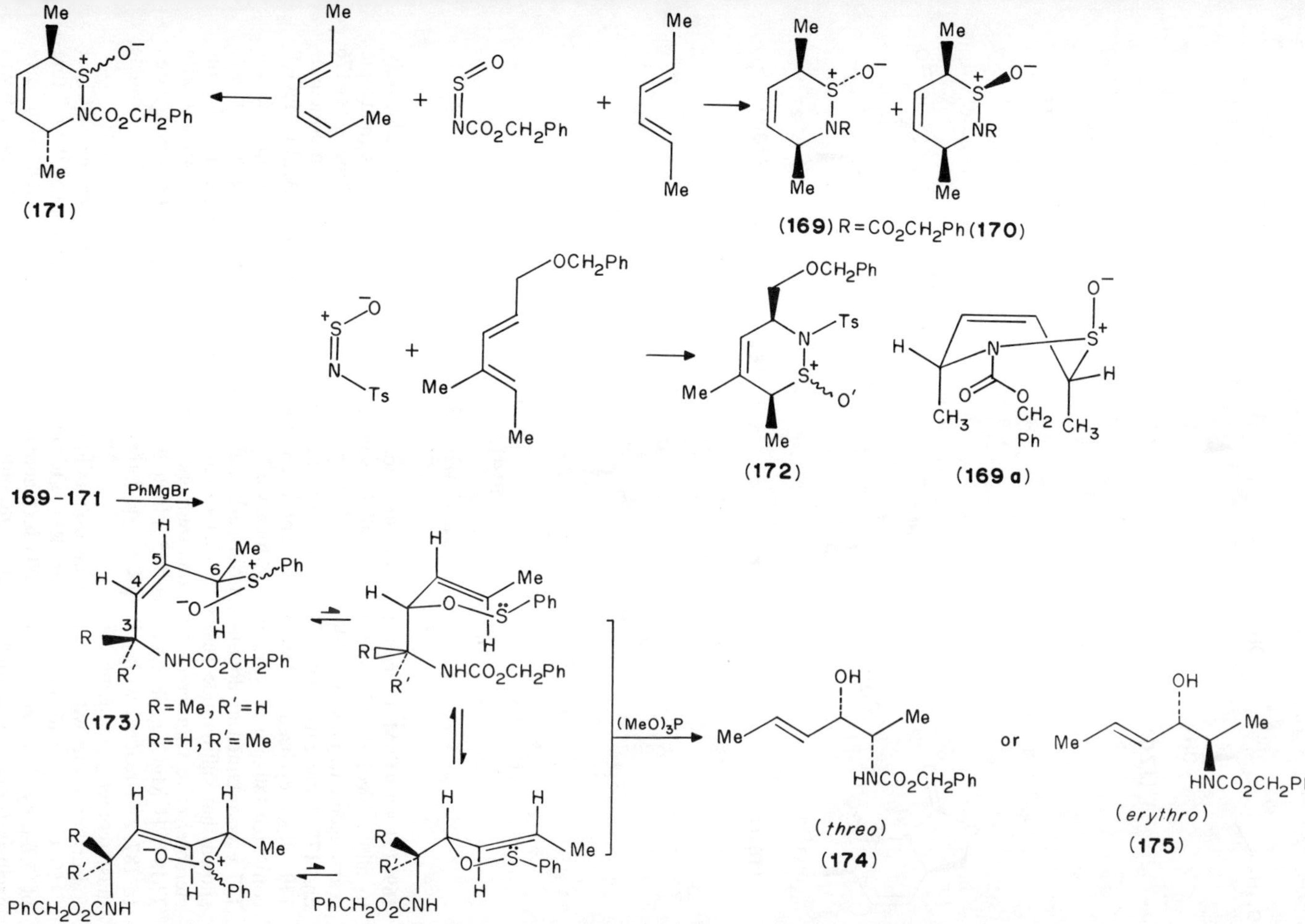


SCHEME 23

*(continued)*

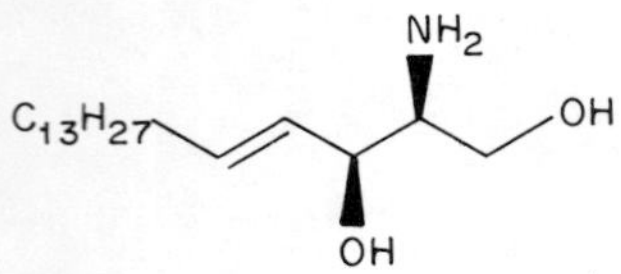
(176)

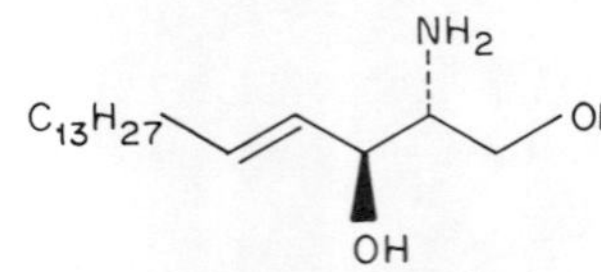
(177)

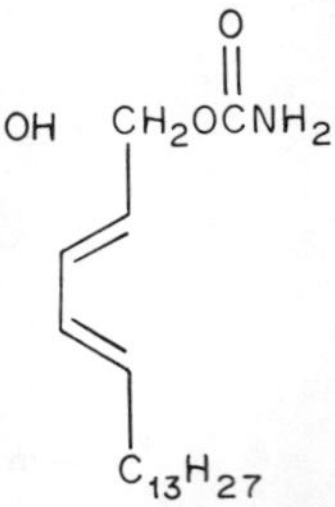
(178)

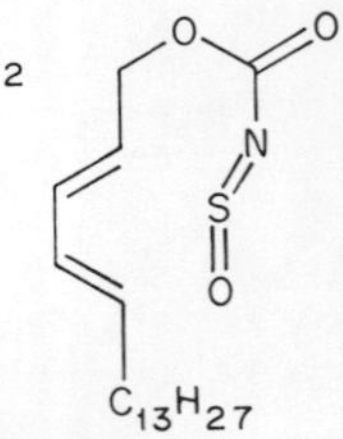
(179)

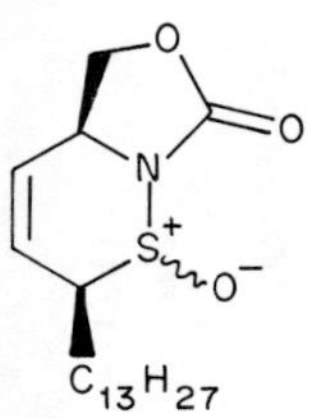
(180)

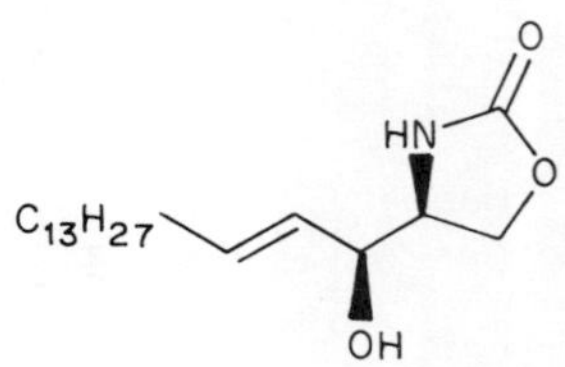
(181)

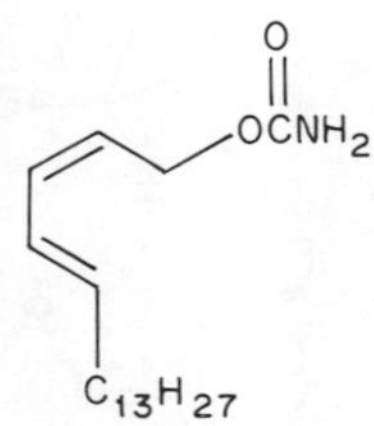
(182)

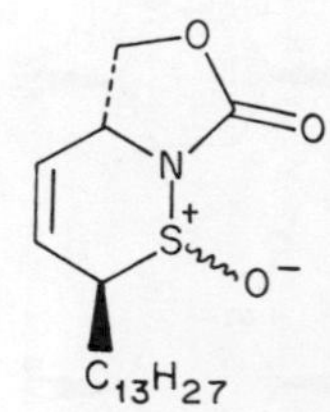
(183)

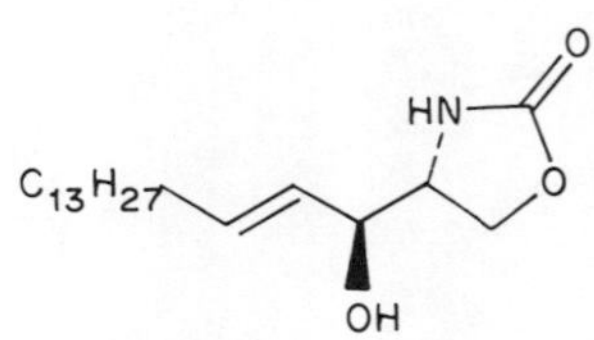
(184)

SCHEME 23

**169–171** with PhMgBr leads to allylic sulfoxides, which were converted stereoselectively to allylic alcohols **174** and **175** via allylic sulfoxide/sulfenate ester [2, 3]-sigmatropic rearrangement. This methodology was then applied to the synthesis of sphingosines **176** and **177**. Thus, carbamate **178** upon treatment with thionyl chloride gave intermediate **179**, which cyclized to **180**. Subsequent stereospecific reaction with PhMgBr/$MeO_3P$ converted **180** into carbamate **181**, which was hydrolyzed to the desired **176**. Sphyngosine **177** was obtained analogously from carbamate **184** (Scheme 23). Subsequently it was shown that sulfinamide **169**, as a 15:1 mixture of sulfur epimers, can be used for the stereoselective transfer of functionalized one- and two-carbon units. Treatment with TMS–$CH_2$MgCl gave sulfoxide **185** which deoxygenated to sulfide **186**, methylated to **187**, and reacted with *t*-BuOK to yield the crystalline silyl sulfide **188** as a single diastereomer[132] (Scheme 24).

Further studies with N-sulfinyl dienophiles have shown that in the presence of $TiCl_4$ at – 50 °C, cyclohexadiene reacted with R*—OCON=S=O (R* = a) to give a 9:1 mixture of 3, 6-dihydrothiazine oxides **189a, b**, whereas R*—OCON=S=O (R* = b) gave exclusively one cycloadduct. The absolute configuration at sulfur in these compounds was

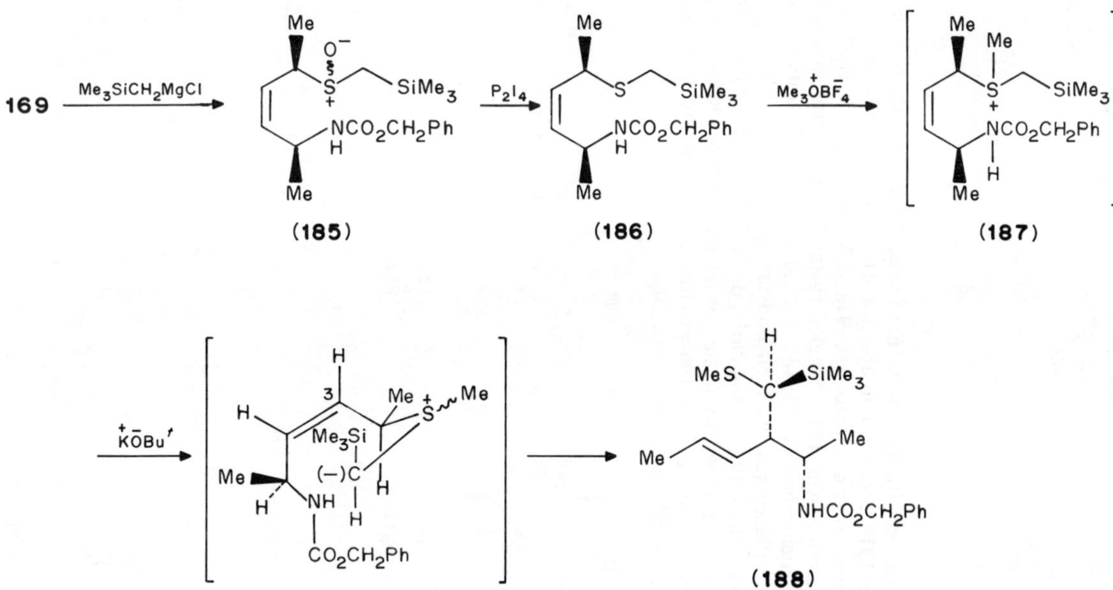
169
$Me_3SiCH_2MgCl$
(185)
$P_2I_4$
(186)
$Me_3\overset{+}{O}\overset{-}{BF_4}$
(187)
$\overset{+}{K}\overset{-}{O}Bu^t$
(188)

SCHEME 24

not determined, however **189a** was shown to be epimeric at sulfur since oxidation gave a single sulfone[133].

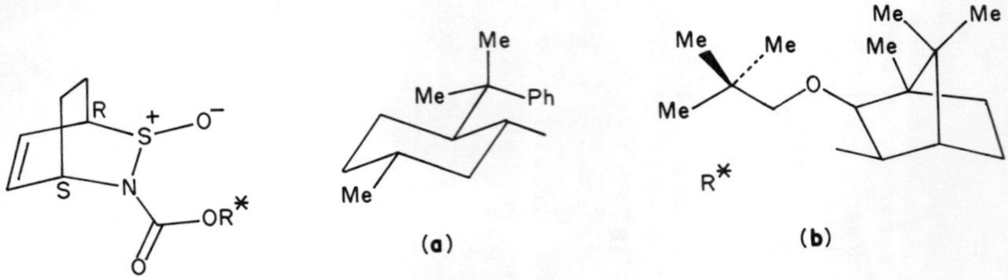


(a) (b)

(189a,b)

When **190** was reacted with the acyclic (*E*, *E*)-hexa-2, 4-diene in the presence of $SnCl_4$, a single diastereomer **191b** formed. In the absence of the Lewis acid, thermal cyclization gave a mixture of all four possible diastereomers **191a–d**. Moreover, the least abundant product in the thermal reaction was **191b**. Single crystal X-ray determination was used to establish the stereochemistry of **191d**. The absolute configurations of the other diastereomers were determined by their conversion to phenyl sulfoxides followed by desulfurization to alcohols **192a, b**. Similarly, diene **193** gave primarily one isomer **194** whose structure was confirmed by facile conversion to **195**. Other reactions of **191d** showed that, under basic conditions, **196** was formed, whereas treatment with MeMgBr gave a single diastereomer **197**[134] (Scheme 25).

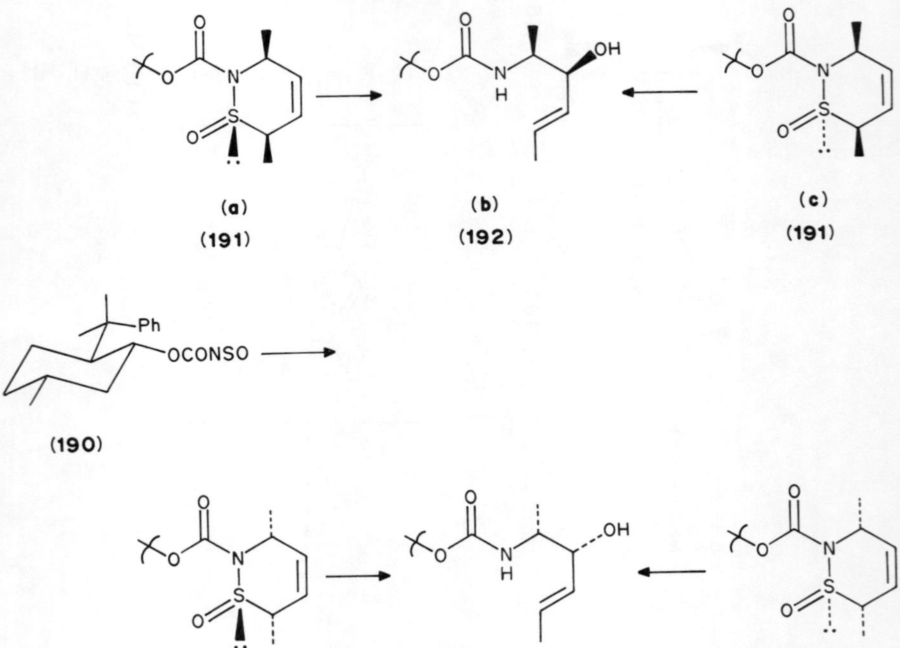


(a) (191) (b) (192) (c) (191)

(190)

(b) (b) (d)

SCHEME 25 *(continued)*

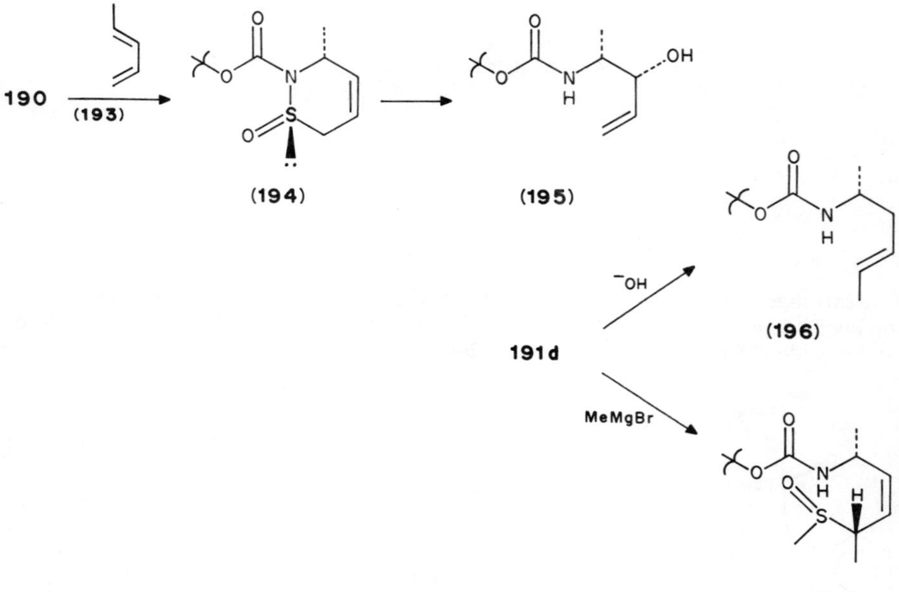


SCHEME 25

Sulfinamides RS(O)NR[1]R[2] (R = Ph, R[1] = H, R[2] = Me; R = *p*-Tol, R[1] = H, R[2] = Cyclohexyl; R[1] = R[2] = Me, have been resolved via cyclodextrin complexes[135].

Stereoselective oxidation of sulfenamidocobalt(III) complexes **198**, prepared from (*R*)-cysteine and ethylenediamine, with NBS, gave a 4:1 ratio of (*R*)- and (*S*)-sulfinamides **199** epimeric at sulfur. The individual isomers were separated by chromatography or fractional crystallization and their absolute configuration was established by X-ray crystallography. The sulfinamides which were found to be optically stable at sulfur, except in 3 M HCl, did not disproportionate, were stable to S—N hydrolysis, upon further oxidation gave a single sulfone **200** and were found to possess bacteriostatic properties[136].

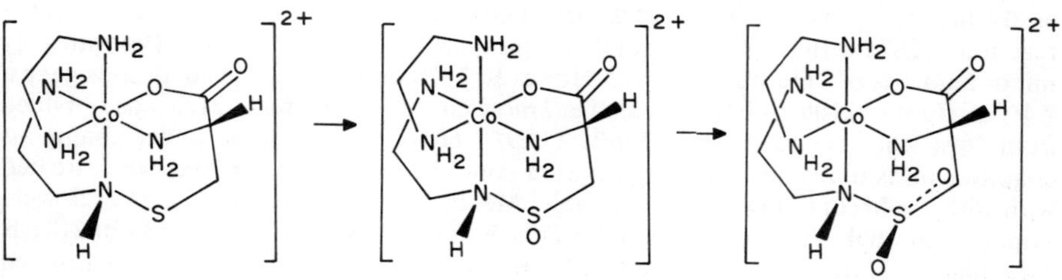

(198) (199) (200)

Anilino sulfinamide **201** underwent acid-catalyzed rearrangement to anilino sulfoxide **202** with complete loss of optical activity. It was not determined whether the racemization took place at the sulfoxide product under the influence of HCl or at some other step in the rearrangement process[137].

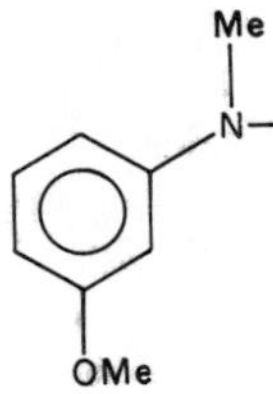

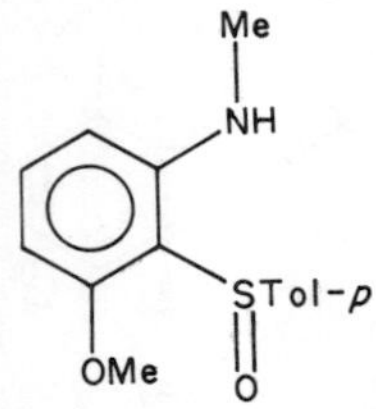

**(201)** **(202)**

Rearrangement of sulfonimidoate **203** in the presence of LDA, by a mechanism involving an elimination–addition, gave[138] sulfonimidamide **204** (45%) accompanied by sulfinamide **205** (38%) and quinonimine **206**.

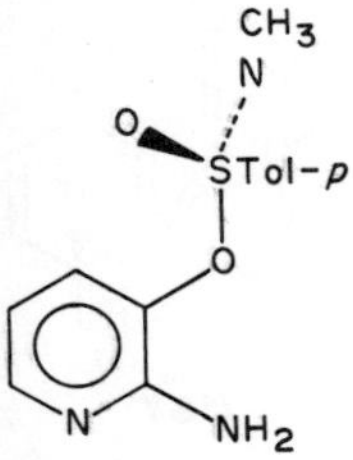

**(203)**

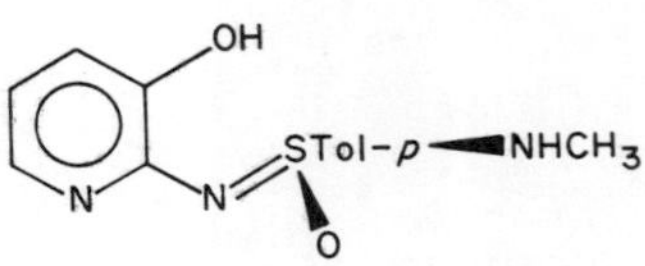

**(204)**

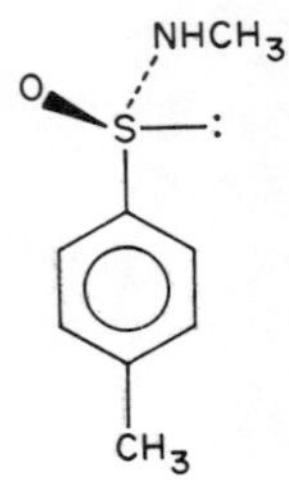

**(205)**

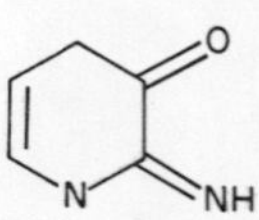

**(206)**

## IV. SULFINIMIDAMIDES AND SULFINIMIDOATES

Sulfur diimide **207a** when condensed in a Diels–Alder fashion with (*E*, *E*)-2, 4-hexadiene gave a 1.1:1 separable mixture of isomeric sulfinimidamides **208a** and **209a**. The absolute configuration of racemic **208a** was established by X-ray crystallography. The analogous reaction with the bis-carbamate **207b** gave a 1:8 mixture of **208b** and **209b.** The *cis* adducts **208** displayed characteristic proton NMR spectra where the olefinic hydrogens appeared as distinct multiplets, whereas in **209b** they appeared as a broad singlet. In the case of the reaction of **207a** with (*E*, *Z*)-2, 4-hexadiene, unexpectedly, a major isomer **210a** with only minor amounts of **211a** was obtained. However, **207b** under the same conditions gave a 2.4:1 mixture of **210b** and **211b**. A single *threo*-vicinal disulfonamide **212a** was obtained from **209a** when treated with PhMgBr/THD/ − 60 °C, whereas **208a** under analogous conditions was unreactive. Compound **212a** could be obtained from **208a** when treated with PhLi or MeLi followed by methanolic $Me_3P$. Analogous results to give **212b** were displayed by **209b**. The *erythro* diamides **213a, b** were respectively prepared from **210a, b**. The epimeric **211b** did not react with PhMgBr, and only low yields of **213b** were isolated from the reaction with PhLi or MeLi. Furthermore, it was shown that adducts **208a, b** when refluxed in benzene underwent a novel [2, 3]-sigmatropic rearrangement to give high yields of stable thiadiazolines **214a, b**. Similarly, **211a, b** gave **215a, b**. The *Z-threo* product **216** formed from **210b** when heated for 9 h in toluene, but under analogous conditions **209b**, underwent extensive decomposition. A parallel sequence of reactions with 1, 3-cyclohexadiene gave **217–220**[139] (Scheme 26).

This methodology, which enables the convenient preparation of vicinal diamines, was

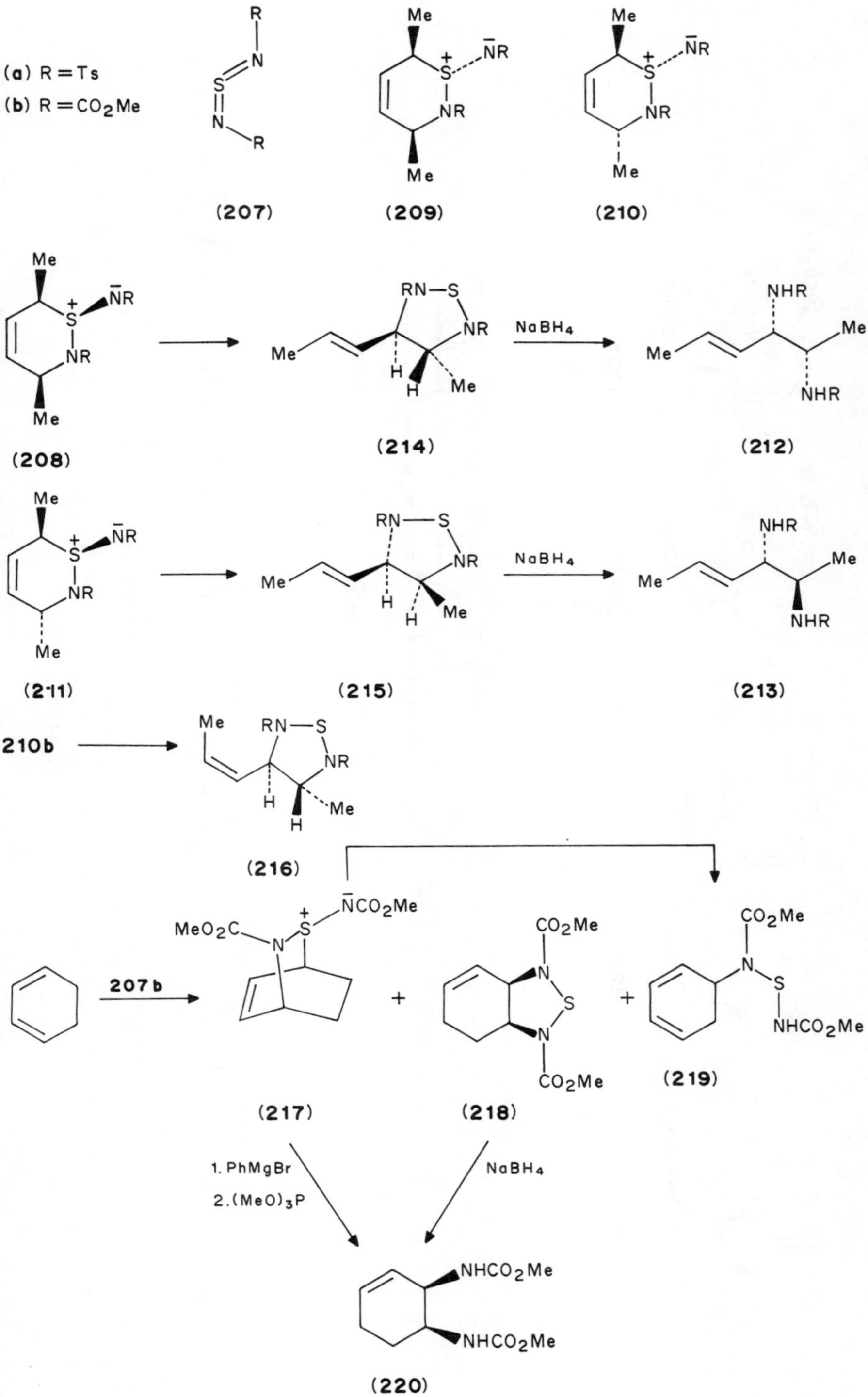
(a) R = Ts
(b) R = CO2Me
(207)
(209)
(210)
(208)
(214)
(212)
(211)
(215)
(213)
210b
(216)
207b
(217)
(218)
(219)
1. PhMgBr
2. (MeO)3P
NaBH4
(220)

SCHEME 26

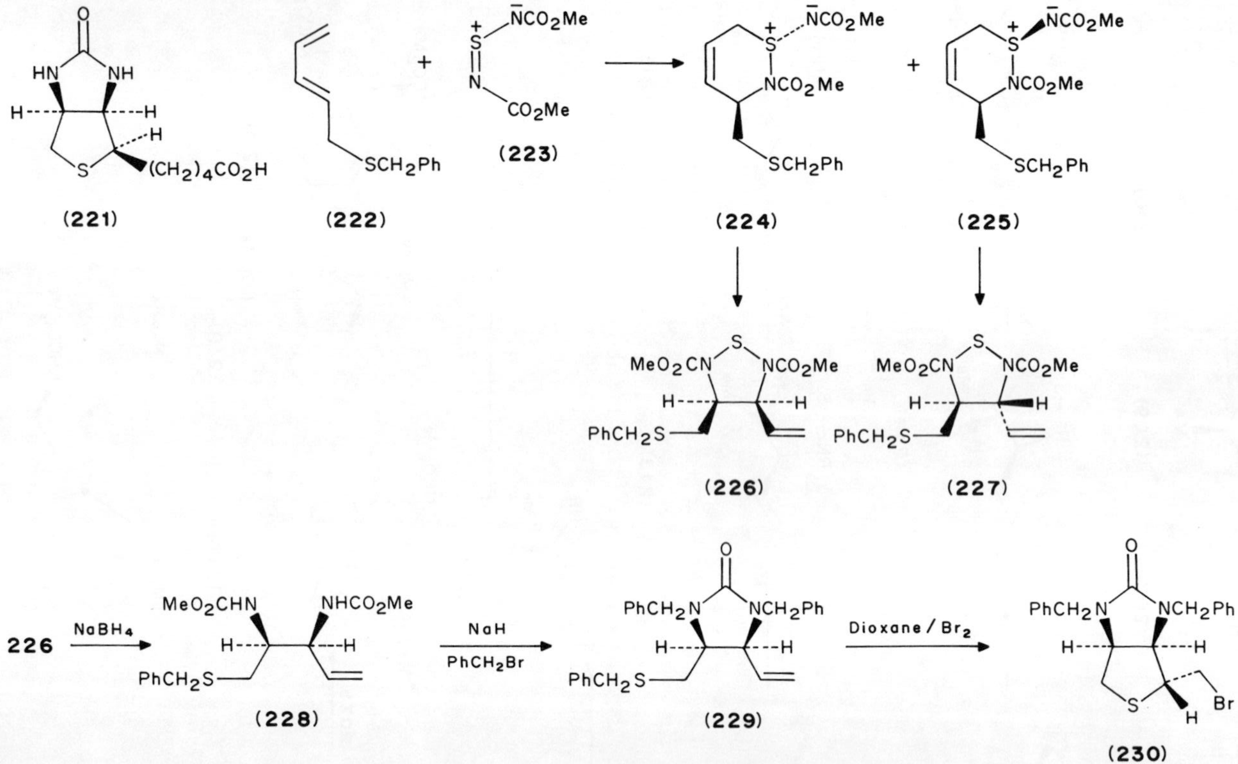
(221)
(222)
(223)
(224)
(225)
(226)
(227)
226
NaBH₄
(228)
NaH
PhCH₂Br
(229)
Dioxane / Br₂
(230)

SCHEME 27

used in an attempted stereospecific synthesis of biotin **221**. Condensation of **222** with diimide **223** gave a mixture of sulfinimidamides **224** and **225** which were readily converted to thiadiazolidines **226** and **227**. Subsequent elaboration of **226**→**230** was successful, however the procedure unfortunately gave product **230** with the undesired epimeric configuration to that of biotin at the tetrahydrothiophene ring[140] (Scheme 27).

Treatment of a disulfide **231** with chloramine T in alcohols gives a mixture of sulfinimidamide **232** and sulfinimidoate **233**. The reaction proceeds via an intermediate which can then react with $TsNH^-$ to give **232** or undergoes alcoholysis to **233**. When the alcohol used is (*l*)-menthol a separate mixture of diastereomeric sulfinimidoates is obtained[141].

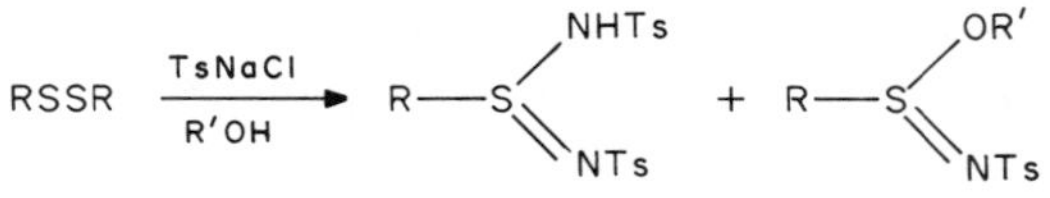


(231) (232) (233)

## V. THIOSULFINATES

The optical stability of thiosulfinates is rather small, since they frequently undergo rapid thermal racemization[1]. Various novel methods have been recently described for the synthesis of optically active sulfinates. Asymmetric oxidation of dibenzyl and di-*t*-butyl disulfides **236** with chiral 2-sulfonyloxaziridine **235** is a convenient procedure for the preparation of the corresponding (*S*)-thiosulfinates **237** in 2.1 and 13.8 e.e.%. Moreover, this procedure enabled the assignment of the absolute configuration of the oxidation products. It was suggested that the reaction proceeds via a chiral recognition model whereby the 2-chloro-5-nitrophenyl group behaves as if it were smaller than the camphorsulfonyl group. Thus, a preferable diastereomeric transition state should be attacked by the enantiotopic electron pair of sulfur on the oxaziridine oxygen in such a way that the large $R_L$ and small $R_S$ groups face the small and large regions of the oxaziridine ring[142] (Scheme 28).

The asymmetric oxidizing reagent Ti(O—*i*-Pr)$_4$/(+)-diethyl tartrate (DET)/$H_2O$ and *t*-BuOOH has been found useful in oxidation of disulfides **238** (X = S) to thiosulfinates **239**, and could also be used in converting sulfenamides (X = NR″) to sulfinamides and sulfenates (S=O) to sulfinates to give optically active products. The thiosulfinates were obtained in up to 40% e.e., whereas a smaller degree of asymmetric induction was detected for the sulfinamides and sulfinates. The simplest thiosulfinate MeSO—SMe was thus prepared for the first time in optically active form, and its absolute configuration was established to be (*S*), upon conversion to methyl *p*-tolyl sulfoxide. The absolute configuration of the N-*i*-propyl *p*-toluenesulfinamide obtained was also established by correlation to (*S*)-methyl *p*-tolylsulfinate. Other steric correlations were also made as indicated[143] (Scheme 29).

Acid-catalyzed displacement of sulfinamides **240** by thiols **241** has been found to proceed with inversion of configuration. This novel reaction is useful in the preparation of chiral thiosulfinates **242** with up to 80% stereospecificity. The thiosulfinates obtained did not racemize, and the sulfinamides underwent slow racemization under the reaction conditions. The reaction mechanism is understood to involve a sulfurane intermediate **A** which gave the product with inverted configuration, however, if the reaction proceeded via three Berry pseudorotations ($\psi$) to give sulfurane **B**, then the product with retained configuration was obtained[144] (Scheme 30).

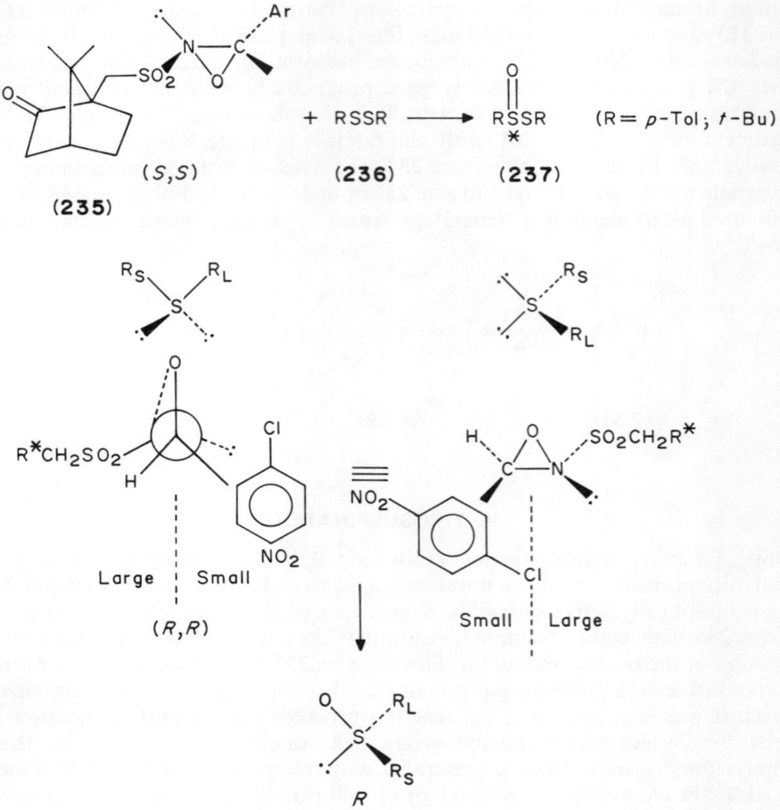


SCHEME 28. Chiral recognition model.

Asymmetric dehydration of *t*-butanesulfenic acid in the presence of optically active amines provides chiral thiosulfinates in up to 26% optical purity. Thermal decomposition of di-*t*-butyl sulfoxide **243** gave the intermediate sulfenic acid **244**, which easily condensed to the thiosulfinate **245**. The sulfenic acid was effectively achiral, although it exists in two tautomeric forms **244a** and **244b**, where the former is chiral. In the presence of chiral amines the amine–sulfenic acid complex underwent condensation in an asymmetric way, leading to the optically active sulfinates[145] (Scheme 31).

The chiral 2-thiacephem methyl ester **246a** when treated with *m*-chloroperbenzoic acid afforded two of the four possible thiosulfinates **247a** and **248a** in a 4:1 ratio. Upon further oxidation both isomers gave a single sulfone **249a** which formed by rearrangement of an intermediate α-disulfoxide **256** (Scheme 32). The formation of two isomeric sulfones (**249** and **250**) from a single α-sulfoxide **256** is understood to involve the formation of two isomeric sulfinate intermediates. Oxidation of **248a** alone gave mostly **249a** and only traces of the sulfone **250a**. Oxidation of the *trans* substrates **246b** and **246c** afforded only the single thiosulfinates **247b** and **247c**, which were respectively oxidized to **249b** and **249c**. From the oxidation of *cis* disulfide **251a**, the thiosulfinates **252a** and **253a** were isolated and these gave upon further oxidation similar mixtures of sulfones **254a** and **250a**. Analogous results were obtained from **251c** but the products were much less stable. The

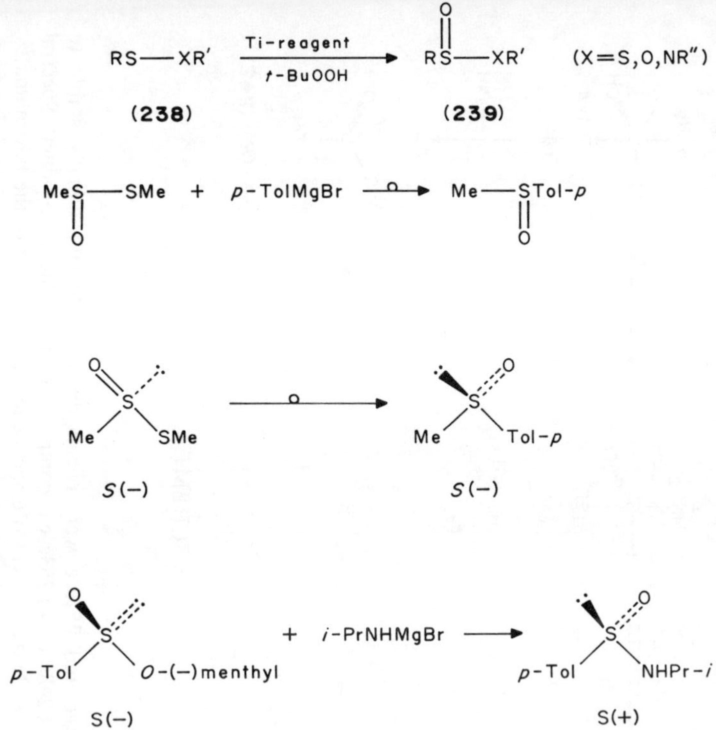

Determination of the absolute configurations of MeS(O)—SMe and p-TolS(O)NHPr-i

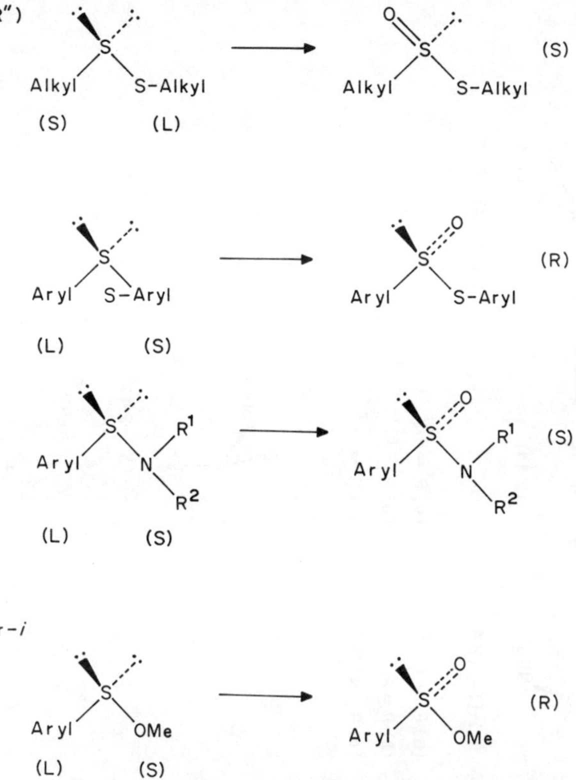

Steric course of the asymmetric oxidation of various prochiral sulfur derivatives by $Ti(Oi\text{-}Pr)_4$/(+) DET/$H_2O$/ t-BuOOH (1:2:1:1)

SCHEME 29

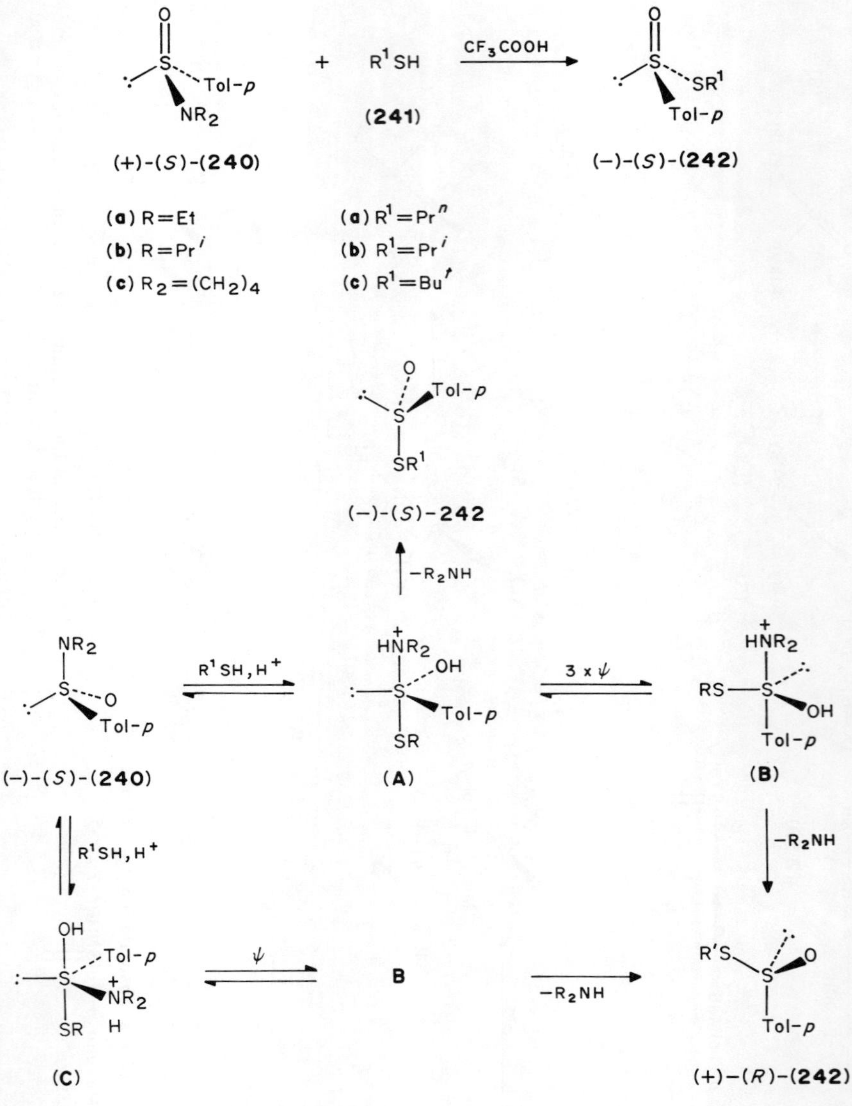


SCHEME 30

thiosulfinates **247a–c** and **252a, c** were chemically and configurationally stable to thermolysis, whereas **248a, c** and **253a, c** decomposed to non-$\beta$-lactam products. Spectral data, in particular $^{13}C$ NMR, provided evidence for the structure of the thiosulfinates, where downfield shifts for $C_6$ of 13.0 ppm and 7.7 ppm were observed in going from **246a** → **247a** and **246a** → **248a**, respectively. The orientation of the SO bond was

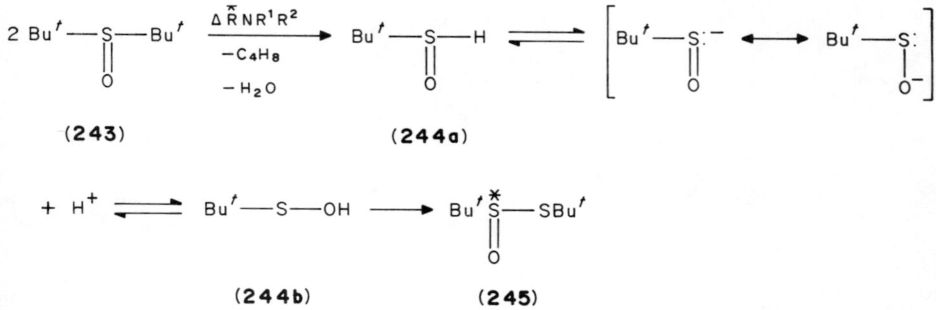

SCHEME 31

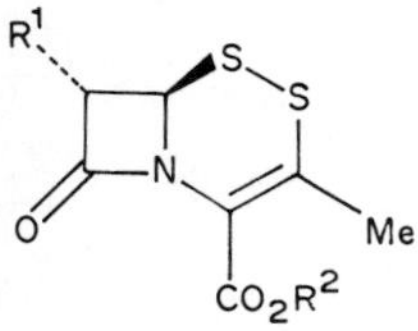

**(246)**

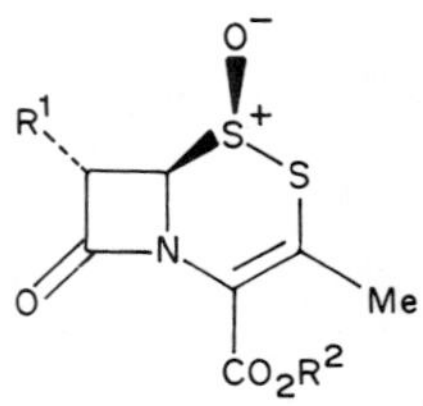

**(247)**

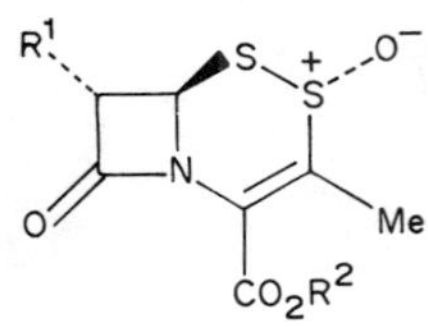

**(248)**

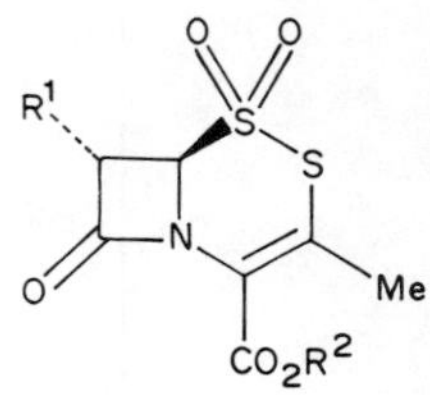

**(249)**

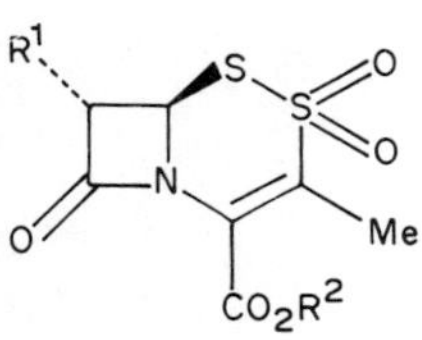

**(250)**

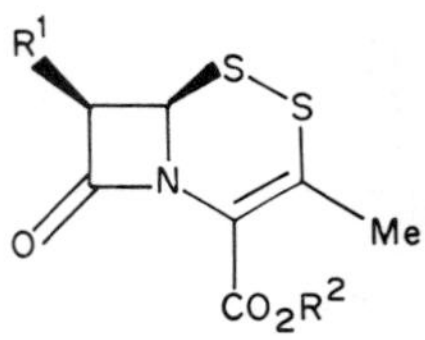

**(251)**

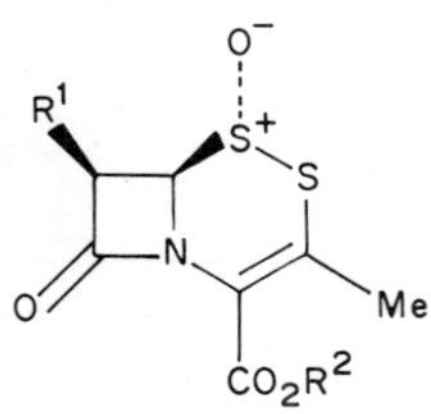

**(252)**

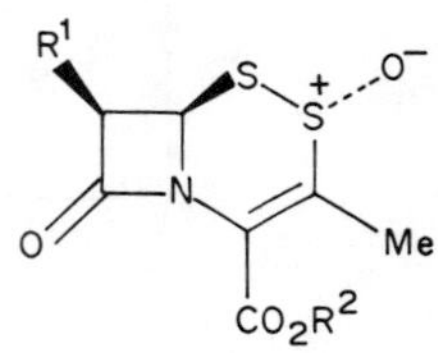

**(253)**

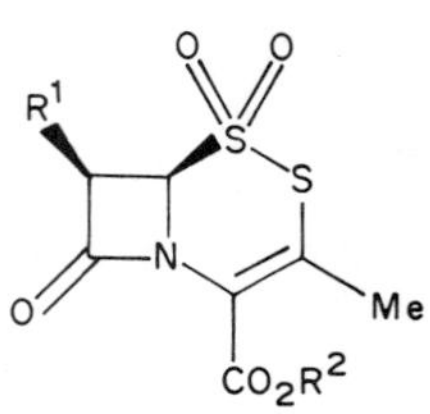

**(254)**

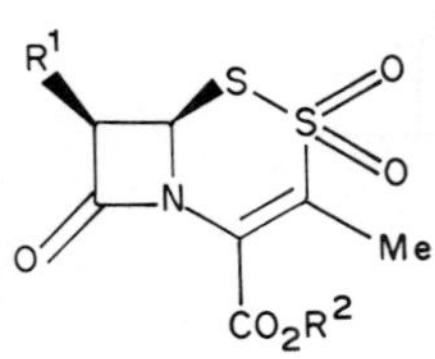

**(255)**

**(a)** $R^1$= 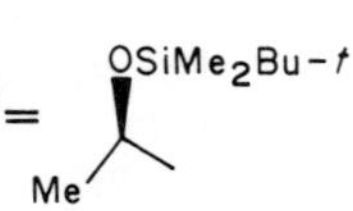 , $R^2$=Me **(b)** $R^1$= 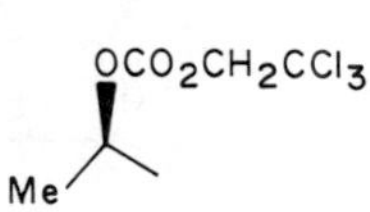 , $R^2$=$CH_2CCl_3$

**(c)** $R^1$= 1-phthalimido, $R^2$= *p*-nitrobenzyl

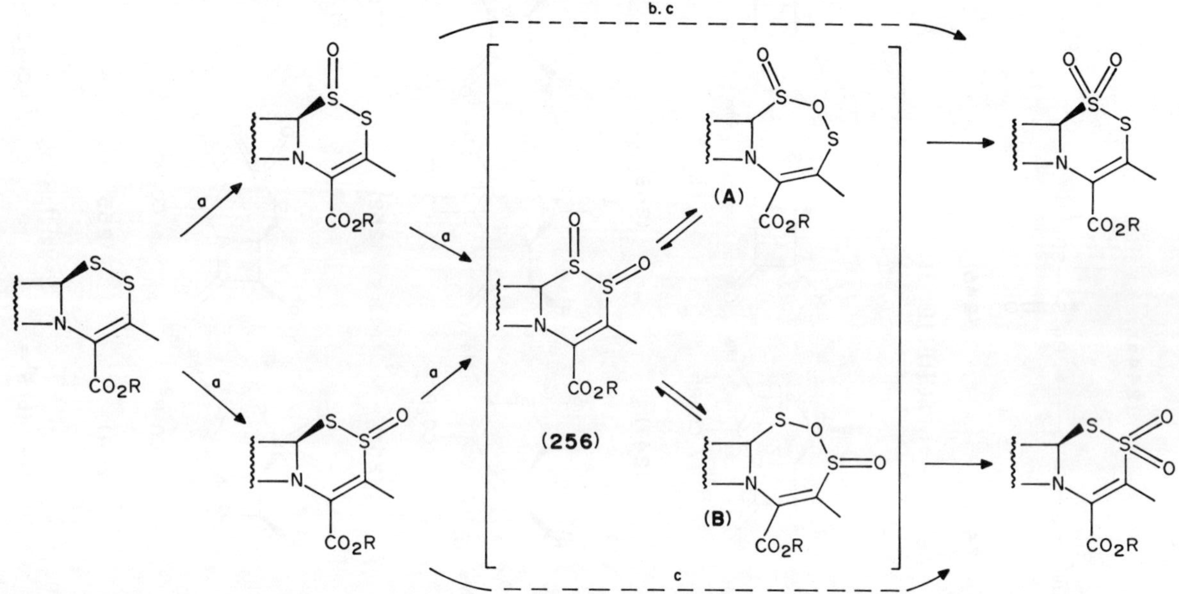


SCHEME 32

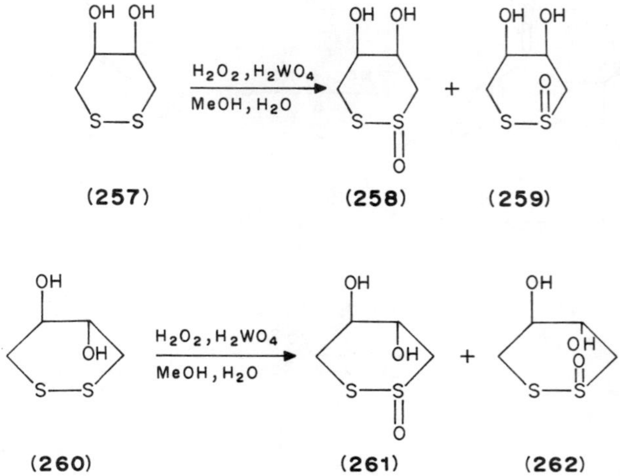


SCHEME 33

obtained by means of analysis of aromatic solvent shifts (ASIS) in the $^1$H NMR spectra determined in $DCCl_3$ and $C_6D_6$[146].

Oxidation of *cis* and *trans* diols **257** and **260**, respectively, gave separable mixtures of geometrically isomeric sulfinates **258** + **259** and **261** + **262** (Scheme 33). The stereochemistry of the products was assigned on the basis of broader IR bands at lower frequency for the hydrogen bonded OH and SO groups[147].

NMR experiments involving variable temperature, double irradiation, solvent effects and use of shift reagent $Eu(fod)_3$ indicate that the axial conformation of the cyclic thiosulfinate **263a** predominated to a great extent over the equatorial conformation[86]. In the case of the 4, 4, 5, 5-tetramethyl-1, 2-dithiane mono-*S*-oxide **263b** the predominance of the axial conformer was absolute[148,149] (Scheme 34).

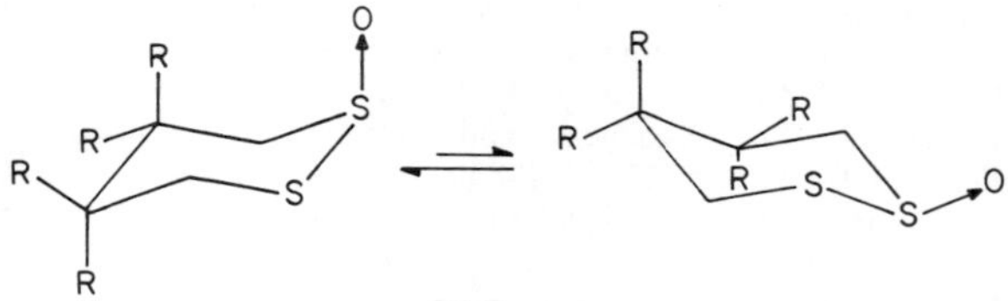

**263**

(**a**) R=H ; (**b**) R=Me

SCHEME 34

## VI. SULFINYL HALIDES

The chiral sulfinyl chlorides **265** obtained from sulfoxides **264** upon treatment with N-chlorosuccinimide (NCS) or N-chlorophthalimide (NCP) were each smoothly converted into a single isomeric chloromethyl sulfoxide **266** when reacted with $CH_2N_2$/excess HCl at $-20\,^{\circ}C$[150].

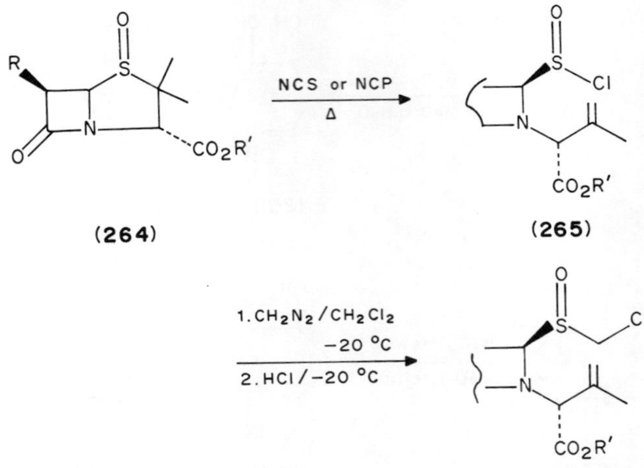


(264) (265) (266)

(**a**) R = phthalimido; R′ = *p*-nitrobenzyl, Me

(**b**) R = $PhOCH_2C(=O)NH$—; R′ = *p*-nitrobenzyl, $Ph_2CH$

## VII. REFERENCES


1. A. Nudelman, *The Chemistry of Optically Active Sulfur Compounds*, Gordon and Breach, New York, 1984.
2. E. Krauthausen, in *Methoden der Organischen Chemie* (*Houben–Weyl*), Vierte Auflage (extension), Vol. E11, (Ed. D. Kalman), Georg Thieme Verlag, Stuttgart, 1985, pp. 632–664.
3. M. Mikolajczyk and J. Drabowicz, in *Topics in Stereochemistry*, Vol. 13, (Eds. N. L. Allinger, E. L. Eliel and S. H. Wilen), Wiley, New York, 1982, p. 333.
4. J. Drabowicz and M. Mikolajczyk, *Org. Prep. Proced. Int.*, **14**, 45 (1982).
5. M. Mikolajczyk, in *Perspectives in Organic Chemistry of Sulfur*, Invited Lectures of the Twelfth International Symposium on the Organic Chemistry of Sulfur, Nijmegen, The Netherlands, 29 June–4 July 1986 (Eds. B. Zwanenburg and A. H. J. Klunder), Elsevier, Amsterdam, 1987, p. 23.
6. E. Wenschuh, K. Dolling, M. Mikolajczyk and J. Drabowicz, *Z. Chem.*, **20**, 122 (1980).
7. J. L. Kice, *Adv. Phys. Org. Chem.*, **17**, 66 (1980).
8. M. Cinquini, S. Colonna and A. Maia, Chim. Ind. (Milan), **62**, 685, 859 (1980).
9. S. Colonna, R. Annunziata and M. Cinquini, *Phosphorus and Sulfur*, **10**, 197 (1981).
10. G. Solladie, *Synthesis*, 185 (1981).
11. G. Solladie, *Chem. Scr.*, **25**, 149 (1985).
12. (a) G. Solladie, *Chimia*, **38**, 233 (1984).
    (b) G. Solladie, in Reference 5, p. 293.
13. G. H. Posner, in *The Chemistry of Sulfones and Sulfoxides* (Eds. S. Patai, Z. Rappoport and C. J. M. Stirling), Wiley, New York, 1987.
14. G. H. Posner, *Chem. Scr.*, **25**, 157 (1985).
15. K. Hiroi, *Yuki Gosei Kagaku Kyokaishi*, **44**, 907 (1986); *Chem. Abstr.*, **107**, 39270r (1986).
16. M. Hulce, J. P. Mallamo, L. L. Frye, T. P. Kogan and G. A. Posner, *Org. Synth.*, **64**, 196 (1986).
17. M. Furukawa, T. Owkawara, Y. Noguchi and M. Nishikawa, *Synthesis*, 441 (1978).
18. M. Furukawa, T. Ohkawara, Y. Noguchi, M. Nishikawa and M. Tomimatsu, *Chem. Pharm. Bull.*, **28**, 134 (1980).
19. R. B. Boar and A. C. Patel, *Synthesis*, 584 (1982).
20. K. Hiroi, R. Kitayama and S. Sato, *Synthesis*, 1040 (1983).

21. R. Kitayama, S. Iwata and K. Hiroi, Annu. Rep. Tohoku Coll. Pharm., 41 (1983); *Chem. Abstr.*, **101**, 170826v (1984).
22. D. D. Ridley and M. A. Smal, *Aust. J. Chem.*, **35**, 495 (1982).
23. M. Mikolajczyk, J. Drabowicz and B. Bujnicki, *Tetrahedron Lett.*, **26**, 5699 (1985).
24. M. Mikolajczyk, *Phosphorus and Sulfur*, **27**, 31 (1986).
25. M. G. Ambrose and R. W. Binkley, *J. Org. Chem.*, **48**, 674 (1983).
26. J. M. Klunder and K. B. Sharpless, *J. Org. Chem.*, **52**, 2598 (1987).
27. D. N. Harp, B. T. Friedlander, C. Larsen, K. Steliou and A. Stockton, *J. Org. Chem.*, **43**, 3481 (1978).
28. J. Drabowicz, S. Legedz and M. Mikolajczyk, *J. Chem. Soc., Chem. Commun.*, 1670 (1985).
29. J. Drabowicz, S. Legedz and M. Mikolaczyk, *Tetrahedron*, in press.
30. J. S. Grossert, P. K. Dubey and T. Elwood, *Can. J. Chem.*, **63**, 1263 (1985).
31. J. Drabowicz and M. Pacholczyk, *Phosphorus and Sulfur*, **29**, 257 (1987).
32. P. Brownbridge and I. C. Jowett, *Synthesis*, 252 (1988).
33. K. Yamanari and Y. Shimura, *Chem. Lett.*, 761 (1984).
34. R. Arad-Yellin, B. S. Green, M. Knossow and G. Tsoucaris, *J. Am. Chem. Soc.*, **105**, 4561 (1983).
35. S. Kokolja, W. A. Spitzer and J. K. Scott, *J. Org. Chem.*, **46**, 1934 (1981).
36. T. W. Hambley, M. C. Harsanyi and R. K. Morris, *J. Org. Chem.*, **53**, 3104 (1988).
37. E. J. Corey, L. O. Weigel, A. R. Chamberlin, H. Cho and D. H. Hua, *J. Am. Chem. Soc.*, **102**, 6613 (1980).
38. P. Magnus, T. Gallagher, P. Brown and J. C. Huffman, *J. Am. Chem. Soc.*, **106**, 2105 (1984).
39. H. Koisugi, H. Konda and H. Uda, *J. Chem. Soc., Chem. Commun.*, 211 (1985).
40. G. Solladie, J. Hutt and A. Girardin, *Synthesis*, 173 (1987).
41. P. Bravo, G. Resnati and F. Viani, *Tetrahedron Lett.*, **26**, 2913 (1985).
42. J. Drabowicz, B. Bujnicki and M. Mikolajczyk, *J. Org. Chem.*, **47**, 3325 (1982).
43. Y. Arai, S. Kuwayama, Y. Takehuchi and T. Koizumi, *Tetrahedron Lett.*, **26**, 6205 (1985).
44. R. Annunziata, M. Cinquini, S. Colonna and F. Cozzi, *J. Chem. Soc., Perkin Trans. 1*, 1005 (1981).
45. D. R. Williams and J. G. Phillips, *J. Org. Chem.*, **46**, 5452 (1981).
46. G. Solladie, R. Zimmermann, R. Bartsch and H. W. Walborsky, *Synthesis*, 662 (1985).
47. G. Solladie and R. Zimmermann, *J. Org. Chem.*, **50**, 4062 (1985).
48. P. Bravo, G. Resnati, F. Viani and A. Arnone, *Tetrahedron*, **43**, 4635 (1987).
49. D. G. Farnum, T. Veysoglu, A. M. Carde, B. Duhl-Emswiler, T. A. Pancost, T. J. Reitz and R. T. Carde, *Tetrahedron Lett.*, 4009 (1977).
50. S. Juge and G. Meyer, *Tetrahedron*, **36**, 959 (1980).
51. R. Annunziata, M. Cinquini, S. Colonna and F. Cozzi, *J. Chem. Soc., Perkin Trans. 1*, 614 (1981).
52. J. Nokami, T. Mandai, A. Nishimura, T. Takeda and S. Wakabayashi, *Tetrahedron Lett.*, **27**, 5109 (1986).
53. K. Ogura, M. Fujita, I. Ueda, H. Fushimi and H. Iida, *Kenkyu Hokoku-Asahi Garasu Kogyo Gijutsu Shoreikai*, **39**, 135 (1981); *Chem. Abstr.*, **97**, 161940z (1982).
54. K. Ogura, M. Fujita, K. Takahashi and H. Iida, *Chem. Lett.*, 1697 (1982).
55. C. Iwata, M. Fujita, K. Hattori, S. Uchida and T. Imanishi, *Tetrahedtron Lett.*, **26**, 2221 (1985).
56. H. Kosugi, M. Kitaoka, K. Tagami, A. Takahashi and H. Uda, *J. Org. Chem.*, **52**, 1078 (1987).
57. J. P. Marino, R. Fernandez de la Pradilla and E. Laborde, *Synthesis*, 1088 (1987).
58. H. Kosugi, M. Kitaoka, K. Tagami and H. Uda, *Chem. Lett.*, 805 (1985).
59. G. H. Posner, T. P. Kogan, S. R. Haines and L. L. Frye, *Tetrahedron Lett.*, **25**, 2627 (1984).
60. G. A. Posner, M. Weitzberg, T. G. Hamill, E. Asirvatham, H. Cun-heng and J. Clardy, *Tetrahedron*, **42**, 2919 (1986).
61. M. A. Brimble and B. R. Davis, *Tetrahedron*, **41**, 4965 (1985).
62. G. Solladie and G. Moine, *J. Am. Chem. Soc.*, **106**, 6097 (1984).
63. T. Koizumi, I. Hakamada and E. Yoshii, *Tetrahedron Lett.*, **25**, 87 (1984).
64. A. Guessous, F. Rouessac and C. Maignan, *Bull. Soc. Chim. Fr.*, 837 (1986).
65. M. Mikolajczyk, W. Midura and M. Kajtar, *Phosphorus and Sulfur*, **36**, 79 (1988).
66. Y. Arai, S. Kuwayama, Y. Takeuchi and T. Koizumi, *Synth. Commun.*, **16**, 233 (1986).
67. M. Cinquini, A. Manfredi, H. Molinari and A. Restelli, *Tetrahedron*, **41**, 4929 (1985).
68. R. Annunziata, M. Cinquini, F. Cozzi, F. Montanari and A. Restelli, *J. Chem. Soc., Chem. Commun.*, 1138 (1983).

69. R. Annunziata, M. Cinquini, F. Cozzi, F. Montanari and A. Restelli, *Tetrahedron*, **40**, 3815 (1984).
70. C. Mioskowski and G. Solladie, *Tetrahedron*, **36**, 227 (1980).
71. C. Papageorgiou and C. Benezra, *Tetrahedron Lett.*, **25**, 1303 (1984).
72. G. Solladie, F. Matloubi-Moghadam, C. Luttmann and C. Mioskowski, *Helv. Chim. Acta*, **65**, 1602 (1982).
73. M. C. Carreno, J. L. Garcia Ruano and A. Rubio, *Tetrahedron Lett.*, **28**, 4861 (1987).
74. G. Posner, K. Miura, J. P. Mallamo, M. Hulce and T. P. Kogan, *Current Trends in Organic Synthesis*, IUPAC, 1983, p. 177.
75. F. Schneider and R. Simon, *Synthesis*, 582 (1986).
76. R. Annunziata, M. Cinquini and F. Cozzi, *Synthesis*, 767 (1982).
77. A. Bernardi, L. Colombo, C. Gennari and L. Prati, *Tetrahedron*, **40**, 3769 (1984).
78. R. Annunziata, S. Cardini, C. Gennari and G. Poli, *Synthesis*, 702 (1984).
79. L. Banfi, L. Colombo, C. Gennari, R. Annunziata and F. Cozzi, *Synthesis*, 829 (1982).
80. R. Annunziata, F. Cozzi, M. Cinquini, L. Colombo, C. Gennari, G. Poli and C. Scolastico, *J. Chem. Soc., Perkin Trans. 1*, 251 (1985).
81. R. Annunziata, M. Cinquini, A. Restelli and F. Cozzi, *J. Chem. Soc., Perkin Trans. 1*, 1183 (1982).
82. R. Herrmann, G. Hubener and I. Ugi, *Tetrahedron*, **41**, 941 (1985).
83. N. Furukawa, T. Shibutani, K. Matsumura, H. Fujihara and S. Oae, *Tetrahedron Lett.*, **27**, 3899 (1986).
84. O. Itoh, T. Numata, T. Yoshimura and S. Oae, *Bull. Chem. Soc. Jpn.*, **56**, 266 (1983).
85. G. D. Figuly and J. C. Martin, *J. Org. Chem.*, **45**, 3728 (1980).
86. R. Annunziata, M. Cinquini, A. Gilardi and F. Cozzi, *Synthesis*, 1016 (1983).
87. M. Cinquini, F. Cozzi and M. Gilardi, *J. Chem. Soc., Chem. Commun.*, 551 (1984).
88. R. Annunziata, M. Cinquini, F. Cozzi and A. Restelli, *J. Chem. Soc., Chem. Commun.*, 1253 (1984).
89. R. Annunziata, M. Cinquini, F. Cozzi and L. Raimondi, *J. Chem. Soc., Chem. Commun.*, 403 (1985).
90. R. Annunziata, M. Cinquini, F. Cozzi, A. Gilardi and A. Restelli, *J. Chem. Soc., Perkin Trans. 1*, 2289 (1985).
91. T. Shimizu and M. Kobayashi, *J. Org. Chem.*, **52**, 3399 (1987).
92. T. W. Hambley, B. Raguse and D. D. Ridley, *Aust. J. Chem.*, **39**, 1833 (1986).
93. B. Raguse and D. D. Ridley, *Aust. J. Chem.*, **37**, 2059 (1984).
94. D. D. Ridley and M. A. Smal, *J. Chem. Soc., Chem. Commun.*, 505 (1981).
95. G. H. Posner, L. L. Frye and M. Hulce, *Tetrahedron*, **40**, 1401 (1984).
96. G. A. Posner, J. P. Mallamo, M. Hulce and L. L. Frye, *J. Am. Chem. Soc.*, **104**, 4180 (1982).
97. G. Demailly, C. Greck and G. Solladie, *Tetrahedron Lett.*, **25**, 4113 (1984).
98. K. Hiroi and N. Matsuyama, *Chem. Lett.*, 65 (1986).
99. K. K. Andersen, B. Bujnicki, J. Drabowicz, M. Mikolajczyk and J. B. O'Brien, *J. Org. Chem.*, **49**, 4070 (1984).
100. N. Kunieda, J. Nokami and M. Kinoshita, *Chem. Lett.*, 871 (1973).
101. P. Vermeer, H. Westmijze, H. Kleijn and L. A. van Dijk, *J. R. Neth. Chem. Soc.*, **97**, 56 (1978).
102. C. Maignan, A. Guessous and F. Rouessac, *Bull. Soc. Chim. Fr.*, 645 (1986).
103. K. Hiroi, R. Kitayama and S. Sato, *Chem. Pharm. Bull.*, **32**, 2628 (1984).
104. K. Hiroi, R. Kitayama and S. Sato, *J. Chem. Soc., Chem. Commun.*, 1470 (1983).
105. K. Hiroi, R. Kitayama and S. Sato, *Chem. Lett.*, 929 (1984).
106. K. Hiroi, R. Kitayama and S. Sato, *J. Chem. Soc., Chem. Commun.*, 303 (1984).
107. K. Hiroi and K. Makino, *Chem. Lett.*, 617 (1986).
108. M. Kobayashi, H. Minato, Y. Miyaji, T. Yoshioka, K. Tanaka and K. Honda, *Bull. Soc. Chem. Jpn.*, **45**, 2817 (1972).
109. S. Braverman and Y. Duar, *J. Am. Chem. Soc.*, **105**, 1061 (1983).
110. R. M. J. Liskamp, H. J. Blom, R. J. F. Nivard and H. C. J. Ottenheijm, *J. Org. Chem.*, **48**, 2733 (1983).
111. J. Drabowicz, *Phosphorus and Sulfur*, **31**, 123 (1987).
112. M. D. M. Gray, D. R. Russell, D. J. H. Smith, T. Durst and B. Gimbarzevsky, *J. Chem. Soc., Perkin Trans. 1*, 1826 (1981).
113. C. A. G. Haasnoot, R. M. J. Liskamp, P. A. W. van Dael, J. H. Noordik and H. C. J. Ottenheijm, *J. Am. Chem. Soc.*, **105**, 5406 (1983).

114. K. Okuma, H. Minato and M. Kobayashi, *Bull. Chem. Soc. Jpn.*, **53**, 435 (1980).
115. R. Annunziata, M. Cinquini and F. Cozzi, *J. Chem. Soc., Perkin Trans. 1*, 339 (1982).
116. K. Hiroi, R. Kitayama, and S. Sato, *Heterocycles*, **15**, 879 (1981).
117. K. Hiroi, S. Sato and R. Kitayama, *Chem. Pharm. Bull.*, **31**, 3471 (1983).
118. R. Bussas, H. Munsterer and G. Kresze, *J. Org. Chem.*, **48**, 2828 (1983).
119. S. G. Pyne, *Tetrahedron Lett.*, **27**, 1691 (1986).
120. S. G. Pyne, *J. Org. Chem.*, **51**, 81 (1986).
121. K. Hiroi, S. Sato and R. Kitayama, *Chem. Lett.*, 1595 (1980).
122. K. H. Pilgram and R. D. Skiles, *J. Org. Chem.*, **47**, 3865 (1982).
123. A. Dondoni, P. Giorgianni, A. Battaglia and G. D. Andreetti, *J. Chem. Soc., Chem. Commun.*, 350 (1981).
124. J. M. Whitesell and J. F. Carpenter, *J. Am. Chem. Soc.*, **109**, 2839 (1987).
125. H. Munsterer, G. Kresze, M. Brechbiel and H. Kwart, *J. Org. Chem.*, **47**, 2677 (1982).
126. S. M. Weinreb, R. S. Garigipati and J. A. Gainor, *Heterocycles*, **21**, 309 (1984).
127. R. S. Garigipati and S. M. Weinreb, *J. Am. Chem. Soc.*, **105**, 4499 (1983).
128. S. W. Remiszewski, R. R. S. Whittle and S. M. Weinreb, *J. Org. Chem.*, **49**, 3243 (1984).
129. S. W. Remiszewski, T. S. Stouch and S. M. Weinreb, *Tetrahedron*, **41**, 1173 (1985).
130. R. S. Garigipati, J. A. Morton and S. M. Weinreb, *Tetrahedron Lett.*, **24**, 987 (1983).
131. R. S. Garipipati, A. J. Freyer, R. R. Whittle and S. M. Weinreb, *J. Am. Chem. Soc.*, **106**, 7861 (1984).
132. R. S. Garigipati, R. Cordova, M. Pervez and S. M. Weinreb, *Tetrahedron*, **42**, 2979 (1986).
133. S. W. Remiszewski, J. Yang and S. M. Weinreb, *Tetrahedron Lett.*, **27**, 1853 (1986).
134. J. K. Whitesell, D. James and J. F. Carpenter, *J. Chem. Soc., Chem. Commun.*, 1449 (1985).
135. J. Drabowicz and M. Mikolajczyk, in *Proc. Int. Symp. Cyclodextrins, 1st, 1981* (Ed. J. Szejtli), Reidel, Dordrecht, Netherlands, 1981, pp. 205–215; *Chem. Abstr.*, **98**, 71128g (1981).
136. G. J. Gainsford, W. G. Jackson and A. M. Sargeson, *J. Chem. Soc., Chem. Commun.*, 875 (1981).
137. K. K. Andersen and O. Malver, *J. Org. Chem.*, **48**, 4803 (1983).
138. K. K. Andersen, S. Chumpradit and D. J. McIntyre, *J. Org. Chem.*, **53**, 4667 (1988).
139. H. Natsugari, R. R. Whittle and S. M. Weinreb, *J. Am. Chem. Soc.*, **106**, 7867 (1984).
140. E. Turos, M. Parvez, R. S. Garigipati and S. M. Weinreb, *J. Org. Chem.*, **53**, 1116 (1988).
141. C. Dell'Erba, G. P. Corallo, M. Novi and G. Leandri, *Phosphorus and Sulfur*, **12**, 123 (1981).
142. F. A. Davis, R. H. Jenkins, Jr., S. B. Awad, O. D. Stringer, W. H. Watson and J. Galloy, *J. Am. Chem. Soc.*, **104**, 5412 (1982).
143. C. Nemecek, E. Dunach and H. B. Kagan, *New J. Chem.*, **10**, 762 (1986).
144. J. Drabowicz and M. Mikolajczyk, *Tetrahedron Lett.*, **26**, 5703 (1985).
145. J. Drabowicz, P. Lyzwa and M. Mikolajczyk, *Phosphorus and Sulfur*, **16**, 267 (1983).
146. E. Perrone, M. Alpegiani, A. Bedeschi, D. Borghi, F. Giudici and G. Franceschi, *J. Org. Chem.*, **51**, 3413 (1986).
147. P. K. Singh, L. Field and B. J. Sweetman, *J. Org. Chem.*, **53**, 2608 (1988).
148. E. Juaristi, J. Guzman, V. V. Kane and R. S. Glass, *Tetrahedron*, **40**, 1477 (1984).
149. E. Juaristi and J. S. Cruz-Sanches, *J. Org., Chem.*, **53**, 3334 (1988).
150. D. O. Spry, *Tetrahedron Lett.*, **22**, 3695 (1981).

The Chemistry of Sulphinic Acids, Esters and their Derivatives
Edited by S. Patai


CHAPTER **4**

# Analytical methods

M. R. F. ASHWORTH

*Organische und Instrumentelle Analytik, Universität des Saarlandes, D-6600 Saarbrücken, FRG*



## I. INTRODUCTION

Few general analytical procedures have been published for sulphinic acids and their derivatives, such as salts, esters, acid amides, acid chlorides and anhydrides. Almost all of these few are for the free acids and their alkali metal salts.

Interest has centred often on some special individual compounds, which may be listed here:

(a) So-called 'Rongalite', $HOCH_2SO_2Na$, variously termed sodium formaldehyde sulphoxylate, sodium hydroxymethyl sulphite and sodium hydroxymethanesulphinate.

(b) Thiourea dioxide, with a tautomeric form of

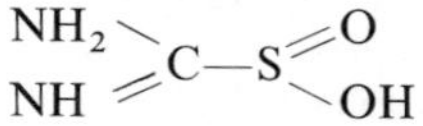

termed also formamidinesulphuric acid or amino-imino-methanesulphinic acid.

(c) Some amino acids, such as cysteinesulphinic acid (3-sulphinoalanine)

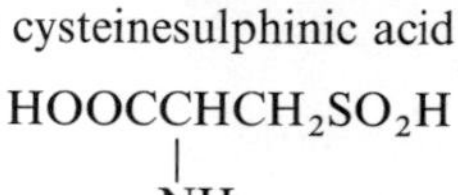

and hypotaurine(2-aminoethanesulphinic acid), $NH_2CH_2CH_2SO_2H$. Considerable analytical work (especially chromatography) has been carried out on the amino acids. The published examples include occasionally one or other of these sulphinyl-amino acids. It is unreasonable to quote any publication unless the sulphinyl compounds form a significant proportion of the whole, say about one quarter, or where these acids receive special mention.

(d) Neoarsphenamine is a derivative of an early chemotherapeutic agent, arsphenamine, in which a $—SO_2Na$ group has been introduced in order to increase solubility in water. The methods for assaying neoarsphenamine are usually based on determination of arsenic or sulphur, or on biological tests of toxicity. It could be argued that the determination of sulphur is indirectly a determination of the sulphinate group, since it is the only source of sulphur in the molecule. However, it cannot be regarded as a determination of the functional group, $—SO_2Na$, and such examples of determination are therefore not included here.

Methods have been given here for some compound classes containing sulphur(IV) which, with some goodwill, can be considered as sufficiently closely related to the sulphinic acids. These include:

sulphilimines or sulphidimines $\mathrm{(R)(R')S{=}NR'}$ or $\mathrm{(R)(R')S{=}NSO_2Tos}$

sulphinylamines $\mathrm{RN{=}S{=}O}$

The analytical methods have been divided into so-called chemical methods (i.e. those based on chemical reaction of the compound or compounds to be detected or determined), physical/instrumental methods (including polarography) and microbiological methods.

## II. CHEMICAL METHODS

The principal chemical methods are based on oxidation to the corresponding sulphonic acid or to sulphate, i.e. to sulphur(VI). About a dozen oxidizing agents have found use. A second group is that of acid/base reactions. These are usually direct titrations, under suitable conditions, of the sulphinic acids with standard bases, or of the salts (mostly alkali metal) of sulphinic acids with a standard acid. Further types of chemical methods are reactions of precipitation or colour change with metal-containing reagents (cations or anions), and a category of assorted procedures under the heading 'miscellaneous'.

There is no reason to doubt that derivatives of sulphinic acids can be determined by standard methods, for example: esters by hydrolysis using excess standard alkali and back titration of the unused amount; amides by hydrolysis with alkali and determination of the ammonia evolved; acid halides and anhydrides by the differential procedure using an alcohol and water, or by reaction with excess primary or secondary amines and determination of the unused amine. However, no published example could be found.

### A. Oxidation to Sulphur(VI)

This oxidation proceeds comparatively easily so that many reagents have been used. In fact, sulphinic acids are highly susceptible to atmospheric oxidation, which endangers the accuracy of oxidative determinations. Most quantitative determinations have been titrimetric. A convenient classification is according to frequency of use.

#### *1. Nitrite*

The reaction of sulphinic acids with nitrous acid was used preparatively by Koenigs[1] as long ago as 1878. His reaction equation was

$$2RSO_2H + HNO_2 \longrightarrow (RSO_2)_2NOH + H_2O$$

This reaction has been the basis of methods of titration. Thus Marvel and coworkers[2] determined the purity of dodecanesulphinic acid by potentiometric titration in acetic acid solution with sodium nitrite. In a study of the reactions of this sulphinic acid, Marvel and Johnson[3] titrated the magnesium salt in acetic acid/hydrochloric acid at 0 °C with sodium nitrite to an external indicator of starch/iodide. They referred to a tendency to formation of the product $(RSO_2)_3NO$ in the presence of excess nitrous acid. This should not affect direct titration in a significant way. Further examples of titration to external starch/iodide can be found in the work of: Ponzini[4], who titrated aromatic sulphinic acids in water cooled to 5 °C and containing hydrochloric acid, taking as end-point a colour persisting for at least several minutes; Kice and Bowers[5] in a study of disproportionation of sulphinic acids; and Danehy and Elia[6] for *p*-chlorobenzenesulphinic acid.

In his study of various methods for determining aromatic sulphinic acids and their salts, Lindberg[7] titrated with sodium nitrite a solution of the sample in dilute sulphuric acid containing potassium bromide and determined the end-point potentiometrically at a platinum electrode. Fleszar[8], likewise in an investigation of various procedures, titrated an acidified solution (hydrochloric acid) of sodium benzenesulphinate with sodium nitrite but used amperometric end-point indication with two platinum electrodes. Matrka and collaborators[9] also employed amperometric or potentiometric end-point determination with platinum and calomel electrodes, or titrated biamperometrically with two platinum electrodes; they titrated aromatic sulphinic acids in dilute hydrochloric acid at 20 °C.

Marek[10] detected some aromatic sulphinic acids and their derivatives on paper chromatograms by exposure for 1 min to the vapours from a mixture of sodium nitrite and hydrochloric acid, followed by spraying on a solution of R-salt (2-naphthol-3,6-disulphonic acid) and exposure to ammonia to yield yellow-green spots. Czerwicz and Malata[11] visualized substituted phenylsulphinylamines Ph—N=S=O, on silica gel thin layer chromatograms, by exposure for 30–60 s to nitrous gases from a sodium nitrite/hydrochloric acid mixture. After 24 h the compounds appeared as yellow, orange or red-brown spots.

### 2. *Hypohalite (and chloramine T)*

After Allen[12] had found low results in oxidative titrations (e.g. with permanganate) of acidified solutions of sodium and magnesium alkanesulphinates, he proposed direct titration with basic calcium hypochlorite [Ca(OCl)Cl, 'bleaching powder'] in alkaline or neutral solution and using an external indicator of starch/iodide. Ackerman[13] titrated sodium benzenesulphinate with standardized sodium hypochlorite, likewise using an external starch/iodide indicator. He gave the reaction equation

$$2\,PhSO_2Na + NaOCl \longrightarrow PhSO_3Na + PhSO_2Cl$$

In order to determine halates, Atkin[14] standardized hypochlorite solution by titrating with sodium benzenesulphinate solution; he, too, used the starch/iodide external indicator.

Coulometric titration with hypochlorite and hypobromite was performed by Liberti and Lazzari[15]. Their examples included sodium benzenesulphinate. They generated chlorine electrolytically at pH 1.3, bromine likewise at pH 5.8. In each case the halogen was led into a 0.6 M solution of hydrogen carbonate of pH 8.3 containing the sample. They used amperometric end-point indication with rotating platinum electrodes at +0.2 V. Hashmi and Ayaz[16] also utilized the reaction of benzenesulphinate (and other reducing agents) with hypochlorite and chlorite, but their aim was the determination of inorganic ions. They titrated the sulphinate with hypochlorite using the starch/iodide external indicator, but also with tartrazine and Bordeaux as internal indicators if sodium hydrogen carbonate and a little potassium bromide were added to the solution to improve the end-point.

In an adaptation of the Ackerman hypochlorite procedure (see above) Uhlenbroek and coworkers[17] used chloramine T for titrating benzenesulphinic acids containing various substituents [e.g. *o*-acetylamino-, *p*-(2-hydroxy)ethoxy-]. They added sodium hydroxide solution to the acid sample until methyl orange indicator just changed colour. Barium chloride was then added and any precipitate filtered off. An aliquot of the filtrate was treated with hydrochloric acid and a measured amount in excess of the chloramine T reagent solution was added. After 2 min reaction time they added solid potassium iodide and titrated with thiosulphate the iodine liberated by unused reagent.

Sumizu[18] determined hypotaurine (2-aminoethanesulphinic acid) by oxidation with excess hypoiodite. Unreacted reagent was then decomposed with phosphoric acid to iodine which was estimated by absorption at 590 nm after treatment with starch.

### *3. Iodine, iodine oxyacids and iodine monochloride*

Iodine has been used to titrate sodium hydroxymethanesulphinate ('Rongalite') according to the reaction

$$HOCH_2SO_2^- + I_2 + H_2O \longrightarrow HOCH_2SO_3^- + 2HI$$

Examples are the work of: Salkin[19], who titrated with iodine but considered it to be less satisfactory than titration with copper(II) because the iodine reacted also with sodium hydrogen sulphite and sodium thiosulphate; Furness[20], who titrated the Rongalite in connection with a polarographic study; Badinand and Rondelet[21], who determined the hydroxymethanesulphinate present as an antioxidant in *p*-aminosalicylate by adding phosphoric acid before titrating with iodine: and Maros[22], who determined it in the presence of formaldehyde bisulphite, $HOCH_2SO_3^-$, by adding formaldehyde and acetate/acetic acid and titrating with iodine. Tsau and Poole[23] commented on the difficult end-point indication in conventional titration of antioxidants including the hydroxymethanesulphinate. They proposed a method involving HPLC of reaction mixtures obtained by adding consecutive amounts of titrant. They used a column of C 18 HL (Alltech) and four mobile phases, monitoring spectrophotometrically.

Thiourea dioxide has also been determined by titration with iodine, e.g. by Wojtasiewicz-Obrzut[24] and Shafran and coworkers[25]. The compound is oxidized to urea and sulphate:

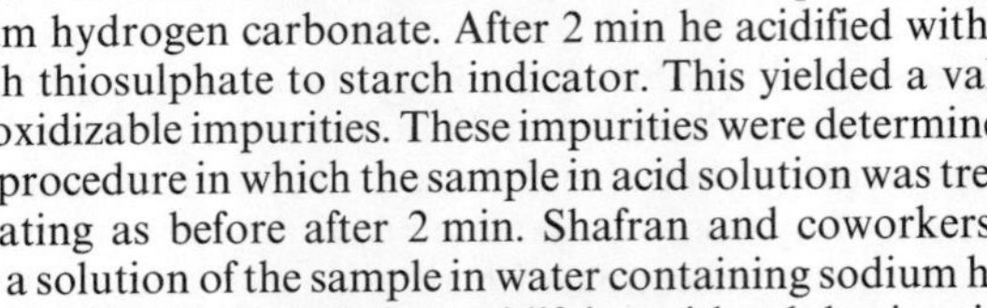

The former added excess standard iodine solution to an aqueous solution of the sample containing sodium hydrogen carbonate. After 2 min he acidified with sulphuric acid and back titrated with thiosulphate to starch indicator. This yielded a value for the thiourea dioxide and any oxidizable impurities. These impurities were determined and corrected for by an analogous procedure in which the sample in acid solution was treated with the excess iodine, back titrating as before after 2 min. Shafran and coworkers also added excess iodine reagent to a solution of the sample in water containing sodium hydrogen carbonate; they, too, used 2 min reaction time before acidifying with sulphuric acid and back titrating with thiosulphate to starch, while impurities were similarly determined in a blank. Mahadevappa[26] used potassium iodate and periodate to titrate sodium hydroxymethanesulphinate, finding also that 4 equivalents of oxidizing agent were consumed.

Iodine monochloride was used by Krishnan Nambisan and Ramachandran Nair[27] to determine some poorly soluble compounds, including Rongalite. They used excess reagent in 5 M hydrochloric acid. After 10 min they added 10% potassium iodide solution and titrated the liberated iodine with thiosulphate. Four equivalents took part in the oxidation:

$$HOCH_2SO_2^- + 4ICl + 2H_2O \longrightarrow HCHO + SO_4^{2-} + 2I_2 + 4Cl^- + 5H^+$$

### 4. Bromine and bromide/bromate

Ramberg[28] titrated ethanesulphinic acid in a hydrochloric acid medium with potassium bromate, taking decolouration of added methyl orange as end-point. He admitted that some atmospheric oxidation took place. Faster titration gave higher values and results were better in the presence of potassium bromide. Fleszar[8] titrated sodium benzenesulphinate with several reagents, including potassium bromate. He dissolved the sample in water, strongly acidified with hydrochloric acid, and added potassium bromide. End-point indication was potentiometric. An example of titrimetric determination of thiosulphinate using bromide/bromate is the work of Ostermayer and Tarbell[29]. They dissolved the sample in 80% acetic acid and took the first permanent appearance of bromine colour in the solution as their end-point. The reaction equation is

$$\underset{\displaystyle SCH_3}{CH_3\overset{}{S}}{=}O + 4Br_2 + 3H_2O \longrightarrow 2CH_3SO_2Br + 6HBr$$

(Siggia and Edsberg[30] used similar conditions for titrating disulphides, which also yield sulphonyl bromides)

Bromine has found some limited use for detecting organic sulphur compounds on chromatograms. Thus Bayfield and collaborators[31] studied the visualization of thiosulphinates and sulphinamides (also thiols and sulphides) by drawing the paper chromatograms through 3% aniline in petrol ether, allowing the petrol ether to evaporate and then exposing to bromine vapour. This gave blue or mauve spots within 30–60 s and was more sensitive for thiosulphinates than for sulphinamides.

### 5. Cerium(IV)

The stoichiometry of the cerium(IV) reaction with sulphinates is controversial. Forrest and Ryan[32] reported titrations with cerium(IV) sulphate of sodium benzenesulphinate, using ferroin as indicator, in which they found a mole ratio of cerium to sulphinate of 1.8 to 1; they quoted diphenyl disulphone, $PhSO_2SO_2Ph$, as a reaction product. Gringras and Sjöstedt[33] also titrated aromatic sulphinic acids with cerium(IV) sulphate, using potentiometric end-point indication with platinum electrodes. They found varying mole ratios, depending on the concentration (within the range $5 \times 10^{-3}$ to $10^{-1}$ M). One equation given by them was

$$4ArSO_2^- + 6Ce^{4+} + 2H_2O \longrightarrow ArSO_2SO_2Ar + 2ArSO_3^- + 6Ce^{3+} + 4H^+$$

Tsaikov[34] reported titrations of benzene- and *p*-toluenesulphinic acids in sulphuric acid solution at pH 1–2, using cerium(IV) sulphate (and also potassium dichromate) and potentiometric end-point determination at a platinum electrode. Grossert and Langler[35] used the 'cerium ammonium nitrate' reagent (ammonium hexanitratocerium), $(NH_4)_2[Ce(NO_3)_6]$, in nitric acid solution as a spray reagent for visualizing some organic sulphur compounds on silica gel thin layers. These compounds included thiols, disulphides and two thiosulphinate esters, the methyl esters of benzenesulphinic and chloromethanesulphinic acids. These esters appeared at room temperature as colourless zones on a yellow background.

### 6. Permanganate

Permanganate has long been known as an oxidizing agent for sulphinates. Reuterskiöld[36] titrated sulphinic acids and sulphinoacetic acid, $HOOCCH_2SO_2H$, with this reagent. As mentioned in Section II.A.2, Allen[12] used potassium permanganate to titrate aliphatic sulphinate salts. He found low results in acid solution but gave two procedures for the use of permanganate in non-acidic solution: (a) direct potentiometric titration in alkaline or neutral solution; (b) the sample together with sodium hydroxide was left in contact with a measured amount of permanganate reagent in excess. After 5 min he added excess standard arsenite solution, some potassium iodate as catalyst, then concentrated hydrochloric acid and finally titrated the unused arsenite with potassium permanganate. Lindberg[7] investigated comprehensively the determination of aromatic sulphinic acids and their sodium salts. His procedures included direct potentiometric titration (platinum indicator electrode) with potassium permanganate, and oxidation with excess permanganate reagent in neutral solution, followed after 10 min by adding acid and potassium iodide and finally titrating the liberated iodine with thiosulphate to a dead-stop end-point. Fleszar and coworkers[37] determined the 4,4′-disulphinic acid of diphenyl ether by precipitation as zinc salt (Section II.C.5) and titrating this in acid solution with permanganate. In an extensive study of the paper chromatography of many compounds Reio[38] detected cysteinesulphinic acid (3-sulphinoalanine) with several reagents which included permanganate.

### 7. Copper(II)

Copper(II) was used as a titrant for hydroxymethanesulphinate in early work. Thus Helwig[39] titrated this sulphinate (sodium salt) in aqueous solution with an ammoniacal copper(II) reagent in a stream of carbon dioxide until decolourization ceased. The solution was then heated over a free flame and titrated further until a faint blue colour persisted for 10 s at the boiling temperature. Salkin[40] also considered copper(II) sulphate in concentrated ammonium hydroxide to be the best titrant for Rongalite, sodium hydroxymethanesulphinate, superior to iodine which reacted with other compounds possibly present, such as hydrogen sulphite and thiosulphate. Spitzer[41] determined the sodium hydroxymethanesulphinate by adding excess concentrated copper(II) sulphate solution, acidified with sulphuric acid. After a short interval he added potassium iodide to yield iodine with the unused copper(II); this was titrated with thiosulphate.

### 8. Mercuric ion

Mercuric chloride has been used in analytical work with sulphinates. Thus Spitzer[42] determined Rongalite by reacting it with mercuric chloride:

$$HOCH_2SO_2^- + 2HgCl_2 + H_2O \longrightarrow HOCH_2SO_3^- + Hg_2Cl_2 + 2HCl$$

He filtered off the precipitated mercurous chloride, dissolved it in excess standard iodine and back titrated the unused part.

Mercuric chloride has also been used to prepare solid crystalline derivatives of sulphinic acids:

$$RSO_2H(Na) + HgCl_2 \longrightarrow RHgCl + H(Na)Cl + SO_2$$

This reaction was described as long ago as 1905 by Peters[43]. Marvel and coworkers[2] were the first to prepare such a derivative of an aliphatic sulphinic acid (dodecanesulphinic acid).

A spot test for aromatic sulphinic acids depends on this reaction yielding sulphur dioxide. Feigl and coworkers[44] heated the sample to 80 °C with mercuric chloride and tested for sulphur dioxide evolved through the blue colour given by Congo paper treated with hydrogen peroxide and held in the issuing gases.

### 9. Chromium(VI)

Tsaikov[34] titrated benzene- and *p*-toluenesulphinic acids in sulphuric acid solution at pH 1–2 with potassium dichromate, using potentiometric end-point indication with a platinum electrode.

Langler[45] visualized some sulphur-containing compound classes (including sulphinate esters) on silica gel $HF_{254}$ thin layers with a chromium trioxide/acetic acid reagent. The sulphinate esters tested (methyl esters of benzenesulphinic and chloromethanesulphinic acids) appeared after 15–30 min as blue-green spots. This detection was considered superior to that with cerium(IV) (see Section II.A.5).

### 10. o- and p-Dinitrobenzene

Feigl and Mainberger[46] distinguished hydroxymethanesulphinate (Rongalite) and dithionite through their reactions with *o*- or *p*-dinitrobenzene in the presence of alkaline ethanol. Both reduce the reagent to violet or orange quinonoid compounds on heating but only dithionite accomplishes this in the cold. Feigl and coworkers[44] detected alkali metal or calcium salts of aromatic sulphinic acids by reaction with *o*-dinitrobenzene to yield nitrite, which is itself detected with the help of the Griess reagent of sulphanilic acid/α-naphthylamine (the former is diazotized in acid solution by the nitrous acid and the

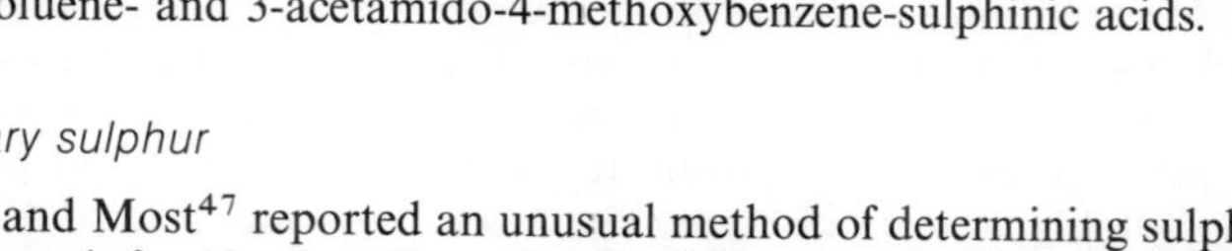


diazonium ion couples with the latter component). Feigl and coworkers tested this with benzene-, *p*-toluene- and 3-acetamido-4-methoxybenzene-sulphinic acids.

### 11. Elementary sulphur

Scandurra and Most[47] reported an unusual method of determining sulphinates. They refluxed the sample for 10 min with a solution of elementary sulphur in 95% ethanol, plus some drops of concentrated ammonium hydroxide. After evaporating to dryness they dissolved the residue in water and removed unused sulphur with chloroform. The aqueous phase was then heated with ammonium hydroxide and potassium cyanide on the water bath for 30 min. After cooling they added a reagent of ferric nitrate in nitric acid, yielding a red colour, evaluated at 470 nm after 10 min. Evidently the sulphinate is first converted by the sulphur into thiosulphinate, ArS(=O)SH, which then reacts with the cyanide to give thiocyanate. This in turn reacts with the ferric ion to give the characteristic red product.

### 12. Vanadate/ferrocyanide

Sodium hydroxymethanesulphinate (also hydrosulphite and stannous chloride) but neither sulphite, hydrogen sulphite nor formaldehyde, respond positively to the test of Gentil and Miranda[48]. They used a reagent of potassium ferrocyanide and vanadate which together yield a green precipitate. This is reduced by the sulphinate to a red product.

*13. Photochemical oxidation*

An unusual example of oxidation of sulphinate to sulphonate come from the work of Kataoka and colleagues[49] on hypotaurine. They treated it first with *i*-butyl chloroformate, then photooxidized this product through irradiation with a 300 W tungsten lamp for 5 min at 20 cm distance. The product was then converted into the sulphonyl chloride with thionyl chloride and the sulphonyl chloride finally reacted with dibutylamine to yield the end-product of 2-(isobutoxycarbonylamino)-*N*,*N'*-dibutylethanesulphonamide.

$$H_2NCH_2CH_2SO_2H \xrightarrow{ClCOOBu\text{-}i} i\text{-}BuOCONHCH_2CH_2SO_2H \xrightarrow{\text{Photooxidation}}$$

$$i\text{-}BuOCONHCH_2CH_2SO_3H \xrightarrow{SOCl_2} i\text{-}BuOCONHCH_2SO_2Cl \xrightarrow{Bu_2NH}$$

$$i\text{-}BuOCONHCH_2CH_2SO_2NHBu$$

## B. Acid/Base Reactions of the Acids and their Salts

*1. Determination and identification of free acids with bases*

Free sulphinic acids can, of course, be directly titrated with bases. They are several powers of ten weaker than the corresponding sulphonic acids so that a differentiating titration of the two, e.g. potentiometrically, is possible. An early reference to an acidimetric determination is that of Krishna and Bhagwan Das[30]. They used the potassium iodide/potassium iodate reagent at 0 °C from which the acid liberated iodine according to the usual reaction:

$$KIO_3 + 5KI + 6H^+ \longrightarrow 3I_2 + 3H_2O$$

They did not determine the iodine by the customary titration with thiosulphate or arsenite. Instead, the reaction mixture was allowed to warm up to room temperature and then treated with alkaline hydrogen peroxide. This yielded oxygen which they determined gas-volumetrically.

Wetzel and Meloan[51] titrated several aromatic sulphinic acids (with methyl, ethyl, butyl and other nuclear substituents) in non-aqueous solution. They tested many solvents, using quaternary ammonium hydroxides in benzene-methanol as titrant. End-point indication was potentiometric. They were able to titrate the sulphinic acids in the presence of the corresponding sulphonic acids in benzene-methanol, t-butanol, DMF–DMSO, tetrahydrofuran and pyridine solution.

The usual organic bases can serve for preparing sulphinate salts for identification of the free sulphinic acids. An early example is the work of Hälssig[52] who prepared salts with phenylhydrazine. Solid derivatives for identification through melting point are, however, generally prepared from alkali metal salts of the acids (see Section II.D.2b below).

*2. Determination of sulphinate salts with acids*

As salts of strong bases and comparatively weak acids, the alkali metal sulphinates can be titrated with strong acids. End-points may not be very sharp in aqueous solution because the sulphinic acids have p*K* values between 2 and 3 (methanesulphinic acid, *ca.* 2.3; benzenesulphinic acid, *ca.* 2.8). Titration is non-aqueous solution is, however, successful. Examples of this are the extensive investigations of Lindberg[7] and Fleszar[8]. Among other procedures, each titrated with perchloric acid. Lindberg titrated *p*-methoxybenzenesulphinate in acetone-methanol or methanol-*i*-butyl acetate, using perchloric acid in dioxan, and some other substituted benzenesulphinic acids (*p*-methyl, *p*-chloro) in acetic acid using perchloric acid in this solvent also. End-point determination was potentiometric with a platinum indicator electrode. Fleszar titrated sodium benzenesulphinate potentiometrically in acetic acid–dioxan (1 + 1) using glass and silver chloride electrodes.

## C. Reactions with Metal-containing Reagents

Reaction with metal-containing cations or anions has been used for quantitative determination of sulphinates via precipitation or colour change, and also for detection and identification.

### *1. Ferric ion*

The earliest reference to the sensibly quantitative reaction of sulphinates with ferric ion appears to be that of Thomas[53]. He obtained orange-yellow ferric salts of aromatic sulphinic acids in practically theoretical yield by adding ferric chloride to the strongly acidified sulphinic acid solution. Krishna and Singh[54] determined a wide range of aromatic sulphinic acids by addition to acidified ferric chloride in excess. After filtration they reduced unused ferric ion in the filtrate with stannous chloride in concentrated hydrochloric acid or zinc dust and sulphuric acid, and titrated the resulting ferrous ion with dichromate to an external indicator of potassium ferrocyanide. Forrest and Ryan[55] undertook a comprehensive study of the reactions between a wide range of metal cations and several aromatic sulphinic acids (benzene-, *p*-toluene-, naphthalene-2-, thiophene-2- and benzyl-). The work was aimed more at detection, identification and determination of the metals. They stated that ferric ion gave at least two products with benzenesulphinic acid. The ferric trisulphinates were soluble in many organic solvents, a property which does not appear to have been exploited analytically. Alimarin and Kuznetsov[56] tested a new reagent for ferric ion, *o*-hydroxybenzenesulphinic acid. At pH values between 1.9 and 7.53 it yielded a violet complex with the ferric ion, evidently in the ratio of 1:1. This could be used for colorimetric estimation of ferric ion. Presumably the method could be used in the reverse sense, to determine the sulphinic acid. However, the colour is probably due to the *o*-hydroxyl group (like salicylic acid).

Ferric chloride has been used for chromatographic visualization of sulphinic acids and related compounds. Thus Mondoví and coworkers[57] used it in the paper electrophoresis of some sulphur-containing compounds of biological interest, inclusing cysteinesulphinic acids and hypotaurine. Fondarai and Richert[58] also used ferric chloride to reveal cysteinesulphinic acid on paper chromatograms. An example of its use in thin-layer chromatography comes from the work of Westley and Westley[59], who visualized organic thiosulphinates (also sulphonates and thiosulphonates) on silica gel layers using ferric chloride in acetone.

### *2. Platinum(IV) (hexachloro, hexaiodo-platinate)*

These reagents with a metal-containing anion have been used in analytical procedures depending on colour change. Thus Winegard and collaborators[60] visualized sulphur-containing acids, including cysteinesulphinic acid, on paper chromatograms by spraying with a hexachloroplatinate-potassium iodide reagent. They then suspended the paper strip in hydrogen chloride fumes, which revealed the amino acids as uncoloured zones on a pink background. Fondarai and Richert[58] also visualized cysteinesulphinic acid and other amino acids with this iodoplatinate reagent, while De Marco and coworkers[61] likewise applied this method of detection to hypotaurine and homohypotaurine (and corresponding selenium compounds) on paper chromatograms. Jolles-Bergeret[62] profited from the fact that cysteine- and homocysteinesulphinic acids decolourize hexachloroplatinic acid in acetic acid solution. Quantitative determination was possible, based on the decrease in light absorption at 500 nm.

### 3. *Palladium(II) (tetrachloropalladate)*

Åkerfeldt and Loevgren[63] reported that various sulphur-containing compound classes (thiols, sulphides, disulphides and sulphinic acids) yielded coloured complexes with palladium(II). They treated the sample with ammonium tetrachloropalladate, $(NH_4)_2[PdCl_4]$, in hydrochloric acid. Colour developed within 5 min with the thiols and sulphinic acids and spectrophotometric determination was possible at wavelengths between 350 and 415 nm.

### 4. *Thallium(III)*

Gilman and Abbott[64] characterized the sodium salt of *p*-toluenesulphinic acid (and many sulphonic acids) by conversion to a derivative of the formula *p*-$TolSOTlCl_2$. This was accomplished by mixing aqueous solutions of the salt(s) and thallium(III) chloride. Other sulphinate salts should react similarly.

### 5. *Zinc(II)*

Fleszar and coworkers[37] precipitated the 4,4′-disulphinic acid of diphenyl ether (as disodium salt) with excess zinc sulphate, centrifuged the product, acidified it and titrated with permanganate (Section II.A.6).

## D. Miscellaneous Other Reactions

### 1. *Reduction*

Methods of reduction apply only in special cases because the sulphinic acids are not normally oxidizing agents. Three examples of reagent are given below:

*a. Iodide.* Compounds with oxidizing properties can convert iodide to the easily detectable and determinable iodine. Thus Bretschneider and Klotzer[65] treated thiosulphinate esters with iodide in sulphuric acid solution:

$$RS(=O)SR + 2I^- + 2H^+ \longrightarrow RS{-}SR + I_2 + H_2O$$

They then titrated the iodine. Barnard and Cole[66] found low results with this method and developed their own procedure for alkyl (ethyl, butyl) esters of aromatic thiosulphinic acids (benzene-, *p*-methoxybenzene-). They dissolved the sample in glacial acetic acid, added aqueous potassium iodide solution in an oxygen-free atmosphere (stream of carbon dioxide) and titrated the iodine yielded with thiosulphate. They also visualized thiosulphinate esters on paper chromatograms as blue-violet spots by exposure for 10 s to hydrogen chloride and then spraying with a starch-iodide reagent. A similar method of detection was used also by Fondarai and Richert[58] for cysteinesulphinic acid on paper chromatograms and by Freytag and Ney[67] for aliphatic sulphilimines on thin-layer chromatograms of silica gel by spraying first with potassium iodide-dilute hydrochloric acid, heating the plate for 8 min at 120 °C and then spraying with starch solution to give violet to brown-violet zones. Bayfield and coworkers[68] visualized aromatic sulphinamides (and sulphenamides) as blue-mauve spots on paper and thin-layer chromatograms by exposure to hydrogen chloride or trichloroacetic acid fumes and then spraying with starch-iodide. De Marco and collaborators[61] visualized hypotaurine and homohypotaurine (and the selenium-containing acids) by spraying with potassium iodide-hydrochloric acid.

*b. Methyl violet.* Yakovleva and coworkers[69] determined thiourea dioxide and hydrogen peroxide through reaction with methyl violet for 10 min at 100 °C and pH 8.5 to 9. The diminution in absorbance of the dye at 590 nm was proportional to the amount of the oxidizing agents [hydrogen peroxide alone was estimated through a colour reaction with titanium(IV), enabling the thiourea dioxide to be determined by the difference].

*c. Reduction with metals.* Feigl[70] gave methods for detecting benzenesulphinic acid by reduction. Zinc-hydrochloric acid or Devarda's alloy reduced it to thiophenol, detected in the vapours through the blackening of paper saturated with lead acetate solution held there. Raney alloy (nickel–aluminium) with some drops of hydrochloric or sulphuric acid reduces the sulphur to hydrogen sulphide, which can be detected in the same way. This test must be applicable to other sulphinic acids but probably also to other sulphur-containing organic compounds.

### 2. Introduction of characterizing organic groups

*a. Diazonium coupling.* Babbs and Gale[71] determined sulphinic acids by reaction with the diazonium salt Fast Blue BB to yield a coloured azo compound:

$$RS(=O)OH + R'\overset{+}{N}\equiv N \longrightarrow RS(=O)ON=NR'$$

They extracted the azo compound into toluene-butanol, stabilized this solution with pyridine and evaluated colorimetrically through the light absorption at 420 nm. It is surprising that no further examples have been found of such coupling procedures. They must be reasonably specific, easy to carry out and sensitive, and there is a wide choice of diazonium salts.

*b. Reaction with active halides.* Jayson and coworkers[72] visualized oxidation products of cysteamine by reaction for 45 min at 100 °C with 1-fluoro-2,4-dinitrobenzene in the presence of sodium hydrogen carbonate. The amino group of the amino acid products reacted with the fluoro compound, but it is of interest that the sulphinic acid group of hypotaurine also reacted, to yield a di-substituted product:

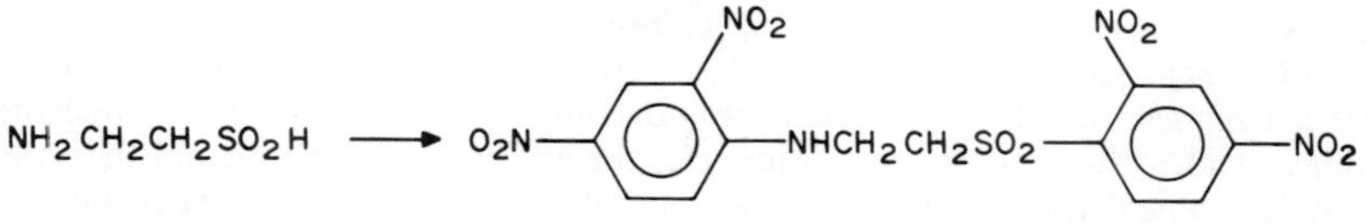


The sulphonic acid group of taurine does not react. This could enable sulphinic acids to be determined in the presence of sulphonic acids, although this idea does not appear to have been tested.

Derivatives for identification of sulphinic acids have been prepared by reaction of their alkali metal salts with suitable halides. Thus Marvel and Johnson[3] used chloroacetic acid:

$$ArSO_2Na + ClCH_2COOH \longrightarrow ArSO_2CH_2COOH + NaCl$$

Douglass and collaborators[73] prepared a derivative of ethanesulphinic acid by reaction with benzyl chloride:

$$EtSO_2^- + PhCH_2Cl \longrightarrow EtSO_2CH_2Ph + Cl^-$$

The well-known reagent benzylthiouronium chloride can be used for characterization, as with carboxylic and sulphonic acids:

$$RSO_2^-Na^+ + [PhCH_2SC(NH_2){=}NH_2]^+Cl^- \longrightarrow [PhCH_2SC(NH_2){=}NH_2]^+[RSO_2]^-$$

*c. Reaction with N-ethylmaleimide.* This reaction, which has been used to determine thiosulphinic acids, is not considered to be addition of the acid to the double bond of the reagent but the formation of an enolic compound of the corresponding diketone[74]; it has an absorption maximum at 515 nm. Carson and Wong[75] mixed the sample and N-ethylmaleimide, both in isopropanol, then added potassium hydroxide; they measured the light absorption at 515 nm after a reaction time of 10 to 18 min from the moment of addition of alkali. They tested the procedure with thiosulphinates containing phenyl, *p*-tolyl, ethyl, propyl and 2-propenyl groups. They used the same colour reaction to visualize thiosulphinates on paper chromatograms. The paper was dipped first into the N-ethylmaleimide solution, dried for 5–10 min, and then dipped into the potassium hydroxide. Nakata and coworkers[76] based their determination on the same principle, mixing sample and reagent in isopropanol with potassium hydroxide but then adding ascorbic acid to stabilize the colour before finally evaluating also at 515 min after 10 min reaction time. According to Watanabe and Komada[77], who worked with the S-propyl ester of propanethiosulphinic acid and the S-allyl ester of 2-propenethiosulphinic acid ('allicin'), the light absorption of the product can be measured also in the ultraviolet region at 355 nm after 20–30 min reaction time.

*d. Pyrolysis.* A spot test of Feigl and Costa Neta[78] to distinguish dithionite and hydroxymethanesulphinate (Rongalite) depends on heating the sample to 200–300 °C. The latter decomposes according to

$$2\,HOCH_2SO_2Na \longrightarrow 2\,HCHO + Na_2SO_4 + H_2S$$

The hydrogen sulphide is detected by the black stain formed on lead acetate paper held in the vapours.

Only in rare cases can sulphinic acids be decomposed with loss of sulphur dioxide:

$$RSO_2H \longrightarrow SO_2 + RH$$

There is thus no example parallel to decarboxylation for quantitative determination.

*e. Miscellaneous colour reactions.* Most colour reactions for sulphinic acids given in the literature are carried out under rather drastic conditions and are therefore probably not specific. Thus Limpricht[79] stated that aromatic sulphinic acids dissolve in concentrated sulphuric acid without colour formation, but the addition of some phenol then yields a blue to deep blue colour. Smiles and Le Rossignol[80] reported a characteristic blue colour given by aromatic sulphinic acids (e.g. *p*-ethoxybenzenesulphinic acid) with anisole or phenetole in concentrated sulphuric acid. The formation of $ArS(=O)SC_6H_4Et$ (or Me) was postulated. Further phenetole led to the disappearance of the colour, ascribed to formation of diphenetyl phenylsulphonium sulphate. Bazlen and Scholz[81] detected sodium and potassium salts of aliphatic and aromatic sulphinic acids through their ability to decolourize indigo solutions in glycerol at 180 °C. Probably many other sulphur-containing compounds are also capable of this.

*f. Addition to double bonds.* Sulphinic acids add comparatively readily to double bonds, usually in a 1,4-addition, e.g. to $\alpha,\beta$-unsaturated carbonyl compounds or quinones. Addition to the last named appears to have been of special interest to Russian teams. Thus

Obtemperanskaya and Zlobin[82] studied the reaction of anthraquinone with thiourea dioxide, which yielded anthrahydroquinone, the absorbance of which at 417 nm was used to determine the anthraquinone in anthracene. Stadnik and coworkers[83] investigated the addition of benzenesulphinic acid to benzoquinone, substituted benzoquinones and 1,4-naphthaquinone; at pH 1–2 this yielded coloured products in a 1:1 ratio. Stom and collaborators[84] reported the chromatographic and spectroscopic determination of quinones in aqueous solution through reaction with benzenesulphinic acid. In all this work the interest has centred on determination of the quinones. The reaction does not appear to have been the basis of determination of a sulphinic acid.

## III. PHYSICAL/INSTRUMENTAL METHODS

Included under this heading are polarographic methods (these concern both reduction and anodic oxidation but it was considered to be more convenient to deal with them in the same section), chromatographic methods of separation and some spectroscopic procedures, of which only a few have been published.

### A. Polarography

Polarographic anodic waves (for oxidized sulphinates) and cathodic waves (for thiosulphinates and thiourea dioxide) have been reported permitting, in some cases, quantitative determination.

The earliest reference to polarographic determination of a sulphinic acid is evidently that of Furness[20], in work on dithionites, hydroxymethane- and hydroxyethane-sulphinates. Horner and Nickel[85] worked on the polarographic reduction of sulphinate esters in 75% dioxan, but their work was not expressly analytical. Several authors have published results from the polarographic determination of sodium hydroxymethanesulphinate (Rongalite). They include Kolthoff and Tamberg[86], who obtained anodic oxidation waves in alkaline solution (pH 9) showing a linear relation between wave height and concentration. Their equations were

$$HOCH_2SO_2^- + 2OH^- \longrightarrow HOCH_2SO_3^- + H_2O + 2e$$
$$HOCH_2SO_3^- + OH^- \longrightarrow HCHO + SO_3^{2-} + H_2O$$

Sokolov and Leonova[87] determined the compound in latex by coagulating with nitric acid–hydrochloric acid, bringing it to pH 1.2–1.5, deaerating and then submitting to polarography in which wave height was proportional to sulphinate concentration. Fernandez-Martin and coworkers[88] found single anodic waves at pH values between 6.92 and 11.6, showing above pH 9 a linear relation between diffusion current and concentration. Edgar[89] studied the polarography of the hydroxymethanesulphinate in a McIlvaine citrate–monohydrogen phosphate buffer of pH 4, using a rotating platinum electrode and obtaining a linear relation between the height of the anodic wave at 0.935 V and concentration in the range from 0.02 to 0.2 mM (the influence of hydrogen sulphite could be eliminated by adding formaldehyde).

Ruff and Kucsman[90] determined polarographically sulphimides, $RR'S{=}NSO_2Ar$, in deaerated Britton–Robinson buffer of pH 4.5, as part of a study of the reaction of organic sulphides with chloramine T.

Gründler and Choschzick[91] made an alternating-current polarographic study of sodium salts of aromatic sulphinic acids (components of photographic emulsions) in borate buffer of pH 6.5 + potassium chloride. They were able to separate the polarographic waves from those of aryl thiosulphinates present (their attention was devoted to these last named). In the cathodic region wave height was proportional to concentration.

Czerwicz and Bogaczek[92] polarographed phenylsulphinylamine, PhN=S=O and the *o*-, *m*- and *p*-substituted (methyl group) compounds in dimethylformamide or benzene-dimethylformamide, using tetrabutylammonium iodide as supporting electrolyte. They obtained two or three reduction waves, the second wave showing a linear relationship between diffusion current and concentration.

Probably classifiable here is the linear sweep voltammetric method of Kirchnerová and Purdy[93], using a vitreous carbon working electrode which they applied to thiourea and thiourea dioxide. They obtained a rectilinear graph up to 0.35 mV. The pH was below 5 in a medium of mineral acid plus alkali metal or ammonium salt. Their sweep rate was 2 mV per second.

## B. Chromatography

### *1. Paper chromatography*

Barnard and Cole[66] separated alkyl and aryl thiosulphinate esters on Whatman No. 1 paper, impregnated with phenoxyethanol, and using heptane as developing solvent. They visualized with hydrochloric acid–iodide (Section II.D.1a). Gringras and Sjöstedt[94] carried out ascending paper chromatography of some aliphatic and (mostly) aromatic sulphinic acids and their salts. They used Whatman No. 1 paper, developing for 16 h at 20 °C with the solvent mixture butanol–propanol–water (1 + 1 + 1) and visualized with tetrazotized *o*-dianisidine (Echtblau B Salt), stabilized with zinc chloride which gave canary yellow dye products, insoluble in water. They were able to determine *p*-toluenesulphinic acid quantitatively via spot size. Fondarai and Richert[58] conducted paper chromatography of sulphur-containing amino acids, using as mobile phase 95% ethanol–chloroform–water (6 + 3 + 1); cysteinesulphinic acid was among the compounds and was visualized with iodoplatinate, ferric chloride and hydrochloric acid–iodide (Sections II.C.1, II.C.2 and II.D.1a). Bayfield and coworkers[68] used conditions similar to those of Barnard and Cole for paper chromatographic separation of some aromatic sulphinamides (and sulphenamides), i.e. with Whatman No. 1 paper, impregnated with phenoxyethanol, and heptane as developing solvent in descending chromatography. They also visualized with acid–iodide (Section II.D.1a). De Marco and collaborators[61] separated sulphur- and selenium-containing amino acids, including hypotaurine and homohypotaurine and their selenium analogues, on Whatman No. 1 paper using three solvents: water-saturated phenol in the presence of ammonia vapour, the upper phase of butanol–acetic acid–water (4 + 1 + 5) and water-saturated 2,4,6-trimethylpyridine-2,6-dimethylpyridine. They visualized with hydrochloric acid–iodide and iodoplatinate (Section II.C.2 and II.D.1a). Marek[10] separated mixtures of sulphinic acids (benzene-, *p*-toluene-, *p*-acetamidobenzene- and *m*-nitrobenzene-) at 24 °C on Whatman No. 1 paper using various solvent systems made up of propanol + ammonium hydroxide and/or butanol and/or water. He visualized by exposure to nitrous oxides, then spraying with R salt solution and finally exposing to ammonia to give yellow-green zones (Section II.A.1).

### *2. Thin-layer chromatography*

In addition to separating some aromatic sulphinamides and sulphenamides by paper chromatography, Bayfield and colleagues[68] also tried thin-layer chromatography on kieselguhr G and cellulose layers, using methanol of various concentrations, of which 65% proved to be the best; however, they found separation to be less satisfactory than with paper chromatography. Detection was carried out with the help of an acid–iodide reagent (Section II.D.1a). Freytag and Ney[67] carried out thin-layer chromatography of *S*,*S*-dialkyl-*N*-(*p*-toluenesulphonyl) sulphilimines (alkyl groups being methyl, ethyl, propyl, butyl and pentyl) on silica gel G layers. Their mobile phase was diethyl ether–ethanol

(4 + 1 for lower alkyl groups, 20 + 1 for higher). Acid–iodide was used for visualization (Section II.D.1a). Czerwicz and Malata[11] used thin-layer chromatography on silica gel layers to separate isomeric sulphinylamines, PhN=S=O, with methyl substituents in the 2,3-,2,4-,2,5-,2,6-,3,4- and 3,5-positions. As solvent they used binary and ternary mixtures of hexane, benzene, chloroform, carbon tetrachloride, amyl alcohol, diethyl ether, acetone and ethyl acetate (also benzene alone). They visualized with nitrous fumes (Section III.A.1). Westley and Westley[59] performed thin-layer chromatography of aliphatic and aromatic thiosulphonates, sulphonates and sulphinates, on silica gel with various mobile phases: isopropanol–0.2 M ammonium hydroxide (3 + 1), ethyl acetate–methanol–water (4 + 1 + 1) and acetone–butanol–water (2 + 2 + 1). They visualized also with ferric chloride (Section II.C.1). The *R* values were in the sequence: $RSO_2S^- > RSO_3^- > RSO_2^-$.

### 3. Gas–liquid chromatography

Block and O'Connor[95] subjected alkyl thiosulphinates (also deuteriated compounds) to gas–liquid chromatography on 10% silicone rubber UCW-98 on 80–100 mesh Chromosorb W at 70–75 °C and using flame ionization detection. The work was primarily to obtain mass spectrometric data. Czerwicz and Markowski[96] carried out gas–liquid chromatography of aromatic sulphinylamines on Chromosorb P + 20% Rheoplex at 170 °C, using hydrogen as carrier gas and a heat conductivity detector. Their samples were phenylsulphinylamine, PhN=S=O, also with *o*-, *m*- and *p*-methyl and chloro substituents. The same authors[97] later used the same principle to separate the six isomeric dimethyl-substituted (2,3-,2,4-,2,5-,2,6-,3,4- and 3,5-)phenylsulphinylamines, injecting 10% solutions of the samples in benzene and using hydrogen as carrier gas and flame ionization detection. Their best results were obtained using Apiezon N on 80–100 mesh Chromosorb G AW-DMCS at 200 °C. Good results were also obtained with the impregnation polyphenyl ether OS 138, also at 200 °C. Other packings were less satisfactory. Silar lOC on 80–100 mesh Chromosorb G AW-DMCS yielded a different elution sequence of the isomers. Czerwicz[98] separated the six dichlorophenylsulphinylamines (substituents in the same positions as in the dimethyl compounds mentioned above) by gas–liquid chromatography, the best column packing being polyphenyl ether OS 138 on Chromosorb G at 180 °C, with hydrogen carrier gas and flame ionization detector.

MacKenzie and Finlayson[99] gas chromatographed cysteic and cysteinesulphinic acids as their N-heptafluorobutyryl isobutyl ester derivatives. Their column was 3% SE-30 on 100–120 mesh Chromosorb W HP. It was kept for 2 min at 100 °C, then warmed to 250 °C at 4°/min. Carrier gases were hydrogen, nitrogen and air with detection by flame ionization.

Gas–liquid chromatography of methyl alkanesulphinates, alkanesulphonates and dialkyl disulphides was carried out by Toro[100] using a column of 10% Ucon on 80–100 mesh Chromosorb HP. Column temperature was programmed, helium was carrier gas and detection was by flame photometry.

Kataoka and colleagues[101] determined cysteinesulphinic acid and hypotaurine in animal tissues by extraction and centrifuging followed by gas–liquid chromatography on a glass column treated with dichlorosilane and packed with 1% silicone OV-17 + 0.2% FFAP at 210–250 °C (programmed at 4°/min) with flame ionization detection.

### 4. High performance liquid chromatography (HPLC)

Some work has been done recently on separating diastereoisomers with the help of HPLC. Thus Szókán and colleagues[102] studied diastereoisomeric sulphoxides and

sulphilimines of the general formula

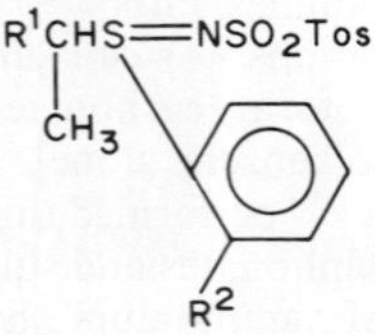


where $R^1$ was hexyl or ethyl and $R^2$ was COOH or H. They used a column of either Hypersil (Shandon GB) or Chromsfer-Sil (Budapest Labor-MIM), eluting with various mixtures of diethyl ether, pentane, methanol and acetic acid. The best eluent for the Chromsfer column proved to be diethyl ether–pentane (80 + 20) and, for the Hypersil column, diethyl ether plus a few percent methanol. Detection was in the ultraviolet at 254 nm. In later work, Jalsovszky and coworkers[103] investigated diastereoisomers of cyclic sulphilimines, e.g.

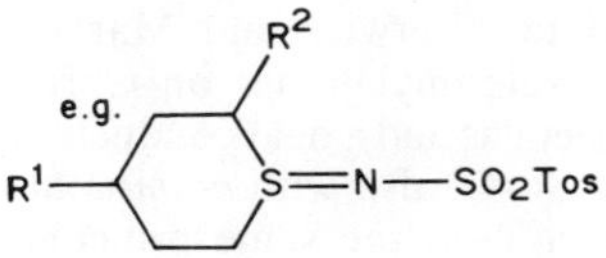


containing alkyl group substituents in the nucleus. They used Hypersil columns, eluting with mixtures of diethyl ether, pentane, methanol, 95% ethanol and 2-octanol and detecting also at 254 nm. Wainer and colleagues[104] used HPLC to separate several classes of compound, including sulphinamides, $R^1SONHR^2$ and sulphilimines, $R^3MeS{=}NR^4$ where $R^1$ was *p*-tolyl or phenyl, $R^2$ was methyl or 2-pyridyl, $R^3$ was phenyl or butyl and $R^4$ was —COPh or —Tos, respectively. [They also worked with sulphoximines, but these contain S(VI) and do not qualify for inclusion here.] Their chiral stationary phase was a cellulose triphenyl carbamate coated onto a macroporous silanized silica gel (*ca* 20–23% by weight). The mobile phase was hexane modified with an alcohol, or an alcohol + acetonitrile; detection was again at 254 nm.

### *5. Ion exchange chromatography*

De Marco and coworkers[105] separated sulphur-containing amino acids, e.g. cysteic and cysteinesulphinic acids, taurine and hypotaurine, on the ion exchange resin 150A, the elution agent being citric acid–sodium chloride. Identification was through light absorption at 440 or 570 nm after colour reaction with ninhydrin. Lombardini and colleagues[106] separated cysteinesulphinic acid from other enzymatic oxidation products of L-cysteine on Dowex-50 at pH 2, ultimately forming and assaying the coloured product with ninhydrin. Purdie and Hanafi[107] separated sulphinic and sulphonic acid derivatives of cysteine and glutathione on Dowex 1X8 anion exchange resin, improved by adding Sephadex G 10 in the ratio 8:5, with the help of a monochloroacetate gradient.

Another example of ion exchange chromatography is the work of Williams[108], who separated some organic sulphur-containing compounds, including aromatic sulphinates. His stationary phase was a monolayer of aminated latex beads, agglomerated to a styrene–divinylbenzene (S/DVB) resin. The eluent system was an ordinary hydrogen carbonate/carbonate one, especially satisfactory being 0.001 M sodium hydrogen carbonate. Detection was by conductivity or, with aromatic compounds, also light absorption in the ultraviolet. He was able to separate benzene- and *p*-toluenesulphinic acids. Ida and Kuriyama[109] determined cysteic and cysteinesulphinic acids in rat brain on the strongly basic anion exchanger ISA-07/S 504. They detected by reacting with *o*-phthalaldehyde in the presence of 2-mercaptoethanol to give fluorescent products.

### C. Spectroscopy

Spectroscopic determination or detection of sulphinic acids or their derivatives have not generally been the subject of special publications. Mostly they have been a tool in a particular study. The best example of this is the monitoring of light absorption in the ultraviolet.

Wojtasiewicz-Obrzut[24] used UV measurements to determine thiourea dioxide and thiourea at 269 and 239 nm, where each has its respective absorption maximum. Mori and Ueda[110] recorded the infrared spectra of 25 aromatic sulphinamides in chloroform solution and in potassium bromide discs. They found characteristic bands at 1070 $cm^{-1}$ (solution) and 1056 $cm^{-1}$ (disc), both of which have undoubtedly found unpublished use for identification. Freeman and McBreen[111] determined thiosulphinate in onions by extraction of the juice plus water with hexane and measuring the ultraviolet absorption at 254 nm. In a study of the reaction of organic sulphides with chloramine T, Ruff and Kucsman[90] determined sulphimides (and sulphoxides) through their absorption at 286 nm.

## IV. MICROBIOLOGICAL METHODS

Some microbiological assays of amino acids containing the sulphinic acid group may be mentioned briefly. Leinweber and Monty[112] determined cysteinesulphinic acid by reaction with α-ketoglutarate in the presence of highly purified glutamic–oxaloacetic acid transaminase. This gave sulphite, $SO_3^{2-}$, which they determined colorimetrically with the Schiff reaction using fuchsine (rosaniline).

Lombardini and coworkers[106] studied the enzymatic oxygenation of L-cysteine to L-cysteinesulphinic acid. They assayed this acid in two ways: (a) separation through ion exchange chromatography (Section III.B.5) followed by colour reaction with ninhydrin and colorimetric evaluation; (b) by transamination with α-ketoglutarate, followed by desulphination using lactate dehydrogenase, yielding pyruvate which they determined spectrophotometrically.

Baba and colleagues[113] assayed cysteinesulphinic acid by enzymatic conversion to lactate, ultimately transforming NAD into NADH, the light absorption of which was found to be proportional to the concentration of the cysteinesulphinic acid.

Soda and coworkers[114] degraded hypotaurine with an aminotransferase, leading to sulphino-acetaldehyde which decomposed spontaneously into acetaldehyde and sulphur dioxide. The former was assayed via aldehyde dehydrogenase, and the latter, more relevant to a sulphinic acid determination, through the Schiff reaction.

## V. REFERENCES


1. C. Koenigs, *Chem. Ber.*, **11**, 615 (1878).
2. C. S. Marvel, G. E. Adams and R. S. Johnson, *J. Am. Chem. Soc.*, **68**, 2735 (1946).
3. C. S. Marvel and R. S. Johnson, *J. Org. Chem.*, **13**, 822 (1948).
4. S. Ponzini, *Farm. Sci. Tec.* (*Pavia*), **2**, 198 (1947).
5. J. L. Kice and K. W. Bowers, *J. Am. Chem. Soc.*, **84**, 605 (1962).
6. J. P. Danehy and V. J. Elia, *Anal. Chem.*, **44**, 1281 (1972).
7. B. Lindberg, *Acta Chem. Scand.*, **17**, 383 (1963).
8. B. Fleszar, *Chem. Anal.* (*Warsaw*), **10**, 49 (1965).

9. M. Matrka, Z. Sagner and A. Spevak, *Chem. Prům.*, **22**, 551 (1972).
10. J. Marek, *Collect. Czech. Chem. Commun.*, **44**, 3357 (1979).
11. Z. Czerwicz and E. Malata, *J. Chromatogr.*, **132**, 168 (1977).
12. P. Allen, *J. Org. Chem.*, **7**, 23 (1942).
13. L. Ackerman, *Ind. Eng. Chem., Anal. Ed.*, **18**, 243 (1946).
14. S. Atkin, *Ind. Eng. Chem., Anal. Ed.*, **19**, 816 (1947).
15. A. Liberti and P. Lazzari, *Ric. Sci.*, **26**, 825 (1956).
16. M. H. Hashmi and A. A. Ayaz, *Anal. Chem.*, **35**, 2194 (1963).
17. J. H. Uhlenbroek, M. J. Koopmans and H. O. Huisman, *Recl. Trav. Chim. Pays-Bas*, **76**, 129 (1957).
18. K. Sumizu, *Anal. Biochem.*, **4**, 378 (1962).
19. B. Salkin, *Ind. Eng. Chem., Ind. Ed.*, **15**, 848 (1927).
20. W. Furness, *J. Soc. Dyers Colour.*, **66**, 270 (1950).
21. A. Badinand and J. Rondelet, *Ann. Pharm. Fr.*, **14**, 691 (1956).
22. L. Maros, *Magy. Kem. Foly.*, **64**, 41 (1958); *Chem. Abstr.*, **52**, 11661 (1958).
23. J. Tsau and J. W. Poole, *Int. J. Pharm.*, **12**, 185 (1982).
24. D. Wojtasiewicz-Obrzut, *Chem. Anal. (Warsaw)*, **10**, 1133 (1965).
25. I. G. Shafran, A. G. Stepanova and L. I. Pankratova, *Tr. Vses. Nauchno-Issled. Inst. Khim. Reaktivov*, No. 25, 215 (1963).
26. D. S. Mahadevappa, *Curr. Sci.*, **35**, 517 (1966).
27. P. N. Krishnan Nambisan and C. G. Ramachandran Nair, *Anal. Chim. Acta*, **52**, 475 (1970).
28. L. Ramberg, *Arkiv Kemi Mineral. Geol.*, **12B**, No. 29 (1936).
29. F. Ostermayer and D. S. Tarbell, *J. Am. Chem. Soc.*, **82**, 3752 (1960).
30. S. Siggia and R. L. Edsberg, *Anal. Chem.*, **20**, 938 (1948).
31. R. E. Bayfield, V. Clarke and E. R. Cole, *J. Chromatogr.*, **19**, 370 (1965).
32. T. P. Forrest and D. E. Ryan, *Can. J. Chem.*, **36**, 1677 (1958).
33. L. Gringras and G. Sjöstedt, *Acta Chem. Scand.*, **15**, 433 (1961).
34. Ts. Tsaikov, *Zh. Prikl. Khim.*, **55**, 475 (1982).
35. J. S. Grossert and R. F. Langler, *J. Chromatogr.*, **97**, 83 (1974).
36. J. A. Reuterskiöld, *J. Prakt. Chem.*, [2] **127**, 271 (1930).
37. B. Fleszar, J. Kowalski, P. Sanecki and J. Plaszynska, *Chem. Anal. (Warsaw)*, **22**, 169 (1973).
38. L. Reio, *J. Chromatogr.*, **47**, 60 (1970).
39. E. L. Helwig, *Am. Dyest. Rep.*, **7**, 12 (1920).
40. B. Salkin, *Ind. Eng. Chem., Ind. Ed.*, **15**, 848 (1927).
41. L. Spitzer, *Ann. Chim. Appl.*, **29**, 184 (1939).
42. L. Spitzer, *Ann. Chim. Appl.*, **28**, 252 (1938).
43. W. Peters, *Chem. Ber.*, **38**, 2567 (1905).
44. F. Feigl, D. Haguenauer-Castro and E. Libergott, *Mikrochim. Acta*, 595 (1961).
45. R. F. Langler, *J. Chromatogr.*, **104**, 228 (1975).
46. F. Feigl and L. Hainberger, *Mikrochim. Acta*, 105 (1955).
47. R. Scandurra and R. Mosti, *Giorn. Biochim.*, **12**, 236 (1963).
48. V. Gentil and D. P. Miranda, *An. Assoc. Bras. Quim.*, **19**, 73 (1960).
49. H. Kataoka, H. Yamamoto, Y. Sumida, T. Hashimoto and M. Makita, *J. Chromatogr.*, **382**, 242 (1986).
50. S. Krishna and Bhagwan Das, *Quart. J. Indian Chem. Soc.*, **4**, 367 (1927).
51. D. L. Wetzel and C. E. Meloan, *Anal. Chem.*, **36**, 2474 (1964).
52. A. Hälssig, *J. Prakt. Chem.*, [2] **56**, 217 (1897).
53. J. Thomas, *J. Chem. Soc.*, **95**, 342 (1909).
54. S. Krishna and H. Singh, *J. Am. Chem. Soc.*, **50**, 792 (1928).
55. T. P. Forrest and D. E. Ryan, *Can. J. Chem.*, **36**, 1674 (1958).
56. I. P. Alimarin and D. I. Kuznetsov, *Dokl. Akad. Nauk SSSR*, **131**, 821 (1960).
57. B. Mondovì, G. Modiano and C. De Marco, *Giorn. Biochim.*, **4**, 324 (1955).
58. J. Fondarai and C. Richert, *J. Chromatogr.*, **9**, 262 (1963).
59. A. Westley and J. Westley, *Anal. Biochem.*, **142**, 163 (1984).
60. H. M. Winegard, G. Toennies and R. J. Block, *Science*, **108**, 506 (1948).
61. C. De Marco, P. Cossu, S. Dernini and A. Rinaldi, *J. Chromatogr.*, **129**, 369 (1976).
62. B. Jolles-Bergeret, *Eur. J. Biochem.*, **10**, 569 (1969).
63. S. Åkerfeldt and G. Loevgren, *Anal. Biochem.*, **8**, 223 (1964).

64. H. Gilman and R. K. Abbott, *J. Am. Chem. Soc.*, **71**, 659 (1949).
65. H. Bretschneider and W. Klotzer, *Monatsh. Chem.*, **81**, 589 (1950).
66. D. Barnard and E. R. Cole, *Anal. Chim. Acta*, **20**, 540 (1959).
67. W. Freytag and K. H. Ney, *J. Chromatogr.*, **27**, 289 (1967).
68. R. F. Bayfield, V. Clarke and E. R. Cole, *J. Chromatogr.*, **26**, 132 (1967).
69. E. N. Yakovleva, N. P. Nikonova and S. U. Kreingold, *Zavod. Lab.*, **49**, 24 (1983).
70. F. Feigl, *Anal. Chem.*, **33**, 1118 (1961).
71. C. F. Babbs and M. J. Gale, *Anal. Biochem.*, **163**, 67 (1987).
72. G. G. Jayson, T. C. Owen and A. C. Wilbraham, *Analyst (London)*, **89**, 788 (1964).
73. I. B. Douglass, K. R. Brower and F. T. Martin, *J. Am. Chem. Soc.*, **74**, 5770 (1952).
74. B. Kakáč and Z. Vejdělek, *Handbuch der photometrischen Analyse*, Band 1, Verlag Chemie 6940 Weinheim, FRG, 1974, p. 207.
75. J. F. Carson and F. F. Wong, *Nature (London)*, **183**, 1673 (1959).
76. C. Nakata, T. Nakata and A. Hishikawa, *Anal. Biochem.*, **37**, 92 (1970).
77. T. Watanabe and K. Komada, *Agric. Biol. Chem. (Tokyo)*, **30**, 418 (1966).
78. F. Feigl and C. Costa Neto, *Chemist-Analyst*, **44**, 91, 93 (1955).
79. H. Limpricht, *Ann. Chem.*, **278**, 243 (1893).
80. S. Smiles and P. Le Rossignol, *J. Chem. Soc.*, **89**, 701 (1906).
81. M. Bazlen and F. Scholz, *Chem. Ber.*, **68**, 2045 (1935).
82. S. I. Obtemperanskaya and V. K. Zlobin, *Zh. Anal. Khim.*, **31**, 1217 (1976).
83. A. S. Stadnik, Yu. Yu. Lur'e and Yu. M. Dedkov, *Zh. Anal. Khim.*, **32**, 1801 (1977).
84. D. I. Stom, S. S. Timofeeva, N. F. Kashina, L. I. Belykh, S. N. Suslov, V. V. Butorov and M. S. Apartsin, *Acta Hydrochim. Hydrobiol.*, **8**, 203 (1980).
85. L. Horner and H. Nickel, *Chem. Ber.*, **89**, 1681 (1956).
86. I. M. Kolthoff and N. Tamberg, *J. Polarogr. Sci.*, **3**, 54 (1958).
87. M. I. Sokolov and R. I. Leonova, *Tr. Probl. Lab. Khim. Vysokomol. Soedin. Voronezh, Gos. Univ.*, No. 4, 234 (1966); *Chem. Abstr.*, **68**, 105867 (1966).
88. R. Fernandez-Martin, R. G. Rinker and W. H. Corcoran, *Anal. Chem.*, **38**, 930 (1966).
89. J. S. Edgar, *J. Soc. Dyers Colour.*, **91**, 149 (1975); *Analyst (London)*, **100**, 735 (1975).
90. F. Ruff and Á. Kucsman, *J. Chem. Soc., Perkin Trans. 2*, 509 (1975).
91. P. Gründler and H. Choschzick, *Z. Chem.*, **12**, 274 (1972).
92. Z. Czerwicz and J. Bogaczek, *Z. Anal. Chem.*, **273**, 415 (1978).
93. J. Kirchnerová and W. C. Purdy, *Anal. Lett., Part A*, **13**, 1031 (1980).
94. L. Gringras and G. Sjöstedt, *Acta Chem. Scand.*, **15**, 435 (1961).
95. E. Block and J. O'Connor, *J. Am. Chem. Soc.*, **97**, 3921 (1974).
96. Z. Czerwicz and J. Markowski, *Chem. Anal. (Warsaw)*, **20**, 213 (1975).
97. Z. Czerwicz and J. Markowski, *J. Chromatogr.*, **124**, 76 (1976).
98. Z. Czerwicz, *J. Chromatogr.*, **139**, 177 (1977).
99. S. L. MacKenzie and A. J. Finlayson, *J. Chromatogr.*, **187**, 239 (1980).
100. P. Toro, *Bol. Soc. Chil. Quim.*, **26**, 59 (1981).
101. H. Kataoka, K. Ohishi, J. Imai and M. Makita, *Bunseki Kagaku*, **35**, 508 (1986).
102. G. Szókán, F. Ruff and Á. Kucsman, *J. Chromatogr.*, **198**, 207 (1980).
103. I. Jalsovszky, G. Szókán, F. Ruff and Á. Kucsman, *J. Chromatogr.*, **389**, 439 (1987).
104. I. W. Wainer, M. C. Alembik and C. R. Johnson, *J. Chromatogr.*, **361**, 374 (1986).
105. C. De Marco, R. Mosti and D. Cavallini, *J. Chromatogr.*, **18**, 492 (1965).
106. J. B. Lombardini, P. Turini, D. R. Biggs and T. P. Singer, *Physiol. Chem. Phys.*, **1**, 1 (1969).
107. J. W. Purdie and D. E. Hanafi, *J. Chromatogr.*, **59**, 181 (1971).
108. R. J. Williams, *J. Chromatogr. Sci.*, **20**, 560 (1982).
109. S. Ida and K. Kuriyama, *Anal. Biochem.*, **130**, 95 (1983).
110. K. Mori and Y. Ueda, *Chem. Pharm. Bull.*, **20**, 829 (1972).
111. G. G. Freeman and F. McBreen, *Biochem. Soc. Trans.*, **1**, 1150 (1973).
112. F. J. Leinweber and K. J. Monty, *Anal. Biochem.*, **4**, 252 (1962).
113. A. Baba, S. Yamagami, H. Mizuo and H. Iwata, *Anal. Biochem.*, **101**, 288 (1980).
114. K. Soda, T. Hirasawa, S. Toyama and K. Yonaha, *Ganryu Aminosan*, **5**, 249 (1982).

The Chemistry of Sulphinic Acids, Esters and their Derivatives
Edited by S. Patai


CHAPTER **5**

# Mass spectra of sulfinic acids, esters and derivatives

KALEVI PIHLAJA

*Department of Chemistry, University of Turku, SF-20500 Turku, Finland*



## I. INTRODUCTION

Fairly limited data are available on the electron-impact-induced mass spectrometric behavior of sulfinic acids, esters and their derivatives. Besides reports dealing with the well-known sulfinate ester rearrangements in sulfones and their implications in mass spectrometry[1,2] only about 40 papers—including at least some useful results to be discussed—could be found.

## II. MASS SPECTRA OF SULFINIC ACIDS, THEIR SALTS AND COMPLEXES

One of the difficulties often met when studying the mass spectrometric behavior of sulfinic acids and their derivatives is their simultaneous thermal decomposition. This was already noted by Wudl and coworkers[3]. They investigated the electron-impact-induced mass spectrum of methanesulfinic acid (**1**), which gave $CS_2$ and MeSSMe via thermal decomposition, when a heated inlet system at 323 and 453 K was applied. Typical mass spectrometric fragments in addition to the molecular ion (50%) for **1** were $SOOH^+$ (100%), $SO_2^{+\bullet}$ (37%), $MeSO^+$ (33%), $CH_2OS^+$ (10%), $OS^{+\bullet}$ and $CS^{+\bullet}$. Practically the same main fragment ions were given by Lorenz and coworkers[4] for silver methanesulfinate.

Filby and coworkers[5] prepared butanesulfinic acid (**2**) but their mass spectrometric data correspond to its methyl ester [*m/z*(%): $M^{+\bullet}$ 136(4), $C_4H_9SO^+$ 105(16), $CH_4SO_2^{+\bullet}$ 80(60), $SOOH^+$ 65(24), $C_4H_9^+$ 57(100)] as can also be seen from their later communication on the mass spectra of methyl alkanesulfinates[6].

Phillips and Deacon[7] prepared thallium(I) 2, 3, 4, 5-tetrafluorobenzenesulfinate (**3**) the electron-impact mass spectrum of which showed the following structurally significant features: *m/z* 418 ($M^{+\bullet}$), 354 ($C_6HF_4Tl^+$), 269 ($O_2STl^+$); rearrangement peaks *m/z* 558 ($C_{24}H_4F_{14}^+$), 540 ($C_{24}H_5F_{13}^+$), 428 ($C_{18}H_3F_{11}^+$), 410 ($C_{18}H_4F_{10}^+$), 298 ($C_{12}H_2F_8^+$), 280 ($C_{12}H_3F_7^+$).

Binder and Schmidt[8] prepared tetrachloroantimony(V) methanesulfinate (**4**) and proved its dimeric structure by mass spectrometry. Furthermore they concluded that the abundant ions $CH_3SO_2$ and $SbCl_3$ originate from a thermal decomposition rather than from a direct electron impact reaction:

$$\underset{(\mathbf{4})}{[MeS(O)OSbCl_4]_2} \longrightarrow \underset{\underset{MeSO_2^+}{\downarrow}}{2MeSO_2Cl} + \underset{\underset{SbCl_3^{+\bullet}}{\downarrow}}{2SbCl_3} \qquad (1)$$

## III. MASS SPECTRA OF SULFINATES AND SULTINES

### A. Sulfinate Esters

Baarschers and Krupay[9] investigated the electron-impact mass spectra of methyl arenesulfinates (**5** and **6**) and aryl methanesulfinates (**7** and **8**) and compared them with those of the corresponding sulfones (**9** and **10**). They found that the spectra of **5** and **6** were distinctly different from the spectra of the isomeric **7** and **8** (Table 1). Compounds **5**–**8** were in turn easy to distinguish from the sulfones with which they are isomeric (Scheme 1 and Table 1).

It is interesting to note that both **7** and **9** gave the ion *m/z* 94, the latter after rearranging to **7**, whereas **8** and **10** gave similarly the ion *m/z* 107 instead of *m/z* 108 (Scheme 1). Obviously the electron-donating methyl group inhibits the necessary hydrogen transfer in the case of **8** and **10**. The data for isopropyl benzenesulfinate (**11**)[10] and ethyl benzenesulfinate (**12**)[11,12] can be explained in agreement with the fragmentations shown in Scheme 1 taking into account that, via a McLafferty-type rearrangement, they gave also the molecular ion of benzenesulfinic acid (*m/z* 142) and this in turn benzene (*m/z* 78). The ion *m/z* 94 could also be found both for **11** (17%) and for **13** (24%).

The electron-impact mass spectra of methyl alkanesulfinates **14a**–**f** are characterized by the presence of molecular ions (1–9% except 31% for **14a**) and by the peaks at *m/z* 80 ($CH_4SO_2^{+\bullet}$: 9–92%), *m/z* 65 ($HSO_2^+$: 11–100%) and *m/z* 50 ($H_2SO^+$: 7–59%). With the exception of **14a** (89%) the hydrocarbon fragment $R^+$ arising by the cleavage of the C—S bond was always the base peak. Filby and coworkers[6] postulated that the fragment *m/z* 80 is formed via a sulfoxyl rearrangement process, which can also account for the presence of alkoxy peaks (up to 26%) and their complements in all spectra:

$$RS(O)OMe^{+\bullet} \xrightarrow[\mathbf{14a–f}]{} ROSOMe^{+\bullet} \begin{cases} \xrightarrow{-\bullet SOMe} RO^+ \\ \xrightarrow{-RO^\bullet} MeSO^+ \end{cases} \qquad (2)$$

**a**, R = Et; **b**, Pr; **c**, i-Pr; **d**, Bu; **e**, *s*-Bu; **f**, t-Bu; **g**, MeOC(O)$CH_2$

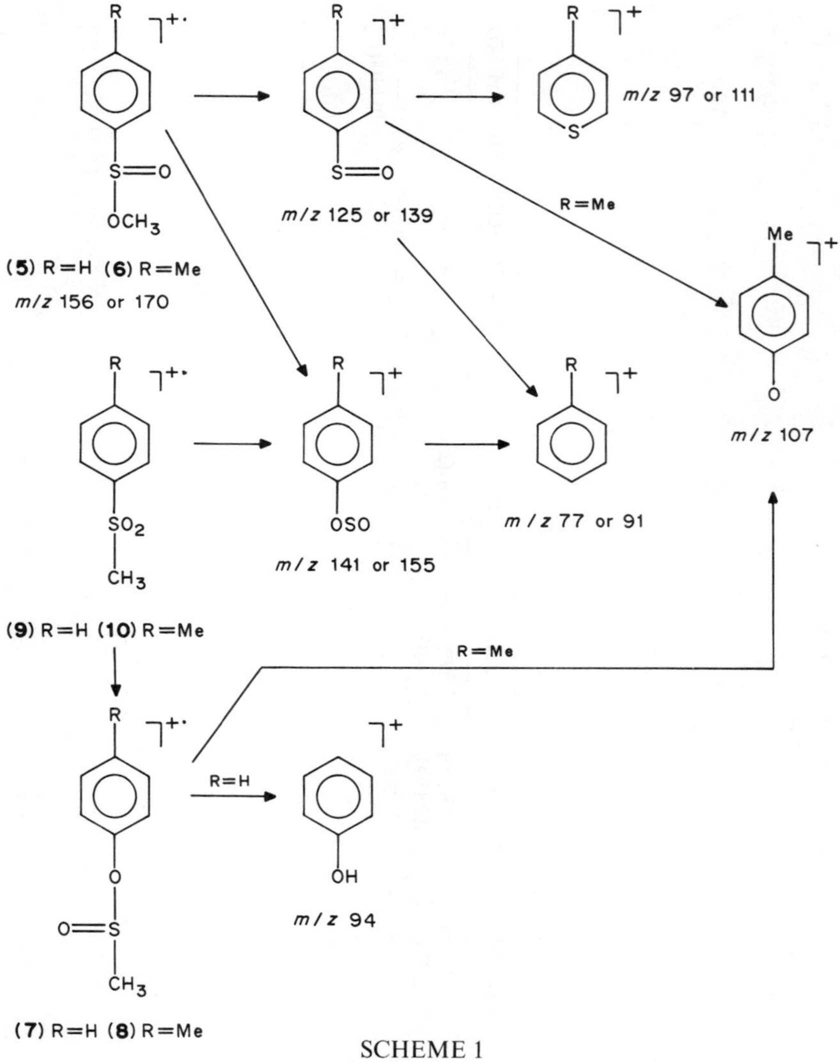


SCHEME 1

Harpp and Back[12], however, stated that **14d** forms ion $m/z$ 80, $CH_4SO_2^{+\cdot}$ (45%) as well as isopropyl (2-methyl)ethanesulfinate (**15**) ion $m/z$ 108, $C_3H_8SO_2^{+\cdot}$ (8%) in a process analogous to the five-center McLafferty rearrangement observed in some sulfinamides (see Section V.A).

The mass spectrum of ethyl (phenyl)methanesulfinate[12] [**15**: $M^{+\cdot}$ 184(2%), 92(10%), 91(100%), $HSO_2^+$ 65 (11%)] is very simple due to overwhelming domination by the tropylium ion. Carbon–sulfur cleavage is similarly important in the spectrum of diester **14g**, although the charge was retained almost exclusively by the sulfur-bearing fragment $MeSO_2^+$, $m/z$ 79[12].

Bell[13] has tabulated the main peaks in the mass spectra of methyl 2-methoxy- (**16**), methyl 4-methoxy- (**17**) and methyl 4, 8-dimethoxynaphthalene-1-sulfinates (**18**). Both ethyl 2-(benzylthio)ethanesulfinate (**19**) and ethyl 2-(*p*-toluenesulfonyl)ethanesulfinate (**20**)

TABLE 1. Major peaks [$m/z$ (rel. int., %)] in the spectra of compounds **5–10**

| Compound | $M^{+\cdot}$ | $[M-Me]^+$ | $[M-OCH_3]^+$ | $Ar^+$ | $[M-OCH_3-CO]^+$ | $[M-CH_2SO]^+$ | $[M-CH_3SO]^+$ |
|---|---|---|---|---|---|---|---|
| (**5**) $PhS(O)OCH_3$ | 156(84) | 141(15) | 125(100) | 77(84) | 97(26) | — | — |
| (**6**) *p*-$TolS(O)OCH_3$ | 170(49) | 155(3) | 139(100) | 91(39) | 111(6) | — | — |
| (**7**) $CH_3S(O)OPh$ | 156(15) | — | — | — | — | 94(100) | — |
| (**8**) $CH_3S(O)OTol$-*p* | 170(31) | — | — | — | — | — | 107(100) |
| (**9**) $PhSO_2CH_3$ | 156(28) | 141(27) | — | 77(100) | — | 94(31) | — |
| (**10**) *p*-$TolSO_2CH_3$ | 170(33) | 155(36) | 139(2) | 91(100) | — | — | 107(27) |

gave a $[M - OEt]^+$ ion in their chemical ionization spectra[14]. The mass spectrum of MeOS(O)C(Me)$_2$C(O)NHTol-*p* (**21**) showed peaks at 255 ($M^{+\cdot}$), 176 (loss of $MeSO_2$), 148 (loss of *p*-Me-aniline) and 107 (*p*-Me-aniline)[15].

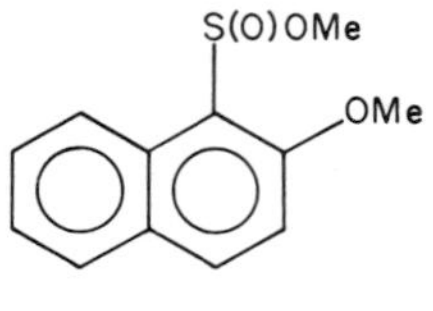


(**16**)

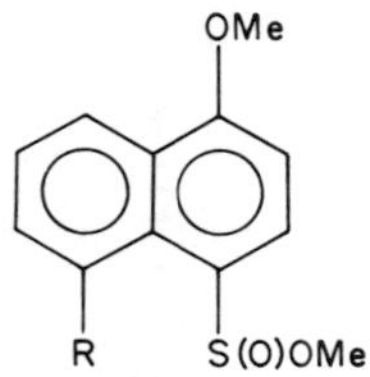


(**17**) R=H
(**18**) R=MeO

## B. Sultines

Also cyclic sulfones can rearrange to sultines–cyclic sulfinate esters[15] as shown in equation 3 for a naphthothiete sulfone (**22**). Similarities exist also between the mass spectra of **22** and 4, 9-diphenyl-3*H*-naphth[2, 3*c*]-2, 1-oxathiole 1-oxide (**23**). The base peak of the

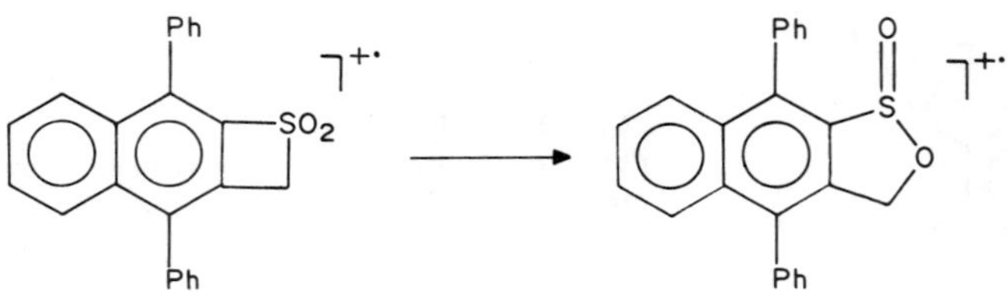


(**22**) (**23**) (3)

latter is the parent ion at *m/z* 356 whereas that of thiete sulfone **22** is ion $C_{23}H_{15}^+$ at *m/z* 291[1]. The sultine **23** shows an intense $[M - SO]^{+\cdot}$ peak at *m/z* 308, which is relatively weak (10%) in the spectrum of sulfone **22**, indicating that the latter does not rearrange at all exclusively to the former (equation 3) prior to fragmentation. Other characteristic peaks in the mass spectrum of **23** correspond to loss of HSO, $SO_2$, $HSO_2$, $H_3SO_2$, $MeSO_2$, $CH_4SO_2$, $PhSO_2$ and $C_7H_8SO_2$.

Scala and coworkers[17] found that the photolysis of sulfolane **24** in solid phase gave a product which, according to its electron-impact mass spectrum, was 1-oxathietane 2-oxide **25** (equation 4).

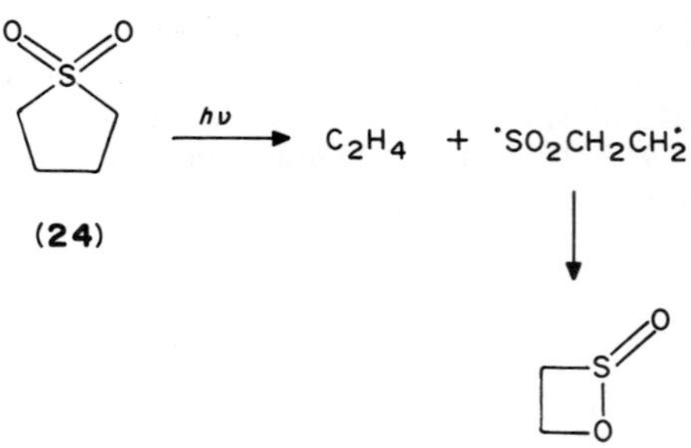


(**25**) (4)

| *m/z*(%) | |
|---|---|
| $M^{+\cdot}$ | 92(79) |
| $SO_2^{+\cdot}$ | 64(29) |
| $CH_2SO^+$ | 62(100) |
| $SO^{+\cdot}$ | 48(44) |
| $CH_2S^{+\cdot}$ | 46(36) |
| $CH_2O^{+\cdot}$ | 30(8) |
| $C_2H_4^{+\cdot}$ | 28(19) |

The major peaks in the mass spectra of α, β-unsaturated γ-sultines (**26**–**31**)[18] correspond to their parent ions (15–39%) and to ions $[M - SO]^{+\cdot}$ (11–50%), $[M - SOH]^+$ (12–40%),

$[M - CHO - SO]^+$ (14–100%), $[M - SO_2]^{+\cdot}$ (10–33%, **28**–**30** only) and $[M - SO - CH_2O]^{+\cdot}$ (8–51%). Fragmentation data for 3, 3, 4, 4-tetraphenyl-1, 2-oxathiolan-5-one 2-oxide (**32**) have also been given[19].

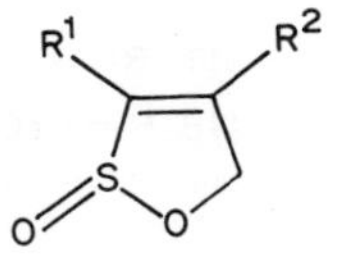

| | $R^1$ | $R^2$ |
|---|---|---|
| (**26**) | H | Et |
| (**27**) | H | Ph |
| (**28**) | H | $CH_2CH{=}CH_2$ |
| (**29**) | Ph | H |
| (**30**) | Ph | Et |
| (**31**) | Ph | Ph |

Durst and coworkers[20] prepared several benz-fused $\delta$-sultines and characterized them (**33**–**37**) by their electron-impact mass spectra. Compounds **33**–**35** gave practically no molecular ion. For all of them the main fragments seem to correspond to loss of MeSO

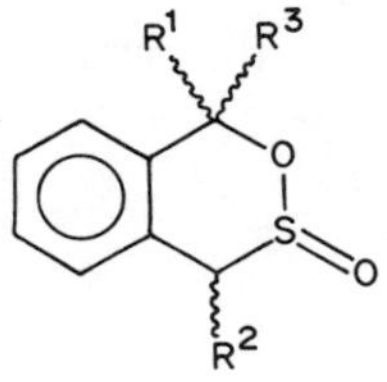

| | $R^1$ | $R^2$ | $R^3$ | |
|---|---|---|---|---|
| (**33**) | H | H | H | |
| (**34**) | H | Ph | H | (obviously *trans*) |
| (**35**) | Me | H | H | (*cis* and *trans*) |
| (**36**) | Me | Ph | Me | (*trans* (?); mp 104 °C |

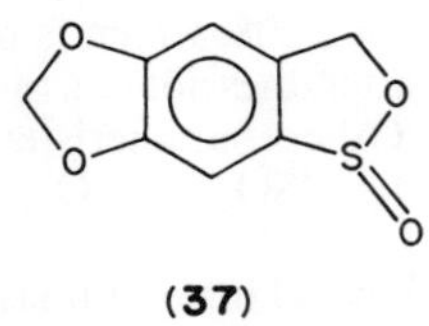

**(37)**

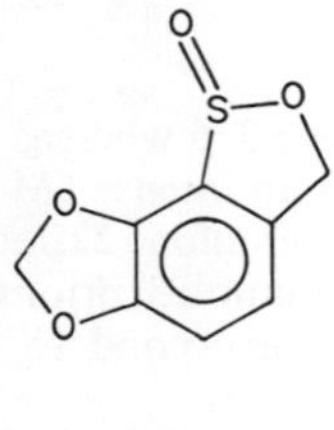

**(38)**

(10–14%), $SO_2$ (93–100%) and $HSO_2$ (42–100%). Furthermore, peaks corresponding to ion $[M - H_2SO_2]^+$ at *m/z* 116 could be found for **34** and for one of the isomers of compound **35** (46 and 16%, respectively). The other isomer of compound **35** gave no peak at *m/z* 116 but ion $[M - H_3SO_2]^+$ at *m/z* 115 (21%) instead. Both compounds **35** gave also peaks at *m/z* 91 (22–25%). The major fragmentations of sultine **36** were due to loss of MeSO, $MeSO_2$ and $C_2H_6SO_2$.

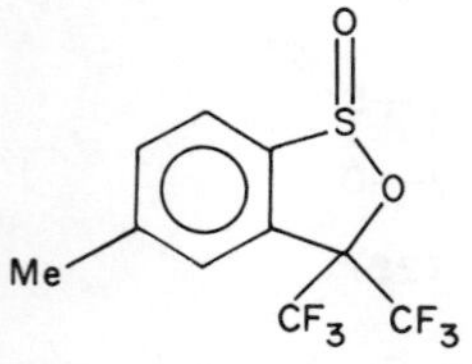

**(39)**

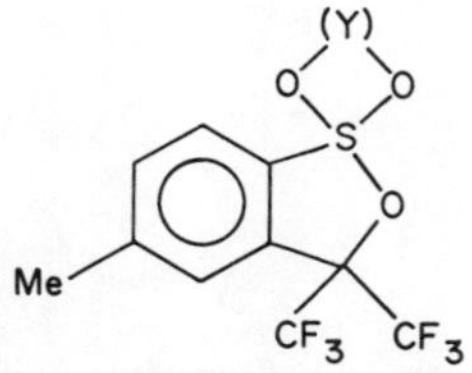

| | Y |
|---|---|
| (**40**) | $—C(Me)_2C(Me)_2—$ |
| (**41**) | $—CH_2CH_2—$ |
| (**42**) | $—CH_2C(Me)_2CH_2—$ |
| (**43**) | $—CH_3 \quad CH_3—$ |

The parent peaks ($m/z$ 198) of compounds **37** and **38** were very strong (75 and 100%, respectively). Only **37** showed loss of $HO^{\bullet}$ (10%). Other major peaks for **37** and **38** seem to correspond to loss of SO (100/20%), $SO_2$ (24/94%) and $CHSO_2$ (21/11%).

The characteristic ions for 5-methyl-3, 3-bis(trifluoromethyl)-3*H*-2, 1-benzoxathiole 1-oxide (**39**)[21] were $m/z$ 304 ($M^{+\bullet}$, 100%), 288 ($[M-O]^+$, 1.4%), 256 ($[M-SO]^{+\bullet}$, 2.7%], 240 ($[M-SO_2]^{+\bullet}$, 13%), 237 (58%), 235 ($[M-CF_3]^+$, 32%) and 166 (74%). In the same report, mass spectrometric data for some orthosulfinates (**40**–**43**) have also been given but they do not resemble those for sulfinate esters.

The mass spectrum of 1, 2-benzenedisulfinic anhydride **44** has been given[22] and its $[M-SO_2]^{+\bullet}$ fragment specified (equation 5).

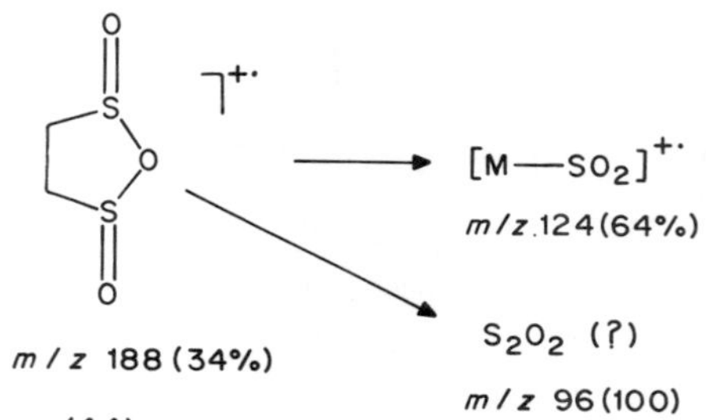


(5)

**(44)**

3′-Phenylfluorene-9-spiro-4′, 1′, 5′, 2′-oxathiazole 5′-oxide (**45**) showed no molecular ion but strong peaks corresponding to ions $m/z$ 267 $[M-SO_2]^{+\bullet}$, 190 $[M-Ph-SO_2]^+$ and 164 $[M-Ph-CN-S_2]^{+\bullet}$ were found[23].

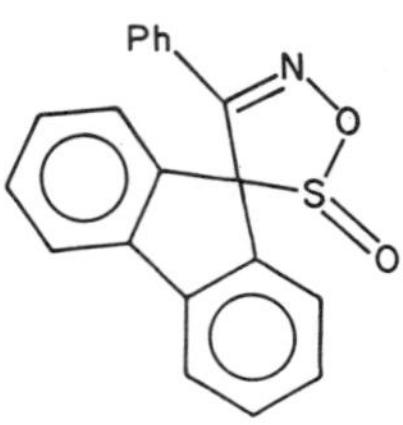

**(45)**

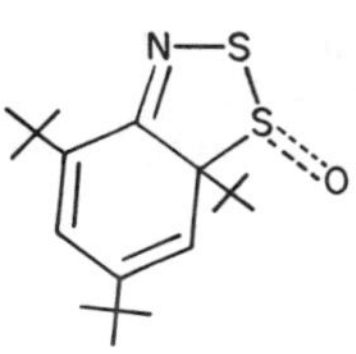

**(46)**

The electron-impact mass spectrum[24] of 2, 4, *r*-6-tri-*t*-butyl-7, 8-dithia-9-aza-bicyclo[4.3.0]nona-2,4,9-triene *t*-7-oxide (**46**) showed the following peaks: 339 ($M^{+\bullet}$, 0.7%), 307 ($[M-S]^+$, 0.4%), 291 ($[M-SO]^{+\bullet}$, 31%), 266 ($[M-C_4H_9O]^+$, 100%) and 250 ($[M-C_4H_9S]^+$, 25%).

## IV. MASS SPECTRA OF THIOSULFINATES

The first paper on the mass spectra of thiosulfinates (**47**–**50**) was published by Oae and coworkers[25]. Very weak or no molecular peaks can be seen in Table 2, where the characteristic fragments for **47**–**50** are also shown. Since thiosulfinates can undergo thermal disproportionation to the corresponding thiosulfonates and disulfides even in mass spectrometric conditions[26], it is possible that some features due to the latter

TABLE 2. Major fragment ions in the spectra of **47–50**

| | | | | |
|---|---|---|---|---|
| $M^{+\cdot}$ | — | 5 | + | + |
| $[M-O]^+$ | 89 | 42 | 66 | 12 |
| $[M-SO]^{+\cdot}$ | 7 | 3 | 3 | 2 |
| $[M-S_2O]^+$ | 5 | 5 | 8 | 4 |
| $XC_6H_4SO^+$ | 23 | 14 | 8 | 10 |
| $YC_6H_4SO^+$ | | 14 | | 6 |
| $XC_6H_4S^+$ | 100 | 100 | 100 | 23 |
| $YC_6H_4S^+$ | | 34 | | 100 |
|  | 4.6 | 4.0 | 5 | 6 |
|  | | | 0.7 | — |
|  | | 3.4 | | — |

compounds can also be found in the spectra of thiosulfinates[25].

| | | X | Y |
|---|---|---|---|
| p—$XC_6H_4S(O)$—$SC_6H_4Y$—p | **(47)** | H | H |
| | **(48)** | H | Me |
| | **(49)** | Me | Me |
| | **(50)** | H | OMe |

Compound **51** showed ions at $m/z$ 240 ($M^{+\cdot}$, 152 ($[M-C_4H_8S]^+$), 103, 102, 77, 54 and 41 (100%) while **52** had ions at $m/z$ 222 ($M^{+\cdot}$), 166 ($[M-C_4H_8]^{+\cdot}$), 134 ($[M-C_4H_8S]^+$), 89, 57 (100%) and 41[26b]. In the mass spectrum of *i*-PrS(O)SMe (**53**) the base peak corresponds to $[M-C_3H_6]^{+\cdot}$ ion, $MeSSOH^{+\cdot}$. Mass spectra of EtS(O)SMe, EtS(O)SEt, *t*-BuS(O)SMe and *t*-BuS(O)SBu-*t* also indicated the occurrence of a similar fragmentation process[26b].

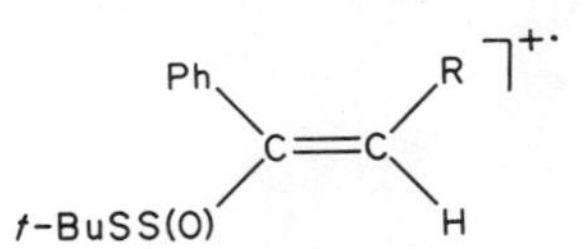

**(51)** R=H

**(52)** R=$CO_2CH_3$

Block and O'Connor[26c] determined the S—S bond energy in MeS(O)SMe by appearance potential methods to be 192 kJ mol$^{-1}$, compared to a corresponding value of *ca* 319 kJ mol$^{-1}$ in MeSSMe. They also solved completely the mass spectral fragmentation pathways of EtS(O)SEt (**54**, equation 6). A unique feature was the formation of the fragment $H_2S_2O$, corresponding to the unknown parent acid of thiosulfinate esters. Another significant observation was the formation of fragments corresponding to EtSOH and EtSSOH. The features of processes included: (i) incomplete site specificity for hydrogen transfer, (ii) persistence of the peaks corresponding to EtSOH, $CH_2CH{=}S$, EtSSOH at low ionizing voltage (8.6 eV), suggesting thermal as well as electron impact derived origins for these products, and (iii) variation of the RSSOH/R'SOH ratio with thiosulfinate structure. Thus for the thiosulfinates RS(O)R' (R, R' = i-Pr, Me; Et, Me; Et, $CD_3$; Et, Et; Et, $CD_2Me$; Et, $CH_2CD_3$) the respective intensity ratios were 25, 1, 3, 2, 5, and 0.7.

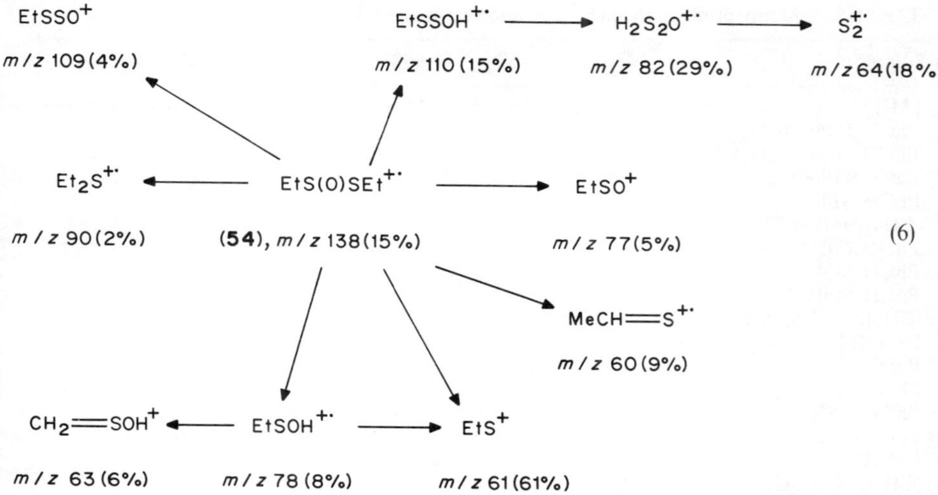


(6)

Block and O'Connor[26c] also compared the mass spectra of **54** and MeS(O)SEt (**55**) with each other. Since the spectra were not identical, the oxygen crossover process proposed by Oae and coworkers[25] to occur during fragmentation of dialkylthiosulfinates was not operative. Freeman and Angletakis[27] came to the same conclusion by comparing the electron-impact mass spectra of *t*-$BuCH_2S(O)SPh$ (**56**) and $PhS(O)SCH_2Bu$-*t* (**57**) with each other. Also the spectra[27] of *t*-$BuCH_2S(O)SCH_2Bu$-*t* (**58**), $PhCH_2S(O)SCH_2Ph$ (**59**), $PhCH_2S(O)SPh$ (**60**) and PhS(O)SPh (**61**) can be explained in accordance with the fragmentations shown in equations 6 and 7.

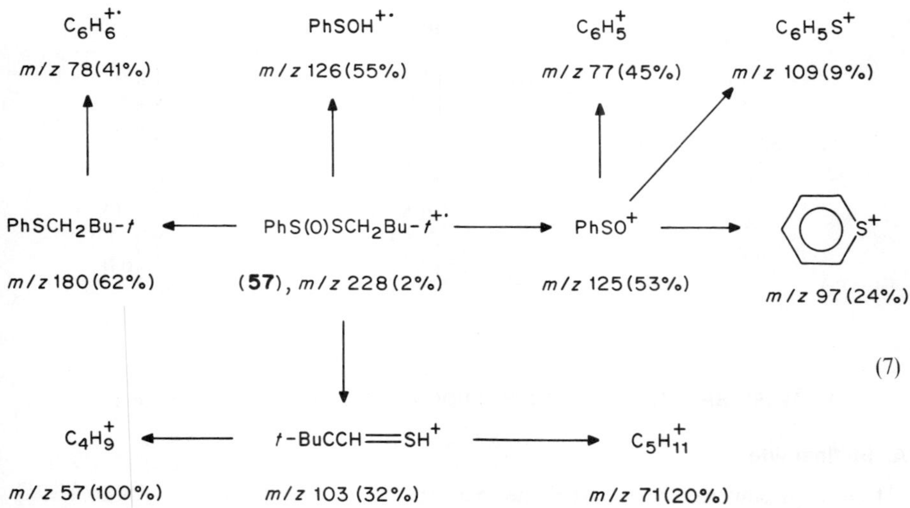


(7)

The 2-methylpropane chemical ionization spectra[27] have also been recorded for compounds **56**–**61**. They are listed in Table 3.

TABLE 3. 2-Methylpropane chemical ionization mass spectra of **56–61** at 100 eV [*m/z* (%)]

| Ion | **56** | **57** | **58** | **59** | **60** | **61** |
|---|---|---|---|---|---|---|
| $[MH]^+$ | 229(100) | 229(100) | 223(100) | 263(18) | 249(100) | 235(100) |
| $PhCH_2S(Ph)C(CH_3)_3^+$ | — | — | — | — | 257(20) | — |
| $PhCH_2S(Ph)C(CH_3)_2^{+\cdot}$ | — | — | — | — | 243(8) | — |
| $PhCh_2S(Ph)C(CH_3)^+$ | — | — | — | — | 229(9) | — |
| $PhCH_2SPh^+$ | — | — | — | — | 200(6) | — |
| $(CH_3)_3S(OH)Ph^+$ | — | — | — | — | — | 183(9) |
| $PhS(H)C(CH_3)_3^+$ | — | — | — | — | — | 167(17) |
| $PhCH_2SOH_2^+$ | — | — | — | 141(43) | 141(45) | — |
| $PhCH_2SOH^{+\cdot}$ | — | — | — | 140(9) | 140(13) | — |
| $(CH_3)_3CCH_2S(O)OH_2^+$ | — | — | — | — | 137(8) | — |
| $PhSOH_2^+$ | — | 127(13) | — | — | — | 127(78) |
| $PhSOH^{+\cdot}$ | — | 126(9) | — | — | — | 126(44) |
| $PhSO^+$ | — | — | — | — | — | 125(19) |
| $PhCH{=}SH_2^+$ | — | — | — | 124(11) | — | — |
| $PhCH{=}SH^+$ | — | — | — | 123(100) | 123(19) | — |
| $PhCHS^{+\cdot}$ | — | — | — | 122(9) | — | — |
| $(CH_3)_3CCH_2SOH_2^+$ | 121(9) | — | 121(19) | — | — | — |
| $PhSH_2^+$ | — | — | — | 111(20) | 111(29) | — |
| $PhSH^{+\cdot}$ | — | — | — | — | 110(9) | 110(17) |
| $PhCH_2O^+$ | — | — | — | 107(15) | 107(51) | — |
| $(CH_3)_3CCH{=}SH^+$ | — | 103(11) | 103(16) | — | — | — |
| $PhCH_3^{+\cdot}$ | — | — | — | 92(7) | — | — |
| $PhCH_2^+$ | — | — | — | 91(36) | 91(40) | — |
| $C_5H_{11}^+$ | 71(8) | — | 71(22) | — | 71(9) | — |

Harpp and Granata[28] reported the electron-impact mass spectrum of benzylsulfinyl methyl thiocarbonate (**62**):

$PhCH_2S(O)SCOOMe^{+\cdot}$
(**62**), *m/z* 230(1%)

| Ion | *m/z* (%) |
|---|---|
| $[M-O]^+$ | 214(3) |
| $[M-SO]^{+\cdot}$ | 182(2) |
| $[M-SO-CO]^{+\cdot}$ (?) | 138(4) |
| $PhCH_2S^+$ | 123(19) |
| $PhCO^+$ (?) | 105(47) |
| $PhMe^{+\cdot}$ | 92(50) |
| $PhCH_2^+$ | 91(100) |
| $Ph^+$ | 77(63) |

## V. MASS SPECTRA OF SULFINAMIDES AND RELATED COMPOUNDS

### A. Sulfinamides

Ueda and coworkers[29] reported the mass spectra of 30 sulfinamide [RS(O)NHR′] derivatives (**63–67**). Most of the spectra had peaks attributable to thermal decomposition products. It was concluded that the above sulfinamides gave the following ions after electron impact: $M^{+\cdot}$, $[M-R]^+$, $[M-R+H]^{+\cdot}$, $[M-SO_2]^{+\cdot}$, $RS^+$, $NHR'^+$,

TABLE 4. Electron-impact mass spectra of sulfinamides **63a**, **63b** and **65d** [$m/z$ (rel. abund., %)]

| Ion | **63a** | **63b** | **65d** |
|---|---|---|---|
| $M^{+\cdot}$ | 149(15) | 175(92) | 237(25) |
| $[M-OH]^+$ | 132(13) | 158(2) | 220(11) |
| $[M-R]^+$ | 120(19) | 146(97) | 146(18) |
| $[M-R+H]^{+\cdot}$ | 121(8) | 147(10) | 147(3) |
| $[RSO]^+$ | 77(3) | 77(2) | 139(100) |
| $[RSO+H]^{+\cdot}$ | 78(3) | 78(2) | 140(50) |
| $[NHR']^+$ | 72(3) | 98(11) | 98(51) |
| $[NHR'+H]^{+\cdot}$ | — | 99(2) | 99(5) |
| $R^+$ | 29(12) | 29(4) | 91(7) |
| $[R+H]^{+\cdot}$ | 30(16) | 30(2) | 92(22) |
| $[R']^+$ | 57(100) | 83(100) | 83(9) |
| $[M-SO]^{+\cdot}$ | — | — | 189(93) |
| $[RS]^+$ | — | — | 123(4) |

$[NHR'+H]^{+\cdot}$, $RSO^+$, $[RSO+H]^{+\cdot}$, $R^+$ and $[M-OH]^+$ and that the thermal decomposition products gave the following ions: $[RSO_2SR]^{+\cdot}$, $[RSSR]^{+\cdot}$, $[M-O]^{+\cdot}$, $[M+O]^{+\cdot}$ and $[RSOC_6H_4NH_2]^{+\cdot}$.

RS(O)NHR′

(**63**): R = Et, **a** R′ = Bu; **b** cyclohexyl; **c** Ph
(**64**): R = Ph, **a** R′ = H; **b** Pr; **c** Ph; **d** *p*-Tol; **e** *p*-An;
**f** *p*-$ClC_6H_4$; **g** p-$MeCOC_6H_4$
(**65**): R = *p*-Tol, **a** R′ = H; **b** Pr; **c** Bu; **d** cyclohexyl; **e** Ph;
**f** *p*-Tol; **g** *p*-An; **h** *p*-$ClC_6H_4$; **i** *p*-$EtCO_2C_6H_4$;
**j** *p*-$MeCOC_6H_4$
(**66**): R = *p*-An, **a** R′ = *p*-Tol; **b** *p*-An; **c** *p*-$ClC_6H_4$
(**67**): R = *p*-$ClC_6H_4$, **a** R′ = Ph; **b** *p*-Tol; **c** *p*-An
**d** *p*-$ClC_6H_4$; **e** *p*-$EtCO_2C_6H_4$; **f** *p*-$MeCOC_6H_4$

Three sulfinamides[29] were shown to be stable on heating at 150 °C for 5 min. Therefore, their mass spectra (Table 4) could be considered to be those derived from the original molecules. Examples of the characteristic fragment peaks for the rest of compounds **63**–**67**, which all underwent the thermal decomposition, are listed in Table 5. In this table, $[M+O]^{+\cdot}$ means the ion $[RSO_2NHR']^{+\cdot}$ and $[M-O]^{+\cdot}$ the ion $[RSNHR']^{+\cdot}$. The ions of the group B can be considered to be formed from artefacts[29].

It can be concluded that (i) $[M-R]^+$ ions are observed, particularly for alkylsulfinylamides, (ii) the $[M-SO]^{+\cdot}$ and $[RS]^+$ ions are observed only for arylsulfinylamides, (iii) the $[NHR']^+$ and $[NHR'+H]^{+\cdot}$ ions are generally observed for all kinds of sulfinamides but are characteristically very abundant (either one or both) for sulfinyl arylamides, (iv) $[RSO]^+$ ion is characteristically abundant in the spectra of arylsulfinyl amides.

Harpp and Back[12] studied the electron-impact mass spectra of seven sulfinamides. None of these compounds has been stated to undergo thermal decomposition in a mass spectrometer. However, at least the spectrum of **68** ($M^{+\cdot}$ at $m/z$ 135) shows a strong artefact peak (54%) at $m/z$ 132. Otherwise **68** was supposed to fragment as shown in equation 8.

TABLE 5. Fragment peaks given by some of the compounds RS(O)NHR′ [$m/z$ (%)]

| Ion | Type | **64a** | **64f** | **65a** | **65h** | **65i** | **66c** | **67d** | **67e**[a] |
|---|---|---|---|---|---|---|---|---|---|
| $M^{+\cdot}$ | — | 141(83) | 251(20) | 155(65) | — | — | — | — | — |
| $[M-OH]^+$ | A | — | 234(3) | 138(1) | 248(6) | 286(1) | — | — | — |
| $[RSO]^+$ | A | 125100) | 125(41) | 139(100) | 139(26) | 139(23) | 155(39) | 159(38) | 159(4) |
| $[RSO+H]^{+\cdot}$ | A | 126(8) | 126(37) | 140(11) | 140(3) | 140(3) | 156(4) | 160(4) | — |
| $[NHR']^+$ | A | — | 126(37) | — | 126(3) | 164(2) | 126(2) | 126(26) | 164(18) |
| $[NHR'+H]^{+\cdot}$ | A | — | 127(100) | — | 127(100) | 165(77) | 127(100) | 127(100) | 165(44) |
| $R^+$ | A | 77(68) | 77(14) | 91(31) | 91(14) | 91(12) | 107(9) | 111(22) | 111(19) |
| $[R+H]^{+\cdot}$ | A | 78(17) | 78(6) | 92(6) | 92(14) | 92(15) | — | 112(5) | 112(2) |
| $[R']^+$ | A | — | — | — | — | — | — | 111(22) | — |
| $[M-SO]^{+\cdot}$ | A | 93(59) | — | 107(48) | — | — | — | — | — |
| $[RS]^+$ | A | 109(26) | 109(9) | 123(4) | 123(17) | 123(24) | 139(87) | 143(31) | 143(8) |
| $[RSO_2SR]^{+\cdot}$ | B | 250(27) | 250(41) | 278(2) | 278(22) | 278(17) | 310(35) | 318(15) | — |
| $[RSSR]^{+\cdot}$ | B | — | 218(16) | — | 246(20) | 246(35) | 278(41) | 286(12) | 286(3) |
| $[M+O]^{+\cdot}$ | B | — | — | — | 281(1) | 319(1) | 297(11) | 301(4) | 339(5) |
| $[M-O]^{+\cdot}$ | B | — | 235(14) | 139(100) | 249(12) | — | 265(2) | 269(7) | 307(10) |

[a] $[H_2NC_6H_4CO]^{+\cdot}$ was the most abundant ion.

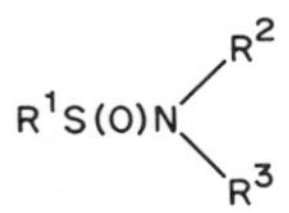

| | $R^1$ | $R^2$ | $R^3$ |
|---|---|---|---|
| (**68**) | Me | Et | Et |
| (**69**) | *i*-Pr | Me | Bu |
| (**70**) | Ph | cyclohexyl | H |
| (**71**) | Me | Ph | H |
| (**72**) | $C_6H_4CH_2$ | $—CH_2CH_2OCH_2CH_2—$ | |
| (**73**) | Et | $—(CH_2)_5—$ | |
| (**74**) | Me | $—CH_2CH_2N(SOCH_3)CH_2CH_2—$ | |

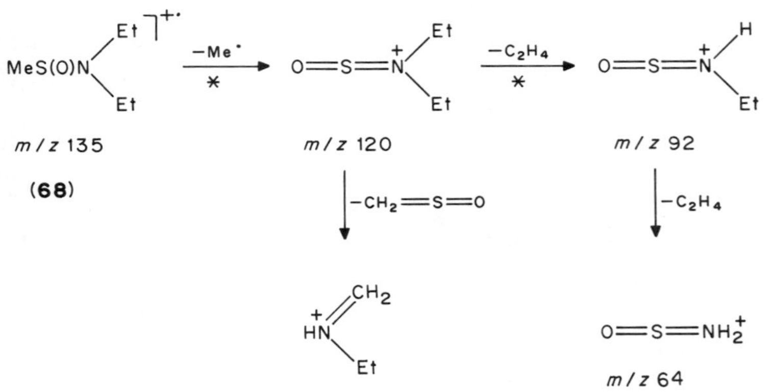


(8)

The spectrum of **69** again reveals carbon–sulfur bond fission to be a key process, although charge retention by the alkyl fragment (*m*/*z* 43, *m*/*z* 57) is now stronger than in **68**. Furthermore **69** can exhibit a five-center McLafferty rearrangement (equation 9) leading

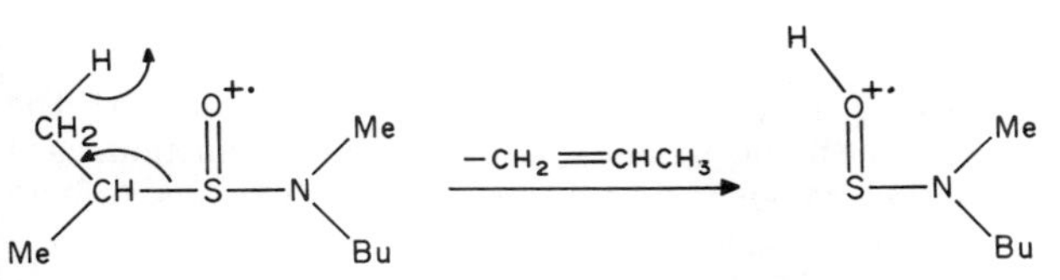


(9)

(**69**)

to ion *m*/*z* 135, a mechanism operative in alkylsulfinylphthalimides containing hydrogen atoms *β* to the sulfur atom[30].

The complexity of the spectrum of sulfinamide **70** is largely due to the presence of the cyclohexyl group (equation 10).

Sulfinanilide **71** gave a very strong molecular ion (60%) and its fragmentation was dominated by formation of aniline and its fragment ions as could be expected[9]. Compound **72** gave only a weak parent ion peak and its fragmentation was dominated by formation of the tropylium ion at *m*/*z* 91 and, via charge retention, that of ion $O{=}S{=}N(CH_2CH_2)_2O^{+\bullet}$ at *m*/*z* 134.

Compound **73**, the piperidine derivative, displayed a strong parent peak (29%) and fragmented predominantly by the alkyl–sulfur bond scission (equation 11). Like **69** also **73**

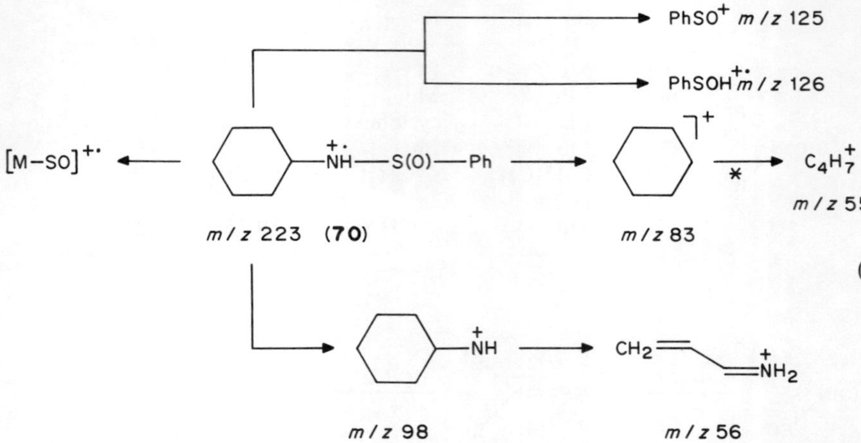


(10)

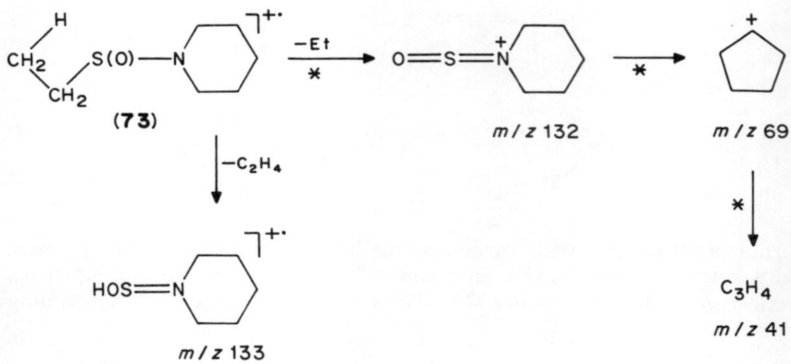


(11)

underwent a McLafferty-type rearrangement giving an ion at $m/z$ 133 (equation 11). The spectrum of **74** was quite complex[12] due to the presence of two sulfinamide functions each of which may undergo fragmentation. Major fragments resulted from cleavage of both S—N bonds are from fission of the piperazine ring.

Bell[13] prepared several sulfinamides (**75**–**86**) and stated their electron-impact mass spectra to be consistent with those reported earlier by Ueda and coworkers[29] and Harpp and Back[12]. The $[M-O]^{+\bullet}$ ion formed the base peak in the mass spectra of compounds **75**, **77**, **78** and **80**–**84** and was very strong also in the spectra of **76** and **79**. Compounds **85** and **86** apparently gave no $[M-O]^{+\bullet}$ ion[13]. Other typical fragments for **75**–**82** were most

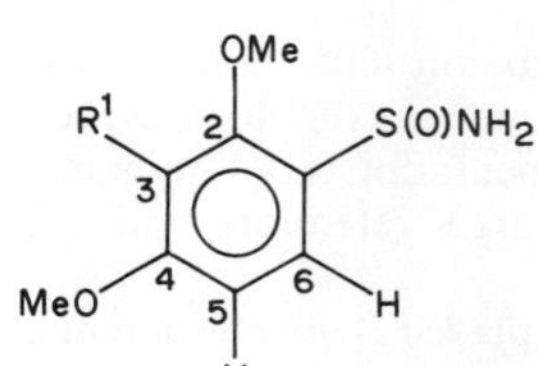


| | $R^1$ | X |
|---|---|---|
| **(75)** | H | H |
| **(76)** | H | Cl |
| **(77)** | Me | H |
| **(78)** | MeO | H |

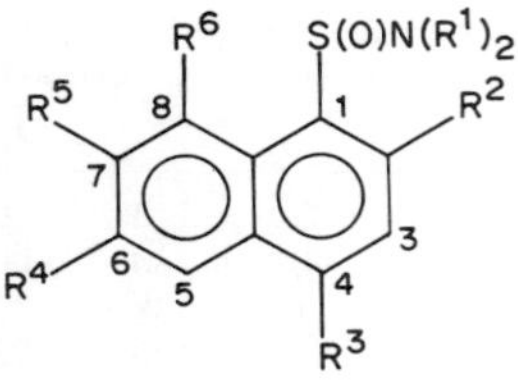

| | $R^1$ | $R^2$ | $R^3$ | $R^4$ | $R^5$ | $R^6$ |
|---|---|---|---|---|---|---|
| **(79)** | H | MeO | H | H | H | H |
| **(80)** | H | H | MeO | H | H | H |
| **(81)** | H | MeO | H | H | MeO | H |
| **(82)** | H | H | MeO | MeO | H | H |
| **(83)** | H | MeO | H | H | MeO | H |
| **(84)** | H | H | MeO | H | H | MeO |
| **(85)** | Et | MeO | H | H | H | H |
| **(86)** | Et | H | MeO | H | H | H |

probably ions corresponding to loss of HS$^\bullet$, SO and NHSO$^\bullet$, although Bell[13] did not report any high resolution data. 4-Methoxy substituted naphthalene sulfinamides **80**, **82** and **84** exhibited also fairly strong $[M - MeO]^+$ ions (14,21 and 30%, respectively). The spectra of **85** and **86** are difficult to explain and deserve a more careful study. Actually a systematic study on the mass spectrometric behavior of various sulfinamides with modern techniques is most desirable in order to understand better the role of their mass spectrometric as well as the possible simultaneous thermal reactions[12,13,29–32].

Gupta and Pizey[31] prepared 2-methyl-3-oxobutane-2-sulfinamide and 2-methyl-3-oxobutane 2-sulfin-*m*-toluidine and gave also the locations of peaks in their mass spectra. The peaks can be postulated to correspond to ions *m*-$XC_6H_4SO^{+\bullet}$ (*m/z* 139 and 153), *m*-$XC_6H_4^+$ (*m/z* 77 and 91), $MeC(O)CH(Me)_2^+$ (*m/z* 86), $MeC(O)CHMe^+$ (*m/z* 71), $SONH_2^{+\bullet}$ (*m/z* 64) and $MeCO^+$ (*m/z* 43).

Mass spectrometric fragmentation of bis(trimethylsilyl)amide of pentafluorobenzenesulfinic acid, $(Me_3Si)_2NS(O)C_6F_5$ was typical for molecules with the $Me_3SiO$ group. Therefore Rinne and Blaschette[32] suggested that this compound—at least in mass spectrometric conditions—has an imidoester structure, $C_6F_5S(SiOMe_3){=}NSiMe_3$.

## B. Sulfinylphthalimides

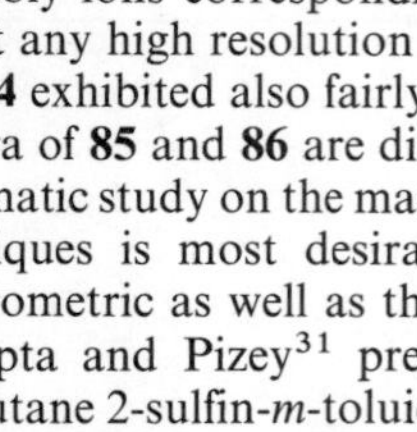

**(87)** (**a**) R = Me; (**b**) Et; (**c**) *i*-Pr; (**d**) Bu; (**e**) *t*-Bu; (**f**) Ph; (**g**) *p*-Tol; (**h**) $C_6H_4CH_2$; (**i**) $CH_3O_2CCH_2$; (**j**) *i*-PrS;

Harpp and Back[30] studied the electron-impact-induced fragmentations of several sulfinyl phthalimides (**87a–j**). All compounds gave a molecular ion of varying abundance (Table 6) and showed intense peaks at *m/z* 147, 130, 104 and 76 likely arising from the fragments shown in equation 12.

Compounds **87**, except **87i** and **j**, exhibited intense peaks for the alkyl or aryl fragments ($R^+$ in Table 6). For **87e** and **87h** it was even the base peak. The charge can also resides on the sulfur-containing fragments (equation 13; cf. **87a**, **f**, **h** and **i** in Table 6). If hydrogen atoms $\beta$ to the sulfinyl group are available, a five-center McLafferty-type rearrangement occurs (cf. equation 11) and the ion *m/z* 195 (equation 14) is obtained. This ion in turn lost $SO_2H$ and gave an ion at *m/z* 130.

Compounds **87f** and **87g** produced additional strong ions at *m/z* 125 and 139, respectively, suggesting formation of *p*-$XC_6H_4SO$ (X = H or Me). Loss of SO occurred directly from compounds **87f, g** and **i**, as usual for many sulfinamides[13,29]. Formation of

TABLE 6. Mass spectra of sulfinyl phthalimides (87)

| Ion | Compound | | | | | | | | | |
|---|---|---|---|---|---|---|---|---|---|---|
| | **a** | **b** | **c** | **d** | **e** | **f** | **g** | **h** | **i** | **j** |
| $M^{+\cdot}$ | 29 | 6 | <1 | <1 | <1 | 14 | 14 | 3 | 2 | 10 |
| *m/z* 147 | 16 | 15 | 10 | 88 | 9 | 100 | 100 | 90 | 80 | 100 |
| *m/z* 130 | 20 | 100 | 100 | 92 | 33 | 26 | 3 | <1 | 38 | 5 |
| *m/z* 104 | 18 | 29 | 13 | 100 | 15 | 58 | 57 | 55 | 100 | 78 |
| *m/z* 76 | 27 | 41 | 18 | 88 | 27 | 60 | 52 | 58 | 96 | 68 |
| $R^+$ | 6 | 19 | 23 | 50 | 100 | 42 | 34 | 100 | — | — |
| *m/z* 194 | 36 | — | — | — | — | 8 | <1 | 8 | — | 37 |
| *m/z* 195 | — | 19 | 66 | 42 | 27 | — | — | — | 10 | — |

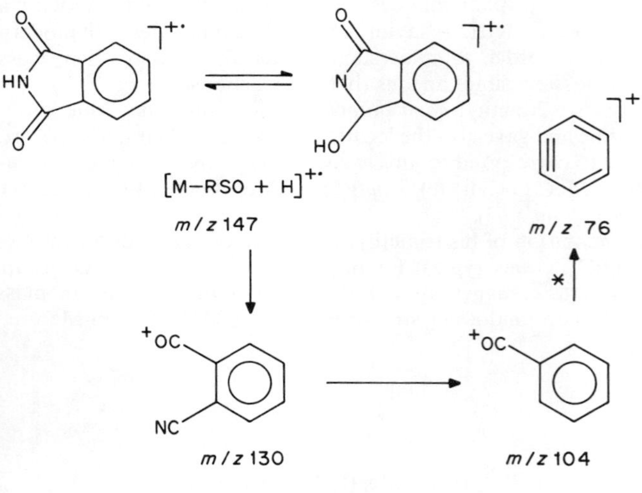


(12)

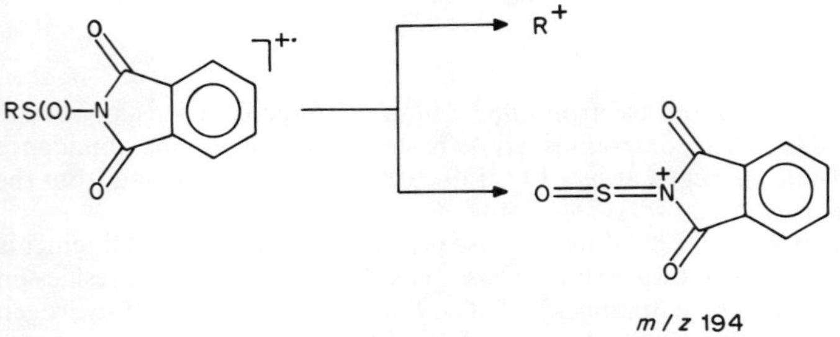


(13)

the abundant ions *m/z* 120 and *m/z* 89 in the spectrum of **87i** can be explained by loss of phthalimide from the molecular ion followed by loss of MeO$^\cdot$. Thiosulfinylphthalimide **87j** underwent a rearrangement process which led to an abundant peak at *m/z* 227 (equation 15).

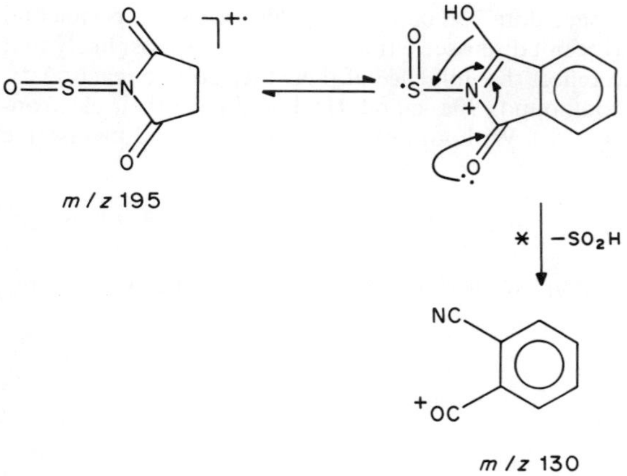


(14)

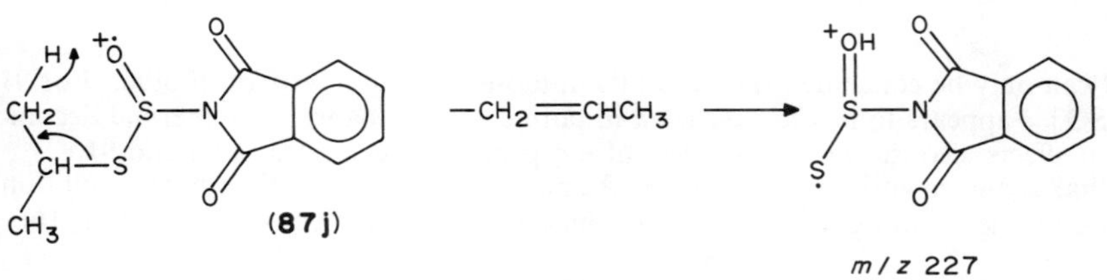


(15)

## C. Cyclic Sulfinamides

No systematic investigation has been carried out on the mass spectrometric behavior of cyclic sulfinamides, although some mass spectrometric data can be found[15,33–36].

The mass spectra of 1, 2-thiazetidin-3-one 1-oxides **88a**–**c** exhibited, in addition to $M^{+\cdot}$, peaks corresponding to ions $[M-SO]^{+\cdot}$ and isocyanates, $RNCO^{+\cdot}$[15]. Jäger and

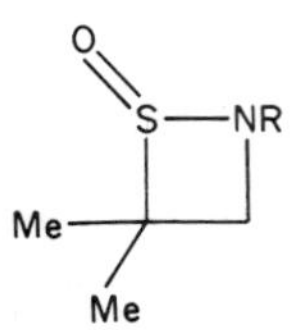


(**88**) (**a**) R = *p*-Tol; (**b**) *p*-An; (**c**) cyclohexyl

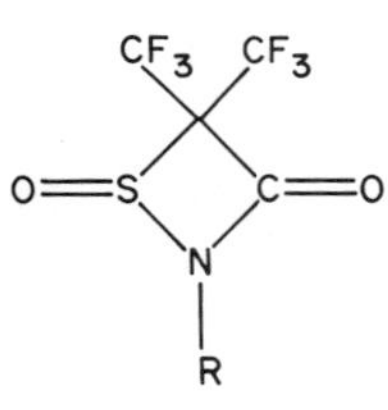


(**89**) (**a**) R = Me; (**b**) cyclohexyl; (**c**) Ph; (**d**) $C_6F_5$; (**e**) $CF_3$

coworkers[33] listed the electron-impact data for compounds **89a–e**. No experimental details were given and the spectra were not discussed either. However, it is most likely that the spectra of these compounds do reflect the influence of their thermolytic reactions[23].

Hanson and Stone[34] prepared compounds **90a–c** and **91a, b** and gave their electron-impact fragmentation patterns as *m/z* (%). Without any high-resolution and metastable

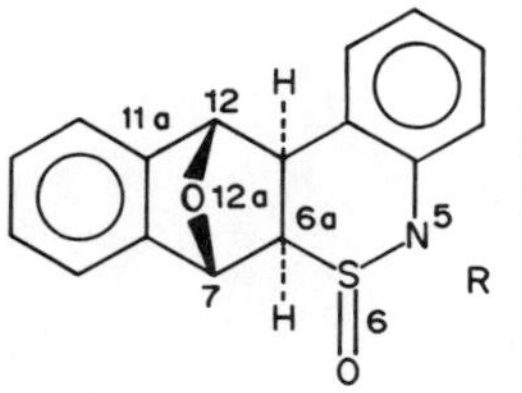


(**90**) (**a**) R = Me; (**b**) $MeSCH_2$; (**c**) EtOCO

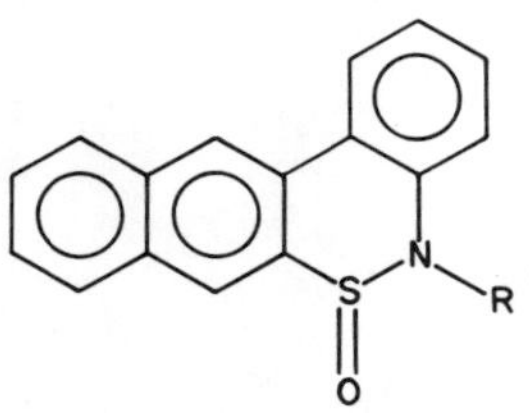


(**91**) (**a**) R = H; (**b**) Me

data it can only be concluded that all of them loose SO in one way or another. For **91** $[M - SO]^{+\cdot}$ appears to be the base peak in both cases. Probably some thermal decomposition affects also the variety of different ion peaks in the spectra of **90** and **91**.

Borthakur and coworkers[35] determined the molecular formula of **92** by high-resolution mass spectrometry and gave the *m/z* (%) values for some fragment ions (equation 16).

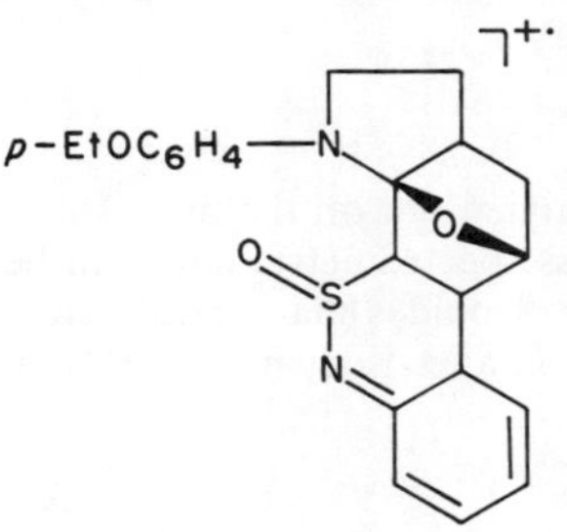


*m/z* 380(28%)
*m/z* 367(24%)
*m/z* 201(20%) (16)
*m/z* 150(16%)
*m/z* 118(28%)

(**92**) *m/z* 396 (100%)

The electron-impact mass spectrum of 4-chloro-2-(1-chloro-2, 2, 2-trifluoromethyl)-1-thia-2, 3-diazaspiro[4, 5]dec-3-ene 1-oxide **93** (both diastereomers) showed the following

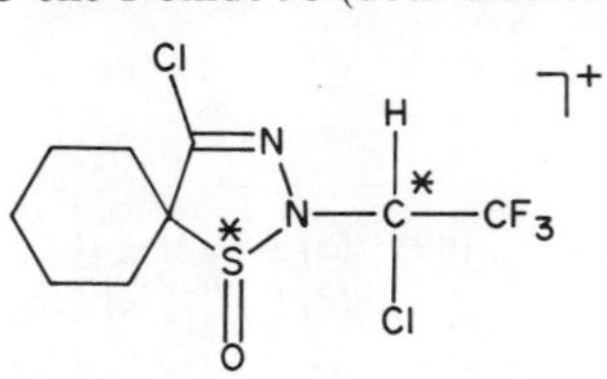


(**93**)

peaks: $m/z$ 322 $M^{+\cdot}$, 287 $[M-Cl]^+$, 274 $[M-SO]^{+\cdot}$, 239 $[M-Cl-SO]^{+\cdot}$, 203 $[M-Cl-SO-HCl]^+$, 108 $C_7H_{10}N^+$ (100%), 69 $CF_3^+$, and 36 $HCl^{+\cdot}$[36].

### D. Sulfinyl Diamines

Neidlein and Walser[37] used 100 eV mass spectra to characterize sulfinyl diamines **94–99**. Inspection of their data allows us to compile the elemental compositions shown in Table 7 for the fragment peaks in the spectra of **94–99**.

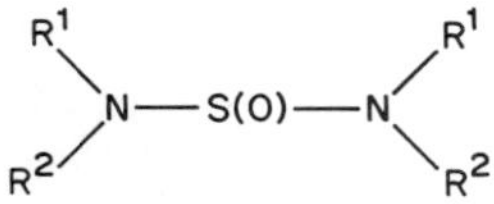

| | $R^1$ | $R^2$ |
|---|---|---|
| **(94)** | *i*-Pr | *i*-Pr |
| **(95)** | —$(CH_2)_4$— | |
| **(96)** | Ph | Me |
| **(97)** | $C_6H_5CH_2$ | Ne |
| **(98)** | $C_6H_5CH_2$ | $C_6H_5CH_2$ |
| **(99)** | —$(CH_2)_2$—NMe—$(CH_2)_2$— | |

### E. 2-Oxo-1, 2, 3-oxathiazolidines

Nishiyama and coworkers[38a–c] investigated the electron-impact mass spectra of several 3-aryl-2-oxo-1, 2, 3-oxathiazolidines (**100–103**). The major fragment ions from com-

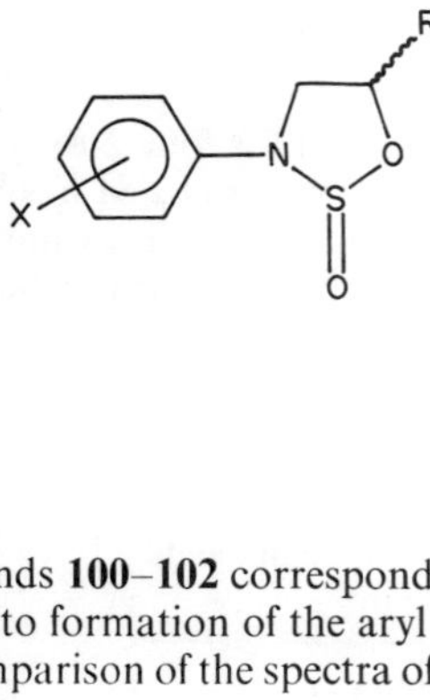

| | X | R |
|---|---|---|
| **(100)** | H | H |
| **(101)** | *p*-Me | H |

**(102)** R = Me: (**a**) X = H, (**b**) *o*-Me (*t*), (**c**) *m*-Me (*t*), (**d**) *p*-Me (*c*), (**e**) *o*-Cl (*c* and *t*), (**f**) *m*-Cl (*t*), (**g**) *p*-Cl (*t*)

**(103)** R = $CH_2Cl$: (**a**) X = H (*c*), (**b**) *m*-Me (*t*), (**c**) *o*-Cl (*c*), (**d**) *p*-Cl (*c*), (**e**) *o*-Me (*c*), (**f**) *m*-Me (*c*), (**g**) *m*-Cl (*t*), (**h**) 2, 4, 6-$Cl_3$ (*t*)

pounds **100–102** corresponded to loss of $SO_2$, $HSO_2$, $RCHSO_2$, $RCH_2SO_2$, $RC_2H_3SO_2$ and to formation of the aryl ions. The configurational isomers gave very similar spectra. Comparison of the spectra of compounds **100** and **101** with those of **102** supports one-step

TABLE 7. Fragment peak compositions for compounds **94–99**

| Compound | $m/z$ (composition, %) |
|---|---|
| **94** | 248($M^{+\cdot}$, 6); 148($C_6H_{14}NSO$, 93); 106($C_3H_8NSO$, 100) |
| **95** | 188($M^{+\cdot}$, 10); 172($C_8H_{16}NS$, 30); 118($C_4H_8NSO$, 100); 102($C_4H_8NS$, 51) |
| **96** | 260($M^{+\cdot}$, 6); 154($C_7H_8NSO$, 95); 138($C_7H_8NS$, 2); 106($C_7H_8N$, 100) |
| **97** | 288($M^{+\cdot}$, 4); 272($C_{16}H_{20}N_2O$, 95); 188(?, 18); 168($C_8H_{10}NSO$, 100); 152($C_8H_{10}NS$, 4); 120($C_8H_{10}N$, 95); 105($C_7H_7N$, 16); 91($C_7H_7$, 99) |
| **98** | 440($M^{+\cdot}$, 3); 424($C_{28}H_{28}N_2S$, 59); 244($C_{14}H_{14}N_2SO$, 95); 196($C_{14}H_{14}N$, 100) |
| **99** | 246($M^{+\cdot}$, 21); 231($C_9H_{19}N_4SO$, 1); 147($C_5H_{11}N_2SO$, 100); 131($C_5H_{11}N_2S$, 40) |

formation of the above ions rather than successive losses of e.g. $SO_2$ and $C_2H_4$, as stated by Nishiyama and his group[38a,b]. As usual **102e** with *o*-chloro substitution gave a much stronger $[M - SO_2 - Cl]^+$ peak than the corresponding *m*- and *p*-derivatives (**102f** and **102g**, respectively).

Bartnik and his group[39] prepared compounds **104–107**. Compounds **104**, **105a**, **106a, b**

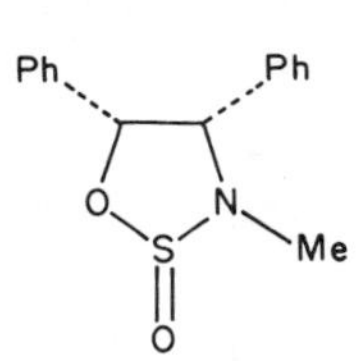


(104)

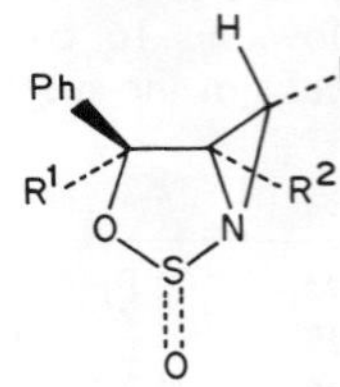


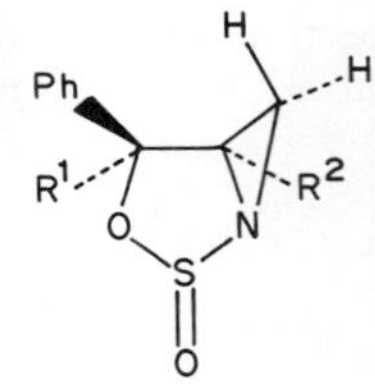


| | $R^1$ | $R^2$ | | $R^1$ | $R^2$ |
|---|---|---|---|---|---|
| (**105a**) | H | H | (**106b**) | Ph | Ph |
| (**106a**) | Ph | Ph | (**107b**) | Ph | Et |
| (**107a**) | Ph | Et | | | |

and **107a, b** all lost $HSO_2$ (100, 100, 100, 73 and 53%, respectively) as their major initial fragmentation. Compound **104** gave 8% of $[M - SO_2]^{+\bullet}$ instead. Its base peak was at *m/z* 118 ($PhCNMe^+$) and those of **107a** and **b** at *m/z* 77 ($Ph^+$).

## F. Octahydro-3, 2, 1-benzoxathiazine 2-Oxides

Some electron-impact-induced fragmentation data can be found for four 2-oxides of octahydro-3, 2, 1-benzoxathiazines, namely for **108–111**[40]. In this paper *m/z* 81 has been printed as the base peak in all cases, but this author believes that it must be tropylium ion *m/z* 91 and this is the value which appears in Table 8. Similarly we believe that **110** did not give ion *m/z* 122 but *m/z* 132 instead.

TABLE 8. Postulated fragment peak compositions for compounds **108–111**

| | | Compound (%) | | | |
|---|---|---|---|---|---|
| Ion | *m/z* | **108** | **109** | **110** | **111** |
| $M^{+\bullet}$ | 265 | 6 | 4 | 6 | 5 |
| $[M - SO_2]^{+\bullet}$ | 201 | 10 | 16 | 11 | 16 |
| $C_6H_{10}SO_2^+$ | 146 | 11 | — | 14 | — |
| $C_5H_9SO_2^+$ | 133 | 54 | 67 | — | 80 |
| $C_5H_8SO_2^+$ | 132 | 14 | — | 82 | — |
| ? | 110 | 31 | 26 | 12 | 16 |
| $C_7H_7^+$ | 91 | 100 | 100 | 100 | 100 |
| $C_5H_5^+$ | 65 | 17 | 17 | 18 | 13 |
| $C_4H_7^+$ | 55 | — | 10 | — | — |

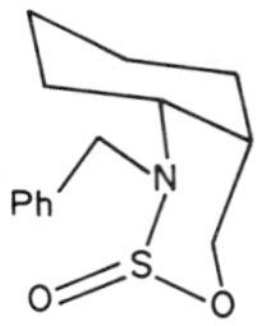

**(108)** *N-out*

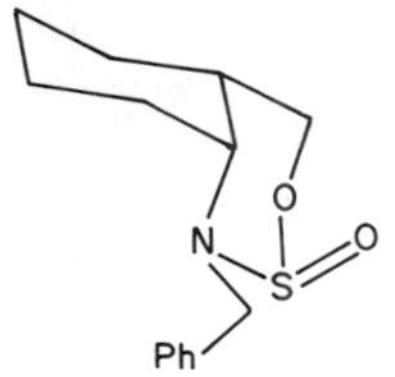

**(109)** *N-in*

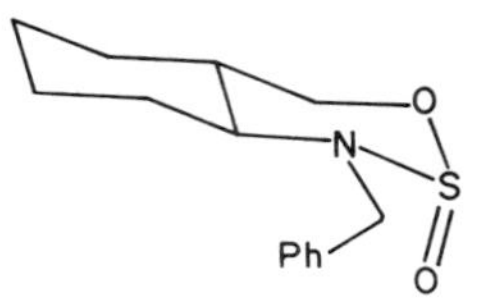

**(110)**

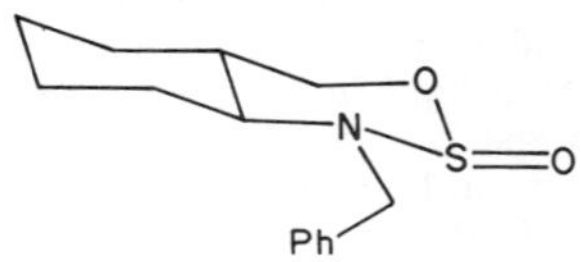

**(111)**

## VI. CONCLUDING REMARKS

Surprisingly little data could be found on the mass spectra of sulfinic acids and their derivatives. Although the mass spectrometric reactions appear to be disturbed by thermal decomposition, one can hope that new, carefully done mass spectrometric studies on them will become available in the near future. At the present, a great deal of the limited results were given without experimental details and any discussion. Therefore this author has been often obliged to postulate the ion structures illustrated in this chapter by a method which in some cases can be called an advanced guess only.

## VII. REFERENCES

1. K. Pihlaja, in *The Chemistry of Sulphones and Sulphoxides*, Chap. 6, Wiley, Chichester, 1988, and references 1–4, 6, 11 cited therein.
2. A. Tangerman and B. Zwanenburg, *J. Chem. Soc., Perkin Trans. 2*, 461 (1973).
3. F. Wudl, D. A. Lightner and D. J. Cram, *J. Am. Chem. Soc.*, **89**, 4099 (1967).
4. I.-P. Lorenz, E. Lindner and W. Reuther, *Chem. Ber.*, **110**, 833 (1977).
5. W. G. Filby, K. Günther and R. D. Penzhorn, *J. Org. Chem.*, **38**, 4070 (1973).
6. W. G. Filby, R. D. Penzhorn and L. Stieglitz, *Org. Mass Spectrom.*, **8**, 409 (1974).
7. R. J. Phillips and G. B. Deacon, *Aust. J. Chem.*, **32**, 2381 (1979).
8. G. E. Binder and A. Schmidt, *Z. Anorg. Allg. Chem.*, **430**, 263 (1977).
9. W. H. Baarschers and B. W. Krupay, *Can. J. Chem.*, **51**, 156 (1973).
10. J. Nokami, Y. Fujita and R. Okawara, *Tetrahedron Lett.*, 3659 (1979).
11. J. S. Grossert, P. K. Dubey and T. Elwood, *Can. J. Chem.*, **63**, 1263 (1985).
12. D. N. Harpp and T. G. Back, *Phosphorus and Sulfur*, **1**, 159 (1976).
13. K. H. Bell, *Aust. J. Chem.*, **38**, 1209 (1985).
14. J. P. Harmon and L. Field, *J. Org. Chem.*, **51**, 5235 (1986).
15. A. Dondoni, P. Giorgianni and A. Battaglia, *J. Chem. Soc., Chem. Commun.*, 350 (1981).
16. D. C. Dittmer, R. S. Henion and N. Takashima, *J. Org. Chem.*, **34**, 1310 (1969).
17. A. A. Scala, I. Colon and W. Rourke, *J. Phys. Chem.*, **85**, 3603 (1981).
18. E. Theumazeau, B. Jousseaume, F. Duboudin and J.-G. Duboudin, *Org. Mass Spectrom.*, **17**, 596 (1982).
19. H. Kohn, P. Charumilind and S. H. Simonsen, *J. Am. Chem. Soc.*, **101**, 5431 (1979).
20. T. Durst, J. L. Charlton and D. B. Mount, *Can. J. Chem.*, **64**, 246 (1986).
21. G. W. Astrologes and J. C. Martin, *J. Am. Chem. Soc.*, **99**, 4390 (1977).

22. H. C. Hansen and J. L. Kice, *J. Org. Chem.*, **48**, 2943 (1983).
23. B. F. Bonini, G. Maccagnani, G. Mazzanti, L. Thijs, H. P. M. M. Ambrosius and B. Zwanenburg, *J. Chem. Soc., Perkin Trans. 1*, 1468 (1977).
24. Y. Inagaki, R. Okazaki and N. Inamoto, *Bull. Chem. Soc. Jpn.*, **52**, 3615 (1979).
25. S. Kozuka, H. Takahashi and S. Oae, *Bull. Chem. Soc. Jpn.*, **43**, 129 (1970).
26. (a) E. Block, *J. Am. Chem. Soc.*, **94**, 642 (1972).
(b) E. Block, *J. Am. Chem. Soc.*, **94**, 644 (1972).
(c) E. Block and J. O'Connor, *J. Am. Chem. Soc.*, **96**, 3921 (1974).
27. F. Freeman and C. N. Angletakis, *Org. Mass Spectrom.*, **17**, 114 (1982).
28. D. N. Harpp and A. Granata, *Synthesis*, 782 (1978).
29. Y. Ueda, K. Mori, K. Kouno and T. Niiya, *Org. Mass Spectrom.*, **11**, 1027 (1976).
30. D. N. Harpp and T. G. Back, *J. Org. Chem.*, **38**, 4328 (1973).
31. R. P. Gupta and J. S. Pizey, *Phosphorus and Sulfur*, **7**, 325 (1979).
32. D. Rinne and A. Blaschette, *Z. Naturforsch.*, **30b**, 323 (1975).
33. U. Jäger, M. Schwab and W. Sundermeyer, *Chem. Ber.*, **119**, 1127 (1986).
34. P. Hanson and T. W. Stone, *J. Chem. Soc., Perkin Trans. 1*, 2429 (1984).
35. D. R. Borthakur, D. Prajapati and J. S. Sandhu, *Heterocycles*, **24**, 2739 (1986).
36. K. H. Pilgram and R. B. Skiles, *J. Org. Chem.*, **47**, 3865 (1982).
37. R. Neidlein and P. Walser, *Chem. Ber.*, **115**, 2428 (1982).
38. (a) F. Yamada, T. Nishiyama, Y. Fujimoto and M. Kinugasa, *Bull. Chem. Soc. Jpn.*, **44**, 1152 (1971).
(b) T. Nishiyama, Y. Fujimoto and F. Yamada, *Bull. Chem. Soc. Jpn.*, **45**, 928 (1972).
(c) T. Nishiyama and Y. Fujimoto, *Bull. Chem. Soc. Jpn.*, **46**, 2166 (1973).
39. R. Bartnik, Z. Cebulska, B. Orlowska, R. Faure, A. Laurent and H. Loiseleur, *Bull. Soc. Chim. Fr.*, 397 (1986).
40. R. J. Goodridge, T. W. Hambley and D. D. Ridley, *Aust. J. Chem.*, **39**, 591 (1986).

The Chemistry of Sulphinic Acids, Esters and their Derivatives
Edited by S. Patai


CHAPTER **6**

# The NMR and ESR spectra of sulphinic acids and their derivatives

ALAN R. BASSINDALE and JAMES N. ILEY

*POCRG, Department of Chemistry, The Open University, Walton Hall, Milton Keynes MK7 6AA, England*

## I. INTRODUCTION

The NMR spectra of sulphinic acids and their derivatives provide a rich source of information on molecular structure and dynamics. In addition to $^1H$ and $^{13}C$ NMR the nuclei $^{33}S$, $^{17}O$, $^{15}N$ and $^{19}F$ have all been used to study these compounds. In principle, if not in practice, all nuclei in most sulphinic acid derivatives are susceptable to NMR analysis. In this review, the NMR sections are concerned with the tabulation and correlation of chemical shifts and coupling constants, where available (Sections II.A and II.B); the dynamic features and consequences of chiral sulphur are covered in Section II.C, and, as a link with ESR spectra, CIDNP is reviewed in Section II.D.

The dynamic processes of sulphinic acid derivatives are particularly interesting and a variety of mechanisms have been proposed to account for the observed phenomena.

In the description of atoms or groups within a sulphinic acid or one of its derivatives the terms α, β, γ, δ and α′, β′, γ′, δ′ have been used consistently throughout in the following way, with the carbon attached to the S(IV) centre always labelled α. Protons attached to $C_\alpha$ are also labelled α. Groups attached to X are always labelled α′, β′ etc.

$$\begin{array}{ccccccccccccccccc} & & & & & & & & O & & & & & & & & \\ & & & & & & & & \| & & & & & & & & \\ C & — & C & — & C & — & C & — & S & — & X & — & C & — & C & — & C & — & C \\ \delta & & \gamma & & \beta & & \alpha & & & & & & \alpha' & & \beta' & & \gamma' & & \delta' \end{array}$$

(X = O, S, NR)

There do not appear to have been any previous reviews on the NMR spectra of sulphinic acids and their derivatives.

For ESR spectroscopic investigations of sulphinic acid derivatives the $RSO_2$ system has provided a fruitful area of study. Less well studied are the sulphinamides. Solid state, solution and theoretical studies have provided a clear understanding of the nature of radicals from sulphinic acid derivatives. There have been no previous extensive reviews of the ESR spectra of sulphinic acids and their derivatives, and here we provide both a detailed tabulation of $g$-values, and coupling constants $a$(H), $a$(Cl) and $a$(F), as well as a discussion of structural and dynamic features of the radical species involved. We follow the terminology described above for identifying C and H atoms attached to the sulphinyl moiety.

## II. THE NMR SPECTRA OF SULPHINIC ACIDS AND DERIVATIVES

### A. Proton and Carbon-13 Chemical Shifts and Coupling Constants

#### *1. Introduction*

Sulphinyl groups —SOX (X = OH, OR, SR, $NR_2$, Cl etc.) are strongly electron-withdrawing, largely by inductive effects, and are therefore expected to deshield strongly

TABLE 1. Some substituent constants for —SOX and related groups

| Group | $\sigma_m$ | $\sigma_p$ | Reference | $\sigma_I$ | $\sigma_R^0$ | Reference |
|---|---|---|---|---|---|---|
| $SON(CH_3)_2$ | | | | 0.3 | 0.03 | 1 |
| SOCl | | | | 0.68 | 0.14 | 1 |
| $SOOCH_3$ | 0.50 | 0.54 | 1, 3 | 0.45 | 0.09 | 1 |
| | 0.66 | | 2 | | | |
| $SOCH_3$ | 0.21 | 0.17 | 1, 3 | 0.25 | −0.08 | 1 |
| $SO_2OCH_3$ | 0.71 | 0.9 | 4, 2, 3 | 0.50 | | 5 |
| $COOCH_3$ | 0.35 | 0.44 | 3, 6 | 0.31 | 0.16 | 5 |

adjacent alkyl groups. Table 1 shows some substituent constants for a few sulphinyl groups. The electron-withdrawing ability falls in the order sulphonyl > sulphinyl > sulphenyl as would be expected. However, it will be shown later that $^1H$ and $^{13}C$ chemical shifts in sulphinyl compounds are sensitive to effects other than simple inductive effects. Conformational preferences, the magnetic anisotropy of the SO bond[7,8] and resonance effects[9] can all influence chemical shifts of adjacent protons and carbon-13 nuclei. The SOX groups are resonance electron-withdrawing with $\sigma_p^-$ for SOOMe being estimated[2] as 0.89, compared with 0.74 for COOMe[10]. Each effect will be discussed as it arises in the interpretation of chemical shifts.

## 2. *Sulphinic acids*

Proton chemical shift data for sulphinic acids are very limited[11]. The proton NMR spectrum of methanesulphinic acid was reported[11a] in 1967, and in $CDCl_3$ at 25 °C consists of two singlets: one at $\delta$ 2.7 ppm for the methyl group, and one at $\delta$ 10.4 ppm for the acidic proton. The spectrum was relatively invariant with temperature, and at −65 °C the only differences were minor chemical shift changes. This was said to suggest that the equilibrium shown in equation 1 lies far to the left[11a].

$$\mathrm{MeSOOH} \rightleftharpoons \mathrm{Me}\overset{\mathrm{O}}{\underset{\mathrm{O}}{\overset{\|}{\underset{\|}{\mathrm{S}}}}}\mathrm{H} \qquad (1)$$

Carbon-13 chemical shifts for a selection of sulphinic acids are available[12,13]. The data obtained by Freeman[13] are shown in Table 2. Freeman also tabulated the $\alpha$, $\beta$, $\gamma$, $\delta$-carbon additivity parameters[14,15] relative to the corresponding thiols. It is clear that the sulphinic acid group does substantially deshield the $\alpha$-carbon nuclei as expected. The $\alpha$-effect is, however, not constant, decreasing with increasing methyl substitution at C-$\alpha$ (Table 2). The deshielding decreases by about 6.5 ppm for each methyl group in the series $R = CH_3CH_2$, $(CH_3)_2CH$ and $(CH_3)C$. This decrease was suggested to be a consequence of increasing polarization of the C—S bond with increasing methyl substitution. Why this should lead to the postulated decrease in electron density at the $\alpha$-substituted carbon nuclei was not fully understood[13].

It was suggested[13] that the shielding or deshielding effects of the sulphinyl group may be rationalized by partition of the group into an S=O component and a S—O— (or S—X) component. For example, the $\alpha$-carbon nuclei in sulphinic acids are deshielded by about 30 ppm relative to the corresponding thiols, which is appropriate for an

TABLE 2. $^{13}C$ NMR chemical shifts[a] and substituent constants[b] for some sulphinic acids, RSOOH[13]

| | C-α | | C-β | | C-γ | | C-δ | |
|---|---|---|---|---|---|---|---|---|
| R | $\delta_C$[a] | $\alpha^b$ | $\delta_C$ | $\beta^b$ | $\delta_C$ | $\gamma^b$ | $\delta_C$ | $\delta^b$ |
| $CH_3$ | 44.30 | 46.4 | | | | | | |
| $CH_3CH_2$ | 51.26 | 45.36 | 5.41 | −0.49 | | | | |
| $CH_3CH_2CH_2$ | 59.65 | 44.05 | 13.26 | −3.84 | 15.32 | −0.28 | | |
| $CH_3CH_2CH_2CH_2$ | 57.53 | 44.3 | 23.56 | −1.4 | 21.8 | −3.1 | 13.71 | 0.5 |
| $(CH_3)_2CH$ | 55.23 | 39.1 | 13.79 | −1.81 | | | | |
| $(CH_3)_3C$ | 56.59 | 32.3 | 21.35 | −3.8 | | | | |
| $(CH_3)_3CCH_2$ | 72.02 | 44.1 | 30.95 | −0.5 | 27.9 | 1.9 | | |
| $C_6H_5CH_2$ | 64.46 | 43.2 | | | | | | |

[a] $^{13}C$ chemical shift ppm in $CDCl_3$ with TMS standard. Measured at 62.89 MHz except for $R = (CH_3)_3CCH_2$ and $C_6H_5CH_2$ which were measured at 22.63 MHz.
[b] Chemical shift differences from the same carbon of the corresponding alkane[14].

approximately 20 ppm deshielding on oxidation of $R_2S$ to $R_2SO$[16] plus an approximately +10 ppm shift for a β-OH group (modelled by the corresponding alcohol shifts[17]). Freeman's[13] analysis of substituent effects of sulphinyl groups, relative to thiols, is empirical, but does have predictive value when the corresponding thiol shifts are known. We shall confine most of our discussion to shifts relative to alkanes as this is more useful in the general case.

Interestingly, oxidation of sulphinic acids to sulphonic acids causes a *shielding* effect of −8.63 to −0.68 ppm at C-α[13] which is the opposite to that expected from electronegativities and substituent constants. This shielding was attributed to a steric compression shift resulting from bond angle widening in the sulphonic acid relative to the sulphinic acid[13,18].

The β-carbon nuclei in sulphinic acids are slightly shielded relative to those in alkanes[13] (Table 2) but not as strongly as in sulphones or sulphoxides[19,20]. There have been many attempts to explain such β-carbon shieldings, but none are completely satisfactory for sulphinic acids and derivatives[13,19,20].

### 3. Sulphinate esters, anhydrides and thioesters

The $^1H$ NMR spectra of sulphinate esters[21–23] and the $^1H$ and $^{13}C$ NMR spectra of anhydrides[24,25] and particularly thiosulphinate esters[9,24,26–32] have been reasonably well reported. The interest in these classes falls into two parts: the effect of the RSOXR′ group on the chemical shifts of R and R′ and the consequences of the magnetic anisotropy of the chiral sulphinate group. In this section we discuss the chemical shift data.

The deshielding effect of the sulphinate group on both α and α′ protons is shown in Table 3, where the $^1H$ methylene shifts for some sulphinate esters are recorded. For comparison, the methylene proton shifts in ethyl ethanesulphinate are δ 2.59 and 3.99 ppm[21] whereas the corresponding protons in ethyl ethanoate have chemical shifts of δ 2.35 and 4.15 ppm[11b]. From the limited data available it appears the —SOO— group is more deshielding for α-$CH_2SO$ protons than the α-COO— group, whereas the —$SOOCH_2$— protons are generally less deshielded than —$COOCH_2$— protons.

TABLE 3. $^{1}H$ NMR chemical shift of the methylene protons in some sulphinates[21] $RCH_2SOOCH_2R^1$

| $R^a$ | R′ | δ/ppm —$CH_2SO$ ($^3J$/Hz) | δ/ppm —$OCH_2$ ($^3J$/Hz) |
|---|---|---|---|
| H | $CH_3$ | 2.48 | 3.98 (7.20) |
| H | $CH_3CH_2$ | 2.59 | 3.93 (7.20) |
| H | $(CH_3)_2CH$ | 2.50 | 3.71 (6.65) |
| H | $(CH_3)_2CHCH_2$ | 2.51 | 3.97 (6.60) |
| $CH_3$ | H | 2.64 (7.5) | 3.69 |
| $CH_3$ | $CH_3$ | 2.59 (7.20) | 3.99 (7.20) |
| $CH_3$ | $CH_3CH_2$ | 2.63 (7.50) | 3.91 (6.50) |

[a]60 MHz spectra, 10% w/w in $CCl_4$, 35 °C.

Conformational effects are important in determining shielding parameters[23] as a consequence of the anisotropy of the SO bond[7,8]. It is well-established that in six-membered rings containing the S=O moiety the oxygen atom is preferentially axial[33–35]; particularly when heteroatoms are adjacent to the SO group[36,37]. Protons in a *syn*-axial relationship with an axial S—O group experience significant deshielding which can be used in conformational and configurational analysis[34–37]. The origin of this so-called *syn*-axial effect is not fully understood but has been attributed to a proximity effect[34] and/or an acetylene-like anisotropy of the S=O bond[7,34,38] or a carbonyl-like anisotropy of the S=O bond[37]. The operation of the *syn*-axial effect is illustrated in Figure 1, which shows the 100 MHz $^{1}H$ NMR spectrum of 1,2-oxathiane-2-oxide[37]. The multiplets assigned to $H_1$, $H_2$ and $H_3$ are indicated on the spectrum and were assigned by double resonance and analysis of coupling constants[37]. Thus the spectrum can be interpreted as that of a single conformer with an axial SO group. The *syn*-axial proton $H_1$ is 0.70 ppm to high frequency of the *syn*-equatorial proton $H_2$. The deshielding of

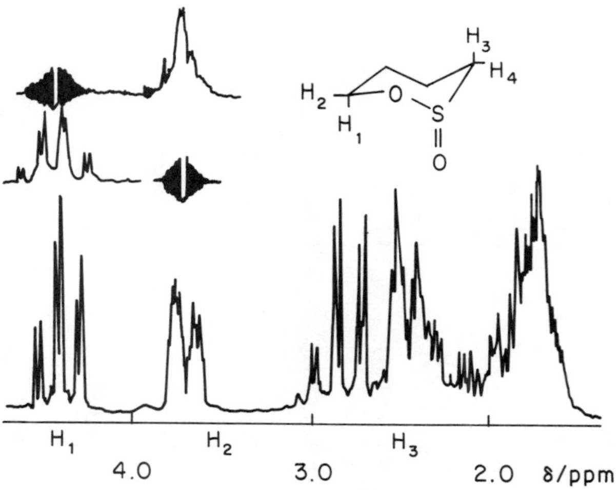


FIGURE 1. 100-MHz $^{1}H$ NMR spectrum of 1,2-oxathiane-2-oxide

TABLE 4. $^1$H NMR chemical shifts of some thiosulphinates RSOSR′, $\delta$/ppm($^2J$/Hz)

| R | R′ | H-α | H-β | H-γ | H-δ | H-α′ | H-β′ | H-γ′ | H-δ′ | Ref. |
|---|---|---|---|---|---|---|---|---|---|---|
| $C_6H_5$ | $CH_3$ | | | | | 2.53 | | | | 28[a] |
| $C_6H_5$ | $CH_2CH_3$ | | | | | {3.13[d], 3.16 | 1.43 | | | 28 |
| $C_6H_5$ | $CH_2CH_2CH_3$ | | | | | {3.12[d], 3.09 | 1.80 | 1.03 | | 28 |
| $C_6H_5$ | $CH_2CH_2CH_2CH_3$ | | | | | 3.14 | | | 0.92 | 28 |
| $CH_3$ | $C_6H_5$ | 2.90 | | | | | | | | 28 |
| $CH_3CH_2$ | $C_6H_5$ | 3.10 | 1.41 | | | | | | | 28 |
| $CH_3CH_2CH_2$ | $C_6H_5$ | 3.09 | 1.86 | 1.08 | | | | | | 28 |
| $CH_3CH_2CH_2CH_2$ | $C_6H_5$ | 3.11 | | | 0.96 | | | | | 28 |
| $(CH_3)_3CCH_2$ | $CH_2C(CH_3)_3$ | {3.06[d], 3.11 (13.2) | | 1.14 | | {3.01[d], 3.16 (4.2)[c] | | 1.03 | | 31[b] |
| $(CH_3)_3CCH_2$ | $C_6H_5$ | {3.04[d], 3.22 (13.2) | | 1.14 | | | | | | 31 |
| $C_6H_5$ | $CH_2C(CH_3)_3$ | | | | | {2.99[d], 3.16 (13.2) | | 1.03 | | 31 |
| $C_6H_5CH_2$ | $CH_2C_6H_5$ | {4.27[d], 4.31 (13.1) | | | | {4.25[d], 4.28 (13.4) | | | | 31 |
| $C_6H_5CH_2$ | $C_6H_5$ | {4.33[d], 4.43 (12.9) | | | | | | | | 31 |
| $C_6H_5$ | $CH_2C_6H_5$ | | | | | {4.22[d], 4.41 (13.2) | | | | 31 |

[a]Reference 28, 27° in $CDCl_3$.
[b]Reference 31, $CDCl_3$ solvent, 60, 90 or 250 MHz.
[c]As reported, suggested to be exceptional[31].
[d]Diastereotopic protons, see Section II.C.

the *anti*-axial proton $H_3$ relative to $H_4$ is not consistent with an acetylenic anisotropy of the S=O bond. A calculation[37] using the McConnell point dipole approximation[39] predicted that $H_3$ should be 0.66 ppm to low frequency of $H_4$ whereas it is actually found as 0.4 ppm to high frequency of $H_4$. This deshielding was said to be more appropriate to a carbonyl-like anisotropy of the S=O bond[37,40]. The marked preference for axial S=O is analogous to the anomeric effect in carbohydrates[41] and has been rationalized[32] in terms of the gauche effect[42], although stereoelectronic effects, as described by Deslongchamps[43], can also be succesfully applied. The stereoelectronic argument is that preferred conformations have lone pairs in an *anti*-relationship to polar bonds.

There do not appear to have been any reports of $^{13}C$ NMR chemical shifts for sulphinic esters. Freeman[30] reported $^{13}C$ NMR shifts of $\delta$21.41 and 60.18 ppm for one pair of stereoisomers (tentatively assigned as *R, R* and *S, S*) of the anhydride $(CH_3)_3CS(O)OS(O)C(CH_3)_3$ and $\delta$21.59 and 60.70 ppm for the other (possibly *R, S*) stereoisomer.

The oxidation of disulphides has received much attention and consequently the NMR spectra of thiosulphinates have been well reported. Proton chemical shifts for a series of thiosulphinates are shown in Table 4. For many $\alpha$ and $\alpha'$ methylene groups the protons are diastereotopic as the sulphinate sulphur atom is chiral. This is discussed in detail in Section II.C.2. There are several interesting aspects to the proton chemical shifts of thiosulphinates. Simple inductive effects would suggest that, in the series illustrated by **1, 2** and **3**, the protons (and carbon nuclei) should be progressively deshielded in the order **3** > **2** > **1**. This general trend is followed adequately except in the case of the $\alpha'$-protons of the thiosulphinates **2** which are significantly more deshielded (by about 0.15 ppm[28]) than those in the thiosulphinates **3**. The $\beta$-protons in **2** and **3** have very similar chemical shifts and are both deshielded relative to **1** by 0.4–0.5 ppm[28,31]. Oae and coworkers[28] rationalized the strong $\alpha'$ deshielding in **2** by proposing a polarization of the $\alpha'$ C—H bond through interactions of the type shown in **4**. A similar effect, but somewhat attenuated, can be envisaged for the $\beta$-proton as shown in **5**.

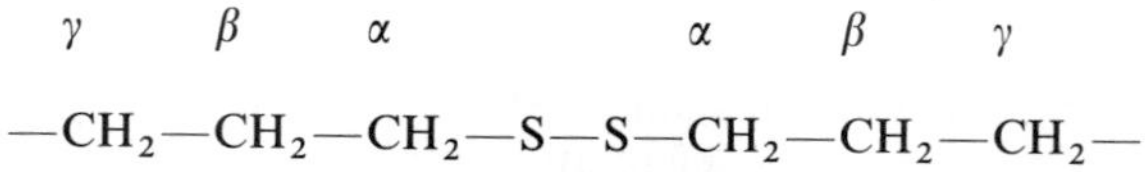


**(1)**

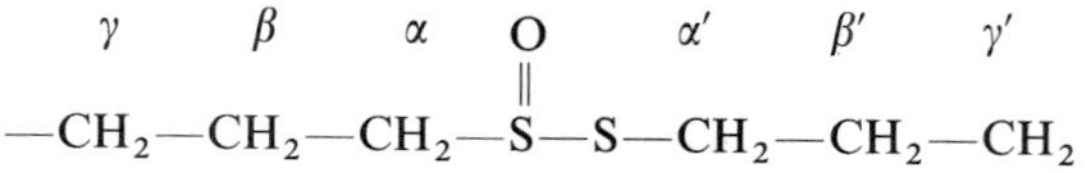


**(2)**

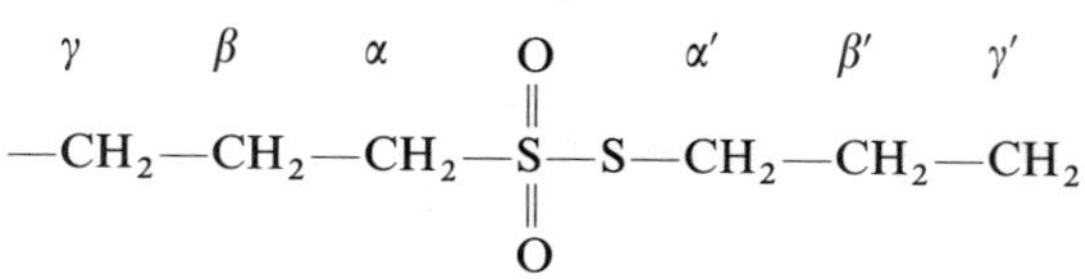


**(3)**

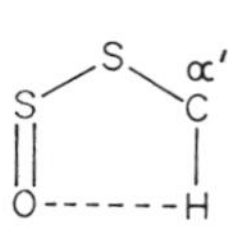


(**4**)

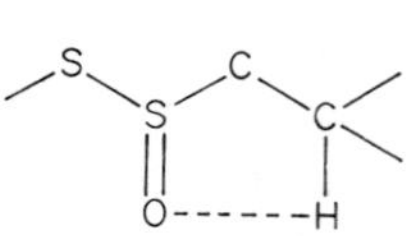


(**5**)

Freeman[31] proposed that structures such as **4** may be partially responsible for the observed effects, but suggested that the data shown in Table 5 is not consistent with this being the only explanation. The chemical shift difference between the α′-protons of **7** and those of **9** is about 0.14 ppm whereas the related protons in **6** and **8** have the same chemical shift.

Complementary phenomena to the anomalous α′- and β-proton chemical shifts are found in the $^{13}$C NMR spectra of thiosulphinates. Table 6 gives $^{13}$C NMR shifts for some thiosulphinates and Table 7 gives substituent effects for thiosulphinates and thiosulphonates relative to the parent disulphides. In these cases the substituent effect is most appropriately referred to the disulphides as they are frequently precursors or found in the same reaction mixtures.

The α-carbons are deshielded in the order $RSO_2 > RSO > RS$, according to the electron-withdrawing properties. The α′-carbon nuclei of thiosulphinates are generally shielded by 6–8 ppm relative to the disulphide (and by about 4 ppm relative to the thiosulphonates). The β-carbon nuclei are also generally shielded by about 6 ppm.

It is generally agreed[9,31,32] that the α-carbon deshielding has a similar origin to the α-proton deshielding, in inductive effects and the β-effect of the sulphinyl oxygen[44].

The strong shielding effect on the α′-carbon in the thiosulphinates has been ascribed[9,31] to some or all of the following: hyperconjugation, sulphur lone-pair donation into the C—S σ* orbital or hydrogen bonding effects as illustrated by **4**[28]. Conformational effects are also important in determining the magnitude of the α′-carbon shielding[32]. It has been observed[9] that as the steric bulk of R and R′ increases, so the magnitude of the α′ shielding decreases. It was suggested[9] that for larger groups, contributions from hyperconjugation and/or back donation increase, with the effect that the inductive effect at C-α′ is increased. Shielding is thereby reduced.

Similar arguments can be made for the shielding effects on C-β. The *syn*-axial effect is also observed in cyclic thiosulphinates[9,28]. The papers by Evans[9] and Freeman[31,32] contain more detailed analyses of thiosulphinate chemical shifts.

### 4. Sulphinamides

The NMR spectroscopy of sulphinamides RSONR′R″ is particularly interesting in view of the chirality at sulphur and the configuration at nitrogen being potentially planar or pyramidal. In terms of chemical shifts, shieldings and coupling constants there are, however, few surprises. Table 8 lists the $^{1}$H chemical shifts of many sulphinamides.

TABLE 5. Comparative $^{1}$H NMR chemical shift data for thiosulphinates and thiosulphonates[31]

| Compound | | $\delta^{1}$H/ppm $\overset{*}{C}H_2$ |
|---|---|---|
| $C_6H_5S(O)CH_2\overset{*}{C}H_2CH_3$ | (**6**) | 1.66 |
| $C_6H_5S(O)S\overset{*}{C}H_2CH_3$ | (**7**) | 3.13, 3.16[a] |
| $C_6H_5SO_2CH_2\overset{*}{C}H_2CH_3$ | (**8**) | 1.66 |
| $C_6H_5SO_2S\overset{*}{C}H_2CH_3$ | (**9**) | 3.0 |

[a]Diastereotopic protons.

TABLE 6. $^{13}C$ NMR shifts for some thiosulphinates RS(O)SR′

| R | R′ | C-α | $\delta^{a}$/ppm C-β | C-γ | C-α′ | C-β′ | C-γ′ | Ref. |
|---|---|---|---|---|---|---|---|---|
| $CH_3$ | $CH_3$ | 42.66 | | | 13.77 | | | 9 |
| | | 42.79 | | | 14.44 | | | 32 |
| $CH_3CH_2$ | $CH_2CH_3$ | 49.88 | 7.67 | | 26.81 | 16.26 | | 9 |
| $(CH_3)_2CH$ | $CH(CH_3)_2$ | 55.26 | {15.70[b] 16.63 | | | 38.27 | {24.57[b] 24.68 | 9 |
| $(CH_3)_3C$ | $C(CH_3)_3$ | 58.81 | 24.01 | | 47.93 | 32.20 | | 9 |
| $(CH_3)_2CHCH_2$ | $CHCH_2(CH_3)_2$ | 65.18 | 24.92 | {21.58[b] 22.52 | 41.54 | 26.69 | 21.66 | 9 |
| $(CH_3)_3CCH_2$ | $CH_2C(CH_3)_3$ | 71.55 | 32.26 | 29.56 | 46.93 | 32.07 | 28.72 | 31 |
| $(CH_3)_3CCH_2$ | $C_6H_5$ | 70.42 | 32.18 | 29.56 | | | | 31 |
| $C_6H_5$ | $CH_2C(CH_3)_3$ | | | | 47.09 | 32.15 | 28.75 | 31 |
| $C_6H_5CH_2$ | $CH_2C_6H_5$ | 62.30 | | | 36.09 | | | 31 |
| $C_6H_5$ | $CH_2CH_3$ | | | | 27.6 | 15.9 | | 28 |
| $CH_3CH_2$ | $C_6H_5$ | 49.8 | 7.6 | | | | | 28 |
| $C_6H_5$ | $CH_2CH_2CH_3$ | | | | 35.1 | 23.9 | 13.1 | 28 |
| $CH_3CH_2CH_2$ | $C_6H_5$ | 57.8 | 17.1 | 13.1 | | | | 28 |
| $CH_3CH_2CH_2$ | $CH_2CH_2CH_3$ | 58.15 | 17.23 | 13.21 | 24.91 | 24.29 | 13.21 | 32 |
| $CH_3CH_2CH_2CH_2$ | $CH_2CH_2CH_2CH_3$ | 56.09 | 25.50 | 21.91[c] | 32.91 | 32.62 | 21.76 | 32 |

[a] 1–15% w/w in $CDCl_3$ on a variety of spectrometers.
[b] Diastereotopic methyl groups.
[c] C-δ 13.53 ppm C-δ′ 13.56 ppm.

TABLE 7. $^{13}C$ NMR substituent effects for thiosulphinates RS(O)SR′ (and thiosulphonates, $RSO_2SR'$)[a]

| R | R′ | α | β | γ | δ | α′ | β′ | γ′ | δ′ | Ref. |
|---|---|---|---|---|---|---|---|---|---|---|
| $CH_3$ | $CH_3$ | 20.75 | | | | −7.60 | | | | 32 |
| | | (26.70) | | | | (−3.81) | | | | |
| | | 20.62 | | | | −8.30 | | | | 9 |
| | | (26.70) | | | | (−3.81) | | | | |
| $CH_3CH_2$ | $CH_2CH_2$ | 17.06 | −6.83 | | | −6.01 | 1.76 | | | 9 |
| | | (24.12) | (−6.19) | | | (−2.28) | (0.62) | | | |
| $CH_3CH_2CH_2$ | $CH_2CH_2CH_3$ | 16.89 | −5.33 | 0.09 | −6.35 | 1.73 | −0.46 | | | 32 |
| | | (23.42) | (−4.93) | (0.24) | (−2.90) | (0.89) | (0.23) | | | |
| $CH_3CH_2CH_2CH_2$ | $CH_2CH_2CH_2CH_3$ | 17.12 | −5.87 | 0.23 | −0.14 | −6.06 | 1.25 | 0.08 | 0.01 | 32 |
| | | (23.56) | (−5.84) | (0.08) | (−0.23) | (−3.00) | (0.34) | (−0.30) | (−0.11) | |
| $(CH_3)_2CH$ | $CH(CH_3)_2$ | 14.12 | −6.43[b] | | | −2.87 | 2.03 | | | 9 |
| | | (22.11) | (−6.34) | | | (1.56) | (1.62) | | | |
| $(CH_3)_3C$ | $C(CH_3)_3$ | 13.18 | −6.50 | | | 2.30 | 1.69 | | | 9 |
| | | (22.39) | (−6.77) | | | (10.66) | (1.01) | | | |
| $(CH_3)_2CHCH_2$ | $CH_2CH(CH_3)_2$ | 16.58 | −3.29 | 0.27[b] | | −7.06 | −1.52 | 0.12 | | 9 |
| | | (21.88) | (−3.00) | (0.70) | | (−4.03) | (0.69) | (−0.10) | | |
| $(CH_3)_3CCH_2$ | $CH_2C(CH_3)_3$ | 15.59 | 1.95 | 0.73 | | −9.03 | 1.76 | −0.11 | | 32 |
| | | (18.99) | (1.81) | (0.93) | | (−6.04) | (3.16) | (0.03) | | |
| $C_6H_5CH_2$ | $CH_2C_6H_5$ | 18.98 | | | | −7.23 | | | | 32 |
| | | (25.69) | | | | (−2.47) | | | | |

[a]The substituent effects are calculated as $\Delta\delta = \delta_C(\text{-S(O)S-}) - \delta_C(\text{-SS-})$ or $\Delta\delta = \delta_C(\text{-SO}_2\text{S-}) - \delta_C(\text{-SS-})$. Values for thiosulphonates are given in parentheses.
[b]Based on the average value for diastereotopic resonances.

TABLE 8. $^1$H NMR chemical shifts of some sulphinamides RSONR′R″

| R | R′ | R″ | δ $^1$H/ppm (J/Hz) H-α | H-β | H-α′ | H-β′ | H-α″ | H-β″ | Ref. |
|---|---|---|---|---|---|---|---|---|---|
| p-$CH_3C_6H_4$ | H | $CH_3$ | | | 4.94(q) | | 2.35(d) | | 45 |
| p-$CH_3C_6H_4$ | H | $CH_2CH_3$ | | | 4.82(t) | | 2.9(m) | 1.05(t) | 45 |
| p-$CH_3C_6H_4$ | H | $CH(CH_3)_2$ | | | 4.62(d) | | 2.35(m) | 1.08(dd) | 45 |
| $CH_3$ | H | $C_6H_{11}$ | | | 4.95(d) | | 3.13(m) | | 45 |
| F | $CH_3$ | $CH_3$ | | | 2.77(d) (4.7) | | *a* | | 46 |
| | | | | | 2.81 (5.0) | | | | 47 |
| F | $CH_2CH_3$ | $CH_2CH_3$ | | | 3.33(qd) (4.5) | 1.16(t) (4.5) | *a* | | 46 |
| Cl | $CH_3$ | $CH_3$ | | | 2.83 | | *a* | | 46 |
| | | | | | 2.90 | | *a* | | 47 |
| Cl | $CH(CH_3)_2$ | $CH(CH_3)_2$ | | | 4.09(sep)[b] | 1.44(d) | *a* | | 46 |
| $CH_3$ | $CH_3$ | $CH_3$ | 2.50[c] | | 2.68[c] | | *a* | | 48 |
| | | | 2.54[d] | | 2.67[d] | | | | 48 |
| $CH_3$ | $CH_2CH_3$ | $CH_2CH_3$ | | | 1.2(t) (7) | | *a* | | 48 |
| $Cl_3C$ | $CH_3$ | $CH_3$ | | | 2.99 | | *a* | | 47 |
| Br | $CH_3$ | $CH_3$ | | | 2.75 | | *a* | | 47 |
| $CCl_2F$ | $CH_3$ | $CH_3$ | | | 2.94(d) (1.3) | | *a* | | 47 |
| $CH_3$ | piperidino | | 2.42 | | 2.98(m) | 1.55[e] | *a* | | 49 |
| $CH_3CH_2$ | piperidino | | 2.59(t) (8) | 1.07(q) (8) | 2.99(m) | 1.56[e] | | | 49 |
| $CH_3$ | morpholino | | 2.47 | | 2.99 | 3.64(t) (5.0) | *a* | | 49 |
| $CH_3CH_2$ | morpholino | | 2.65 | 1.09 | 2.99 | 2.64(t) (5.0) | *a* | | 49 |

[a]Spectra of RSONR′R″ run at room temperature generally show only one set of signals for R′ and R″ when R′ = R″.
[b]Septet.
[c]In $CDCl_3$.
[d]Neat liquid.
[e]Multiplet, includes H-$\gamma^1$.

The chemical shifts of the methyl groups in $CH_3SON(CH_3)_2$ are about $\delta 2.5$ ppm for the $CH_3$—SO protons and about $\delta 2.68$ ppm for the $NCH_3$ protons[48]. For $CH_3CON(CH_3)_2$ the $^1H$ chemical shifts are $\delta 2.10$ ppm for the $CH_3CO$ protons and the mean value of the two $NCH_3$ singlets is at $\delta 3.0$ ppm[11b]. So, as with the sulphinates, the α-protons are more shielded than in the carbon analogue, but the α′ protons are less shielded. In $RSONHR^2$ compounds coupling between the NH proton and the H-α′ protons was observed[45], as is often the case with amides[11b].

In compounds $RSONR^1R^2$, where R′ = R″, the protons in $R^1$ and $R^2$ are isochronous at ambient temperature[46–49] owing to rapid exchange of environment between $R^1$ and $R^2$. This is discussed in detail later.

There have been two studies using proton NMR spectroscopy to probe the transmission of substituent effects across the N—S bond in sulphinamides[50,51].

Mori and Ueda[51] examined the $^1H$ NMR spectra of some *para*-substituted derivatives of PhSONHPh. There was no variation in the aromatic proton chemical shifts of the phenyl ring adjacent to sulphur when the N-phenyl ring bears the substituents, but the N-proton shift was affected. This was taken to suggest that there is little double-bond character in the N—S bond, but there is significant double-bond character in the bond between sulphur and its phenyl substituent.

Davis and coworkers[50] measured the $^1H$ NMR spectra of compounds of the general formula shown in **10**.

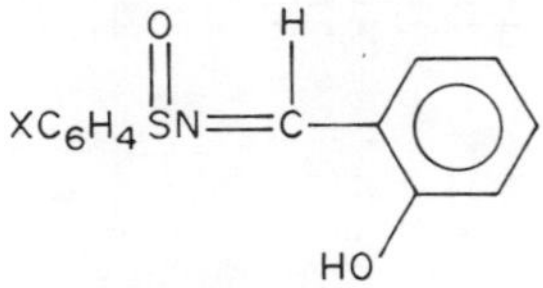


**(10)**

The Hammett constant, ρ, was measured by plotting the value of the hydroxyl proton chemical shift against the σ-values for the various substituents X. The value of ρ thus obtained was suggested to be a measure of the transmission of electronic effects through the N—S bond. The effect of the substituents, X, on the imidoyl proton chemical shift was also measured. The related sulphenamides did show transmission of electronic effects across the N—S bond, possibly through d orbital involvement. There was no effective transmission of substituent effects in the sulphinamides[50]. The results are compatible with those of Mori and Ueda[51].

A limited amount of $^{13}C$ NMR shift data are available for sulphinamides[52–54] (Table 9). In aromatic sulphinamides $C_6H_5SONR_2$ the aromatic C-1 is about 9 ppm

TABLE 9. $^{13}C$ NMR chemical shifts of some sulphinamides, RSONR′R″

| R | R′ | R″ | H-α | H-α′ | H-α″ | Ref. |
|---|---|---|---|---|---|---|
| $CH_3$ | H | H | 48.9[a] | — | — | 52 |
| $CH_3$ | $CH_3$ | H | 40.0[a] | 25.8 | *c* | 52 |
| $CH_3$ | $CH_3$ | $CH_3$ | 39.0[b] | 36.1 | | 53 |
| $C_6H_5$ | $CH_3$ | $CH_3$ | *d* | 36.7 | | 54 |

[a]Acetone solvent.
[b]Neat liquid.
[c]Both methyl carbons isochronous at ambient temperature.
[d]Neat liquid; aromatic ring resonances, C-1, 144.3; C-2, 6, 125.9; C-3, 5, 128.9; C-4, 130.8 ppm.

TABLE 10. One-bond carbon–proton coupling constants [$^{1}J$(CH)/Hz] for some sulphinamides[52]

| Compound | Solvent | Aliphatic S—$CH_3$ | Aromatic C-2,6 | Aromatic C-3,5 | Aromatic C-4 | Aliphatic N—$CH_3$ |
|---|---|---|---|---|---|---|
| $CH_3SONH_2$ | Acetone | 139.1 | | | | |
| $CH_3SONHCH_3$ | Neat | 137.3 | | | | 137.3 |
| | Acetone | 137.3 | | | | 137.3 |
| $CH_3SON(CH_3)_2$ | Neat | 137.1 | | | | 137.2 |
| | Acetone | 137.1 | | | | 137.2 |
| $C_6H_5SON(CH_3)_2$ | Neat | | 164.2 | 162.5 | 161.0 | |

more deshielded than in the equivalent sulphonamide[52,54] and is said to be largely an inductive deshielding[54]. Mesomeric effects may be more important in sulphonamides. On the other hand, the N-methyl carbon nuclei in sulphinamides are less deshielded than those of the corresponding sulphonamides[54]. The one- bond C—H coupling constants in sulphinamides are given in Table 10. The S—CH coupling constants are nearly the same in sulphinamides and sulphonamides, but those for the N-methyl groups are slightly smaller (1–2 Hz) in sulphinamides than in sulphonamides[52]. The $^{13}$C NMR spectra of sulphinamides differ sufficiently from sulphonamides and sulphenamides to allow identification, but are not sufficiently well defined for complete structure determinations[52–54].

### 5. *Sulphinyl chlorides*

The main interest in the NMR spectra of sulphinyl chlorides is again the effect of the chiral sulphur atom on neighbouring groups. There are some scattered $^{1}$H NMR chemical shifts[55–59] but no systematic study.

In $CDCl_3$ the methylene protons in $CH_3CH_2SOCl$ are isochronous and appear at $\delta$3.30 ppm with the methyl resonance at $\delta$1.39 ppm. The additional deshielding effect of the chlorine atom is clearly observed, when compared with the $\delta$2.6 ppm typical for $RCH_2SOOR'$ in sulphinates (see Table 3). Other $^{1}$H NMR shifts will be shown later in the discussion of dynamic effects in sulphinyl chlorides.

TABLE 11. $^{13}$C NMR chemical shifts and substituent constants for some sulphinyl chlorides[13] (RSOCl)

| | C-$\alpha$ | | C-$\beta$ | | C-$\gamma$ | | C-$\delta$ | |
|---|---|---|---|---|---|---|---|---|
| R | $\delta_C$[a] | $\alpha$[b] | $\delta_C$ | $\beta$[b] | $\delta_C$ | $\gamma$[b] | $\delta_C$ | $\delta$[b] |
| $CH_3$ | 52.42 | 54.5 | | | | | | |
| $CH_3CH_2$ | 58.44 | 52.5 | 5.72 | −0.2 | | | | |
| $CH_3(CH_2)_2CH_2$ | 64.35 | 51.2[c] | 24.26 | −0.2 | 21.68 | −3.3 | 13.65 | 0.4 |
| $(CH_3)_2CH$ | 62.17 | 46.1 | 14.46 | −1.1 | | | | |
| $(CH_3)_3C$ | 64.41 | 40.1 | 22.47 | −2.7 | | | | |
| $(CH_3)_3CCH_2$ | 79.24 | 51.3 | 32.73 | 1.2 | 29.55 | 1.6 | | |
| $C_6H_5CH_2$ | 71.12 | 49.8 | | | | | | |

[a] $^{13}$C chemical shift/ppm in $CDCl_3$ with TMS standard. Measured at 62.89 MHz, except for R = $(CH_3)_3CCH_2$ and $CH_3CH_2$ which were measured at 22.63 MHz.
[b] Chemical shift differences from the same carbon of the corresponding alkane[14].
[c] Reported as 57.9 in Reference 13; recalculated from the original data to be 51.2 ppm.

Some systematic data are available for the $^{13}C$ NMR parameters of sulphinyl chlorides[13]. The data are given in Table 11. The pattern in shielding effects is very similar to that for the parent sulphinic acids. The α-carbon nuclei are more deshielded than those in the equivalent sulphinic acid by about 6–8 ppm, and this can be understood in terms of the greater electron-withdrawing ability of the —SOCl group. As with the sulphinic acids, increased alkyl substitution on the α-carbon diminishes the deshielding effect, by about 8.1 ppm for each additional methyl group.

### B. Multinuclear Studies of Sulphinic Acids and their Derivatives

In addition to carbon and hydrogen, the element in common to all sulphinic acids and their derivatives is sulphur. The isotope $^{33}S$ is present in 0.76% natural abundance and has spin $\frac{3}{2}$, and a receptivity of 0.097 relative to $^{13}C$. Some hundred or so $^{33}S$ chemical shifts have been reported[60] but the only compound related to sulphinic acids is $SOCl_2$, which has a $^{33}S$ chemical shift[61] of 210 ppm relative to $SO_4^{2-}$. In general, the linewidths of sulphinic acid derivatives are too broad for measurement, owing to the unsymmetrical electron distribution around sulphur. The large $^{33}S$ NMR linewidths of RSOX compounds compared with the narrow $^{33}S$ NMR linewidths in $RSO_2X$ have been used diagnostically[62] to distinguish between the two possible structures **11** and **12**.

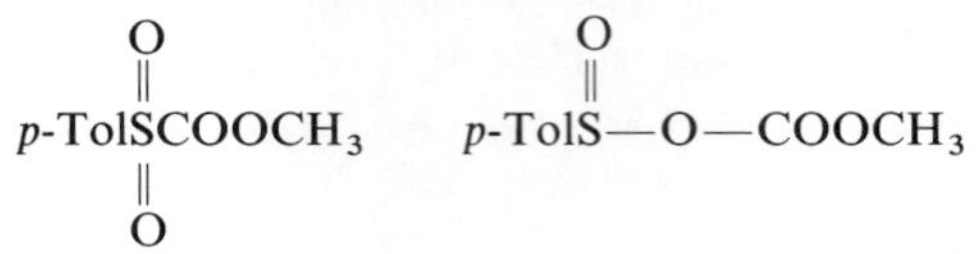

**(11)** **(12)**

The $^{33}S$ NMR spectrum gave a signal of linewidth 283 Hz which could not correspond to **12**[62]. The chemical shift δ 314 ppm, relative to carbon disulphide (−20.2 w.r.t. $SO_4^{2-}$), is also in the same region as other sulphonyl derivatives, confirming **11** as the correct structure.

After sulphur the most common 'other nucleus' in sulphinic acid derivatives is oxygen. The nucleus $^{17}O$ is now relatively commonly used in NMR studies, despite having a natural abundance of 0.037%, a receptivity of 0.61 compared to carbon and being quadrupolar with spin $\frac{5}{2}$. The usual standard for $^{17}O$ NMR is $H_2O$, and here all shifts are referenced to water, usually as an external standard. The available $^{17}O$ shifts are given in Table 12. Sulphinyl oxygen atoms are considerably shielded relative to sulphonyl oxygen atoms (δ 150–170 ppm)[52]. The $^{17}O$ chemical shifts of sulphinamides

TABLE 12. $^{17}O$ NMR chemical shifts of some sulphinic acid derivatives

| Compound | $\delta\,^{17}O$/ppm$^a$ | Solvent | Ref. |
|---|---|---|---|
| $CH_3SON(CH_3)_2$ | 79 | neat | 54 |
| | 78.5 | acetone | 52 |
| $CH_3SONHCH_3$ | 92.2 | acetone | 52 |
| $C_6H_5SON(CH_3)_2$ | 65 | neat | 52 |
| $CH_3SOSCH_3$ | 73 | | 63 |
| $CH_3CH_2SOSCH_2CH_3$ | 64 | | 63 |
| $(CH_3)_2CHSOSCH(CH_3)_2$ | 57 | | 63 |

$^a$Relative to external $H_2O$.

TABLE 13. $^{15}N$ chemical shifts of some sulphinamides

| Compound | $\delta\,^{15}N$/ppm[a] | Solvent | Ref. |
|---|---|---|---|
| $CH_3SON(CH_3)_2$ | −308.9 | neat | 54 |
| | −309.2 | acetone | 54 |
| $C_6H_5SON(CH_3)_2$ | −305.1 | neat | 54 |
| $CH_3SONH_2$ | −285.4 | acetone | 52 |
| $CH_3SONHCH_3$ | −302.7 | neat | 52 |
| | −303.4 | acetone | 52 |
| $ClSON(CH_3)_2$ | −261 | neat | 65 |
| $ClSON(CH_2CH_3)_2$ | −235.4 | neat | 65 |
| $ClSON(CH(CH_3)_2)_2$ | −216.0 | neat | 65 |
| $ClSON(CHCH_3(C_2H_5))_2$ | {−218.5[b] / −219.9 | neat | 65 |

[a]Relative to external $CH_3NO_2$, negative values to low frequency.
[b]Two diastereomers present.

(65–79 ppm) are greater than those of the corresponding sulphoxides (2–13 ppm)[64]. There are too few $^{17}O$ shifts available to comment on any other trends or special effects associated with sulphinic acid derivatives.

The $^{15}N$ chemical shifts for some sulphinamides have been measured[52–54,65] and are reported in Table 13. It is interesting to note in Table 14 the relative orders of $^{15}N$ chemical shifts for the two series $ClSO_xN(CH_3)_2$ and $CH_3SO_xN(CH_3)_2$. The α-chloro atom is significantly deshielding as expected on electronegativity arguments. However, the more interesting observation is that the order of shifts is different for each series. From electronegativity values alone the order of chemical shifts (from low to high frequency) is expected to be sulphonamide > sulphinamide > sulphenamide. That is the order observed for the S—$CH_3$ series[54], but for the S—Cl series the order sulphinamide > sulphonamide > sulphenamide is observed[65]. It may be that pπ–dπ interactions are stronger in $ClSON(CH_3)_2$ than in the corresponding sulphonamide, with the effect being enhanced by the α-chloro atom. The geometry around the sulphonamide nitrogen is said to have more $sp^3$ character than that in the sulphinamides (see later) but this cannot provide a complete explanation[65]. It is difficult to rationalize the differences in $^{15}N$ chemical shifts for the compounds with different oxidation states of sulphur and between the two series. More work is required in this area.

The $^{15}N$ shifts in the $CH_3SONH_n(CH_3)_{2-n}$ series show an increased shielding of 17 ppm on going from $n = 2$ to $n = 1$ and a further 6 ppm shielding for $n = 1$ to $n = 0$. In the series $ClSO_xN(CH_3)_2$, $ClSO_xN(CH_2CH_3)_2$ and $ClSO_xN[CH(CH_3)_2]_2$ the

TABLE 14. The $^{15}N$ chemical shifts for two series of sulphinamides, sulphonamides and sulphenamides

| Compound | $\delta\,^{15}N$/ppm | Ref. | Compound | $\delta\,^{15}N$/ppm | Ref. |
|---|---|---|---|---|---|
| $ClSO_2N(CH_3)_2$ | −273 | 65 | $CH_3SO_2N(CH_3)_2$ | −300.7 | 54 |
| $ClSON(CH_3)_2$ | −261 | 65 | $CH_3SON(CH_3)_2$ | −308 | 54 |
| $ClSN(CH_3)_2$ | −304 | 65 | $CH_3SN(CH_3)_2$ | −355[a] | 54 |

[a]Extrapolated; $CH_3SN(CH_2CH_3)_2$ has $\delta\,^{15}N$, −335 ppm and substitution of each methyl group on the α-carbon is generally deshielding by about 10 ppm[65].

TABLE 15. Differences $\Delta\delta$ $^{15}N$ between the chemical shifts of some sulphur amides and the corresponding secondary amines[65]

| X | $(CH_3)_2NX$ $\Delta\delta$ $^{15}N$/ppm | $(\beta)^a$ | $(C_2H_5)_2NX$ $\Delta\delta$ $^{15}N$/ppm | $(\beta')^a$ | $(i\text{-}C_3H_7)_2NX$ $\Delta\delta$ $^{15}N$/ppm |
|---|---|---|---|---|---|
| SCl | 65.3 | (14.8) | 59.0 | (12.5) | 56.8 |
| SOCl | 108.4 | (12.9) | 98.3 | (9.7) | 90.5 |
| $SO_2Cl$ | 96.1 | (10.1) | 80.5 | (6.7) | 66.5 |

$^a\beta$ and $\beta'$ correspond respectively to the effect on $\delta$ $^{15}N$ of the first and second substitution on the $\alpha$-carbon to nitrogen.

variations in $^{15}N$ chemical shift from those of the parent secondary amines[66,67], $HN(CH_3)_2$ etc., are shown in Table 15.

The relative values of $\beta$ and $\beta'$ (the effect on $\delta^{15}N$ of increasing substitution at the $\alpha$-carbon) are claimed[68] to reflect the degree of planarity at the nitrogen atom. Small values of $\beta$ and $\beta'$ are said to indicate $sp^3$ nitrogen. Hence the order of $sp^3$ character is suggested to be sulphonamides > sulphinamides > sulphenamides. So sulphinamides are not planar at nitrogen according to this analysis[65] but are somewhat flattened relative to a purely $sp^3$ hybridized nitrogen atom. Some corroboration of this view may be found in the observation that the $^1J(NH)$ coupling constant in sulphinamides is about 80 Hz[52]. The $^1J(NH)$ coupling in amides, where N is planar, is about 90–100 Hz and that in alkylamines is about 65 Hz[69].

The $^{19}F$ NMR spectra of a few fluorosulphinates FSOOR have been reported[70] and have quite different chemical shifts from fluorosulphenates FSOR.

Given the variety of NMR active nuclei that can be present in sulphinic acids and their derivatives, the multinuclear NMR data are rather limited; however, the reported data show that there is scope for further interesting work.

## C. Dynamic NMR of Sulphinic Acids and their Derivatives, and the Effect of Chiral Sulphur on NMR Spectra

### 1. Introduction

The sulphur atom in sulphinic acids is effectively tetrahedral as shown in **13**.

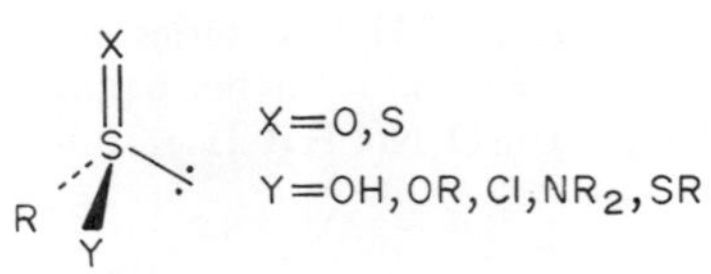


**(13)**

The effect of a chiral centre on the NMR spectra of neighbouring groups within a molecule has long been recognized and the principles elucidated[71]. There are however certain misconceptions that occasionally impede the interpretation of NMR spectra of sulphinyl compounds, and in particular the interpretation of temperature-dependent NMR spectra. The general principles of the effect of chiral groups on NMR spectra as they apply to sulphinic acids and their derivatives are outlined briefly below.

In molecules such as **14** and **15** the groups A and B are diastereotopic and therefore magnetically anisochronous, so the nuclei comprising A and B are expected to have different chemical shifts. A and B *cannot* be made isochronous simply by rotation about chemical bonds. If we take the simple example of an ethyl group attached to S in **14** (A = B = H, R′ = $CH_3$), the methylene protons are diastereotopic and the three staggered conformations are shown in **16–18**.

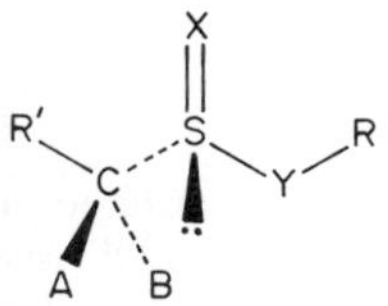

**(14)**

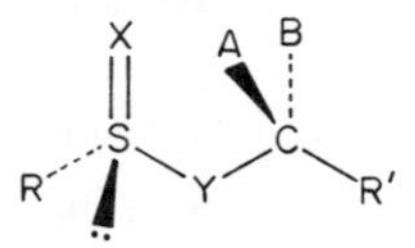

**(15)**

A=B=H,$CH_3$ etc.
R′ ≠ A,B
R,R′=H,$CH_3$,alkyl, aryl

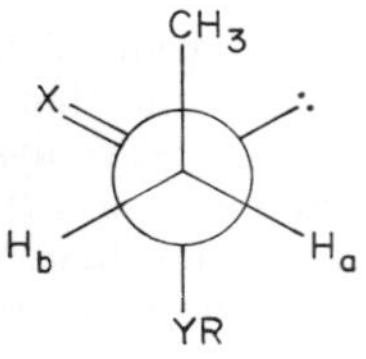

**(16)**

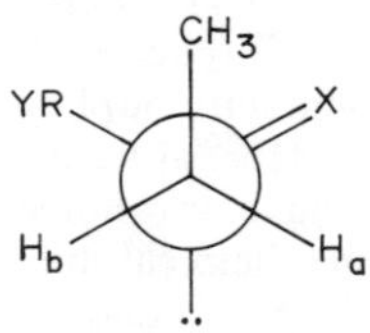

**(17)**

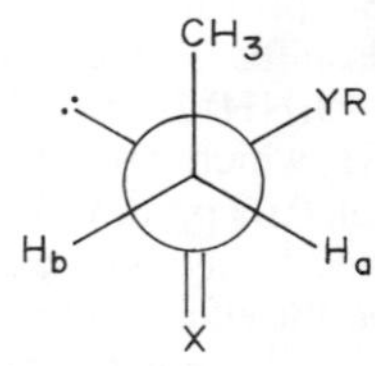

**(18)**

Regardless of the conformational preference, if any, for **16**, **17** or **18**, $H_a$ and $H_b$ can never be in the same magnetic environment, nor can rapid rotation make the average environment identical, except through accidental equivalence. The integration ratio of $H_a$:$H_b$ must always be 1:1 if they have different shifts. There can only be two possible explanations if $H_a$ and $H_b$ (or other diastereotopic groups) have the same chemical shifts: either their chemical shifts are fortuitously or accidentally equivalent, or the sulphur atom is undergoing inversion of configuration rapidly on the NMR time scale. It is possible that in one or more conformations the protons $H_a$ and $H_b$ could be accidentally equivalent, but anisochronous in other conformations, in which case if the conformational population changes with temperature so will the NMR spectrum. This is perhaps a rather unlikely combination of circumstances to account for apparent inversion of sulphur.

When there are two chiral centres within a molecule there are four stereoisomers possible, if each chiral centre is different.

The compounds **19** and **20** are enantiomers and will have identical NMR spectra in achiral solvents. The compounds **21** and **22** are also enantiomers, with identical NMR spectra, but they differ from **19** and **20** as they are diastereoisomers. It is therefore expected for a mixture of **19–22** that two sets of resonances will be observed in the NMR spectra. If the two sets of resonances coalesce to one set as the temperature is increased, then the *configuration* at either S or Y is being inverted fast on the time scale. As **19–22** are all different compounds, conformational changes by rotations about bonds cannot inter-

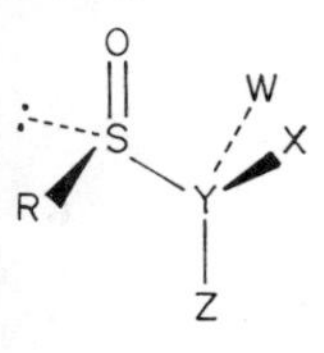

**(19)**

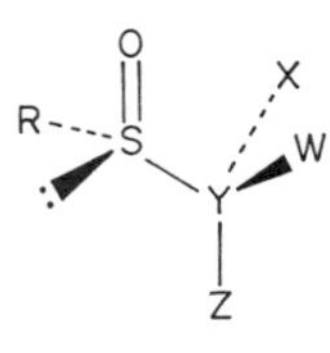

**(20)**

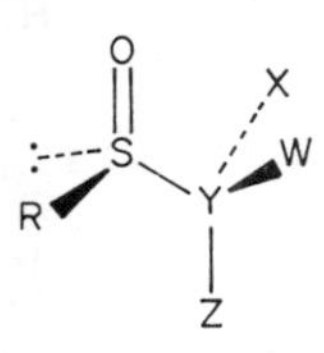

**(21)**

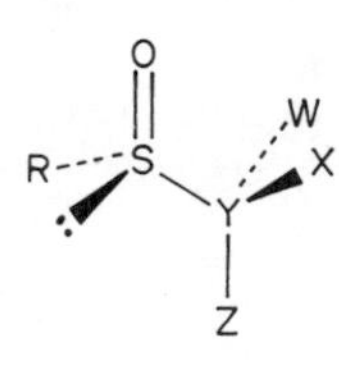

**(22)**

change one compound with another. The molar ratio of **19** and **20** to **21** and **22** is not necessarily, or even usually, 1:1. The ratio depends on the internal energies of the diastereoisomers, their method of preparation and whether their populations are equilibrating under measurement conditions.

The only set of compounds where these principles need any modification is the sulphinamides and these are discussed later in the section.

*2. Diastereotopism in sulphinates and thiosulphinates*

We have been unable to find examples of magnetic inequivalence of diastereotopic groups in sulphinic acids, although there does not appear to be any fundamental reason why this should be so. There are however very limited NMR data for sulphinic acids (see Section II.A.2).

As early as 1961 Waugh and Cotton[72] and Kaplan and Roberts[73] showed that the two methylene protons in $C_6H_5SOOCH_2CH_3$ were anisochronous and have different chemical shifts. The chemical shift difference is 0.434 ppm and the geminal coupling constant $^2J(HH)$ is 10.0 Hz. This geminal coupling constant is similar to that in FSO $OCH_2CH_3$ which is reported to be 9.8 Hz[70] although the chemical shift non-equivalence is very small, 0.09 ppm. Other studies have also noted such geminal non-equivalence[21,74]. Norton and Douglass[21] examined the chemical shift non-equivalence for sulphinates, and compared the effect on the $^1H$ NMR spectra of diastereotopic groups adjacent to oxygen with those adjacent to sulphur. The isopropyl esters of a variety of sulphinic acids all showed chemical shift non-equivalence of the diastereotopic methyl groups as shown in Table 16. The non-equivalence is particularly pronounced for the ester of phenylsulphinic acid; the phenyl group appears specifically to enhance shielding of the more shielded methyl group ($\delta$ 1.15 ppm) while having almost no effect on the other methyl group ($\delta$ 1.33 ppm).

Surprisingly, esters of 2-propanesulphinic acid show little or no intrinsic non-equivalence in the $^1H$ NMR in most cases[21]. The isopropyl methyl groups in $(CH_3)_2CHSOOCH_3$ show no difference in chemical shift in 10% $CCl_4$. In benzene solution the isopropyl methyl groups in $(CH_3)_2CHSOOCH_3$ do show some difference (0.05 ppm) attributed to a specific interaction with the phenyl ring[21]. Figure 2 shows the effect of temperature on the non-equivalent methyl groups in neat $CH_3SOOCH(CH_3)_2$ (**23**) and $(CH_3)_2CHSOOCH_3$ (**24**) in benzene solution. The $^1H$ NMR spectrum of **23** is essentially invariant to 102 °C, after which some coalescing is observed. By contrast the spectrum of **24** is much more temperature dependent, and this is accounted for by progressive decomplexation of the benzene and **24**[21]. There does not appear to be any further systematic study on the anisotropic effects of the sulphur atom on diastereotopic groups. It is possible that the equivalence of the diastereotopic protons bonded to S in

TABLE 16. Chemical shift non-equivalence in some sulphinates $RSOOCH(CH_3)_2$[21]

| R | $\delta CH_3$/ppm[a] | $\Delta\delta$/ppm | $\delta CH$/ppm |
|---|---|---|---|
| $CH_3$ | 1.28, 1.32 | 0.04 | 4.42 |
| $CH_3CH_2$ | 1.25, 1.31 | 0.06 | 4.42 |
| $CH_3CH_2CH_2$ | 1.27, 1.33 | 0.06 | 4.39 |
| Cl | 1.47, 1.52 | 0.05 | 5.46 |
| $C_6H_5$ | 1.15, 1.33 | 0.18 | 4.50 |

[a]10% w/w in $CCl_4$ 35 °C, 60MHz, $^1H$ NMR spectra.

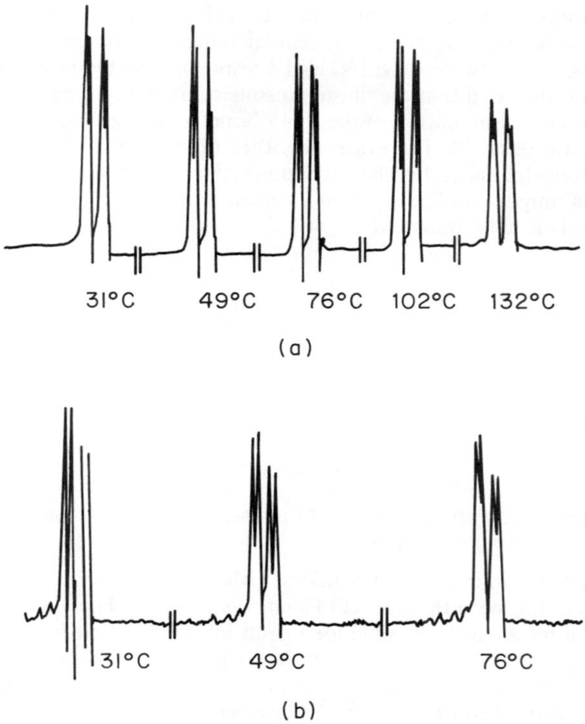


FIGURE 2. The $^{1}H$ NMR spectra of the diastereotopic methyl groups in (a) $CH_3SOOCH(CH_3)_2$ and (b) $(CH_3)_2CHSOOCH_3$ in $C_6H_6$ at a variety of temperatures

sulphinates is a very limited effect with no general significance. In the thioester series compound **25** shows magnetic non-equivalence for both sets of diastereotopic methylene protons, α and α′, in $CDCl_3$ with Δδ 0.05 ppm for the α-protons and Δδ 0.15 ppm for the α′ protons[31]. For compound **26** both the β and β′ methyl carbon nuclei show magnetic non-equivalence[9] with the difference being 0.9 ppm for the β carbons but only 0.1 ppm for the β′ carbon nuclei (Table 6). Additionally, the γ-methyl groups are non-equivalent in compound **27** (Δδ, 1.0 ppm) while the γ′ groups are isochronous[9] (Table 6).

$(CH_3)_3CH_2S(=O)SCH_2(CH_3)_3$ (α, α′)

**(25)**

$(CH_3)_2CHS(=O)SCH(CH_3)_2$ (β, β′)

**(26)**

$(CH_3)_2CHCH_2S(=O)SCH_2CH(CH_3)_2$ (γ, γ′)

**(27)**

Oae and coworkers[28] observed only one $^{13}C$ NMR methyl resonance for **28** and **29**. Each of these compounds has two chiral centres and can therefore exist as two pairs of diastereoisomers, so that two methyl $^{13}C$ NMR resonances would be expected for each of **28** and **29**. It was suggested that the diastereoisomers are in rapid equilibrium through a rapid cleavage and recombination of the S—S bond, as had already been suggested for ArSOSR racemizations[75,76]. These are no other reports of dynamic phenomena for sulphinates or thiosulphinates, but the indication is that in these compounds the sulphur is generally maintaining its configuration at ambient temperature or above[21] as magnetic non-equivalence is usually observed.

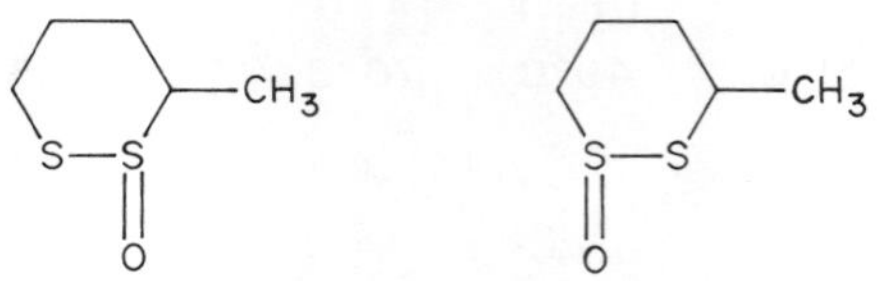

(**28**) (**29**)

*3. Diastereotopism in sulphinamides and the mechanism of exchange of magnetic environment of the nitrogen ligands*

The dynamic behaviour of sulphinamides, although superficially simple, is actually quite complex and has been the subject of some controversy. There are three related but different types of mechanism possible for ligand interconversion in the sulphinamides, shown in Figure 3–5. If the nitrogen atom has a planar configuration, then the ligand interconversion is directly analogous to that in amides[77], but the available evidence does suggest that the nitrogen is not planar[52,65]. The mechanism for planar nitrogen is shown in Figure 5, and those for ligand interconversion with $sp^3$-type nitrogen are given in Figure 3 and 4.

Initially[48] it seemed that there was fast exchange between the groups on nitrogen through fast rotations about the N—S bond coupled with fast inversion of configuration at nitrogen, or possibly fast rotation about N—S with a planar nitrogen configuration.

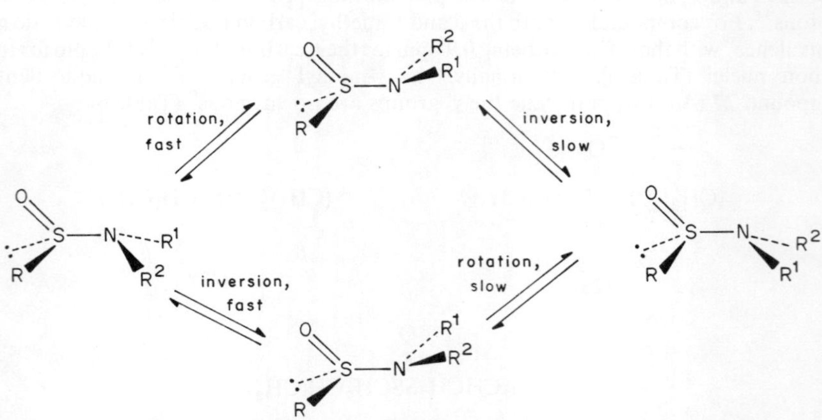


FIGURE 3. Ligand interconversion in sulphinamides with either rotation or inversion rate-limiting

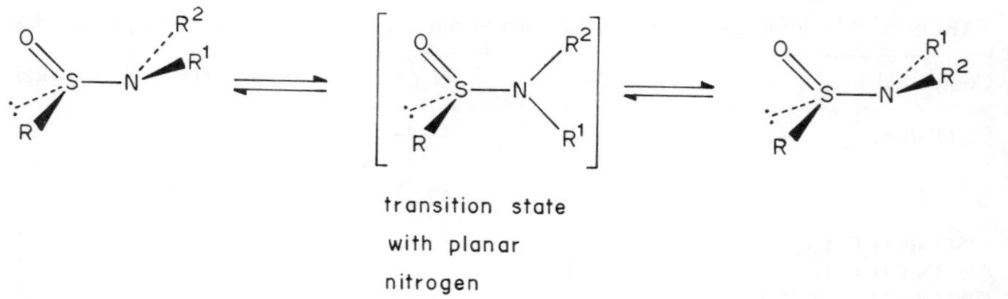

FIGURE 4. Concerted rotation-inversion in sulphinamides

FIGURE 5. Slow rotation about planar nitrogen in sulphinamides

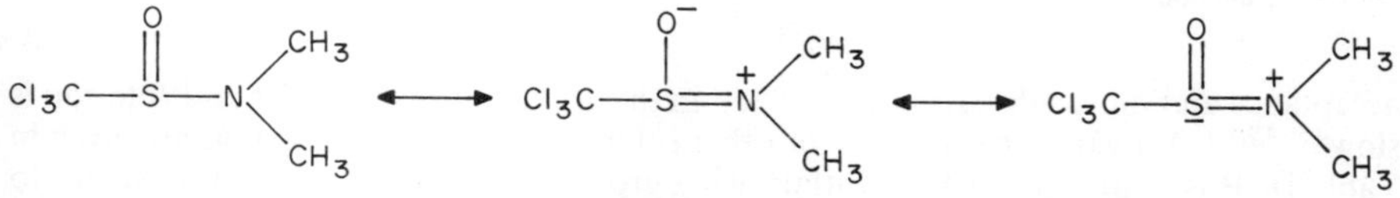

FIGURE 6. Resonance canonicals for a sulphinamide

This was based on the observation[48] that in $CH_3SON(CH_3)_2$ only one N-methyl resonance was observed, even at low temperature (−60 °C). The configuration at sulphur in sulphinamides is stable, as shown by the diastereotopic methyl groups in $(CH_3)_2CHSON(CH_3)_2$ giving separate signals in the $^1H$ NMR[48]. Jakobsen and Senning[78] subsequently observed non-equivalence in the methyl groups of $Cl_3CSON(CH_3)_2$ at temperatures below −46 °C (60 MHz) which they attributed to restriction of rotation about the N—S bond through (p–d)$\pi$ bonding as shown in Figure 6. A line-shape analysis of the temperature variation of the $^1H$ NMR spectrum of $Cl_3CSON(CH_3)_2$ gave the activation parameters shown in Table 17. Raban[79] disputed the interpretation involving (p–d)$\pi$ bonds on the grounds that there is little geometric requirement for such bonds. He proposed[79] that the temperature dependence of the NMR spectrum of $Cl_3CSON(CH_3)_2$ could be explained if nitrogen were undergoing slow inversion (see Figure 3), presumably with fast rotation about the N—S bond. This suggestion has not met with great

TABLE 17. Activation parameters[78] for $CH_3$ interconversion in $Cl_3CSON(CH_3)_2$

| $T_c$/°C | $\Delta\nu_{AB}{}^a$/Hz | $E_a$/kJ mol$^{-1}$ | $\Delta G^‡$/kJ mol$^{-1}$ | $\Delta H^‡$/kJ mol$^{-1}$ | $\Delta S^‡$/J K$^{-1}$ mol$^{-1}$ |
|---|---|---|---|---|---|
| $-46 \pm 2$ | $19.5 \pm 0.2$ | $39.2 \pm 4$ | $49.7 \pm 4.2$ | $31.8 \pm 4$ | $-79 \pm 20$ |

[a]The separation in Hz between the N—$CH_3$ resonances in the absence of exchange at −82 °C.

TABLE 18. Activation parameters for nitrogen ligand interconversion in some sulphinamides

| Compound | $\Delta\nu$/Hz[a] | $T_c$/°C[c] | $\Delta G^{\ddagger}$/kJ mol$^{-1}$ | Ref. |
|---|---|---|---|---|
| $ClSON(CH_3)_2$ | 3.8 | −44 | 50.6 | 80 |
| | 3.6 | −35 | 53.5 | 47 |
| | 3.3[d] | −48 | 47.6 | 46 |
| | 3.9 | −39 | 52.2 | 57 |
| $ClSON(CH_2CH_3)_2$ | 5.5 | −37[d] | 52.4 | 82 |
| $ClSON(CH_2C_6H_5)_2$ | 10[b] | −29 | 53.1 | 80 |
| $ClSON(CH_2CH(CH_3)_2)_2$ | 7.35[b] | −19 | 56.0 | 80 |
| $ClSONCH(CH_3)_2$ | 1.8 | −14 | 61.5 | 80 |
| | 3.6 | −17 | 54.3 | 57 |
| $FSON(CH_2CH_3)_2$ | | −105 | ≈35 | 80 |
| $FSON(CH_3)_2$ | 7.5[d] | −99 | ≈35 | 46 |
| $FSON(C(CH_3)_3)_2$ | | −102 | ≈35 | 80 |
| $FSON(CH(CH_3)_2)_2$ | 7.0 | −82 | 41.8 | 80 |
| $BrSON(CH_3)_2$ | 3.5 | −27 | 56.5 | 80 |
| | 3.2 | −27 | 55.6 | 47 |
| $FCl_2CSON(CH_3)_2$ | 18.8 | −59 | 45.1 | 47 |

[a]Difference in chemical shift between diastereotopic methyl protons unless otherwise stated.
[b]For methylene protons.
[c]$T_c$-Coalescence temperature at 60 MHz in $CH_2Cl_2$.
[d]In $CFCl_3$ solution.

acceptance and most subsequent papers retain the argument that N—S bond rotation is slow[45–47,80]. Activation parameters for other sulphinamide interconversions are given in Table 18. It is clear that electron-withdrawing groups on sulphur increase the barrier to ligand interconversion, and the barrier increases with increasing bulk of the ligand on N (as would be appropriate for a rotational process)[46,80]. The origin of the barriers to rotation (or inversion) is still a matter for debate. For rotation it has been suggested[46] that lone-pair–lone-pair interactions could be important. Subsequent to the published work on sulphinamide dynamic processes, Cowley, Wolfe and coworkers[81] published a theoretical interpretation of the N–P torsional process in aminophosphines, which is analogous in some ways to the sulphinamide torsional process. They suggested that in the N–P torsion the geometry of the nitrogen atom changed *as a function of the dihedral angle* so that a 'rotation' in fact incorporates both inversion and rotation. An extension of this mechanism to sulphinamides is shown in Figure 4. This concerted rotation/inversion does seem quite well suited to the sulphinamide dynamic processes, but no computational studies have been carried out on this system. The mechanism of ligand interconversion in sulphinamides has not been completely elucidated and could be fruitfully restudied using modern NMR techniques.

One further complication arises in the NMR spectra of halosulphinamides. In compounds such as **30**, at the low-temperature limit it would be expected from symmetry arguments that each proton $H_a$–$H_d$ would give rise to a separate resonance. However, in **30** and other derivatives such as the N-benzyl derivative only one signal was observed for each methylene group[80,57]. It was suggested that this may result from fast inversion of configuration at sulphur through halide exchange[80]. Halide exchange was demonstrated[80] (but not shown to be fast on the NMR time scale) and has precedent in other studies of sulphinyl halide racemization[82]. Rapid inversion at N would render $H_a$ and $H_b$ ($H_c$ and $H_d$) enantiotopic rather than diastereotopic while maintaining the diastereotopic relationship between $H_a$ and $H_c$ (and so on). For further comment on this see the next section.

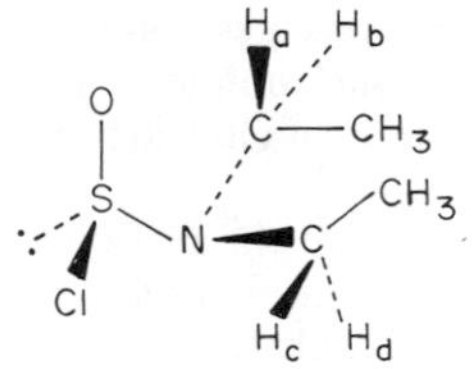

(30)

The question of silatropism has been investigated for monosilylsulphinamides[83] and bissilylsulphinamides. The monosubstituted compounds have the N-silyl form[83], whereas the bissilyl compounds, by $^{29}$Si NMR analysis, appear to have the NO bisimido form, with the possible exception of $C_6H_5SON(SiMe_3)_2$ which may have the *N, N* bissilyl form[84].

*4. Diastereotopism in sulphinyl halides*

In contrast to the chlorosulphinamides, the sulphur in sulphinyl chlorides is frequently configurationally stable, on the NMR time scale, at ambient temperature. King and Beatson[55] observed separate $^1$H NMR signals for the two diastereoisomers of $CH_3CHClSOCl$. The diastereoisomers were present in unequal amounts showing some asymmetric induction at S.

Three groups independently published data[56–58] on isopropylsulphinyl chloride showing that the two methyl groups are diastereotopic in several solvents. The data are given in Table 19. Taddei and coworkers[56] showed that as the temperature of the $CS_2$ solution was decreased, the methyl groups in $(CH_3)_2CHSOCl$ became anisochronous and $\Delta\delta$ increased to 0.03 ppm at $-80\,°C$. This behaviour was attributed to a conformational effect, but inversion at sulphur appears to be a more likely explanation. Rinne and Blaschette[58] found that, on heating, the diastereotopic methyl resonances in $(CH_3)_2CHSOCl$ in benzene solution coalesced at 54 °C. They suggested that inversion at sulphur was responsible for this dynamic behaviour. A rough estimate of $\Delta G^{\ddagger}$ from $\Delta\delta$/Hz and $T_c$ for the dynamic process in $(CH_3)_2CHSOCl$ is $75\,kJ\,mol^{-1}$ in $C_6H_6$ and $72\,kJ\,mol^{-1}$ in $CS_2$. The $\Delta G^{\ddagger}$ values are in good agreement and suggest that the same process, inversion at sulphur is occurring in each case. Thus it may be that for $(CH_3)_2CHSOOCH_3$ the temperature dependence[21] (see Section II.C.2) is not a result of benzene–sulphinate complexation, but is an inversion at sulphur.

A plausible explanation for the optical stability order of sulphinic acid derivatives is that sulphur is susceptible to nucleophile-induced racemization, either by halogen–halogen exchange[80] or by a more direct route. Silicon compounds are well known for their facile ligand exchanges in the presence of nucleophiles[85]. As sulphinamides contain the $NR_2$

TABLE 19. The 60 MHz $^1$H NMR spectra of the diastereotopic methyl groups of $(CH_3)_2CHSOCl$ in various solvents at ambient temperature

| Solvent | $\delta CH_3{}^a$/ppm | $\delta CH_3{}^b$/ppm | $\Delta\delta$/ppm | Ref. |
|---|---|---|---|---|
| $CDCl_3$ | 1.462 | 1.482 | 0.02 | 46 |
| $CS_2$ | 1.430 | 1.430 | 0.00 | 56 |
| $(CD_3)_2CO$ | 1.450 | 1.450 | 0.00 | 56 |
| $C_6D_6$ | 0.951 | 0.983 | 0.032 | 56 |
| $C_6H_6$ | 1.01 | 1.04 | 0.03 | 58 |
| $C_6H_6$ | 1.015 | 1.04 | 0.025 | 57 |

group which is nucleophilic, these molecules could be expected to be more labile than sulphinyl chlorides, for which no nucleophile is present. In such cases adventitious moisture could provide trace amounts of chloride ions to catalyze the ligand exchange process.

Taddei and coworkers[56] reported that the methylene protons in $CH_3CH_2SOCl$ were anisochronous (solvent not reported), Mikolajczyk[57] could not see separate signals for the same protons, whereas Rinne and Blaschette[58] reported that the methylene protons in $CH_3CH_2SOCl$ were anisochronous in benzene but not in $CHCl_3$. As $CH_3CH_2SOCl$ is the least hindered compound in which protons can be diastereotopic, it may be that inversion at sulphur is most facile. Differences in spectra may be due to differences in amounts of $Cl^-$ available, as well as to anisotropic and other solvent effects.

Pizey and coworkers[59] studied the $^1H$ NMR of some isopropyl-$\beta$-ketosulphinyl chlorides, and conformational effects were said to be important in determining spectral features.

*5. The use of NMR spectroscopy in the measurement of enantiomeric excess and in determining the absolute configuration of sulphinic acid derivatives*

The enantiomeric excess of a number of sulphinates has been measured[86] using tris-[3-(trifluoromethyl-hydroxymethylene-(+)-camphorato]europium (TFMC-Eu)[87]. No difference, however, was observed[88] between the enantiotopic methyl groups in racemic methyl *p*-toluenesulphinate in the presence of tris-[3-(*t*-butylhydroxymethylene)-(+)-camphorato]europium.

Pirkle and coworkers[89,90] have developed the use of the chiral alcohols **31** and **32** as solvents for use in both optical purity and configurational assignments in sulphinates and thiosulphinates. Specific solvation models are used[90] to predict the configuration at sulphur.

$$C_6H_5CH(OH)CF_3 \qquad \alpha\text{-NaphCH(OH)}CF_3$$

**(31)** **(32)**

### D. CIDNP

*N,N*-Disubstituted hydroxylamines such as oximes[91], ketoximines[92], *N*-alkylhydroxamic acids[93,94], *N*-alkyl-*N*-hydroxycarbamates[95] and the hydroxylamines themselves[96] react with sulphinyl chlorides at low temperatures (*ca* $-70\,^{\circ}C$) to form sulphinate esters of structure **33** (equation 2). Above approximately $0\,^{\circ}C$, these sulphinate

$$R^1R^2NOH + R^3S(O)Cl \longrightarrow R^3S(O)ONR^1R^2 \qquad (2)$$

**(33)**

esters undergo a facile rearrangement to form, amongst other products, the isomeric sulphonamides **34** (equation 3). Ketoximines appear to give the highest yields of the sulphonamide product **34**, *ca* 85%[92]. The reactions can be monitored using $^{13}C$ NMR spectroscopy, and the spectra display significant polarization effects. Therefore, part of the

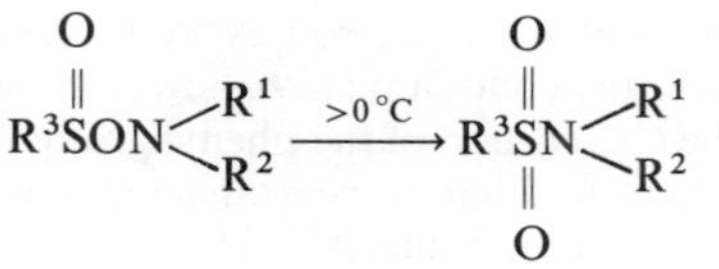


(34)

(3)

reaction at least proceeds via a radical process. For compounds **35** and **36**, the imine carbon atoms are observed in emission, as are the carbon atoms attached to sulphur. In both compounds, the carbon atoms attached to the imine group exhibit enhanced absorbtion signals. Analysis of the spectra using a radical pair model and Kaptein's equation for the net effect (equation 4) predicts precisely this pattern of polarization if the

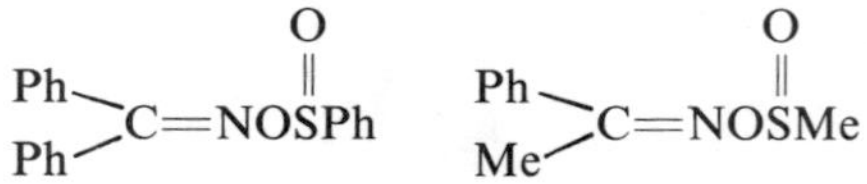

(**35**) (**36**)

$$\Gamma_{\text{ne}} = \mu\varepsilon\Delta g\, A_i \qquad (4)$$

reaction proceeds via an in-cage recombination of radicals. Such an analysis requires (i) the assumption that the reaction involves a singlet state precursor (therefore $\mu$ is negative) and (ii) the sign of $A_i$ to be correctly calculated by INDO MO calculations. Both requirements are reasonable. The sign of $\Delta g$ can be determined from the ESR spectra of the radicals involved (see Section III). Thus, the mechanism of the rearrangement involves homolysis of the N—O bond to form iminyl and sulphonyl radicals (equation 5). Recombination of these two radicals at the sulphur atom of the sulphonyl radical, at which there is significant spin density (see Section III.B.2), yields the sulphonamide product[92].

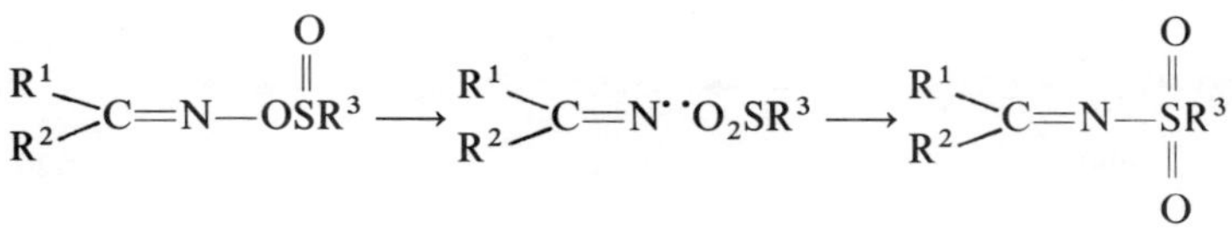

(5)

For aldoximes the position is less clear cut[91]. The deuterio compound **37** behaves in an analogous fashion to ketoximes, forming the sulphonamide **38** (and also benzonitrile) by an in-cage process (equation 6). Interestingly, the *E*- and *Z*-isomers of compound **37** give

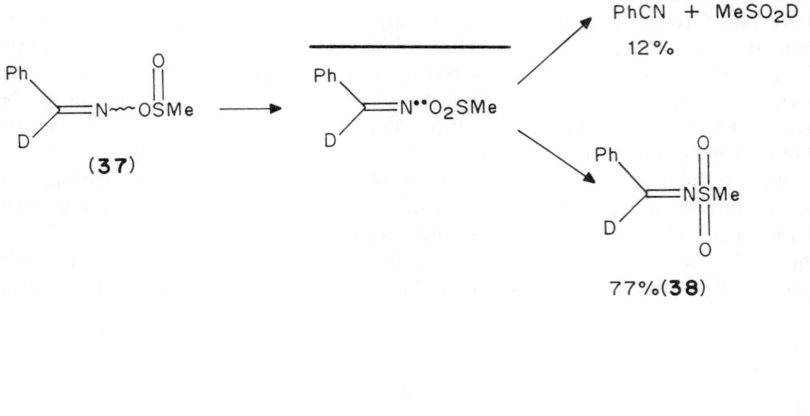


(6)

identical results, thus ruling out a six-electron cyclic pathway for the *E*-isomer[91]. The corresponding protio substituted compound (**39**), however, displays polarizations for the imine carbon atom, and the C-1 carbon of the phenyl group in the opposite sense. These observations lead to the conclusion that the rearrangement of compound **39** must involve an out-of-cage process from escaped radicals[91]. No explanation for this switch from an incage to an out-of-cage process on isotopic substitution has been forwarded.

*N*-Hydroxycarbamates **40** rearrange via an in-cage recombination of the radical pair to form both the sulphonamide and sulphonate products **41** and **42** (equation 7)[95]. This

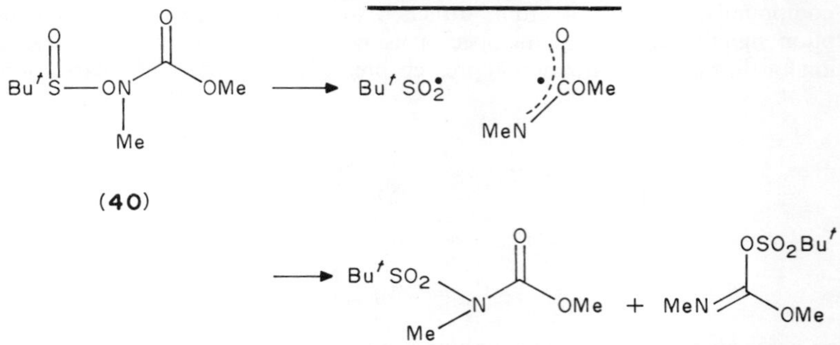


(**40**) (**41**) (**42**) (7)

reflects the ambident nature of the amidyl radical. Polarization effects can be observed in the $^1$H NMR for both products; thus, **41** and **42** display emission for the protons of the Bu$^t$ group and enhanced absorption for the *N*-Me group, whereas the *O*-Me remains unpolarized. A major product observed in this reaction is the carbamate **43** (equation 8),

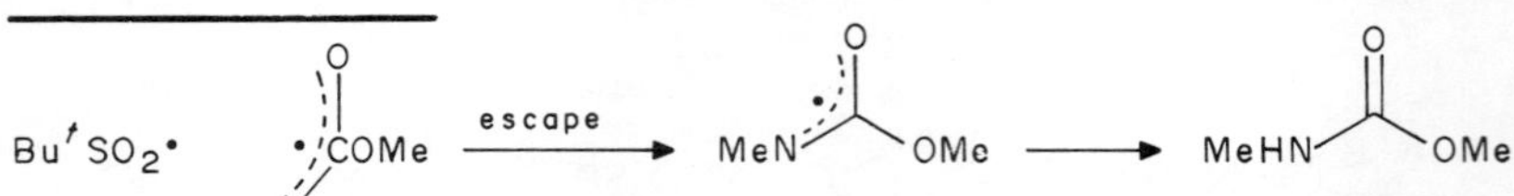


(**43**) (8)

formed by escape of the amidyl radical followed by hydrogen atom abstraction. Polarization in the *N*-Me group of **43** is observed as emission, i.e. the opposite sense to that observed for **41** and **42**, as expected for an out-of-cage process.

The related sulphinyl esters of *N*-alkylhydroxamic acids **44** generate the *N*-sulphonylamide **45** via an in-cage recombination of radicals, and the parent amide **46** via radical escape (equation 9). In this case, no recombination of the caged pair was observed to take place at the oxygen atom of the amidyl radical. For compounds **45** and **46** polarization was observed in both $^1$H and $^{13}$C NMR spectra: for **45**, the aromatic C-1 carbon exhibits emission, and the carbonyl carbon and *N*-Me carbon atoms enhanced absorption in the $^{13}$C spectrum, while the *N*-Me group exhibits emission in the $^1$H spectrum; for **46**, the carbonyl and *N*-Me carbon atoms appear in emission, and the *N*-Me protons in enhanced absorption[93]. A further product, accounting for 16–32% of the total products of the rearrangement of **44**, is the *O*-sulphonylhydroxamic acid **47**. This, too, exhibits polarization; the aromatic C-1 carbon appears in enhanced absorption, and the carbonyl and *N*-Me carbon atoms appear in emission while the *N*-Me signal in the $^1$H NMR spectrum displays enhanced absorption. These polarizations are in the opposite sense to those observed for **45** and indicate recombination of escaped radicals. A

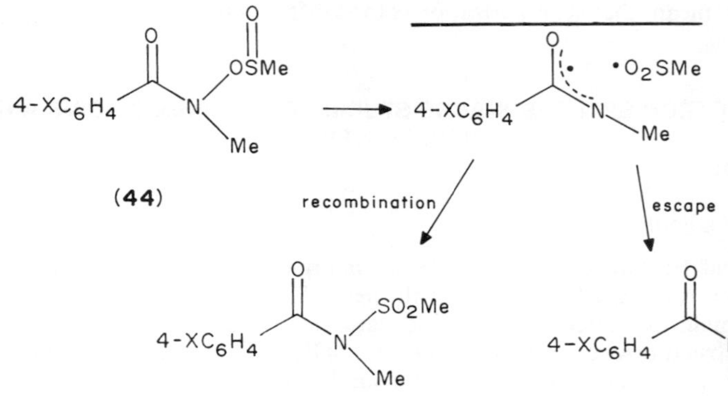

(44) (45) (46)

(9)

mechanism that accounts for the formation of compound **47**[93] is shown in equation 10. That both N—O and S—O bond cleavages occur in these reactions is verified by the observation of the acylnitroxyl radical in the ESR spectrum.

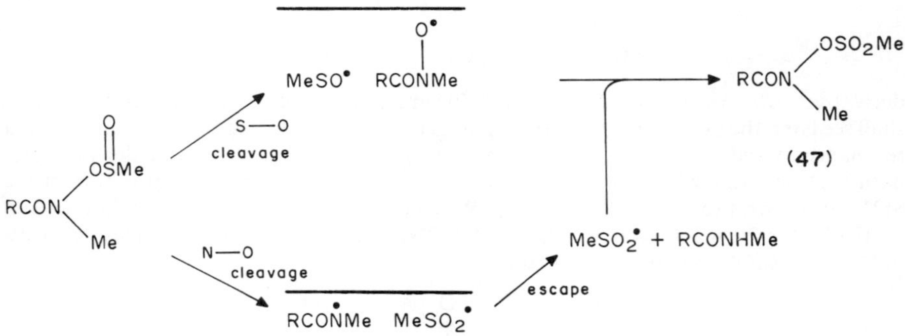

(47)

(10)

Sulphinyl esters of *N*-arylhydroxamic acids behave similarly[94], except that delocalization of the spin density in the amidyl radical onto the *N*-aryl ring enable further products, resulting from recombination of the $RSO_2^{\cdot}$ with the *ortho*- and *para*-positions of the aryl ring, to be isolated. Strong $^{13}C$ polarization is observed, with the carbonyl carbon atom of the products formed by in-cage recombination appearing in emission while the carbonyl carbon atom of products formed by out-of-cage processes appear in absorption.

Alkyl- and arylsulphinate esters of *N,N*-dialkylhydroxylamines also undergo in-cage radical recombination to form the corresponding sulphonamides. For *N,N*-dimethylmethanesulphonamide **48**, the *N*-Me and *S*-Me carbon atoms appear in emission in the $^{13}C$ spectrum, while in the $^{1}H$ spectrum of the *N*-Me signal appears as enhanced absorption. The small coupling of the unpaired electron to the *S*-Me protons (see

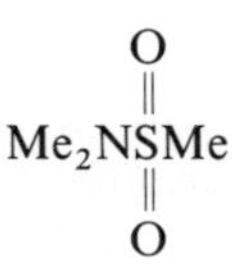

(**48**)

Section III.B.1) means that net polarization of the *S*-Me signal in the $^1H$ NMR spectrum is not observed[92].

## III. ELECTRON SPIN RESONANCE STUDIES OF SULPHINIC ACIDS AND DERIVATIVES

### A. Introduction

#### *1. Radicals of sulphinic acid derivatives*

Sulphinic acid derivatives of structure **49** can, in principle, exhibit tautomeric equilibria involving structures **49a**, **49b** and **49c**. For the neutral molecules of sulphinic acid (X=O) and sulphinamide (X=NR), spectroscopic data clearly identify tautomer **49a** as the predominant structure (of course, for X=O, **49a** ≡ **49b**)[97–99]. The sulphinate anion $RSO_2^-$ is best described as a resonance hybrid in a similar way to the analogous carboxylate anion, $RCO_2^-$ [98]. A strictly parallel situation arises for radical species of sulphinic acid

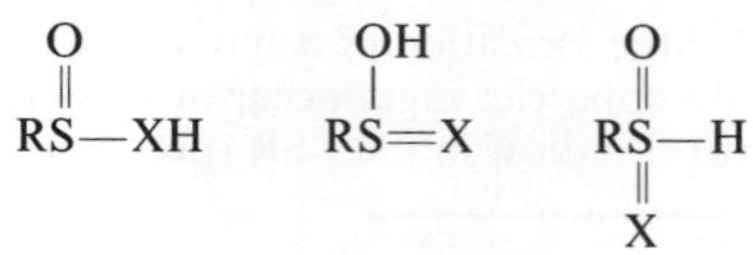

**(49a)** **(49b)** **(49c)**

derivatives, where structures **50a**, **50b** and **50c** are all plausible candidates. Indeed, as we shall see later, the experimental evidence points to a structure which can be thought of as a resonance hybrid of **50a–c**. ESR spectroscopy has been used to determine the structural nature of such radicals ($\sigma$ or $\pi$), the atomic spin densities, and the conformation of the $SOX^\bullet$ group with respect to the R group. We shall describe each of these studies here, in particular for the $RSO_2^\bullet$ (sulphonyl) and $R^1SONR^{2\bullet}$ (sulphinylamidyl) radicals. Radicals of the type $RSOS^\bullet$ remain unreported.

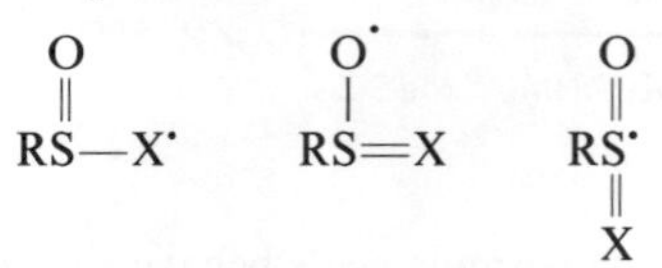

**(50a)** **(50b)** **(50c)**

Depending on the conditions, sulphinic acid derivatives can undergo a range of homolytic bond cleavage reactions to generate a variety of radical species:

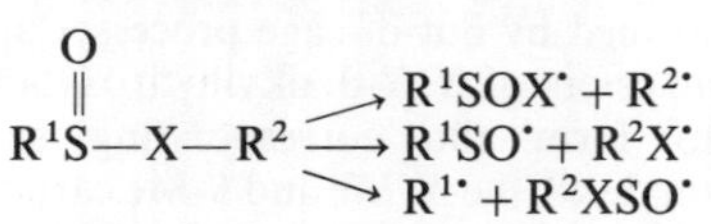

We shall only consider the formation of $RSOX^\bullet$ in this section, leaving the formation of $R^1SO^\bullet$ and $R^2XSO^\bullet$ to be properly discussed in a future companion volume on the chemistry of sulfenic acid derivatives.

#### *2. Formation of radicals of sulphinic acid derivatives*

A useful, recent review detailing the formation of $RSO_2^\bullet$ radicals is that of Freeman and Keindl[100]. We shall therefore only provide an outline of the methods involved. The most obvious route to $RSOX^{\bullet}$ radicals is direct H-atom abstraction from RSOXH (equation 11), and this is known for both $RSO_2^{\bullet}$[101] and $R^1SONR^{2\bullet}$[102]. Indeed, the formation of $RSO_2^{\bullet}$ from the reaction of dialkylsulphoxides, RSOR, with hydroxyl radicals, $OH^{\bullet}$, ultimately involves a similar process (equation 12)[103]. However, other

$$\mathrm{RS(=O)XH} + \mathrm{Bu}^t\mathrm{O}^{\bullet} \longrightarrow \mathrm{RSOX}^{\bullet} + \mathrm{Bu}^t\mathrm{OH} \tag{11}$$

$$\mathrm{RS(=O)R} + \mathrm{OH}^{\bullet} \longrightarrow \mathrm{RS(=O)OH} + \mathrm{R}^{\bullet} \longrightarrow \mathrm{RSO}_2^{\bullet} + \mathrm{RH} \tag{12}$$

simple, though equally direct, routes to $RSOX^{\bullet}$ from a range of compounds that contain sulphur in various oxidation states are more commonly employed for ESR studies. These include halogen abstraction from sulphonyl halides (X=Cl, F) by trialkylsilyl radicals (equation 13)[101,104,105], radical addition to $SO_2$[101,103b,106,107] and *N*-sulphinylamines (RNSO)[108] [but not to sulphurdiimides (RNSNR)[109]] (equation 14), photochemical cleavage of alkyl alkanesulphinates (but not arenesulphinates[110]) (equation 15), and the oxidation of thiols, RSH[111], and disulphides,

$$\mathrm{RSO_2X} + \mathrm{Et_3Si}^{\bullet} \longrightarrow \mathrm{RSO}_2^{\bullet} + \mathrm{Et_3SiX} \tag{13}$$
$$\mathrm{X=F,\ Cl}$$

$$\mathrm{R}^{\bullet} + \mathrm{O{=}S{=}X} \longrightarrow \mathrm{RSOX}^{\bullet} \tag{14}$$

$$\mathrm{R^1SO}^{\bullet} + {}^{\bullet}\mathrm{OR}^2 \xleftarrow[\mathrm{R^1=aryl}]{h\nu} \mathrm{R^1S(=O)OR^2} \xrightarrow[\mathrm{R^1=alkyl}]{h\nu} \mathrm{R^1SO_2^{\bullet}} + \mathrm{R}^{\bullet} \tag{15}$$

$RSSR$[111], by the Ti(III)–$H_2O_2$ couple at pH 1–2, or of thiols[111] and arenesulphinic acids[112] by Ce(IV) (equations 16 and 17). A somewhat less direct method involves the thermal rearrangement of the sulphinate ester **51** formed by the reaction of an oxime with a sulphinyl chloride (equation 18)[91,92]. X-[113,114] and $\gamma$-irradiation[115] of sulphones and $\gamma$-irradiation of the sulphonic acid taurine[116] have also been used, but these clearly are not of general utility.

$$\mathrm{RSH\ or\ RSSR} \xrightarrow[\mathrm{H_2O_2}]{\mathrm{Ti(III)}} \mathrm{RSO}_2^{\bullet} \tag{16}$$

$$\mathrm{RSH\ or\ RSO_2H} \xrightarrow{\mathrm{Ce(IV)}} \mathrm{RSO}_2^{\bullet} \tag{17}$$

$$\mathrm{R^1R^2C{=}NOH} + \mathrm{R^3S(=O)Cl} \xrightarrow{-\mathrm{HCl}} \underset{(\mathbf{51})}{\mathrm{R^1R^2C{=}NOS(=O)R^3}} \longrightarrow \mathrm{R^1R^2C{=}N}^{\bullet} + \mathrm{R^3SO_2^{\bullet}} \tag{18}$$

### B. The Sulphonyl Radical, $RSO_2^{\bullet}$

#### 1. g-Values and hyperfine coupling constants

Since the first ESR observation of an $RSO_2^{\bullet}$ radical in 1959[117], there have been a series of quite detailed studies, both experimental and theoretical, which enable the nature of such radicals to be described. Table 20 contains *g*-values and hyperfine coupling constants for alkyl and aryl sulphonyl radicals in solution. (For reasons that will become apparent, the $RSO_2^{\bullet}$ radical is generally referred to as a sulphonyl radical, despite it being formally related to sulphinic acid.) Some typical spectra of various $RSO_2^{\bullet}$ radicals are shown in Figure 7, and these clearly demonstrate the coupling between the unpaired electron to α-CH, β-CH, γ-CH and aromatic protons. The analysis of these spectra is contained in Table 20.

Careful inspection of Table 20 enables some general observations to be made regarding *g*-values and the hyperfine coupling constants *a*(H) and *a*(X) (where X = F, Cl etc.). These observations are listed below and we shall refrain from detailed analysis here, leaving such discussions to Section III.B.3 where the structure of $RSO_2^{\bullet}$ is described.

(i) *g*-Values are largely structure independent, ranging from 2.0041–2.0055 for both alkyl and aryl sulphonyl radicals. Only for those arylsulphonyl radicals which contain more than one heavy element substituent, i.e. Br, does the *g*-value exceed 2.0055, presumably due to spin–orbit coupling. At the lower end of the range, sulphonyl radicals in which the sulphur atom is directly bonded to a heteroatom appear to have, with the exception of Cl, *g*-values *ca* 2.0035. *g*-Values are also solvent and temperature independent.

(ii) Hyperfine coupling *a*(H) to the α-, β- and γ-CH protons of alkylsulphonyl radicals follows the trend: β-CH > α-CH ≈ γ-CH. Indeed, the hyperfine couplings observed for the α-CH protons are remarkably low (see later). A similar trend is noted for the alkenyl analogues. In contrast, the hyperfine coupling to fluorine *a*(F), cf. $CF_3SO_2^{\bullet}$ and $CF_3CH_2SO_2^{\bullet}$, follows the more usual trend α-CF > β-CF. However, a small coupling to β-Cl has been observed, e.g. $ClCH_2CH_2SO_2^{\bullet}$, whereas no coupling to an α-Cl, e.g. $ClCH_2SO_2^{\bullet}$, has been reported.

(iii) For the aromatic sulphonyl radicals, $ArSO_2^{\bullet}$, coupling to the protons of the aryl ring is observed. Thus, unpaired spin density is being transferred to the aryl ring. Originally, the size of *a*(H) was ordered *ortho*-H > *para*-H > *meta*-H[101,112]. This order has been corrected by the later work of Gilbert and colleagues[104] who showed that *meta*-H > *para*-H ⩾ *ortho*-H. This has important structural consequences which will be discussed later. This correction is based on an analysis of the spectra of 3-Me-4-Br(or Cl)-$C_6H_3SO_2^{\bullet}$ and 3,5-$(CF_3)_2C_6H_3SO_2^{\bullet}$[104]. For the first two compounds, the spectra obtained exhibit a doublet of triplets. The larger splitting, *a*(H) 1 G, arises from coupling to only *one* proton which must be the *meta* -CH. For the third compound, the spectrum revealed a small quartet splitting, *a*(H) 0.55 G, from which it follows that *ortho* and *para* couplings are small and of comparable size. Further confirmation of this effect can be found by analysis of the data for other substituted arylsulphonyl radicals, in particular the isomeric 2,4-$Br_2C_6H_3SO_2^{\bullet}$ and 2,5-$Br_2C_6H_3SO_2^{\bullet}$. The 2,4-isomer exhibits only one large doublet coupling, *a*(H) 1.80 G, whereas the 2,5-analogue only exhibits couplings < 0.3 G. Thus, the large coupling in the 2,4-disubstituted radical must be due to the *meta*-5H hydrogen atom, since the 3-H and 6-H hydrogen atoms are common to both radicals.

(iv) In certain circumstances, equivalent atoms or groups, i.e. those in *ortho*- or *meta*-positions, can display different hyperfine coupling constants (see later). The 2,4-$Br_2C_6H_3SO_2^{\bullet}$ radical mentioned above is an example, where the *meta*-3H coupling is too small to be resolved and the *meta*-5H coupling is 1.8 G. At 193 K, the 2,4,6-$Cl_3C_6H_2SO_2^{\bullet}$ radical displays an analogous effect, where only *one ortho*-Cl, *a*(Cl) 1.4 G, and *one meta*-H,

TABLE 20. ESR parameters for sulphonyl radicals, $RSO_2^{\bullet}$, in solution

| Radical | Solvent | $T$/K | $g$-Value | Hyperfine coupling/G $a$(H) | Ref. |
|---|---|---|---|---|---|
| $MeSO_2^{\bullet}$ | $H_2O$ | | 2.0050 | 0.95(3H) | 103b, 118 |
| | $H_2O$(pH 1–2) | | 2.0049 | 0.8(3H) | 111 |
| | $H_2O$ | | 2.0055 | 0.9(3H) | 107 |
| | $CCl_2FCClF_2$ | 248 | 2.0052 | 0.58(3H) | 119 |
| | toluene | 233 | 2.0049 | 0.58(3H) | 101, 120 |
| | cyclopropane | 148 | 2.0049 | 0.76(3H) | 106, 121 |
| $EtSO_2^{\bullet}$ | $H_2O$(pH 1–2) | | 2.0050 | 0.95(2H), 1.95(3H) | 103b, 111 |
| | toluene | 233 | | 0.90(2H), 1.74(3H) | 101 |
| | cyclopropane | 223 | 2.0050 | 0.75(2H), 1.73(3H) | 121 |
| $Pr^nSO_2^{\bullet}$ | toluene | 233 | | 0.70($\alpha$-2H), 2.12($\beta$-2H), 0.70(3H) | 101 |
| | cyclopropane | 233 | 2.0051 | 0.70($\alpha$-2H), 2.12($\beta$-2H), 0.70(3H) | 121 |
| $Pr^iSO_2^{\bullet}$ | toluene | 233 | | 0.40(1, H), 1.90(6H) | 101 |
| | cyclopropane | 213 | 2.0052 | 0.40(1H), 1.90(6H) | 121 |
| $Bu^nSO_2^{\bullet}$ | toluene | 233 | | 0.47($\alpha$-2H), 2.09($\beta$-2H) 0.47($\gamma$-2H) | 101 |
| $Bu^tSO_2^{\bullet}$ | $H_2O$(pH 1–2) | | 2.0053 | 2.60(9H) | 103b, 111 |
| | toluene, $CCl_4$ | 233 | 2.0054 | 2.10(9H) | 101 |
| n-$C_{16}H_{33}SO_2^{\bullet}$ | toluene | 233 | | 0.50($\alpha$-2H), 1.89($\beta$-2H), 0.50($\gamma$-2H) | 101 |
| $ClCH_2SO_2^{\bullet}$ | cyclopropane | 195 | 2.0055 | 1.29(2H) | 121 |
| $CF_3SO_2^{\bullet}$ | cyclopropane | 173 | 2.0052 | 15.5(3F) | 122 |
| $ClCH_2CH_2SO_2^{\bullet}$ | oxirane-cyclopropane | 240 | 2.0054 | 0.65($\alpha$-2H), 2.67($\beta$-2H), 0.65(Cl) | 121 |
| $HOCH_2CH_2SO_2^{\bullet}$ | $H_2O$(pH 1–2) | | 2.0050 | 3.90($\beta$-2H) | 103b, 111 |
| | cyclopropane | 226 | 2.0052 | 1.30($\alpha$-2H), 2.75($\beta$-2H), 0.65(OH) | 121 |
| $EtOCH_2CH_2SO_2^{\bullet}$ | $H_2O$(pH 1–2) | | 2.0050 | 3.80($\beta$-2H) | 111 |
| $H_3\overset{+}{N}CH_2CH_2SO_2^{\bullet}$ | $H_2O$(pH 1–2) | | 2.0049 | 2.60($\beta$-2H) | 111 |
| $HSCH_2CH_2SO_2^{\bullet}$ | $H_2O$(pH 1–2) | | 2.0052 | 2.60($\beta$-2H) | 111 |
| $CF_3CH_2SO_2^{\bullet}$ | cyclopropane | 179 | 2.0047 | 0.35(2H), 3.83(3F) | 121 |
| $HO_2CCH_2CH_2SO_2^{\bullet}$ | $H_2O$(pH 1–2) | | 2.0047 | 1.25($\alpha$-2H), 2.50($\beta$-2H) | 103b, 111 |
| $HO_2CCH_2CH(CO_2H)SO_2^{\bullet}$ | $H_2O$(pH 1–2) | | 2.0050 | 3.0($\beta$-2H) | 111 |
| $HO_2CCH(NH_3)CH_2SO_2^{\bullet}$ | | | 2.0048 | 2.1($\beta$-1H) | 111 |
| $HOCMe_2CH_2SO_2^{\bullet}$ | $H_2O$(pH 1) | | 2.0051 | | 103b |
| $HS(CH_2)_4SO_2^{\bullet}$ | $H_2O$(pH 1) | | 2.0050 | 0.75($\alpha$-2H), 2.3($\beta$-2H) | 111 |
| $EtCOCH_2CH_2SO_2^{\bullet}$ | $H_2O$(pH 1) | | 2.0050 | 1.3($\alpha$-2H), 2.6($\beta$-2H) | 103b |

(continued)

TABLE 20. (*continued*)

| Radical | Solvent | $T$/K | $g$-Value | Hyperfine coupling/G $a$(H) | Ref. |
|---|---|---|---|---|---|
| $CH_2{=}CHSO_2^{\bullet}$ | cyclopropane | 153 | 2.0045 | 0.50(α-H), 0.85(*Z*-β-H), 5.20(*E*-β-H) | 121 |
| $CH_2{=}C(Me)SO_2^{\bullet}$ | cyclopropane | 153 | 2.0046 | 0.72(3H), 0.95(*Z*-β-H), 5.61(*E*-β-H) | 121 |
| *E*-$MeCH{=}CHSO_2^{\bullet}$ | cyclopropane | 153 | 2.0045 | 0.15(1H), 0.25(1H), 1.85(3H) | 121 |
| *E*-$PhCH{=}CHSO_2^{\bullet}$ | cyclopropane | 153 | 2.0045 | 0.15(1H), 0.35(1H) | 121 |
| $PhSO_2^{\bullet}$ | 1 M $H_2SO_4$ | | 2.0044 | 1.23(2H, *ortho*), 0.17(2H, *meta*), 0.6(1H, *para*) | 112 |
| | toluene | 233 | 2.0046 | 1.06(2H, *ortho*), 0.33(2H, *meta*), 0.5(1H, *para*) | 101 |
| | toluene | *ca* 200 | 2.0045 | 0.33(2H, *ortho*), 1.13(2H, *meta*), 0.52(1H, *para*) | 104 |
| 4-$MeC_6H_4SO_2^{\bullet}$ | 1 M $H_2SO_4$ | | 2.0041 | 0.98(2H, *ortho*), 0.98(3H) | 112 |
| | toluene | 233 | | 1.18(2H, *ortho*), 0.3(2H, *meta*), 0.63(3H) | 101 |
| | toluene | *ca* 200 | 2.0045 | 0.32(2H, *ortho*), 1.18(2H, *meta*), 0.65(3H) | 104 |
| 4-$BrC_6H_4SO_2^{\bullet}$ | toluene | 233 | | 0.95(2H, *ortho*), 0.31(2H, *meta*) | 101 |
| | toluene | *ca* 200 | 2.0047 | 0.31(2H, *ortho*), 0.95(2H, *meta*) | 104 |
| 2-$ClC_6H_4SO_2^{\bullet}$ | 1 M $H_2SO_4$ | | | 1.94(1H, *ortho*) | 112 |
| 4-$ClC_6H_4SO_2^{\bullet}$ | toluene | 233 | | 0.90(1H, *ortho*), 1.20(1H, *ortho*), 0.30(2H, *meta*) | 101 |
| | toluene | 213 | 2.0045 | 0.32(2H, *ortho*), 0.96(2H, *meta*), 0.12($^{35}$Cl), 0.10($^{37}$Cl) | 104 |
| 4-$FC_6H_4SO_2^{\bullet}$ | 1 M $H_2SO_4$ | | | 1.03(2H, *ortho*), 1.96(F) | 112 |
| | toluene | 233 | | 0.30(1H, *ortho*), 1.75(1H, *ortho*), 0.30(2H, *meta*), 1.75(F) | 101 |
| | toluene | *ca* 200 | 2.0044 | 0.31(2H, *ortho*), 0.90(2H, *meta*), 1.75(F) | 104 |
| 4-$EtC_6H_4SO_2^{\bullet}$ | 1 M $H_2SO_4$ | | | 0.91(2H, *ortho*), 0.91(2H, *meta*) | 112 |
| 4-$Bu^tC_6H_4SO_2^{\bullet}$ | toluene | *ca* 200 | 2.0045 | 0.28(2H, *ortho*), 1.03(2H, *meta*) | 104 |
| 4-$MeOC_6H_4SO_2^{\bullet}$ | 1 M $H_2SO_4$ | | | 0.91(2H, *ortho*) | 112 |
| | toluene | 233 | | 0.78(2H, *ortho*), 0.16(2H, *meta*), 0.16(3H) | 101 |
| 4-$NO_2C_6H_4SO_2^{\bullet}$ | 1 M $H_2SO_4$ | | | 1.17(2H, *ortho*) | 112 |
| 4-$HOOCC_6H_4SO_2^{\bullet}$ | 1 M $H_2SO_4$ | | | 1.12(2H, *ortho*), 0.2(2H, *meta*) | 112 |
| 2,4-$Cl_2C_6H_3SO_2^{\bullet}$ | toluene | *ca* 200 | 2.0050 | 0.50(2H, *ortho* and 3-*meta*), 1.5(1H, 5-*meta*) | 104 |
| 3,4-$Cl_2C_6H_3SO_2^{\bullet}$ | toluene | 233 | | 1.08(1H, *ortho*), 0.52(1H, *ortho*), 0.52(1H, *meta*) | 101 |
| 2,5-$Cl_2C_6H_3SO_2^{\bullet}$ | toluene | *ca* 200 | 2.0050 | 0.75(2H, *ortho* and *meta*) | 104 |
| 3-Me-4-$ClC_6H_3SO_2^{\bullet}$ | toluene | *ca* 200 | 2.0046 | 0.30(2H, *ortho*), 1.00(1H, *meta*) | 104 |
| 3-Me-4-$BrC_6H_3SO_2^{\bullet}$ | toluene | *ca* 200 | 2.0046 | 0.30(2H, *ortho*), 1.00(1H, *meta*) | 104 |
| 3,5-$(CF_3)_2C_6H_3SO_2^{\bullet}$ | toluene | *ca* 200 | 2.0043 | 0.55(2H, *ortho*), 0.55(1H, *para*) | 104 |

| | | | | | |
|---|---|---|---|---|---|
| $C_6F_5SO_2^{\bullet}$ | toluene | 233 | | 0.85(2F), 1.67(1F), 0.30(2F, *meta*) | 101 |
| | | | | 1.75(2F, *meta*), 0.3(3F, *ortho* and *para*) | 104 |
| 2,3,4-$Cl_3C_6H_2SO_2^{\bullet}$ | toluene | 233 | | 1.63(1H, *ortho*), 0.53(1H, *meta*) | 101 |
| 2,4,6-$Cl_3C_6H_2SO_2^{\bullet}$ | toluene | 240 | 2.0054 | 0.70(2Cl, *ortho*), 0.70(2H, *meta*) | 104 |
| 2,4,5-$Cl_3C_6H_2SO_2^{\bullet}$ | toluene | *ca* 200 | 2.0050 | 0.75(2H, *ortho* and *meta*) | 104 |
| 2,4,6-$Me_3C_6H_2SO_2^{\bullet}$ | toluene | *ca* 200 | 2.0049 | 0.68(7H), 1.07(2H, *meta*) | 104 |
| 2,4,5,6-$Me_4C_6HSO_2^{\bullet}$ | toluene | *ca* 200 | 2.0049 | 0.68(6H, *ortho* $CH_3$), 1.35(1H) | 104 |
| 2,3,5,6-$Me_4C_6HSO_2^{\bullet}$ | toluene | 250 | 2.0049 | 0.6(6H, *ortho* $CH_3$) | 104 |
| $Me_5C_6SO_2^{\bullet}$ | toluene | 250 | 2.0049 | 0.6(6H, *ortho* $CH_3$) | 104 |
| 2,4-$Cl_2$-5-$MeC_6H_2SO_2^{\bullet}$ | toluene | *ca* 200 | 2.0051 | 0.55(2H, *ortho* and *meta*) | 104 |
| 2,4-$Cl_2$-3-$MeC_6H_2SO_2^{\bullet}$ | toluene | *ca* 200 | 2.0051 | 0.55(1H, *ortho*), 1.70(1H, *meta*) | 104 |
| 2,4-$Br_2C_6H_3SO_2^{\bullet}$ | toluene | *ca* 200 | 2.0068 | 1.80(1H, 5-*meta*) | 104 |
| 2,5-$Br_2C_6H_3SO_2^{\bullet}$ | toluene | *ca* 200 | 2.0069 | *a* | 104 |
| 2,4-$Me_2$-5-Br-$C_6H_2SO_2^{\bullet}$ | toluene | *ca* 200 | 2.0050 | 0.4(4H, *ortho* and $CH_3$), 1.2(1H, *meta*) | 104 |
| $C_4H_3S$-2-$SO_2^{\bullet}$ | toluene | 230 | 2.0042 | 0.53(1H, 3-H), 0.74(2H, 4-H and 5-H) | 123 |
| 5-$BrC_4H_2S$-2-$SO_2^{\bullet}$ | toluene | 230 | 2.0047 | 0.67(1H, 3-H), 0.80(1H, 4-H) | 123 |
| 3-$BrC_4H_2S$-2-$SO_2^{\bullet}$ | toluene | 230 | 2.0054 | 0.65(1H, 4-H) | 123 |
| 4,5-$Br_2C_4HS$-2-$SO_2^{\bullet}$ | toluene | 230 | 2.0048 | 0.56(1H) | 123 |
| 3,4-$Br_2C_4HS$-2-$SO_2^{\bullet}$ | toluene | 230 | 2.0057 | | 123 |
| 3,5-$Br_2C_4HS$-2-$SO_2^{\bullet}$ | toluene | 230 | 2.0058 | 0.87(1H, 4-H) | 123 |
| 3-Br-5-D-$C_4H$-2-$SO_2^{\bullet}$ | toluene | 230 | 2.0054 | 0.66(1H) | 123 |
| 3-Br-4-D-$C_4H$-2-$SO_2^{\bullet}$ | toluene | 230 | 2.0053 | | 123 |
| $ClSO_2^{\bullet}$ | toluene | *ca* 200 | 2.0046 | | 104, 123 |
| $HOSO_2^{\bullet}$ | $H_2O$ | | 2.0033 | | 107 |
| $SO_3^{\bullet -}$ | | | 2.0036 | | 124 |
| $EtOSO_2^{\bullet}$ | hexane | 233 | | 1.43(2H), 0.43(3H) | 101 |
| $Bu^tOSO_2^{\bullet}$ | $CF_2ClCFCl_2$ | 233 | 2.0036 | 0.28(9H) | 101 |
| | MeOH | | 2.0034 | 0.28(9H) | 107 |
| $H_2NSO_2^{\bullet}$ | $Et_2O$ | 170 | 2.0036 | 5.00(N), 5.00(2H) | 125 |
| $D_2NSO_2^{\bullet}$ | $Et_2O$ | 170 | 2.0036 | 5.00(N), 0.77(2D) | 125 |
| $MeNHSO_2^{\bullet}$ | $Et_2O$ | 170 | 2.0036 | 5.80(N), 5.80(1H), 4.50(3H) | 125 |
| $Me_2NSO_2^{\bullet}$ | $Et_2O$ | 170 | 2.0036 | 6.90(N), 5.22(6H) | 125 |

[a]Too small to be assigned unambiguously.

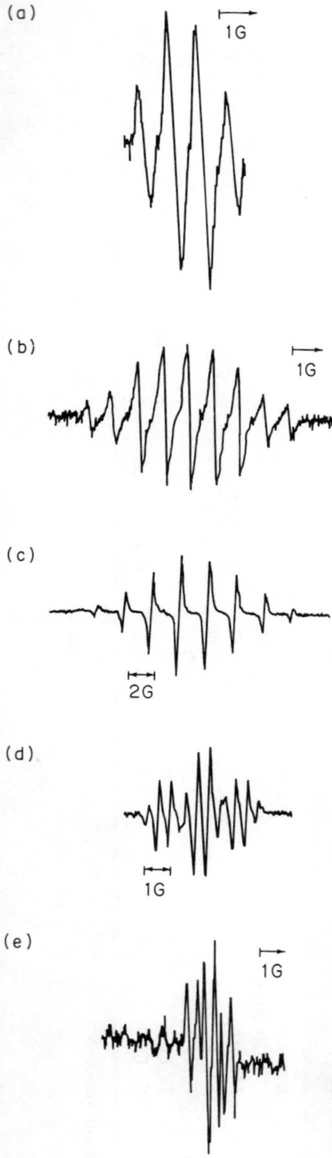


FIGURE 7. ESR spectra of (a) $MeSO_2^{\bullet}$, (b) $EtSO_2^{\bullet}$, (c) $Bu^tSO_2^{\bullet}$, (d) $EtOSO_2^{\bullet}$ and (e) $C_4H_3S$-2-$SO_2^{\bullet}$ (adapted from References 101, 121 and 123). Reproduced by permission of The Royal Society of Chemistry

$a$(H) = 1.4 G, are discernible. Also at 193 K, the spectrum of 2,4,5,6-$Me_4C_6HSO_2^{\bullet}$ shows only *one ortho*-$CH_3$ coupling. These observations point to an asymmetric distribution of spin density.

(v) Coupling to F substituents attached to the *S*-aryl ring is also observable, but the relative size of the hyperfine couplings $a$(F) for *ortho*-, *meta*- and *para*-F are not known with certainty. In some circumstances, coupling to ring substituted chlorine atoms is observed, but the value of $a$(Cl) is very small.

(vi) For alkenesulphonyl radicals, a large coupling of *ca* 5 G is attributable to the hydrogen atom *E*- to the sulphonyl centre and couplings < 1 G to both the *Z*- and C-1 hydrogen atoms.

(vii) Not apparent from Table 20 is the temperature dependence of *a*(H) hyperfine couplings for both alkyl and aryl sulphonyl radicals. This results from conformational effects and will be discussed in detail later.

### 2. $RSO_2^{\bullet}$ radicals in solid matrices

Most studies have employed observation of $RSO_2^{\bullet}$ radicals in solution. In conjunction with MO calculations, the data obtained by such studies (Section III.B.1) enable the structure of the $RSO_2^{\bullet}$ radical to be described. However, before embarking upon such a description, we shall outline the few solid state studies that have been performed because, uniquely, the results of these studies enable the atomic spin density at the sulphur atom to be defined.

The spectrum of $PhCH_2SO_2^{\bullet}$ trapped in $(PhCH_2)_2SO_2$ (interestingly, $PhCH_2SO_2^{\bullet}$ has not been observed in solution) is shown in Figure 8[114]. The triplet structure $a(H) = 5.0$ G,

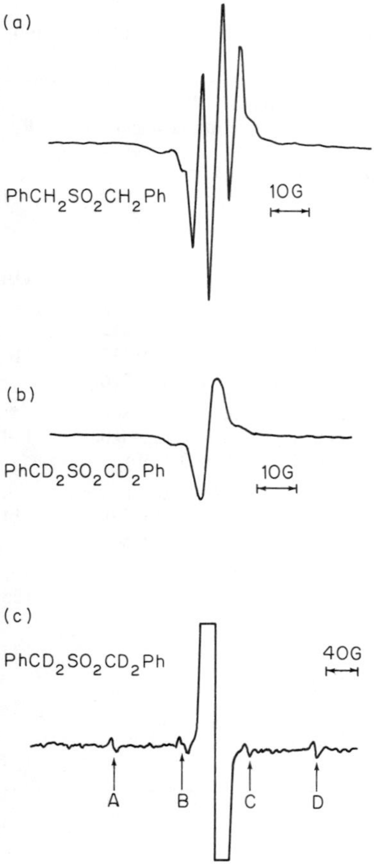


FIGURE 8. ESR spectrum of (a) $PhCH_2SO_2^{\bullet}$, (b) $PhCD_2SO_2^{\bullet}$ and (c) $PhCD_2SO_2^{\bullet}$ at high gain in a dibenzyl sulphone crystal. Reproduced with permission from Reference 114

due to coupling to the $CH_2$ protons, is clearly discernible. The deuterated analogue, $PhCD_2SO_2^{\cdot}$, exhibits a singlet which at high gain is accompanied by a 1:1:1:1 quartet each line of which is $1.8 \times 10^{-3}$ times as intense as the central line (Figure 8). These peaks are due to coupling to $^{33}S$ ($I = 3/2$, natural abundance 0.74%). Confirmation of this assignment comes from the spectrum of $PhCH_2SO_2^{\cdot}$, for which the $^{33}S$ lines are also triplets[114]. Accurate measurements for $SO_3^{\cdot -}$ show that the $^{33}S$ spectrum is shifted to lower field than that of the $^{32}S$ spectrum, though to a first approximation the $^{32}S$ spectrum lies in the centre of the $^{33}S$ spectrum[124]. For $PhCD_2SO_2^{\cdot}$, both the $g$-values of the $^{32}S$ spectrum and the $^{33}S$ hyperfine coupling are anisotropic[114]. The principal components of the $g$-tensor and the $a(^{33}S)$ tensor are given in Table 21 together with those from other sulphonyl radicals. It would appear that whereas for some radicals the $g$-value is essentially isotropic, for others there is significant anisotropy with $g$ reaching values (*ca* 2.01) seen for sulphinyl radicals $RSO^{\cdot}$. Interestingly, the average $g$-value observed for $PhSO_2^{\cdot}$, 2.0045, is identical to the solution value; however, that for $H_3NCH_2CH_2SO_2^{\cdot}$ is significantly different, 2.0059 as compared with 2.0049. It appears to be the case that, for those radicals with anisotropic $g$ and $a(^{33}S)$ tensors, the smallest principal $g$-value lies close to the largest principal $^{33}S$ coupling tensor. This direction is presumably that of a sulphur $3p$ orbital containing the unpaired electron.

The principal values of the $a(^{33}S)$ tensor given in Table 21 can be analyzed to give the

TABLE 21. Principal values of the $g$ tensor and $a(^{33}S)$ tensor and their direction cosines for various sulphonyl radicals

| Radical | Principal value of $a(^{33}S)$/G (direction cosines) | $g$-Value of $^{32}S$ isotopomer (direction cosines) | Ref. |
|---|---|---|---|
| $SO_3^{\cdot -}$ | 152.6(−0.778, ±0.384, −0.496) | 2.0036[c] | |
| | 112.7(−0.211, ±0.584, 0.784) | | 124 |
| | 112.0(0.591, ±0.715, −0.373) | | |
| | 135.2(0.516, ±0.540, 0.665) | 2.0035[c] | |
| | 99.2(0.714, ±0.700, 0.014) | | 116 |
| | 97.9(0.473, ±0.468, −0.747) | | |
| $H_3NCH_2CH_2SO_2^{\cdot}$ | 49.6(0.593, ±0.454, 0.666) | 2.0024(0.527, ±0.504, 0.684) | |
| | 9.1(0.577, ±0.815, 0.042) | 2.0056(0.586, ±0.798, 0.137) | 116 |
| | 1.2(0.562, ±0.360, −0.745) | 2.0097(0.615, ±0.329, 0.716) | |
| $PhCD_2SO_2^{\cdot}$ | 96.3(±0.53, 0.26, 0.80) | 2.0027(±0.33, −0.37, 0.86) | |
| | 56.2(±0.64, 0.74, 0.17) | 2.0056(±0.76, −0.64, −0.06) | 114 |
| | 63.4(±0.55, 0.60, −0.57) | 2.0094(±0.60, 0.60, 0.50) | |
| $PhSO_2^{\cdot}$ | 107.1(1.0, 0, 0) | 2.0023(0.86, 0.46, ±0.19) | |
| | 71.3(0, 1.0, 0) | 2.0044(0.44, −0.50, ±0.73) | 114 |
| | 71.3(0, 0, 1.0) | 2.0069(0.24, −0.72, ±0.64) | |
| $H_2NSO_2^{\cdot}$ | 118[a] | 2.0035[c] | |
| | 80[b] | | 125 |
| $MeHNSO_2^{\cdot}$ | 120[a] | 2.0035[c] | |
| | 84[b] | | 125 |
| $Me_2NSO_2^{\cdot}$ | 118[a] | 2.0035[c] | |
| | 76[b] | | 125 |
| $CH_2(CH_2)_4NSO_2^{\cdot}$ | 120[a] | 2.0035[c] | |
| | 78[b] | | 125 |

[a] $a_{\parallel}$.
[b] $a_{\perp}$.
[c] Isotropic.

TABLE 22. Isotropic and anisotropic components of the $a(^{33}S)$ hyperfine coupling and orbital spin densities in $RSO_2^{\cdot}$

| Radical | Matrix | $a_{iso}$/G | $a_{aniso}$/G | %s | %p | %(s+p) | p(s+p) |
|---|---|---|---|---|---|---|---|
| $H_2NSO_2^{\cdot}$ | THF/77K | 92.7 | 12.7 | 9.6 | 45.2 | 54.8 | 0.82 |
| $MeNHSO_2^{\cdot}$ | THF/77K | 96 | 12 | 9.9 | 42.9 | 52.8 | 0.81 |
| $Me_2NSO_2^{\cdot}$ | THF/77K | 90 | 14 | 9.3 | 50.0 | 59.3 | 0.84 |
| $\overline{CH_2(CH_2)_4N}SO_2^{\cdot}$ | THF/77K | 92 | 14 | 9.5 | 50.0 | 59.5 | 0.84 |
| $SO_3^{\cdot-}$ | taurine crystal | 110.8 | 12.3 | 11.4 | 43.3 | 55.2 | 0.79 |
| | $K_2CH_2(SO_3)_2$ crystal | 125.8 | 13.5 | 13.0 | 48.2 | 61.2 | 0.79 |
| $H_3\overset{+}{N}CH_2CH_2SO_3^{\cdot}$ | taurine crystal | 20 | 14.9 | 2.1 | 53.2 | 55.3 | 0.96 |
| $MeSO_2^{\cdot}$ | $Me_2SO_2$ crystal | 71.5 | 9.3 | 7.4 | 33.2 | 40.6 | 0.82 |
| $PhSO_2^{\cdot}$ | $PhSO_2CH_2CO_2H$ crystal | 83.2 | 11.9 | 8.6 | 42.6 | 51.2 | 0.83 |
| $PhCD_2SO_2^{\cdot}$ | $(PhCD_2)_2SO_2$ crystal | 72.0 | 12.2 | 7.4 | 43.6 | 51.0 | 0.85 |

isotropic and anisotropic components in Table 22. The isotropic coupling results from spin density in the sulphur 3*s* orbital and the anisotropic coupling from the 3*p* orbital. Since the isotropic coupling of an electron in a pure 3*s* orbital is 970 G and the anisotropic coupling in a pure 3*p* orbital is 28 G[126], it follows that the ratios of the observed isotropic and anisotropic components to these theoretical values yield the unpaired electron spin density in the sulphur atomic orbitals of the $RSO_2^{\cdot}$ radicals. These values are also contained in Table 22, together with the ratio $p/(s+p)$ which can vary from 0 for a pure *s* orbital to 1 for a pure *p* orbital. An $sp^3$ hybrid orbital has a value of 0.75. Several observations are worthy of note, the most important being that the total unpaired spin density at sulphur is *ca* 50–60%. This implies that *ca* 40–50% of the unpaired spin density must reside on the two oxygen atoms. Thus structures (**50a**–**c**) all contribute significantly to the overall structure of the radical. Further, it appears that alkyl and aryl sulphonyl radicals tend to have less unpaired spin density at sulphur than sulphonyl radicals bonded to electronegative elements. Moreover, it has been suggested on the basis of the $p/s$ ratio that the more electronegative the atom bonded to the sulphonyl group the more pyramidal the radical centre[114,125]. However, close scrutiny of Table 22 does not provide definitive substantiation for this trend. Certainly, the $SO_3^{\cdot-}$ radical is essentially pyramidal and the unpaired spin density in $H_3NCH_2CH_2SO_2^{\cdot}$ resides in an almost pure sulphur 3*p* orbital. However, this would appear to be an exception since the other radicals involving a C—S bond have a $p/(p+s)$ ratio of *ca* 0.83. This ratio appears to be identical to that for sulphonyl radicals involving a S—N bond, which implies that both types are nearly pyramidal but have somewhat more *p* character and are therefore flatter than $SO_3^{\cdot-}$. The generally quoted $p/s$ ratio must therefore be used with caution.

### *3. The structure and conformation of RSO˙ radicals in solution: empirical observations and molecular orbital calculations*

*Ab initio* molecular orbital calculations at the STO-3G* level theory have been performed for various sulphonyl radicals, including $MeSO_2^{\cdot}$[127]. The optimized geometry for $MeSO_2^{\cdot}$ is **52**; the OSO and OSC angles are 122.8° and 106.5°, respectively, from which It follows that, consistent with solid state studies, the sulphur atom is pyramidal. The S—O and S—C bond lengths are 147.9 and 181.8 pm, respectively. Almost identical configurations of the sulphonyl group in other radicals $XSO_2^{\cdot}$ (X = H, N, O, F, Cl) are found, i.e. OSO *ca* 123 (±2)°, XSO *ca* 106 (±1.5)°, and S—O 147 (±1) pm. Again, this

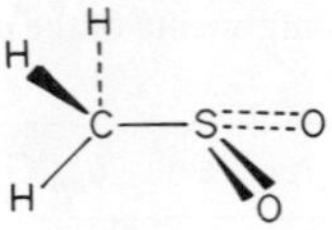

(52)

appears consistent with the *s* and *p* spin densities calculated for the sulphur atom from the solid state spectra. Significantly, for the $HSO_2^{\bullet}$ radical a planar arrangement (OSO 133.2° and HSO 113.4°) of the sulphonyl radical is found to be *ca* 100 kJ mol$^{-1}$ *less* stable than the pyramidal arrangement[127]. In spite of this, atomic spin densities for pyramidal $HSO_2^{\bullet}$ are reported to be 0.23 for sulphur and 0.42 for oxygen. These are considerably different from those calculated from solid state observations, viz. 0.5–0.06 for sulphur and 0.4–0.5 for oxygen. However, spin densities for planar $HSO_2^{\bullet}$ are 0.62 and 0.39, which appear consistent with the experimentally determined values. This anomaly remains unresolved, but the commonly accepted structure for the sulphonyl radical is pyramidal.

The Newman projection **53** of structure **52** reveals that one proton of the methyl group is magnetically distinct from the other two.

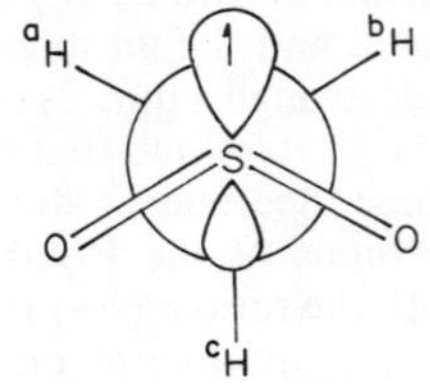

(53)

The doublet of triplets expected from such a radical, if rotation about the C—S bond is frozen out, has never been observed. However, the quartet seen for $MeSO_2^{\bullet}$ at 200 K, due to averaging of the proton environments by rapid C—S bond rotation, is temperature dependent (Figure 9), the central lines showing significant broadening at 153 K[121]. Moreover, the size of the hyperfine couplings *a*(H) increase at lower temperatures; thus at 223 K *a*(H) is 0.55 G and at 148 K it is 0.76 G[121]. INDO MO calculations for $MeSO_2^{\bullet}$ are able to reproduce these averaged couplings, 0.52–0.77 G, for a pyramidal structure in which the OSO angle is 105°, the C—S and S—O bond lengths are 188 and 143 pm, and the plane containing the OSO atoms and the normal to the S—C bond subtends an angle, α, between 130° and 135° (**54**)[121]. The sulphur 3*s* spin density in such a structure is *ca* 0.06, in line with the values discussed in Section III.B.2.

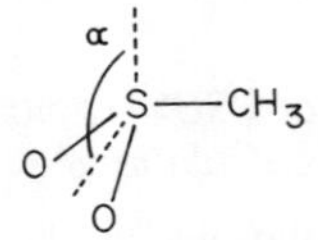

(54)

The individual couplings for the three protons are conformationally dependent. Thus for structure **53**, the hyperfine couplings for *a*($^a$H, $^b$H) are −2.36 G and for *a*($^c$H), 7.04 G[121]. Further, while *a*(H) is *ca* 7 G for a proton *trans*, i.e. $^c$H, to the orbital containing the unpaired spin density (e.g. **53**) for a proton *cis* to this orbital, *a*(H) is approximately

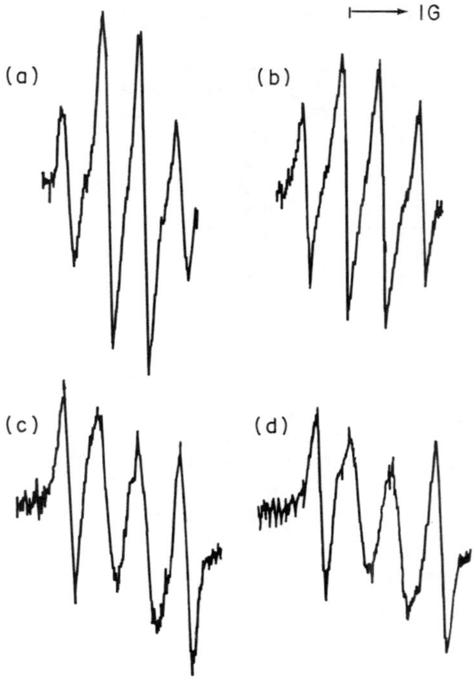


FIGURE 9. ESR spectrum of $MeSO_2^{\bullet}$ in diethyl ether: (a) 200 K, (b) 176 K, (c) 163 K, (d) 153 K (from Reference 121). Reproduced by permission of The Royal Society of Chemistry

0[121]. Thus, in part, the low values of *a*(H) for α-CH protons remarked upon in Section III.B.1. are due to averaging of the conformationally dependent hyperfine couplings. However, the degree of bending at sulphur may also contribute, since it is found that as the angle α becomes larger the α-CH proton splittings diminish[121]. Both effects can explain the observed temperature variation in the magnitude of *a*(α-CH). Interestingly, the hyperfine splittings for $a(^{a}H, ^{b}H, ^{c}H)$ above have been used to simulate the line broadening seen in the spectrum for $MeSO_2^{\bullet}$. The rate constant $k$ at 163 K is $0.8 \times 10^9\,s^{-1}$ and the activation energy for rotation about the C—S bond is calculated to be *ca* 15 kJ mol$^{-1}$ [121]. Despite the uncertainties involved, this correlates nicely with the value of 9.6 kJ mol$^{-1}$ calculated by STO-3G* for the rigid rotor barrier in $MeSO_2^{\bullet}$ [127].

Other alkanesulphonyl radicals behave similarly[101,121]. For example, the ethanesulphonyl radical $EtSO_2^{\bullet}$ exhibits temperature-dependent hyperfine coupling to the α-CH but not the β-CH protons (Table 23)[101]. This has been analyzed by INDO MO calculations, and the average values of *a* (α-CH) 1.6 G and *a*(β-CH) 1.1 G obtained for a structure in which S—O and C—S bond lengths are 140 and 188 pm, respectively, and the angles OSO and α (see **54**) 105° and 130°, respectively[121]. These are reasonably close to the observed values. Indeed, the β-CH hyperfine couplings depend on the conformation about the C—C bond and can achieve values as high as 2.5 G[121]. It is clear, therefore, that a pyramidal structure for the sulphur atom, and restricted rotation about the C—S bond, can satisfactorily account for the low values of *a*(α-CH), the relative order *a*(β-CH) > *a*(α-CH) and the temperature dependence of *a*(α-CH). The low value of *a*(α-H)

TABLE 23. Temperature dependence of $a$(H) for $EtSO_2^{\bullet}$[101]

| $T/K$ | $a(\alpha\text{-CH})/G$ | $a(\beta\text{-CH})/G$ |
|---|---|---|
| 263 | 0.71 | 1.74 |
| 253 | 0.80 | 1.73 |
| 243 | 0.84 | 1.74 |
| 233 | 0.90 | 1.74 |
| 223 | 0.96 | 1.73 |
| 203 | 1.10 | 1.73 |
| 183 | 1.25 | 1.71 |

for $Me_2CHSO_2^{\bullet}$, for example, suggests that the preferred conformation of this radical is **55** rather than **56**.

Arenesulphonyl radicals can, in principle, adopt structures in which the orbital at sulphur containing the unpaired electron is either in the plane of the aryl ring **57** or, as in **58**, co-planar with the $\pi$-system. These are, of course, rotamers of the type discussed above

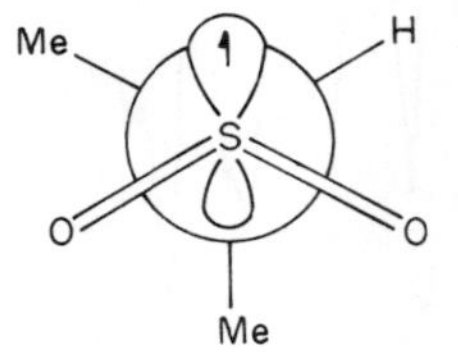


(**55**)

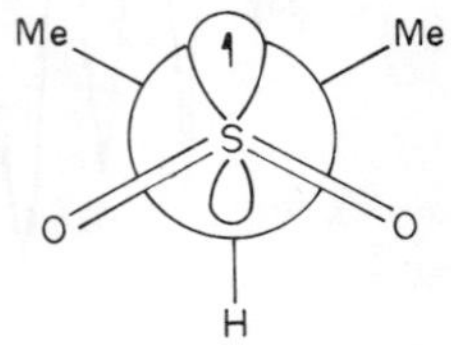


(**56**)

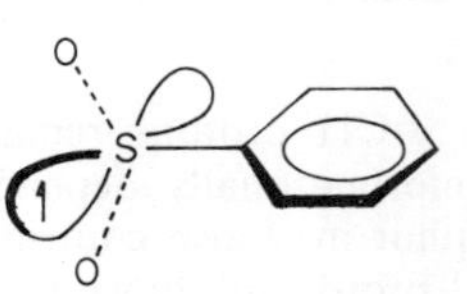


(**57**)

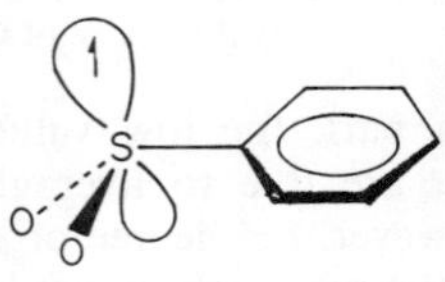


(**58**)

for alkanesulphonyl radicals. Thermochemical arguments have been used to calculate a stabilization energy of $58 \pm 5\,\text{kJ mol}^{-1}$ for $PhSO_2^{\bullet}$[128], thus implying a structure such as **58**. However, for reasons we shall now discuss, the preferred structure is **57**. First, it has already been noted that the *meta*-protons display the largest coupling. This points to a structure in which the unpaired electron resides in an orbital co-planar with the aryl ring. Indeed, for the alternative $\pi$-type structure, the magnitude of the hyperfine splittings would be expected to follow the order $a$(*para*-H) > $a$(*ortho*-H) > $a$(*meta*-H) as has been observed for $PhO^{\bullet}$ and $PhCH_2^{\bullet}$[112]. Second, the $\pi$-type structure **58** implies that the spin density is symmetrical. Thus, the two *ortho*- and *meta*-couplings should always be identical. While this is experimentally observed for many aryl radicals, there are several examples where the two *ortho*- or the two *meta*-couplings are different. Thus, at temperatures >240 K, the 2,3,5,6-tetramethylbenzenesulphonyl and 2,3,4,5,6-pentamethylbenzenesulphonyl radicals exhibit coupling to both *ortho*-Me groups, $a$(6H) 0.6 G, whereas at 193 K coupling to only one *ortho*-Me group, $a$(3H) 1 G, is observed[104]. Further, the 2,4,6-trichlorobenzenesulphonyl radical exhibits coupling to two chlorine atoms and two protons $a(2Cl) = a(2H) = 0.7$ G at 240 K, but at 193 K coupling to only one chlorine atom and one proton $a(Cl) = a(H) = 1.4$ G is detectable with line broadening at intermediate temperatures[104]. Moreover, analysis of the spectra of the 2,4- and 2,5-dibromobenzenesulphonyl radicals reveals that the *meta*-3H proton has a coupling $< 3$ G, whereas the *meta*-5H proton has a coupling of 1.8 G[104]. The inevitable conclusion to be drawn from these observations is that arenesulphonyl radicals are $\sigma$ radicals of structure **57**. The averaged coupling of *ortho*- and *meta*-substituents observed for most arenesulphonyl radicals must therefore be due to rapid rotation around the C—S bond. Spectrum simulation enables the kinetic data for C—S bond rotation that are presented in Table 24 to be obtained[104,121]. Clearly, rotation about an aryl C—S bond is less facile than about an alkyl C—S bond, though the activation energies for both are of similar magnitude to the 9.6 kJ mol$^{-1}$ calculated by *ab initio* methods[127]. It should be noted that the restricted rotation process has been observed only when the aryl ring contains two *ortho*-substituents.

INDO MO calculations for the $\sigma$-type benzenesulphonyl radical **59** identifiy the likely magnitude of the hyperfine proton couplings. Such calculations nicely reproduce the experimental data[104]. Interestingly, it is the *meta*-proton *anti* to the sulphur orbital containing the unpaired spin density that exhibits the largest proton coupling. This appears to be observed in practise for the 2,4-dibromobenzenesulphonyl radical **60** where the only resolvable coupling of 1.8 G is attributed to the *meta*-proton *anti* to the orbital of the unpaired electron. Such a conformation, which is fixed for this radical, is ascribed to unfavourable interactions between the sulphonyl oxygen atoms and the *ortho* bromine atom[104].

The structure of the sulphonyl group used to calculate the range of hyperfine couplings in **59** is similar to that for $MeSO_2^{\bullet}$, i.e. S—O 141 pm, C—S 182 pm, OSO 105–120° and $\alpha$ (the angle between the OSO plane and the normal to the C—S bond) 100–110°. All

TABLE 24. Kinetic data for rotation about the C—S bond in sulphonyl radicals

| Radicals | $T/K$ | $10^6\,k/s^{-1}$ | $E_a$/kJ mol$^{-1}$ | Ref. |
|---|---|---|---|---|
| 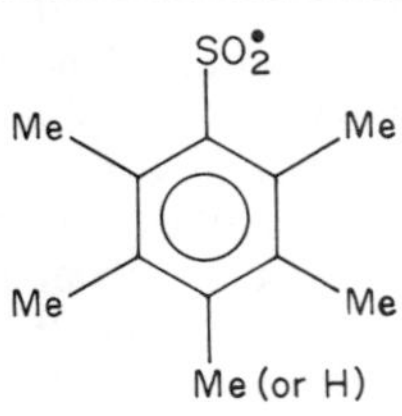 | 193<br>238 | 0.9<br>120 | 22.2 | 104 |
| 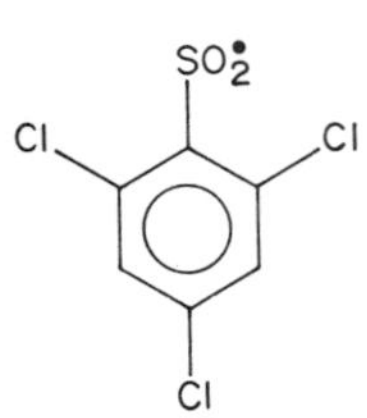 | 223 | 10 | | 104 |
| $MeSO_2^{\bullet}$ | 163 | 800 | 15 | 121 |

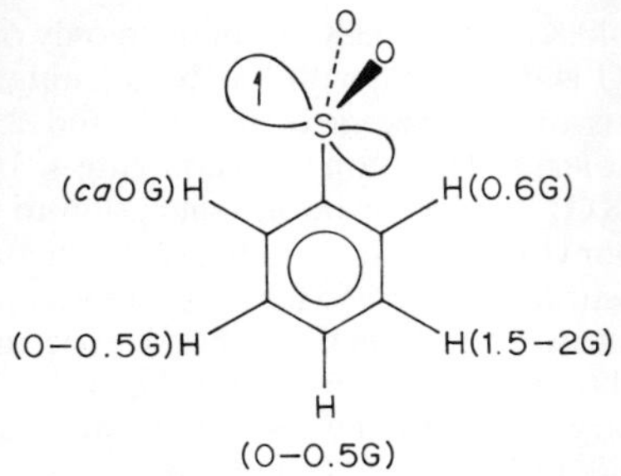


**(59)**

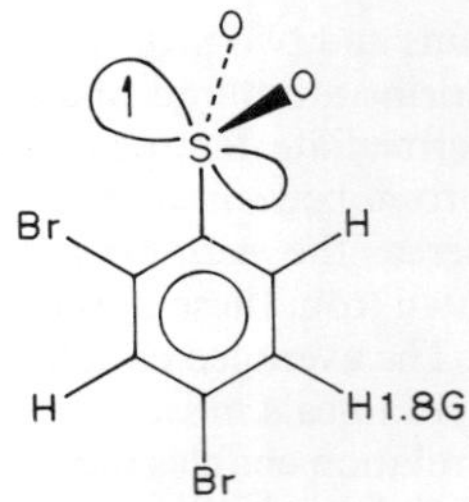


**(60)**

protons are particularly sensitive to the extent of bending of the sulphonyl group as expressed by $\alpha$, and, as for $MeSO_2^{\cdot}$, when the angle $\alpha$ increases the hyperfine couplings diminish[104]. The sulphur 3*s* spin density calculated for the above structure for $PhSO_2^{\cdot}$is *ca* 0.04–0.06[104], consistent with the 0.086 calculated from the $^{33}S$ hyperfine splittings (Table 22).

In view of the above discussion for $PhSO_2^{\cdot}$, it is interesting to note that the thiophene-2-sulphonyl radical **61** is suggested to prefer a more co-planar arrangement of the $\pi$-system and the orbital containing the unpaired electron[123].

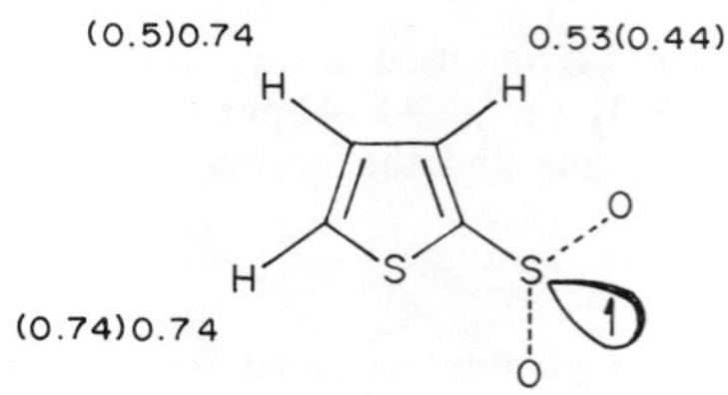


**(61)**

The observed hyperfine couplings *a*(H) (see **61**) are reasonably matched (values in parentheses) when the orbital containing the unpaired electron and the plane of the ring subtend a dihedral angle of 45°. Moreover, for the 3-bromothiophene-2-sulphonyl radical the 5-H coupling is *ca* 0 G, which according to INDO calculations corresponds to a dihedral angle of 90°[123]. However, it is known that the extent of bending of the $SO_2$ group can exert a significant effect on *a*(H) values[104,121]. Since this was not examined for the thiophene radicals, it may be that $SO_2$ bending could account for such an anomaly. Alternatively, the conformation **62** (disregarded because of oxygen–bromine interactions)

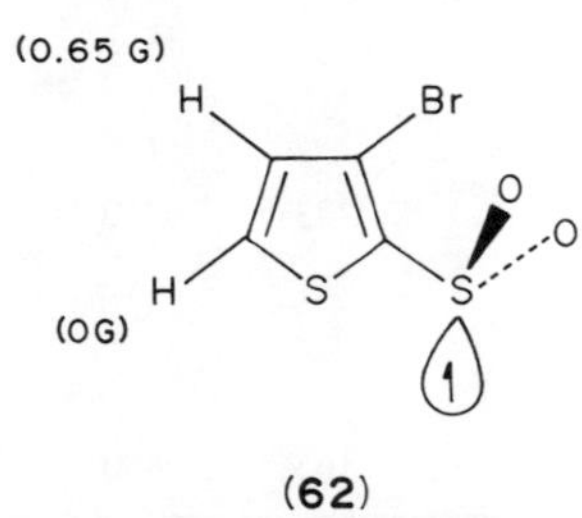


**(62)**

could account for the relative magnitude of the *a*(H) couplings. The H-4 atom is *anti* to the orbital containing the unpaired spin density and is therefore expected to exhibit a large coupling, whereas H-5 is *syn* and expected to have hyperfine coupling close to 0[121]. Indeed, this is verified by INDO calculations[123]. Thus, it remains unclear whether or not the sulphur orbital containing the unpaired spin density thiophenesulphonyl radicals is coplanar with the $\pi$-system.

A similar uncertainty exists for alkenesulphonyl radicals. For the ethenesulphonyl radical, INDO calculations suggest that the in-plane structure **63a** probably reflects the most likely structure, though **63b** is also in reasonable accord[121] [observed values of *a*(H) in G are shown in parentheses].

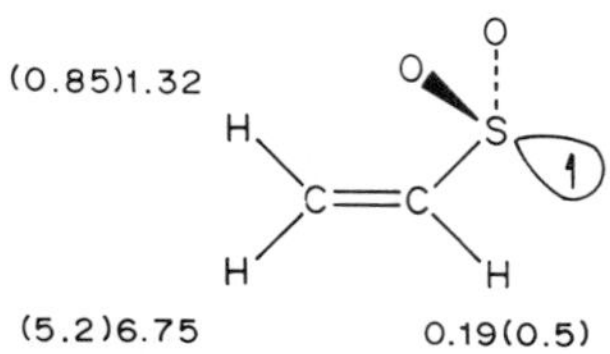


**(63a)**

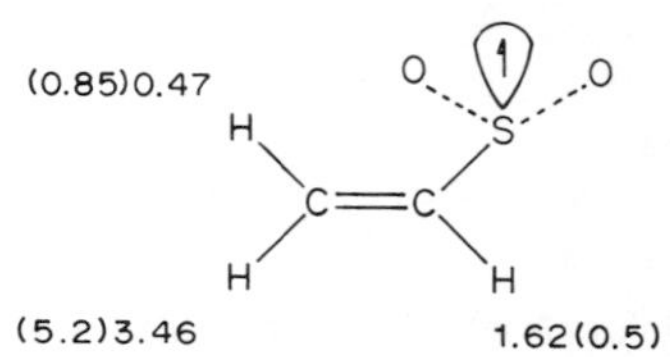


**(63b)**

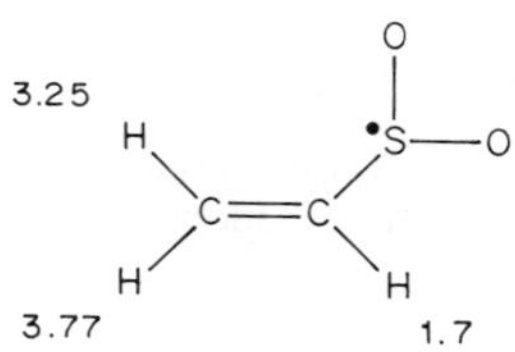


**(63c)**

The structure of the sulphonyl group in **63a** is C—S 182 pm, S—O 141 pm, OŜO 120° and the angle $\alpha$, 130°. It is noteworthy that the allyl-type structure **63c**, in which the sulphonyl group is planar and the electron therefore in a sulphur 3*p* orbital, is unable both to distinguish between the *cis*- and *trans*-2-H protons and reproduce the hyperfine splittings[121]. This is consistent with the observation that the planar configuration of the $SO_2$ group for $MeSO_2^{\bullet}$ is 100 kJ mol$^{-1}$ less stable than the pyramidal configuration[127]. For the 1-Me and 2*E*-2-Me substituted ethenesulphonyl radicals, the structures **64** and **65** appear to be the most likely on the basis of INDO calculations[121] [observed values of

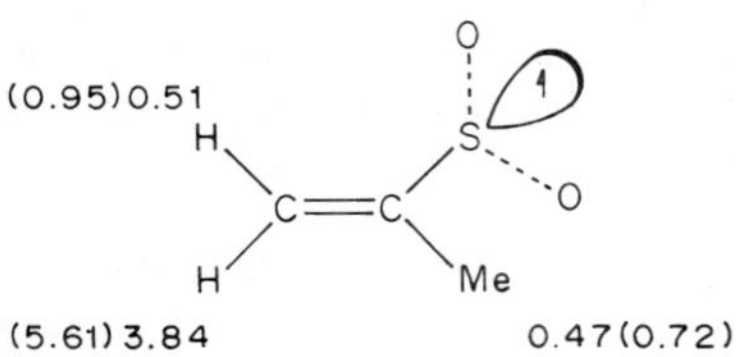


**(64)**

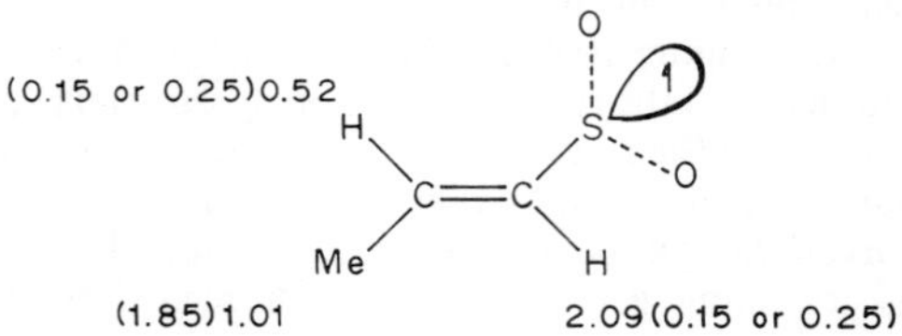


**(65)**

*a*-(H) in parentheses]. It is unclear why alkenesulphonyl radicals prefer to adopt such conformations whereas arenesulphonyl radicals prefer an in-plane $\sigma$-type structure.

Aminosulphonyl radicals, $R_2NSO_2^{\bullet}$, in which a heteroatom is directly bonded to the sulphonyl sulphur, are thought to adopt a conformation, e.g. **66**, in which the orbital containing the nitrogen lone pair of electrons is co-planar with the sulphur orbital containing the unpaired spin density[125]. Conformation **66**, but not the twisted conformation **67**, can explain the observation that *a*(H) for a *N*-Me group is of similar magnitude to

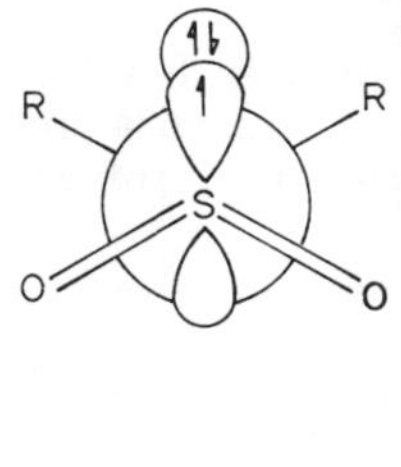


**(66)**

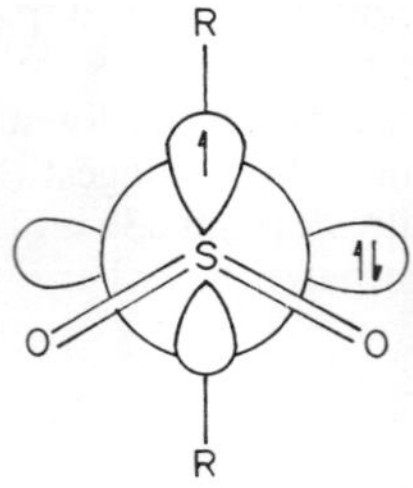


**(67)**

*a*(H) for *N*-H[125]; for **67**, *a*(Me) would be expected to be much smaller than the corresponding *a*(H). Structure **66** is that predicted by *ab initio* calculations, from which a barrier to S—N bond rotation of 19.2 kJ mol$^{-1}$ is found[127].

## C. The Sulphinylaminyl Radical, $R^1SONR^{2\cdot}$

In contrast to the extensive investigations of the sulphonyl radical, the sulphinylaminyl system remains little studied. The two reports that have been made are, however, complementary[102,108].

### 1. g-Values and hyperfine coupling constants

Sulphinylaminyl radicals have *g*-values in the range 2.0035–2.0044 (Table 25). This is somewhat lower than the corresponding $RSO_2^{\cdot}$ radicals'(*g*-values *ca* 2.005) and similar to the simple aminyls[129]. The spectrum of $MeSO\dot{N}Bu^t$ is shown in Figure 10a[108]. The notable feature of this spectrum is the twelve-line 1:1:1 triplet of quartets. This is easily interpreted in terms of a larger coupling to the nitrogen nucleus and a smaller coupling due to the S—$CH_3$ group. Coupling to the $Bu^t$ protons is clearly too small to be observed. The spectrum of $MeSO\dot{N}C_6H_3$-3, 5-$Bu^t$ (Figure 10b) is much more complicated but can be interpreted as coupling to nitrogen, the $CH_3$ protons and the two *ortho* and one *para* protons in the *N*-aryl ring (Table 25). The quartet arising from the $SCH_3$ group is quite clearly discernible in the wing lines. Interestingly, the spectrum of 4-$MeC_6H_4SO\dot{N}C_6H_3$-3, 5-$Bu^t_2$ (Figure 10c) is a much simpler eighteen-line spectrum arising from coupling to nitrogen, and the two *ortho* and one *para* *N*-aryl protons. The most striking feature of the spectrum is the lack of any coupling to the protons in the *S*-aryl ring. Table 25 indicates that this appears to be a general phenomenon.

Even though the data set is rather limited, further inspection of Table 25 enables us to discern some general trends with regard to both *g*-values and hyperfine coupling constants:

(i) *N*-alkyl groups result in *g*-values 2.0041–2.0046 regardless of the substituent at sulphur (alkyl, aryl, alkoxy).

(ii) *N*-aryl groups result in somewhat lower *g*-values, 2.0034–2.0035.

(iii) The nitrogen hyperfine coupling constant *a*(N) is larger for *N*-alkyl substituents, 8.5–10.3 G, than for *N*-aryl substituents, 8.13–8.18 G.

TABLE 25. ESR parameters for various sulphinylaminyl radicals, $R^1SO\dot{N}R^2$

| Radical | g-Value | Hyperfine coupling/G | | Ref. |
|---|---|---|---|---|
| | | a(N) | a(H) | |
| $MeSO\dot{N}Et$ | 2.0043[a] | 9.3 | 22.2(1H), 18.7(1H), 1.3(3H) | 108 |
| $MeSO\dot{N}Pr^i$ | 2.0041[a] | 9.0 | 9.4(1H) | 108 |
| $MeSO\dot{N}Bu^t$ | 2.0044[a] | 8.4 | 1.1(3H) | 108 |
| | 2.0042[b] | 8.52 | 1.0(3H) | 102 |
| 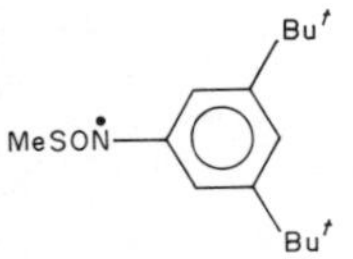 | 2.0034[b] | 8.16 | 4.89(2H, *ortho*), 6.38(1H, *para*) 0.85(3H) | 102 |
| $Bu^tSO\dot{N}Et$ | 2.0042[a] | 9.3 | 18.5(1H), 17.1(1H), 1.1(9H) | 108 |
| $Bu^tSO\dot{N}Pr^i$ | 2.0041[a] | 9.0 | 9.0(1H) | 108 |
| $Bu^tSO\dot{N}Bu^t$ | 2.0042[a] | 8.7 | 1.0(9H) | 108 |
| $4\text{-}Bu^tC_6H_4SO\dot{N}Bu^t$ | 2.0041[b] | 8.58 | | 102 |
| 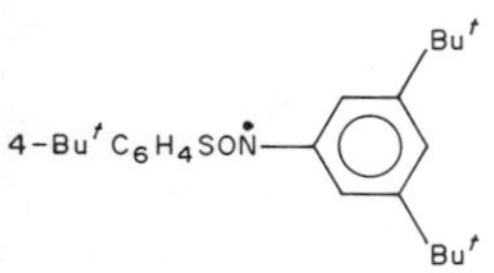 | 2.0035[c] | 8.18 | 4.87(2H, *ortho*), 6.44(1H, *para*) | 102 |
| 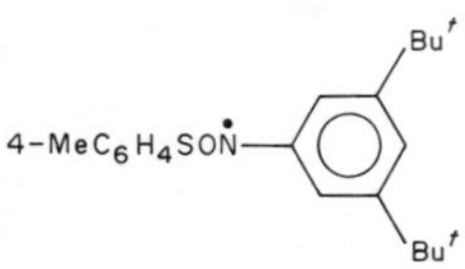 | 2.0035[c] | 8.18 | 4.86(2H, *ortho*), 6.44(1H, *para*) | 102 |
| 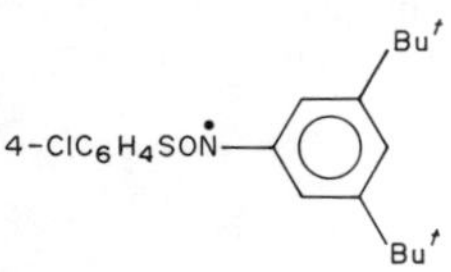 | 2.0034[c] | 8.13 | 4.82(2H, *ortho*), 6.43(1H, *para*) | 102 |
| $Bu^tOSO\dot{N}Et$ | 2.0041[a] | 10.3 | 29.2(1H), 25.2(1H) | 108 |
| $Bu^tOSO\dot{N}Pr^i$ | 2.0041[a] | 9.6 | 10.0(1H) | 108 |
| $Bu^tOSO\dot{N}Bu^t$ | 2.0042[a] | 8.7 | | 108 |
| $Me_3SiOSO\dot{N}Et$ | 2.0041[d] | 10.0 | 31.1(1H), 26.0(1H) | 108 |

[a] In cyclopropane at −73 °C.
[b] In benzene at 40 °C.
[c] In benzene at 21 °C.
[d] In cyclopropane at −111 °C.

(iv) *a*(N) decreases across the series *N*—Et > *N*—$Pr^i$ > *N*—$Bu^t$.
(v) *a*(N) appears to be larger for *S*-alkoxy substituents than for *S*-alkyl substituents.
(vi) The proton hyperfine splitting, *a*(H), of the *N*—CH protons of an *N*-alkyl group is large, whereas *a*(H) for *N*—C—C*H* is difficult to observe and/or assign.
(vii) *a*(H) for both the *ortho*- and *para*-protons is significant.

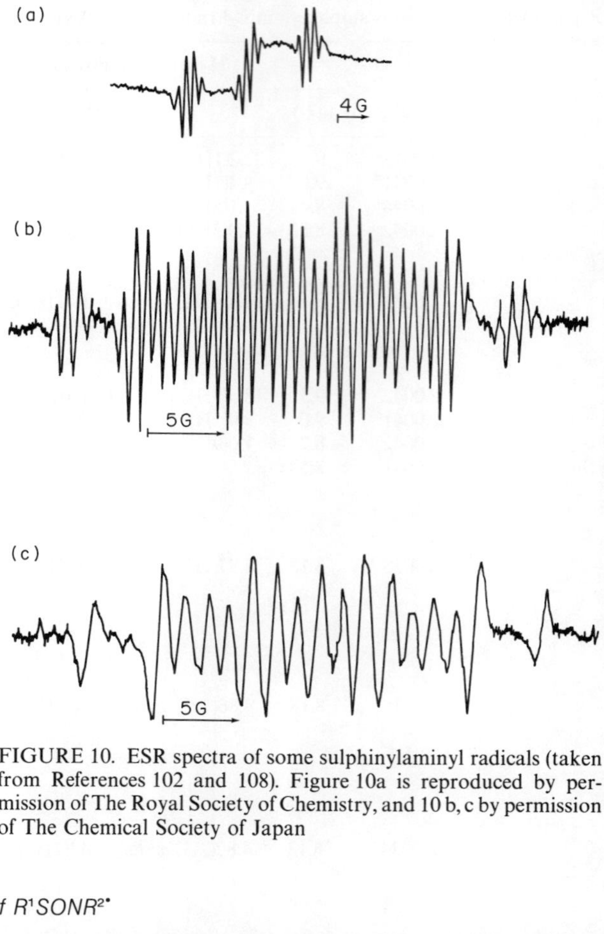


FIGURE 10. ESR spectra of some sulphinylaminyl radicals (taken from References 102 and 108). Figure 10a is reproduced by permission of The Royal Society of Chemistry, and 10 b, c by permission of The Chemical Society of Japan

### 2. *Structure of* $R^1SONR^{2\bullet}$

Structures **68a–c** may all make a contribution to the structure of sulphinylaminyl radicals. However, the *g*-value of sulphinylaminyls as compared to sulphonyl (*g ca* 2.005) and sulphinyl (*g ca* 2.009)[130] radicals on the one hand, and simple aminyls[129] on the other, together with the lack of any hyperfine splitting due to the protons of *S*-aryl substituents (in contrast to arylsulphonyl radicals)[101,102,121], all point to the relative unimportance of structure **68c**. This is a clear difference between $R^1SO\dot{N}R^2$ and $RSO_2^{\bullet}$ and presumably

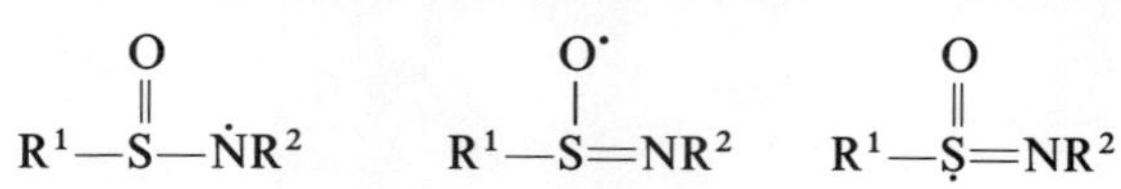

**(68a)** **(68b)** **(68c)**

results from the greater electronegativity of oxygen compared with that of nitrogen. Moreover, for the *N*-Et and *N*-Pr$^i$ derivatives, the high values of the hyperfine coupling constants for the α-C*H* protons are of similar magnitude to those for $Et_2N^{\bullet}$ and $Pr^i_2N^{\bullet}$ [129], from which it appears that *ca* 70% of the unpaired electron density is localized on the nitrogen atom. Again, this is consistent with the greater electronegativity of oxygen. Thus

**68a** makes the major contribution to the overall structure of the sulphinylaminyl radical.

Simple dialkylaminyls are π-radicals. The similar ESR parameters (*g*-values, *a*(N) and *a*(H) for the *N*—*CH* protons) of $R^1SO\dot{N}R^2$ to $R_2N^{\bullet}$ imply the sulphinylaminyls are also π-radicals. This is further supported by the significant coupling observed to the ring protons of the *N*-aryl group. Such coupling implies a transfer spin density from the nitrogen atom to the *N*-aryl ring, which can only occur via overlap of the nitrogen *p*-orbital containing the unpaired electron with the aromatic π-system. This would accord with both the lower *g*-values and the smaller *a*(N) constants observed for the *N*-arylsulphinylaminyls compared with the corresponding *N*-alkylsulphinylaminyls.

One feature of the ESR spectra of $R^1SO\dot{N}R^2$ that we have yet to remark upon is the non-equivalence of the hyperfine coupling constants of the two *N*—$CH_2$ protons. Table 25 shows that, for MeSO$\dot{N}$Et, Bu$^t$SO$\dot{N}$Et, Bu$^t$OSO$\dot{N}$Et and $Me_3$SiOSO$\dot{N}$Et, *a*(H) for the $CH_2$ protons is clearly different. Significantly, the magnitude of the hyperfine coupling is temperature dependent (Table 26)[108] and, more importantly, the difference, Δ, between

TABLE 26. Temperature dependence of *a*(H) for the $NCH_2$ protons in $RSO\dot{N}CH_2CH_3$ radicals

| Radical | *T*/K | *a*(H)/G | | Δ/G |
|---|---|---|---|---|
| Bu$^t$SO$\dot{N}$Et | 200 | 18.5 | 17.1 | 1.4 |
| | 188 | 18.9 | 17.3 | 1.6 |
| | 173 | 19.5 | 17.6 | 1.9 |
| | 163 | 20.0 | 18.0 | 2.0 |
| Bu$^t$OSO$\dot{N}$Et | 268 | 26.9 | 24.8 | 2.1 |
| | 245 | 27.6 | 24.9 | 2.7 |
| | 223 | 28.2 | 25.2 | 3.0 |
| | 200 | 29.2 | 25.2 | 4.0 |
| | 186 | 30.4 | 26.2 | 4.2 |
| | 168 | 32.0 | 26.8 | 5.2 |
| | 151 | 34.2 | 27.8 | 6.4 |

the two *a*(H) values decreases at higher temperatures. The most likely explanation for this phenomenon is restricted rotation about the C—N bond. The non-equivalence of the two C—H protons arises not from a higher energy barrier to S—N rotation as the authors suggest[108], but from the chiral sulphur atom.

Structure **69** reveals that radicals such as Bu$^t$SO$\dot{N}$Et, etc., are enantiomeric. Thus, $H_a$ and $H_b$ are diastereotopomeric and magnetically non-equivalent. One would therefore

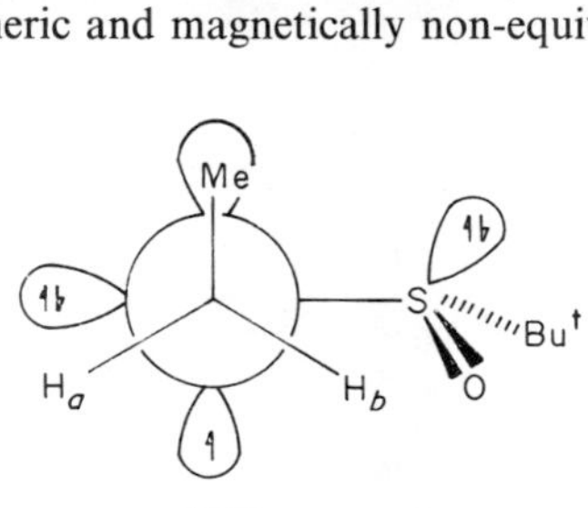


(**69**)

expect their hyperfine couplings to be of a different magnitude. The size of such coupling is presumably determined by the dihedral angle between the C—H bond and the singly occupied *p*-orbital on nitrogen. It would appear from Table 26, however, that $H_a$ and $H_b$

have not reached their average hyperfine coupling values since Δ is clearly diminishing at the highest temperature of experimental observation. Both $H_a$ and $H_b$ must reach their average values when rotation about the C—N bond is rapid on the ESR time scale, and this will result in Δ reaching a limiting value. It would, however, appear that such rotation has a remarkably high barrier, since a limiting value of Δ has not been reached even at −5 °C.

### D. α-Sulphinyl Radicals, RĊHSOX

Carbon-centred α-sulphinyl radicals of the type $R^1\dot{C}HSOR^2$ have been well characterized for sulphoxides ($R^1, R^2$ = alkyl or aryl). Such radicals are generated directly from the parent sulphoxides by hydrogn atom abstraction from an α-CH group by the phenyl radical (equation 19)[131].

$$Ph^{\cdot} + R^1CH_2SOR^2 \longrightarrow PhH + R^1\dot{C}HSOR^2 \qquad (19)$$

The phenyl radical was chosen because it has a greater tendency to abstract an α-CH hydrogen atom than to react at sulphur. This contrasts with the hydroxyl radical[103], which only reacts with sulphoxides at sulphur to generate, after the loss of an alkyl radical, a sulphonyl radical (*cf.* equation 12) and also with simple alkyl radicals which do not react with sulphoxides. The *tert*-butoxyl radical does, however, abstract an α-hydrogen atom from dialkyl sulphoxides to generate α-sulphinyl radicals[119]. An alternative route to the formation of α-sulphinyl radicals is via halogen atom abstraction using the $HPO_2^{\cdot-}$ radical anion (equation 20). However, clever use of the propensity of $OH^{\cdot}$ to react at sulphur has been made to generate α-sulphinyl radicals from cyclic 1,3-bis-sulphoxides (equation 21)[131].

$$RSOCH_2Br \xrightarrow{HPO_2^{\cdot-}} RSOCH_2^{\cdot} \qquad (20)$$

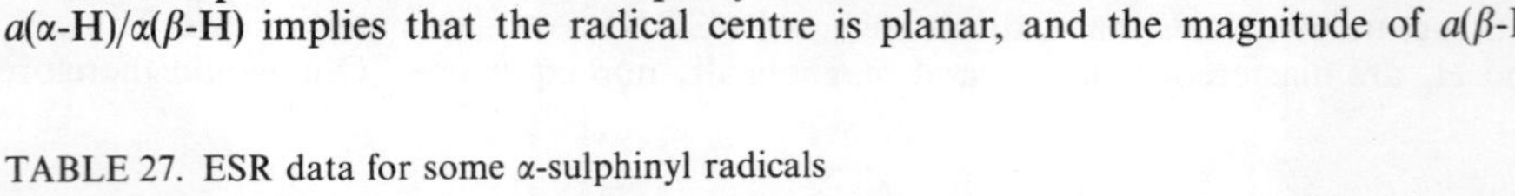

(21)

The ESR parameters of some α-sulphinyl radicals are contained in Table 27. The ratio $a(\alpha\text{-H})/a(\beta\text{-H})$ implies that the radical centre is planar, and the magnitude of $a(\beta\text{-H})$

TABLE 27. ESR data for some α-sulphinyl radicals

| Radical | *g*-Value | Hyperfine couplings/G | Ref. |
|---|---|---|---|
| $MeSOCH_2^{\cdot}$ | 2.0025 | 20.0(2H) | 119, 131 |
| $EtSOCHMe^{\cdot}$ | 2.0025 | 20.2(2H), 25.3(3H) | 119, 131 |
| $Pr^iSOCMe_2^{\cdot}$ | 2.0026 | 22.4(6H) | 119 |
| $Pr^nSOCMe_2^{\cdot}$ | 2.0026 | 22.5(6H) | 119 |
| $EtSOCMe_2^{\cdot}$ | 2.0026 | 22.5(6H) | 119 |
| $HO_2S(CH_2)_3SOCH_2^{\cdot}$ | 2.0025 | 20.0(2H) | 131 |
| $HO_2S(CH_2)_2SOCH_2^{\cdot}$ | 2.0025 | 20.0(2H) | 131 |
| $HO_2S(CH_2)_2SOCHMe^{\cdot}$ | 2.0025 | 20.1(1H), 25.2(3H) | 131 |
| $HO_2S(CH_2)_2SOCMe_2^{\cdot}$ | 2.0025 | 23.3(6H) | 131 |

suggests that the sulphinyl group removes only *ca* 6% of the unpaired spin density from the carbon-centred radical[131]. This contrasts with an adjacent carbonyl group, $\dot{C}H_2COR$, which is able to remove *ca* 16% and a thioether, $\dot{C}HMeSEt$, which removes *ca* 22% of the spin density[131]. Canonical forms such as **70b, c** thus contribute little to the structure of such radicals.

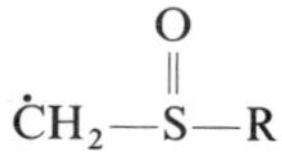

**(70a)**

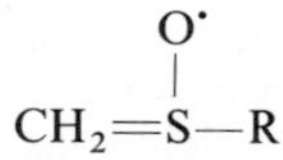

**(70b)**

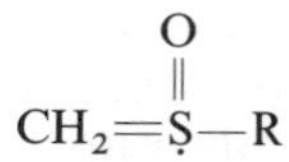

**(70c)**

α-Sulphinyl radicals of sulphinic acid derivatives, i.e. $R^1\dot{C}HSOX$ (X = O, N etc.), have yet to be reported, even under conditions [e.g. $Bu^tO^{\bullet}$ with $MeSO_2Me$, $MeSO_2Bu$ and $(CH_2)_2SOOCH_2$] where analogous radicals from dialkyl sulphoxides are observed[19]. The only radicals detected in these reactions are those from dealkylation and dealkoxylation processes (equation 22). Both of these involve $S_H2$ attack of the alkoxyl radical at the

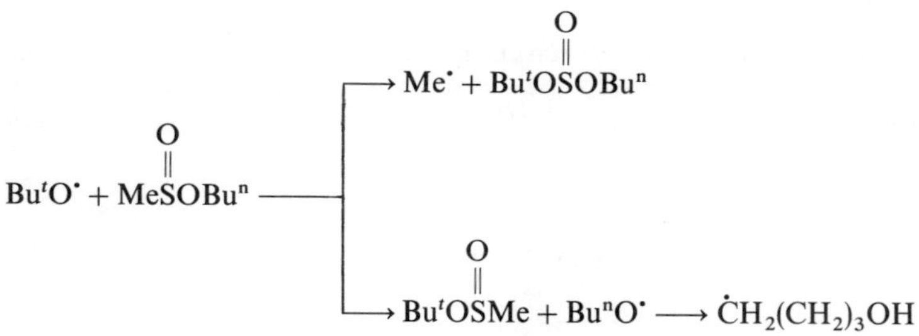

(22)

sulphur centre, and it appears that the replacement of Me (in DMSO) by RO (in the sulphinate esters) preferentially increases the rate of the $S_H2$ reaction as compared to α-hydrogen atom abstraction. The sulphuranyloxyl radical **71**, which is presumably an intermediate in these processes, has not been detected for alkyl alkanesulphinates $(R^1SO_2R^2)$[119]. However, the cyclic sulphite, $CH_2OSOOCH_2$, yields a spectrum on reaction with $Bu^tO^{\bullet}$ [*g* 2.0044; *a*(H) 2.38 G(1H), 0.37 G(2H) at 163 K] which has been assigned to the sulphuranyloxyl radical **72**[132,133]. This observation is likely to stem from

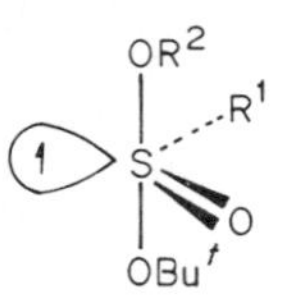

(71)

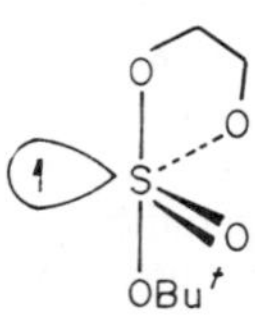

(72)

the known propensity for dealkylation versus dealkoxylation seen in cyclic sulphinic acid esters[119]. Similar sulphuranyloxyl radicals derived from acyclic sulphites have not been confirmed, but a signal at *g ca* 2.0053 could possibly be due to such a species. However, the lack of hyperfine coupling makes such an assignment tentative[133].

A further complication in the attempt to observe α-sulphinyl radicals results from abstraction of hydrogen atom in the alkoxy group (equation 23) to form radicals of structure **73**. These can fragment to form the $R^1SO^{\bullet}$ radical[133], but are observable if $R^2$ can provide stabilization, e.g. $R^2 = —CH{=}CH_2$[110,133].

$$R^1\overset{\overset{\displaystyle O}{\|}}{S}OCH_2R^2 + RO^{\bullet} \longrightarrow R^1\overset{\overset{\displaystyle O}{\|}}{S}O\dot{C}HR^2 + ROH \qquad (23)$$

**(73)**

Thus, it would appear that the most likely route to successfully observing α-sulphinyl radicals may lie with the hydrogen abstraction of an α-H from a methanesulphinate with the phenyl radical. Alternatively, the reaction between an alkoxy radical and a sulphine (equation 24), which is analogous to that of an alkoxy radical with a sulphinylamine[108], may also provide a useful approach. Both methods are, as yet, untried.

$$RCH{=}S{=}O + Bu^tO^{\bullet} \longrightarrow R\dot{C}H{-}\overset{\overset{\displaystyle O}{\|}}{S}{-}OBu^t \qquad (24)$$

### E. Spin Trapping of $RSO_2^{\bullet}$

Despite the fact that $RSO_2^{\bullet}$ radicals have been extensively studied by direct observation, there have been several reports where such radicals have been trapped by reaction with $Bu^tNO$[134–137], nitrosodurene[140], nitrones[134–136] and thioketones[120] (equations 25, 26 and 27).

$$RSO_2^{\bullet} + Bu^tN{=}O \longrightarrow RSO_2\overset{\overset{\displaystyle O^{\bullet}}{|}}{N}Bu^t \qquad (25)$$

$$RSO_2^{\bullet} + PhCH{=}\overset{+}{N}(Bu^t)O^- \longrightarrow Ph(RSO_2)CH\overset{\overset{\displaystyle O^{\bullet}}{|}}{N}Bu^t \qquad (26)$$

$$RSO_2^{\bullet} + Ph_2C{=}S \longrightarrow Ph_2\dot{C}SSO_2R \qquad (27)$$

The sulphonamide-based nitroxyl radicals formed in reaction 25 appear as 1:1:1 triplets due to coupling to the nitrogen atom, with *a*(N) *ca* 12.5 G (Table 28). Interestingly, the nitroso spin traps, $Bu^tNO$ and 2,3,5,6-$Me_4C_6HNO$, appear to be the most effective; it has been reported that $PhCH{=}\overset{+}{N}(Bu^t)O^-$ fails to trap $MeSO_2^{\bullet}$[44] and various $ArSO_2^{\bullet}$ radicals[136] under conditions where $Bu^tNO$ does. Moreover, $Bu^tNO$ is capable of trapping $PhCH_2SO_2^{\bullet}$, a radical that has been observed only in the solid state[110,114]. It has been proposed, on the basis of smaller *g*-values and larger *a*(N) values for the sulphonylnitroxyl radicals as compared with acylnitroxyl radicals, that the nitrogen atom is pyramidal and that the orbital containing the spin density has greater *s* character than dialkylnitroxides[139].

The nitroxyl radicals formed in reaction 26 also exhibit hyperfine coupling to nitrogen, but couple further to the β-CH protons. The low value of the hyperfine coupling to these protons is interpreted in terms of an eclipsing of the C—H bond and the π-orbital containing the unpaired electron, **74**.

The radicals formed in reaction 27 have *g*-values *ca* 2.0025–2.0028, typical of the unpaired electron residing on a carbon atom. The spin density is, however, delocalized over the C-aryl group, as demonstrated by hyperfine coupling to the *ortho*-, *meta*- and *para*-protons.

Interestingly, whereas both diphenyl thioketone and phenyl(triphenylsilyl) thioketone are able to spin trap sulphonyl radicals $RSO_2^{\bullet}$ (R = Me and 4-$MeC_6H_4$), di-*tert*-butyl

TABLE 28. ESR parameters for spin-trapped sulphonyl radicals

| Sulphonyl radical | Spin trap | Radical detected | $g$-Value | Hyperfine coupling $a$(N) | Hyperfine coupling $a$(H) | Ref. |
|---|---|---|---|---|---|---|
| $MeSO_2^{\bullet}$ | $Bu^tNO$ | $MeSO_2N(Bu^t)O^{\bullet}$ | 2.0060 | 12.5 | | 138 |
| | $Ph_2C{=}S$ | $MeSO_2S\dot{C}Ph_2$ | 2.0025 | | 2.92 (*ortho* H)<br>1.21 (*meta* H)<br>3.28 (*para* H) | 120 |
| | $Ph(Ph_3Si)C{=}S$ | $MeSO_2S\dot{C}Ph(Ph_3Si)$ | 2.0027 | | 4.02 (*ortho* H)<br>1.41 (*meta* H)<br>4.74 (*para* H)<br>0.21 (3H) | 120 |
| $EtSO_2^{\bullet}$ | $Bu^tNO$ | $EtSO_2N(Bu^t)O^{\bullet}$ | | 12.2 | | 134, 135 |
| $PhCH_2SO_2^{\bullet}$ | $Bu^tNO$ | $PhCH_2SO_2N(Bu^t)O^{\bullet}$ | 2.0060 | 12.5 | | 136 |
| $PhSO_2^{\bullet}$ | $Bu^tNO$ | $PhSO_2N(Bu^t)O^{\bullet}$ | 2.0061 | 12.2 | | 134–136, 139 |
| | $PhCH{=}\overset{+}{N}(Bu^t)O^-$ | $PhSO_2CHPhN(Bu^t)O^{\bullet}$ | 2.0062 | 13.5 | 1.5 (1H) | 136 |
| | $AcOCH_2Me_2CNO$ | $PhSO_2N(CMe_2CH_2OAc)O^{\bullet}$ | | 11.9 | | 139 |
| | $Pr^iNO$ | $PhSO_2N(Pr^i)O^{\bullet}$ | | 11.5 | 2.3 (1H) | 139 |
| | $c\text{-}C_5H_9NO$ | $PhSO_2N(c\text{-}C_5H_9)O^{\bullet}$ | | 11.7 | 2.6 (1H) | 139 |
| | $c\text{-}C_6H_{11}NO$ | $PhSO_2N(c\text{-}C_6H_{11})O^{\bullet}$ | | 11.4 | 2.5(1H) | 139 |
| | PhMeCHNO | $PhSO_2NCHPhMeO^{\bullet}$ | | 11.6 | 3.4 (1H) | 139 |
| | $Ph_2CHNO$ | $PhSO_2NCHPh_2O^{\bullet}$ | | 11.2 | 2.0 (1H) | 139 |
| | $PhCH_2NO$ | $PhSO_2NCH_2PhO^{\bullet}$ | | 11.4 | 6.3 (2H) | 139 |
| | PhNO | $PhSO_2NPhO^{\bullet}$ | | 11.6 | 1.7 (*o*/*p*, Ph)<br>0.8 (*m*, *Ph*) | 139 |
| $4\text{-}MeC_6H_4SO_4^{\bullet}$ | $Bu^tNO$ | $ArSO_2N(Bu^t)O^{\bullet}$ | | 12.2 | | 136 |
| | $2,3,5,6\text{-}Me_4C_6HNO$ | $ArSO_2N(Me_4C_6H)O^{\bullet}$ | 2.0061 | 11.3 | | 140 |
| | $PhCH{=}\overset{+}{N}(Bu^t)O^-$ | $ArSO_2CHPhN(Bu^t)O^{\bullet}$ | 2.0062 | 13.6 | 1.5 (1H) | 136 |
| | $CH_2{=}\overset{+}{N}(Bu^t)O^-$ | $ArSO_2CH_2N(Bu^t)O^{\bullet}$ | 2.0062 | 12.7 | 6.25 (2H) | 136 |
| | $Ph_2C{=}S$ | $ArSO_2S\dot{C}Ph_2$ | 2.0025 | | 2.90 (*ortho* H)<br>1.23 (*meta* H)<br>3.31 (*para* H) | 120 |

(*continued*)

Table 28. (*Continued*)

| Sulphonyl radical | Spin trap | Radical detected | *g*-Value | Hyperfine coupling *a*(N) | Hyperfine coupling *a*(H) | Ref. |
|---|---|---|---|---|---|---|
| | $Ph(Ph_sSi)C{=}S$ | $ArSO_2S\dot{C}Ph(Ph_3Si)$ | 2.0028 | | 3.81 (*ortho* H)<br>1.39 (*meta* H)<br>4.48 (*para* H) | 120 |
| $4\text{-}ClC_6H_4SO_2^{\cdot}$ | $Bu^tNO$ | $ArSO_2N(Bu^t)O^{\cdot}$ | 2.0061 | 12.2 | | 136 |
| | $PhCH{=}\overset{+}{N}(Bu)O^-$ | $ArSO_2CHPhN(Bu^t)O^{\cdot}$ | 2.0061 | 13.2 | 1.5 (1H) | 136 |
| $4\text{-}BrC_6H_4SO_2^{\cdot}$ | $Bu^tNO$ | $ArSO_2N(Bu^t)O^{\cdot}$ | 2.0061 | 12.3 | | 136 |
| $4.MeOC_6H_4SO_2^{\cdot}$ | $Bu^tNO$ | $ArSO_2N(Bu^t)O^{\cdot}$ | 2.0061 | 12.5 | | 136 |
| | PhNO | $ArSO_2N(Ph)O^{\cdot}$ | | 11.8 | 1.7 (*o*, *p* Ph)<br>0.8 (*m*, Ph) | 139 |
| $4\text{-}NO_2C_6H_4SO_2^{\cdot}$ | $Bu^tNO$ | $ArSO_2N(Bu^t)O^{\cdot}$ | 2.0059 | 12.0 | | 136 |
| | PhNO | $ArSO_2N(Ph)O^{\cdot}$ | | 11.4 | 1.7 (*o*, *p*, Ph)<br>0.8(*m*, Ph) | 139 |
| $3,5\text{-}Bu^t_2\text{-}4\text{-}HOC_6H_2SO_2^{\cdot}$ | $Bu^tNO$ | $ArSO_2N(Bu^t)O^{\cdot}$ | 2.0060 | 12.5 | | 136 |
| | $PhCH{=}\overset{+}{N}(Bu^t)O^-$ | $ArSO_2CHPhN(Bu^t)O^{\cdot}$ | 2.0063 | 13.35 | 1.3(1H) | 136 |
| | $CH_2{=}\overset{+}{N}(Bu^t)O^-$ | $ArSO_2CH_2N(Bu^t)O^{\cdot}$ | 2.0064 | 12.7 | 6.25 (2H) | 136 |
| $FSO_2^{\cdot}$ | $Bu^tNO$ | $FSO_2N(Bu^t)O^{\cdot}$ | | 11.7 | | 134, 135 |
| $ClSO_2^{\cdot}$ | $Bu^tNO$ | $ClSO_2N(Bu^t)O^{\cdot}$ | | 11.7 | | 134, 135 |
| $BrSO_2^{\cdot}$ | $Bu^tNO$ | $BrSO_2N(Bu^t)O^{\cdot}$ | | 11.9 | | 134, 135 |

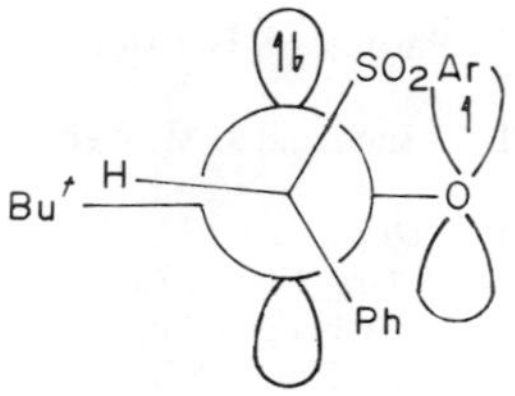


**(74)**

thioketone is not[120]. The authors forward no explanation for this observation, but it must relate to the stabilization afforded to the radical by the aryl group(s) attached to the thiocarbonyl carbon. This is reflected in the observed hyperfine coupling to the aryl protons.

## IV. REFERENCES


1. W. A. Sheppard and R. W. Taft, *J. Am. Chem. Soc.*, **94**, 1919 (1972).
2. B. J. Lindbergh, *Arkiv Kemi*, **32**, 317 (1970).
3. O. Exner, in *Correlation Analysis in Chemistry* (Eds. N. B. Chapman and J. Shorter), Chap. 10, Plenum Press, New York, 1978.
4. G. P. Schiemenz, *Angew. Chem.*, **80**, 559 (1968).
5. R. T. C. Brownlee, R. E. J. Hutchinson, A. R. Katritzky, T. T. Tidwell and R. D. Topsom, *J. Am. Chem. Soc.*, **90**, 1757 (1968).
6. M. Sjöström and S. Wold, *Chem. Scr.*, **9**, 200 (1976).
7. J. G. Pritchard and P. C. Lauterbur, *J. Am. Chem. Soc.*, **83**, 2105 (1961).
8. P. Albriktsen, *Acta. Chem. Scand.*, **29**, 824 (1975).
9. S. W. Bass and S. A. Evans Jr., *J. Org. Chem.*, **45**, 710 (1980).
10. A. J. Hoefnagel and B. M. Wepster, *J. Am. Chem. Soc.*, **95**, 5350 (1973).
11. a. F. Wudl, D. A. Lightner and D. J. Cram, *J. Am. Chem. Soc.*, **89**, 4099 (1967). b. C. Pouchert, *The Aldrich Library of NMR Spectra*, Vol. 2, Aldrich Chemical Co. Inc. Milwaukee, 1983.
12. S. Sato, C. Nagata and S. Tanaka, *Bunseki Kagaku*, **34**, 392 (1985).
13. F. Freeman and C. N. Angeletakis, *Org. Magn. Reson.*, **21**, 86 (1983).
14. J. B. Stothers, *Carbon-13 NMR* Spectroscopy, Academic Press, New York, 1972.
15. D. M. Grant and E. G. Paul, *J. Am. Chem. Soc.*, **86**, 2984 (1964).
16. G. Barbarella, P. Dembeck, A. Garbesi and A. Fava, *Org. Magn. Reson.*, **8**, 108 (1976).
17. J. D. Roberts, F. J. Weigert, J. I. Kroschwitz and H. J. Reich, *J. Am. Chem. Soc.*, **92**, 1338 (1970).
18. S. S. McCrachren and S. A. Evans Jr., *J. Org. Chem.*, **44**, 3551 (1979).
19. J. C. Batchelor, *J. Am. Chem. Soc.*, **97**, 3410 (1975).
20. E. L. Eliel, W. F. Bailey, L. D. Kopp, R. L. Willer, D. M. Grant, R. Bertrand, K. A. Christensen, P. K. Dalling, M. W. Duch, E. Wenkert, F. M. Schell and D. W. Cochran, *J. Am. Chem. Soc.*, **97**, 322 (1975).
21. R. V. Norton and I. B. Douglass, *Org. Magn. Reson.*, **6**, 89 (1974).
22. J. S. Waugh and F. A. Cotton, *J. Phys. Chem.*, **65**, 562 (1961).
23. K. S. Dhami, *Indian J. Chem.*, **12**, 278 (1974).
24. F. Freeman and C. N. Angeletakis, *J. Am. Chem. Soc.*, **105**, 4039 (1983).
25. F. Fzeeman, C. N. Angeletakis and M. C. Keindl, *J. Org. Chem.*, **49**, 454 (1984).
26. A. Kato and M. Numata, *Tetrahedron Lett.*, 203 (1972).
27. L. E. Legler, S. L. Jindal and R. W. Murry, *Tetrahedron Lett.*, 3907 (1972).
28. T. Takata, Y. H. Kim and S. Oae, *Tetrahedron Lett.*, 4303 (1978).
29. S. Oae and T. Takata, *Chem. Lett.*, 845 (1981).
30. F. Freeman and C. N. Angeletakis, *J. Am. Chem. Soc.*, **103**, 6232 (1981).
31. F. Freeman, C. N. Angeletakis and T. J. Maricich, *Org. Magn. Reson.*, **17**, 53 (1981).
32. F. Freeman and C. N. Angeletakis, *J. Org. Chem.*, **47**, 4194 (1982).
33. K. Mislow, M. Axelrod, D. R. Rayner, H. Gotthardt, L. M. Coyne and G. S. Hammond, *J. Am. Chem. Soc.*, **87**, 4958 (1965).

34. K. W. Buck, A. B. Foster, W. D. Pardoe, M. H. Qadir and J. M. Webber, *J. Chem. Soc., Chem Commun.*, 759 (1966).
35. A. B. Foster, J. M. Duxbury, T. D. Inch and J. M. Webber, *J. Chem. Soc., Chem Commun.*, 881 (1967).
36. J. A. Deyrup and C. L. Moyer, *J. Org. Chem.*, **34**, 175 (1969).
37. D. N. Harpp and J. G. Gleason, *J. Org. Chem.*, **36**, 1314 (1971).
38. R. D F. Cooper, P. V. DeMarco, J. C. Cheng and N. D. Jones, *J. Am. Chem. Soc.*, **91**, 1408 (1969).
39. H. M. McConnell, *J. Chem. Phys.*, **37**, 226 (1957).
40. M. Nishio, *J. Chem. Soc., Chem. Commun.*, 560 (1969).
41. R. U. Lemieux and S. Koto, *Tetrahedron*, **30**, 1933 (1974).
42. S. Wolfe, *Acc. Chem. Res.*, **5**, 102 (1972).
43. P. Deslongchamps, *Stereoelectronic Effects in Organic Chemistry*, Pergamon Press, Oxford, 1983.
44. J. B. Lambert, D. A. Netzel, H. Sun and K. K. Lilianstrom, *J. Am. Chem. Soc.*, **98**, 3778 (1976).
45. A. Kolbe and E. Wenschuh, *J. Mol. Struct.*, **28**, 359 (1975).
46. R. Keat, D. S. Ross and D. W. A. Sharp, *Spectrochim. Acta*, **27A**, 2219 (1971).
47. H. J. Jakobsen, A. Senning and S. Kaae, *Acta Chem. Scand.*, **25**, 3031 (1971).
48. R. M. Moriarty, *J. Org. Chem.*, **30**, 600 (1965).
49. E. Wenschuh, U. Kühne, M. Mikolajczyk and B. Bujnicki, *Z. Chem.*, **21**, 217 (1981).
50. F. A. Davis, J. M. Kaminski, E. W. Kluger and H. S. Freilich, *J. Am. Chem. Soc.*, **97**, 7085 (1975).
51. K. Mori and Y. Ueda, *Yakugaku Zasshi*, **91**, 940 (1971); *Chem. Abstr.*, **75**, 150938 (1971).
52. P. Ruostesuo, A.-M. Häkkinen and T. Mattila, *Magn. Reson. Chem.*, **25**, 189 (1987).
53. A-M. Häkkinen, P. Ruostesuo and S. Kirkisuo, *Magn. Reson. Chem.*, **23**, 311 (1985).
54. A-M. Häkkinen and R. Ruostesuo, *Magn. Reson. Chem.*, **23**, 424 (1985).
55. J. F. King and R. P. Beatson, *J. Chem. Soc., Chem Commun.*, 663 (1970).
56. G. Canalini, G. Maccagnani and F. Taddei, *Tetrahedron Lett.*, 3035 (1971).
57. M. Mikolajczyk and J. Drabowicz, *Z. Naturforsch. B*, **26b**, 1372 (1971).
58. D. Rinne and A. Blaschette, *Z. Anorg. Allg. Chem.*, **428**, 237 (1977).
59. R. P. Gupta, J. S. Pizey and K. Symeonides, *Tetrahedron*, **32**, 1917 (1976).
60. H. C. E. McFarlane and W. McFarlane, in *Multinuclear NMR* (Ed. J. Mason), Chap. 15, Plenum, New York, 1987, p. 417.
61. R. E. Wasylishen, C. Connor and J. R. Friedrich, *Can. J. Chem.*, **62**, 981 (1984).
62. T. C. Farrar, B. M. Trost, S. L. Tang and S. E. Springer-Wilson, *J. Am. Chem. Soc.*, **107**, 262 (1985).
63. S. A. Evans Jr., *NATO Adv. Study Inst. Ser., Ser. C*, **124**, 757 (1984).
64. J. C. Dyer, D. L. Harris and S. A. Evans Jr., *J. Org. Chem.*, **47**, 3660 (1982).
65. J. Dorie and J. P. Gouesnard, *J. Chem. Phys. Phys.-Chim. Biol.*, **81**, 15 (1984).
66. R. O. Duthaler and J. D. Roberts, *J. Am. Chem. Soc.*, **100**, 3889 (1978).
67. R. O. Duthaler and J. D. Roberts, *J. Am. Chem. Soc.*, **101**, 3706 (1979).
68. J. P. Gouesnard and J. Dorie *Nouv. J. Chim.*, **6**, 143 (1982).
69. J. Mason, in *Multinuclear NMR* (Ed J. Mason), Chap. 12, Plenum, New York, 1987, p. 335.
70. F. Seel, R. Budenz and W. Gombler, *Z. Naturforsch. B*, **25**, 885 (1970).
71. K. Mislow and M. Raban, *Top. Stereochem.*, **1**, 1 (1967).
72. J. S. Waugh and F. A. Cotton *J. Phys. Chem.*, **65**, 562 (1961).
73. F. Kaplan and J. D. Roberts, *J. Am. Chem. Soc.*, **93**, 4666 (1961).
74. J. W. Wilt and W. J. Wagner, *Chem. Ind.*, 1389 (1964).
75. P. Koch and A. Fava, *J. Am. Chem. Soc.*, **90**, 3867 (1968).
76. P. C. Turley and P. Haake, *J. Am. Chem. Soc.*, **89**, 4617 (1967).
77. L. M. Jackman, in *Dynamic NMR Spectroscopy* (Eds. L. M. Jackman and F. A. Cotton), Chap. 7, Academic Press, New York, 1975.
78. H. J. Jakobsen and A. Senning, *J. Chem. Soc., Chem. Commun.*, 617 (1967).
79. M. Raban, *J. Chem. Soc., Chem. Commun.*, 1017 (1967).
80. W. R. Jackson, T. G. Kee and W. B. Jennings, *J. Chem. Soc., Chem. Commun.*, 1154 (1972).
81. A. H. Cowley, M. W. Taylor, M-H. Whangbo and S. Wolfe, *J. Chem. Soc., Chem Commun.*, 838, (1976).
82. H. Herbrandson and R. T. Dickerson Jr., *J. Am. Chem. Soc.*, **81**, 4102 (1959).
83. L. Golebiowski and Z. Lasocki, *Bull. Acad. Pol. Sci., Ser. Sci. Chim.*, **24**, 439 (1975).

84. V. A. Blaschette, D. Rinne and H. C. Marsmann, *Z. Anorg. Allg. Chem.*, **420**, 55 (1976).
85. A. R. Bassindale, J. C-Y. Lau and P. G. Taylor, *J. Organomet. Chem.*, **341**, 213 (1988).
86. M. Mikolajczyk and J. Drabowicz, *J. Chem. Soc., Chem. Commun.*, 547 (1974).
87. A. F. Cockerill, G. L. O. Davies, R. C. Harden and D. M. Rackham, *Chem. Rev.*, **73**, 553 (1973).
88. H. Nozaki, K. Yoshino, K. Oshima and Y. Yamamoto, *Bull. Soc. Chem. Jpn.*, **45**, 3495 (1972).
89. W. H. Pirkle, S. D. Beare and R. L. Muntz, *J. Am. Chem. Soc.*, **91**, 4575 (1969).
90. W. H. Pirkle and M. S. Hoekstra, *J. Am. Chem. Soc.*, **98**, 1832 (1976).
91. M. R. Banks, C. Brown, R. F. Hudson and K. A. F. Record, *J. Chem. Soc., Perkin Trrans. 2*, 1501 (1986).
92. C. Brown, R. F. Hudson and K. A. F. Record, *J. Chem. Soc., Perkin Trans. 2*, 822 (1978).
93. M. R. Banks and R. F. Hudson, *J. Chem. Soc., Perkin Trans. 2*, 822 (1978).
94. A. Heesing, W. K. Homann and W. Müllers, *Chem. Ber.*, **113**, 152 (1980).
95. W. J. Bouma and J. B. F. N. Engberts, *J. Org. Chem.*, **41**, 143 (1976).
96. M. R. Banks and R. F. Hudson, *J. Chem. Soc., Perkin Trans. 2*, 151 (1986).
97. J. L. Kice, *Adv. Phys. Org. Chem.*, **17**, 65 (1980).
98. C. J. M. Stirling, *Int. J. Sulfur Chem. B*, **6**, 277 (1971).
99. Y. H. Chiang, J. S. Luloff and E. Schipper, *J. Org. Chem.*, **34**, 2397 (1969).
100. F. Freeman and M. C. Keindl, *Sulfur Rep.*, **4**, 231 (1985).
101. A. G. Davies, B. P. Roberts and B. R. Sanderson, *J. Chem. Soc., Perkin Trans. 2*, 626 (1973).
102. Y. Miura, Y. Nakamura and M. Kinoshita, *Chem. Lett.*, 521 (1978).
103. B. C. Gilbert, R. O. C. Norman and R. C. Sealy, *J. Chem. Soc., Perkin Trans. 2*, (a) 303, (b) 308 (1975).
104. C. Chatgilialoglu, B. C. Gilbert and R. O. C. Norman, *J. Chem. Soc., Perkin Trans. 2*, 770 (1979).
105. C. Chatgilialoglu, L. Lunazzi and K. U. Ingold, *J. Org. Chem.*, **48**, 3588 (1983).
106. T. Kawamura, P. J. Krusic and J. K. Kochi, *Tetrahedron Lett.*, 4075 (1972).
107. B. D. Flockhart, K. J. Ivin, R. C. Pink and B. D. Sharma, *J. Chem. Soc., Chem. Commun.*, 339 (1971).
108. J. A. Baban and B. P. Roberts, *J. Chem. Soc., Perkin Trans. 2*, 678, (1978).
109. G. Brunton, J. F. Taylor and K. U. Ingold, *J. Am. Chem. Soc.*, **98**, 4879 (1976).
110. C. Chatgilialoglu, B. C. Gilbert and R. O. C. Norman, *J. Chem. Soc., Perkin Trans. 2*, 1084 (1979).
111. B. C. Gilbert, H. A. H. Laue, R. O. C. Norman and R. C. Sealy, *J. Chem. Soc., Perkin Trans. 2*, 892, (1975).
112. M. McMillan and W. A. Walters, *J. Chem. Soc. (B)*, 422 (1966).
113. R. S. Andersen, *J. Chem. Phys.*, **66**, 5610 (1977).
114. M. Geoffroy and E. A. C. Lucken, *J. Chem. Phys.*, **55**, 2719 (1971).
115. P. B. Ayscough, K. J. Ivin and J. H. O'Donnell, *Trans. Faraday Soc.*, **61**, 1110 (1965).
116. G. Lind and R. Kewley, *Can. J. Chem.*, **50**, 43 (1972).
117. Z. Kuri, H. Ueda and S. Shida, *J. Chem. Phys.*, **32**, 371 (1960).
118. W. Damerau, G. Lassmann and K. Lohs, *Z. Chem.*, **9**, 343 (1969).
119. W. B. Gara and B. P. Roberts, *J. Chem. Soc., Perkin Trans. 2*, 1708, (1977).
120. A. Alberti, B. F. Bonini and G. F. Pedulli, *Tetrahedron Lett.*, **28**, 3737 (1987).
121. C. Chatgilialoglu, B. C. Gilbert and R. O. C. Norman, *J. Chem. Soc., Perkin Trans. 2*, 1429 (1980).
122. C. Chatgilialoglu, B. C. Gilbert, C. M. Kirk and R. O. C. Norman, *J. Chem. Soc., Perkin Trans. 2*, 1084 (1979).
123. A. Alberti, C. Chatgilialoglu and M. Guerra, *J. Chem. Soc., Perkin Trans. 2*, 1179 (1986).
124. G. W. Chantry, A. Horsfield, J. R. Morton, J. R. Rowlands and D. H. Whiffen, *Mol. Phys.*, **5**, 233 (1962).
125. C. Chatgilialoglu, B. C. Gilbert, R. O. C. Norman and M. C. R. Symons, *J. Chem. Res. (M)*, 2610 (1980).
126. J. R. Bolton and J. E. Wertz, *Electron Spin Resonance: Elementary Theory and Applications*, McGraw-Hill, New York, 1972.
127. R. J. Boyd, A. Gupta, R. F. Langler, S. P. Lownie and J. A. Pincock, *Can. J. Chem.*, **58**, 331 (1980).
128. S. W. Benson, *Chem. Rev.*, **78**, 23 (1978).
129. W. C. Danen and F. A. Neugebauer, *Angew. Chem., Int. Ed. Engl.*, **14**, 783 (1975).

130. B. C. Gilbert, C. M. Kirk, R. O. C. Norman and H. A. H. Laue, *J. Chem. Soc., Perkin Trans. 2*, 497 (1977).
131. P. M. Carton, B. C. Gilbert, H. A. H. Laue, R. O. C. Norman and R. C. Sealy, *J. Chem. Soc., Perkin Trans. 2*, 1245 (1975).
132. W. B. Gara, B. P. Roberts, C. M. Kirk, B. C. Gilbert and R. O. C. Norman, *J. Magn. Reson.*, **27**, 509 (1977).
133. B. C. Gilbert, C. M. Kirk and R. O. C. Norman, *J. Chem. Res. (M)*, 1974 (1977).
134. I. I. Kandror, R. G. Gasanov and R. K. Freidlina, *Tetrahedron Lett.*, 1075 (1976).
135. I. I. Kandror, R. G. Gasanov and R. K. Freidlina, *Izv. Akad. Nauk SSSR, Ser. Khim.*, 550 (1977).
136. C. Chatgilialoglu, B. C. Gilbert, B. Gill and M. D. Sexton, *J. Chem. Soc., Perkin Trans. 2*, 1141 (1980).
137. M. Kobayashi, E. Akiyama, H. Minato and N. Kito, *Bull. Chem. Soc. Jpn.*, **47**, 1504 (1974).
138. M. Kobayashi, M. Gotoh and H. Minato, *J. Org. Chem.*, **40**, 140 (1975).
139. T. A. J. W. Wajer, H. W. Geluk, J. B. F. N. Engberts and T. J. De Boer, *Recl. Trav. Chim. Pays-Bas*, **89**, 696 (1970).
140. T. L. Hall, M. F. Lappert and P. W. Lednor, *J. Chem. Soc., Dalton Trans.*, 1448 (1980).

The Chemistry of Sulphinic Acids, Esters and their Derivatives
Edited by S. Patai


CHAPTER **7**

# Syntheses of sulfinic acids

URI ZOLLER

*Division of Chemical Studies, Haifa University—Oranim, P.O. Kiryat Tivon 36910, Israel*

## I. INTRODUCTION

The structure, physical and chemical properties as well as the unique features of the sulfinyl functional group form the subject matter of this volume.

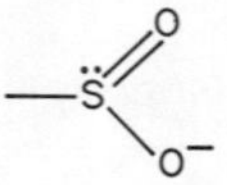

The following three features are of particular significance from the point of view of the methods and strategies for the syntheses of sulfinic acids and esters:

(a) The intermediate position of the sulfinyl group (**1**) on the oxidation coordinate of the divalent sulfur in **2** (i.e. thiols, sulfides, sulfenyl halides, sulfenic acids and esters) to the hexavalent sulfur(VI) in **3** (i.e. sulfones, sulfonyl chlorides, sulfonic acids and esters).

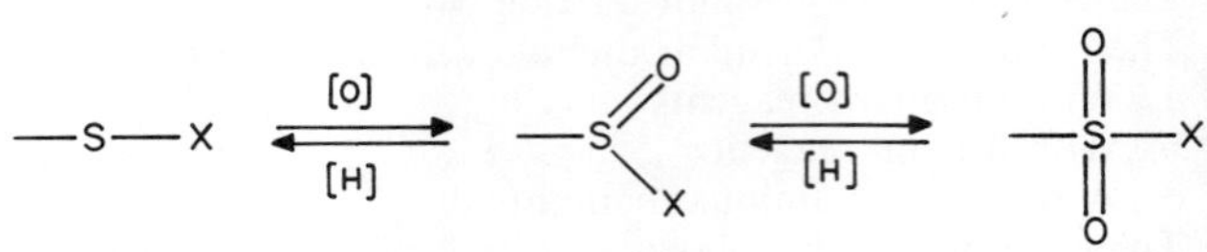


**(2)** **(1)** **(3)**

$X = H, R, Hal, OH, OR, O^-M^+$

(b) The presence of a lone electron pair on the sulfur atom of the sulfinyl group not only makes the latter prone to easy oxidation to the corresponding sulfonyl group, but also facilitates its function as a nucleophile.

(c) The facile hydrolysis, disproportionation and reduction which the sulfinic esters and acids undergo under various reaction conditions is reminiscent, in some respects, of the behaviour of carboxylic compounds.

These factors determine the feasibility of all synthetic methods for the preparation of sulfinic acids and esters. No single general method is available for their synthesis in high yields and purity, due to the instability of the desired product(s) under the reaction conditions employed, and the formation of undesirable byproducts. Furthermore, some of the most obvious starting materials, the sulfinyl chlorides (the alkanesulfinyl chlorides in particular), are not readily available and are rather unstable, giving various side-reactions and therefore low yields.

In spite of the interesting chemistry associated with them, the syntheses of sulfinic acids and their esters are relatively unexplored. Their preparations were previously reviewed by several authors[1-6], but the method of choice is still an open question. Recently modified strategies of convenient and general syntheses of sulfinic acids[7] and their chiral esters[8] appeared in the literature. If this renewal of interest will gain momentum, hopefully the full scope of the chemical behaviour of the sulfinyl functionality will be uncovered, like that of the corresponding sulfones and sulfoxides[9].

## II. THE SYNTHESIS OF SULFINIC ACIDS

Sulfinic acids are formed readily via the following methodologies:

(a) Reduction of sulfonyl halides.

(b) Alkaline hydrolysis of sulfinic acid derivatives.

(c) Nucleophilic cleavage of the sulfur–sulfur bond in thiosulfonates.

(d) Oxidation of thiols and thioureas.

(e) Formation of the C—S bond by (i) the reaction of sulfur dioxide with alkanes, organometallics, allenes, alkynes and arenediazonium salts; (ii) the reaction of thionyl chloride with olefins; and (iii) the transfer of a sulfinyl group from other sulfinic acids.

(f) Cleavage of the C—S bond of sulphones and sulfoxides, and other sulfinic acid derivatives by bases or electrochemically.

(g) Cleavage of the sulfur–oxygen and sulfur–nitrogen bonds of sulfonic acids and esters, sulfonamides and sulfonyl hydrazines.

(h) Miscellaneous.

The first three appear to be the methods of choice from the point of view of availability of starting materials, generality, conveniency, efficiency and the ease of work-up procedures.

The descriptions of the above methods are accompanied by some very abbreviated illustrative experimental procedures.

### A. Reduction of Sulfonyl Halides

The reduction of sulfonyl chlorides was for many years the most important route to obtain sulfinic acids[6]. The most commonly used procedure is the treatment of sulfonyl chlorides with zinc dust[10] or iron[11] in aqueous caustic alkali solution.

$$2RSO_2Cl + 2Zn \longrightarrow (RSO_2)_2Zn + ZnCl_2 \qquad (1)$$

$$(RSO_2)_2Zn + NaCO_3 \xrightarrow{NaOH} 2RSO_2Na + ZnCO_3 \qquad (2)$$

The mechanism of the above reduction was studied and discussed[12].

This reductive method is illustrated in the following procedure[10a].

*p-Toluenesulfinic acid* (**4**): To a stirred suspension of zinc dust (40 g) in warm water (300 ml, 70°), *p*-toluenesulfonyl chloride (50 g) is added over about 10 min. The temperature is raised to 90 °C. Sodium hydroxide (25 ml, 12 N) is then added, followed by 5 g portions of sodium carbonate until the mixture becomes strongly alkaline. The precipitate is filtered and washed. Evaporation of the filtrates to about 100 ml and thorough cooling affords 36 g (64%) of the sodium salt **5** ($p$-$CH_3C_6H_4SO_2Na{\cdot}2H_2O$). The free sulfinic acid **4** may be prepared by careful acidification of a cold aqueous solution of **5** with hydrochloric acid. The drying of the acid without its partial conversion to sulfonic acid and thiosulfonic ester is difficult.

A comparably convenient method of preparing sulfinic acids is the reduction of sulfonyl halides with sodium sulfite[13,14]:

$$RSO_2X + NaSO_3 + H_2O \longrightarrow RSO_2H + NaX + NaHSO_4 \qquad (3)$$

R = Ar (mainly) (**6**)

X = Cl or F (**a**) R = $p$-$ClC_6H_4$

Since the lower aliphatic alkanesulfinic acids are unstable and easily disproportionate[1,15], this route is mainly applicable for the preparation of either the arenesulfinic acids[5,13] or the aliphatic disulfinic acids[15]. Two illustrative procedures follow.

*p-Chlorobenzenesulfinic acid* (**6a**)[13]: To a stirred solution of sodium sulfite (30 g) in water at 70 °C is added *p*-chlorobenzenesulfonyl fluoride (10 g). The mixture is stirred for 5 h at 70–80 °C and then heated for a few minutes at 100 °C, followed by acidification with concentrated hydrochloric acid, cooling and filtration. The yield of **6a** is 7.28 g (81%).

The reduction of *p*-chlorobenzenesulfonyl chloride with sodium sulfite afforded 80% yield of **6a**.

*1,4-Butanedisulfinic acid* (**7**)[15]: To a warm stirred solution of sodium sulfite and sodium bicarbonate in water the 1,4-disulfonyl chloride (prepared by oxidative chlorination of the corresponding diisothiuronium salt) is added over a period of 1 h. The mixture is stirred at 70–80 °C for an additional 2 h, cooled and filtered. The filtrate is acidified by hydrochloric acid. The resulting precipitate is recrystallized from water and gives the acid **7** (60.3%).

Only in the late forties was the rather stable and crystalline 1-dodecanesulfinic acid prepared for the first time[16]. However, even this long-chain sulfinic acid undergoes, on heating or long standing (two months), the disproportionation typical for aliphatic sulfinic acids[17].

$$3RSO_2H \longrightarrow RSO_2SR + RSO_3H + H_2O \tag{4}$$

The preparation of perhaloalkanesulfinic acids can be achieved by reduction of the corresponding perhalosulfonyl halides with sodium sulfite[18], hydrazine[19] or hydrogen sulfide[20]. An illustrative procedure of the use of $H_2S$ is given in equation 5.

$$Cl_3CSO_2Cl + H_2S \longrightarrow Cl_3CSO_2H + HCl + S_{(8)} \tag{5}$$

*Trichloromethanesulfinic acid* (**8**)[20]: Hydrogen sulfide is bubbled into a solution of trichloromethanesulfenyl chloride in methanol. Loss of the color of the solution and formation of coagulated sulfur show the end of the reaction. Filtration under reduced pressure affords 99% of the crude product. Distillation *in vacuo*, accompanied by partial decomposition, gives a colorless oil which turns after a while into hygroscopic crystals.

The reduction of sulfonyl chlorides to the corresponding sulfinic acids can be accomplished in good yields by the use of lithium aluminum hydride[21]. The reduction was suggested to proceed through the nucleophilic displacement of chloride ion from the sulfur atom by a complex hydride ion, followed by attack of another complex hydride ion on the hydrogen of the resulting sulfinic acid with the formation of a sulfinate salt and hydrogen[21]:

$$RSO_2Cl \xrightarrow{AlH_4^-} RSO_2H + Cl^- \xrightarrow{AlH_4^-} RSO_2^- + H_2 \tag{6}$$

By this experimental procedure, benzenesulfinic acid (**9**)[21] could be obtained in 89% yield, while *p*-toluenesulfonyl chloride affords 93% of *p*-toluenesulfinic acid (**4**). Slow addition of the hydride to the sulfonyl halide at low temperatures avoids partial reduction beyond the sulfinic acid stage and gives high yields of the desired product.

Although the reaction of selected diarylcadmium compounds with both arenesulfonyl halides[22a] and alkanesulfonyl halides[22b] gave the corresponding sulfinic acids in fair yields (30–45% in most cases), the produced acids were not isolated, but rather were immediately converted to the corresponding benzyl sulfones. Also, the reaction mixture contained both reduced and oxidized products. Thus, the practicality of this method for the synthesis of sulfinic acids on a preparative scale is questionable and further experimentation and process optimization are still needed.

Both aliphatic and aromatic sulfonyl chlorides were reported to be easily reduced to sulfinic acids by either triethylaluminum or ethylaluminum-sesquichloride in 1:1 mole ratio[23]. Although the reported yields are high (see Table in Section III at the end), handling of these reducing agents and work-up procedures of the mixtures obtained do not suggest any particular advantage.

An interesting convenient method of preparing the potassium salt of trifluoromethanesulfinic acid (potassium triflinate; $CF_3SO_2K$, **10**) is the reduction of the corresponding sulfonyl chloride with potassium iodide[24]. This method appears to be limited in scope and applicable only to low molecular weight perhalosulfonyl halides.

The preparation of sulfinic acids by electrolytic reduction of sulfonyl chlorides was also reported[25].

## B. Alkaline Hydrolysis of Sulfinic Acid Derivatives

Although the hydrolysis of sulfinic acid derivatives such as sulfinic esters, sulfinamides and sulfinyl halides is, supposedly, an effective, straightforward and easy process, it constitues in fact a rather indirect strategy. The functional group is already attached to the starting material at a certain site to begin with and, consequently, the versatility with respect to the final product is rather limited.

The alkaline hydrolysis of sulfinic esters to afford the corresponding sodium sulfinate is a straightforward, fast process which is completed within minutes in dilute sodium hydroxide solution even at 0 °C[26]. Methyl methane sulfinate is miscible with water, but the other alkanesulfinates show a greater tendency to dissolve water than to dissolve in water themselves. Furthermore, the base catalyzed solvolysis of several arenesulfinates was shown to involve sulfur–oxygen bond fission giving rise—in aqueous ethanol—to ester interchange rather than to sulfinic acids. With certain arenesulfinic esters, however, one can control the reaction to give carbon–oxygen bond fission by choice of an appropriate base[27]. The acidic hydrolysis of sulfinic esters is a much slower process than that of the alkaline one[26,28] (less than 5% hydrolysis of methyl methanesulfinate in 3 h in 0.06 N hydrochloric acid at 25 °C; 95% in 45 min at 100 °C)[26].

High yields of arenesulfinic acids are obtained when sulfinamides are hydrolyzed in basic aqueous ethanol[29] (equation 7). The reaction was shown to be first order in base and first order in sulfinamide, with the expected substitution effect on the relative rate of the

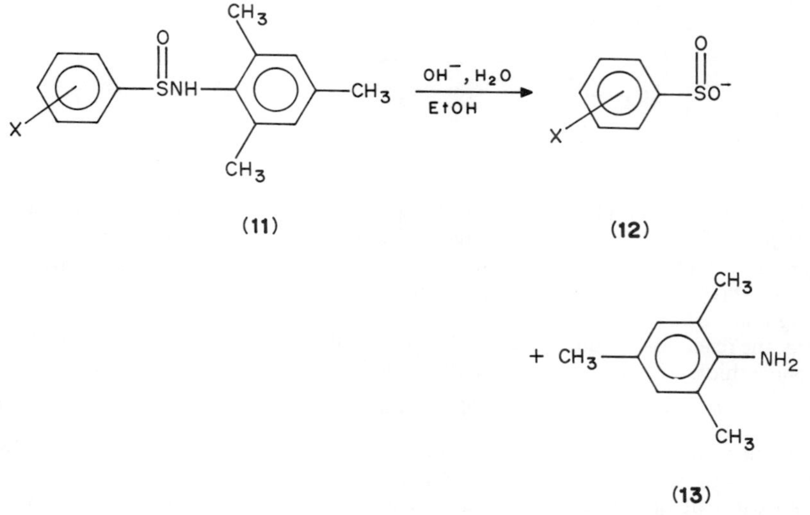


X= *p*-$CH_3O$, $CH_3$, H, Cl, $NO_2$ or *m*-$CH_3$, Cl, $CF_3$, $NO_2$ (7)

hydrolysis. From a practical point of view, the overall reflux time required is rather long, e.g. 3 days in the preparation of **6a** from *N*-mesityl-*p*-chlorobenzenesulfinamide with sodium hydroxide in refluxing ethanol.

Hydrolysis of sulfinyl chlorides is a very effective way of synthesizing the corresponding sulfinic acids[5,12b].

$$RSOCl \xrightarrow{H_2O} RSO_2H + HCl \qquad (8)$$

The major advantage of this method is the easy accessibility of the sulfinyl chlorides via direct chlorination of the readily available thiols or disulfides with chlorine in aqueous media[26,31,32]. The chlorination is probably the entry of choice to sulfinic acids and esters series starting with simple sulfur-containing materials. Also, the hydrolysis of sulfinyl chlorides is useful for the preparation of $^{18}O$-labeled sulfinic acids which, in turn, are used as starting materials for $^{18}$-O-labeled trivalent and tetravalent oxygen-containing organosulfur derivatives (e.g. sulfoxides, sulfinic esters, sulfoximines, etc.)[5]

Thus, methanesulfinic acid (**14**) is obtained[33] by addition of water dropwise with stirring under a dry nitrogen atmosphere at $-30\,°C$ over *ca* 5 min to methanesulfinyl chloride[34]. The oily reaction mixture, when mixed with anhydrous ether and stored at $-15\,°C$ for 24 h, affords long transparent, colorless hygroscopic needles.

### C. Nucleophilic Cleavage of the Sulfur–Sulfur Bond in Thiosulfonates

The nucleophilic cleavage of the sulfur–sulfur bond in thiosulfonates[35] and in cyclic thiosulfonates (e.g. 1,2-dithiane 1,1-dioxides[36] and 1,2-dithiole 1,1-dioxides[37]) is closely related to the alkaline hydrolysis route described above. However, in contrast to the hydrolysis process in which there is no change in the trivalent sulfur(IV) of the starting sulfinic derivative, the reductive cleavage of the S—S bond is accompanied here by an S(VI) → S(IV) sulfonate → sulfinate transformation.

$$R^1SO_2SR^2 + Nu^- \longrightarrow R^2SNu + R^2SO_2^- \qquad (9)$$

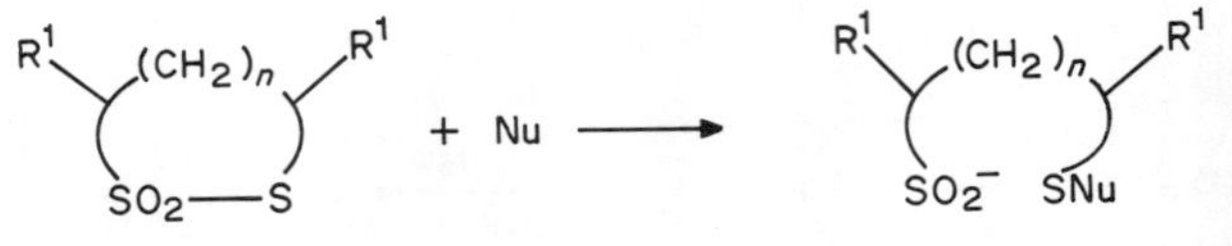


(10)

**(15)** **(16)**

In both acyclic and cyclic systems, the particular sulfinic acid obtained is contingent on the structure of $R^1$ attached to the sulfinic sulfur atom. In the cyclic thiosulfonates [for the preparation of **15a** ($R^1 = H$; $n = 4$) see Reference 36a)] the final product is necessarily bifunctional (i.e. **16**) due to the second sulfur atom contained in the cyclic array of the starting material.

Thus, the reaction of sodium dialkyl phosphites, $(RO)_2PONa$, with thiosulfonates gives phosphorothiolates and sodium sulfinates in high yields.

$$R^1SO_2SR^2 + (R^3O)_2PONa \longrightarrow R^1SO_2Na + R^2SP(O)(OR^3)_2 \qquad (11)$$

As an illustrative example[35], butanesulfinic acid (**17**) is obtained from ethyl butanethiosulfonate with sodium dibutyl phosphite in 86% yield.

The treatment of the aliphatic or aromatic cyclic thiosulfonates with thiolate salts, sodium polysulfide or sodium amide provides the corresponding alkyldithioalkane sulfinates $R^1 = H$, Nu = Et, Pr, Bu, *t*-Bu, pentyl) or the arenesulfinate salts, 4,4′-polythiobis(butanesulfinates), $NaO_2S(CH_2)_4S_m(CH_2)_4SO_2Na$ (**18**; $m = 3$–6), and the disodium salt of 4-mercaptobutanesulfinic acid, $NaO_2S(CH_2)_4SNa$ (**19**), respectively[36,37].

Detailed procedures for the synthesis of the sulfide–sulfinate salts **16** [e.g. sodium 4-(*tert*–butyldithio)butanesulfinate[36a]] and of the *4-mercaptobutanesulfinic acid, disodium salt*[36b] (**19**) have been described.

The alkaline pyrolysis of sulfonyl hydrazones (the Bamford–Stevens reaction) yields diazoalkanes and arenesulfinates[6,38]. However, this method has no preparative value as far as the sulfinates are concerned[6]. Similarly, the alkaline pyrolysis of *N*-acyltosylhydrazides[39] as well as the alkaline reduction of sulfonamides[40] also yield sulfinates, but have no preparative value with respect to sulfinic acids.

Treatment of alkanesulfonyl-hydrazides with alkali also provides sulfinic acid salts and free acids in moderate to good yields (see Section II.G).

## D. Oxidation of Thiols and Thioureas

All previously described methods for the preparation of sulfinic acids are indirect routes, suffering often from complications and competing side-reactions, in addition to the instability of the free alkanesulfinic acids themselves. Consequently, the direct oxidation of thiols, under relatively mild conditions, appears to be the method of choice for the one-step synthesis of alkanesulfinic acids.

$$RSH \xrightarrow{[O]} RSOH \xrightarrow{[O]} RSO_2H \qquad (12)$$

Direct oxidation of aliphatic thiols with 2 equivalents of *m*-chloroperbenzoic acid (MCPBA) in methylene chloride yields sulfinic acids in a high state of purity and good yield. The experimental procedure is rather simple and applicable to all paraffinic $C_2$—$C_4$ thiols as well as to thiophenol[41].

Neither disulfides (disproportionation products of sulfinic acids) nor sulfonic acids have been observed to accompany the freshly isolated sulfinic acids via this procedure. Apparently, the *m*-chloroperbenzoic acid is too mild to further oxidize the sulfinic acid formed under these conditions. Thus, *n*-butanesulfinic acid (**17**)[41] could be obtained from the corresponding thiol by MCPBA 'interval' oxidation (at $-30\,°C$) in 81.5% yield and can be preserved *in vacuo* at $-30\,°C$ for months without noticeable decomposition.

In an attempt to avoid completely the further oxidation of the sulfinic acid to the sulfonic acid in the oxidation of thiols, a new synthetic method is based upon protection of thiols and subsequent deprotection[42] using the 2-benzothiazolyl protecting group. The protected thiols (i.e. **20**), easily prepared by alkylation of 2-mercaptobenzothiazole with alkyl halides (or by substitution of 2-chlorobenzothiazole with sodium alkane thiolates), are oxidized to the corresponding sulfone **21** which is cleanly cleaved to the targeted sodium sulfinate and benzothiazole with sodium borohydride (equation 13). This method

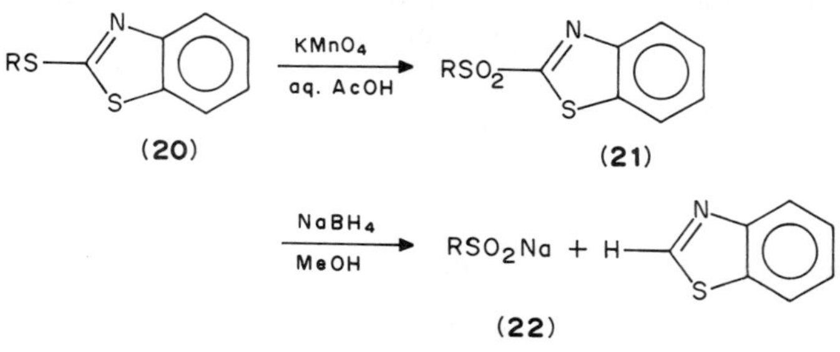


(13)

was shown to be applicable for the synthesis of alkanesulfinic acids (e.g. methane-, hexane- and cyclohexanesulfinic acids **22a–c**), α-toluenesulfinic acid (**22d**, $R=C_6H_5CH_2$) and 2-pyridinesulfinic acid (**22e**) in overall yields of 64–81%[42]. Significantly, the method is applicable for the preparation of optically pure chiral sulfinic acids through a slight modification in the preparation of the protected starting thiols (equation 14).

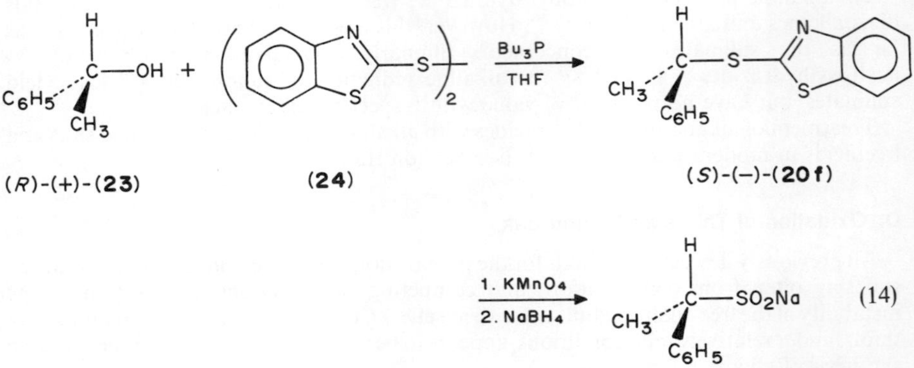


(*R*)-(+)-(**23**) (**24**) (*S*)-(–)-(**20f**)

(14)

(*S*)-(–)-(**22f**)

A typical procedure[42] for the synthesis of the optically active (*S*)-(–)-α-methylbenzylsulfinic acid (**22f**) gives the corresponding sulfone (**21f**) in 77% yield, and the final product **22f** in an overall yield of < 66%, based on the optically active alcohol **23**.

It is possible to oxidize thiols with air[43,44] as well as with the superoxide anion[45]. The air autoxidations of octanethiol and of thiophenol in nonaqueous alkaline media ($t$-BuO$^-$K$^+$ in $t$-BuOH) at atmospheric pressure were shown to yield mixtures of the sodium sulfinate, sodium sulfonate and the corresponding disulfides[43], the formation of which is catalyzed by unavoidable traces of metal ions. This oxidation route is thus, probably, more important mechanistically than preparatively.

Organic sulfur compounds such as disulfides, thiosulfinates, thiosulfonates, sodium thiolates, sodium sulfinates and thiols were readily oxidized under mild conditions with superoxide anion generated from potassium superoxide and 18-crown-6 ether[45]. Although thiols were easily oxidized with $O_2^-$ at room temperature to the corresponding disulfides and further oxidized to the corresponding sulfinic acids at 60 °C, this oxidation was accompanied by oxidation to the sulfonic acids so that mixtures have been obtained. In most of the above oxidations the sulfonic acids predominate in the mixtures and the yields of the sulfinic acids are rather low. Consequently, this method (equation 15) appears to have no preparative potential as far as the sulfinic acids are concerned.

$$\left.\begin{array}{l} RSSR \\ RSOSR \\ RSO_2SR \\ RS^-Na^+ \end{array}\right\} \xrightarrow[\text{18-Crown-6-ether, Pyridine}]{KO_2} RSO_2^- + RSO_3^- \qquad (15)$$

The oxidation of thioureas to the amino-iminomethanesulfinic acids is a well-known process which has been executed for many years on an industrial scale with air/ozone mixtures in water or acetone at 1–3 °C[46], or with 50% $H_2O_2$[47], particularly in the preparation of the formamidine sulfinic acid (**25a**; R = H)—an industrial reducing agent and stabilizer.

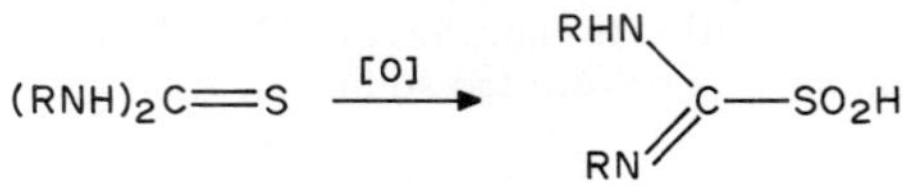

(16)

**(25)**

The experimental procedure for the preparation of amino-iminoalkane sulfinic acids is based on the treatment of the starting thiourea with an aqueous solution of hydrogen peroxide as applied for the preparation of thiourea *S,S*-dioxides[48]. A modified version of this procedure[49] yields amino-imino-n-butanesulfinic acid (*N,N'*-di-n-butylthiourea *S,S*-dioxide (**25b**)[50] from *N,N'*-di-n-butylthiourea and $H_2O_2$ in 38% yield.

Although both IR and NMR (two separated triplets for the two $NCH_2$ protons 0.45 ppm apart) are in accord with the amino-iminosulfinic acid representation of **25b**, the chemical behaviour of **25b** on thermolysis[49,50] suggests the *S,S*-dioxide structure, at least on heating.

## E(a). Sulfinations with Sulfur Dioxide (formation of a C—S bond)

### *1. The condensation of sulfur dioxide with alkanes*

Irradiation of sulfur dioxide in the gas phase with UV light in the presence of gaseous alkanes leads to the formation of alkanesulfinic acids[51], apparently via excited $SO_2$[52]. Similarly, the biradicals formed by $\gamma$-hydrogen abstraction on photoexcitation of *t*-butyl-*p*-benzoquinones add to $SO_2$ to give the benzoquinonyl-alkanesulfinic acids (**26**)[53].

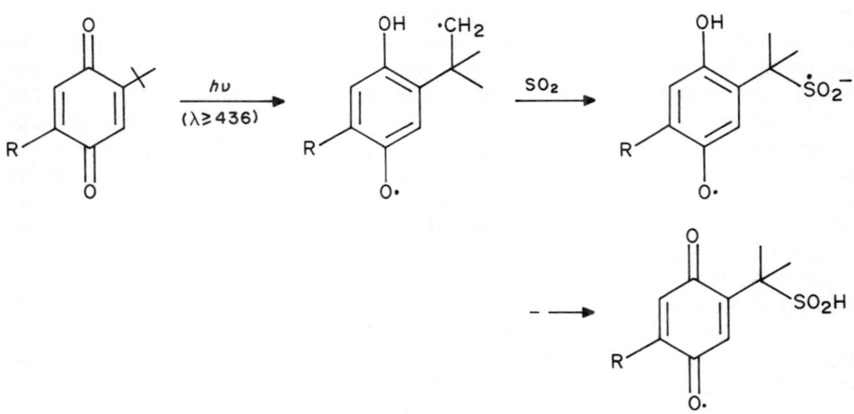

(17)

**(26)**

These methods have no practical value since the yields of the sulfinic acids obtained are relatively low and they are accompanied by other photoproducts and, in most cases, they undergo further reactions under the reaction conditions and cannot therefore be isolated as such.

Alcohols, ethers, sulfides, chloroalkanes, dimethylformamide and even isobutane were shown to give α-substituted alkanesulfinic acids on reaction in the liquid phase and at low temperatures with photoexcited $SO_2$, albeit in small yields[54]. Only in the case of ethers as the starting materials are the yields within the fair range of 43–55%. Thus, this method is a facile one for producing *solutions* of α-substituted sulfinic acids.

$$\text{RX or } R_2X \xrightarrow[-75\,^\circ\text{C, 7–18 h}]{SO_2} \text{R}(SO_2\text{H})\text{X or HOOSRXR} \tag{18}$$

$$\text{R} = \text{alkyl; X} = \text{OH, Cl, O, S}$$

For example, α-hydroxyethanesulfinic acid (**27**)[54] is obtained from ethanol in liquid $SO_2$ by irradiation with a UV lamp and its sodium salt is stable at room temperature.

Tetrahydrofuran gives a low yield of the tetrahydrofuran-2-sulfinic acid by the same method[55].

### 2. *The condensation of sulfur dioxide with organometallics*

*a. With Grignard reagents.* The classical reaction of Grignard reagents with sulfur dioxide is probably one of the most known and commonly used reactions for the formation of the carbon–sulfur bond in organic synthesis. The synthesis of sulfinic acids for their own sake or as intermediates in the preparation of sulfones, sulfinic esters and other oxidized sulfur-containing compounds is quite useful. It suffers, however, from competing side-reactions[1,16].

$$\text{RX} \xrightarrow{\text{Mg}} \text{RMgX} \xrightarrow{SO_2} (\text{RSO}_2)_2\text{Mg} \xrightarrow{\text{H}^+} \text{RSO}_2\text{H} \tag{19}$$

1-Dodecanesulfinic acid (**28**)[16] was prepared from the Grignard reagent of 1-bromododecane in ether, treated with sulfur dioxide at −40 to −35 °C[56]. The overall yield of the free sulfinic acid (**28**) was 64%.

The condensation of sulfur dioxide with organometallic reagents is a very promising method of sulfinic acid synthesis, albeit with various shortcomings: The formation of the corresponding sulfoxides (presumably by reaction of the sulfinate salt with the unconsumed organometallic reagents) appears to be the main undesired side-reaction. The 'reverse' addition, i.e. addition of the organometallic reagent to excess sulfur dioxide, should eliminate this problem, and indeed, the addition of either Grignard reagents or organolithium reagents to roughly 10 equivalents of $SO_2$ in ether gives a nearly quantitative yield of the corresponding sulfinate salts[57]. For instance, both $C_6H_5CH_2SO_2MgCl$ (**29a**)[57] and *sec*-$BuSO_2MgCl$ (**29b**) could be prepared in 97% yield by using the reverse addition procedure.

Interestingly, the treatment of the Grignard reagent prepared from 3-chloro-2-methylpropene with liquid sulfur dioxide in ether (equation 20) produces the isolable magnesium salt **29c** of the corresponding allylic sulfinic acid. Acidification of an ether suspension of this magnesium salt in a reaction flask protected with a dry ice condenser led to the instantaneous liberation of the sulfur dioxide.

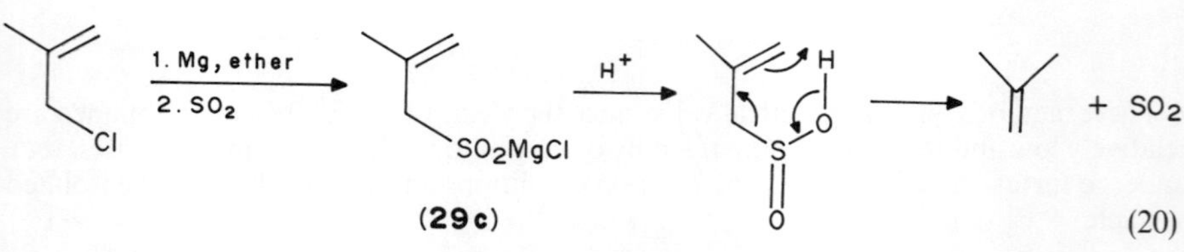


(20)

Magnesium salts of other allylic sulfinic acids have been prepared by this method, which is applicable also for synthesis on a preparative scale. On acid hydrolysis all these magnesium salts gave instantaneously the corresponding olefins and sulfur dioxide[58]. The instability of the allylic sulfinic acids may be rationalized in terms of the retro-ene reaction available to them.

*b. With lithium reagents.* In principle, the condensation of lithium reagents with sulfur dioxide[57,59] is analogous in all respects to the condensation of the latter with Grignard reagents. The method is applicable to both aliphatic and aromatic sulfinic acids.

$$RLi + SO_2 \longrightarrow RSO_2^- Li^+ \xrightarrow{H^+} RSO_2H \qquad (21)$$

High yields of lithium sulfinates can be obtained on adding the organolithium reagents into a large excess of sulfur dioxide, since by using this procedure undesired side-reactions are avoided[57]. For instance, lithium butanesulfinate (**30**) is obtained in quantitative yield by dropwise addition of an hexane solution of n-butyllithium to liquid $SO_2$ at $-78\,°C$.

The lithium salts of the sulfinic acids derived from trithiane, tetrathiocane and pentathiecane, namely $C_3H_5S_3SO_2Li$ (**31**), $C_4H_7S_4SO_2Li$ and $C_5H_9S_5SO_2Li$, have been prepared[60] by the addition of sulfur dioxide to the corresponding lithium derivatives of the parent *s*-trithiane, tetrathiocane and pentathiecane, respectively. The slightly impure lithium sulfinates precipitated from THF solutions at room temperature, although on standing they were converted to the sulfones. Acidification led to decomposition with liberation of sulfur dioxide[60].

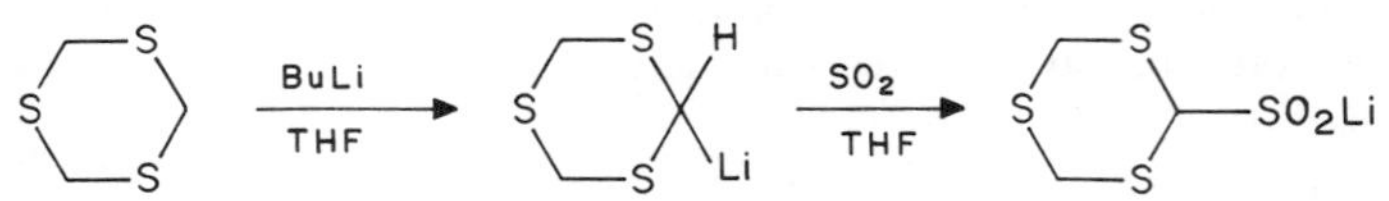


(22)

(**31**)

Aromatic sulfinic acids may be prepared by the same method, i.e. by treatment of the lithiated aromatic ring with liquid sulfur dioxide. Thus, *p*-*n*-dodecylbenzenesulfinic acid (**32**) is obtained[59] from lithium and *p*-bromo-*n*-dodecylbenzene and sulfur dioxide, in 63% yield.

*c. With organoaluminum compounds.* The patent literature describes several procedures for the synthesis of alkanesulfinic acids by treating trialkyl—or triaryl—aluminum compounds with sulfur dioxide at low temperatures[61]. For instance, *n*-octanesulfinic acid (**33**) is obtained from tri-*n*-octylaluminum and $SO_2$ in THF, in 94% yield[61a].

It is also possible to obtain arenesulfinic acids by the $AlCl_3$ catalyzed sulfination of aromatic hydrocarbons with sulfur dioxide[62].

$$ArH + SO_2 \xrightarrow{AlCl_3} ArSO_2H \qquad (23)$$

### 3. *The reaction of sulfur dioxide with allenes and alkynes*

Methyl-substituted allenes undergo ene addition to sulfur dioxide to give vinylic/allylic sulfinic acids which possess stability with respect to the reagents[63] (in contrast to the case with allylic sulfinic acids[58]). The sulfinic acids **35** (equation 24) can be isolated by distillation of the excess $SO_2$ at $-78\,^{\circ}C$ *in vacuo*. However, since they decompose at room temperature, they have been converted to their corresponding sulfinic esters via their sodium salts[63].

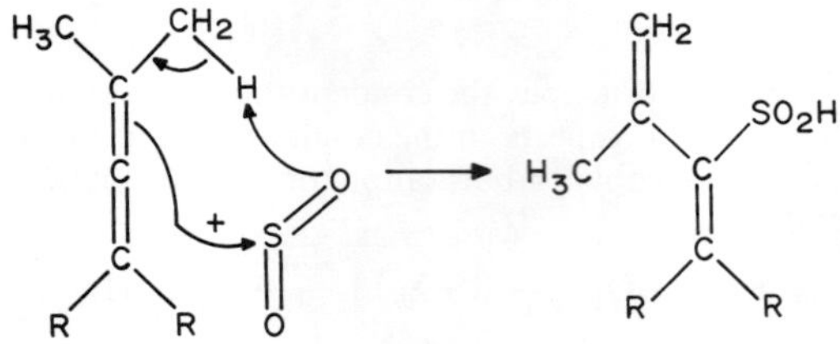


(24)

**(34)** **(35)**

(R=Me or H)

An interesting entry into the diaryl-substituted ethenesulfinic acid series, the synthetic potential of which has not as yet been explored, is the reaction of alkynes with benzene, antimony pentafluoride and liquid sulfur dioxide according to the suggested mechanism[64] in equation 25. Although the sulfinylations of aromatic and unsaturated systems by sulfur dioxide are known to be facilitated by antimony pentafluoride[65] and the somewhat similar reactions of alkyl or aryl halides with alkynes under Friedel–Crafts conditions are also known[66], it is not yet known how to control the selectivity of the above reaction (e.g. equation 25). On the other hand, the sulfinic acids **37** are not easily available by other routes. Some more work in this direction is clearly needed.

$$C_6H_5C{\equiv}CX \xrightarrow[SbF_5]{SO_2} C_6H_5\overset{+}{C}{=}CXSO_2SbF_5^- \xrightarrow[(-H^+)]{C_6H_6}$$

**(36)**

**(a)** X = H

**(b)** X = Br

$$(C_6H_5)_2C{=}CXSO_2SbF_5^- \xrightarrow[-SbF_5]{H^+} (C_6H_5)_2C{=}CXSO_2H \qquad (25)$$

**(37)**

**(a)** X = H

**(b)** X = Br

The procedure for the preparation of 1-bromo-2,2-diphenylethenesulfinic acid (**37b**)[64] according to equation 25 gave a yield of 67%, which could be easily separated from the accompanying 2-bromo-3-phenylbenzothiophene sulfoxide (16%) by-product.

### *4. The reaction of sulfur dioxide with arenediazonium salts*

Arenesulfinic acids can be obtained by the direct reaction of sulfur dioxide with arenediazonium salts in the presence of cuprous salts[5].

$$ArNH_2 \longrightarrow ArN_2^+HSO_3^- \xrightarrow[2.\ H_2O]{1.\ SO_2/Cu} ArSO_2H + N_2 + H_2SO_4 \qquad (26)$$

Although the yields are generally high and the method has been known for many years[67], its scope is somewhat limited since the presence of the amino group in the aromatic ring of the starting material is essential and, in turn, determines the possible substitution pattern of the aromatic portion of the target molecule.

All three isomers of chlorobenzenesulfinic acid (**6a**) were prepared[67a,c] starting from the corresponding chloroanilines in excellent yields for the *ortho* and *para* isomers, but in poor yield for the *meta* isomer.

### E(b). Sulfination of Olefins with Thionyl Chloride

2,2-Diarylethylene-1-sulfinic acids can be prepared by the reaction of 1,1-diarylethylenes, with $SOCl_2$, followed by hydrolysis of the initially formed $Ar_2C{=}CHSOCl$ in 28–33% yield[68a].

The reaction of olefins with thionyl chloride in the presence of aluminum chloride followed by hydrolysis leads to the formation of the corresponding 2-chloroalkanesulfinic acids by a very simple procedure[68b].

$$H_2C{=}CH_2 + SOCl_2 \xrightarrow{AlCl_3} ClCH_2CH_2\overset{O}{\overset{\|}{S}}{}^+AlCl_3^- \xrightarrow{H_2O} ClCH_2CH_2SO_2H \qquad (27)$$

The simplest such derivative, 2-chloroethanesulfinic acid (**38**), is obtained from thionyl chloride, aluminum chloride and ethylene, in 98% yield.

## F. Cleavage of the Carbon–Sulfur Bond

### 1. *Reductive cleavage of sulfones*

The base-induced reductive cleavage of sulfones appears to be by far the most extensively explored method for the synthesis of sulfinic acids.

This strategy was applied in various modifications in the preparation of all types of both aromatic and aliphatic sulfinic acids[5,6], the common feature being the cleavage of one of the carbon–sulfur bonds of the sulfone group with the concomitant reduction of the sulfur atom. Thus, in principle, this method capitalizes on the ready availability of the sulfones as starting materials. Since the key step in the synthesis is the sulfonyl–sulfinyl reduction, the reducing agents and methods which are also responsible for the carbon–sulfur bond cleavage can vary widely.

*a. Electrochemical reduction of sulfones to sulfinic acids.* Sulfones containing all combinations of alkyl, aryl and benzyl groups undergo electrochemical reduction on the mercury cathode with the aid of tetramethyl ammonium ions, leading to sulfinic acids in high yields (50–90%)[69].

$$\left.\begin{array}{ll}
\text{(i)} & 2[N(CH_3)_4]^+ + 2e^- + n\text{Hg} \longrightarrow 2[N(CH_3)_4]Hg_n \equiv \text{TMA} \\
\text{(ii)} & \underset{\textbf{(39a–d)}}{R^1SO_2R^2} + 2\text{TMA} + H^+ \longrightarrow R^1SO_2^- + R^2H + 2\text{TMA}^+ \\
\text{(iii)} & RSO_2^- \xrightarrow{H^+} \underset{\textbf{(40a–d)}}{R^1SO_2H}
\end{array}\right\} \qquad (28)$$

The alkyl–sulfur bonds are cleaved in the alkyl aryl sulfones, while the alkyl vinyl sulfones gave the 1-alkenesulfinic acids on reduction[70]. Aliphatic sulfones do not undergo this reductive cleavage.

The sulfinic acids **40a–d** [(**a**) $R^1 = p\text{-}CH_3C_6H_5$; (**b**) $R^1 = C_6H_5$; (**c**) $R^1 = C_3H_7$; (**d**) $R^1 = C_6H_5CH_2$], as illustrative examples, were obtained[69] starting from **39a–d**, respectively [(**a**) $R^1 = p\text{-}CH_3C_6H_4$, $R^2 = CH_3$ or $CH_2{=}CHCH_2$; (**b**) $R^1 = C_6H_5$, $R^2 = C_6H_5CH_2$ or $C_6H_5$; (**c**) $R^1 = C_3H_7$, $R^2 = C_6H_5CH_2$; (**d**) $R^1 = R^2 = C_6H_5CH_2$]. In all cases, the sulfone and tetramethylammonium chloride in methanol on electrochemical reduction at a mercury cathode gave the acids $R^1SOOH$ in 75–90% yield (as the corresponding sodium salts).

*b. Sodium amalgam reduction.* The use of sodium amalgam for the reduction of sulfones is closely related to the electrochemical synthesis of sulfinic acids. Thus, diaryl and alkyl aryl sulfones were found to undergo reductive fission upon treatment with sodium amalgam to the corresponding arenesulfinic acid[71]:

$$ArSO_2R \xrightarrow{\text{Na–Hg}} ArSO_2H + RH \qquad (29)$$

In view of the several alternative available methods for the preparation of arenesulfinic acids, there appears to be no particular advantage in using this method.

### 2. *Reductive fission of sulfones with alkaline metal amides*

It is possible to cleave aryl–S bonds and benzyl–S bonds of alkyl aryl sulfones[72] and alkyl benzyl sulfones[73] by using metallic lithium in methylamine or metallic sodium in liquid ammonia as the reducing agents. The carbon–sulfur bond in dialkyl sulfones can only be cleaved with the lithium amide, whereas in diaryl sulfones the lithium amide may further reduce the sulfinic acids formed to the corresponding thiols.

$$R^1SO_2R^2 + 2M + RNH_2 \xrightarrow[(-MNH_2)]{} R^1SO_2^-M^+ + R^2H \qquad (30)$$

$$M = \text{Li or Na}$$

$$R = \text{Me or H}$$

Whether by simultaneous addition of two electrons, or stepwise addition of two single electrons, the cleavage of the carbon–sulfur bond in the sulfones occurs by release of electrons from the dissolving metal eventually to the compound being cleaved.

According to the above method, n-decanesulfinic acid (**41**) was obtained[72] from n-decyl phenyl sulfone in methylamine with lithium metal, in 95% yield.

The reduction of phenyl *o*-tolyl sulfone (**42**) with sodium in liquid ammonia followed by

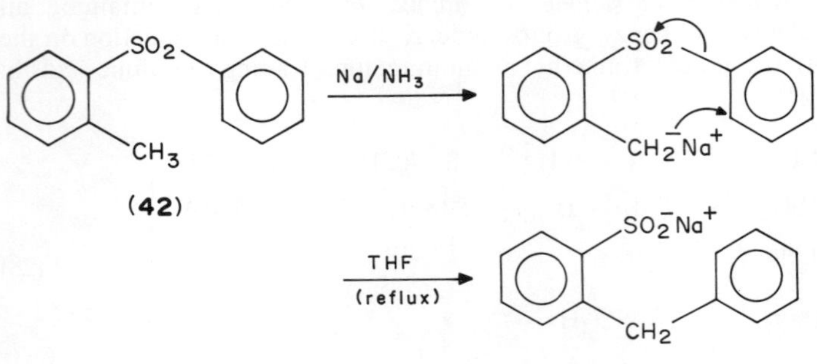


(**42**)

(**43**)

(31)

reflux in THF affords the Truce–Smiles rearrangement product (**43**) in good yields[74].

This constitutes an interesting entry into some *ortho*-substituted arenesulfinic acids (see Section II.F.5 below).

Reduction of the toluenesulfonyl derivatives of cysteamine, L-cysteine, and L-homocysteine with metallic sodium in liquid ammonia gives the corresponding sulfinic acids[73] as in the case of 'ordinary' sulfones. Thus, L-cysteine sulfinic acid (**44**) was obtained[73] from *S*-benzyl-L cysteine sulfone in liquid ammonia with sodium ($[\alpha]_D^{23} = +24.0°$; 77% yield).

### 3. *The base-induced cleavage of β-substituted sulfones*

The reductive, nucleophilic cleavage of sulfones with bases/nucleophiles is best achieved with sulfones substituted in the β-position with electron-withdrawing groups. No β-nucleophilic substitution accompanies the predominant elimination–addition sequence which leads to the formation of the sulfinic acid[75]. The method is applicable for the synthesis of both aliphatic and aromatic sulfinic acids, and is based on cleavage of the sulfones by thiolate[76] or cyanide[77] ions.

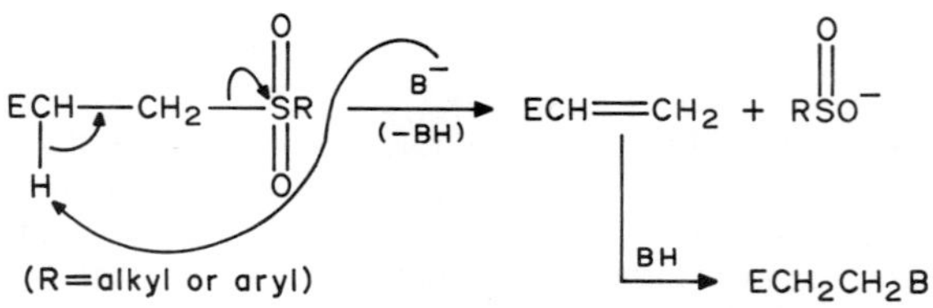


(32)

The sodium salt of n-propanesulfinic acid (**40c**) was prepared[76] from β-propylsulfonylpropionitrile with 1-propanethiol, sodium and ethanol, in 91% yield. The free sulfinic acid was obtained by acidification in *ca* 60% yield.

Similar high yields of n-butane— as well as *p*-toluene— and *o*-toluene-sulfinic acids were also obtained through this procedure.

The same principle is applied to symmetrical β-substituted sulfones, the cleaving base in the elimination–addition sequence being again either the cyanide or the thiophenolate ion. The resulting sulfinic acids are substituted at the β-position, and the yields are good whenever the β-substituent is a good electron-withdrawing group[77].

$$\underset{(\mathbf{45})}{(C_6H_5COCH_2CH_2)_2SO_2} \xrightarrow[2.\ H_3O^+]{1.\ CN^-\ \text{or}\ RS^-} \underset{(\mathbf{46})}{C_6H_5COCH_2CH_2SO_2H} + C_6H_5COCH_2CH_2CN \quad (33)$$

3-Oxo-3-phenylpropanesulfinic acid (**46**) was prepared[77] from the sulfone **45** and sodium cyanide in refluxing ethanol, in a yield of 81%.

In quite an analogous fashion, the basic ($K_2CO_3$–$CH_3CN$) β-elimination of γ-keto triflones (i.e. the β-substituted sulfone **47**) removes the triflyl group and thus provides the trifluoromethanesulfinic acid (as its potassium salt **10**)[24].

$$\underset{(\mathbf{47})}{C_2H_5COCH_2CH(C_6H_5)SO_2CF_3} \xrightarrow[CH_3CN]{K_2CO_3} C_6H_5CH{=}CHC_6H_5 + \underset{(\mathbf{10})}{CF_3SO_2^-K^+} \quad (34)$$

*4. The base-induced cleavage of phthalimidomethyl sulfones and sulfonylpyridines*

Although the alkaline cleavage of alkyl sulfones to alkali metal alkanesulfinates has been improved by using rather reactive sulfones such as 1,2-dialkyl-sulfonylethane[16] and alkyl, 3-alkylsulfonylacrylates[76] (see Section II.F.3 above), the yields and purity of the products are not always sufficiently high. An attractive alternative route for the preparation of sulfinic acids is the cleavage of phthalimidomethyl sulfones with sodium ethoxide in ethanol. The procedure is simple and the yields are high (92–100%)[78].

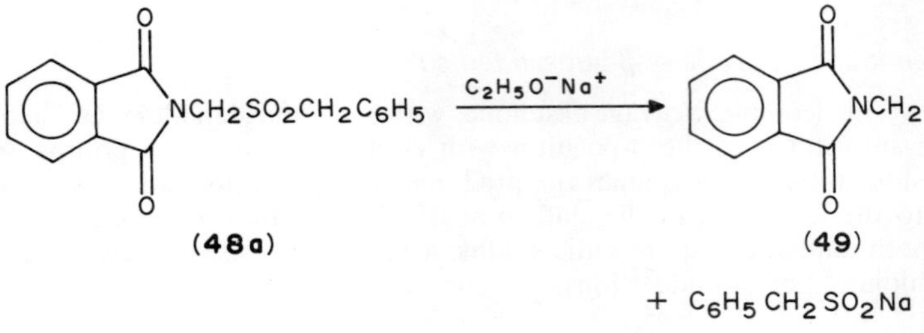


(48a) (49) (40d–Na) (35)

The use of a slight excess of ethoxide facilitates the clear separation between the two products so that both **49** and **40d–Na** are obtained in nearly quantitative yields. Whereas sulfones **50** are ordinarily cleaved with sodium ethoxide followed by work-up of the resulting mixture, in the case of higher alkanesulfinates (n-$C_{12}H_{25}SO_2H$, n-$C_{14}H_{29}SO_2H$, etc.), the crystalline products precipitate from the reaction solution and could thus be isolated in pure state by simple filtration. If the cleavage of **50** is carried out with the appropriate alkanethiolates in ethanol, then the released sulfide **52** may be recycled upon oxidation with potassium permanganate[78] (equation 36).

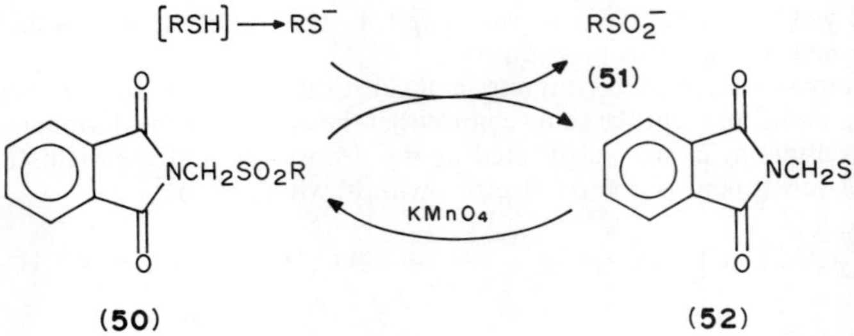


(50) (52) (36)

The cleavage reaction with thiolates is much faster than the cleavage reaction by the ethoxide due, primarily, to the difference in nucleophilicity between these two nucleophiles.

Two general procedures are used[78] for the cleavage of alkyl phthalimidomethyl sulfones **50** to sodium alkanesulfinates **51** [R = n-$C_{5-16}H_{11-33}$ and R = $(C_2H_5)_2CH$- or *i*-$C_3H_7$—$CH(CH_3)$—]. In method A, sodium ethoxide is used to cleave the phthalimido sulfone and the alkanesulfinate is obtained as a powder in 92–100% yield. In method B, sodium alkanethiolates are the reagents and the alkanesulfinates are obtained as a powder in essentially quantitative yields.

In a similar manner, *ipso*-substitution reaction of 2-sulfonylpyridines and their *N*-oxides with alkoxide or thiolate anions affords the sodium salts of sulfinic acids in high yields (38–92% with the pyridines, and 70%-quantitative with the pyridine-*N*-oxides)[79]. In this method the sulfone of the pyridine-*N*-oxide (i.e. **53b**) may work as a mediator in a cycle for the preparation of various kinds of sulfinic acids by using the thiolate ion as the cleaving nucleophile, as illustrated below[79] (Scheme **37a**). Thus, in comparison with other methods for the synthesis of sulfinic acids, this procedure has the advantages of (a) not requiring tedious processes, and (b) the mediated sulfone can be easily prepared from the commercially available 2-mercaptopyridine *N*-oxide.

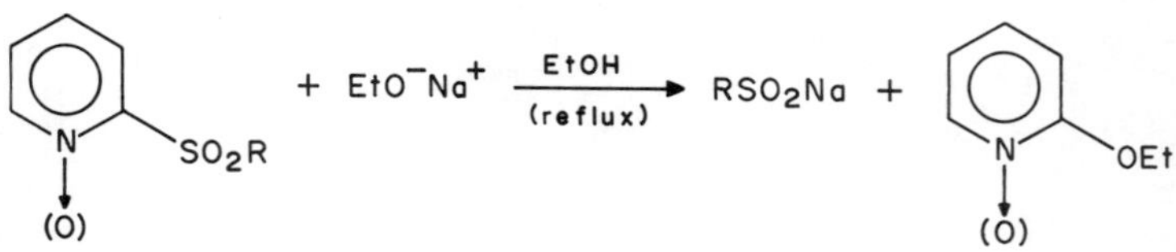


(37)

**(53a,b)** **(54a,b)**

**(a)** pyridine; **(b)** pyridine-*N*-oxide

(R=alkyl or aryl)

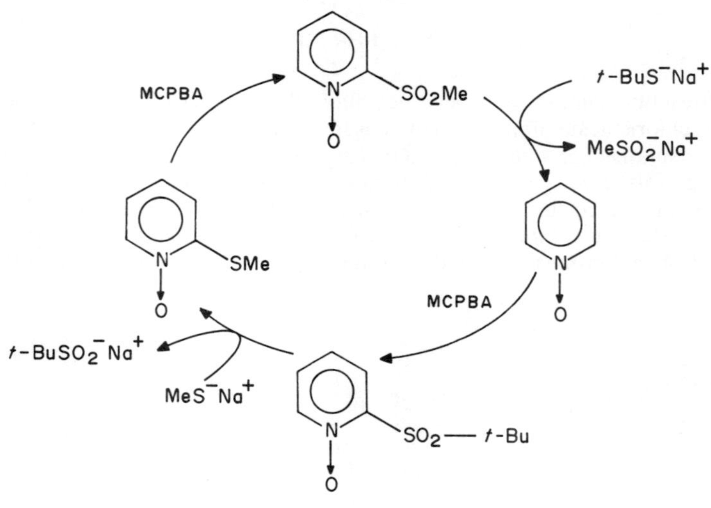


(37a)

## *5. Base-promoted Smiles rearrangement of aryl sulfones and of benzylically lithiated sulfones*

Several 2-substituted diaryl or alkyl aryl sulfones were found to undergo base-induced rearrangement to give sulfinic acids in high yields[6,80] (equation 38). Thus, various substituted arenesulfinic acids can be prepared by dissolving *ortho*-substituted aryl sulfones in aqueous sodium hydroxide and letting the two react. Extracting the $CO_2$-

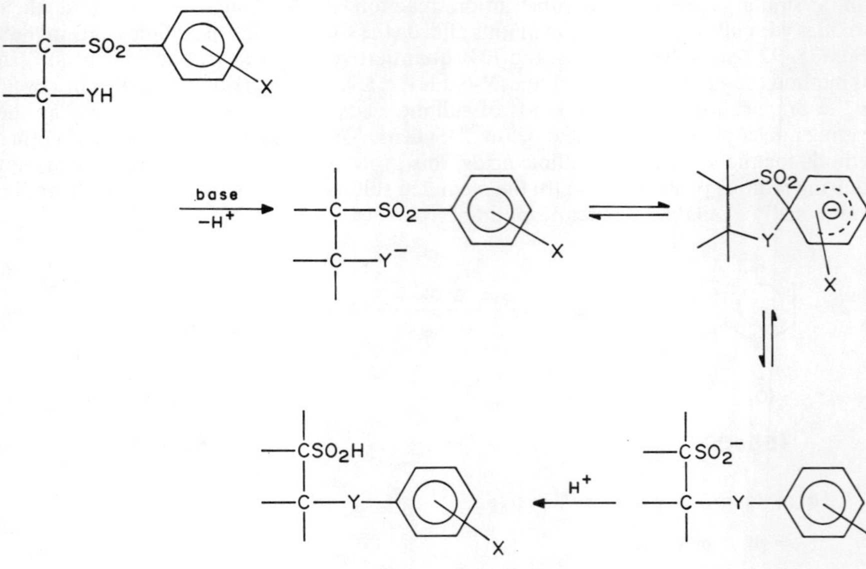

$X = 2$ or $4\text{-}NO_2$; $4\text{-}SO_2R$; 2-Hal

$YH = OH$; $NH_2$; NHR; $CH_3$ (38)

saturated reaction solution with benzene, followed by acidification of the aqueous layer with hydrochloric acid, affords the rearranged sulfinic acid[80c].

Aromatic sulfones containing an *o*-methyl group were found by Truce and coworkers[81] to rearrange to the *o*-arylmethylated arenesulfinic acids upon treatment with n-butyllithium as illustrated in equation 39. This route enabled the synthesis of some interesting *o*-substituted arenesulfinic acids on a preparative scale. Not only phenyl and substituted phenyl groups are capable of migrating (see equations 38 and 39), but also

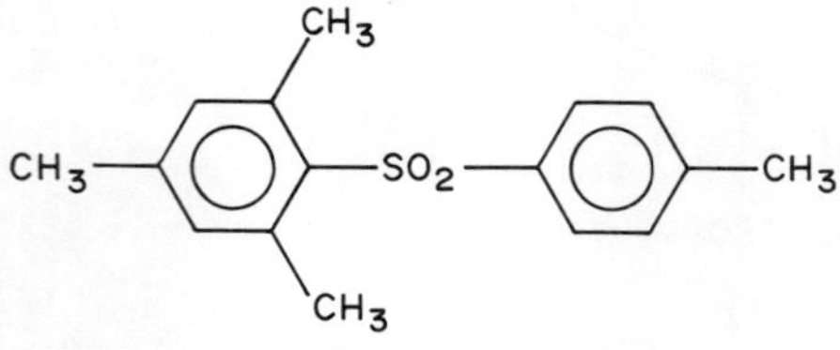

**(55)**

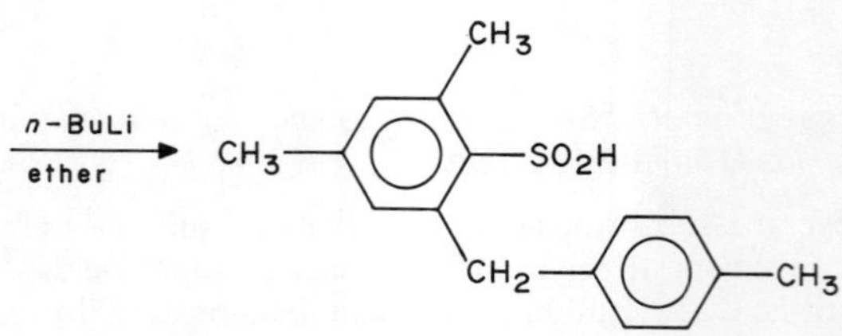

(39)

**(56)**

naphthyl groups. Thus, both mesityl, 1-naphthyl and mesityl 2-naphthyl sulfones rearrange in the presence of butyllithium in ether or potassium *t*-butoxide in dimethyl sulfoxide to yield the *ortho*-naphthylmethylsulfinic acids. The yields with potassium *t*-butoxide are substantially higher[82].

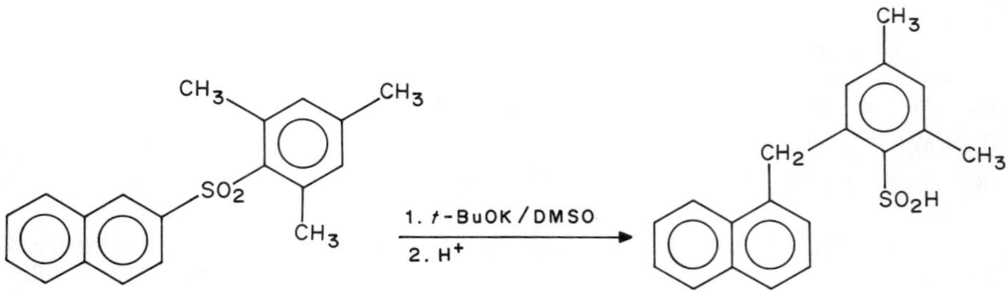


**(57)** **(58)** (40)

The products from the rearrangement of the naphthyl sulfones indicate that the nucleophilic displacement has occurred at the ring carbon α to the point of attachment of the sulfone, rather than directly at the ring carbon bearing the sulfone group as in the case of the phenyl sulfone series[82].

Recently, it was shown that lithiation of the benzylic carbon of *o*- and *p*-tolyl *tert*-butyl sulfones is followed by migration of the *alkyl* (*t*-Bu) group from sulfur to the benzylic carbon[83]. This further extended the scope of this rearrangement and its potential in the synthesis of sulfinic acids.

The rearrangement of mesityl 1-naphthyl sulfone[82] in the presence of butyllithium (Method A) yielded 42% of the 2-(2′-naphthyl-methyl)-4,6-dimethylbenzenesulfinic acid (**59**) while 51% of the starting sulfone could be recovered. In Method B, the potassium *t*-butoxide-dimethyl sulfoxide induced rearrangement yielded the sulfinic acid **59** in 84% yield, while 12% of the sulfone was recovered. By using procedure B the mesityl *p*-tolyl sulfone **55** yields the sulfinic acid **56** in 74% yield and in Method A the yield of the sulfinic acid **56** is 88%[82].

### 6. Base-induced cleavage of SO- and $SO_2$-containing heterocycles

The base-induced cleavage of the carbon–sulfur bonds in cyclic sulfoxides and sulfones is well known and thoroughly studied[84], particularly as far as the three-membered sulfoxides and sulfones are concerned[85]. However, the practicality of the use of SO- and $SO_2$-containing 3–6 membered ring heterocycles for the synthesis of acyclic sulfinic acids has to be carefully assessed in each case. Only if the starting heterocycle is readily available or if the alternative strategies are very difficult, should this methodology be applied.

*a. Cleavage of six-membered ring sulfones.* Thioxanthen-9-one, 10,10-dioxide derivatives readily react with methoxide ion, resulting in the displacement of the sulfone linkage to give the corresponding methoxybenzophenonesulfinic acids (**61**) in high yields (> 90% in most cases)[86]. The displacement reaction occurs exclusively on the more electrophilic aromatic ring (equation 41). For instance, methoxybenzophenonesulfinic acid **61e** can be prepared[86] from the sulfone[87] **60e** in 94% yield.

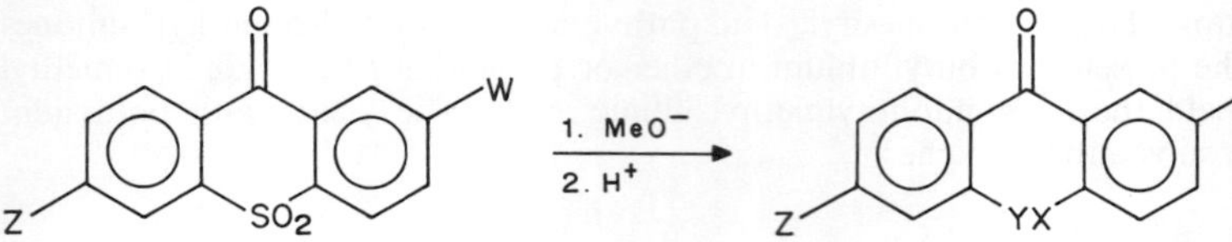


(41)

**(60)** **(61)**

| | | | |
|---|---|---|---|
| **(a)** | $W = Z = H$ | **(a)** | $W = Z = H$; $X = OCH_3$; $Y = SO_2H$ |
| **(b,c)** | $W = Cl, NO_2$; $Z = H$ | **(b,c)** | $W = Cl, NO_2$; $X = OCH_3$; $Y = SO_2H$; $Z = H$ |
| **(d,e)** | $W = CH_3, OCH_3$; $Z = H$ | **(d,e)** | $W = CH_3, OCH_3$; $X = SO_2H$; $Y = OCH_3$; $Z = H$ |

*b. Cleavage of five-membered rings: dihydrothiophene dioxides and 1,3-oxathiolane S-dioxides.* 1,3-Alkadienesulfinates can be obtained by ring cleavage at the carbon–sulfur bond of several 2,5-dihydrothiophene 1,1-dioxides with Grignard reagents[88], as shown in equation 42. The claimed yields of the bromomagnesium salts obtained are high (98%) and they appear to be quite stable to light at room temperature. However, the actual preparation of the corresponding sulfinic acids is not reported.

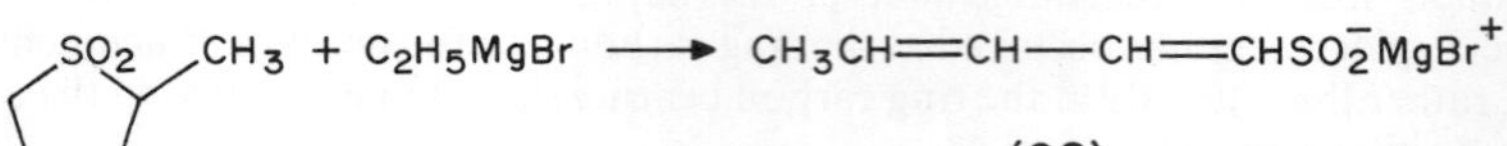


(42)

**(62)**

The 1,3-oxathiolane *S*-dioxide (obtained by oxidation of the parent oxathiolane) undergoes cyclofragmentation when treated with strong bases. Thus unstable ethenesulfinates (as well as formaldehyde) are formed as shown below[88]; in 80% yield.

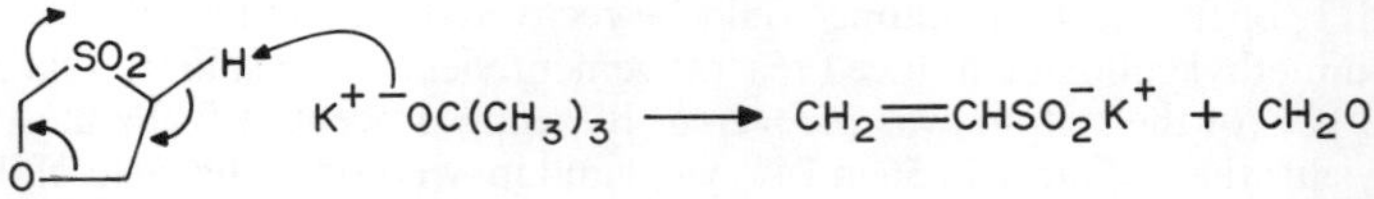


(43)

**(63)**

On acidification, the free vinyl sulfinic acid (30%) is obtained as a viscous colorless oil which polymerizes after a short time in the presence of aqueous acids.

*c. Cleavage of (four-membered) thietane oxides and dioxides.* Reaction of either *cis*- or *trans*-2,4-diphenylthietane 1-oxide (**64a**, **65a**) with potassium *tert*-butoxide yielded a mixture of *cis*-1,2-diphenylcyclopropanethiol (**66**) and *cis*-1,2-diphenylcyclopropanesulfinic acid[89] (**67**). Since (a) the starting thietane oxides are not readily available; (b) the reaction provided a mixture of products; and (c) the yield of the ultimately isolated three-membered ring sulfinic acids is rather low (10–20%), this strategy has no preparative value.

However, *cis*- and *trans*-2,4-diphenylthietane 1,1-dioxides (**64b**, **65b**), when treated with ethylmagnesium bromide, rearrange apparently through a mechanism resembling that of the Stevens rearrangement[89] to *trans*-1,2-diphenylcyclopropanesulfinic acid (**68**) in a highly stereoselective manner and in good yields[90] as shown in equation 44. Starting from *trans*-2,4-diphenylthietane 1,1-dioxide (**65b**) the yield is 77%, while *cis*-1,2-diphenylthietane (**64b**) yields the same acid (**68**), in 74% yield.

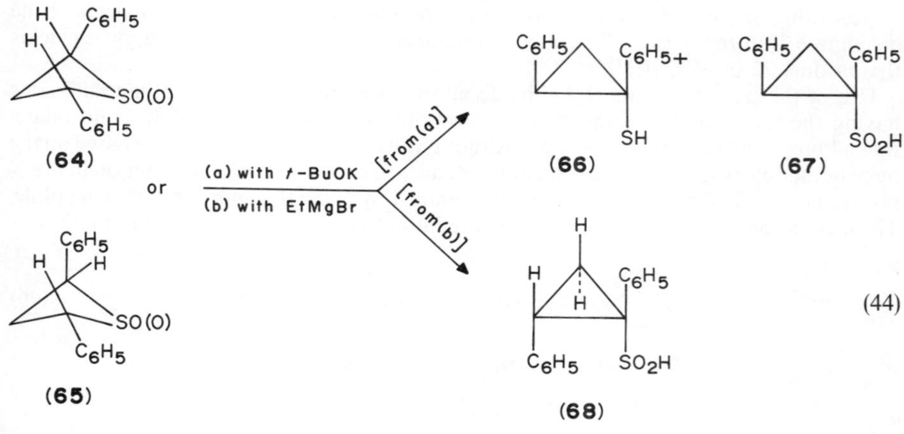


(44)

(a) sulfoxide ; (b) sulfone

*d. Cleavage of the carbon–sulfur bond in thiirane dioxides.* The cleavage of the carbon–sulfur bond of thiirane-1,1-dioxides, with soft bases—thiolate, sulfide and thiourea—or with organometallic reagents to yield the corresponding sulfinic acids is a facile process[91,92]. The yields are good and the bifunctional products obtained may serve as versatile intermediates for further transformations (equation 45). It should be pointed out that in the case of the alkyl-organometallic reagents (route d), the sulfur and not carbon is the site of initial attack. Consequently, the thiirane dioxide serves as an $SO_2$ transfer agent to the metal of the organometallic reagent, the result being the extrusion of the $SO_2$ moiety from the three-membered ring[85] with concomitant formation of the sulfinate of the alkyl group. This may constitute an easy route to alkanesulfinic acids, but since several other easy methods are available, its preparative usefulness is questionable.

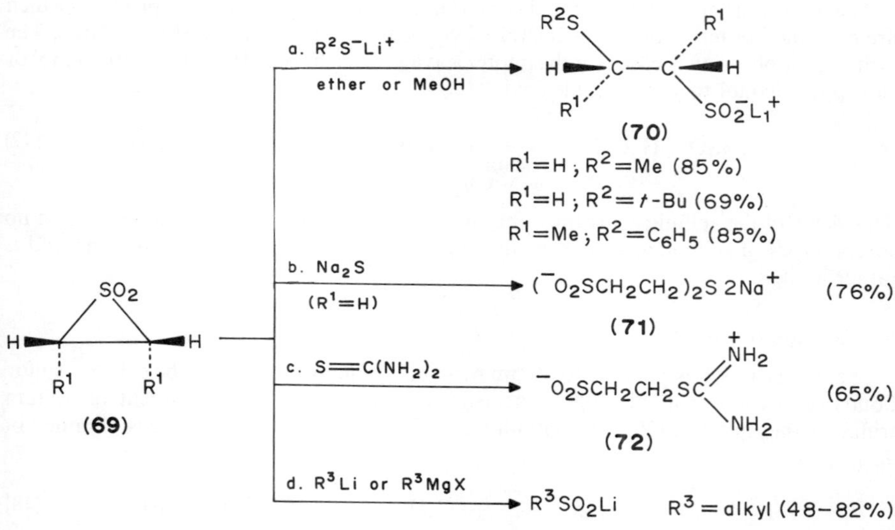


(73)

(45)

According to method b in equation 45, the parent sulfone **69** with $Na_2S$ yields of the disulfinate **71** in 76% yield[91]. The same sulfone according to method c with thiourea yields the product **72** in 65% yield[91].

Use of the parent thiirane 1,1-dioxide affords a useful synthesis of ethanesulfinates having the following functions in the 2-position: thiol, disulfide, trisulfide, thiosulfate, thiosulfonate and phosphorothioate[93]. Although all these transformations are based on the nucleophilic cleavage of the three-membered thiirane ring, the particular function in the 2-position of the final product can be predetermined by using the appropriate nucleophile. Thus, equation 46 is an extension of the method already described in equation 45.

(46)

(**69**; R=H) (**74**) (**75a–e**)

(**a**) Nu = p-$MeC_6H_4S^-$; M = Na (73%)

(**b**) Nu = $C_6H_5CH_2S^-$; M = Na (93%)

(**c**) Nu = $HS^-$; M = Na (> 67%)

(**d**) Nu = $NaO_3SS^-$; M = Na (72%)

(**e**) Nu = $Li_2O_2(O)PS^-$; M = Li (100%)

Interestingly, the sulfinic esters corresponding to **75** are rather stable. For instance, the Bunte salt, disodium *S*-(2-sulfinoethyl)thiosulfate (**75d**)[93], could be prepared from thiirane 1,1-dioxide and $Na_2S_2O_3$ in 72% yield. In several instances, however, the salts gave unexpected results due to the involvement of the sulfinate function with the sulfur atom of the 2-substituent[93].

### *7. Photochemical cleavage of benzylic sulfones*

The photolysis of benzylic sulfonyl compounds ($RSO_2CH_2C_6H_5$) gives products which are explained in terms of the chemistry of the benzyl and sulfonyl ($RSO_2^{\bullet}$) radicals. The synthetic application is based on the photocleavage of such benzyl sulfones in methanol or isopropyl alcohol to yield sulfinic acids[94].

$$RSO_2CH_2C_6H_5 \xrightarrow[\textit{i}\text{-PrOH (or MeOH)}]{h\nu} RSO_2H + C_6H_5CH_2CH_2C_6H_5 \quad (47)$$

The yields of the sulfinic acids thus obtained are not particularly high (26–63%) and no special types of sulfinic acids can be thus synthesized. Hence, the use of this method is rather limited.

### *8. Cleavage of the carbon–sulfur bond of sulfinic acids*

In certain cases, it is possible to obtain new sulfinic acids by cleaving the carbon–sulfur bonds of already available sulfinic acids. A case in point is the treatment of lithium trifluoroethensulfinate (**76**) with butyllithium. The lithium butanesulfinate is obtained in 90% yield[95].

$$F_2C{=}CF{-}SO_2^-Li^+ \xrightarrow{\text{BuLi}} C_4H_9{-}SO_2^-Li^+ + F_2C{=}CFLi \quad (48)$$

(**76**) (**17–Li**)

## G. Cleavage of Sulfur–Nitrogen and Sulfur–Oxygen Bonds

### 1. Sulfonamides

Aromatic sulfonamides can be cleaved electrochemically at the mercury cathode with trimethylammonium (TMA) ions to produce the corresponding sulfinic acids on a preparative scale[96].

$$\mathrm{ArSO_2NHR} \xrightarrow[\mathrm{TMA}]{+2e+2\mathrm{H}} \mathrm{ArSO_2H + RNH_2} \tag{49a}$$

Only the aromatic sulfonamides (e.g. benzene- and toluene-sulfonamides) undergo readily this reductive cleavage to provide the arylsulfinic acids in excellent yields (86–97%). Although the scope of this method is limited to the aromatic series and is contingent on the availability of the corresponding sulfonamides, the actual procedure is simple and straightforward (see Section II.G.2 below).

An additional fringe benefit of this method is that, if the amine portion of the starting sulfonamide is optically active, this configuration is preserved in the liberated amine during and after the cleavage process of the sulfur–nitrogen bond.

### 2. Sulfonic esters

Aromatic sulfonic esters readily undergo electrochemical, reductive fission of the sulfur–alkoxy oxygen bond leading to aromatic sulfinic acids and alcohols in high yields and purity[96,97].

$$\mathrm{ArSO_2{-}OR} \xrightarrow[\mathrm{TMA}]{+2e+2\mathrm{H}} \mathrm{ArSO_2H + ROH} \tag{49b}$$

This process is completely analogous to the electrochemical, reductive cleavage of the sulfur–nitrogen bond described in the previous subsection. In this case too, both the original configuration and optical activity of the alcohol portion in the molecule (if chiral) are maintained.

The following is an illustrative example of the actual procedure which can be applied equally to both aromatic sulfonamides and sulfonic esters[97]. The tosyl esters containing 1-menthyl, 1-bornyl, or methylbenzylcarbinyl group, on electrolysis (15–18 V and 0.5–0.75 A) in the presence of tetramethylammonium chloride in ethanol, yield, after the usual work-ups and final acidification, 91–99% yield of *p*-toluenesulfinic acid (**4**). The yields of the alcohols are: 1-menthol—73%; 1-borneol—95%; and menthylbenzylcarbinol—85%.

Arenesulfonates of mandelonitrile (i.e. **77**) eliminate the arenesulfinate ion in high yields when treated with sodium ethoxide in ethanol[98].

$$\mathrm{EtO^- + H{-}\underset{\displaystyle CN}{\underset{|}{C}}R{-}O{-}SO_2Ar \longrightarrow ArSO_2^- + RCOCN + EtOH} \tag{50}$$

(**77**)

Although the starting material is readily available[98], this method is not attractive for the synthesis of arenesulfinic acids, since easier and simpler methods are available, Thus, the sulfonyl chloride itself, which is required for the synthesis of the ester, can be transformed directly to the corresponding sulfinic acid.

### 3. Sulfonyl hydrazines

On heating, the cross-linked polystyrene–divinylbenzene matrix which is functionalized with sulfonylhydrazine groups (i.e. **78**) evolves diimide and a sulfinic acid–functionalized resin is obtained[99].

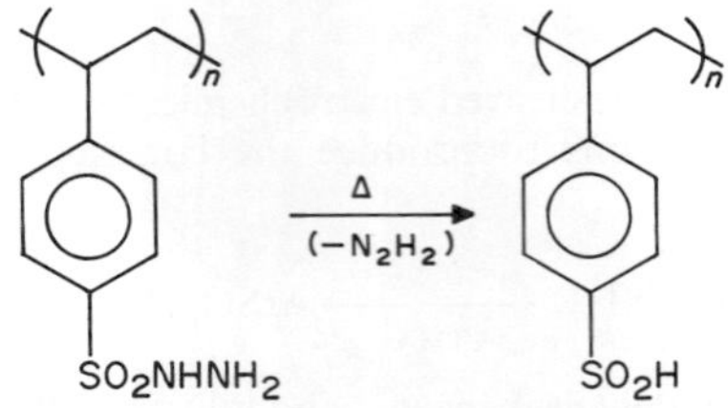

(78)

(51)

This reaction demonstrates the potential of sulfonylhydrazines for the synthesis of sulfinic acids via the cleavage of the sulfur–nitrogen bonds.

## H. Miscellaneous

A few more routes leading to sulfinic acids will be just mentioned briefly for the sake of completeness.

The following equilibrium should be considered in order to find the best conditions for obtaining the product:

$$M^+ + RSO_2H \rightleftharpoons RSO_2^- M^+ + H^+ \qquad (52)$$

Ammonium and alkali metal salts are most frequently used for preparative purposes[100], and sulfuric acid is preferred to hydrochloric acid for acidification[101].

Thiolates generated electrochemically from disulfides in dry *N*,*N*-dimethylformamide (DMF) react with oxygen to give a mixture of disulfide and sulfinate. If oxygen is present during the electrochemical reduction, sulfinate derivatives are obtained in good yield[102].

$$\begin{aligned} &RSSR \longrightarrow 2RS^- \\ &RS^- + O_2 \rightleftharpoons RS\cdot + O_2^- \\ &RS\cdot + O_2^- \longrightarrow RSO_2^- \\ &2RS\cdot \longrightarrow RSSR \end{aligned} \qquad (53)$$

Since the sulfinic acids have not been isolated as such from the reaction mixture, more work is still needed before the preparative value of this method can be assessed.

A recent paper[103] describes the preparation of disulfide sulfinates $RSS(CH_2)_nSO_2Na$ (**79**) $[NaO_2S(CH_2)_nS]_2S$ ($n = 3$–5) (**80**). This method has been already described and discussed in Section II.E(a).2.c. Equation 54 shows an extension of the possibilities available. The disulfide sulfinate with $n = 3$ and $R = HO_2CCH_2$, i.e. sodium 3-(2-carboxyethyldithio) propanesulfinate (**79a**), was obtained[103] in 79% yield from 3-mercaptopropionic acid and 1,2-dithiolane 1,1-dioxide[36b] (**81**, $n = 3$).

$$RS{-}S(CH_2)_n\ddot{S}O_2Na \rightleftharpoons RSNa + \overline{S(CH_2)_nSO_2} \xrightarrow{0.5Na_2S} (\mathbf{80})$$

(**79**) (**81**)

$$\updownarrow \; {+79}, \; {-RSSR} \qquad (54)$$

$$NaS(CH_2)_nSO_2Na \xrightarrow[(or\ +\mathbf{81})]{+\mathbf{79}\text{-}RSNa} [S(CH_2)_nSO_2Na]_2$$

(**82**)

$R = HO_2CCH_2$; $p$-$CH_3C_6H_4$
$C_6H_5$; 2,4,6-$(CH_3O)_3C_6H_2$

## III. TABLE

Synthesis of selected sulfinic acids $RSO_2H$ (or their corresponding salts): Starting materials, methods, yields and references

| R | Starting materials/reagents | Method* | Yield(%) | Reference |
|---|---|---|---|---|
| $CH_3$ | $CH_3SOCl$; $H_2O$ | II.B | 49–61[a] | 33 |
| $CH_3$ | 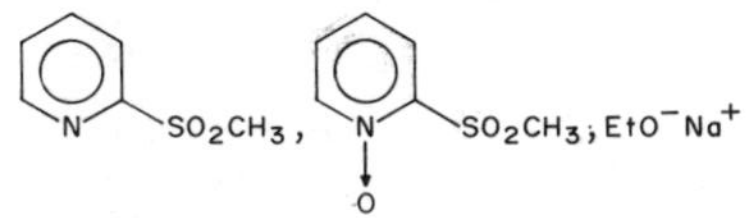 | II.F.4 | 58; 100 | 79 |
| $CH_3$ | $CH_3S(O)OCH_3$; NaOH | II.B. | ~100[b] | 26 |
| $CH_3$ | 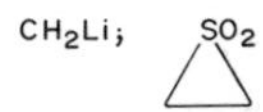 | II.F.6 | 69[c] | 92 |
| $CH_3$ | 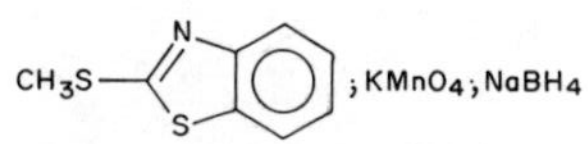 | II.D. | ~70[b] | 42 |
| $ClCH_2CH_2$ | $CH_2{=}CH_2$; $AlCl_3$; $SOCl_2$ | II.E(b) | 98 | 68 |
| $CH_3CH(OH)$ | $CH_3CH_2OH$; $SO_2$ | II.E(a).1 | ~45[b] | 54 |
| $NaO_3SSCH_2CH_2$ | 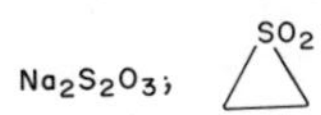 | II.F.6 | 72 | 93 |
| $(CH_3)_2C(OCH_3)$ | $(CH_3)_2CHOCH_3$; $SO_2$ | II.E(a).1 | 55 | 54 |
| $C_6H_5COCH_2CH_2$ | $(C_6H_5COCH_2CH_2)_2SO_2$; $CN^-$ | II.F.3 | 81 | 77 |
| $CH_3CH_2CH_2$ | $CH_3CH_2CH_2SO_2CH_2C_6H_5$; $(CH_3)_4\overset{+}{N}Cl^-$ | II.F.1 | 90[b] | 69 |
| $CH_3CH_2CH_2$ | $CH_3(CH_2)_2SO_2CH_2CH_2CN$; $CH_3(CH_2)_2S^-Na^+$ | II.F.3 | ~60 (>90)[b] | 76 |
| $(CH_3)_2CH$ | 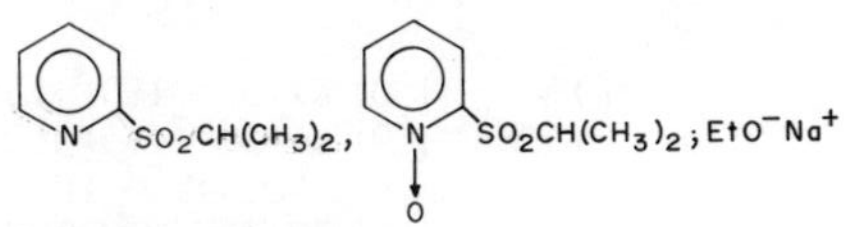 | II.F.4 | 38; 100 | 79 |
| $(CH_3)_3C$ | $(CH_3)_3CLi$; $SO_2$ | II.E(a).2 | 92[c] | 57 |
| Bu | BuSH; $m$-$ClC_6H_4CO_3H$ | II.D | 81 | 41 |
| Bu | $BuSO_2Cl$; $(C_2H_5)_3Al$ | II.A | 99 | 23 |

*(continued)*

III. **TABLE** *(continued)*

| R | Starting materials/reagents | Method* | Yield (%) | Reference |
|---|---|---|---|---|
| Bu | $BuSO_2SC_2H_5$; $Bu_2PONa$ | II.B | 86[b] | 35 |
| $CH_3(CH_2)_7$ | $(C_8H_{17})_3Al$; $SO_2$ | II.E(a).2 | 94 | 61a |
| $CH_3(CH_2)_7$ | $CH_3(CH_2)_7SO_2Cl$; $(C_2H_5)_3Al$ | II.A | 93 | 23 |
| $CH_3(CH_2)_7$ | $CH_3(CH_2)_7SO_2Cl$; $(C_2H_5)_{1.5}AlCl_{1.5}$ | II.A | 95 | 23 |
| $CH_3(CH_2)_7$ | 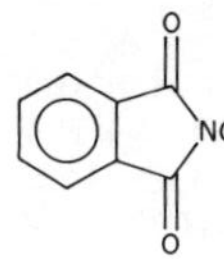 $NCH_2SO_2C_8H_{17}$; $C_2H_5O^-Na^+$ | II.F.4 | 95 | 78 |
|  | 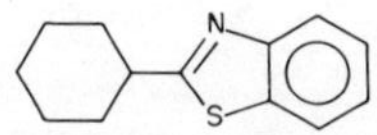 ; $KMnO_4$, $NaBH_4$ | II.D | 68 | 42 |
| $CH_3(CH_2)_9$ | $(CH_3(CH_2)_9SO_2C_6H_5$; Li, $MeNH_2$ | II.F.2 | 95 | 72 |
| $CH_3(CH_2)_{11}$ | $CH_3(CH_2)_{11}Br$, Mg; $SO_2$ | II.E(a).2 | 80 | 16 |
| $CH_3(CH_2)_{11}$ | 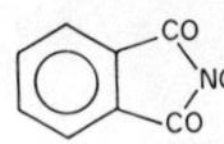 $NCH_2SO_2C_{12}H_{25}$; $C_{12}H_{25}S^-Na^+$ | II.F.4 | 100 | 78 |
| $C_6H_5\overset{*}{C}H(CH_3)$ | $C_6H_5\overset{*}{C}H(CH_3)$—S— 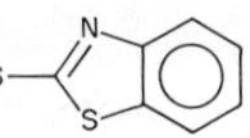 ; $KMnO_4$; $NaBH_4$ | II.D | <66 | 42 |
| $Cl_3C$ | $Cl_3CSO_2Cl$; $H_2S$ | II.A | 99[d] | 20 |
| $CF_3(CF_2)_3$ | $CF_3(CF_2)_3SO_2F$; $H_2NNH_2$ | II.A | 13[e] | 19b |
| $CH_2N(CH_3)CHO$ | $(CH_3)_2NCHO$; $SO_2$ | II.E(a).1 | 35 | 54 |
| $CH_2{=}CH$ | 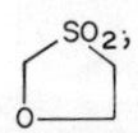 ; $t$-$BuO^-K^+$ | II.F.6 | 30(80)[b] | 88 |
| $CH_2{=}C(CH_3)C{=}C(CH_3)_2$ | $(CH_3)_2C{=}C{=}C(CH_3)_2$; $SO_2$ | II.E(a).3 | *f* | 63 |
| $[Bu_3N(H)]_2C$ | $(BuNH)_2C{=}S$; $H_2O_2$ | II.D | 38 | 49 |
| $[C_2H_5CH(CH_3)N(H)]_2C$ | $[C_2H_5CH(CH_3)NH]_2C{=}S$; $H_2O_2$ | II.D | 37 | 48 |
| 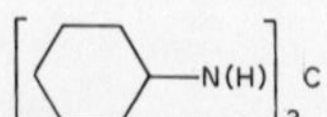 | 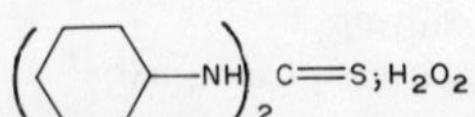 ; $H_2O_2$ | II.D | 90 | 48 |

| | | | | |
|---|---|---|---|---|
| $HO_2S(CH_2)_4$ | $(CH_2)_4(SO_2Cl)_2$; $Na_2SO_3$ | II.A | 60–66 | 15 |
| $HO_2S(CH_2)_5$ | $(CH_2)_5(SO_2Cl_2)_2$; $Na_2SO_3$ | II.A | *f* | 15 |
| $HO_2S(CH_2)_{10}$ | $(CH_2)_{10}(SO_2Cl_2)_2$; $Na_2SO_3$ | II.A | 77[d] | 15 |
| $C_6H_4SS(CH_2)_3$ | $\overline{S—(CH_2)_3—SO_2}$; $C_6H_5S^-Na^+$ | II.C | 86[b] | 103 |
| $(CH_3)_3CSS(CH_2)_4$ | $\overline{S—(CH_2)_4—SO_2}$; $(CH_3)_3CSH$ | II.C | 62[b] | 36a |
| $HO(CH_2)_2SS(CH_2)_4$ | $\overline{S—(CH_2)_4—SO_2}$; $HO(CH_2)_2SH$ | II.C | 81[b] | 36a |
| $NaS(CH_2)_4$ | $\overline{S—(CH_2)_4—SO_2}$; Na, $NH_3$(liq.) | II.C | 99[b] | 36 |
| —$(CH_2)_2S(CH_2)_2$— |  ; $Na_2S$ | II.F.6 | 76 | 91 |
| $C_6H_5$ | $C_6H_5SO_2Cl$; $LiAlH_4$ | II.A | 89 | 21 |
| $C_6H_5$ | $C_6H_5SO_2Cl$; $(C_2H_5)_3Al$ | II.A | 98 | 23 |
| $C_6H_5$ | $C_6H_5SO_2CH_2C_6H_5$; $(CH_3)_4\overset{+}{N}Cl^-$ | II.F.1 | 90 | 69 |
| $C_6H_5$ | $C_6H_5SO_2SC_2H_5$; $(EtO)_2PONa$ | II.B | 85[b] | 35 |
| $C_6H_5$ | 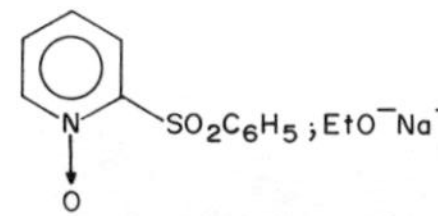 ; $EtO^-Na^+$ | II.F.4 | 89[b] | 79 |
| $p$-$CH_3C_6H_4$ | $p$-$CH_3C_6H_4SO_2Cl$; Zn | II.A | 68 | 10a |
| $p$-$CH_3C_6H_4$ | $p$-$CH_3C_6H_4SO_2Cl$; $LiAlH_4$ | II.A | 93 | 21 |
| $p$-$CH_3C_6H_4$ | $p$-$CH_3C_6H_4SO_2Cl$; $(C_2H_5)_3Al$ | II.A | 95 | 23 |
| $p$-$CH_3C_6H_4$ | $p$-$CH_3C_6H_4SO_2NH—C_6H_{13}$; $(CH_3)_4\overset{+}{N}Cl^-$ | II.G.1 | 97 | 96 |
| $p$-$CH_3C_6H_4$ | $p$-$CH_3C_6H_4SO_2CH_2CN$; $RS^-Na^+$ | II.F | 91[b] | 76 |
| 2,6-$(CH_3)_2C_6H_3$ | 2,6-$(CH_3)_2C_6H_3NH_2$; $SO_2$ | II.E(a).4 | 75 | 27 |
| $p$-$ClC_6H_4$ | $p$-$ClC_6H_4SO_2F(Cl)$; $Na_2SO_3$ | II.A | 81; 80 | 13 |
| $p$-$ClC_6H_4$ | $p$-$ClC_6H_4SONH_6H_2(CH_3)_3$; NaOH | II.B | 99 | 29 |
| $o$-$ClC_6H_4$ | $o$-$ClC_6H_4NH_2(\overset{+}{N}_2)$; Cu; $SO_2$ | II.E(a).4 | 86 | 67a |
| $p$-$FC_6H_4$ | $C_6H_5F$; $AlCl_3$; $SO_2$ | II.E(a).2 | 75 | 62 |
| $p$-$CH_3OC_6H_4$ | $(p$-$CH_3OC_6H_4)_2SO_2$; $(CH_3)_4\overset{+}{N}Cl^-$ | II.F.1 | 65[b] | 69 |
| $p$-$CH_3CONHC_6H_4$ | $p$-$CH_3CONHC_6H_4SO_2Cl$; $SO_2$ | II.A | 43–47 | 14 |
| $C_6H_5CH_2$ | 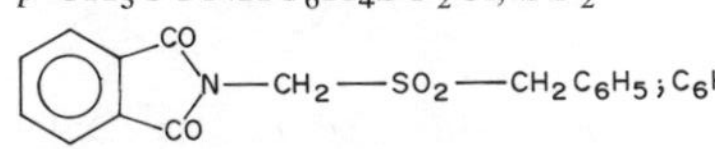 ; $C_6H_5O^-Na$ | II.F.4 | 98 | 78 |
| $C_6H_5CH_2$ | $C_6H_5CH_2Cl$; Mg; $SO_2$ | II.E(a).2 | 97[a] | 57 |
| $C_6H_5CH_2$ | $(C_6H_5CH_2)_2SO_2$; $(CH_3)_4\overset{+}{N}Cl^-$ | II.F.1 | 83 | 69 |
| $C_6H_5CH_2$ | 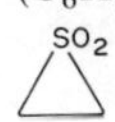 ; $C_6H_5CH_2MgX$ | II.F.6 | 82[h] | 92 |

(*continued*)

**III. TABLE** *(continued)*

| R | Starting materials/reagents | Method* | Yield (%) | Reference |
|---|---|---|---|---|
| $C_6H_5CH{=}CH$ | $C_6H_5CH{=}CHSO_2Cl$; $LiAlH_4$ | II.A | 78[b,d] | 21 |
| $(C_6H_5)_2C{=}CH$ | $C_6H_5C{\equiv}CH$; $SbF_5$; $SO_2$; $C_6H_6$ | II.E(a).3 | 64 | 64 |
| $(C_6H_5)_2C{=}C(Br)$ | $C_6H_5C{\equiv}CBr$; $SbF_5$; $SO_2$; $C_6H_6$ | II.E(a).3 | 67 | 64 |
| $p$-n-$C_{12}H_{25}C_6H_5$ | $p$-n-$C_{12}H_{25}{-}C_6H_4{-}Br$; Li; $SO_2$ | II.E(a).2 | 60 | 59 |
| 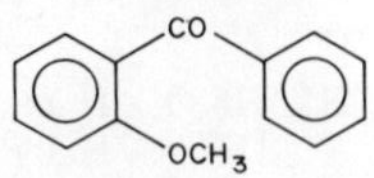 | 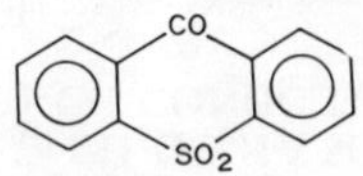 ; $CH_3O^-Na^+$ | II.F.6 | 97 | 86 |
| 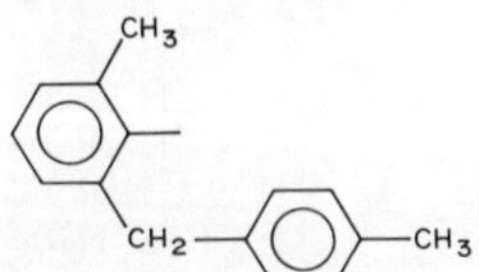 | 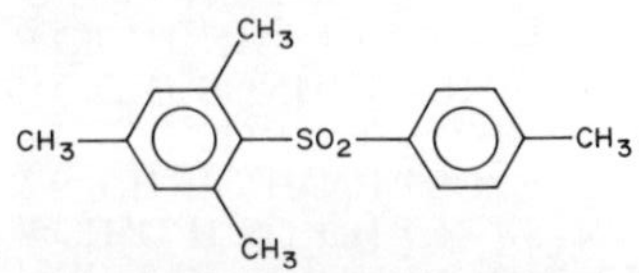 ; $n$-BuLi or $t$-$BuO^-K^+$/DMSO | II.F.5 | 88; 74 | 82 |
| 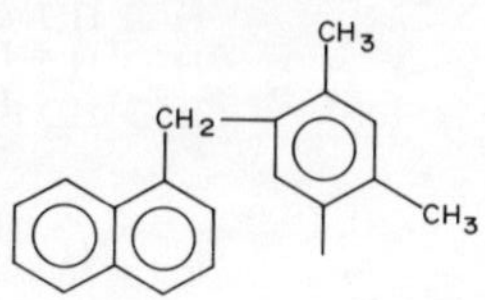 | 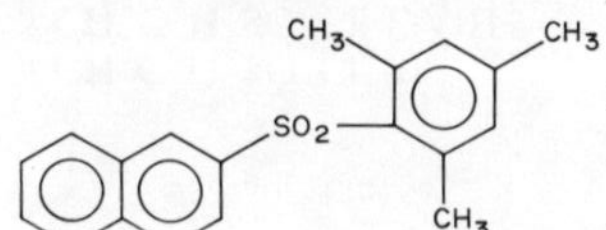 ; $n$-BuLi or $t$-$BuO^-K^+$/DMSO | II.F.5 | 5; 84 | 82 |
| 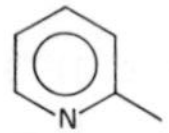 | 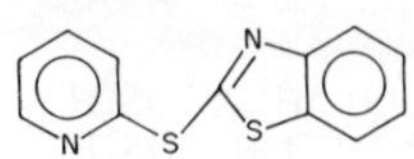 ; $KMnO_4$; $NaBH_4$ | II.D | 67[b] | 42 |
| 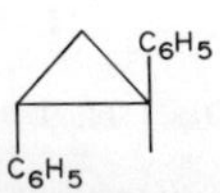 | 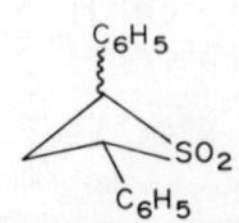 ; EtMgBr (*cis* & *trans*) | II.F.6 | 74–77 | 90 |

*The entries in this column correspond to the number of the section and subsection where the method is described.
[a]In a form of pure, transparent colorless needles.
[b]As the sodium salt.
[c]As the lithium salt.
[d]Yield of the crude product.
[e]After vacuum distillation of the crude product.
[f]Not specified.

## IV. REFERENCES


1. W. E. Truce and A. M. Murphy, *Chem. Rev.*, **48**, 69 (1951).
2. C. J. M. Stirling, in *The Chemistry of Organic Sulfur Compounds* (Eds. N. Kharasch and C. Y. Meyers), Vol. 3, Pergamon Press, New York, 1966.
3. C. J. M. Stirling, *Int. J. Sulfur Chem.*, Part B, **6**, 277 (1971).
4. I. Field, *Synthesis*, 101 (1972).
5. S. Oae and N. Kunieda, in *Organic Chemistry of Sulfur* (Ed. S. Oae), Plenum Press, New York, 1977, pp. 604–606, and references cited therein.
6. E. Krauthausen, in *Methoden Der Organischen Chemie* (Houben-Weyl) (Ed. D. Klamann), Georg Thieme Verlag, New York, 1985, pp. 618–631.
7. Y. Ueno, A. Kojima and M. Okawara, *Chem. Lett.*, 2125 (1984).
8. J. M. Klunder and K. B. Sharpless, *J. Org. Chem.*, **52**, 2598 (1987).
9. S. Patai, Z. Rappoport and C. J. M. Stirling (Eds.), *The Chemistry of Sulfones and Sulfoxides*, Wiley, Chichester, 1988.
10. a. F. C. Whitmore and F. H. Hamilton, *Org. Synth., Coll.*, **1**, 492 (1941).
    b. H. Gilman, E. W. Smith and H. J. Oatfield, *J. Am. Chem. Soc.*, **56**, 1413 (1934).
11. German Patent 912,091 (May 24, 1954).
12. a. F. Klivényi, E. Vinkler and J. Lázár, *Acta Chim. Acad. Sci. Hung.*, **46**, 357 (1965); *Chem. Abstr.*, **64**, 8066 (1966).
    b. N. Kunieda, K. Sakai and S. Oae, *Bull. Chem. Soc. Jpn.*, **41**, 3015 (1968).
13. M. Kulka, *J. Am. Chem. Soc.*, **72**, 1215 (1950).
14. S. Smiles and C. M. Bere, *Org. Synth., Coll.*, **1**, 7 (1941).
15. M. T. Beachem, J. T. Shaw, G. D. Sargent, R. B. Fortenbaugh and J. M. Salsburg, *J. Am. Chem. Soc.*, **81**, 5430 (1959).
16. C. S. Marbel and R. S. Johnson, *J. Org. Chem.*, **13**, 822 (1948).
17. J. Von Braun and K. Weissbach, *Chem. Ber.*, **63**, 2836 (1930).
18. U. S. Patent 3420877 (1961), Minnesota Mining and Manufacturing Co.; *Chem. Abstr.*, **70**, 67609 (1969).
19. a. C. Harzdorf, J.-N. Meubdoerffer, H. Niederprüm and M. Wechsberg, *Justus Liebigs Ann. Chem.*, 33 (1973).
    b. H. W. Roesky, *Angew. Chem., Int. Ed. Engl.*, **10**, 810 (1971).
20. U. Schöllkopf and P. Hilbert, *Justus Liebigs Ann. Chem.*, 1061 (1973).
21. L. Field and F. A. Grunwald, *J. Org. Chem.*, **16**, 964 (1951).
22. A. I. Khodair, A. Swelim and A. A. Abdel-Wahab, *Phosphorus and Sulfur*, **2**, (a) 165, (b) 169 (1976).
23. H. Reinheckel and D. Jahnke, *Chem. Ber.*, **99**, 1718 (1966).
24. J. B. Hendrickson, A. Giga and J. Wareing, *J. Am. Chem. Soc.*, **96**, 2275 (1974).
25. S. G. Mairanovski and M. B. Neiman, *Dokl. Akad. Nauk SSSR*, **79**, 85 (1951).
26. I. B. Douglass, *J. Org. Chem.*, **30**, 633 (1965).
27. D. Darwish and J. Noreyko, *Can. J. Chem.*, **43**, 1366 (1965).
28. C.A. Bunton and B. W. Hendy, *Chem. Ind.* (*London*), 466 (1960).
29. J. B. Biasotti and K. K. Andersen, *J. Am. Chem. Soc.*, **93**, 1178 (1971).
30. C. C. Price and S. Oae, *Sulfur Bonding*, Ronald Press, New York, 1962.
31. I. B. Douglass and D. R. Poole, *J. Org. Chem.*, **23**, 330 (1958).
32. S. Oae and K. Ikura, *Bull. Chem. Soc. Jpn.*, **39**, 1309 (1966).
33. F. Wudl, D. A. Lightner and D. J. Cram, *J. Am. Chem. Soc.*, **89**, 4099 (1967).
34. I. B. Douglass, B. S. Farah and E. G. Thomas, *J. Org. Chem.*, **26**, 1996 (1961).
35. J. Michalski, J. M. Modro, and J. Wieczorkowski, *J. Chem. Soc.*, 1665 (1960).
36. a. R. Chandra and L. Field, *Phosphorus and Sulfur*, **27**, 247 (1986).
    b. L. Field and V. Eswarakrishnan, *J. Org. Chem.*, **46**, 2025 (1981).
37. B. Boduszek and J. L. Kice, *J. Org. Chem.*, **47**, 2055 (1982).
38. D. Farnum, *J. Org. Chem.*, **28**, 870 (1963).
39. H. Paulsen and D. Stoye, in *The Chemistry of Amides* (Ed. J. Zabicky), Interscience, London, 1970, pp. 557 and 569.
40. D. Klamann and G. Hofbauer, *Chem. Ber.*, **86**, 1246 (1953).
41. W. G. Filby, K. Günther and R. D. Penzhorn, *J. Org. Chem.*, **38**, 4070 (1973).
42. Y. Ueno, A. Kojima and M. Okawara, *Chem. Lett.*, 2125 (1984).

43. H. Berger, *Recl. Trav. Chim. Pays-Bas*, **82**, 773 (1963).
44. NL Patent 287952 (1963); *Chem. Abstr.*, **63**, 9815 (1965).
45. a. S. Oae, T. Takata and Y. H. Kim, *Tetrahedron*, **37**, 37 (1981).
    b. T. Takata, Y. H. Kim and S. Oae, *Tetrahedron Lett.*, 821 (1979).
46. U. S. Patent 2872482 (1959); *Chem. Abstr.*, **53**, 15994 (1959).
47. U. S. Patent 3355486 (1965); *Chem. Abstr.*, **68**, 49061 (1968).
48. W. Walter and G. Randau, *Justus Liebigs Ann. Chem.*, **722**, 80 (1969).
49. E. M. Burges, U. Zoller and R. L. Burger Jr., *J. Am. Chem. Soc.*, **106**, 1128 (1984).
50. U. Zoller, *Tetrahedron*, **44**, 7413 (1988).
51. F. S. Dainton and K. J. Ivin, *Trans. Faraday Soc.*, **46**, 374,382 (1950).
52. Y. Ogata, Y. Izawa and K. Takagi, *Bull Chem. Soc. Jpn.*, **39**, 2438 (1966).
53. S. Farid, *Chem. Commun.*, 73 (1971).
54. J. R. Nooi, P. C. van der Hooven and W. P. Haslinghuis, *Tetrahedron Lett.*, 2531 (1970).
55. H. Takeuchi, T. Nagai and T. Tokura, *Bull. Chem. Soc. Jpn.*, **46**, 695 (1973).
56. a. H. G. Houlton and H. V. Tartar, *J. Am. Chem. Soc.*, **60**, 544 (1938).
    b. P. Allen, Jr., *J. Org. Chem.*, **7**, 23 (1942).
57. H. W. Pinnick and M. A. Reynolds, *J. Org. Chem.*, **44**, 160 (1979).
58. D. Masilamani and M. M. Rogic, *J. Am. Chem. Soc.*, **100**, 4634 (1978).
59. W. E. Truce and J. F. Lyons, *J. Am. Chem. Soc.*, **73**, 126 (1951).
60. M. E. Peach, M. Schmidt and E. Weisflog, *Can. J. Chem.*, **53**, 3720 (1975).
61. a. GB Patent 916751 (1960); *Chem. Abstr.*, **58**, 12422 (1963).
    b. DE Patent 1050762 (1957); *Chem. Abstr.*, **55**, 2483 (1961).
62. R. M. Hann, *J. Am. Chem. Soc.*, **57**, 2166 (1935).
63. G. Capozzi, V. Lucchini, F. Marcuzzi and G. Melloni, *Tetrahedron Lett.*, **21**, 3289 (1980).
64. R.-L. Fan, J. I. Dickstein and S. I. Miller, *J. Org. Chem.*, **47**, 2466 (1982).
65. a. G. G. Yakobson and G. G. Furin, *Synthesis*, 345 (1980).
    b. G. A. Olah and T. E. Kiovsky, *J. Am. Chem. Soc.*, **90**, 2583 (1968).
66. M. I. Kanischev, A. A. Schegolev, W. A. Smit, R. Caple and M. J. Kelner, *J. Am. Chem. Soc.*, **101**, 5660 (1979).
67. a. H. R. Todd and R. L. Shriner, *J. Am. Chem. Soc.*, **56**, 1382 (1934).
    b. W. A. Silvester and W. P. Wynne, *J. Chem. Soc.*, 693 (1936).
    c. M. E. Hanke, *J. Am. Chem. Soc.*, **45**, 1325 (1923).
68. a. S. Patai and F. Bergmann, *J. Am. Chem. Soc.*, **72**, 1034 (1950); S. Patai and A. Patchornik, *J. Am. Chem. Soc.*, **74**, 4494 (1952).
    b. A. J. Titov and A. N. Baryshnikova, *Dokl. Akad. Nauk SSSR*, **157**, 139 (1964); *Chem. Abstr.*, **61**, 9396 (1964).
69. L. Horner and H. Neumann, *Chem. Ber.*, **98**, 1715 (1965).
70. B. A. Arabuzov and E. A. Berdnikov, *Dokl. Akad. Nauk SSSR*, **171**, 860 (1966); *Chem. Abstr.*, **66**, 61267 (1967).
71. R. E. Dabby, J. Kenyon and R. E. Mason, *J. Chem. Soc.*, 4881 (1952).
72. W. E. Truce, D. P. Tate and D. N. Burdge, *J. Am. Chem. Soc.*, **82**, 2872 (1960).
73. D. B. Hope, C. D. Morgan and M. Wälti, *J. Chem. Soc.* (*C*), 270 (1970).
74. G. P. Crowther and C. R. Houser, *J. Org. Chem.*, **33**, 2228 (1968).
75. A. T. Kader and C. J. M. Stirling, *J. Chem. Soc.*, 3425, 3686 (1962).
76. W. E. Truce and F. E. Roberts, Jr., *J. Org. Chem.*, **28**, 593 (1963).
77. P. Messinger and H. Greve, *Justus Liebigs Ann. Chem.*, 1457 (1977).
78. M. Uchino, K. Suzuki and M. Sekiya, *Synthesis*, 794 (1977).
79. N. Furukawa, M. Tsuruoka, and H. Fujihara, *Heterocycles*, **24**, 3019 (1986).
80. a. C. S. McClement and S. Smiles, *J. Chem. Soc.*, 1016 (1937).
    b. W. E. Truce, E. M. Kreider and W. W. Brand, *Org. React.*, **18**, 99 (1970).
    c. T. Okamoto and J. F. Bunnet, *J. Am. Chem. Soc.*, **78**, 5357 (1956).
81. W. E. Truce, W. J. Ray, Jr., O. L. Norman and D. B. Eickemeyer, *J. Am. Chem. Soc.*, **80**, 3625 (1958).
82. W. E. Truce, C. R. Robbins and E. M. Kreider, *J. Am. Chem. Soc.*, **88**, 4027 (1966).
83. a. D. M. Snider and W. E. Truce, *J. Am. Chem. Soc.*, **101**, 5432 (1979).
    b. E. J. Madaj, Jr., D. M. Snider and W. E. Truce, *J. Am. Chem. Soc.*, **108**, 3466 (1986).
84. U. Zoller, in *Cyclic Sulfones and Sulfoxides* (Eds. S. Patai, Z. Rappoport and C. J. M. Stirling), *The Chemistry of Sulfones and Sulfoxides*, Wiley, Chichester, 1988, pp. 783–885.

85. U. Zoller, *Three-membered rings containing sulfur*, in *Small Ring Heterocycles* (Ed. A. Hassner), Wiley, New York, 1983, pp. 333–660.
86. O. F. Bennett, G. Saluti and F. X. Quinn, *Synth. Commun.*, **7**, 33 (1977).
87. O. F. Bennett, M. J. Bouchard, R. Malloy, P. Dervin and G. Saluti, *J. Org. Chem.*, **37**, 1356 (1972).
88. K. Schank, R. Wilmes and G. Ferdinand, *Int. J. Sulfur Chem.*, **8**, 397 (1973).
89. R. M. Dodson, P. D. Hammen and J. Yu. Fan, *J. Org. Chem.*, **36**, 2703 (1971).
90. R. M. Dodson, P. D. Hammen, E. H. Jancis and G. Klose, *J. Org. Chem.*, **36**, 2698 (1971).
91. E. Vilsmaier and G. Becker, *Synthesis*, 55 (1975).
92. E. Vilsmaier, R. Tropitzsch and O. Vostrowsky, *Tetrahedron Lett.*, 3987 (1974).
93. J. P. Harmon and L. Field, *J. Org. Chem.*, **51**, 5235 (1986).
94. R. F. Langler, Z. A. Marini and J. A. Pincock, *Can. J. Chem.*, **56**, 903 (1978).
95. R. Sauvétre, D. Masure, L. Chuit and J.-F. Normant, *C.R. Acad. Sci. Paris, Ser. C*, 335 (1979).
96. L. Horner and H. Neuman, *Chem. Ber.*, **98**, 3462 (1965).
97. L. Horner and R.-J. Singer, *Chem. Ber.*, **101**, 3329 (1968).
98. J. D. Loudon and I. Wellings, *J. Chem. Soc.*, 1780 (1959).
99. D. W. Emerson, R. R. Emerson, S. C. Joshi, E. M. Sorensen and J. E. Turek, *J. Org. Chem.*, **44**, 4634 (1979).
100. a. H. Bredereck and E. Bäder, *Chem. Ber.*, **87**, 129 (1954).
b. B. Lindberg, *Acta. Chem. Scand.*, **17**, 377 (1963).
101. H. Bredereck, A. Wanger, R. Blaschke, G. Demetriades and K.-G. Kottenhahn, *Chem. Ber.*, **92**, 2628 (1959).
102. C. Degrand and H. Lund, *Acta Chem. Scand., Ser. B*, **33**, 512 (1979).
103. J. D. Macke and L. Field, *J. Org. Chem.*, **53**, 396 (1988).

The Chemistry of Sulphinic Acids, Esters and their Derivatives
Edited by S. Patai


CHAPTER **8**

# Syntheses of sulfinic esters

URI ZOLLER

*Division of Chemical Studies, Haifa University—Oranim, P.O. Kiryat Tivon 36910, Israel*



## I. INTRODUCTION

The sulfinic esters and sulfinic acids constitute an interesting case study in organic chemistry in that the chemistry and synthetical methodologies of the offspring derivatives (i.e. the sulfinic esters) appear to have been much more explored and developed compared with that of the 'parent' compounds (i.e. the sulfinic acids). The reason for that is probably twofold:

First, the sulfinic acids, the low molecular weight aliphatic ones in particular, are not stable at room temperature, and readily undergo disproportionation and oxidation reactions (under mild reaction conditions and even spontaneously under certain circumstances). In addition, most of the acids are very hygroscopic and thus resist isolation, purification and handling.

Second, the sulfinic esters are configurationally stable which makes them easier and more convenient to work with (compared with their acid counterparts). The stability of the *pyramidal* structure[1,2] of the central sulfur(IV) atom facilitates the synthesis of stable, optically active sulfinic esters[3]. The importance of sulfinic esters in establishing the absolute configuration of sulfur(IV) compounds is well recognized[4]. In fact, optically active sulfinic esters have long been known and active $\beta$-octyl and menthyl *p*-toluenesulfinates were synthesized about sixty-five years ago[5]. However, in spite of such a long history and 'recognition', many papers, including recent ones, complain about the limited applicability, low yields, hard-to-obtain precursors, the occurrence of side-reactions and by-products, laborious and expensive experimental procedures encountered in these syntheses.

Apparently, in spite of all the effort in the development of new methodologies for convenient, straightforward and general syntheses of sulfinic esters, there still remains much scope for further exploration and innovation. A review of all main synthetic routes reported for obtaining sulfinic esters will follow.

## II. THE SYNTHESIS OF SULFINIC ESTERS

Several methods are available to the organic chemist for the synthesis of sulfinic esters on a preparative scale. These can be divided into the following categories:

A. Direct synthesis from sulfinic acids or their salts.
B. Esterification of sulfinyl halides (mainly chlorides) with alcohols.
C. Esterification of sulfinic acids and sulfinate salts by hard alkylating agents.
D. Cleavage of the S—S and S—N bonds in thiosulfinic S-esters and sulfinamides.
E. Oxidation of disulfides, thiols and sulfenic esters.
F. Reaction of sulfenyl derivatives with oxiranes.
G. Reduction of sulfonyl derivatives.
H. S—S and S—N bond cleavage.
I. Carbon–sulfur bond formation.
J. Carbon–sulfur bond cleavage.
K. Miscellaneous.

The first two appear to be the methods of choice, the second being the most commonly used.

### A. Synthesis of Sulfinic Esters Directly from Sulfinic Acids (and their salts)

In contrast to carboxylic acids, sulfinic acids do not undergo acid catalyzed esterification with alcohols but rather disproportionate into sulfonates and thiosulfonates. However, through appropriate activation of the sulfinyl group with various activating reagents (or a combination of reagents), the direct esterification of sulfinic acids with alcohols can be realized[6–8]. This is illustrated in equation 1 by two different combinations of reagents in which the nucleophilic cleavage of the sulfur–oxygen bond by the alcohol is facilitated by the enhanced electrophilicity of the sulfinyl sulfur atom in the intermediate salts (i.e. **4** and **5**)[6,7]. Using this sulfur–oxygen bond activation principle, sulfinic acids have been directly transformed into their corresponding esters (e.g. **1** $\rightarrow$ **2**) by using the following activating coupling reagents (or combination of reagents): Dicyclohexylcarbodiimide[6,9,8a], diethyl diazodicarboxylic ester/triphenylphosphine[6,8a], 2-chloro-1-methylpyridinium iodide/triethylamine[6,8a] (equation 1a), 3-chloro-1,2-benzothiazole-1,1-dioxide/triethylamine[8a], 3-chloro-1,2-benzothiazole-1,1-dioxide/*N*-hydroxyphthalimide[8b], phenyl phosphorodichloridate/pyridine[7,8b] (equation 1b), diphenyl phosphorochloridate/pyridine[8b] and saccharine chloride[8a].

The yields obtained are generally fair to good (52–85%) and since this is a one-pot

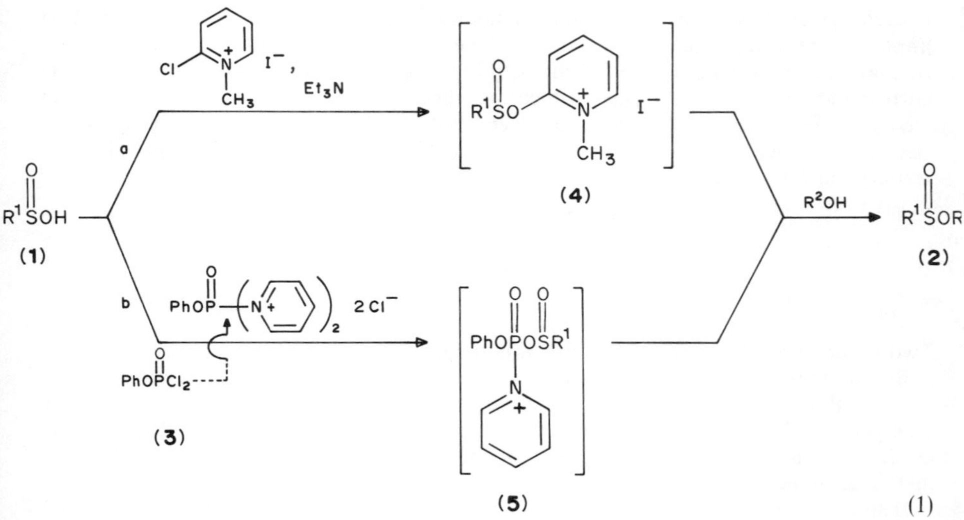


(1)

synthesis of sulfinates, it is quite convenient provided, of course, that the starting sulfinic acids are readily available (see Table in Section III).

As an example, *p*-toluenesulfinic acid butyl ester (**2a**)[7] was obtained from *p*-toluenesulfinic acid in a dichloromethane pyridine solvent with phenyl phosphorodichloridate (**3**) and n-butanol at $-15\,^{\circ}C$ under a nitrogen atmosphere. The yield was 85%.

The above procedure is not only applicable for the preparation of other sulfinic esters (including *t*-butyl,cyclohexyl, vinyl and benzyl esters[6,7]), but also for the preparation of the corresponding sulfinamides[8b,10] as well as S-alkyl and S-aryl thiosulfinates[8b] by using amines and thiols (rather than alcohols), respectively[7]. Several sulfinic acid derivatives are known to be useful as starting materials and versatile synthetic intermediates[11], and some have attracted interest on account of their biological activities, particularly in the case of thiosulfinates[12].

An alternative direct preparation of sulfinic esters from sulfinic acids can be realized by the reaction of sulfinate salts with alkyl halides in sufficiently polar solvents[13]. This method was applied successfully for the preparation of the easily rearrangeable vinylic sulfonic esters[14] as shown in equation 2. However, there are only a few reports in the literature about the use of this method. Probably, the use of phase transfer agents might prove useful in soving the problem of mixing the reagents in the above synthetic strategy.

$$\underset{(6)}{CH_2{=}C(CH_3)\overset{\overset{SO_2^-N_a^+}{|}}{C}{=}CMe_2} \xrightarrow[EtOH]{EtBr} \underset{(7)}{CH_2{=}C(Me)\overset{\overset{\overset{O}{\|}}{S-OEt}}{\overset{|}{C}}{=}CMe_2} \qquad (2)$$

### B. Esterification of Sulfinyl Halides with Alcohols

This is by far the most popular and commonly used method for the preparation of both aliphatic and aromatic sulfinic esters, mainly because of the availability of the starting materials, including the necessary sulfinyl halides[15] (albeit the latter are accompanied by impurities resulting from side-reactions). The sulfinyl halides (usually chlorides) are obtained through the oxidation of thiols, thiophenols or disulfides with chlorine to the corresponding sulfinyl chlorides which, in turn, are reacted with the appropriate alcohol[16]. The hydrogen chloride can be removed by applying vacuum[16,17], heating[18], a stream of nitrogen or, alternatively, by adding equivalent or excess amounts of either tertiary amines or alkali metal carbonates to the reaction mixture.

$$R^1\overset{\overset{\displaystyle O}{\|}}{S}Cl + HOR^2 \longrightarrow \underset{\textbf{(2)}}{R^1\overset{\overset{\displaystyle O}{\|}}{S}OR^2} \qquad (3)$$

Two procedures of this sulfinic ester synthesis methodology are given below.

Butanesulfinic acid methyl ester (**8**)[4] was obtained from (−) menthol in pyridine and butanesulfinyl chloride[19] in anhydrous ether at −78 °C under a nitrogen atmosphere as a pale yellow oil in essentially quantitative yield (based on the menthol used). Similarly, crude methanesulfinic acid (−) menthyl ester (**9**) was obtained in 93% yield[3], and methanesulfinic acid methyl ester[16] in 71% yield.

It should be pointed out that sulfinyl chlorides tend to disproportionate into sulfonyl and sulfenyl chlorides. Sulfonyl chlorides may also be formed as by-products during the chlorination of disulfides. On the other hand, the reaction of the alcohol with the sulfinyl chloride to form the sulfinic ester rapidly consumes the sulfinyl chloride, but leaves any sulfonyl chloride present essentially unchanged[16]. It is not surprising, therefore, that even the most careful syntheses of sulfinic esters via this route are accompanied by traces of by-products or impurities, mainly the sulfonyl chloride. However, the latter can be removed by adding an aromatic amine to the reaction mixture following the esterification procedure to form the sulfonamide from which the liquid sulfinic esters can be separated by distillation.

Various types of sulfinic esters may be conveniently prepared by this method[20] and representative examples[21–29] are presented in the Table at the end.

Most optically active sulfinates have been obtained by the reaction of racemic sulfinyl chlorides with optically active alcohols, affording a mixture of diasteromeric sulfinic esters in unequal amounts[30]. The synthetic utility of the transesterification of menthyl sulfinates[5] and the asymmetric oxidation of sulfenates[31] to give optically active sulfinic esters is limited by the low stereoselectivity (1–2%). It is possible, however, to achieve an asymmetric synthesis of optically active sulfinic esters by the reaction o sulfinyl chlorides with achiral alcohols in the presence of optically active tertiary amines (i.e. **10**) as asymmetric reagents[32]:

$$R^1\overset{\overset{\displaystyle O}{\|}}{S}Cl + R^2OH \xrightarrow[\textbf{(10)}]{Me_2N\overset{*}{R}_3} R^1\underset{*}{\overset{\overset{\displaystyle O}{\|}}{S}}OR^2 \qquad (4)$$

The optical purity of the sulfinic esters thus obtained strongly depends on the reaction temperature and, to some extent, on the structure of all reaction components. Typically, the optical purity is within the range of 10–29%. For instance, the reaction of methanesulfinyl chloride with propanol in the presence of (+)-*N*,*N*-dimethyl-α-phenylethylamine (+**10**) at −60 °C gave (*S*)-(−)-methanesulfinic acid propyl ester in 19.3% optical purity; at −12 °C, the same reaction gave an ester having only 6.2% optical purity. Ethanesulfinyl chloride with propyl alcohol at *ca* −70 °C and in the presence of the (−)-*N*,*N*-dimethyl-α-phenylamine (−**10**) afforded the *S*-(+)-ethanesulfinic acid propyl ester in 23.9% optical purity[32]. Several other optically active esters have been synthesized via this method (see Table).

It was later shown[30] that in the methanolysis of primary and secondary alkanesulfinyl chlorides, the methyl sulfinates are formed by nucleophilic reaction at the electrophilic sulfur atom of the sulfinyl chloride. Preliminary dehydrochlorination followed by addition of methanol to the sulfines thus formed was excluded. This implies a nucleophilic attack of the oxygen atom of alkoxytrimethylsilanes on the electrophilic sulfinyl sulfur which should facilitate the synthesis of sulfinyl esters by reaction of the former with sulfinyl halides[18,34].

$$\mathrm{R\overset{O}{\overset{\|}{S}}Cl + Me_3SiOR^2 \longrightarrow R^1\overset{O}{\overset{\|}{S}}OR^2 + Me_3SiCl} \qquad (5)$$

Indeed, alkoxytrimethylsilanes and sulfinyl chlorides have been shown of couple efficiently to afford sulfinic esters apparently via a nonionic four-center transition state as suggested by kinetic data[34].

This method is equivalent, in principle, to the 'classical' direct esterification of sulfinyl chloride with alcohols already described above, but it has some advantages in certain cases. First, the reaction proceeds smoothly at room temperature and its progress can be conveniently followed by $^1$H NMR spectroscopy (the singlet for chlorotrimethylsilane increasing at the expense of the peak of the trimethylsilyl group of the alkoxytrimethylsilane). Second, the chlorosilane may be easily removed at the end of the reaction by evaporation, saving the need for distillation of the ester product. The precursor alcohols may be conveniently silylated[35] with hexamethyldisilazane using imidazole as catalyst.

Although the silylation adds one extra step in the synthesis, a useful application of this method is the synthesis of methyl sulfinates, precursors of chiral sulfoxides[36]. For instance, (−)-methyl(−)-(*S*)-benzenesulfinate was prepared from benzenesulfinyl chloride and 1-menthoxytrimethylsilane in 91% yield. The final pure product melted at 37–40 °C and had $[\alpha]_D - 195.3°$ ($c = 2.0$, acetone)[34].

Ethyl and benzyl methane-, benzene- and α-toluene-sulfinic esters have been prepared in good to fair yields by using this procedure[34].

A simple, convenient modification of the above method for the synthesis of sulfinic esters in good yields (73–87%) in netural conditions has been recently reported[37]. Typically, a few drops of dimethyl sulfoxide are added to an equimolar mixture of sulfinyl chloride, chlorosulfite and hexamethylsiloxane, and the reaction is allowed to proceed at room temperature without any solvent.

$$\mathrm{R^1\overset{O}{\overset{\|}{S}}Cl + R^2O\overset{O}{\overset{\|}{S}}Cl + (Me_3Si)_2O \xrightarrow{(CH_3)_2SO} R^1\overset{O}{\overset{\|}{S}}OR^2 + 2MeSiCl + SO_2} \qquad (6)$$

Since the mechanism of this reaction is not yet clear and the availability of the chlorosulfites may cause a problem, the scope and preparative potential of this method are still open questions.

The use of metal alkoxides rather than the free alcohols in the reaction with sulfinyl chlorides to yield the esters has been reported[38]. Thus, according to equation 7, trichloromethanesulfinic acid methyl ester (**12**)[38a] was obtained from trichloromethanesulfinyl chloride, methanol and anhydrous potassium carbonate in 73% yield.

$$\underset{\textbf{(11)}}{\mathrm{Cl_3CSOCl}} \xrightarrow{\mathrm{CH_3OK}} \underset{\textbf{(12)}}{\mathrm{Cl_3CSOOCH_3}} \qquad (7)$$

Sulfinyl fluorides react with alcohols to give sulfinic esters[39,40]. Thus, reactions of trifluoromethane sulfinyl fluoride with fluoro alcohols in the presence of sodium and cesium fluorides were used to prepare fluorosulfinic esters containing diastereoisomers due to the chiral sulfur center[41].

$$\underset{(\mathbf{13})}{CF_3\overset{\overset{O}{\|}}{S}F} + R_fOH \xrightarrow[25\,^\circ C]{NaF/CsF} \underset{(\mathbf{14})}{CF_3\overset{\overset{O}{\|}}{S}OR_f} \qquad (8)$$

$R_f =$ (**a**) $CF_3(CH_3)CH$; (**b**) $CF_3(CH_3)_2C$; (**c**) $(CF_3)_2CH$; (**d**) $(CF_3)_2C(CH_3)$

The reactions of trifluoromethanesulfinyl fluoride (**12**) with secondary and tertiary alcohols are slow compared to its reactions with hydrogenated and partially fluorinated primary alcohols (several days versus several hours). For example, trifluoromethanesulfinyl fluoride (**12**) and the alcohol [$R_f =$ (a)] in the presence of anhydrous. NaF were shaken in a closed vessel at 25 °C for 5–7 days, occasionally adding more $CF_3SOF$ in order to maintain high pressure which enhances the rate of the reaction. The yield of the product **14a** was 91%.

### C. Esterification of Sulfinic Acids and Sulfinate Salts by Hard Alkylating Agents

The direct esterification of sulfinic acids or their metal (sodium) salts by hard alkylating agents[42–44] appears—at first glance—to be quite attractive. However, in view of the relative instability of the starting sulfinic acids (the aliphatic ones in particular) and the hygroscopicity of the sulfinate salts, the problems associated with the generation and/or handling of the necessary alkylating agents (not to mention their toxicity and/or sensitivity to moisture and/or light) the application of this method is rather limited. Moreover, with the exception of the diazomethane alkylating agents in highly polar media (e.g. ether/MeOH; DMSO) which provide the sulfinic esters exclusively and in high yields, all other alkylating agents employed (e.g. dimethyl sulfate, alkyl halides and methyl sulfonates) afforded mixtures of sulfinic esters together with the corresponding sulfoxides[42,43], as also did triethyloxonium tetrafluoroborate in dichloromethane[44,45].

$$R^1\overset{\overset{O}{\|}}{S}OH(Na) + \left\{\begin{array}{l} R^1CN_2 \\ \text{or:} \\ (MeO)_2SO_2 \\ \text{or:} \\ MeOSO_2C_6H_4CH_3\text{-}p \end{array}\right. \longrightarrow \underset{(\mathbf{15})}{R^1\overset{\overset{O}{\|}}{S}OCR^1} \quad \underset{(\mathbf{16})}{(+R^1\overset{\overset{O}{\|}}{S}CR^1_2)} \qquad (9)$$

(**a**) $R^1 = H$

(**b**) $R^1 = C_6H_5$

In conclusion, these esterifications of sulfinic acids are preparatively useful only in some special cases, when the starting sulfinic acid is readily available, when the methyl (or ethyl) sulfinic ester is the target compound and when the alternative available synthetic routes are laborious or provide low yields of the desired products.

### D. Cleavage of the Sulfur–Sulfur and Sulfur–Nitrogen Bonds in Thiosulfinic S-Esters and Sulfinamides

Thiosulfinic S-esters are readily converted to the corresponding sulfinic esters in fair to good yields by treating them with alcohols using a catalytic amount of $I_2$, $Br_2$ or HCl in the presence of $H_2O_2$[46]. The sulfenyl group in the thiosulfinates is thus replaced by the $OR^2$ group of the employed alcohol (equation 10). Apparently, the reaction involves formation of a sulfinyl halide [$R^1S(O)Y$] which is derived by the reaction with the catalyst XY, followed by the reaction of the resulting mixture $R^1S(O)Y$ and $R^2SX$ with the excess of the alcohol $R^3OH$ to afford (in the presence of the oxidizing $H_2O_2$) the observed products[46,47].

The difficult availability of thiosulfinic S-esters on the one hand, and the need to supress the formation of the disulfide by-products to a minimum on the other hand, constitute limiting factors in using this method.

Benzenesulfinic acid methyl ester could be prepared from the thiosulfinic S-ester **17a** ($R^1 = R^2 = C_6H_5$)[48] in methanol, 30% $H_2O_2$ and traces of bromine stirred at room temperature for 1–2 h. The pure sulfinic ester was obtained in 91% yield[46].

$$\underset{(\mathbf{17})}{R^1\overset{O}{\overset{\|}{S}}SR^2} + R^3OH \xrightarrow[20\,^{\circ}C,\ 1\text{–}2\,h]{XY/30\%\ H_2O_2} R^1\overset{O}{\overset{\|}{S}}OR^3 + R^2SSR^2 + R^1SSR^2 \qquad (10)$$

$$XY = I_2 \text{ or } Br_2 \text{ or } HCl$$

The presence of the bulky *t*-butyl group in thiosulfinic S-esters (i.e. **17**; $R^2 = t$-Bu) prevents nucleophilic substitution on sulfur and thus increases the chemical and optical stability of chiral thiosulfinates. However, the reaction of (−)-(*S*)-**17b** ($R^1 = p\text{-}CH_3C_6H_4$; $R^2 = t$-Bu) with MeOH in the presence of *N*-bromosuccinimide affords (+)-(*R*)-methyl toluene-*p*-sulfinate (**2**; $R^1 = p\text{-}CH_3C_6H_4$, $R^2 = CH_3$), the optical purity of the product being 5.8% (compared with 11.3% in **17b**)[49]. The reaction most likely proceeds with inversion at the sulfinyl center providing a useful synthetic entry to relatively stable, optically active sulfinic esters.

Similarly to the above method, the nucleophilic cleavage of the sulfur–nitrogen bond in sulfinamides by alcohols or alkoxides also leads to the formation of sulfinic esters.

Sulfinylphthalimides are excellent precursors of sulfinate esters: they react with solutions of alkoxides in alcohol at room temperature (method A) to provide the esters in high yield. Alternatively, the alcoholysis may be accomplished by simply refluxing the sulfinylphthalimides in the appropriate alcohol (method B) to yield the sulfinate esters in nearly quantitative yields and in a high state of purity[50]. So, the limiting factor of this method is the availability of the appropriate sulfinylphthalimide.

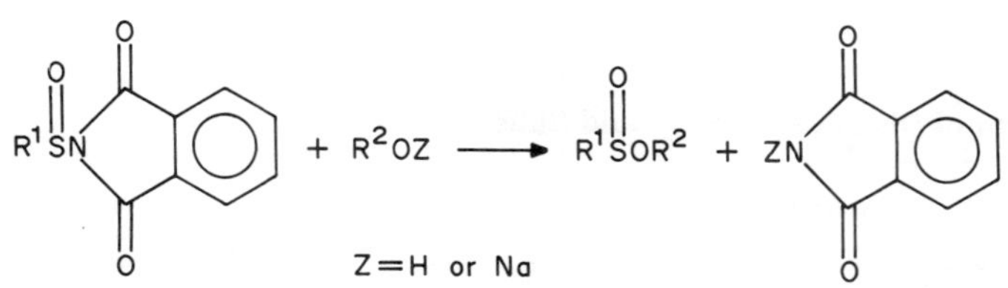


(11)

**(18)** **(19)**

As an illustrative example, the benzenesulfinic acid methyl ester (**2b**) was prepared according to method A with sodium methoxide in methanol and *N*-(phenylsulfinyl)phthalimide[57] (**18**, $R^1 = Ph$). After 0.5 h of stirring at room temperature and the usual purification procedure, 90% of the product was obtained. According to method B, **18** ($R^1 = Ph$) was refluxed for 2 h in methanol and the yield of the product was 95%.

Several aliphatic and aromatic sulfinic esters can be synthesized by this route in good to high yields (see Table).

Treatment of sulfinamides with alcohols in the presence of strong acids results in the formation of sulfinic esters in good yields. This reaction was shown to proceed with inversion of configuration at the sulfinyl center and with high stereospecificity which is dependent on the structure of the alcohol used[52].

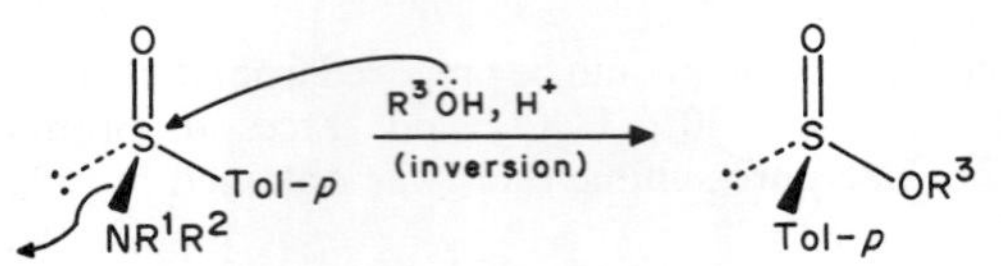


(12)

**(20)** **(21)**

The acid catalyzed alcoholysis of sulfinamides (+)-**20** ($R^1$, $R^2 = Et$) with primary alcohols (e.g. MeOH and EtOH) proceeds with complete or almost complete inversion. The stereospecificity is lowered with *i*-PrOH and is further lowered with *t*-BuOH[52]. Of the several acid catalysts used ($CF_3CO_2H$, $PhSO_2H$, $HSbF_6$ and others), the mixture $CF_3CO_2H$–$AgClO_4$ was proved to be the best as far as the stereospecificity is concerned. The yields in the transformation **20** ⟶ **21** are within the range of 53–95%; the enantiomeric excess is 50–88%, the percent of inversion is 64–100 and the stereospecificity is 58–98%, except in the case of *t*-BuOH where the values are significantly lower.

The alcoholysis of sulfinamides is carried out successfully by heating a solution of equivalent amounts of *N*-benzylsulfinamides and alcohols in benzene containing one equivalent of sulfuric acid[8a]. Although the yields of the resulting sulfinic esters are rather low (23–58%) this modified version of the acid catalyzed alcoholysis can provide sulfinic esters having sterically bulky groups, such as *tert*-butyl or menthyl *p*-toluenesulfinate. The latter is obtained from the corresponding sulfinamide (**20**; $R^1 = H$, $R^2 = CH_2C_6H_5$), menthol, and concentrated sulfuric acid in benzene by refluxing for 3 h. After work-up, **21** ($R^3$ = menthyl) is obtained in 23% yield, $[\alpha]_D^{75} = -196.8°$ ($c = 1.0$, acetone).

Other *p*-toluenesulfinic esters are obtained in the same manner by using different alcohols for the alcoholysis[8a].

The photolysis of *p*-toluenesulfinamides with a low-pressure mercury lamp in methanol also resulted in sulfur–nitrogen bond cleavage yielding methyl sulfinates in 30–40% yield[53]. Since more efficient methods are available for the sulfur–nitrogen bond cleavage of sulfinamides to form sulfinic esters, the photolytic method has no preparative value at present.

## E. Oxidation of Disulfides, Thiols and Sulfenic Esters

### 1. *Oxidation of disulfides*

Low-temperature oxidation of disulfides with chlorine in the presence of an alcohol leads directly to the corresponding sulfinic esters in good yields[54,55].

$$R^1SSR^1 + 4R^2OH + 3Cl_2 \xrightarrow{-20\,^\circ C} 2R^1S(O)OR^2 + 2R^2Cl + 4HCl \quad (13)$$

Quite often this is the method of choice for the preparation of sulfinic esters, since the starting disulfides are readily available and so are the other reactants, and it is a one-pot reaction leading directly to the desired target compound[56].

The problem is that, although the crude sulfinic esters are produced in 60–85% yield by this route, they are contaminated with chlorine-containing by-products, mainly $R^1SOCl$, $R^1SO_2Cl$ and also others[57] such as the corresponding thiosulfonic ester $R^1SO_2SR^1$.

The thiosulfonic esters boil higher than the sulfinic esters and can be removed by distillation, whereas the sulfinyl and sulfonyl chloride impurities may be removed by treatment with additional alcohol or with a high boiling primary amine (such as *p*-toluidine), and subsequent distillation[54]. All these required procedures complicate the synthesis, turning the work-up of the mixture into a laborious process.

Methanesulfinic acid methyl ester (**2**, $R^1 = R^2 = Me$) is prepared from a mixture of methyl disulfide, methanol and chlorine at $-20$ to $-25\,^\circ C$. A transient reddish-orange color caused by the formation of the methanesulfenyl chloride is observed. The work-up involves distillation, vacuum distillation, removal of various acylchlorides by *p*-toluidine and redistillation in vacuo. The final yield of the ester is 54%.

The oxidation of disulfides to the corresponding sulfinic esters can also be accomplished by lead tetraacetate[58].

$$R^1SSR^1 + 3Pb(OAc)_4 + R^2OH \longrightarrow 2R^1S(O)OR^2 + 3Pb(OAc)_2(+4AcOH + 2AcOCH_3) \quad (14)$$

This method affords a successful one-step synthesis of a variety of aromatic sulfinic esters from the readily available disulfides but is rather unattractive when starting from aliphatic disulfides, due to the by-products formed which make purification of the sulfinic esters impractical[59]. The yields of the arenesulfinic esters are 62–68%. Thus benzenesulfinic acid methyl ester (**2**, $R^1 = Ph$, $R^2 = Me$) is easily obtained[58] in the form of a pure oil in 62–68% yield.

Methyl and ethyl alkanesulfinic esters have been prepared in 68–79% yields via the oxidation of disulfides with singlet oxygen[60].

$$R^1SSR^1 + 2R^2OH + O_2 \xrightarrow{h\nu,\ \text{Rose Bengal}} 2R^1S(O)OR^2 \quad (15)$$

This reaction was performed on a low scale and the presence of the sulfinic ester products was determined without actually isolating them from the mixtures obtained; the preparative value of this method is doubtful.

### *2. Oxidation of thiols*

Electrolysis of thiophenol in acetic acid with aliphatic alcohols in the presence of sodium acetate leads to the formation of alkyl benzenesulfinic esters in satisfactory yields[61]. The electrolysis of the corresponding disulfides also affords the sulfinic esters, albeit in lower yields.

$$\text{PhSH} \xrightarrow[\text{CH}_3\text{CO}_2\text{H; R}^2\text{OH; 15–45 °C}]{\text{electrolysis/CH}_3\text{CO}_2\text{Na}} \text{PhS(=O)OR}^2 \qquad (16)$$

(83–95%)

The procedure is rather simple and the scale of the reported syntheses is satisfactory from a preparative point of view. However, the generality of this method has not yet been established.

Thus, to obtain benzenesulfinic acid methyl ester (**2**, $R^1 = Ph$, $R^2 = Me$) a mixture of thiophenol and sodium acetate in acetic acid and methanol is electrolyzed in an undivided cell by using two platinium electrodes at 15–45 °C. After passing about 6 F $mol^{-1}$ of electricity at 0.01–0.04 A $cm^{-2}$ of current density (30 V), the solvent is removed and the residue worked-up as usual, to give 95% yield.

The 'classical' oxidation of thiols with chlorine in the presence of acetic acid[5] leads to the corresponding sulfinyl chloride[56] which is then esterified by added alcohol[62], thus merging with the method described in Section II.B.

### 3. Oxidation of sulfenic esters (and sulfenyl chlorides)

The synthesis of sulfinic esters by oxidation of sulfenic esters has been known for quite some time[63]. Both sulfenic esters and sulfenyl chlorides are obtained from the corresponding sulfides and disulfides, and the chlorides have been most commonly used for the preparation of sulfinic esters. However, in certain cases the use of sulfenic esters as starting materials is advantageous. Such a case is the preparation of methyl trichloromethane sulfinate[64]: benzyl and variously *p*-substituted benzyl trichloromethane sulfinates have been synthesized in almost quantitative yields by oxidation of the appropriate sulfenates with *m*-chloroperbenzoic acid in methylene chloride. All the products were crystalline solids. No further oxidation to the sulfonates took place.

$$p\text{-RC}_6\text{H}_4\text{CH}_2\text{OSCCl}_3 \xrightarrow[\text{CH}_2\text{Cl}_2/0\,°\text{C}]{\text{MCPBA}} p\text{-RC}_6\text{H}_4\text{CH}_2\text{OS(=O)CCl}_3 \qquad (17)$$

**(22)** **(23)**

$R = H; CH_3; Cl; NO_2$

Methanesulfenyl chloride reacts with an equimolar or greater ratio of lithium pentyloxide, 4-methyl-2-pentyloxide or *t*-butoxide in 1,2-dimethoxyethane at −40 to −60 °C to give the corresponding methanesulfenate esters $R^1SOR^2$ in moderate yields. However, when the reaction is run using 1.5:1 or a greater ratio of methanesulfenylchloride to lithium pentyloxide or 4-methyl-2-pentyloxide, the corresponding sulfinic esters are formed in good yields[65].

$$3\text{CH}_3\text{SCl} + 2\text{LiOR} \longrightarrow \text{CH}_3\text{S(=O)OR} + \text{CH}_3\text{SSCH}_3 + \text{RCl} + 2\text{LiCl} \qquad (18)$$

**(24)** **(25)**

This procedure could have been a convenient method for preparing primary and secondary sulfinic esters. However, as one can see from the stoichiometry of the reaction (equation 18), one half of the alkoxide and two thirds of the methanesulfenyl chloride are

$$\underset{(24)}{CH_3SCl} \xrightarrow{ROLi} \underset{(26)}{CH_3SOR} \xrightarrow{CH_3SCl} \left[\begin{array}{c} CH_3 \\ | \\ CH_3SS^{+}ORCl^{-} \end{array}\right]$$

$$\longrightarrow \underset{(27)\ +RCl}{CH_3\overset{\overset{O}{\|}}{S}SCH_3} \xrightarrow{CH_3SOR} \underset{(25)}{CH_3\overset{\overset{O}{\|}}{S}OR} + CH_3SSCH_3 \tag{19}$$

converted to other products, most probably through the mechanism indicated[65] in equation 19. The process involves the reaction of sulfenyl chlorides with sulfenyl esters to give sulfinate esters (**25**) and disulfides as shown, although the thiosulfinate **27** could also react with the sulfenyl chloride **24** to form methanesulfinyl chloride, which inturn could then react with alcohol or alkoxide to give the sulfinic ester **25**[57b]. Regardless of which alternative is actually operating, the ultimate result is the same.

If the starting sulfenyl chloride is easily available and the alcohol to be used is not very expensive, this method might be considered, its drawbacks notwithstanding.

## F. Reaction of Sulfenyl Derivatives with Oxiranes

The reaction of methanesulfenyl chloride and ethylene oxides results in the formation of methanesulfinic esters alongside with dimethyl disulfide and ethylene dichloride[66].

$$3CH_3SCl + 2\underset{\diagdown O \diagup}{CH_2{-}CH_2} \longrightarrow \underset{(28)}{CH_3\overset{\overset{O}{\|}}{S}OCH_2CH_2Cl} + CH_3SSCH_3 + ClCH_2CH_2Cl \tag{20}$$

Here, again, the methanesulfenate initially formed by cleavage of the epoxide ring reacts with the excess of sulfenyl chloride to give a thiosulphinate which, in turn, reacts further to give the sulfinic ester **28**.

In view of the other convenient alternatives available for the synthesis of sulfinic esters, this route does not appear to be attractive. Even if the starting sulfenyl chloride and oxirane are easily available, only one-third of the first and one-half of the second is actually used for the formation of the sulfinic ester.

## G. Reduction of Sulfonyl Derivatives

Reaction of trialkyl phosphites $(R^3O)_3P$ with alkyl esters of aliphatic and aromatic thiosulfonic acids $R^1SO_2SR^2$ leads to the formation of sulfinic esters in high yield accompanied by *O,O,S*-trialkyl phosphorothiolates[67]. However, with *aryl* esters of aromatic thiosulfonic acids, sulfinic acids are not formed, and reduction of the thiosulfonates to disulfides occurs. Similarly, reaction of sodium dialkyl phosphites $(R^3O)_2PONa$ with thiosulfonates gives sodium sulfinates.

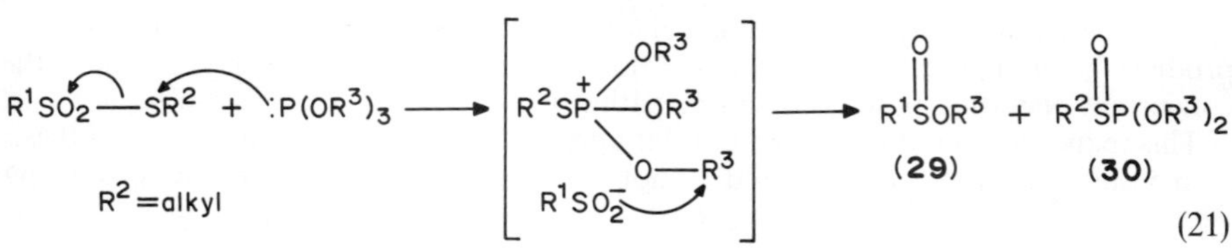


(21)

These reactions are carried out without a solvent, at 20–30 °C. The products are isolated by distillation *in vacuo* in 80–90% yield. In some cases the isomeric sulfone (i.e. $R^1SO_2R^3$) is also formed.

In this facile and convenient one-step synthesis of sulfinic esters, the availability of the starting alkyl thiosulfonic esters is the main limiting factor for wide application. Also, the separation between the sulfinic ester and the phosphorothiolate **30** by distillation is not complete in some cases. An example of this procedure[67], butanesulfinic acid butyl ester (**29**, $R^1 = R^3 = Bu$) was prepared from ethyl butanethiosulfonate[68] with tributyl phosphite in 90% yield.

The reaction of trimethyl phosphite and 1,2-di-*p*-toluenesulfonylethene to give methyl *p*-toluenesulfinate almost exclusively[69] is closely related and probably also involves a thiosulfonate–sulfinic ester transformation as described above.

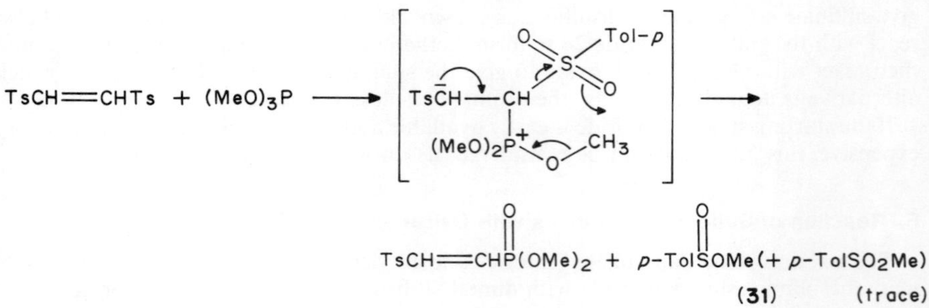


(22)

Based on the observation (a) that *in situ* sulfonylations of epoxy alcohols following asymmetric epoxidation persistently afforded sulfinate esters as significant by-products[70] and also that the isolation of *O,O*-dimethyl *S*-*p*-tolyl phosphorothiolate [*p*-TolSP(O)OMe$_2$] provides an additional by-product[71], and (b) that the sequential deoxygenation of sulfonyl chlorides in the reaction between triethyl phosphite and sulfonyl chlorides afforded the corresponding phosphorothiolates and triethyl phosphate[72], an extremely convenient, one-step synthesis of sulfinic esters from readily available sulfonyl chlorides proceeding by *in situ* reduction has been developed[73].

$$R^1SO_2Cl + R^2OH \xrightarrow[CH_2Cl_2]{(MeO)_3P,\ Et_3N} [R^1SO_2^- Cl\overset{+}{P}(OMe)_3] \longrightarrow [Cl^- R^1\overset{O}{\overset{\|}{S}}O\overset{+}{P}(OMe)_3]$$

$$\longrightarrow \overset{O}{\overset{\|}{P}}(OMe)_3 + R^1\overset{O}{\overset{\|}{S}}Cl \xrightarrow{R^2OH} R^1\overset{O}{\overset{\|}{S}}OR^2 \qquad (23)$$

**(2)**

Apparently, the intermediate sulfinyl chloride is intercepted by the alcohol present to produce the sulfinate ester **2**. It is possible, however, that the sulfinic ester is formed by the concurrent operation of more than one pathway[73].

This reduction method has been thus far applied primarily for the successful synthesis of menthyl sulfinates. The method is not suitable for sulfonyl chlorides possessing α-hydrogens, which give rise to the formation of sulfonate esters via sulfene intermediates. Phosphorothiolate $RSP(O)(OMe)_2$ is a consistent by-product resulting in reduction in the sulfinic ester yield. Given the experimental simplicity and the low cost of the reagents, the method promises to find widespread application in the preparation of sulfinate esters, especially when the corresponding sulfinic acid is not commercially available. Thus, a 0.2-mol-scale preparation of (−)-menthyl *p*-toluenesulfinate compared favorably (66% yield of the pure diasteromer) with previously reported methods[74], while the chiral sulfinic ester (+)-[(−) menthyl 2-naphthalenesulfinate[73] was obtained from 1-(−)-menthol and 2-naphthalenesulfonyl chloride in $CH_2Cl_2$ under nitrogen, with triethylamine and trimethyl posphite in 96% crude yield and 28% yield after two recrystallizations.

The conversion of 1,2-dithiolane-1,1-dioxides to the corresponding 1,2-oxathiolane-2-oxides has been achieved by using hexaethylphosphoramide as the reducing agent[75].

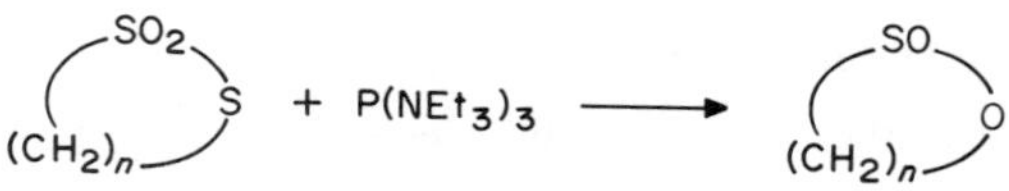

(24)

n = 3, 4

For more details, the interested reader is referred to the chapter on cyclic sulfinic esters in this book.

## H. Sulfur–Sulfur and Sulfur–Nitrogen Bond Cleavage

Both aryl α-disulfones and sulfinyl sulfones[76–78] are very reactive towards nucleophiles, which cleave the sulfur–sulfur bond resulting in the displacement of an aryl sulfinate ion[77]:

$$ArSO_nSO_2Ar + Nu^- \longrightarrow ArSO_nNu + ArSO_2^- \qquad (25)$$

$n = 1$ or 2

Thus, the reaction of sulfinyl sulfones **32a, b** with a solution of sodium methoxide in methanol[77] or with various alcohols (methanol, borneol, adamantanol, *t*-butanol) in the presence of pyridine[78] provides the corresponding sulfinic esters, most probably as a result of direct substitution[77].

$$\underset{\mathbf{(32)}}{R^1\overset{O}{\overset{\|}{S}}SO_2R^1} + R^2OH \xrightarrow[CH_2Cl_2]{R^2O^- \text{ or pyridine}} \underset{\substack{\mathbf{(2)}\\(62\text{–}90\%)}}{R^1\overset{O}{\overset{\|}{S}}OR^2} + R^1SO_2^- \qquad (26)$$

(**a**) $R^1 = Bu$

(**b**) $R^1 = p\text{-Tol}$

The procedure is simple and convenient and the yields are good. The required sulfinyl sulfones **32** should be prepared just before their use since they are relatively unstable.

The preparation of 1-adamantyl *p*-toluenesulfinate (**2**, $R^1 = p\text{-Tol}$, $R^2$ = adamantyl) is achieved[78] from freshly prepared sulfinyl sulfone **32b** with 1-adamantanol and pyridine in dichloromethane at room temperature. After about one hour the mixture is worked-up to afford, after chromatography on silicagel, the product in 79% yield.

The oxidation of $N$-alkyl-$N^1$-tosylhydrazines with $SeO_2$, $CrO_3$ and HgO results in the isolation of $p$-toluenesulfinic esters and olefins[79], probably via a radical pathway.

$$p\text{-TolSO}_2\text{NHNHCH(R}^1\text{)CH}_2\text{R}^2 \xrightarrow[(-\text{H}_{2j}\text{–N}_2)]{\text{SeO}_2\text{ or CrO}_3\text{ or HgO}} p\text{-TolS(=O)OCH(R}^1\text{)CH}_2\text{R}^2 \quad (27)$$

Although the yields of the sulfinic esters produced are within the respectable range of 60–80% for long-chain alkyl sulfinic esters and 40–70% for cycloalkyl sulfinic esters, the synthetic potential of this route cannot be assessed at this state in view of the limited relevant data available.

## I. Carbon–Sulfur Bond Formation

Early attempts to synthesize simple sulfinic esters by the reaction of dialkyl sulphites and Grignard reagents were unsuccessful, and symmetrical sulfoxides were always obtained probably from the sulfinates first formed and the Grignard reagents present in the reaction mixture. Later it was found that this undesired reaction can be blocked at the stage of the sulfinate formation if tertiary Grignard reagents are employed. Thus, a one-step synthesis of alkyl $t$-alkanesulfinates from dialkyl sulfites and tertiary alkyl magnesium chlorides is feasible, the yields being fair to good[80].

$$\underset{\textbf{(33)}}{\text{R}^1\text{MgCl}} + \underset{\textbf{(34)}}{\text{R}^2\text{OS(=O)OR}^2} \longrightarrow \underset{\textbf{(2)}}{\text{R}^1\text{S(=O)OR}^2} \quad (28)$$

$R^1 = t\text{-}C_4H_9$; $t\text{-}C_5H_{11}$
$R^2 = CH_3$; $C_2H_5$; n-$C_3H_7$; $i\text{-}C_3H_7$; n-$C_4H_9$

Although two molar equivalents of the tertiary alkyl magnesium chloride (**33**) are used in boiling tetrahydrofuran for 4–8 h, only one of the sulfur–oxygen bonds of the sulfite is being cleaved and only one new carbon–sulfur bond is being formed in the process.

Interestingly, unlike the dialkyl derivatives, alkylchlorosulfites do give sulfinic esters, though in moderate yields, even in their reactions with nonbulkyl alkyl magnesium chlorides[81].

The reaction of 4-hydroxy 1-butene (or 1-pentene) with thionyl chloride in the presence of pyridine provides an entry into cyclic sulfinic esters (1,2-oxathiane-2-oxides) through a carbon-1–sulfur bond formation[82].

The reaction of dialkyl esters of sulfoxylic acid (i.e. **35**) with low molecular weight alkyl iodides to give sulfinic esters[83] represents another sulfinic ester synthesis via a carbon–sulfur bond formation.

$$\text{R}^1\text{I} + \underset{\textbf{(35)}}{\text{R}^2\text{—O—S—O—R}^2} \xrightarrow[10\text{–}15\,\text{h};\ 70\text{–}100\,^\circ\text{C}]{\text{reflux}} \underset{(44\text{–}79\%)}{\text{R}^1\text{S(=O)OR}^2} + \text{R}^2\text{I} \quad (29)$$

$R^1 = CH_3$ or $C_2H_5$; $R^2 = C_3H_7$, $C_4H_9$, $C_5H_{11}$

Since comparable or higher yields of the same sulfinic esters can be achieved by using more readily available starting materials than the diesters **35** (see Section II.B, II.E and Table), the use of this method is not expected to gain much ground.

### J. Carbon–Sulfur Bond Cleavage

Both benzyl and *t*-butyl alkyl or aryl sulfoxides react with *N*-bromo- and *N*-chlorosuccinimide to provide the relatively stable benzylic and *t*-butyl carbocations via the cleavage of the carbon–sulfur bond. In the presence of alcohol sulfinic esters are obtained in good yields (60–95%)[84].

$$\underset{\textbf{(36)}}{R^1\overset{\overset{\displaystyle O}{\|}}{S}R^3} + R^2OH \xrightarrow[\text{CHCl}_3]{\text{NBS(NCS)}} \underset{\textbf{(2)}}{R^1\overset{\overset{\displaystyle O}{\|}}{S}OR^2} \qquad (30)$$

$R^1 = t\text{-alkyl}; C_6H_5CH_2; C_6H_5CH(CH_3)$

$R^3 = t\text{-Bu}; \quad C_6H_5CH_2$

$R^2 = Et$

Since *t*-butyl alkyl sulfoxides can be conveniently prepared by treatment of the lithioderivatives of simpler sulfoxides with electrophilic reagents (equation 31), the formation of sulfinic esters by the cleavage of the carbon–sulfur bond of the *t*-butyl sulfoxides represents a general synthesis. *t*-Butyl 3-,4-,5- and 6-hydroxyalkyl sulfoxides can be transformed into the corresponding cyclic sulfinic esters by the same method[84,85].

$$t\text{-Bu}\overset{\overset{\displaystyle O}{\|}}{S}CH_2R^4 \xrightarrow[2.\ R^5X]{1.\ \text{MeLi}} t\text{-Bu}\overset{\overset{\displaystyle O}{\|}}{S}CHR^4R^5 (\equiv 36;\ R^1 = CHR^4R^5) \qquad (31)$$

$$\xrightarrow[R^2OH]{\text{NCS}} \underset{\textbf{(2)}}{R^1\overset{\overset{\displaystyle O}{\|}}{S}OR^2}$$

### K. Miscellaneous

In certain cases one may be interested in using a readily available sulfinic ester for the preparation of another sulfinic ester in which the group $R^2$ is replaced by another group. This can be done by either acid- or base-catalyzed alcoholysis of the sulfinic ester[28,86].

$$\underset{\textbf{(2)}}{R^1\overset{\overset{\displaystyle O}{\|}}{S}OR^2} + R^3OH \xrightarrow[\Delta]{\text{Acid or base catal.}} R^1\overset{\overset{\displaystyle O}{\|}}{S}OR^3 \qquad (32)$$

In the base-catalyzed alcoholysis, the stronger the base the higher the rate of the sulfur–oxygen bond fission[28].

Methyl sulfinates are often used to produce higher molecular weight esters[86]. Thus, methanesulfinic acid butyl ester (**37**) is prepared from methanesulfinic acid methyl ester and excess 1-butanol and concentrated sulfuric acid catalyst by refluxing for 45 min. Distillation of the remaining liquid under reduced pressure resulted in 79% yield.

Germanium mono-, di- and trisulfinic esters, $R^1S(O)OGeR_3$, $[R^1S(O)O]_2GeR_2$ and

# III. TABLE

Synthesis of selected sulfinic esters $R^1SOOR^2$:
Starting materials, methods, yields and references

| $R^1$ | $R^2$ | Starting materials reagents | Method* | Yield(%) | Reference |
|---|---|---|---|---|---|
| $CH_3$ | $CH_3$ | $CH_3SSCH_3$; $CH_3OH$; $Cl_2$ | II.E | 54 | 54 |
| $CH_3$ | $CH_3$ | $CH_3SOCl$; $CH_3OH$ | II.B | 71 | 16 |
| $CH_3$ | $C_2H_5$ | $CH_3SOCl$; $C_2H_5OSiMe_3$ | II.B | 81 | 34 |
| $CH_3$ | $C_4H_9$ | $CH_3SOCl$; $C_4H_9OH$; (+)-(**10**) | II.B | *a,b* | 32 |
| $CH_3$ | $C_4H_9$ | $CH_3I$; $C_4H_9{-}O{-}S{-}O{-}C_4H_9$ | II.I | 79 | 83 |
| $CH_3$ | $C_4H_9$ | $CH_3SSCH_3$; $C_4H_9OH$; $Cl_2$ | II.E | 84 | 54 |
| $CH_3$ | $C_5H_{11}$ | $CH_3SCl$; $C_5H_{11}OH$; MeLi | II.E.3 | 84 | 65 |
| $CH_3$ | menthyl | $CH_3SOCl$; (−)-menthol | II.B | 93 | 3 |
| $C_2H_5$ | $C_2H_5$ | $C_2H_5SSC_2H_5$; $C_2H_5OH$; $Cl_2$ | II.E.1 | 75 | 54 |
| $C_2H_5$ | $i$-$C_3H_7$ | $C_2H_5S(O)SC_2H_5$; $i$-$C_3H_7OH$; $H_2O_2$ | II.D | 45 | 46 |
| $C_3H_7$ | $CH_3$ | $CH_3SOCl$; $CH_3OH$ | II.B | 82 | 33 |
| $C_3H_7$ | $CH_2{=}CHCH_2{-}$ | $C_3H_7SOCl$; $CH_2{=}CHCH_2OH$ | II.B | 54 | 17 |
| $i$-$C_3H_7$ | $C_2H_5$ | $i$-$C_3H_7SOBu$-$t$; NCS(NBS); $C_2H_5OH$ | II.J | 85[c] | 84 |
| $i$-$C_3H_7$ | $i$-$C_3H_7$ | 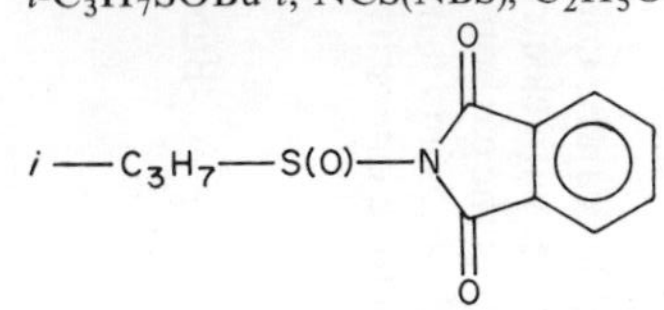 ; $i$-$C_3H_7O^-Na^+$ | II.D | 63 | 50 |
| $C_4H_9$ | $CH_3$ | $C_4H_9SOCl$; $CH_3OSOCl$; $(Me_3Si)_2O$ | II.B | 81 | 37 |
| $C_4H_9$ | $C_4H_9$ | $BuSO_2SC_2H_5$; $(n$-$C_4H_9O)P$ | II.G | 90 | 67 |
| $C_4H_9$ | menthyl | $C_4H_9SOCl$; (−)menthol | II.B | quant.[c] | 4 |
| $t$-$C_4H_9$ | n-$C_4H_9$ | n-$C_4H_9OS(O)OC_4H_9$-n; $t$-$C_4H_9MgCl$ | II.I | 62 | 80 |
| $t$-$C_5H_{11}$ | n-$C_3H_7$ | n-$C_3H_7OS(O)OC_3H_7$-n; $t$-$C_5H_{11}MgCl$ | II.I | 51 | 80 |
| $CH_3OCOCH_2$ | $CH_3$ | 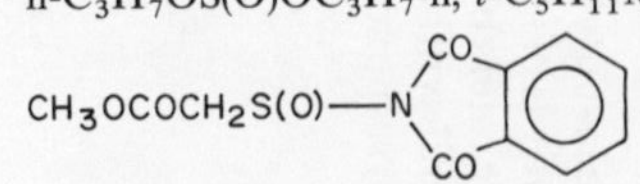 ; $CH_3OH$ | II.D | 93 | 50 |
| n-$C_3H_7CH(CH_3)$ | $C_2H_5$ | n-$C_3H_7CH(CH_3)SO$-$t$-$C_4H_9$; NCS(NBS); $C_2H_5OH$ | II.J | 94[c] | 84 |
|  | $CH_3$ | 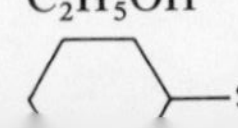 ; $CH_3OH$ | II.B | 73 | 33 |

| | | | | | |
|---|---|---|---|---|---|
| 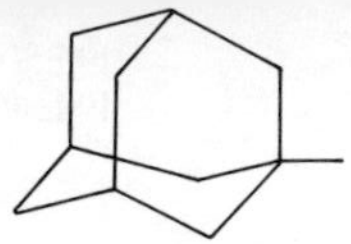 | $CH_3$ | 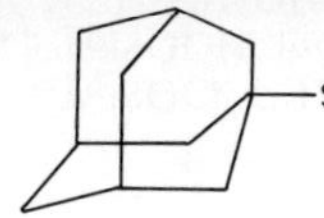 –SOCl; $CH_3OH$ | II.B | 83 | 20, 21 |
| $C_{12}H_{25}$ | $C_2H_5$ | $C_{12}H_{25}SO_2H$; 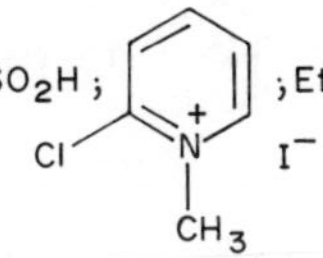 ; $Et_3N$; $C_2H_5OH$ | II.A | 62 | 6 |
| $C_{12}H_{25}$ | $C_2H_5$ | $C_{12}H_{25}S(O)NHCH_2C_6H_5$; $C_2H_5OH$; $H_2SO_4$ | II.D | 58 | 8a |
| $Cl_3C$ | $CH_3$ | $Cl_3CSOCl$; $CH_3OH/K_2CO_3$ | II.B | 73 | 38a |
| $Cl_3C$ | $p$-$RC_6H_4CH_2$<br>R = H; $CH_3$; Cl; $NO_2$ | $Cl_3CS(O)CH_2C_6H_4R$; MCPBA | II.E.3 | quant. | 64 |
| $CF_3$ | $CF_3(CH_3)_2C$ | $CF_3SOCl$; $CF_3(CH_3)_2COH$; NaF | II.B | 67 | 41 |
| Ph, $CH_2$ | $C_2H_5$ | 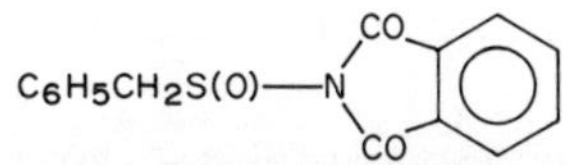 ; $C_2H_5OH$ | II.D | 97 | 50 |
| $PhCH_2$ | $C_4H_9$ | $C_6H_5CH_2SOCl$; $C_4H_9OH$ | II.A | 66 | 22 |
| $PhCH_2$ | $C_3H_7$ | $C_6H_5CH_2SOCl$; $C_3H_7OH$; (−)-(**10**) | II.B | *a,d* | 32 |
| $PhCH_2$ | $C_6H_5CH_2$ | $C_6H_5CH_2SOCl$; $C_6H_5CH_2CH_2OSi(CH_3)_3$ | II.B | 48 | 34 |
| Ph | $CH_3$ | $C_6H_5S(O)SC_6H_5$; MeOH; $H_2O_2$ | II.D | 91 | 46 |
| Ph | $CH_3$ | 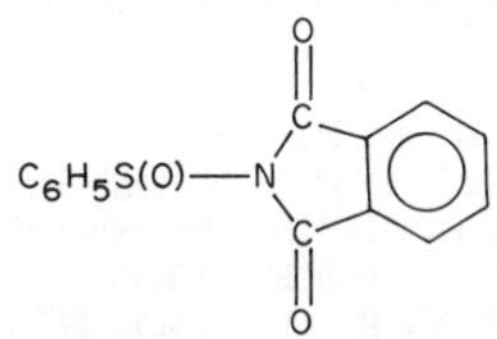 ; $CH_3OH$ | II.D | 95 | 50 |
| Ph | $CH_3$ | $C_6H_5SOCl$; $CH_3OSOCl$; $(Me_3Si)_2O$ | II.B | 86 | 37 |
| Ph | $CH_3$ | $C_6H_5SOCl$ (i.e. $C_6H_5SH + Cl_2$); $CH_3OH$ | II.B | 65[e] | 16 |
| Ph | $CH_3$ | $C_6H_5SSC_6H_5$; $CH_3OH$; $Pb(OAc)_4$ | II.E.1 | 62–68 | 58a |
| Ph | $CH_3$ | $C_6H_5SH$; $CH_3OH$; $CH_3CO_2H(Na)$ | II.E.2 | 95 | 61 |
| Ph | $C_2H_5$ | $C_6H_5SO_2SC_2H_5$; $(C_2H_5O)_3P$ | II.G | 74 | 67 |
| $C_6H_5$ | $(CH_3)_2CH$ | $C_6H_5SH$; $(CH_3)_2CHOH$; $CH_3SO_2H(Na)$ | II.E.2 | 85 | 61 |
| $C_6H_5$ | $C_6H_5CH_2$ | $C_6H_5SOCl$; $C_6H_5CH_2OSi(CH_3)_3$ | II.B | 95 | 34 |

(*continued*)

**III. TABLE** *(continued)*

| $R^1$ | $R^2$ | Starting materials reagents | Method* | Yield(%) | Reference |
|---|---|---|---|---|---|
| $C_6H_5$ | 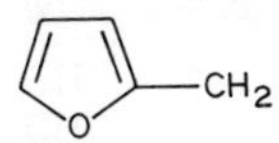 | $C_6H_5SOCl$; 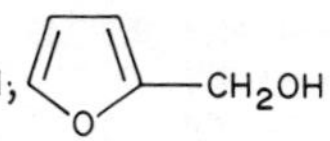 | II.B | 71 | 29 |
| $C_6H_5$ | $HC{\equiv}C{-}C(CH_3)_2$ | $C_6H_5SOCl$; $-C(CH_3)_2C{\equiv}CH$ | II.B | 66 | 23 |
| $C_6F_5$ | $CH_3$ | $C_6F_5SOCl$; $CH_3OH$ | II.B | 75 | 27 |
| *p*-Tol | $CH_3$ | *p*-TolS(O)S$C_6H_5$; MeOH; $H_2O_2$ | II.D | 81 | 46 |
| *p*-Tol | $CH_3$ | *p*-Tol$SO_2^-Na^+.2H_2O$; $(CH_3O)_2SO_2$ | II.C | $70^f$ | 42 |
| *p*-Tol | $CH_3$ | *p*-Tol$SO_2^-Na^+.2H_2O$; $CH_2N_2$ | II.C | 100 | 42 |
| *p*-Tol | $CH_3$ | *p*-TolSOCl; $CH_3OH$ | II.B | 65 | 24 |
| [(−)-(*S*)]-*p*-$CH_3C_6H_4$ | $CH_3$ | (+)-(*S*)-*p*-TolSON$Et_2$; MeOH; $CF_3CO_2H$ | II.D | 94 | 52 |
| [(−)-(*S*)]-*p*-$CH_3C_6H_4$ | $C_2H_5$ | (+)-(*S*)-*p*-TolSON$Et_2$; EtOH; $C_6H_5SO_3H$ | II.D | 76.5 | 52 |
| *p*-Tol | $C_2H_5$ | *p*-Tol$SO_2H$; $C_6H_5O{-}P(=O)Cl_2$; $P_y$; $C_2H_5OH$ | II.A | 79 | 7 |
| *p*-Tol | *i*-$C_3H_7$ | *p*-TolSOCl; *i*-$C_3H_7OSOCl$; $(Me_3Si)_2O$ | II.B | 87 | 37 |
| *p*-Tol | $C_4H_9$ | *p*-Tol$SO_2H$; $C_6H_5O{-}P(=O)Cl_2$; $P_y$; BuOH | II.B | 85 | 7 |
| *p*-Tol | *t*-$C_4H_9$ | *p*-Tol$SO_2H$; $C_6H_5O{-}P(=O)Cl_2$; $P_y$; *t*-BuOH | II.B | 81 | 7 |
| *p*-Tol | *t*-$C_4H_9$ | *p*-Tol$SO_2H$; $(C_6H_5O)_2P(=O)Cl_2$; $P_y$; *t*-BuOH | II.B | 87 | 8b |
| *p*-Tol | *t*-$C_4H_9$ | *p*-TolS(O)NH$CH_2C_6H_5$; *t*-BuOH; $H_2SO_4$ | II.D | 42 | 8a |
| *p*-Tol | $C_6H_5CH_2$ | *p*-Tol$SO_2H$; Cl 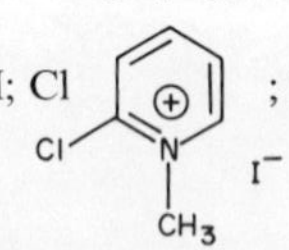 ; $Et_3N$; $C_6H_5CH_2OH$ | II.B | 65 | 6 |
| *p*-Tol | $C_6H_5CH_2$ | *p*-TolSOCl; $C_6H_5CH_2OH$ | II.B | 52 | 25 |
| *p*-Tol | $C_6H_5CH_2$ | *p*-TolSOOH; $EtO_2CN{=}NCO_2Et$; $(C_6H_5)_3P$; $C_6H_5CH_2OH$ | II.A | 35 | 6 |

| | | | | | |
|---|---|---|---|---|---|
| p-Tol | menthyl | p-TolSOO$_2$H; 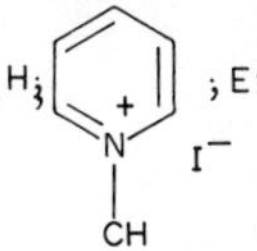 ; Et$_3$N; menthol | II.A | 35 | 6 |
| p-Tol | 1-damantyl | p-TolSOSO$_2$—$C_6H_4CH_3$-p; 1-adamantanol | II.H | 79 | 78 |
| p-Tol | menthyl | p-TolSOSO$_2$—$C_6H_4CH_3$-p; menthol | II.H | 82 | 78 |
| p-Tol | CH≡CCH$_2$ | p-TolSO$_2$H; 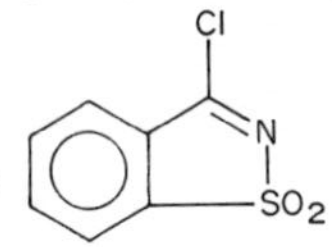 ; Et$_3$N; CH≡CCH$_2$OH | II.A | 57 | 8a |
| p-Tol | (−)-$C_6H_{13}CH(CH_3)$ | p-TolSOCl(−)$C_6H_{13}CH(CH_3)$; pyridine | II.B | 56 | 25 |
| 2,6-$(CH_3)_2C_6H_3$ | $CH_3CH(C_6H_5)$ | 2,6-$(CH_3)_2C_6H_3SOCl$; $CH_3CH(C_6H_5)OH$ | II.B | *c* | 26 |
| 2,6-$(CH_3)_2C_6H_3$ | t-$C_4H_9$ | 2,6-$(CH_3)_2C_6H_3SOCl$; t-$C_4H_9OH$ | II.B | *c* | 26 |
| 2,6-$(CH_3)_2C_6H_3$ | p-$CH_3OC_6H_4C(CH_3)_2CH_2$ | 2,6-$(CH_3)_2C_6H_3SOCl$; p-$CH_3OC_6H_4C(CH_3)_2CH_2OH$ | II.B | 49 | 2 |
| p-$NO_2C_6H_4$ | $C_2H_5$ | p-$NO_2C_6H_4SO_2H$; Py; $(C_2H_5)_3\overset{+}{O}BF_4^-$ | II.C | 37 | 44 |
| 2,4,5-$Cl_3C_6H_2$ | menthyl | 2,4,5-$Cl_3C_6H_2SO_2Cl$; 1-menthol | II.G | 75[g] | 73 |
| p-(t-Bu)$C_6H_4$ | menthyl | p-(t-Bu)$C_6H_4SO_2Cl$; 1-menthol | II.G | 87[h] | 73 |
| 2-naphthalenyl | menthyl | 2-naphthalenesulfonyl chloride; 1-menthol | II.G | 96[i] | 73 |
| $C_6H_5CH_2S(CH_2)_2$- | $C_2H_5$ | $C_6H_5CH_2S(CH_2)_2SO_2Na$; $(C_2H_5)_3\overset{+}{O}BF_4^-$ | II.C | 41 | 45 |

*The entries in this column correspond to the number of the section and subsection where the method is described.

[a]Not specified.
[b]14.3% optical purity.
[c]Before final distillation.
[d]21% optical purity.
[e]Based on thiophenol.
[f]Contaminated with some 4-menthylphenyl-methylsulfone.
[g]Diastereomeric selectivity, 2.1:1.
[h]Diastereomeric selectivity, 1.4:1.
[i]Diastereomeric selectivity, 1.5:1.

$[R^1S(O)]_3GeR$, have been synthesized from the reaction of anhydrous $R^1SO_2Ag$ with $R_3GeCl$, $R_2GeCl_2$ and $RGeCl_3$, respectively[87].

## IV. REFERENCES


1. S. Oae and N. Kunieda, in *Organic chemistry of Sulfur* (Ed. S. Oae), Plenum Press, New York, 1977, pp. 610–612.
2. O. Bastiansen and H. Viervoll, *Acta Chem. Scand.*, **2**, 702 (1948).
3. K. K. Andersen, *J. Org. Chem.*, **29**, 1953 (1964).
4. K. Mislow, M. M. Green, P. Laur, J. T. Melillo, T. Simmons and A. L. Ternay, Jr., *J. Am. Chem. Soc.*, **87**, 1958 (1965).
5. H. Phillips, *J. Chem. Soc.*, **127**, 2552 (1925).
6. M. Furukawa, T. Ohkawara, Y. Noguchi and M. Nishikawa, *Synthesis*, 441 (1978).
7. M. Furukawa, T. Ohkawara, Y. Noguchi, M. Isoda and T. Hitoshi, *Synthesis*, 937 (1980).
8. a. M. Furukawa, T. Ohkawara, Y. Noguchi, M. Nishikawa and M. Tominatsu, *Chem. Pharm. Bull.*, **28**, 134 (1980).
   b. Y. Noguchi, M. Isoda, K. Kuroki and M. Furukawa, *Chem. Pharm. Bull.*, **30**, 1646 (1982).
9. Y. Miyaji, H. Minato and M. Kobayashi, *Bull. Chem. Soc. Jpn.*, **44**, 862 (1971).
10. M. Furukawa and T. Ohawara, *Synthesis*, 339 (1976).
11. L. Field, *Synthesis*, 713 (1978).
12. A. S. Weisberger and J. Pensky, *Science*, **126**, 1112 (1957).
13. K. Schank and H. G. Schmitt, *Chem. Ber.*, **107**, 3026 (1974).
14. G. Capozzi, V. Luccini, F. Marcuzzi and G. Melloni, *Tetrahedron Lett.*, **21**, 3289 (1980).
15. I. B. Douglass, B. S. Farah and E. G. Thomas, *J. Org. Chem.*, **26**, 1996 (1961); *Org. Synth.*, **40**, 62 (1960).
16. I. B. Douglass, *J. Org. Chem.*, **30**, 633 (1965).
17. De. OS. 2257050 (1972); *Chem. Abstr.*, **79**, 41926 (1973).
18. E. Wenschuh, C. Steyer and G. Bär, *Z. Chem.*, **19**, 211 (1979).
19. I. B. Douglass and D. R. Poolee, *J. Org. Chem.*, **22**, 536 (1957).
20. E. Krauthausen, in *Methoden Der Organischen Chemie* (Houben-Weyl) (Ed. D. Klamann), Georg Thieme Verlag, New York, 1985, pp. 643–647, and references cited therein.
21. H. Stetter, M. Krause and W. D. Last, *Chem. Ber.*, **102**, 3357 (1969).
22. E. Wenschuh and H. Lankau, *Z. Chem.*, **13**, 427 (1973).
23. S. Braverman and H. Mechoulam, *Isr. J. Chem.*, **5**, 71 (1967).
24. A. H. Wragg, J. S. McFadyen and T. S. Stevens, *J. Chem. Soc.*, 3603 (1958).
25. J. W. Wilt, R. G. Stein and W. J. Wagner, *J. Org. Chem.*, **32**, 2097 (1967).
26. D. Darwish and R. McLaren, *Tetrahedron Lett.*, 1231 (1962).
27. W. A. Sheppard and S. S. Foster, *J. Fluorine Chem.*, **2**, 53 (1972).
28. D. Darwish and J. Noreyko, *Can. J. Chem.*, **43**, 1366 (1965).
29. S. Braverman and T. Globerman, *Tetrahedron*, **30**, 3873 (1974).
30. I. D. Morrison and H. S. Mosher, *Asymmetric Organic Reactions*, Prentice-Hall, Englewood Cliffs, New Jersey, 1971, pp. 362–365.
31. E. Ciuffarin, M. Isoda and A. Fava, *J. Am. Chem. Soc.*, **90**, 3594 (1968).
32. M. Mikolajczyk and J. Drabowicz, *Chem. Commun.*, 547 (1974).
33. A. Hessig, M. Jaspers and J. Schwermann, *Chem. Ber.*, **112**, 2903 (1979).
34. D. N. Harpp, B. T. Friedlander, C. Larsen, K. Steliou and A. Stockton, *J. Org. Chem.*, **43**, 3481 (1978).
35. S. H. Langer, S. Connel and I. Wender, *J. Org. Chem.*, **23**, 50 (1958).
36. M. Axelrod, P. Bickart, J. Jacobus, M. M. Green and K. Mislow, *J. Am. Chem. Soc.*, **90**, 4835 (1968).
37. J. Drabowicz, *Chem. Lett.*, 1753 (1981).
38. a. U. Schöllkopf and P. Hilbert, *Justus Liebigs Ann. Chem.*, 1061 (1973).
    b. W. H. Mueller and M. B. Dines, *Chem. Commun.*, 1205 (1969).
39. D. Sianesi, G. C. Berardi and G. Moggi, *Tetrahedron Lett.*, 1313 (1970).
40. D. T. Sauer and J. M. Shreeve, *Inorg. Chem.*, **10**, 358 (1971).
41. A. Majid and J. M. Shreeve, *Inorg. Chem.*, **13**, 2710 (1974).
42. J. S. Meek and J. S. Fowler, *J. Org. Chem.*, **33**, 3422 (1968).
43. M. Kobayashi, H. Minato and H. Fukuda, *Bull. Chem. Soc. Jpn.*, **46**, 1266 (1973).

44. M. Kobayashi, M. Terao and A. Yamamoto, *Bull. Chem. Soc. Jpn.*, **39**, 1296 (1966).
45. J. P. Harmon and L. Field, *J. Org. Chem.*, **51**, 5235 (1986).
46. T. Takata and S. Oae, *Bull. Chem. Soc. Jpn.*, **55**, 3937 (1982).
47. Y. H. Kim, T. Takata and S. Oae, *Tetrahedron Lett.*, 2305 (1978).
48. S. Oae, T. Takata and J. H. Kim, *Tetrahedron*, **37**, 37 (1981).
49. M. Mikolajczyk and J. Drabowicz, *Chem. Commun.*, 220 (1976).
50. D. N. Harpp and T. G. Back, *J. Org. Chem.*, **38**, 4328 (1973).
51. D. N. Harpp and T. G. Back, *Tetrahedron Lett.*, 5313 (1972).
52. M. Mikolajczyk, J. Drabowicz and B. Bujnicki, *Chem. Commun.*, 568 (1976).
53. H. Tsuda, H. Minato and M. Kobayashi, *Chem. Lett.*, 149 (1976).
54. I. B. Douglass, *J. Org. Chem.*, **39**, 563 (1974).
55. E. N. Givens and L. A. Hamilton, *J. Org. Chem.*, **32**, 2857 (1967).
56. I. B. Douglass and R. V. Norton, *J. Org. Chem.*, **33**, 2104 (1968).
57. a. M. L. Kee and I. B. Douglass, *Org. Prep. Proceed. Int.*, **2**, 235 (1970).
b. I. B. Douglass and D. A. Koop, *J. Org. Chem.*, **27**, 1398 (1962).
58. a. L. Field and J. M. Locke, *Org. Synth.*, **46**, 62 (1966).
b. L. Field, C. B. Hoelzel and J. M. Locke, *J. Am. Chem. Soc.*, **84**, 847 (1962).
59. L. Field, J. M. Locke, C. B. Hoelzel and J. E. Lawson, *J. Org. Chem.*, **27**, 3313 (1962).
60. W. Ando, J. Suzuki, T. Arai and T. Mijita, *Tetrahedron*, **29**, 1507 (1973).
61. J. Nokami, Y. Fujita and R. Okawara, *Tetrahedron Lett.*, 3659 (1979).
62. C. J. M. Stirling, *J. Chem. Soc.*, 5741 (1963).
63. D. Barnard, *J. Chem. Soc.*, 4547 (1957).
64. S. Braverman and Y. Duar, *Tetrahedron Lett.*, 343 (1975).
65. T. L. Moore and D. E. O'Connor, *J. Org. Chem.*, **31**, 3587 (1966).
66. I. B. Douglass and J. A. Douville, *J. Org. Chem.*, **25**, 2221 (1960).
67. J. Michalski, T. Mordo and J. Wieczorkowski, *J. Chem. Soc.*, 1665 (1960).
68. B. J. Boldyrev and A. K. Litkovets, *Zh. Obshch. Khim.*, **16**, 3360 (1956).
69. J. S. Meek and J. S. Fowler, *J. Org. Chem.*, **33**, 985 (1968).
70. R. M. Hanson, S. Y. Ko, Y. Gao, H. Masamune, J. M. Klunder and K. B. Sharpless, *J. Am. Chem. Soc.*, in press.
71. A. H. Lee and R. L. Metcalf, *Pestic. Biochem. Physiol.*, **2**, 408 (1973).
72. R. W. Hoffmann, T. R. Moore and B. Kagan, *J. Am. Chem. Soc.*, **78**, 6413 (1956).
73. J. M. Klunder and K. B. Sharpless, *J. Org. Chem.*, **52**, 2598 (1987).
74. H. Hulce, J. P. Mallamo, L. L. Frye, T. P. Kogan and G. H. Posner, *Org. Synth.*, **64**, 196 (1986).
75. a. D. N. Harpp, J. G. Gleason and D. K. Ash, *J. Org. Chem.*, **36**, 322 (1971).
b. D. N. Harpp and J. G. Gleason, *J. Org. Chem.*, **36**, 1314 (1971).
76. L. Kice, G. Guaraldi and C. G. Venier, *J. Org. Chem.*, **31**, 3561 (1961).
77. J. L. Kice and S.-M. Wu, *J. Org. Chem.*, **46**, 3913 (1981).
78. R. B. Boar and A. C. Patel, *Synthesis*, 584 (1982).
79. L. Caglioti and F. Gasparrini, *Chem. Commun.*, 138 (1974).
80. M. Mikolajczyk and J. Drabowicz, *Synthesis*, 124 (1974).
81. P. Carre and D. Libermann, *C.R. Acad. Sci. Paris*, **200**, 2086 (1935).
82. K. S. Dhami, *Indian J. Chem.*, **12**, 278 (1974).
83. E. Wenschuh, R. Fahsl and R. Höhne, *Synthesis*, 829 (1976).
84. F. Jung and T. Durst, *Chem. Commun.*, 4 (1973).
85. N. K. Sharma, F. de Reinbach-Hirtzbach and T. Durst, *Can. J. Chem.*, **54**, 3012 (1976).
86. I. B. Douglass, F. J. Ward and R. V. Norton, *J. Org. Chem.*, **32**, 324 (1967).
87. E. Lindner and K. Schardt, *J. Organomet. Chem.*: a. **81**, 145 (1974). b. **82**, 73 (1974). c. **82**, 81 (1974).

The Chemistry of Sulphinic Acids, Esters and their Derivatives
Edited by S. Patai


CHAPTER **9**

# Cyclic sulphinic acid derivatives (sultines and sulphinamides)

DONALD C. DITTMER and MICHAEL D. HOEY
*Department of Chemistry, Syracuse University, Syracuse, New York 13244-1200, USA*

## I. INTRODUCTION

Both cyclic and acyclic sulphinate esters and amides are covered to some extent in the Houben-Weyl series[1] and in *Comprehensive Organic Chemistry*[2a]. Discussion of these compounds is scattered throughout *Comprehensive Heterocyclic Chemistry*[2b]. Cyclic sulphinates (sultines)[3] and aspects of cyclic sulphinamide chemistry[4] have been reviewed. The sultine **1** and its substituted derivatives are useful precursors of *o*-quinodimethanes (*o*-xylylenes), and dihydrothiazine 1-oxides, **2**, have been shown to be versatile intermediates in synthesis[4d]. Ring sizes from four up to eight are known, although the four-membered $\beta$-sultines lose sulphur dioxide readily. In this review, structures are drawn with sulphur–oxygen double bonds for the sake of convenience in representing these polar bonds that are more correctly written as $>S^+—O^-$. The reader should bear in mind that the tetrahedral sulphur atom is a chiral centre.

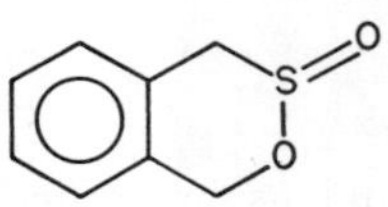

(1)

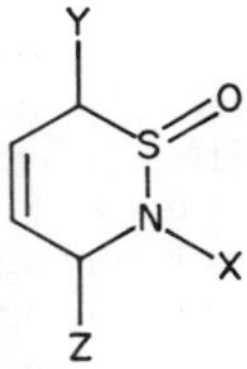

(2)

## II. CYCLIC SULPHINATES (SULTINES)

### A. Synthesis of Sultines

The first sultine, reported in 1893, was obtained by a dehydration reaction of a sulphinic acid and an alcohol[5]. Although a sultine had been suggested in 1966 as an intermediate species in the mass spectrum of dibenzothiophene sulfone[6], the next isolated sultines were described in 1967[7,8].

#### *1. Cyclizations involving an alcohol and a sulphur function*

An alcohol may attack the positive sulphur atom derived from a sulphinic acid[5,9,10,14a], a sulphinyl halide[11], a sulphenyl halide or a related species (followed by oxidation under the reaction conditions[12]) or an oxosulphonium ion[11,13]. These syntheses are exemplified by equations 1–7. Four- to eight-membered sultines can be obtained by the method illustrated in equation 5. A useful method for preparation of $\alpha,\beta$-unsaturated $\gamma$-sultines is the reaction of sulphur dioxide with vinyl Grignard reagents substituted with a hydroxymethyl group (equation 4)[14a]. Substituents $\alpha$ to the sultine oxygen can be introduced by treatment of **3** (equation 1) with alkyllithium reagents. Grignard reagents are unsatisfactory for this purpose.

Hydrolysis of 1,2-ethanedisulphinyl chloride is said to give the anhydride **4** (equation 8)[14], but the mixed sulphinic–carboxylic structural analogue reported in

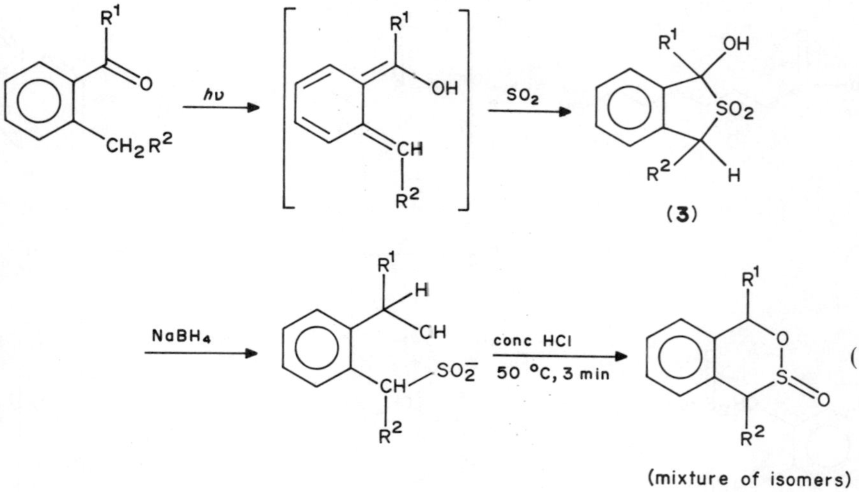


(1)[9a]

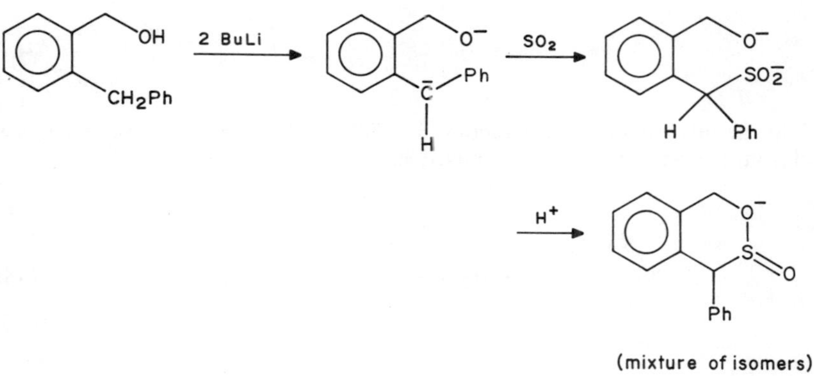


(2)[9a]

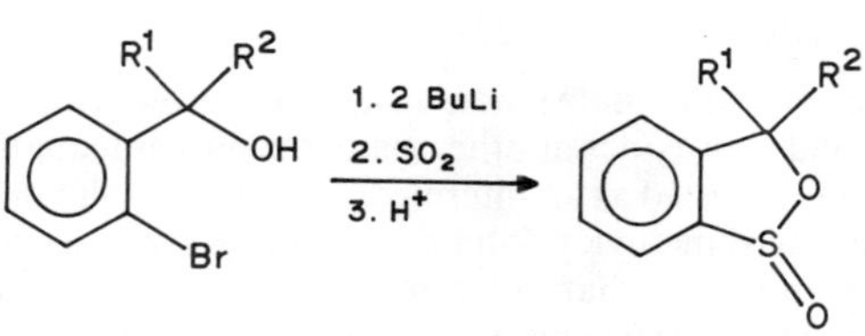


(3)[9a]

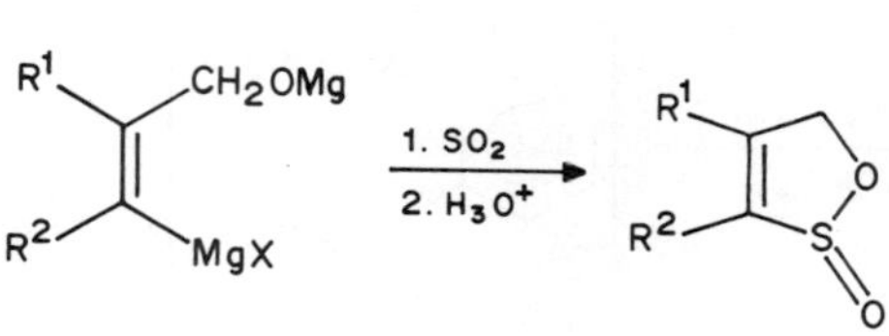


(4)[14a]

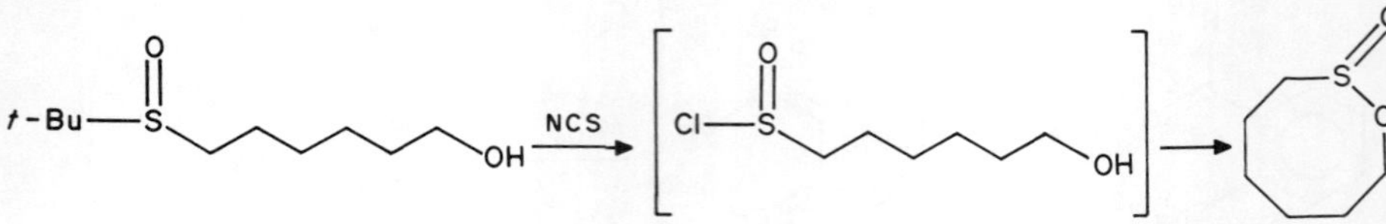

(5)[11]

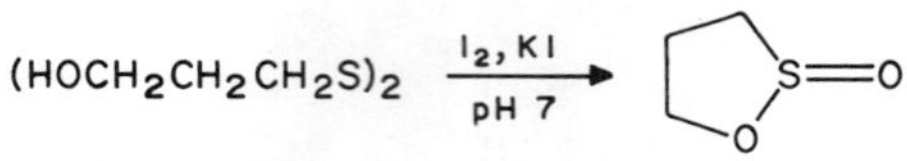

(6)[12c]

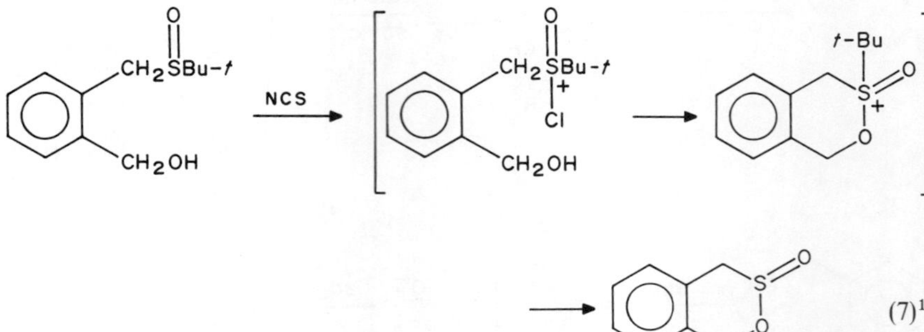

(7)[13e]

1969[15a] apparently has an acyclic structure[15b]. 3-Methyl-1, 2-oxathiolan-5-one-2-oxide is derived from the acid chloride of the mixed sulphinic–carboxylic diacid[15c].

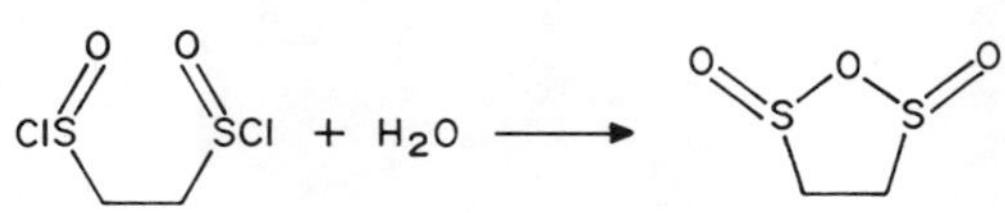

**(4)**

(8)[14b]

*2. Cyclizations involving nucleophilic attack by sulphinates*

Cyclization via attack of a sulphinate anion on a carbon atom with a leaving group[16] or on a reactive double bond[17] provides another route to sultines (equations 9–12). A four-membered $\beta$-sultine was suggested as an intermediate but not isolated in the treatment of the trichloroaluminate-tetramethylcyclobutenyl cation zwitterion with sulphur dioxide[18a]. Presumably an intermediate $\alpha$-chlorosulphinate underwent cyclization to the sultine. Similarly the stable, bicyclic sultine formed by reaction of the sulphur dioxide–antimony pentafluoride complex with 1, 3-cyclohexadiene arises from a step-wise process

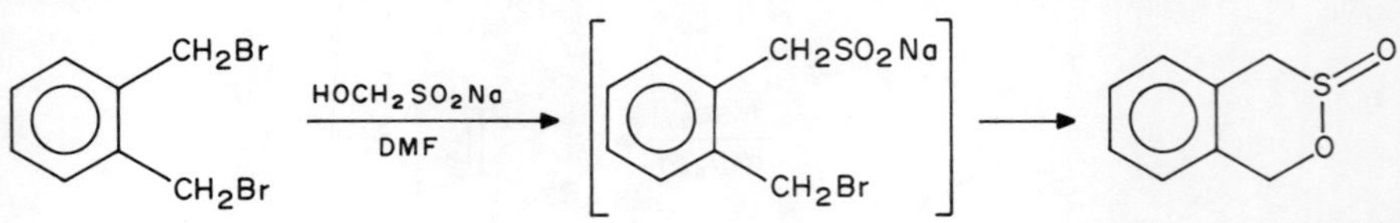

(9)[16a]

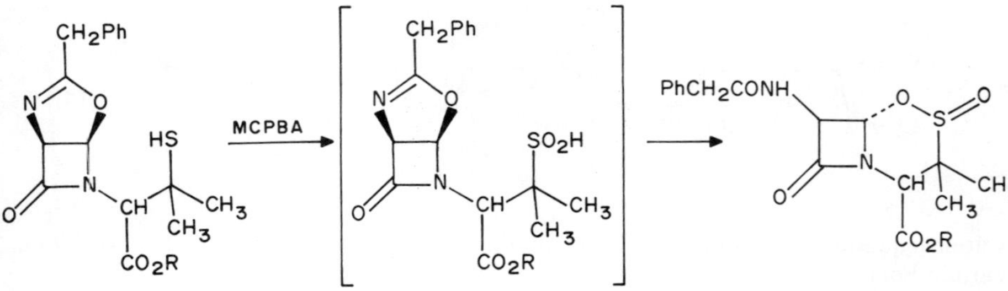

(10)[16b]

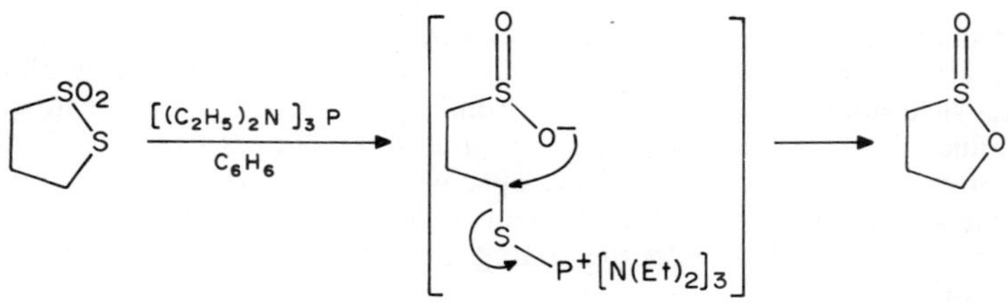

(11)[16c]

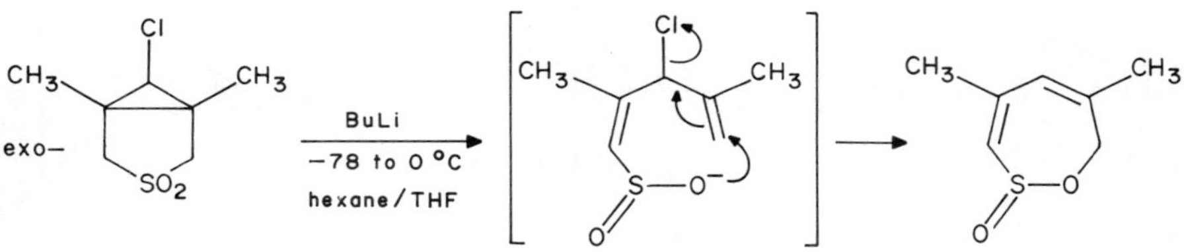

(12)[17]

involving a sulphinate-cyclohexenyl (allyl) cation which cyclizes following a hydride rearrangement[18b].

### 3. *Cyclizations involving electrophilic attack on multiple bonds*

Conversion of an hydroxyl function to a chlorosulphite intermediate, ROS(O)Cl, can dispose a neighbouring multiple bond to effect a displacement of chloride ion to form a

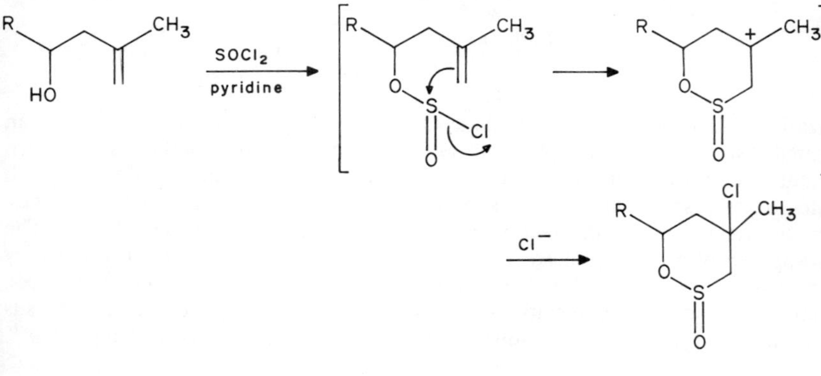

(13)[19a]

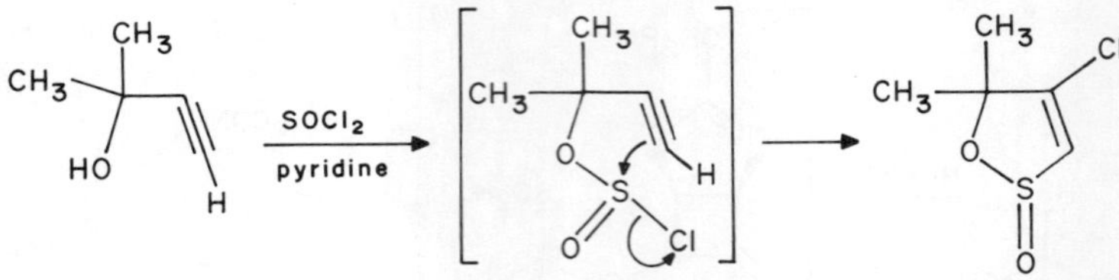


(14)[19c]

sultine (equation 13)[19]. In one case, the multiple bond is apparently an enolate carbon–carbon bond[20].

*4. Miscellaneous cyclizations*

If a sulphinate precursor group (sulphinate ester, sulphone) is situated where an oxygen atom can attack an electrophilic site, cyclization can occur. Allene sulphones and sulphinates yield sultines on treatment with bromine (equation 15)[21]. Allene stereochemistry determines the product stereochemistry and chiral sultines can be obtained[21c]. A neighbouring cyclopropyl group in (2-cyclopropylphenyl) benzyl sulphone provides the electrophilic site when the compound is treated with sulphuric acid. Loss of the benzyl cation (as benzyl alcohol) provides the driving force for sultine formation in good yields (equation 16)[22].

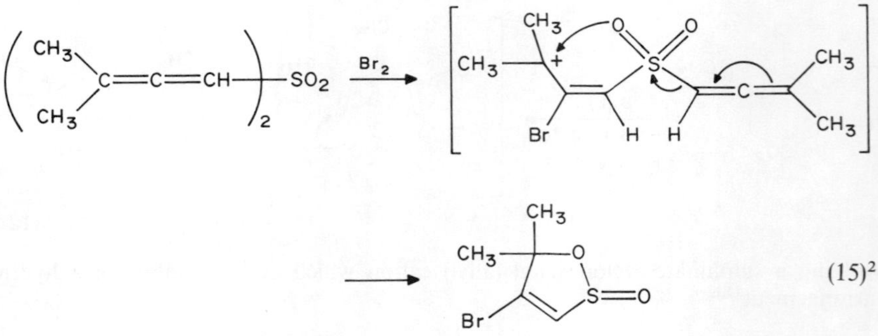


(15)[21a]

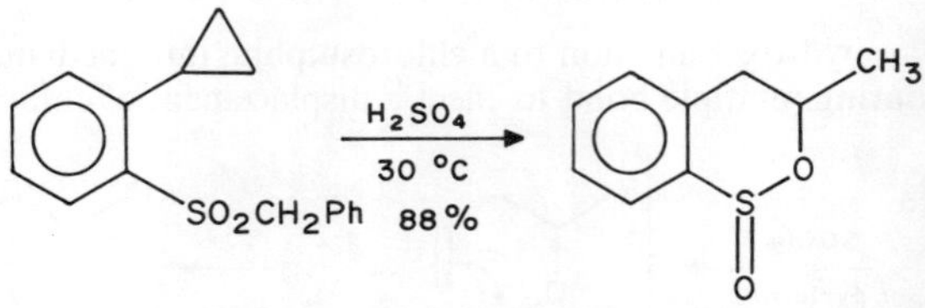


(16)[22]

Cyclization by addition of sulphur dioxide via both an oxygen and a sulphur atom to an unsaturated system can, in principle, yield sultines. In practice, this method is not much used, since sulphones are usually the major product, although selenium dioxide reacts with 1,3-dienes to give selenium analogues of sultines[23a]. Several highly reactive dienes, however, do yield sultines on reaction with sulphur dioxide (equation 17)[23b,24]. The sultines apparently are the preferred product with 1,3-dienes[24], but on heating they rapidly rearrange to the cyclic dihydrothiophene sulphones via a cycloreversion process.

Formation of four-membered sultines via the addition of sulphur dioxide to alkenes is rare[25], although a number of $\beta$-sultines have been suggested as intermediates[26].

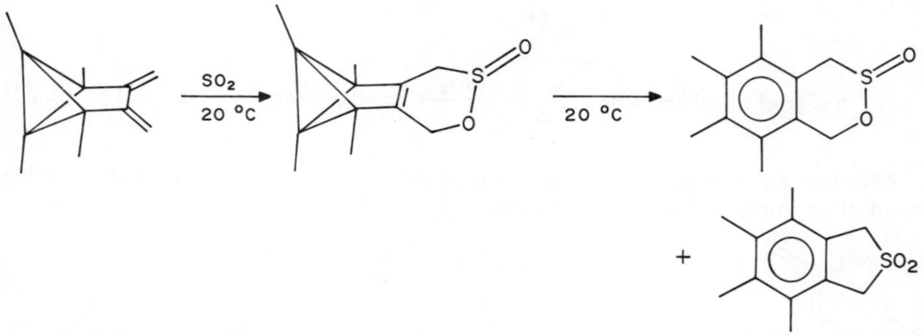


(17)[23b]

Somewhat more common is the addition of sulphur dioxide to cyclopropanes to give mainly $\gamma$-sultines[27]. An acid catalyst is apparently required for the reaction with simple cyclopropanes[27f]. The insertion of sulphur dioxide into a silicon[28a] or germanium bond[28b] of sila- or germacyclobutanes or a stannocyclopentane[28c] is analogous to the insertions into carbon–carbon bonds shown in equations 18 and 19. In the sultines that are formed, the oxygen atom is attached to silicon, germanium or tin. Sulphur dioxide is said to add to $\beta$-thiopropiolactone to give a six-membered sultine[28d].

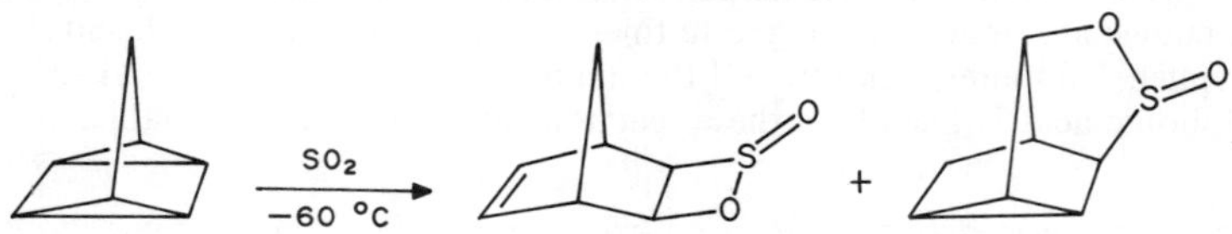


(18)[27a]

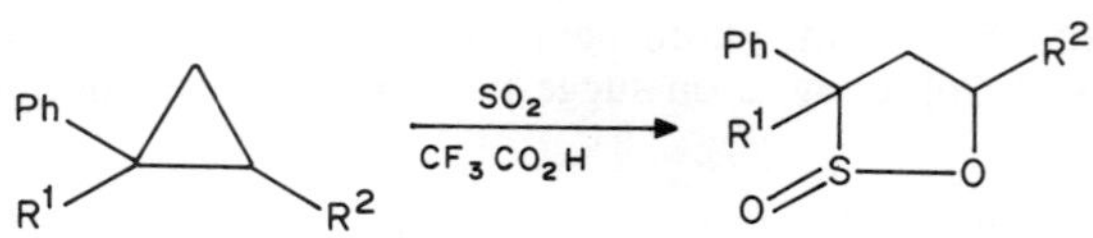


(19)[27f]

Various sultines are obtained by addition of sulphur dioxide to a mixture of ketenes and ketimines[29], to an $\alpha$-chloroimine[30] and to $\alpha$-alkynyl transition metal derivatives[31] (equations 20–22).

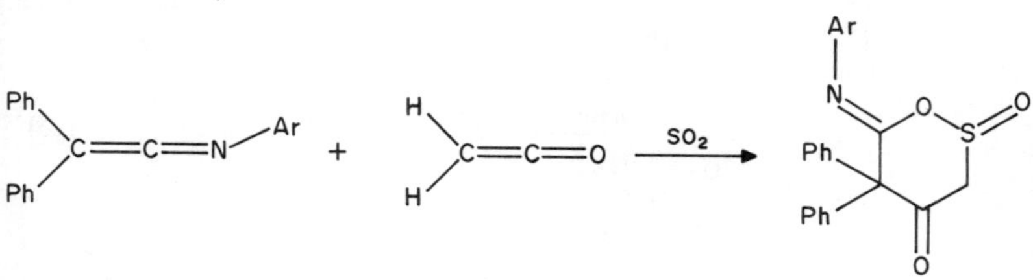


(20)[29b]

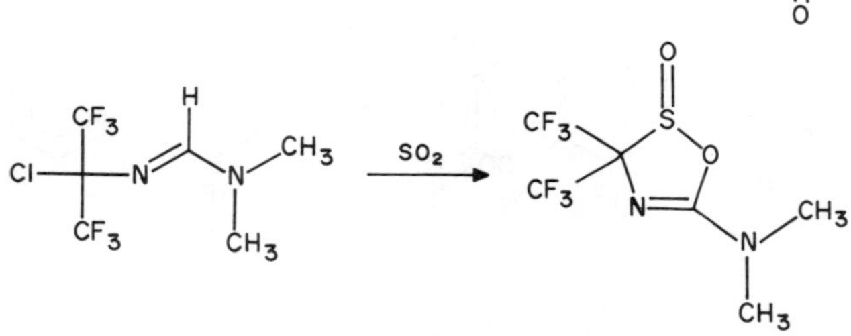


(21)[30]

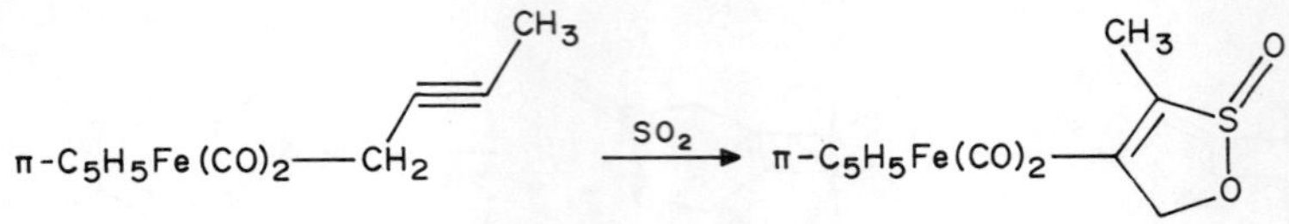

(22)[31c]

Addition of *o*-chloranil[32] or imine or nitrile oxides[33] across the carbon–sulphur double bond of sulphines yields sultines (equation 23).

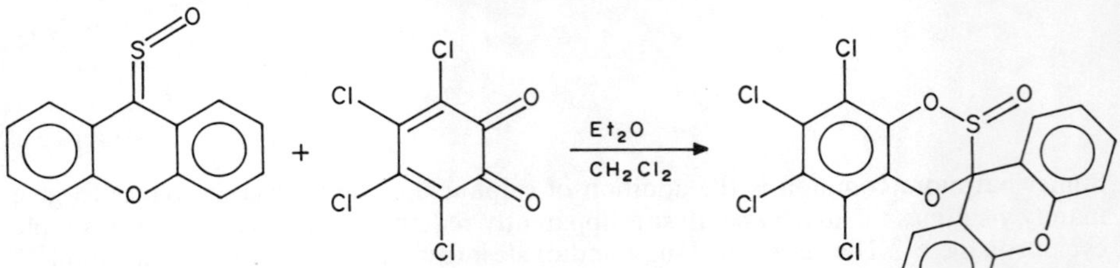

(23)[32]

Sultines were obtained by cyclization of sulphinyl diradicals, $RR^1CCH_2CBr_2SO_2$. Somewhat unusual reactions leading to sultines are the cyclization of the diacid chloride of *o*-carboxyphenyl sulphinic acid[34b], and the treatment of 1, 3, 5-triisopropylbenzene with chlorosulphonic acid[35]. Details of these reactions are not readily available.

*5. Sultines from cyclic derivatives by ring expansion, contraction or rearrangement*

Successful thermal[7,8,27d,36] and photochemical[37] ring expansions of cyclic sulphones to sultines (equations 24–26) followed on suggestions of such rearrangements observed in mass spectra[6,38].

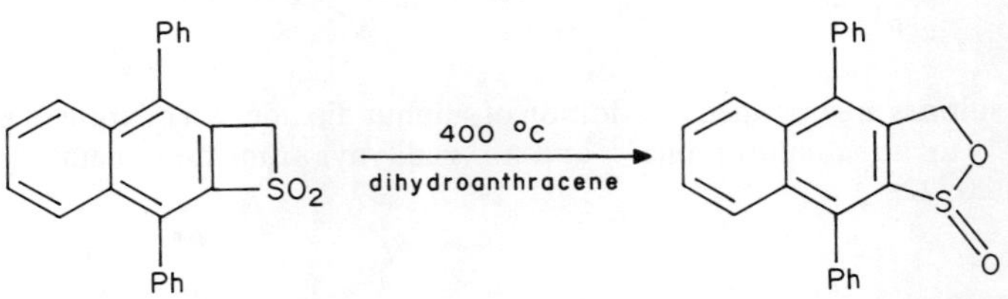

(24)[8]

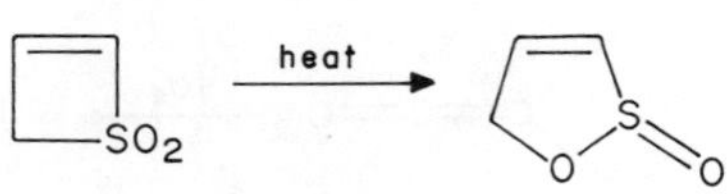

(25)[36b]

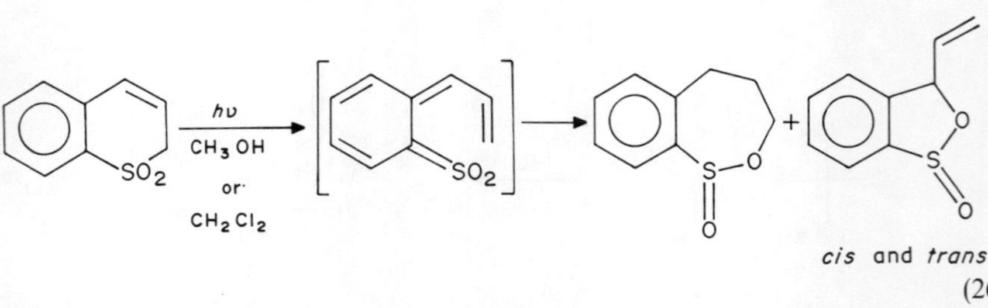

(26)[37a]

The photolysis of saturated cyclic sulphones has been little investigated. In one case[37b], thiolane 1, 1-dioxide goes to the six-membered sultine, but ethylene may be lost with formation of a transient, four-membered sultine[39]. The ketosulphone, **5**, smoothly photoisomerizes via a diradical to sultine, **6** (equation 27)[40].

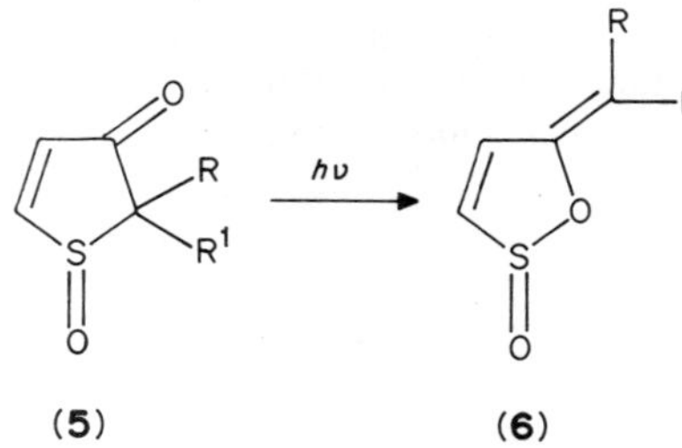


(27)[40]

Treatment of four-membered cyclic sulphones with *tert*-butoxymagnesium bromide yields the five-membered sultines (equation 28)[12h,41]. The 2,4-diphenyl-substituted compounds preserve the stereochemical integrity (*cis* or *trans*) of the substituents on going to the sultine.

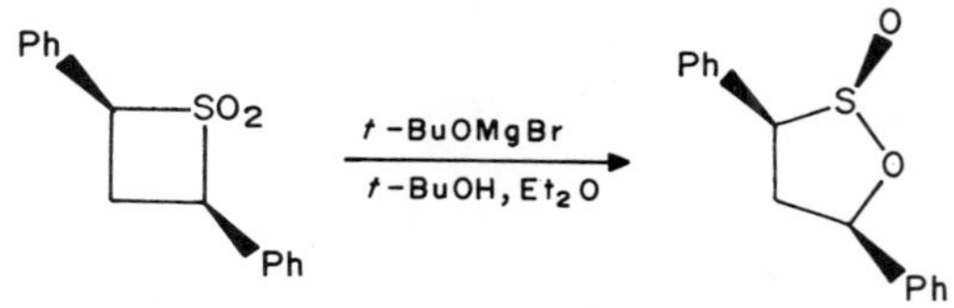


(28)[41a]

Thermolysis or chlorination of the benzothiadiazine sulphone **7** yields five-membered sultines (equation 29)[42].

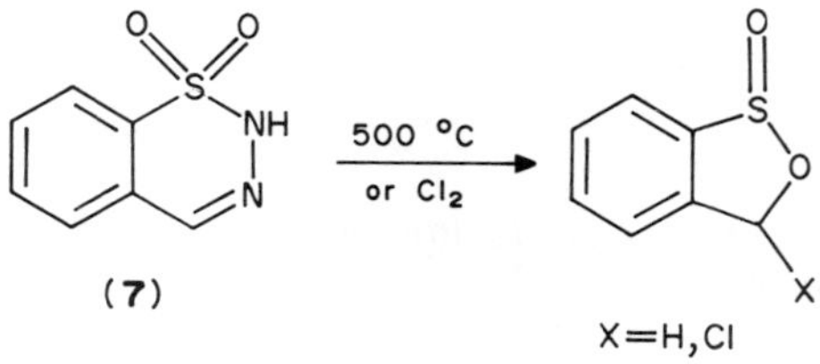


(29)[42a]

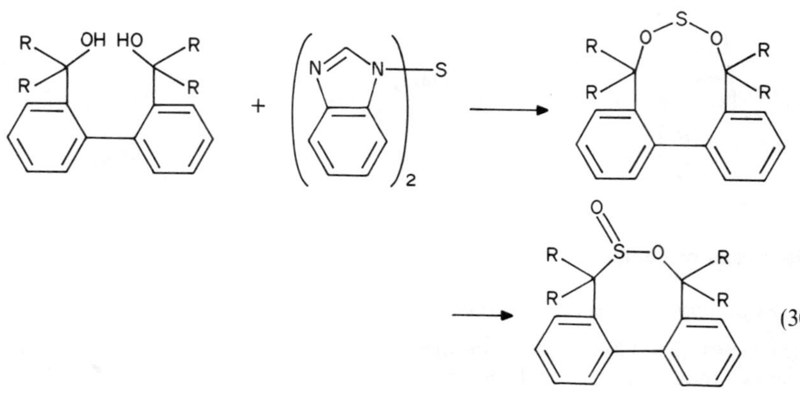


(30)[43]

Cyclic sulphoxylate esters, formed by reaction of diols with sulphur transfer agents, rearrange to the sultines (equation 30)[43].

*6. Sultines by oxidation or reduction*

Oxidation of cyclic sulphenates yields sultines (equation 31)[44], but the method is limited by the availability of the starting materials. Similar is the oxidation via a sulphurane (equation 32)[44c,45]. A β-thiolactone affords the cyclic mixed carboxylic–sulphinic anhydride, **8** (equation 33)[46].

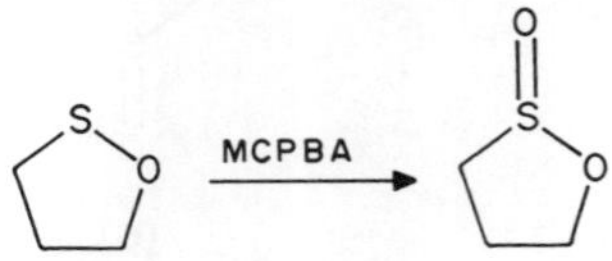


(31)[44d]

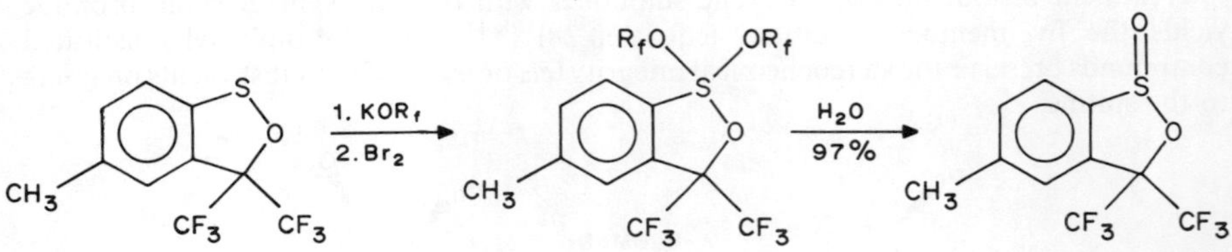


(32)[44c]

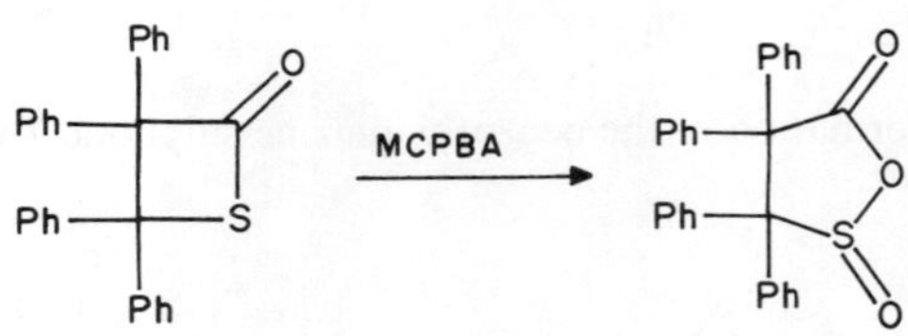


(33)[46]

**(8)**

Three examples of the reduction of a sultone to a sultine have been reported to proceed in moderate yields (equation 34)[47]. Ring-opened products also are obtained.

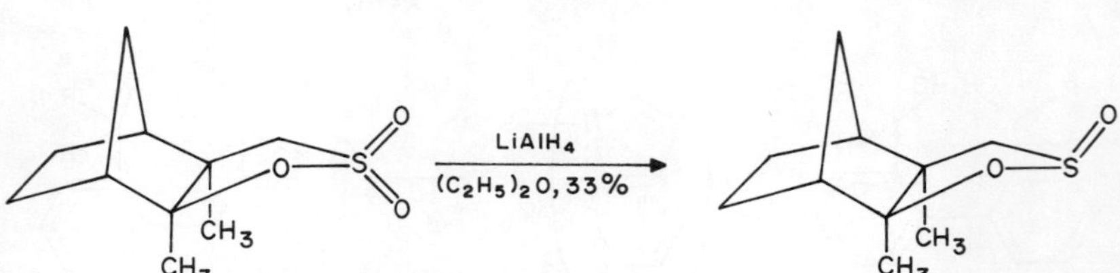


(34)[47]

## B. Reactions of Sultines

*1. Ring opening by nucleophiles, bases and electrophiles*

The base-catalysed hydrolysis of sultines occurs readily[14b,19d,20,29b,36b,47,48] and it has been shown that oxygen-18 exchange between the sultine and $^{18}OH^-$ does not occur in the hydrolysis of the five-membered sultine, **9** (equation 35)[48]. Trigonal bipyrimidal intermediates that do not permit this oxygen exchange were considered. The sulphinate ion may be alkylated *in situ*[47]. Ammonia[34b], a neighbouring hydroxyl group

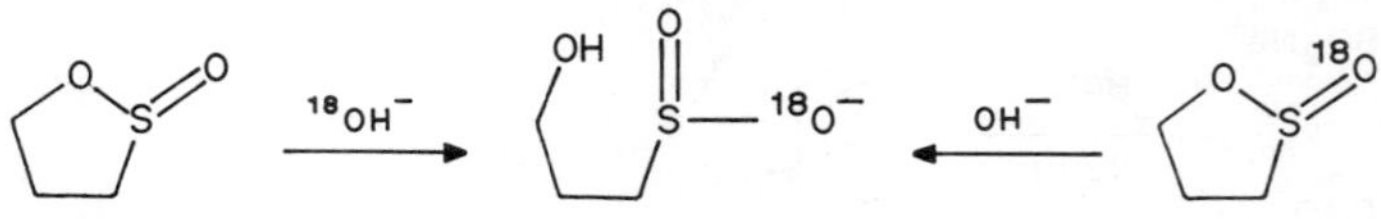

(35)[48]

**(9)**

(equation 36)[12e] and organometallic reagents[49] attack sultines. Incomplete reaction of 3*H*-2, 1-benzoxathiole 1-oxide with (*S*)-2-methyl-1-butyl- or (*S*)-2-phenyl-1-butylmagnesium chloride yields recovered sultine enriched in the (*S*) enantiomer (8–64% ee)[49b]. The simple saturated five- and six-membered sultines react with Grignard reagents to give complex mixtures of sulphides and sulphoxides, but dialkyl cuprate reagents cleanly give the sulphoxides (equation 37)[49a]. Treatment of sultines, **10**, with

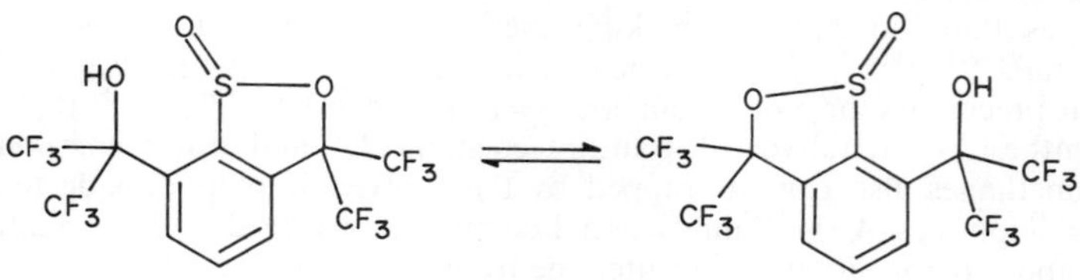

(36)[12e]

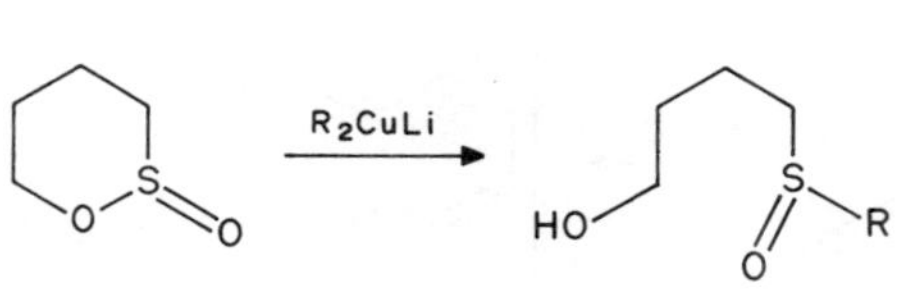

(37)[49a]

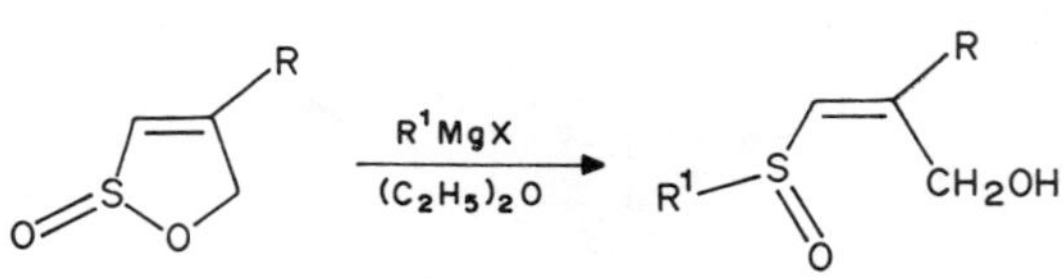

(38)[49c]

**(10)**

hydride ion followed by alkylation with methyl iodide yields $\alpha,\beta$-unsaturated aldehydes[49c]. A mechanism involving formation of a sulphenate anion was suggested.

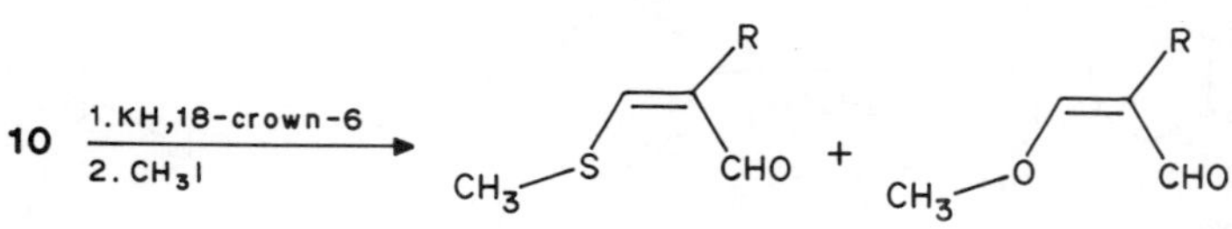

Several sultines undergo chlorination with ring opening to give sulphonyl chlorides[11,42b,c]. The hydroxysultine **12** is in equilibrium with the acyclic sulphinate-aldehyde, **11** (equation 39).

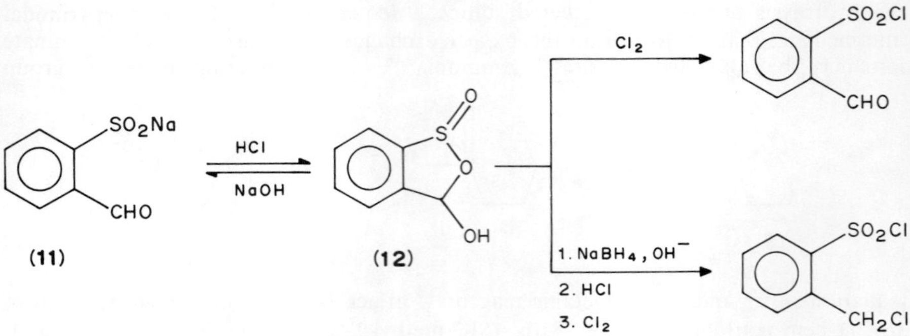


(11) (12)

(39)[42b,c]

*2. Extrusion of sulphur dioxide or sulphur monoxide*

The $\beta,\gamma$-unsaturated six-membered sultines lose sulphur dioxide at much lower temperatures than do sulfones, making them more suitable for generating dienes (equation 40)[9a–c,13e,16a,23b,50]; in particular, the benzo-fused six-membered sultines are convenient precursors for *o*-quinodimethanes (*o*-xylylenes)[9a–c,13e,16a,50a] that are useful in the synthesis of tetrahydro-1,4-anthracenediones[50a] and other compounds. The *o*-quinodimethanes also can be trapped by the evolved sulphur dioxide to give five-membered sulphones. A rhodium carbonyl complex is involved in the formation of the sulphone and a benzo-fused cyclopentenone from sultine, **13** (equation 41)[50b].

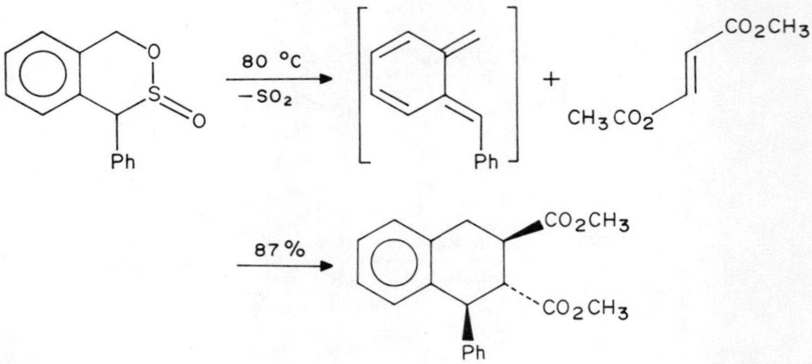


(40)[9c]

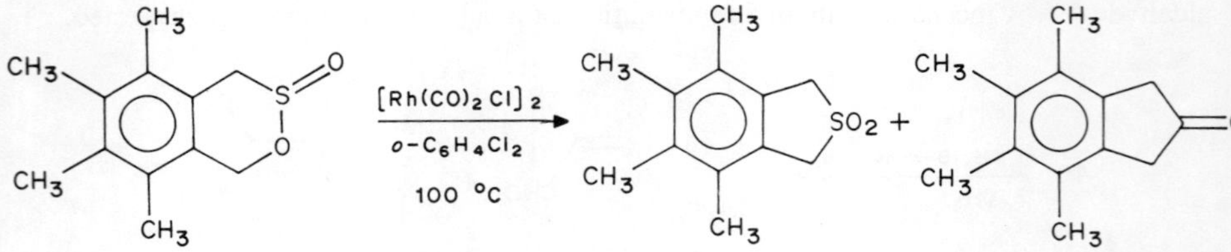


(13) (41)[50b]

The cyclopropane-substituted sultine, **14**, undergoes a stereospecific $\pi_{2s}+\pi_{2s}+\pi_{2s}$ cycloreversion (equation 42)[13f].

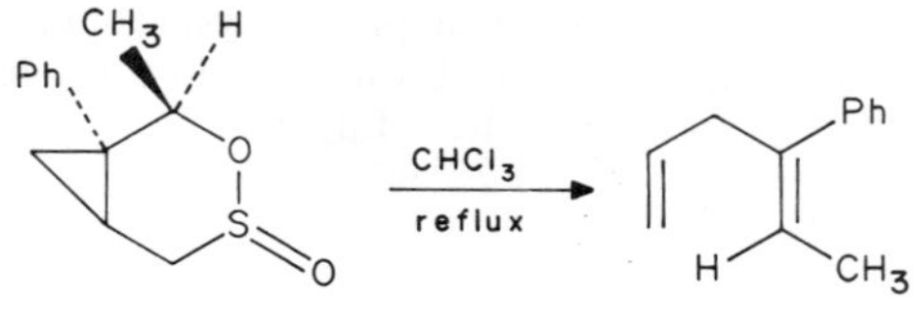


(14)

(42)[13f]

β-Sultines are analogues of the intermediate in the Wittig olefin synthesis and lose sulphur dioxide via a *cis* elimination process even at room temperature to give alkenes (equation 43)[13d,j,k]. The process has been discussed from a theoretical viewpoint which shows that a $\sigma_{2s} + \sigma_{2a}$ process is not obligatory[51]. In a flash vacuum thermolysis experiment in which a β-sultine was suggested as an intermediate in the reaction of sulphur dioxide with ethylene, the products were ethylene oxide and elemental sulphur[26g]. Thermolysis of the γ-sultine, 1,2-oxathiolane 2-oxide, gave a mixture of products including cyclopropane, ethylene, 1,2-oxathiolane, formaldehyde, sulphur dioxide and sulphur monoxide[52a]. Substituted γ-sultines give analogous products formed via diradicals (equation 44)[52b]. Other ring sizes behave similarly, diradical intermediates being formed[43b]. Both sulphur dioxide and sulphur monoxide are apparently produced in the thermolysis of certain sultines[6,8b,38,43b,53]. Sulphur monoxide apparently is lost in the thermolysis of **15** (equation 45)[30] and from a tetracyclic sultine[27b]. Desulphinylation of a tungsten-substituted sultine occurs on alumina[32d]. Photochemically, sultines also extrude sulphur dioxide[52b,54], but 1,2-oxathietan-4-one 2-oxide photochemically loses carbon dioxide[25a]. A 4-imino-β-sultine intermediate, believed formed in the reaction of a ketimine with sulphur dioxide, does not revert to starting materials by extrusion of sulphur dioxide

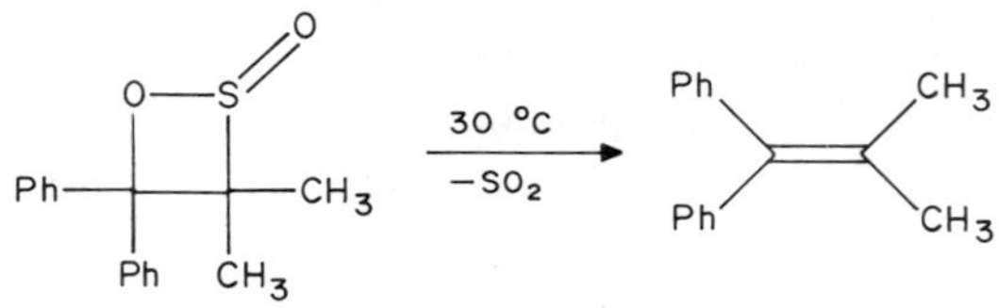


(43)[13j]

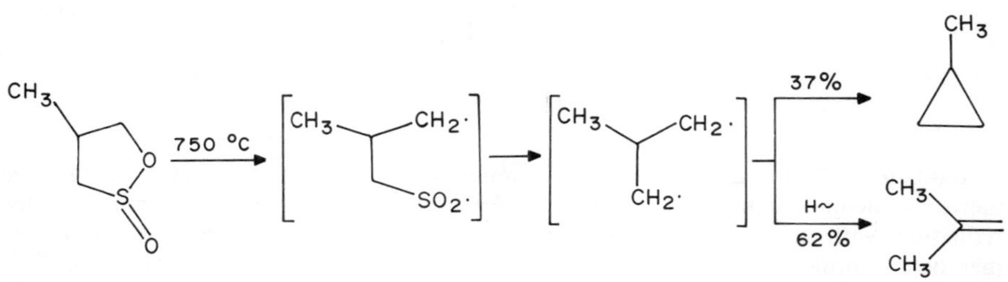


(44)[52b]

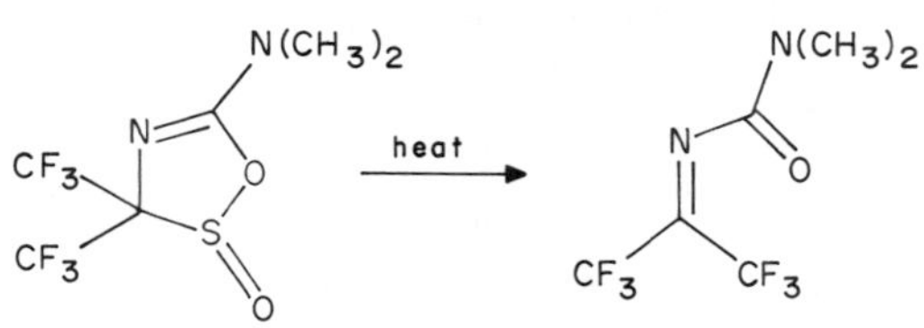


(15)

(45)[30]

but instead rearranges via cleavage of the sulphur–oxygen bond (equation 46)[26f]. The isomerizations of sultines to sulphones probably involves loss of sulphur dioxide followed by its recombination with a diradical or diene fragment[9c,13e,43b].

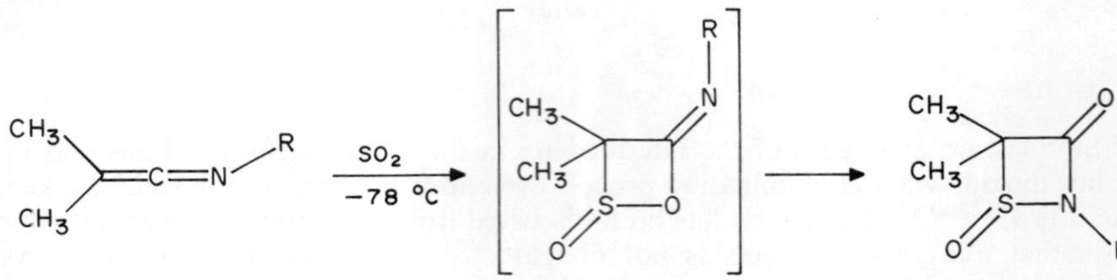


(46)[26f]

### 3. Rearrangements

The conversion of sultines to cyclic sulphones has been described in the previous section[9c,13e,43b] as has the rearrangement of a *β*-sultine (equation 46)[26f]. The diradicals formed by loss of sulphur dioxide may give a variety of products[8b,43b]. A naphtho-fused sultine (equation 24) undergoes extensive rearrangement at 380–400 °C to give low yields of a fluorenone and a fluorene, the former presumably being formed by loss of sulphur monoxide[8b]. This reaction may be related to the fragmentations seen in the mass spectra of sulphones[6,8b,38,43b].

### 4. Oxidation and reduction

Sultines are oxidized to sultones in good yields. Oxidants include *m*-chloroperbenzoic acid[11,14a,21c,27g,41a], hydrogen peroxide[16f,19a,36b], potassium hydrogen persulphate[14a], potassium permanganate[16d,f,32a] and positive halogen (NCS, NBS, $Cl_2$, $I_2$) followed by hydrolysis[12b–d,13a,41a].

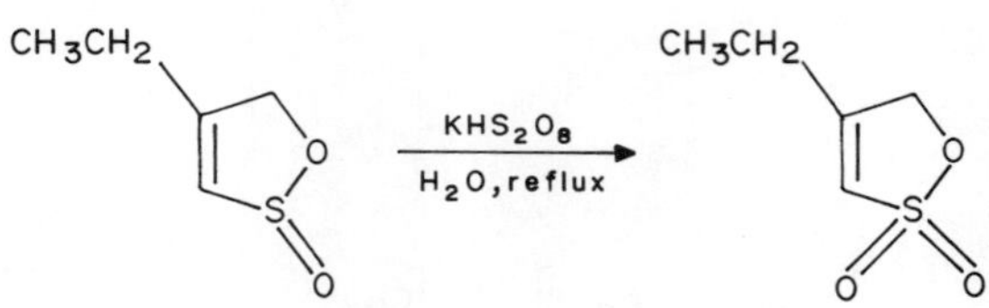


(47)[14a]

Reductions of sultines with lithium aluminium hydride involve ring opening to a mercapto-alcohol (equation 48)[8b,12f,g,i,47]. Symmetrical disulphides are obtained by reduction with trichlorosilane-tripropylamine (equation 49)[55]. Reduction of a six-membered sultine with hydrogen is said to give the oxathiane without ring opening[29a,b], and reduction of a cyclic five-membered thiolsulphinate gave the cyclic disulphide[56]. Sultines were not reduced by treatment with $P_4S_{10}$[57].

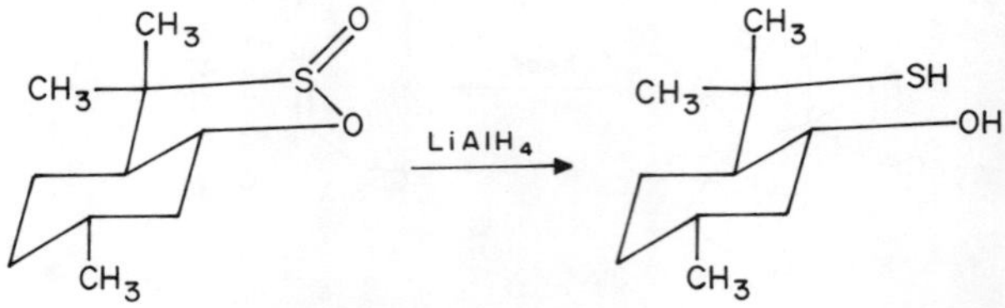


(48)[12g]

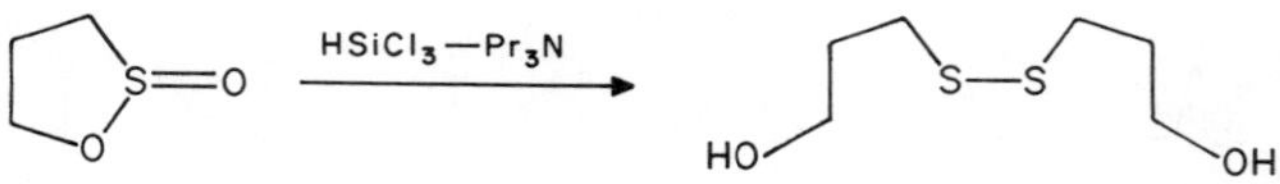


(49)[55]

*5. Reactions involving ring substituents*

Hydrolysis of the 5-chloro-3,4-benzosubstituted $\gamma$-sultine, **16**, results in replacement of the chlorine atom by hydroxyl, the new sultine in basic medium being in equilibrium with the acyclic sulphinate-aldehyde (equation 50)[42b,c]. The unsaturated chlorosultine, **17**, undergoes addition–elimination reactions (equation 51)[19d]. A $\beta$-chloro-$\delta$-sultine undergoes elimination of chlorine to give mainly the unconjugated $\beta,\gamma$-sultine when treated with diethylamine[19a].

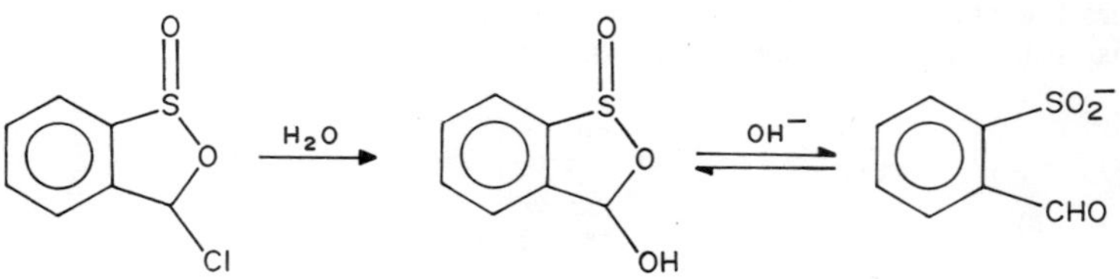


(50)[42c]

(**16**)

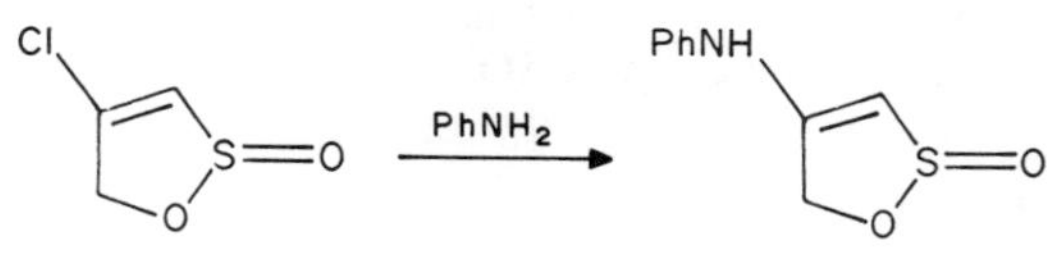


(51)[19d]

(**17**)

## C. Physical Properties of Sultines

The structures of sultines **18**[13k], **19**[27g], **20**[58], **21**[59], **22**[60] have been established by X-ray analysis. The four-membered ring in **18** is puckered with a dihedral angle of 20.3° between the O—S—C and C—C—O planes of the ring. The exocyclic sulphinyl oxygen in the five-membered rings is axial, unlike the situation in thietane oxides where the oxygen is equatorial[61]. An anomeric effect was suggested[13k,16f]. The four ring atoms exclusive of sulphur in **20** are in essentially one plane; the sulphur atom is displaced toward the cyclopentadienyl ring which lies above the plane and is *syn* to the sulphur–oxygen bond. In **21**, the oxygen atom is *exo*.

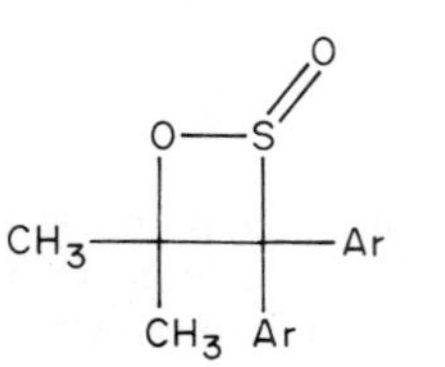


(**18**) Ar= $p$-$FC_6H_4$

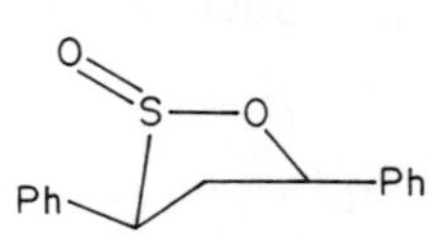


(**19**) *cis* Ph

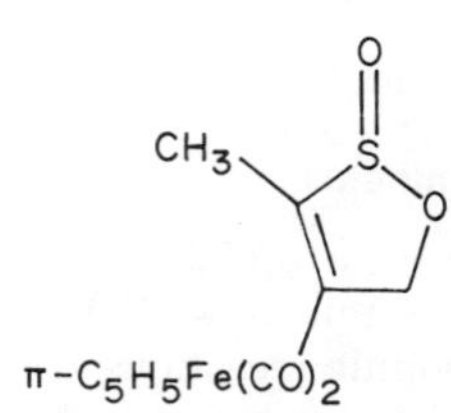


(**20**)

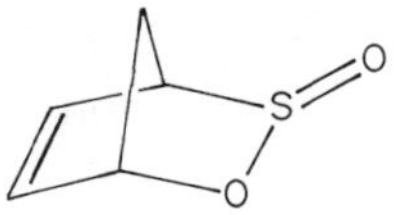

(21)

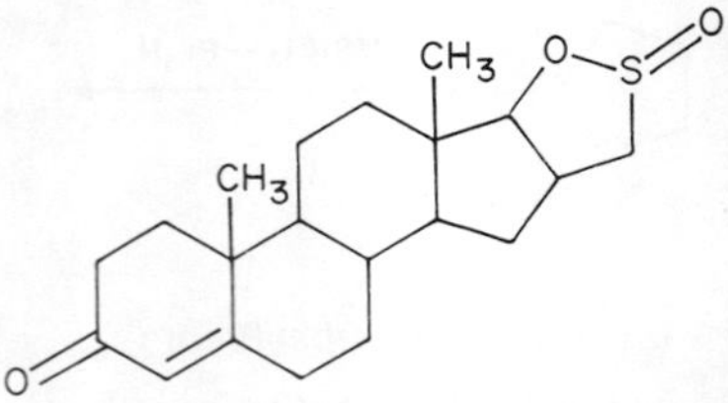

(22) two isomers

Other investigations involving NMR[11,16f,19b,27g,62], IR[19b] and dipole moments[63] indicate that the exocyclic sulphinyl oxygen is axial in six-membered sultines as well. In the proton NMR, the well-known *syn* axial deshielding effect is observed. The carbon-13 NMR chemical shifts for 1,2-oxathiane-2-oxide are given in structure **23**[62]. This sultine undergoes fast chair–chair interconversions even at $-90\,°C$, and the axial–equatorial barrier is estimated to be in excess of 2 kcal $mol^{-1}$ [16f].

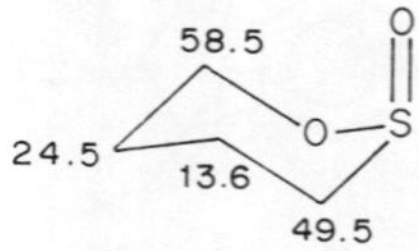


(23)

Theoretical treatments of α- and β-sultines have been reported[51,64]. The α-sultine is a suggested valence isomer of sulphene, $CH_2SO_2$.

### D. Uses of Sultines

The usefulness of sultines in olefin and *o*-quinodimethane syntheses have been described in Section II.B. Compound **22** and its analogues are competitive with steroids for binding to receptor proteins[13g–i,65]: one is an inhibitor of aldosterone acetate in rats[13h] and another is a diuretic[13i]. The sultine **24** and its six-membered analogue are chiral NMR resolving agents[49b].

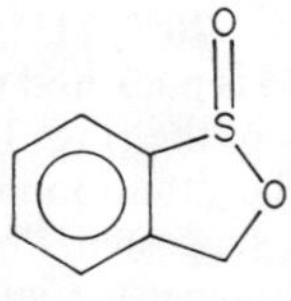

(24)(*S*)(+)

## III. CYCLIC SULPHINAMIDES

### A. Synthesis

The most useful intermediates in the synthesis of cyclic sulphinamides are the *N*-sulphinylamines, RNSO, which are imines of sulphur dioxide. These undergo a variety of cycloaddition reactions to give the compounds under discussion. Numerous reviews attest to the importance and interest in these reactions[4a–d,66–68]. In particular, the addition of 1,3-dienes to *N*-sulphinylamines has considerable potential in organic synthesis[67].

*1. 2 + 2 Cycloadditions with N-sulphinylamines*

Treatment of ketenes with *N*-sulphinylamines gives the four-membered mixed sulphinyl–carboxylic imides, **25** (equation 52)[29b,69–72a]. The adduct with ketene itself is not very stable, but diphenylketene gives good yields of isolable products. Additions to vinyl

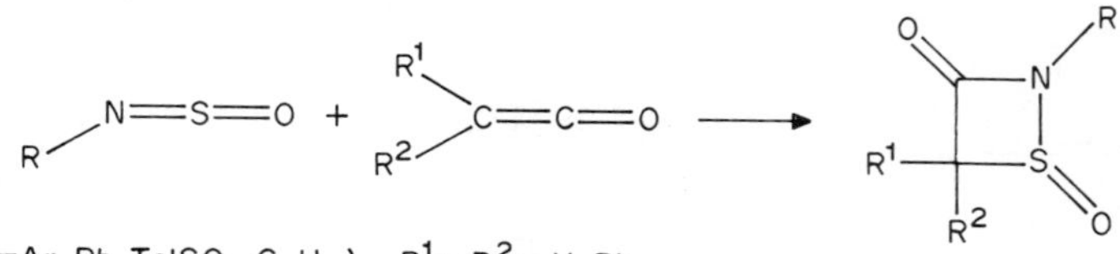

(52)

(R = Ar, Ph, TolSO$_2$, C$_6$H$_{11}$) R$^1$ = R$^2$ = H, Ph;

R$^1$ = Ar, R$^2$ = PhCO, PhSO$_2$

(**25**)

ethers give good yields of cyclic products when a sulphinyl sulphonamide is used (equation 53)[73], and an adduct is reported from 9-ethylidene fluorene[74]. The cationic sulphinylamine derivative shown in equation 54 reacted smoothly with ketenimines[75]; the initially formed adduct or adducts undergo rearrangement[75]. Dipolar intermediates have been suggested in some of these reactions. Adducts with the carbon–oxygen double bond

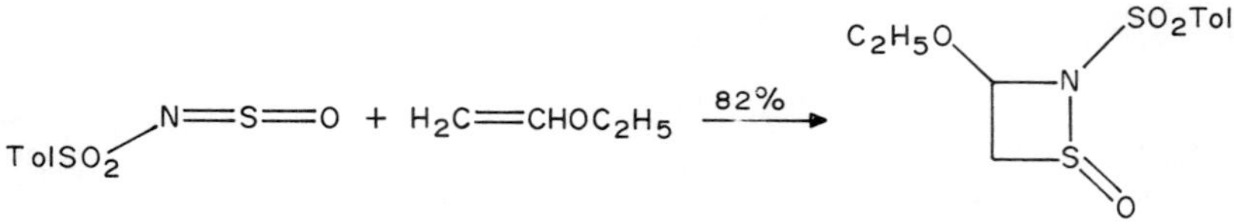

(53)[73a]

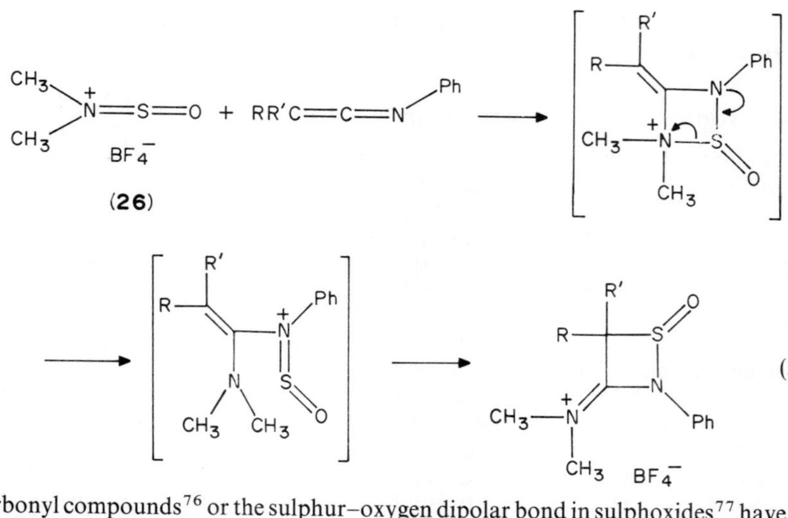

(**26**)

(54)[75]

of carbonyl compounds[76] or the sulphur–oxygen dipolar bond in sulphoxides[77] have been proposed as intermediates.

*2. 3 + 2 Cycloaddition with N-sulphinylamines*

1,3-Dipolar species add across the sulphur–nitrogen double bond (equation 55)[78]. The behaviour of sulphinylamines with diphenylcyclopropenone[79] and with transition metal

2-alkynyl, cyclopropylmethyl, or $\eta^1$-allyl complexes[80] appear to fit this pattern (equation 56).

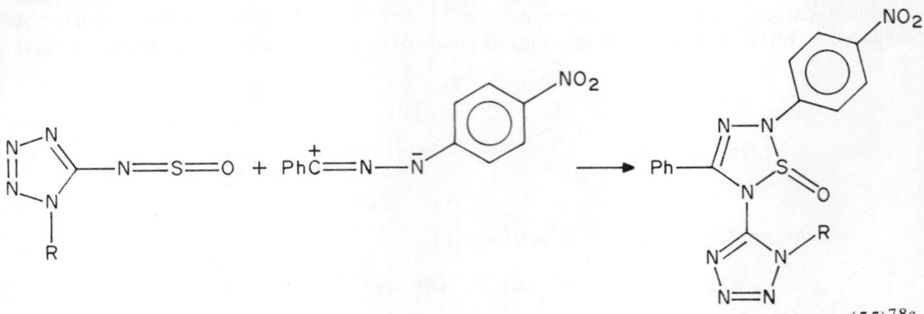

(55)[78c]

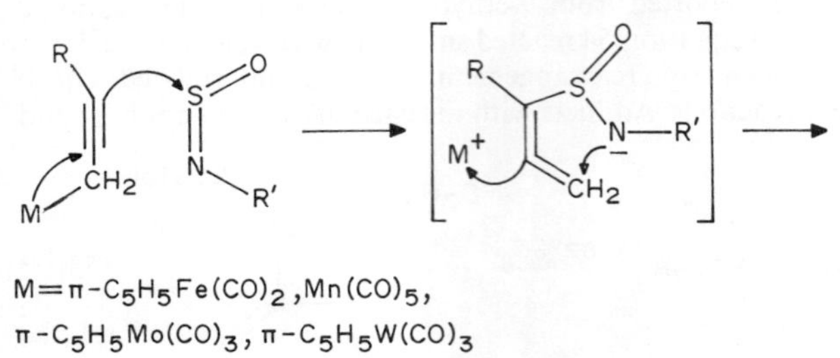

(56)[80]

A reaction in which the sulphinylamine functions as the 1,3-dipolar species and an *N*-substituted maleimide as the dipolarophile has been observed[81].

*3. 4+2 Cycloadditions with N-sulphinylamines as dienophiles*

The addition of a variety of 1,3-dienes to the S=N bond of *N*-sulphinylamines to give 3,6-dihydro-1,2-thiazine-1-oxides has been widely investigated since it was first reported in 1953[82]. The addition is *syn* and is reversible in some cases. Early reports did not recognize the possibility that the sulphinylamines could function as dienes also, and the cyclopentadiene adduct structure has been corrected[83a]. The substituents of the *N*-sulphinylamines, RNSO, are as follows: aryl[82,83], heteroaryl[78c,84], RCO[76f,85], ROCO-[86], $RSO_2$[73c,87], $R_2PO$[86f,88], $(CH_3)_2S^+$[89] and CN[85b,90]. An ammonium salt, $(CH_3)_2N^+SO$, also has been used[91]. Those compounds with electron-withdrawing groups, such as *N*-sulphinylamides and *N*-sulphinylsulphonamides, are the most reactive. Instead of the usual 1,3-dienes, a 1,2,4,5-tetraene also has been employed[83e]. 1,4-Substitution in the dienes hinders the addition[85i]. The regioselectivity for TolCONSO is such that, when possible, a 1-substituted diene yields the thiazine oxide with the substituent *ortho* to nitrogen and such that a 2-substituted diene gives the adduct with the substituent *para* to the nitrogen[85h,86m]. The same pattern is followed with disubstituted 1,3-dienes, except that the orienting power of a phenyl group is greater than that of a methyl group[85a,i]. The geometry of adducts of *N*-sulphinyl-*p*-toluenesulphonamide with (*Z,Z*)-, (*E,Z*)- and (*E,E*)-2,4-hexadienes has been elucidated and interpreted on the basis of a dipolar mechanism of addition[87l], although a concerted process is in agreement with frontier molecular orbital theory[86m]. The stereochemistry at the sulphur atom is variable. The regiochemistry reported in earlier work has been corrected[86m].

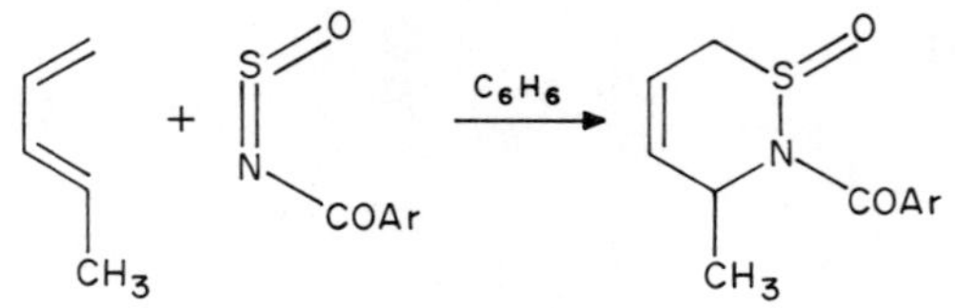

(57)[85h]

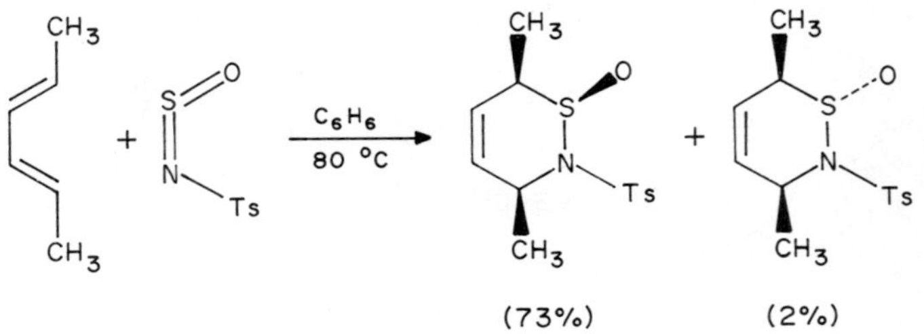

(58)[87l]

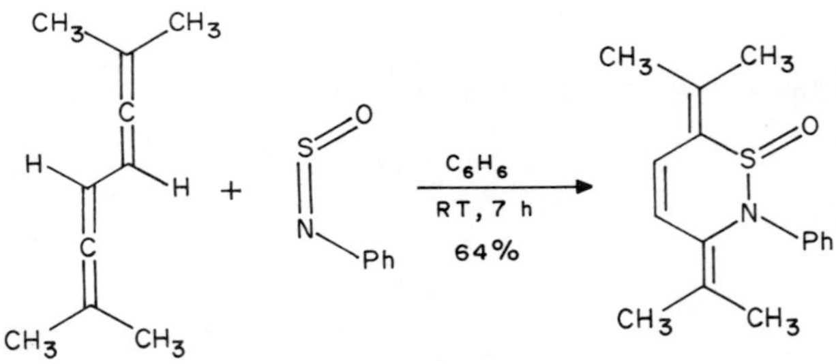

(59)[83e]

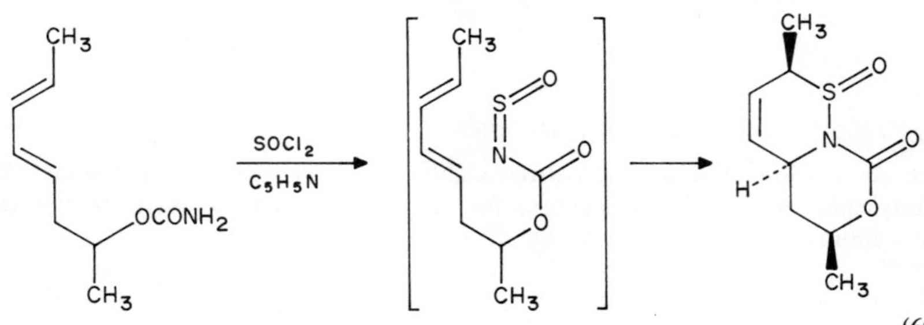

(60)[86h]

*4. 4 + 2 Cycloadditions with N-sulphinylamines as the diene components*

Aromatic *N*-sulphinylamines react as dienes with alkynes or with alkenes possessing somewhat strained double bonds[92]. Analogous reactions occur with *N*-sulphinylamides and related compounds[86a,n,93]. These reactions are exemplified by equations 61–65.

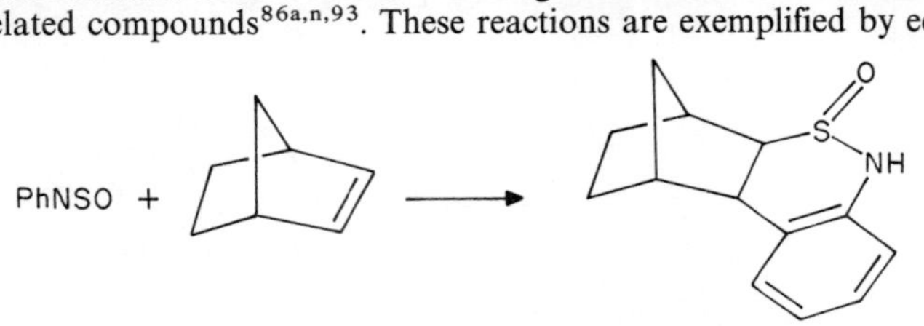

(61)[92a]

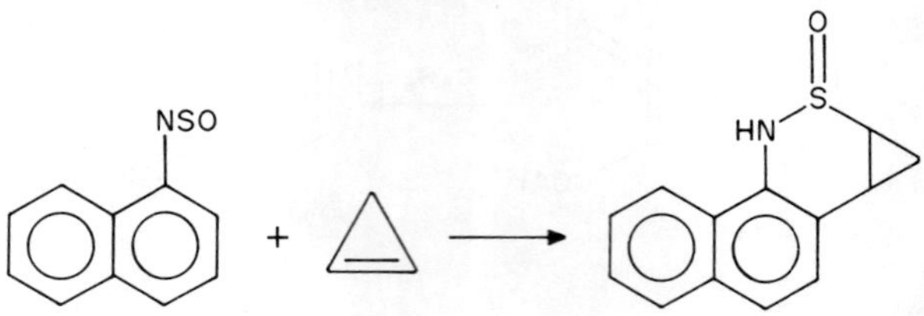

(62)[94h]

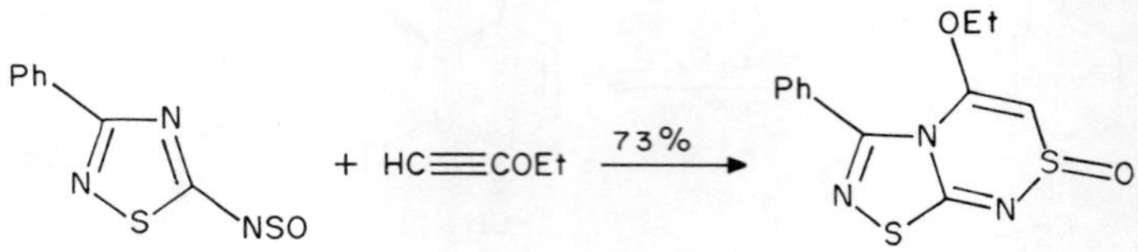

(63)[92d]

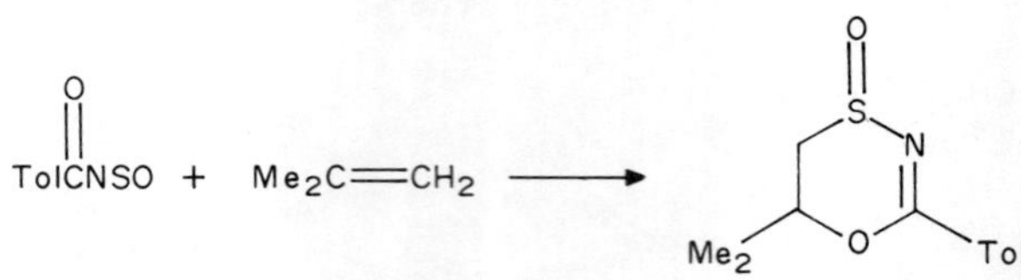

(64)[93b]

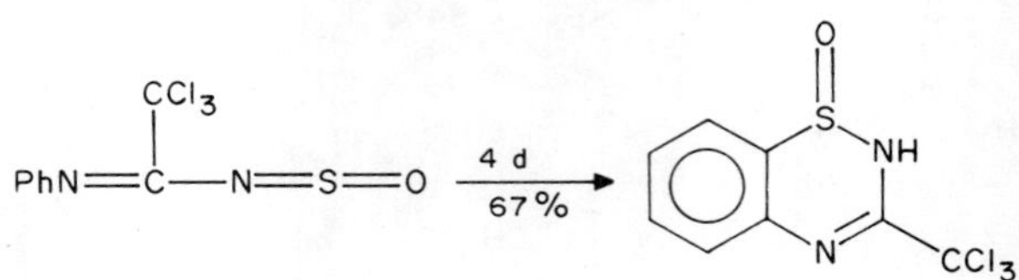

(65)[93d]

*5. Cyclizations involving sulphinic acid derivatives and amines*

The elimination of water or hydrogen chloride between a sulphinic acid[29b,94] or sulphinyl chloride[34b,72a,95] and an amine function has found limited use in the synthesis of cyclic sulphinamides (equations 66–68).

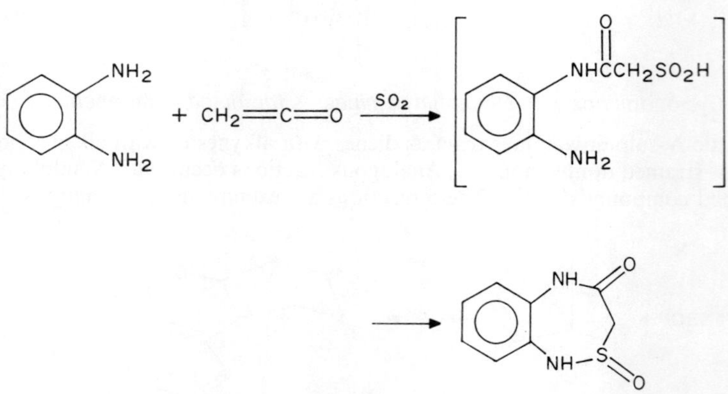

(66)[94]

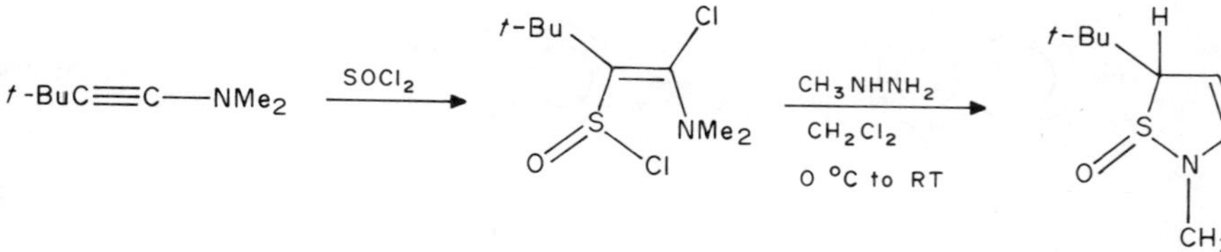


(67)[95c]

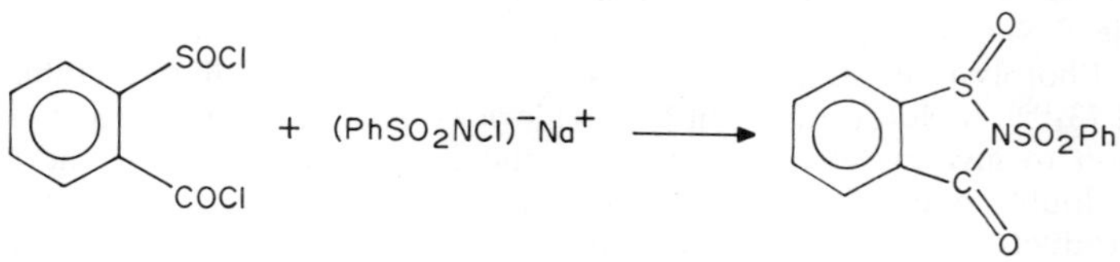


(68)[95e]

### 6. Oxidation of cyclic sulphenamides

Mild oxidation of cyclic sulphenamides with peracids[72,96], hydrogen peroxide[97], dilute nitric acid[96b,97c,98], dinitrogen tetroxide[96b], chromium trioxide[96b], bromine[96f], iodine[99] or chlorine[97c] yields the desired sulphinamides (equations 69 and 70). Overoxidation to the dioxide is the principal side-reaction[96c,i,j] or even the major reaction[96i]. Nitrogen functionality if present elsewhere in the sulphenamide may be converted to an *N*-oxide function[96g,j].

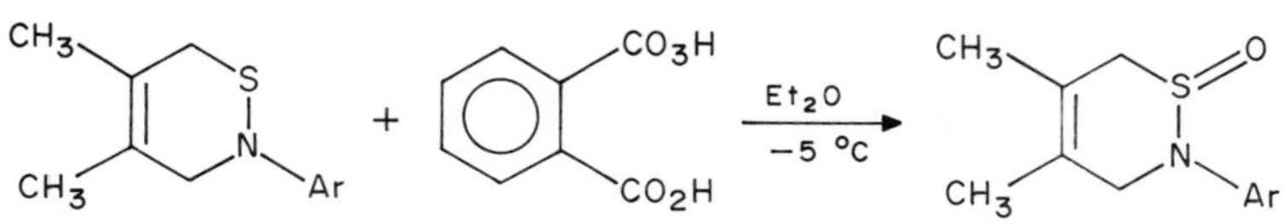


(69)[96a]

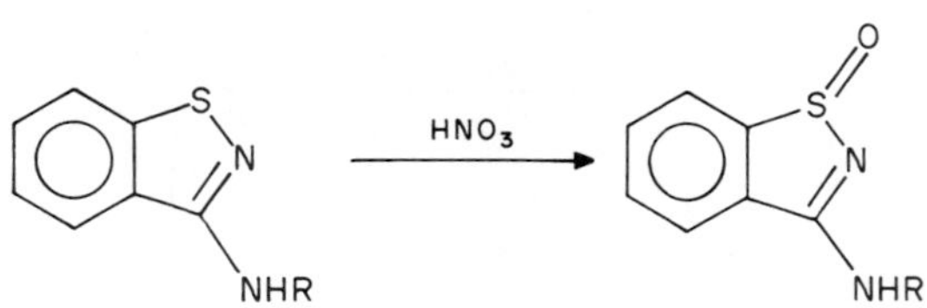


(70)[97c]

### 7. Miscellaneous methods

Addition of water across sulphur–nitrogen double bonds provides the sulphinyl functionality in the preparation of several cyclic sulphenamides (equations 71 and 72)[95e,100].

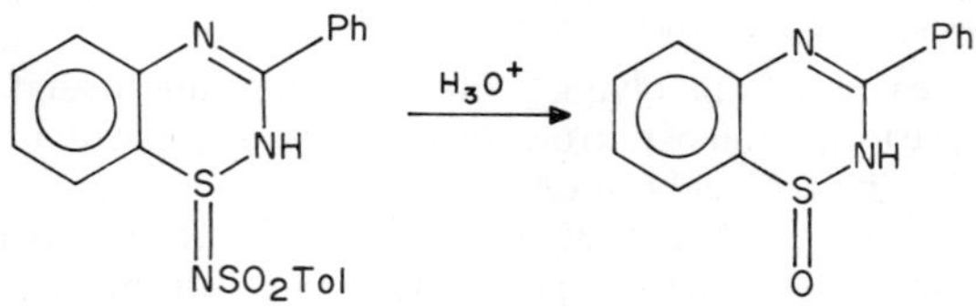


(71)[100b]

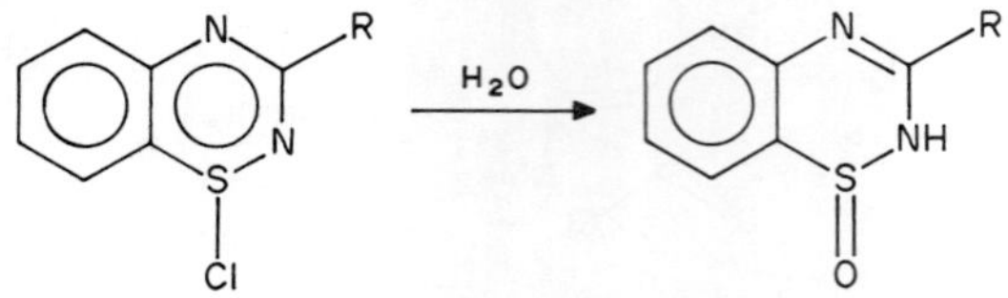


(72)[100h]

Several conversions of sultines to cyclic sulphinamides have been reported[26f,34b,101], and an $S$-oxide of a penicillin derivative is rearranged by base to the sulphinamide[96l]. An isothiazole $S,S$-dioxide may have been reduced in an unspecified manner to the $S$-oxide[102]. Photolysis of sultam **26** yields the $N$-hydroxy- or alkoxysulphinamide, **27** (equation 73)[103]. $N$-Substituted sulphinyl chlorides may undergo aromatic electrophilic substitution to give cyclic products[95i,104], and an attack of a sulphenic acid on an activated double bond gives a cyclic sulphinamide[105]. Other reactions that give the desired products are the cyclization of an acetylenic acyclic sulphinamide through the nitrogen atom (equation 74)[106], addition of sulphinylaniline to a cyclic nitrone[107], the cyclization of an ylide derived from a sulphoximine[108], rearrangement of the initial adduct of thiofluorenone with a nitrone[96m] and cyclization of an $o$-amino-substituted benzenesulphinamide with an *ortho* ester or dimethylformamide dimethyl acetal[109].

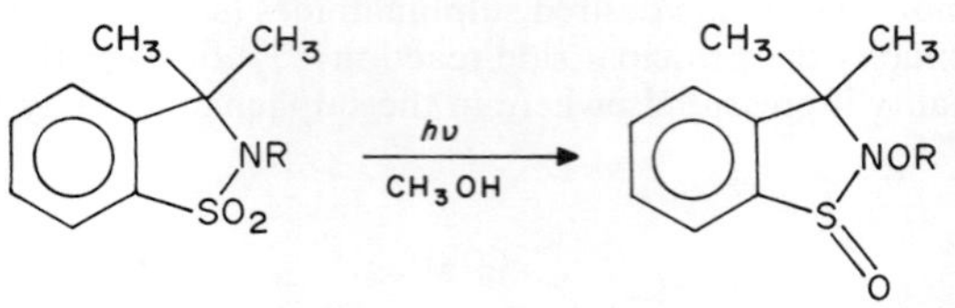


(73)[103]

**(26)** **(27)**

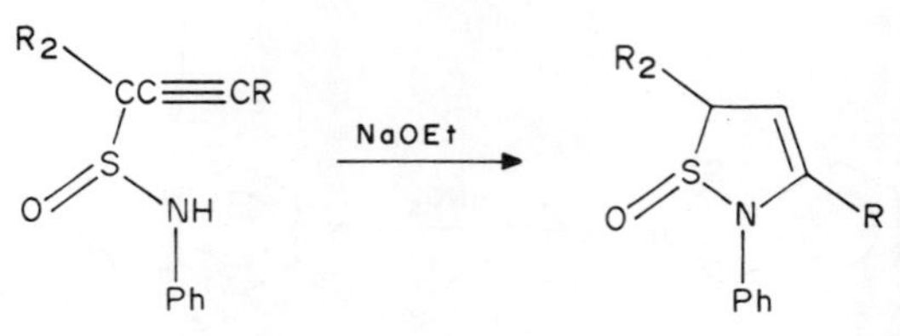


(74)[106]

## B. Reactions

### *1. Ring-opening reactions—hydrolysis and nucleophilic attack*

Hydrolysis of cyclic sulphinamides under either acidic or basic conditions ordinarily proceeds with loss of sulphur dioxide to give amine derivatives, the process with diene adducts of sulphinylamines apparently involving a retro-ene reaction[110a] as shown in equation 75[4a,4c,29b,67,68,75,83a,c,85a,b,d,e,h,86b,87e,g,i,89,91a,b,97c,110]. Under acidic conditions further reactions involving carbocation intermediates may occur (equation 77)[110b]. The formation of pyrroles by ring contraction with loss of sulphur from six-membered sulphinamides has been observed[83b,86f].

Alcohol[89,91b,110d], amine[71,110d] and sulphur[96f,110d] nucleophiles may attack the sulphur atom of cyclic sulphinamides with ring opening. Grignard reagents react similarly to give

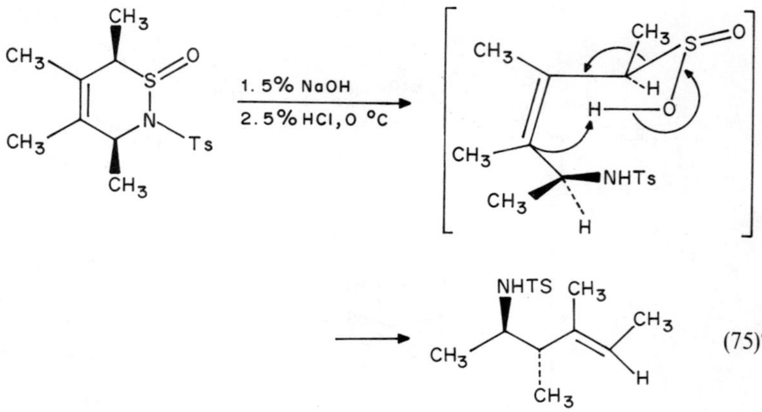


(75)[67,110c]

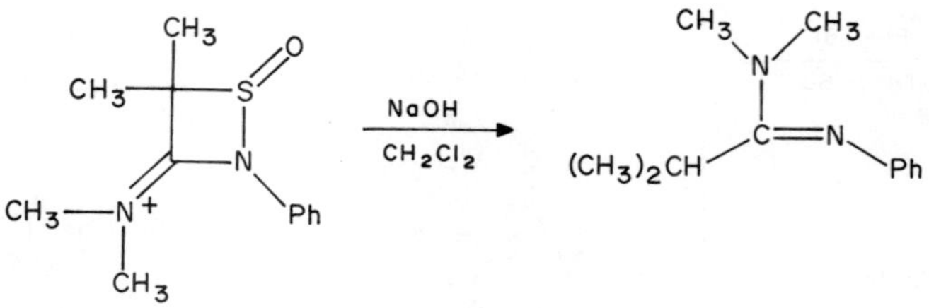


(76)[75]

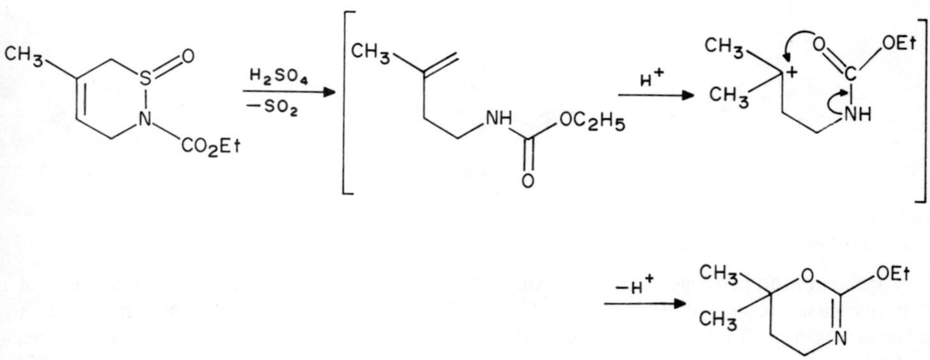


(77)[110b]

sulphoxides[73b,86g–k,o] which, in unsaturated systems, can be utilized in sigmatropic rearrangements[4c,d,67,86g–k,o].

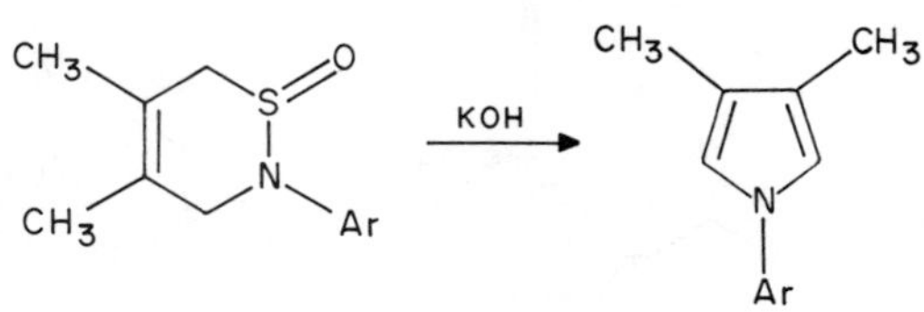


(78)[83b]

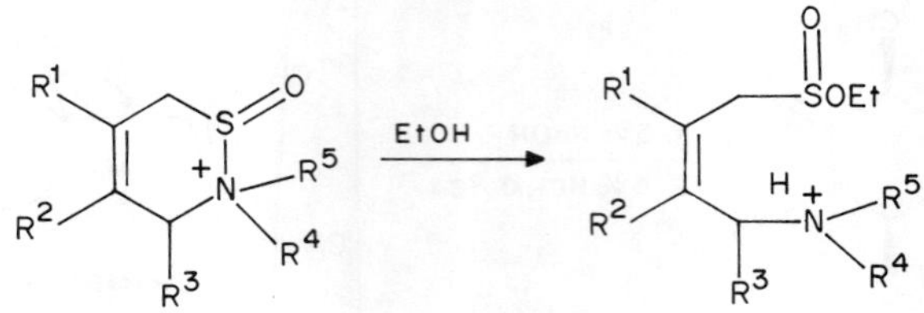


(79)[91b]

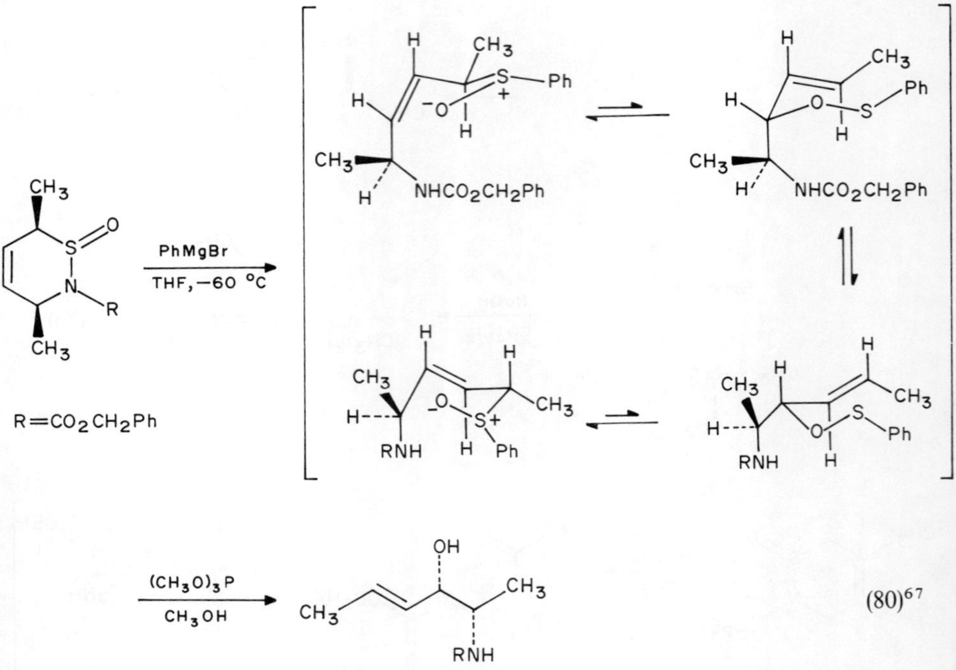


(80)[67]

## 2. Oxidation and reduction

Peracids and hydrogen peroxide oxidize cyclic sulphinamides to *S*-dioxides (sultones) in many cases (equation 81)[67,68,86l,87l,92a,f,93b–d,95i,96c,j,m,100a–c,104]. In one instance, a tertiary amine function is oxidized, thus precipitating the elimination of the benzoyl derivative of 1-hydroxypyrrolidine (equation 82)[96g], and in another case (equation 83) oxidation of the sulphinyl sulphur atom is accompanied by ring opening[86n]. Ozone is reported to oxidize dihydro-1, 2-thiazine 1-oxides to formaldehyde[87g]. The epoxidation of the double bond in this latter class of sulphinamides is directed by the SO group[86l].

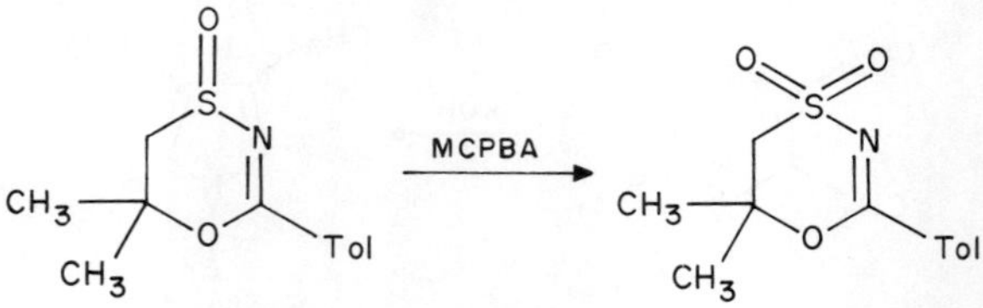


(81)[93b]

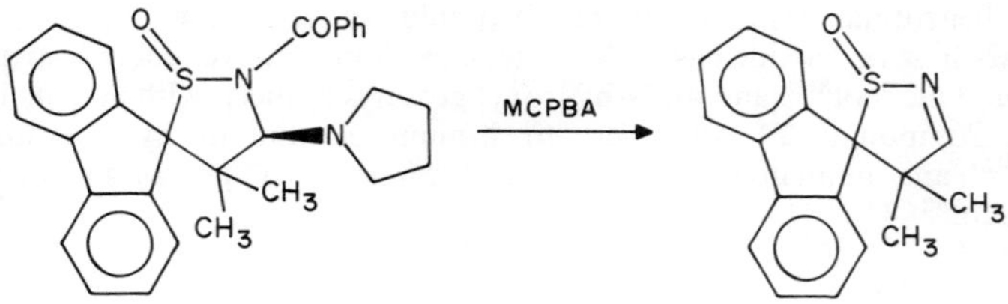


(82)[96g]

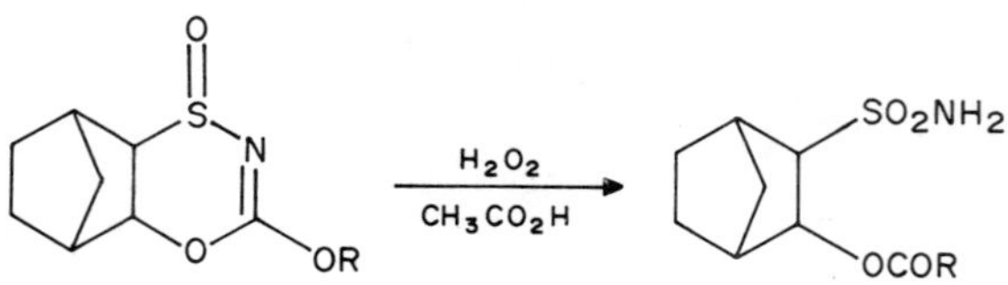


(83)[86n]

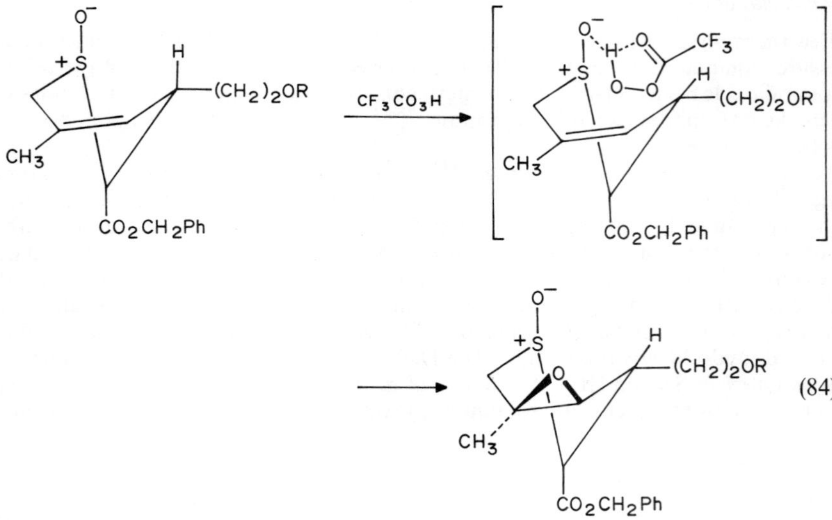


(84)[86l]

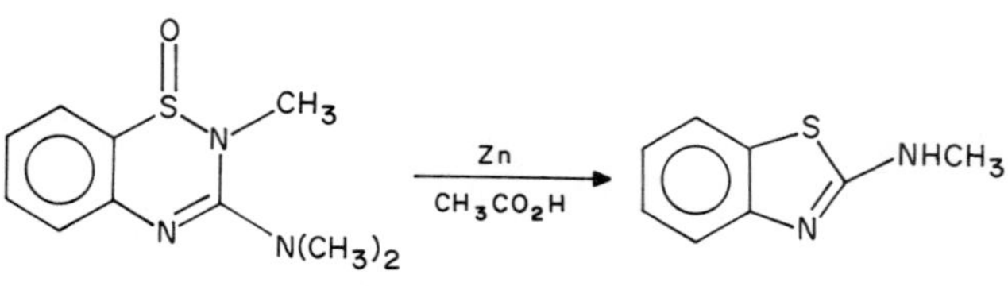


(85)[104]

**(28)**

Reduction of **28** with zinc-acetic acid gave a benzothiazole (equation 85)[100a,104], and treatment of several other cyclic sulphinamides (e.g. **29**) with Raney nickel eliminates sulphur (equation 86)[92a,b,d]. Deoxygenation of benzothiadiazine 1-oxides has been observed with thionyl chloride[100b,109] and with tributylphosphine[109]; with excess phosphine, benzothiazoles were obtained. A trichloroethyl carbamate derivative of a 3, 6-dihydro-1, 2-thiazine 1-oxide has its N-protecting group removed on treatment with zinc and *tert*-butyl alcohol[86d], and an N-hydroxyl group is reduced with acidified potassium iodide[103]. Compound **29** was inert to lithium aluminium hydride and also to hydrolysis[92a], and an analogous compound gave a green to blue colour in concentrated sulphuric acid[92e].

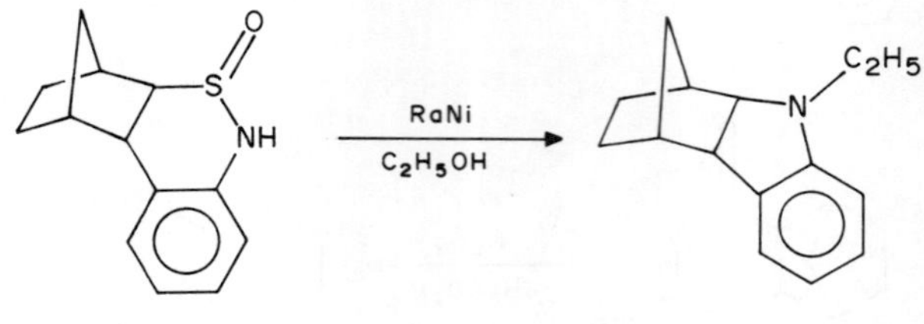


(29)

(86)[92a]

*3. Miscellaneous reactions*

Few thermolyses of cyclic sulphinamides have been reported. At 600 °C under reduced pressure, compound **27** (equation 73) is converted to **26**[103]. The thermal extrusion of sulphur dioxide from four-membered intermediates in the reaction of *N*-tosylamines with aldehydes or sulphoxides has been proposed[76a,77], and an extrusion of sulphur monoxide has been observed[95d].

Acylation or sulfenylation of nitrogen[111,112] and alkylation of nitrogen[92b] or carbonyl oxygen[96k] atoms have been reported. An *N*-chloroimino substituent is reduced to the imino derivative by hydrogen chloride[97c]. A possible aldol condensation with *p*-nitrobenzaldehyde may occur with a four-membered mixed sulphinic–carboxylic imide[70]. An electrophilic attack on the sulphur atom by elemental chlorine has been reported[113], and dehydration involving oxygen loss from sulphur occurs with an unstable uracil derivative of a five-membered sulphinamide[95g]. Diels–Alder reactions are successful with the mixed imide **30** (equation 87)[114]. The Diels–Alder adducts may undergo cycloreversion as noted in Section III.A.3[87l]. Removal of the dimethylsulphide group from **31** is accomplished by treatment with sodium methoxide (equation 88)[89], and the intermediate,

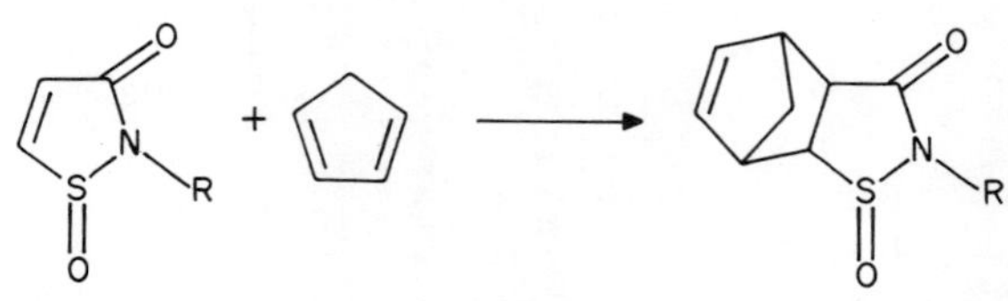


(30)

(87)[114a]

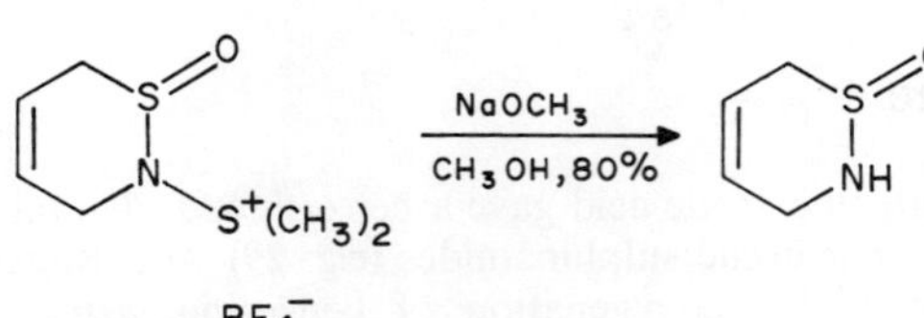


(31)

(88)[89]

**32**, reacts with *N*-sulphinyl carbamates with loss of carbon dioxide (equation 89)[86n]. Sulphur dioxide is evolved on similar treatment of analogues of **28**[4b]. A selenium analogue of a cyclic sulphinamide transfers its oxygen atom to phosphines and oxidizes hydroquinone[115], a reaction that has also been reported for the sulphur compounds[109].

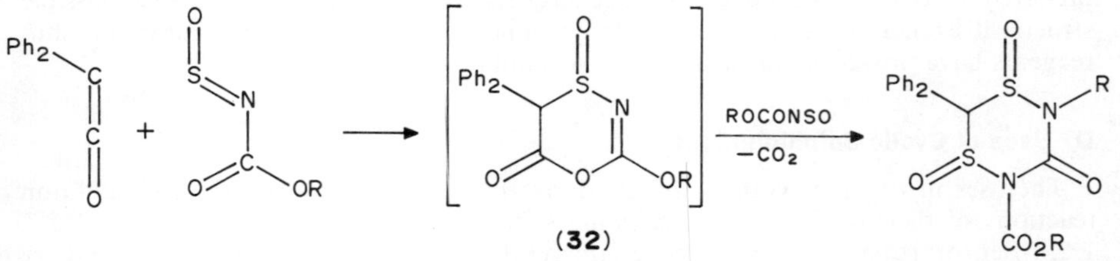


(**32**)

(89)[86n]

### C. Physical Properties of Cyclic Sulphinamides

Structures **33**–**37** have been established by X-ray analysis. The five-membered rings are non-planar with intermolecular hydrogen bonding being observed in **35**. The stereochemistry of **36** is *exo*, and enantiomeric molecules in the crystal are paired by

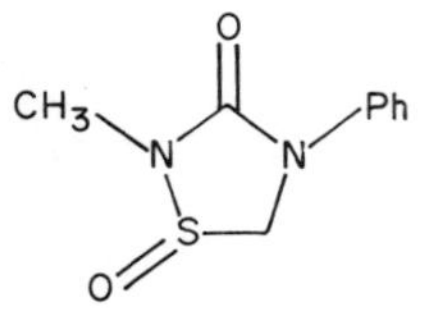

(**33**)[116]

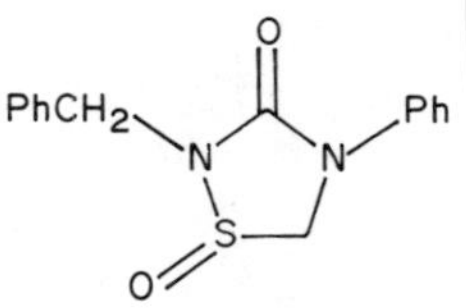

(**34**)[116]

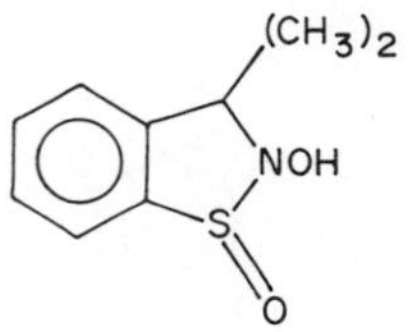

(**35**)[103]

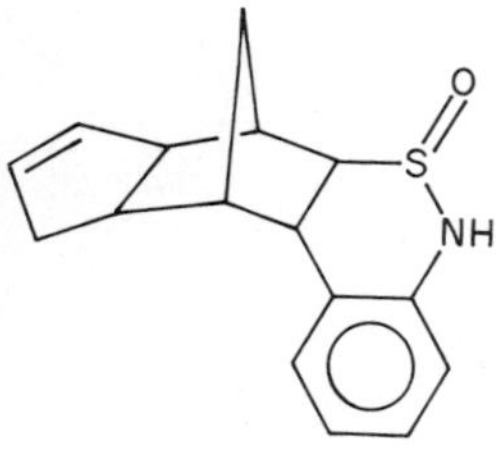

(**36**)[117]

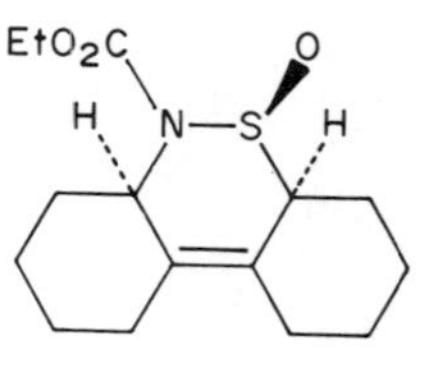

(**37**)[118]

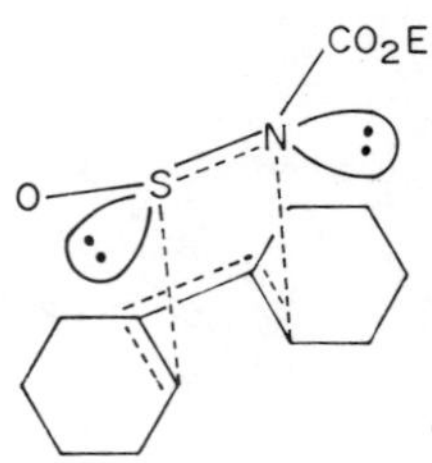

(**38**)

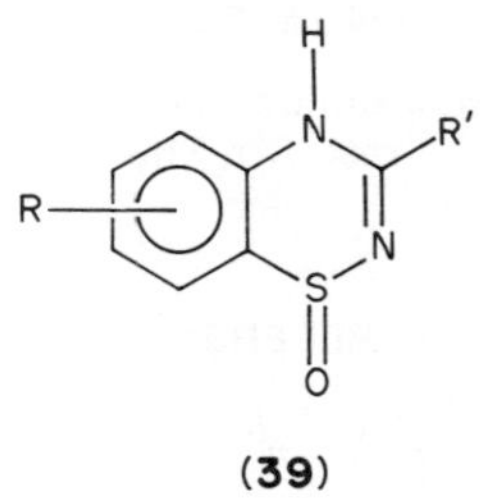

(**39**)

NH···O=S hydrogen bonds[117]. From considerations of structure **37** and theory[86m], the transition state for cycloaddition of the diene to the ethyl *N*-sulphinylcarbamate to give **37** has the geometry shown in **38**, involving the *Z* configuration of the sulphinylamine.

Molecular orbital calculations on the tautomers of **39** show that the 4*H* tautomer is preferred[119]. Nuclear magnetic resonance has been used to differentiate between possible structural isomers of adducts of sulphinylamines with dienes[120], and europium shift reagents have proved useful with the cyclic sulphinamides[86l].

### D. Uses of Cyclic Sulphinamides

The uses in organic synthesis of the derivatives obtained by [4 + 2] cycloaddition reactions of dienes with *N*-sulphinylamines have already been mentioned[4c,67,68,86i]. 1, 2, 4-Benzothiadiazine 1-oxides have apparently shown little antihypertensive activity compared to the 1, 1-dioxides[100i,109]. Five-membered mixed sulphinic–carboxylic imides, e.g. **40**, show antifungal activity[95b,112,121a], and one 3, 6-dihydro-1, 2-thiazine 1-oxide demonstrated fungicidal and antibacterial properties[121b]. Compound **40** is said to be an anti-inflammatory agent and a central nervous system depressant[95b], and a benziso-thiazole *S*-oxide is claimed to have antipsychotic activity[98]. The dihydrobenzothiadiazine *S*-oxides, **41**, are claimed to be pesticides[96d], and **42** is related to compounds that inhibit 3′, 5′-nucleotide-phosphodiesterases[95d]. A related compoud, **43**, stabilizes photographic emulsions[122]. A diamino-isothiazole *S*-oxide inhibits the secretion of stomach acid[102].

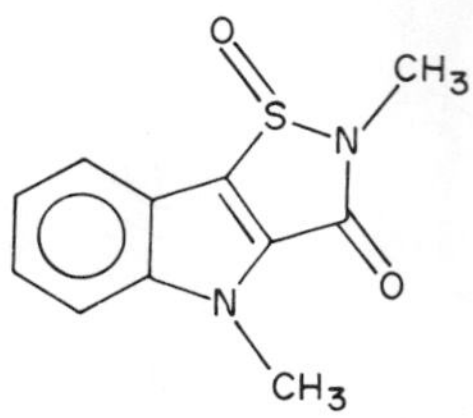


**(40)**

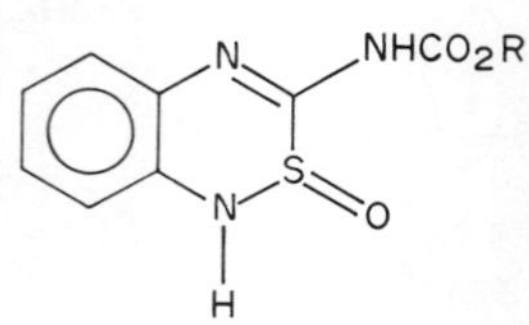


**(41)**

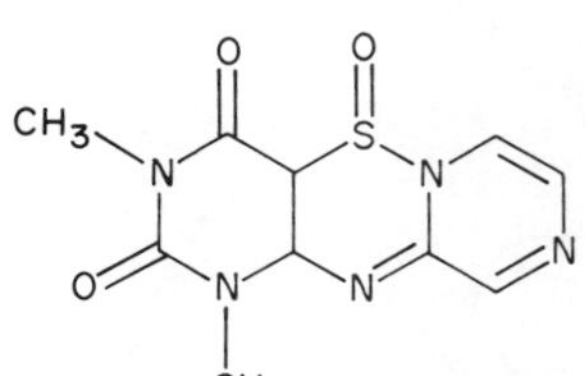


**(42)**

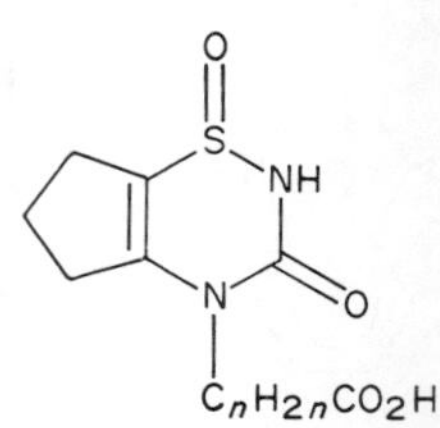


**(43)**

## IV. ACKNOWLEDGEMENTS

The authors would like to thank Mrs. Sandra Thompson for typing this manuscript. They would also like to thank Mr. James Grace, Mr. Martin Kociolek and Mr. Brian Peters for helping with the literature search.

## V. REFERENCES

1. E. Krauthausen, in *Methoden der Organischen Chemie* (Ed. D. Klamann), Vol. E II, George Thieme, Stuttgart, 1985, pp. 640, 655.

2. (a) K. K. Andersen, in *Comprehensive Organic Chemistry* (Eds. D. Barton and W. D. Ollis), Vol. 3, Pergamon, Oxford, 1979, pp. 322, 326.
(b) A. R. Katritzky and C. W. Rees (Eds.), *Comprehensive Heterocyclic Chemistry*, Vols. 1–8, Pergamon, Oxford, 1984.
3. R. S. Henion, *Eastman Org. Chem. Bull.*, **41**, No. 3 (1969).
4. (a) G. Kresze, A. Maschke, R. Albrecht, K. Bederke, H. P. Patzschke, H. Smalla and A. Trede, *Angew. Chem., Int. Ed. Engl.*, **1**, 89 (1962).
(b) G. Kresze and W. Wucherpfennig, *Angew. Chem., Int. Ed. Engl.*, **6**, 149 (1967).
(c) S. M. Weinreb, *Acc. Chem. Res.*, **21**, 313 (1988).
(d) D. L. Boger and S. M. Weinreb, *Hetero Diels–Alder Methodology in Organic Synthesis*, Vol. 47, *Organic Chemistry* (Ed. H. Wasserman), Chap. 1, Academic Press, New York, 1987.
5. E. Baumann and G. Walter, *Chem. Ber.*, **26**, 1124 (1893).
6. E. K. Fields and S. Meyerson, *J. Am. Chem. Soc.*, **88**, 2836 (1966).
7. R. W. Hoffmann and W. Sieber, *Justus Liebigs Ann. Chem.*, **703**, 96 (1967).
8. (a) D. C. Dittmer, R. S. Henion and N. Takashina, 153rd National Meeting, American Chemical Society, Miami, Florida, April, 1967, Abstract O-101.
(b) D. C. Dittmer, R. S. Henion and N. Takashina, *J. Org. Chem.*, **34**, 1310 (1969).
9. (a) T. Durst, J. L. Charlton and D. B. Mount, *Can. J. Chem.*, **64**, 246 (1986).
(b) J. L. Charlton and T. Durst, *Tetrahedron Lett.*, **25**, 5287 (1984).
(c) T. Durst, E. C. Kozma and J. L. Charlton, *J. Org. Chem.*, **50**, 4829 (1985).
10. (a) P. Messinger and F. Von Vietinghoff-Scheel, *Arch. Pharm.* (*Weinheim Ger.*), **318**, 813 (1985).
(b) P. Messinger, *Arch. Pharm.* (*Weinheim Ger.*), **318**, 950 (1985).
11. N. K. Sharma, F. de Reinach-Hirtzbach and T. Durst, *Can. J. Chem.*, **54**, 3012 (1976).
12. (a) E. N. Givens and L. A. Hamilton, *J. Org. Chem.*, **32**, 2857 (1967).
(b) J. F. King, S. Skonieczny, K. C. Khemani and J. B. Stothers, *J. Am. Chem. Soc.*, **105**, 6514 (1983).
(c) J. T. Doi, G. W. Luehr and W. K. Musker, *J. Org. Chem.*, **50**, 5716 (1985).
(d) C. W. Perkins and J. C. Martin, *J. Am. Chem. Soc.*, **105**, 1377 (1983).
(e) C. W. Perkins, S. R. Wilson and J. C. Martin, *J. Am. Chem. Soc.*, **107**, 3209 (1985).
(f) M. Ohwa, T. Kogure and E. L. Eliel, *J. Org. Chem.*, **51**, 2599 (1986).
(g) J. E. Lynch and E. L. Eliel, *J. Am. Chem. Soc.*, **106**, 2943 (1984).
(h) R. M. Dodson, P. D. Hammen and R. A. Davis, *J. Chem. Soc., Chem. Commun.*, 9 (1968).
(i) K.-Y. Ko, W. J. Frazee and E. L. Eliel, *Tetrahedron*, **40**, 1333 (1984).
13. (a) F. Jung and T. Durst, *J. Chem. Soc., Chem. Commun.*, 4 (1973).
(b) N. K. Sharma, F. Jung and T. Durst, *Tetrahedron Lett.*, 2863 (1973).
(c) T. Durst, K.-C. Tim and J. M. V. Marcil, *Can. J. Chem.*, **51**, 1704 (1973).
(d) F. Jung, N. K. Sharma and T. Durst, *J. Am. Chem. Soc.*, **95**, 3420 (1973).
(e) F. Jung, M. Molin, R. Van Den Elzen and T. Durst, *J. Am. Chem. Soc.*, **96**, 935 (1974).
(f) F. Jung, *J. Chem. Soc., Chem. Commun.*, 525 (1976).
(g) G. Rousseau and V. Torelli, German Patent 2, 711, 516, 22 Sept. 1977; *Chem. Abstr.*, **88**, 23268 (1978).
(h) V. Torelli and D. Philbert, Ger. Offen, 2, 748, 250, 11 May 1978; *Chem. Abstr.*, **89**, 110126 (1978).
(i) Roussel-UCLAF, Fr. Demande, 2, 421, 913, 2 Nov. 1979; *Chem. Abstr.*, **93**, 26659 (1980).
(j) T. Durst and B. P. Gimbarszevsky, *J. Chem. Soc., Chem. Commun.*, 724 (1975).
(k) M. D. M. Gray, D. R. Russell, D. J. H. Smith, T. Durst and B. Gimbarszevsky, *J. Chem. Soc., Perkin Trans. 1*, 1826 (1981).
14. (a) E. Thoumazean, B. Jousseaume, F. Tiffon and J.-G. Durboudin, *Heterocycles*, **19**, 2247 (1982).
(b) W. H. Mueller and M. B. Dines, *J. Chem. Soc.* (*D*), 1205 (1969).
15. (a) Y. H. Chiang, J. S. Luloff and E. Schipper, *J. Org. Chem.*, **34**, 2397 (1969).
(b) J. G. Kasperek and G. J. Kasperek, *J. Org. Chem.*, **43**, 3393 (1978).
(c) T. P. Vasil'eva, N. V. Kalyuzhnaya, V. M. Bystrova, M. G. Lin'kova, O. V. Kildisheva and I. L. Knunyants, *Izv. Akad. Nauk SSSR, Ser. Khim.*, 2187 (1980); *Chem. Abstr.*, **94**, 47192 (1981).
16. (a) W. F. Jarvis, M. D. Hoey, A. L. Finocchio and D. C. Dittmer, *J. Org. Chem.*, **53**, 5750 (1988).
(b) W. Baker, C. M. Pant and R. J. Stoodley, *J. Chem. Soc., Perkin Trans. 1*, 668 (1978).
(c) D. Knittel, *Monatsh. Chem.*, **113**, 37 (1982).
(d) D. N. Harpp and J. G. Gleason, *Tetrahedron Lett.*, 1447 (1969).
(e) D. N. Harpp, J. G. Gleason and D. K. Ash, *J. Org. Chem.*, **36**, 322 (1971).

(f) D. N. Harpp and J. G. Gleason, *J. Org. Chem.*, **36**, 1314 (1971).
(g) H. Inoue, N. Inoguchi and E. Imoto, *Bull. Chem. Soc. Jpn.*, **50**, 197 (1977).
17. Y. Gaoni, *J. Org. Chem.*, **46**, 4502 (1981).
18. (a) P. B. J. Priessen and H. Hogeveen, *J. Organomet. Chem.*, **156**, 265 (1978).
(b) K. S. Fongers and H. Hogeveen, *Tetrahedron Lett.*, 275 (1979).
19. (a) K. S. Dhami, *Chem. Ind.* (*London*), 1004 (1968).
(b) K. S. Dhami, *Indian J. Chem.*, **12**, 278 (1974).
(c) K. Henrick and B. L. Johnson, *Aust. J. Chem.*, **25**, 2263 (1972).
(d) M. A. Torosyan, R. S. Mirzoyan, S. S. Isakhanyan and A. M. Arutyunyan, *Arm. Khim. Zh.*, **39**, 124 (1986); *Chem. Abstr.*, **106**, 138299 (1987).
20. W. Barbieri, L. Bernardi, S. Coda and A. Vigevani, *Tetrahedron Lett.*, 4913 (1971).
21. (a) S. Braverman and D. Reisman, *J. Am. Chem. Soc.*, **99**, 605 (1977).
(b) S. Braverman and D. Reisman, *Tetrahedron Lett.*, 1753 (1977).
(c) S. Braverman and Y. Duar, *J. Am. Chem. Soc.*, **105**, 1061 (1983).
22. S. V. Veselovskaya, L. G. Saginova and Y. S. Shabarov, *J. Org. Chem. USSR*, **23**, 115 (1987).
23. (a) W. L. Mock and J. H. Mc Causland, *Tetrahedron Lett.*, 391 (1968).
(b) R. Heldeweg and H. Hogeveen, *J. Am. Chem. Soc.*, **98**, 2341 (1976).
24. T. Durst and L. Tetreault-Ryan, *Tetrahedron Lett.*, 2353 (1978). See also Reference 9c.
25. (a) I. R. Dunkin and J. G. Mac Donald, *J. Chem. Soc., Chem. Commun.*, 1020 (1978).
(b) N. B. Kaz'mina, I. L. Knunyants, E. I. Mysov and G. M. Kuz'yants, *Bull. Acad. Sci. USSR, Div. Chem. Sci.*, **27**, 142 (1978).
26. (a) H. Staudinger and F. Pfenninger, *Chem. Ber.*, **49**, 1941 (1916).
(b) E. Tempesti, L. Giuffre, M. Fornaroli and G. Airoldi, *Chem. Ind.* (*London*), 183 (1973).
(c) H. W. Sidebottom, C. C. Badcock, J. G. Calvert, B. R. Rabe and E. K. Damon, *J. Am. Chem. Soc.*, **93**, 3121 (1971).
(d) G. C. Bernardi, G. Moggi and D. Sianesi, *Ann. Chim.* (*Rome*), **62**, 95 (1972).
(e) D. Sianesi, G. C. Bernardi and G. Moggi, *Tetrahedron Lett.*, 1313 (1970).
(f) A. Dondoni, P. Giorgianni, A. Battaglia and G. D. Andreetti, *J. Chem. Soc., Chem. Commun.*, 350 (1981).
(g) A. V. Devekki, Yu. N. Koshelev and Yu. I. Malov, *J. Org. Chem. USSR*, **19**, 829 (1983).
27. (a) O. De Lucchi and V. Lucchini, *J. Chem. Soc., Chem. Commun.*, 464 (1982).
(b) H. Hogeveen and L. Zwart, *J. Am. Chem. Soc.*, **104**, 4889 (1982).
(c) A. Cutler, R. W. Fish, W. P. Giering and M. Rosenblum, *J. Am. Chem. Soc.*, **94**, 4354 (1972).
(d) M. Christl, E. Brunn and F. Lanzendorfer, *J. Am. Chem. Soc.*, **106**, 373 (1984).
(e) Yu. S. Shabarov, L. G. Saginova and O. B. Bondarenko, USSR SU 1, 209, 694, 7 Feb. 1986; *Chem. Abstr.*, **106**, P 5002 (1987).
(f) L. G. Saginova, O. B. Bondarenko and Yu. S. Shabarov, *J. Org. Chem. USSR*, **21**, 608 (1985).
(g) O. B. Bondarenko, T. I. Voevodskaya, L. G. Saginova, V. A. Tafeenko and Yu. S. Shabarov, *J. Org. Chem. USSR*, **23**, 1736 (1987).
28. (a) J. Dubac, P. Mazerolles, M. Joly, W. Kitching, C. W. Fong and W. H. Atwell, *J. Organomet. Chem.*, **25**, C 20 (1970).
(b) J. Dubac and P. Mazerolles, *C. R. Acad. Sci. Paris, Ser. C*, **267**, 411 (1968).
(c) E. J. Bulten and H. A. Budding, *J. Organomet. Chem.*, **166**, 339 (1979).
(d) M. Luria, R. G. DePena, K. J. Olszyna and J. Heicklen, *J. Phys. Chem.*, **78**, 325 (1974).
29. (a) J. M. Bohen and M. M. Joullie, *Tetrahedron Lett.*, 1815 (1971).
(b) J. M. Bohen and M. M. Joullie, *J. Org. Chem.*, **38**, 2652 (1973).
30. H. Gruetzmacher and H. W. Roesky, *Chem. Ber.*, **120**, 995 (1987).
31. (a) D. W. Lichtenberg and A. Wojcicki, *J. Organomet. Chem.*, **33**, C 77 (1971).
(b) R. N. Haszeldine, W. D. Bannister, B. L. Booth and P. L. Loader, *J. Chem. Soc.* (*A*), 930 (1971).
(c) J. E. Thomasson, P. W. Robinson, D. A. Ross and A. Wojcicki, *Inorg. Chem.*, **10**, 2130 (1971).
(d) J. O. Kroll and A. Wojcicki, *J. Organomet. Chem.*, **66**, 95 (1974).
(e) M. R. Churchill, J. Wormald, D. A. Ross, J. E. Thomasson and A. Wojcicki, *J. Am. Chem. Soc.*, **92**, 1795 (1970).
32. J. Strating, L. Thijs and B. Zwanenburg, *Recl. Trav. Chim. Pays-Bas*, **86**, 641 (1967).
33. (a) B. F. Bonini, G. Maccagnani, G. Mazzanti, P. Pedrini and B. Zwanenburg, *Gazz. Chim. Ital.*, **107**, 283 (1977).

(b) B. F. Bonini, G. Maccagnani, G. Mazzanti, L. Thijs, H. P. M. M. Ambrosius and B. Zwanenburg, *J. Chem. Soc., Perkin Trans. 1*, 1468 (1977).
34. (a) C. J. Kelley and M. Carmack, *Tetrahedron Lett.*, 3605 (1975).
(b) V. N. Klyuev, A. B. Korzhenevski and B. D. Berezin, *Izv. Vyssh. Uchebn. Zaved., Khim. Khim. Tekhnol.*, **21**, 31 (1978); *Chem. Abstr.*, **88**, 170041 (1978).
35. J. Pedain, O. Bayer and G. Oertel, Ger. Patent 1, 643, 341, 20 Feb. 1975; *Chem. Abstr.*, **83**, 131333 (1975).
36. (a) J. F. King, K. Piers, D. J. H. Smith, C. Mc Intosh and P. de Mayo, *J. Chem. Soc., Chem. Commun.*, 31 (1969).
(b) J. F. King, P. De Mayo, C. L. Mc Intosh, K. Piers and D. J. H. Smith, *Can. J. Chem.*, **48**, 3704 (1970).
(c) D. C. Dittmer and T. R. Nelsen, *J. Org. Chem.*, **41**, 3044 (1976).
37. (a) C. R. Hall and D. J. H. Smith, *Tetrahedron Lett.*, 3633 (1974).
(b) H. P. Schuckmann and C. Von Sonntag, *J. Photochem.*, **22**, 55 (1983).
38. R. D. Chambers and J. A. Cunningham, *J. Chem. Soc., Chem. Commun.*, 583 (1967).
39. A. A. Scala, I. Colon and W. Rourke, *J. Phys. Chem.*, **85**, 3603 (1981).
40. R. Kowalewski and P. Margarethe, *Angew. Chem.*, **100**, 1437 (1988).
41. (a) R. M. Dodson, P. D. Hammen and R. A. Davis, *J. Org. Chem.*, **36**, 2693 (1971).
(b) R. M. Dodson, P. D. Hammen and J. Y. Fan, *J. Org. Chem.*, **36**, 2703 (1971).
42. (a) J. F. King, B. L. Huston, A. Hawson, J. Komery, D. M. Deaken and D. R. K. Harding, *Can. J. Chem.*, **49**, 936 (1971).
(b) J. F. King, A. Hawson, B. L. Huston, L. J. Danks and J. Komery, *Can. J. Chem.*, **49**, 943 (1971).
(c) J. F. King, A. Hawson, D. M. Deaken and J. Komery, *J. Chem. Soc., Chem. Commun.*, 33 (1969).
43. (a) D. N. Harpp, K. Steliou and T. H. Chan, *J. Am. Chem. Soc.*, **100**, 1222 (1978).
(b) D. N. Harpp and D. F. Mullins, *Can. J. Chem.*, **61**, 757 (1983).
44. (a) C. Kemal, T. W. Chan and T. C. Bruice, *Proc. Natl. Acad. Sci. U.S.A.*, **74**, 405 (1977).
(b) G. W. Astrologes and J. C. Martin, *J. Am. Chem. Soc.*, **97**, 6909 (1975).
(c) G. W. Astrologes and J. C. Martin, *J. Am. Chem. Soc.*, **99**, 4390 (1977).
(d) A. P. Davis and G. H. Whitham, *J. Chem. Soc., Chem. Commun.*, 741 (1981).
(e) E. Block and A. Wall, *J. Org. Chem.*, **52**, 809 (1987).
45. (a) L. D. Martin and J. C. Martin, *J. Am. Chem. Soc.*, **99**, 3511 (1977).
(b) G. W. Astrologes and J. C. Martin, *J. Am. Chem. Soc.*, **99**, 4400 (1977).
46. (a) H. Kohn, P. Charumilind and S. H. Simonsen, *J. Am. Chem. Soc.*, **101**, 5431 (1979).
(b) H. Kohn and P. Charumilind, *J. Org. Chem.*, **45**, 4359 (1980).
47. J. Wolinsky and R. L. Marhenke, *J. Org. Chem.*, **40**, 1766 (1975).
48. A. A. Najam and J. G. Tillett, *J. Chem. Soc. Perkin Trans. 2*, 858 (1975).
49. (a) D. N. Harpp, S. M. Vines, J. P. Montillier and T. H. Chan, *J. Org. Chem.*, **41**, 3987 (1976).
(b) W. H. Pirkle and M. S. Hoekstra, *J. Am. Chem. Soc.*, **98**, 1832 (1976).
(c) J. G. Duboudin, B. Jousseaume and E. Thoumazeau, *Bull. Soc. Chim. Fr.*, II 105 (1983).
(d) W. E. Truce and B. Van Gemert, *J. Am. Chem. Soc.*, **100**, 5525 (1978).
50. (a) S. Askari, S. Lee, R. R. Perkins and J. R. Scheffer, *Can. J. Chem.*, **63**, 3526 (1985).
(b) R. F. Heldeweg and H. Hogeveen, *J. Am. Chem. Soc.*, **98**, 6040 (1976).
51. L. Carlsen and J. P. Snyder, *Tetrahedron Lett.*, 2045 (1977).
52. (a) L. Carlsen, H. Egsgaard and D. N. Harpp, *J. Chem. Soc., Perkin Trans. 2*, 1166 (1981).
(b) T. Durst, J. D. Finlay and D. J. H. Smith, *J. Chem. Soc., Perkin Trans. 1*, 950 (1979).
53. E. Thoumazeau, B. Jousseaume, F. Duboudin and J. G. Duboudin, *Org. Mass Spectrom.*, **17**, 596 (1982).
54. T. Durst, J. C. Huang, N. K. Sharma and D. J. H. Smith, *Can. J. Chem.*, **56**, 512 (1978).
55. T. H. Chan, J. P. Montillier, W. F. Van Horn and D. N. Harpp, *J. Am. Chem. Soc.*, **92**, 7224 (1970).
56. S. Tamagaki, H. Hirota and S. Oae, *Bull. Chem. Soc. Jpn.*, **47**, 2075 (1974).
57. I. W. J. Still, S. K. Hasan and K. Turnbull, *Can. J. Chem.*, **56**, 1423 (1978).
58. M. R. Churchill and J. Wormald, *J. Am. Chem. Soc.*, **93**, 354 (1971).
59. A. L. Spek, *Cryst. Struct. Commun.*, **8**, 127 (1979).
60. E. Surcouf, *Acta Crystallogr., Sect. B*, **B35**, 1922, 1925 (1979).

61. D. C. Dittmer and T. C. Sedergran, in *Small Ring Heterocycles* (A. Hassner, Ed.), Vol. 42, Part 3, Wiley, New York, 1985, p. 476.
62. G. W. Buchanan, N. K. Sharma, F. de Reinach-Hirtzbach and T. Durst, *Can. J. Chem.*, **55**, 44 (1977).
63. O. Exner, D. N. Harpp and J. G. Gleason, *Can. J. Chem.*, **50**, 548 (1972).
64. L. Carlsen and J. P. Synder, *J. Org. Chem.*, **43**, 2216 (1978).
65. J. Delettre, J. P. Mornon, G. Lepicard, T. Ojasoo and J. P. Raynaud, *J. Steroid Biochem.*, **13**, 45 (1980).
66. S. M. Weinreb and R. R. Staib, *Tetrahedron*, **38**, 3087 (1982).
67. S. M. Weinreb, R. S. Garigipati and J. A. Gainor, *Heterocycles*, **21**, 309 (1984).
68. S. M. Weinreb, J. A. Gainor and R. P. Joyce, *Bull. Soc. Chim. Belg.*, **95**, 1021 (1986).
69. H. Beecken and F. Korte, *Tetrahedron*, **18**, 1527 (1962).
70. J. C. Martin, U.S. Patent 3, 271, 417, Sept. 6, 1966; *Chem. Abstr.*, **65**, 18731 (1966).
71. J. E. Semple and M. M. Joullie, *J. Org. Chem.*, **43**, 3066 (1978).
72. (a) L. Capuano, G. Urhahn and A. Willmes, *Chem. Ber.*, **112**, 1012 (1979).
(b) T. Minami, K. Yamataka, Y. Ohshiro, T. Agawa, N. Yasuoka and N. Kasai, *J. Org. Chem.*, **37**, 3810 (1972).
73. (a) G. Maccagnani, *Boll. Sci. Fac. Chim. Ind. Bologna*, **23**, 399 (1965).
(b) F. Effenberger and R. Gleiter, *Chem. Ber.*, **99**, 3903 (1966).
(c) W. Wucherpfennig, *Tetrahedron Lett.*, 1891 (1971).
74. R. Bussas and G. Kresze, *Justus Liebigs Ann. Chem.*, 545 (1982).
75. R. Lux and G. Kresze, *Justus Liebigs Ann. Chem.*, 527 (1987).
76. (a) G. Kresze and R. Albrecht, *Angew Chem., Int. Ed. Engl.*, **1**, 595 (1962).
(b) C. Carpanelli and G. Gaiani, *Gazz. Chim. Ital.*, **100**, 689 (1970).
77. G. Schultz and G. Kresze, *Angew. Chem., Int. Ed. Engl.*, **2**, 736 (1963).
78. (a) P. Rajagopalan and H. U. Daeniker, *Angew. Chem., Int. Ed. Engl.*, **2**, 46 (1963).
(b) K. Burger, H. Goth and E. Daltrozzo, *Z. Naturforsch., B*, **37B**, 473 (1982).
(c) R. N. Butler and G. A. O'Halloran, *Chem. Ind. (London)*, 750 (1986).
79. A. Baba, Y. Ohshiro and T. Agawa, *Chem. Lett.*, 11 (1976).
80. (a) L. S. Chen, D. W. Lichtenberg, P. W. Robinson, Y. Yamamoto and A. Wojcicki, *Inorg. Chem. Acta*, **25**, 165 (1977).
(b) T. W. Leung, G. G. Christoph, J. Gallucci and A. Wojcicki, *Organometallics*, **5**, 846 (1986).
81. R. Grigg, M. Dowling and V. Sridharan, *J. Chem. Soc., Chem. Commun.*, 1777 (1986).
82. (a) O. Wichterle and J. Rocek, *Chem. Listy*, **47**, 1768 (1953).
(b) O. Wichterle and J. Rocek, *Coll. Czech. Chem. Commun.*, **19**, 282 (1954).
(c) J. Rocek and O. Wichterle, *Czech. Patent 83770*, Jan. 3, 1955; *Chem. Abstr.*, **50**, 7152 (1956).
83. (a) E. G. Kataev and V. V. Plemenkov, *J. Gen. Chem. USSR*, **32**, 3745 (1962).
(b) R. A. Jones, *Aust. J. Chem.*, **19**, 289 (1966).
(c) T. Sasaki, S. Eguchi and T. Ishii, *J. Org. Chem.*, **34**, 3749 (1969).
(d) E. S. Levchenko and E. I. Slyusarenko, *Fiziol. Akt. Veshchestva*, **8**, 36 (1976); *Chem. Abstr.*, **87**, 23185 (1977).
(e) C. Boan and L. Skattebol, *J. Chem. Soc., Perkin Trans. 1*, 1568 (1978).
84. (a) H. P. Malach, R. Bussas and G. Kresze, *Justus Liebigs Ann. Chem.*, 1384 (1982).
(b) R. N. Butler, D. A. O'Donoghue and G. A. O'Halloran, *J. Chem. Soc., Chem. Commun.*, 800 (1986).
85. (a) W. Wucherpfennig, *Tetrahedron Lett.*, 3235 (1967).
(b) O. J. Scherer and R. Schmitt, *Chem. Ber.*, **101**, 3302 (1968).
(c) E. S. Levchenko and E. M. Dorokhova, *J. Org. Chem. USSR*, **8**, 2573 (1972).
(d) E. M. Dorokhova and E. S. Levchenko, *Tezisy Dokl. Nauchn. Sess. Khim. Tekhnol. Org. Soedin, Sery. Sernistykh Neftei 13th*, 207 (1974); *Chem. Abstr.*, **85**, 160005 (1976).
e) E. M. Dorokhova, E. S. Levchenko and T. Ya. Lavrenyuk, *J. Org. Chem. USSR*, **10**, 1877 (1974).
(f) G. S. Borovikova, E. S. Levchenko and E. I. Borovik, *J. Org. Chem. USSR*, **15**, 2095 (1979).
(g) E. M. Dorokhova, E. S. Levchenko and N. P. Pel'kis, *J. Org. Chem. USSR*, **11**, 755 (1975).
(h) C. Carpanelli and G. Gaiani, *Gazz. Chim. Ital.*, **112**, 187 (1982).
(i) C. Carpanelli and G. Gaiani, *Gazz. Chim. Ital.*, **112**, 191 (1982).
86. (a) H. Hoerhold, *Angew. Chem., Int. Ed. Engl.*, **6**, 357 (1967).

(b) E. S. Levchenko, Ya. G. Bal'on and A. V. Kirsanov, *J. Org. Chem. USSR*, **3**, 1838 (1967).
(c) E. S. Levchenko and Ya. G. Bal'on, *Metody Poluch. Khim. Reactivov Prep. No. 19*, 153 (1969); *Chem. Abstr.*, **74**, 125599 (1971).
(d) L. Wald and W. Wucherpfennig, *Justus Liebigs Ann. Chem.*, **746**, 28 (1971).
(e) K. Ichimura, Japan Kokai 75 18, 462, 26 Feb. 1975; *Chem. Abstr.*, **83**, 131448 (1975).
(f) K. Ichimura, S. Ichikawa and K. Imamura, *Bull. Chem. Soc. Jpn.*, **49**, 1157 (1976).
(g) R. S. Garigipati, A. J. Freyer, R. R. Whittle and S. M. Weinreb, *J. Am. Chem. Soc.*, **106**, 7861 (1984).
(h) S. W. Remiszewski, R. R. Whittle and S. M. Weinreb, *J. Org. Chem.*, **49**, 3243 (1984).
(i) S. M. Weinreb, R. S. Garigipati and J. A. Gainor, *Heterocycles*, **21**, 309 (1984).
(j) S. W. Remiszewski, T. R. Stouch and S. M. Weinreb, *Tetrahedron*, **41**, 1173 (1985).
(k) R. S. Garigipati, R. Cordova, M. Parvez and S. M. Weinreb, *Tetrahedron*, **42**, 2979 (1986).
(l) R. P. Joyce, J. A. Gainor and S. M. Weinreb, *J. Org. Chem.*, **52**, 1177 (1987).
(m) P. Hanson and W. A. Stockburn, *J. Chem. Soc., Perkin Trans. 2*, 589 (1985).
(n) H.-H. Hoerhold and H. Eibisch, *Chem. Ber.*, **101**, 3567 (1968).
(o) R. S. Garigipati and S. M. Weinreb, *J. Am. Chem. Soc.*, **105**, 4499 (1983).
87. (a) G. Kresze and A. Maschke, Ger. Patent 1, 117, 566, Nov. 23, 1961; *Chem. Abstr.*, **57**, 11110 (1962).
(b) E. S. Levchenko and A. V. Kirsanov, *J. Gen. Chem. USSR*, **32**, 157 (1962).
(c) E. S. Levchenko, Ya. G. Bal'on and A. V. Kirsanov, *J. Gen. Chem. USSR*, **33**, 1540 (1963).
(d) E. S. Levchenko, Ya. G. Bal'on and A. V. Kirsanov, *J. Gen. Chem. USSR*, **33**, 1540 (1963).
(e) E. S. Levchenko, Ya. G. Bal'on and A. A. Kisilenko, *J. Org. Chem. USSR*, **1**, 151 (1965).
(f) A. Macaluso and J. Hamer, *J. Org. Chem.*, **31**, 3049 (1966).
(g) Ya. G. Bal'on, E. S. Levchenko and A. V. Kirsanov, *J. Org. Chem. USSR*, **3**, 2009 (1967).
(h) E. S. Levchenko and Ya. G. Bal'on, *Metody Poluch. Khim. Reaktivov Prep., No. 19*, 143 (1969); *Chem. Abstr.*, **76**, 3775 (1972).
(i) G. Kresze and U. Wagner, *Justus Liebigs Ann. Chem.*, **762**, 93 (1972).
(j) R. Bergamasco and Q. N. Porter, *Aust. J. Chem.*, **30**, 1061 (1977).
(k) J. W. McFarland, D. Schut and B. Zwanenburg, *Tetrahedron*, **37**, 389 (1981).
(l) W. L. Mock and R. M. Nugent, *J. Am. Chem. Soc.*, **97**, 6521 (1975).
(m) W. L. Mock and R. M. Nugent, *J. Am. Chem. Soc.*, **97**, 6526 (1975).
88. (a) E. G. Kataev, V. V. Plemenkov and V. V. Markin, *Proc. Acad. Sci. USSR, Chem. Sect.*, **165**, 1208 (1965).
(b) E. G. Kataev and V. V. Plemenkov, *Khim. Primen, Fosfororg. Soedin., Tr. Vses. Konf 3rd, 1965*, 268 (1972); *Chem. Abstr.*, **77**, 88415 (1972).
89. M. Roessert, W. Kraus and G. Kresze, *Tetrahedron Lett.*, 4669 (1978).
90. O. J. Scherer and R. Schmitt, *Angew. Chem., Int. Ed. Engl.*, **6**, 701 (1967).
91. (a) G. Kresze and M. Roessert, *Angew. Chem., Int. Ed. Engl.*, **17**, 63 (1978).
(b) G. Kresze and M. Roessert, *Justus Liebigs Ann. Chem.*, 58 (1981).
92. (a) G. R. Collins, *J. Org. Chem.*, **29**, 1688 (1964).
(b) G. R. Collins, U.S. Patent 3, 285, 911, Nov. 15, 1966; U.S. Patent 3, 337, 581, Aug. 22, 1967; *Chem. Abstr.*, **66**, P37932 (1967); **68**, P69012 (1968).
(c) A. Macaluso and J. Hamer, *J. Org. Chem.*, **32**, 506 (1967).
(d) H. Beecken, *Chem. Ber.*, **100**, 2159 (1967).
(e) H. Beecken, *Chem. Ber.*, **100**, 2164 (1967).
(f) E. G. Kataev, V. V. Plemenkov and V. K. Pluzhnov, *J. Org. Chem. USSR*, **3**, 319 (1967).
(g) E. G. Kataev and V. V. Plemenkov, *J. Org. Chem. USSR*, **4**, 1056 (1968).
(h) V. V. Plemenkov and L. A. Yanykina, *J. Org. Chem. USSR*, **6**, 2049 (1970).
(i) D. R. Borthakur, D. Prajapati and J. S. Sandhu, *Heterocycles*, **24**, 2739 (1986).
93. (a) C. Carpanelli, G. Gaiani and F. Sancassan, *Gazz. Chim. Ital.*, **112**, 469 (1982).
(b) C. Carpanelli, G. Gaiani and F. Sancassan, *Gazz. Chim. Ital.*, **114**, 399 (1984).
(c) C. Carpanelli, G. Gaiani and F. Sancassan, *Gazz. Chim. Ital.*, **115**, 265 (1985).
(d) L. I. Samarai, V. P. Belaya and G. I. Derkach, *Chem. Heterocycl. Compd.*, **4**, 271 (1968).
94. A. de Souza Gomes and M. M. Joullie, *J. Heterocycl. Chem.*, **6**, 729 (1969).
95. (a) J. Szmuszkovicz, *J. Org. Chem.*, **29**, 178 (1964).
(b) J. Szmuszkovicz, U.S. Patent 3, 147, 273, Sept. 1, 1964; *Chem. Abstr.*, **61**, P14676 (1964).
(c) R. Buyle and H. G. Viehe, *Tetrahedron*, **24**, 3987 (1968).

(d) I. M. Goldman, Ger. Offen. 1, 809, 013, 7 Aug. 1969; *Chem. Abstr.*, **72**, P12771 (1970).
(e) E. S. Levchenko and I. N. Berzina, *J. Org. Chem. USSR*, **6**, 2281 (1970).
(f) T. J. Maricich, R. A. Jourdenais and T. A. Albright, *J. Am. Chem. Soc.*, **95**, 5831 (1973).
(g) K. Senga, M. Ichiba and S. Nishigaki, *Tetrahedron Lett.*, 1129 (1976).
(h) E. Schaumann, J. Ehlers and H. Mrotzek, *Justus Liebigs Ann. Chem.*, 1734 (1979).
(i) G. Kresze and C. Seyfried, Ger. Patent 1, 620, 459, 12 Apr. 1973; *Chem. Abstr.*, **78**, 159688 (1973).

96. (a) P. Tavs, *Angew. Chem., Int. Ed. Engl.*, **5**, 1048 (1966).
(b) S. N. Lewis, G. A. Miller, M. Hausman and E. C. Szamborski, *J. Heterocycl. Chem.*, **8**, 591 (1971).
(c) S. N. Lewis and G. A. Miller, U.S. Patent 3, 562, 283, 9 Feb. 1971; *Chem. Abstr.*, **74**, 100025 (1971).
(d) J. R. Hadfield, F. L. C. Baranyovits and J. I. Masters, Ger. Offen. 2, 050, 022, 13 May 1971; *Chem. Abstr.*, **75**, P63842 (1971).
(e) E. M. Burgess and H. R. Penton, Jr., *J. Am. Chem. Soc.*, **95**, 279 (1973).
(f) R. B. Morin, E. M. Gordon and J. R. Lake, *Tetrahedron Lett.*, 5213 (1973).
(g) E. M. Burgess and H. R. Penton, Jr., *J. Org. Chem.*, **39**, 2885 (1974).
(h) T. Minami and T. Agawa, *J. Org. Chem.*, **39**, 1210 (1974).
(i) H. P. Braun and H. Meier, *Tetrahedron*, **31**, 637 (1975).
(j) H. Meier, G. Trickes and H. P. Braun, *Tetrahedron Lett.*, 171 (1976).
(k) T. Gilchrist, C. W. Rees and I. W. Southon, *J. Chem. Res.* (*S*), 214 (1979).
(l) T. Kamiya, T. Teraji, M. Hashimoto, O. Nakaguchi and T. Oku, *J. Am. Chem. Soc.*, **97**, 5020 (1975).
(m) G. Mazzanti, G. Maccagnani, B. F. Bonini, P. Pedrini and B. Zwanenburg, *Gazz. Chim. Ital.*, **110**, 163 (1980).
(n) T. Kamiya, T. Teraji, M. Hashimoto, O. Nakaguchi and T. Oku, Japan Kokai 76 39, 667, 2 Apr. 1976; *Chem. Abstr.*, **85**, 192710 (1976).

97. (a) R. J. A. Walsh and K. R. H., Wooldridge, *J. Chem. Soc., Perkin Trans. 1*, 1247 (1972).
(b) M. Davis, M. C. Dereani, J. L. McVicars and I. J. Morris, *Aust. J. Chem.*, **30**, 1815 (1977).
(c) H. Boeshagen, W. Geiger and H. Medenwald, *Chem. Ber.*, **103**, 3166 (1970).

98. J. P. Yevich and W. G. Lobeck, Ger. Offen. DE 3, 613, 739, 20 Nov. 1986; *Chem. Abstr.*, **106**, 102322 (1987).

99. J. T. Doi and W. K. Musker, *J. Org. Chem.*, **50**, 1 (1985).

100. (a) G. Kresze, C. Seyfried and A. Trede, *Tetrahedron Lett.*, 3933 (1965).
(b) G. Kresze, C. Seyfried and A. Trede, *Justus Liebigs Ann. Chem.*, **715**, 223 (1968).
(c) G. Kresze, A. Trede and C. Seyfried, Ger. Patent 1, 470, 316, 7 Dec. 1972; *Chem. Abstr.*, **78**, 72227 (1973).
(d) H. Beecken, *Chem. Ber.*, **100**, 2151 (1967).
(e) L. N. Markovskii, E. A. Darmokhval and E. S. Levchenko, *J. Org. Chem. USSR*, **9**, 2055 (1973).
(f) V. W. Boehnisch, T. L. Gilchrist and C. W. Rees, *J. Chem. Soc., Perkin Trans. 1*, 2851 (1979).
(g) T. L. Gilchrist, C. W. Rees and D. Vaughan, *J. Chem. Soc., Perkin Trans. 1*, 49 (1983).
(h) E. S. Levchenko, G. S. Borovikova, E. I. Borovik and V. N. Kalinin, *J. Org. Chem. USSR*, **20**, 176 (1984).
(i) D. D. Ross and D. Lednicer, *J. Heterocycl. Chem.*, **19**, 975 (1982).

101. (a) W. Ried, O. Moesinger and W. Schuckmann, *Angew. Chem., Int. Ed. Engl.*, **15**, 103 (1976).
(b) T. Kamiya, T. Teraji, M. Hashimoto, O. Nakaguchi and T. Oku, Japan Kokai 76 39, 666, 2 Apr. 1976; *Chem. Abstr.*, **85**, 192709 (1976).

102. S. F. Britcher, W. C. Lumma, Jr., C. N. Habecker, J. J. Baldwin and J. M. Hoffman, *Eur. Pat. Appl. EP 67,436*, 22 Dec. 1982; *Chem. Abstr.*, **98**, 160577 (1983).

103. D. Doepp, C. Krueger, P. Lauterfeld and E. Raabe, *Angew. Chem., Int. Ed. Engl.*, **26**, 146 (1987).

104. G. Kresze and A. Hatjiissaak, *Phosphorus Sulfur*, **29**, 41 (1987).

105. H. Wittman, E. Ziegler, H. Stark and G. Dworak, *Monatsh. Chem.*, **100**, 959 (1969).

106. K. Ruitenberg and P. Vermeer, *J. Organomet. Chem.*, **256**, 175 (1983).

107. O. Tsuge, M. Tashiro and S. Mataka, *Tetrahedron Lett.*, 3877 (1968).

108. K. Okuma, N. Higuchi, H. Ohta and M. Kobayashi, *Chem. Lett.*, 1503 (1980).

109. N. Finch, S. Ricca, Jr., L. H. Werner and R. Rodebaugh, *J. Org. Chem.*, **45**, 3416 (1980).
110. (a) W. L. Mock and R. M. Nugent, *J. Org. Chem.*, **43**, 3433 (1978).
(b) E. I. Slyusarenko, N. P. Pel'kis and E. S. Levchenko, *J. Org. Chem. USSR*, **14**, 1016 (1978).
(c) R. S. Garigipati, J. A. Morton and S. M. Weinreb, *Tetrahedron Lett.*, **24**, 987 (1983).
(d) W. Wucherpfennig, *Justus Liebigs Ann. Chem.*, **746**, 16 (1971).
(e) E. S. Levchenko and E. M. Dorokhova, *J. Org. Chem. USSR*, **12**, 432 (1976).
111. P. Stoss, G. Satzinger and M. Herrmann, Ger. Offen. 2, 340, 784, 20 Feb. 1975; *Chem. Abstr.*, **83**, 58797 (1975).
112. S. Watanabe, K. Matsumoto and M. Iwata, Japan Kokai, 74 36, 673; *Chem. Abstr.*, **83**, 10058 (1975).
113. L. N. Markovskii and E. A. Darmokhval, *J. Org. Chem. USSR*, **9**, 664 (1973).
114. (a) E. D. Weiler and J. J. Brennan, *J. Heterocycl. Chem.*, **15**, 1299 (1978).
(b) G. A. Miller and E. D. Weiler, U.S. Patent 4, 199, 591, 22 Apr. 1980; *Chem. Abstr.*, **93**, 186337 (1980).
115. N. Kamigata, M. Takata, H. Matsuyama and M. Kobayashi, *Sulfur Lett.*, **5**, 1 (1986).
116. W. Schuckmann, H. Fuess, O. Moesinger and W. Ried, *Acta Crystallogr., Sect. B*, **B34**, 1516 (1978).
117. D. J. Pointer, J. B. Wilford and K. M. Chui, *J. Chem. Soc., Perkin Trans. 2*, 1134 (1972).
118. Z. Dauter, P. Hanson, C. D. Reynolds and W. A. Stockburn, *Acta Crystallogr., Sect. C*, **C40**, 521 (1984).
119. A. J. Wohl, in *Molecular Orbital Studies in Chemical Pharmacology Symposium* (Ed. L. B. Kier), Springer, New York, 1970, p. 262; *Chem. Abstr.*, **74**, 40780 (1971).
120. Ya. G. Bal'on, V. F. Bystrov and A. U. Stepanyants, *Chem. Heterocycl. Compd.*, **4**, 50 (1968).
121. (a) R. J. Lukens and J. G. Horsfall, *Phytopathology*, **57**, 876 (1967).
(b) V. V. Stopkan, T. T. Cherepenko, E. S. Levchenko and Ya. G. Bal'on *Fiziol. Akt. Veshchestva*, No. 2, 111 (1969); *Chem. Abstr.*, **73**, 13452 (1970).
122. T. I. Abbott, Def. Publ. U.S. Pat. Off. 882, 019, 26 Jan. 1971; *Chem. Abstr.*, **74**, 93463 (1971).

The Chemistry of Sulphinic Acids, Esters and their Derivatives
Edited by S. Patai


CHAPTER **10**

# Acidity, hydrogen bonding and complexation

HISASHI FUJIHARA and NAOMICHI FURUKAWA

*Department of Chemistry, University of Tsukuba, Tsukuba, Ibaraki 305, Japan*



## I. INTRODUCTION

Sulphinic acids (**A**) particularly the alkanesulphinic acids, are unstable and disproportionate on standing to the thiolsulphonates and sulphonic acids. They are usually handled as their stable sodium salts. The free acids are liberated from aqueous solutions of these salts upon careful acidification by hydrochloric acid. Few sulphinic acids are formed in nature, doubtless due to their instability, but they exist as intermediates in the oxidation of thiols[1,2].

$$\underset{\textbf{(A)}}{RSO_2H} \qquad \underset{\textbf{(B)}\ M = \text{metal}}{RSO_2M}$$

Both sulphinic acids and sulphinatometal complexes (**B**) are often used in industrial processes. Copolymerization of butadiene and styrene is promoted by adding sulphinic acids and transition metal salts[3]. Silver benzenesulphinate is bactericidal and prevents the growth of skin fungi. Even during the short duration of normal washing, soaps containing arenesulphinato complexes of silver have a bactericidal effect[4]. The study of sulphinato complexes with regard to their structural diversity and their wide range of industrial applications are important.

TABLE 1. p$K_a$ Values of several sulphinic acids in water

| Sulphinic acid | p$K_a$ | Reference |
|---|---|---|
| $CH_3SO_2H$ | 2.28 | 5 |
| $CH_3(CH_2)_3SO_2H$ | 2.11 | 6 |
| $PhCH_2SO_2H$ | 1.45 | 6 |
| $Ph(CH_2)_2SO_2H$ | 1.89 | 6 |
| $Ph(CH_2)_3SO_2H$ | 2.03–2.05 | 6 |
| $Ph(CH_2)_4SO_2H$ | 2.23 | 6 |
| *p*-$AnSO_2H$ | 1.70 | 7 |
| *p*-$TolSO_2H$ | 1.55 | 7 |
| $PhSO_2H$ | 1.29 | 6 |
| | 1.45 | 7 |
| *p*-$ClC_6H_4SO_2H$ | 1.15 | 7 |
| *m*-$NO_2C_6H_4SO_2H$ | 0.55 | 7 |

In this chapter we describe the acidity, the hydrogen bonding of sulphinic acids and the property and structure of sulphinato metal complexes.

## II. ACIDITY OF SULPHINIC ACIDS

The p$K_a$ values of several sulphinic acids are listed in Table 1[5–7]. Sulphinic acids are stronger acids than the structually similar carboxylic acids, i.e. benzenesulphinic acid is stronger than benzoic acid and as acidic as dichloroacetic acid.

## III. THE HYDROGEN BONDS OF SULPHINIC ACIDS

Since sulphinic acids are labile and readily undergo disproportionation or decomposition, their isolation and characterization are generally performed after converting the acids to the corresponding salts. Only a few studies on the physical properties, particularly the hydrogen bonds, of sulphinic acids have been reported.

Reinhoudt and coworkers[8,9] reported the isolation of 2-sulphino-1,3-xylyl crown ethers **1** by the reaction of the 2-lithio-1,3-xylyl crown ethers of **2** with $SO_2$ at $-78\,^{\circ}C$. The 2-sulphino-1,3-xylyl crown ethers **1** were obtained in good yields (Scheme 1). The macrocyclic ring has some stabilizing effect on the sulphino group. However, long-term storage of **1**, even in the dark and under argon, results in decomposition. The structure of **1** was determined by X-ray diffraction. The crystal structures of **1a** and **1b** are given in Figures 1 and 2. The macrocyclic cavity of **1b** is filled by the sulphino OH group, which is engaged in a bifurcated hydrogen bond with two ether oxygens. The O ··· O distances and

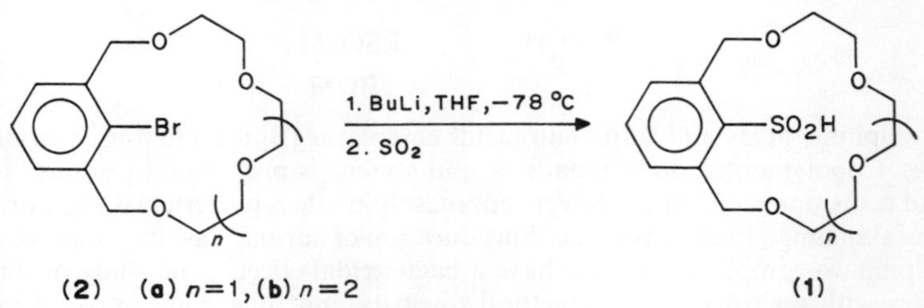


SCHEME 1

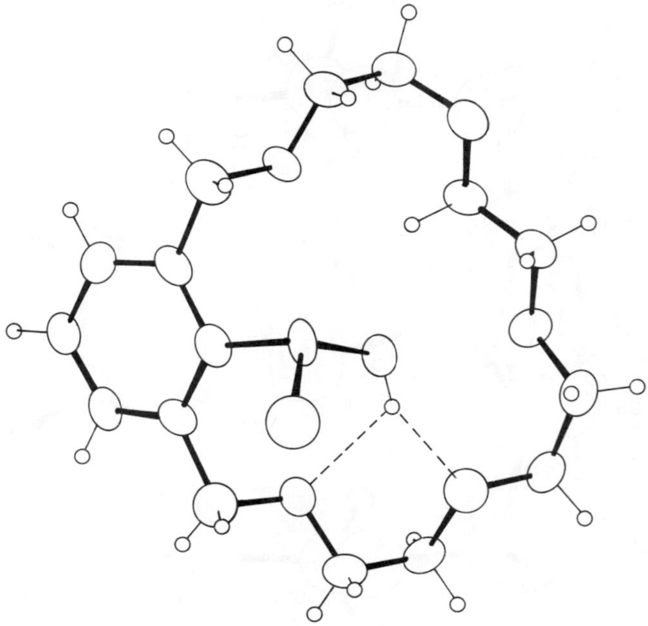

FIGURE 1. Crystal structure of **1b**. From Reference 9

the OH···O angles are 2.79 and 2.88 Å and 155° and 126°, respectively. Compound **1a** adopts the structure with the aromatic ring perpendicular to the mean macrocyclic plane (Figure 2). The hydrogen atom in the $SO_2H$ group is not located on O(21) [S(19—O(21): 1.54 Å vs. S(19)—O(20): 1.44 Å]. However, short distances between this sulphinyl oxygen atom and two oxygen atoms in the crown ether [O(21)···O(10): 2.89 Å and O(21)···O(13): 2.81 Å] indicate the presence of an intramolecular interaction between the proton in the sulphinic acid and the oxygen atoms of the macrocyclic ring.

Cram and coworkers reported the existence of monomer–dimer equilibrium of methanesulphinic acid in chloroform by osmometric molecular-weight determination (equation 1)[5].

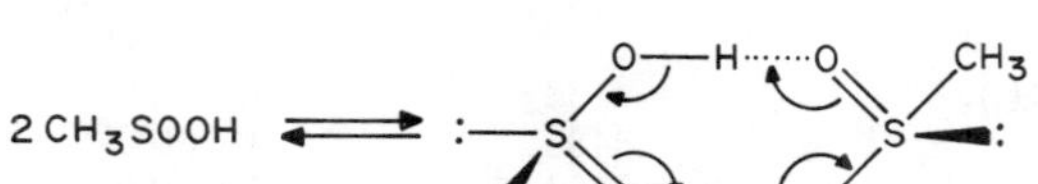


(1)

Engberts and Zuidema[10] studied the intramolecular hydrogen bonding between phenol and several sulphinic esters using IR spectroscopy. The stretching frequency shift ($\Delta\nu$) towards lower values for the OH stretching frequency of phenol observed upon addition of sulphinic esters indicates that these sulphinyl compounds are proton acceptors in hydrogen bonding. The results are shown in Table 2. Variations in either the concentration of phenol (0.005–0.02 mol liter$^{-1}$) or the concentration of the acceptor molecule (0.007–0.1 mol liter$^{-1}$) resulted in no significant changes of $\Delta\nu$. This suggests the formation of a 1:1 hydrogen bonding complex at infinite dilution. From the magnitude of the $\Delta\nu$ values of sulphinic esters it is concluded that in all cases the sulphinyl oxygen atom

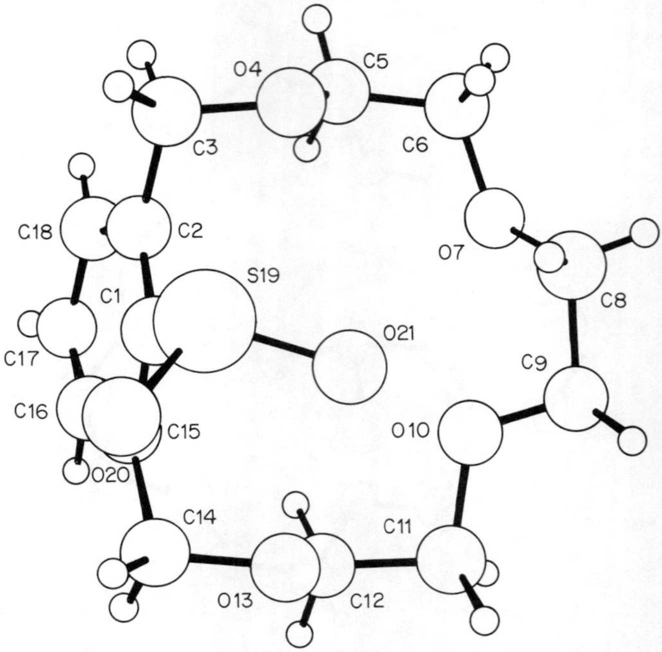


FIGURE 2. Structure of 2-sulphinyl-1,3-xylyl-15-crown-4 (**1a**). From Reference 8

TABLE 2. Intermolecular hydrogen bonding of phenol[a] with sulphinic esters in $CCl_4$ at 40 °C

| Sulphinic esters | $\nu_{S=O}$ ($cm^{-1}$) | $\Delta\nu^b$ ($cm^{-1}$) |
|---|---|---|
| *p*-AnS(O)$OCH_3$ | 1134 | 230 (0.007) |
| *p*-TolS(O)$OCH_3$ | 1137 | 225 (0.007)[c], 220 (0.004)[d] |
| | | 222 (0.100) |
| *p*-TolS(O)$OCH_2C(CH_3)_3$ | 1138 | 251 (0.007) |
| PhS(O)$OCH_3$ | 1136 | 218 (0.007) |
| *p*-$ClC_6H_4S$(O)$OCH_3$ | 1139 | 206 (0.007) |
| *p*-$NO_2C_6H_4S$(O)$OCH_3$ | 1142 | 187 (0.007), 185 (0.100) |
| $CH_3S$(O)$OCH_3$ | 1141 | 243 (0.007) |
| $CH_3S$(O)$OC_2H_5$ | 1140 | 268 (0.007), 265 (0.070) |
| $CH_3S$(O)$OC_4H_9$ | 1141 | 264 (0.007) |
| $CH_3S$(O)$OCH_2C(CH_3)_3$ | 1142 | 260 (0.007) |
| $C_2H_5S$(O)$OCH_3$ | 1136 | 250 (0.007) |
| $CH_3S$(O)$OC_6H_5$ | 1146 | 220 (0.007), 220 (0.100) |

[a]0.005 M phenol in $CCl_4$.
[b]Molarity of proton acceptor in parentheses.
[c]0.02 M phenol: $\Delta\nu = 220\,cm^{-1}$.
[d]0.01 M phenol: $\Delta\nu = 220\,cm^{-1}$.

is the proton acceptor site and not the oxygen atom of the alkoxy group attached to the sulphur. The IR data show that the structures of the groups R and R′ in the sulphinic esters RS(O)OR′ influence the $\Delta\nu$ values markedly, although the position of the S═O stretching band at about 1140 cm$^{-1}$ is only slightly affected. For five methyl *p*-substituted benzenesulphinates the $\Delta\nu$ values correlate linearly with Hammett's $\sigma_p$ constants.

## IV. SULPHINATO METAL COMPLEXES

### A. General

Generally sulphinato complexes, $RSO_2M$ (R = organic residue, M = central ion), may be classified into the five possible structures depending on the bonding mode of the $RSO_2^-$ ligand to the coordination centres[11–13], namely as S-sulphinate (I), an O-sulphinate (II), an O,O′-sulphinate (III), an O,S-sulphinate (IV) and an O-alkyl-S-sulphoxylate (V). Although complexes involving dinuclear and polynuclear structures with bridged $RSO_2$ groups as ligands are possible, those are little known and will not be described here.

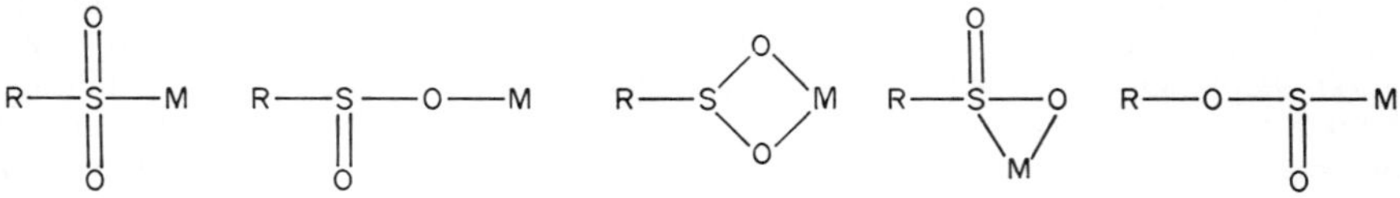

(I) (II) (III) (IV) (V)

The structure of the $SO_2$ insertion products is generally determined on the basis of infrared and proton NMR spectroscopic analysis, and X-ray crystallography. Structures (I) usually exhibit the sulfur–oxygen stretching frequencies in the ranges 1250–1100 cm$^{-1}$ as $\nu_{as}(SO_2)$ and 1100–1000 cm$^{-1}$ as $\nu_s(SO_2)$, which are shifted to higher wave numbers as compared with the absorptions at 1085–1050 cm$^{-1}$ and 1000–800 cm$^{-1}$ of the structures (II)–(V). The four structures (II)–(V) are normally difficult to distinguish from each other by IR.

The sulphinate ion may act as an electron donor via one of two ways, namely either as a 'soft' donor via the S-atom or as a 'hard' donor by one or both O-atoms. For preparation of *S-sulphinato* complexes, the central ion must have a soft character, such as the lowest possible oxidation number, a low positive charge, a large ionic radius, occupied outer orbitals and a high polarizability. On the other hand, for obtaining *O-sulphinato* complexes the central ion should have a hard character, i.e. a higher oxidation number, a higher effective charge, a small ionic radius, a low polarizability and a low oxidizability.

### B. Preparation of Sulphinato Metal Complexes

The general preparative methods for sulphinato complexes are as follows:

(1) Insertion of $SO_2$ into the metal–carbon bond of organometallic compounds. Insertion of sulphur dioxide is shown by equation 2, where M stands for a metal together with its ancillary ligands and R is an alkyl or a related $\sigma$-bonded carbon group. In this reaction two processes are conceivable, since $SO_2$ may attack the central ion either by the sulphur or by the oxygen atom, resulting, in the first case, in formation of an S-sulphinato (equation 3) and, in the second case, of an O-sulphinato complex (equation 4). Detailed studies on the mechanisms of $SO_2$ insertions into metal–carbon bonds have been reported[13–20,84–88].

$$R{-}M + SO_2 \longrightarrow RSO_2{-}M \qquad (2)$$

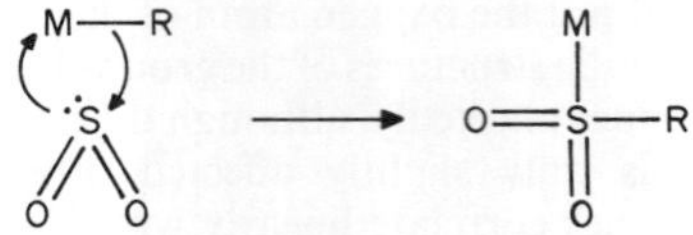

(3)

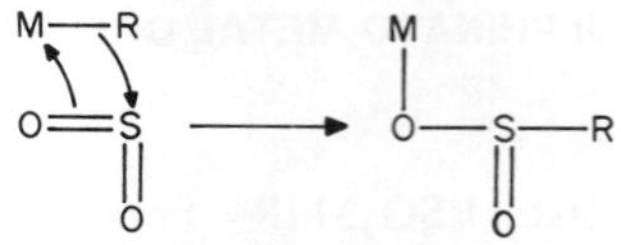

(4)

(2) Reaction of sulphonyl chlorides and sulphonic anhydrides with organometallic Lewis bases or neutral carbonyl–metal compounds. Treatment of a nucleophilic agent [Lewis base]$^-$ with sulphonyl halides or anhydrides in polar solvents gave the corresponding metal sulphinates[21–28] (equation 5). For example, the bis(trifluoromethanesulphinato)metal complexes of molybdenum, iron and nickel are obtained by treating the corresponding bis($\pi$-cyclopentadienylmetalcarbonyls) with $CF_3SO_2Cl$ in tetrahydrofuran[26] (equation 6). Monosulphinato, bis(sulphinato), tris(sulphinato) and tetrakis(sulphinato) complexes are obtained by reactions of soluble metal halides or acetates with sodium sulphinates in water, ethanol or THF at 25–80 °C (equation 7)[32]. Sodium sulphinates are easily obtained by *ipso* substitutions of 2-sulphonylpyridines and their N-oxides with alkoxides[82] (equation 8).

$$RSO_2X + [\text{Lewis base}]^- \xrightarrow[\text{THF}]{-80\,^\circ\text{C}} RSO_2[\text{Lewis base}] + X^- \quad (5)$$

$$(R = CF_3,\ p\text{-Tol};\ X = Cl)$$

$$[\text{Lewis base}]^- = [Mn(CO)_5]^-,\ [Re(CO)_5]^-,$$

$$[\pi\text{-}C_5H_5Fe(CO)_2]^-,\ [Co(CO)_3P(C_6H_5)_3]^-$$

$$2\,CF_3SO_2Cl + [\pi\text{-}C_5H_5M(CO)n]_2 \longrightarrow MCl_2 + (CF_3SO_2)_2M + 2n\,CO + 2\,C_5H_5 \quad (6)$$

$$(M = Mo,\ Fe,\ Ni;\ n = 1, 2, 3)$$

$$MX_n + n\,RSO_2Na \longrightarrow n\,NaX + M(O_2SR)_n \quad (7)$$

$$(M = Ag,\ Hg,\ Co,\ Ni;\ R = p\text{-Tol},\ Ph,\ CH_3)$$

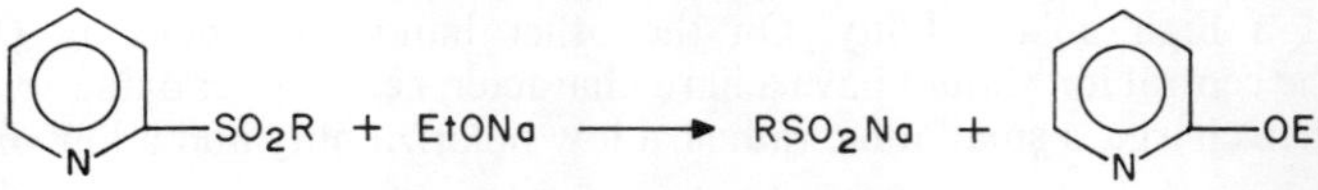

(8)

$$(R = Me,\ Octyl,\ i\text{-Pr},\ t\text{-Bu},\ PhCH_2,\ Ph)$$

(3) Reaction of metal halides with alkyl metal sulphides

$$R'_nMX_m + mRSO_2^- \longrightarrow R'_nM(SO_2R)_m + mX^- \quad (9)$$

$$[Bu_4N][M(CO_5)I] + RSO_2^- \longrightarrow [M(CO)_5\text{—}SO_2R]^- + Bu_4NI$$

$$(M = Cr,\ W;\ R = C_6H_5)^{29}$$

### C. Characterization and Properties of Sulphinatometal Complexes

#### *1. S-Sulphinato complexes*

The infrared spectroscopic data of various S-sulphinato complexes are listed in Table 3.

TABLE 3. IR spectra ($cm^{-1}$) of S-sulphinato complexes[a]

| Compound | $\nu_{as}(SO_2)$ | $\nu_s(SO_2)$ | Reference |
|---|---|---|---|
| $PhSO_2HgPh$ | 1175 vs, b | 1049 s | 30, 31 |
| $(TS)_2Hg$ | 1229 m, 1203 vs | 1040 vs | 32, 33 |
| $(CH_3SO_2)_2Hg$ | 1177 vs | 1061 vs | 32 |
| $(BS)_2Pd(OH_2)_2$ | 1195 s 1103 s | 1057 s | 34 |
| $Na[(BS)_2PdCl(OH_2)]$ | 1189 s 1099 s | 1051 s | 34 |
| $Li_2[(BS)_2PdCl_2]$ | 1200 s 1103 s | 1060 s | 34 |
| $TSIr(CO)[P(C_6H_5)_3]_2Cl_2$ | 1240 1220 | 1065 1055 | 35 |
| $TSPt[P(C_6H_5)_3]_2Cl$ | 1205 | 1043 | 36 |
| $CH_3SO_2Mn(CO)_3bipy$ | 1148 s | 1044 s | 37 |
| $CH_3SO_2Mn(CO)_4P(C_6H_5)_3$ | 1145 s | 1035 s | 37 |
| $(TS)_2Fe(bipy)_2$ | 1219 vs 1199 vs | 1034 m 1012 m | 38 |
| $(TS)_2Ni(bipy)_2$ | 1219 vs 1204 vs | 1035 s 1013 s | 38 |
| $(BS)_2Pt(NH_2C_6H_4CH_3)_2$ | 1164 | 1040 | 39 |
| $BSPdCl(py)_2$ | 1184 | 1032 | 39 |
| $MeSO_2Ir(CO)(Cl)_2P(C_6H_5)_3$ | 1220 | 1070 | 35 |

[a] TS = *p*-$CH_3C_6H_4SO_2^-$, BS = $C_6H_5SO_2^-$, bipy = bipyridine, py = pyridine.

The preparation, characterization, physical properties and structure of several S-sulphinato complexes are described below.

A mixture of the two isomers of phenylmercuric benzenesulphinate is obtained from the reaction of liquid sulphur dioxide with diphenylmercury in acetone (equation 10). The S-sulphinato complex is obtained from cold acetone, methyl ethyl ketone or methanol. Isomerization is readily effected by evaporating a solution of the S-sulphinato isomer in chloroform or acetone to dryness at room temperature, when the O-sulphinato isomer is obtained[30,31].

$$(C_6H_5)_2Hg + SO_2 \longrightarrow C_6H_5HgSO_2C_6H_5 \qquad (10)$$

Bis(sulphinato) complexes are formed in the reaction of sodium sulphinates with soluble mercuric halides in water or alcohol (equation 11)[32,33]. The bonding mode of the ligands to the Hg ion is particularly dependent on the water content of the compounds. The IR spectra indicate that the anhydrous mercuric complexes $Hg(SO_2R)_2$ probably have the S-bonded form.

$$2RSO_2Na + HgCl_2 \longrightarrow Hg(SO_2R)_2 + 2NaCl \qquad (11)$$
$$R = p\text{-Tol}, CH_3$$

Various sulphinato complexes of iridium(III) have been prepared by the oxidative addition of sulphonyl chlorides to square-planar iridium compounds[35,40]. In these compounds, the sulphinato group is connected to the metal by the sulphur atom. For example, alkyl- and arylsulphonyl chlorides were found to combine readily with **3** to afford a new type of complex formulated as iridium(III) sulphinate derivatives **4** (equation 12)[35]. The alkyl iridium sulphinates are thermally stable and remain unchanged after boiling for

24 h in solvents such as chloroform or toluene. However, some arylsulphinates undergo extrusion of $SO_2$ upon heating for 3 h in boiling toluene. For instance, the *p*-tolylsulphinate **4** is smoothly and quantitatively transformed into the *p*-tolyl derivative **5** (equation 13), in which the infrared absorptions corresponding to the $SO_2$ group are completely absent. Oxygen-bonded sulphinates or alkoxysulphinato complexes of iridium(III) are also known[41,42].

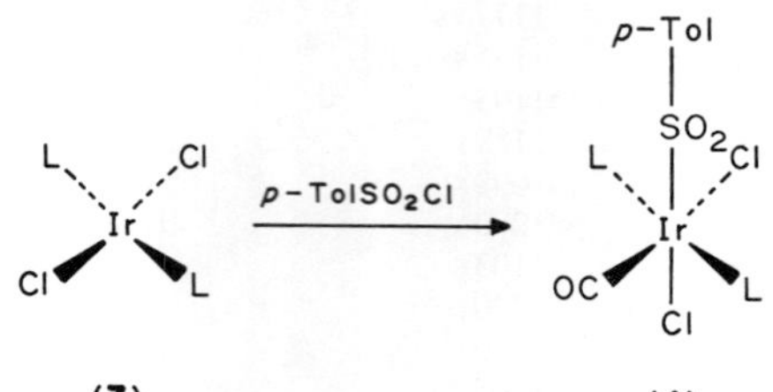

(12)

(3) (4)

$L = P(C_6H_5)_3$

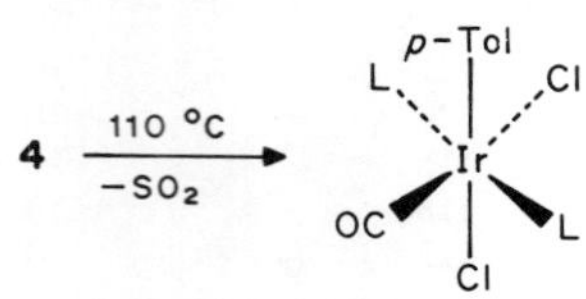

(13)

(5)

Tetrakis(triphenylphosphine)platinum(0) reacts smoothly with *p*-toluenesulphonyl chloride to give the complex **6** (equation 14)[36]. When the S-sulphinate (**6**) was heated, the aryl platinum complex was formed by losing $SO_2$.

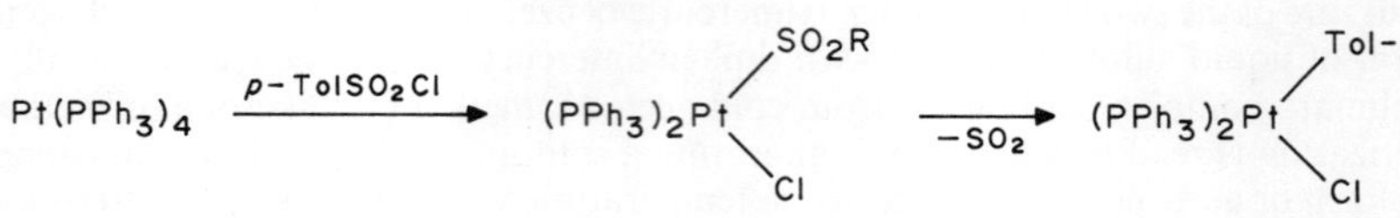

(6)

(14)

Formation of the complexes between $Fe(O_2SR)_2(OH)_2$ and 2,2′-bipyridine (=bipy) depends markedly on the solvent used, for example, in pyridine $Fe(O_2SR)_2(bipy)_2$ (**7**) is obtained. On heating to 115 °C in pyridine or at its melting point (175 °C), **7** is converted irreversibly into the thermodynamically more stable **8** as the S-sulphinato complex (equation 15)[38]. The splitting frequency in IR of $\nu_{as}$ and $\nu_s(SO_2)$ observed in **8** indicates *cis* bonding of the sulphinato ligand.

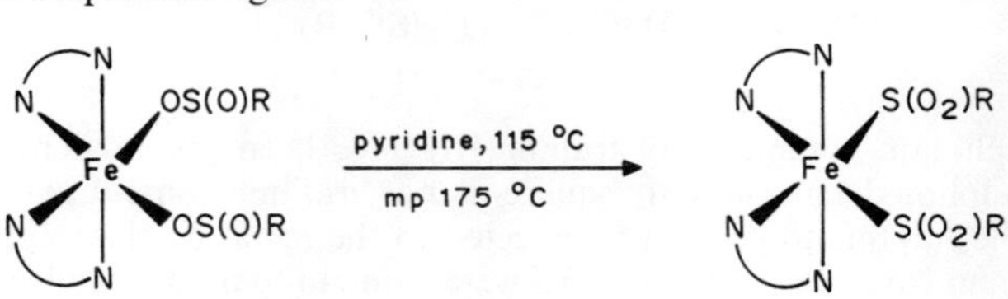

(15)

(7) (8)

Wojcicki and coworkers[43] studied the mechanism of sulphur dioxide insertion between the transition metal and alkyls and/or aryls of the type $h^5$-$C_5H_5Fe(CO)_2R$, $h^5$-$C_5H_5Mo(CO)_3R$, $Mn(CO)_5R$ and $Re(CO)_5R$ in liquid $SO_2$, and in organic solvents containing $SO_2$, using $^1H$ NMR spectroscopy. $^1H$ NMR spectroscopic data (see Table 4) indicate that the reactions of these compounds with $SO_2$ proceed via the intermediacy of the oxygen-bonded sulphinates, which subsequently rearrange to the thermodynamically stable and isolable sulphur-bonded sulphinates (equation 16). These O-sulphinato complexes are stable in the presence of $SO_2$, with stability being highest when $R=CH_3$. However, upon complete removal of $SO_2$ during their isolation, the O-sulphinato complexes immediately isomerize to the corresponding S-sulphinates. For example, in the $^1H$ NMR spectrum of $h$-$C_5H_5Fe(CO)_2CH_2Ph$ in liquid $SO_2$ recorded at $-18\,°C$, an AB quartet of the $CH_2$ resonance in the parent alkyl at $\tau$ 7.31 diminishes in intensity and an AB quartet and two new signals appear at $\tau$ 6.49 and 5.79 and grow, respectively. The signal at $\tau$ 5.79 is assigned to the $CH_2$ protons of the isolable $h^5$-$C_5H_5Fe(CO)_2SO_2CH_2Ph$. After storage of the solution for *ca* 24 h at $-20\,°C$, the $CH_2$ resonance at $\tau$ 7.31 disappeared and the quartet at $\tau$ 6.49 was barely discernible, while the intensity of the peak at $\tau$ 5.79 increased considerably. Furthermore, changes of the IR spectra are also consistent with these observations. IR spectra show lines at 1118 s, 828 m $\nu$(SO) for the O-sulphinate and at 1174 s, 1054 s, 1034 s $\nu$(SO) for the S-sulphinate complex.

$$R{-}M + SO_2 \longrightarrow [M]{-}OS(O)R \longrightarrow [M]{-}S(O)_2R \qquad (16)$$

$$[M] = h^5\text{-}C_5H_5Fe(CO)_2,\ \text{etc.}$$

Although it is well known that $SO_2$ inserts into the metal–alkyl or metal–aryl (M–R) bond to form an S-sulphinate, the reaction of $SO_2$ with $\sigma$-allyls is rather complicated. After observing that $(OC)_5MnCH_2CH{=}CH_2$ inserts $SO_2$ much faster than the analogous methyl or benzyl derivatives[45], Wojcicki and coworkers studied $SO_2$ insertion into a number of manganese carbonyl complexes containing unsymmetrically substituted allyl

TABLE 4. $^1H$ NMR spectra of metal alkyls and their $SO_2$-insertion products in liquid $SO_2$[a]

| | Chemical shift, $\tau$ | | |
|---|---|---|---|
| Compound | $CH_2(CH_3)$ | $h^5$-$C_5H_5$ | $C_6H_5$ |
| $h^5$-$C_5H_5Fe(CO)_2OS(O)CH_2Ph$ | 6.57, 6.41 AB ($J = 12.6$ Hz) | 4.91 s | 2.65 m |
| $h^5$-$C_5H_5Fe(CO)_2SO_2CH_2Ph$ | 5.79 s | 4.91 s | 2.53 s |
| $h^5$-$C_5H_5Mo(CO)_3OS(O)CH_2Ph$ | 6.49, 6.31 AB ($J = 12.5$ Hz) | 4.29 s | 2.75 m |
| $h^5$-$C_5H_5Mo(CO)_3SO_2CH_2Ph$ | 5.78 s | 4.32 s | 2.57 m |
| $Mn(CO)_5OS(O)CH_2Ph$ | 6.41, 6.23 AB ($J = 12.5$ Hz) | | 2.65 m |
| $Mn(CO)_5SO_2CH_2Ph$ | 5.70 s | | 2.55 s |
| $h^5$-$C_5H_5Fe(CO)_2OS(O)CH_3$ | 7.85 s | 4.75 s | |
| $h^5$-$C_5H_5Fe(CO)_2SO_2CH_3$ | 6.95 s | | |
| $Mn(CO)_5OS(O)CH_3$ | 7.72 s | | |
| $Mn(CO)_5SO_2CH_3$ | 6.91 s | | |
| $Re(CO)_5OS(O)CH_3$ | 7.73 s | | |
| $Re(CO)_5SO_2CH_3$ | 6.79 s | | |

[a]The intermediates are designated as the O-sulfinates.
Key: S, singlet; m, multiplet; AB, AB quartet.

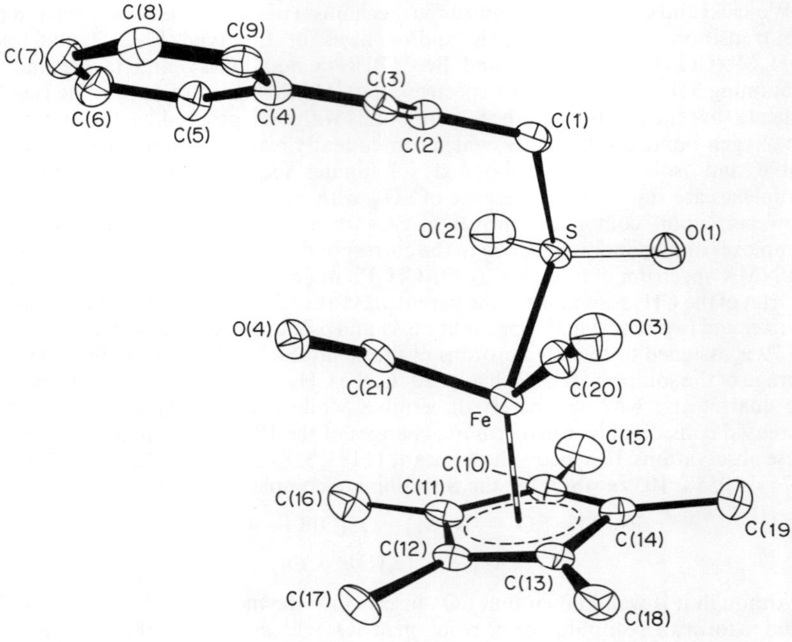


FIGURE 3. Crystal structure of $[\pi\text{-}C_5(CH_3)_5]Fe(CO)_2SO_2CH_2CH{=}CH(C_6H_5)$ (**9**). From Reference 44

groups[14,15]. Thus, $(OC)_5MnCH_2CH{=}CHCH_3$ and $(OC)_5MnCH_2CH{=}C(CH_3)_2$ were found to undergo rearrangement upon insertion of $SO_2$, yielding the products $(OC)_5MnSO_2CH(CH_3)CH{=}CH_2$ and $(OC)_5MnSO_2C(CH_3)_2CH{=}CH_2$, respectively. Churchill and Wormald[44] reported the results of an X-ray diffraction study on $[\pi\text{-}C_5(CH_3)_5]Fe(CO)_2SO_2CH_2CH{=}CH(C_6H_5)$ (**9**) prepared from $[\pi\text{-}C_5(CH_3)_5]Fe(CO)_2CH_2CH{=}CH(C_6H_5)$ and $SO_2$. The crystallographic analysis has confirmed that this molecule is formed by insertion of an $SO_2$ molecule into the iron-($\sigma$-allyl) bond without rearrangement of the allyl fragment (Figure 3). The X-ray structure indicates that the formallyl $d^6$ Fe(II) ion achieves the expected noble gas configuration by the donation of six electrons from the $\pi$-pentamethylcyclopentadienyl ion, two electrons from each of the carbonyl ligands and two electrons from the S-bonded sulphinate moiety. The iron atom is regarded as a pseudo-octahedrally coordinated form, since it is linked to three monodentate ligands and to a formally tridentate $\pi\text{-}C_5(CH_3)_5$ ligand.

On the other hand, Churchill and Wormald[46] reported an X-ray crystal structure of the product from $SO_2$ and 2-alkynyl complex of a transition metal, $\pi\text{-}C_5H_5Fe(CO)_2C_4H_5SO_2$ (**10**) [prepared from $\pi\text{-}C_5H_5Fe(CO)_2CH_2C{\equiv}CCH_3$ and $SO_2$]. The results demonstrate that the incoming $SO_2$ molecule does not insert into the iron–($\sigma$-alkynyl) linkage, but rather is involved in a sultine ring which is bonded to the iron atom via an iron–($\sigma$-vinyl) linkage to form the system $Fe{-}C{=}C(CH_3){-}S({=}O){-}O{-}CH_2$ (Figure 4).

Bruce and Redhouse[59] reported that the reaction of $C_6F_5SO_2Cl$ with the anion $[(\pi\text{-}C_5H_5)Fe(CO)_2)]^-$ gave the S-bonded sulphinato complex $C_6F_5SO_2Fe(CO)_2(\pi\text{-}C_5H_5)$, which was characterized by X-ray diffraction (Figure 5).

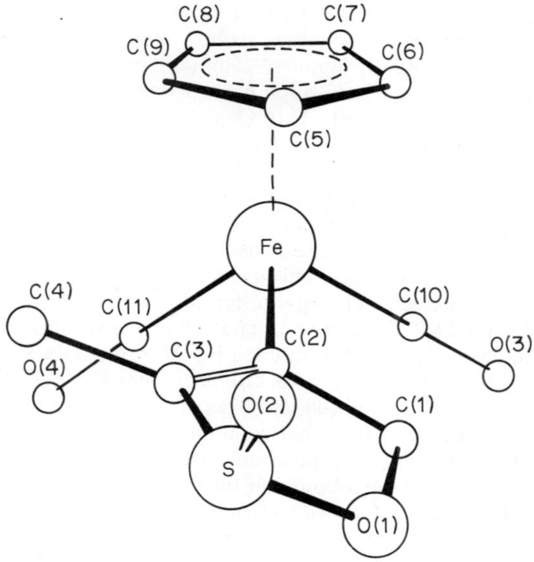


FIGURE 4. Crystal structure of $\pi$-$C_5H_5Fe(CO)_2$-$C_4H_5SO_2$ (**10**). From Reference 46

It is well established that in cobalt(III) chemistry S-bonded sulphite produces a specific and dramatic labilization of the ligand situated *trans* to it[47]. The chemistry of octahedral complexes containing S-sulphinato ligands is little known[12], but an S-bonded benzene-sulphinate has been shown to exhibit a *trans* effect in a Pd(II) complex[34].

Elder and coworkers[52] reported the synthesis and detailed X-ray structural characterization of an S-bonded sulphinic acid complex of cobalt(III). A bis(ethylenediamine)-

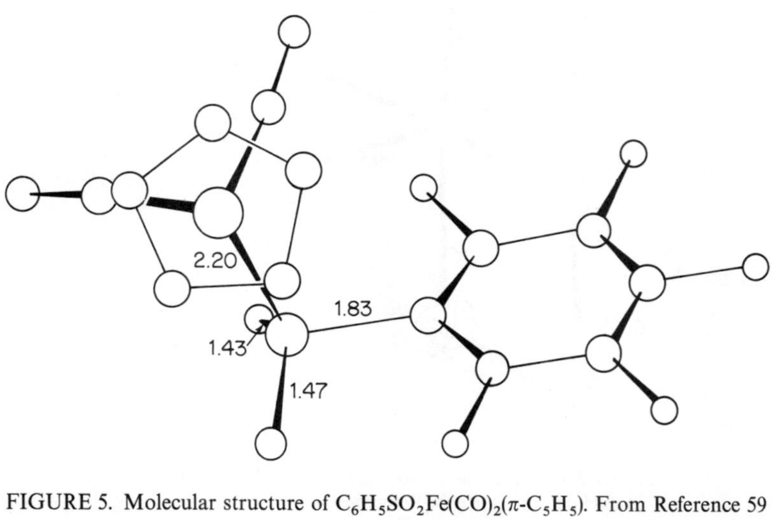


FIGURE 5. Molecular structure of $C_6H_5SO_2Fe(CO)_2(\pi$-$C_5H_5)$. From Reference 59

cobalt(III) complex of cystein shows that of the three potential donor groups N, O, S, both the amino and thiolate functional groups coordinate to the Co(III)[48–51]. Treatment of (2-mercaptoethylamine-*N*,*S*)bis(ethylenediamine)cobalt(III) with excess hydrogen peroxide provides (2-sulphinatoethylamine-*N*,*S*)bis(ethylenediamine)-cobalt(III), $[Co(en)_2(S(O)_2CH_2CH_2NH_2)]\ (NO_3)(ClO_4)$ (**11**) in good yield.

The visible–UV absorption spectrum of $[Co(en)_2(S(O)_2CH_2CH_2NH_2)]^{2+}$ ion observed at $\lambda_{max}(\varepsilon)$ 432 (220), 288 (14.200) exhibits d–d bands characteristic of cobalt(III) complexes as well as an intense ligand-to-metal charge transfer (LTMCT) band characteristically arising from coordination of the sulphur atom to the potentially oxidizing Co(III) centre. The nearly identical positions of the UV–LTMCT bands indicate that in this system $RSO_2^-$ works as an as effective reductant as $RS^-$, which in turn implies that for both complexes the electron being transferred is the one involved in the Co—S $\sigma$-bond (i.e. a $\sigma_L$–$\sigma_M^*(e_g)$ LTMCT process[53]. The relative positions of the visible d–d absorption bands indicate that the coordinated $RSO_2^-$ provides a stronger ligand field than the coordinated $RS^-$. The infrared spectrum of $[(en)_2Co(S(O)_2(CH_2CH_2NH_2)]I_2$ exhibits bands at 1220 as $\nu_{as}(SO_2)$ and 1080 cm$^{-1}$ as $\nu_s(SO_2)$, respectively. The shift of these bands to higher frequencies, relative to their positions in the unbound sulphinato anion, confirms that sulphur–oxygen band positions may be used to indicate whether a coordinated sulphinic acid is in the S-bound or the O-bound form. A single-crystal X-ray structure analysis of **11** shows the following features (Figure 6): (1) The sulphur atom of **11** has an oxidation number by four units higher than in the starting thiolate complex. (2) The primary coordination sphere (octahedral, one sulphur and five nitrogen atoms) of the cobalt atom remains intact throughout the oxidation process. (3) The $RSO_2^-$ group induces a ground-state *trans* effect of 0.049(5) Å [average *cis* Co—N bond length

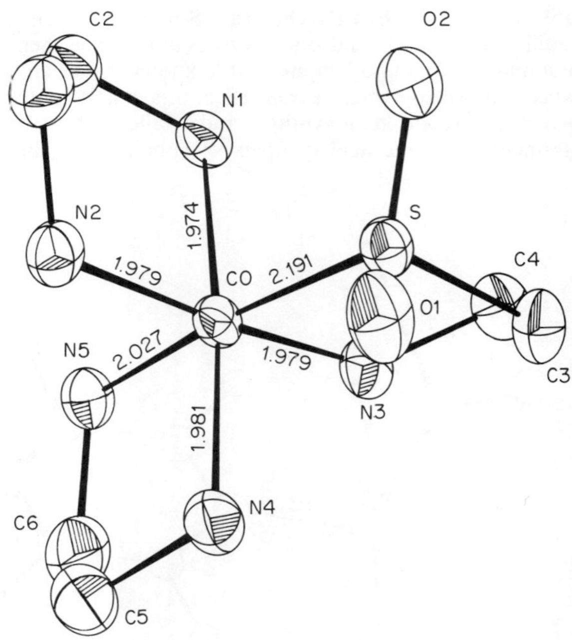


FIGURE 6. Molecular structure of $[(en)_2Co(S(O)_2$-$CH_2NH_2)]^{2+}$ (**11**). From Reference 52

1.978(3) Å; *trans* Co—N bond length 2.027(4) Å]. (4) The hydrogen bonds are formed between the coordinated amino hydrogen atoms and the oxygen atoms of the nitrate anions. One of the sulphinato oxygens [O(2)] forms a hydrogen bond with an amino hydrogen atom on an adjacent cation, which may be reflected in the fact that the S—O(2) distance [1.476(4) Å] is longer than the S—O(1) distance [1.456(4) Å].

Two recent investigations indicate that in cobalt(III) complex the extent of the sulphur-induced kinetic *trans* effect (KTE) is correlated with the extent of the concomitant ground-state structural *trans* effect (STE). Thus, S-bonded sulphinic acids ($RSO_2^-$) exert a kinetic *trans* effect (KTE) in bis(dimethylglyoximato)cobalt(III) complexes which is smaller than that exerted by $SO_3^{2-}$ in the same complexes[47,55], while the S-bonded sulphinic acid in $[(en)_2Co(S(O)_2CH_2CH_2NH_2)]^{2+}$ exerts a ground-state structural *trans* effect (STE) which is smaller than that exerted by $SO_3^{2-}$ in $[(NH_3)_5\text{-}CoSO_3]^{+}$[52].

Deutsch and coworkers reported the single-crystal X-ray structure analysis of (*p*-toluenesulphinato-S)pentaamminecobalt(III) perchlorate monohydrate (**12**), $[(NH_3)_5CoS(O)_2C_6H_4CH_3](ClO_4)_2{\cdot}H_2O$[56]. An X-ray structure (Figure 7) shows that the central cobalt(III) is ligated by one sulphur five nitrogen atoms in a closely octahedral arrangement. The salient structural feature of the complex is that the Co—N bond *trans* to the sulphur is significantly longer than the average of the *cis* Co—N bonds in the same complex. In $[(NH_3)_5CoS(O)_2C_6H_4CH_3]^{2+}$, the structural *trans* effect (STE) is 0.054(12) Å. Two other sulphinato complexes have STEs of 0.049(5) and 0.060() Å, suggesting that 0.054(6) Å is the best estimation of the sulphinato STE.

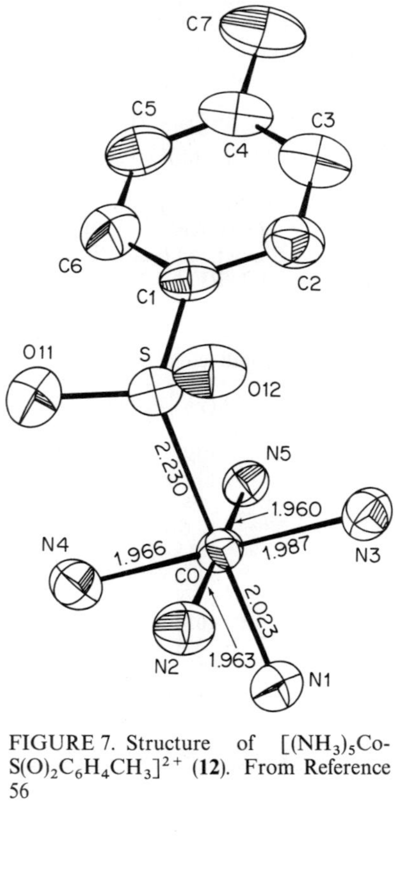


FIGURE 7. Structure of $[(NH_3)_5Co\text{-}S(O)_2C_6H_4CH_3]^{2+}$ (**12**). From Reference 56

Furthermore, a kinetic *trans* effect (KTE) and a structural *trans* effect (STE) of sulphinato–cobalt(III) complexes are discussed in detail[57,58].

### 2. O-Sulphinato complexes

Infrared spectroscopic data of several O-sulphinato complexes are listed in Table 5. The characterization and structure of O-sulphinato complexes are described below.

Langs and Hare reported the crystal structure of bis(*p*-toluenesulphinato)copper(II) tetrahydrate, **13**[65]. An X-ray structure of $Cu(CH_3C_6H_4SO_2)_2 \cdot 4H_2O$, **13**, indicates that the ligand field of the copper ion is nearly a tetragonally elongated octahedron. A water molecule and the sulphinate oxygen form the approximate tetragonal plane of the centrosymmetric complex with bond distances of 2.020(4) and 1.973(4) Å, respectively, and the apical position is occupied by a water molecule at 2.347(4) Å. The toluenesulphinate group is non-planar and the bonds about the S atom are disposed in a trigonal pyramidal array. The S—O bond length directed to the copper ion is 1.541(4) Å. The X-ray structure is illustrated in Figure 8.

Edmondson and Newlands reported the $SO_2$ insertion reaction into a tin–carbon bond[62]. The compound $[C_5H_5Fe(CO)_2Sn(SO_2Ph)_2]$ was obtained as orange-yellow crystals by passing $SO_2$ into bis($\pi$-cyclopentadienyldicarbonyl-iron)diphenyltin in benzene at room temperature. The structure of the insertion product was confirmed by an independent synthesis from bis($\pi$-cyclopentadienyldicarbonyliron)dichlorotin and sodium benzenesulphinate in methanol (equation 17). From the X-ray study[63], it is clear that insertion of $SO_2$ into the parent compound takes place in the Sn—C bonds to give an Sn–O–(SO)–C unit. The geometry at the S atom is approximately tetrahedral in each case with a lone pair of electrons presumably occupying the fourth arm of the tetrahedron.

$$[C_5H_5Fe(CO)_2]_2SnCl_2 + 2PhSO_2Na \longrightarrow [C_5H_5Fe(CO)_2]_2Sn(SO_2Ph)_2 + 2NaCl \qquad (17)$$

The insertion of $SO_2$ into the W—R bond of $\eta^5$-$C_5H_5W(CO)_3R$ (R = $CH_3$, **14**) in liquid $SO_2$ is markedly promoted by the Lewis acids $BF_3$ and $SbF_5$. The promoted reaction proceeds to give the corresponding Lewis acid stabilized O-sulphinato complexes, $\eta^5$-$C_5H_5W(CO)_3[OS(OBF_3$ or $OSbF_5)R]$ ($BF_3$: R = $CH_3$, **16**, $SbF_5$: R = $CH_3$, **17**), which

TABLE 5. IR spectra ($cm^{-1}$) of O-sulphinato–metal complexes[a]

| Compound | $\nu_{as}(SO_2)$ or $\nu(SO)$ | $\nu_s(SO_2)$ or $\nu_{as}(SOM)$ | Reference |
|---|---|---|---|
| $C_6H_5CH_2SO_2HgCH_2C_6H_5$ | 1050 s | 878 s | 16, 60 |
| $C_6H_5SO_2HgC_6F_5$ | 1035 m | 828 vs | 31 |
| $(TS)_2Cu(OH_2)_4$ | 998 | 938 | 65 |
| $(TS)_2Cd$ | 1031 s | 924 m | 32 |
| | 1015 s | 904 m | |
| $[\pi$-$C_5H_5Fe(CO)_2]_2Sn(O_2SC_6H_5)_2$ | 1103 | 869 | 62, 63 |
| | 1088 | 853 | |
| $(TS)_2Fe(bipy)_2$ | 1054 vs | 918 vs | 38 |
| $(TS)_2Ni(bipy)_2$ | 1055 vs | 958 m | 11 |
| | | 943 m-s | |
| $(CF_3SO_2)_2Ni(bipy)_2$ | 1180 sh | 985 m-s | 11 |
| | 1164 s-vs | | |
| | 1145 vs | | |

[a] TS = $p$-$CH_3C_6H_4SO_2^-$, bipy = bipyridine.

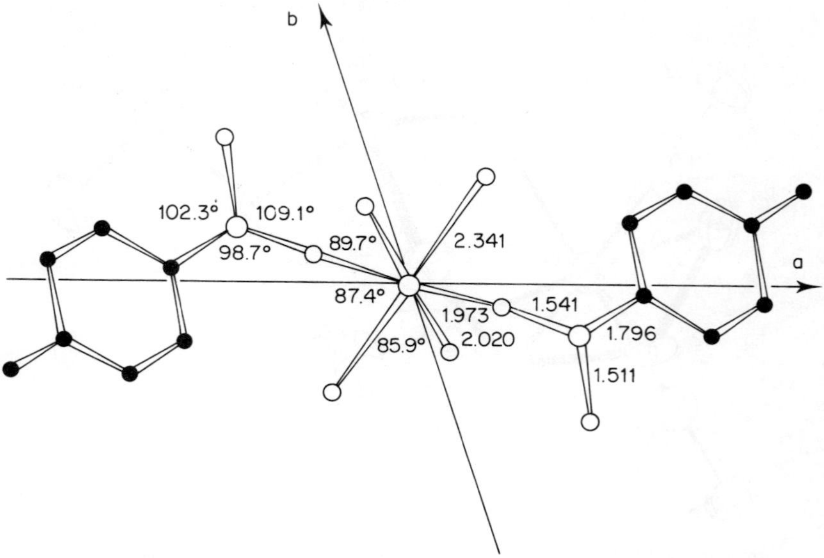


FIGURE 8. Structure of $Cu(CH_3C_6H_4SO_2)_2{\cdot}4H_2O$ (**13**). From Reference 65

were characterized by elemental analysis and infrared and $^1H$ NMR spectroscopy; by contrast, the insertion of $SO_2$ alone continues to yield the S-sulphinato complex, $\eta^5$-$C_5H_5W(CO)_3S(O)_2R$ ($R = CH_3$, **15**)[83].

| | |
|---|---|
| (**15**) $\eta^5$-$C_5H_5W(CO)_3S(O)_2CH_3$ | 1188, 1054 |
| (**16**) $\eta^5$-$C_5H_5W(CO)_3[OS(OBF_3)CH_3]$ | 870 |
| (**17**) $\eta^5$-$C_5H_5W(CO)_3[OS(OSbF_5)CH_3]$ | 990, 870 m<br>845 sh |

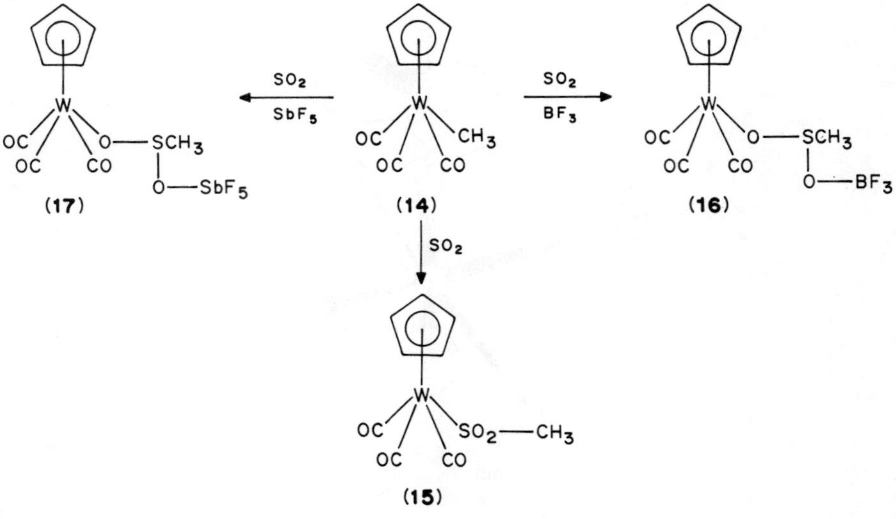

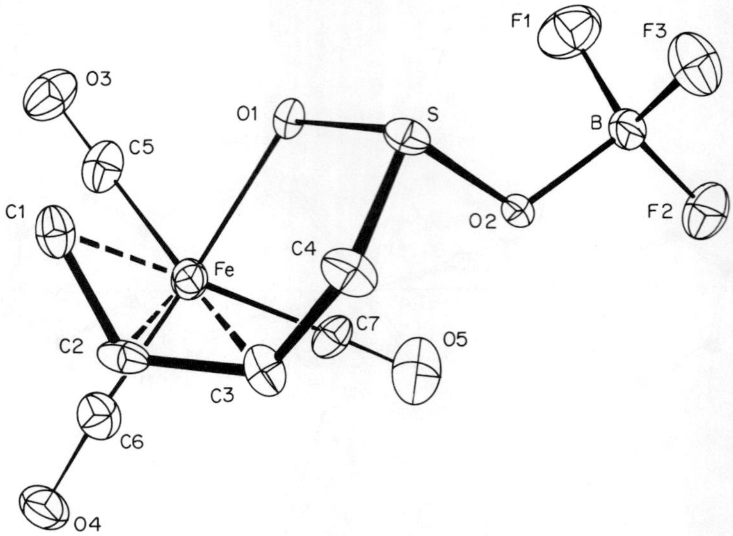


FIGURE 9. Molecular structure of $C_4H_6Fe(CO)_3SO_2BF_3$ (**18**). From Reference 75

Two examples of Lewis acid enhancement of electrophilic reactivity of $SO_2$ have been reported. Sulphur dioxide adds to cyclooctatetraene in the presence of $SbF_5$ at $-70\,^{\circ}C$[61,64]; in contrast, $SO_2$ alone appears unreactive[61]. In the presence of $BF_3$, $SO_2$ undergoes addition with $\eta^4$-$C_4H_4Fe(CO)_3$ to afford ($\eta^3$-$CH_2CHCHCH_2$)$Fe(CO)_3[OS(OBF_3)]$ (**18**)[69], which was characterized by X-ray crystallography[69,75]. An X-ray crystal structure of $C_4H_6Fe(CO)_3{\cdot}SO_2{\cdot}BF_3$ (**18**) is illustrated in Figure 9. The central iron atom is linked to three carbonyl ligands, a $\pi$-allyl system and an

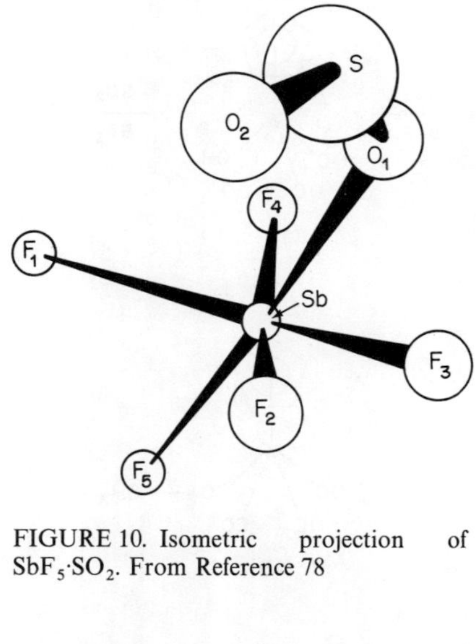


FIGURE 10. Isometric projection of $SbF_5{\cdot}SO_2$. From Reference 78

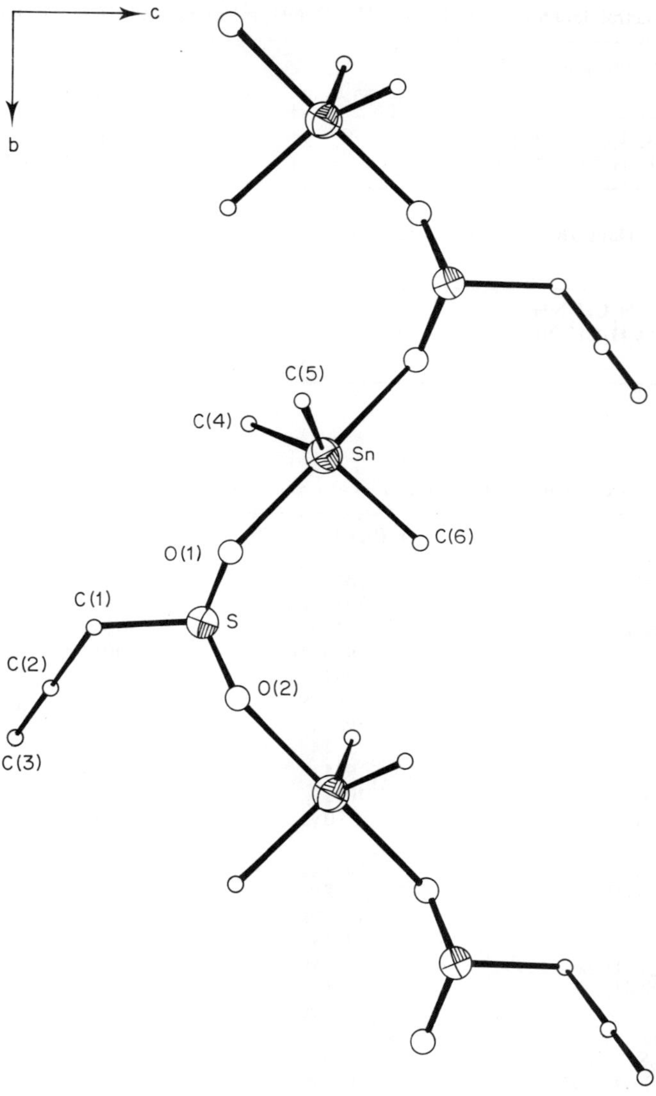


FIGURE 11. Structure of $C_6H_{12}SnSO_2$ polymer. From Reference 66

oxygen atom of the inserted sulphur dioxide molecule [Fe–O(1) = 2.00 ± 0.01 Å]. Sulphur dioxide reacted with antimony(V) fluoride to produce the 1:1 adduct, $SbF_5{\cdot}SO_2$, which was characterized by X-ray diffraction (Figure 10)[78]. Numerous studies were performed on the chemistry of sulphur dioxide complexes, $SO_2M$[81], which are not described here.

Ginderow and Huber[66] reported that the crystal structure of $C_6H_{12}SnSO_2$ consists of infinite chains. These chains are formed by oxygen bridges linking the tin and the sulphur atoms. The tin atoms are five-coordinated as shown in Figure 11.

TABLE 6. Spectral data for $Co(en)_2[S(O)_2CH_2CH_2NH_2]^{2+}$ and $Co(en)_2[OS(O)CH_2CH_2NH_2]^{2+}$

A. Electronic Absorption ($H_2O$, 25 °C)

| Complex | $\lambda_{max}(\varepsilon)$ | |
|---|---|---|
| $Co(en)_2(S(O)_2CH_2CH_2NH_2)(ClO_4)_2$ | 432 (220) | 288 (14,200) |
| $Co(en)_2(OS(O)CH_2CH_2NH_2)(ClO_4)_2$ | 512 (134) | 326 (4,100) |

B. Vibrational Data (IR; KBr pellet; Raman, $H_2O$, 647.1 nm)

| Complex | IR[a] ($cm^{-1}$) | Raman[b] ($cm^{-1}$) |
|---|---|---|
| $Co(en)_2(S(O)_2CH_2CH_2NH_2)^{2+}$ | 1190 s | 1204 w |
| $Co(en)_2(OS(O)CH_2CH_2NH_2)^{2+}$ | 950 m | 950 w |
| | 1030–1037 s | 1038 w |

[a]Halide salts.
[b]$ClO_4^-$ salts.

TABLE 7. IR spectra ($cm^{-1}$) of metal–sulphinato complexes[a]

| Compound | $\nu_{as}(SO_2)$ or $\nu(SO)$ | $\nu_s(SO_2)$ or $\nu_{as}(SOM)$ | Reference |
|---|---|---|---|
| $CH_3SO_2ZnCH_3$ | 1005 s | 955 m | 68 |
| $PhSO_2Ag$ | 1027 vs | 956 s | 12 |
| $PhSO_2Ag(OH_2)_n$ | 1020 vs | 970 s | 12 |
| $(TS)_2Cr(OH_2)_2$ | 970 vs,b | 940 vs,b | 26 |
| $(TS)_2Mn(OH_2)_2$ | 994 s | 952 s | 26 |
| $(TS)_2Fe(OH_2)_2$ | 994 vs | 963 s | 26 |
| $(TS)_2Co(OH_2)_2$ | 983 s | 963 m-s | 26 |
| | | 938 vs | |
| $(TS)_2Ni(OH_2)_2$ | 984 s | 945 vs | 26 |
| $(TS)_2Cu(OH_2)_2$ | 1011 | 953 | 70 |
| $(TS)_2Cd(OH_2)_2$ | 990 vs | 965 sh | 32, 71 |
| | | 947 vs | |
| $(TS)_2Hg(OH_2)_2$ | 1037 vs | 980 s | 32 |
| $(CF_3SO_2)_2Ni(THF)_2$ | 1192 vs | 1053 s | 26 |
| | 1166 vs | 1040 s | |
| $(PhSO_2)_2Pb$ | 991 s | 932 vs | 72 |
| $[(CH_3)_3SnO_2SCH_3]_4$ | 993 vs | 945 m-s | 73,74 |
| $(CH_3)_3PbO_2SCH_3$ | 1001 vs | 935 s | 54 |
| | 987 vs | 916 s | |
| | 970 vs | | |
| $[(CH_3)_3SnO_2SC_6H_5]_n$ | 994 vs | 957 vs | 76 |
| $(CH_2{=}CH)_3SnO_2SCH{=}CH_2$ | 1001 vs | 936 vs | 12 |
| $Ph_2Sn(O_2SPh)_2$ | 958 s | 936 vs | 72 |
| | 945 sh | | |
| $Ph_2Pb(O_2SPh)_2$ | 958 | 937 | 77 |
| $(CH_3)_2Sn(O_2SCH_3)_2$ | 974 s | 941 m | 72 |
| $[(PhCH_2SO_2Mn(CO)_3py]_n$ | 1020 s | 958 s | 37 |
| | 1013 s | | |
| $[TSNi(bipy)_2]Cl$ | 1032 s | 957 m | 79 |
| | 1019 s | | |
| $(TS)_2Mn(en)_2$ | 1012 vs | 977 m | 80 |
| $(TS)_2Febipy$ | 1025 sh | 972 s-vs | 11 |
| | 1015 vs | | |

[a]TS = $p$-$CH_3C_6H_4SO_2^-$, bipy = bipyridine.

The ability of sulphinic acids, $RSO_2H$, to coordinate to main-group elements and to transition-metal ions is well known, and their structures and bonding modes have been studied in some detail[12].

The bonding is performed by one or both oxygen atoms, by sulphur or by a sulphur–oxygen $\pi$-system. However, all the reported Co(III) complexes containing sulphito, sulphinato or sulphenato ligands show only Co—S bonding[48].

Recently, Adamson and coworkers[67] obtained by a photochemical method a *robust* Co(III) complex containing an O-bonded sulphinato ligand and characterized it (Table 6). The visible–UV absorption spectra and infrared and Raman spectra indicate the following characteristic properties: (1) The photoproduct is confirmed to be the O-bonded isomer by analyzing the absorption spectrum. The position of the first ligand field absorption band of CoOSON is characteristic of a Co(III) complex having one oxygen and five nitrogens coordinated. The lack of a characteristic intense charge-transfer (CT) band at ~285 nm confirms that the sulphur is *not* coordinated, but a new CT band at 326 nm indicates that the sulphinate moiety can still interact with the Co(III) centre. This would be true for the O-bonded isomer. (2) The vibrational data obtained from the infrared and Raman spectra of the two complexes are too complicated to be analyzed due to the presence of a large number of ligand vibrations. It can be seen that the strong vibrational band at 1190 $cm^{-1}$, likely due to the asymmetric O═S═O stretching, is not present in the product. However, two new vibrational bands appear in the product at ~1035 and 950 $cm^{-1}$, attributable to the S═O stretching mode and to the asymmetric Co–O–S stretching mode of an O-sulphinato ligand, respectively.

### 3. *O,O′-Sulphinato complexes*

The structures of O,O′-sulphinato complexes are not discussed in detail, because it is difficult to distinguish between the structures **C** and **D**. Only IR spectral data are listed in Table 7.

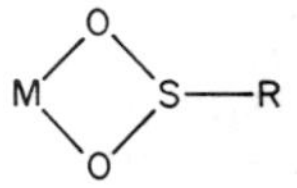


**(C)**

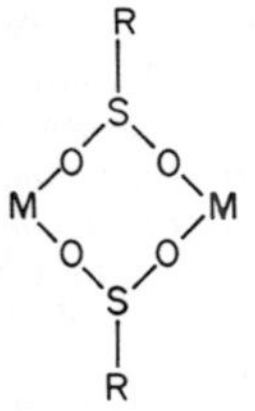


**(D)**

## V. REFERENCES


1. S. Oae and N. Kunieda, in *Organic Chemistry of Sulfur* (Ed. S. Oae), Chap. 11, Plenum, New York, 1977.
2. W. E. Truce and A. M. Murphy, *Chem. Rev.*, **48**, 69 (1951).
3. C. S. Marvel and N. A. Meinhardt, *J. Polym. Sci.*, **6**, 733 (1951).
4. Permachem Corp., *Chem. Abstr.*, **53**, 12716c (1959).
5. F. Wudl, D. A. Lightner and D. J. Cram, *J. Am. Chem. Soc.*, **89**, 4099 (1967).
6. P. Rumpf and J. Sadet, *Bull. Soc. Chim. Fr.*, 447 (1958).
7. Y. Ogata, Y. Sawaki and M. Isono, *Tetrahedron*, **26**, 731 (1970).
8. M. S.-Ptasinska, P. Telleman, V. M. L. J. Aarts, P. D. J. Grootenhuis, J. v. Eerden, S. Harkema and D. N. Reinhoudt, *Tetrahedron Lett.*, **28**, 1937 (1987).
9. M. S.-Ptasinska, V. M. L. J. Aarts, R. J. M. Egberink, J. v. Eerden, S. Harkema and D. N. Reinhoudt, *J. Org. Chem.*, **53**, 5484 (1988).
10. J. B. F. N. Engberts and G. Zuidema, *Recl. Trav. Chim. Pays-Bas*, **89**, 1202 (1970).

11. E. Lindner and G. Vitzthum, *Chem. Ber.*, **102**, 4062 (1969).
12. G. Vitzthum and E. Lindner, *Angew. Chem., Int. Ed. Engl.*, **10**, 315 (1971).
13. A. Wojcicki, *Adv. Organomet. Chem.*, **12**, 31 (1974).
14. F. A. Hartman, P. J. Pollick, P. L. Downs and A. Wojcicki, *J. Am. Chem. Soc.*, **89**, 2493 (1967).
15. F. A. Hartman and A. Wojcicki, *Inorg. Chim. Acta*, **2**, 289 (1968).
16. W. Kitching, B. Hegarty, S. Winstein and W. G. Young, *J. Organomet. Chem.*, **20**, 253 (1969).
17. W. Kitching, C. W. Fong and A. J. Smith, *J. Am. Chem. Soc.*, **91**, 767 (1969).
18. C. W. Fong and W. Kitching, *J. Organomet. Chem.*, **22**, 95 (1970).
19. C. W. Fong and W. Kitching, *J. Organomet. Chem.*, **22**, 107 (1970).
20. C. W. Fong and W. Kitching, *J. Organomet. Chem.*, **21**, 365 (1970).
21. E. Lindner, H. Weber and G. Vitzthum, *J. Organomet. Chem.*, **13**, 431 (1968).
22. E. Lindner and H. Weber, *Angew. Chem., Int. Ed. Engl.*, **5**, 727 (1966).
23. E. Lindner and H. Weber, *Z. Naturforsch.*, **22b**, 1243 (1967).
24. E. Lindner, *Angew. Chem., Int. Ed. Engl.*, **9**, 114 (1970).
25. E. G. W. Hieber, W. Beck and E. Lindner, *Z. Naturforsch.*, **16b**, 229 (1961).
26. E. Lindner, G. Vitzthum and H. Weber, *Z. Anorg. Allg. Chem.*, **373**, 122 (1970).
27. R. B. King, S. L. Stafford, P. M. Treichel and F. G. A. Stone, *J. Am. Chem. Soc.*, **83**, 3604 (1961).
28. H. Alper, *Tetrahedron Lett.*, **16**, 1239 (1969).
29. F. A. Hartman and A. Wojcicki, *Inorg. Chem.*, **7**, 1504 (1968).
30. G. B. Deacon and P. W. Felder, *J. Am. Chem. Soc.*, **90**, 493 (1968).
31. G. B. Deacon and P. W. Felder, *Aust. J. Chem.*, **22**, 549 (1969).
32. E. Lindner, G. Vitzthum, D. Langner and I.-P. Lorenz, *Angew. Chem., Int. Ed. Engl.*, **9**, 160 (1970).
33. G. B. Deacon, *Aust. J. Chem.*, **20**, 1367 (1967).
34. B. Chiswell and L. M. Venanzi, *J. Chem. Soc. (A)*, 1246 (1966).
35. J. P. Collman and W. R. Roper, *J. Am. Chem. Soc.*, **88**, 180 (1966).
36. C. D. Cook and G. S. Jauhal, *Can. J. Chem.*, **45**, 301 (1967).
37. F. A. Hartman and A. Wojcicki, *Inorg. Chim. Acta*, **2**, 351 (1968).
38. E. Lindner, I.-P. Lorenz and G. Vitzthum, *Angew. Chem., Int. Ed. Engl.*, **10**, 193 (1971).
39. C. W. Dudley and C. Oldham, *Inorg. Chim. Acta*, **3**, 3 (1969).
40. A. J. Deeming and B. L. Shaw, *J. Chem. Soc. (A)*, 1128 (1969).
41. C. A. Reed and W. R. Roper, *J. Chem. Soc., Chem. Commun.*, 1556 (1971).
42. T. A. George and D. D. Watkins Jr., *Inorg. Chem.*, **12**, 398 (1973).
43. S. E. Jacobson, P. R. Rohrwig and A. Wojcicki, *Inorg. Chem.*, **12**, 717 (1973).
44. M. R. Churchill and J. Wormald, *Inorg. Chem.*, **10**, 572 (1971).
45. F. A. Hartman and A. Wojcicki, *J. Am. Chem. Soc.*, **88**, 844 (1966).
46. M. R. Churchill and J. Wormald, *J. Am. Chem. Soc.*, **93**, 354 (1971).
47. J. M. Palmer and E. Deutsch, *Inorg. Chem.*, **14**, 17 (1975).
48. C. P. Sloan and J. H. Krueger, *Inorg. Chem.*, **14**, 1481 (1975).
49. M. P. Schubert, *J. Am. Chem. Soc.*, **55**, 3336 (1933).
50. L. S. Dollimore and R. D. Gillard, *J. Chem. Soc., Dalton Trans.*, 933 (1973).
51. L. S. Dollimore and R. D. Gillard, *J. Chem. Soc., Dalton Trans.*, 369 (1975).
52. B. A. Lange, K. Libson, E. Deutsch and R. C. Elder, *Inorg. Chem.*, **15**, 2985 (1976).
53. V. Balzani and V. Carassiti, *Photochemistry of Coordination Compounds*, Academic Press, New York, 1970, p. 57.
54. U. Stahlberg, R. Gelius and R. Muller, *Z. Anorg. Allg. Chem.*, **355**, 230 (1967).
55. L. Seibles and E. Deutsch, *Inorg. Chem.*, **16**, 2273 (1977).
56. R. C. Elder, M. J. Heeg, M. D. Payne, M. Trkula and E. Deutsch, *Inorg. Chem.*, **17**, 431 (1978).
57. R. C. Elder, G. J. Kennard, M. D. Payne and E. Deutsch, *Inorg. Chem.*, **17**, 1296 (1978).
58. J. N. Cooper, J. D. McCoy, M. G. Kats and E. Deutsch, *Inorg. Chem.*, **19**, 2265 (1980).
59. M. I. Bruce and A. D. Redhouse, *J. Organomet. Chem.*, **30**, C78 (1971).
60. P. J. Pollick, J. P. Bibler and A. Wojcicki, *J. Organomet. Chem.*, **16**, 201 (1969).
61. J. Gasteiger and R. Huisgen, *J. Am. Chem. Soc.*, **94**, 6541 (1972).
62. R. C. Edmondson and M. J. Newlands, *J. Chem. Soc., Chem. Commun.*, 1219 (1968).
63. R. F. Bryan and A. R. Manning, *J. Chem. Soc., Chem. Commun.*, 1220 (1968).
64. L. A. Paquette, U. Jacobsson and M. Oku, *J. Chem. Soc., Chem. Commun.*, 115 (1975).
65. D. A. Langs and C. R. Hare, *J. Chem. Soc., Chem. Commun.*, 853 (1967).
66. P. D. Ginderow and M. Huber, *Acta Cryst.*, **B29**, 560 (1973).
67. H. Mäcke, V. Houlding and A. W. Adamson, *J. Am. Chem. Soc.*, **102**, 6888 (1980).

68. N. A. D. Carey and H. C. Clark, *Can. J. Chem.*, **46**, 649 (1968).
69. M. R. Churchill, J. Wormald, D. A. T. Young and H. D. Kaesz, *J. Am. Chem. Soc.*, **91**, 7201 (1969).
70. C. W. Dudley and C. Oldham, *Inorg. Chim. Acta*, **2**, 199 (1968).
71. G. B. Deacon and P. G. Cookson, *Inorg. Nucl. Chem. Lett.*, **5**, 607 (1969).
72. E. Lindner, U. Kunze, G. Vitzthum, G. Ritter and A. Haag, *J. Organomet. Chem.*, **24**, 131 (1970).
73. E. Lindner, U. Kunze, G. Ritter and A. Haag, *J. Organomet. Chem.*, **24**, 119 (1970).
74. G. Vitzthum, U. Kunze and E. Lindner, *J. Organomet. Chem.*, **21**, P38 (1970).
75. M. R. Churchill and J. Wormald, *Inorg. Chem.*, **9**, 2430 (1970).
76. M. Pang and E. I. Becker, *J. Org. Chem.*, **29**, 1948 (1964).
77. E. Lindner and U. Kunze, *J. Organomet. Chem.*, **23**, C53 (1970).
78. J. W. Moore, H. W. Baird and H. B. Miller, *J. Am. Chem. Soc.*, **90**, 1358 (1968).
79. E. Lindner and G. Vitzthum, *Chem. Ber.*, **102**, 4053 (1969).
80. E. Lindner and G. Vitzthum, *Angew. Chem., Int. Ed. Engl.*, **9**, 308 (1970).
81. W. A. Schenk, *Angew. Chem., Int. Ed. Engl.*, **26**, 98 (1987).
82. N. Furukawa, M. Tsuruoka and H. Fujihara, *Heterocycles*, **24**, 3019 (1986).
83. R. G. Severson and A. Wojcicki, *J. Am. Chem. Soc.*, **101**, 877 (1979).
84. T. G. Attig and A. Wojcicki, *J. Am. Chem. Soc.*, **96**, 262 (1974).
85. T. C. Flood and D. L. Miles, *J. Am. Chem. Soc.*, **95**, 6460 (1973).
86. U. Kunzo, E. Lindner and J. Koola, *J. Organomet. Chem.*, **40**, 327 (1972).
87. A. Dormond, C. Moise, A. Dahchour and J. Tirouflet, *J. Organomet. Chem.*, **177**, 181 (1979).
88. A. E. Crease and M. D. Johnson, *J. Organomet. Chem.*, **100**, 8013 (1978).

The Chemistry of Sulphinic Acids, Esters and their Derivatives
Edited by S. Patai


CHAPTER **11**

# Rearrangements

SAMUEL BRAVERMAN

*Department of Chemistry, Bar-Ilan University, Ramat Gan 59100, Israel*

## I. INTRODUCTION

Rearrangements involving sulfinic acids and their derivatives, especially esters, have played a significant role in the development of the chemistry of these functional groups. It is therefore not surprising that all major literature surveys on sulfinic acids[1-6] or sulfones[7-13] also include a discussion of this subject. However, while excellent and detailed coverage exists for certain rearrangements of general mechanistic and synthetic interest, such as, for example, the Smiles[14,15] or Truce–Smiles[16] rearrangement, the treatment of all other rearrangements is usually brief and partial. An attempt has therefore been made to provide the reader with a comprehensive and systematic survey of the literature dealing with rearrangements involving sulfinic acids and their derivatives, some of which have never been reviewed before. An exception to this statement are the rearrangements of sulfinate esters to sulfones and the reverse rearrangements which have been extensively reviewed by the present author, as part of a chapter on rearrangements involving sulfones in a recent volume of this series[17]. An effort has also been made to scan the literature through 1988, as far as possible, and to cover the most significant aspects and most important advances, particularly work of the last two decades.

Rearrangements have been included in which sulfinic acids and their derivatives participate not only as reactants but also as products. Reactions have been classified according to mechanism, but although the main emphasis has been on mechanism and stereochemistry, special attention to synthetic applications has also been given, wherever appropriate. Obviously, due to space limitations, only selected and representative results of general importance, as judged by the concern of the reviewer, are presented below. Thus, the exclusion of a particular piece of work in no way passes judgement on its scientific value.

## II. REARRANGEMENTS INVOLVING SULFINIC ACIDS

### A. Pericyclic Rearrangements

Pericyclic reactions involving sulfur dioxide constitute a fascinating chapter in organosulfur chemistry[18]. One of the best studied pericyclic reactions of sulfur dioxide is its facile cheletropic 1,4-cycloaddition reaction with a variety of conjugated dienes to give the corresponding 2,5-dihydrothiophene 1,1-dioxide[19], which dates back to the discovery of sulfolene in 1914[20]. Contrary to previous reports that cycloaddition does not occur with 1,4-dienes[19], Rogic and Vitrone[21] observed that sulfur dioxide reacts with a mixture of isomeric 4-alkoxy-1,3- and 1,4-dienes to give an essentially quantitative yield of the corresponding 1,4-adduct (equation 1). The authors concluded that the cycloaddition was clearly preceded by a facile isomerization of the 1,4- to 1,3-diene, and decided to investigate the scope and mechanism of this isomerization in more detail[22-24].

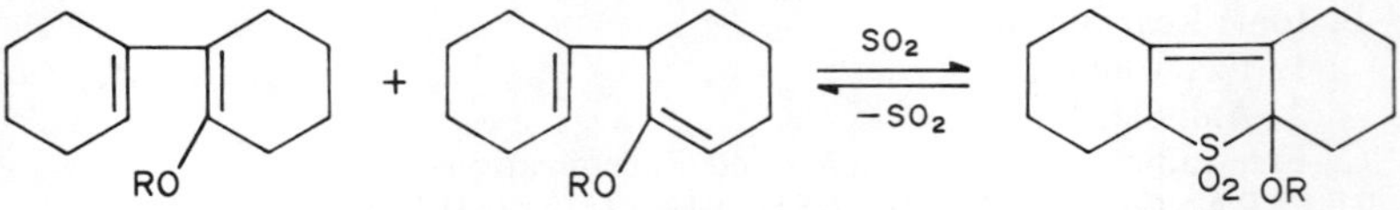


(1)

These studies revealed that sulfur dioxide indeed catalyzed a facile and regiospecific isomerization of a variety of olefins to the thermodynamically more stable isomers at room temperature. Based on kinetic and deuterium labeling studies the authors suggested that the isomerization proceeds by a sequence of reversible reactions (Scheme 1) that involves formation of a dipolar olefin–sulfur dioxide adduct, which in an ene reaction provides the corresponding allylic sulfinic acid as a reactive intermediate. The 1,3-rearrangements of the allylic sulfinic acid followed by retro-ene reaction and elimination of sulfur dioxide provides the isomerized olefin[23].

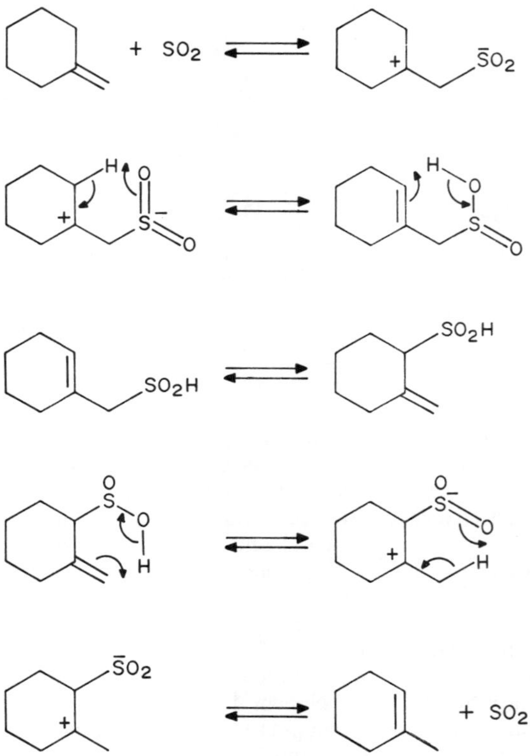

SCHEME 1

Evidence that allylic sulfinic acids are indeed very unstable and undergo smooth decomposition to sulfur dioxide and olefin has also been provided by the same authors[23]. Thus, magnesium salts of the allylic sulfinic acids prepared by reaction of sulfur dioxide with the Grignard reagents derived from 1-chloro-3-methyl-2-butene and 2-chloromethylenecyclohexane on acid hydrolysis gave the olefin and sulfur dioxide (equations 2 and 3, respectively). The deuterolyses of the magnesium salt of α,α-dimethylallylsulfinic acid in the presence of deuteriosulfinic acid gave 4-deuterio-2-methyl-2-butene (equation 2), as expected from the retro-ene mechanism. Hydrolysis of the chloromagnesium salt of the sulfinic acid derived from 2-chloromethylenecyclohexane afforded an approximately 1:1 mixture of 1-methylcyclohexene and methylenecyclohexane and sulfur dioxide. This result suggests that in this case the generated allylic sulfinic acid had a sufficiently long lifetime to undergo the 1,3-rearrangement before the retro-ene reaction occurred.

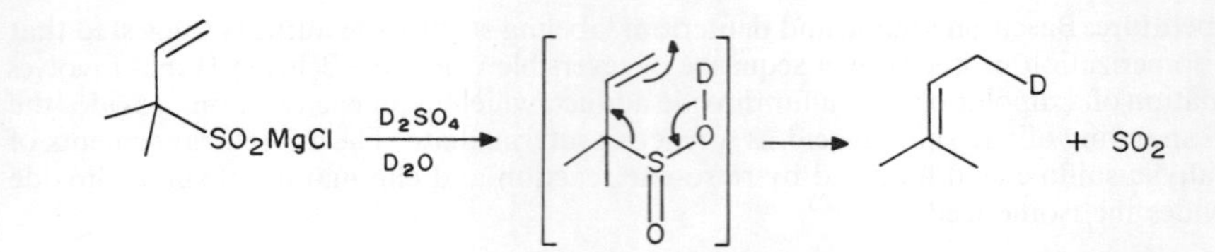

(2)

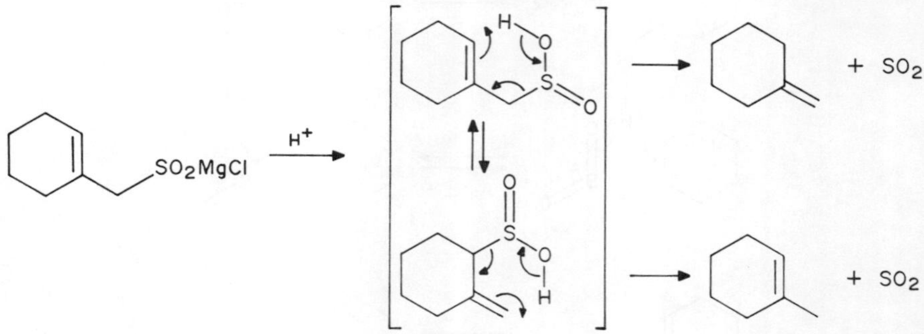

(3)

The authors[23] also pointed out that the 1,3-allylic sulfur migration is not well understood, but may involve a four-membered cyclic dipolar intermediate with the negative charge localized on the sulfinyl oxygen and the positive charge on the tertiary carbon atom. The effectiveness with which various allylic sulfur compounds such as sulfides[25,26], sulfoxides[27,28] and sulfones[17] undergo the 1,3-rearrangement may depend on the ability of the corresponding sulfur centers to open up new coordination sites. Thus Kwart and coworkers[25,26] have discussed the thiaallylic rearrangement as a well-characterized process involving a dipolar trigonal bipyramid intermediate.

Interestingly, the fragmentation of allylsulfinic acids[29–33] had been known long before the studies by Rogic described above, but the stereochemical course of this reaction was reported subsequently by Mock and Nugent[34a]. These authors reported evidence of a stereochemical nature which tends to support a cyclic mechanism for this transformation. Thus, treatment of cyclic sulfinamides **1** and **2**, prepared by cycloaddition of *N*-sulfinyl *p*-toluenesulfonamide to (*E*,*E*) and (*E*,*Z*)-2,4-hexadiene, respectively[35], with aqueous sodium hydroxide (scission of the S—N bond), followed by acidification of the sulfinate

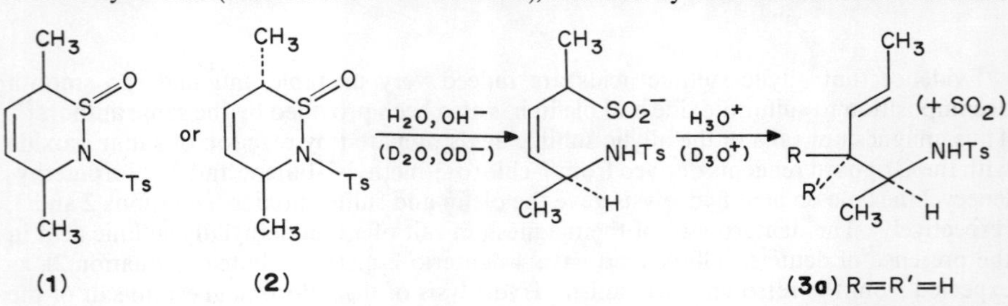

(**1**) (**2**)

(**3a**) R=R′=H
(**3b**) R=D; R′=H
(**3c**) R=H; R′=D

(4)

salts with hydrochloric acid yielded 5-(*p*-tolylsulfonamide)-(*E*)-2-hexene (**3a**) as the only isolable product (equation 4). Hydrolysis of **1** and **2** in a deuterated medium proceeded stereospecifically and afforded diastereomers **3b** and **3c**, respectively. The authors[34] suggested that formation of diastereomeric products **3b, c** implies diastereomeric transition states. Configurational control was rationalized by a cyclic retro-ene mechanism (equation 5). Strong preference for a chair configuration in the transition state could explain both predominant (*E*)-alkene formation as well as diastereomeric induction at the 4-position as a result of 1,3-transfer of chirality. This reaction has been applied in the synthesis of homoallylic amine derivatives having predictable stereochemistry and double bond geometry[34b].

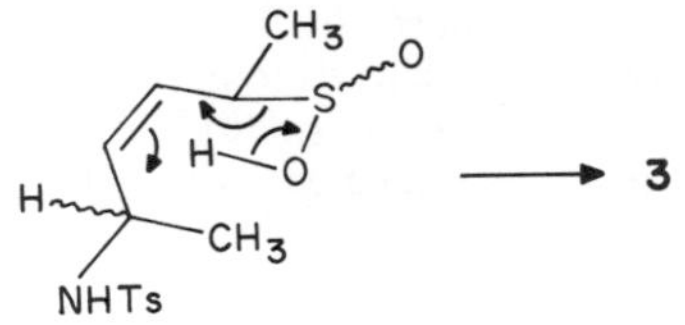


(5)

Following the proposal of the mechanism of sulfur dioxide isomerization of olefins involving allylic sulfinic acids as intermediates (Scheme 1)[22,23], and the various reports on the spontaneous fragmentation of the latter, direct evidence for their involvement was provided by Raasch and Smart[36]. These workers reported the isolation of a stable sulfinic acid in an ene reaction of an olefin with sulfur dioxide. Thus, on passing sulfur dioxide into a solution of the cyclic olefin **4** in methylene chloride, an ene reaction occurs and the sulfinic acid **5** precipitates in 76% yield.

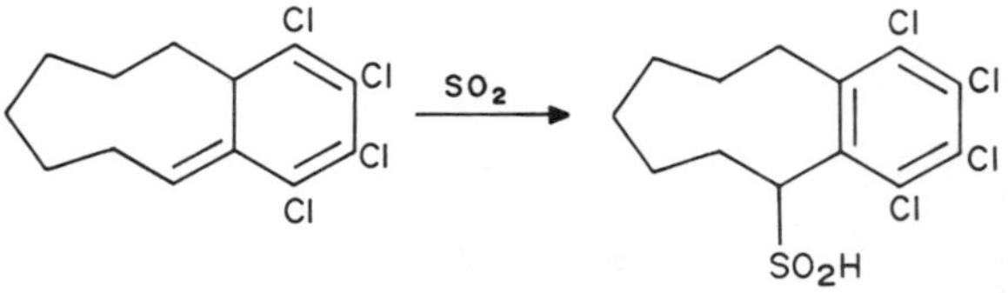


(**4**) (**5**)

In this case, the benzylic sulfinic acid **5** is isolable because the benzene ring does not participate in the rearrangement and retro-ene reactions characteristic of allylic sulfinic acids. Subsequently, several other examples of relatively stable and isolable sulfinic acids generated by ene reactions with sulfur dioxide have been published[37–39]. Thus, Lucchini and coworkers[37] have found that cycloheptatriene (**6**) is converted to α-toluenesulfinic acid (**7**) in quantitative yield on standing for three days in liquid sulfur dioxide at room temperature. Interestingly, while 1,2- and 3,4-benzocycloheptatriene are unreactive in liquid $SO_2$, except for a slow isomerization of the latter to the former, more stable isomer, another valence isomer, benzonorcaradiene (**8**), is converted to a mixture of α- and β-naphthylmethanesulfinic acids (**9** and **10**) in a 45:55 ratio after only 2 hours, under the same conditions.

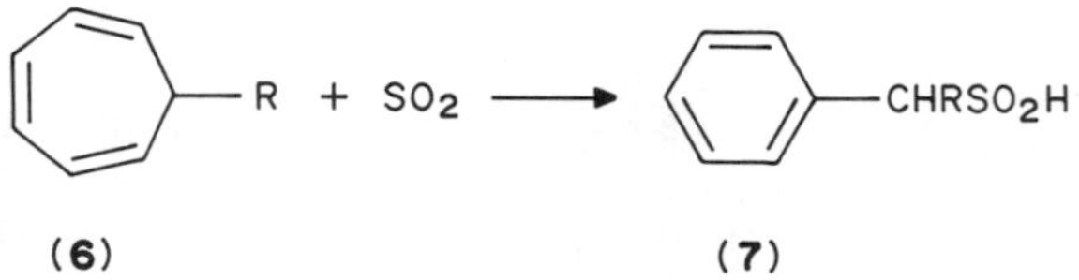


(**6**) (**7**)

(**a**) R=H; (**b**) R=Me; (**c**) R=Ph

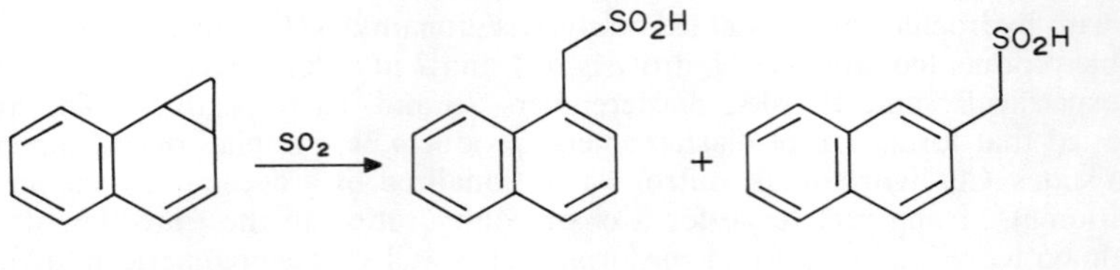


(8) (9) (10)

The following two mechanisms were suggested by the authors. One mechanism (a) involves an electrophilic attack on the cyclopropane ring of norcaradiene (**11**), similar to the mechanism proposed for the thermally induced reaction between homocycloheptatriene (**12**) and sulfur dioxide, which leads to the formation of the bicyclic sulfone **13**, as shown in equation 7[38]. The other mechanism (b) is an ene reaction, implying attack of $SO_2$ on the 1,7 or 6,7 bond in **11**, as indicated by the arrow in equation 6.

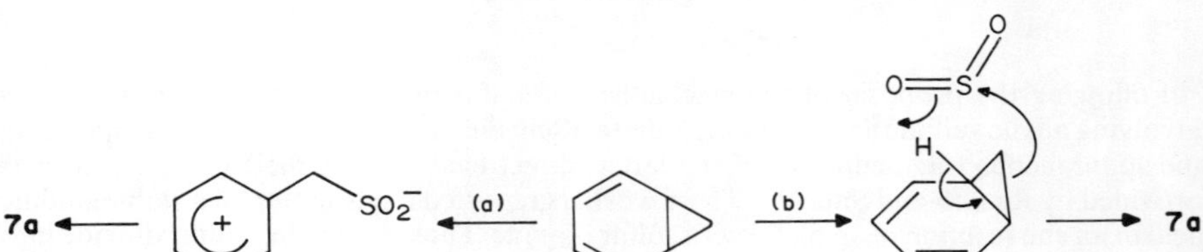


(11)

(6)

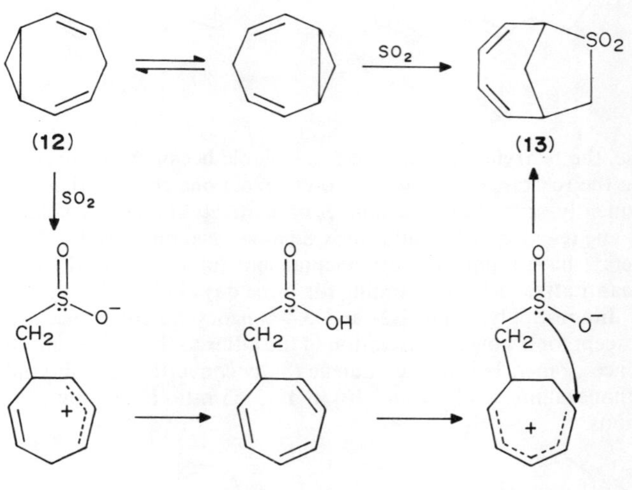


(12) (13)

(7)

The formation of stable sulfinic acids from the ene reaction of methyl substituted allenes with sulfur dioxide has also been reported[39]. For example, instantaneous formation of the allylic-vinylic sulfinic acid **15** from the reaction of tetramethylallene (**14**) and $SO_2$ at $-60\,°C$ was observed by NMR. Although **15** decomposes at room temperature as do sulfinic acids in general, it may be isolated as the corresponding sulfone by reaction of its sodium salt with ethyl bromide[40]. More recently, the utility of the $SO_2$ isomerization for the stereospecific synthesis of *cis* or *trans* hydrindanones[40a], and some more examples of stable allylic sulfinic acids[40b] have been described.

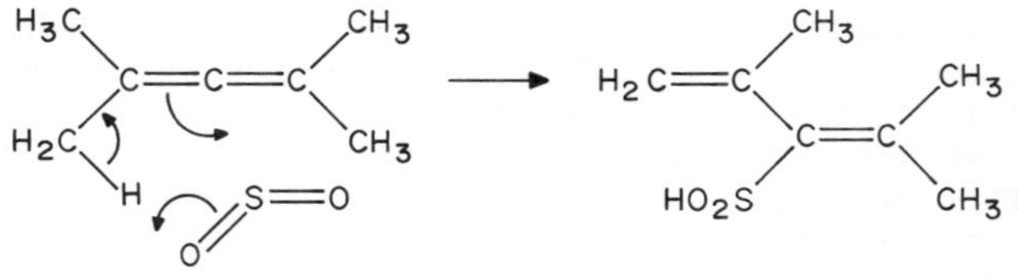

(14) (15)

**B. Anionic and Nucleophilic Rearrangements**

In a re-examination of the base catalyzed rearrangement of 2-nitrobenzenesulfenanilides, Cava and Blake[41] showed that the product from the reaction of **16a** with base was not the sodium salt of an aminothiol, as previously reported[42], but the azosulfinate **17a**. This assignment was confirmed by methylation using methyl iodide to the corresponding sulfone **18a**. The same workers[41] proposed a mechanism for the transformation which involved attack of hydroxide anion on sulfur, ultimately to form S═O bonds, and loss of hydroxide ions from the acid form of the nitro group. However, the X-ray structural determination of the related sulfenate ester **19** indicated a strong interaction between one of the oxygen atoms of the $NO_2$ group and the sulfur atom[43]. This observation has led Brown[44] to assume that such an interaction might well be involved in the conversion of **16** to **17**. Accordingly, experiments were performed by the author to check on the origin of the sulfinate oxygen in $^{18}O$-labeled material, together with some kinetic and product studies, pertinent to the mechanism. The rearrangement of 2-nitrobenzenesulfenanilide (**16a**) and

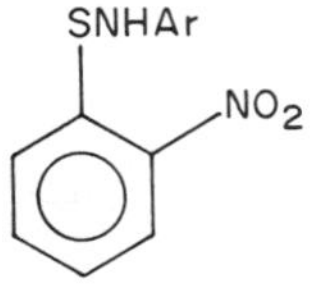

(**16a**) Ar═$C_6H_5$–
(**16b**) Ar═4-$MeOC_6H_4$–
(**16c**) Ar═4-$NO_2C_6H_4$–

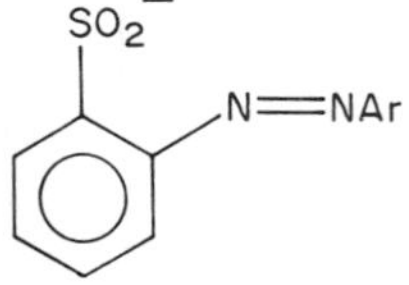

(**17a**) Ar═$C_6H_5$–
(**17b**) Ar═4-$MeOC_6H_4$–

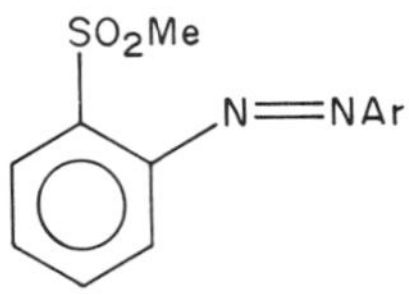

(**18a**) Ar═$C_6H_5$–
(**18b**) Ar═4-$MeOC_6H_4$–

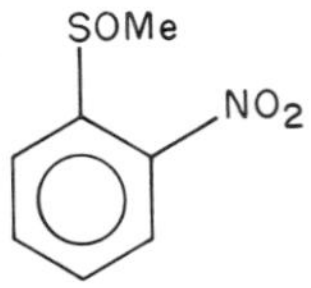

(**19**)

its 4-methoxy derivative **16b** to the azobenzenesulfinates **17a** and **17b**, respectively, in aqueous alcoholic sodium hydroxide have thus been examined. The reactions were first order in sulfenanilide and in hydroxide ion, and **16b** rearranged at a slightly faster rate than **16a**. When the rearrangement of **16a** was conducted using $^{18}O$-labeled sodium hydroxide solution, essentially zero incorporation of label into the sulfinate was observed. These results rule out any mechanism involving oxygenation of sulfur by attack of hydroxide, and clearly show that both oxygens of the $NO_2$ group are transferred to sulfur. The author[44] has also shown that hydroxide ions as such are not essential since the rearrangement of **16a** to **17a** can be promoted by any comparable strong base, such as for example dry ethanolic NaOEt. Based on these results and the failure of 4-nitrobenzenesulfenanilide (**16c**) to rearrange, an alternative mechanism for the rearrangement of **16a** to **17a** has been suggested as shown in Scheme 2.

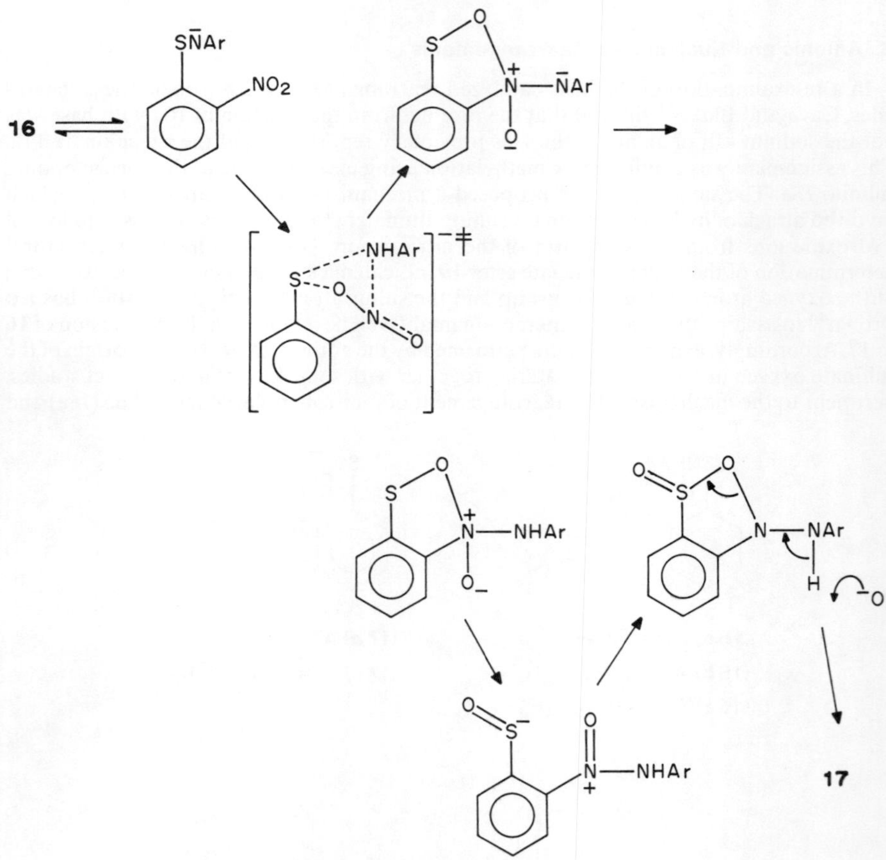


SCHEME 2

A variety of anionic or nucleophilic rearrangements of sulfones to sulfinate salts, including the well-known Smiles rearrangement described in the following section, have been reported. For example, nucleophiles such as cyanide and benzenesulfinate ions in DMF add across the carbon–carbon double bond in 2,3-diphenylthiirene 1,1-dioxide (**20**) to give an intermediate which undergoes electrocyclic ring-opening to vinylsulfinates **21** and **22**, respectively. These sulfinate anions were trapped with methyl iodide and isolated as their respective methyl sulfones **23** and **24**[45].

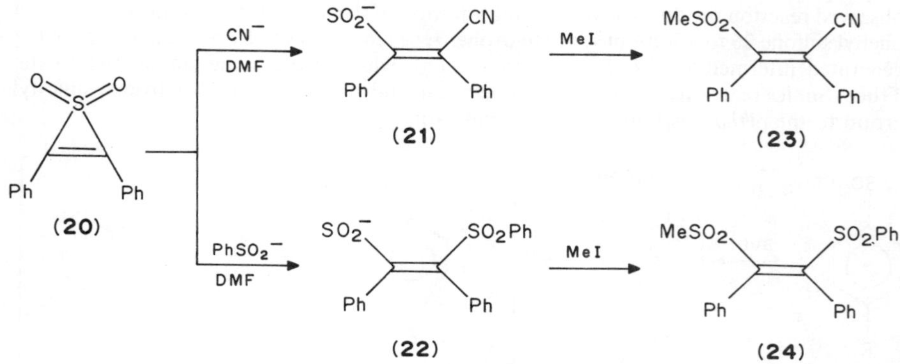

Under certain basic conditions 2,5-dihydrothiophene 1,1-dioxide also undergoes ring-opening reactions[46,47] and the resulting 1,3-butadienyl sulfinate anions may be alkylated to the corresponding esters or sulfones (equation 8)[48].

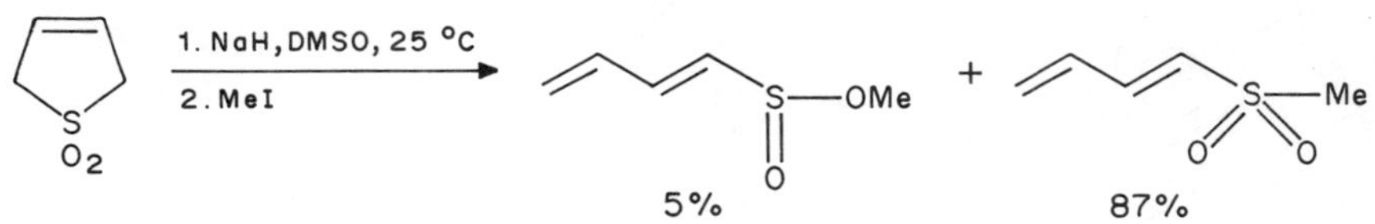

(8)

Dodson and coworkers[49] have observed ring contraction in the rearrangement of *cis*- or *trans*-2,4-diphenylthietane 1,1-dioxide to *trans*-1,2-diphenylcyclopropanesulfinic acid upon treatment with ethylmagnesium bromide (equation 9).

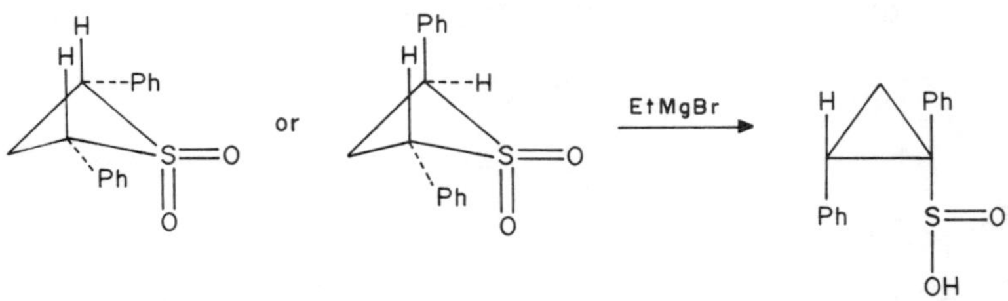

(9)

Directed lithiation of aromatic compounds is a reaction of broad scope and considerable synthetic utility[50]. The metalation of arenesulfonyl systems was first observed by Gilman and Webb[51] and by Truce and Amos[52] who reported that diphenyl sulfone is easily metalated at an *ortho* position by butyllithium.

Following earlier observations by Stoyanovich and coworkers[53,54] that the action of three or more moles of alkyllithium with one mole of *t*-butyl phenyl sulfone (**25**) proceeds with elimination of lithium *t*-butylsulfinate and formation of 2,6-dilithium-1-alkylbenzene, the same authors[55] attempted to clarify the mechanism of this reaction, and to detect the intermediacy of a 2,6-dilithium derivative of *t*-butyl phenyl sulfone (**27**, Scheme 3), by lowering the reaction temperature. Thus, treatment of *t*-butyl phenyl sulfone (**25**) with butyllithium at − 20 to − 30 °C in THF–ether unexpectedly leads to lithium (2′-*t*-butyl-6′-lithium-biphenyl) sulfinate (**30**). Hydrolysis of the latter gave 2′-*t*-butylbiphenyl-2 sulfinic acid (**31**), whereas carboxylation led to 6′-carboxyl-2′-*t*-butylbiphenyl-2 sulfinic acid (**32**). A mechanism was suggested for the formation of observed reaction products, involving the addition of the *ortho*-lithio derivative of *t*-butyl phenyl sulfone **26** to 3-lithium 1,2-dehydrobenzene (**28**), followed by rearrangement of the generated intermediate **29**. This rearrangement, which bears some similarities to the Truce–Smiles rearrangement described below, includes *t*-butyl migration from a sulfonyl group to the *ortho* position of the adjacent aromatic ring.

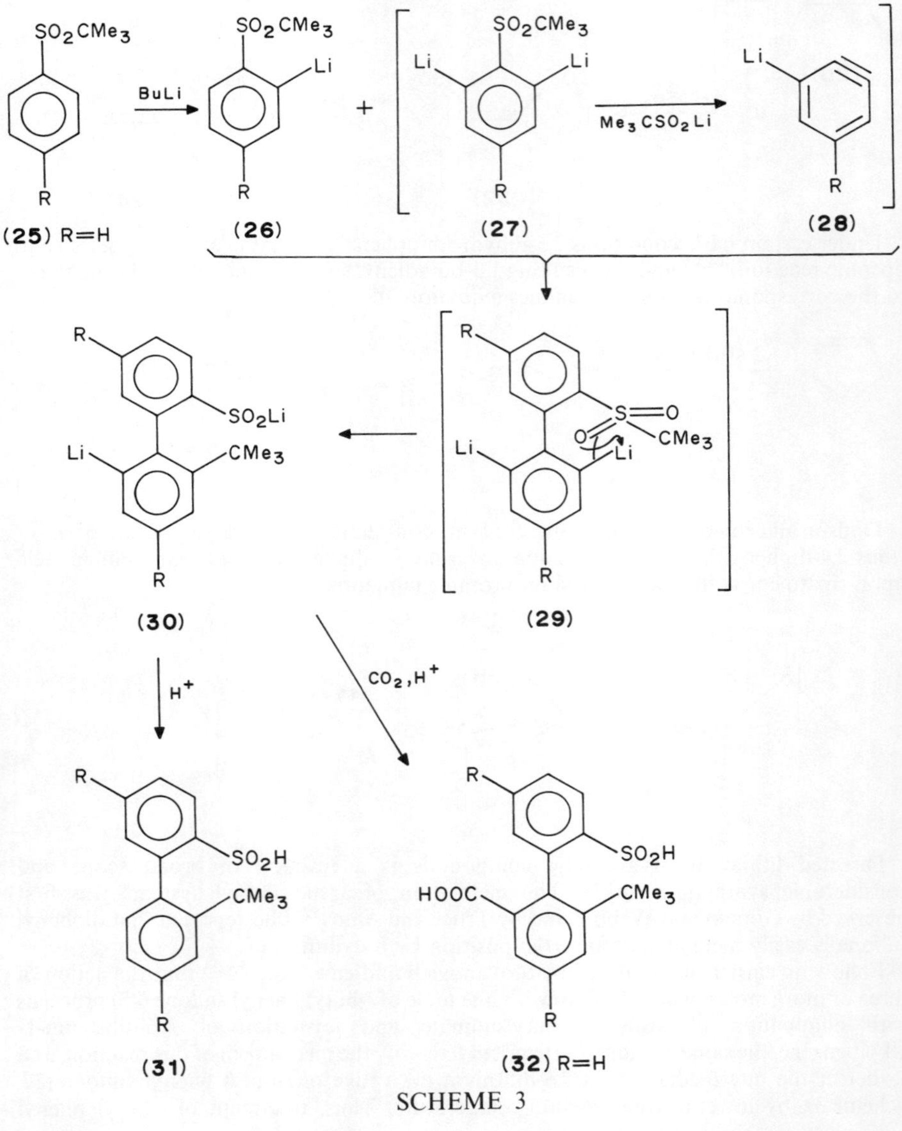


SCHEME 3

## C. The Smiles Rearrangement

The Smiles rearrangement is one of the oldest and best studied rearrangements of sulfones. Although first reported by Henrique[56] and by Hinsberg[57], the rearrangement is named after Smiles[58–61], who has not only established the correct structure of the products, but also recognized the occurrence of a novel rearrangement and developed its chemistry. The rearrangement involves the isomerization of a sulfone to a sulfinic acid, and can be described as an intramolecular aromatic substitution of a sulfonyl group initiated by a nucleophilic group attached to the sulfonyl group through two atoms, which may also be part of an aromatic system. Several typical examples of this rearrangement, which is usually catalyzed by base, are given in equations 10–13.

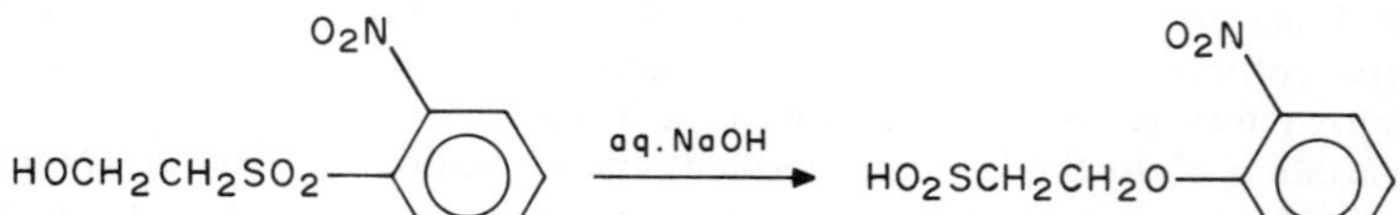


(10)

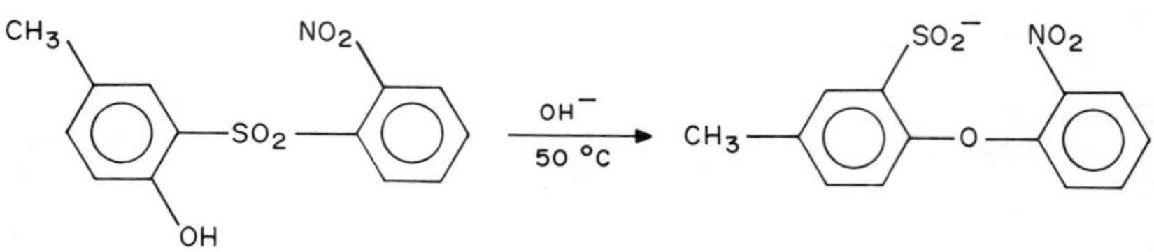


(11)

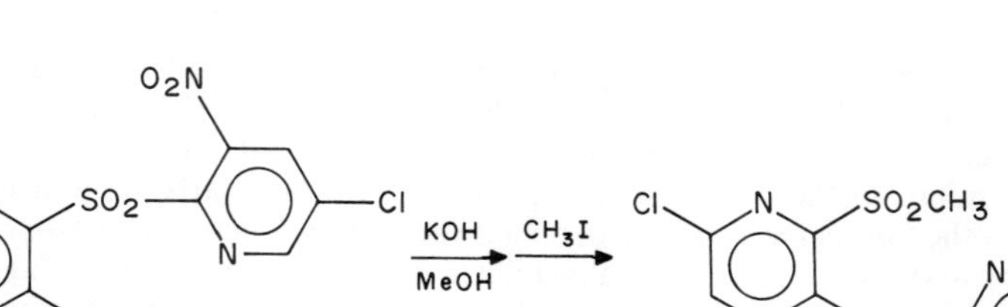


(12)

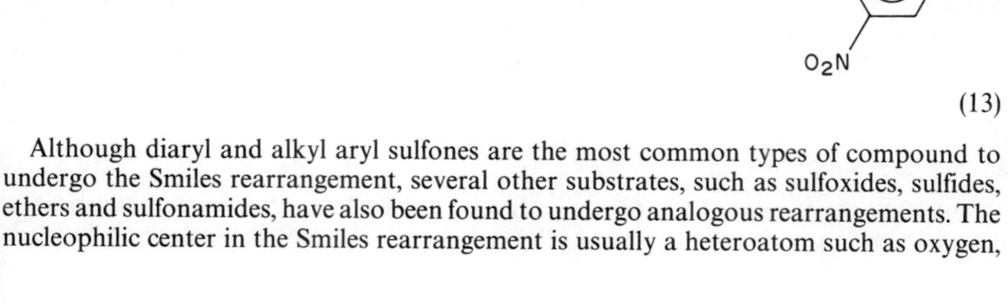


(13)

Although diaryl and alkyl aryl sulfones are the most common types of compound to undergo the Smiles rearrangement, several other substrates, such as sulfoxides, sulfides, ethers and sulfonamides, have also been found to undergo analogous rearrangements. The nucleophilic center in the Smiles rearrangement is usually a heteroatom such as oxygen, nitrogen or sulfur, while in the Truce–Smiles modification it may also be a carbanion. If one or both aromatic rings are pyridine, the rearrangement may also be catalyzed by acid[63]. Because of the considerable interest in the Smiles rearrangement several excellent and comprehensive reviews have also been published in the past[15,64–66], as well as a very recent brief survey by the present author[17]. The reader is therefore directed to these sources for further details.

### D. The Truce–Smiles Rearrangement

#### 1. *Diaryl sulfones*

In 1958, Truce and coworkers[67] discovered that metalation of mesityl phenyl sulfone (**33**) occurred entirely at an *ortho*-methyl group and not at a ring carbon, as expected[51,52]. Furthermore, refluxing an ether solution of the lithiated species resulted in a novel and unusual variation of the Smiles rearrangement and formation of 2-benzyl-4,6-dimethylbenzenesulfinic acid (**34**) in almost quantitative yield (equation 14). Several other *o*-methyldiaryl sulfones have also been shown to rearrange to *o*-benzylbenzenesulfinic acids when heated in ether solution with BuLi[68–70].

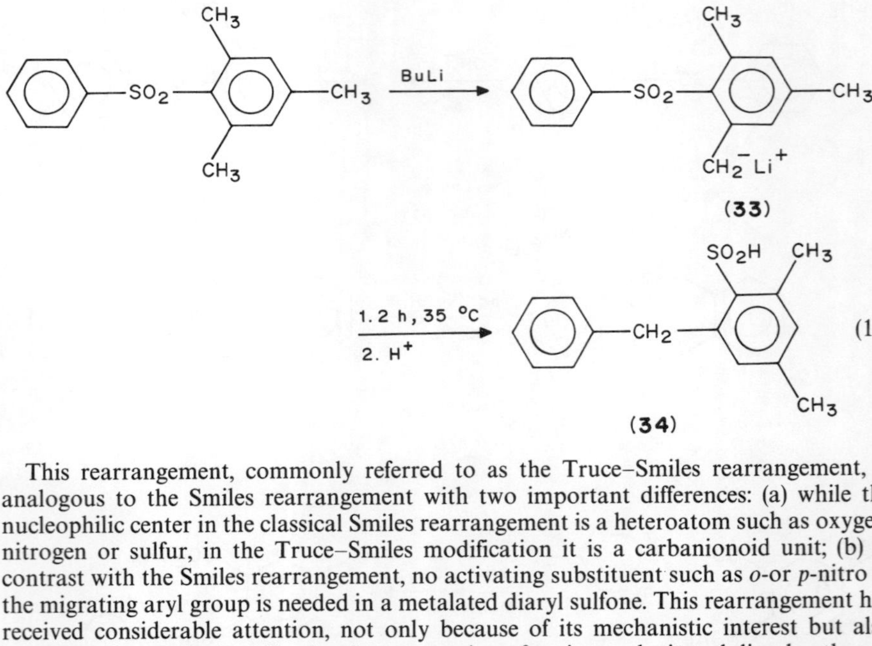


(**33**)

(**34**) (14)

This rearrangement, commonly referred to as the Truce–Smiles rearrangement, is analogous to the Smiles rearrangement with two important differences: (a) while the nucleophilic center in the classical Smiles rearrangement is a heteroatom such as oxygen, nitrogen or sulfur, in the Truce–Smiles modification it is a carbanionoid unit; (b) in contrast with the Smiles rearrangement, no activating substituent such as *o*-or *p*-nitro in the migrating aryl group is needed in a metalated diaryl sulfone. This rearrangement has received considerable attention, not only because of its mechanistic interest but also because of its synthetic utility for the preparation of various substituted diarylmethanes. However, two excellent and comprehensive reviews have been published by Truce[16] and Drozd[15], and therefore the reader is referred to these sources as well as to a more recent survey by the present author[17].

#### 2. *Alkyl aryl sulfones*

One of the most recent and interesting extensions of the Truce–Smiles rearrangement is the analogous rearrangement of aryl *t*-alkyl sulfones. For example, Snyder and Truce[71] reported facile metalation of *o*-tolyl *t*-butyl sulfone (**35**) with butyllithium in THF to yield the benzyllithium species **35a**, which was stable at room temperature or below. However, refluxing the solution for several hours resulted in the formation of the salt of *o*-neopentylbenzenesulfinic acid (**36**) in good yield (equation 15). This reaction, which constitutes a Truce–Smiles rearrangement with an alkyl group as the migrating unit, has also been observed with other *o*-methylaryl *t*-alkyl sulfones. Subsequently, and unexpectedly, Truce and coworkers[72] observed that this rearrangement can also be extended to *p*-tolyl *t*-alkyl sulfones. Thus, an attempt to metalate *p*-tolyl *t*-butyl sulfone (**37**) with butyllithium resulted in metalation at an *ortho* position, but metalation with lithium diisopropylamide in THF resulted in benzylic metalation. Furthermore, the resulting metalated sulfone **37a** was found to rearrange readily to lithium *p*-neopentylbenzenesulfinate (**38**) even at room temperature (equation 16).

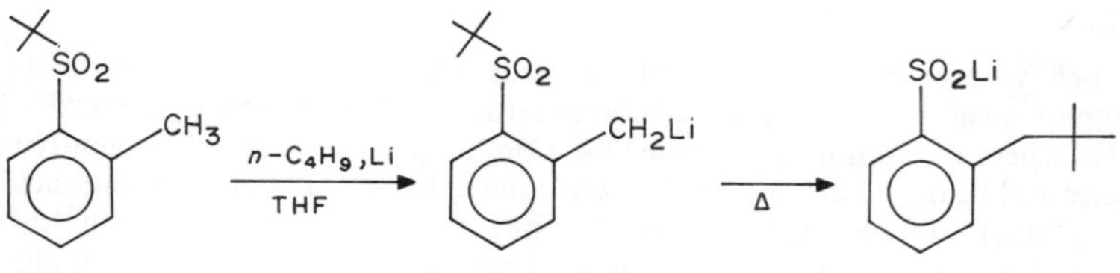


(15)

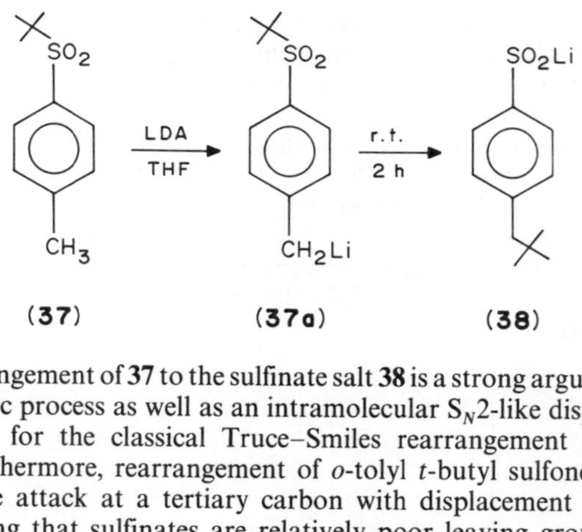


(16)

The facile rearrangement of **37** to the sulfinate salt **38** is a strong argument against both a concerted pericyclic process as well as an intramolecular $S_N2$-like displacement mechanism, as suggested for the classical Truce–Smiles rearrangement of *o*-methyl diaryl sulfones[16,17]. Furthermore, rearrangement of *o*-tolyl *t*-butyl sulfone (**35**) via an intramolecular $S_N2$-like attack at a tertiary carbon with displacement of sulfinate is also unlikely considering that sulfinates are relatively poor leaving groups in nucleophilic displacements and few documented examples exist of $S_N2$-type reactions at tertiary carbons, even with good leaving groups[73,74]. On the other hand, considerable evidence for a free radical mechanism was obtained[71]. For example, the rearrangement of *o*-tolyl cumyl sulfone yielded radical coupling products such as bicumyl, in addition to the normal rearrangement product. Consequently, an electron-transfer radical-anion chain mechanism has been suggested[72,16,17].

## III. REARRANGEMENTS INVOLVING SULFINIC ANHYDRIDES

Although acid chlorides acylate sulfinate ions at oxygen[75,76] rather than at sulfur (equation 17), with sulfinyl chlorides, the isolated product of their reaction with sulfinate ions is not the corresponding sulfinic anhydride (**39**), as originally assumed[77], but rather the sulfinyl sulfone **40**, as demonstrated by Bredereck and coworkers[78] many years later (equation 18).

$$RSO_2^- + R\overset{O}{\overset{\|}{C}}Cl \longrightarrow Cl^- + R\overset{O}{\overset{\|}{S}}O\overset{O}{\overset{\|}{C}}R \qquad (17)$$

$$\underset{(\mathbf{39})}{R\overset{O}{\overset{\|}{S}}O\overset{O}{\overset{\|}{S}}R} \longleftarrow RSO_2^- + R\overset{O}{\overset{\|}{S}}Cl \longrightarrow \underset{(\mathbf{40})}{R\overset{O}{\overset{\|}{S}}\underset{\underset{O}{\|}}{\overset{O}{\overset{\|}{S}}}R} \qquad (18)$$

This behavior is reminiscent of sulfenic acids whose anhydrides have the analogous thiosulfinate structure[5]. However, although structure **40** is thermodynamically favored, two cases where the sulfinic anhydride (**39**) is preferred have been subsequently reported. Thus Kice and Ikura[79] have shown that while the anhydride of butanesulfinic acid has the normal sulfinyl sulfone structure (**40**, R = Bu), the anhydride of 2-methylpropane-2-sulfinic acid has the sulfinic anhydride structure (**39**, R = *t*-Bu). These findings were explained by the relief in steric interference between the bulky *t*-butyl groups by the oxygen bridge of the anhydride isomer. The authors also estimated that the decrease in free energy associated with releasing the interference between two *t*-butyl groups on going from **40** to **39** should be of the order of only a few kilocalories, and concluded that the difference in free energy between the sulfinic anhydride and sulfinyl sulfone functionalities should also be of the same order of magnitude. This is reminiscent of the thermodynamics for sulfenate sulfoxide isomerizations[5,28].

The second successful preparation of a sulfinic anhydride was reported by Mueller and Dines[80], who obtained the anhydride of ethane-1,2-disulfinic acid (**41**) by carefully controlled hydrolysis of the corresponding dichloride in tetrahydrofuran at room temperature. Similar to the previous case, absence of sulfone-like bands in the IR spectrum of the product provided compelling evidence against the isomeric sulfinyl sulfone (**42**). The diminished stability of the latter is easily explained by the strain associated with the four-membered ring in **42**.

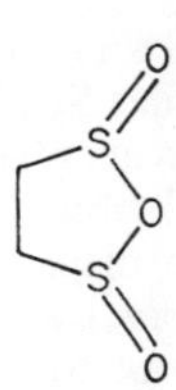

**(41)**

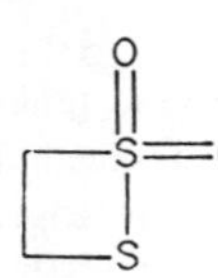

**(42)**

Support of this explanation may be found in the unusual stability of five- and six-membered sulfinyl sulfones not only with regard to rearrangement to the corresponding anhydrides, but even with respect to their generally facile hydrolysis. Thus, while the equilibrium constant for a sulfinic acid–sulfinyl sulfone equilibrium (equation 19) in a medium containing much water is too small to be detected, it can be measured spectroscopically in a low-water-content solvent, such as 1% $H_2O$ in AcOH, though the concentration of the sulfinyl sulfone is still very small. In contrast, in a disulfinic acid such as **43** where sulfinyl sulfone formation can be an intramolecular reaction, this percent increases dramatically. In the same solvent, the sulfinyl sulfone **44** is present at equilibrium to the extent of 88%[81]. Even more remarkable is the behavior of naphthalene-1, 8-disulfinic acid **45**, with which even in 60% aqueous dioxane almost 75% of the disulfinic acid is present at equilibrium as the cyclic sulfinyl sulfone **46**[82].

$$2\mathrm{ArSO_2H} \rightleftharpoons \mathrm{Ar}\overset{\overset{\mathrm{O}}{\|}}{\mathrm{S}}\underset{\underset{\mathrm{O}}{\|}}{\overset{\overset{\mathrm{O}}{\|}}{\mathrm{S}}}\mathrm{Ar} + \mathrm{H_2O} \tag{19}$$

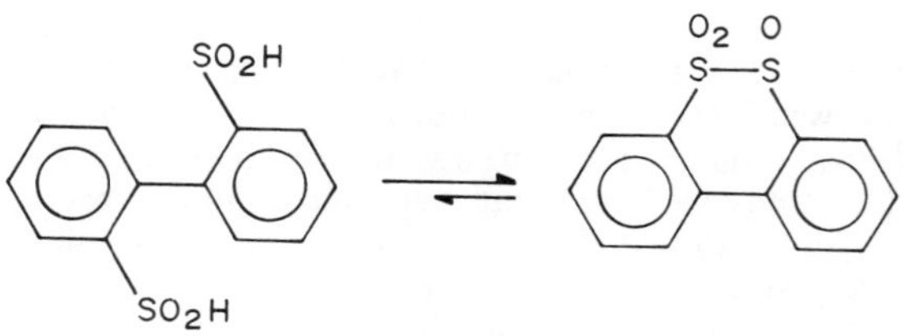


**(43)** **(44)**

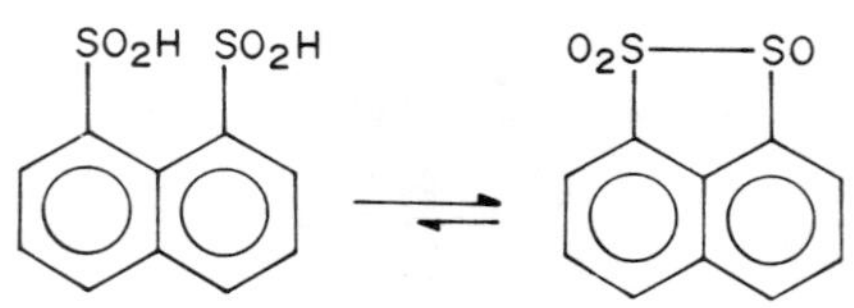


**(45)** **(46)**

Kice[5] has noted that the question, whether the formation of sulfinyl sulfones under normal conditions (equation 18) is because *S*-sulfinylation is kinetically preferred, or rather results because the sulfinic anhydride generated by initial *O*-sulfinylation is readily converted by some of the remaining sulfinate ions to the thermodynamically more stable sulfinyl sulfone (equation 20), has not been definitely established. Alternatively, one could also suggest a free radical mechanism for the isomerization of the two species, especially in view of the facile homolytic dissociation of the S—S bond (equation 21). For example, Kice and Pawlowski[83,84] have found that, when heated in anhydrous dioxan, aryl sulfinyl sulfones $ArS(O)SO_2Ar$ undergo thermal decomposition very readily ($t_{1/2} = 30$ min at 50 °C for Ar = *p*-tolyl) with clean first-order kinetics. Evidence for the intermediacy of free radicals by trapping experiments was obtained. The rate-determining step (equation 21) of the decomposition is the dissociation of the sulfinyl sulfone into a pair of free radicals. Interestingly, the $\Delta H^{\ddagger}$ for this reaction is only 28 kcal mol$^{-1}$, which is 13 kcal mol$^{-1}$ less than $\Delta H^{\ddagger}$ for homolysis of the S—S bond in the corresponding α-disulfones. This result shows that ArSO· radicals enjoy particular stability and are easier to form compared to ArS· and $ArSO_2$· free radicals.

$$\mathrm{RSO_2^-} + \mathrm{R}\overset{\overset{\mathrm{O}}{\|}}{\mathrm{S}}\mathrm{O}\overset{\overset{\mathrm{O}}{\|}}{\mathrm{S}}\mathrm{R} \longrightarrow \mathrm{R{-}}\underset{\underset{\mathrm{O}}{\|}}{\overset{\overset{\mathrm{O}}{\|}}{\mathrm{S}}}\mathrm{{-}}\overset{\overset{\mathrm{O}}{\|}}{\mathrm{S}}\mathrm{{-}R} + \mathbf{RSO_2^-} \tag{20}$$

$$\mathrm{ArS(O)_2S(O)Ar \longrightarrow ArSO\cdot + ArSO_2\cdot} \qquad (21)$$

$$\mathrm{3ArSO_2H \longrightarrow ArS(O)_2SAr + ArSO_3H} \qquad (22)$$

**(47)**

Following a proposal[78,85] that direct reaction of sulfinic acid with sulfinyl sulfone may be the key step in the well-known disproportionation of sulfinic acids to thiosulfonates (**47**) and sulfonic acids (equation 22), the same authors[83,84] performed a kinetic study to test this hypothesis. This study indicated that there was no direct reaction between **40** and sulfinic acid in anhydrous dioxan. Rather, consumption of the sulfinic acid under such conditions occurs as a result of its reaction with an intermediate sulfenyl sulfonate (**48**) formed in the rate-determining unimolecular decomposition of **40**, as illustrated in Scheme 4.

$$\mathrm{ArS(O)_2S(O)Ar \xrightarrow{slow} ArSO_2\cdot + ArSO\cdot}$$

$$\mathrm{ArSO_2\cdot + ArSO\cdot \longrightarrow ArS(O)_2OSAr}$$

$$\mathrm{\underset{(48)}{ArS(O)_2OSAr} + ArSO_2H \xrightarrow{fast} ArS(O)_2SAr + ArSO_3H}$$

SCHEME 4

## IV. REARRANGEMENTS INVOLVING SULFINYL HALIDES

Following the observation[86] of a facile isomerization of 1-(aziridine) carbonyl and thiocarbonyl chlorides to 2-chloroethyl isocyanate and isothiocyanate, respectively, under extremely mild conditions, Tomalia[87] reported the analogous rearrangement of 1-(aziridine)sulfinyl chloride to 2-chloro-*N*-sulfinylethylamine, which readily occurred at room temperature (equation 23).

An interesting and useful rearrangement of penicillin sulfoxides (**49**) to 3-methylenecephams (**51**) via a sulfinyl chloride intermediate (**50**) has been reported by Kukolja and coworkers[88,89]. Thus, treatment of **49** with *N*-chlorosuccinimide in refluxing $CCl_4$ for

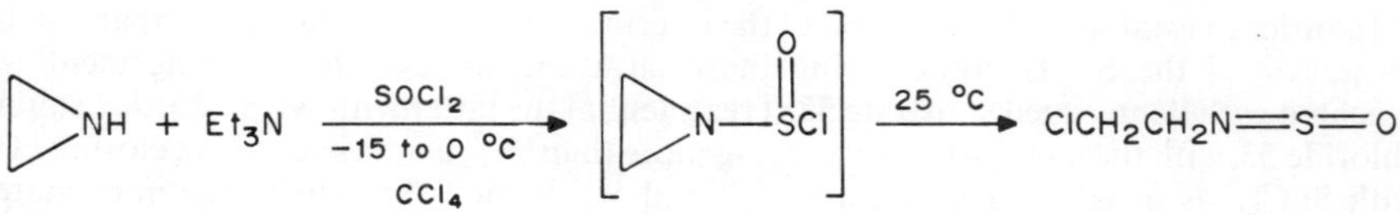


(23)

70 min gave in almost quantitative yield a mixture of the sulfinyl chlorides **50** which are epimeric at sulfur. Ring closure of **50** with various Lewis acids occurred readily at room temperature and gave a mixture of *R* and *S* sulfoxides **51** in the ratio 2:1, separable by chromatography. The latter was considered to be a very versatile intermediate for the synthesis of a wide variety of commercially significant cephalosporins. The highly desirable exomethylene moiety located at the 3-position offers the opportunity to functionalize that group and to prepare various 3-substituted cephalosporins.

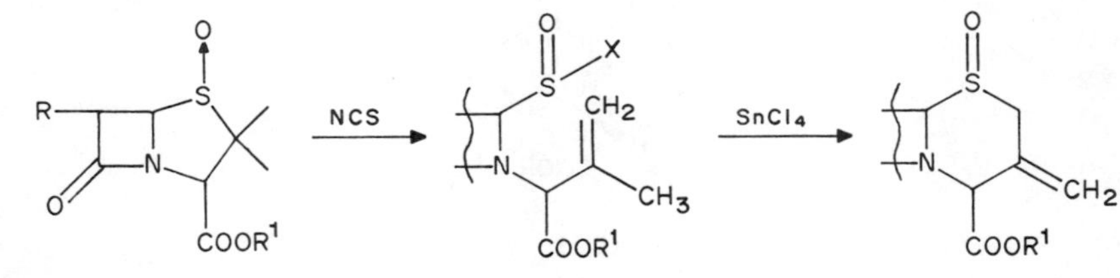


**(49)** **(50)** **(51)**

R = phthalimido (Ft), phenoxyacetamido
$R^1 = CH_3$, *p*-nitrobenzyl (pNB)
NCS = *N*-chlorosuccinimide

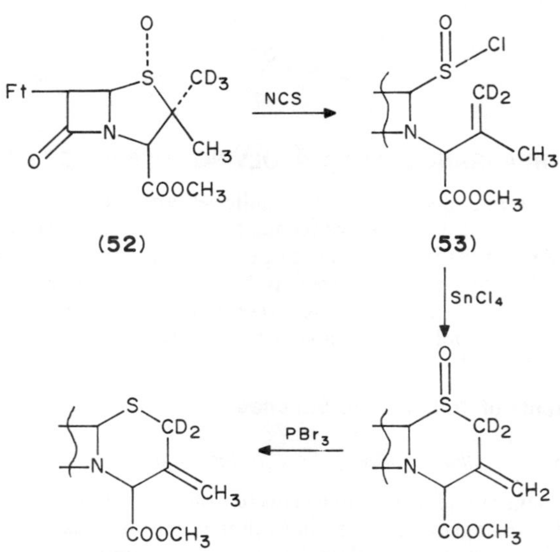


**(52)** **(53)**

**(55)** **(54)**

In order to establish which carbon of the intermediate sulfinyl halide participates in the formation of the S—C bond during the cyclization process, the rearrangement was repeated with deuterated substrate **52**. Treatment of the latter with NCS afforded sulfinyl chloride **53**, with the methylene group being more than 95% deuteriated. Ring closure of **53** with $SnCl_4$, as usual, gave a mixture of *R* and *S* sulfoxides **54**, which was immediately reduced with $PBr_3$ to methyl 2-dideuterio-3-methylene-7-phthalimidocepham-4-carboxylate (**55**).

A sulfinium cation, **56**, was suggested as a probable intermediate in the ring closure of sulfinyl chloride **50**, with Lewis acids, and the mechanism visualized as an intramolecular ene reaction. In support of this suggestion it was found that other sulfinic acid derivatives capable of forming a sulfinium cation, including sulfinic acids themselves[90], have also been found to cyclize under the same conditions.

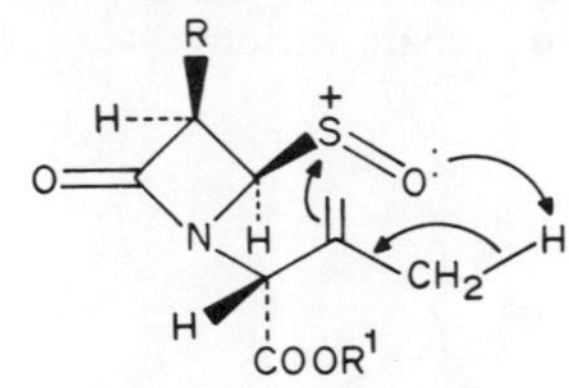

**(56)**

Formation of α-fluorosulfinylacyl fluorides by rearrangement of a postulated β-sultine intermediate, generated by photoreaction of sulfur dioxide with perfluoroolefins in the condensed phase under UV irradiation, has also been reported (equation 24)[91].

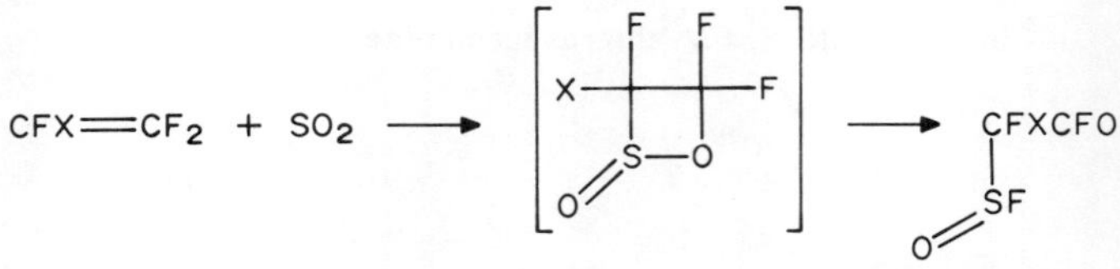

(24)

## V. REARRANGEMENTS INVOLVING SULFINATE ESTERS

Rearrangements of esters of sulfinic acids to sulfones, and the reverse process, are among the best studied and most useful rearrangements of organosulfur compounds in general, and sulfinic acid derivatives in particular. The extent of interest in these rearrangements is reflected by more than one hundred papers published on this subject. However, since this literature was recently reviewed by the present author in considerable detail[17], the following discussion will only present a summary of the main features of these reactions.

### A. Rearrangements of Sulfinates to Sulfones

#### *1. Rearrangements of alkyl and benzyl sulfinates to sulfones*

The rearrangement of esters of sulfinic acids to sulfones (equation 25) is one of the oldest and best studied rearrangements involving sulfones, dating back to 1930, when Kenyon and Phillips[92] first reported that α-phenylethyl *p*-toluenesulfinate rearranged on standing to α-phenylethyl *p*-tolyl sulfone. Subsequently, Kenyon and coworkers[93] observed that this rearrangement was favored by an increase in solvent polarity and that in formic acid the optically active ester was converted to completely racemic sulfone. These results were considered as consistent with an ionic mechanism. Similarly, Stevens and coworkers[94] investigated the rearrangement of a number of sulfinates to sulfones and suggested an ionic mechanism. It should be pointed out that the driving force for the sulfinate to sulfone isomerization is the formation of the strong sulfur oxygen bond in the sulfonyl group (112 kcal mol$^{-1}$)[95], a result of back donation of a pair of nonbonding electrons from the oxygen atom into empty *d* orbitals of the sulfur atom, with consequent $p_\pi$–$d_\pi$ overlap[96].

$$\mathrm{Ar}\overset{\mathrm{O}}{\overset{\|}{\mathrm{S}}}\mathrm{OR} \longrightarrow \mathrm{Ar}\underset{\underset{\mathrm{O}}{\|}}{\overset{\overset{\mathrm{O}}{\|}}{\mathrm{S}}}\mathrm{R} \tag{25}$$

Neither of these reports provides any information with regard to the type of ionization, i.e. ionization to free ions or ion pairs. The more recent, and mechanistically detailed investigations by Darwish and coworkers[97–102] prove quite useful in this respect, and may be regarded as the most important contribution in this field. These authors have examined the rearrangement of *t*-butyl[97], α-phenylethyl[97,98], α-(*p*-methoxyphenyl)ethyl[97,98], benzhydryl[97,99], cumyl[100] and trityl[101,102] 2,6-dimethylbenzenesulfinates under a variety of conditions. The main findings revealed by these investigations are as follows. The rate of rearrangement and solvolysis of these esters showed a high sensitivity to the ionizing power of the solvent and to the introduction of a *para*-methoxy group into the aromatic group indicative of an ionization mechanism.

However, several pieces of evidene indicate that sulfone is not formed by recombination of dissociated ions. For example, when the reactant was optically active, diastereomerically pure α-phenylethyl 2,6-dimethylbenzenesulfinate, the sulfone which was produced was also optically active and of over 95% retained configuration, but the ester recovered after partial reaction was a mixture of diastereomers[103]. Furthermore, addition of 2,6-dimethylbenzenesulfinate anion to any of the systems did not increase the fraction of sulfone formed, as would have been expected if the ions were competing with solvent for the cation. Even with the highly stable trityl cation produced during the rearrangement of trityl 2-methylbenzenesulfinate in acetonitrile, only 45% at most could be diverted to form trityl azide by the addition of tetrabutylammonium azide. On the basis of these observations and other pertinent data, the mechanism which was proposed included the intermediacy of ion pairs, as shown in equation 26, where $ArSO_2^-R^+$ is a noncapturable intimate ion pair, and $ArSO_2^- \| R^+$ is a capturable solvent-separated ion pair. This mechanism has received further support from related studies conducted by several other investigators[104–108].

$$\mathrm{Ar}\overset{\mathrm{O}}{\overset{\|}{\mathrm{S}}}\mathrm{OR} \rightleftharpoons \mathrm{Ar}\overset{\mathrm{O}}{\overset{\|}{\mathrm{S}}}\mathrm{O^-R^+} \rightleftharpoons \mathrm{Ar}\overset{\mathrm{O}}{\overset{\|}{\mathrm{S}}}\mathrm{O^-} \| \mathrm{R^+} \longrightarrow \mathrm{Ar}\underset{\underset{\mathrm{O}}{\|}}{\overset{\overset{\mathrm{O}}{\|}}{\mathrm{S}}}\mathrm{R} \tag{26}$$

$$\mathrm{R} = t\text{-Bu, } \mathrm{Ph_2CH, PhCHCH_3, Ph_3C}$$

The rearrangement of furfuryl benzenesulfinate (**57**) appears of special interest. In contrast with the corresponding benzyl ester, this sulfinate was found to undergo a facile rearrangement to sulfone. Furthermore, in nonhydroxylic solvents a mixture of furfuryl phenyl sulfone (**58**) and 2-methyl-3-furyl phenyl sulfone (**59**) is obtained (equation 27)[106].

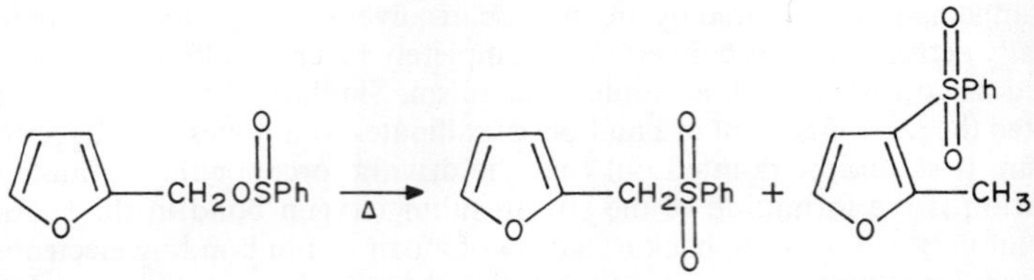

(57) (58) (59) (27)

Of special interest are also the rearrangements of benzylic trichloro- and trifluoromethanesulfinates, which are easily prepared by MCPBA oxidation of the corresponding sulfenate esters, and which rearrange to sulfones orders of magnitude faster than the corresponding arenesulfinates, a consequence of the high leaving-group ability of the $X_3CSO_2^-$ anion[107,108]. Similar observations have been made by Hendrickson[109], who has also demonstrated the synthetic utility of the rearranged trifluoromethyl sulfones, so-called triflones, in the variety of ways they facilitate carbon–carbon bond construction[110].

The rearrangement of several cyclic benzylic sulfinates have also been described in the literature by the groups of Durst[111a] and of Hogeveen[111b] and seem to proceed by a special two-step mechanism: retro Diels–Alder extrusion of $SO_2$, followed by its cheletropic addition to the unstable quinodimethane intermediate (equation 28).

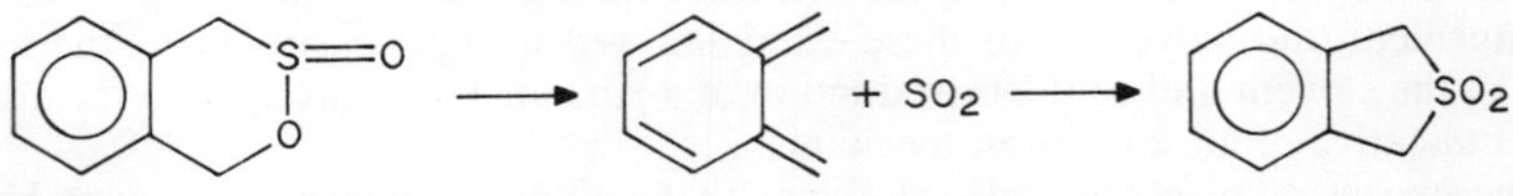

(28)

Finally, Kornblum and coworkers[112] reported a particularly interesting and efficient isomerization of *p*-nitrocumyl arenesulfinates to the corresponding sulfones which occurs at room temperature in quantitative yield. This rearrangement is believed to occur by the general and well-known electron-transfer chain-substitution mechanism of ambident anions introduced by Kornblum (Scheme 5)[112b].

*2. The [2,3]sigmatropic rearrangement of allylic sulfinates to sulfones*

In 1950, Cope and coworkers[113] have examined the thermal stability of allylic arenesulfinates. They found that allyl, crotyl and α-methylallyl benzenesulfinates on heating underwent rearrangement to sulfones in low yields, but were unable to reach a decision with regard to the reaction mechanism, mainly because the last two esters gave the same product: crotyl phenyl sulfone.

Some ten years later, Darwish and Braverman[114,115] undertook a more extensive study of this rearrangement, which has revealed some unique features. These investigators examined the behavior of six different esters, namely allyl, crotyl, α-methylallyl, racemic and optically active α,γ-dimethylallyl, cinnamyl and α-phenylallyl 2,6-dimethylbenzenesulfinates under various reaction conditions.

All these esters have been found to undergo rearrangement to sulfones in high yields even under solvolytic conditions (equation 29). This result seems of interest in view of the fact that under such conditions sulfinates in general undergo solvolysis, with relatively little sulfone formation[97,104–107]. The second point of interest is that the rearrangement of unsymmetrically substituted allylic esters involves simultaneous isomerization of the allylic group. Similarly, optically active α,γ-dimethyl 2,6-dimethylbenzenesulfinate rearranged to the corresponding optically active sulfone with practically complete inversion of configuration (equation 30)[114,115].

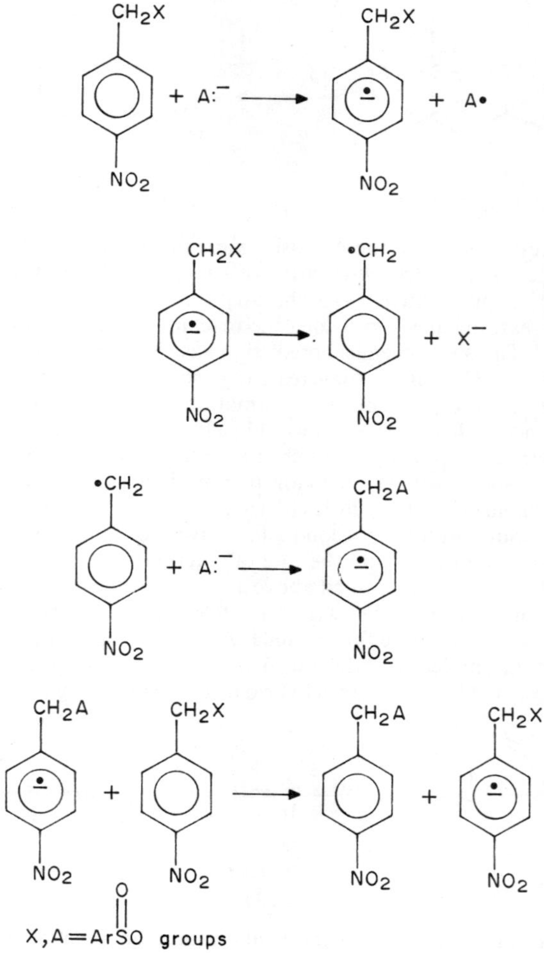


SCHEME 5

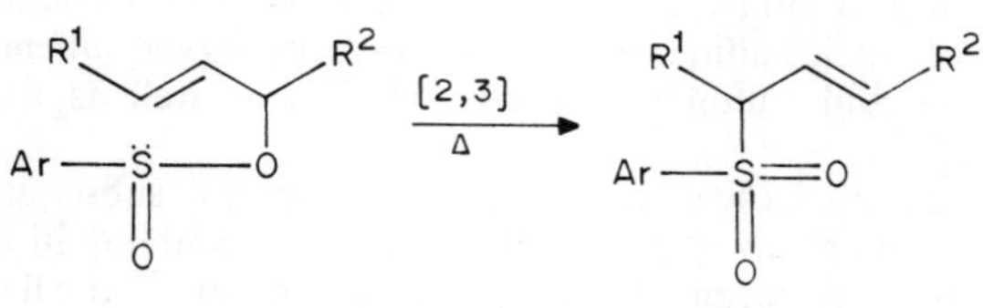


| | a | b | c | d | e | f |
|---|---|---|---|---|---|---|
| $R^1$= | H | Me | H | Me | Ph | H |
| $R^2$= | H | H | Me | Me | H | Ph |

Ar=2,6-$Me_2C_6H_3$ (29)

**(60)** **(61)**

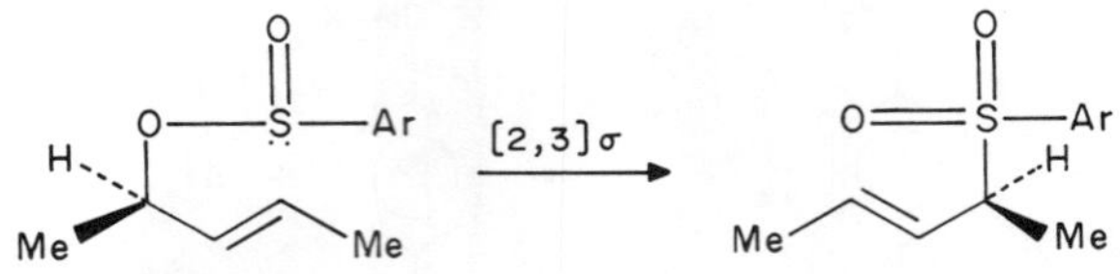

(30)

(**60d**) $[\alpha]_D^{25}$ + 3.86°   (**61d**) $[\alpha]_D^{25}$ + 7.02°

It may be of interest to note that the stereospecific transformation shown in equation 30 has been cited as the first reported observation of a 1 → 3 chirality transfer[116]. As pointed out by Hoffmann[116], quantitative 1 → 3 chirality transfer will follow from the suprafacial[117] course of rearrangement, provided the reactant has a uniform configuration at the $\beta,\gamma$-double bond. This stereochemical prediction has also been confirmed by the results obtained in several other [2,3]sigmatropic rearrangements, subsequently reported[118–120].

The evidence presented so far excludes the formation of dissociated ions as the principal route to sulfones, since such a mechanism would yield a mixture of two isomeric sulfones. Similarly, in the case of an optically active ester a racemic product would be formed. The observed data are consistent with either an ion-pair mechanism or a more concerted cyclic intramolecular mechanism involving little change between the polarity of the ground state and transition state. Support for the second alternative was found from measurements of the substituent and solvent effects on the rate of reaction[17].

On the basis of the evidence presented above as well as some other pertinent data (e.g. negative entropies of activation), Darwish and Braverman[114] have suggested that the rearrangement of allylic arenesulfinates (**60a–f**) to corresponding sulfones (**61a–f**) proceeds by a cyclic intramolecular mechanism involving a five-membered transition state which may be represented by a resonance hybrid **62** of the following resonance structures.

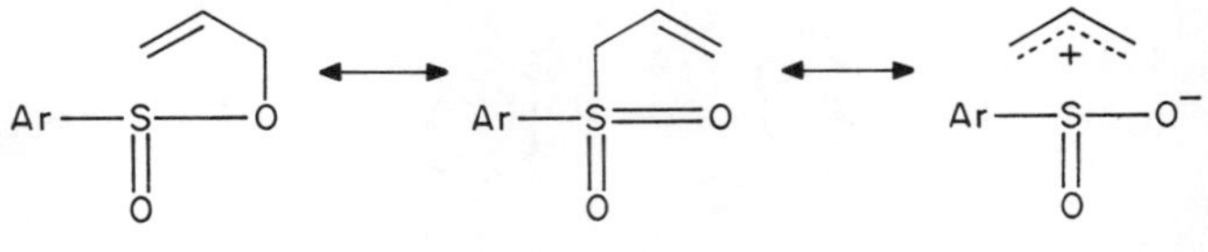

(**62**)

One would expect a graded sequence of transition states between the covalent and ionic structures. It is conceivable that with allyl 2,6-dimethylbenzenesulfinate in nonpolar solvents the covalent resonance structure of the transition state is the major contributor. It is also probable that replacement of a hydrogen of the allyl group by a carbenium ion-stabilizing substituent such as alkyl and phenyl groups and the use of solvents of high ionizing power will enhance the contribution of the ionic resonance structure. It should be added that this is not only one of the first and best studied [2,3]sigmatropic rearrangements[121], but it has also been used as a model for the prediction of the closely related [2,3]sigmatropic rearrangements of allylic sulfinates to sulfoxides, propargylic sulfenats and sulfinates to allenic sulfoxides and sulfones, respectively[28,115], as well as their corresponding selenium analogues[122].

Further support of the proposed mechanism can be found in several subsequent studies[123–129], including the most elegant investigation of oxygen-18 scrambling in the rearrangement of allylic arenesulfinates performed by Darwish and Armour[123], the flash pyrolysis of deuterated allylic sulfinates[124] and the transfer of chirality from sulfur to carbon in the rearrangement of optically active *cis* and *trans* $\gamma$-substituted allylic *p*-toluenesulfinates to optically active chiral sulfones[125,126] (equation 31).

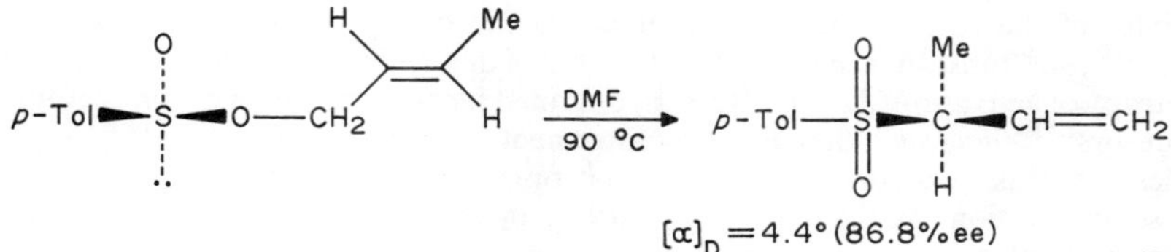

(31)[126]

Some useful synthetic applications of the allylic sulfinate–sulfone rearrangement have also been reported[128,129] (equation 32).

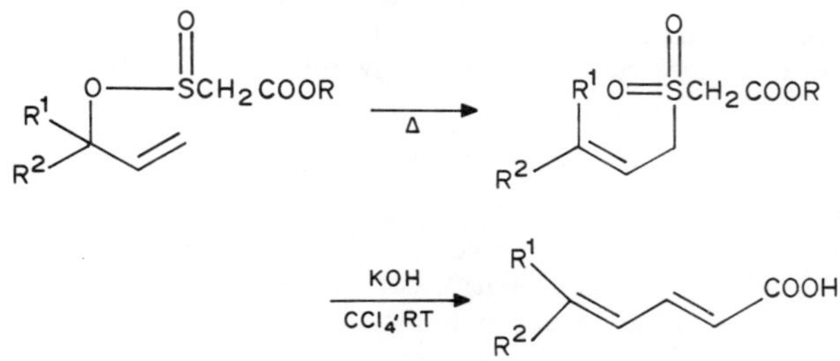

(32)[128]

*3. The [2,3]sigmatropic rearrangement of propargylic arenesulfinates to allenyl aryl sulfones*

Following studies on the rearrangement of allylic arenesulfinates, Braverman and coworkers have investigated a number of natural extensions of this unique transformation including the predictable [2,3]sigmatropic rearrangement of propargylic sulfinates to allenic sulfones, described in equation 33.

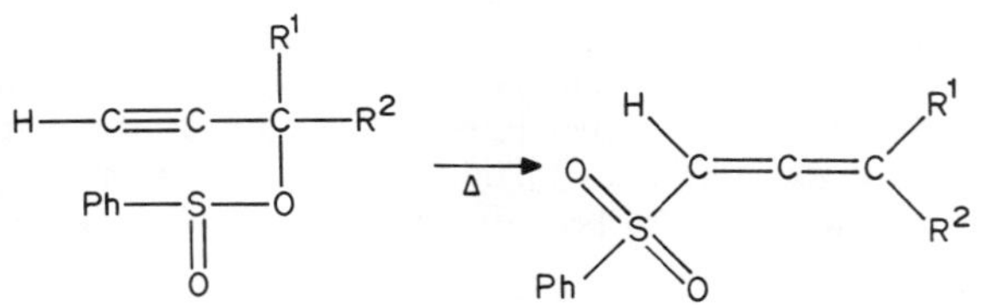

(33)

**(63)** **(64)**

(**a**) $R^1 = R^2 = Me$
(**b**) $R^1 = R^2 = H$
(**c**) $R^1 = H, R^2 = Me$
(**d**) $R^1 = H, R^2 = Ph$
(**e**) $R^1 = Me, R^2 = Ph$

Rearrangements of propargylic systems to allenes in general have been widely studied and are well documented[130–133]. In 1966. Braverman and Mechoulam[134a] first reported the facile thermal rearrangement (equation 33) of α,α-dimethylpropargyl benzenesulfinate (**63a**) to γ,γ-dimethylallenyl phenyl sulfone (**64a**) thus indicating the occurrence of an 'allylic shift' for this system as well, in spite of certain geometrical differences. The analogy between the rearrangement of allylic arenesulfinates, as described in the previous section, and the corresponding propargylic esters was further demonstrated by the almost exclusive rearrangement to sulfone even under solvolytic conditions, as well as by a low sensitivity of the rate of the rearrangement to the change in ionizing power of the solvent[134b], and substituent effect[134c]. In the light of this evidence and the negative value of the entropy of activation ($\Delta S^{\ddagger} = -12.8$ eu) obtained for the reaction of **63a** in acetonitrile, the authors[134] suggested that the rearrangement of propargylic sulfinates to allenyl sulfones proceeds by a concerted [2,3]sigmatropic shift mechanism.

These observations on the rearrangement of propargylic arenesulfinates are confirmed by the work of Stirling and Smith[135] performed contemporaneously with that by Braverman and Mechoulam[134]. These authors reported that $\gamma$-deuteriopropargyl *p*-toluenesulfinate rearranged to $\gamma$-deuterioallenyl *p*-tolyl sulfone on heating at 130 °C, and that under similar conditions *R*-(+)-$\alpha$-methylpropargyl *p*-toluenesulfinate rearranged to (−)-$\gamma$-methylallenyl *p*-tolyl sulfone whose absolute configuration, predicted on the basis of a cyclic intramolecular mechanism, agrees with that calculated from the polarizability sequence of substituents attached to the allene system.

### 4. *The double [2,3]sigmatropic rearrangement of allylic and propargylic sulfoxylates*

An interesting extension of the [2,3]sigmatropic rearrangements of allylic and propargylic sulfinates, discussed in the preceding two subsections, as well as the analogous rearrangements of allylic[136] and propargylic[28,137] sulfenates, is the double [2,3]sigmatropic rearrangement of the corresponding sulfoxylate esters.

Braverman and Segev[138] first reported a convenient method for the preparation of conjugated diallenyl sulfones, involving a double [2,3]sigmatropic shift of propargylic sulfoxylates, as illustrated in equation 34. While the rearrangement of sulfinate **65** requires moderate heating for several hours, the rearrangement of its sulfoxylate precursor proceeds spontaneously at low temperature. Diallenyl sulfone **66** was found to undergo some interesting thermal and ionic rearrangements to cyclic products[138]. Subsequently, Büchi and Freidinger[139] have reported the analogous rearrangement of allylic sulfoxylates to diallylic sulfones, as illustrated in equation 35. In this case too, the rearrangement of sulfinate **67** to the bisallyl sulfone **68** is best accomplished by brief reflux in toluene, while the formation of **67** itself takes place below room temperature. The analogous rearrangement of allyl thiosulfoxylate to allyl thiosulfinate (equation 36), believed to represent the antibacterial principle of Allium sativum (garlic), has also been reported[140]. In this case, however, the rearrangement can be reversed by $\alpha$-substitution in the product. A useful conversion of diallylic sulfones to the corresponding trienes, by way of the Ramberg–Bäcklund reaction, has also been described and applied for the synthesis of various natural products.

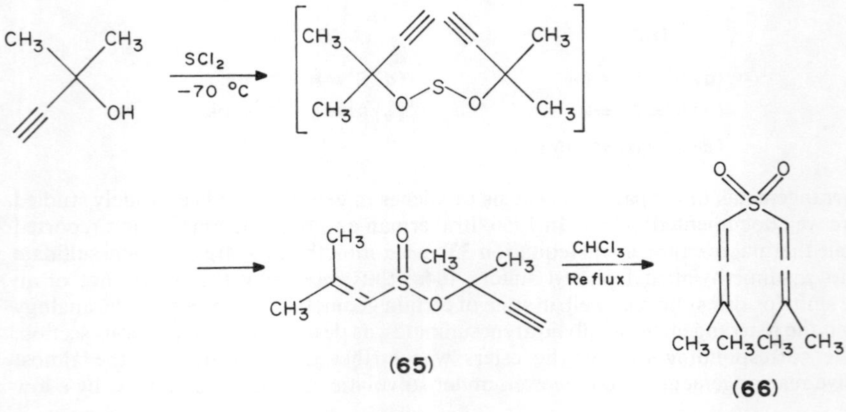


(65) (66) (34)

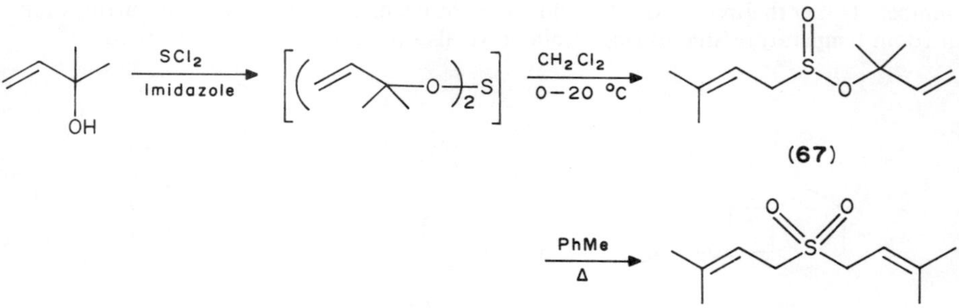

(**67**)

(**68**) (35)

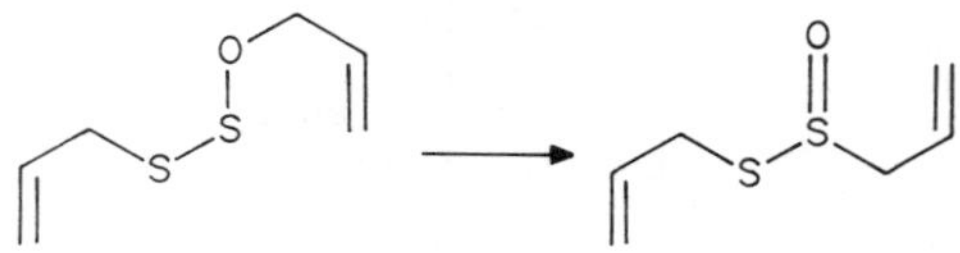

(36)

## B. Rearrangements of Sulfones to Sulfinates

### 1. *Thermal rearrangements*

In contrast with the relatively facile thermal rearrangement of sulfinates to sulfones discussed in the preceding section, the reverse process is, relatively, rarely encountered and is usually observed only at elevated temperatures. One of the first thermal sulfone to sulfinate isomerizations has been invoked by Fields and Meyerson[141] to occur during the pyrolysis of dibenzothiophene *S,S*-dioxide (**69**) to dibenzofuran, through elimination of sulfur monoxide from the sultine intermediate **70** (equation 37).

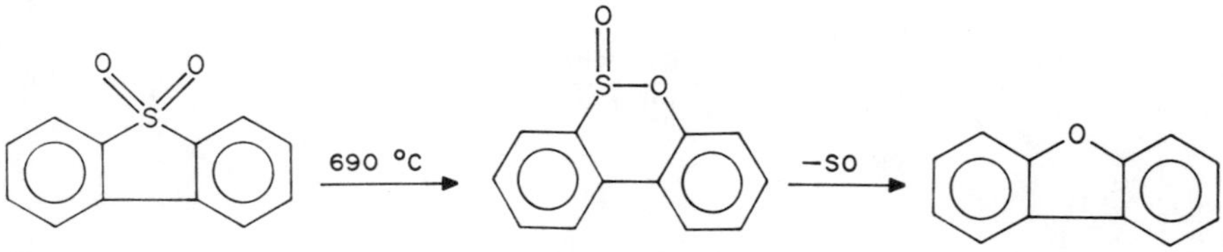

(**69**) (**70**) (37)

Several reports involving the rearrangement of cyclic four-membered $\alpha,\beta$-unsaturated sulfones to the corresponding five-membered cyclic sulfinates ($\gamma$-sultines) were observed. For example, Dittmer and coworkers[142] have observed the rearrangement of benzothiete 1,1-dioxide (**71**) to benzosultine (**72**) in 90% yield at 210 °C, while Hoffman and Sieben[143] reported the gas-phase rearrangement of **73** to **74** at 300 °C. Contemporaneously, King and coworkers[144–146] have studied the thermal rearrangement of the parent molecule thiete 1,1-dioxide (**75**) to $\gamma$-sultine **76**, and rationalized their results in terms of a mechanism involving vinyl sulfene as a reactive intermediate, which is formed and reacts in a concerted manner (equation 38). This intermediate could be trapped by reaction with phenol, and the release of ring strain during its formation provides the driving force for the reaction. A number of other thermal sulfone to sulfinate rearrangements, some of them occurring even at room temperature, due to ring strain, have also been reported and reviewed[17].

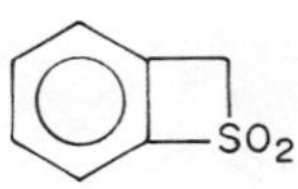


**(71)**

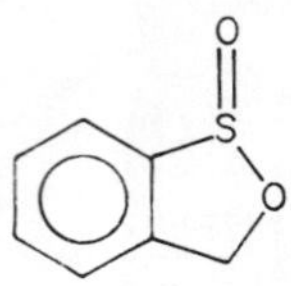


**(72)**

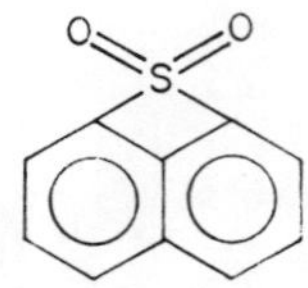


**(73)**

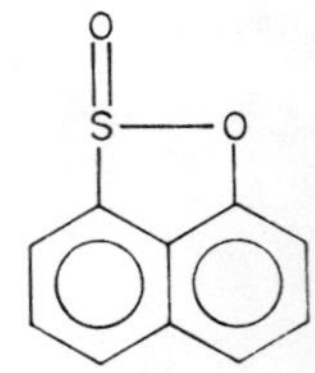


**(74)**

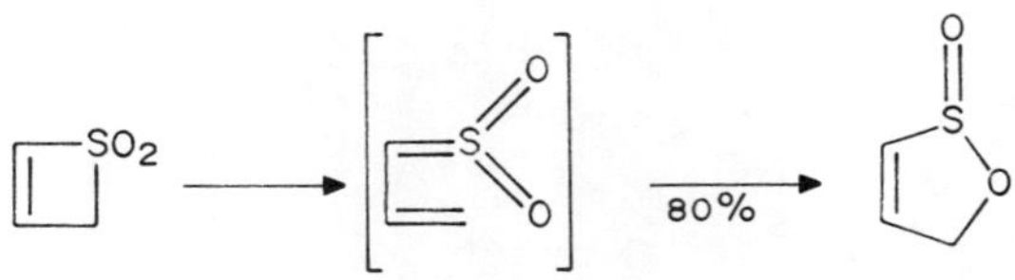


**(75)** **(76)**

(38)

2. *Ionic rearrangements*

Similarly to the thermal rearrangements discussed in the previous subsection, the base-catalyzed rearrangements of cyclic four-membered sulfones to five-membered sulfinates have also been reported. For example, Dodson and coworkers[147] have observed the rearrangement of *cis*- and *trans*-2,4-diphenylthietane 1,1-dioxides to *cis*- and *trans*-3,5-diphenyl-1,2-oxathiolane (2,3)-*cis*-oxides, respectively (equation 39), on treatment with *t*-butoxymagnesium bromide, which is stereospecific with respect to the phenyl group but stereoselective with respect to the oxygen atoms on sulfur.

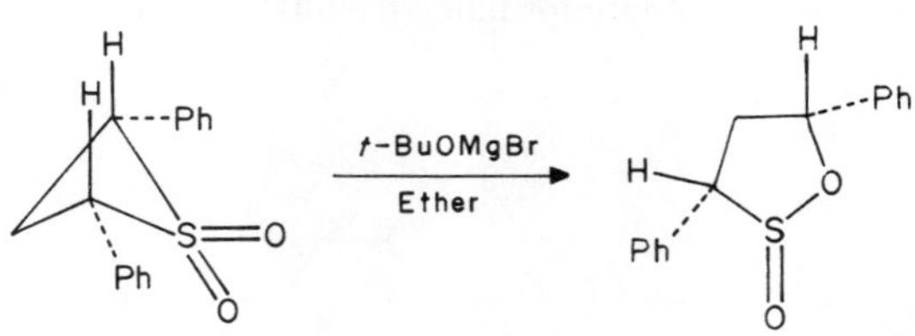


(39)

More recently, Braverman and Reisman[148] have found that addition of a carbon tetrachloride solution of bromine to bis-$\gamma,\gamma$-dimethylallenyl sulfone (**66**) at room temperature unexpectedly resulted in spontaneous and quantitative fragmentation of the sulfone, with formation of the $\alpha,\beta$-unsaturated $\gamma$-sultine **77a** and the tribromo products **78** and **79** (equation 40). Analogously, treatment of the same sulfone with trifluoroacetic acid gives rise to $\gamma$-sultine **77b**. It is interesting to note that from a synthetic point of view it is not even necessary to prepare the diallenyl sulfone **66**, since one can use its sulfinate precursor (equation 34) to obtain exactly the same results, under the same conditions[149]. The authors suggested that the fragmentation cyclization of sulfone **66** may take place by the mechanism depicted in equation 41.

The conversion of sulfones to sulfinates under electrophilic conditions such as those described above appears to be unique. In continuation, a stereochemical study of the reaction has also been performed[149–151]. Racemic $\gamma$-methyl and $\gamma$-*t*-butyl allenyl *t*-butyl

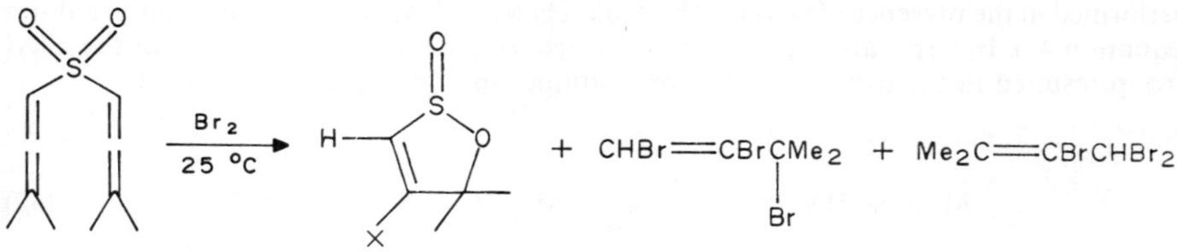

(**66**) (**77a**) X=Br (**78**) (**79**)
(**77b**) X=H

(40)

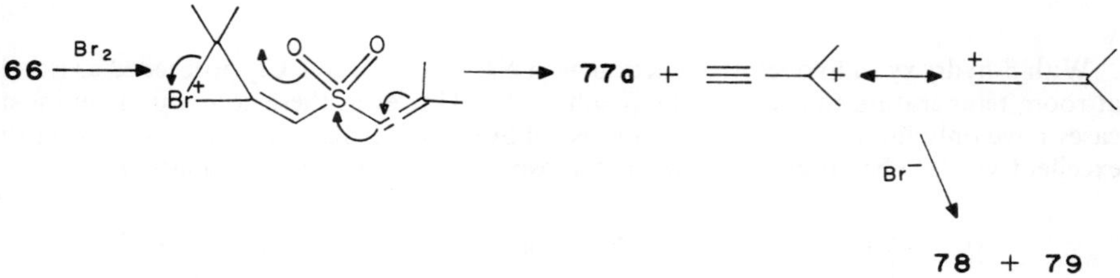

(41)

sulfones (**80**,**81**) were prepared by [2,3]sigmatropic rearrangements of the corresponding α-alkylpropargyl *t*-butylsulfinate. Optically active sulfones (−)-**80** ($[\alpha]^{25}_{D}$—47.5°) and (−)-**81** ($[\alpha^{25}_{D}$—58.5°) were obtained by the elegant method of kinetic resolutions[152], and were assigned the *R* absolute configuration. Treatment of these sulfones with bromine and methanesulfenyl chloride gave optically active γ-sultines **82**–**85**. A stereoselective synthesis of optically active chiral α,β-unsaturated γ-sultines of known absolute configuration has thus been achieved[17].

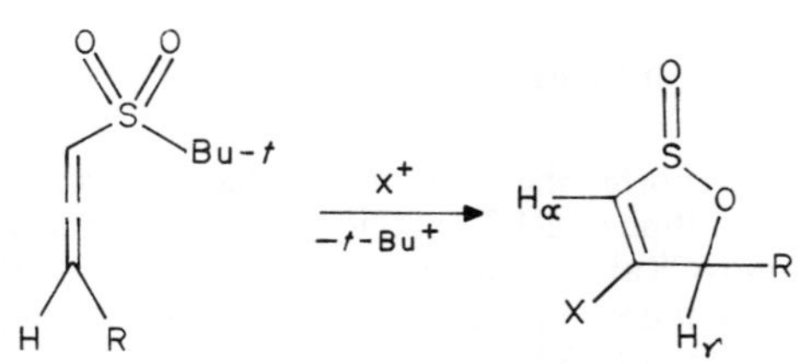

(*R*)-(−)-(**80**) (**81**)
(**80**) R=Me
(**81**) R=*t*-Bu

| | $[\alpha]^{25}_{D}$ (deg) |
|---|---|
| (**82**) X=Br, R=Me | + 16.9 |
| (**83**) X=Br, R=*t*-Bu | + 15.6 |
| (**84**) X=MeS, R=Me | + 23.7 |
| (**85**) X=MeS, R=*t*-Bu | + 15.9 |

### C. Rearrangements of Sulfoxides to Sulfinates

Jung and Durst[153] reported that certain sulfoxides such as *t*-butyl alkyl or aryl sulfoxides undergo carbon–sulfur bond cleavage upon treatment with positive halogen species such as *N*-bromo or *N*-chlorosuccinimide. When the cleavage reaction was performed in the presence of alcohol, the products were alkyl sulfinates and *t*-butyl halides (equation 42). Incorporation of the hydroxy group into the $\gamma$ or $\delta$ position of the alkyl group resulted in the formation of $\gamma$- or $\delta$-sultines in high yields (equation 43).

$$\mathrm{Me_3C\overset{\overset{\displaystyle O}{\|}}{S}CHR^1R^2 \xrightarrow[CH_2Cl_2,\ ROH]{NCS} Me_3CCl + R^1R^2CH\overset{\overset{\displaystyle O}{\|}}{S}OR} \qquad (42)$$

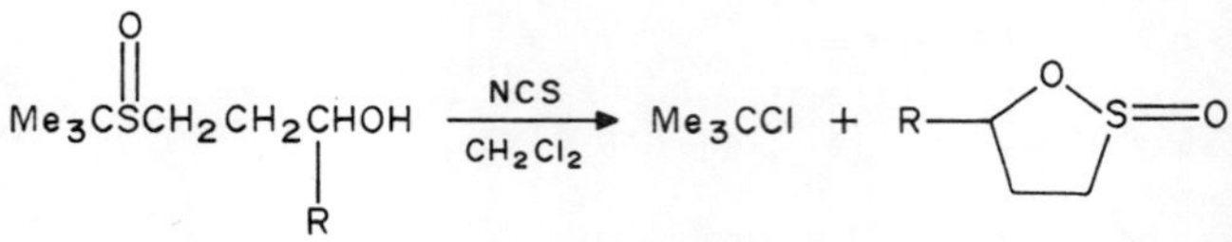

(43)

With $\beta$-hydroxy sulfoxides, the reaction with NBS, NCS or $SO_2Cl_2$ proceeded as usual at room temperature to give initially $\beta$-sultines[154]. However, these compounds in most cases have only limited stability, due to loss of $SO_2$ and formation of olefins in good to excellent yields. The latter occurs by a stereospecific *cis* elimination (equation 44).

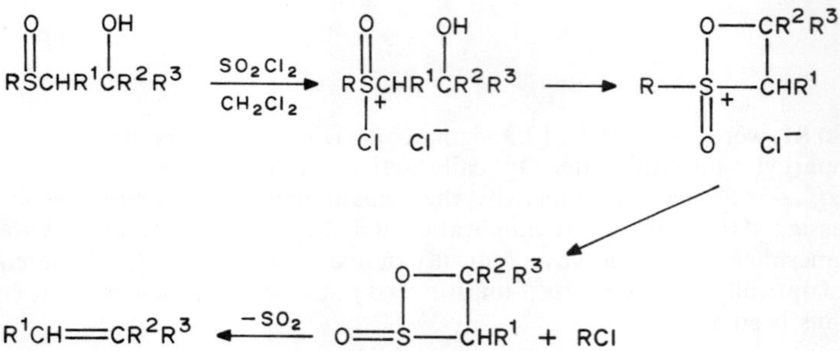

(44)

R = *tert*-butyl, $R^1, R^2, R^3$ = alkyl or aryl

Subsequently, the same group[155] reported the isolation and characterization of a stable crystalline $\beta$-sultine, 3,3-dimethyl-2,2-diphenyl-1,2-oxathiethan 2-oxide, from the reaction of the $\beta$-hydroxy sulfoxide **85a** with NCS at room temperature. This compound is stable at room temperature for several days but decomposes quantitatively into 1,1-diphenyl-2,2-dimethylethene and $SO_2$, when warmed to 30 °C in $CH_2C_2$ ($t_{1/2} \approx 24$ h).

## VI. REARRANGEMENTS INVOLVING SULFINAMIDES

### A. Pericyclic Reactions

#### 1. *Cycloaddition and electrocyclization reactions*

Maricich and Hoffman[156] observed that, unlike other azides, the reaction of benzenesulfinyl azide (**86**) with sulfoxides gave *N*-benzenesulfonyl sulfimides (**88**), instead of the expected sulfoximide, adduct **87**. The results were interpreted by a two-step 1,2-dipolar cycloaddition of a delocalized sulfinyl nitrene intermediate **89** with sulfoxide, followed by an electrocyclic ring opening of the cyclic sulfurane **90**.

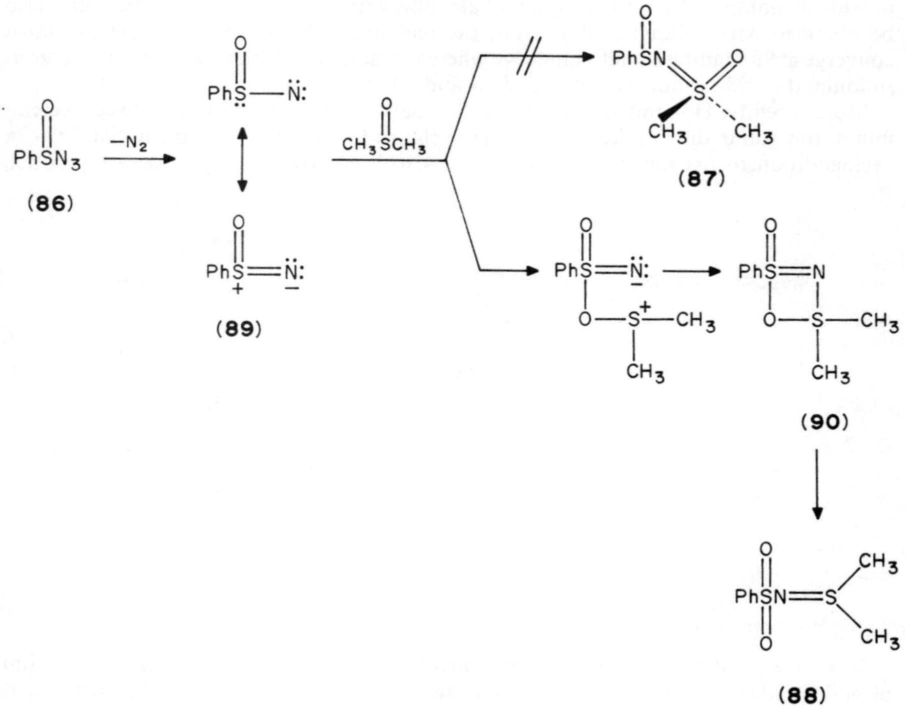

In a related study, the same authors[157] reported that certain *N*-alkoxybenzene-sulfinamides (**91**) reacted with alcohols in a way that implicated a sulfinyl nitrenium ion intermediate (**92**). The latter is generated by an unprecedented dissociative rearrangement from **91** to **93** (equation 45) involving migration of an alkoxy group from nitrogen to adjacent sulfur, which can also exchange with the alcohol solvent. The sulfonimidate intermediates **93** also alkylate the solvent.

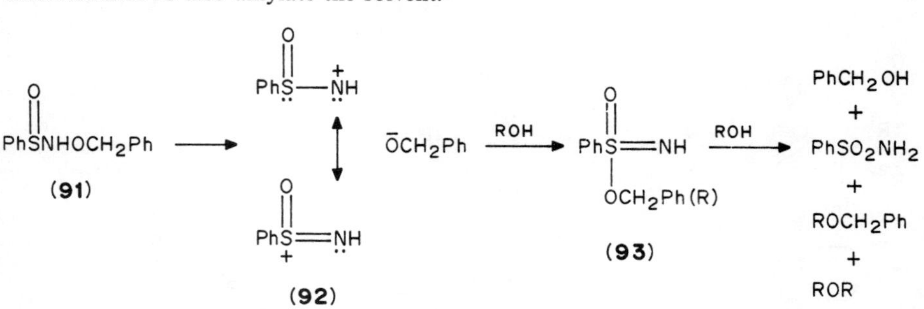

(45)

Subsequently, evidence supporting the convergence of mechanisms between the sulfinyl azide and *N*-alkoxysulfinamide reactions with both sulfoxides and alcohols has also been presented[158]. The major products previously obtained from the reaction of sulfoxides with sulfinyl azides were also obtained with *N*-alkoxysulfinamides. Likewise, major products previously obtained from the reaction of alcohols with $N$-alkoxysulfinamides could also be obtained with sulfinyl azides. Thus, the reaction of **86** and **91** with DMSO must converge at **90**, leading to sulfinimide **88**, whereas reaction with alcohols must converge at sulfonimidate **93**, leading to sulfonamides and ethers.

More recently, Dondoni and coworkers[159] have found that reaction between ketenimines and sulfur dioxide leads to stable cyclic sulfinimides **95**, via an initial $2\pi + 2\pi$ cycloaddition, to give the unstable $\beta$-sultines **94**, which rearrange to the observed product.

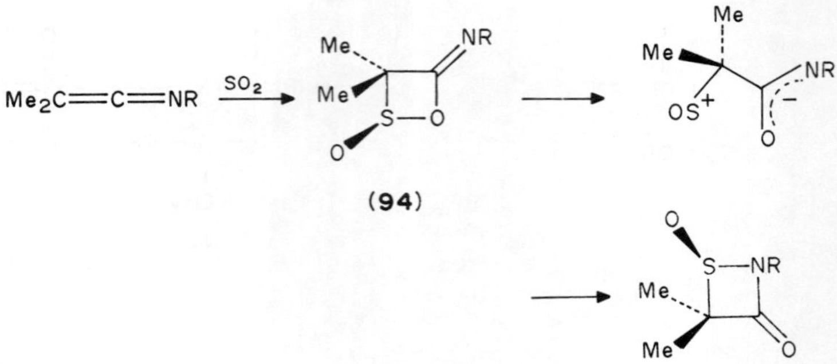


(94)

(95)

2. *Sigmatropic rearrangements*

Several sigmatropic rearrangements involving the formation or transformation of sulfinamides have been reported. For example, Scherer and Schmitt[160] reported that $N,N$-bis-trimethylsilylmethanesulfinamide (**96**), undergoes tautomerization from an amido to imido ester (**97**, equation 46) by a [1,3]sigmatropic rearrangement of a trimethylsilyl group.

$$\underset{(\mathbf{96})}{Me_3Si{-}N(SiMe_3){-}S(=O)CH_3} \rightleftharpoons \underset{(\mathbf{97})}{Me_3Si{-}N{=}S(OSiMe_3)CH_3} \qquad (46)$$

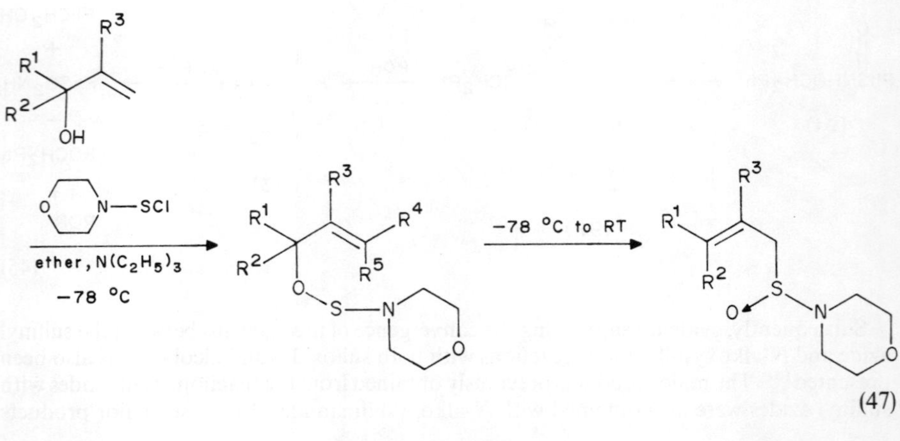


(47)

(48)

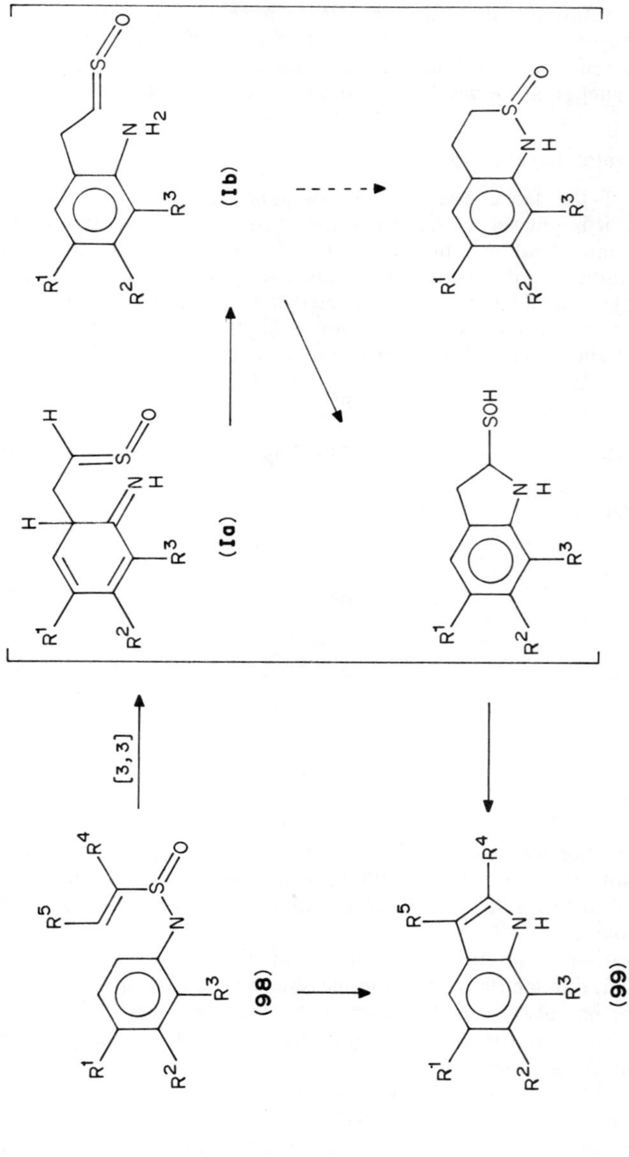

An interesting application of the well-known [2,3]sigmatropic rearrangement of allylic sulfenates to sulfoxides[28] to the preparation of allylic sulfinamides was recently reported by Baudin and Julia[161]. Thus, reaction of the easily available 4-morpholinesulfenyl chloride[162] with various allylic alcohols in the presence of triethylamine at low temperature readily affords allylic sulfinamides via the intermediacy of the unstable 4-morpholinesulfenate esters (equation 47).

The same authors[163] have also reported a novel synthesis of indoles (**99**) through a [3,3]sigmatropic rearrangement of *N*-aryl-1-alkenylsulfinamides **98**, which are smoothly obtained by reaction of vinylic Grignard reagents with *N*-sulfinylanilines (Ar—N=S=O), and suggested the mechanism shown in equation 48.

### 3. Ene and retro-ene reactions

In view of the facile ene reaction of sulfur dioxide with olefins discussed in Section II.A, it is not surprising that some of its nitrogen analogs, in particular sulfur diimides[164] and *N*-sulfinylamilines[165], would also react analogously. These reactions have been studied by three different groups led by Kresze, Sharpless and Deleris.

Schönberger and Kresze[166] first reported that propene derivatives react with *N*,*N*′-ditosyl sulfur diimide (**100**) with formation of allylsulfinamidines (**101**), which are rather unstable and undergo a [2,3]sigmatropic rearrangement to *N*-allyl-disulfenamides (**102**).

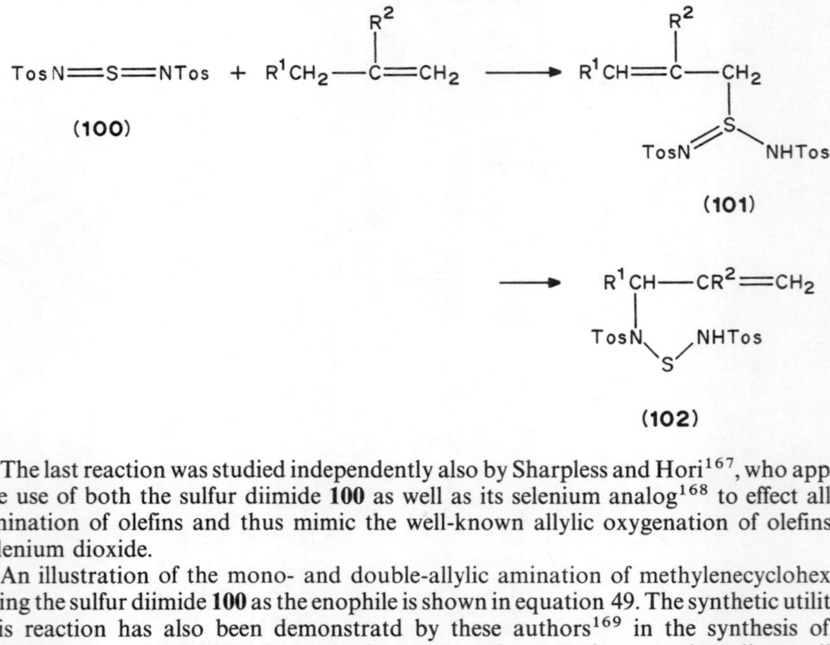


The last reaction was studied independently also by Sharpless and Hori[167], who applied the use of both the sulfur diimide **100** as well as its selenium analog[168] to effect allylic amination of olefins and thus mimic the well-known allylic oxygenation of olefins by selenium dioxide.

An illustration of the mono- and double-allylic amination of methylenecyclohexane using the sulfur diimide **100** as the enophile is shown in equation 49. The synthetic utility of this reaction has also been demonstratd by these authors[169] in the synthesis of *dl*-gabuculine, an inhibitor of *γ*-aminobutyrate amino-transferase, using direct allylic amination as the key step.

Subsequently, Sharpless and coworkers[170] reported that the ene reaction of the related monooxo compounds, the *N*-sulfinylsulfonamides, is reversible under mild conditions, and that this reversibility can be exploited to specifically introduce deuterium or tritium into the allylic position of the substrate. Thus, when *N*-sulfinyl-*p*-toluenesulfonamide (**103**) was stirred with *β*-pinene in benzene for 3 h at 25 °C, a 1:1 adduct, the *N*-tosylsulfinamide

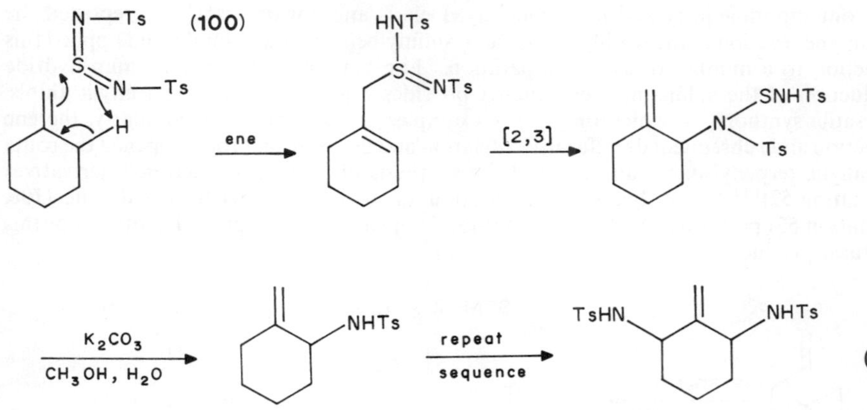

(63%)

(49)

**104**, was isolated in 89% yield (equation 50). However, upon standing in moist air at room temperature for a few days or upon heating above 150 °C, **104** was found to decompose with the liberation of $\beta$-pinene. The same result was also observed when **104** was refluxed in benzene with an excess of $H_2O$. The observed behavior is consistent with a reversible ene reaction, although initial hydrolysis of the allylic sulfinamide adduct **104** to the corresponding sulfinic acid which then undergoes retroene reaction is another likely possibility.

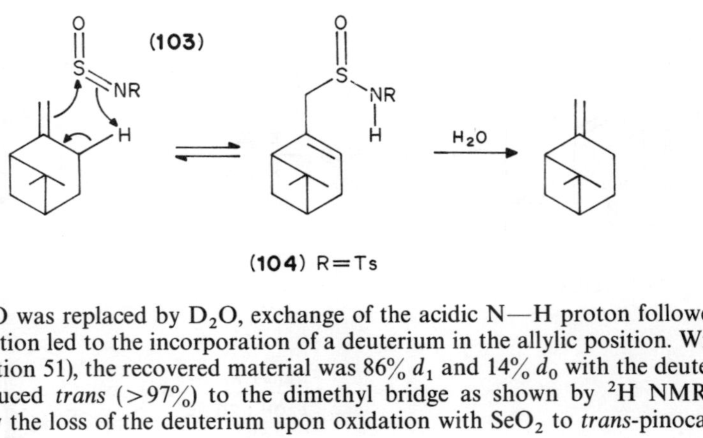

(50)

(**104**) R=Ts

When $H_2O$ was replaced by $D_2O$, exchange of the acidic N—H proton followed by retroene reaction led to the incorporation of a deuterium in the allylic position. With $\beta$-pinene (equation 51), the recovered material was 86% $d_1$ and 14% $d_0$ with the deuterium being introduced *trans* (>97%) to the dimethyl bridge as shown by $^2H$ NMR and confirmed by the loss of the deuterium upon oxidation with $SeO_2$ to *trans*-pinocarveol (**105**).

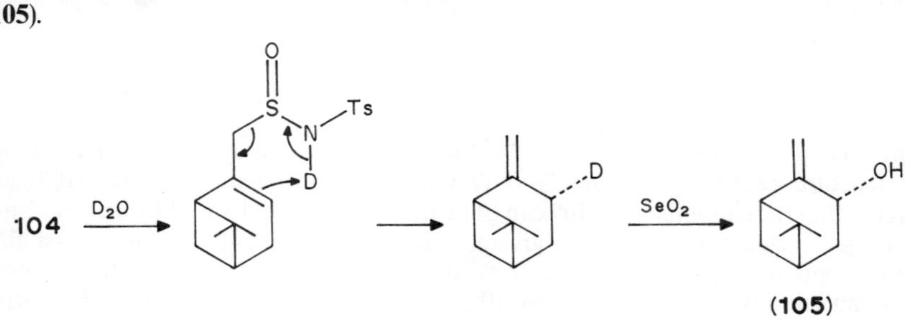

(51)

(**105**)

Contemporaneously and independently, Deleris and coworkers[171–174] reported the facile ene reaction of the readily available *N*-sulfinylbenzenesulfonamide, and applied this reaction to a number of useful preparations. For example, lithium aluminum hydride reduction of the sulfinamide ene-adduct provides a general synthesis of allylic thiols, versatile synthons, of which only a few examples are known[171,172]. Similarly, the ene reaction and subsequent desulfurative silylation and desilylation process opened the route to allylic terpenylsilanes and allowed the synthesis of terpenoid functional derivatives (equation 52)[171,173]. A two-step preparation of aristolene (**107**) from calorene (**106**, equation 53) performed by the same authors[174] provides an improved synthesis of this natural product.

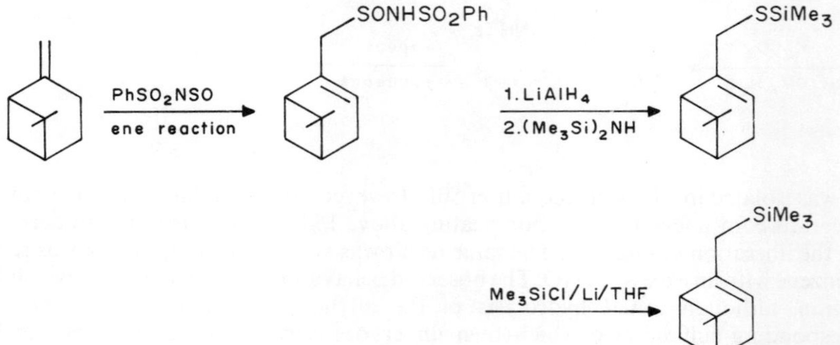


(52)

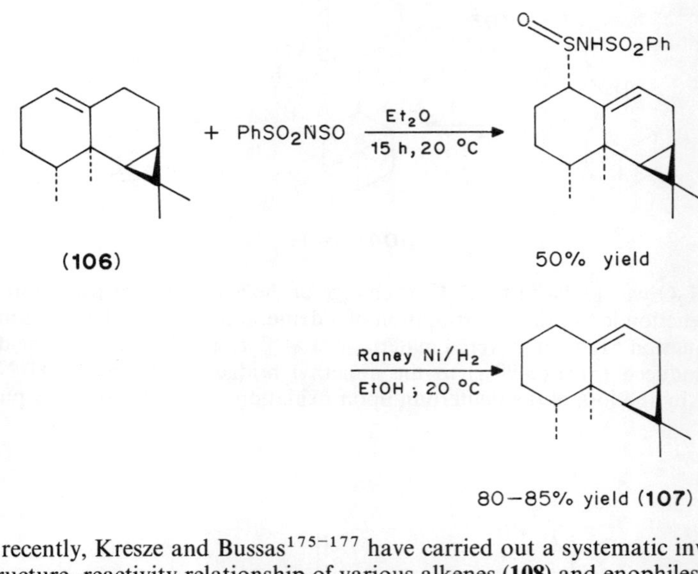


(53)

More recently, Kresze and Bussas[175–177] have carried out a systematic investigation of the structure–reactivity relationship of various alkenes (**108**) and enophiles (**109**), and discovered that the enophilicity of **109** can be further markedly enhanced by incorporation of strongly electron-attracting groups $R^6$ (equation 54). Thus, *N*-sulfinyl nonafluorobutanesulfonamide (**109**, $R^6 = n\text{-}C_4F_9$), easily obtained from mona fluorobutanesulfonyl fluoride, is about $10^3$–$10^4$ times more reactive than the corresponding *p*-tolyl derivative (**103**)[175]. This so-called 'superenophile' reacts almost instantaneously at room temperature even with electron-deficient and slow-reacting alkenes.

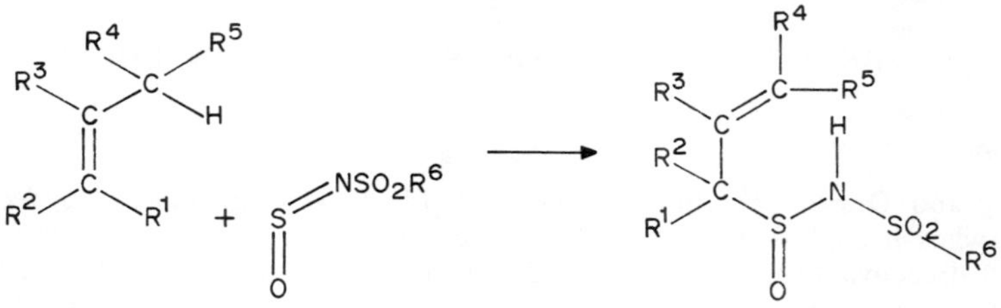

(**108**) (**109**) (54)

## B. Ionic Rearrangements

### 1. Electrophilic

The rearrangement of various *N*-methyl-*N*-aryl-*p*-toluenesulfinanilides to anilino sulfoxides upon treatment with gaseous HCl in chloroform at room temperature was observed by Andersen and Malver[178]. For example, the rearrangement of sulfinamide **110** gave a mixture of the two anilino sulfoxides **111** and **112** in 70 and 26% yield, respectively. This reaction is believed to proceed by electrophilic aromatic substitution, with the HCl aiding in formation of the electrophile. Preequilibrium protonation on nitrogen, followed by nucleophilic attack by chloride anion on sulfur, would give the corresponding *N*-methylaniline and *p*-toluenesulfinyl chloride. The latter could act as an electrophilic sulfinylating agent on the highly reactive aniline ring.

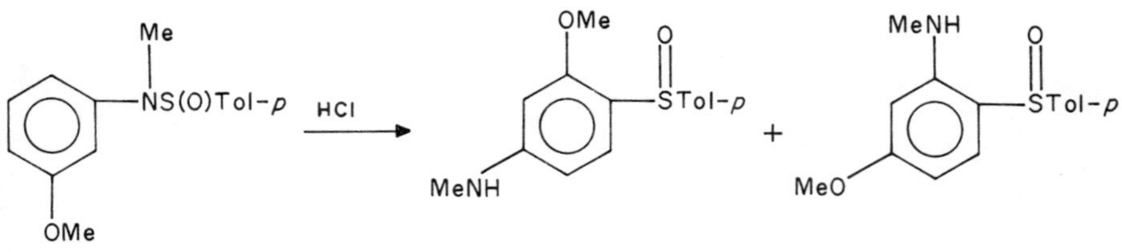

(**110**) (**111**) (**112**)

Apparently, the first example of a Pummerer-type reaction of a sulfinamide was observed during a synthesis of 7α-methoxycephalosporins. Thus, compound **114** was obtained from sulfinamide **113** by treatment with thionyl chloride and quinoline at 0 °C in 54% yield[179].

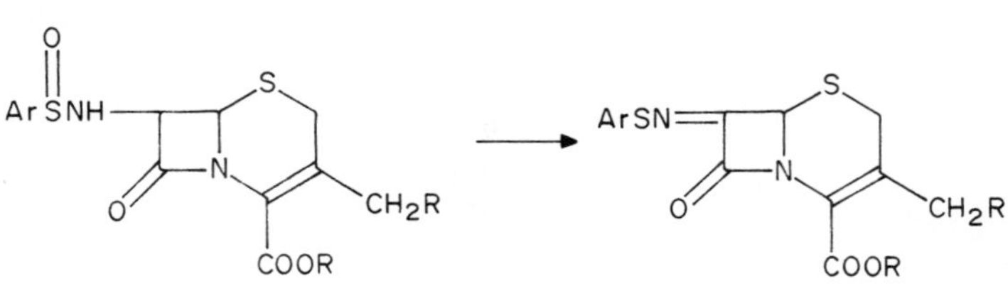

(**113**) (**114**)

Subsequently, a variety of secondary sulfinamides bearing one hydrogen atom at the α-carbon to nitrogen have been found to undergo reaction with electrophilic reagents such is acetic anhydride, leading to the formation of *N*-sulfenylimines via a Pummerer-type rearrangement (equation 55)[180,181].

$$\mathrm{ArSONHCHRR'} \xrightarrow[25\,^{\circ}\mathrm{C}]{\mathrm{Ac_2O}} \mathrm{ArSN{=}CRR' + AcNHCHRR'} \qquad (55)$$

*2. Anionic*

Corey and Durst[182,183] have shown that β-hydroxysulfinamides **115** decompose cleanly when heated alone at melting point or on refluxing in dry benzene for 5 h to form 1,1-diphenylethene, *p*-toluidine and $SO_2$, in quantitative yield. The reaction has been tentatively suggested to proceed via intermediates of typpe **116** and **117**, the former being easily accessible because of the enhancement of basicity of the sulfinyl group by nitrogen (equation 56). A similar intermediate has also been suggested for the rearrangement of β-ketosulfinamides to azomethines on treatment with a secondary base such as diethylamine[184].

$$\underset{(\mathbf{115})}{\mathrm{R^1R^2C(OH)CH_2S(O)NHTol\text{-}}p} \xrightarrow{\Delta} \underset{(\mathbf{116})}{\mathrm{R'C(O^-)(R^2)CH_2S(OH){=}\overset{+}{N}HTol\text{-}}p}$$

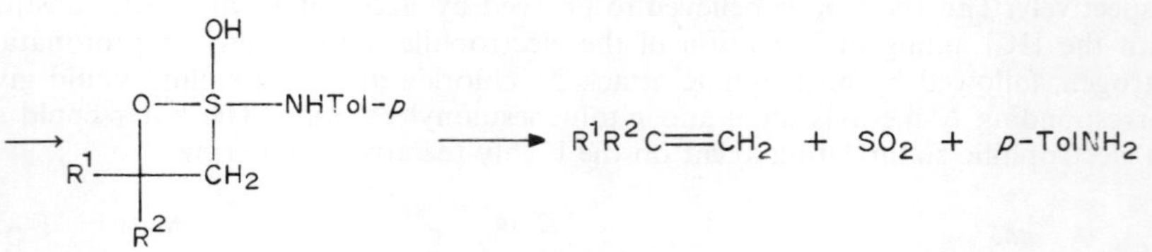


**(117)**

(56)

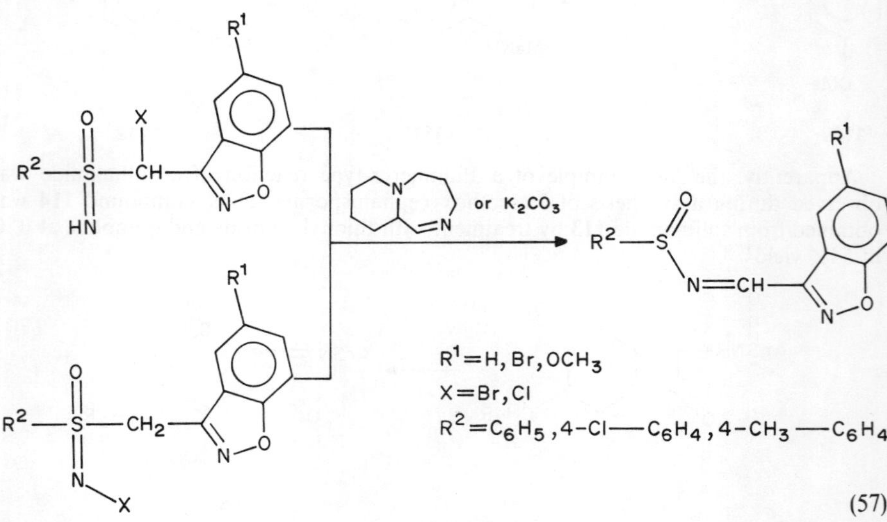


(57)

The rearrangement of several *N*-halo or 1-haloalkyl sulfoximides on treatment with 1,5-diazobicyclo[5.4.0]undec-5-ene (DBU), or simply potassium carbonate, to alkylidene arenesulfinamides (equation 57) has been recently reported[185].

## C. Free-radical Rearrangements and Racemizations

Kobayashi and coworkers[186] have observed a photoisomerization of aromatic sulfinamides to *p*-anilino sulfoxides upon irradiation in aprotic solvents such as benzene or acetonitrile with a low-pressure mercury lamp at room temperature. This rearrangement, which is similar to the rearrangement of **110** to **111** under electrophilic conditions, has been suggested to take place by a free radical mechanism as shown in equation 58, and is accompanied by the formation of some other products arising by different radical recombinations.

$$p\text{-TolS(O)NHPh} \xrightarrow{h\nu} [p\text{-Tol}\overset{\overset{\displaystyle O}{\|}}{S}\cdot\ \cdot\text{NHPh}] \longrightarrow p\text{-TolS(O)C}_6\text{H}_4\text{NH}_2\text{-}p \qquad (58)$$

A free-radical mechanism has also been suggested by Booms and Cram[187] for the racemization of optically active sulfinamides. Following a previous observation on the optical lability of sulfinamides in the solid state in the presence of sunlight[188], these authors found that optically active arenesulfinanilides **118** (R = H or $CH_3$) racemize very readily even in the absence of light at room temperature, and that this thermal racemization is the result of a free-radical chain reaction that is initiated by the dissociation of some of the sulfinamide into an ArSO· and a PhṄR radical (equation 59).

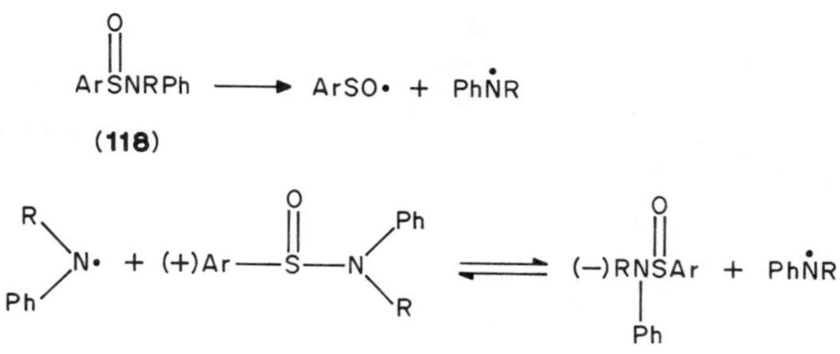

(**118**)

(59)

Racemization of optically active sulfinamides is not limited to homolytic S–N fission, but may also occur by ionic processes as well. For example, Nudelman and Cram[189] reported that when (−)-(*S*)-methyl *p*-toluenesulfinate (**119**) was treated with one mole of lithium anilide in ether at 0 °C (inverse addition), (+)-(*S*)-*N*-phenyl-*p*-toluenesulfinamide (**118**) was produced in 41% yield. However, when reaction is carried out with excess lithium anilide by addition of ester to anilide salt, totally racemic **118** was obtained. The results provide strong evidence that the substitution reaction occurred essentially stereospecifically with inversion, and that in the presence of excess anilide ion the optically active sulfinanilide was converted to racemic material by multiple substitutions of anilide ions by anilide ion with inversion. This conclusion is confirmed by the contemporaneous observations on the reaction of optically active sulfinamides with methyllithium[190] and the well-known synthesis of optically active sulfoxides by reaction of **119** with organometallic reagents[191].

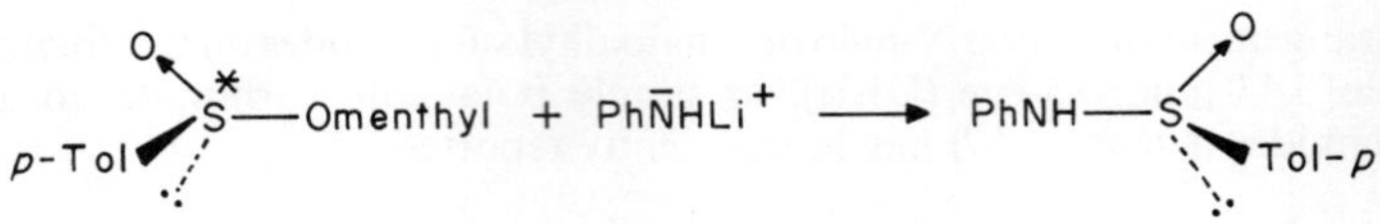


(−)-(*S*)-(**119**) (+)-(*S*)-(**118**)

More recently, a new asymmetric synthesis of chiral sulfoxides, based on the conversion of a 1,2,3-oxathiazolidine 2-oxide (**120**) derived from *l*-ephedrine, to methyl aryl sulfoxides via sulfinamides (**121**) has been reported by Wudl and Lee[192,193].

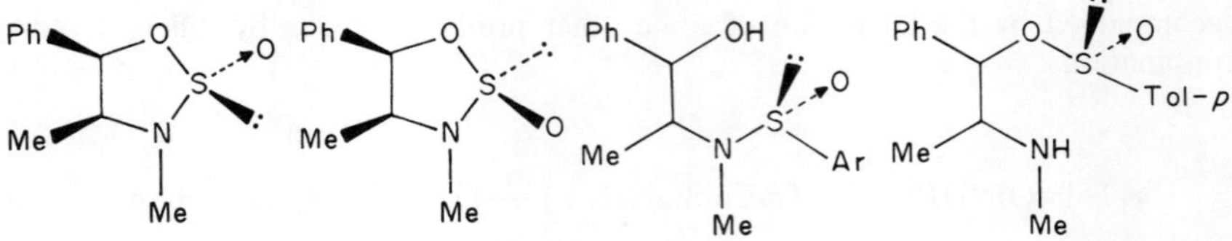


(**120**) (**120a**) (**121**) (**122**)

The same authors[193] also investigated the stereochemistry of sulfinyl transfer in the *O*-sulfinylated ethanolamine **122**. As shown in equations 60 and 61, this rearrangement proceeds via two competitive paths: intramolecular and intermolecular O–N sulfinyl migrations. The intramolecular path yields a sulfinamide (**121**) with retention of configuration at sulfur, whereas the intermolecular path results in inversion of configuration, as expected.

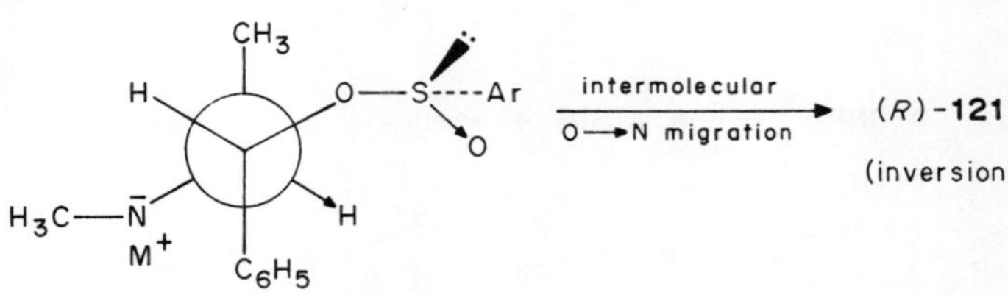


(60)

(*S*)-(**122**)

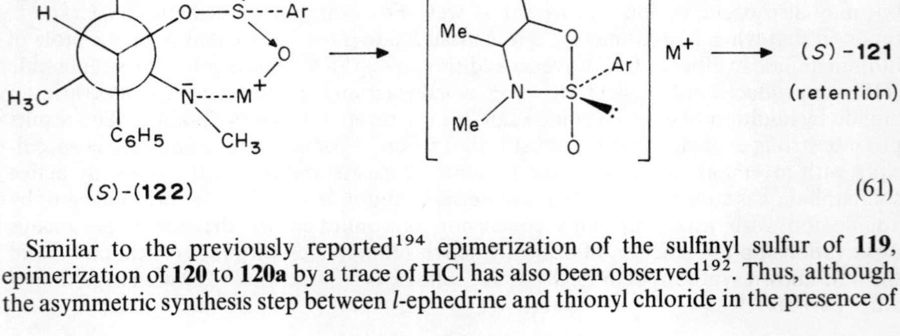


(*S*)-(**122**) (61)

Similar to the previously reported[194] epimerization of the sulfinyl sulfur of **119**, epimerization of **120** to **120a** by a trace of HCl has also been observed[192]. Thus, although the asymmetric synthesis step between *l*-ephedrine and thionyl chloride in the presence of triethylamine occurs with high efficiency (80%), the overall stereochemical efficiency of this reaction may be boosted to 100% of one diastereomer, due to solubility differences of the two diastereomers.

## VII. REARRANGEMENTS OF *O*-SULFINYL OXIMES AND HYDROXYLAMINES

The reaction of a variety of *N*-hydroxy compounds with alkyl or arylsulfinyl chlorides has been studied in recent years. This reaction is usually performed at low temperatures and leads to the expected *O*-sulfinylated products, which can be isolated and characterized only in certain cases, and which are thermally unstable and rearrange to the corresponding *N*-sulfonyl products when warmed to room temperature.

For example, Hudson and coworkers[195–197] have found that oximes are easily sulfinylated with sulfinyl chlorides in the presence of a molar equivalent of triethylamine in ether at −30 °C. Although the sulfinyl derivatives of ketoximes (**123**) can be isolated in the solid state at *ca* 0 °C, they have limited stability at room temperature. When the sulfinylation reaction is performed at room temperature, the thermally stable *N*-sulfonyl imines **124** are produced directly in high yield.

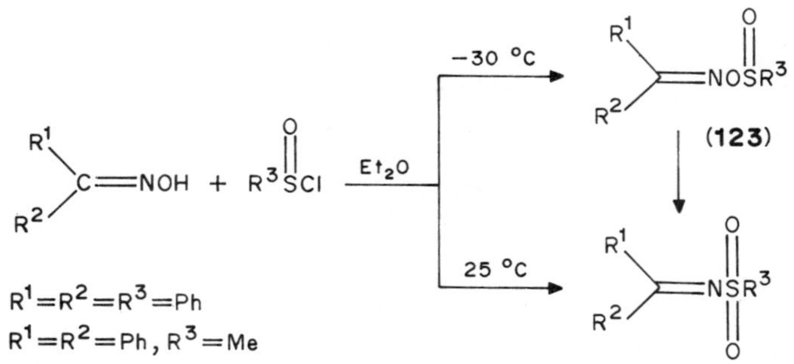


(**123**)

(**124**)

The sulfinylated ketoximes **123** readily rearrange on warming either in the solid state or in solution to give the corresponding *N*-sulfonylimines **124** also in high yields. Although an intramolecular cyclic 1,2-shift was also considered by the authors[197], a dissociative mechanism involving homolytic N—O bond cleavage to give iminyl and sulfonyl radicals, followed by radical recombination with N—S bond formation, was proposed (equation 62). Evidence for this mechanism by spectroscopic and kinetic studies has been obtained. Thus, when a solution of **123**, prepared below room temperature, was warmed at *ca* 35 °C in the probe of an ESR spectrometer, strong signals due to both iminyl[198] and sulfonyl[199] radicals were detected. More direct evidence for the participation of these radicals in the formation of the product was obtained from $^{13}$C CIDNP effects[200] detected in the $^{13}$C NMR spectra of the reaction products when the above experiments were repeated in the probe of an NMR spectrometer.

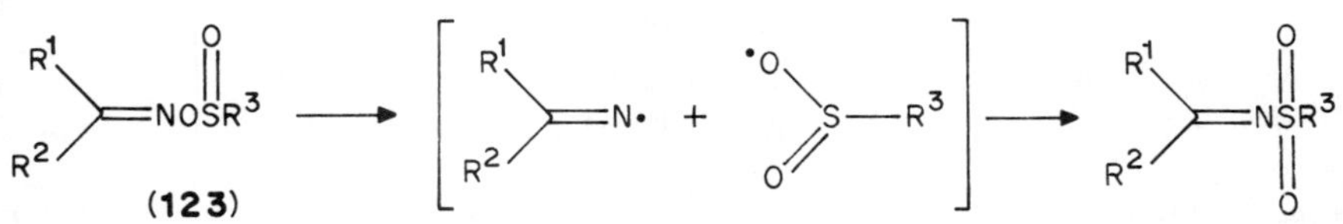

(**123**)

(62)

Interestingly, the rearrangement of *N*-sulfinyl aldoximes (**125**) also gives the corresponding sulfonyl imines (**126**) under similar conditions. However, in this case the reaction is accompanied by the formation of aryl nitriles and aldehydes, which can also be explained by the free-radical mechanism (equation 63).

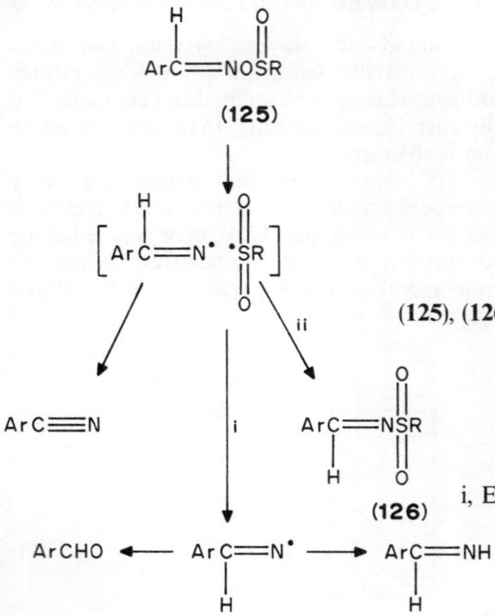


(63)

(**125**), (**126**) (**a**) Ar = R = Ph
(**b**) Ar = Ph, R = $C_6H_4Me$-*p*
(**c**) Ar = Ph, R = Mc
(**d**) Ar = $C_6H_4NO_2$-*p* R = $C_6H_4Mc$-*p*
(**e**) Ar = $C_6H_4NO_2$-*p*, R = $C_6H_4Mc$-*p*

i, Escape; ii, In-cage recombination.

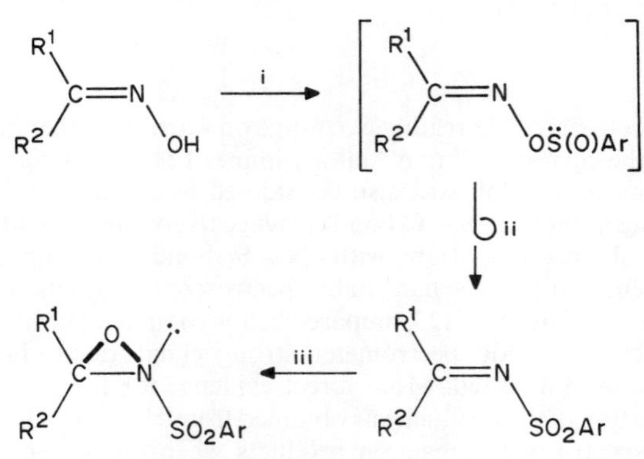


(64)

Ar = $C_6H_4Me$

(**a**) $R^1 = R^2 = [CH_2]_4$ (**d**) $R^1$ = Et, $R^2$ = Me
(**b**) $R^1 = R^2$ = Me (**e**) $R^1 = PhCH_2$, $R^2$ = Me
(**c**) $R^1 = R^2 = PhCH_2$ (**f**) $R^1$ = Ph, $R^2$ = Me

*Reagents and conditions*: i. 4-$MeC_6H_4SOCl$, $Et_3N$, $Et_2O$, −15 °C, 10 min; ii. stir at ambient temperature for 1 h: iii. *N*-sulphonyl imine (10 mmol), 3-$ClC_6H_4CO_3H$ (22 mmol), $NaHCO_3$ (25 mmol), $CH_2Cl_2/H_2O$, 0 °C, 4 h.

Furthermore, compounds **125** decompose explosively at room temperature but are quite stable in solution or as solids at −30 °C. Recently, the rearrangement of *O*-sulfinyl ketoximes to n-sulfonylimines has been applied for the preparation of 3,3-disubstituted 2-sulfonyloxaziridines (equation 64)[201]; some of these compounds provide the first reported examples of *cis–trans* isomerism in *N*-sulfonyloxaziridines. Similar to the reaction of sulfinyl chlorides with oximes, the reaction of thiocarbonyl chlorides[202a], chlorophosphines and chlorophosphites[202b] also produce reactive esters with oximes which undergo facile rearrangement of the acidic group, involving homolytic N—O bond fission. Besides the reactions of oximes with sulfinyl chlorides discussed above, the reactions of hydroxylamines, *N*-hydroxyureas, *N*-hydroxycarbamates and *N*-phenylhydroxamic acids with sulfinyl chlorides have also attracted considerable attention. The reactions of hydroxylamine and its *N*-substituted derivatives with sulfinyl chlorides to give sulfonamides directly were first reported in 1925 by Whalen and Jones[203]. Subsequently, Hovius and Engberts[204] have found that these reactions are rapid at room temperature and can be carried out under similar conditions irrespective of the extent of substitution in the amino group and the nature of the sulfinyl chlorides. No intermediates were detected by these authors, but since *O*-methyl-*N*-alkylhydroxylamines gave *N*-sulfinylated derivatives under the same conditions[204], the reaction was assumed to proceed by attack of nitrogen on the sulfinyl chloride, followed by rapid rearrangement (equation 65). No polarization of the $^1$H NMR spectrum was observed during the reaction of 1,1-dimethylethanesulfinyl chloride with *N*-substituted hydroxylamines[205] carried out under CIDNP conditions, and there was no evidence for a radical mechanism.

$$\mathrm{RNHOH + R'S(=O)Cl \longrightarrow R'S(=O)N(OH)R \longrightarrow R'S(=O)_2NHR} \qquad (65)$$

More recently, however, it has been shown by Banks and Hudson[206,207] that the reactions of several *N*,*N*-dialkylhydroxylamines with methane- and benzenenesulfinyl chlorides below 0 °C give *O*-sulfinylated intermediates, which have been isolated and characterized by NMR spectroscopy. These compounds rearrange at ambient temperatures to give the corresponding sulfonamides (equation 66), and in some cases the imines and products derived from decomposition of the accompanying sulfinic acids. In addition, $^1$H and $^{13}$C NMR spectra show strong polarization in the sulfonamides, indicating a radical-cage mechanism. Apparently, this is the first recorded example of a homolytic process involving an aminyl radical. Since no CIDNP signals were observed in the imines, a six-electron symmetry-allowed cyclic elimination has been suggested to be responsible for their formation.

$$\mathrm{R_2NOH + R'S(=O)Cl \xrightarrow[CH_2Cl_2,\ Et_3N]{-70\,^{\circ}C} R_2NOS(=O)R' \xrightarrow{25\,^{\circ}C} R_2NS(=O)_2R'} \qquad (66)$$

R = Me, Et or $PhCH_2$
R′ = Me or Ph

Although not isolated, the formation of several *O*-sulfinylated *N*-hydroxycarbamates has been observed by NMR, before their room-temperature rearrangement to the corresponding sulfonyl carbamates (equation 67)[208]. This rearrangement is also believed

$$t\text{-BuS(O)Cl} + \text{HONR}^2\text{COOR}^1 \xrightarrow[\text{pyr.}]{\text{CHCl}_3} [t\text{-BuS(O)ONR}^2\text{COOR}^1]$$

$$\longrightarrow t\text{-Bu}\overset{\text{O}}{\underset{\text{O}}{\overset{\|}{\underset{\|}{\text{S}}}}}\text{NR}^2\text{COOR}^1 + \text{other products} \qquad (67)$$

to proceed by homolytic nitrogen–oxygen bond cleavage followed by recombination of the radical pair, as evidenced by the observation of pronounced proton CIDNP effects during conversion of the *O*-sulfinyl esters and the formation of some escape-type products. Subsequently, the same authors reported the related rearrangement of *O*-sulfinylated *N*-hydroxyureas (equation 68)[209] and *N*-hydroxysulfonamides (equation 69)[210], which occur readily at room temperature.

$$t\text{-BuS(O)Cl} + \text{HONH}\overset{\text{O}}{\overset{\|}{\text{C}}}\text{NR}_2 \longrightarrow [t\text{-BuS(O)ONH}\overset{\text{O}}{\overset{\|}{\text{C}}}\text{NR}_2] \longrightarrow t\text{-Bu}\overset{\text{O}}{\underset{\text{O}}{\overset{\|}{\underset{\|}{\text{S}}}}}\text{NH}\overset{\text{O}}{\overset{\|}{\text{C}}}\text{NR}_2 \qquad (68)$$

$$t\text{-BuS(O)Cl} + \text{R}'\overset{\text{O}}{\underset{\text{O}}{\overset{\|}{\underset{\|}{\text{S}}}}}\text{—}\underset{\text{OH}}{\underset{|}{\text{NR}^2}} \longrightarrow [t\text{-BuS(O)O—NR}^2\overset{\text{O}}{\underset{\text{O}}{\overset{\|}{\underset{\|}{\text{S}}}}}\text{—R}'] \longrightarrow t\text{-Bu}\overset{\text{O}}{\underset{\text{O}}{\overset{\|}{\underset{\|}{\text{S}}}}}\text{NR}^2\overset{\text{O}}{\underset{\text{O}}{\overset{\|}{\underset{\|}{\text{S}}}}}\text{R}' \qquad (69)$$

Observation of pronounced $^1$H and $^{13}$C CIDNP effects and the formation of both free-radical recombination and escape products provide clear evidence for a homolytic cleavage mechanism of the N—O bond in this case as well. Heesing and coworkers[211]

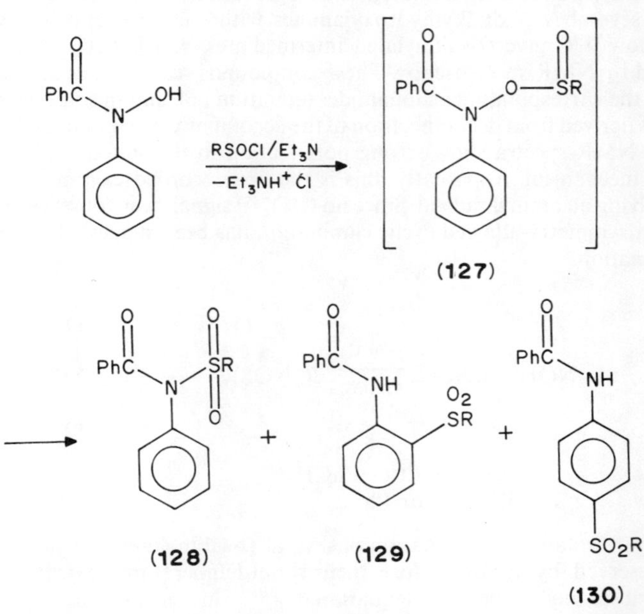


(127)

(128) (129) (130)

have reported that *N*-phenylbenzohydroxamic acid reacts with various alkanesulfinyl chlorides to give the corresponding *O*-sulfinyl esters **127** which rearrange at $-70\,^{\circ}\mathrm{C}$ to yield the corresponding sulfonamide **128** together with the *o*- and *p*-alkylsulfonyl derivatives **129** and **130**. The reaction has been suggested to proceed by an intramolecular radical-pair mechanism, as evidenced by experiments with $^{18}O$ labeling and $^{13}C$-CIDNP effects. A similar mechanism has been proposed for the rearrangement of the corresponding thionocarbamates[212].

## VIII. REARRANGEMENTS INVOLVING THIOLSULFINATES

One of the characteristic features of thiolsulfinates is their relatively low thermal stability. Backer and Kloosterziel[213] first reported the occasional spontaneous disproportionation of thiolsulfinates into thiolsulfonates and disulfides. Subsequently, Barnard[214] has noticed that aryl thiolsulfinates are stable for months under normal atmospheric conditions but undergo rapid decomposition in vacuum, and suggested a free-radical mechanism (equation 70). Homolytic fission of the S(O)—S bond to give sulfinyl and thiyl radicals is followed by dimerization to the observed products. The dimerization of sulfinyl radicals, perhaps through the intermediacy of mixed sulfenic sulfinic anhydrides (equation 70a), is well established[215] and the dimerization of thiyl radicals to disulfides is self-evident. More recently, a mechanistic study of the thermal disproportionation of arylarenethiosulfinates was performed by Fava and coworkers[216], who concluded that their data may be interpreted in terms of a radical process: a unimolecular decomposition along with an induced decomposition. These processes may be facilitated by the unusually weak S—S bond of about 35 kcal mol$^{-1}$ [216].

$$2\mathrm{Ar}\overset{\mathrm{O}}{\overset{\|}{\mathrm{S}}}\mathrm{SAr'} \longrightarrow [2\mathrm{Ar}\overset{\mathrm{O}}{\overset{\|}{\mathrm{S}}}\cdot + 2\mathrm{Ar'S}\cdot] \longrightarrow \mathrm{Ar}\underset{\mathrm{O}}{\underset{\|}{\overset{\mathrm{O}}{\overset{\|}{\mathrm{S}}}}}\mathrm{SAr} + \mathrm{Ar'SSAr'} \qquad (70)$$

$$2\mathrm{ArSO}\cdot \longrightarrow \mathrm{ArSO}\overset{\mathrm{O}}{\overset{\|}{\mathrm{S}}}\mathrm{Ar} \longrightarrow \mathrm{Ar}\underset{\mathrm{O}}{\underset{\|}{\overset{\mathrm{O}}{\overset{\|}{\mathrm{S}}}}}\mathrm{SAr} \qquad (70a)$$

Of particular interest are the detailed and systematic studies by Blcok and coworkers[217–220] on the pyrolysis of alkyl thiosulfinates. Thus, the pyrolysis of dialkyl thiosulfinates has been shown to afford alkanesulfenic or alkanethiosulfoxylic acids by the two thermal cycloelimination pathways shown in equation 71. Path a should be favored in view of the weakness of the S—S bond and the enhanced acidity of the α-sulfenyl protons. The sulfenic and thiosulfoxylic acids arising from pyrolysis could be trapped in good yields with acetylenes giving α,β-unsaturated sulfoxides or thiosulfinates, respectively. In the absence of trapping agents, the sulfenic acids can undergo a variety of reactions, including dehydration to thiosulfinates and exchange with thiolsulfinate via nucleophilic displacement leading to a scrambling process, if two different thiosulfinates are involved (Scheme 6).

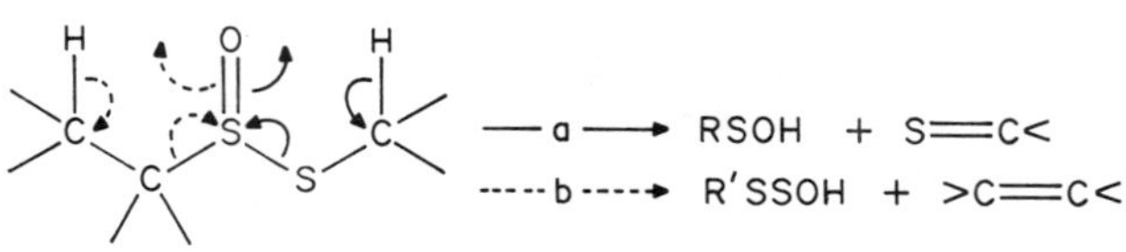


(71)

$$\mathrm{MeS(O)SMe} \xrightarrow{k_1} \mathrm{MeSOH + CH_2{=}S} \qquad \text{Initiation}$$

$$\mathrm{EtS(O)SEt} \xrightarrow{k_1'} \mathrm{EtSOH + CH_3CH{=}S}$$

$$\mathrm{MeSOH + Et(O)SEt} \underset{}{\overset{k_2}{\rightleftharpoons}} \mathrm{EtSOH + MeS(O)SEt} \qquad \text{Propagation}$$

$$\mathrm{EtSOH + MeS(O)SMe} \overset{k_2'}{\rightleftharpoons} \mathrm{MeSOH + EtS(O)SMe}$$

$$\mathrm{RSOH + RS(O)SR} \xrightarrow{k_3} \mathrm{RSO_2H + RSSR} \qquad \text{Termination}$$

SCHEME 6 Proposed mechanism for thiolsulfinate scrambling

The thermal stability of optically active aryl arenethiolsulfinates, prepared by asymmetric oxidation of the corresponding diaryl disulfides with percamphoric acid[221–225], has also received considerable attention. Fava and coworkers[221–223] have reported the unusually facile thermal racemization of optically active aryl arenethiolsulfinates, obtaining rate constants ranging from *ca* $2 \times 10^{-5}\,s^{-1}$ (in benzene saturated with water) to $46 \times 10^{-5}\,s^{-1}$ (in dry benzene) at 50 °C (Ar = $p$-$ClC_6H_4$), with $\Delta H^{\neq} = 23$ kcal mol$^{-1}$ [223].

After excluding an intramolecular oxygen transfer between the two sulfur atoms, a slow homolytic fission of the S—S bond and a normal pyramidal inversion of the sulfinyl group, the authors[223] suggested the unusual cyclic intramolecular mechanism shown in equation 72.

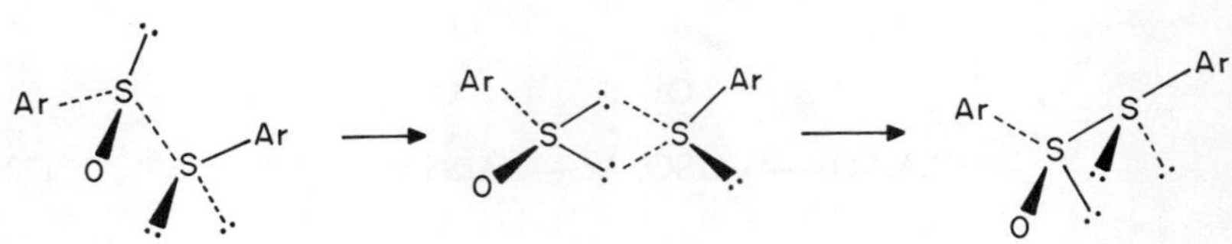


(72)

Ar = $p$-$CH_3C_6H_4$, $p$-$ClC_6H_4$

The same authors[221] also reported that the rearrangement of ArS(O)SAr in pyridine takes place spontaneously at 25 °C. In this case, the reaction is initiated by nucleophilic attack on the sulfenyl sulfur to give an ion pair, which could lead to racemization by its collapse (equation 73). A detailed kinetic and mechanistic study by Kice and Large[225] on the combined nucleophile- and acid-catalyzed racemization of optically active phenyl benzenethiolsulfinate also supports a rate-determining attack of the nucleophilic catalyst on the sulfenyl sulfur of the sulfinyl-protonated thiolsulfinate. In acidic aqueous dioxane in

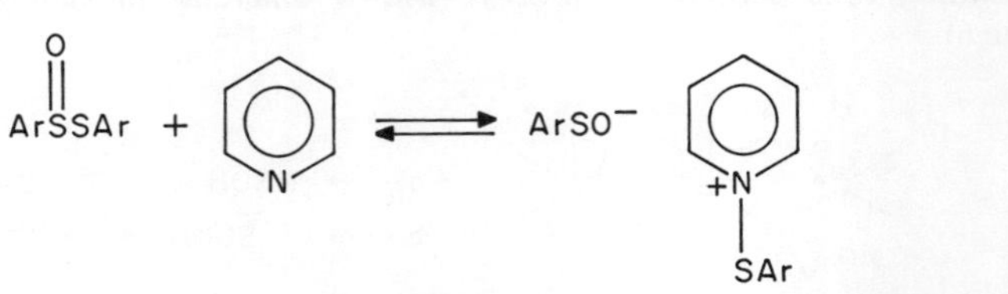


(73)

the absence of added nucleophile, optically active ArS(O)SAr racemizes only very slowly, but the addition of small amounts of alkyl sulfides, halide ions or thiocyanate ion leads to quite rapid racemization of the substrate.

In contrast to the relatively facile racemization of aryl arenethiolsulfinates described above, the cyclic thiolsulfinate **131** was found by Padwa and coworkers[226] to be configurationally stable up to 166 °C.

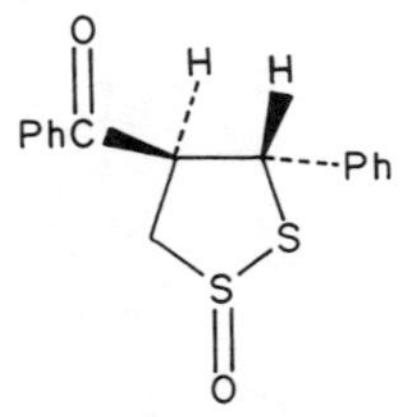

**(131)**

An explanation which removes the puzzling inconsistency between the mechanism offered by Fava[223] for the rapid racemization of diaryl thiolsulfinates and the lack of isomerization of **131** has been provided by Block[220], based on the observation of scrambling of thiolsulfinates (Scheme 6). According to Block, the facile reaction of arenesulfenic acid (generated perhaps from the thiolsulfinate by reaction with traces of

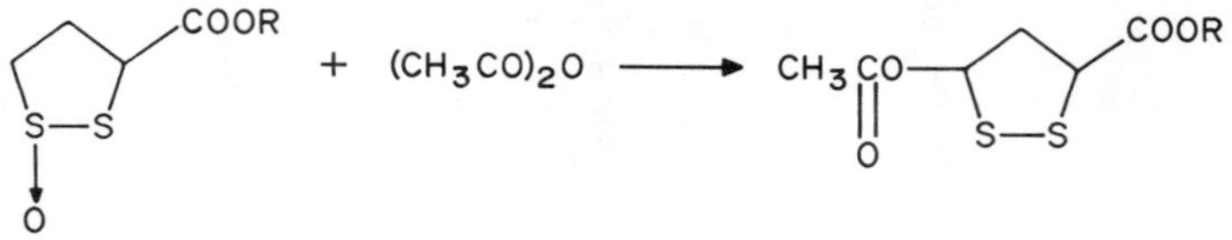

(74)

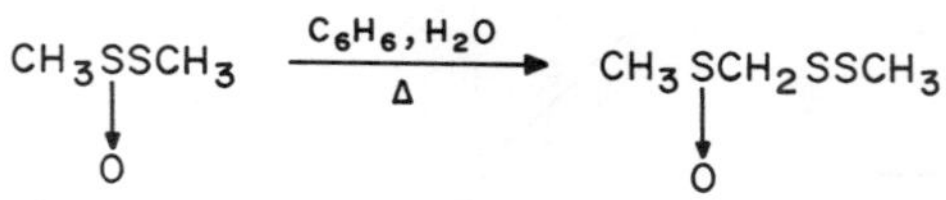

(75)

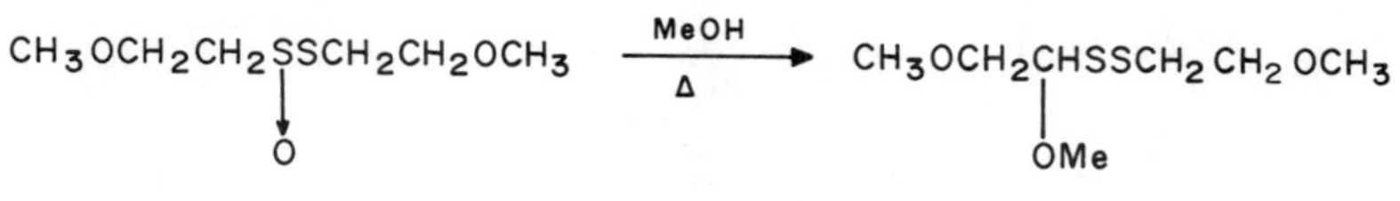

(76)

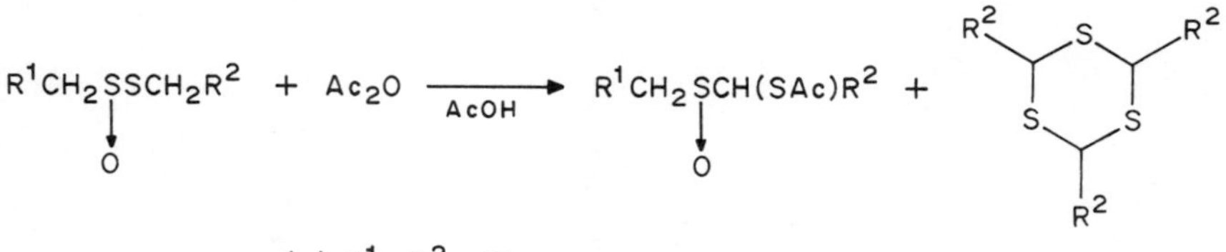

(a) $R^1=R^2=Ph$
(b) $R^1=R^2=p\text{-Tol}$
(c) $R^1=Ph, R^2=p\text{-Tol}$
(d) $R^1=R^2=H$
(e) $R^1=R^2=Me$
(f) $R^1,R^2=[CH_2]_2$

(77)

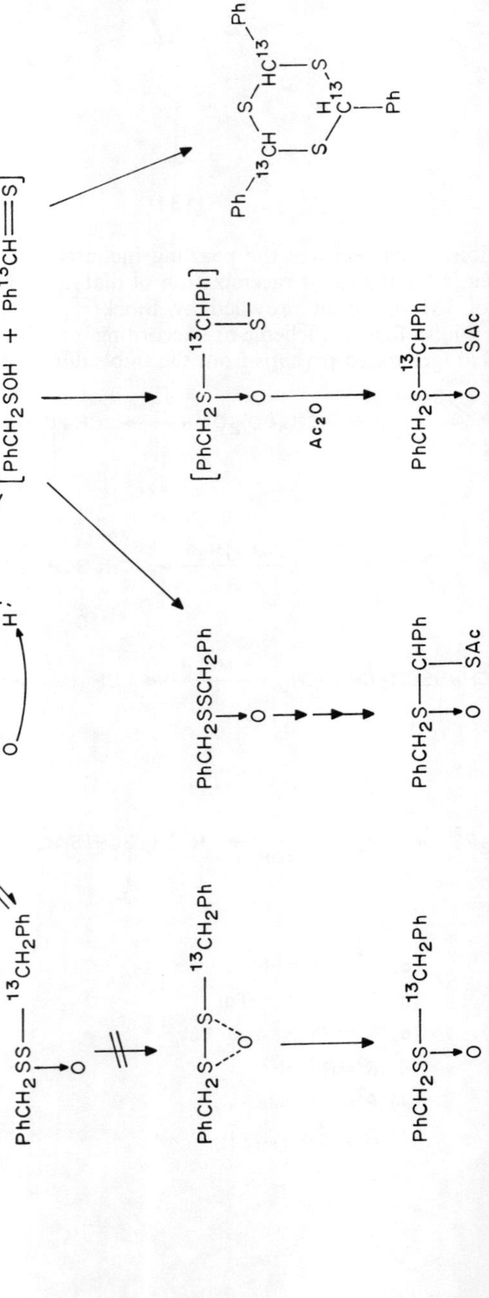

(78)

water) with optically active thiolsulfinate may be responsible for the low optical stability of these esters. Similarly, the decreased rate of racemization in protic solvents compared to aprotic solvents may possibly be due to hydrogen-bonding effects by the protic solvent, which interferes with the transition state of the reaction of sulfenic acid with thiolsulfinate. On the other hand, a coplanar Cope elimination of the type indicated in equation 71a is impossible for **131**, as is the chain-type scrambling of Scheme 6.

Although the Pummerer rearrangement of sulfoxides has been investigated extensively[227], the analogous reaction of thiolsulfinates has received relatively little attention. Thus, Saito and Fukui[228] reported that the reaction of α-lipoic acid monoxide with acetic anhydride in acetonitrile gives the normal Pummerer-type product, but only in 6% yield (equation 74)! Although Block[229] could not observe the same type of reaction on treatment of *t*-butyl methanethiolsulfinate with acetic anhydride, he found, however, a Pummerer-type rearrangement when a few thiolsulfinates were pyrolyzed in benzene saturated with water (equation 75)[219b]. Similarly, Kondo and Negishi[230] reported that α-methoxydisulfides can be obtained in excellent yields when heated in methanol at 90 °C (equation 76).

More recently, Oae and coworkers[231–233] reported a new type of rearrangement of thiolsulfinates under Pummerer reaction conditions. Thus, thiolsulfinates with at least one proton on the carbon adjacent to the sulfenyl sulfur react with acetic anhydride containing acetic acid to afford the corresponding α-acetylthiosulfoxides (equation 77). Although the authors[231] first considered a Pummerer-type mechanism, they preferred the mechanism shown in equation 78. This mechanism, which is inspired by the Block mechanism of thiosulfinate disproportionation, involves an initial thermal cycloelimination to form the corresponding sulfenic acid and thioaldehyde, followed by attack of the former on the latter, and formation of the observed product after acetylation of the thiolate intermediate. The suggested mechanism is supported by $^{2}H$, $^{13}C$ and $^{18}O$ labeling experiments and by trapping of the sulfenic acid intermediate with methyl acrylate.

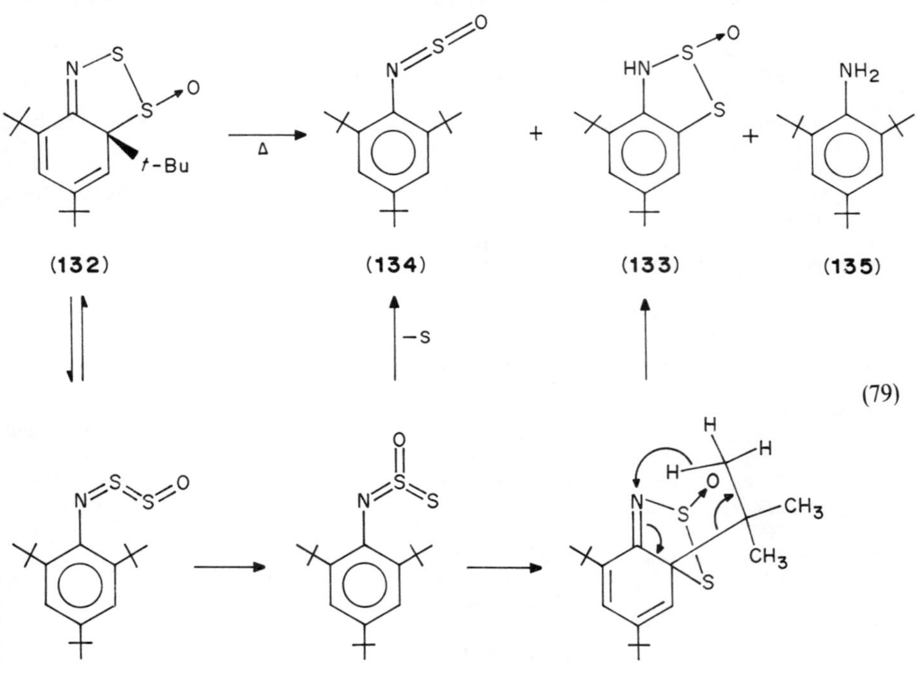


(79)

An interesting reaction of thiolsulfinates is their oxidation with peracids to thiolsulfinates. Although the oxidation of symmetrical thiolsulfinates RS(O)SR affords only one product as expected, the oxidation of unsymmetrical thiolsulfinates may lead to the formation of four different thiolsulfonates ($RSO_2SR$, $RSO_2SR'$, $R'SO_2SR$ and $R'SO_2SR'$)[234], and the oxidation of cyclic thiolsulfinates leads to the formation of two different thiolsulfonates[235]. *A priori*, these reactions may be explained by an oxygen migration $RS(OSR' \rightleftarrows RSS(O)R'$. However, since such a process is believed not to occur[220], and in view of some other data, the reaction has been suggested to occur by initial oxidation of the thiolsulfinate ester to an α-disulfoxide RS(O)S(O)R′, followed by homolytic fission of the sulfur–sulfur bond to generate two different sulfinyl radicals, and reaction of the latter as shown in equation 70a.

However, the occurrence of oxygen migration has been detected during the thermolysis of the unusual thiolsulfinate **132** in refluxing benzene, which affords the isomeric compound **133** together with two other products[236]. The mechanism suggested for this reaction is shown in equation 79.

## IX. ACKNOWLEDGMENT

The excellent hospitality offered by Professor S. Gronowitz at the University of Lund during the preparation of the final parts of this chapter is gratefully acknowledged.

## X. REFERENCES

1. (a) M. Quadeavlieg, in *Houben-Weyl, Methoden der Organischen Chemie* (Ed. E. Müller), Vol. 9, Chap. 12, G. Thieme Verlag, Stuttgart, 1955.
   (b) F. Muth, in Reference 1a, Chap. 13.
2. C. J. M. Stirling, *Int. J. Sulfur Chem. (B)*, **6**, 277 (1971).
3. S. Oae and N. Kumeda, in *Organic Chemistry of Sulfur* (Ed. S. Oae), Chap. 11, Plenum Press, New York, 1977.
4. K. K. Andersen, in *Comprehensive Organic Chemistry* (Eds. D. H. R. Barton and W. D. Ollis), Vol. 3 (Ed. D. N. Jones), Pergamon Press, Oxford, 1979, p. 317.
5. J. L. Kice, *Adv. Phys. Org. Chem.*, **17**, 65 (1980).
6. E. Krauthausen, in *Houben-Weyl, Methoden der Organischen Chemie* (Ed. D. Klamann), E11, Part 1, G. Thieme Verlag, Stuttgart, 1985, p. 614.
7. A. Schöberl and A. Wagner, in Reference 1, p. 223.
8. J. Strating, in *Organic Sulfur Compounds* (Ed. N. Kharasch), Vol. 1, Pergamon Press, Oxford, 1961, p. 146.
9. W. E. Truce, T. C. Klingler and W. W. Brand, in Reference 3, p. 527.
10. P. D. Magnus, *Tetrahedron*, **33**, 2019 (1977).
11. L. Field, *Synthesis*, 713 (1978).
12. T. Durst, in Reference 4, p. 171.
13. K. Schank, in Reference 6, Part 2, p. 1129.
14. W. E. Truce, E. M. Kreider and W. W. Brand, *Org. React.*, **18**, 19 (1970).
15. V. N. Drozd, *Int. J. Sulfur Chem.*, **8**, 443 (1973).
16. W. E. Truce and E. J. Madaj, Jr., *Sulfur Rep.*, **3**, 259 (1983).
17. S. Braverman, in *The Chemistry of Sulfones and Sulfoxides* (Eds. S. Patai, Z. Rappoport and C. J. M. Stirling), Chap. 13, Wiley, Chichester, 1988.
18. E. Block, *Reactions of Organosulfur Compounds*, Chap. 7, Academic Press, New York, 1978.
19. S. D. Turk and R. L. Cobb, in *1,4-Cycloaddition Reactions* (Ed. J. Hamer), Academic Press, New York, 1967, p. 13.
20. (a) G. de Bruin, *Konikl. Ned. Akad. Witenschap*, **17**, 585 (1914).
    (b) H. Staudinger and F. Pfenninger, *Chem. Ber.*, **49**, 1446 (1916).
21. M. M. Rogic and J. Vitrone, *J. Am. Chem. Soc.*, **94**, 8642 (1972).
22. M. M. Rogic and D. Masilamani, *J. Am. Chem. Soc.*, **99**, 5219 (1977).
23. D. Masilamani and M. M. Rogic, *J. Am. Chem. Soc.*, **100**, 4634 (1978).

24. (a) D. Masilamani and M. M. Rogic, *Tetrahedron Lett.*, 3785 (1978).
(b) D. Masilamani, E. H. Manahan, J. Vitrone and M. M. Rogic, *J. Org. Chem.*, **48**, 4918 (1983).
25. (a) H. Kwart, N. A. Johnson, T. Eggericks and T. J. George, *J. Org. Chem.*, **42**, 172 (1977).
(b) H. Kwart and T. C. Stanulonis, *J. Am. Chem. Soc.*, **98**, 4009 (1976).
(c) P. Brownbridge and S. Warren, *J. Chem. Soc., Perkin Trans. 1*, 1131 (1977).
26. H. Kwart and K. King, *d-Orbitals in the Chemistry of Silicon, Phosphorus and Sulfur*, Springer-Verlag, New York, 1977.
27. D. A. Evans and G. C. Andrews, *Acc. Chem. Res.*, **7**, 147 (1974).
28. S. Braverman, in Reference 17, Chap. 14.
29. O. Wichterle and J. Rocek, *Chem. Listy*, **47**, 1768 (1953).
30. O. Wichterle and J. Rocek, *Collect. Czech. Chem. Commun.*, **19**, 282 (1954).
31. G. Kresze, A. Maschpe, R. Albrecht, K. Bederke, H. P. Patzschke, H. Smalla and A. Trede, *Angew. Chem., Int. Ed. Engl.*, **1**, 89 (1962).
32. W. Wucherpfennig, *Tetrahedron Lett.*, 3235 (1967).
33. W. Wucherpfennig, *Justus Liebigs Ann. Chem.*, **746**, 16 (1971).
34. (a) W. L. Mock and R. M. Nugent, *J. Org. Chem.*, **43**, 3434 (1978).
(b) R. S. Garigipati, J. A. Morton and S. M. Weinreb, *Tetrahedron Lett.*, **24**, 987 (1983).
35. W. L. Mock and R. M. Nugent, *J. Am. Chem. Soc.*, **97**, 6521 (1975).
36. M. S. Raasch and B. E. Smart, *J. Am. Chem. Soc.*, **101**, 7734 (1979).
37. O. De Lucchi, F. Filipuzzi and V. Lucchini, *Tetrahedron Lett.*, **25**, 1407 (1984).
38. J. Dalling, J. H. Gall and D. D. MacNicol, *Tetrahedron Lett.*, 4789 (1979).
39. G. Capozzi, V. Lucchini, F. Marcuzzi and G. Melloni, *Tetrahedron Lett.*, **21**, 3289 (1980).
40. (a) E. J. Corey and T. A. Engler, *Tetrahedron Lett.*, **25**, 149 (1984).
(b) J.-B. Boudin and S. A. Julia, *Tetrahedron Lett.*, **29**, 3255 (1988).
41. M. P. Cava and C. E. Blake, *J. Am. Chem. Soc.*, **78**, 5444 (1956).
42. M. L. Moore and T. B. Johnson, *J. Am. Chem. Soc.*, **57**, 2235 (1935).
43. W. J. Hamilton and S. J. La Placa, *J. Am. Chem. Soc.*, **86**, 2289 (1964).
44. C. Brown, *J. Am. Chem. Soc.*, **91**, 5832 (1969).
45. B. B. Jarvis, W. P. Tong and H. L. Ammon, *J. Org. Chem.*, **40**, 3188 (1975).
46. H. Kloostterziel, J. A. H. van Drunen and P. Galama, *J. Chem. Soc., Chem. Commun.*, 885 (1969).
47. E. V. Polanin, I. M. Zaks, A. M. Moisernkov and A. V. Semenovskii, *Izv. Akad. Nauk SSSR, Ser. Khim.*, 641 (1979); *Chem. Abstr.*, **91**, 5005 (1979).
48. R. L. Crumbie and D. D. Ridley, *Aust. J. Chem.*, **34**, 1017 (1981).
49. R. M. Dodson, P. D. Hammen, E. H. Jancis and G. Klose, *J. Org. Chem.*, **36**, 2698 (1971).
50. P. Beak and V. Snieckus, *Acc. Chem. Res.*, **15**, 306 (1982).
51. H. Gilman and J. F. Webb, *J. Am. Chem. Soc.*, **71**, 4062 (1949).
52. W. E. Truce and F. M. Amos, *J. Am. Chem. Soc.*, **73**, 3013 (1951).
53. F. M. Stoyanovich and B. P. Feodorov, *Angew. Chem.*, **78**, 116 (1966).
54. R. G. Karpenko, F. M. Stoyanovich, S. P. Raputo and Ya. L. Goldfarb, *Zh. Org. Khim.*, **6**, 112 (1970).
55. F. M. Stoyanovich, R. G. Karpenko, G. I. Gorushkina and Ya. L. Goldfarb, *Tetrahedron*, **28**, 5017 (1972).
56. B. Henrique, *Chem. Ber.*, **27**, 2993 (1894).
57. O. Hinsberg, *J. Prakt. Chem.*, **90**, 345 (1914); **91**, 307 (1913); **93**, 277 (1918).
58. L. A. Warren and S. Smiles, *J. Chem. Soc.*, 956, 1327 (1930); 914, 2207 (1931); 1040 (1932).
59. B. A. Keat and S. Smiles, *J. Chem. Soc.*, 422 (1934).
60. W. J. Evans and S. Smiles, *J. Chem. Soc.*, 181 (1935).
61. A. A. Levi, H. C. Rains and S. Smiles, *J. Chem. Soc.*, 3264 (1931).
62. T. Takahashi and Y. Maki, *Chem. Pharm. Bull.* (Tokyo), **6**, 369 (1958); *Chem. Abstr.*, **53**, 9228 (1959).
63. O. R. Rodig, R. E. Collier and R. K. Schlatzer, *J. Org. Chem.*, **29**, 2652 (1964).
64. J. F. Bunnett and R. E. Zahler, *Chem. Rev.*, **49**, 362 (1951).
65. W. E. Truce, T. C. Klingler and W. W. Brand, in Reference 14, p. 564.
66. S. W. Schneller, *Int. J. Sulfur Chem.*, **8**, 569 (1976).
67. W. E. Truce, W. J. Ray, Jr., O. L. Norman and D. B. Eickemeyer, *J. Am. Chem. Soc.*, **80**, 3625 (1958).
68. W. E. Truce and W. J. Ray, Jr., *J. Am. Chem. Soc.*, **81**, 481 (1959).
69. W. E. Truce and W. J. Ray, Jr., *J. Am. Chem. Soc.*, **81**, 484 (1959).

70. W. E. Truce and M. M. Guy, *J. Org. Chem.*, **26**, 4331 (1961).
71. D. M. Snyder and W. E. Truce, *J. Am. Chem. Soc.*, **101**, 5432 (1979).
72. E. J. Madaj, Jr., D. M. Snyder and W. E. Truce, *J. Am. Chem. Soc.*, **108**, 3466 (1986).
73. S. Winstein, S. Smith and D. Darwish, *Tetrahedron Lett.*, 24 (1959).
74. F. G. Bordwell, P. F. Wiley and T. G. Mecca, *J. Am. Chem. Soc.*, **97**, 132 (1975).
75. K. Schank, *Justus Liebigs Ann. Chem.*, **702**, 75 (1967).
76. M. Kobayashi, *Bull. Chem. Soc. Jpn.*, **39**, 967 (1966).
77. E. Knoevenagel and L. Pollack, *Chem. Ber.*, **41**, 3323 (1908).
78. H. Bredereck, A. Wagner, H. Beck and R. J. Klein, *Chem. Ber.*, **93**, 2736 (1960).
79. J. L. Kice and K. Ikura, *J. Am. Chem. Soc.*, **90**, 7378 (1968).
80. W. A. Mueller and M. B. Dines, *J. Chem. Soc., Chem. Commun.*, 1205 (1969).
81. M. M. Chau and J. L. Kice, *J. Org. Chem.*, **42**, 3265 (1977).
82. J. L. Kice and H. C. Margolis, *J. Org. Chem.*, **40**, 3623 (1975).
83. J. L. Kice and N. E. Pawlowski, *J. Org. Chem.*, **28**, 1162 (1963).
84. J. L. Kice and N. E. Pawlowski, *J. Am. Chem. Soc.*, **86**, 4898 (1964).
85. H. Bredereck, A. Wagner, E. H. Beck, H. Berlinger and K. G. Kottenhahn, *Angew. Chem.*, **70**, 268 (1958).
86. D. A. Tomalia, *J. Heterocycl. Chem.*, **3**, 384 (1966).
87. D. A. Tomalia, *Tetrahedron Lett.*, 2559 (1967).
88. S. Kukolja, S. R. Lammert, M. R. B. Gleissner and A. I. Ellis, *J. Am. Chem. Soc.*, **98**, 5040 (1976).
89. S. Kukolja, in *Recent Advances in the Chemistry of β*-L actam Antibiotics (Ed. J. Elka), The Chemical Society, Burlington House, London, 1977, p. 181.
90. W. A. Spitzer, T. Goodson, S. R. Lammert and S. Kukolja, *J. Org. Chem.*, **46**, 3568 (1981).
91. D. Sianesi, G. C. Bernardi and G. Moggi, *Tetrahedron Lett.*, 1313 (1970).
92. J. Kenyon and H. Phillips, *J. Chem. Soc.*, 1676 (1930).
93. C. L. Arcus, M. P. Balfe and J. Kenyon, *J. Chem. Soc.*, 485 (1938).
94. A. H. Wragg, J. S. McFadyen and T. S. Stevens, *J. Chem. Soc.*, 3603 (1958).
95. H. Mackle, *Tetrahedron*, **19**, 1159 (1963).
96. A. B. Burg, in Reference 8, pp. 35–37.
97. D. Darwish and R. A. McLaren, *Tetrahedron Lett.*, 123 (1962).
98. R. A. McLaren, Ph.D. Thesis, University of Alberta, 1964.
99. J. Grover, Ph.D. Thesis, University of Alberta, 1969.
100. R. Mermelstein, Ph.D. Thesis, University of Alberta, 1966.
101. D. Darwish and E. A. L. Preston, *Tetrahedron Lett.*, 113 (1964).
102. H. H. Persad, Ph.D. Thesis, University of Alberta, 1966.
103. D. Darwish and R. A. McLaren, 148th National Meeting of the American Chemical Society, Chicago, Ill., Sept. 1964, Abstracts 44S.
104. E. Cuiffarin, M. Isola and A. Fava, *J. Am. Chem. Soc.*, **90**, 3594 (1968).
105. S. Braverman and S. Steiner, *Isr. J. Chem.*, **5**, 267 (1967).
106. S. Braverman and T. Globerman, *Tetrahedron*, **30**, 3873 (1974).
107. S. Braverman anf Y. Duar, *Tetrahedron Lett.*, 343 (1975).
108. S. Braverman and H. Manor, *Phosphorus and Sulfur*, **2**, 213 (1976).
109. J. B. Hendrickson and P. L. Skipper, *Tetrahedron*, **32**, 1627 (1976).
110. J. B. Hendkrickson, D. D. Sternbach and K. W. Bair, *Acc. Chem. Res.*, **10**, 306 (1977).
111. (a) F. Jung, M. Molin, R. Van Den Elzen and T. Durst, *J. Am. Chem. Soc.*, **96**, 935 (1974).
(b) R. F. Heldeweg and H. Hogeveen, *J. Am. Chem. Soc.*, **98**, 2341 (1976).
112. (a) N. Kornblum, P. Ackermann and R. T. Swiger, *J. Org. Chem.*, **45**, 5294 (1980).
(b) N. Kornblum, *Angew. Chem., Int. Ed. Engl.*, **14**, 734 (1975).
113. A. C. Cope, D. E. Morrison and L. Field, *J. Am. Chem. Soc.*, **72**, 59 (1950).
114. (a) S. Braverman, Ph.D. Thesis, University of Alberta, 1963.
(b) D. Darwish and S. Braverman, 48th Chemical Institute of Canada Conference, Montreal, Canada, May 30–June 2, 1965.
115. (a) S. Braverman, 3rd Symposium on Mechanism and Structure in Sulfur Chemistry, Cork, Ireland, Sept. 29–Oct. 3, 1969.
(b) S. Braverman, *Int. J. Sulfur Chem.* (*C*), **6**, 149 (1971).
116. R. W. Hoffman, *Angew. Chem., Int. Ed. Engl.*, **18**, 563 (1979).
117. J. E. Baldwin and J. E. patrick, *J. Am. Chem. Soc.*, **93**, 3556 (1971).
118. K. K. Chan and G. Saucy, *J. Org. Chem.*, **42**, 3828 (1977).

119. Y. Yamamoto, J. Oda and Y. Ynouye, *J. Org. Chem.*, **41**, 303 (1976).
120. D. N. Jones, J. Blekinsopp, A. F. C. Edwards, E. Helms and R. J. K. Taylor, *J. Chem. Soc., Perkin Trans. 1*, 2602 (1973).
121. (a) A. Jefferson and F. Scheinmann, *Quart Rev.*, **22**, 391, 420 (1968).
(b) J. E. Baldwin, R. E. Hackler and D. P. Kelly, *J. Chem. Soc., Chem. Commun.*, 538 (1968).
122. H. J. Reich, I. L. Reich and S. Wollowitz, *J. Am. Chem. Soc.*, **100**, 5981 (1978).
123. (a) D. Darwish and A.-M. Armour, unpublished results.
(b) A.-M. Armour, Ph.D. Thesis, University of Alberta, 1970.
124. S. O. Myong, L. W. Lindler, Jr., S. C. Seike and R. D. Little, *J. Org. Chem.*, **50**, 2244 (1985).
125. K. Hiroi, R. Kitayama and S. Sato, *Synthesis*, 1040 (1983).
126. K. Hiroi, R. Kitayama and S. Sato, *J. Chem. Soc., Chem. Commun.*, 1470 (1983); *Chem. Pharm. Bull.*, **32**, 2628 (1984).
127. D. J. Knight, G. H. Whithan and J. G. Williams, *J. Chem. Soc., Perkin Trans. 1*, 2149 (1987).
128. P. A. Grieco and D. Boxler, *Synth. Commun.*, **5**, 315 (1975).
129. J. E. Baldwin, O. W. Lever, Jr. and N. R. Tzadikov, *J. Org. Chem.*, **41**, 2312 (1976).
130. J. D. Roberts and C. M. Sharts, *Org. React.*, **12**, 1 (1962).
131. D. R. Taylor, *Chem. Rev.*, **67**, 317 (1967).
132. W. D. Huntsman, in *The Chemistry of Ketenes, Allenes and Related Compounds*, (Ed. S. Patai), Wiley, Chichester, 1980, p. 521.
133. S. Braverman, in *The Chemistry of Double-Bonded Functional Groups; Supplement A2* (Ed. S. Patai), Wiley, Chichester, 1989.
134. (a) S. Braverman and H. Mechoulam, *Isr. J. Chem.*, **4**, 17 (1966).
(b) S. Braverman and H. Mechoulam, *Isr. J. Chem.*, **5**, 71 (1967).
(c) S. Braverman and H. Mechoulam, *Tetrahedron*, **30**, 3883 (1974).
135. (a) C. J. M. Stirling, *J. Chem. Soc., Chem. Commun.*, 131 (1967).
(b) G. Smith and C. J. M. Stirling, *J. Chem. Soc. (C)*, 1530 (1971).
136. S. Braverman and Y. Stabinsky, *J. Chem. Soc., Chem. Commun.*, 270 (1967).
137. S. Braverman and Y. Stabinsky, *Isr. J. Chem.*, **5**, 125 (1967).
138. S. Braverman and D. Segev, *J. Am. Chem. Soc.*, **96**, 1245 (1974).
139. G. Büchi and R. M. Freidinger, *J. Am. Chem. Soc.*, **96**, 3332 (1974).
140. J. E. Baldwin, G. Hölfe and S. C. Choi, *J. Am. Chem. Soc.*, **93**, 2810 (1971).
141. E. K. Fields and S. Meyerson, *J. Am. Chem. Soc.*, **88**, 2836 (1966).
142. (a) D. C. Dittmer, R. S. Henion and N. Takashima, *J. Org. Chem.*, **34**, 1310 (1969).
(b) D. C. Dittmer and T. R. Nelsen, *J. Org. Chem.*, **41**, 3044 (1976).
143. R. W. Hoffman and W. Sieben, *Justus Liebigs Ann. Chem.*, **703**, 96 (1967).
144. J. F. King, K. Piers, D. J. H. Smith, C. L. McIntosh and P. de Mayo, *J. Chem. Soc., Chem. Commun.*, 31 (1969).
145. J. F. King, P. de Mayo, C. L. McIntosh, K. Piers and D. J. H. Smith, *Can. J. Chem.*, **48**, 3704 (1970).
146. J. F. King, *Acc. Chem. Res.*, **8**, 10 (1975).
147. R. M. Dodson, P. D. Hammen and R. A. Davis, *J. Chem. Soc., Chem. Commun.*, 9 (1968); *J. Org. Chem.*, **36**, 2693 (1971).
148. S. Braverman and D. Reisman, *J. Am. Chem. Soc.*, **99**, 605 (1977).
149. S. Braverman and D. Reisman, *Tetrahedron Lett.*, 1753 (1977).
150. S. Braverman and Y. Duar, *J. Am. Chem. Soc.*, **105**, 1061 (1983).
151. S. Braverman, *Phosphorus and Sulfur*, **23**, 297 (1985).
152. M. Cinquini, S. Colonna and F. Cozzi, *J. Chem. Soc., Perkin Trans. 1*, 247 (1978).
153. F. Jung and T. Durst, *J. Chem. Soc., Chem. Commun.*, 4 (1973).
154. F. Jung, N. K. Sharma and T. Durst, *J. Am. Chem. Soc.*, **95**, 3420 (1973).
155. T. Durst and B. P. Gimbarzewsky, *J. Chem. Soc., Chem. Commun.*, 724 (1975).
156. T. J. Maricich and V. L. Hoffman, *J. Am. Chem. Soc.*, **96**, 7770 (1974).
157. T. J. Maricich, R. A. Jourdenais and T. A. Albright, *J. Am. Chem. Soc.*, **95**, 5831 (1973).
158. T. J. Maricich, S. Medhusoodanan and C. A. Kapfer, *Tetrahedron Lett.*, 983 (1977).
159. A. Dondoni, P. Giorgianni and A. Battaglia, *J. Chem. Soc., Chem. Commun.*, 350 (1981).
160. O. J. Scherer and R. Schmitt, *Tetrahedron Lett.*, 6235 (1968).
161. J.-B. Baudin and S. A. Julia, *Tetrahedron Lett.*, **29**, 3251 (1988).
162. J.-B. Baudin, S. A. Julia and O. Ruel, *Tetrahedron*, **43**, 881 (1987).
163. J.-B. Baudin and S. A. Julia, *Tetrahedron Lett.*, **27**, 837 (1986).

164. G. Kresze and W. Wucherpfennig, *Angew. Chem., Int. Ed. Engl.*, **6**, 149 (1967).
165. G. Kresze and W. Wucherpfennig, in *Newer Methods of Preparative Organic Chemistry* (Ed. W. Foerst), **5**, 109 (1968).
166. (a) N. Schönberger and G. Kresze, *Justus Liebigs Ann. Chem.*, 1725 (1975).
(b) G. Kresze, in *Organic Sulfur Chemistry* (Ed. C. J. M. Stirling), Butterworths, London, 1975, p. 65.
167. K. B. Sharpless and T. Hori, *J. Org. Chem.*, **41**, 176 (1976).
168. K. B. Sharpless, T. Hori, L. K. Truesdale and C. O. Dietrich, *J. Am. Chem. Soc.*, **98**, 269 (1976).
169. S. P. Singer and K. B. Sharpless, *J. Org. Chem.*, **43**, 1448 (1978).
170. T. Hori, S. P. Singer and K. B. Sharpless, *J. Org. Chem.*, **43**, 1456 (1978).
171. G. Deleris, J. Kowalski, J. Dunogues and R. Calas, *Tetrahedron Lett.*, 4211 (1977).
172. A. Laporterie, J. Dubac, G. Manuel, G. Deleris, J. Kowalski, J. Dunogues and R. Calas, *Tetrahedron*, **34**, 2669 (1978).
173. J.-P. Pillot, G. Deleris, J. Dunogues and R. Calas, *J. Org. Chem.*, **44**, 3397 (1979).
174. G. Deleris, J. Dunogues and R. Calas, *Tetrahedron Lett.*, 4835 (1979).
175. R. Bussas and K. Kresze, *Angew. Chem., Int. Ed. Engl.*, **19**, 732 (1980).
176. G. Kresze and R. Bussas, *Angew. Chem., Int. Ed. Engl.*, **19**, 737 (1980).
177. R. Bussas and G. Kresze, *Justus Liebigs Ann. Chem.*, 843 (1980).
178. K. K. Andersen and O. Malver, *J. Org. Chem.*, **48**, 4803 (1983).
179. T. Kobayashi, K. Iino and T. Hiraoka, *J. Am. Chem. Soc.*, **99**, 5506 (1977).
180. M. Isola, E. Ciuffarin, L. Sagramora and L. Niccolai, *Tetrahedron Lett.*, **23**, 1381 (1982).
181. K. Okuma, K. Nakanishi and H. Ohta, *J. Org. Chem.*, **49**, 1402 (1984).
182. E. J. Corey and T. Durst, *J. Am. Chem. Soc.*, **88**, 5656 (1966).
183. E. J. Corey and T. Durst, *J. Am. Chem. Soc.*, **90**, 5548 (1968).
184. R. P. Gupta and J. P. S. Pizey, *Phosphorous and Sulfur*, **7**, 325 (1979).
185. T. Yoshida, S. Narato, H. Ano and H. Nishimura, *Chem. Pharm. Bull.*, **30**, 2820 (1982).
186. H. Tsuda, H. Minato and M. Kobayashi, *Chem. Lett.*, 149 (1976).
187. R. E. Booms and D. J. Cram, *J. Am. Chem. Soc.*, **94**, 5438 (1972).
188. S. Colonna, R. Giovini and F. Montanari, *J. Chem. Soc., Chem. Commun.*, 865 (1968).
189. A. Nudelman and D. J. Cram, *J. Am. Chem. Soc.*, **90**, 3869 (1968).
190. J. Jacobus and K. Mislow, *J. Chem. Soc., Chem. Commun.*, 253 (1968).
191. (a) K. K. Andersen, *Tetrahedron Lett.*, 93 (1962).
(b) O. N. Sorensen, *Int. J. Sulfur Chem.* (*B*), 6, 321 (1971).
192. F. Wudl and T. B. K. Lee, *J. Chem. Soc., Chem. Commun.*, 61 (1972).
193. F. Wudl and T. B. K. Lee, *J. Am. Chem. Soc.*, **95**, 6349 (1973).
194. (a) H. E. Herbrandson and R. T. Dickerson, *J. Am. Chem. Soc.*, **81**, 4102 (1959).
(b) K. Mislow, T. Simons, J. T. Melillo and A. L. Ternay, Jr., *J. Am. Chem. Soc.*, **86**, 1452 (1964).
195. R. F. Hudson and K. A. F. Record, *J. Chem. Soc., Chem. Commun.*, 831 (1976).
196. C. Brown, R. F. Hudson and K. A. F Record, *J. Chem. Soc., Chem. Commun.*, 540 (1977).
197. C. Brown, R. F. Hudson and K. A. F Record, *J. Chem. Soc., Perkin Trans. 2*, 822 (1978).
198. R. F. Hudson, A. J. Lawson and K. A. F Record, *J. Chem. Soc., Chem. Commun.*, 488 (1974).
199. A. G. Davies, P. D. Roberts and B. R. Sanderson, *J. Chem. Soc., Perkins Trans. 2*, 626 (1973).
200. (a) A. R. Lepley and G. L. Closs (Eds.), *Chemically Induced Magnetic Polarization*, Wiley, New York, 1973.
(b) R. Kaptein, *Adv. Free-Radical Chem.*, **5**, 319 (1975).
201. W. B. Jennings, S. P. Watson and D. R. Boyd, *J. Chem. Soc., Chem. Commun.*, 931 (1988).
202. (a) R. F. Hudson, A. J. Lawson and K. A. F. Record, *J. Chem. Soc., Perkin Trans. 2*, 869 (1974).
(b) C. Brown, R. F. Hudson, A. Maron and K. A. F. Record, *J. Chem. Soc., Chem. Commun.*, 663 (1973).
203. H. F. Whalen and L. F. Jones, *J. Am. Chem. Soc.*, **47**, 1353 (1925).
204. K. Hovius and J. B. F. N. Engberts, *Tetrahedron Lett.*, 181 (1972).
205. I. P. Bleeker, Ph.D. Thesis, Groningen, 1981.
206. M. R. Banks and R. F. Hudson, *J. Chem. Soc., Chem. Commun.*, 799 (1985).
207. M. R. Banks and R. F. Hudson, *J. Chem. Soc., Perkin Trans. 2*, 151 (1986).
208. W. J. Bouma and J. B. F. N. Engberts, *J. Org. Chem.*, **41**, 143 (1976).
209. I. P. Bleeker and J. B. F. N. Engberts, *Recl. Trav. Chim. Pays-Bas*, **98**, 120 (1979).
210. I. P. Bleeker and J. B. F. N. Engberts, *J. Org. Chem.*, **46**, 1012 (1981).
211. A. Heesing, W. K. Homann and W. Müller, *Chem. Ber.*, **113**, 152 (1980).

212. W. B. Ankers, R. F. Hudson and A. J. Lawson, *J. Chem. Soc., Perkin Trans. 2*, 1826 (1974).
213. H. J. Backer and H. Kloosterziel, *Recl. Trav. Chim. Pays-Bas*, **73**, 129 (1954).
214. D. Barnard, *J. Chem. soc.*, 4675 (1957).
215. (a) R. M. Topping and N. Kharasch, *J. Org. Chem.*, **27**, 4353 (1962).
(b) C. M. M. da Silva Correa and W. Waters, *J. Chem. Soc. (C)*, 1874 (1968).
216. P. Koch, E. Ciufarin and A. Fava, *J. Am. Chem. Soc.*, **92**, 5971 (1970).
217. E. Block, *J. Am. Chem. Soc.*, **94**, 644 (1972).
218. E. Block and S. Weidman, *J. Am. Chem. Soc.*, **95**, 5046 (1973).
219. E. Block and J. O'Connor, *J. Am. Chem. Soc.*, **95**, 5048 (1973).
220. E. Block and J. O'Connor, *J. Am. Chem. Soc.*, **96**, 3929 (1974).
221. W. E. Savige and A. Fava, *J. Chem. Soc., Chem. Commun.*, 417 (1965).
222. L. Sagramora, P. Koch, A. Garbesi and A. Fava, *J. Chem. Soc., Chem. Commun.*, 985 (1967).
223. P. Koch and A. Fava, *J. Am. Chem. Soc.*, **90**, 3867 (1968).
224. J. L. Kice and G. B. Large, *Tetrahedron Lett.*, 3537 (1965).
225. J. L. Kice and G. B. Large, *J. Am. Chem. Soc.*, **90**, 4069 (1968).
226. F. Wudl, R. Gruber and A. Padwa, *Tetrahedron Lett.*, 2133 (1969).
227. (a) G. A. Russell and G. J. Mikol, in *Mechanisms of Molecular Migrations* (Ed. B. S. Thayagarajan), Vol. 1, Interscience, New York, 1968, p. 157.
(b) S. Oae and T. Numata, *Isot. Org. Chem.*, **5**, 45 (1980).
228. I. Saito and S. Fukui, *J. Vitaminol. (Kyoto)*, **12**, 244 (1966).
229. E. Block, *J. Org. Chem.*, **39**, 734 (1974).
230. K. Kondo and A. Negishi, *Chem. Lett.*, 1525 (1974).
231. N. Furukuwa, T. Morishita, T. Akasaka and S. Oae, *Tetrahedron Lett.*, 1653 (1977).
232. N. Furukawa, T. Morishita, T. Akasaka and S. Oae, *Tetrahedron Lett.*, 1567 (1978).
233. N. Furukawa, T. Morishita, T. Akasaka and S. Oae, *J. Chem. Soc., Perkin Trans. 2*, 432 (1980).
234. (a) M. M. Chan and J. L. Kice, *J. Am. Chem. Soc.*, **98**, 7711 (1976).
(b) S. Oae, Y. H. Kim, T. Takata and D. Fukushima, *Tetrahedron Lett.*, 1195 (1977).
235. S. Oae and T. Takata, *Tetrahedron Lett.*, **21**, 3213 (1980).
236. Y. Inagaki, R. Okazaki and N. Inamato, *Chem. Lett.*, 1095 (1978).

The Chemistry of Sulphinic Acids, Esters and their Derivatives
Edited by S. Patai


CHAPTER **12**

# Sulphinic acids and esters in synthesis

JOZEF DRABOWICZ, PIOTR KIEŁBASIŃSKI AND MARIAN MIKOŁAJCZYK

*Centre of Molecular and Macromolecular Studies, Polish Academy of Sciences, Department of Organic Sulphur Compounds, Sienkiewicza 112, 90-363, Łódź, Poland*

## I. INTRODUCTION

Sulphinic acids, as a result of their high reactivity and ambident character of the sulphinate anion, are very convenient starting materials for the synthesis of a variety of organosulphur compounds with the same, lower or higher oxidation state. Although the tetrahedral-like configuration around the sulphur atom in sulphinic acids is stable and one can formally write two enantiomeric structures (**A** and **B**) of the acid, sulphinic acids are effectively achiral. This is due to a fast proton exchange between the **A** and **B** forms via the achiral sulphinic acid anion.

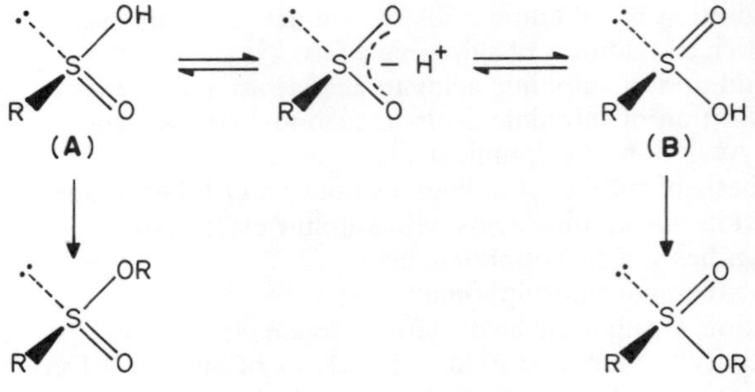

On the contrary, sulphinic acid esters, which may be formally derived from sulphinic acids by replacement of the hydroxy by the alkoxy group, are chiral and were obtained in optically active forms. For this reason, all the reactions of sulphinic acids usually result in the formation of achiral or racemic products while the transformations of chiral, optically active sulphinates are in the majority of cases highly or fully stereoselective and give optically active products.

Cursory discussions on reactivity and properties of sulphinic acids and sulphinates may be found in many review articles and books devoted to sulphur chemistry. This review represents an attempt, perhaps the first one, to summarize in a systematic and comprehensive way various synthetic applications of both classes of compounds with emphasis on the most recent findings. Therefore, an effort has been made to cover the results published up to the end of 1988.

## II. SULPHINIC ACIDS

Sulphinic acids and sulphinate anions exhibit a typical ambident reactivity and react with a variety of electrophilic reagents to form a new bond either by means of the sulphur or oxygen atoms (equation 1). In the former case sulphonyl derivatives are produced and this can be considered as a formal oxidation of S(IV) in the substrate to S(VI) in the product. The electrophilic attack at the oxygen atom affords sulphinyl derivatives which often undergo subsequent reactions or rearrangements. The direction of the attack depends on the nature of the electrophile and reaction conditions. Sometimes competition between O and S attack is observed which results in the formation of a mixture of both types of product.

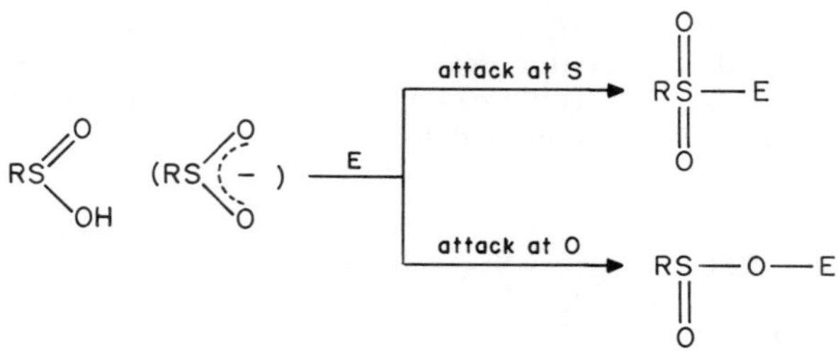


(1)

A large majority of the reactions of sulphinic acids or their salts, and hence their applications to the synthesis of other organic sulphur compounds, may be classified according to this scheme. For this reason the discussion on the application of sulphinic acids and sulphinate anions in organic synthesis will be divided into three parts: (A) S reactivity of sulphinic acids—synthesis of sulphonyl derivatives; (B) O reactivity of sulphinic acids—synthesis of sulphinyl derivatives; (C) other applications of sulphinic acids.

### A. S Reactivity of Sulphinic Acids—Synthesis of Sulphonyl Derivatives

#### *1. Synthesis of sulphones*

The reaction of sulphinic acids or their salts with C electrophiles, resulting in the formation of a new C—S bond, constitutes the most important method for the synthesis of sulphones. Although it has been investigated and used for over 100 years, new reports concerning modification and optimization of this reaction are still appearing in the chemical literature. Since two comprehensive reviews by Schank devoted to the synthesis of sulphones have recently been published[1,2], only selected examples of general importance or interesting from the synthetic and mechanistic points of view will be presented here. In this section alkylation, condensation, ring-opening reactions, alkenylation and arylation, addition to non-activated alkenes, Michael addition, addition to acetylenes and allenes, addition to carbonyl compounds and S acylation of sulphinic acids will be discussed.

*a. Alkylation of sulphinic acids and their anions.* S alkylation of sulphinic acids belongs to the classical methods of the synthesis of sulphones. The reaction has a very wide scope. Only tertiary alkyl halides do not alkylate sulphinate anions, instead olefins are produced[3]. All the examples of alkylation by means of alkyl halides or sulphates published until 1942 have been listed by Suter[4] (equation 2). This method has been used for the synthesis of some special types of sulphones. Thus, long-chain ($C_6$—$C_{14}$) dialkyl sulphones were obtained in 27 to 42% yield[5]. Cycloalkyl sulphones were also synthesized in this way in yields from 11% (for dicyclohexyl sulphone) to 82.5% (for cyclohexyl methyl sulphone)[6]. Schank investigated the reaction of sodium arenesulphinates with α-halogenoethers **1** and found that a mixture of S— and O-alkylation products (**2** and **3**, respectively) was always formed[7]. S alkylation was found to prevail when X = Cl, while O alkylation predominated when X = Br (equation 3). Better yields of sulphones were obtained in apolar solvents, such as benzene and petroleum ether (40–66% when X = Cl; for X = Br the yields were much lower)[8]. α,α-Dihalogenoethers do not give sulphones at all, instead arenesulphonyl halides are formed among other products[9]. α-Methoxy-α-halogenoketones **4** behave similarly and form upon treatment with metal arenesulphinates α-methoxy-α-ketomethyl sulphones **5** and α-methoxy-α-ketomethyl sulphinates **6**. The latter undergo decomposition to form S-aryl arenethiosulfonates **7** (equation 4). Detailed investigations on the influence of external (solvent, concentration) and internal (substituents, cation, alkylating reagent) factors showed that the optimal yield of **5** (Ar = *p*-Tol, R = Me) of 81% could be obtained when M = Na, X = Br and in the presence of catalytic amounts of sodium iodide[10].

$$\mathrm{RSO_2H + Alk{-}X \longrightarrow R\overset{\overset{\displaystyle O}{\|}}{\underset{\underset{\displaystyle O}{\|}}{S}}{-}Alk} \qquad (2)$$

$$\underset{(\mathbf{1})}{\mathrm{ArSO_2Na + XCH_2OR}} \longrightarrow \underset{(\mathbf{2})}{\mathrm{ArSO_2CH_2OR}} + \underset{(\mathbf{3})}{\mathrm{ArS\underset{\underset{\displaystyle O}{\|}}{O}CH_2OR}} \qquad (3)$$

$$\mathrm{ArSO_2M} + \underset{(\mathbf{4})}{\mathrm{XCH\overset{OMe}{|}\underset{\underset{\displaystyle O}{\|}}{C}R}} \longrightarrow \underset{(\mathbf{5})}{\mathrm{ArSO_2\overset{OMe}{\overset{|}{C}H}\underset{\underset{\displaystyle O}{\|}}{C}R}} + \underset{(\mathbf{6})}{\left[\mathrm{Ar\underset{\underset{\displaystyle O}{\|}}{S}O\overset{OMe}{\overset{|}{C}H}\underset{\underset{\displaystyle O}{\|}}{C}R}\right]} \qquad (4)$$

$$(\mathbf{6}) \longrightarrow \underset{(\mathbf{7})}{\mathrm{ArSO_2{-}SAr}}$$

α-Haloalkyl carboxylic esters **8** react with sulphinates to give acyloxyalkyl sulphones **9** in low yields. The presence of water was found to be crucial[11] (equation 5). However, 3-arylsulphonyl iso-chromane-1,4-diones **11** could be obtained from the bromolactone **10** in the yields of 40 to 54% (equation 6)[12].

$$p\text{-}\mathrm{TolSO_2Na{\cdot}2H_2O} + \underset{(\mathbf{8})}{\mathrm{ClCH_2O\underset{\underset{\displaystyle O}{\|}}{C}Ph}} \xrightarrow{18\%} \underset{(\mathbf{9})}{p\text{-}\mathrm{TolSO_2CH_2O\underset{\underset{\displaystyle O}{\|}}{C}Ph}} \qquad (5)$$

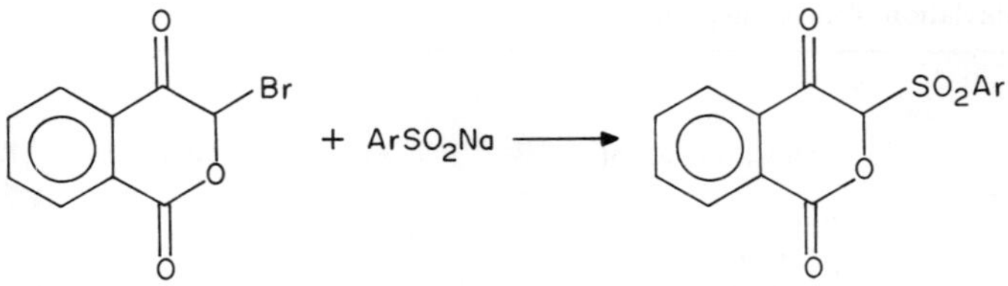


(10) (11) (6)

α-Halogenoketones **12** react very easily with metal sulphinates to afford β-oxo sulphones **13** in high yields[13,14] (equation 7). Other examples are spread over many references; see also Tables 1–4 below. On the other hand, α-chloro-1,3-dicarbonyl compounds **14** do not give the expected β, β′-dioxo sulphones **15** but sulphones **16** which are the products of elimination (equation 8)[15]. Moreover, in the case of α-halogeno-β, β,β-tricarbonyl compounds the halogen atom becomes so electropositive that it is first attacked by a sulphinate anion to form a sulphonyl halide[16].

$$\mathrm{R^1\underset{\underset{\displaystyle O}{\|}}{C}\overset{\overset{\displaystyle R^2}{|}}{\underset{\underset{\displaystyle R^3}{|}}{C}}X + R^4SO_2M \longrightarrow R^1\underset{\underset{\displaystyle O}{\|}}{C}\overset{\overset{\displaystyle R^2}{|}}{\underset{\underset{\displaystyle R^3}{|}}{C}}SO_2R^4} \qquad (7)$$

(12) (13)

$$\mathrm{Ar^1SO_2Na + ClCH}\begin{cases}\mathrm{C(O)Me}\\ \mathrm{C(O)NHAr^2}\end{cases} \longrightarrow \left[\mathrm{Ar^1SO_2CH}\begin{cases}\mathrm{C(O)Me}\\ \mathrm{C(O)NHAr^2}\end{cases}\right]$$

(14) (15)

$$\xrightarrow[\mathrm{-\,MeCO_2Et}]{\mathrm{EtOH}} \mathrm{Ar^1SO_2CH_2\underset{\underset{\displaystyle O}{\|}}{C}NHAr^2} \qquad (8)$$

(16)

An interesting example of alkylation is the reaction of 2-chloro-1, 2-dithiolium perchlorates **17** with sodium benzenesulphinate which leads to the formation of 2-benzenesulphonyl-1, 2-dithiolium perchlorates **18**[17] (equation 9).

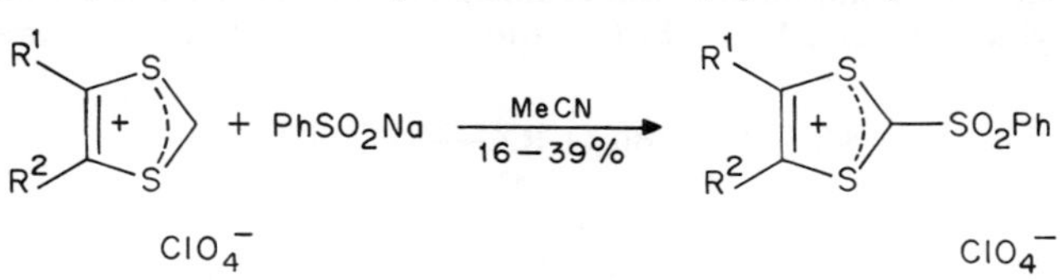


(17) (18) (9)

Silver salts of sulphinic acids were reported long ago to undergo alkylation on the oxygen atom and to produce sulphinic esters[3,18]. More recently, Meek and Fowler demonstrated that when the reaction of silver *p*-toluenesulphinate with methyl iodide is performed in methanol, the sulphone to sulphinate ratio is 98:2, the overall yield being 77%[19]. Later on, Russian workers investigated in detail the bidirectional course of the

TABLE 1. Alkylation of silver sulphinates

| | | Yield (%) | | |
|---|---|---|---|---|
| Silver salt | Alkylating agent | ester | sulphone | Reference |
| $MeSO_2Ag$ | $PhCH_2I$ | 24 | 76 | 20 |
| $MeSO_2Ag$ | $p$-$NO_2C_6H_4CH_2I$ | | 80 | 20 |
| $PhSO_2Ag$ | $PhCH_2I$ | 36 | 63 | 20 |
| $PhSO_2Ag$ | $p$-$NO_2C_6H_4CH_2I$ | 14 | 80 | 20 |
| $CF_3SO_2Ag$ | $PhCH_2I$ | 56.5 | 41.3 | 20 |
| $CF_3SO_2Ag$ | $p$-$NO_2C_6H_5CH_2I$ | 50 | 40 | 20 |
| $CF_3SO_2Ag$ | $PhC(O)CH_2Br$ | 26 | 26 | 20 |
| $CF_3SO_2Ag$ | $p$-$ClC_6H_4SCH_2Cl$ | 0 | 57 | 20 |
| $p$-$TolSO_2Ag$ | MeI/DMF | 6.6 | 66.5 | 19 |
| $p$-$TolSO_2Ag$ | MeI/MeOH | 1.5 | 75.5 | 19 |

alkylation of silver sulphinates in anhydrous acetonitrile and found that the product ratio depends on both the kind of acid and alkylating agents (Table 1)[20]. No reaction was observed when α-chlorosulphones were used as alkylating agents.

An attempt to prepare 4-sulphonyl derivatives of isophorone by treatment of 4-bromoisophorone **19** with sodium arenesulphinates led to the isomeric 2-sulphonyl derivatives **20**, most probably via an $S_N2'$ mechanism[21] (equation 10).

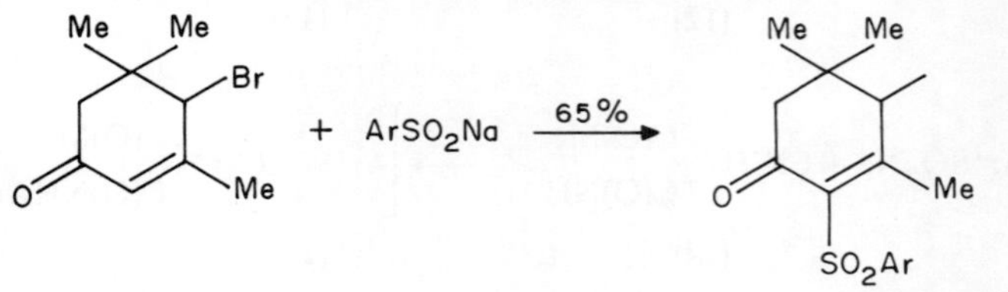


(19) (20)

(10)

Several general improvements of the experimental methodology of alkylation of sulphinic acids with alkyl halides have been published recently. The first one is based on the application of tetraalkylammonium (mainly tetrabutylammonium) sulphinates obtained by the ion-pair extraction method[22]. The reaction is performed in a THF solution at 10–40 °C using equimolar amounts of alkyl halides and gives sulphones in yields usually higher than those obtained by other methods (equation 11, Table 2)[22]. In a similar way tetrabutylammonium trifluoromethanesulphinate **21** is alkylated to give trifluoromethyl sulphones **22** and **23** (equation 12)[23]. An interesting variation of this

$$p\text{-TolSO}_2^- \overset{+}{N}Bu_4 + RX \longrightarrow p\text{-TolSO}_2R \tag{11}$$

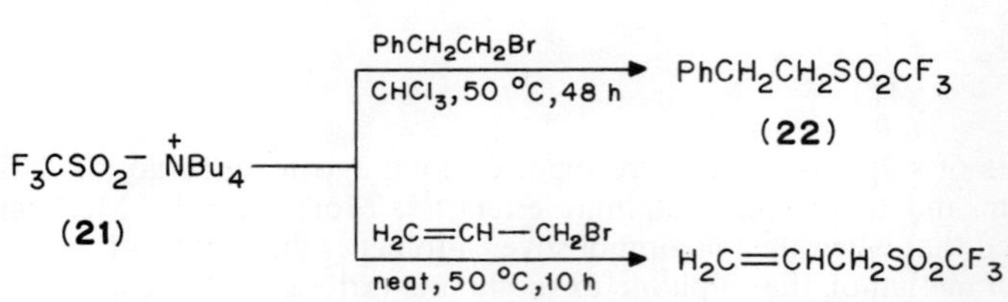


(12)

TABLE 2. Alkylation of tetrabutylammonium *p*-toluenesulphinate

| Alkyl halide | Reaction temperature (°C) | Reaction time (h) | Yield of sulphone (%) |
|---|---|---|---|
| MeI | 20 | 3 | 93 |
| *p*-$ClC_6H_4CH_2Br$ | 20 | 2 | 93 |
| $H_2C{=}CHCH_2Br$ | 20 | 2.5 | 80 |
| *i*-PrBr | 40 | 4 | 63 |
| $MeOCH_2Cl$ | 40 | 4 | 59 |
| $N{\equiv}CCH_2Cl$ | 30 | 3 | 85 |
| $EtOC(O)CH_2Br$ | 20 | 2 | 80 |
| $PhC(O)CH_2Br$ | 20 | 2 | 81 |
| $ClCH_2C(O)CH_2Cl$ | 20 | 2 | 75[a] |
| $PhCH{=}CHC(O)CH_2Cl$ | 20 | 4 | 85 |
| $BrCH_2Br$ | 20 | 4 | 89[a] |

[a] Monosulphone.

method has been described which consists in the alkylation of benzenesulphinate anion supported on Amberlyst A-26, a macroreticular anion exchange resin containing quaternary ammonium groups (equation 13, Table 3)[24].

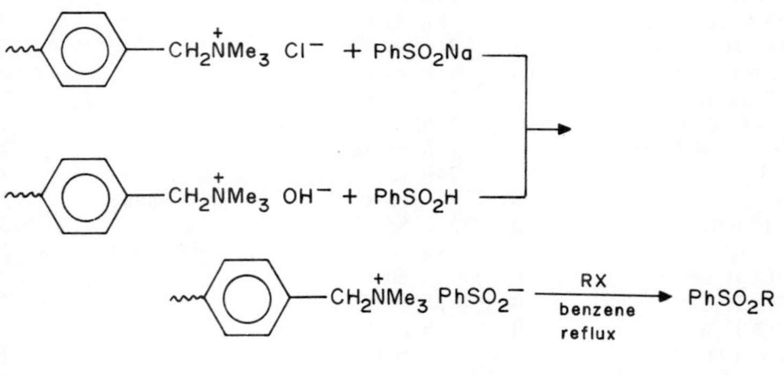

(13)

TABLE 3. Alkylation of benzenesulphinic acid using Amberlyst A-26

| Alkyl halide | Reaction time (h) | Yield (%) |
|---|---|---|
| MeI | 3 | 95 |
| PrI | 3 | 94 |
| Hexyl-Br | 3 | 92 |
| Octyl-Br | 3 | 92 |
| $C_6H_{13}CH(Me)Br$ | 3 | 60 |
| $PhCH_2Cl$ | 2 | 93 |
| $Me_2C{=}CHCH_2Br$ | 1.5 | 95 |
| (*E*)-$EtO_2CCH{=}C(Me)CH_2Br$ | 1.5 | 92 |
| (*E*)-$MeO_2CC(Me){=}CHCH_2Br$ | 1.5 | 93 |
| $EtO_2CCH_2Cl$ | 2 | 91 |
| $N{\equiv}CCH_2Cl$ | 2 | 95 |

Another improvement is based on the application of phase-transfer catalysis (PTC) conditions. Three different approaches have been described: (A) a solid–liquid PTC method, using DME as the solvent and tetrabutylammonium bromide as the catalyst[25]; (B) a solid–liquid PTC method with neat alkylating agents playing the role of the organic phase and "Aliquot 336" as the catalyst[26]; (C) a liquid–liquid PTC method using water, benzene and acetone (4:3:3) as the solvent system and tetrabutylammonium bromide or iodide as catalysts[27]. These methods were applied for the synthesis of a very broad variety of differently substituted sulphones (equation 14), and the results are collected in Table 4.

$$p\text{-}R^1C_6H_4SO_2Na + R^2X \longrightarrow p\text{-}R^1C_6H_4SO_2R^2 \tag{14}$$

TABLE 4. Synthesis of sulphones by PTC methods

| $R^1$ | $R^2$—X | Method | Reaction conditions | Yield (%) | Reference |
|---|---|---|---|---|---|
| Me | EtI | C | 6 h, reflux | 89 | 27 |
| Me | *i*-PrI | C | 12 h, reflux | 65 | 27 |
| Me | *i*-PrBr | A | 48 h, 85 °C | 68 | 25 |
| Me | BuBr | A | 4 h, 85 °C | 94 | 25 |
| Me | HexBr | C | 8 h, reflux | 85 | 27 |
| Me | $PhCH_2Cl$ | A | 30 min, 85 °C | 98 | 25 |
| Me | $PhCH_2Br$ | A | 15 min, 85 °C | 96 | 25 |
| H | $PhCH_2Br$ | B | 24 h, 60 °C | 95 | 26 |
| Me | $p$-$NO_2C_6H_4CH_2Br$ | B | 2 h, 85 °C | 93 | 26 |
| H | $o$-$MeC_6H_4CH_2Br$ | A | 24 h, 20 °C | 97 | 25 |
| H | 3-Me-Naph-2-$CH_2Br$ | A | 5 h, 20 °C | 97 | 25 |
| Me | $o$-$N{\equiv}CC_6H_4CH_2Br$ | A | 24 h, 20 °C | 96 | 25 |
| Me | $CH_2{=}C(Me)CH_2Cl$ | C | 12 h, reflux | 71 | 27 |
| Me | $Me_2C{=}CHCH_2Br$ | C | 12 h, reflux | 87 | 27 |
| Me | $ICH_2I$ | C | 24 h, reflux | 55 (monosulphone) | 27 |
| H | $ClCH_2Br$ | B | 24 h, 85 °C | 47 (monosulphone) | 26 |
| H | ClCH(Me)Br | B | 24 h, 85 °C | 35 (monosulphone) | 26 |
| Me | $Cl(CH_2)_3Br$ | C | 8 h, reflux | 93 (monosulphone) | 27 |
| H | $N{\equiv}CCH_2Br$ | B | 2 h, 60 °C | 91 | 26 |
| H | $N{\equiv}CCH_2Cl$ | B | 2 h, 85 °C | 93 | 26 |
| Me | $N{\equiv}CCH_2Br$ | B | 2 h, 60 °C | 85 | 26 |
| Me | $MeC(O)CH_2Cl$ | C | 6 h, reflux | 94 | 27 |
| Me | $PhC(O)CH_2Cl$ | A | 30 min, 85 °C | 96 | 25 |
| H | $MeO_2CCH_2Br$ | B | 4 h, 60 °C | 95 | 26 |
| H | $MeO_2CCH_2Cl$ | B | 4 h, 60 °C | 53 | 26 |
| Me | $EtO_2CCH_2Cl$ | A | 30 min, 85 °C | 90 | 25 |
| H | $MeO_2CCH(Me)Br$ | B | 4.5 h, 60 °C | 88 | 26 |
| Me | $EtO_2CCH(Me)Br$ | C | 8 h, reflux | 79 | 27 |
| H | $H_2NC(O)CH_2Cl$ | B | 3 h, 120 °C | 73 | 26 |
| Me | 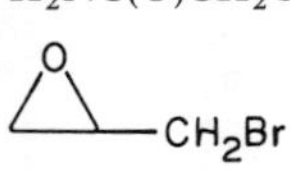 | C | 12 h, reflux | 81[a] | 27 |
| Me | 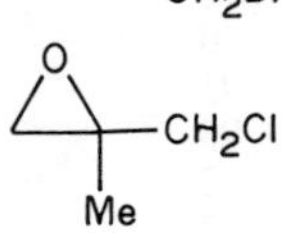 | C | 12 h, reflux | 95[b] | 27 |

[a]Sulphone of the structure (*E*)-$HOCH_2CH{=}CH{-}SO_2$-Tol-*p* is produced.
[b]Mixture of (*E*) and (*Z*) $HOCH_2CH{=}C(Me){-}SO_2$-Tol-*p*, ($E/Z = 65{:}35$) is obtained.

Reactions of sulphinic acids with alkyl halides bearing electron-withdrawing groups sometimes require special conditions, since in some cases they proceed according to a radical mechanism. Thus, the synthesis of trifluoromethyl sulphones **24** with iodotrifluoromethane was achieved in liquid ammonia under UV irradiation[28] (equation 15). Alkylation with the very hindered *p*-nitro-1, 1-dimethylbenzyl chloride **25** proceeds according to the electron-transfer mechanism to give the sulphone **26** in high yield[29] (equation 16). 1-Iodo-1-nitroalkanes **27** react with sodium sulphinates via a radical intermediate **28** to give α-nitroalkyl sulphones **29** (equation 17)[30]. The latter were also obtained in an analogous reaction from bromonitromethane[31] and 1-methyl-1-nitroethylthiocyanate[32]. Another approach to **29** involves the coupling of nitroparaffin salts with benzenesulphinate anions by the agency of an oxidizing agent, e.g. potassium ferricyanide. The reaction proceeds according to the anion-radical mechanism (equation 18)[33,34].

$$ArSO_2H + ICF_3 \xrightarrow[\text{liq. } NH_3]{h\nu} \underset{\textbf{(24)}}{ArSO_2CF_3} \qquad (15)$$

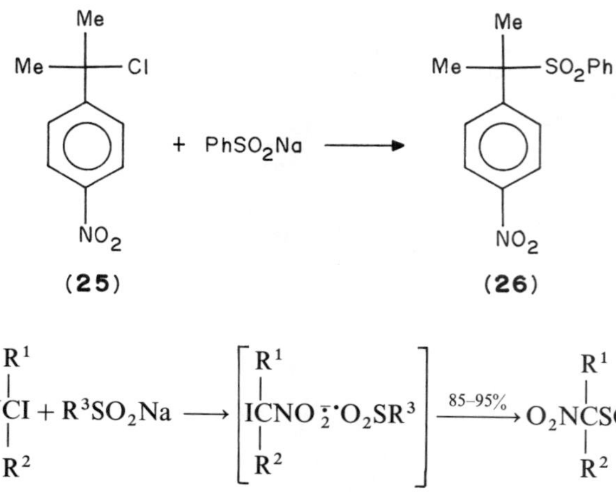

**(25)** **(26)** (16)

$$O_2N\overset{R^1}{\underset{R^2}{C}}I + R^3SO_2Na \longrightarrow \left[ I\overset{R^1}{\underset{R^2}{C}}NO_2^{\bullet-}\ {}^{\bullet}O_2SR^3 \right] \xrightarrow{85\text{–}95\%} O_2N\overset{R^1}{\underset{R^2}{C}}SO_2R^3 \qquad (17)$$

**(27)** **(28)** **(29)**

$$\begin{matrix}R^1\\R^2\end{matrix}\!\!>\!C{=}NO_2^- + R^3SO_2^- \xrightarrow[\substack{K_3Fe(CN)_6\\60\text{–}86\%}]{-2e} O_2N\overset{R^1}{\underset{R^2}{C}}SO_2R^3 \qquad (18)$$

**(29)**

There are several other alkylating agents which give sulphones with sulphinic acids (those which alkylate the oxygen atom of the sulphinate anion leading to sulphinates will be presented in Section II.B.1). For example, allylic acetates proved to be very useful in the synthesis of allylic sulphones. Fischli and Mayer took advantage of this reaction in their synthesis of apocarothenoids **31** from vitamin A acetate **30** (equation 19)[35]. Later on, this reaction was performed in the presence of palladium complexes [e.g. $Pd(PPh_3)_4$[36,38,39], Pd on graphite, carbon or $Al_2O_3$[37]], which made it possible to direct the relative ratio of the isomeric sulphones **32a** and **32b** (equation 20)[36]. In a similar way, 4-sulphonyl azetidinones **33** were obtained from the corresponding 4-acyloxy derivatives in 61–95% yields (equation 21)[38]. An efficient, one-pot procedure for the preparation of methylthiomethyl *p*-tolyl sulphone **34** was accomplished by the Pummerer reaction of dimethyl sulphoxide with acetic anhydride, followed by treatment of the resulting acetoxymethyl methyl sulphide with sodium *p*-toluenesulphinate (equation 22)[39].

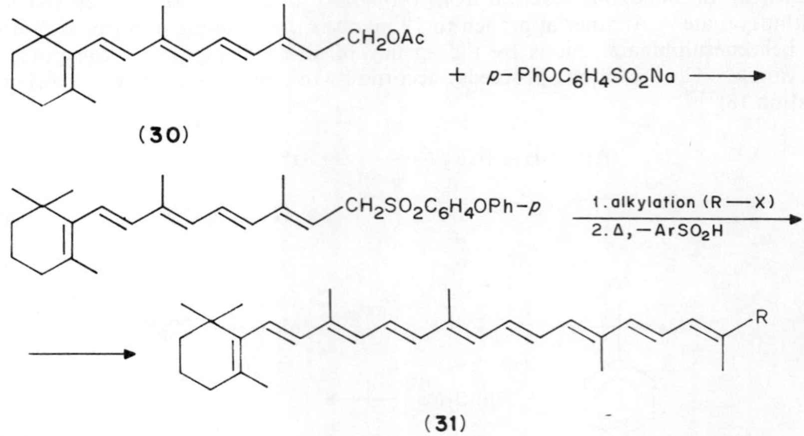


(19)

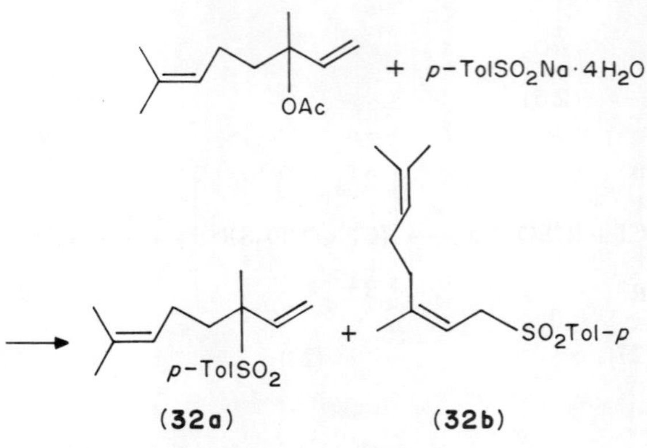


(20)

| Conditions | Ratio | |
|---|---|---|
| 0 °C, 3.5 h | 78% | 9% |
| r.t., 1 min | 62% | 23% |
| r.t., overnight | — | 84% |

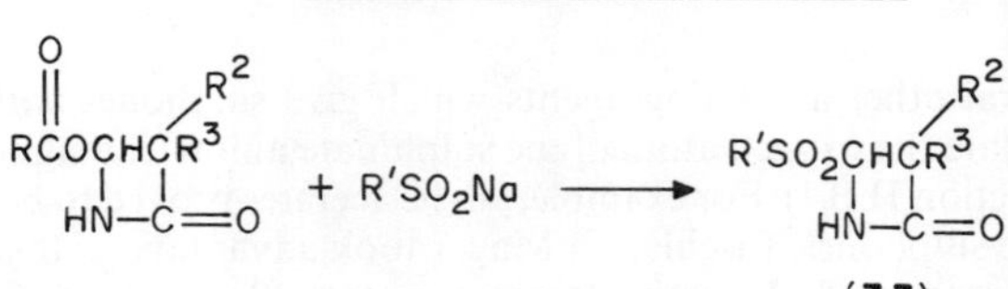


(21)

$$\underset{\substack{\| \\ \mathrm{O}}}{\mathrm{MeSMe}} \xrightarrow{\mathrm{Ac_2O}} \mathrm{MeSCH_2OAc} \xrightarrow[\mathrm{AcOH,\ AcONa}]{p\text{-}\mathrm{TolSO_2Na}} \underset{(\mathbf{34})}{\mathrm{MeSCH_2SO_2Ph}} \qquad (22)$$

Acetals react with sulphinic acids (or their sodium salts) in the presence of $BF_3{\cdot}Et_2O$ to give, depending on the reaction conditions, either sulphinic esters **35** or alkoxysulphones **36** (equation 23)[40]. It is interesting to note that only these two products (of four possible) are produced. To achieve a selective sulphone synthesis the following procedure must be applied: first the acetal and $BF_3{\cdot}Et_2O$ are mixed in a 6:2 ratio, then 1 equivalent of a sulphinic acid is added. The yields of **36** are 82–91%[40].

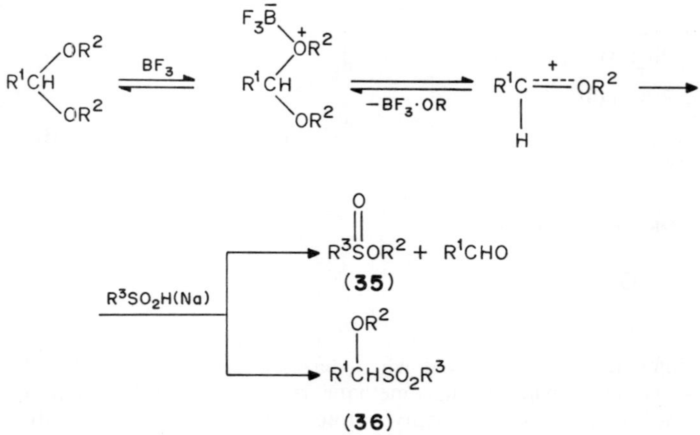


(23)

High yields of aryl methyl sulphones (~95%) may also be obtained when dimethyl methanephosphonate **37** is used as an alkylating agent[41] (equation 24).

$$\mathrm{ArSO_2Na} + \underset{(\mathbf{37})}{\underset{\substack{\| \\ \mathrm{O}}}{\mathrm{MeP(OMe)_2}}} \xrightarrow[170\,^{\circ}\mathrm{C}]{\mathrm{B}} \mathrm{ArSO_2Me} + \underset{\substack{\| \\ \mathrm{O}}}{\mathrm{MeP}}\begin{matrix}\mathrm{OMe} \\ \mathrm{ONa}\end{matrix} \qquad (24)$$

There are two different reports concerning alkylation of sulphinic acids by sulphonium salts. Julia and coworkers obtained 3-methyl-2-butenyl phenyl sulphone **39** in 78% yield using a long-chain sulphonium salt **38** under PTC conditions (equation 25)[42]. On the other hand, Kobayashi and Toriyabe investigated the alkylating properties of diphenylmethylsulphonium perchlorate **40** and found that a mixture of O- and S-alkylation products was always formed (equation 26, Table 5)[43].

$$\mathrm{PhSO_2H} + \underset{(\mathbf{38})}{\mathrm{Me_2C{=}CHCH_2\overset{+}{S}(C_{12}H_{25})_2ClO_4^-}} \longrightarrow \qquad (25)$$

$$\xrightarrow[\mathrm{CH_2Cl_2,\ r.t}]{\mathrm{K_2CO_3}} \underset{(\mathbf{39})}{\mathrm{PhSO_2CH_2CH{=}CMe_2}}$$

TABLE 5. Alkylation of *p*-$TolSO_2K$ with sulphonium and oxosulphonium salts **40** and **41**

| Alkylating agent | Solvent | Crown ether | Time (h) | Product ratio | |
|---|---|---|---|---|---|
| | | | | sulphinate | sulphone |
| **40** | $CH_2Cl_2$ | none | 24 | 44 | 56 |
| **40** | $CH_2Cl_2$ | 18-crown-6 | 2 | 40 | 60 |
| **41** | $CH_2Cl_2$ | none | 24 | 56 | 44 |
| **41** | $CH_2Cl_2$ | 18-crown-6 | 2 | 25 | 75 |
| **41** | DMF | none | 26 | 24 | 76 |

$$Ph_2\overset{+}{S}Me\ ClO_4^- \ \textbf{(40)} \quad \text{or} \quad Ph_2\overset{+}{S}(=O)Me\ ClO_4^- \ \textbf{(41)} \xrightarrow{p\text{-}TolSO_2K} p\text{-}TolS(=O)Me + p\text{-}TolSO_2Me \qquad (26)$$

Similar differences were observed when diazoalkanes were used for alkylation of sulphinic acids. Thus, whereas diazomethane reacts with *p*-toluenesulphinic acid in ether/methanol to give 100% of methyl *p*-toluenesulphinate[19], diphenyldiazomethane gives upon treatment with the same acid a mixture of the sulphinate and sulphone, the ratio of which depends on the solvent used (equation 27, Table 6)[44,45].

$$p\text{-}TolSO_2H + Ph_2C{=}N_2 \longrightarrow p\text{-}TolS(=O){-}OCHPh_2 + p\text{-}TolSO_2CHPh_2 \qquad (27)$$

Application of dimethyl sulphate (or other sulphates[4]) usually leads to the predominant formation of sulphinates[1,19]. However, magnesium trimethylsilylmethanesulphinate **42** reacts with dimethyl sulphate to afford trimethylsilylmethyl methyl sulphone **43** in 78.9%

TABLE 6. Reaction of *p*-$TolSO_2H$ with diphenyldiazomethane[44]

| Solvent | Product ratio | | Total yield |
|---|---|---|---|
| | sulphinate | sulphone | |
| $CH_2Cl_2$ | 0 | 100 | 80 |
| benzene | 20 | 80 | 96 |
| MeCN | 81 | 19 | 100 |
| dioxane | 83 | 17 | 100 |
| DMSO | 100 | 0 | 82 |

yield (equation 28)[46]. The sulphone **43** is also produced in 50.1% yield, together with 15.4% of methyl trimethylsilylmethanesulphinate **45**, when dimethyl sulphate is replaced by trimethyloxonium tetrafluoroborate in nitromethane and the sodium sulphinate **44** is used instead of **42** (equation 29)[46].

$$\underset{\textbf{(42)}}{(Me_3SiCH_2SO_2)\ Mg_{1/2}} + (MeO)_2SO_2 \longrightarrow \underset{\textbf{(43)}}{Me_3SiCH_2SO_2Me} \qquad (28)$$

$$\underset{\textbf{(44)}}{Me_3SiCH_2SO_2Na} + Me_3\overset{+}{O}\overset{-}{B}F_4 \longrightarrow \underset{\textbf{(43)}}{Me_3SiCH_2SO_2Me} + \underset{\textbf{(45)}}{Me_3SiCH_2\overset{\overset{O}{\|}}{S}OMe} \qquad (29)$$

The usefulness of carbenes as alkylating agents in the synthesis of sulphones from sulphinic acids is strongly dependent on their structure. Thus, phenylcarbene[47] and methoxy-carbomethoxycarbene[48] are not suitable for these purposes since they produce the corresponding sulphones in the yield of only 11 and 20%, respectively. On the other hand, methoxy-*p*-toluenesulphonylcarbene **47**, formed by an α-elimination of HCl from the chlorosulphone **46**, reacts with sodium *p*-toluenesulphinate to give the disulphone **48** in 63% yield (equation 30)[49]. Similarly, chloroform and bromoform react with sodium sulphinates in the presence of aqueous base (in the conditions enabling dichloro- and dibromocarbene formation) to give the dichloromethyl sulphones **49** and dibromomethyl sulphones **50**, respectively (equation 31, Table 7)[50,51]. It should be added that in the case of sodium phenylmethanesulphinate (R = $PhCH_2$ in Table 7) the major product was (*E*)-PhCH=CHSO$_3$H, produced as a result of the Ramberg–Bäcklund reaction of the initially formed sulphone. Therefore, this reaction may be used for preparation of α,β-unsaturated sulphonic acids[50].

$$\underset{\textbf{(46)}}{p\text{-}TolSO_2\overset{\overset{OMe}{|}}{C}HCl} \xrightarrow[\text{DMF, 100\%}]{t\text{-BuOK}} \underset{\textbf{(47)}}{\left[p\text{-}TolSO_2\overset{\overset{OMe}{|}}{C}:\right]} \xrightarrow{p\text{-TolSO}_2\text{Na}} \underset{\textbf{(48)}}{p\text{-}TolSO_2\underset{\underset{OMe}{|}}{C}HSO_2Tol\text{-}p} \qquad (30)$$

TABLE 7. Reaction of sodium sulphinates with haloforms

| R | Yield 49 (%) | Yield 50 (%) | Reference |
|---|---|---|---|
| Ph | 87 | 77 | 50 |
| *p*-Tol | 81 | 75 | 50 |
| *p*-$ClC_6H_4$ | 63 | | 51 |
| 2-Naph | 70 | 48 | 50 |
| $Me_3C$ | 55 | 7 | 50 |
| $PhCH_2$ | 5 | | 50 |

$$RSO_2Na \xrightarrow[NaOH]{CHCl_3} RSO_2CHCl_2 \ \textbf{(49)} \qquad RSO_2Na \xrightarrow[NaOH]{CHBr_3} RSO_2CHBr_2 \ \textbf{(50)} \tag{31}$$

Very recently, certain nitroalkanes have been found to be also good alkylating agents[52,53]. For example, cyclic allylic nitro compounds **51**, which are readily prepared by the amine-catalysed reaction of nitroalkanes with cycloalkanones, react with sodium benzenesulphinate in the presence of 5% mol. of $Pd(PPh_3)_4$ to give allylic sulphones **52** with predominance of the endo form **52a** (equation 32, Table 8)[52]. In the case of acylic nitro compounds **53** the ratio of regioisomeric sulphones **54a** and **54b** depends on R and reaction conditions (equation 33, Table 9)[53].

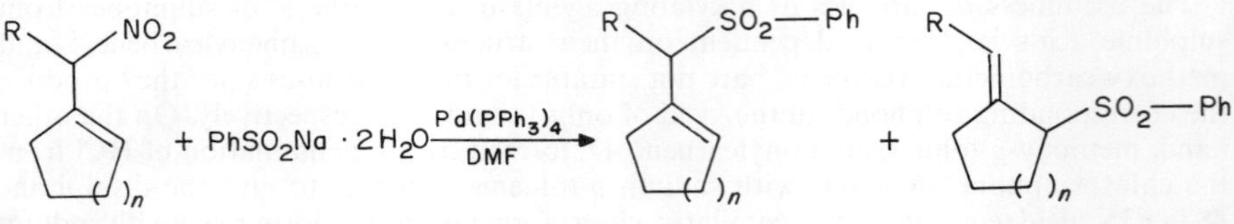

**(51)** **(52a)** **(52b)**

(32)

TABLE 8. Synthesis of cyclic allylic sulphones by denitrosulphonylation

| R | *n* | Temp. (°C) | Time (h) | Product | Yield (%) |
|---|---|---|---|---|---|
| H | 1 | 20 | 10 | **52a** | 70[a] |
| H | 2 | 70 | 1 | **52a** | 85[a] |
| H | 2 | 20 | 10 | **52a** | 70 |
| Me | 2 | 20 | 15 | **52a** | 75 |
| H | 3 | 70 | 1 | **52a** | 92 |
| H | 4 | 70 | 1 | **52a** | 76 |

[a] 2–3% of **52b** was also formed.

TABLE 9. Synthesis of acyclic allylic sulphones by denitrosulphonylation

| R | Time (h) | Product | Yield (%) | **54a/54b** |
|---|---|---|---|---|
| Me | 10 | **54a** | 75 | 100/0 |
| $C_6H_{13}$ | 15 | **54a** | 96 | 95/5 |
| $MeO_2CCH_2CH_2$ | 10 | **54a** | 79 | 95/5 |
| $MeC(O)CH_2CH_2$ | 15 | **54a** | 80 | 95/5 |
| Ph | 15 | **54a + 54b** | 76 | 53/57 |
| $AcOCH_2$ | 15 | **54a + 54b** | 95 | 51/49 |
| $EtO_2C$ | 15 | **54b** | 87 | 0/100 |
| H | 15 | **54a + 54b** | 60 | 25/75 |

$$RC(Me)(NO_2)CH{=}CH_2 + PhSO_2Na \xrightarrow[20\,^{\circ}C]{Pd(PPh_3)_4} RC(Me)(SO_2Ph)CH{=}CH_2 + C(Me)(R){=}CHCH_2SO_2Ph \qquad (33)$$

(53) (54a) (54b)

*b. Condensation of sulphinic acids (or anions) with alcohols, Mannich bases and related compounds.* Aryl alkyl and diaryl carbinols react with sodium *p*-toluenesulphinate in aqueous methanol or ethanol[54] or with *p*-toluenesulphinic acid in a formic acid/sulphuric acid solution[55] to afford the corresponding sulphones in very high yields (examples are shown in equations 34 and 35). Benzyl alcohol does not react with arenesulphinic acids in 100% formic acid, but *p*-methoxybenzyl alcohol gives good yields of methoxybenzyl sulphones[56,57], as do flavanols **55** (equation 36)[56]. Introduction of the methoxy group at the 7- and at the 5- and 7-positions in **55** allows the reaction to occur even in 8% acetic acid, as expected for an $S_N1$ mechanism. Allylic alcohols are also suitable substrates for this reaction, though rearranged products are sometimes formed (equation 37)[58].

$$p\text{-}Me_2NC_6H_4CH(OH)Ph + p\text{-}TolSO_2Na \xrightarrow[-\,NaOH]{EtOH,\ H_2O} p\text{-}Me_2NC_6H_4CH(SO_2Tol\text{-}p)Ph \qquad (34)$$

$$Ph_2CHOH + p\text{-}TolSO_2H \xrightarrow{HCO_2H/H_2SO_4} Ph_2CHSO_2Tol\text{-}p \qquad (35)$$

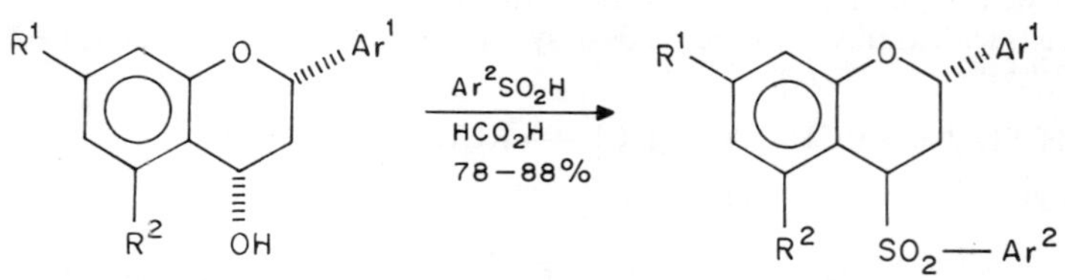


(55) (56) (36)

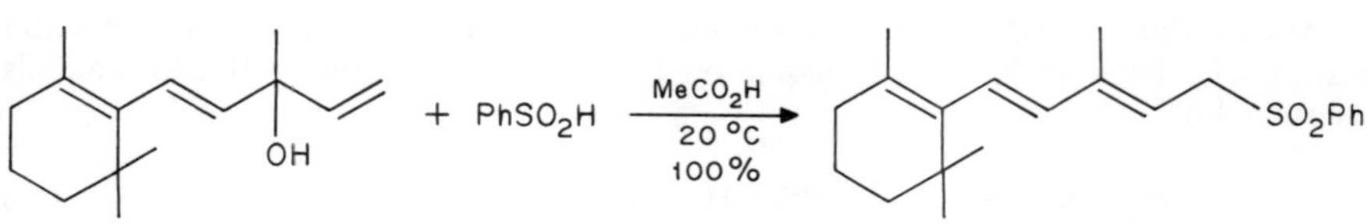


(37)

Condensation of sulphinic acids or their salts with Mannich bases[59,60], their hydrochlorides[61] or quaternary ammonium salts[62] gives sulphones. In this way indolemethyl sulphones[60], aroylethyl sulphones[61,62], e.g. **57** (equation 38)[61] and quinonylmethyl sulphones **58** (equation 39)[59], have been obtained in reasonable to good yields. The mechanism is assumed to be either a direct $S_N$ substitution of the amine by the sulphinate anion or an E–A mechanism, shown in equation 40[59].

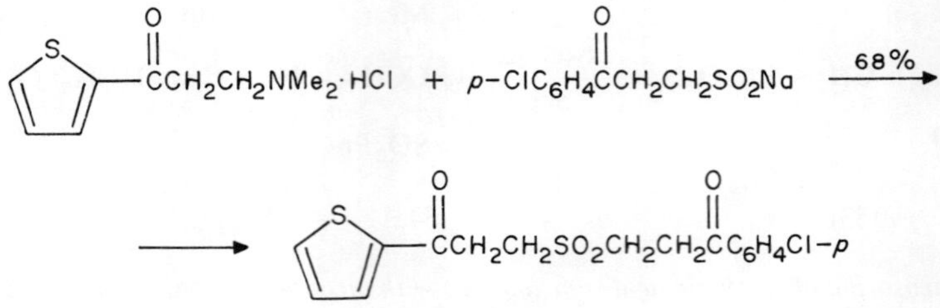


(57) (38)

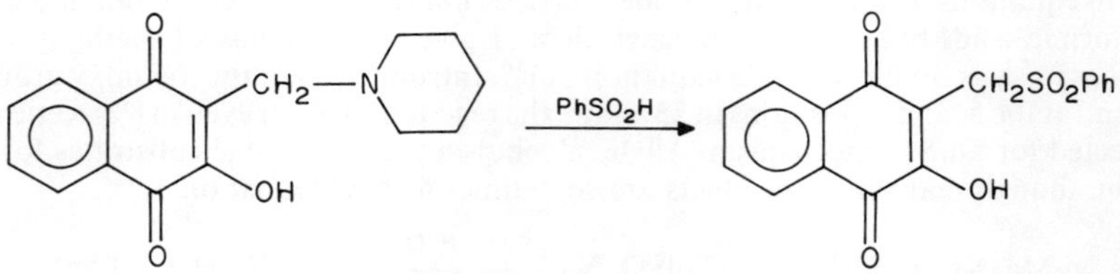


(39)

(58)

$$ZCH_2CH_2NR_2 + R'SO_2H \longrightarrow (ZCH_2CH_2NHR_2)^+R'SO_2^-$$
$$\longrightarrow ZCH{=}CH_2 + R_2\overset{+}{N}HR'SO_2^- \longrightarrow ZCH_2CH_2SO_2R' + R_2NH \quad (40)$$

Reaction of sodium hydroxymethanesulphinate **59** with Mannich bases **60** does not lead to the expected hydroxymethyl sulphones **61** but to the symmetrical sulphones **62**. One of the possible ways of this reaction is shown in equation 41[63]. In the same way the alkylation of **59** proceeds with aralkyl halides; yields of symmetrical sulphones of the type **62**, R = Ar, are 25–47%[64].

$$\underset{(59)}{HOCH_2SO_2Na} + \underset{(60)}{RCH_2NMe_2{\cdot}HCl} \longrightarrow \underset{(61)}{RCH_2SO_2CH_2OH} \quad (41)$$

$$\xrightarrow[-CH_2O]{} RCH_2SO_2H \xrightarrow{60} \underset{(62)}{RCH_2SO_2CH_2R} \text{ yields 14–66\%}$$

α-Aminosulphones **63** and **64** and α-amidosulphones **65** are formed in the reaction of sulphinic acids with aminals (equation 42), triazines (equation 43) and amidals (equation 44)[65].

$$H_2C(NMePh)_2 + p\text{-TolSO}_2H \xrightarrow{51\%} \underset{(63)}{p\text{-TolSO}_2CH_2NMePh} \quad (42)$$

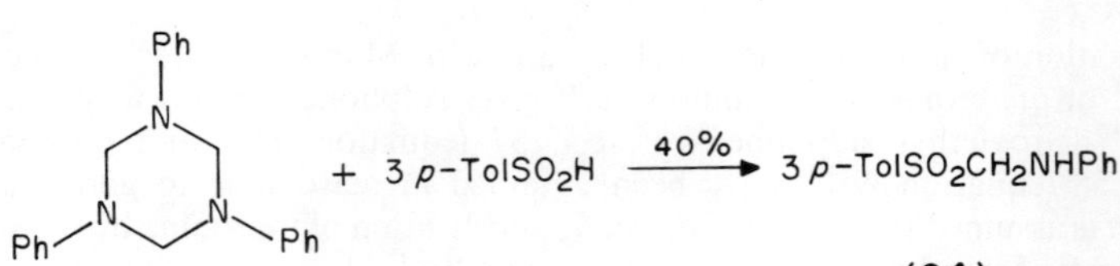


(43)

(64)

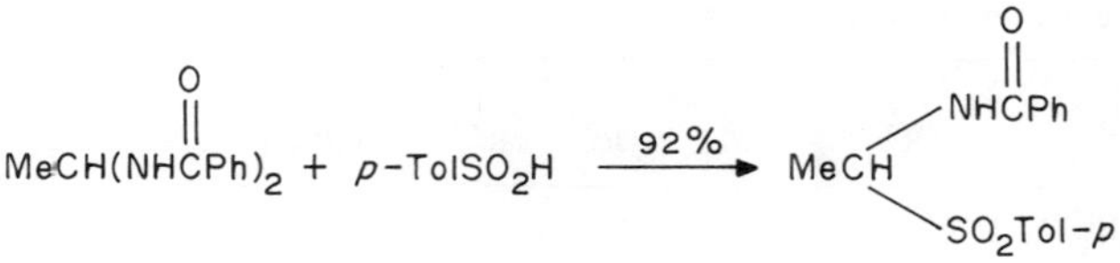

(65)

(44)

*c. Ring-opening reactions.* Oxiranes undergo ring-opening on treatment with sulphinate salts to give 2-hydroxyalkyl sulphones, e.g. **66** (equation 45)[27,66,67]. ω-Carboxyalkyl sulphones (e.g. **67**) are obtained in the reaction of sulphinate salts with lactones (equation 46)[68] and 4-arenesulphonylsulphonic acids from sultones[3].

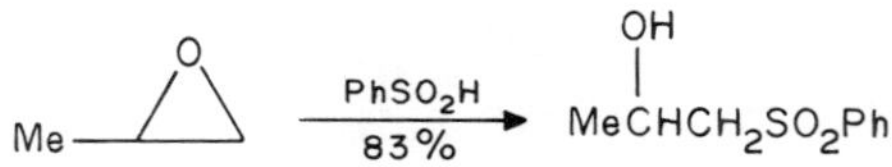

(66)

(45)

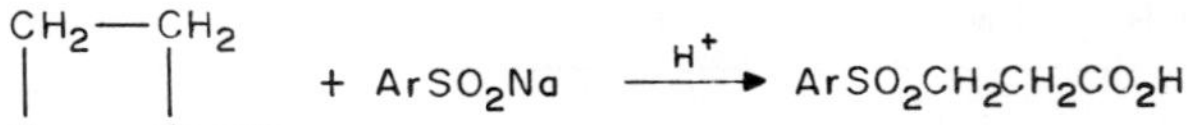

(67)

yields up to 58%

(46)

*d. Alkenylation and arylation of sulphinic acids.* A direct attachment of an alkenyl moiety to sulphinic acids to form α,β-unsaturated sulphones **69** was achieved either by a photostimulated coupling of 1-alkenylmercury halides **68** with sodium sulphinates (method A, equation 47)[69] or by the reaction of alkenes[70,71] or conjugated dienes[72] with mercury(II) chloride and sodium sulphinates followed by base-catalysed eliminative demercuration (methods B and C, equation 48). It is interesting to note that treatment of alkenes with sodium sulphinates and iodine ('iodosulphonylation') followed by basic hydroiodide elimination produces sulphones which are regioisomers of those obtained by the previous method (an example is shown in equation 49)[71]. The detailed results of the above approaches are collected in Table 10.

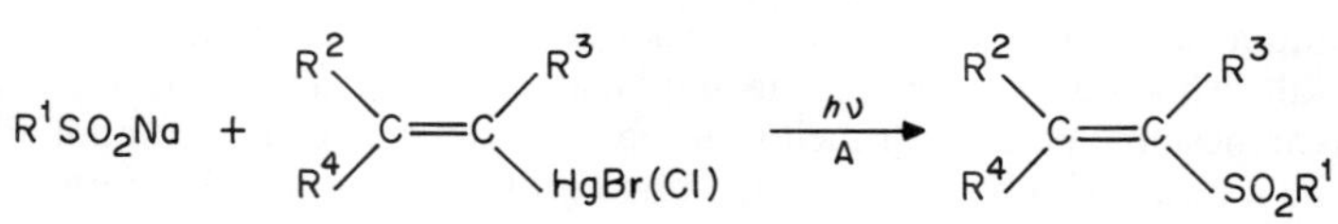

(68) (69)

(47)

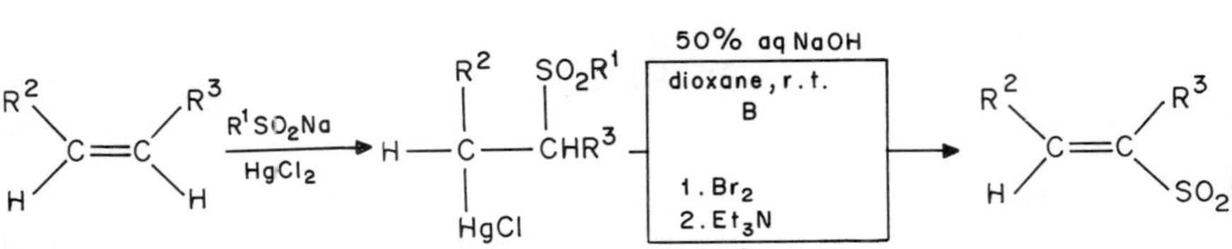

(48)

TABLE 10. Alkenylation of sulphinic acids

| $R^1$ | $R^2$ | $R^3$ | $R^4$ | Method | Isolated yield (%) | Reference |
|---|---|---|---|---|---|---|
| *p*-Tol | *t*-Bu | H | H | A | 68 | 69 |
| *p*-Tol | Pr | H | H | A | 63 | 69 |
| *p*-Tol | H | Me | Me | A | 67 | 69 |
| *p*-Tol | Ph | H | H | A | 77 | 69 |
| *p*-Tol | Ph | H | Ph | A | 61 | 69 |
| $c$-$C_6H_{11}$ | *t*-Bu | H | H | A | 66 | 69 |
| $c$-$C_6H_{11}$ | Ph | H | H | A | 65 | 69 |
| *t*-Bu | Ph | H | H | A | 55 | 69 |
| Pr | *t*-Bu | H | H | A | 69 | 69 |
| Ph | *t*-Bu | H | H | A | 85 | 69 |
| Ph | H | H | H | C | 77.5 | 70 |
| Ph | $—(CH_2)_3—$ | | H | B | 79 | 70 |
| Ph | $—(CH_2)_4—$ | | H | B | 79 | 70 |
| *p*-Tol | $—(CH_2)_4—$ | | H | B | 84 | 70 |
| Ph | $CO_2Me$ | Ph | H | C | 75 | 70 |
| Ph | COMe | Ph | H | C | 59.5 | 70 |
| Ph | $—(CH_2)_2—CH=CH—$ | H | | | 97 | 72 |
| Ph | $—(CH_2)_2—CMe=CH$ | H | | | 82 | 72 |

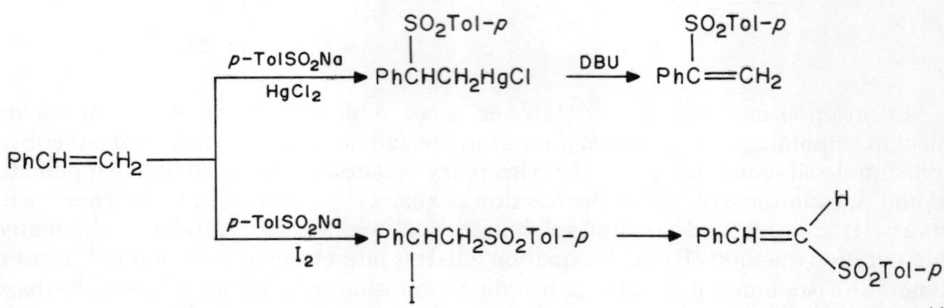


(49)

A direct arylation of sodium sulphinates was accomplished by their reaction with diaryliodonium salts. The yields of sulphones were up to 56%, based on the consumed iodonium salt, while 40% of the starting sulphinates was left unreacted (equation 50)[73]. Anhydrous arenesulphinic acids or their salts react with aromatic nitro compounds to give sulphones as a result of substitution of the nitro group (equation 51)[74]. Organometallic compounds have recently also been used for arylation of sulphinic acids. Thus, pentaphenylbismuth **70** or the trifluoroacetate **71** derived from it react with *p*-toluenesulphinic acid at 80 °C to give phenyl *p*-tolyl sulphone in 87 and 76% yield,

$$Ar_2I^+X^- + Ar^1SO_2Na \longrightarrow ArSO_2Ar^1 \qquad (50)$$

$$p\text{-}ClC_6H_4\overset{\overset{\displaystyle O}{\|}}{C}C_6H_4NO_2\text{-}p + PhSO_2Na \xrightarrow[150\,°C,\ 24\,h]{DMSO} p\text{-}ClC_6H_4\overset{\overset{\displaystyle O}{\|}}{C}C_6H_4(SO_2Ph)\text{-}p \qquad (51)$$

respectively (equation 52)[75]. Another example of arylation is the oxidative sulphonylation of azulene with copper(II) arenesulphinates. The reaction takes place only in the five-membered ring giving equimolar amounts of both regioisomeric sulphones (equation 53)[76]. For earlier examples of arylation see References 1 and 3.

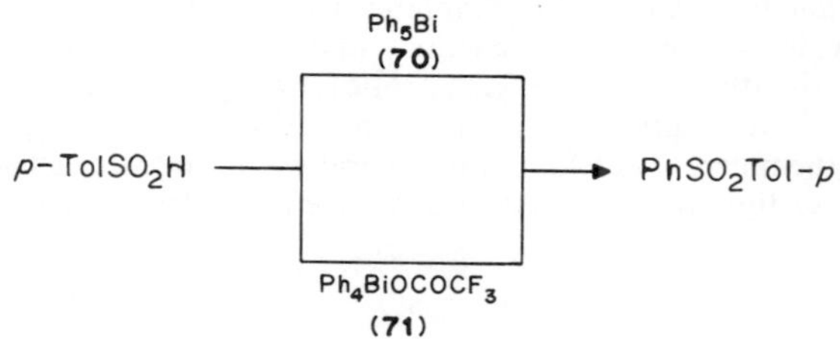


(52)

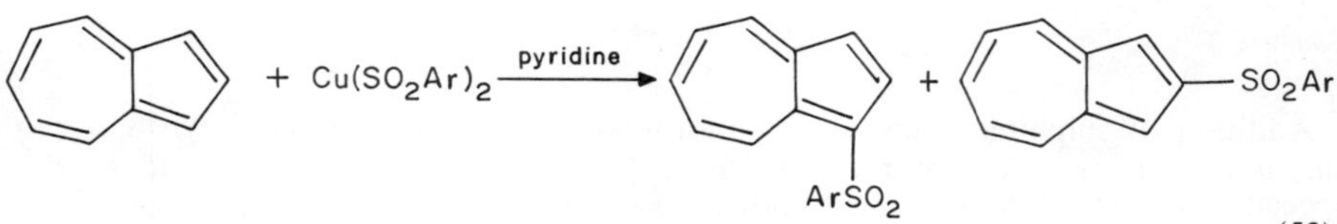


(53)

*e. Addition of sulphinic acids to non-activated alkenes.* Sulphinic acids react with non-activated alkenes only in the presence of catalysts, the exception being the reaction of very strong, perfluoroalkylsulphinic acids with conjugated dienes (equation 54)[77]. Russian workers investigated in detail the reaction of sulphinic acids with butadiene in the presence of palladium complexes as catalysts [e.g. $Pd(acac)_2$–$PPh_3$–$AlEt_3$, 1:3:4] and found that the products **72**, **73** and **74** were formed usually in the ratio 51:34:15, which is independent of the substituents in the acid used (equation 55)[78]. The cyclopentadiene dimer reacts similarly in the presence of palladium(II) chloride[79]. Telomerization of *p*-toluenesulphinic acid and butadiene in the presence of nickel catalysts gives a mixture of telomers, the yield and composition of which depend on the ligands in the catalyst molecule and the catalyst/substrates ratio[80].

$$C_nF_{2n+1}SO_2H + H_2C{=}C(Me)CH{=}CH_2 \longrightarrow C_nF_{2n+1}SO_2CH_2CH{=}CMe_2 \quad (54)$$

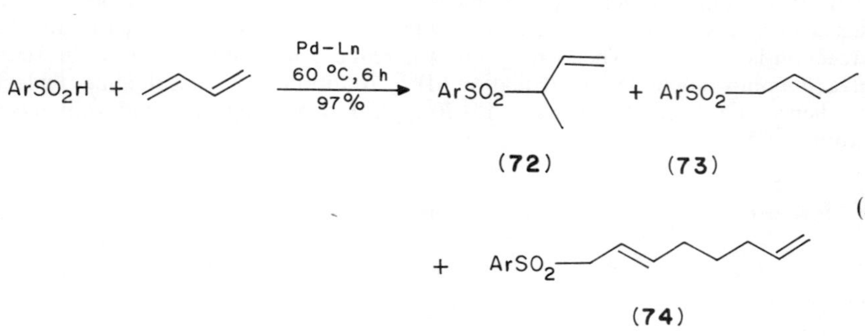


(55)

Sulphinic acids catalyze the $Z$–$E$ equilibration of disubstituted olefins, the equilibrium being obtained in less than 15 min in refluxing dioxane. The yields are high and no migration of the double bond is observed. The highest yields ($\sim 95\%$) and $E$:$Z$ ratio (81:19) was obtained by using 10 mol% of *p*-chlorobenzenesulphinic acid as a catalyst[81].

*f. Michael addition of sulphinic acids.* Sulphinic acids add very easily to olefins bearing electron-withdrawing groups. The reaction may be carried out under various conditions—from slightly acidic to basic, in protic (also aqueous) and aprotic media. The number of $\beta$-substituted sulphones obtained in this way is so huge that it is quite impossible to list them here. Therefore only general groups of Michael acceptors will be mentioned in this section and several representative examples will be given.

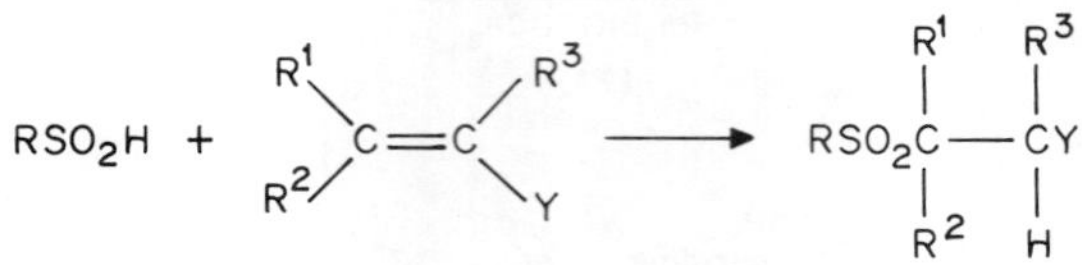
(56)

Addition of sulphinic acids to $\alpha,\beta$-unsaturated carbonyl compounds gives $\beta$-oxo sulphones. Among the acceptors are chalcones (17–94% yield, slightly acidic conditions[82]; special activated sulphinic acids[83]), other $\alpha,\beta$-unsaturated aldehydes and ketones[84,85], esters, amides[84–86], imides[87] and acids[86]. An interesting example is the addition of arenesulphinic acids to 3-aryl-3-cyclobutene-1,2-dione **75** leading to 4-arenesulphonyl-2-hydroxy-3-aryl-2-cyclobuten-1-ones **76** in 52–80% yield (equation 57)[88]. Of some

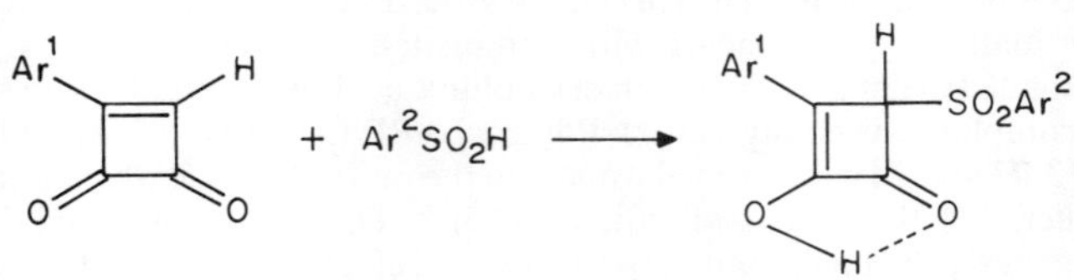
(57)

**(75)** **(76)**

synthetic importance are 2-sulphonylethyl-1,3-dioxolanes **77** obtained as a result of a Michael addition of sulphinic acids to $\alpha,\beta$-unsaturated aldehydes, followed by acetalization with ethanediol (equation 58)[89,90]. These compounds were successfully used for the synthesis of 4-hydroxycyclopentenones **78** (equation 59)[91] and 2-methyl-4-hydroxybut-2-enolide **79** (equation 60)[90].

Addition of sulphinic acids to nitroolefins affords $\beta$-nitro sulphones **80**[84,92]. The adducts can be easily split into substrates by treatment with NaOH, which means that this reaction is reversible. The nitro group was reduced with $SnCl_2$/HCl or Zn/AcOH to give $\beta$-amino sulphones **81** (equation 61)[92]. Addition of arenesulphinic acids to 2-nitrodienes affords sulphones **82** in 13–70% yield, depending on the substituents R (equation 62)[93].

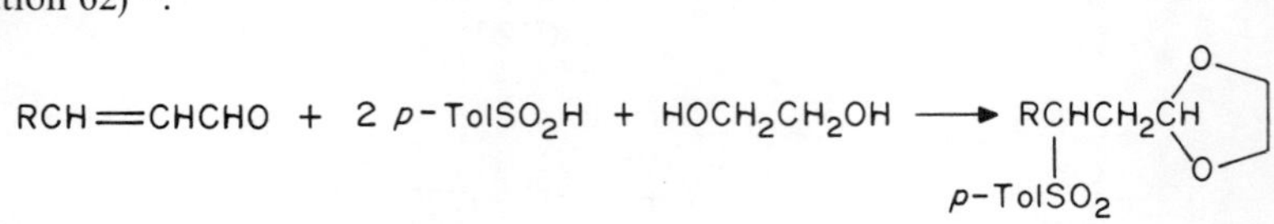
(58)

**(77)**

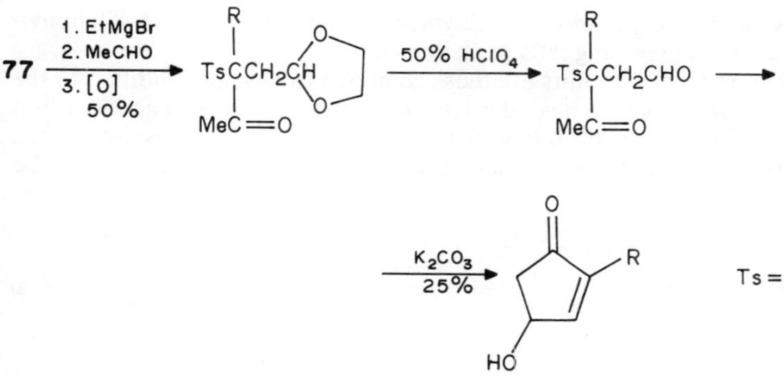

**(78)** $Ts = p\text{-}TolSO_2$ (59)

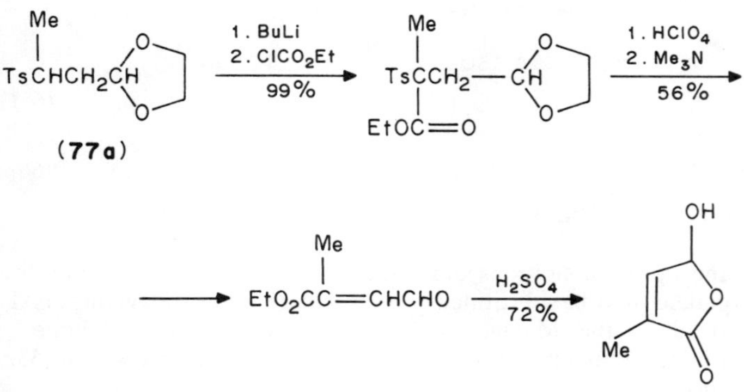

**(77a)** **(79)** $Ts = p\text{-}TolSO_2$ (60)

$$RCH{=}CHNO_2 + R'SO_2H \xrightarrow[18\text{–}95\%]{} R'SO_2\overset{\displaystyle R}{\overset{|}{C}}HCH_2NO_2 \xrightarrow{\text{red.}} R'SO_2\overset{\displaystyle R}{\overset{|}{C}}HCH_2NH_2 \quad (61)$$

**(81)**

$$CH_2{=}C(NO_2)CMe{=}CHR + ArSO_2H \longrightarrow ArSO_2CH_2C(NO_2)CMe{=}CHR \quad (62)$$

**(82)**

Addition of sulphinic acids to $\alpha,\beta$-unsaturated nitriles produces $\beta$-cyano sulphones[84,86]. However, the reaction of acrylonitrile with sodium hydroxymethanesulphinate **59** affords $\beta,\beta'$-dicyano sulphone **84** (equation 63)[94]. It is formed via the intermediate **83** which has been isolated (compare equation 41, Reference 63).

$$HOCH_2SO_2Na + H_2C{=}CHCN \longrightarrow NaOCH_2SO_2CH_2CH_2CN \quad (63)$$

**(59)**

$$\xrightarrow[-CH_2O]{} NaO_2SCH_2CH_2CN \xrightarrow{H_2C{=}CHCN} N{\equiv}CCH_2CH_2SO_2CH_2CH_2C{\equiv}N$$

**(83)** **(84)**

Sulphinic acids undergo addition to 2-vinylpyridine to give 2-$\beta$-sulphonylethylpyridines **85** in yields exceeding 90% (equation 64)[95].

A special group of Michael acceptors are those compounds which, in addition to the electron-withdrawing group, possess a good leaving group attached to an $\alpha$ and $\beta$ carbon atom. The primary addition is usually followed by an elimination process resulting in the formation of $\beta$-substituted alkenyl sulphones (equations 65 and 66)[96–98]. It should be

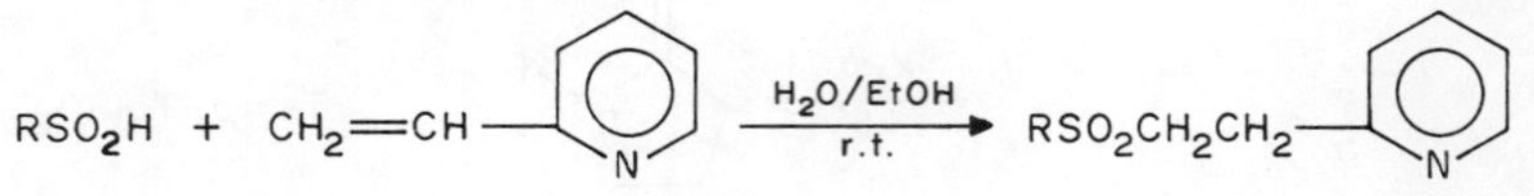

(64)

(**85**)

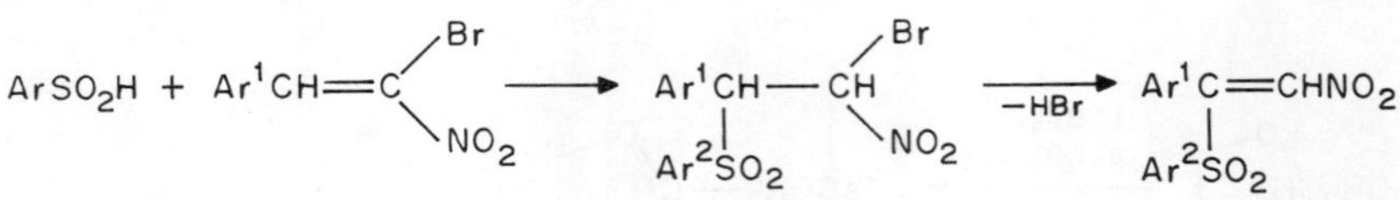

(65)

$$PhCH{=}CHSO_2Na + ClCH{=}CHY \longrightarrow PhCH{=}CHSO_2CH{=}CHY \qquad (66)$$

$Y = C(O)Me, CN, NO_2$ Yield 50–65%

mentioned that, under appropriate conditions and in the presence of palladium catalysts, the nitro group in a nitroalkene may be substituted by a sulphinate anion, thus giving alkyl sulphones, e.g. **86** instead of the Michael addition product, $\beta$-nitro sulphone[99] (equation 67) (for substitution of the nitro group see equations 32, 33, References 52, 53, and equation 51, Reference 74).

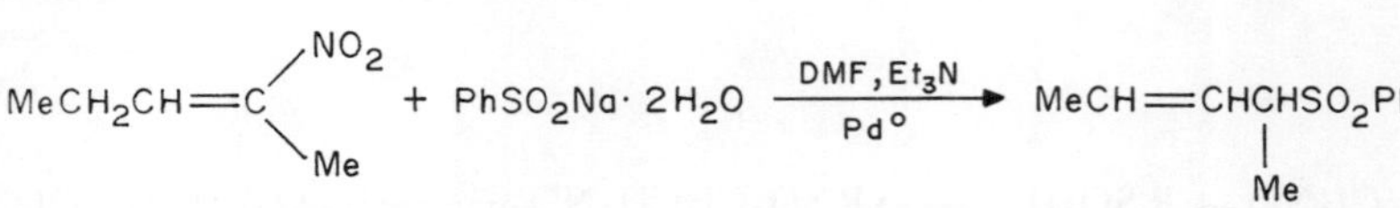

(67)

(**86**)

Addition of sulphinic acids to quinones and the closely related 1,4-benzoquinonedibenzenesulphonimides has been known for a long time. The addition step is usually followed by enolization and the overall process results in the formation of 2,5-dihydroxyaryl sulphones **87** and 2,5-disulphonamidoaryl sulphones **88**, respectively (equations 68 and 69). Nevertheless, some newer reports concerning this subject

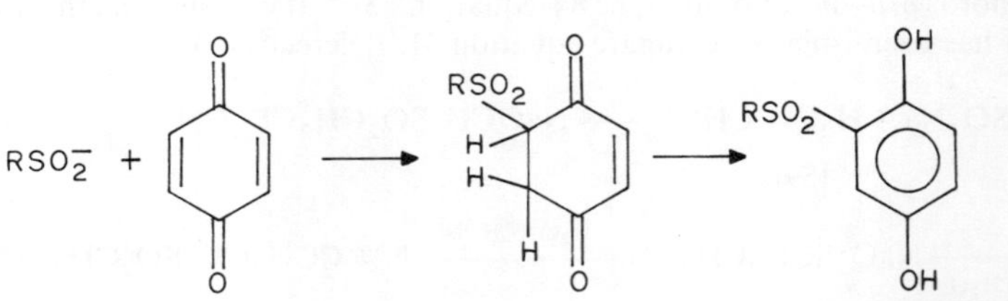

(68)

(**87**)

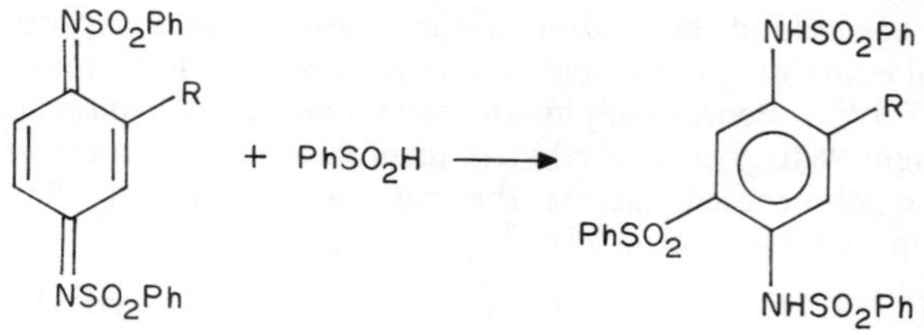

(88)

69)

appeared[100–103]. Reaction of *N*-phenyl-1,4-benzoquinonimine **89** with sulphinic acid gave (2-hydroxy-5-phenylamino)phenyl sulphones **90**. Their oxidation with lead tetraacetate leads to the corresponding sulphonylquinonimines **91** (equation 70). When hydroxymethanesulphinic acid was used in this reaction, bis-quinonimine **92** was formed (equation 71)[104].

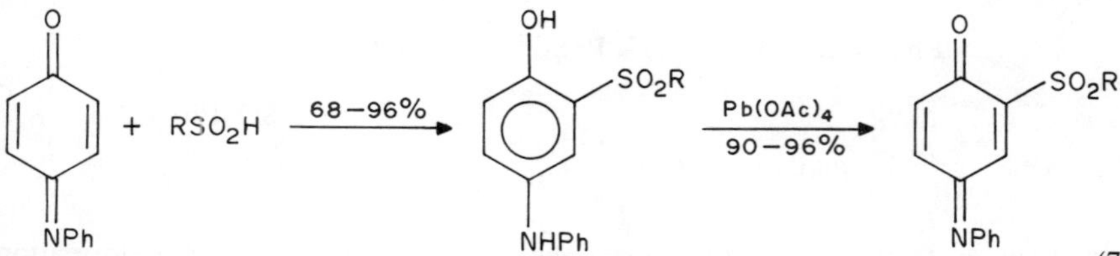

(89) (90) (91)

(70)

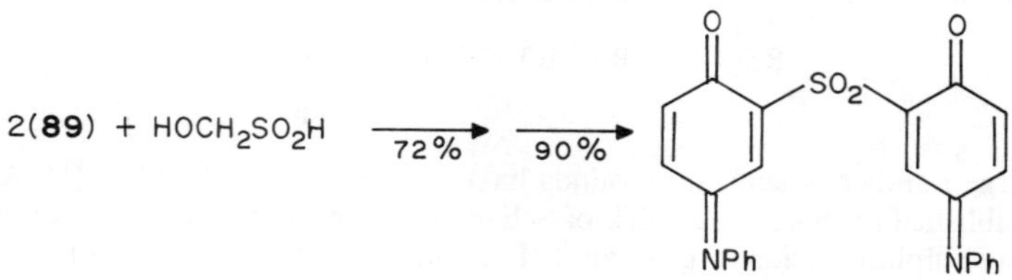

(92)

(71)

*g. Addition of sulphinic acids to acetylenes and allenes.* Addition of free sulphinic acids or metal sulphinates to α-acetylenic ketones was investigated as early as in 1924 by Kohler and Barrett. These authors found that *p*-toluenesulphinic acid combines with phenyl benzoyl acetylene to give a mixture of both diastereomeric alkenyl sulphones **93**. Moreover, they found that the process is stopped at the mono-addition stage (equation 72)[105]. The adducts were later investigated to prove their configuration[106].

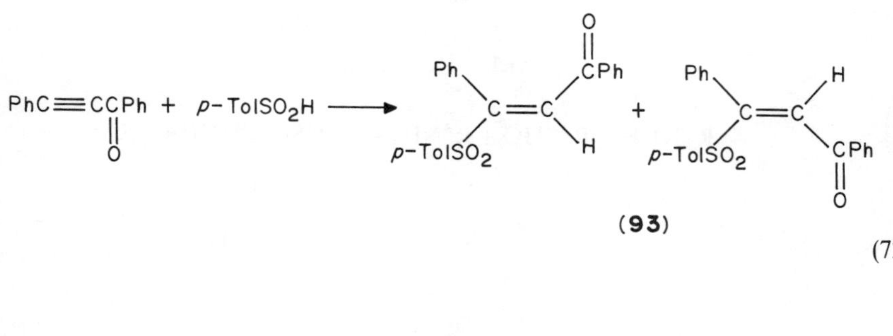

(93)

(72)

Recently, it has been found that sodium arenesulphinates add to propiolamide **94** in an aqueous ethanol containing boric acid to give $\beta$-sulphonyl acrylamides **95** in 55 to 58% yield (equation 73)[107]. Benzenesulphinate salts undergo a *trans*-specific addition to sulphonyl acetylene **96** to give the *Z* ethylene disulphone **97** (equation 74). The addition to sulphonyl allene **98** proceeds across the $\alpha$-double bond and affords 2,3-diphenylsulphonyl-1-propene **99** (equation 75)[108].

$$\underset{(\mathbf{94})}{HC{\equiv}CCONH_2} + ArSO_2Na \longrightarrow \underset{(\mathbf{95})}{ArSO_2CH{=}CHCONH_2} \qquad (73)$$

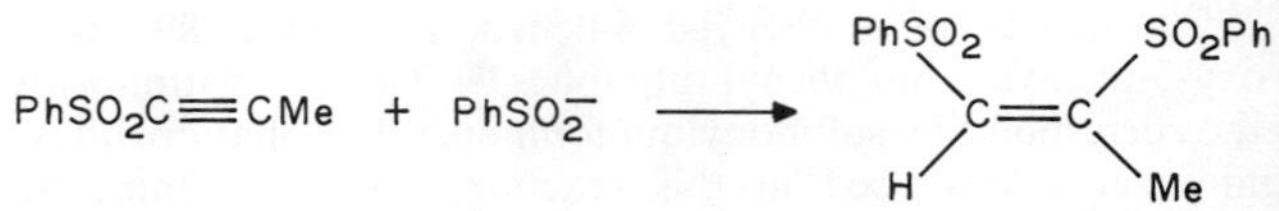

(74)

**(96)** **(97)**

$$\underset{(\mathbf{98})}{PhSO_2CH{=}C{=}CH_2} + PhSO_2^- \longrightarrow \underset{(\mathbf{99})}{PhSO_2CH_2\underset{\displaystyle SO_2Ph}{\overset{|}{C}}{=}CH_2} \qquad (75)$$

*h. Addition of sulphinic acids to carbonyl compounds.* The first report on the formation of adducts between sulphinic acids and aldehydes appeared in 1901[109,110]. The reaction resembles the bisulphite addition to aldehydes. The structure of those adducts has been established fifty years later as $\alpha$-hydroxyalkyl sulphones **100** (equation 76)[111]. Since that

$$RSO_2H + R'CHO \longrightarrow \underset{(\mathbf{100})}{RSO_2CHR'OH} \qquad (76)$$

time a large number of such compounds have been synthesized[110,112–114]. Among the papers published in this area a work of Schank deserves mentioning. He found that the addition of sulphinic acids to glyoxals **101** results in the formation of $\alpha$-hydroxy-$\beta$-oxo sulphones **102**, which may be considered as sulphonyl analogues of aci-reductones (equation 77)[113]. $\alpha$-Hydroxy sulphones **100** were used as substrates in reactions with amines to give $\alpha$-amino sulphones **103**[112]. Later on, it was found that these compounds

$$\underset{(\mathbf{101})}{R'\underset{\displaystyle O}{\underset{\|}{C}}\underset{\displaystyle O}{\underset{\|}{C}}H} + HSO_2R \rightleftharpoons \underset{(\mathbf{102})}{R'\underset{\displaystyle O}{\underset{\|}{C}}\underset{\displaystyle OH}{\underset{|}{\overset{\displaystyle H}{\overset{|}{C}}}}SO_2R} \qquad (77)$$

yields 56.7–82.7%

$$R^1SO_2H + R^2CHO + R^3NH_2 \longrightarrow \underset{(\mathbf{103})}{R^1SO_2\underset{\displaystyle R^2}{\underset{|}{C}}HNHR^3} \qquad (78)$$

can be synthesized by reacting in one pot all three compounds, i.e. a sulphinic acid, an aldehydes and an amine (equation 78). Most probably the condensation does not proceed via the intermediary formation of **100**[115], but it resembles the Mannich reaction with sulphinic acid as an acidic component. However, the dialkylamino derivatives could not be obtained in this way unless the amine possessed an electron-withdrawing group. For example, the reaction proceeds only with piperazine derivatives which bear a substituent like $CO_2Et$ or $p$-$O_2NC_6H_4$ on the nitrogen atom[116].

The Mannich-type condensation has been performed with a broad variety of amino compounds. Thus, the following substrates were used as the amino compounds in these reactions: hydroxylamine **104** (equation 79)[117], benzohydroxamic acid **105**[117], *N*-hydroxybenzenesulphonamide **106**[117], carboxylic amides **107**[118,119], sulphonamides **108**[119] and lactams **109**[119] (equation 80). The Mannich condensation of sulphinic acids, formaldehyde and ethyl carbamate is of special synthetic value, since it gives *N*-sulphonylmethyl urethanes **110** which are convenient intermediates in the synthesis of α-sulphonyl diazomethanes **111** (equation 81)[120–122].

$$2RSO_2H + 2CH_2O + \underset{\textbf{(104)}}{NH_2OH} \xrightarrow[38\text{–}75\%]{} (RSO_2CH_2)_2N\text{—}OH \qquad (79)$$

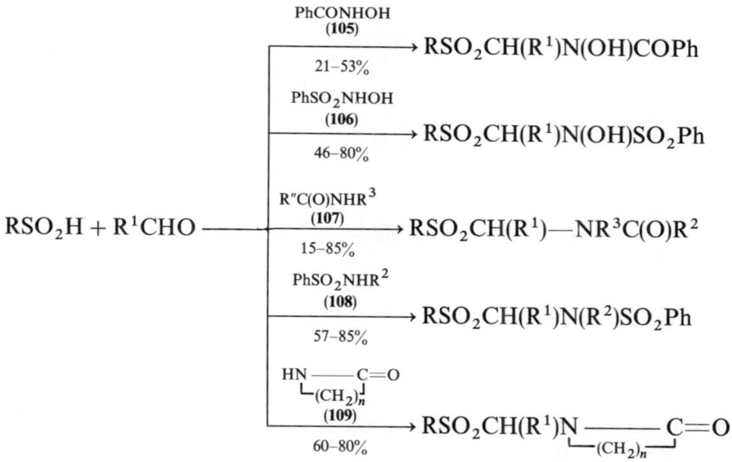
(80)

$$RSO_2H + CH_2O + H_2NCO_2Et \xrightarrow[50\text{–}75\%]{\text{r.t.}} \underset{\textbf{(110)}}{RSO_2CH_2NHCO_2Et} \xrightarrow{\text{1. NOCl; 2. Base}} \underset{\textbf{(111)}}{RSO_2CHN_2} \qquad (81)$$

α-Amino sulphones and derivatives have also been obtained by addition of sulphinic acids to compounds containing the C═N bond such as azomethines **112**, arylsulphonyl imines **113** and azodicarboxylates **114** (equation 82)[123] and 3-*p*-toluenesulphonyl-3*H*-azirine **115**, formed *in situ* by the irradiation of *trans*-β-azidovinyl *p*-tolyl sulphone (equation 83)[124]. Another interesting condensation has been described by Hellmann and Müller[125], who found that aromatic sulphinic acids react in a one-pot procedure with formaldehyde and C—H acids, such as indole **116**, β-naphthol or dimedone, to give unsymmetrical sulphones, e.g. 3-benzenesulphonylmethyl-indole **117** (equation 84).

$$R^2SO_2H \xrightarrow[28\text{–}92\%]{RCH{=}NAr\ (112)} RCH(SO_2R^2)NHAr$$

$$R^2SO_2H \xrightarrow{RSO_2N{=}CHR^1\ (113)} RSO_2NHCH(R^1)SO_2R^2 \qquad (82)$$

$$R^2SO_2H \xrightarrow[66\%]{EtO_2CN{=}NCO_2Et\ (114)} EtO_2CNHN(SO_2R^2)CO_2Et$$

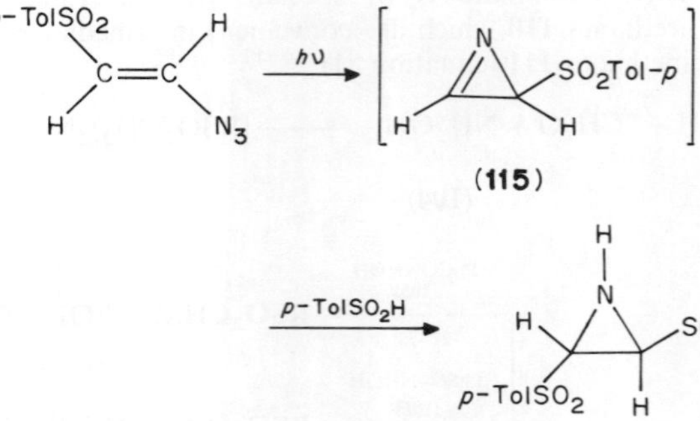


(83)

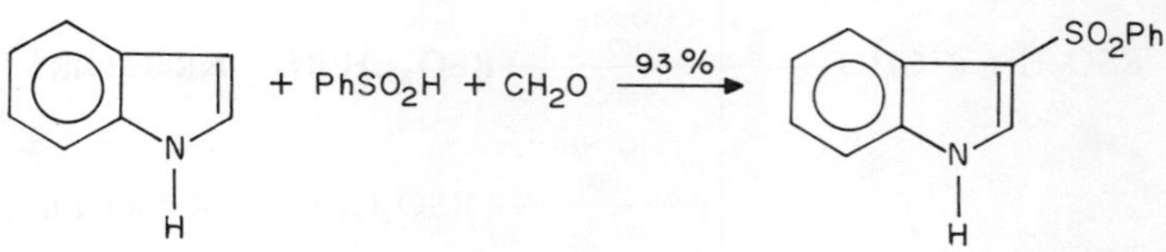


(116) (117)

(84)

*i. S Acylation of sulphinic acids.* Acylation of sulphinic acids proceeds either on the sulphur or on the oxygen atom, the direction being dependent on the acylating agent used. This section is limited to the discussion of the acylations and related reactions which lead to the formation of sulphonyl derivatives. The reactions taking place at the oxygen atom will be presented in Section II.B.2.

Dithiochloroformates **118** react with metal arenesulphinates to give trithiocarbonate *S,S*-dioxides **119** as strongly coloured (deep-red, violet) compounds. Chemical yields of this reaction vary from 7 to 70%, and depend on the nature of the substituents $R^1$ and $R^2$ (equation 85)[126,127]. In the same way thiocarbamoyl chlorides **120** react with sulphinate anions affording dithiocarbamate *S,S*-dioxides **121** in 10 to 70%

$$R^1SO_2^- + ClC({=}S)SR^2 \longrightarrow R^1SO_2C({=}S)SR^2 + Cl^- \qquad (85)$$

(118) (119)

yield (equation 86)[128,129]. However, in the case of thiophosgene the reaction does not give the expected thiocarbonyl bis(methyl sulphone) **122** but affords bis(methylsulphonyl)methyl methanethiosulphonate **123**. The latter is formed by the subsequent thiophilic addition of the sulphinic acid anion to the thiocarbonyl group in **122** as shown in equation 87[130]. The reaction of chlorosulphines **124** with *p*-toluenesulphinate anion leads to the products in which the C=S=O function is replaced by a $CH_2$ group via the transiently formed *p*-tolylsulphonyl sulphine **125** (equation 88)[131].

$$RSO_2^- + R^1R^2NC(=S)Cl \xrightarrow[\text{reflux}]{H_2O/\text{benzene}} RSO_2C(=S)NR^1R^2 \qquad (86)$$

**(120)** **(121)**

$$ClC(=S)Cl \xrightarrow{2MeSO_2^-} MeSO_2C(=S)SO_2Me \xrightarrow{MeSO_2^-}$$

**(122)**

$$\longrightarrow (MeSO_2)_2\bar{C}SSO_2Me \xrightarrow{H^+} (MeSO_2)_2CHSSO_2Me \qquad (87)$$

**(123)**

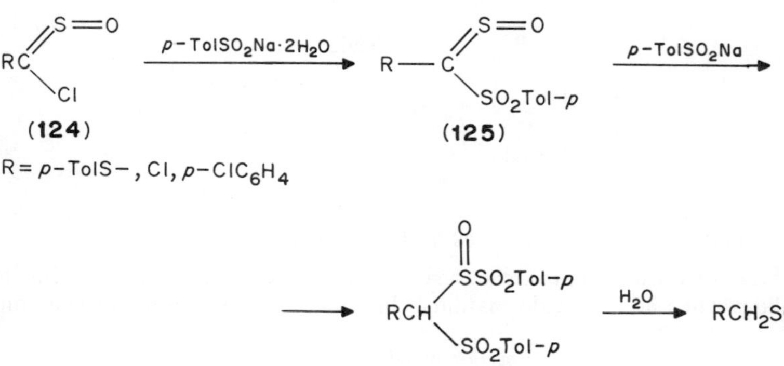

**(124)** **(125)**

R = *p*-TolS–, Cl, *p*-$ClC_6H_4$

(88)

Imino analogues of acyl halides belong also to these compounds which react with sulphinic acids at sulphur to yield α-iminosulphones. For example, the *p*-chlorophenylhydrazone of *p*-chlorobenzoyl chloride (**126**) gives on treatment with methanesulphinate anion the sulphone **127** in 33% yield (equation 89)[132]. Nitro analogues of imidoyl chlorides, e.g. **128**, behave similarly (equation 90)[133]. Another interesting example

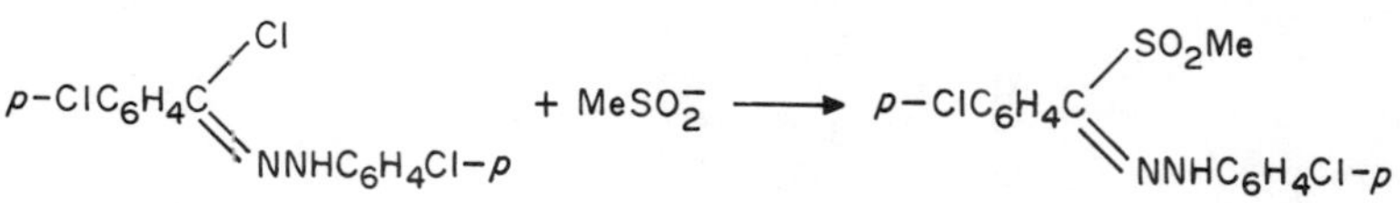

**(126)** **(127)** (89)

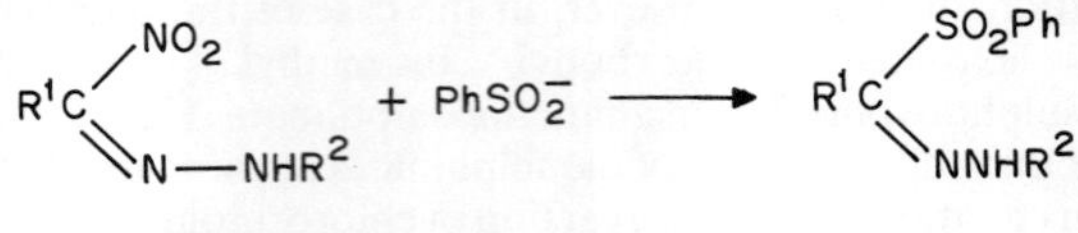


(128)

(90)

of S acylation involves the chemoselective reaction of the aldehyde-imidoyl bromides **129** with sodium arenesulphinates leading to the sulphones **130** arising from replacement of bromide anion (equation 91, Table 11)[134].

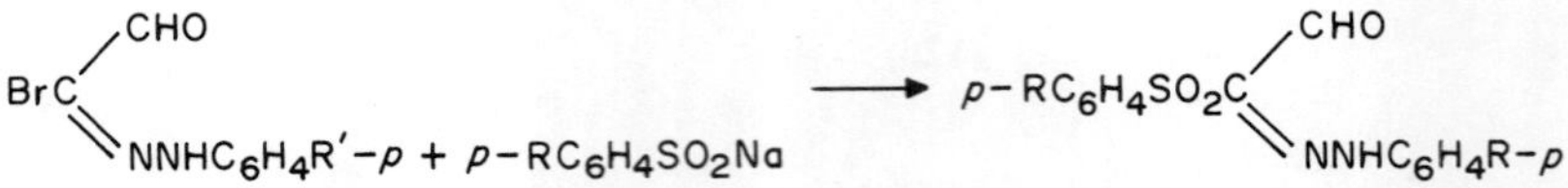


(129) (130)

(91)

Sulphinic acid anhydrides, obtained in the reaction of sodium sulphinates with sulphinyl chlorides, exist in the form of sulphinyl sulphone **131** and not bis-sulphinyl oxide (equation 92)[135] (for examples illustrating O acylation and O sulphinylation see Section II.B.2).

$$RSO_2Na + RS(O)Cl \longrightarrow RSO_2S(O)R \qquad (92)$$

**(131)**

| R | Yield |
|---|---|
| *p*-Tol | 81% |
| *p*-$ClC_6H_4$ | 51% |
| *β*-naphthyl | 60% |

*2. Synthesis of sulphonyl halides, cyanides and thiocyanates*

Relatively little is known about the reaction of sulphinic acids with halogens leading to sulphonyl halides. Thus, aqueous chlorination or bromination of arenesulphinate anions

TABLE 11. Reaction of aldehyde-imidoyl bromides **129** with sulphinates

| R | R′ | Yield of **130** (%) |
|---|---|---|
| H | H | 60 |
| H | Me | 70 |
| H | Br | 60 |
| H | $NO_2$ | 50 |
| Me | Me | 52 |
| Me | Cl | 65 |
| Me | Br | 65 |
| Me | $NO_2$ | 60 |
| Cl | Br | 56 |

affords sulphonyl chlorides in yields up to 81% and sulphonyl bromides in yields about 47%[136]. ω-Hydroxy-1-alkanesulphonyl chlorides have recently been obtained by chlorination of a dichloromethane suspension of sodium ω-hydroxy-1-alkanesulphinates. The yields were practically quantitative, though the propane and butane derivatives ($n = 3$ and 4) were accompanied by about 17% of sultones **132** as products of cyclization (equation 93)[137].

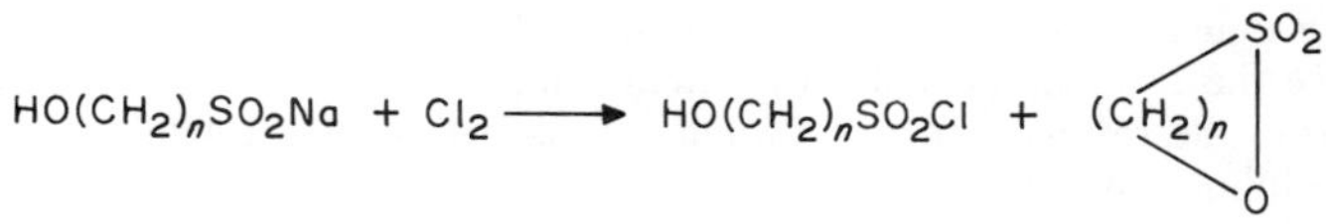

(93)

**(132)**

Sulphonyl iodides have been prepared by treatment of alkali sulphinates with iodine in alcohol solution[136]. In aqueous solution there is an equilibrium, the position of which depends on the substituent in the sulphinic acid (equation 94)[138].

$$RSO_2^- + I_3^- \rightleftarrows RSO_2I + 2I^- \qquad (94)$$

Treatment of dirhodane with sodium or silver arenesulphinates produces the corresponding sulphonyl thiocyanates **133** in 48 to 84% yield (equation 95)[139].

$$ArSO_2M + (SCN)_2 \xrightarrow{CH_2Cl_2} ArSO_2SCN + MSCN \qquad (95)$$

$M = Na, Ag$ **(133)**

When sodium sulphinates react with cyanogen bromide **134** only sulphonyl bromides are formed due to the highly electropositive character of bromine (equation 96)[140] (compare Section II.A.1.a, References 9 and 16). However, cyanogen chloride **135** gives in an analogous reaction sulphonyl cyanides **136**[140] (equation 97). The latter have been reacted with a variety of nucleophiles to give cyanates, thiocyanates and cyanamides (equation 98)[141].

$$RSO_2Na + BrCN \longrightarrow RSO_2Br \qquad (96)$$

**(134)**

$$RSO_2Na + ClCN \longrightarrow RSO_2CN \qquad (97)$$

**(135)** **(136)**

$$ArSO_2CN + Nu \longrightarrow \left[ ArSO_2C{=}N^- \atop \quad | \atop \quad Nu^+ \right] \longrightarrow NuC{\equiv}N + ArSO_2^- \qquad (98)$$

**(136)**
$Nu = {}^-OR,\ {}^-SR,\ NR_3$

*3. Reaction of sulphinic acids with sulphur electrophiles*

*a. Synthesis of thiosulphonic acids.* Nucleophilic attack of a sulphinate ion on elemental sulphur resembles the reaction with sulphites and yields salts of thiosulphonic acids[142]. The formation of thiosulphonic acid salts may be accelerated by addition of sodium sulphide or polysulphide[143]. Recently, sodium arenethiosulphonates **137** have been obtained in quantitative yields by reacting sodium arenesulphinates with elemental sulphur in the presence of amines ($BuNH_2$, *i*-$PrNH_2$, $Et_2NH$, $Et_3N$, morpholine, piperidine, ammonia) (equation 99)[144a] (and sodium methanethiosulphonate by simple heating of sodium methanesulphinate with sulphur in methanol[144b]).

$$p\text{-}RC_6H_4SO_2Na \xrightarrow[20\,^\circ C\ (\text{or } 60\,^\circ C)]{S_8,\ \text{amine}} p\text{-}RC_6H_4SO_2SNa \qquad (99)$$

**(137)**

Sodium *p*-toluenethiosulphonate, prepared in this way, was treated with Amberlyst A-26 to give the resin-supported *p*-toluenethiosulphonate. The same product was also obtained from the resin-supported *p*-toluenesulphinate on treatment with an excess of elemental sulphur in boiling toluene. It was then used as a source of thiosulphonate moiety for the simple conversion of alkyl halides into S-alkyl *p*-toluenethiosulphonates[145].

*b. Synthesis of thiosulphonic S esters.* In the first instance, the title compounds may be prepared by simple disproportionation of sulphinic acids. However, they are contaminated with the corresponding sulphonic acids as a second disproportionation product (equation 100)[3].

$$3RSO_2H \longrightarrow RSO_2SR + RSO_3H + H_2O \qquad (100)$$

Sulphenyl halides react with sulphinic acids[136] or their silver[146] or sodium salts[147] to give thiosulphonates **138** in moderate yields (equation 101). Aminosulphenyl chlorides **139** can also be used for this reaction which results in the formation of S-sulphonyl sulphenamides **140** (equation 102)[148]. Dialkyl disulphides may serve as a source of sulphenyl moiety to give **138**[149]. However, the use of diamino disulphides **141** produces amino sulphonyl disulphides **142** (equation 103)[150]. On the other hand, diamino sulphides **143** react smoothly with sulphinic acids to give **140** (equation 104)[150–152].

$$R^1SO_2M + XSR^2 \longrightarrow R^1SO_2SR^2 \qquad (101)$$

**(138)**

$M = H, Ag, Na \qquad X = Cl, Br$

$$RSO_2K + ClSNR^1_2 \xrightarrow{50\text{–}95\%} RSO_2SNR^1_2 \qquad (102)$$

**(139)** **(140)**

$NR^1_2 = \text{morpholyl}$

$$p\text{-}TolSO_2H + R_2NSSNR_2 \longrightarrow p\text{-}TolSO_2SSNR_2 + (p\text{-}TolSO_2S)_2 \qquad (103)$$

**(141)** **(142)**

$NR_2 = \text{morpholyl}$ yield 62%

$NR_2 = \text{piperidyl}$ yield 32%

$$RSO_2H + R^1_2NSNR^1_2 \longrightarrow RSO_2SNR^1_2 + (RSO_2)_2S \qquad (104)$$

**(143)** **(140)**

$NR^1_2 = \text{morpholyl}$, yield 60%[150]

$NR^1_2 = \text{piperidyl}$, yield 25%[150]

$NR^1_2 = \text{phtalimidyl}$, yields 40–60%[151,152]

Another simple method of preparation of thiosulphonates from sulphinic acids consists in the reaction of the latter with thionitrites **144**, obtained from thiols and dinitrogen tetroxide (equation 105)[153,154], or with thiols in the presence of alkyl nitrites **145** (equation 106)[155].

$$RSH + N_2O_4 \longrightarrow \underset{\mathbf{(144)}}{RSNO} \xrightarrow{R^1SO_2H} R^1SO_2SR \qquad (105)$$

Yields up to 95%

$$R^1SO_2H + RSH + \underset{\mathbf{(145)}}{2EtONO} \longrightarrow R^1SO_2SR + 2NO + 2EtOH \qquad (106)$$

Yields 48–80%

*4. Reaction of sulphinic acids with nitrogen electrophiles*

Sulphinic acids react with nitrous acid to give *N,N*-disulphonyl hydroxylamine **146** or trisulphonylamine oxide **147**, depending on the substrates ratio (equation 107)[156]. In the reaction of sulphinic acids with alkyl nitrites **145** bis-sulphonyl-hydroxylamines **146** are also produced. It is assumed that sulphonyl nitrites **148** are formed as intermediates, which react quickly with the next sulphinic acid molecule to yield **146**[155]. Sulphonyl nitrites **148** have been obtained independently in the reaction of sulphinic acids with dinitrogen tetroxide (equation 108)[157].

$$RSO_2H + HONO \longrightarrow \underset{\mathbf{(146)}}{(RSO_2)_2NOH} + \underset{\mathbf{(147)}}{(RSO_2)_3N \rightarrow O} \qquad (107)$$

$$RSO_2H + N_2O_4 \xrightarrow[38\text{–}55\%]{\text{ether, } -20 \text{ to } 0\,^\circ\text{C}} \underset{\mathbf{(148)}}{RSO_2NO} + RSO_3H \qquad (108)$$

Arenesulphinic acids have been used in acidic media as protecting reagents for the C-nitroso group against reduction and condensation. The method is based on the formation of stable non-reducible N-substituted hydroxylamines **149** and it can be applied to nitrosoarenes. However, this method is not applicable to the N-nitroso group. Deprotection is effected by basic hydrolysis (equation 109)[158,159].

$$Ar^1NO + Ar^2SO_2H \xrightarrow{\text{pH } 0\text{–}3} \underset{\mathbf{(149)}}{Ar^1N(OH)SO_2Ar^2} \xrightarrow{HO^-} Ar^1NO + ArSO_2^- + H_2O \qquad (109)$$

Hydroxylamine-O-sulphonic acid **150** converts sulphinic acids directly to sulphonamides **151**. The reaction is performed in acetate-buffered water as a solvent (equation 110)[160].

$$RSO_2Na + \underset{\mathbf{(150)}}{H_2NOSO_3H} \xrightarrow{\text{acetate buffer}} \underset{\mathbf{(151)}}{RSO_2NH_2} \qquad (110)$$

R = Ph, *p*-Tol, $PhCH_2$, *p*-An, Bu, 2-Thienyl, 5-Me-fur-2-yl; yield 50–94%

## B. O Reactivity of Sulphinic Acids—Synthesis of Sulphinyl Derivatives

*1. Synthesis of sulphinic esters by O alkylation of sulphinic acids*

For many years it has been thought that the alkylation of sulphinic acids or sulphinate anions proceeds only on the sulphur atom, the only exception being the reaction of silver sulphinates with alkyl halides[3,18]. In fact, even silver sulphinates proved to undergo S alkylation (see Section II.B.1.a, Table 1, References 19 and 20). Later on, however, after some early findings that both O- and S-alkylation products may be formed[161], it turned out that the use of proper alkylating agents, namely those bearing a greater positive charge on carbon ('hard' in the HSAB sense), may lead to the prevailing or exclusive formation of O-alkylation products, i.e. sulphinates. Kobayashi was the first to obtain exclusively ethyl sulphinates by using triethyloxonium tetrafluoroborate as an alkylating agent (equation 111)[162]. Later on, Meek and Fowler found that the use of 'hard' alkylating agents and highly polar solvent leads almost exclusively to the formation of sulphinate esters[19]. Recently, Kobayashi and Toriyabe have reported that the contribution of O alkylation may be increased by addition of crown ethers or cryptands, but only in the case when 'hard' alkylating agents are used[43]. Selected examples of the reactions (equation 112) leading to the predominant formation of sulphinic esters are collected in Table 12.

$$RSO_2H + Et_3O^+BF_4^- \xrightarrow{\text{pyridine}} RS(=O)OEt \qquad (111)$$

R = Ph yield 95%
R = $p$-$O_2NC_6H_4$ yield 33%

$$RSO_2M \xrightarrow{\text{Alkyl. agent}} \underset{(152)}{RS(=O)OAlk} + \underset{(153)}{RSO_2Alk} \qquad (112)$$

Methoxymethyl sulphinates **154** have been obtained by alkylation of sulphinic acids with bromomethyl methyl ether (chloromethyl methyl ethers yield predominantly methoxymethyl sulphones; compare equation 3, Reference 8). Blowing dry nitrogen through the reaction mixture is necessary to remove the hydrogen bromide formed, since it has been found that the undesired methoxymethyl sulphones **155** are formed mainly as a result of a proton-catalysed rearrangement of **154** and not via the direct S alkylation[164] (equation 113).

The predominant formation of sulphinic esters has also been found when O alkylisoureas **156** were used as alkylating agents (equation 114)[165]. The sulphinate to

$$RSO_2H + BrCH_2OMe \longrightarrow \underset{(154)}{RS(=O)OCH_2OMe} + \underset{(155)}{RSO_2CH_2OMe} \qquad (113)$$

Ratio 86:14

$$\underset{(156)}{R^1N{=}C(OR)NHR^1} + PhSO_2H \xrightarrow[-(R^1NH)_2CO]{} \underset{(152a)}{PhS(=O)OR} + \underset{(153a)}{PhSO_2R} \qquad (114)$$

TABLE 12. Predominant O alkylation of sulphinic acids

| Substrate R | Substrate M | Alkylating agent | Solvent, conditions | Products: Ov. yield (%) | Products: Ratio 152 | Products: Ratio 153 | Ref. |
|---|---|---|---|---|---|---|---|
| *p*-Tol | H | $CH_2N_2$ | MeOH/ether | 100 | 100 | 0 | 19 |
| *p*-Tol | H | $Ph_2CN_2$ | MeCN | 100 | 81 | 19 | 44 |
| *p*-Tol | H | $Ph_2CN_2$ | dioxane | 100 | 83 | 17 | 44 |
| *p*-Tol | H | $Ph_2CN_2$ | DMSO | 82 | 100 | 0 | 44 |
| *p*-Tol | Na | $TsCH{=}P(OMe)_3$ | none | 100 | 95 | 5 | 19 |
| *p*-Tol | Na | $(MeO)_2SO_2$ | DMF | 80 | 88 | 12 | 19 |
| *p*-Tol | K | $(MeO)_2SO_2$ | $CH_2Cl_2$ | > 90 | 50 | 50 | 43 |
| *p*-Tol | K | $(MeO)_2SO_2$ | $CH_2Cl_2$, 18-cr.-6 (80 mol%) | > 90 | 58 | 42 | 43 |
| *p*-Tol | Na | $MeOSO_2$Tol-*p* | DMF | 66 | 77 | 23 | 19 |
| *p*-Tol | K | $MeOSO_2F$ | $CH_2Cl_2$, 18-cr.-6 (87 mol%) | > 90 | 70 | 30 | 43 |
| *p*-Tol | K | $MeOSO_2F$ | DMF | > 90 | 77 | 23 | 43 |
| *p*-Tol | K | $MeOSO_2F$ | DMF, 18-cr.-6 (87 mol%) | > 90 | 82 | 18 | 43 |
| *p*-Tol | K | $MeOSO_2F$ | HMPA | > 90 | 100 | 0 | 43 |
| *p*-Tol | K | $MeOSO_2F$ | HMPA, 18-cr.-6 (87 mol%) | > 90 | 94 | 6 | 43 |
| 1-Adamantyl | Na | $MeOSO_2F$ | $CH_2Cl_2$, 18-cr.-6 | > 90 | 62 | 38 | 43 |
| *p*-Tol | K | $MeOSO_2F$ | $CH_2Cl_2$, 15-cr.-6 (200 mol%) | > 90 | 81 | 19 | 43 |
| *p*-Tol | K | $MeOSO_2F$ | $CH_2Cl_2$, kryptifix(2,2,2) (100 mol%) | > 90 | 83 | 17 | 43 |
| 1-Adamantyl | Na | $MeOSO_2F$ | $CH_2Cl_2$, kryptifix(2,2,2) | > 90 | 64 | 36 | 43 |
| $Me_3SiCH_2$ | Na | $Et_3\overset{+}{O}BF^-$ | $CH_2Cl_2$ | 82.1 | 100 | 0 | 46 |
| $CF_3$ | H | $Me_3SiCl$ | | 92 | 100 | 0 | 163 |
| n-$C_4F_9$ | H | $Me_3SiCl$ | | 83 | 100 | 0 | 163 |
| $CF_3$ | H | $Me_3SnCl$ | | 66 | 100 | 0 | 163 |

sulphone ratio appeared to be strongly dependent on the nature of the alkyl groups in O alkylisoureas and to some extent on the solvent used. Thus, among the primary alkyl groups, ethyl was found to give the highest **152a**:**153a** ratio of 90:10 in THF. In the case of secondary alkyl groups (*i*-Pr, *sec*-Bu, 2-hexyl) only sulphinates **152a** were formed. An attempt to synthesize optically active sulphinates by using O alkylisoureas bearing optically active substituents at the nitrogen atom ($R^1 = \alpha$-phenylethyl, myrtanyl) gave products with very low e.e. values (up to 8.1%). Two facts may be responsible for the predominant formation of sulphinates in this reaction—the relatively 'hard' character of the alkylating agent (though the $S_N1$ mechanism has been excluded) and the steric effect exerted by the large electrophile which makes the alkyl group more susceptible to nucleophilic attack by the oxygen atoms[165].

For similar reasons the alkylation of sulphinic acid salts with chlorides **157** gives an unusually high proportion of sulphinic esters **158** in addition to the expected sulphones. It should be added, however, that this reaction, performed in DMF, proceeds according to the $S_N1$ mechanism (equation 115)[166].

Finally, two papers of Kobayashi and coworkers should be mentioned which describe alkylation of sulphinic acid salts with alkyl chlorosulphites **159**[167] and alkyl chlorocarbo-

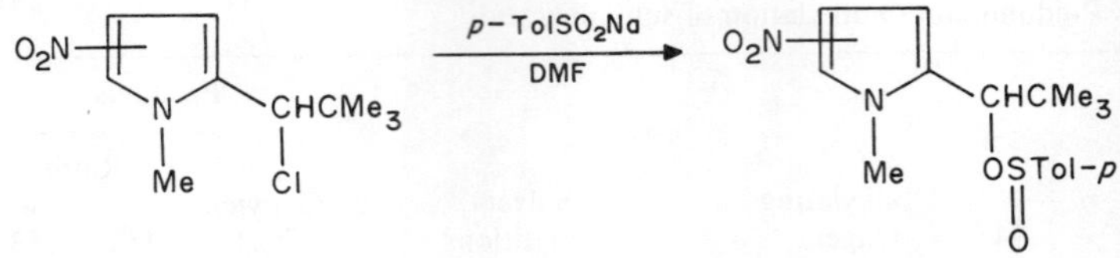


(157) (158)

(115)

nates **161**[168]. In the former case alkyl sulphinates are formed in yields up to 59% and the reaction is assumed to proceed via the intermediary mixed anhydride **160** (equation 116)[167]. The reaction of sodium arenesulphinates with alkyl chlorocarbonates **161** in various alcohols as solvents gives alkyl sulphinates in which the alkoxy group originates from the solvent alcohol and not from **161**. On the basis of experiments with $^{18}O$-labelled sodium sulphinates, the mixed anhydride **162** is postulated as an intermediate whose alcoholysis gives the product (equation 117)[168]. When the reaction is performed in pyridine without addition of an alcohol, sulphinates containing the alkoxy group that originated from **161** are obtained in 30–42% yields[168].

$$RSO_2Na + ClS(=O)OR^1 \longrightarrow RS(=O)OS(=O)OR^1 \longrightarrow RS(=O)OR^1 + SO_2 \qquad (116)$$

(159) (160)

$$RS^{18}O_2Na + ClCO_2R^1 \longrightarrow RS(={}^{18}O){-}^{18}OC(=O)OR^1 \xrightarrow{R^2OH}\ \xrightarrow{36\text{–}69\%} RS(={}^{18}O)OR^2 + CO^{18}O + R^1OH \qquad (117)$$

(161) (162)

*2. O Acylation and O sulphinylation of sulphinic acids*

The carbonyl group of acid chlorides is a hard electropositive centre and acylation of a sulphinate ion may be expected to occur at the oxygen rather than at the sulphur atom (for a different reactivity of the thiocarbonyl and imidoyl groups see Section II.A.1.i, equations 85–91). Indeed, mixed carboxylic–sulphinic anhydrides are produced in this way[11,169]. The acyclic analogues are very unstable (they survive for some time at $-68\,^{\circ}C$)[169] and break down in various ways. However, the cyclic mixed sulphinyl–carboxylic anhydride, namely 3-methyl-1,2-oxathiolan-5-one-2-oxide **163**, has been prepared in 50% yield and proved to be stable (equation 118)[170]. The analogous

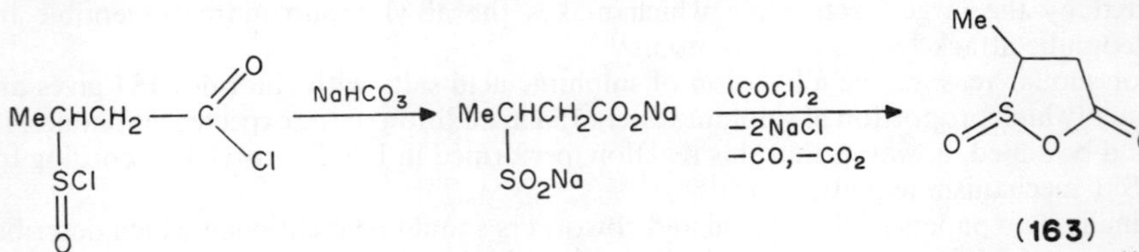


(163)

(118)

compounds **164** have been obtained by the oxidation of benzoxathioles (equation 119)[171].

In contrast to earlier findings that sulphinylation of sulphinic acids leads to the formation of sulphinyl sulphones **131**[135] (equation 92), Kice and Ikura succeeded in the preparation of the sulphinyl anhydride **165**. They reacted silver *tert*-butanesulphinate with *tert*-butanesulphinyl chloride and obtained in 50% yield the product whose structure was univocally proven by spectroscopic methods and by kinetic investigations of its hydrolysis (equation 120)[172]. Later on, the first cyclic sulphinic anhydride **166** was obtained in good yield (equation 121)[173] by carefully controlled hydrolysis of ethanebissulphinyl chloride.

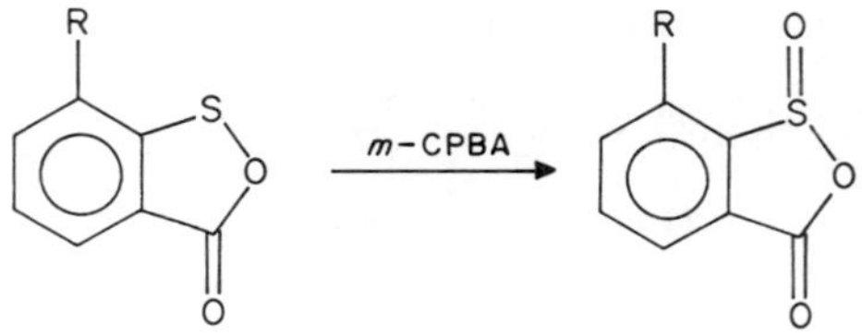


**(164)** (119)

$$t\text{-BuSO}_2\text{Ag} + t\text{-Bu}\overset{\text{O}}{\overset{\|}{\text{S}}}\text{Cl} \longrightarrow t\text{-Bu}\overset{\text{O}}{\overset{\|}{\text{S}}}\text{O}\overset{\text{O}}{\overset{\|}{\text{S}}}\text{Bu-}t \qquad (120)$$

**(165)**

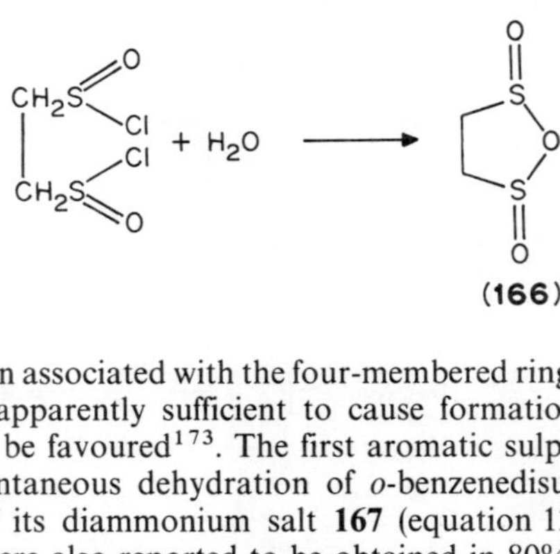


**(166)** (121)

In this instance the strain associated with the four-membered ring in the eventual isomeric sulphinyl sulphone is apparently sufficient to cause formation of the five-membered sulphinic anhydride to be favoured[173]. The first aromatic sulphinic anhydride **168** has been prepared by spontaneous dehydration of *o*-benzenedisulphinic acid formed by careful acidification of its diammonium salt **167** (equation 122)[174]. Perfluoroalkanesulphinic anhydrides were also reported to be obtained in 80% yield but no discussion concerning their structure was presented[163].

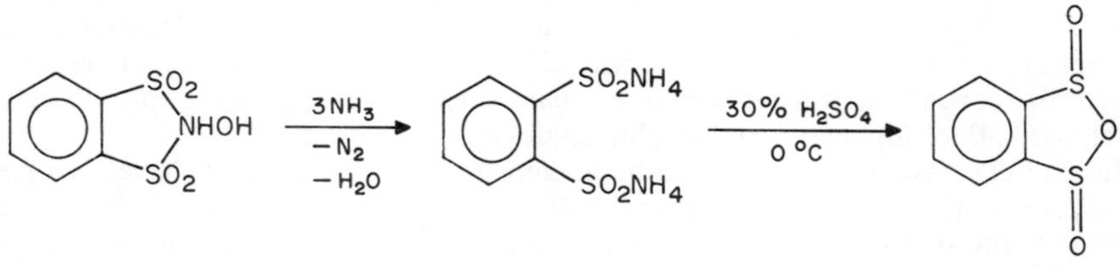


**(167)** **(168)**

(122)

*3. Synthesis of sulphinic esters, sulphinamides, thiosulphinates and sulphoxides by using coupling reagents*

Field and Srivastava have developed several methods for simple esterification of sulphinic acids[175]. Thus, a sodium sulphinate may be reacted with one equivalent of MeOH·HCl in the presence of two equivalents of $BF_3 \cdot Et_2O$ or with methanol itself in the presence of three equivalents of $BF_3 \cdot Et_2O$. The free sulphinic acids with $BF_3 \cdot Et_2O$ in MeOH gave still better results and this would be a method of choice in those cases when free sulphinic acids are readily available and relatively stable. The former methods have been used for esterification of sensitive trisulphide sulphinate salts **169** which were tested for antiradiation properties (equation 123)[175].

$$\underset{\textbf{(169)}}{NaO_2S(CH_2)_4SSS(CH_2)_4SO_2Na} \xrightarrow[BF_3 \cdot Et_2O]{ROH\text{-}HCl} \qquad (123)$$

$$RO_2S(CH_2)_4SSS(CH_2)_4SO_2R$$

$$R = Me \quad \text{yield } 75\text{–}85\%$$

$$R = Et \quad \text{yield } 60\text{–}65\%$$

Kobayashi and coworkers have reported that arenesulphinic acids can be easily converted into the corresponding sulphinates **170** by treatment with an equimolar amount of dicyclohexylcarbodiimide (DCC) as a dehydrating agent and an excess of an appropriate alcohol[176]. Recently, this procedure has been extended to the synthesis of sulphinamides **171** and thiolsulphinates **172**[177,178] (equation 124). The reaction involves

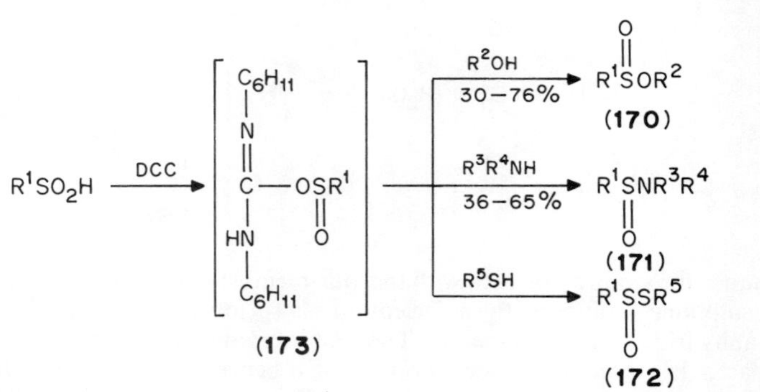


(124)

the primary formation of the intermediary O-sulphinyl *N,N'*-dicyclohexylisourea **173** and the subsequent attack of a nucleophile on the sulphinyl sulphur atom (an alternative mechanism assumes the attack of a second acid molecule on sulphur in **173** to form a sulphinic anhydride which is the real sulphinylating agent). Recently Drabowicz and Pacholczyk treated arylsulphinic acids with alcohols, thiols and secondary amines in the presence of optically active carbodiimides and obtained the corresponding optically active sulphinates, thiolsulphinates and sulphinamides with e.e. up to 10%[178].

Instead of carbodiimides, several other coupling (dehydrating) agents have been used which also made it possible to prepare the sulphinyl derivatives mentioned above. In all cases the crucial step consists in the formation of a bond between the sulphinyl oxygen atom and the coupling reagent. The following reagents have been described: 2-chloro-1-methylpyridinium iodide **174**[179,180] (**170** obtained in 30–76% yield; **171**, 39–52%), $\gamma$-

saccharine chloride **175**[180] (**170**, 24–69%; **171**, 26–75%), diethyl azodicarboxylate **176** and triphenylphosphine (the Mitsunobu reaction cannot be used for the preparation of **171**)[180], phenyl phosphorodichloridate **177** and pyridine (**170**, 50–85% yield; **171**, 15–75%; **172**, 41–82%)[181], diphenyl phosphorochloridate **178** (**170**, 87–97% yield; **171**, 0–36%; **172**, 30–89%)[182], *N*-chlorosuccinimide and triphenylphosphine (**170**, 40–88% yield; **171**, 5–32%; **172**, 22–77%)[182] and 3-(phthalimidoxy)-1,2-benzoisothiazole 1,1-dioxide **179** (**170**, 27–81% yield; **171**, 0–20%; **172**, 22–70%)[182]. Some of these coupling reagents have also

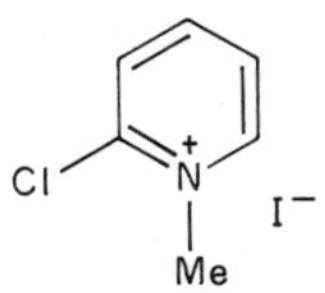

**(174)**

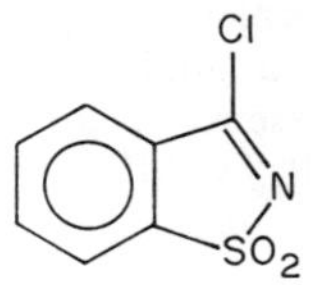

**(175)**

$EtO_2CN{=}NCO_2Et$

**(176)**

$PhOP(=O)Cl_2$

**(177)**

$(PhO)_2P(=O)Cl$

**(178)**

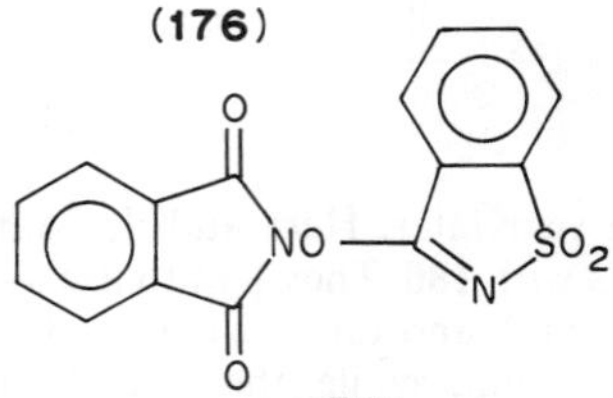

**(179)**

been used for the synthesis of sulphoxides from sulphinic acids. The adducts of sulphinic acids with **177**, **178** or **179** were treated with Grignard reagents or enamines to give sulphoxides in 11–53% yield[183] (e.g. equation 125).

$$p\text{-TolSO}_2\text{H} \xrightarrow{\text{178, pyridine}} \left[ p\text{-TolS(=O)OP(=O)(OPh)}_2 \right] \xrightarrow[\text{or enamine}]{\text{RMgBr}} p\text{-TolS(=O)R} \qquad (125)$$

*4. Synthesis of sulphinyl chlorides from sulphinic acids*

Reaction of sulphinic acids or their salts with an excess of thionyl chloride gives sulphinyl chlorides in good yields[184]. Perfluoromethanesulphinyl chloride and perfluorobutanesulphinyl chloride have been obtained from the corresponding perfluoroalkanesulphinic acids when reacted with thionyl chloride, phosphorus trichloride and phosphorus pentachloride (equation 126)[163]. This method of synthesis of sulphinyl chlorides is of limited value and is used only in special instances, the most useful method being the oxidative chlorination of thiols, disulphides and thioloesters[184].

$$R_fSO_2H \xrightarrow[\text{or PCl}_5\ \ 80\text{–}95\%]{\text{SOCl}_2\ \text{or PCl}_3} R_fS(=O)Cl + R_fS(=O)OS(=O)R_f \qquad (126)$$

(up to 12% when $PCl_3$ or $PCl_5$ are used)

## C. Other Applications of Sulphinic Acids

### *1. Formamidinesulphinic acid as a reducing agent*

Formamidinesulphinic acid **180**, called sometimes thiourea dioxide, is a commercially available and easy to handle reagent, which is used for the reduction of a variety of organic compounds.

Among organic nitrogen compounds, aromatic nitro-, azoxy-, azo- and hydrazo-derivatives are reduced by **180** to give the corresponding amines in high yields[185].

Nakagawa and Minami reported in 1972 that aliphatic, aromatic and heteroaromatic ketones can be easily reduced by **180** in the presence of caustic alkali in ethanolic solution to give the corresponding alcohols in 74–100% yield[186].

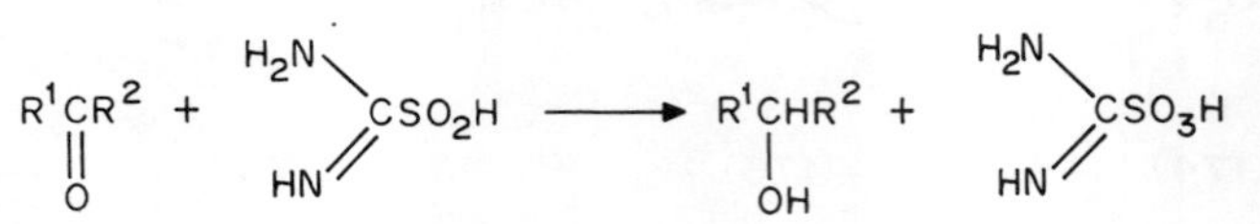

(180) (127)

One year later, Herz and de Marquez described a successful reduction of steroidal ketones with **180**. They had to use a stronger alkaline reagent, namely sodium propoxide in propanol, and under such conditions they were able to reduce a 3-keto group and a 6-keto group, while the reduction of a 20-oxo group could not be achieved[187]. Shanker[188] succeeded in the preparation of α-D-fluoren-9-ol containing at least 90% of deuterium at C-9 from fluorenone using the deuterated **180** in the presence of sodium deuteroxide in deuteroethanol.

The above results, however, have been disputed by Italian workers[189] who have found that ketones are reduced under the alkaline conditions applied even without addition of **180** and that the yields of alcohols are only slightly lower in this case. Thus, their conclusion is that formamidinesulphinic acid **180** does not play a major role in this reaction and therefore it cannot be considered as a useful reducing agent for ketones[189].

There is no doubt, however, that **180** has been successfully applied to the reduction of a variety of organic sulphur, selenium and tellurium compounds. Thus, **180** reduces disulphides to thiols (equation 128) and *N*-tosylsulphimines **181** to sulphides (equation 129) when the reaction is carried out under phase-transfer catalytic conditions in the presence of a catalyst, such as (hexadecyl)tributylphosphonium bromide, in an aqueous–organic two-phase system[190].

$$\mathrm{RSSR} \xrightarrow{\mathbf{180},\ \mathrm{NaOH_{aq}}\ \mathrm{catalyst}} 2\,\mathrm{RSH} \tag{128}$$

R = alkyl, aryl yields 62–90%

$$\underset{\textbf{(181)}}{\mathrm{R^1}\overset{\overset{\mathrm{NTos}}{\|}}{\mathrm{S}}\mathrm{R^2}} \xrightarrow{\mathbf{180},\ \mathrm{NaOH_{aq}}/i\text{-}\mathrm{Pr_2O},\ \mathrm{catalyst}} \mathrm{R^1SR^2} \tag{129}$$

$R^1$ = aryl, $R^2$ = alkyl, aryl; yields 26–100%

To achieve reduction of sulphoxides to sulphides, different conditions were applied, namely the reaction was performed in boiling acetonitrile in the presence of iodine as a catalyst (equation 130)[191]. Finally, aryltellurium trihalides **182**, arylselenium trihalides

$$R^1\underset{\underset{O}{\|}}{S}R^2 + \mathbf{180} \xrightarrow[80\%]{I_2,\,MeCN} R^1SR^2 \qquad (130)$$

$$R^1,R^2 = \text{alkyl, aryl;} \quad \text{yields } 89\text{–}95\%$$

**187**, organyltellurium dichlorides **184** and organylselenium dichlorides **189** and organyl selenoxides **190** and telluroxides **185** are reduced in high yields to the corresponding ditellurides **183**, tellurides **186**, diselenides **188** and selenides **191** with **180** in a two-phase system[192]. Some representative examples are shown in equations 131–134 and in Table 13.

$$\underset{(\mathbf{182})}{ArTeX_3} \xrightarrow[r.t.]{NaOH/H_2O} ArTeO_2^- \xrightarrow[\text{petr. ether}]{\mathbf{180}} \underset{(\mathbf{183})}{ArTeTeAr} \qquad (131)$$

$$\underset{(\mathbf{184})}{R^1RTeCl_2} \xrightarrow[r.t.]{NaOH/H_2O} \underset{(\mathbf{185})}{R\overset{\overset{O}{\|}}{Te}R^1} \xrightarrow[\text{petr. ether}]{\mathbf{180}} \underset{(\mathbf{186})}{RTeR^1} \qquad (132)$$

$$\underset{(\mathbf{187})}{RSeX_3} \xrightarrow{NaOH/H_2O} RSeO_2^- \longrightarrow \underset{(\mathbf{188})}{RSeSeR} \qquad (133)$$

$$\underset{(\mathbf{189})}{R^1RSeCl_2} \xrightarrow{NaOH/H_2O} \underset{(\mathbf{190})}{R\underset{\underset{O}{\|}}{Se}R^1} \longrightarrow \underset{(\mathbf{191})}{RSeR^1} \qquad (134)$$

TABLE 13. Reduction of organoselenium and organotellurium compounds with formamidine-sulphinic acid **180**

| Starting materials | Product | Reaction time (min) | Yield (%) |
|---|---|---|---|
| $PhTeBr_3$ | $Ph_2Te_2$ | 30 | 92 |
| *p*-$AnTeCl_3$ | (*p*-$An)_2Te_2$ | 30 | 94 |
| $PhSeBr_3$ | $Ph_2Se_2$ | 30 | 93 |
| PhCH=CHTe($Cl_2$)Bu | PhCH=CHTeBu | 30 | 90 |
| (*p*-$An)_2TeO$ | (*p*-$An)_2Te$ | 30 | 95 |
| PhSe($Cl_2$)Bu | PhSeBu | 45 | 89 |
| $Ph_2SeO$ | $Ph_2Se$ | 30 | 90 |

*2. Reductive transformations of sulphinic acids*

Sulphinic acids when reduced electrolytically or with lithium aluminium hydride[3,193] give disulphides. Unsymmetrical disulphides **192** are formed in high yields by the reduction of sulphinic acids with a thiol–chlorotrimethylsilane system (equation 135)[194].

$$RSO_2H + 3R^1SH + 4Me_3SiCl \xrightarrow{\text{r.t.}} \underset{\textbf{(192)}}{RSSR^1} + R^1SSR^1 + 2(Me_3Si)_2O \quad (135)$$

R = alkyl, aryl, $R^1$ = alkyl, aryl; yields 80–100%

Two different methods of reductive cyanation of sulphinic acids leading to the formation of aryl or alkylthiocyanates **194** have recently been reported. The first consists of a reaction of sodium sulphinates with diethyl phosphorocyanidate **193** in boiling tetrahydrofuran (equation 136)[195]. Although the mechanism of this transformation has not been elucidated, the intermediacy of sulphinyl cyanide has been proven. The second approach is based on the similar reaction of sodium arenesulphinates with trimethylsilyl cyanide **195** and affords arenethiocyanates in 18–80% yield (equation 137)[196].

$$RSO_2Na + \underset{\textbf{(193)}}{(EtO)_2P(=O)CN} \xrightarrow{\text{THF}} \underset{\textbf{(194)}}{RSCN} \quad (136)$$

R = aryl, benzyl, adamantyl; yields 15–79%

$$ArSO_2Na + \underset{\textbf{(195)}}{3Me_3SiCN} \xrightarrow{\text{HMPA}} ArSCN \quad (137)$$

*3. Condensations of sulphinic acids leading to sulphoxides and sulphinamides*

*p*-Aminobenzenesulphinic acid **196** undergoes condensation on heating to give 4,4′-diaminophenyl sulphoxide **197** (equation 138)[197]. In the same way *p*-acetamidobenzenesulphinic acid **198** reacts with aryl amines to give unsymmetrical *p*-aminophenyl sulphoxides **199** (equation 139)[198].

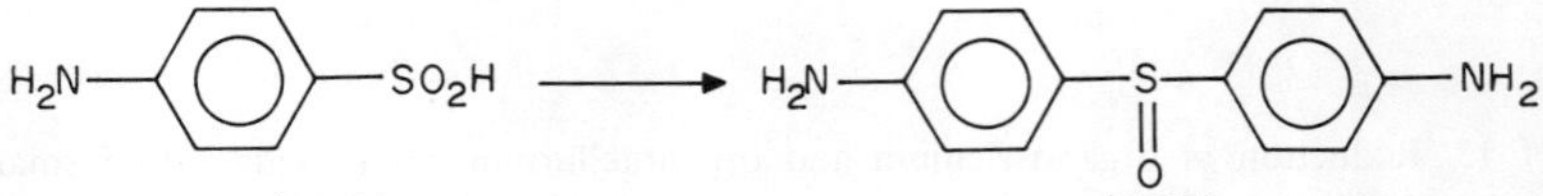


**(196)** **(197)**

(138)

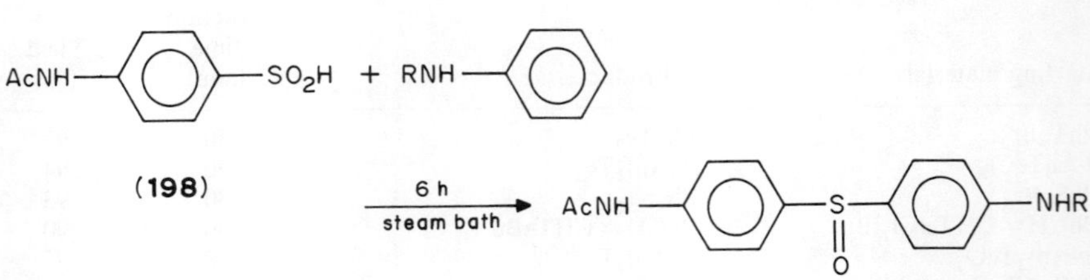


**(198)** **(199)**

(139)

Reaction of isocyanates with sulphinic acids in the presence of 4-dimethylamino-pyridine or 1-methylimidazole results in the formation of sulphinamides in high yields (equation 140)[199]. Reaction of *p*-toluenesulphinic acid with dimorpholino sulphoxide **200** proceeds with evolution of $SO_2$ and produces *p*-toluenesulphinylmorpholine **201**. Most probably the process involves an attack of the sulphinate oxygen atom on the sulphoxide sulphur atom (equation 141)[150].

$$R^1SO_2H + R^2NCO \longrightarrow R^1\overset{\displaystyle O}{\overset{\|}{S}}NHR^2 + CO_2 \qquad (140)$$

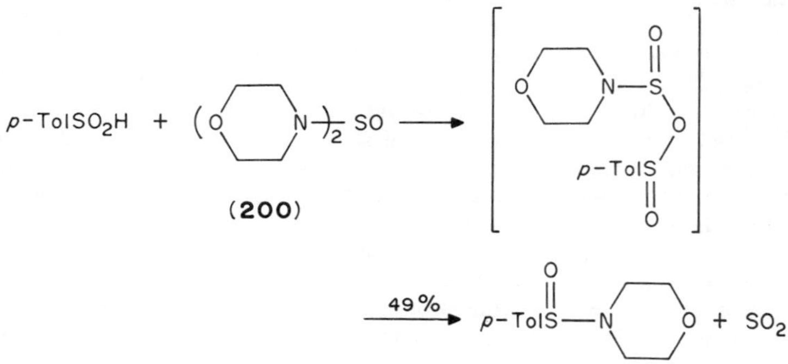


(200)

(201)

(141)

*4. Miscellaneous*

Sulphinic acids can serve as substrates for the synthesis of organometallic compounds, including organomercuric and organothallium derivatives[200–202].

Sodium sulphinates are used to generate arylpalladium complexes serving as reagents for arylation of diolefins[203] and take the role of co-catalysts in $PdCl_2$-catalysed dimerization of butadiene in alcoholic solvents[204].

Fragmentation of the homoconjugated sulphinic acids **202** gives α-methylene carbonyl compounds (equation 142)[205].

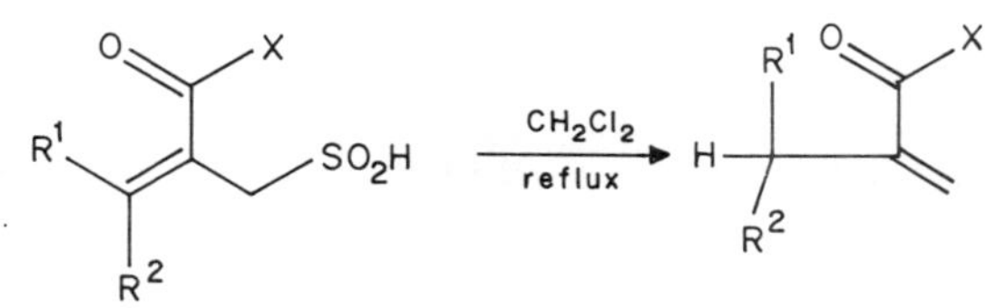


(142)

(202)

X = Me, OMe; yields 50–60%

## III. SULPHINATE ESTERS

### A. Synthetic Applications of Sulphinate Esters Based on Nucleophilic Exchange at the Sulphinyl Sulphur Atom

Although the reactions of sulphinic acid esters with nucleophilic reagents may occur in different ways depending on the site attacked, the most important, from both synthetic and mechanistic points of view, is the nucleophilic substitution that occurs at the electron-deficient sulphinyl sulphur atom, with the alkoxy group being the leaving group.

$$\mathrm{NuX} + \mathrm{R\underset{\displaystyle O}{\overset{}{\underset{\|}{S}}}OR'} \longrightarrow \mathrm{Nu\underset{\displaystyle O}{\underset{\|}{S}}R} + \mathrm{R'OX} \qquad (143)$$

In this section a summary of such reactions will be given.

### 1. Transesterification

Transesterification of sulphinates (equation 144) has only limited applicability as a synthetic method. However, this reaction plays an important role in stereochemical studies as a simple model of the nucleophilic substitution at the sulphinyl sulphur atom[206] (equation 144). Historically, the thermal transesterification of racemic O-alkyl *p*-toluenesulphinate **203** with optically active alcohols (equation 145) described by Phillips[207] in 1925 may be considered as the first example of the application of sulphinic acid esters in organic synthesis. Later it was reported[208] that diastereoisomerically pure (−)menthyl (−)arenesulphinates **204** are converted into the corresponding racemic O-ethyl arenesulphinates **205** in ethanol solution in the presence of sodium ethoxide (equation 146). More recently, it was found that transesterification of sulphinates **203** and **204** proceeds at room temperature in the presence of strong acids[209] or N-bromosuccinimide (NBS)[210] giving products which were isolated by distillation in 50–80% yield (see Table 14). It was also found that transesterification of optically active sulphinates catalysed by NBS is not stereospecific and takes place with predominant inversion of configuration or with racemization when acid catalysts were used.

$$\mathrm{R\underset{\displaystyle O}{\underset{\|}{S}}OR^1} + \mathrm{R^2OH} \longrightarrow \mathrm{R\underset{\displaystyle O}{\underset{\|}{S}}OR^2} + \mathrm{R^1OH} \qquad (144)$$

$$(\pm)2\ p\text{-}\mathrm{Tol\underset{\displaystyle O}{\underset{\|}{S}}OR}\ \mathbf{(203)} \xrightarrow{(-)\text{-}2\text{-}\mathrm{C_8H_{17}OH}} (+)p\text{-}\mathrm{Tol\underset{\displaystyle O}{\underset{\|}{\overset{*}{S}}}OR}\ \mathbf{(203)} + p\text{-}\mathrm{Tol\underset{\displaystyle O}{\underset{\|}{\overset{*}{S}}}O\overset{\displaystyle CH_3}{\overset{|}{C}}HC_6H_{13}} + \mathrm{ROH} \qquad (145)$$

$$(-)\ \mathrm{Ar\underset{\displaystyle O}{\underset{\|}{S}}OMenthyl}\ \mathbf{(204)} \xrightarrow[\mathrm{EtONa}]{\mathrm{EtOH}} (\pm)\ \mathrm{Ar\underset{\displaystyle O}{\underset{\|}{S}}OEt}\ \mathbf{(205)} + (-)\ \mathrm{Menthol} \qquad (146)$$

### 2. Reactions with organometallic reagents

The reaction of organometallic reagents with sulphinate esters consists in the replacement of the sulphur-oxygen bond by a sulphur–carbon or sulphur–nitrogen bond. In the first case, when the organometallic reagent is a carbon nucleophile, sulphoxides are formed (path A) and this reaction is the most important method of their preparation. When the organometallic reagent is a nitrogen nucleophile, the starting sulphinates are converted into sulphinamides (path B); see equation 147.

TABLE 14. Transesterification of sulphinate esters, $RS(O)OR^1$, with alcohols, $R^2OH$

| Starting R | Sulphinate $R^1$ | $R^2OH$ | Conditions | Time (h) | Sulphinate formed Yield (%) | Reference |
|---|---|---|---|---|---|---|
| *p*-Tol | Et | $2\text{-}C_8H_{17}$ | | 54 | 90 | 207 |
| *p*-Tol | Bu | $2\text{-}C_8H_{17}$ | | 54 | 9 | 207 |
| *p*-Tol | Et | Bu | | 18 | 10 | 207 |
| *p*-Tol | Me | *i*-Pr | $CF_3SO_3H$ | 75 | | 209 |
| *p*-Tol | $CH_2CH{=}CH_2$ | *i*-Pr | $CF_3SO_3H$ | 98 | | 209 |
| *p*-Tol | $CH_2C{\equiv}CH$ | *i*-Pr | $CF_3SO_3H$ | 18 | | 209 |
| *p*-Tol | Me | *i*-Pr | NBS | 10 | 82 | 210 |
| *p*-Tol | Me | Pr | NBS | 10 | 80 | 210 |
| *p*-Tol | Me | Bu | NBS | 15 | 79 | 210 |
| *p*-Tol | $CH_2CH{=}CH_2$ | *i*-Pr | NBS | 20 | 51 | 210 |
| *p*-Tol | $CH_2C{\equiv}CH$ | *i*-Pr | NBS | 20 | 52 | 210 |
| Ph | Me | Et | NBS | 30 | 83 | 210 |
| Ph | Et | *i*-Pr | NBS | 15 | 72 | 210 |
| *p*-Tol | Menthyl | Et | EtONa | | | 208 |
| *p*-Tol | Menthyl | Et | EtONa | | | 208 |

$$R^1S(O)NR_2 \xleftarrow[\text{B}]{R_2N^-M^+} R^1S(O)OR^2 \xrightarrow[\text{A}]{R_3C^-M^+} R^1S(O)CR_3 \qquad (147)$$

*a. Reactions with carbon nucleophiles.* Gilman and coworkers[211] were the first to report that the reaction of sulphinate esters **203** with Grignard reagents (equation 148) affords racemic sulphoxides in moderate yields (see Table 15). Much later, a detailed study of the reaction between acyclic and cyclic sulphinate esters **206–210** and various Grignard reagents was carried out by Harpp and coworkers[212]. They found that it is possible to isolate sulphoxides from the reaction mixture but the yield (see Table 15) varied greatly with the structure of both sulphinate and Grignard reagent. Moreover, these authors recommended the use of organocopper reagents in place of the Grignard compounds since sulphoxides were obtained in higher yields.

$$p\text{-TolS(O)OR} + R^1MgX \longrightarrow p\text{-TolS(O)}R^1 + ROMgX \qquad (148)$$

**(203)**

$R = Et, Bu$
$R^1 = Ph, CH_2Ph$

PhS(O)OMe **(206)**

BuS(O)OMe **(207)**

PhS(O)OTol-*p* **(208)**

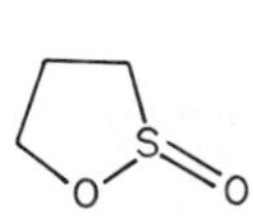

**(209)**

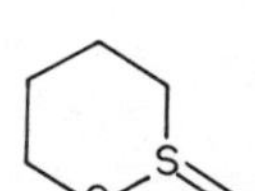

**(210)**

TABLE 15. Formation of racemic sulphoxides, $R^1S(O)R^2$, from the reaction of organometallic reagents, $R^2M$, with sulphinates, $R^1S(O)OR^3$

| $R^1$ | $R^3$ | $R^2M$ | Yield (%) | Reference |
|---|---|---|---|---|
| Ph | Me | MeMgBr | 27 | 211 |
| Ph | Me | EtMgBr | 32.0 | 211 |
| Ph | Me | PhMgBr | 55.0 | 211 |
| Ph | Me | $Me_2CuLi$ | 59.0 | 212 |
| Ph | Me | $Et_2CuLi$ | 36.0 | 212 |
| Ph | Me | $PH_2CuLi$ | 50.0 | 212 |
| Ph | Tol-*p* | $Me_2CuLi$ | 22.0 | 212 |
| *p*-Tol | Et | $PhCH_2MgBr$ | 57.2 | 211 |
| *p*-Tol | Bu | PhMgBr | 46.0 | 211 |
| Bu | Me | $Me_2CuLi$ | 50.0 | 212 |
| Bu | Me | $Bu_2CuLi$ | 52.0[a] | 212 |
| $PhCH_2$ | Et | BuLi | | 213 |
| $PhCH_2$ | *i*-Pr | BuLi | | 213 |
| $PhCH_2$ | Bu | BuLi | | 213 |

[a]Reaction was carried out at −78 °C; at 0 °C no sulphoxide was isolated.

The reaction of alkyl phenylmethanesulphinates **211** with butyllithium in tetrahydrofuran at −80 °C afforded benzyl butyl sulphoxide[213] (equation 149).

$$\mathrm{PhCH_2\underset{\underset{O}{\|}}{S}OR + BuLi \longrightarrow PhCH_2\underset{\underset{O}{\|}}{S}Bu} \qquad (149)$$

**(211)** R = Et, i-Pr, Bu

The hydrolytically and thermally unstable α-silylmethyl sulphoxides **212** were prepared in high yield by the treatment of methyl arenesulphinates with the Grignard reagent obtained from halomethyltrialkylsilanes[214] (equation 150). It is interesting to note that trimethylgermylmethyl phenyl sulphoxide **213**, prepared in a similar way in 78% yield (equation 151), was found to be thermally stable[214]. The Claisen-type condensation between ketone enolate anions **214** and arenesulphinates provides an interesting synthetic approach to β-ketosulphoxides **215** (equation 152)[215,216]. An extension of this procedure[217] was found to be very useful in the synthesis of 16-phenylsulphinyl 17-ketones **216a**–**c** (yields around 90%), pyrolysis of which provides a short and inexpensive route to α,β-unsaturated ketones **217a**–**c** starting from saturated ketones **218a**–**c** (equation 153).

$$\mathrm{Ar\underset{\underset{O}{\|}}{S}OMe + R_3SiCH_2MgX \longrightarrow Ar\underset{\underset{O}{\|}}{S}CH_2SiR_3} \qquad (150)$$

**(212)**

(**a**) Ar = Ph
(**b**) Ar = *p*-Tol

$$\mathrm{Ph\underset{\underset{O}{\|}}{S}OMe + Me_3GeCH_2MgCl \longrightarrow Ph\underset{\underset{O}{\|}}{S}CH_2GeMe_3} \qquad (151)$$

**(213)**

$$\underset{\substack{\| \\ \mathrm{O}}}{\mathrm{RC}}\overset{-}{\cdots}\mathrm{CHR^1}\ \mathrm{Na^+} + \mathrm{Ar}\underset{\substack{\| \\ \mathrm{O}}}{\mathrm{S}}\mathrm{OR} \longrightarrow \mathrm{R}\underset{\substack{\| \\ \mathrm{O}}}{\mathrm{C}}\mathrm{CHR^1}\underset{\substack{\| \\ \mathrm{O}}}{\mathrm{S}}\mathrm{Ar} \qquad (152)$$

(**214**) (**215**)

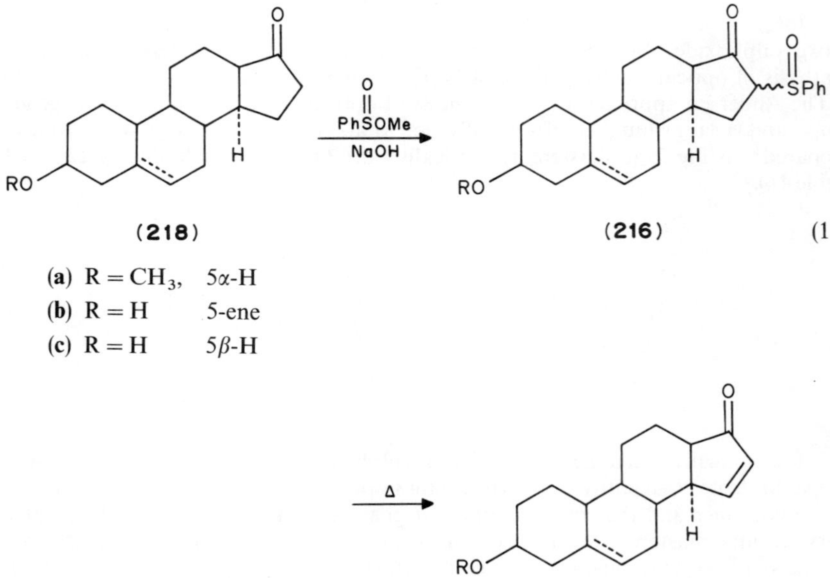

(**218**) (**216**) (**217**) (153)

(**a**) R = $CH_3$, 5α-H
(**b**) R = H 5-ene
(**c**) R = H 5β-H

A few racemic sulphinylsulphones **219** were prepared analogously by reacting arenesulphinates with the carbanions generated from dimethyl[216] or methyl *p*-tolyl sulphones[218] (equation 154).

$$\mathrm{Ar}\underset{\substack{\| \\ \mathrm{O}}}{\mathrm{S}}\mathrm{OEt} + {}^{-}\mathrm{CH_2}\overset{\substack{\mathrm{O} \\ \|}}{\underset{\substack{\| \\ \mathrm{O}}}{\mathrm{S}}}\mathrm{R} \longrightarrow \mathrm{Ar}\underset{\substack{\| \\ \mathrm{O}}}{\mathrm{S}}\mathrm{CH_2}\overset{\substack{\mathrm{O} \\ \|}}{\underset{\substack{\| \\ \mathrm{O}}}{\mathrm{S}}}\mathrm{R} + \mathrm{EtO^-} \qquad (154)$$

(**219**)

R = Me or *p*-Tol

Ar = Ph or *p*-Tol

The highly stereoselective synthesis of optically active sulphoxides, developed by Andersen[219] in 1962, is based on the reaction of the diastereoisomerically pure (or strongly enriched in one diastereoisomer) O-menthyl arene(alkane)sulphinates with Grignard reagents. (+)-(*R*)-Ethyl *p*-tolyl sulphoxide **220** prepared from (−) (*S*)-O-menthyl *p*-toluenesulphinate **221** and ethylmagnesium iodide (equation 155) was the first optically

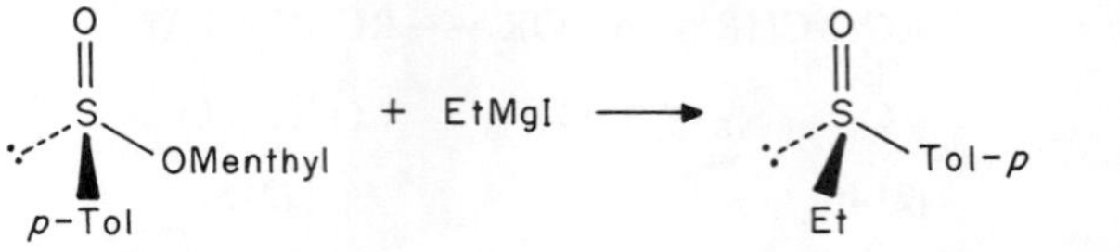


(155)

(−)-(*S*)-(**221**) (+)-(*R*)-(**220**)

active sulphoxide obtained by this, still most important and widely used, method of synthesis of optically active sulphoxides of very high optical purity.

The Andersen approach to the synthesis of optically active sulphoxides is general in scope and a large number of optically active alkyl aryl and diaryl sulphoxides were prepared starting from diastereoisomerically pure O-menthyl sulphinates **221**–**224** (see Table 16).

(**221**) Ar = *p*-Tol[207]
(**222**) Ar = Ph[208]
(**223**) Ar = *p*-An[220]
(**224**) Ar = 1-Naph[220]

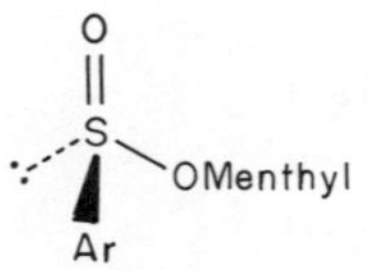


(−)-(*S*)-(**221**)-(**224**)

A few optically active benzyl alkyl(aryl) sulphoxides **225** of high optical purity were prepared from *O*-menthyl phenylmethanesulphinate **226** strongly enriched in one diastereoisomer and the corresponding Grignard reagents[221] (equation 156, Table 16). Very recently, diastereoisomerically pure (+)-(*R*)-mesitylenesulphinic ester **227** was used for the synthesis of optically active mesityl alkyl sulphoxides **228**[222] (equation 157). The Andersen approach to the synthesis of chiral dialkyl sulphoxides of high optical purity starting from diastereoisomeric alkanesulphinates has a serious limitation, because they are not diastereoisomerically pure at sulphur. For example, all known diastereoisomeric O-menthyl alkanesulphinates are oils which cannot be separated into pure diastereoisomers. It was found, however, that O-cholesteryl methanesulphinate **229** is a crystalline compound which, after separation by crystallization into pure diastereoisomers and upon treatment with alkyl Grignard reagents, affords alkyl methyl sulphoxides **230** of high optical purity[223] (euqation 158, Table 16).

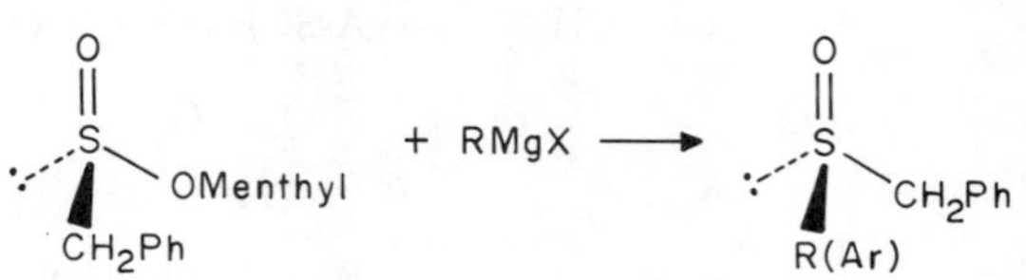


(156)

(−)-(*S*)-(**226**) (**225**)

Usually, the reaction of arenesulphinates with Grignard reagents is carried out in ethyl ether solution. However, in this solvent (see Table 16) chiral sulphoxides are formed in moderate to low yields, depending on the structure of both reaction components. Harpp and coworkers found[212] that the use of lithium-copper reagents ($R_2CuLi$) instead of Grignard reagents gives a cleaner conversion of optically active sulphinates to chiral

TABLE 16. Synthesis of optically active sulphoxides, $RS(O)R^1$, from diastereoisomerically pure aryl(alkyl) sulphinates, $RS(O)OR^2$, and organometallic reagents, $R^1M$

| Sulphinate | | | | Sulphoxide | | |
|---|---|---|---|---|---|---|
| R | $R^2$ | $[\alpha]_{589}$ | $R^1M$ | yield (%) | $[\alpha]_{589}$ | Reference |
| Ph | Men | −206.1 | MeMgI | 60 | +178.3 | 231 |
| | | −205.5 | $Me_2CuLi$ | 16 | +133.9 | 212 |
| | | *a* | EtMgI | 72 | +176.2 | 232 |
| | | −206.1 | *i*-PrMgCl | 60 | +170.0 | 233 |
| | | −206.1 | *t*-BuMgCl | 60 | +180.0 | 231 |
| | | | $C_5H_{11}MgBr$ | | 199.6 | 234 |
| | | | $C_6H_{13}MgBr$ | | 184.0 | 234 |
| *p*-Tol | Men | −198.0 | $MeMgI/Et_2O$ | *a* | +145.5 | 235 |
| | | −195.0 | MeMgI/PhH | 82 | +150.0 | 224 |
| | | −210.0 | $Me_2CuLi$ | 55 | +143.2 | 212 |
| | | −198.0 | $EtMgBr/Et_2O$ | *a* | +187.5 | 235 |
| | | −195.0 | EtMgBr/PhH | 92 | +198.0 | 224 |
| | | *a* | n-PrMgBr | *a* | +201.0 | 236 |
| | | −198.0 | *i*-$PrMgBr/Et_2O$ | 22 | | 235 |
| | | −195.0 | *i*-PrMgBr/PhH | 40 | +173.2 | 224 |
| | | −195.0 | n-BuMgBr/PhH | 73 | +186.0 | 224 |
| | | −198.0 | *t*-BuMgCl | *a* | +190.0 | 235 |
| | | −198.0 | *t*-$BuCH_2MgBr$ | *a* | +220.0 | 237 |
| | | −198.0 | $C_6H_{13}MgI$ | *a* | +176.0 | 238 |
| | | −198.0 | $CH_2{=}CHCH_2MgBr$ | *a* | +212.0 | 236 |
| | | −198.0 | PhMgBr | *a* | + 27.0 | 235 |
| | | −198.0 | *o*-TolMgBr | *a* | + 75.6 | 235 |
| | | −198.0 | *m*-TolMgBr | *a* | + 24.4 | 235 |
| | | −198.0 | 2,4,6-$Me_3C_6H_2MgBr$ | *a* | −259.0 | 235 |
| | | −198.0 | 9-AnthrylMgBr | *a* | −309.0 | 235 |
| | | −198.0 | 4-$CF_3C_6H_4MgBr$ | *a* | + 57.0 | 237 |
| | | −198.0 | 3-$CF_3C_4H_6MgBr$ | *a* | + 58.0 | 237 |
| | | −198.0 | 4-$ClC_6H_4MgBr$ | *a* | + 25.0 | 237 |
| | | −198.0 | 2-$ClC_6H_4MgBr$ | *a* | −120.0 | 237 |
| | | −198.0 | 2-AnMgBr | *a* | −221.0 | 237 |
| | | −192.2 | 4-AnMgBr | *a* | − 25.1 | 220 |
| | | −199.2 | 4-$Me_2NC_6H_4MgBr$ | *a* | + 85.2 | 220 |
| | | −199.2 | 1-NaphMgBr | *a* | −414.2 | 220 |
| | | −210.0 | 4-Me-*c*-$C_6H_{10}CH_2MgBr$ | 61 | +204.0 | 239 |
| | | −210.0 | 4-$MeOCH_2$-*c*-$C_6H_{10}CH_2MgBr$ | 70 | +182.0 | 239 |
| | | −210.0 | 4-$ClCH_2$-*c*-$C_6H_{10}CH_2MgBr$ | 17 | +173.0 | 239 |
| | | −210.0 | 4-$C_5H_{11}$-*c*-$C_6H_{10}CH_2MgBr$ | 72 | +169.0 | 239 |
| | | −210.0 | 4-*t*-Bu-*c*-$C_6H_{10}CH_2MgBr$ | 41 | +155.0 | 239 |
| 4-An | Men | −189.1 | 2-MeO-$C_6H_4MgBr$ | *a* | −217.2 | 220 |
| | | −189.1 | *p*-TolMgBr | *a* | + 24.2 | 220 |
| 1-Naph | Men | −433.2 | *p*-TolMgBr | *a* | +416.2 | 220 |
| Mesityl | Dcg[b] | + 28.8 | MeMgI | 93 | −200.1 | 222 |
| | | | *i*-PrMgI | 71 | −176.9 | 222 |
| Me | Cholesteryl | + 77.35 | n-PrMgBr | 32 | −139.0 | 223 |
| | | + 77.35 | *p*-TolMgBr | 35 | +148.0 | 223 |
| | | −113.0 | n-BuMgBr | 52 | +110.3 | 223 |
| | | −113.0 | *i*-BuMgBr | 50 | +138.0 | 223 |
| | | −111.85 | $PhCH_2MgBr$ | 36 | +106.0 | 223 |
| $PhCH_2$ | Men | +105.0 | MeMgI | *a* | + 96.0 | 221 |
| | | +105.0 | EtMgI | *a* | + 47.0 | 221 |
| | | +123.3 | n-PrMgI | *a* | + 55.0 | 240 |
| | | +105.0 | *i*-PrMgI | *a* | +119.0 | 221 |
| | | +105.0 | n-BuMgI | *a* | + 16.0 | 221 |
| | | +123.3 | *i*-BuMgI | *a* | −110.0 | 240 |
| | | +105.0 | *t*-BuMgCl | *a* | +281.0 | 221 |

[a]Not given.
[b]See equation 157.

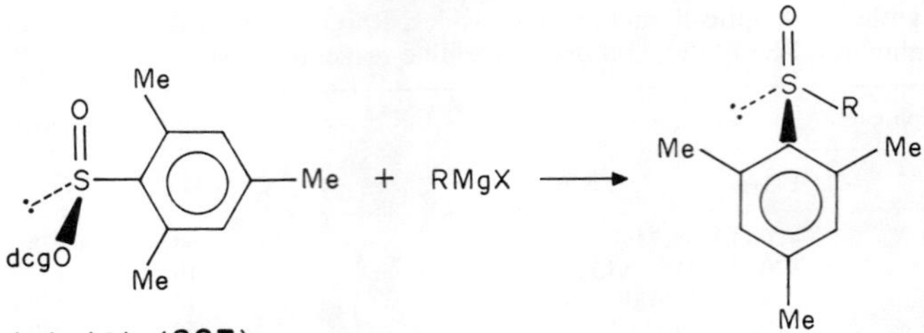

(+)-(*R*)-(**227**) (−)-(*S*)-(**228**) (157)

dcg =

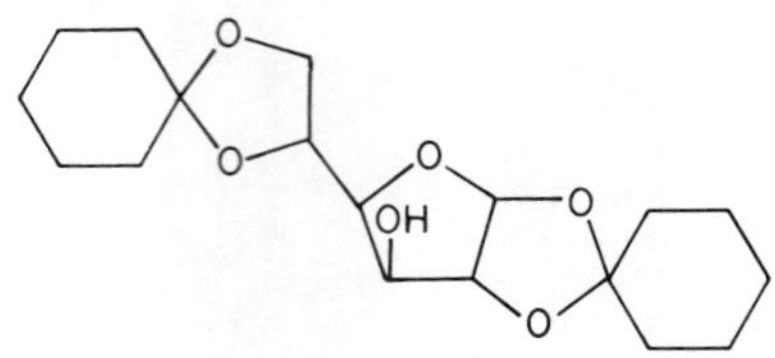

$$\underset{\underset{\displaystyle O}{\|}}{\text{MeSOCholesteryl}} + \text{RMgX} \longrightarrow \text{R}\overset{*}{\text{S}}\text{Me} \quad (\text{S}=\text{O}) \qquad (158)$$

(**229**) (**230**)

(+)(*R*) or (−)(*S*)

sulphoxides. However, in this case also the yields of sulphoxides were in the range between 16 and 59%. Chiral sulphoxides of greater chemical and optical purity and in higher chemical yields are obtained when the reactions of O-menthyl sulphinates with Grignard reagents are carried out in a benzene solution[224].

In his original papers on the synthesis of optically active sulphoxides Andersen[219,220] assumed that the reactions of Grignard reagents with arene(alkane)sulphinates proceed with a full inversion of configuration at the sulphinyl sulphur atom. This steric course was firmly established by Mislow[225] and other investigators[226,227]. However, it was recently found that the reactions of O-alkyl *t*-butanesulphinates with methylmagnesium iodide and O-alkyl methanesulphinates with *t*-butylmagnesium chloride are not fully stereoselective and the reactions of O-alkyl *t*-butanesulphinates with ethylmagnesium halides and O-alkyl ethanesulphinate with *t*-butylmagnesium chloride proceed with predominant retention of configuration at the sulphinyl sulphur atom[228].

In a few cases the highly stereoselective conversion of O-menthyl arenesulphinates into chiral aryl methyl sulphoxides was also achieved by means of methyllithium[229,230].

Utility of the Andersen sulphoxide synthesis is demonstrated by the preparation of optically active sulphoxides **231** and **232** where chirality is due to isotopic substitution (H → D and $^{12}C \rightarrow {}^{13}C$, respectively). The synthesis of **231** involves the reaction of non-labelled O-menthyl methanesulphinate **233** with fully deuteriated methylmagnesium iodide[241] (equation 159). In the second case, non-labelled O-menthyl phenylmethanesulphinate **234** was allowed to react with benzylmagnesium chloride prepared from benzyl chloride labelled with carbon $^{13}C$ (equation 160)[242]. Starting from sulphinate (−)-(*S*)-**221** a series of podands **235** possessing the chiral sulphur atom was prepared[243]. Thus, the compound **235a** was obtained in 34% yield when sulphinate **221** was treated with the Grignard reagent prepared from α, ω-biaryl ether **236** in benzene solution. When this reaction was carried out in THF, none of the desired products was isolated. The only product isolated was 1,7-diphenyl-1,4,7-trioxaheptane. This indicates that the expected Grignard reagent was formed, but it was unreactive towards sulphinate **221** in refluxing THF.

$$\underset{\substack{(-)(R)\\ \mathbf{(233)}}}{CH_3S(=O)OMenthyl} + CD_3MgI \longrightarrow \underset{\mathbf{(231)}}{CH_3\overset{*}{S}(=O)CD_3} \qquad (159)$$

$$\underset{(-)\text{-}(S)\text{-}\mathbf{(234)}}{Ph^{12}CH_2S(=O)OMenthyl} + Ph^{13}CH_2MgCl \longrightarrow \underset{\mathbf{(232)}}{Ph^{12}CH_2S(=O)^{13}CH_2Ph} \qquad (160)$$

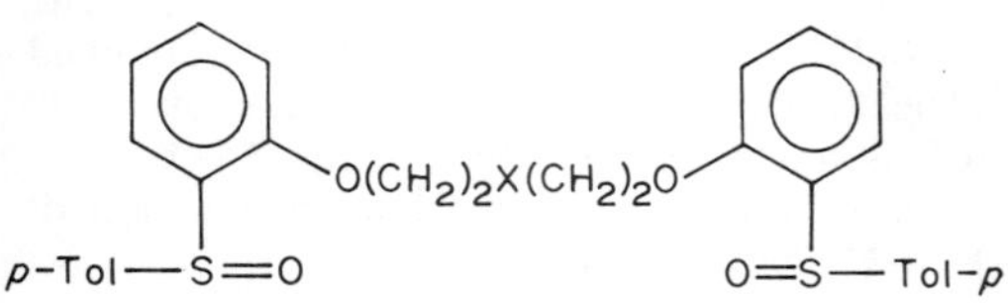

**(235)**

(**a**) X = O
(**b**) X = $NCH_3$
(**c**) X = S

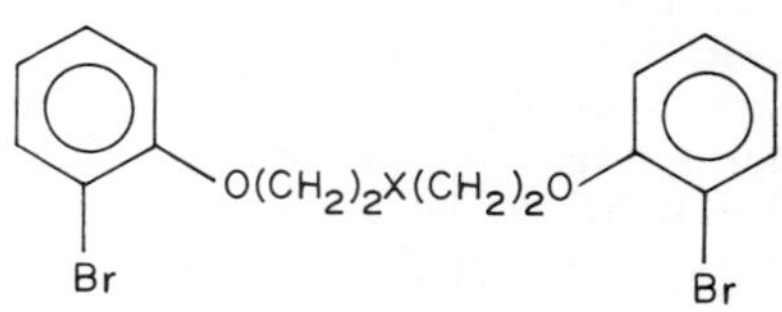

**(236)**

This approach was also applied successfully for the preparation of optically active unsaturated sulphoxides. The first example reported by Stirling and coworkers[244] involved the reaction of sulphinate **221** with the appropriate vinylic Grignard reagents. Later on, Posner and Tang[245] reported a similar preparation of a series of (*E*)-1-alkenyl *p*-tolyl sulphoxides. Posner's group performed also the synthesis of (+)-(*S*)-2-(*p*-tolylsulphinyl)-2-cyclopentanone **237**, which is a key compound in the asymmetric synthesis of various natural products[246] (equation 161).

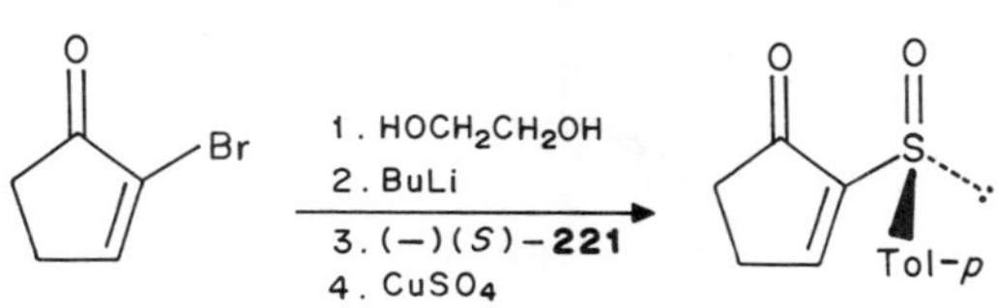

(161)

(+)-(*S*)-(**237**)

Reaction of (−)-(*S*)-**221** with allyl Grignard reagents gives optically active allylic sulphoxides **238** in which the allylic group is rearranged[247]. Mislow and his collaborators[236] evidenced that this conversion involves an allylic rearrangement via transition state **239** (equation 162).

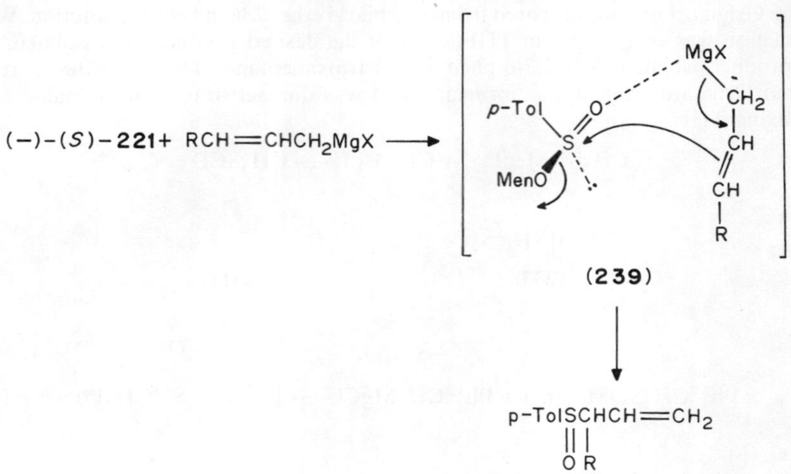


(239)

(238)

(162)

The formation of a mixture of acetylenic sulphoxides **240** and allenic sulphoxides **241** was found to occur by treatment of (−)-(*S*)-**221** with the Grignard reagents obtained from α-acetylenic halides (equation 163). The allenic sulphoxides **241** are most probably formed via the transition state **242** which is analogous to **239**. On the other hand, hex-1-ynyl *p*-tolyl sulphoxide **243** is the only product isolated from the reaction of hex-1-ynylmagnesium bromide with (−)-(*S*)-**221** (equation 164).

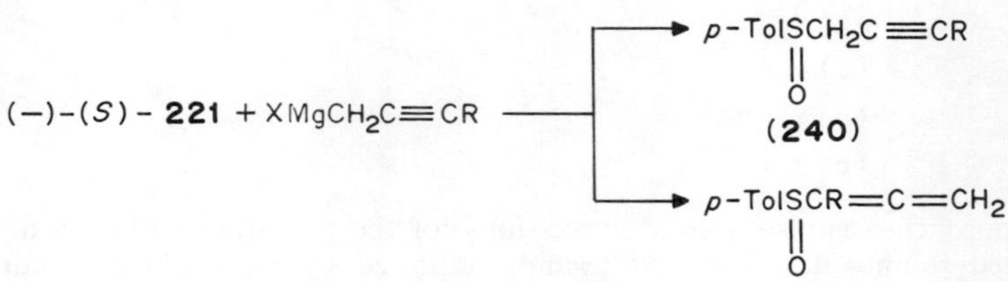


(240)

(241)

(163)

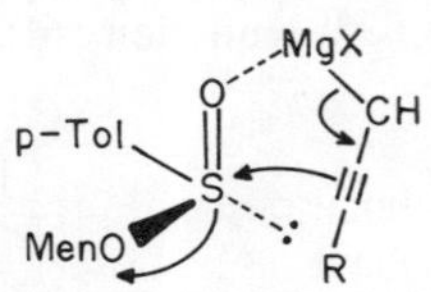


(242)

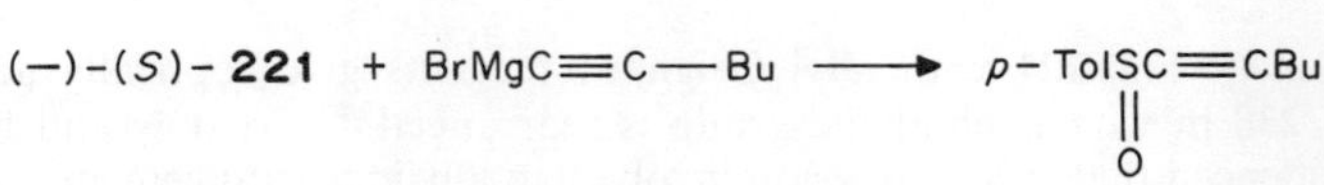


(164)

(243)

The Andersen method allows one to synthesize a variety of α-heteroatom substituted sulphoxides using α-heteroatom stabilized carbanions as nucleophiles in the reaction with (−)-(*S*)-**221**. The selected examples shown in Scheme 1 are the best illustration of the generality of this approach.

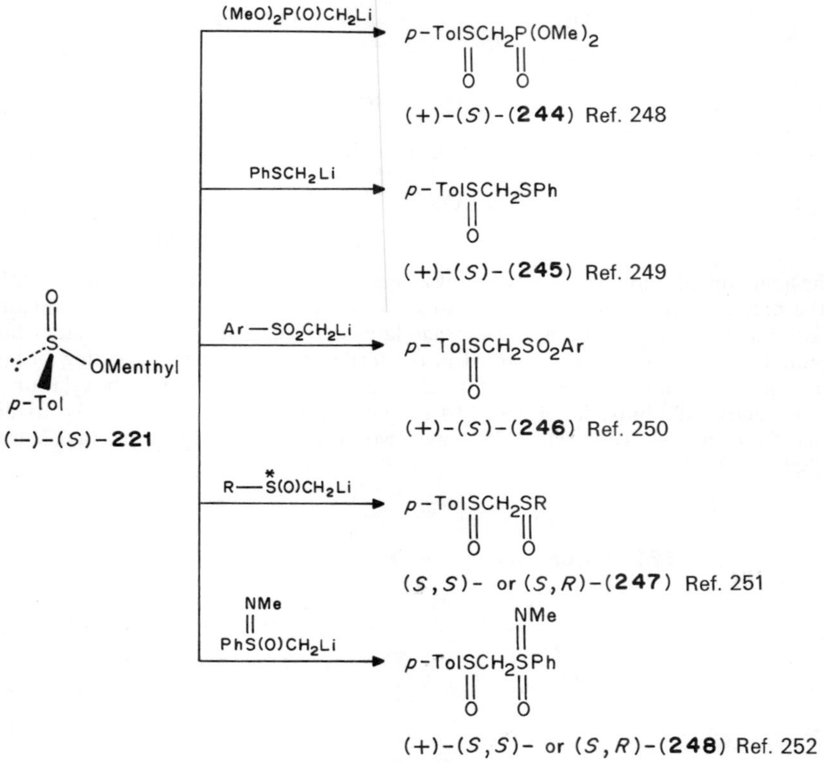

SCHEME 1

The reaction of α-cyano carbanions with (−)-(*S*)-**221** gave the corresponding α-cyanoalkyl *p*-tolyl sulphoxides (+)-(*R*)-**249** in high chemical yield and optical purity[253] (equation 165).

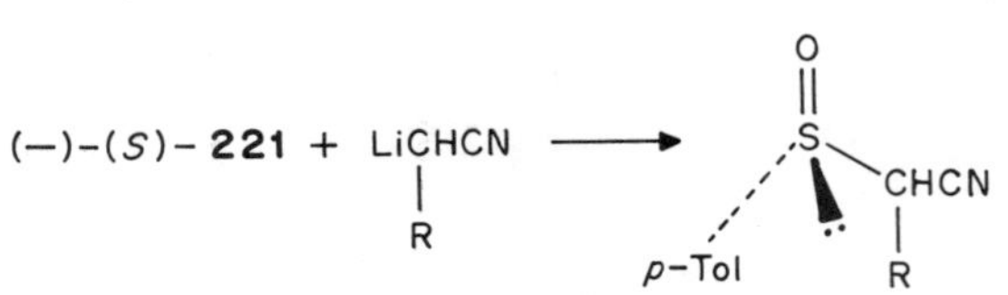

(165)

(+)-(*R*)-(**249**)

Optically active β-enamino sulphoxides **250** and/or β-iminosulphoxides **251** were found to be formed by treatment of (−)-(*S*)-**221** with α-lithiated imines. In an analogous way, optically active α-sulphinylhydrazones **252** were prepared from

(−)-(*S*)-**221** and α-metallated *N,N*-dimethylhydrazones[254].

**(250)** **(251)**

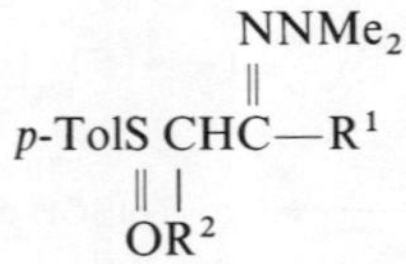

**(252)** $R^1, R^2$ = H, Alkyl, Aryl

The reaction of enolates or enolate-like species with (−)-(*S*)-**221** has been applied for the preparation of optically active α-carboalkoxy sulphoxides. Thus, treatment of this sulphinate with the halomagnesium enolates derived from *t*-butyl acetate, *t*-butyl propionate or *t*-butyl butyrate resulted in the formation of the corresponding (+)-(*R*)-*t*-butyl *p*-toluenesulphinylcarboxylates **253** (equation 166)[255]. Decarboxylation of optically active sulphinyl ketoesters **254** prepared from (−)-(*S*)-**221** and the dianion derived from methyl acetoacetate **255** gave two chiral *p*-tolylsulphinylmethyl ketones **256** (Scheme 2)[256].

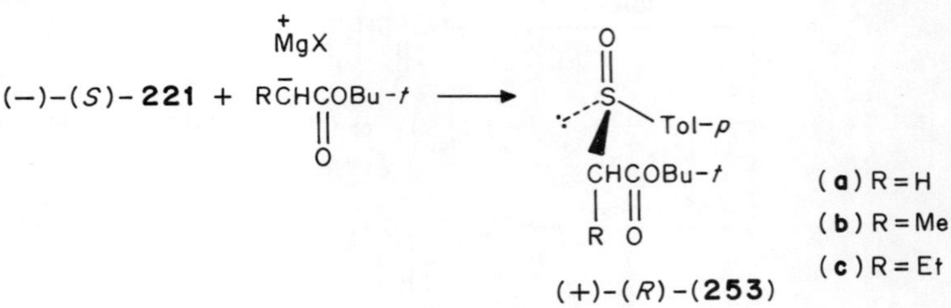


(166)

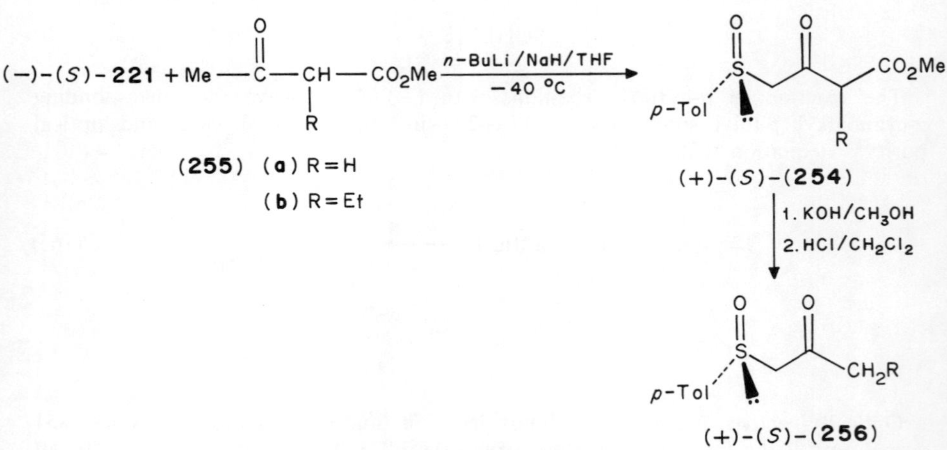


SCHEME 2

The acid-catalysed reaction of enol silyl ethers of cyclic ketones **257** with optically active O-methyl *p*-toluenesulphinate (−)-(*S*)-**258** was reported[257] as a very general entry to chiral α-sulphinylketones **259** (equation 167). It was found that the highest chemical and optical yields were obtained with boron trifluoride etherate as acidic catalyst. It should be noted that the reproducibility of this procedure has recently been questioned by a Spanish group[258]. These authors simultaneously reported[258] a new and efficient one-pot synthesis of chiral sulphinyl cyclohexanones **259b**. They found that the reaction of cyclohexanone with (−)-(*S*)-**221** in benzene, at 0 °C for 16 h, in the presence of *i*-$Pr_2NMgBr$, yielded a 75:25 mixture of both 2-*p*-tolylsulphinylcyclohexanone diastereoisomers **259b** (epimers at C-2) in 70% yield (equation 168).

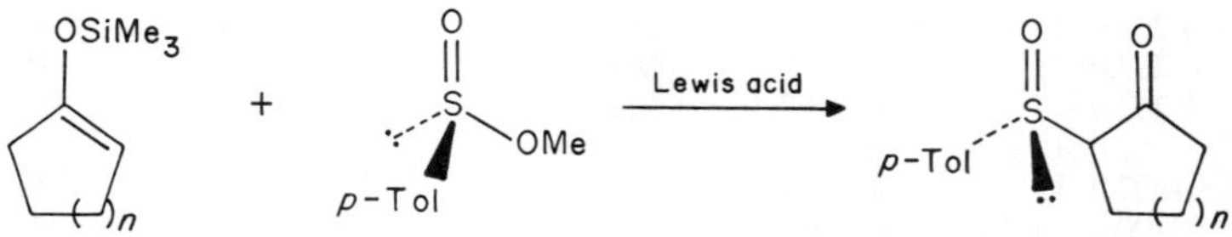


(167)

**(257)** (−)-(*S*)-**(258)** (+)-(*R*)-**(259)**

(**a**) *n* = 1
(**b**) *n* = 2
(**c**) *n* = 3

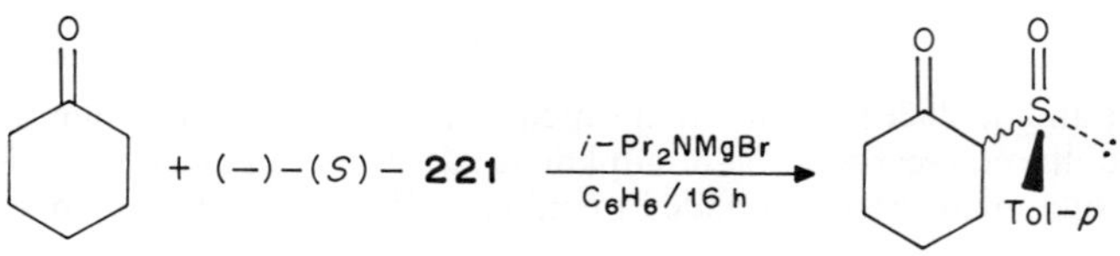


(168)

(75:25 mixture)

**(259b)**

*b. Reactions with nitrogen nucleophiles.* Direct sulphinylation of 1-trimethylsilyl-2-pyrrolidone **260** with methyl benzenesulphinate **206** was found to give the sulphoxide **261** in 67% yield[259] (equation 169). The reaction of sulphinate esters with organometallics containing the nitrogen–metal bond has no synthetic value as a method of the synthesis of racemic sulphinamides. However, this reaction has been applied successfully for the preparation of optically active compounds. Montanari and coworkers[230] showed that the reaction of (−)-(*S*)-**221** with dialkylaminomagnesium halides carried out at 0 °C in ethyl ether solution gives the corresponding optically active sulphinamides (+)-(*S*)-**262** in yields around 60% (equation 170, Table 17). This reaction is highly stereoselective and proceeds with inversion of configuration at the sulphinyl sulphur atom. Similarly, treatment of (−)-(*S*)-**221** with lithium anilide (equation 171) results in the stereospecific formation of the

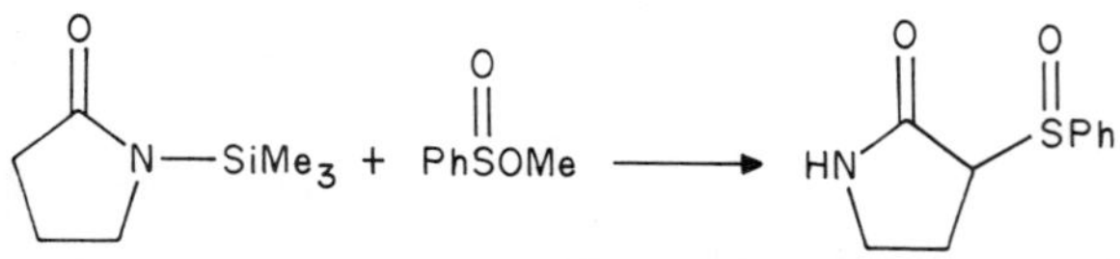


(169)

**(260)** **(206)** **(261)**

TABLE 17. Synthesis of optically active aryl sulphinamides, $ArS(O)NR^1R^2$, from diastereoisomerically pure O-menthyl arenesulphinate **221** or **224** and nitrogen containing organometallic reagent, $R^1R^2NX$

| Sulphinate | | X | Sulphinamide | | | | | | |
|---|---|---|---|---|---|---|---|---|---|
| Ar | $[\alpha]_{589}$ | | No | $R^1$ | $R^2$ | Yield (%) | $[\alpha]_{589}$ | e.e (%) | Ref. |
| *p*-Tol | −196.0 | MgBr | **a** | Me | Me | ~60 | +157.0 | ~100 | 230 |
| | −196.0 | MgBr | **b** | Et | Et | ~60 | +110.0 | ~100 | 230 |
| | −196.0 | MgBr | **c** | *i*-Pr | *i*-Pr | ~60 | +205.0 | ~100 | 230 |
| | −202.0 | Li | **d** | Ph | Me | 27 | −5.64 | 4 | 261 |
| | −202.0 | Li | **e** | *i*-Pr | Me | 26 | +85.9 | 31 | 261 |
| | −202.0 | Li | **f** | Allyl | Me | 62 | +57.3 | 45 | 261 |
| | −202.0 | Li | **g** | Allyl | Ph | 91 | −164.0 | *a* | 261 |
| | −202.0 | Li | **h** | Allyl | Allyl | 89 | +49.2 | 93 | 261 |
| | −200.6 | MgBr | **d** | Ph | Me | 70 | −109.3 | 98 | 260 |
| | −202.0 | Li | **i** | Ph | H | 60 | +199.0 | 89 | 261 |
| | −197.3 | Li | **i** | Ph | H | 41 | +216.9 | 100 | 260 |
| | −202.0 | Li | **j** | *i*-Pr | H | 66 | +167.4 | 100 | 260 |
| | −202.0 | Li | **k** | Allyl | H | 55 | +145.5 | 100 | 261 |
| | | MgBr | | $(CH_2)_5$ | | | | | 264 |
| | | Li | **l** | $c$-$C_6H_{11}$ | H | | | | 263 |
| 1-Naph | −426.5 | Li | **m** | 1-Naph | H | 34 | +561.0 | *a* | 262 |

[a]Not given.

corresponding anilide **262i** provided that an equivalent amount of lithium anilide was slowly added to the ethereal solution of sulphinate **221**. When an excess of lithium anilide was used and sulphinate ester was added to anilide salt, racemic sulphinamide **262i** was obtained[260].

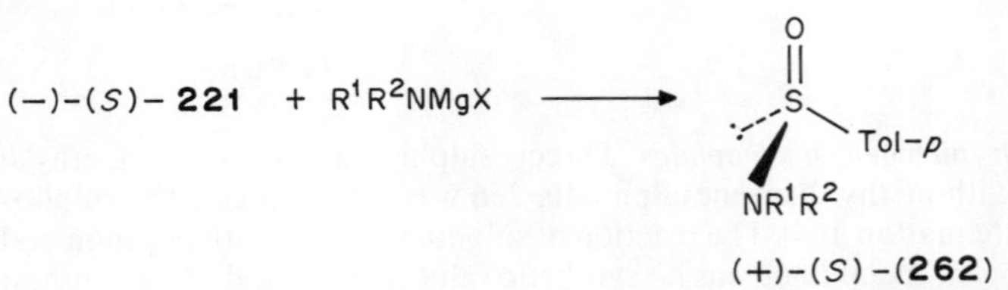

(170)

(+)-(*S*)-(**262**)

(**a**) $R^1 = R^2 = Me$

(**b**) $R^1 = R^2 = Et$

(**c**) $R^1 = R^2 = i$-Pr

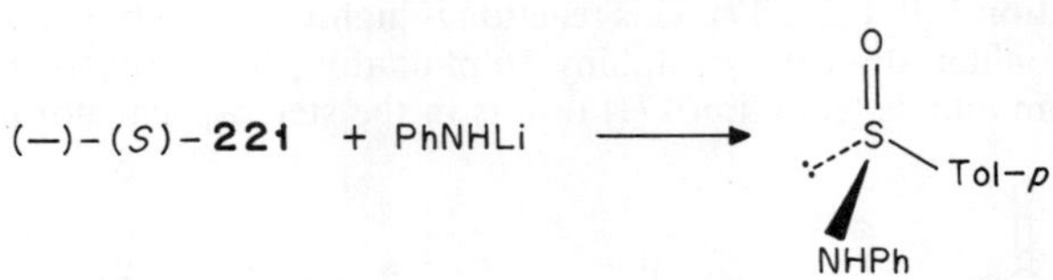

(171)

(+)-(*S*)-**262i**

The results presented above provide strong evidence that racemization of the optically active sulphinanilide **262i** is due to the anilide ion–anilide ion exchange taking place with inversion of configuration at sulphur. In this context, it is interesting to note that the reaction of two molar equivalents of the more hindered lithium salt of (−)-(*S*)-α-phenylethylamine with sulphinate **221** at 25 °C in ether solution (equation 172) affords the corresponding diastereoisomerically pure sulphinamide (+) (*S*,*S*)-**263** in 70% yield[260]. Apparently, the more hindered anion derived from α-phenylethylamine does not displace the amide ion from the sulphinamide **263** formed to give its diastereoisomer. Very recently, lithium amides were applied successfully by Colonna and Stirling[261] for the preparation of a series of optically active primary and secondary sulphinamides **262** (equation 173, Table 17). They found that stereoselectivity of this conversion is strongly influenced by the nature of substituents connected with the nitrogen atom. For example, whereas the reaction of *N*-methyllithium anilide with (−)-(*S*)-**221** gave the corresponding sulphinamide **262** with 5% e.e. only, *N*,*N*-diallylaminolithium afforded the corresponding sulphinamide **262h** having optical purity as high as 93%. The reaction of the imino-Grignard reagents, prepared *in situ* from an alkyl or aryl Grignard reagent and benzonitrile, with sulphinate (−)-(*S*)-**221** was applied for the synthesis of optically active *N*-alkylidenesulphinamides **264** (equation 174)[265]. They are formed in moderate to excellent yields and in very high stereoselectivity (Table 18). The absolute configuration at sulphur in **264** was assigned on the basis of the reasonable assumption that the reaction shown above proceeds with inversion of configuration at sulphur as had been established

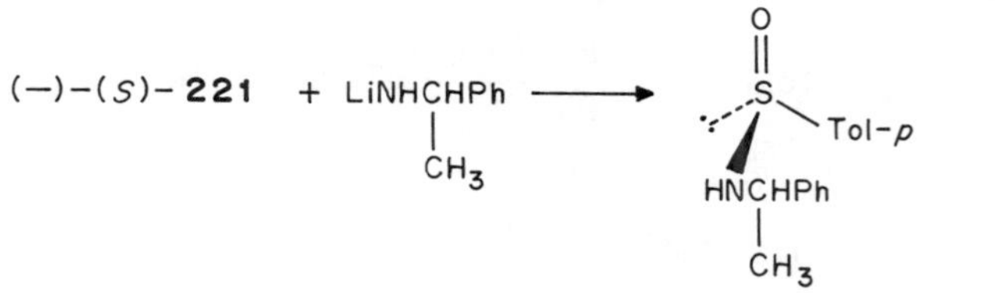

(172)

(+)-(*S*,*S*)-(**263**)

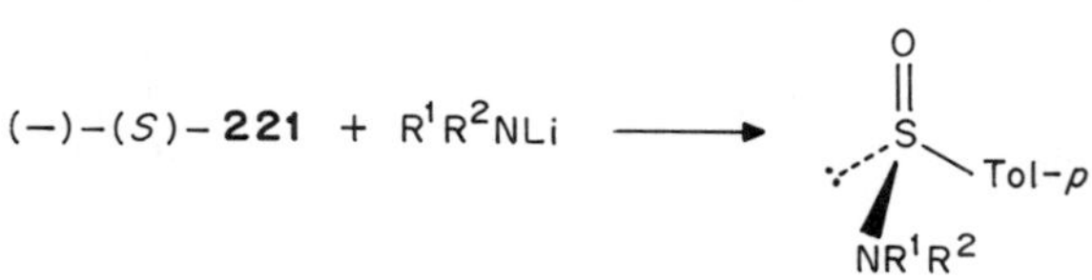

(173)

(+)- or (−)-(*S*)-(**262**)

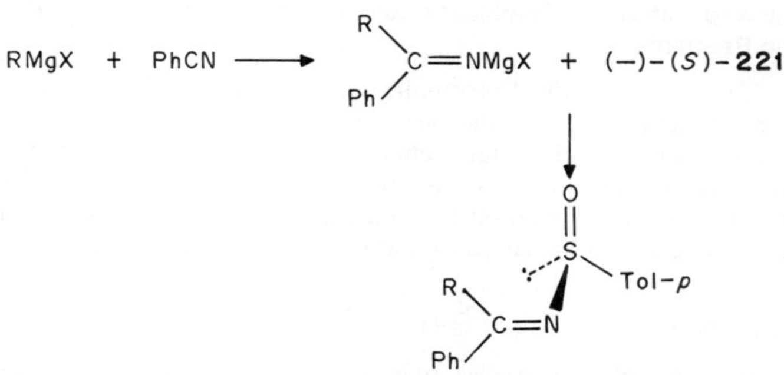

(174)

(−)-(*S*)-(**264**)

TABLE 18. Synthesis of optically active *N*-alkylidenesulphinamides, *p*-TolS(O)N=CRPh, from O-menthyl *p*-toluenesulphinate (−)-(*S*)-**221** and imino-Grignard reagents Ph(R)C=NMgX

| No. | R | $[\alpha]_{589}$ | Absilute configuration | Reference |
|---|---|---|---|---|
| **a** | Me | +98.0 | *S* | 265 |
| **b** | Et | +26.0 | *S* | 265 |
| **c** | *i*-Pr | −288.0 | *S* | 265 |
| **d** | Ph | −56.2 | *S* | 265 |
| **e** | α-Naph | | *S* | 266 |

for the reaction of Grignard reagent with sulphinate esters[207,219]. From the synthetic point of view it is interesting to note that substantial asymmetric induction is observed when optically active *N*-alkylidenesulphinamides **264** are reduced by $LiAlH_4$ (equation 175)[266]. The reduction products **265** obtained were easily cleaved to optically active amines **266** by treatment with methanol in the presence of trifluoroacetic acid[267].

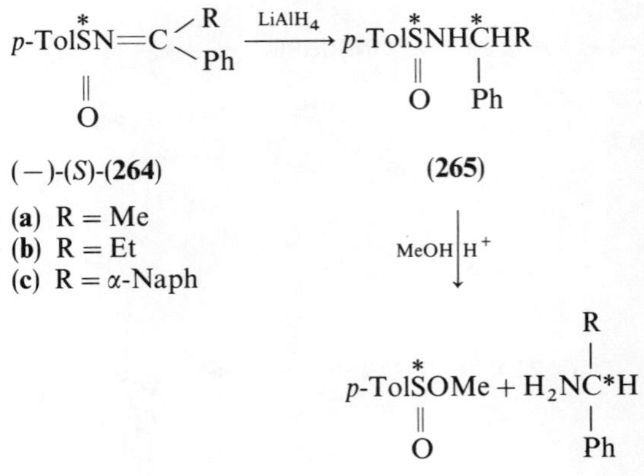

(175)

(−)-(*S*)-(**264**) (**265**) (**266**)

(**a**) R = Me
(**b**) R = Et
(**c**) R = α-Naph

## B. Synthetic Applications of Sulphinate Esters Based on Reactions with Electrophilic Reagents

Reactions of sulphinates with electrophilic reagents may be categorized in two groups depending on the site attacked by the electrophile.

Pathway A shows the reaction that occurs at the electron-rich sulphinyl oxygen atom leading to the corresponding dialkoxy sulphonium salts as final products. In pathway B the electrophilic attack is directed on the lone electron pair on sulphur, resulting in the formation of products with higher coordination number (Scheme 3).

### 1. *Synthesis of dialkoxysulphonium salts*

The first conversion of sulphinic esters into dialkoxysulphonium salts was described by Kobayashi and coworkers[268]. They were able to synthesize stable dialkoxyethane-

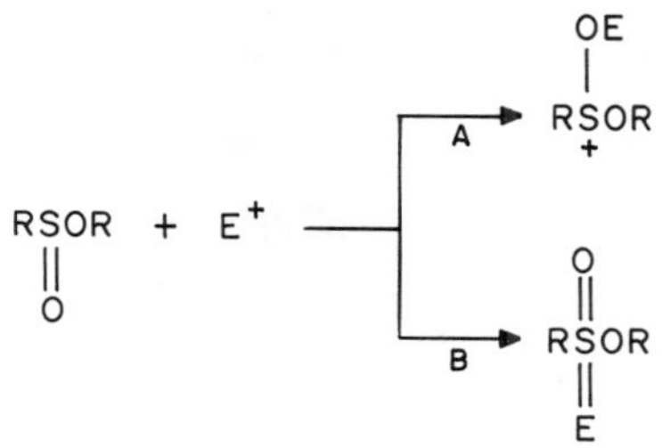


SCHEME 3

sulphonium salts **267** in quantitative yields ($^1$H-NMR assay) by alkylation of alkyl ethanesulphinates **268** with alkyl triflate **269** in nitromethane-$D_3$ (equation 176). They also found that, among the dialkoxyarylsulphonium ions **270a–c** prepared, methoxyneopentyloxy-*p*-tolylsulphonium triflate **270a** and methoxybenzyloxy-*p*-tolylsulphonium triflate **270b** decompose in solution whereas dimethoxy-*p*-tolylsulphonium triflate **270c** is stable. This method was recently employed[269] to convert a series of optically active O-alkyl isopropanesulphinates **271** into the corresponding optically active dialkoxy isopropyl sulphonium triflates **272** (equation 177). Interestingly, their hydrolysis gave a mixture of two optically active sulphinic esters **271** and **273**, both of which were formed with inversion of configuration with respect to that of the starting sulphonium salts **272** (equation 178), the first by substitution of the OR group, the second, of the OMe group.

$$\underset{\mathbf{(268)}}{\mathrm{EtS(=O)OR}} \xrightarrow[\mathbf{(269)}]{\mathrm{CF_3SO_2OR^1}} \underset{\mathbf{(267)}}{\mathrm{Et\overset{+}{S}(OR)OR^1}} \quad \mathrm{CF_3SO_3^-} \qquad (176)$$

(**a**) $R = CH_2Bu\text{-}t$ $R^1 = Et$
(**b**) $R = CH_2Bu\text{-}t$ $R^1 = Me$
(**c**) $R = Et$ $R^1 = Me$

$$\underset{\mathbf{(270)}}{p\text{-}\mathrm{Tol\overset{+}{S}(OR)OR^1}} \quad \mathrm{CF_3SO_3^-}$$

(**a**) $R = Me$ $R^1 = CH_2Bu\text{-}t$
(**b**) $R = Me$ $R^1 = CH_2Ph$
(**c**) $R = Me$ $R^1 = Me$

*2. Oxidation*

The arenesulphonate (e.g. tosylate) and alkanesulphonate (e.g. mesylate) groups are generally considered as particularly useful leaving groups in the initiation of carbonium

$$\underset{\substack{\| \\ \mathrm{O}}}{i\text{-PrSOR}} \xrightarrow{\mathrm{CF_3SO_3R^1}} \underset{\substack{| \\ \mathrm{OR}}}{i\text{-Pr}\overset{+}{\mathrm{S}}\mathrm{OR^1}} \quad \mathrm{CF_3SO_3^-} \qquad (177)$$

**(271)** **(272)**

(**a**) $R = CD_3$ $R^1 = Me$
(**b**) $R = Et$ $R^1 = Me$
(**c**) $R = Pr$ $R^1 = Me$
(**d**) $R = i\text{-}Pr$ $R^1 = Me$
(**e**) $R = Bu$ $R^1 = Me$
(**f**) $R = CH_2Bu\text{-}t$ $R^1 = Me$

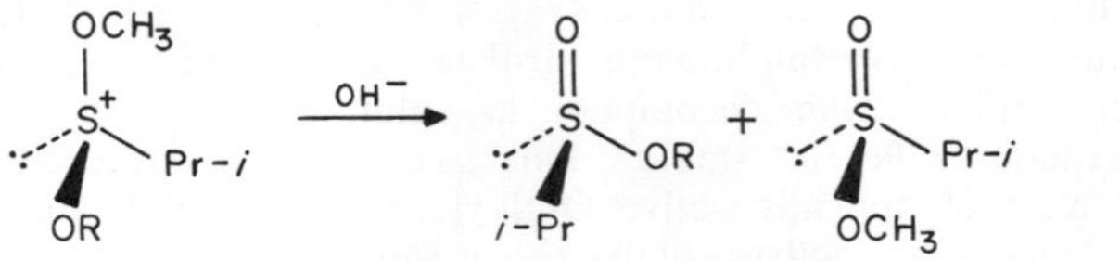


(178)

(*R*)-(**272**) (−)-(*S*)-(**271**) (+)-(*R*)-(**273**)

ion reactions. Their reactivity is comparable to the corresponding halide, and in addition the sulphonate derivative has the advantage of generally being prepared from an alcohol without affecting the C—O bond, thus preserving the stereochemistry at the carbinyl position. Unfortunately, the synthetic method most frequently employed for the preparation of sulphonates, namely the reaction of alcohols with sulphonyl chloride in pyridine, is generally not suitable for either reactive or hindered structures. For this reason, the oxidation of sulphinic acid esters, which in many cases are more easily made and handled, may be considered as a useful synthetic procedure for the preparation of the corresponding sulphonate esters starting from the alcohol (equation 179). For the first time sulphinate esters were oxidized to the corresponding sulphonic derivatives by Phillips and coworkers[270] as early as 1933. These authors found that diastereoisomeric O-α-phenylethyl *p*-toluenesulphinate **274**, although relatively unstable, underwent oxidation with ease, when treated with urea hydroperoxide (equation 180). However, the resulting sulphonate **275** was so reactive that its isolation in the pure state was not possible. They also found that oxidation of sulphinate **274** to sulphonate **275** by air in benzene solution was accompanied by its isomerization into the corresponding *p*-tolyl α-phenylethyl sulphone. Later they also reported[271] oxidation of ethyl (+)-α-*p*-toluenesulphinoxypropionate **276** to the corresponding sulphonate **277** with anhydrous potassium permanganate (equation 181). For stereochemical studies, of great importance was the oxidation of diastereoisomerically pure O-menthyl phenylmethanesulphinate **226** with *R* chirality at sulphur by potassium permanganate containing 90.2% of oxygen $^{18}O$. This reaction was found to be stereospecific and gave the corresponding (−)-*O*-menthyl phenylmethanesulphonate **278** in which the sulphonyl group is chiral due to the presence of two different isotopes of oxygen[272] (equation 182).

$$\underset{\substack{\| \\ \mathrm{O}}}{\mathrm{RSCl}} + \mathrm{R^1OH} \longrightarrow \underset{\substack{\| \\ \mathrm{O}}}{\mathrm{RSOR^1}} \xrightarrow{\text{oxidation}} \overset{\substack{\mathrm{O} \\ \|}}{\underset{\substack{\| \\ \mathrm{O}}}{\mathrm{RSOR^1}}} \qquad (179)$$

$$p\text{-TolS(O)CH(Me)Ph} \xrightarrow[\text{hydroperoxide}]{\text{urea}} p\text{-TolS(O)}_2\text{CH(Me)Ph} \qquad (180)$$

(274) (275)

$$p\text{-TolS(O)CH(Me)COOEt} \xrightarrow{\text{KMnO}_4} p\text{-TolS(O)}_2\text{CH(Me)COOEt} \qquad (181)$$

(276) (277)

$$\text{PhCH}_3\text{S(=O}^{16}\text{)OMenthyl} \xrightarrow{\text{KMn}^{18}\text{O}_4} \text{PhCH}_2\text{—S(=O}^{16}\text{)(=O}^{18}\text{)—OMenthyl} \qquad (182)$$

(226) (−)-(278)

$[\alpha]_{589} = +105$

Coates and Chen[273] have found that *m*-chloroperbenzoic acid in methylene chloride converts *p*-toluenesulphinates into tosylates and that this oxidation procedure allows one to prepare a variety of unstable and hindered tosylates (see Table 19). A good example is the preparation of *p*-toluenesulphonate ester **279** derived from longicamphenylol **280**, a case in which tosyl chloride in pyridine (and several other modifications) gave only hydrocarbon products (equation 183). This oxidation method also has limitations and failed in the case of reactive sulphinate derivatives such as *t*-butyl, *p*-methoxybenzyl and benzhydryl. In the latter case the corresponding sulphone was obtained in 93% yield, thus the internal rearrangement apparently exceeded the rate of oxidation.

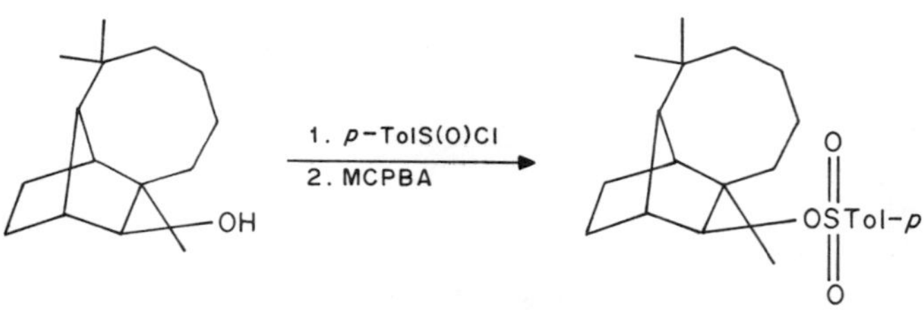


(280) (279) (183)

A few mesylates **281** containing the diethyl phosphonate substituents, for which the direct mesylation procedure using mesyl chloride and triethylamine was not successful, could be prepared by this two-step procedure[275] (equation 184). Treatment of alcohols **282** with methanesulphinyl chloride and triethylamine led to the sulphinate esters **283**, which were oxidized *in situ* to the corresponding mesylates **281** by *m*-chloroperbenzoic acid. The yields of sulphonates **281a–c** were 91, 73 and 94%, respectively.

Oxidation of sulphinic acid esters **284** and **285** containing a bulky alkyl group with perfluoroacetic acid in methylene chloride was found (equations 185 and 186) to be the effective way (see Table 20) for the preparation of the corresponding sulphonic esters **286**

TABLE 19. Oxidation of *p*-toluenesulphinates, *p*-TolS(O)OR, into the corresponding *p*-toluenesulphonates, *p*-TolSO$_2$OR, with *m*-chloroperbenzoic acid

| R | Time (h) | Temperature (°C) | Yield (%) | Reference |
|---|---|---|---|---|
| *c*-$C_6H_{11}$ | 1.5 | 0 | 84 | 273 |
| $CF_3$(Me)CH | 2.4 | 0 | 75 | 273 |
| $PhCH_2$ | 1.5 | 0 | 87 | 273 |
| PhCHMe | 1.5 | 0 | 72 | 273 |
| 2-Bicyclo[3.1.1]heptyl | 1.5 | 0 | 82 | 273 |
| Adamantyl | 1.5 | 0 | *a* | 274 |

[a]Not given.

and **287**, which could not be prepared by direct esterification of sulphonyl chloride[276] due to steric hindrance.

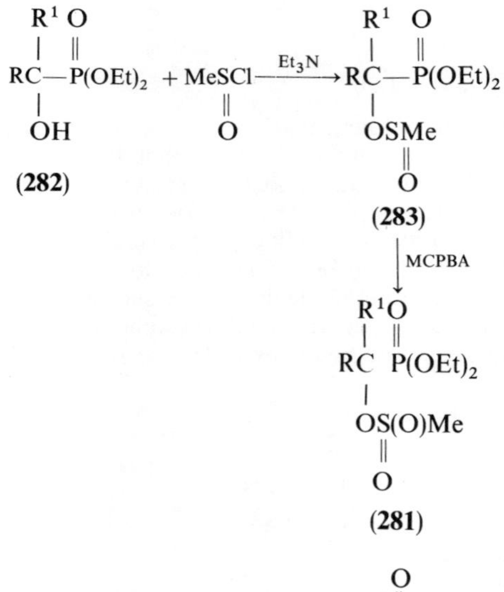

(184)

(**282**) (**283**) (**281**)

(**a**) R = Ph, R$^1$ = Me
(**b**) R = R$^1$ = Me
(**c**) R + R$^1$ = $(CH_2)_5$

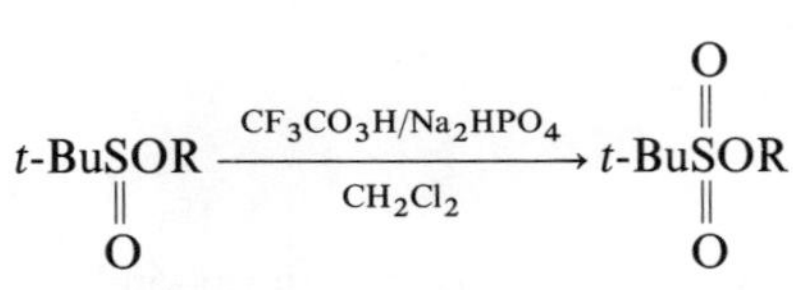

(185)

(**284**) (**286**)

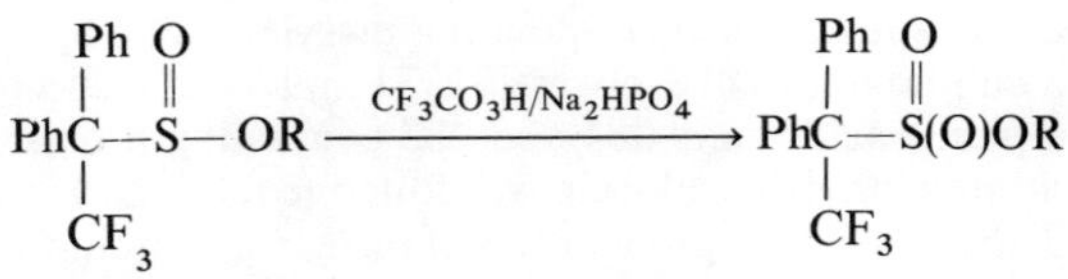

(186)

(**285**) (**287**)

TABLE 20. Oxidation of sulphinates **284** and **285** to the corresponding sulphonates **286** and **287** with perfluoroacetic acid[a]

| No. | R | Yield of **286** or **287** (%) |
|---|---|---|
| **284a** | *t*-$BuCH_2$ | 65 |
| **284b** | $PhCH_2OCH_2CH_2$ | 84 |
| **284c** | 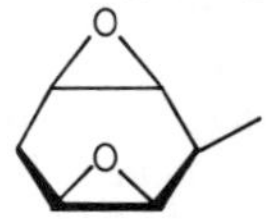 | 89 |
| **284d** | 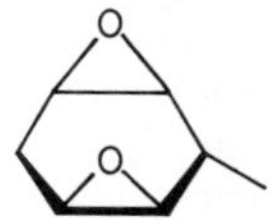 | 90 |
| **284e** | 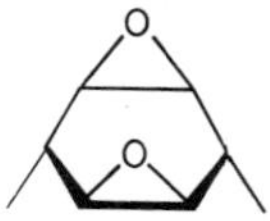 | 85 |
| **284f** | 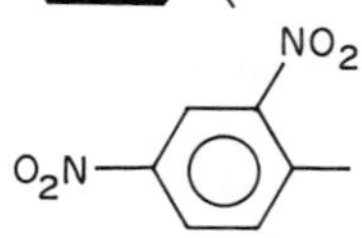 | 79 |
| **285a** | *t*-$BuCH_2$ | 98 |
| **285b** | $PhCH_2OCH_2CH_2$ | 100 |
| **285c** | 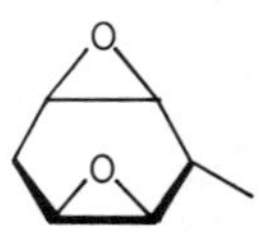 | 94 |

[a]Taken from Reference 276.

Oxidation of the unsaturated cyclic sulphinate ester **288** formed by the thermal rearrangement of thiet-1,1-dioxide **289** using hydrogen peroxide in glacial acetic acid gave the corresponding sulphonate ester **290** in high yield[277] (equation 187).

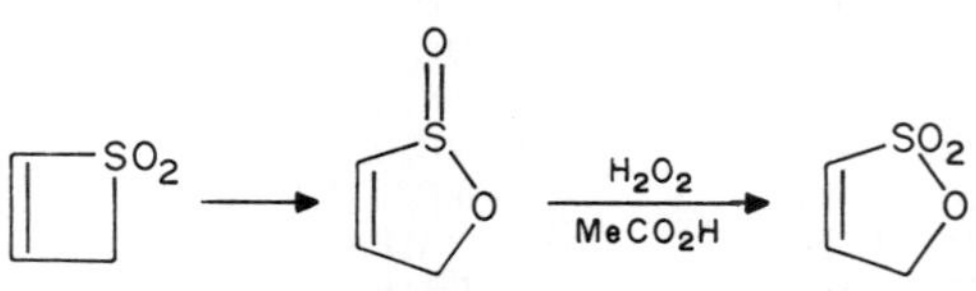


(289) (288) (290) (187)

## C. Synthetic Applications of Sulphinate Esters Based on Rearrangements

### 1. *Rearrangements of alkyl and benzyl sulphinates to sulphones*

Kenyon and Phillips[278] in 1930 first reported that *O*-α-phenylethyl *p*-toluenesulphinate **274** rearranged on standing to the corresponding α-phenylethyl *p*-tolyl sulphonate **291** (equation 188).

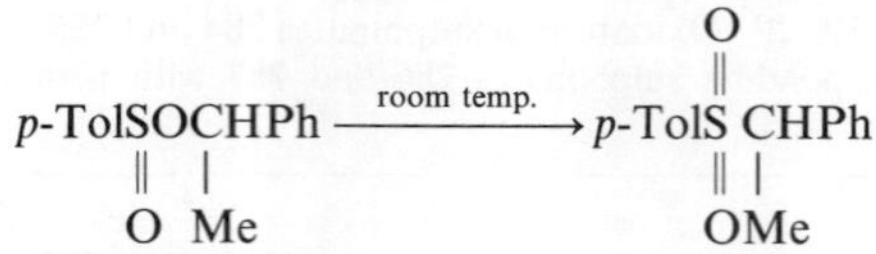

(274) (291) (188)

Soon after, Kenyon and coworkers[279] and later on Stevens and coworkers[280] investigated the rearrangement of a number of sulphinates to sulphones (see Table 21) and suggested an intermolecular ionic mechanism for this reaction. More recently, Darwish and coworkers[281] carried out more detailed studies on the rearrangement of *t*-butyl, α-phenylethyl, α-(*p*-methoxyphenyl)ethyl, benzhydryl, 2-aryl-2-propyl and trityl 2,4-dimethylbenzenesulphinates under a variety of conditions (see Table 21) and have shown that the important route to the sulphone formation involves ion-pair recombination of free ions as shown in equation 189, where $R^1SO_2^- R^+$ is a non-capturable intimate ion pair and $R^1SO_2^- \| R^+$ is a capturable solvent separated ion pair. This mechanistic proposal was supported later by related studies carried out by Fava and coworkers[283] on isomerization of optically active benzhydryl *p*-toluenesulphinates and by Braverman and

**TABLE 21.** Rearrangement of sulphinic esters, $RSOCR^1R^2R^3$, to Sulphones, $RSCR^1R^2R^3$

| Sulphinate | | | | | | |
|---|---|---|---|---|---|---|
| R | $R^1$ | $R^2$ | $R^3$ | Conditions | Yield of sulphone (%) | Reference |
| *p*-Tol | H | H | H | neat/160 °C | 0 | 280 |
| *p*-Tol | H | Ph | Ph | neat/160 °C | 0 | 280 |
| *p*-Tol | H | H | *o*-$NO_2C_6H_4$ | neat/160 °C | 0 | 280 |
| *p*-Tol | H | Me | Ph | neat/160 °C | 20 | 280 |
| *p*-Tol | Me | Me | Ph | neat/160 °C | 87 | 280 |
| *p*-Tol | H | Ph | Ph | neat/160 °C | 100 | 280 |
| *p*-Tol | Ph | Ph | *p*-$ClC_6H_4$ | neat/160 °C | 95 | 280 |
| *p*-Tol | Ph | H | *o*-Tol | neat/160 °C | 95 | 280 |
| *p*-Tol | Ph | H | *p*-Tol | neat/160 °C | 80 | 280 |
| *p*-$ClC_6H_4$ | H | Ph | Ph | neat/160 °C | 100 | 280 |
| *o*-Tol | Ph | Ph | Ph | $CHCl_3$/reflux | ~100 | 282 |
| 2,4-$Me_2C_6H_3$ | Me | Me | Me | 60% EtOH/70 °C | 1 | 281 |
| 2,4-$Me_2C_6H_3$ | H | Ph | Ph | 90% Dioxane/70°C | 68 | 281 |
| 2,4-$Me_2C_6H_3$ | H | Ph | Me | 60% EtOH/70 °C | 12 | 281 |
| 2,4-$Me_2C_6H_3$ | H | Me | *p*-An | 60% EtOH/70 °C | 27 | 281 |
| $CF_3$ | H | H | $C_6H_{13}$ | HMPA/145 °C | 87 | 286 |
| $CF_3$ | H | H | $CH_2Ph$ | DMF/155 °C | 56 | 286 |
| $CF_3$ | H | H | $CH_2CH_2Ph$ | HMPA/145 °C | 71 | 286 |
| $CF_3$ | H | H | $CH_2CH_2CH{=}CH_2$ | HMPA/145 °C | 78 | 286 |
| $CCl_3$ | H | H | Ph | $CH_3CN$/100 °C | H[a] | 285 |
| $CCl_3$ | H | H | *p*-Tol | $CH_3CN$/100 °C | H[a] | 285 |
| $CCl_3$ | H | H | *p*-$ClC_6H_4$ | $CH_3CN$/100 °C | H[a] | 285 |
| $CCl_3$ | H | H | *p*-$NO_2C_6H_4$ | $CH_3CN$/100 °C | H[a] | 285 |

[a]Exact value not given; H denotes high yield.

coworkers[284], who studied the solvolysis and rearrangement of several benzyl and furfuryl sulphinates. An inspection of the results summarized in Table 21 clearly indicates that no sulphone was obtained from either methyl or benzyl *p*-toluene sulphinates. On the other hand, O-benzyl trichloromethanesulphinates **292** undergo a facile and efficient rearrangement to sulphones **293** on heating in strongly polar solvents such as acetonitrile or nitromethane (equation 190)[285].

$$\mathrm{R^1\overset{\overset{\displaystyle O}{\|}}{S}OR} \rightleftharpoons \mathrm{R^1\overset{\overset{\displaystyle O}{\|}}{S}O^-\,R^+} \rightleftharpoons \mathrm{R^1\overset{\overset{\displaystyle O}{\|}}{S}O^-\,\|\,R^+} \longrightarrow \mathrm{R^1\underset{\underset{\displaystyle O}{\|}}{\overset{\overset{\displaystyle O}{\|}}{S}}R} \qquad (189)$$

$$\mathrm{ArCH_2O\underset{\underset{\displaystyle O}{\|}}{S}CCl_3} \xrightarrow{\mathrm{CH_3Cn}} \mathrm{ArCH_2\underset{\underset{\displaystyle O}{\|}}{\overset{\overset{\displaystyle O}{\|}}{S}}CCl_3} \qquad (190)$$

**(292)** **(293)**

Hendrickson and Skipper[286] reported independently that simple primary trifluoromethanesulphinates **294** rearrange cleanly to the corresponding sulphones **295** (equation 191) on heating at 145 °C for 4 h in hexamethylphosphoramide or in dimethylformamide at 155 °C. However, in a less polar solvent like diglyme or sulpholane these sulphinates do not undergo conversion to sulphones.

$$\mathrm{CF_3\underset{\underset{\displaystyle O}{\|}}{S}OR} \xrightarrow[\text{or HMPT}]{\text{DMF}} \mathrm{CF_3\underset{\underset{\displaystyle O}{\|}}{\overset{\overset{\displaystyle O}{\|}}{S}}R} \qquad (191)$$

**(294)** **(295)**

It is interesting to note that benzyl ester **296** rearranges to sulphone **297** during its preparation at 0 °C[284b,c] by the oxidation of the corresponding sulphenyl derivative **298** (equation 192).

$$\mathrm{CF_3SOCH_2Ph} \xrightarrow{\text{MCPBA}} \left[\mathrm{CF_3\overset{\overset{\displaystyle O}{\|}}{S}OCH_2Ph}\right] \longrightarrow \mathrm{CF_3\underset{\underset{\displaystyle O}{\|}}{\overset{\overset{\displaystyle O}{\|}}{S}}CH_2Ph} \qquad (192)$$

**(298)** **(296)** **(297)**

In this context, the rearrangement of furfuryl benzenesulphinate **299** is of special interest. In contrast to the corresponding benzyl ester, sulphinate **299** undergoes thermal rearrangement under buffered non-solvolytic conditions to a mixture of furfuryl phenyl sulphone **300** and 2-methyl-3-furyl phenyl sulphone **301** (equation 193)[284a].

Other types of rearrangements of cyclic benzylic sulphinates have also been described in the literature. Durst and coworkers[287] have found that sultines **302**, when heated in

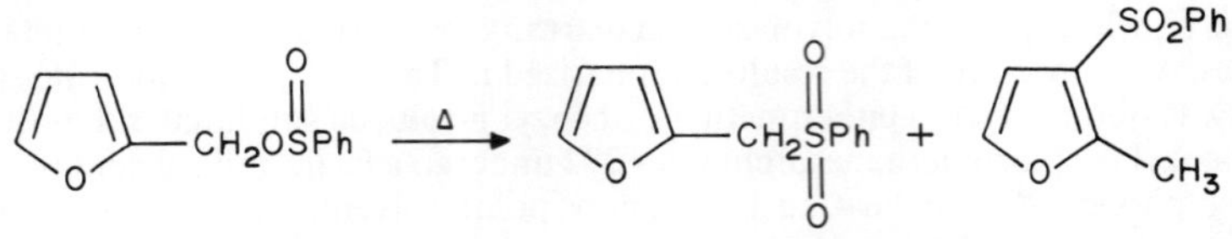

(193)

(299) (300) (301)

refluxing benzene, undergo isomerization to 1,3-dihydrobenzo[*c*]thiophene-2,2-dioxide **303**. In this case a two-step mechanism was proposed which involves retro Diels–Alder extrusion of $SO_2$ from **302** followed by a typical cycloaddition of $SO_2$ to the 1,3-diene **305** (Scheme 4). When a very reactive dienophile such as maleic anhydride was present in the reaction mixture, the tetrahydronaphthalene derivative **304a** was obtained in *ca* 95% yield.

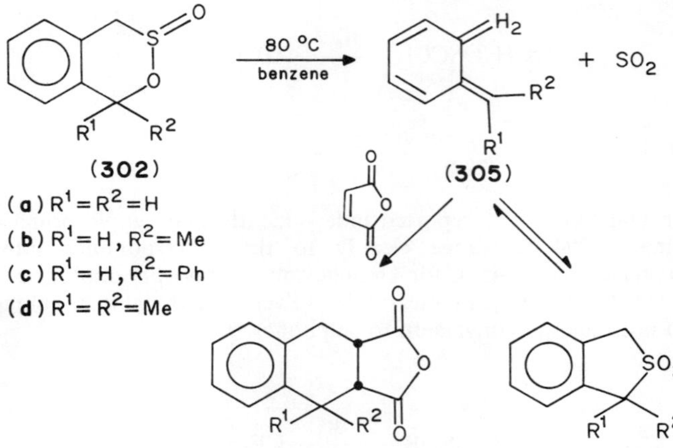

SCHEME 4

Another interesting benzylic-type rearrangement of sulphinates was reported by Heldeweg and Hogeveen[288], who found that cyclic sulphinate **306** [formed *in situ* by the kinetically-controlled [(2 + 4)($\pi + \pi\pi$)] cycloaddition of sulphur dioxide to tricyclic diene **307**] rearranges thermally in two different ways. The first preferred direction leads to aromatic cyclic sulphinate ester **308** while the second leads to sulphone **309** as shown in Scheme 5.

*2. [2,3] Sigmatropic rearrangements of allylic and propargylic sulphinates to sulphones*

This topic has been discussed exhaustively by Braverman[289] in his excellent review on rearrangements involving sulphones. Therefore, only selected examples of general importance illustrating the synthetic utility of the title reactions will be presented here.

The first attempts to convert thermally allylic sulphinates into sulphones were described by Cope and collaborators[290] in 1950. However, the group of Braverman[291] demonstrated that the rearrangement of allylic arenesulphinates **312** to sulphones **313** is a general reaction (equation 194) and clarified the mechanistic and stereochemical features of this reaction. First, Darwish and Braverman[291] found that the above rearrangement may occur thermally or under solvolytic conditions. Secondly, they found that the rearrangement of unsymmetrically substituted allylic esters involves simultaneous isomerization of

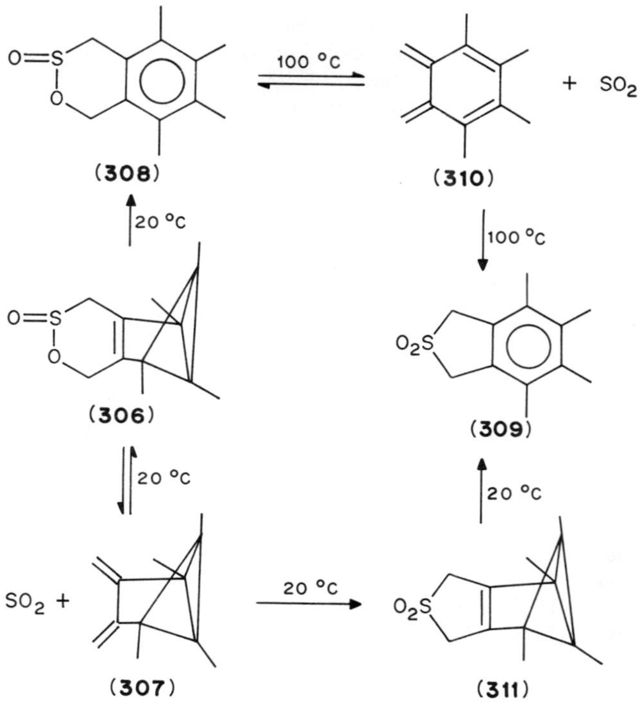


SCHEME 5

the allylic group. Moreover, these authors showed that the conversion of optically active α,γ-dimethylallyl 2,4-dimethylbenzenesulphinate **314** to the corresponding sulphone **315** is accompanied by a full inversion of configuration (equation 195). Interestingly, this reaction represents also the first example of a 1,3-chirality transfer from carbon to carbon.

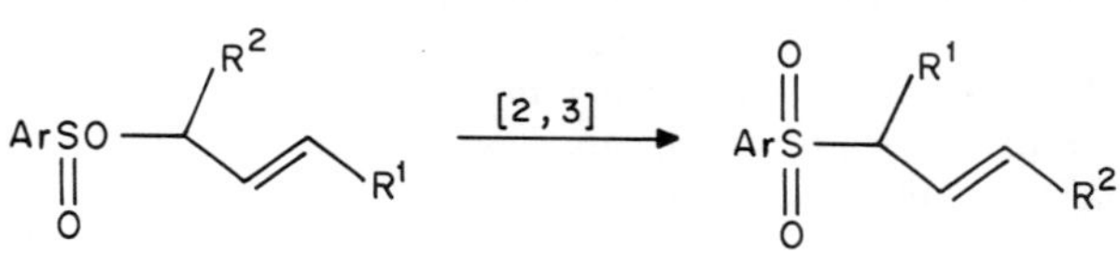


(**312**) (**313**) (194)

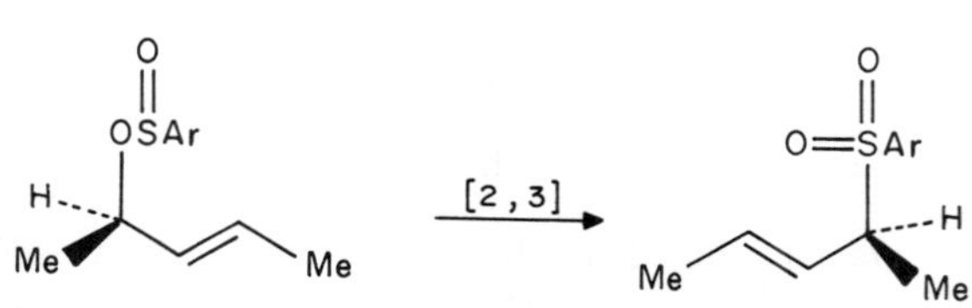


(**314**), $[\alpha]_D = +3.86°$ (**315**), $[\alpha]_D = +7.02°$ (195)

Hiroi and coworkers[292] were the first to report the chirality transfer from sulphur to carbon in the allylic sulphinate to sulphone rearrangement. They prepared a series of optically active *trans*- and *cis*-allyl *p*-toluenesulphinates **316** and investigated their conversion to the corresponding sulphones **317** (equation 196). It was found that all the *trans*-allyl sulphinates gave, on heating at 90–100 °C in *N,N*-dimethylformamide, the corresponding sulphones with the *S*-absolute configuration at the α-carbon atom while, from *cis*-isomers, the sulphones with the *R*-absolute configuration were formed. The stereoselectivity of the rearrangement was observed to be higher than 80%.

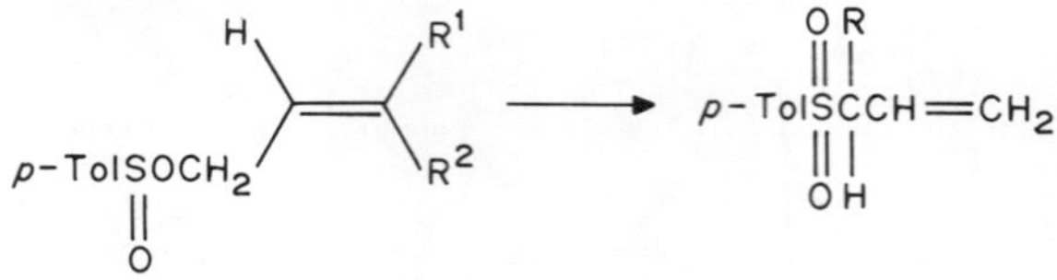


(196)

(−)-(*S*)-(**316**) (**317**)

| | $R^1$ | $R^2$ |
|---|---|---|
| **(a)** | Me | H |
| **(b)** | H | Me |
| **(c)** | Pr | H |
| **(d)** | H | Pr |
| **(e)** | $C_5H_{11}$ | H |
| **(f)** | H | $C_5H_{11}$ |
| **(g)** | Ph | H |

| | R |
|---|---|
| **(a)** | Me |
| **(b)** | Pr |
| **(c)** | $C_5H_{11}$ |
| **(d)** | Ph |

Later, Hiroi's group[293] found that sulphinates **316** undergo very easy conversion to sulphones in the presence of catalytic amounts of palladium complexes. Thus, heating *trans*-(−)-(*S*)-**316a** at 50 °C for 10 h in THF in the presence of tetrakis(triphenylphosphine)palladium **318** (0.15 molar equivalent) gave (+)-(*S*)-sulphone **317a** in 74% yield and with stereoselectivity higher than 90%. The rearrangement of **316b** under the same conditions resulted in the formation of the sulphone (−)-(*R*)-**317a** in 69% yield and with stereoselectivity equal to 86.4%. In both cases the α-rearranged sulphones were also formed. Much lower regioselectivity and stereoselectivity of the palladium-catalysed rearrangement was observed with other sulphinates **316**, which have bulky substituents connected with the α-carbon atom of the allyl moiety. These differences were rationalized in terms of the transition state having more ionic character than that for the typical [2,3]-sigmatropic rearrangement.

Taking into account the well-known fact that the stereospecific replacement of the carbon–sulphur bond in allyl sulphones by the carbon–carbon bond is catalysed by palladium complexes[294], Hiroi and coworkers[295] extended their studies on palladium-catalysed isomerization of chiral allylic sulphinates to the corresponding sulphones and found proper conditions for the direct conversion of optically active allyl sulphinates to optically active dimethyl 1-buten-3-yl-malonates (equation 197). Thus, the reaction of the sulphinate (−)-(*S*)-**316a** with sodium dimethyl malonate carried out in the presence of **318** in refluxing tetrahydrofuran for 10 h gave a 1:1 mixture of (+)-(*S*)-dimethyl 2-buten-3-ylmalonate **319a** and dimethyl 2-butenylmalonate **320** in 75% yield. The stereoselectivity of the **316a**–**319a** conversion was estimated as 83%. When the *cis*-sulphinate **316b** was reacted with sodium dimethyl malonate under the same conditions, the enantiomeric (−)-(*R*)-**319a** was formed with 75% stereoselectivity. With other sulphinates **316** the palladium-catalysed allylation occurred with much lower stereoselectivity. It is obvious that the first step of the reaction under discussion involves rearrangement of allyl sulphinate to allyl sulphone which undergoes, in turn, alkylation.

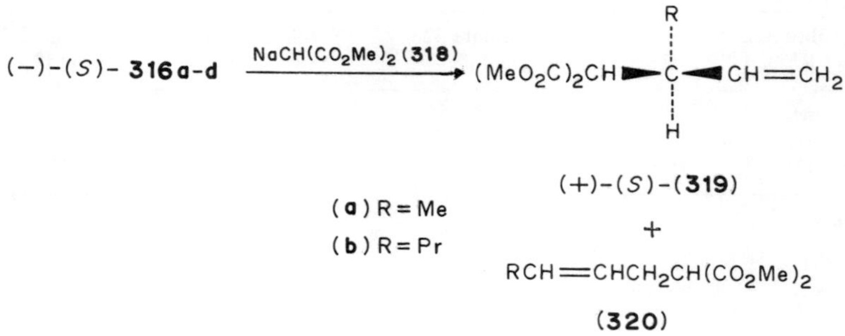

(197)

Grieco and Boxler[296] utilized the allylic sulphinate–sulphone rearrangement in their synthesis of conjugated dienoic acids shown in Scheme 6.

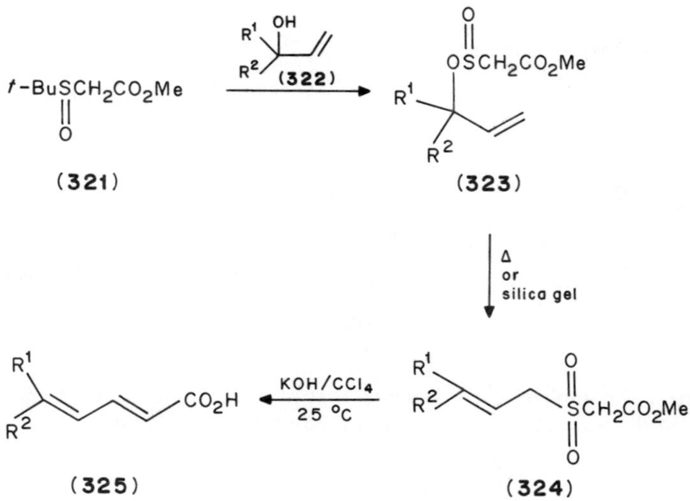

SCHEME 6

The starting allyl sulphinate **323** was prepared from methyl *t*-butylsulphinylacetate **321** according to the procedure elaborated by Jung and Durst[297]. Grieco and Boxler found that the isomerization of **323** to sulphone **324** may be effected by heating at 100 °C or by stirring methylene chloride or benzene–ethyl acetate solution over silica gel at room temperature. The Ramberg–Bäcklund reaction of **324** afforded dienoic acids **325**. The experimental data are summarized in Table 22.

The [2,3]-sigmatropic rearrangement of allylic sulphinates to sulphones is a key step in the synthesis of the dihydroxyacetone derivatives developed by Baldwin and collaborators[298]. Scheme 7 illustrates the most important features of this approach.

Braverman and Mechoulam[299] continued studies on the sulphinate rearrangements and found that propargylic arenesulphinates **326** undergo isomerization to allenyl aryl sulphones **327** (equation 198, Table 23). Independently, this type of rearrangement has also been reported by Stirling and Smith[300]. The mechanism and stereochemistry of this

TABLE 22. Synthesis of dienoic acid **325** via the allylic sulphinate **323** to allylic sulphone **324** rearrangement[a]

| No. | Alcohol **322** $R^1$ | $R^2$ | Sulphinate **323** yield (%) | Sulphone **324** yield (%) | Dienoic acid **325** yield (%) |
|---|---|---|---|---|---|
| **a** | Me | Me | 43 | 50 | 80 |
| **b** | H | Me | 97 | 93 | 63 |
| **c** | H | $C_5H_{11}$ | 92 | 83 | 82 |

[a]Taken from Reference 296.

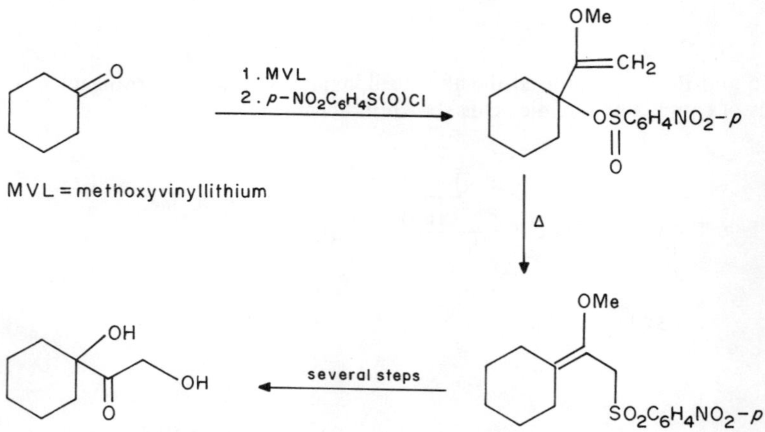


SCHEME 7

rearrangement is completely analogous to that of allylic sulphinates. A more detailed discussion on this subject may be found in the review by Braverman[289].

$$\underset{(326)}{\mathrm{ArS(O)OC(R^1)(R^2)C{\equiv}CH}} \longrightarrow \underset{(327)}{\mathrm{ArS(O)_2CH{=}C{=}C(R^1)(R^2)}} \qquad (198)$$

## D. Miscellaneous Synthetic Applications of Sulphinate Esters

Pyrolysis of sulphinates, like that of carboxylic acid esters[301], xanthates[302], amine-oxides[303] and sulphoxides[304], results in the formation of products of *cis*-elimination (equation 199). Jones and Higgins[305] found that diastereomeric (at sulphur) methyl 5α-cholestane-6 β-sulphinates **328** undergo pyrolytic *syn*-elimination, the rate of which depends upon the chirality at sulphur. Thus, the sulphinate **328** with the *R*-chirality at sulphur gave 5α-cholest-6-ene **329** in 70% yield on boiling in decalin for 16 h, whereas its diastereomer (*S*)-**328** under the same experimental conditions gave only 5% of the olefin, the remainder being starting material (equation 200).

TABLE 23. Rearrangement of propargyl arenesulphinates **326**, $RC{\equiv}CCR^1R^2OS(O)Ar$, to allenyl aryl sulphones **327**, $R^1R^2C{=}C{=}C(R)SO_2Ar$

| No. | Ar | R | R¹ | R² | Solvent | Time (h) | Temp. (°C) | Yield of **327** (%) | Reference |
|---|---|---|---|---|---|---|---|---|---|
| **a** | Ph | H | Me | Me | $CH_3CO_2H$ | 21 | 70 | 91 | 299b |
| **a** | Ph | H | Me | Me | EtOH | 19 | 78 | 85 | 299b |
| **a** | Ph | H | Me | Me | MeCN | 12 | 80 | 88 | 299b |
| **a** | Ph | H | Me | Me | $CHCl_3$ | 23 | 75 | 100 | 299b |
| **a** | Ph | H | Me | Et | EtOH | 14 | 75 | 100 | 299b |
| **c** | Ph | H | H | Ph | $CH_3CN$ | 4.7 | 75 | 80 | 299b |
| **d** | Ph | H | H | Me | $CH_3CN$ | 8.5 | 90 | 100 | 299b |
| **e** | *p*-Tol | H | H | H | $C_6H_5Cl$ | 6 | 130 | 80 | 300 |
| **f** | *p*-Tol | D | H | H | $C_6H_5Cl$ | 6 | 130 | *a* | 300 |
| **g** | *p*-Tol | H | Me | H | $C_6H_5Cl$ | 6 | 130 | *a* | 300 |

[a] Not given.

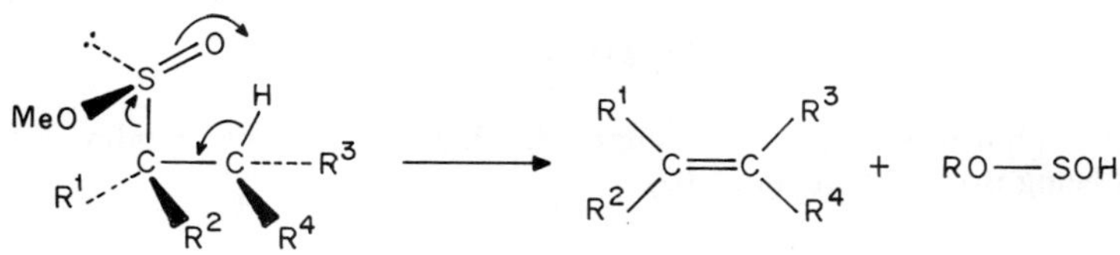


(199)

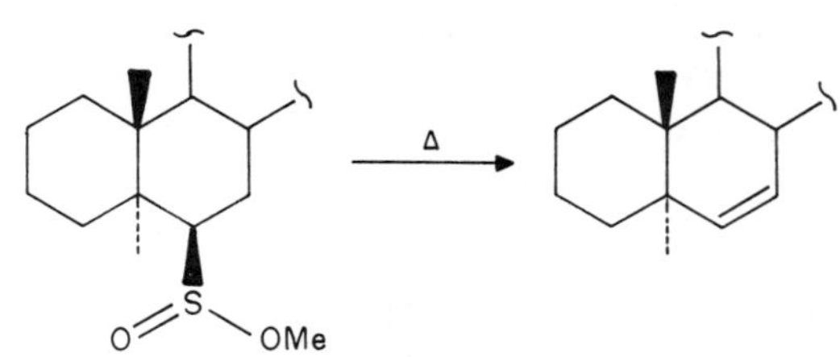


(200)

(**328**) (**329**)

Another interesting application of thermal decomposition of sulphinates was reported by Müller and Schank[306]. Desyl sulphinates **330** prepared as shown in Scheme 8 were found to decompose thermally or at room temperature under basic conditions to yield α-diketones **331** in yields above 90% together with a mixture of disulphide **333** and thiosulphonate **332**. The latter two sulphur-containing products arise undoubtedly from the sulphenic acid **334**, which is formed as the primary pyrolysis product of sulphinates **330**.

Recently, Vermeer and coworkers[307] devised a new synthesis of allenic hydrocarbons starting from 2-propynyl sulphinates **335**. Their reaction with organoheterocuprates of the type (RCuBr)MgX in THF was found to give the desired hydrocarbons **336** in yields over 90% (equation 201). However, when diethyl ether was used as the solvent, substitution

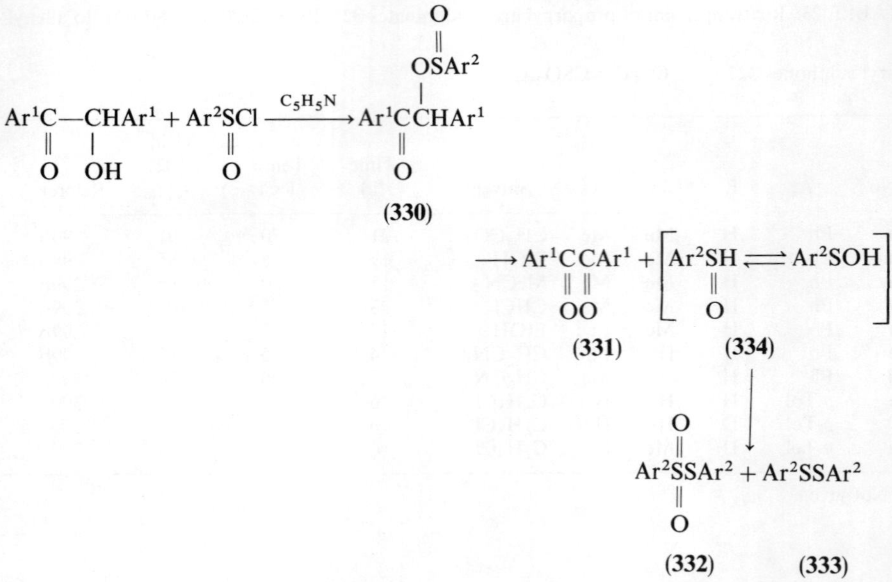

SCHEME 8

reaction at sulphur was observed, the sulphoxide **337** and the corresponding alcoholate anion **338** being isolated (equation 202).

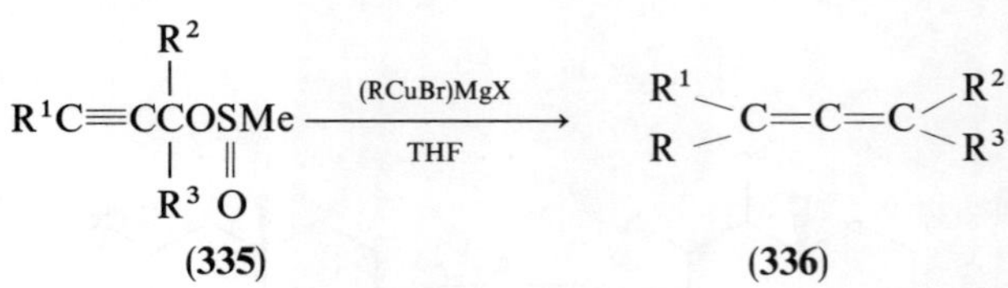

(201)

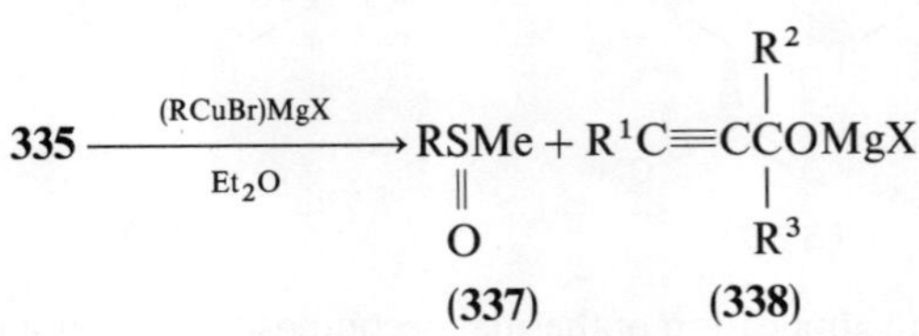

(202)

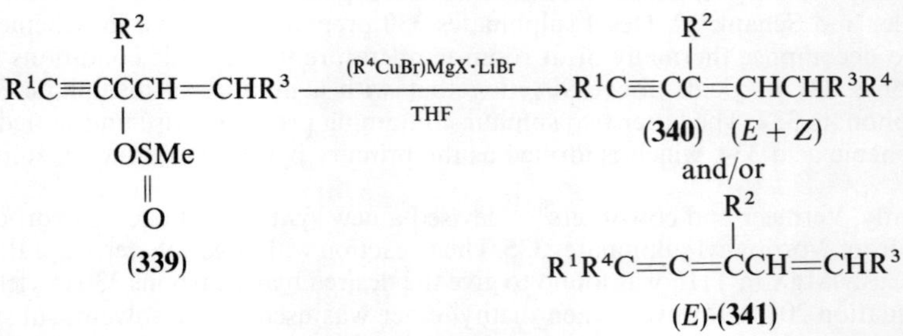

(203)

The reaction of organocuprates with a number of methanesulphinates **339** derived from 3-hydroxy-1-penten-4-ynes was found to give the products of allylic 1,3-substitution and/or propargylic 1,3-substitution[308] (equation 203). This method was applied for the synthesis of pure α- or β-allenic alcohols **342**[309], which can be prepared in yields above 90% as shown in equation 204 (Table 24). Treatment of the epimeric methanesulphinates **345**, derived from epimestranol, with equimolar amounts of the heterocuprate for 1 h at 0 °C afforded the corresponding allenes **346** in nearly quantitative yields[310] (equation 205).

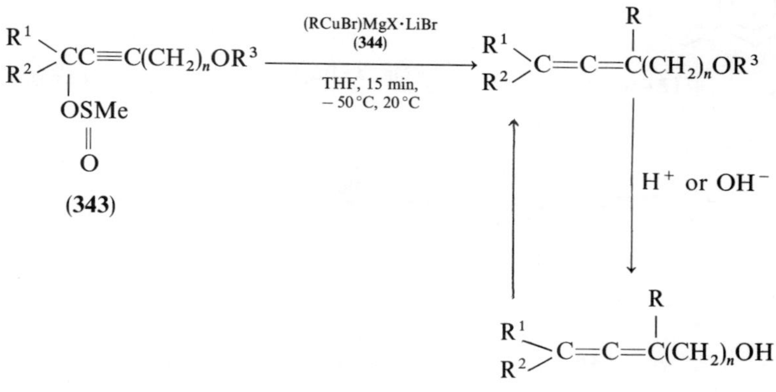

(343) (204) (342)

Vermeer and his group[311] reported also the highly stereoselective synthesis of optically active non-steroidal allenes **347**, which is based on the preferred *anti*-1,3-substitution by an attack of organocuprates on the methanesulphinates **348** derived from (−)-(*R*)-3-hydroxy-3-phenyl-propyne (equation 206).

Finally, it should be mentioned that optically active (+)-α-*p*-toluenesulphinyl-oxypropionate **276** may be converted into a mixture of ethyl (−)-α-chloropropionate **349** and ethyl (−)-lactate **350** when treated with hypochloric acid or with a water/chlorine system[271] (equation 207). Bromination of this sulphinate leads to optically active bromopropionate and *p*-toluenesulphonyl bromide.

TABLE 24. Synthesis of α- and β-allenic alcohols **342** from the hydroxy protected methanesulphinate **343** and organocuprates **344**

| Sulphinate **343** | | | | Cuprate **344** | | |
|---|---|---|---|---|---|---|
| $R^1$ | $R^2$ | $R^3$ | *n* | R | X | Yield of **342** (%) |
| H | H | EtOCHMe | 1 | Et | Br | 60 |
| H | H | EtOCHMe | 1 | *t*-Bu | Cl | 73 |
| Pr | H | EtOCHMe | 2 | Me | Cl | 77 |
| Pr | H | EtOCHMe | 2 | Et | Br | 75 |
| —$(CH_2)_4$— | | $Me_3Si$ | 1 | Me | Cl | 70 |
| —$(CH_2)_4$— | | $Me_3Si$ | 1 | Ph | Br | 90 |

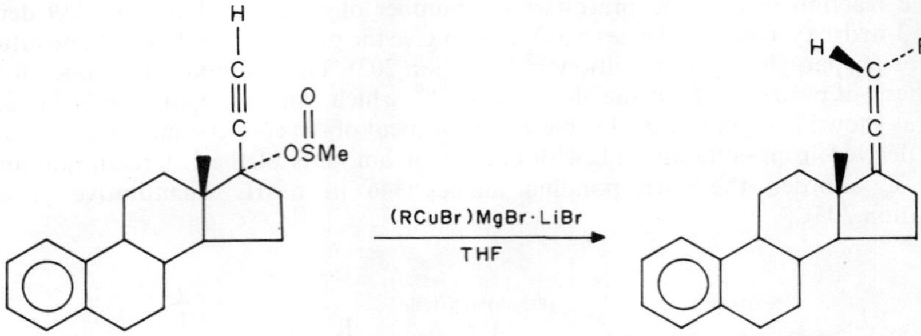


(345) R = Me, t-Bu, Ph (346) (205)

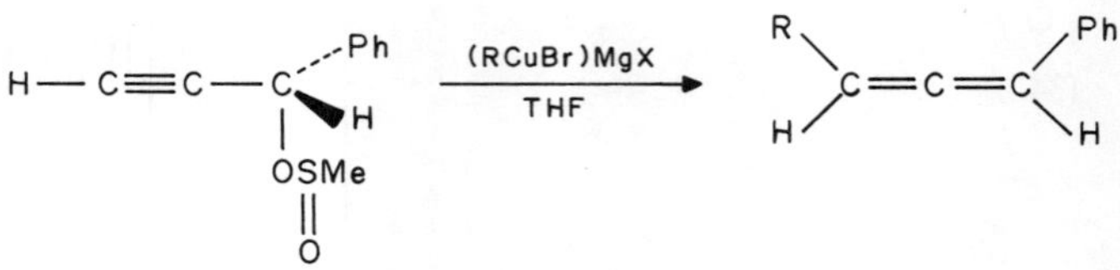


(206)

(348) R=Me,Ph (347)

$$p\text{-TolS(O)OCH(Me)CO}_2\text{Et} \xrightarrow{\text{HClO}} \text{MeCH(Cl)CO}_2\text{Et} + \text{MeCH(OH)CO}_2\text{Et} \quad (207)$$

**(276)** $[\alpha]_{546} + 31.5°$ **(349)** $[\alpha]_{579} - 9.1°$ **(350)** $[\alpha]_{579} - 4.8°$

## IV. REFERENCES


1. K. Schank, in *Methoden der Organischen Chemie (Houben-Weyl)*, Vol. E 11 (Ed. D. Klamann), G. Thieme Verlag, Stuttgart, New York, 1985, p. 1132.
2. K. Schank, in *The Chemistry of Sulphones and Sulphoxides* (Eds. S. Patai, Z. Rappoport and C. J. M. Stirling), Wiley, Chichester, 1988, p. 165.
3. C. J. M. Stirling, *Int. J. Sulfur Chem., B*, **6**, 277 (1971).
4. C. M. Suter, *The Organic Chemistry of Sulfur*, Wiley, New York, London, 1944; reprinted edn., Intra-Science Research Foundation, Santa Monica, 1969, p. 667.
5. P. Allen, Jr., L. S. Karger, J. D. Haygood, Jr. and J. Shrensel, *J. Org. Chem.*, **16**, 767 (1951).
6. W. E. Truce and J. P. Milionis, *J. Org. Chem.*, **17**, 1529 (1952).
7. K. Schank, *Ann. Chem.*, **714**, 117 (1968).
8. K. Schank and A. Weber, *Synthesis*, 367 (1970).
9. K. Schank and F. Schröder, *Phosphorus and Sulfur*, **1**, 307 (1976).
10. K. Schank and A. Weber, *Chem. Ber.*, **105**, 2188 (1972).
11. K. Schank, *Ann. Chem.*, **702**, 75 (1967).
12. M. Adler and K. Schank, *Chem. Ber.*, **111**, 2859 (1978).
13. G. A. Russell and F. Ros, *J. Am. Chem. Soc.*, **107**, 2506 (1985).
14. J. V. Weber, M. Schneider, D. Paquer and P. Faller, *Sulfur Lett.*, 45 (1985).

15. C. Kowal, I. Czernicka and Z. Eckstein, *Przem. Chem.*, **62**, 401 (1983).
16. I. Fleming and C. R. Owen, *J. Chem. Soc., Chem. Commun.*, 1402 (1970).
17. E. Fanghänel, K. H. Kühnemund and A. M. Richter, *Synthesis*, 319 (1984).
18. A. Rosenheim and L. Singer, *Chem. Ber.*, **37**, 2152 (1904).
19. J. S. Meek and J. S. Fowler, *J. Org. Chem.*, **33**, 3422 (1968).
20. V. N. Kondratenko, V. P. Sambur and L. M. Yagupol'skii, *Zh. Org. Khim.*, **7**, 2382 (1971).
21. L. D. Markley, *J. Org. Chem.*, **38**, 3417 (1973).
22. G. E. Veenstra and B. Zwanenburg, *Synthesis*, 519 (1975).
23. J. Hendrickson, D. A. Judelson and T. Chancellor, *Synthesis*, 320 (1984).
24. F. Manescalchi, M. Orena and D. Savoia, *Synthesis*, 445 (1979).
25. J. Wildman and A. M. van Leusen, *Synthesis*, 733 (1979).
26. G. Bram, A. Loupy, M. C. Roux-Schmitt, J. Sansoulet, T. Strazałko and J. Seyden-Penne, *Synthesis*, 56 (1987).
27. J. K. Crandall and C. Pradat, *J. Org. Chem.*, **50**, 1327 (1985).
28. N. V. Kondratenko, V. I. Popov, N. Y. Boyko and L. M. Yagupol'skii *Zh. Org. Khim.*, **13**, 2235 (1977).
29. N. Kornblum, P. Ackermann and R. T. Swiger, *J. Org. Chem.*, **45**, 5294 (1980).
30. N. Kornblum, M. M. Kestner, S. D. Boyd and L. C. Cattran, *J. Am. Chem. Soc.*, **95**, 3356 (1973).
31. B. R. Fishwick, D. K. Rowles and C. J. M. Stirling, *J. Chem. Soc., Chem. Commun.*, 835 (1983).
32. S. I. Al-Khalil and W. R. Bowman, *Tetrahedron Lett.*, **24**, 2517 (1983).
33. Z. Matacz, H. Piotrowska and T. Urbański, *Pol. J. Chem.*, **53**, 187 (1979).
34. N. Kornblum, H. K. Singh and W. J. Kelly, *J. Org. Chem.*, **48**, 332 (1983).
35. A. Fischli and H. Mayer, *Helv. Chim. Acta*, **58**, 1492 (1975).
36. (a) K. Inomata, T. Yamamoto and H. Kotake, *Chem. Lett.*, 1357 (1981).
(b) M. Julia, M. Nel and L. Saussine, *J. Organomet. Chem.*, **181**, C17 (1979).
(c) M. Julia, D. Lave, M. Mulhauser and M. Ramirez-Muñoz, *Tetrahedron Lett.*, **24**, 1783 (1983).
37. G. P. Boldrini, D. Savoia, E. Tagliavini, C. Trombini and A. Umani-Ronchi, *J. Organomet. Chem.*, **268**, 97 (1984).
38. K. Clauss, O. Grimm and G. Prossel, *Ann. Chem.*, 539 (1974).
39. K. Ogura, N. Yahata, J. Watanabe, K. Takahashi and H. Iida, *Bull. Chem. Soc. Jpn.*, **56**, 3543 (1983).
40. K. Schank and H.-G. Schmitt, *Chem. Ber.*, **110**, 3235 (1977).
41. P. Sutter and C. D. Weis, *Phosphorus and Sulfur*, **4**, 335 (1978).
42. B. Badet, M. Julia and M. Ramirez-Muñoz, *Synthesis*, 926 (1980).
43. M. Kobayashi and K. Toriyabe, *Sulfur Lett.*, **3**, 117 (1985).
44. M. Kobayashi, H. Minato and H. Fukuda, *Bull. Chem. Soc. Jpn.*, **46**, 1266 (1973).
45. H. Dorn and H. Graubaum, *Z. Chem.*, **15**, 437 (1975).
46. E. Wenschuh, W. Redeck, A. Porzel, A. Kolbe and S. Edelman, *Z. Anorg. Allg. Chem.*, **528**, 138 (1985).
47. H. Nozaki, R. Noyori and K. Sisido, *Tetrahedron*, **20**, 1125 (1964).
48. J. Gehlhaus and R. W. Hoffmann, *Tetrahedron*, **26**, 5901 (1970).
49. K. Schank, F. Schroeder and A. Weber, *Ann. Chem.*, **553** (1973).
50. W. Middelbos, J. Strating and B. Zwanenburg, *Tetrahedron Lett.*, 351 (1971).
51. E. Gradoń, Z. Ejmocki and Z. Eckstein, *Pol. J. Chem.*, **55**, 469 (1981); *Chem. Abstr.*, **95**, 219 834 (1981).
52. N. Ono, I. Hamamoto, T. Kawai, A. Kaji, R. Tamura and M. Kakihana, *Bull. Chem. Soc. Jpn.*, **59**, 405 (1986).
53. R. Tamura, K. Hayashi, M. Kakihana, M. Tsuji and D. Oda, *Tetrahedron Lett.*, **26**, 851 (1985).
54. M. P. Balfe, E. A. W. Downer, A. A. Evans, J. Kenyon, R. Poplett, S. E. Searle and A. L. Tarnoky, *J. Chem. Soc.*, 797 (1946).
55. M. P. Balfe, J. Kenyon and C. E. Searle, *J. Chem. Soc.*, 3309 (1950).
56. B. R. Brown and M. R. Shaw, *J. Chem. Soc., Perkin Trans. 1*, 2036 (1974).
57. K. Schank, in Reference 1, p. 1149.
58. M. Julia and D. Arnoult, *Bull. Soc. Chim. Fr.*, 746 (1973).
59. P. Messinger, *Arch. Pharm. (Weinheim)*, **306**, 603 (1973).
60. J. J. Licari and G. Dougherty, *J. Am. Chem. Soc.*, **76**, 4039 (1954).
61. P. Messinger and H. Greve, *Arch. Pharm. (Weinheim)*, **311**, 280 (1978).
62. O. R. Hansen and R. Hammer, *Acta Chem. Scand.*, **7**, 1331 (1953).

63. P. Messinger and H. Greve, *Synthesis*, 259 (1977).
64. A. R. Harris, *Synth. Commun.*, **18**, 659 (1988).
65. P. Messinger and J. Gompertz, *Arch. Pharm. (Weinheim)*, **307**, 653 (1974).
66. C. C. J. Culvenor, W. Davies and N. S. Heath, *J. Chem. Soc.*, 278 (1949).
67. I. W. J. Still and F. J. Ablenas, *Synth. Commun.*, **12**, 1103 (1982).
68. T. L. Gresham, J. E. Jansen, F. W. Shaver, M. R. Frederick, F. T. Fiedorek, R. A. Bankert, J. R. Greory and W. L. Beears, *J. Am. Chem. Soc.*, **74**, 1323 (1952).
69. J. Hershberger and G. A. Russell, *Synthesis*, 475 (1980).
70. W. Sas, *J. Chem. Soc., Chem. Commun.*, 862 (1984).
71. K. Inomata, T. Kobayashi, S. Sasaoka, H. Kinoshita and H. Kotake, *Chem. Lett.*, 289 (1986).
72. O. S. Andell and J.-E. Bäckvall, *Tetrahedron Lett.*, **26**, 4555 (1985).
73. F. M. Beringer, A. Brierley, M. Drexler, E. M. Gindler and C. C. Lumpkin, *J. Am. Chem. Soc.*, **75**, 2708 (1953).
74. K. Reuter, Ger. Offen. 3, 616, 065; *Chem. Abstr.*, **108**, 151118 (1988); K. Reuter, Ger. Offen. 3, 616, 066; *Chem. Abstr.*, **108**, 168 124 (1988).
75. D. H. R. Barton, J.-C. Blazejewski, C. Charpiot and W. B. Motherwell, *J. Chem. Soc., Chem. Commun.*, 503 (1981).
76. V. A. Nefedov, L. V. Kryuchkova and R. K. Torygina, *Zh. Org. Khim.*, **13**, 1735 (1977).
77. D. Arlt, M. Jautelat and R. Lantzsch, *Angew. Chem.*, **93**, 719 (1981); *Angew. Chem., Int. Ed. Engl.*, **20**, 703 (1981).
78. U. M. Dzhemilev, R. V. Kunakova and R. L. Gaysin, *Izv. Akad. Nauk SSSR, Ser. Khim.*, 2337 (1983); *Chem. Abstr.*, **100**, 51176 (1984).
79. Y. Tamaru and Z. Yoshida, *J. Org. Chem.*, **44**, 1188 (1979).
80. R. V. Kunakova, R. L. Gaysin, G. A. Tolstikov, L. M. Zelenova and U. M. Dzhemilev, *Izv. Akad. Nauk SSSR, Ser. Khim.*, 2610 (1980).
81. T. W. Gibson and P. Strassburger, *J. Org. Chem.*, **41**, 791 (1976).
82. H. Gilman and L. F. Cason, *J. Am. Chem. Soc.*, **72**, 3469 (1950).
83. P. Messinger and H. Greve, *Arch. Pharm. (Weinheim)*, **311**, 827 (1978).
84. O. Achmatowicz and J. Michalski, *Rocz. Chem.*, **30**, 243 (1956).
85. P. Messinger, *Arch. Pharm (Weinheim)*, **306**, 458 (1973).
86. H. Kamogawa, H. Kusaka and M. Nanasawa, *Bull. Chem. Soc. Jpn.*, **53**, 3379 (1980).
87. I. Matsuda, K. Akijama, T. Toyoshima, S. Kato and M. Mizuta, *Bull. Chem. Soc. Jpn.*, **48**, 3675 (1975).
88. W. Ried, A. H. Schmidt and H. Knorr, *Chem. Ber.*, **108**, 538 (1975).
89. C. G. M. Janssen, P. M. van Lier, H. M. Buck and E. F. Godefroi, *J. Org. Chem.*, **44**, 4199 (1979).
90. G. L. Cooper and L. J. Dolby, *J. Org. Chem.*, **44**, 3414 (1979).
91. G. K. Cooper and L. J. Dolby, *Tetrahedron Lett.*, 4675 (1976).
92. L. F. Cason and C. C. Wanser, *J. Am. Chem. Soc.*, **73**, 142 (1951).
93. D. I. Aleksiev, *God. Vissh. Khim.-Tekhnol. Inst. Burgas, Bulg.*, **12**, Pt. 2, 31 (1977); *Chem. Abstr.*, **90**, 137 400 (1979).
94. R. Kerber and J. Starnik, *Chem. Ber.*, **104**, 2035 (1971).
95. O. Achmatowicz, E. Maruszewska-Wieczorkowska and J. Michalski, *Rocz. Chem.*, **29**, 1029 (1955).
96. D. I. Aleksiev, *Zh. Org. Khim.*, **11**, 211 (1975).
97. D. I. Aleksiev, *Zh. Org. Khim.*, **12**, 2038 (1976).
98. D. I. Aleksiev and I. T. Mladenov, *Zh. Org. Khim.*, **13**, 2005 (1977).
99. R. Tamura, K. Hayashi, M. Kakihana, M. Tsuji and D. Oda, *Chem. Lett.*, 229 (1985).
100. K. Bailey, B. R. Brown and B. Chalmers, *J. Chem. Soc., Chem. Commun.*, 618 (1967).
101. Y. Ogata, Y. Sawaki and M. Isono, *Tetrahedron*, **26**, 731 (1970).
102. A. Etienne and G. Lonchambon, *C. R. Acad. Sci. Paris, Ser. C*, **275**, 375 (1972).
103. H. Maruyama and T. Hiraoka, *J. Org. Chem.*, **51**, 399 (1986).
104. S. I. Burmistrov, N. V. Toropin and K. S. Burmistrov, *Vopr. Khim., Khim. Tekhnol.*, **61**, 36 (1980); *Chem. Abstr.*, **96**, 8296 (1982).
105. E. P. Kohler and G. R. Barrett, *J. Am. Chem. Soc.*, **46**, 747 (1924).
106. K. Bowden, E. C. Braude and E. R. H. Jones, *J. Chem. Soc.*, 945 (1946).
107. R. A. Araviiskii, V. I. Veksler, V. N. Mikhailova and V. V. Yakovlev *Zh. Obshch. Khim.*, **57**, 2574 (1987); *Chem. Abstr.*, **109**, 128 509 (1988).
108. C. J. M. Stirling, *J. Chem. Soc.*, 5856, 5863 (1964).

109. E. von Meyer, *J. Prakt. Chem.*, **63**, 167 (1901).
110. C. M. Suter, in Reference 4, p. 691.
111. L. Field and P. H. Settlage, *J. Am. Chem. Soc.*, **73**, 5870 (1951).
112. H. Bredereck and E. Bäder, *Chem. Ber.*, **87**, 129 (1954).
113. K. Schank, *Chem. Ber.*, **99**, 48 (1966).
114. W. Löwe, G. Eggersmann and A. Kennemann, *Arch. Pharm. (Weinheim)*, **317**, 15 (1984).
115. E. Bäder and H. D. Hermann, *Chem. Ber.*, **88**, 41 (1955).
116. H. Hellmann and G. Opitz, *Chem. Ber.*, **90**, 8 (1957).
117. G. Rawson and J. B. F. N. Engberts, *Tetrahedron*, **26**, 5653 (1970).
118. T. Olijnsma, J. B. F. N. Engberts and J. Strating, *Rec. Trav. Chim. Pays-Bas*, **91**, 209 (1972).
119. T. Olijnsma, J. B. F. N. Engberts and J. Strating, *Recl. Trav. Chim. Pays-Bas*, **86**, 463 (1967).
120. J. Strating and A. M. van Leusen, *Recl. Trav. Chim. Pays-Bas*, **81**, 966 (1962).
121. J. B. F. N. Engberts and J. Strating, *Recl. Trav. Chim. Pays-Bas*, **83**, 733 (1964).
122. J. B. F. N. Engberts and J. Strating, *Recl. Trav. Chim. Pays-Bas*, **84**, 942 (1965).
123. P. Messinger, *Arch. Pharm (Weinheim)*, **307**, 348 (1974).
124. J. S. Meek and J. S. Fowler, *J. Org. Chem.*, **33**, 3418 (1968).
125. H. Hellmann and K. Müller, *Chem. Ber.*, **98**, 638 (1965).
126. N. H. Nilsson, C. Jacobsen and A. Senning, *J. Chem. Soc., Chem. Commun.*, 314 (1971).
127. N. H. Nilsson and A. Senning, *Chem. Ber.*, **107**, 2345 (1974).
128. A. Senning, O. N. Sørensen and Ch. Jacobsen, *Angew. Chem., Int. Ed. Engl.*, **9**, 737 (1968).
129. N. H. Nilsson, C. Jacobsen, O. N. Sørensen, N. H. Hanusøe and A. Senning, *Chem. Ber.*, **105**, 2854 (1972).
130. N. H. Nilsson, C. Jacobsen and A. Senning, *J. Chem. Soc., Chem. Commun.*, 658 (1970).
131. G. E. Veenstra and B. Zwanenburg, *Recl. Trav. Chim. Pays-Bas*, **95**, 28 (1976).
132. J. Perronnet and P. Girault, *Bull. Soc. Chim. Fr.*, 2843 (1973).
133. A. S. Shawali, H. M. Hassaneen and S. M. Sherif, *J. Heterocycl. Chem.*, **17**, 1745 (1980).
134. I. M. Bazavova, V. M. Nepluyev and R. G. Debenko, *Zh. Org. Khim.*, **11**, 2388 (1975).
135. H. Bredereck, A. Wagner, H. Beck and R.-J. Klein, *Chem. Ber.*, **93**, 2736 (1960); J. L. Kice and G. Guaraldi, *J. Am. Chem. Soc.*, **89**, 4113 (1967).
136. C. J. M. Stirling, *J. Chem. Soc.*, 3597 (1957); C. M. Suter, in Reference 4, pp. 460 and 513 and earlier references cited therein.
137. J. F. King and V. Rathore, *Phosphorus and Sulfur*, **33**, 165 (1987).
138. O. Foss, *Kgl. Norske Videnskab. Selskabs. Forh.*, **19**, 68 (1946); *Chem. Abstr.*, **42**, 19f (1948).
139. J. Goerdeler and P. Rosenthal, *Tetrahedron Lett.*, 3665 (1964).
140. J. M. Cox and R. Ghosh, *Tetrahedron Lett.*, 3351 (1969).
141. A. M. van Leusen and J. C. Jagt, *Tetrahedron Lett.*, 967 (1970).
142. F. Kurzer and J. R. Powell, *J. Chem. Soc.*, 3728 (1952).
143. I. L. Khmel'nitskaya and F. M. Kogan, *Zh. Prikl. Khim.*, **25**, 1004 (1952).
144. (a) R. Sato, T. Goto, Y. Takikawa and S. Takizawa, *Synthesis*, 615 (1980).
(b) J. D. Macke and L. Field, *J. Org. Chem.*, **53**, 396 (1988).
145. S. Takano, K. Hiroya and K. Ogasawara, *Chem. Lett.*, 255 (1983).
146. Th. Zincke and F. Farr, *Ann. Chem.*, **391**, 57 (1912).
147. T. A. Parsons, J. D. Buckman, D. E. Pearson and L. Field, *J. Org. Chem.*, **30**, 1923 (1965).
148. L. D. Markley and J. E. Dunbar, *J. Org. Chem.*, **37**, 2512 (1972).
149. M. D. Bentley, I. B. Douglass and J. A. Lacadie, *J. Org. Chem.*, **37**, 333 (1972).
150. M. Furukawa, K. Sato and T. Okawara, *Chem. Lett.*, 2007 (1982).
151. Y. Abe and J. Tsurugi, *Chem. Lett.*, 811 (1972).
152. S. Pawlenko, in Reference 1, p. 1112.
153. S. Oae, Y. H. Kim and D. Fukushima, Jpn. Kokai Tokyo Koho, 78, 127, 402; *Chem. Abstr.*, **90**, 12291 (1979); Jpn. Kokai Tokyo Koho, 78, 108, 906; *Chem. Abstr.*, **90**, 87011 (1979).
154. S. Oae, Y. H. Kim, D. Fukushima and T. Takata, *Chem. Lett.*, 893 (1977).
155. G. Kresze and W. Kort, *Chem. Ber.*, **94**, 2624 (1961).
156. C. S. Marvel and R. S. Johnson, *J. Org. Chem.*, **13**, 822 (1948).
157. S. Oae, K. Shinhama and Y. H. Kim, *Tetrahedron Lett.*, 3307 (1979).
158. E. Yu. Belyaev, L. M. Gornostaev and G. A. Suboch, *Khim. Tekhnol. Polim.*, **4**, 60 (1975); *Chem. Abstr.*, **86**, 189 390 (1976).
159. A. Darchen and C. Moinet, *J. Chem. Soc., Chem. Commun.*, 1031 (1986).
160. S. L. Graham and T. H. Scholz, *Synthesis*, 1031 (1986).

161. J. Michalski, J. Wieczorkowski and T. Modro, *Rocz. Chem.*, **32**, 1409 (1958); J. Michalski, T. Modro and J. Wieczorkowski, *J. Chem. Soc.*, 1665 (1960).
162. M. Kobayashi, *Bull. Chem. Soc. Jpn.*, **39**, 1296 (1966).
163. H. W. Roesky and S. Tutkunkardes, *Chem. Ber.*, **107**, 508 (1974).
164. K. Schank and H.-G. Schmitt, *Chem. Ber.*, **107**, 3026 (1974).
165. P. Kiełbasiński, R. Zurawiński, J. Drabowicz and M. Mikołajczyk, *Tetrahedron*, **44**, 6687 (1988).
166. T. W. Hambley, M. C. Harsanyi and R. K. Norris, *J. Org. Chem.*, **53**, 3104 (1988).
167. M. Kobayashi, M. Terao and A. Yamamoto, *Bull. Chem. Soc. Jpn.*, **39**, 802 (1966).
168. M. Kobayashi and M. Terao, *Bull. Chem. Soc. Jpn.*, **39**, 1292 (1966).
169. M. Kobayashi, *Bull. Chem. Soc. Jpn.*, **39**, 967 (1966).
170. T. P. Vasil'eva, N. V. Kaluzhnaya, V. M. Bystrova, M. G. Lin'kova, O. V. Kil'bisheva and I. L. Knunyants, *Izv. Akad. Nauk SSR, Ser. Khim.*, 2187 (1980).
171. W. Walter, B. Rische and G. Adiwidjaja, *Ann. Chem.*, 14 (1980).
172. J. L. Kice and K. Ikura, *J. Am. Chem. Soc.*, **90**, 7378 (1968).
173. W. H. Mueller and M. B. Dines, *J. Chem. Soc., Chem. Commun.*, 1205 (1969).
174. J. L. Kice and S. Liao, *J. Org. Chem.*, **46**, 2691 (1981).
175. P. K. Srivastava and L. Field, *Phosphorus and Sulfur*, **25**, 161 (1985).
176. Y. Miyaji, H. Minato and M. Kobayashi, *Bull. Chem. Soc. Jpn.*, **44**, 862 (1971).
177. M. Furukawa, T. Ohkawara, Y. Noguchi and M. Nishikawa, *Synthesis*, 441 (1978).
178. J. Drabowicz and M. Pacholczyk, *Phosphorus and Sulfur*, **29**, 257 (1987).
179. M. Furukawa and T. Ohkawara, *Synthesis*, 339 (1976).
180. M. Furukawa, T. Ohkawara, Y. Noguchi, M. Nishikawa and M. Tomimatsu, *Chem. Pharm. Bull.*, **28**, 134 (1980).
181. M. Furukawa, T. Ohkawara, Y. Noguchi, M. Isoda and T. Hitoshi, *Synthesis*, 937 (1980).
182. Y. Noguchi, M. Isoda, K. Kuroki and N. Furukawa, *Chem. Pharm. Bull.*, **30**, 1646 (1982).
183. Y. Noguchi, K. Kuroki, M. Sekioka and M. Furukawa, *Bull. Chem. Soc. Jpn.*, **56**, 349 (1983).
184. E. Krauthausen, in Reference 1, p. 635.
185. P. H. Gore, *Chem. Ind. (London)*, 1355 (1954).
186. K. Nakagawa and K. Minami, *Tetrahedron Lett.*, 343 (1972).
187. J. E. Herz and L. A. de Marquez, *J. Chem. Soc., Perkin Trans.1*, 2633 (1973).
188. R. Shanker, *Chem. Ind. (London)*, 76 (1974).
189. R. Caputo, L. Mangoni, P. Monaco, G. Palumbo and L. Previtera, *Tetrahedron Lett.*, 1041 (1975).
190. G. Borgogno, S. Colonna and R. Fornasier, *Synthesis*, 529 (1975).
191. J. Drabowicz and M. Mikołajczyk, *Synthesis*, 542 (1978).
192. E. S. Lang and J. V. Comasseto, *Synth. Commun.*, **18**, 301 (1988).
193. J. Strating and H. J. Backer, *Recl. Trav. Chim. Pays-Bas*, **69**, 638 (1950).
194. S. Oae, H. Togo, T. Numata and K. Fujimori, *Chem. Lett.*, 1193 (1980).
195. S. Hurusawa and T. Shoiri, *Tetrahedron Lett.*, **23**, 447 (1982).
196. S. Kagabu, M. Maehara, K. Sawahara and S. Saito, *J. Chem. Soc., Chem. Commun.*, 1485 (1988).
197. P. Lapape, *Ann. Pharm. Fr.*, **28**, 181 (1970); *Chem. Abstr.*, **73**, 98 541 (1970).
198. C. Khosla and N. Anand, *J. Sci. Ind. Res.*, **16B**, 69 (1957); *Chem. Abstr.*, **51**, 13 804d (1957).
199. P. W. Henniger and J. K. van der Drift, Ger. Offen. 2, 235, 390; *Chem. Abstr.*, **78**, 124608j (1973).
200. E. Wenshuh, K. Dölling, M. Mikołajczyk and J. Drabowicz, *Z. Chem.*, **20**, 122 (1980) and references cited therein.
201. J. R. Brush, P. G. Cookson and G. B. Deacon, *J. Organomet. Chem.*, **34**, C1 (1972) and references cited therein.
202. G. B. Deacon and I. K. Johnson, *J. Organomet. Chem.*, **122**, 123 (1976).
203. Y. Tamaru and Z. Yoshida, *Tetrahedron Lett.*, 4527 (1978).
204. Y. Tamaru, M. Kagotani, R. Suzuki and Z. Yoshida, *Chem. Lett.*, 1329 (1978).
205. J.-B. Baudin and S. A. Julia, *Tetrahedron Lett.*, **29**, 3255 (1988).
206. M. Mikołajczyk and J. Drabowicz, *Top. Stereochem.*, **13**, 333 (1982).
207. H. Phillips, *J. Chem. Soc.*, **127**, 2552 (1925).
208. H. F. Herbrandson and C. M. Cusano, *J. Am. Chem. Soc.*, **83**, 2124 (1961).
209. M. Mikołajczyk, J. Drabowicz and H. Ślebocka-Tilk, *J. Am. Chem. Soc.*, **101**, 1302 (1979).
210. J. Drabowicz, *Phosphorus and Sulfur*, **31**, 123 (1987).
211. H. Gilman, J. Robinson and N. Beaber, *J. Am. Chem. Soc.*, **48**, 2715 (1926).

212. D. N. Harpp, S. M. Vines, J. P. Montillier and T. H. Chan, *J. Org. Chem.*, **41**, 3987 (1976).
213. E. Wenschuh and H. Lankau, *Z. Chem.*, **13**, 427 (1973).
214. A. G. Brook and D. G. Anderson, *Can. J. Chem.*, **46**, 2115 (1968).
215. R. M. Coates and H. D. Pigott, *Synthesis*, 319 (1975).
216. H. J. Monteiro and J. P. de Souza, *Tetrahedron Lett.*, 921 (1975).
217. G. Groszek, M. M. Kabat, A. Kurek, M. Masnyk and J. Wicha, *Bull. Acad. Polon. Sci., Ser. Chim.*, **34**, 305 (1986).
218. H. Bohme and B. Clement, *Tetrahedron Lett.*, 1737 (1979).
219. K. K. Andersen, *Tetrahedron Lett.*, 93 (1962).
220. K. K. Andersen, W. Gaffield, N. E. Papanikolaou, J. W. Foley and R. I. Perkins, *J. Am. Chem. Soc.*, **86**, 5637 (1964).
221. K. Mislow, M. M. Green and M. Raban, *J. Am. Chem. Soc.*, **87**, 2761 (1965).
222. D. D. Ridley and M. A. Smal, *Aust. J. Chem.*, **35**, 495 (1982).
223. K. K. Andersen, B. Bujnicki, J. Drabowicz, M. Mikołajczyk and J. B. O'Brien, *J. Org. Chem.*, **49**, 4070 (1984).
224. J. Drabowicz, B. Bujnicki and M. Mikołajczyk, *J. Org. Chem.*, **47**, 3325 (1982).
225. M. Axelrod, P. Bickart, J. Jacobus, M. M. Green and K. Mislow, *J. Am. Chem. Soc.*, **90**, 4835 (1968).
226. H. Hope, U. de la Camp, G. Homer, A. W. Messing and L. H. Sommer, *Angew Chem.*, **81**, 619 (1969); *Angew. Chem., Int. Ed. Engl.*, **8**, 612 (1969).
227. U. de la Camp and H. Hope, *Acta Cyrstallogr., Sect. B*, **26**, 846 (1970).
228. J. Drabowicz and M. Mikołajczyk, unpublished results.
229. J. Jacobus and K. Mislow, *J. Am. Chem. Soc.*, **89**, 5228 (1967).
230. S. Colonna, R. Giovini and F. Montanari, *J. Chem. Soc., Chem. Commun.*, 865 (1968).
231. U. Folli, D. Iarrosi, F. Montanari and U. Torre, *J. Chem. Soc. (C)*, 1317 (1968).
232. K. Komiyama, H. Minato and M. Kobayashi, *Bull. Chem. Soc. Jpn.*, **46**, 3895 (1973).
233. P. Bonvicini, A. Levi and G. Scorrano, *Gazz. Chim. Ital.*, **102**, 621 (1972).
234. J. Drabowicz, B. Dudziński and M. Mikołajczyk, unpublished results.
235. K. Mislow, M. M. Green, P. Laur, J. P. Melillo, T. Simons and A. L. Ternay Jr., *J. Am. Chem. Soc.*, **87**, 1958 (1965).
236. P. Bickart, F. W. Carson, J. Jacobus, E. G. Miller and K. Mislow, *J. Am. Chem. Soc.*, **90**, 4869 (1968).
237. D. R. Rayner, A. J. Gordon and K. Mislow, *J. Am. Chem. Soc.*, **90**, 4854 (1968).
238. P. Bravo, G. Resnati, F. Viani and A. Arone, *Tetrahedron*, **43**, 4635 (1987).
239. G. Solladie, R. Zimmermann, R. Bartsch and H. M. Walborsky, *Synthesis*, 662 (1985).
240. M. Mikołajczyk and J. Drabowicz, *J. Am. Chem. Soc.*, **100**, 2518 (1978).
241. W. H. Pirkle and S. D. Beare, *J. Am. Chem. Soc.*, **90**, 6250 (1968).
242. K. K. Andersen, S. Colonna and C. J. M. Stirling, *J. Chem. Soc., Chem. Commun.*, 645 (1973).
243. B. Raguse and D. D. Ridley, *Aust. J. Chem.*, **37**, 2059 (1984).
244. D. J. Abbott, S. Colonna and C. J. M. Stirling, *J. Chem. Soc., Perkin Trans. 1*, 492 (1976).
245. G. H. Posner and P. W. Tang, *J. Org. Chem.*, **43**, 4131 (1978).
246. G. H. Posner, J. P. Mallamo, M. Hulce and L. L. Frye, *J. Am. Chem. Soc.*, **104**, 4180 (1982).
247. M. Cinquini, S. Colonna, F. Cozzi and C. J. M. Stirling, *J. Chem. Soc., Perkin Trans. 1*, 2061 (1976).
248. M. Mikołajczyk, W. Midura, S. Grzejszczak, A. Zatorski and A. Chefczyńska, *J. Org. Chem.*, **43**, 478 (1978).
249. G. L. Colombo, C. Gennari and E. Narisano, *Tetrahedron Lett.*, 3861 (1978).
250. R. Annunziata, M. Cinquini and F. Cozzi, *Synthesis*, 535 (1979).
251. N. Kunieda, J. Nokami and M. Kinoshita, *Bull. Chem. Soc. Jpn.*, **49**, 256 (1976).
252. R. Annunziata, M. Cinquini and F. Cozzi, *Synthesis*, 767 (1982).
253. R. Annunziata, M. Cinquini, S. Colonna and F. Cozzi, *J. Chem. Soc., Perkin Trans. 1*, 614 (1981).
254. L. Banfi, G. Colombo, C. Gennari, R. Annunziata and F. Cozzi, *Synthesis*, 829 (1982).
255. G. Mioskowski and G. Solladie, *Tetrahedron*, **36**, 227 (1980); G. Solladie, F. Matloubi-Moghadam, G. Luttmann and G. Mioskowski, *Helv. Chim. Acta*, **65**, 1602 (1982).
256. F. Schneider and R. Simon, *Synthesis*, 582 (1986).
257. K. Hiroi and N. Matsuyama, *Chem. Lett.*, 65 (1986).
258. M. C. Carreno, J. L. Garcia Ruano and A. Rubio, *Tetrahedron Lett.*, **28**, 4861 (1987).
259. P. A. Zoretic, P. Soja and N. D. Sinha, *J. Org. Chem.*, **43**, 1379 (1978).

260. A. Nudelman and D. J. Cram, *J. Am. Chem. Soc.*, **90**, 3869 (1968).
261. S. Colonna, G. Germinario, A. Manfredi and C. J. M. Stirling, *J. Chem. Soc., Perkin Trans. 1*, 1695 (1988).
262. R. E. Booms and D. J. Cram, *J. Am. Chem. Soc.*, **94**, 5438 (1972).
263. E. Wenschuh, H. Winter, G. Mendel and A. Kolbe, *Phosphorus and Sulfur*, **7**, 321 (1979).
264. M. Mikołajczyk, B. Bujnicki, J. Drabowicz, *Bull. Acad. Pol. Sci., Ser. Chem.*, **25**, 267 (1977).
265. M. Cinquini and F. Cozzi, *J. Chem. Soc., Chem. Commun.*, 502 (1977).
266. M. Cinquini and F. Cozzi, *J. Chem. Soc., Chem. Commun.*, 723 (1977).
267. M. Mikołajczyk, J. Drabowicz and B. Bujnicki, *J. Chem. Soc., Chem. Commun.*, 568 (1976).
268. (a) Minato, K. Yamaguchi, K. Okuma and M. Kobayashi, *Bull. Chem. Soc. Jpn.*, **49**, 2590 (1976).
(b) H. Minato, K. Yamaguchi and M. Kobayashi, *Chem. Lett.*, 991 (1975).
269. M. Mikołajczyk, B. Bujnicki and J. Drabowicz, unpublished results.
270. J. Kenyon, H. Phillips and F. M. H. Taylor, *J. Chem. Soc.*, 173 (1933).
271. W. Gerrard, J. Kenyon and H. Phillips, *J. Chem. Soc.*, 153 (1937).
272. M. A. Sabol and K. K. Andersen, *J. Am. Chem. Soc.*, **91**, 3603 (1969).
273. R. M. Coates and J. E. Chen, *Tetrahedron Lett.*, 2705 (1969).
274. J. C. Martin and B. R. Ree, unpublished results cited in Reference 273.
275. X. Creary, C. C. Geiger and K. Hilton, *J. Am. Chem. Soc.*, **105**, 2851 (1983).
276. T. Netscher and H. Prinzbach, *Synthesis*, 683 (1987).
277. J. F. King, K. Piers, D. J. H. Smith, C. L. McIntosh and P. deMayo, *J. Chem. Soc., Chem. Commun.*, 31 (1969).
278. J. Kenyon and H. Phillips, *J. Chem. Soc.*, 1676 (1930).
279. C. L. Arcus, M. P. Balfe and J. Kenyon, *J. Chem. Soc.*, 485 (1938).
280. A. H. Wragg, J. S. McFadyen and T. S. Stevens, *J. Chem. Soc.*, 3603 (1958).
281. D. Darwish and R. A. McLaren, *Tetrahedron Lett.*, 123 (1962).
282. D. Darwish and E. A. L. Preston, *Tetrahedron Lett.*, 113 (1964).
283. E. Ciuffarin, M. Isola and A. Fava, *J. Am. Chem. Soc.*, **90**, 3594 (1968).
284. (a) S. Braverman and T. Globerman, *Tetrahedron*, **30**, 3873 (1974).
(b) S. Braverman and Y. Duar, *Tetrahedron Lett.*, 343 (1975).
(c) S. Braverman and H. Manor, *Phosphorus and Sulfur*, **2**, 213 (1976).
285. Y. Duar, M.Sc. Thesis, Bar-Ilan University, 1975. This reference is cited by S. Braverman, in *The Chemistry of Sulphones and Sulphoxides* (Eds. S. Patai, Z. Rappoport and C. J. M. Stirling), Wiley, Chichester, 1988, p. 665.
286. J. B. Hendrickson and P. L. Skipper, *Tetrahedron*, **32**, 1627 (1976).
287. F. Jung, M. Malin, R. Van Den Elzen and T. Durst, *J. Am. Chem. Soc.*, **96**, 935 (1974).
288. R. F. Heldeweg and H. Hogeveen, *J. Am. Chem. Soc.*, **98**, 2341 (1976).
289. S. Braverman, in *The Chemistry of Sulphones and Sulphoxides* (Eds. S. patai, Z. Rappoport and C. J. M. Stirling), Wiley, Chichester, 1988, p. 665.
290. A. C. Cope, D. E. Morrison and L. Field, *J. Am. Chem. Soc.*, **72**, 59 (1950).
291. (a) S. Braverman, *Int. J. Sulfur Chem. (C)*, **6**, 149 (1971).
(b) D. Darwish and A.-M. Armour, unpublished results cited by S. Braverman in Reference 289.
292. K. Hiroi, R. Kitayama and S. Sato, *J. Chem. Soc., Chem. Commun.*, 1470 (1983).
293. K. Hiroi, R. Kitayama and S. Sato, *Chem. Pharm. Bull.*, **32**, 2628 (1984).
294. B. Trost, N. R. Schmull and M. J. Miller, *J. Am. Chem. Soc.*, **102**, 5979 (1980).
295. K. Hiroi, R. Kitayama and S. Sato, *Chemistry Lett.*, 929 (1984).
296. P. A. Grieco and D. Boxler, *Synth. Commun.*, **5**, 315 (1975).
297. F. Jung and T. Durst, *J. Chem. Soc., Chem. Commun.*, 4 (1973).
298. J. E. Baldwin, O. W. Lewer, Jr., and N. R. Tzodikow, *J. Org. Chem.*, **41**, 2312 (1976).
299. (a) S. Braverman and H. Mechoulam, *Isr. J. Chem.*, **5**, 71 (1967).
(b) S. Braverman and H. Mechoulam, *Tetrahedron*, **30**, 3883 (1974).
300. (a) C. J. M. Stirling, *J. Chem. Soc., Chem. Commun.*, 135 (1967).
(b) G. Smith and C. J. M. Stirling, *J. Chem. Soc. (C)*, 1530 (1971).
301. D. H. R. Barton, *J. Chem. Soc.*, 2174 (1949).
302. W. Huckel, W. Tappe and G. Luttke, *Ann. Chem.*, **543**, 191 (1940).
303. A. C. Cope, N. A. LeBel, H. H. Lee and W. R. Moove, *J. Am. Chem. Soc.*, **79**, 4720 (1957).
304. C. A. Kingsbury and D. J. Cram, *J. Am. Chem. Soc.*, **82**, 1810 (1960).
305. D. N. Jones and W. Higgins, *J. Chem. Soc. (C)*, 81 (1969).
306. W. Müller and K. Schank, *Chem. Ber.*, **111**, 2870 (1978).

307. P. Vermeer, H. Westmijze, H. Kleijn and L. A. van Dijck, *Recl. Trav. Chim. Pays-Bas*, **97**, 56 (1978).
308. H. Kleijn, H. Westmijze, K. Kruithof and P. Vermeer, *Recl. Trav. Chim. Pays-Bas*, **98**, 27 (1979).
309. H. Kleijn, C. J. Elsevier, H. Westmijze, J. Meijer and P. Vermeer, *Tetrahedron Lett.*, 3101 (1979).
310. H. Westmijze and P. Vermeer, *Tetrahedron Lett.*, 410 (1979).
311. G. Tadema, R. H. Everhardus, H. Westmijze and P. Vermeer, *Tetrahedron Lett.*, 3935 (1978).

The Chemistry of Sulphinic Acids, Esters and their Derivatives
Edited by S. Patai


CHAPTER **13**

# Photochemistry of sulphinic acid derivatives

GIUSEPPE CAPOZZI and PIERO SARTI-FANTONI

*Centro C. N. R. 'Chimica e Struttura dei Composti Eterociclici, c/o Dipartimento di Chimica Organica, Universita' di Firenze, 50121 Firenze, Italy*



## I. INTRODUCTION

In this chapter we have separated the photochemical synthesis and the photochemical properties of sulphinic acids and sulphinic acid derivatives for the sake of clarity. However, these two aspects are strongly connected since they involve the formation of sulphinyl and/or sulphonyl radicals at a certain stage of the reaction under study. The fate of these radicals then depends on the particular substrate and on the reaction conditions.

The structure of the sulphinyl[1] and sulphonyl[2] radicals has been recently reviewed in two chapters of a book of this series and therefore this subject will be not discussed again here.

Although the number of papers dealing with the photochemistry of sulphinic acids and their derivatives is limited, the use of these substances for industrial applications is quite extensive. This includes mainly the use of sulphinic acids or salts as photopolymerization initiators for the synthesis of polymers which can be used in odontology, photography, radar technology and other important technological fields.

## II. PHOTOCHEMICAL SYNTHESIS OF SULPHINIC ACIDS AND SULPHINIC ACID DERIVATIVES

Three general methods for the photochemical synthesis of sulphinic acid derivatives can be envisaged: the photolysis of sulphonyl compounds[3,4], the insertion of sulphur dioxide into a carbon–hydrogen or carbon–carbon bond[5–8], and the photooxidation of suitable compounds[9–14], mainly disulphides[9–12]. The second method has been applied by using excited sulphur dioxide or by trapping photo-generated radicals by ground-state sulphur dioxide.

### A. Photolysis of Sulphonyl Compounds

Photolysis in methanol of α-ketosulphones **1a–c** gave complex mixtures of products containing the sulphinic acids **2a–c** in variable yields (22–44%)[3]. The yields of the products obtained in the photolysis of **1a–c** are listed in Table 1.

TABLE 1. Product composition (% yields) for the photolysis of ketosulphones **1a–c** in methanol

| $RSO_2CH_2COR^1$ (**1**) | $RSO_2H$ (**2**) | R—R (**3**) | $RCH_2COR^1$ (**4**) | $MeCOR^1$ (**5**) | $RSO_2O(R^1)C{=}CH_2$ (**6**) |
|---|---|---|---|---|---|
| (**1a**) $R = PhCH_2$, $R^1 = Me$ | 44 | 22 | 13 | 59 | 6 |
| (**1b**) $R = PhCH_2$, $R^1 = Ph$ | 22 | 24 | — | 78 | — |
| (**1c**) $R = t\text{-Bu}$, $R^1 = Me$ | 28 | — | — | 56 | 16 |

All the reaction products may derive from the radicals formed by the fission of the $SO_2$—$CH_2$ bond. Hydrogen abstraction from the solvent by the sulphonyl radical forms the sulphinic acid.

The formation of the other products has been explained by simple transformations of the initially formed radicals. Loss of sulphur dioxide from the sulphonyl radicals and coupling gives **3**; the ketones **4** and **5** are formed from the $R^1COCH_2$ radicals by hydrogen abstraction (**5**) or coupling with the R radicals (**4**). The vinyl sulphonates **6** are formed by a different recombination of the two primarily formed radicals.

The photolysis of benzyl sulphones **7** in methanol or 2-propanol[4] represents a general and easy synthesis of sulphinic acids. When sulphones **7a–f** were photolyzed the main reaction products were the sulphinic acids **8a–f** (equation 1).

$$\underset{(\mathbf{7})}{PhCH_2SO_2R} \xrightarrow{h\nu} \underset{(\mathbf{8})}{RSO_2H} \qquad (1)$$

(**7a, 8a**) R = Me (**7d, 8d**) R = —$(CH_2)_2Cl$
(**7b, 8b**) R = Et (**7e, 8e**) R = —$(CH_2)_2Ph$
(**7c, 8c**) R = n-Pr (**7f, 8f**) R = Ph

The yields of sulphinic acids are higher in 2-propanol than in methanol. Indeed, in this case also the homolysis of the sulphur–benzylic carbon bond is the primary reaction path. This process generates benzyl radicals and sulphonyl radicals; for the latter there is competition between hydrogen abstraction from the solvent and loss of sulphur dioxide. The easier the hydrogen abstraction, the higher the yield of the sulphinic acid. The two reaction paths have been disclosed in the photolysis of the sulphone **7e** which gives the sulphinic acid **8e** (54%) and phenylethane (36%). It is noteworthy that these two products account for about 90% of the original sulphone; this indicates a very high selectivity of the photolytic fission of sulphur–carbon bonds of **7e** (Scheme 1).

$$\underset{\textbf{(7e)}}{PhCH_2SO_2CH_2CH_2Ph} \xrightarrow[\text{2-propanol}]{h\nu} PhCH_2\cdot + \underset{\textbf{(9)}}{PhCH_2CH_2SO_2\cdot}$$

$$PhCH_2CH_3 \xrightarrow{\text{2-Propanol}} PhCH_2CH_2\cdot \xrightarrow{-SO_2} \textbf{(9)} \xrightarrow{\text{2-propanol}} \underset{\textbf{(8e)}}{Ph_2CH_2CH_2SO_2H}$$

SCHEME 1

### B. Photoinitiated Insertion of Sulphur Dioxide

The insertion of photoexcited sulphur dioxide into a carbon–hydrogen bond was first discovered in the gas phase many years ago[5]. Simple alkanes like methane, ethane and propane gave the corresponding sulphinic acids which, in some cases, were characterized as 2,4-dinitrophenyl sulphones. Unidentified mixtures of isomeric sulphinic acids were obtained in the case of propane and butane.

More recently, a similar study on the photoreaction of sulphur dioxide with hydrocarbons has been carried out and the product analysis (GC-MS) performed after treatment of the reaction mixture with diazomethane[8]. In the case of the reaction of butane, besides butanesulphinic acids (detected as methyl esters), several other sulphur-containing products were identified.

Photoexcited sulphur dioxide reacts in the gas phase also with alkenes[5] to give low boiling products, which are believed to be sulphinic acids.

It is also possible to obtain a variety of functionalized sulphinic acids by low-temperature (−75 °C) irradiation of alcohols, ethers, sulphides, alkyl halides and *N,N*-dimethylformamide[6]. The substances irradiated and the yields of the sulphinic acids obtained are listed in Table 2. Under the same reaction conditions, 2-methylpropane gives a mixture of the two isomeric sulphinic acids derived from sulphur dioxide insertion into the two different carbon–hydrogen bonds.

TABLE 2. Reaction of photoexcited $SO_2$ with various substrates

| Substrate | Product | Yields (%) |
|---|---|---|
| MeOH | $CH_2(OH)SO_2H$ | 5 |
| $PhCH_2OH$ | $PhCH(OH)SO_2H$ | 14 |
| *i*-PrOPr-*i* | *i*-$PrOCMe_2SO_2H$ | 43 |
| MeOPr-*i* | $MeOCMe_2SO_2H$ | 55 |
| EtSEt | $EtSCHMeSO_2H$ | 20 |
| $Me_3CH$ | $Me_3CSO_2H$ | 23 |
| | $Me_2CHCH_2SO_2H$ | 17 |
| EtCl | $MeCH(Cl)SO_2H$ | 5 |
| $Me_2NCHO$ | $HO_2SCH_2MeNCHO$ | 35 |

Although the yields are not high, this reaction represents an easy entry into the class of α-substituted sulphinic acids.

The second way to obtain the photoinduced insertion of sulphur dioxide into an organic molecule is the reaction of ground-state sulphur dioxide with photoexcited substrates.

When diradicals of type **10**, formed by irradiation of **11**, were trapped with sulphur dioxide[7], many compounds containing a sulphonyl functionality were obtained. The formation of most of them can be explained by the intermediacy of the sulphinic acid **12** deriving from **13**, the primarily formed adduct of sulphur dioxide with the diradical **10** (Scheme 2).

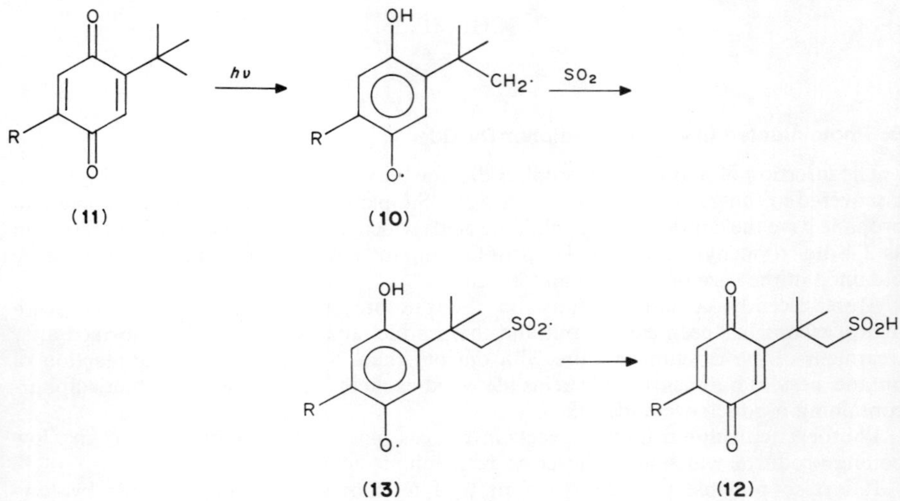


SCHEME 2

Compounds like **14**, **15** and **16** are formed in the photolysis of **11** in sulphur dioxide. Their formation can be explained by intramolecular cyclization of **12** (**14** and **15**) or by intermolecular addition of **12** to the unchanged starting material **11** to yield **16**.

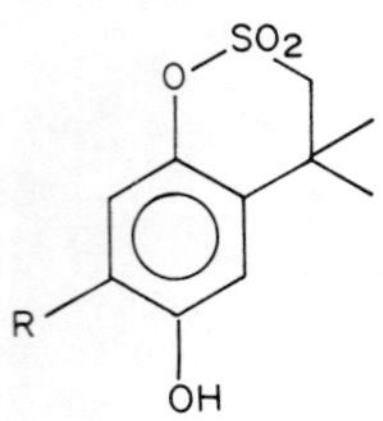


**(14)**

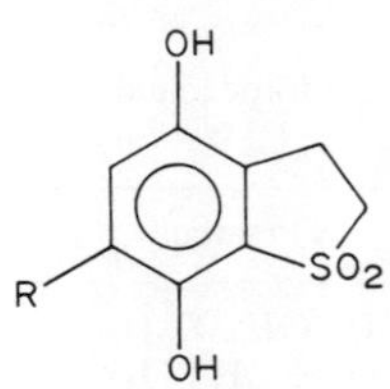


**(15)**

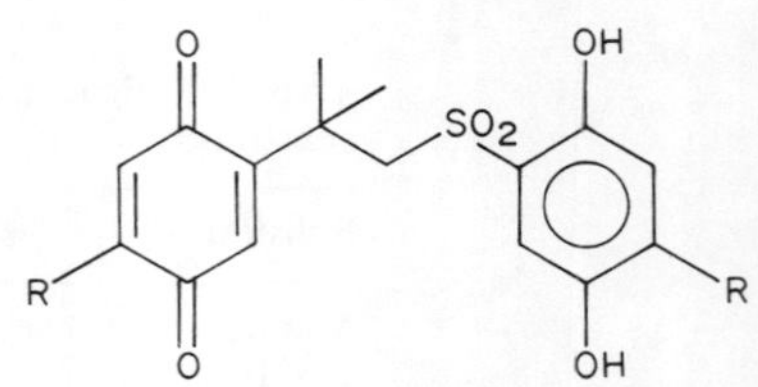


**(16)**

It should be pointed out that the insertion of sulphur dioxide into a carbon–hydrogen bond in this reaction is quite different, on a mechanistic point of view, from the insertions of photoexcited sulphur dioxide described above[5,6]. In fact, in the case of photoexcited organic substrates, there is no absorption of light by sulphur dioxide and, moreover, energy transfer from the excited starting material to the sulphur dioxide seems an energetically unfavoured process[7].

Sultines of type **17** have been synthesized by sulphur dioxide insertion into a carbon–carbon bond of the cyclopropane ring of **18** (equation 2)[15].

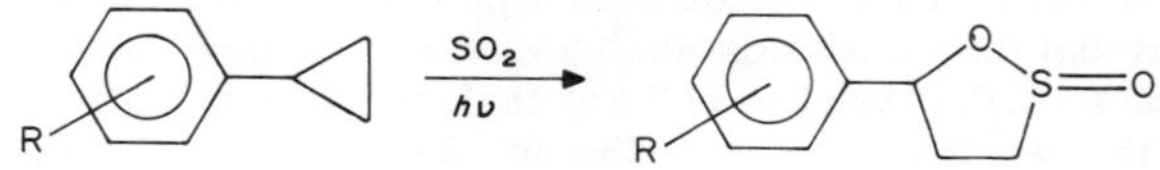


(2)

**(18)** **(17)** *cis* + *trans*

**(17a, 18a)** R = *p*-CN
**(17b, 18b)** R = *p*-MeCO

The reaction produces a mixture of diastereomeric sultines. In fact the oxidation of the mixture of *cis* and *trans* **17a** gave the sultone **19** as a single product. In addition, the insertion of sulphur dioxide is regiospecific; the orientation of the insertion has been demonstrated by further oxidation of the sultone **19** to the ketosulphonic acid **20**.

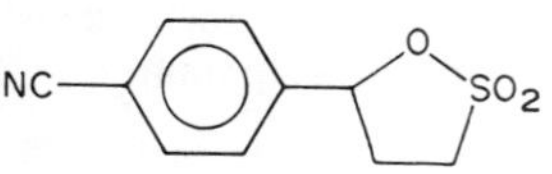


**(19)**

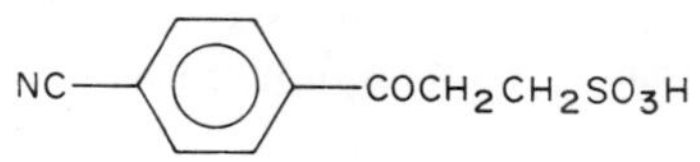


**(20)**

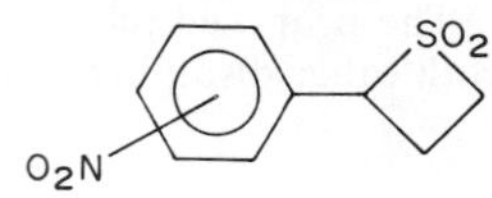


**(21)**

The formation of sultines from the cyclopropanes **18a–b** seems controlled by rather peculiar factors linked to substituent effects and intrinsic stability of the products. Thus *p*-nitro- and *o*-nitro-phenylcyclopropane under the same reaction conditions gave the sulphone **21**; the phenylcyclopropane itself and other derivatives, bearing in the phenyl ring a variety of substituents like *p*-phenyl, *p*-chloro, *p*-bromo and *p*-iodo, do not give photochemical insertion of sulphur dioxide, probably due to the photolability of the corresponding sultines[16,17].

### C. Photooxidation of Disulphides

The photooxidation of dialkyl disulphides **22a–d** by molecular oxygen in methanol using methylene blue as sensitizer gave the corresponding thiosulphinates **23a–d** in fairly good yields (60–75%) (equation 3)[9]. Diphenyl disulphide, under the same reaction conditions, was found to be unreactive[9,18].

$$\underset{(22)}{\mathrm{RSSR}} \xrightarrow[\mathrm{MeOH}]{h\nu/\mathrm{O_2}/\text{methylene blue}} \underset{(23)}{\mathrm{RS{-}\overset{\displaystyle O}{\overset{\|}{S}}R}} \qquad (3)$$

**(22a, 23a)** R = Me **(22d, 23d)** R = *t*-Bu
**(22b, 23b)** R = Et **(22e, 23e)** R = Ph
**(22c, 23c)** R = *i*-Pr

In the photooxidation of **22a** and **22b**, traces of the corresponding thiosulphonates were also found; the presence of thiosulphonates has been explained by a non-photochemical disproportionation of the thiosulphinates eventually catalysed by acidic impurities[19–23] present in the reaction mixtures.

The photooxidation of the disulphides **22a–d** can be strongly retarded by the presence of equimolar amounts of 1,4-diazabicyclo[2,2,2]octane (DABCO). This observation is good evidence for the intervention of singlet oxygen in the oxidation reaction.

Other information on the mechanism of this reaction has been obtained from cooxidation reactions. Firstly, it was found that when a mixture of di-*t*-butyl disulphide and di-iso-propyl disulphide was photooxidized, no mixed thiosulphinates were detected; this demonstrates that the photooxidation does not involve the cleavage of the sulphur–sulphur bond. Secondly, the oxidation of the diethyl disulphide in the presence of diphenyl disulphide gave the two thiosulphinates **23b** and **23e**. Since diphenyl disulphide alone was not oxidized in the same reaction conditions, it was suggested that the zwitterion **24a** (R = Et), formed from the diethyl disulphide and singlet oxygen, was the oxidizing agent of the diphenyl disulphide. Intermediates **25** and **26**, similar to **24**, have been proposed in the photosensitized oxidation of sulphides[24,25].

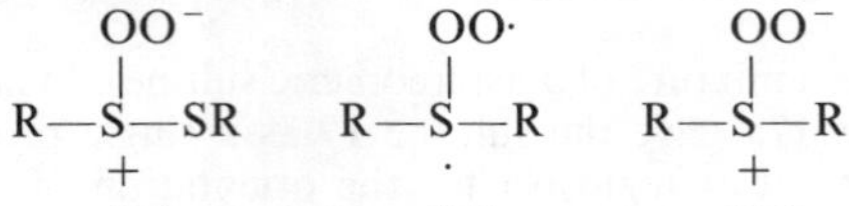

**(24)** **(25)** **(26)**

The proposed mechanism (Scheme 3) suggests the formation of **24** which then reacts with other disulphides present in solution to give two molecules of thiosulphinates.

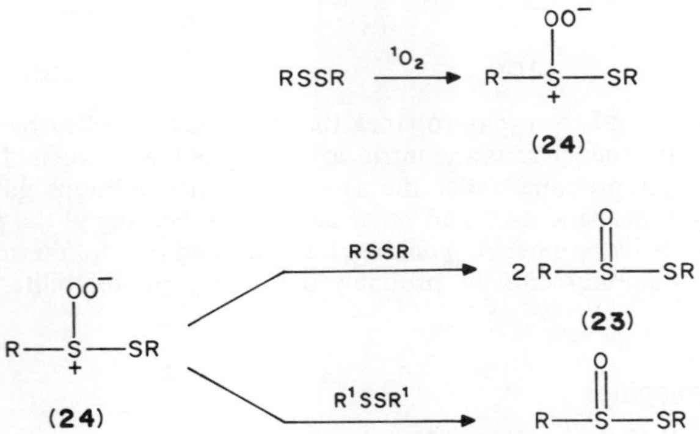

SCHEME 3

The photooxidation of dialkyl disulphides in methanol or ethanol, using Rose Bengal as sensitizer, follows a different course[11,12]. Under these reaction conditions, the dialkyl disulphides **22a** and **22b** gave the alkyl sulphinates **27a–d** (equation 4) while the di-*t*-butyl disulphide **22d** did not give the corresponding sulphinate. It has been suggested that the sulphinates **27a–d** could be obtained from the initially formed thiosulphinates by reaction with the solvent.

$$\text{RSSR} \xrightarrow[\text{R}^1\text{OH}]{h\nu/\text{O}_2/\text{Rose Bengal}} \text{R—S(=O)—OR}^1 \qquad (4)$$

**(22a)** or **(22b)** **(27)**

**(27a)** R = Me, R′ = Me **(27c)** R = Me, $R^1$ = Et

**(27b)** R = Et, $R^1$ = Me **(27d)** R = Et, $R^1$ = Et

Photosensitized oxidation has been also attempted with cyclic disulphides; however, the extension of the reaction to this class of disulphides has some limitations depending on the structure of the disulphide. In fact the 1,2-dithiolane **28** gave[10] the corresponding cyclic thiosulphinate **29** (equation 5), whereas other disulphides like **30** and **31** were found to be inert to photooxidation[18].

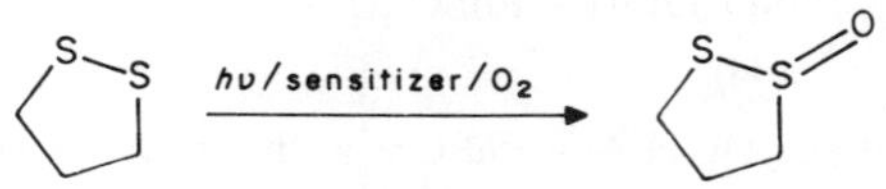


(5)

**(28)** **(29)**

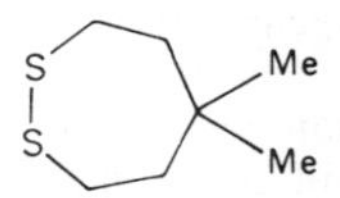

**(30)**

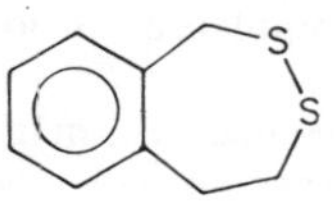

**(31)**

## III. PHOTOCHEMICAL REACTIVITY

### A. Sulphinic Acids

The autooxidation of benzenesulphinic acids to sulphonic acids is quite a slow process which can be made much faster by UV irradiation[26] (Figure 1). The proposed mechanism for this reaction implies the formation of benzenesulphonyl radicals **32**, which react with oxygen to give peroxybenzenesulphonic radicals **33** and the peroxybenzenesulphonic acid **34** (equations 6 and 7).

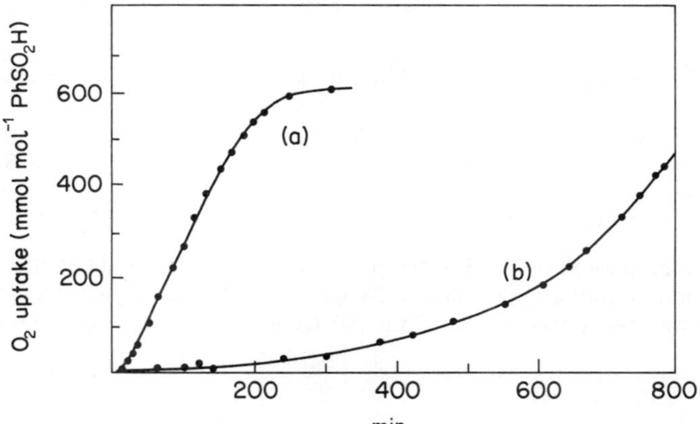


FIGURE 1. Autooxidation of benzenesulphinic acid: (a) with UV light, at 60 °C; (b) without UV light, at 60 °C. Reproduced by permission of VCH Verlagsgesellschalt from Ref. 26

$$PhSO_2\cdot + O_2 \longrightarrow PhSO_2OO\cdot \qquad (6)$$

**(32)** **(33)**

$$PhSO_2OO\cdot + PhSO_2H \longrightarrow PhSO_2OOH + PhSO_2\cdot \qquad (7)$$

**(33)** **(34)** **(32)**

The formation of the sulphonic acid arises from the oxidation of the sulphinic acid by **34** (equation 8) and/or from the disproportionation of **34** which gives benzenesulphonic radicals and **33** (equation 9).

$$\underset{(\mathbf{34})}{PhSO_2OOH} + PhSO_2H \longrightarrow 2\,PhSO_3H \tag{8}$$

$$2\,\underset{(\mathbf{34})}{PhSO_2OOH} \longrightarrow PhSO_3\cdot + \underset{(\mathbf{33})}{PhSO_2OO\cdot} + H_2O \tag{9}$$

The photolysis of 2,4,6-tri-isopropylbenzenesulphinic acid **35** has been carefully studied in the presence or with complete exclusion of oxygen[27] using 2,2′-azobis-(2-methylpropionitrile) as initiator.

In the absence of oxygen the main reaction product of the photolysis was the thiosulphonate **36**; however, small quantities of the disulphide **37** were also formed (equation 10).

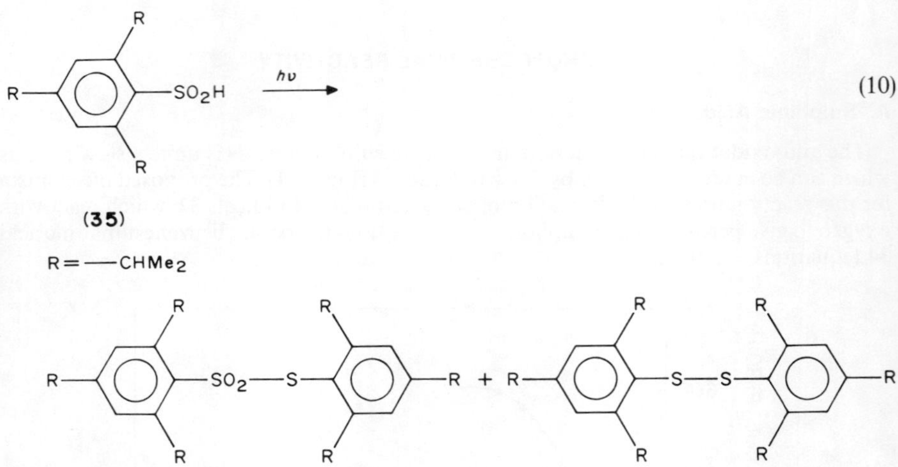


(10)

(**35**)

R = —CHMe$_2$

(**36**) 58%  (**37**) 6%

The initiated photolysis of **35** in the presence of oxygen gave different results. Under these reaction conditions, the sultone **38** was formed in 34% yield (equation 11). The sultone **38** was also obtained from **35** under oxygen in the dark or without initiation.

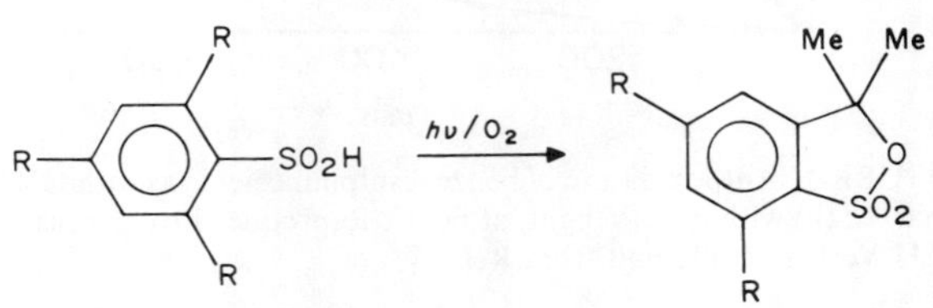


(11)

(**35**)  (**38**)

R = —CHMe$_2$

The yield of **38** in this reaction is similar to that of the photochemical reaction; however, the photochemical transformation is much faster than the dark reaction.

It has been suggested that the thiosulphonate **36** might be an intermediate in the reaction leading to **38**; indeed, when **36** was photolysed in the presence of oxygen and the initiator, good yields of **38** were obtained (equation 12).

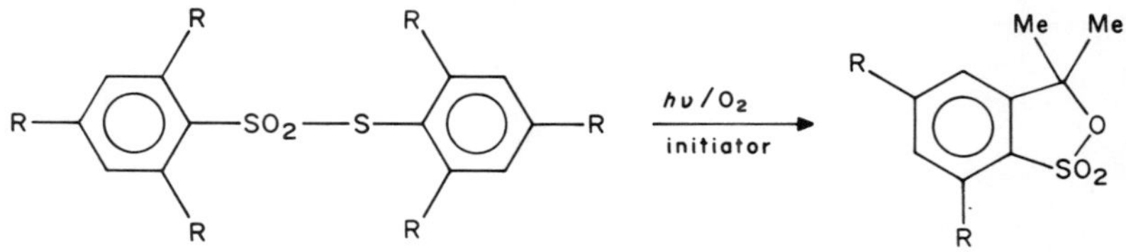


(12)

**(36)** **(38)** 51%

The formation of **38** during the photolysis of **35** under oxygen has been explained by a radical mechanism as indicated in Scheme 4. According to this reaction scheme, the sulphonyl radical **39** is the key intermediate; the reaction of **39** with oxygen gives the peroxysulphonic radical **40**, which generates the new radical **41** by hydrogen abstraction from an *ortho* isopropyl group. The sultone **38** is then formed by cyclization and hydroxy radical elimination.

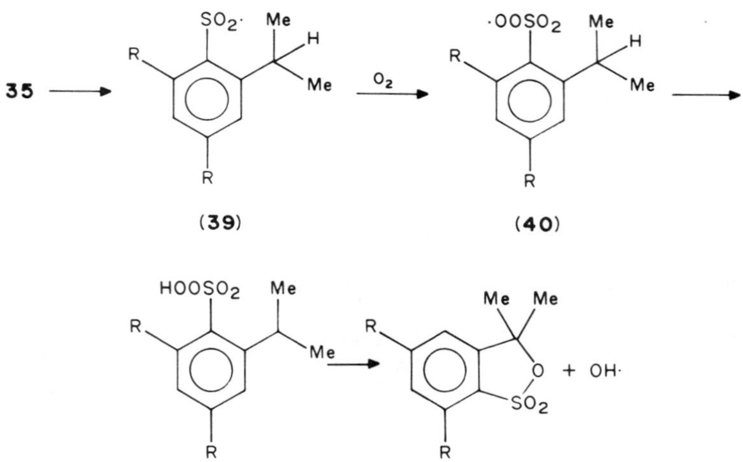


**(39)** **(40)**

**(41)** **(38)**

SCHEME 4

The photochemical behaviour of sulphinic acid salts of type **42** and **43** containing a remote disulphide or trisulphide functionality has been recently reported[28] and the results compared with those of the thermal reaction.

$$RSS(CH_2)_nSO_2^- \qquad {}^-O_2S(CH_2)_nSSS(CH_2)_nSO_2^-$$

**(42)** **(43)**

**(42a)** $n = 3$, R = *p*-Tol **(42d)** $n = 3$, R = $-(CH_2)_4COOH$ **(43a)** $n = 3$
**(42b)** $n = 4$, R = *p*-Tol **(42e)** $n = 4$, R = $-(CH_2)_4COOH$ **(43b)** $n = 4$
**(42c)** $n = 5$, R = *p*-Tol **(42f)** $n = 5$, R = $-(CH_2)_4COOH$ **(43c)** $n = 5$

The disulphides **42d–f** were found to be quite stable to heating: only 5% decomposition was observed after 80 min at 68 °C. The nature of the decomposition products was not investigated; however, since it was reported that under less vigorous conditions **42e** gave the cyclic thiosulphonate **44**[29], it is possible that also the products of the decomposition of **42d–f** have similar structures.

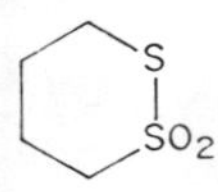

(**44**)

Indeed, the general scheme for the reaction of compounds of type **42** implies the cyclization to the thiosulphonate **45** and the formation of the two disulphides **46** and **47** via **48** as intermediate (Scheme 5).

$$\underset{(\mathbf{42})}{RSS(CH_2)_nSO_2^-} \rightleftarrows \underset{(\mathbf{45})}{\overline{S{-}(CH_2)_n{-}SO_2}} + RS^-$$

$$RS^- + \mathbf{42} \rightleftarrows \underset{(\mathbf{46})}{RSSR} + \underset{(\mathbf{48})}{{}^-S(CH_2)_nSO_2^-}$$

$$\mathbf{48} \xrightarrow{\mathbf{42}} \underset{(\mathbf{47})}{{}^-O_2S(CH_2)_nSS(CH_2)_nSO_2^-} + RS^-$$

$$\mathbf{48} \xrightarrow{\mathbf{45}} \mathbf{47}$$

SCHEME 5

Compounds **42d–f** under UV irradiation decompose at a much faster rate (10% after 10 min). It should be pointed out that apparently the different chain length does not effect the thermal or the photochemical decomposition of **42d–f**. On the other hand, compounds **42a–c** show an evident effect of the chain length on the thermal decomposition. Figure 2 shows this effect for the decomposition of **42a–c** at 25 °C.

From these data it is evident that among the three disulphides **42a–c**, **42c** is by far the most stable compound; this is in line with a heterolytic mechanism involving intramolecular cyclization which is expected to be easier for **42a** ($n=3$) and **42b** ($n=4$) than for **42c** ($n=5$).

The rates of the photochemical decomposition of **42a–c** are very close to each other and the chain-length effect is suppressed. This feature supports a homolytic mechanism involving cleavage and intermolecular reactions, which are not expected to be strongly dependent on the chain length.

The photochemical or the thermal reaction of the trisulphides **43a–c** gave as unique product the disulphides **49a–c**, respectively.

$$\underset{(\mathbf{49})}{{}^-O_2S(CH_2)_nSS(CH_2)_nSO_2S^-}$$

(**49a**) $n=3$

(**49b**) $n=4$

(**49c**) $n=5$

The formation of **49** is not easy to rationalize on the basis of a simple mechanism. The reaction sequence reported in Scheme 6 has been tentatively suggested to account for this unusual rearrangement.

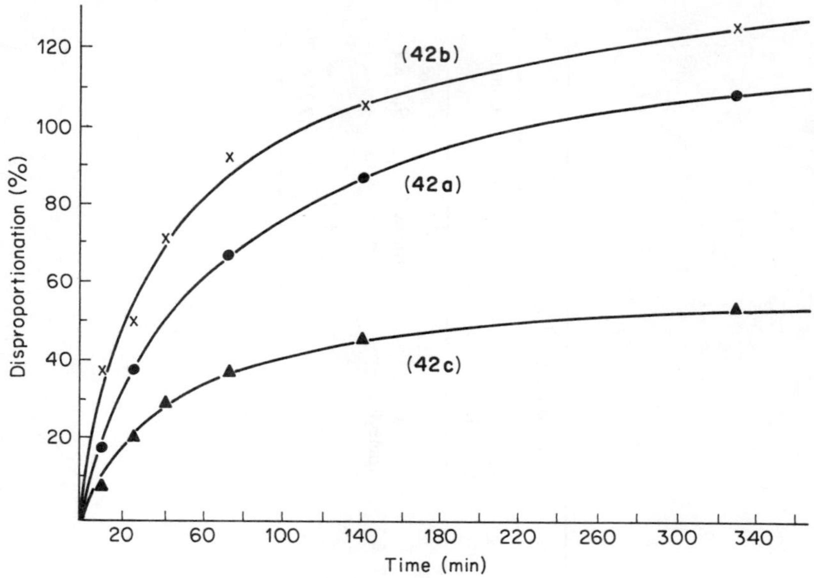


FIGURE 2. Plot of percent diproportionation at 25 °C vs time for the disulphides **42a**, **42b**, and **42c**. Reprinted with permission from Macke and Field, *J. Org. Chem.*, **53**, 396. Copyright (1988) American Chemical Society (Ref. 28)

$$^{-}O_2S(CH_2)_nSSS(CH_2)_nSO_2^{-} \rightleftarrows {}^{-}O_2S(CH_2)_nSS^{-} + \overline{\underline{\;(CH_2)_n\;}}\;SO_2{-}S$$

(**43**) (**50**) (**51**)

$$\mathbf{50} + \mathbf{51} \rightleftarrows {}^{-}O_2S(CH_2)_nSSSO_2(CH_2)_nS^{-}$$

(**52**)

$$\mathbf{52} \rightleftarrows {}^{-}S(CH_2)_nSO_2S^{-} + \mathbf{51} \rightleftarrows {}^{-}O_2S(CH_2)_nSS(CH_2)_nSO_2S^{-}$$

(**53**) (**49**)

SCHEME 6

The rate of the thermal rearrangement of **43a** ($n = 3$) and **43b** ($n = 4$) are similar and faster than that of **43c**. Under UV conditions the three compounds react at a comparable rate, thus suggesting in this case the intervention of radical mechanisms.

### B. Sulphinate Esters

#### *1. Open-chain sulphinates*

The photochemical behaviour of alkyl *p*-toluenesulphinates **54** has been studied in hexane as solvent using a high-pressure mercury lamp[30]. The reaction conditions used and the products of the photolysis of **54a–e** are summarized in Table 3.

The data of Table 3, although not completely homogeneous, show that the chain length of the alkyl residue of **54** has the effect of decreasing the reactivity of the sulphinates. The

TABLE 3. Photolysis of alkyl *p*-toluenesulphinates (*p*-TolS(O)OR) (**54**)

| Sulphinate Ester | Concentration ($mol\,l^{-1}$) | Time (h) | Unreacted **54** (mol%) | Products (mol %) | | | |
|---|---|---|---|---|---|---|---|
| | | | | *p*-TolSO$_2$STol-*p* (**55**) | *p*-TolSSTol-*p* (**56**) | *p*-TolSO$_2$OH | Others |
| (**54a**) R = Me | 0.41 | 130 | — | 19.8 | 8.2 | 3.9 | 7.6[a] |
| (**54b**) R = Et | 0.29 | 100 | 17.6 | 22.2 | 4.8 | traces | 9.2[b], 8.9[c] |
| (**54c**) R = Bu | 0.23 | 107 | 35.6 | 12.5 | detected | — | — |
| (**54d**) R = Oct | 0.40 | 117 | 100 | — | — | — | — |
| (**54d**) R = Oct | neat | 123 | 44.5 | 20.8 | 0.5 | detected | — |
| (**54e**) R = Allyl | 0.27 | 111 | — | 7.0 | detected | — | 12.2[d], 40.7[b] |

[a] *p*-TolSO$_2$OMe.
[b] A sulphone of unknown structure; yield in weight%.
[c] MeCHO.
[d] *p*-TolSO$_2$Allyl.

octyl derivative **54d** does not react under conditions that cause extensive photolysis of **54a–c**.

As far as the product composition is concerned, alkyl *p*-toluenethiosulphonates and *p*-tolyl disulphide are always the main reaction products. However, their ratio varies with the individual reaction, the sulphonate being always present in much greater amount.

A closer inspection of the nature of the products detected in the photolysis of **54a–e** gives some insight into the mechanism of this reaction.

The formation of acetaldehyde, as well as that of the *p*-tolyl *p*-toluenethiosulphonate **55** and di-*p*-tolyl disulphide **56** in the photolysis of **54b** might be explained by the non-radical mechanism depicted in Scheme 7. However, this reaction scheme requires the formation of **55** and **56** in equal amounts since they would be generated by the disproportionation of the thiosulphinate **57**.

All the data fit better a radical mechanism with the initial homolytic fission of the SO—O bond; this is shown in Scheme 8 for the ethyl derivative **54b**[30].

$$p\text{-Tol}\overset{\overset{\displaystyle O}{\|}}{S}\text{OEt} \longrightarrow p\text{-TolSOH} + \text{MeCHO}$$

$$2\,p\text{-TolSOH} \longrightarrow p\text{-Tol}\overset{\overset{\displaystyle O}{\|}}{S}\text{STol-}p + H_2O$$

$$\underset{(\mathbf{57})}{2p\text{-Tol}\overset{\overset{\displaystyle O}{\|}}{S}\text{STol-}p} \longrightarrow \underset{(\mathbf{55})}{p\text{-TolSO}_2\text{STol-}p} + \underset{(\mathbf{56})}{p\text{-TolSSTol-}p}$$

SCHEME 7

The alkoxy radical generates the aldehyde through hydrogen abstraction by the sulphinyl radical; the acetaldehyde then photolyses giving rise to the gaseous products[31]. The formation of the sulphur-containing products can be explained by the intermediacy of sulphonyl radicals which have been shown to generate sulphonic acids, thiosulphonates and disulphides[31]. The radical mechanism with cleavage of the SO—O bond is also supported by the photolysis of ethyl *p*-toluenesulphinate selectively labelled with $^{18}O$ (0.37 atom%) at the sulphinyl oxygen which gives enriched *p*-tolyl *p*-toluenethiosulphonate (0.54 atom%). This result indicates that, although not exclusively, the homolytic cleavage of the SO—O bond predominates over other modes of bond breaking[30].

The presence of *p*-tolyl allyl sulphone in the photolysis of the allyl sulphinate **54e** is noteworthy. The proposed mechanism for its formation is shown in Scheme 9.

The cyclic mechanism has been preferred to the simpler mechanism which implies formation of allyl and sulphonyl radicals and recombination to give *p*-tolyl allyl sulphone, on the basis of the observation that, in the photolysis of benzyl *p*-toluenesulphinate **54f** (**54f**; $R = CH_2Ph$), the corresponding sulphone is not formed. In fact, the great stability of the benzyl radical would favour the dissociative mechanism. Therefore the behaviour of the allyl sulphinate **54e** must be considered as a special case due to the structure of the allyl residue.

Optically active *l*-menthyl *l*-*p*-toluenesulphinate partially loses optical activity upon irradiation. This process is much faster than photolysis; in fact irradiation for short times leads to toluenesulphinates with about 60% of retention of optical activity while no appreciable photolysis is observed. This result has been explained assuming the formation of sulphinyl and alkoxy radicals followed by recombination in the solvent cage. This hypothesis also provides an explanation for the slower photolysis of the octyl sulphinate

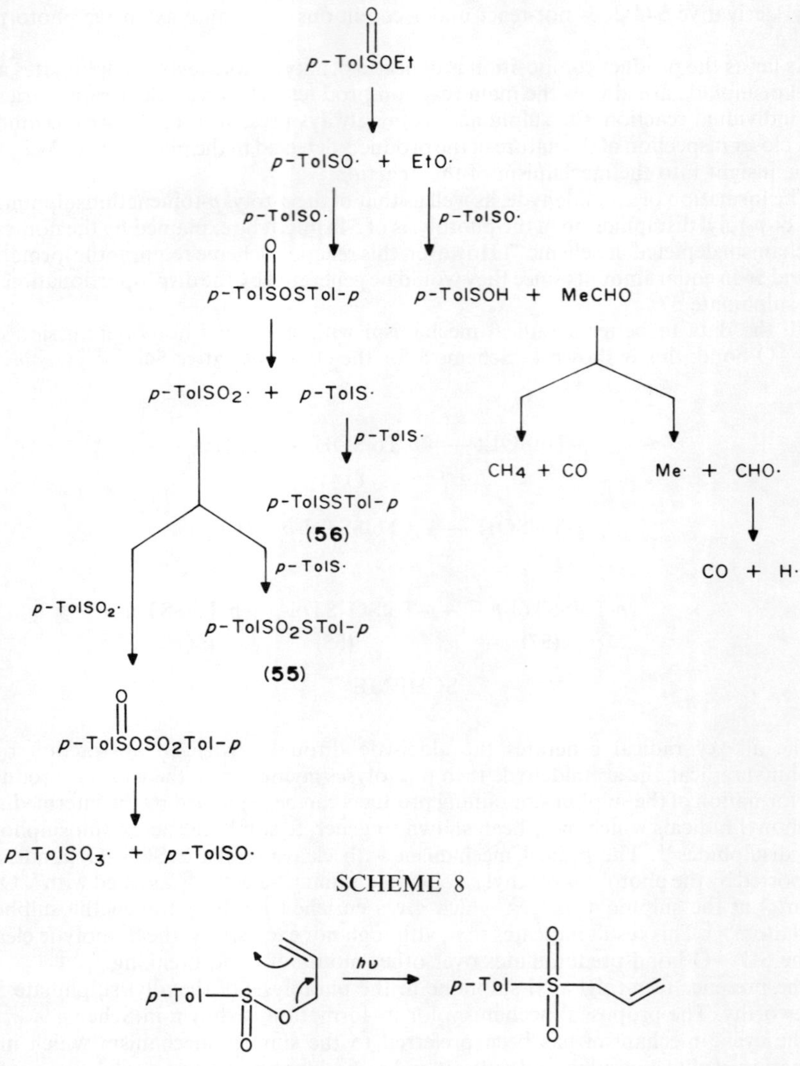


SCHEME 8

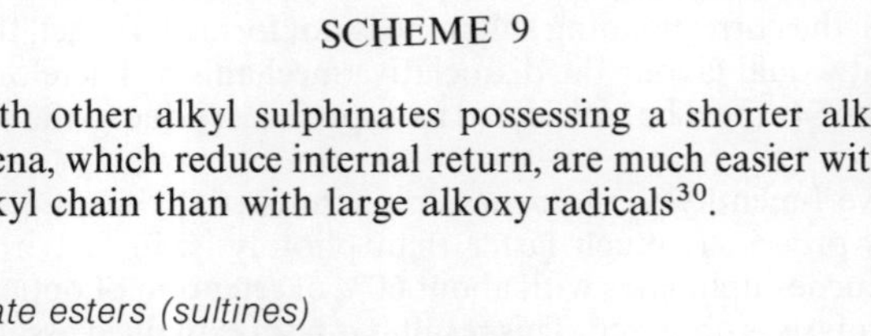


(**54**e)

SCHEME 9

**54d** compared with other alkyl sulphinates possessing a shorter alkyl chain. In fact diffusion phenomena, which reduce internal return, are much easier with alkoxy radicals having a short alkyl chain than with large alkoxy radicals[30].

*2. Cyclic sulphinate esters (sultines)*

Cyclic sulphinate esters (sultines) basically undergo two types of photoreaction: sulphur dioxide extrusion or a sultine-to-sulphone rearrangement.

5-Phenyl-1,2-dioxolan-2-oxides derivatives **58a** and **58b** give, upon irradiation at 253 nm in benzene–acetone solvent, very high yields (>90%) of the phenylcyclopropenes **59a** and **59b**, respectively (equation 13)[16,17].

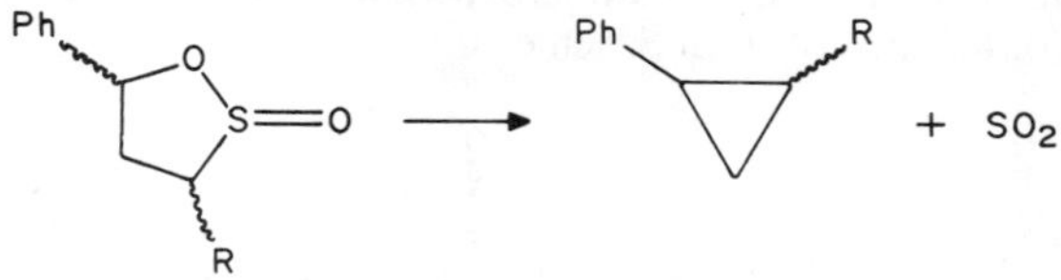

(13)

**(58)** **(59)**

**(58a, 59a)** R=H

**(58b, 59b)** R=Me

The presence of the phenyl substituent at the 5-position of the γ-sultines seems to have a very important role in determining the photoreactivity of γ-sultines; in fact the sultines **60** and **61** do not react under reaction conditions similar to those in which the cyclopropanes **59** are formed[17]. Moreover also the reverse reaction, namely the photochemical insertion of sulphur dioxide into a carbon–carbon bond of cyclopropane derivatives[15] (see Section II.B), shows a strong dependence on the substituent at the ring.

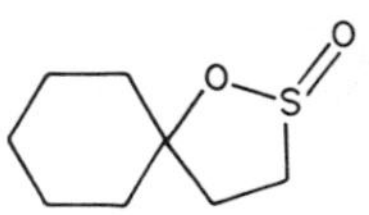

**(60)**

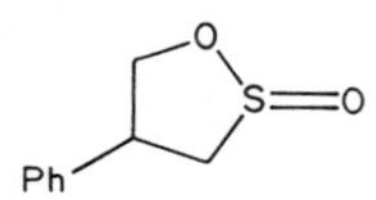

**(61)**

The photolysis of benzo-fused γ-sultines **62b** and **62c** in methanol gave fluorene **63a** and 9-phenylfluorene **63b**, respectively[17] (equation 14).

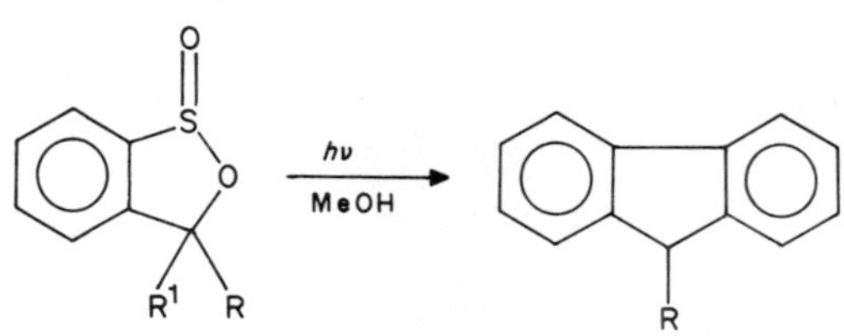

(14)

**(62)** **(63)**

**(62a)** R = R$^1$ = H
**(62b)** R = H, R$^1$ = Ph
**(62c)** R = R$^1$ = Ph
**(62d)** R = R$^1$ = Me

**(63a)** R = H
**(63b)** R = Ph

In the case of the photolysis of **62b**, the methyl ether **64** was also obtained as secondary product.

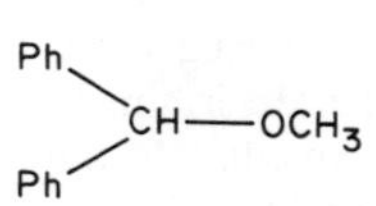

**(64)**

Quite interesting is also the observation that, when a single isomer of **62b** was irradiated for a short time, almost equal amounts of *cis* and *trans* **62b** were detected[17]. This indicates that a photochemical epimerization process is operative and that this process is faster than the process leading to the fluorene. The whole experimental evidence led to rationalization of the photolysis of **62b** described in Scheme 10.

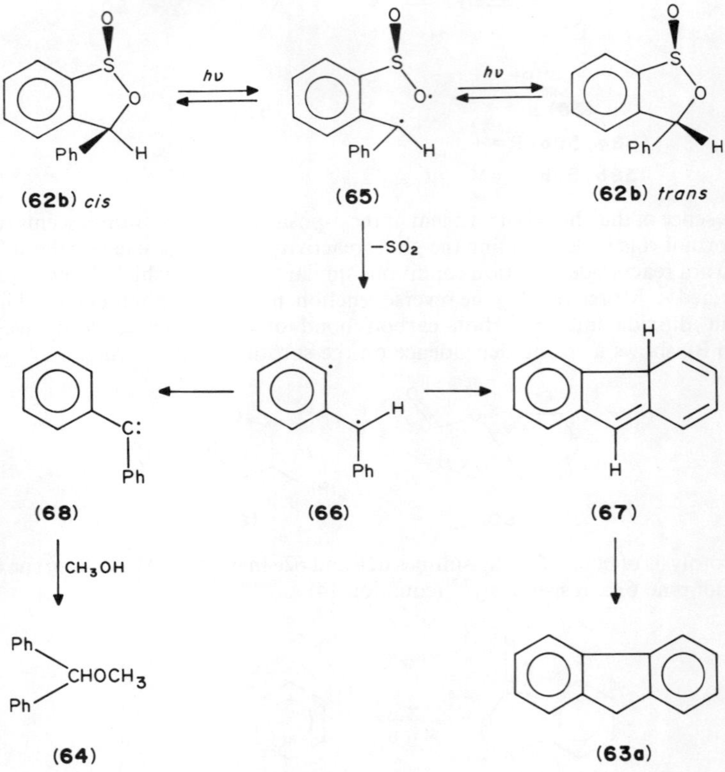


SCHEME 10

Reversible bond breaking of the carbon–oxygen bond causes isomerization of **62b**. The isolated products arise from the diradical **65** that loses sulphur dioxide to give **66**; ring closure to **67** and hydrogen migration give the fluorene **63a**. The ether **64** is also formed from **66** by hydrogen migration to give the carbene **68**, which is then trapped by the solvent.

The sultines **62a** and **62d**, which do not have phenyl substituents on the heterocycle, are less reactive than the phenyl-substituted derivatives **62b** and **62c** and react photolytically only after prolonged irradiation; however, the reaction products have not been identified.

The six-membered ring sultine **69** and the benzo-fused δ-sultine **70** behave differently than the corresponding five-membered ring derivatives[17]. The δ-sultine **69** gives the sulphinic acid **71** as a single isomer of undetermined stereochemistry. It has been suggested that this reaction occurs via the intermediate diradical **72** (equation 15).

Photolysis of **70** gave the sulphone **73** in quantitative yield (equation 16).

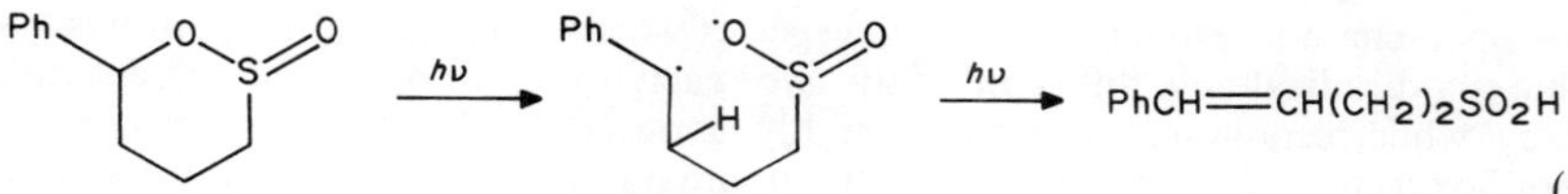

(69) (72) (71)

(15)

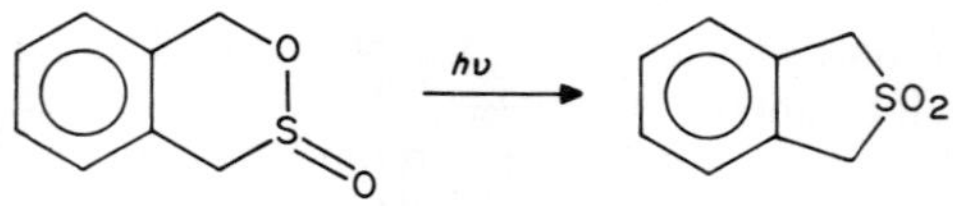

(70) (73)

(16)

Seven- and eight-membered ring sultines **74** and **75** have also been photolyzed[17,33]. The sultine **74** gave the hydrocarbon **76**, the product of sulphur dioxide extrusion and carbon–carbon bond formation[33] (equation 17). A similar behaviour is also shown by the sultine **75** that gave the 9,10-dihydrophenantrene **77** together with some phenantrene in ratios depending on the reaction conditions[17,33] (equation 18).

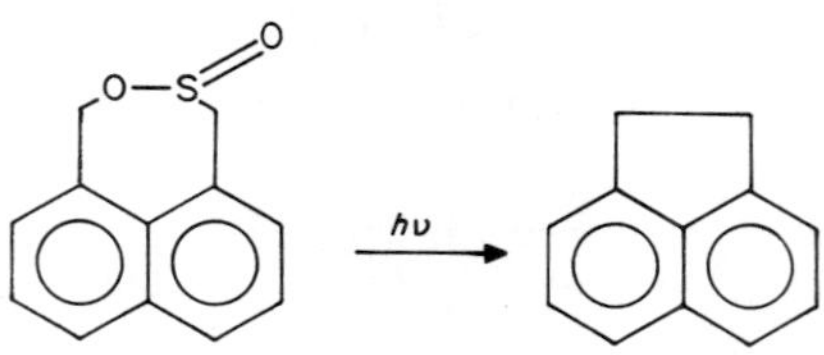

(74) (76)

(17)

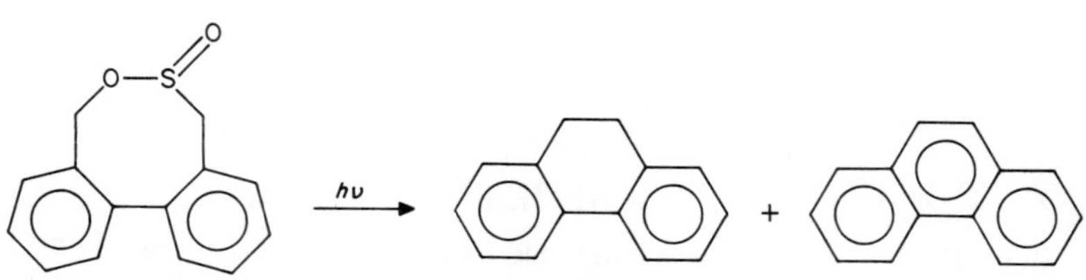

(75) (77)

(18)

The photolysis of sultines of various ring sizes has been explained assuming the homolytic cleavage of the carbon–oxygen bond of the heterocycle[16,17,33]. On the other hand, compelling evidence for the preferred sulphur–oxygen bond homolysis has been inferred for the photolysis of open-chain sulphinates[30]. This might be only an apparent discrepancy if one considers that both mechanisms have been shown to be reversible. Therefore a general scheme for the photolysis of both open-chain and cyclic sulphinates can be drawn (Scheme 11).

$$R\cdot + \cdot O\overset{\overset{O}{\|}}{S}R^1 \rightleftarrows RO\overset{\overset{O}{\|}}{S}R^1 \rightleftarrows RO\cdot + \cdot \overset{\overset{O}{\|}}{S}{-}R^1$$

SCHEME 11

This simple scheme might explain why the sultines do not give reactions similar to those of the open-chain sulphinates. In fact the species which is formed in the photolysis of a sultine is a diradical and 'diffusion' of the two reactive centres out of the solvent cage is unlikely while recombination becomes highly favoured. Under these circumstances the carbon–oxygen bond cleavage with sulphur dioxide elimination or sultine–sulphone rearrangement may take place.

### C. Sulphites, Chlorosulphites and Sulphinamides

Limited information only is available on the photochemical behaviour of sulphites, chlorosulphites and sulphinamides.

Dibutyl sulphite **78** has been photolysed at room temperature using a high-pressure mercury lamp[30]. Many products were formed and identified in the photolysis of **78**; among them butanol, butyraldehyde, 1-butene, butane, propane, propene, ethylene, ethane, methane, sulphur dioxide and carbon monoxide were found. This result points to a homolytic cleavage of the sulphur–oxygen bond with formation of the two radicals **79** and **80**; the formation of the observed products has been explained as shown in Scheme 12.

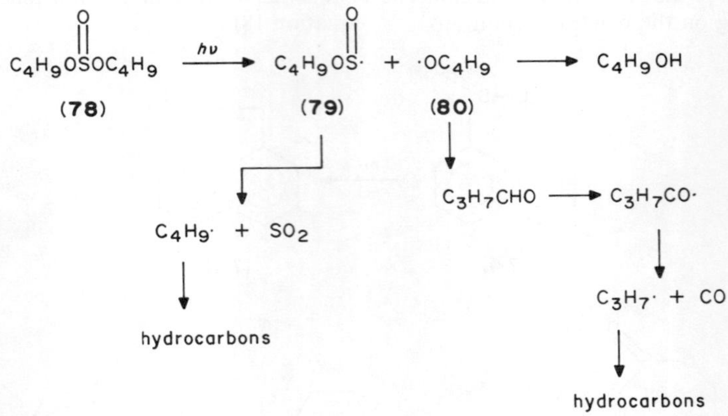


SCHEME 12

When butyl chlorosulphite **81** was irradiated under the same conditions as **78**, butyl chloride, sulphur dioxide, hydrochloric acid and some dibutyl sulphite **78** were formed[30]. The presence of butyl chloride suggests that the fission of the chlorine–sulphur bond is the preferred initial cleavage (Scheme 13); however, the presence of **78** might indicate that sulphur–oxygen bond fission also occurs.

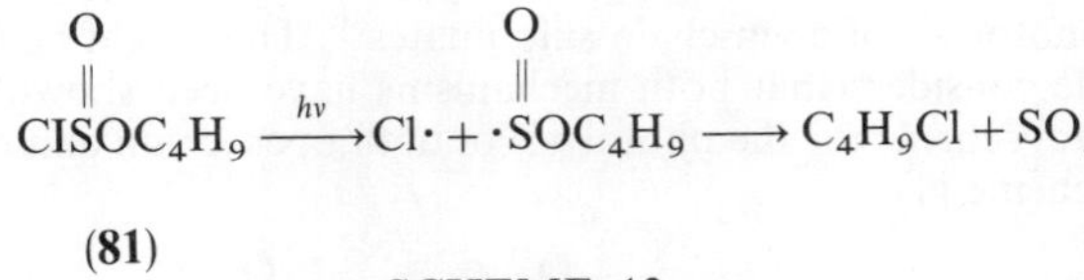


SCHEME 13

Photolysis of methanol solutions of *p*-toluenesulphinamides gave substitution of the amino group with the alkoxy group producing the methyl *p*-toluenesulphinate ester **54a**[34].

### D. Sulphinic Acid Derivatives in Photopolymerization

The importance of arylsulphinic acids and arylsulphinate ions in photopolymerization is well established on the basis of a large number of patents and papers reported in the literature.

In some cases the mechanism involving sulphinate ions in the photopolymerization reactions was elucidated[35–37]. In particular, the use of *para*-substituted benzenesulphinate ions **82a**–**f** in the presence of methylene blue as sensitizer for the photopolymerization of acrylamide monomer in aqueous solution (pH = 7, $\lambda = 666$ nm) received much attention, since the rapid polymer formation under irradiation conditions is an interesting entry in the field of the imaging process formations[35].

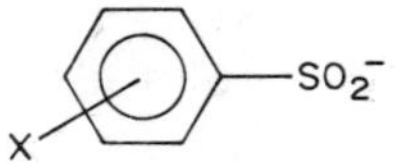


**(82)**

(**82a**) X = H
(**82b**) X = 4-Me
(**82c**) X = 4-MeCONH
(**82d**) X = 4-Cl
(**82e**) X = 4-Br
(**82f**) X = 4-$NO_2$
(**82g**) X = 4-$NH_2$
(**82h**) X = 2-$NH_2$
(**82i**) X = 2-Me
(**82j**) X = 2-$NO_2$
(**82k**) X = 2, 4, 6-Me

The results obtained from flash photolysis, fluorescence measurements and quantum yields of monomer polymerization were used to clarify the role of the *para*-substituted benzenesulphinate ions **82** and that of methylene blue ($D^+$) which acted as sensitizer.

The reactions involved during the photopolymerization of acrylamide under the above conditions are summarized in equations 19–28.

$$D^+ \xrightarrow{h\nu} {}^1D^+ \tag{19}$$

$$^1D^+ \longrightarrow {}^3D^+ \tag{20}$$

$$^1D^+ + RSO_2^- \longrightarrow D^+ + RSO_2^- \tag{21}$$

$$^3D^+ + RSO_2^- \longrightarrow D\cdot + RSO_2\cdot \tag{22}$$

$$D\cdot + RSO_2\cdot \longrightarrow D^+ + RSO_2^- \tag{23}$$

$$2D\cdot + H_2O \longrightarrow D^+ + DH + OH^- \tag{24}$$

$$RSO_2\cdot + M \longrightarrow RSO_2M\cdot \tag{25}$$

$$RSO_2M\cdot + xM \longrightarrow RSO_2\dot{(M)}_{x+1} \tag{26}$$

$$RSO_2(M)_x\cdot + D\cdot \longrightarrow RSO_2(M)_xD \tag{27}$$

$$2RSO_2(M)_x\cdot \longrightarrow RSO_2(M)_x(M)_xSO_2R \tag{28}$$

The dye quenching was observed when the concentration of benzenesulphinate ions was $10^{-3}$–$10^{-2}$ molar or higher, whereas at lower concentrations no quenching was found. In the first case the ion reacted with the excited single state of the methylene blue ($^1D^+$) according to equation 21.

The effect of the *para* substituents of the sulphinate salts on the quenching constants of the dye was also studied and found to be in the order $NO_2 >$ MeCONH > Br > Cl > Me > H.

When the concentration of benzenesulphinate ions was low, the triplet state ($^3D^+$)

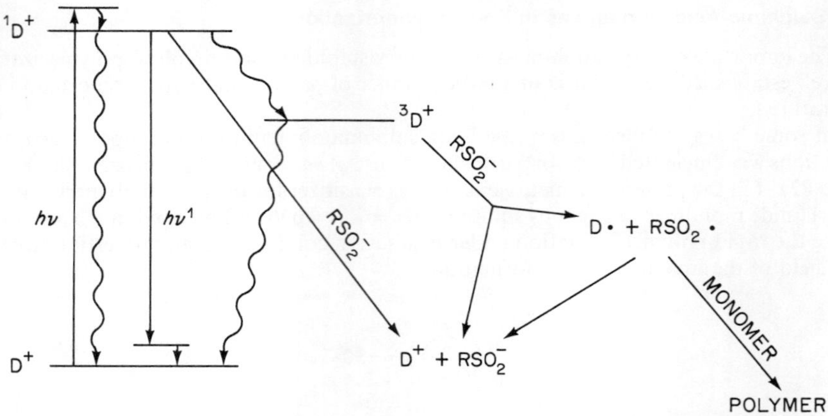


FIGURE 3. Schematic diagram for reaction mechanisms in dye-sensitized photopolymerization with sulphinate ions. Reprinted with permission from Margerum *et al.*, *J. Phys. Chem.*, **75**, 3066. Copyright (1971) American Chemical Society (Ref. 35)

obtained from $^1D^+$ (equation 20) reacted with the sulphinate ion to give D· and sulphinate radicals (equation 22). The initiators of the polymerization reaction of acrylamide are the sulphinate radicals rather than D· (equation 25). In fact sulphonyl residues were found bonded to the polymers.

The electron-withdrawing effect of the *para* substituents on the benzenesulphinate ions reduces the reaction rates of equation 22. The reaction of $ArSO_2$· and D· to give $D^+$ and $RSO_2^-$ (equation 23) was slower in the presence of acrylamide.

The reaction of the sulphonylated monomer radicals ($RSO_2M$·) with acrylamide monomers (M) to give the new radicals $RSO_2(M)\cdot_x$ is the propagation step (equation 26).

The dye radical (D·) may react with polymer radicals ($RSO_2(M)\cdot_x$) according to equation 27 or give rise to a disproportionation reaction (equation 24) leading to leucomethylene blue (DH) and methylene blue ($D^+$). The alternative final stage of polymerization is shown in equation 28.

The schematic diagram of the mechanism of the dye-sensitized photopolymerization is shown in Figure 3.

The dye-sensitized polymerization of acrylamide was also studied with *ortho*-substituted benzenesulphinate ions **82h**–**k** and the results compared with those obtained from the same substituents in the *para* position[36]. In particular, the methyl and the amino substituents at the *para* or at the *ortho* position of the benzenesulphinate ion do not affect the rate of photopolymerization, whereas in the case of the nitro substituent, the rate for the *ortho* derivative **82j** was found to be three time slower than that of the *p*-nitro derivative **82f**.

The thermal stability and the effect of pH on photopolymerization reactions of barium acrylate in the presence of *p*-toluenesulphinate ions and methylene blue was also investigated[37]. Heating of photopolymerizable solutions at pH 6 resulted in an initial increase of the photosensitivity whereas, after longer time, a desensitization process occurred. The latter effect, which is more evident at low pH values, is due to the concomitant addition reaction of the sulphinic acid, or the sulphinate salt, to the acrylic species which leads to photostable sulphones (equation 29).

$$RSO_2^- + CH_2{=}CHCOO^- + H_2O \longrightarrow RSO_2CH_2CH_2COO^- + OH^- \qquad (29)$$

Finally, it should be mentioned that sulphinate derivatives have been largely employed in several technological fields like imaging[38–65], odontology[66,67] and photopolymerization processes, mainly for photographic applications[68–79].

## IV. REFERENCES


1. C. Chatgilialoglu, in *The Chemistry of Sulphones and Sulphoxides* (Eds. S. Patai, Z. Rappoport and C. Stirling), Wiley, London, 1988, p. 1081.
2. C. Chatgilialoglu, in *The Chemistry of Sulphones and Sulphoxides* (Eds. S. Patai, Z. Rappoport and C. Stirling), Wiley, London, 1988, p. 1089.
3. C. L. McIntosh, P. de Mayo and R. W. Yip, *Tetrahedron Lett.*, 37 (1967).
4. R. F. Langler, Z. A. Marini and J. A. Pincock, *Can. J. Chem.*, **56**, 903 (1978).
5. F. S. Dainton and K. J. Ivin, *Trans. Faraday Soc.*, **46**, 374 (1950).
6. J. R. Nooi, P. C. van der Hoeven and W. P. Haslinghuis, *Tetrahedron Lett.*, 2531 (1970).
7. S. Farid, *J. Chem. Soc., Chem. Commun.*, 73 (1971).
8. R. D. Penzhorn, L. Stieglitz, W. G. Filby and K.'Guenther, *Chemosphere*, **2**, 111 (1973).
9. R. W. Murray and S. L. Jindal, *J. Org. Chem.*, **37**, 3516 (1972).
10. J. A. Barltrop, P. M. Hayes and M. Calvin, *J. Am. Chem. Soc.*, **76**, 4348 (1954).
11. W. Ando, J. Suzuki, T. Arai and T. Migita, *J. Chem. Soc., Chem. Commun.*, 477 (1972).
12. W. Ando, J. Suzuki, T. Arai and T. Migita, *Tetrahedron*, **29**, 1507 (1973).
13. A. Schonberg, G. O. Schenck and O. Neumuller, *Preparative Organic Photochemistry*, Springer-Verlag, New York, 1968, p. 444.
14. R. S. Goudie and P. N. Preston, *J. Chem. Soc. (C)*, 3081 (1971).
15. D. E. Applequist and L. F. McKenzie, *J. Org. Chem.*, **42**, 1251, (1977).
16. N. K. Sharma, F. Jung and T. Durst, *Tetrahedron Lett.*, 2863 (1973).
17. T. Durst, J. C. Huang, N. K. Sharma and D. J. H. Smith, *Can. J. Chem.*, **56**, 512 (1978).
18. K. Gollnick, *Adv. Photochem.*, **6**, 110 (1968).
19. H. J. Backer and H. Kloosterziel, *Recl. Trav. Chim. Pays-Bas*, **73**, 129 (1954).
20. D. Barnard, *J. Chem. Soc.*, 4675 (1957).
21. C. J. Cavallito, J. S. Buck and C. M. Suter, *J. Am. Chem. Soc.*, **66**, 1952 (1944).
22. P. Koch, E. Ciuffarin, and A. Fava, *J. Am. Chem. Soc.*, **92**, 5971 (1970).
23. J. L. Kice, G. C. Venier, G. B. Large and L. Heasley, *J. Am. Chem. Soc.*, **91**, 2028 (1969).
24. G. O. Schenck and C. H. Krauch, *Chem. Ber.*, **96**, 517 (1963).
25. L. Weil, *Arch. Biochem. Biophys.*, **110**, 57 (1965).
26. L. Horner and O. H. Basedow, *Justus Liebigs Ann. Chem.*, **612**, 108 (1958).
27. R. M. Adlington and A. G. M. Barrett, *J. Chem. Soc., Perkin Trans. 1*, 1076 (1980).
28. J. D. Macke and L. Field, *J. Org. Chem.*, **53**, 396 (1988).
29. P. K. Srivastava and L. Field, *J. Chem. Eng. Data*, **31**, 252 (1986).
30. M. Kobayashi, H. Minato, Y. Miyaji and T. Yoshioka, *Bull. Chem. Soc. Jpn.*, **45**, 2817 (1972).
31. J. G. Calvert and J. N. Pitts *Photochemistry*, Wiley, New York, 1966, p. 371.
32. C. M. M. da Silva Correa and W. A. Waters, *J. Chem. Soc. (C)*, 1874 (1968).
33. D. N. Harpp and D. F. Mullins, *Can. J. Chem.*, **61**, 757 (1983).
34. H. Tsuda, H. Minato and M. Kobayashi, *Chem. Lett.*, 149 (1976).
35. J. D. Margerum, A. M. Lackner, M. J. Little and C. T. Petrusis, *J. Phys. Chem.*, **75**, 3066 (1971).
36. J. D. Margerum, R. G. Brault, A. M. Lackner and L. J. Miller, *J. Phys. Chem.*, **77**, 2720 (1973).
37. L. J. Miller, J. D. Margerum, J. B. Rust, R. G. Brault and A. M. Lackner, *Macromolecules*, **7**, 179 (1974).
38. A. H. Herz, *U.S. patent*, 3, 144, 336; *Chem. Abstr.*, **61**, 11520e (1964).
39. O. Gorgon, *Czech. patent*, 122, 710; *Chem. Abstr.*, **68**, 64598n (1968).
40. J. D. Margerum, L. J. Miller and J. B. Rust, *Photogr. Sci. Eng.*, **12**, 177 (1968); *Chem. Abstr.*, **69**, 40084m (1968).
41. E. Klein and E. Moisar, *Belg. patent*, 722, 149; *Chem. Abstr.*, **71**, 118343h (1969).
42. Y. Z. Zaidenberg, K. M. Ginzburg, E. S. Shvaishtein, L. V. Krasnyi-Admoni and L. N. Fialka, *Usp. Nauch. Fotogr.*, **14**, 104 (1970); *Chem. Abstr.*, **74**, 81741v (1971).
43. K. Ohkubo, T. Masuda, K. Yamasue K. Hayashi, *German patent*, 2, 052, 698; *Chem. Abstr.*, **75**, 93004r (1971).
44. J. B. Rust, *U.S. patent*, 3, 642, 487; *Chem. Abstr.*, **77**, 20412d (1972).

45. J. D. Margerum and A. D. Jacobson, *U.S. patent*, 3, 694, 218; *Chem. Abstr.*, **78**, 50599h (1973).
46. J. B. Rust, *U.S. patent*, 3, 726, 688; *Chem. Abstr.*, **79**, 12004y (1973).
47. R. G. Brault, J. A. Jenney, A. M. Lackner and J. D. Margerum, *Unconventional Photogr. Syst., Symp., 3rd*, 68 (1971); *Chem. Abstr.*, **79**, 72261q (1973).
48. E. Inoue, H. Kokado and T. Yamase, *German patent*, 2, 215, 474; *Chem. Abstr.*, **80**, 54530t (1974).
49. E. Inoue, H. Kokado and T. Yamase, *German patent*, 2, 333, 793; *Chem. Abstr.*, **80**, 102309b (1974).
50. J. D. Margerum, *U.S. Natl. Tech. Inform. Serv.*, AD Rep. No. 770068/5GA (1973); *Chem. Abstr.*, **80**, 151081e (1974).
51. T. Yamase, T. Ikawa, H. Kokado and E. Inuoe, *Photogr. Sci. Eng.*, **19**, 57 (1975); *Chem. Abstr.*, **82**, 92001v (1975).
52. E. S. Almazov, V. V. Andreyanov, V. A. Bekunov, M. S. Portnova, V. P. Protsenko and L. V. Grechko, *USSR patent*, 480, 039; *Chem. Abstr.*, **84**, 24385q (1976).
53. N. Suzuki, Y. Noguchi and T. Masuda, *German patent*, 2, 511, 361; *Chem. Abstr.*, **84**, 52121u (1976).
54. R. G. Brault, C. C. DeAnda, J. A. Jenney and J. D. Margerum, *U.S. patent*, 4, 036, 647; *Chem. Abstr.*, **88**, 43768r (1978).
55. T. Masuda, *Japan patent*, 78 28, 417; *Chem. Abstr.*, **89**, 83003q (1978).
56. K. Frank, *German patent*, 2, 700, 290; *Chem. Abstr.*, **89**, 171825q (1978).
57. S. Ikegami and T. Masuda, *Japan patent*, 78 20, 923; *Chem. Abstr.*, **89**, 155609s (1978).
58. R. J. LeStrange, *U.S. patent*, 4, 175, 970; *Chem. Abstr.*, **92**, 102289t (1980).
59. J. Yamaguchi, T. Naoi, H. Sera, K. Ishigaki and M. Ogawa, *German patent*, 3, 023, 112; *Chem. Abstr.*, **94**, 217545g (1981).
60. Konishiroku Photo Industry Co. Ltd., *Japan patent*, 81, 67, 842; *Chem. Abstr.*, **96**, 13595e (1982).
61. Fuji Photo Film Co. Ltd., *Japan patent*, 81, 151, 937; *Chem. Abstr.*, **97**, 31213d (1982).
62. Process Shizai Co. Ltd and E. Inoue, *Japan patent*, 82 07, 423; *Chem. Abstr.*, **97**, 136648c (1982).
63. A. Nagashima, *Japan patent*, 60, 138, 545 [85, 138, 545]; *Chem. Abstr.*, **104**, 99547j (1986).
64. H. Hirai, Y. Yabuki and K. Sato, *Japan patent*, 61, 51, 139 [86, 51, 139]; *Chem. Abstr.*, **105**, 235893u (1986).
65. T. Ishizuka and A. Yagishita, *Japan patent*, 60, 227, 280 [85, 227, 280]; *Chem. Abstr.*, **105**, 88718k (1986).
66. T. Sakashita and S. Arata, *Japan patent*, 62, 27, 403 [87, 27, 403]; *Chem. Abstr.*, **107**, 60242t (1987).
67. T. Sakashita, *Japan patent*, 61, 296, 002 [86, 296, 002]; *Chem. Abstr.*, **106**, 210779s (1987).
68. J. N. Rust, L. J. Miller and J. D. Margerum, *Polym. Eng. Sci.*, **9**, 40 (1969); *Chem. Abstr.*, **70**, 68782g (1969).
69. J. B. Rust, *British patent*, 1, 153, 813; *Chem. Abstr.*, **71**, 50700q (1969).
70. K. Iwasaki, K. Yamaguchi, J. Inomata, Y. Otsuka and K. Kasahara, *Japan patent*, 70 13, 584; *Chem. Abstr.*, **73**, 77829s (1970).
71. L. J. Miller and J. B. Rust, *U.S. patent*, 3, 531, 282; *Chem. Abstr.*, **73**, 121175b (1970).
72. J. N. Rust, L. J. Miller and J. D. Margerum, *Polym. Prepr., Am. Chem. Soc., Div. Polym. Chem.*, **10**, 294 (1969); *Chem. Abstr.*, **73**, 120978d (1970).
73. L. J. Miller and J. B. Rust, *U.S. patent*, 3, 627, 656; *Chem. Abstr.*, **76**, 113938z (1972).
74. J. D. Margerum, *U.S. patent*, 3, 788, 858; *Chem. Abstr.*, **81**, 56666m (1974).
75. T. Yamase, T. Sumi, T. Ikawa, H. Kokado and E. Inoue, *Photogr. Sci. Eng.*, **18**, 25 (1974); *Chem. Abstr.*, **81**, 26015f (1974).
76. N. Baumann and H. P. Schlunke, *German patent*, 2, 360, 350; *Chem. Abstr.*, **81**, 170266f (1974).
77. N. Yang and S. Wong, *Polym. Prepr., Am. Chem. Soc., Div. Polym. Chem.*, **14**, 1301 (1973); *Chem. Abstr.*, **84**, 31569f (1976).
78. Hughes Aircraft Co., *Neth. patent*, 75, 06, 973; *Chem. Abstr.*, **85**, 6396k (1985).
79. W. Weissflog, P. Moeckel, K. Kock, H. Schlegel, G. Griebel, P. Veckenstedt and H. Schuelert, *German (East) DD patent*, 152, 349; *Chem. Abstr.*, **97**, 24742x (1982).

The Chemistry of Sulphinic Acids, Esters and their Derivatives
Edited by S. Patai


CHAPTER **14**

# The oxidation and reduction of sulphinic acids and their derivatives

JEFFREY HOYLE

*Chemistry-Soils Department, Nova Scotia Agricultural College, P.O. Box 550, Truro, Nova Scotia, Canada B2N 5E3*



## I. INTRODUCTION

Sulphinic acids (**1**) possess a single tri-coordinate sulphur atom that is bonded to two oxygen atoms and a carbon atom. The sulphur atom in this moiety is at the sulphur(IV) oxidation state.

$$\mathrm{R{-}\overset{\overset{\displaystyle O}{\|}}{S}{-}OH}$$

**(1)**

Oxidation of sulphinic acids, and their derivatives, results in the formation of a sulphur(VI)-containing moiety as is present in a sulphonic acid, and exemplified in equation 1. Such oxidations are the subject of Section II in this chapter. It should be carefully noted that sulphones, which are also sulphur(VI)-containing compounds, are formed from sulphinic acids by a nucleophilic displacement reaction and these reactions are the subject of another chapter in this volume.

$$RSO_2H \longrightarrow RSO_3H \tag{1}$$

Reduction of sulphinic acids, and their derivatives, results in the formation of a sulphur(II)-containing species such as a thiol or disulphide, as exemplified in equation 2. These reactions are covered in Section III of this chapter.

$$RSO_2H \longrightarrow RSH|RSSR \tag{2}$$

Disproportionation, the concomitant oxidation and reduction, of sulphinic acids and sulphinic acid derivatives, to a sulphur(VI)-containing moiety and a sulphur(II)-containing moiety, is covered in Section IV. Such processes are extremely important for many sulphur compounds at the sulphur(IV) oxidation level.

This chapter covers the oxidation, reduction and disproportionation of sulphinic acids and their derivatives up to the middle of 1988. The derivatives, other than sulphinic acids and their salts, discussed here are sulphinic acid esters (**2**), sulphinyl halides (**3**), sulphinamides (**4**) and thiosulphinates (**5**). Whenever the terms sulphinate ester or sulphinic acid ester are used in this work they refer to O-esters only. The S-esters are referred to as thiosulphinates in all cases.

Neither oxidative nor reductive desulphurization are covered in this review.

$$\mathrm{R{-}\overset{\overset{\displaystyle O}{\|}}{S}{-}OR'} \qquad \mathrm{R{-}\overset{\overset{\displaystyle O}{\|}}{S}{-}X} \qquad X = F, Cl, Br, I$$

**(2)** **(3)**

$$\mathrm{R{-}\overset{\overset{\displaystyle O}{\|}}{S}{-}NH_2} \qquad \mathrm{R{-}\overset{\overset{\displaystyle O}{\|}}{S}{-}SR'}$$

**(4)** **(5)**

## II. OXIDATION

The oxidation of sulphinic acids and their derivatives have been studied by many workers for at least the last hundred years. Much of the very early work concentrated on the formation of sulphonic acids from the corresponding sulphinic acids. Work in the early part of this century also began to consider the oxidation of thiosulphinates, as part of the study of the oxidation of disulphides and their derivatives. These latter studies have generated some controversy concerning the structure of the initial oxidation product. Early work tended to favour the thiosulphonate structure (**6**) whilst later workers have been increasingly in favour of the α-disulphoxide (**7**). There is now very strong evidence to show that the α-disulphoxide is initially formed and this will be discussed in more detail

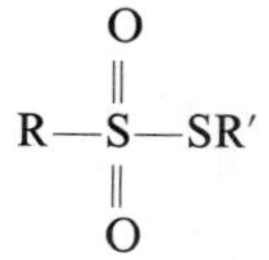

(6)

(7)

where appropriate. An excellent *ab initio* study[1] has been published which details the energetics involved in the conversion of α-disulphoxides to thiosulphonates.

## A. General Oxidation Methods

### 1. Nitric acid and nitrogen oxides

Nitric acid is one of the most common, and cheaper, oxidants used in organic chemistry, which produces few by-products. It is thus not surprising that nitric acid was one of the earliest oxidizing agents used for the conversion of sulphinic acids into their sulphonic acid analogues. Thus benzenesulphinic acid is converted into benzenesulphonic acid in good yield[2]. Other aromatic sulphinic acids undergo a similar conversion although ring nitration occurs in the presence of excess oxidant[3], as exemplified in equation 3. Both

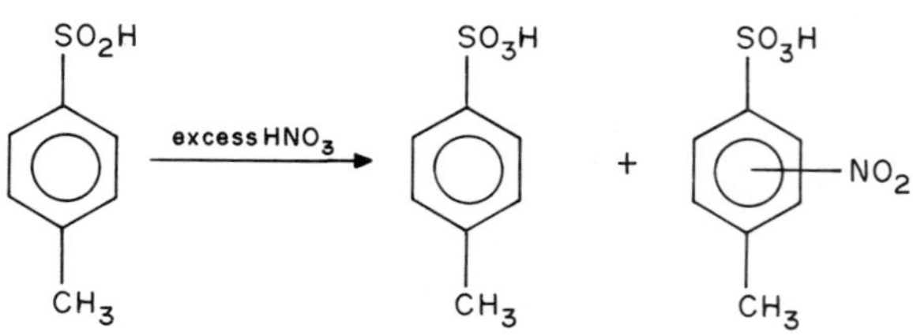


(3)

aliphatic sulphinic acids and thiosulphinates are unstable in the presence of nitric acid and so no synthetically useful reactions have been reported.

The only oxide of nitrogen that has been reported to oxidize sulphinic acids and their derivatives in dinitrogen tetraoxide. In the presence of dinitrogen tetraoxide aromatic sulphinic acids are converted to novel sulphonyl nitrites and sulphonic acids[4] (equation 4).

$$\mathrm{Ar}\overset{\overset{\displaystyle O}{\|}}{\mathrm{S}}\!-\!\mathrm{OH} \xrightarrow{N_2O_4} \mathrm{Ar}\overset{\overset{\displaystyle O}{\|}}{\mathrm{S}}\!-\!\mathrm{NO} + \mathrm{ArSO_3H} \qquad (4)$$

Dinitrogen tetraoxide may also be used to oxidize alkyl aryl thiosulphinates[5] into either a mixture of sulphonic acids (equation 5), or the corresponding thiosulphonate and the alkyl sulphonic acid (equation 6). In both cases a small quantity of sulphonic anhydrides ($RSO_2OSO_2R$) are formed.

$$\mathrm{PhSOSCH_3} \xrightarrow{N_2O_4} \mathrm{PhSO_3H} + \mathrm{CH_3SO_3H} \qquad (5)$$

$$\mathrm{PhSSOCH_3} \xrightarrow{N_2O_4} \mathrm{PhSO_2SPh} + \mathrm{CH_3SO_3H} \qquad (6)$$

*2. Oxygen and ozone*

Oxygen, in the air, is probably the cheapest, most readily available oxidizing agent and may be used to convert sulphinic acids into sulphonic acids by an autocatalytic, radical chain mechanism. Such a reaction has been reported[6,7], and a mechanism, based on careful kinetic studies in many solvents, has been proposed, as detailed in equation 7.

$$\begin{aligned} PhSO_2H + In^{\bullet} &\longrightarrow PhSO_2^{\bullet} \\ PhSO_2^{\bullet} + O_2 &\longrightarrow PhSO_2OO^{\bullet} \\ PhSO_2OO^{\bullet} + PhSO_2H &\longrightarrow PhSO_2OOH + PhSO_2^{\bullet} \\ PhSO_2OOH + PhSO_2H &\longrightarrow 2\,PhSO_3H \end{aligned} \qquad (7)$$

Phenylpersulphonic acid ($PhSO_2OOH$), implicated as the key intermediate in the mechanism, has precedence in other reactions. This autocatalytic oxidation of sulphinic acids can be easily prevented by the addition of an antioxidant, such as benzaldehyde, which is a better radical scavenger than sulphinic acids.

Oxygen has also been used for the synthetic formation of a sultone (a cyclic sulphonate ester) from a sulphinic acid[8], as shown in equation 8. In addition, oxygen has been used to oxidize sulphinate ligands in iron(III) and indium(III) sulphinato porphryns to the sulphonate oxidation level[9,10].

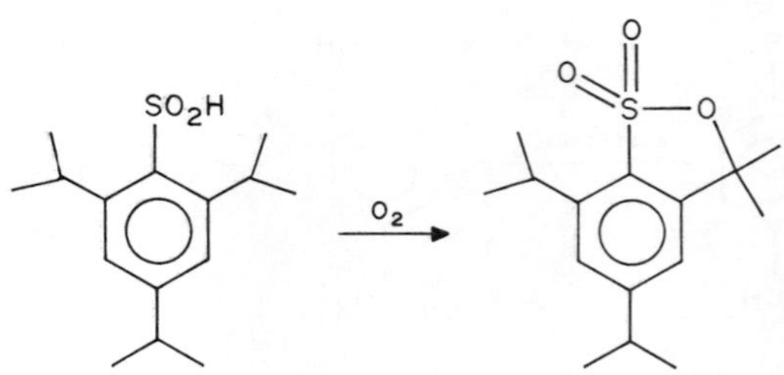


(8)

Reports of the oxidation of sulphinic acids, and their derivatives, employing ozone as oxidant are surprisingly scarce. One notable exception is the preparation of sulphonic anhydrides from thiosulphinates in 85–100% yields[11], as indicated in equation 9.

$$ArSOSAr + O_3 \longrightarrow ArSO_2OSO_2Ar \qquad (9)$$

Superoxide ion, generated *in situ* by the reaction of potassium superoxide with a crown ether, has been successfully employed for the oxidation of sulphinic acids, sulphinyl chlorides and thiosulphinates to the sulphur(VI) oxidation level under mild, inert conditions[12–14]. It is rather surprising that these reactions all proceed so readily when it is usually considered that superoxide ion is a rather weak oxidizing agent. In fact, superoxide ion may act as a reductant or an oxidant depending on the reaction conditions[15] and in some cases reduction products are evident.

Using this reagent, sodium arylsulphinates are converted, in good yields, to sulphonic acids using one equivalent of potassium superoxide at 25 °C in 2.5 h. The first step of the reaction is a one-electron process of the type initially proposed by Berger[16]. This is then followed by superoxide uptake giving the peroxysulphonate which decomposes to give the sulphonic acid and oxygen, as shown in equation 10.

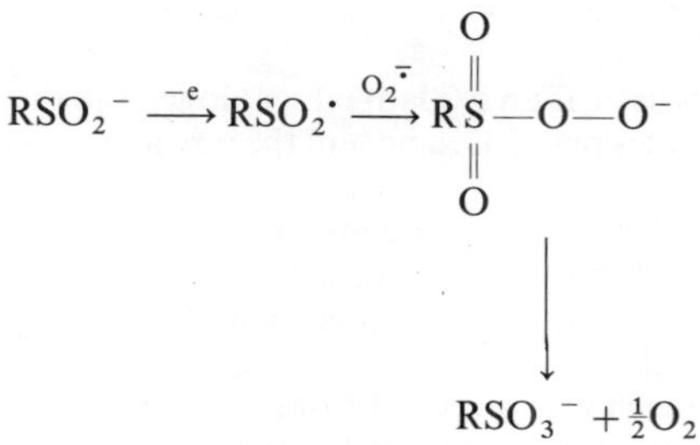


(10)

Aromatic sulphinyl chlorides are oxidized to sulphonic acids in 90 min at 20 °C using excess potassium superoxide. In this case the reaction is initiated by the nucleophilic attack of superoxide on the sulphinyl chloride. A subsequent one-electron transfer from superoxide followed by rearrangement gives the sulphonic acid in good yield, as indicated in equation 11.

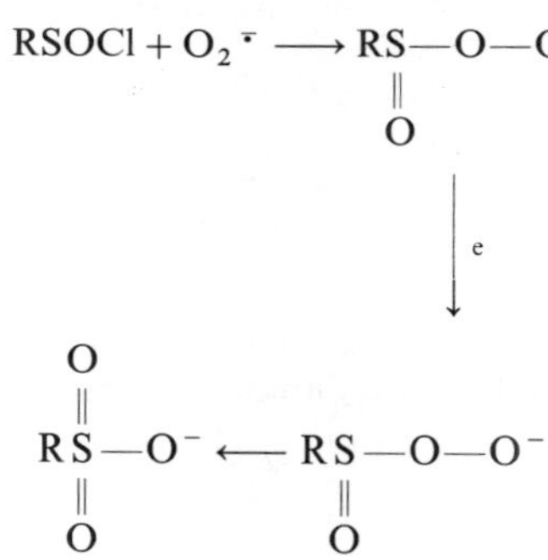


(11)

Thiosulphinates are even more easily oxidized by superoxide. The reaction occurs even at −40 °C in about 30 min using excess superoxide. The products formed are a disulphide, derived from the sulphenyl side of the thiosulphinate, and a sulphonic acid from the sulphinyl side of the thiosulphinate. In this case there are two postulated routes to the sulphonic acid, one involving a sulphonyl radical which presumably proceeds to the acid as shown in equation 10 above. The second route is via a peroxysulphinate which would form the sulphonic acid by rearrangement. The postulated pathway for the oxidation of thiosulphinate by superoxide is shown in equation 12.

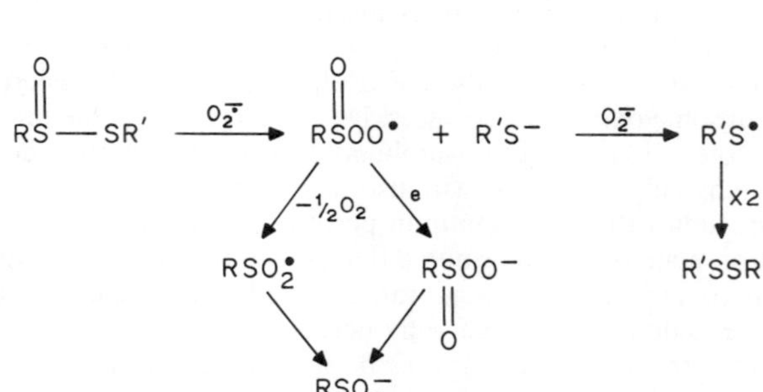


(12)

### 3. Hydrogen peroxide

Hydrogen peroxide is used as an oxidant either alone or in the presence of acetic acid. In the latter case, the oxidant is peracetic acid and these reactions will be dealt with in the next part of this section.

Hydrogen peroxide has been used to convert sulphinic acids and thiosulphinates into products containing a sulphur(VI) moiety under a variety of conditions. On the other hand, sulphinamides are apparently unaffected by this reagent[17]. In 1935, Hann synthesized a series of chemotherapeutic agents, one of which was *p*-fluorophenylsulphonic acid which was prepared from the sulphinic acid using excess hydrogen peroxide at room temperature[18]. Other workers have also oxidized barium salts of aromatic sulphinic acids to the corresponding sulphonic acids in 31–60% yields using the same procedure[19].

Two careful kinetic studies[20,21] have shown that the oxidation of alkali metal salts of arylsulphinic acids proceeds via a second-order reaction, over a wide pH range. It was concluded in one of these studies[21] that the rate-determining step involves the nucleophilic attack of the sulphinate ion on the neutral hydrogen peroxide molecule with a rate constant of $0.02\,M^{-1}s^{-1}$ at 40 °C. The overall mechanism proposed is shown in equation 13.

$$ArSO_2^- + HO{-}OH \xrightarrow{\text{slow}} ArSO_3H + OH^- \xrightarrow{\text{fast}} ArSO_3^- + H_2O \qquad (13)$$

One hydrogen peroxide oxidation of a sulphinic acid (equation 14) has been used in a commercial pilot plant[22] and this procedure was apparently the best method available.

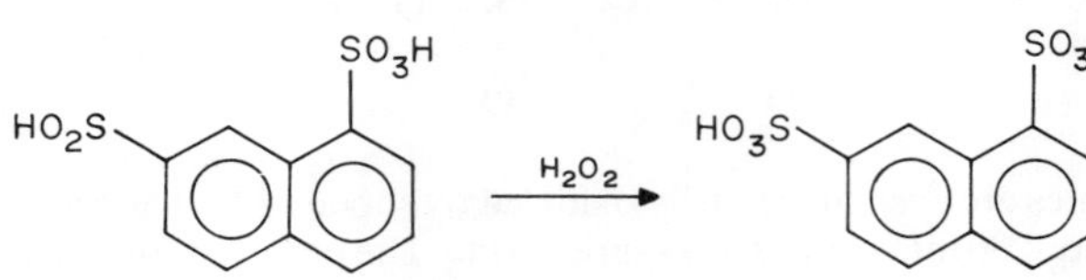


(14)

Sulphinic acid esters have also been oxidized, to the sulphonic acid ester, with hydrogen peroxide although the reaction proceeds in poor yield[23].

### 4. Other peroxy species

Peroxy species, other than hydrogen peroxide, have been widely used for the study of the oxidation of disulphides, and related compounds. These studies have been performed because of the importance of the oxidation of disulphides *in vivo* where peroxy species have been implicated in some oxidative processes. Due to this interest, the present section will deal mainly with the oxidation of thiosulphinates (disulphide monoxides), although the literature concerning sulphinic acids will also be covered.

In all of these studies the most common peroxy species used are peracetic acid and *m*-chloroperbenzoic acid. It should be noted that peracetic acid is usually generated *in situ* from hydrogen peroxide and acetic acid, rather than being purchased from commercial sources, since fewer side-reactions generally occur.

An early study reported the use of barium peroxide for the conversion of 3,4-dimethylphenylsulphinic acid to the corresponding sulphonic acid[24]. However, the synthetic utility of this reaction has not been reported to date.

An innovative procedure for the preparation of unstable tosylates which relies on preparing the much more stable sulphinate has been reported[25]. The tosylate is formed, when required, by oxidation of the sulphinate with *m*-chloroperbenzoic acid, as shown in equation 15. The synthetic utility of this method lies in the fact that the reaction producing

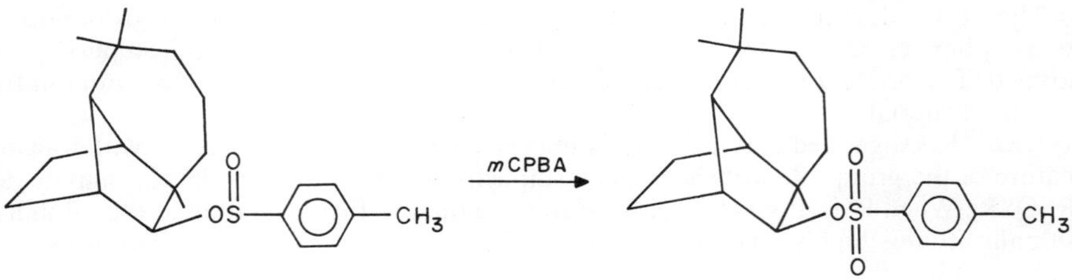


(15)

the tosylate (*p*-toluenesulphonic acid ester) occurs under extremely mild conditions (0 °C in methylene chloride). Other more normal methods of preparing tosylates were found to produce little or none of the required product.

Both cyclic[26] and acyclic[17] sulphinamides may also be oxidized to the sulphur(VI) level with *m*-chloroperbenzoic acid. However this reagent will also oxidize alkenic double bonds, present in the substrate, to the epoxide, as shown in equation 16. The best yields for

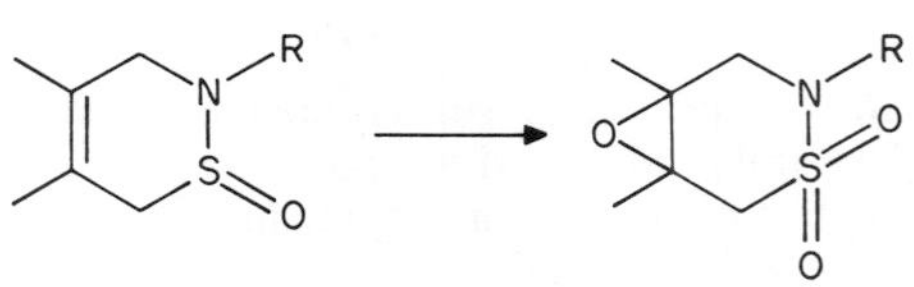


(16)

this oxidation occur when a large excess of the peracid is used under carefully buffered conditions.

Thiosulphinates are oxidized by peracid to various multi-oxygenated species[5,27–41] as shown in equation 17. The α-disulphoxide **7** and the sulphinyl sulphone **8** are not stable species although both have been identified, at low temperatures, by the use of NMR spectroscopy (for example, Reference 42). The final product formed depended upon the initial ratio of substrate to oxidizing agent. The reaction has been shown to be catalysed by vanadium pentoxide[43] and by tungsten trioxide[44].

$$\underset{}{\text{RSOSR}} \longrightarrow \underset{(\mathbf{7})}{\text{RSOSOR}} \longrightarrow \text{RSO}_2\text{SR} \longrightarrow \underset{(\mathbf{8})}{\text{RSO}_2\text{SOR}} \downarrow \text{RSO}_2\text{SO}_2\text{R} \qquad (17)$$

When the thiosulphonate is produced from an unsymmetrical thiosulphinate, there are four different products that may be formed as shown in equation 18. Which of these products is formed seems to depend upon the structure of the thiosulphinate. For example, Barnard and Percy[45] have shown that peracid oxidation of alkyl arylthiosulphinates produces all four possible thiosulphonate products. They have suggested that this is due to the initial formation of the α-disulphoxide which subsequently undergoes S—S bond scission. This process would then be followed by radical–radical combinations and rearrangement of the resulting sulphinyl radicals, as discussed in a review published in a previous volume of the present series[46].

$$RSOSR' \xrightarrow{[O]} RSO_2SR' + RSO_2SR + R'SO_2SR + R'SO_2SR' \quad (18)$$

The above described behaviour is in contrast with that seen for the oxidation of thiosulphinates as reported by other workers[34,38], where only one thiosulphonate is formed. The product is that expected by oxidation of the sulphinyl sulphur atom in the starting material.

Kice[27] has suggested that this contrasting behaviour is due to the electron-withdrawing nature of the group, R′, attached to the sulphenyl sulphur atom in the thiosulphinate, RSOSR′. When R′ is more electron-withdrawing than R, then oxidation at the sulphinyl sulphur atom is highly favoured and thus a single product is formed (equation 19).

$$RSOSR' \xrightarrow{[O]} RSO_2SR' \quad (19)$$

Contrarywise, when R′ is more electron-donating than R, then the sulphenyl sulphur atom is the preferential site for initial oxidation. In this case, the α-disulphoxide is formed in the first instance. Subsequently S—S bond scission and radical recombination allow the four thiosulphonates to be formed (equation 20).

$$RSOSR' \xrightarrow{[O]} RSOSOR' \longrightarrow RSO^{\bullet} + R'SO^{\bullet}$$

$$\longrightarrow \text{Products of equation 18} \quad (20)$$

It is perhaps useful to note that studies concerning the oxidation of proteins, and other disulphide-containing natural products, with peracids have also reported the formation of a dioxide species when starting from a thiosulphinate. This dioxide has been shown to contain a thiosulphonate group in some cases[47–49]. α-Disulphoxides have also been implicated in these oxidation reactions both *in vitro* and *in vivo*[50,51].

There are also a few examples where an unexpected product, containing three sulphur atoms, is formed on oxidation of a thiosulphinate at −40 °C with peracetic acid[30,52], as indicated in equation 21.

$$t\text{-BuSOSBu-}t \longrightarrow t\text{-BuSO}_2\text{SBu-}t + t\text{-BuSO}_2\text{SSBu-}t \quad (21)$$

*5. Chlorine-containing reagents*

There is a wide range of chlorine-containing oxidants available to organic chemists. However, only a few of these have been utilized for the oxidation of sulphinic acids, the main exceptions being chlorine and hypochlorite ions.

Arylsulphinic acids, and their alkali metal salts, have long been used as precursors for the preparation of sulphonyl chlorides. This interconversion has most often been performed with chlorine as the oxidizing agent in either water or acetic acid solvent, as shown in equation 22[2,53–61]. In some cases chlorination of the aromatic ring also occurs[62].

$$ArSO_2H + Cl_2 \longrightarrow ArSO_2Cl \quad (22)$$

Arylsulphinyl chlorides may also be converted to the sulphonyl chloride by a similar process in 80% yield[63] (equation 23). This reaction also occurs for sulphinate esters. For example, methyl methanesulphinate is converted to methanesulphonyl chloride in excellent yield at 0 °C[64], as shown in equation 24.

$$\mathrm{RSOCl + Cl_2 \longrightarrow RSO_2Cl} \quad (23)$$

$$\mathrm{CH_3\overset{\overset{O}{\|}}{S}OCH_3 \xrightarrow[Cl_2]{0\,^\circ C} CH_3\underset{\underset{O}{\|}}{\overset{\overset{O}{\|}}{S}}{-}Cl + CH_3Cl} \quad (24)$$

The only generally useful method for the preparation of tertiary alkyl sulphonyl chlorides is by the oxidation of the sulphinate formed on reaction of a Grignard reagent, from a tertiary alkyl bromide, with sulphur dioxide as shown in equation 25. The method provides a rapid, clean and simple route for the preparation of sulphonyl chlorides in good yields and high purity. This method has also been used for the preparation of arylsulphonyl chlorides and the sulphinate salt may be isolated prior to oxidation, or used *in situ*[65–67]. A similar method has been patented, using trialkyl organo-aluminium compounds in place of the Grignard reagent (equation 26)[68].

$$\mathrm{RMgBr + SO_2 \longrightarrow RSO_2^-\,MgBr^+ \xrightarrow{Cl_2} RSO_2Cl + MgBrCl} \quad (25)$$

$$\mathrm{R_3Al + 3SO_2 \longrightarrow (RSO_2)_3Al \xrightarrow{Cl_2} 3RSO_2Cl + AlCl_3} \quad (26)$$

An interesting variation of this oxidation procedure leads to an arylsulphonamide by the reaction of chlorine with an ammonium arylsulphinate in aqueous solution (equation 27)[69,70].

$$\mathrm{ArSO_2^-NH_4^+ \xrightarrow{Cl_2} ArSO_2NH_2 + 2HCl} \quad (27)$$

This reaction also takes place with hypochlorite ion, as oxidant[69]. Hypochlorite has also been used for the oxidation of a pyrazolophenanthridine sulphinate salt, under basic conditions[71]. In aqueous solutions, arylsulphinates react with hypochlorite to give sulphonate salts[72]. The mechanism for this reaction involves nucleophilic attack by hypochlorite ion on the sulphinate salt to give a sulphurane-like intermediate, which then decomposes to give the products as shown in equation 28.

$$\mathrm{ArSO_2^- + OCl^- \longrightarrow \left[ ArS(O)_2(OCl) \right]^{2-} \longrightarrow ArSO_3^- + Cl^-} \quad (28)$$

Secondary sulphinamides undergo a rather novel oxidation reaction with chlorine, either in benzene at room temperature or in ether at $-78\,^\circ$C, as indicated in equation 29. In this reaction, the oxosulphonium salt is assumed to be the intermediate[73]. The product, a sulphonimidoyl chloride, may also be prepared if the chlorine is replaced by N-chlorobenzotriazole or by tertiary butyl hypochlorite[74]. These latter oxidants are chosen when other groups, that are sensitive to chlorine oxidation, are present in the sulphinamide. The use of N-chlorobenzotriazole has been shown to undergo reaction, but not oxidation of the sulphur(IV) moiety, under other conditions[75].

Thiosulphinates are also oxidized, by chlorine, to the sulphur(VI) level. The thiosulphonate products formed are produced by chlorination which involves a S—S bond scission, as shown in equation 30[76]. It seems rather surprising that only a single product has been reported. Under anhydrous conditions the reaction is stopped at the sulphinyl chloride stage[77].

$$\mathrm{RS(O)NHR' + Cl_2 \longrightarrow \left[R{-}\overset{\overset{\displaystyle O}{\|}}{\underset{\underset{\displaystyle NHR'}{|}}{S^{+}}}{-}ClCl^{-}\right]^{2-} \longrightarrow R{-}\overset{\overset{\displaystyle O}{\|}}{\underset{\underset{\displaystyle NR'}{\|}}{S}}{-}Cl + HCl} \qquad (29)$$

$$\mathrm{RS(O)SR' + Cl_2 \longrightarrow RS(O)Cl + R'SCl \longrightarrow RSO_2SR' + 2Cl^{-}} \qquad (30)$$

### *6. Bromine- and iodine-containing reagents*

A wide range of bromine- and iodine-containing reagents have been used as oxidants in many areas of organic chemistry for decades. The oxidation of sulphinic acids and their derivatives has been performed using these species, but their use has been surprisingly infrequent.

As early as 1893, Limpricht[78] showed that sulphinic acids may be oxidized to the corresponding sulphonyl bromides using bromine. Other authors have also reported this reaction[53,57,58,79,80] which is shown in equation 31. The product from this reaction is either a sulphonyl bromide or a sulphonic acid, depending upon the reaction conditions. Methyl methanesulphinate has also been oxidized with bromine. In this reaction, at 0 °C, the products are methyl bromide and methanesulphonyl bromide (equation 32)[64].

$$\mathrm{RSO_2H \xrightarrow{Br_2} RSO_2Br \quad or \quad RSO_2OH} \qquad (31)$$

$$\mathrm{CH_3S(O)OCH_3 + Br_2 \longrightarrow CH_3Br + CH_3SO_2Br} \qquad (32)$$

Alkyl magnesium bromides have been used to prepare alkyl sulphinate salts, which have then been oxidized to their sulphonyl bromides in high yields[81]. As with the similar reaction involving chlorine, described in the last section, this is an excellent route to sulphur(VI)-containing compounds that are not easily obtainable by other routes.

Iodine has also been used to oxidize sulphuric acids, and their salts, to sulphonyl iodides or sulphonic acids, equation 33[8,57,58,82–84]. Indeed, this was the first method by which sulphonyl iodides were prepared and isolated.

$$\mathrm{RSO_2H \,|\, RSO_2^{-} \xrightarrow{I_2} RSO_2I \quad or \quad RSO_3H} \qquad (33)$$

Periodate has also been used successfully for the oxidation of thiosulphinates to thiosulphonates[35,85], although the use of this oxidant with sulphinamides produced a complex mixture of products[17].

Sodium periodate oxidation of (2,2-dimethylpropyl)benzenethiosulphinate produces the thiosulphonate in quantitative yield (equation 34) whilst attempted oxidation of phenyl 2,2-dimethylpropanethiosulphinate with the same reagent was unsuccessful after 48 h (equation 35)[33]. It has been found that for most unsymmetrical thiosulphinates the thiosulphonate is produced in good yield by this method and is catalysed by iodine or acid[85]. This result should be contrasted with the oxidation of the two above-mentioned thiosulphinates with *m*-CPBA, which yielded a complex mixture[33,39].

$$PhSOSCH_2Bu^t \xrightarrow{IO_4^-} PhSO_2SCH_2Bu^t \quad (34)$$

$$^tBuCH_2SOSPh \xrightarrow{IO_4^-} \text{no reaction} \quad (35)$$

*7. Metal ion oxidants*

There are many transition-metal-ion oxidants currently available to organic chemists. However, there have been very few metal ion oxidants used for the conversion of sulphinic acid derivatives into sulphur(VI)-containing compounds, that have been reported in the literature, the main exception being the use of permanganate ion, under a variety of conditions. In the early 1900s, Borsche and Lange[86–88] converted cyclic alkylsulphinate salts into sulphonic acids using aqueous potassium permanganate. These reactions have been pursued by some workers to apparent synthetic advantage[89,90]. Other workers, however, have reported that α-disulphones are produced as unfortunate by-products[91–93], or as the only product[22,94,95]. In addition, permanganate oxidation of the sulphinate salts, prepared by reaction of Grignard reagents with sulphur dioxide, proceeds to the sulphonic acid in low yield[96].

A review of these reports suggests that either the α-disulphone, or the sulphonic acid, may be produced free of the other if the conditions are carefully controlled. For example, Allen and coworkers[94,95] have shown that the α-disulphone is the only product if aqueous, acidic potassium permanganate is employed as oxidant, as shown in equation 36. On the other hand, when cold, glacial acetic acid, or a buffered system (pH 7.2–7.5), is used as solvent, then the sulphonic acid is the major product (equation 37).

$$2RSO_2H \xrightarrow[H^+|H_2O]{KMnO_4} RSO_2SO_2R + H_2O \quad (36)$$

$$RSO_2H \xrightarrow[\text{acetic acid}]{KMnO_4} RSO_3H \quad (37)$$

Cobalt(III) sulphate has also been used to oxidize both alkyl and arylsulphinic acids. In this case, only the α-disulphone was produced, with yields ranging from 35–56%[97], and it has been suggested that the reaction occurs via a one-electron oxidation process. The sulphonyl radical, thus formed, then undergoes further reaction to give the α-disulphone.

Aromatic sulphinate esters undergo oxidation to the sulphonate ester with permanganate in aqueous solution[98,99], in good yields, as shown in equation 38.

$$ArSO_2R + [O] \longrightarrow ArSO_3R \quad (38)$$

Thiosulphinates are oxidized to thiosulphonates in poor yields using permanganate[35]. Other products are also formed in this reaction. If selenium dioxide is used as the oxidant, then synthetically useful yields of the thiosulphonate result, as shown in equation 39.

$$RSOSR' \xrightarrow{SeO_2} RSO_2SR' \quad (39)$$

The oxidation of sulphinamides to sulphonamides, with permanganate, has met with varying degrees of success. In some cases the reaction is totally unsuccessful[100], or multiple products are formed[17]. On the other hand, other reports have indicated that excellent yields of the sulphonamide are produced[101,102]. Excess manganese dioxide, suspended in dry benzene, has also been used for this oxidation reaction[103]. In this case, the sulphonamide was produced in quantitative yield, at 70 °C (equation 40).

$$C_6F_5SONH_2 \xrightarrow{MnO_2} C_6F_5SO_2NH_2 \qquad (40)$$

*8. Other oxidations*

There have been few reports of other oxidations of sulphinic acid derivatives. However, reactions such as enzymatic oxidation and the use of oxygen-transfer reagents (like N-oxides) have been carried out and these are discussed here.

An enzymatic preparation, extracted from rabbit liver microsomes, has been shown[5] to oxidize a thiosulphinate to the thiosulphonate, as shown in equation 41. It was shown that in order for the reaction to occur, to any significant extent, the correct co-factors and minerals must be present.

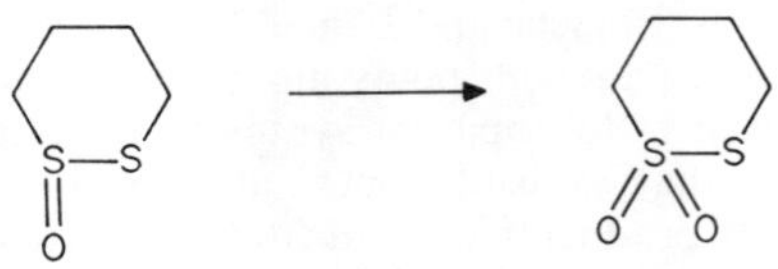

(41)

Electrochemical oxidation of thiosulphinates leads cleanly to the corresponding thiosulphonate in reasonable yields with no observed side-products[104]. It is rather surprising that this method has apparently not been used to synthetic advantage.

Tertiary amine N-oxides have been shown to oxidize arylsulphinyl chlorides to sulphonic acids, albeit in low yields[105,106]. In this reaction other products, such as thiosulphonates, are also produced.

Methanesulphinyl chloride and *p*-nitrobenzenesulphinyl chloride have been used to reduce sulphoxides to sulphides[107]. During this process, the sulphonyl chloride is produced by direct oxygen transfer. It is difficult to see how this reaction could be synthetically useful for the preparation of sulphonyl chlorides.

Finally, aromatic sulphinic acids have been shown to react rapidly with benzeneseleninic acid (the selenium equivalent of a sulphinic acid) in a range of solvents, at 0 °C, producing a sulphonate salt and a selenosulphonate[108], as shown in equation 42. Benzeneseleninic anhydride (PhSe(O)OSe(O)Ph) may be used in this reaction in place of the seleninic acid.

$$ArSO_2H + 2PhSeO_2H \longrightarrow ArSO_3^-(PhSeO_2H_2)^+ + PhSeSO_2Ar \qquad (42)$$

## B. Oxidative Analytical Methods

Sulphinic acids are usually determined analytically using either the iron(III) salt method or oxidative methods. The latter methods are relevant to the present work and one of these has been discussed in an excellent review of the methods available for the determination of organic sulphur-containing functional groups[109].

Probably the best oxidative method of analysis involves the oxidation of a sulphinate salt with hypochlorite to the sulphonate as depicted in equation 43. This method has been recommended, by several groups[72,110,111], for the determination of either hypochlorite or sulphinate.

$$RSO_2^- + OCl^- \longrightarrow RSO_3^- + Cl^- \qquad (43)$$

Lindberg[112,113] has indicated that oxidative determination by potassium permanganate in neutral solution can be used. In addition, Allen[95] has reported that either calcium hypochlorite or potassium permanganate can be used, but the solution must be alkaline if quantitative results are to be expected.

It has also been reported[113] that oxidations with bromine, iodine and cerium(IV) salts have been attempted, but these oxidants have proved to be unreliable for quantitative analysis.

## III. REDUCTION

In comparison with oxidation reactions, the reduction of sulphinic acids, and their derivatives, has been little studied. Indeed, there has been no report of a systematic study of the reduction of these compounds with the usual range of reducing agents available to organic chemists. There are, however, some reduction reactions that have been studied in some detail and these are reviewed in this section.

### A. Hydride-transfer Reagents

Hydride-containing reagents, such as sodium borohydride and lithium aluminium hydride, are the reagents of choice in many reductions in organic chemistry. These reagents have been used rarely for the reduction of sulphinic acid derivatives.

Reduction of sulphinic acids and sulphinyl chlorides, with lithium aluminium hydride, leads to disulphides[114]. The reaction is thought to occur by initial reduction of the sulphur(IV) moiety to the sulphur(II) level, as shown in equations 44 and 45. These initially formed products then undergo further reaction to form disulphides as the final product (equation 46). If excess lithium aluminium hydride is used, then a thiol is the final product.

$$RSO_2H \xrightarrow{LiAlH_4} RSOH \tag{44}$$

$$RSOCl \xrightarrow{LiAlH_4} RSCl \tag{45}$$

$$RSOH \quad \text{or} \quad RSCl \longrightarrow RSSR \tag{46}$$

Cyclic sulphinate esters (sultines) may also be reduced by lithium aluminium hydride[115]. In this case, the product is a thioalcohol, as shown in equation 47. Sulphinamides are not reduced by lithium aluminium hydride[116].

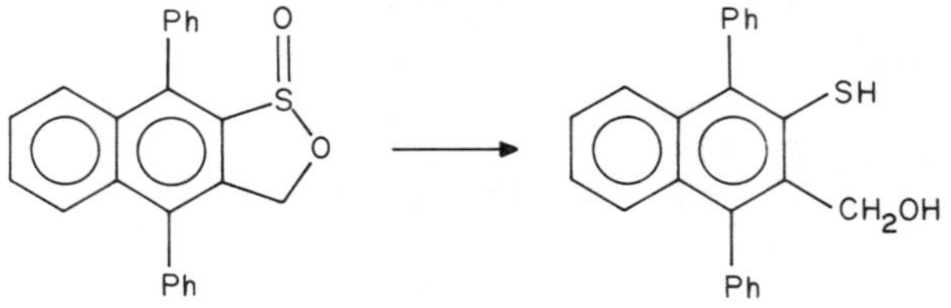


(47)

### B. Silicon-containing Reagents

Since silicon forms strong bonds with oxygen, organosilicon-containing reagents are ideal candidates for reducing agents. Halogenated methylsilanes and trichlorosilane have been reported as successful reducing reagents for sulphinic acid derivatives.

Oae and coworkers[117] have found that both alkyl and arylsulphinic acids are reduced by chlorotrimethylsilane, in the presence of thiols, to produce disulphides, as shown in equation 48. When the reaction is carried out at room temperature, in chloroform with excess silane present, then the yields are nearly quantitative. However, when the reaction is performed under refluxing conditions, then only aromatic sulphinic acids give good yields of disulphides (70%). On the other hand, yields for this reaction with alkyl sulphinic acids are less than 40%. In the latter case, the major product is the thiosulphonate which is formed by nucleophilic attack by the thiol on the sulphinic acid.

$$RSO_2H + 3R'SH + 4Me_3SiCl \longrightarrow RSSR' + R'SSR' + 2(Me_3Si)_2O \qquad (48)$$

Olah and coworkers[118] have shown that iodotrimethylsilane may be used to reduce sulphinic acids, their salts and esters and sulphinyl chlorides, to disulphides in yields varying from 75–96% (equation 49). These reactions are performed in methylene chloride, at room temperature, for 16 h. The silicon-containing reagent can either be purchased ready-for-use, or generated *in situ* from chlorotrimethylsilane/sodium iodide or from hexamethyldisilane/iodine. This reaction probably occurs by a mechanism involving the formation of a sulphenyl iodide, as shown in equation 50.

$$RS(=O){-}X \xrightarrow{Me_3SiI} RSSR \qquad (49)$$

$$X = OH,\ O^-Na^+,\ OR',\ Cl$$

$$RS(=O){-}X \longrightarrow RS^+(OSiMe_3){-}X\ \ I^- \longrightarrow RSOSiMe_3 \longrightarrow RSI \longrightarrow RSSR \qquad (50)$$

Trichlorosilane and tripropylamine have also been used to reduce sulphinyl chlorides and sulphinate esters to disulphides, in benzene solution[119] (equation 51). In this reaction, the amine acts as a proton sponge, removing the HCl produced in the reaction. With cyclic sulphinate esters, the $\alpha,\omega$-diol of the disulphide is the product, as shown in equation 52; such compounds are fairly difficult to prepare by other simple routes.

$$RS(=O){-}X + HSiCl_3 + Pr_3N \longrightarrow RSSR + (SiCl_2O)_n + Pr_3\overset{+}{N}HCl^- \qquad (51)$$

$$X = Cl,\ OR'$$

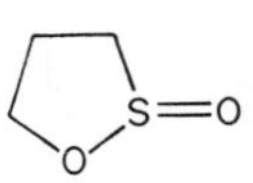 $\longrightarrow HO(CH_2)_3SS(CH_2)_3OH$ (52)

## C. Phosphorus-containing Reagents

As reducing agents, phosphorus-containing compounds usually act by an oxygen-transfer mechanism whereby a phosphorus(V) species ($POX_3$) is formed from the phosphorus(III)-containing reducing agent. However, phosphorus(V)-containing reagents are also able to reduce organic compounds. Both types of reagent are exemplified below.

Triphenylphosphine reacts rapidly with either aryl or alkylthiosulphinates to produce disulphides, even at $-25\,^{\circ}C$[120]. Triphenylphosphine oxide is formed as a by-product, as indicated by equation 53, and is easily removed from the product by its differential solubility in organic solvents. It should be noted that triphenylarsine and triphenylstibine may also be used to reduce thiosulphinates to dysulphides but, in these cases, more forcing conditions are required. In addition, alkylthiosulphinates are unaffected by both of these reagents.

$$\mathrm{RSOSR + Ph_3P \longrightarrow RSSR + Ph_3PO} \qquad (53)$$

Ethyl hypophosphite has been used to reduce the salts of aromatic sulphinic acids to disulphides[121,122], as shown in equation 54. This reaction produces good yields of the disulphide and the best solvent is dimethylsulphoxide.

$$\mathrm{ArSO_2^- \xrightarrow{EtO\overset{O}{\overset{\|}{P}}H_2} ArSSAr} \qquad (54)$$

A rather novel reduction reaction involves the conversion of sulphinic acid salts into thiocyanates[123]. The reaction is performed in refluxing tetrahydrofuran, using diethyl phosphorocyanidate as the reducing agent (equation 55). The reduction of sulphinic acids using this reagent produces poor yields.

$$\mathrm{RSO_2^-Na^+ \xrightarrow{(EtO)_2\overset{O}{\overset{\|}{P}}CN} RSCN} \qquad (55)$$

Sulphinic acid groups are also reduced, to acetylthio groups, on reaction with thionyl chloride, thioacetic acid and triphenylphosphine, in succession (equation 56). Such a sequence has been used in the synthesis of novel antibiotics[124].

$$\mathrm{RSO_2H \xrightarrow[1.\,SOCl_2\quad 2.\,CH_3COSH\quad 3.\,Ph_3P]{} RS\overset{O}{\overset{\|}{C}}CH_3} \qquad (56)$$

## D. Sulphur-containing Reagents

Thiols may be used as reducing agents for sulphinyl chlorides[125] and thiosulphinates[126–129]. In both cases the yields of disulphides are good. Methanesulphinyl chloride is reduced by excess ethanethiol, producing dimethyl disulphide, as shown in equation 57. However, diethyl disulphide is also formed and thus this reaction is unlikely to be of synthetic utility. Cyclic thiosulphinates are reduced to produce *bis* disulphides, as shown in equation 58.

$$\mathrm{2CH_3SOCl + 6EtSH \longrightarrow CH_3SSCH_3 + 3EtSSEt + 2HCl + 2H_2O} \qquad (57)$$

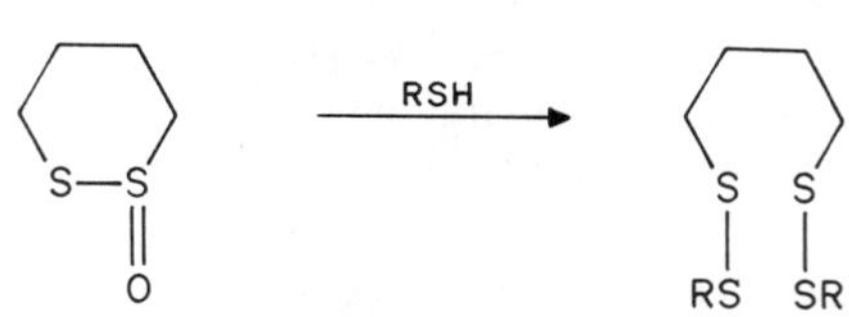


(58)

It should be noted that aromatic seleninic acids are reduced to aromatic selenenic acids by thiols under similar conditions, as shown in equation 59[130]. Perhaps sulphinic acids react initially by a similar route, but then further reaction of the sulphenic acid produces the disulphide.

$$ArSeO_2H + 2RSH \longrightarrow RSSR + ArSeOH + H_2O \qquad (59)$$

## E. Electrochemical Methods

Electrochemical methods of reduction usually yield clean products. It is thus surprising that, to date, these methods have been used very infrequently in the field of organo-sulphur chemistry. Perhaps this is, in part, due to the fact that sulphur has a 'poisoning effect' on mercury, which is often used as an electrode.

Electrochemical reduction of sulphinic acids, in acidic ethanol, yields thiols[131]. This reaction occurs through the intermediacy of thiosulphonates which are presumably formed by disproportionation of the sulphinic acid (see Section IV). The thiosulphonate is then reduced to sulphinic acid and thiol. The overall reaction is thus a reduction, as shown in equation 60. The latter step of this process involves a two-electron transfer, as detailed in equation 61.

$$RSO_2H \longrightarrow RSO_2SR \longrightarrow RSO_2H + RSH \qquad (60)$$

$$ArSO_2SAr + 2H^+ + 2e \longrightarrow ArSO_2H + ArSH \qquad (61)$$

*β*-Lipoic acid (a thiosulphinate) has been reduced polarographically, in a four-electron process, yielding dihydrolipoic acid[132]. Sulphinate esters may also be reduced in a similar fashion[133,134]. In this case, polarographic reduction is also a four-electron process which yields a thiol and an alcohol, in buffered aqueous solution, as shown in equation 62.

$$R\overset{\overset{\Large O}{\|}}{S}OR' + 4e + 4H^+ \longrightarrow RSH + R'OH + H_2O \qquad (62)$$

In a similar process, diseleninic acids are reduced to diselenols[134], as shown in equation 63.

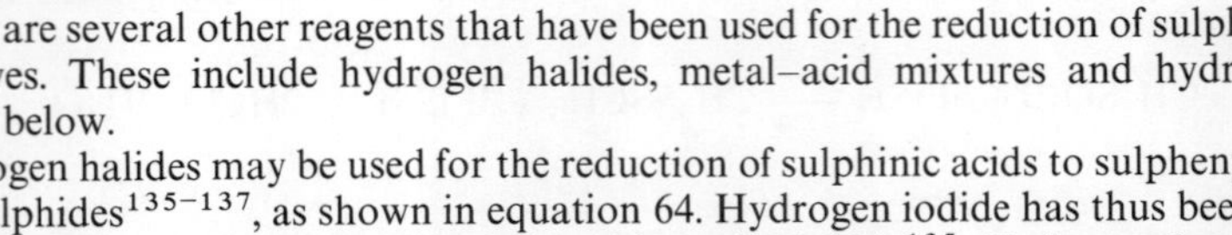


(63)

## F. Other Reductions

There are several other reagents that have been used for the reduction of sulphinic acid derivatives. These include hydrogen halides, metal–acid mixtures and hydrazine, as detailed below.

Hydrogen halides may be used for the reduction of sulphinic acids to sulphenyl halides and disulphides[135–137], as shown in equation 64. Hydrogen iodide has thus been used to reduce glutathione sulphinic acid to glutathione disulphide[135]. Hydrogen bromide has been used to prepare a range of disulphides or sulphenyl bromides from sulphinic acids[136,137].

$$\begin{gathered} RSO_2H + 3HX \longrightarrow RSX + X_2 + 2H_2O \\ 2RSX \longrightarrow RSSR + X_2 \\ X = Br, I \end{gathered} \qquad (64)$$

Zinc with sulphuric acid[138–140] and tin(II) chloride with hydrochloric acid[141] have also been used to reduce sulphinic acids. In the former case thiols are formed (equation 65). In the latter case, the thiolate ion, that is initially formed, reacts further with sulphinic acid to produce the thiosulphonate. In this case the overall reaction appears to be a disproportionation reaction, as shown in equation 66. However, the tin(IV) chloride, isolated at the completion of the reaction, is evidence for the reduction process.

$$ArSO_2H \xrightarrow[H_2SO_4]{Zn} ArSH \tag{65}$$

$$ArSO_2H \xrightarrow[HCl]{SnCl_2} ArS^- \xrightarrow{ArSO_2H} ArSO_2SAr \tag{66}$$

Anhydrous hydrazine has been shown to be a useful reagent for the reduction of sulphinate esters, sulphinyl chlorides and thiosulphinates giving disulphides as products, under mild conditions[142]. Under more forcing conditions thiols are formed. If carbon–carbon multiple bonds are present in the substrate molecule, these too are reduced. Sulphinic acids, on the other hand, are seemingly unaffected by this reduction process.

Finally, an interesting reduction of sulphinyl chlorides to disulphides has been reported by Harpp and MacDonald[143]. These workers found that benzenesulphinyl chloride reacts with a molybdenum–persulphide complex to produce a 68% yield of diphenyl disulphide, as shown in equation 67.

$$RS(=O)\text{—}Cl \xrightarrow{Mo_2S_{12}} RSSR \tag{67}$$

## IV. DISPROPORTIONATION

Sulphinic acids, and their derivatives, have a propensity to disproportionate and this process has been found to be catalysed by various species, such as acids and iodide ions. Disproportionation results in an overall oxidation and a concomitant reduction of the sulphur(IV) moiety [i.e. sulphur(VI)- and sulphur(II)-containing species are formed] and so is discussed in the present chapter.

It has been known since as early as 1868[54,144] that sulphinic acids disproportionate, although in the nineteenth century there was some dispute concerning this matter[145]. Perhaps this discrepancy was due, at least in part, to the widely different rates of disproportionation. There have been reports of essentially instantaneous reactions[146] whilst others have reported that the reaction takes about twenty months to complete[144]. The former example was for the intramolecular disproportionation of the 2,2′-disulphinic acid derivative of biphenyl.

There have been three different overall stoichiometries described for this reaction, as shown in equations 68–70. The first two of these equations have been used to describe disproportionation in earlier times and the second was even used relatively recently to describe the disproportionation of phenylsulphinic acid[147]. During the past few decades it is the last of these stoichiometries (equation 70) that has become accepted as the norm for the disproportionation of sulphinic acids.

$$5RSO_2H \longrightarrow 3RSO_3H + RSSR + H_2O \tag{68}$$

$$3RSO_2H \longrightarrow RSH + 2RSO_3H \tag{69}$$

$$3RSO_2H \longrightarrow RSO_3H + RSO_2SR + H_2O \tag{70}$$

Kinetic and mechanistic studies have resulted in two conflicting views of the reaction mechanism, one involving radicals and the other not. Horner and Basedow[148] have suggested that the mechanism for disproportionation of sulphinic acids does not involve radicals, but is as shown in equation 71. A similar mechanism was also supported by the work of Allen and Reich[149].

$$\begin{array}{r}2RSO_2H \longrightarrow RSO_3H + RSOH\\ RSOH + RSO_2H \longrightarrow RSO_2SR + H_2O\end{array} \qquad (71)$$

This rather simple mechanism has now been replaced, by a process involving radicals, mainly due to the extensive kinetic studies of Kice and coworkers[150–154] and others[155]. In this mechanism, the key intermediate is the sulphinyl sulphone. This species undergoes rate-limiting S—S bond homolysis to form both sulphinyl and sulphonyl radicals. Recombination and further reaction with sulphinic acid then occurs to produce the thiosulphonate and the sulphonic acid, as shown in equation 72. This mechanism gives a much better account of the experimental evidence, compared with the non-radical process discussed above, and it is thus the currently accepted mechanism.

$$\begin{array}{l}2ArSO_2H \rightleftharpoons ArSO_2SOAr + H_2O\\ \qquad\qquad\qquad\qquad \downarrow\\ ArSO_2{-}O{-}SAr \longleftarrow ArSO_2^{\bullet} + ArSO^{\bullet}\\ \quad \downarrow ArSO_2H\\ ArSO_2SAr + ArSO_3H\end{array} \qquad (72)$$

Aromatic thiosulphinates also disproportionate and in this case the products are a disulphide and a thiosulphonate, as indicated in equation 73. This process seems to be more rapid under anhydrous conditions[156]. The disproportionation of thiosulphinates has not received as much attention as the similar reaction of sulphinic acids. Notwithstanding this, Koch and coworkers[157] have proposed a mechanism involving radicals for the reaction, as shown in equation 74. The careful work of Koch's group[157] and of others[156,158,159] has indicated that the reaction is not as simple as described by equations 73 and 74, since both sulphonic acid and sulphonyl anhydride ($RSO_2OSO_2R$) are also isolated, albeit in low yields. Kice and coworkers[158,159] have also shown the reaction to be catalyzed by both acids and nucleophiles, and these workers have described more complex reaction schemes for the catalyzed reactions.

$$2RSOSR \longrightarrow RSO_2SR + RSSR \qquad (73)$$

$$\begin{array}{r}ArSOSAr \longrightarrow ArSO^{\bullet} + ArS^{\bullet}\\ 2ArSO^{\bullet} \longrightarrow ArS(=O)SAr \longrightarrow ArSO_2SAr\\ 2ArS^{\bullet} \longrightarrow ArSSAr\end{array} \qquad (74)$$

Other derivatives of sulphinic acids are also known to disproportionate. Trifluoromethanesulphonyl chloride and bromide disproportionate, especially in the presence of a catalytic amount of water, to give the sulphonyl and the sulphenyl halides (equation 75)[154]. Methanesulphinyl chloride also undergoes a similar disproportionation rather readily[160]. In the case of sulphinyl fluorides, it seems that disproportionation yields the disulphide rather than the sulphenyl fluoride. Alkyl thiosulphinates also undergo a similar disproportionation process[161].

$$2CF_3SOX \longrightarrow CF_3SO_2X + CF_3SX \qquad (X = Cl, Br) \qquad (75)$$

Sulphinate esters have also been reported to disproportionate[162]. In this case a thiosulphonate and a sulphonate ester are formed, as shown in equation 76. There has also been a report that arylsulphinyl nitrates ($RSONO_3$) disproportionate quantitatively in the presence of water or alcohols[163].

$$\mathrm{ArS(=O)-OCH_3 \longrightarrow ArS(=O)_2-OCH_3 + ArSSCH_3} \qquad (76)$$

## V. ACKNOWLEDGEMENTS

I would like to thank the library staffs at the Nova Scotia Agricultural College, the Atlantic Regional Laboratory of the National Research Council of Canada, Halifax and Dalhousie University for all their kind assistance. In addition, I thank my students for leaving me to get at the task of writing this chapter when I asked. Finally, and most important of all, I wish to thank my wife, Niki, for her patience and understinding when I needed it most.

## VI. REFERENCES

1. F. Freeman, C. N. Angeletakis, W. J. Pietro and W. J. Hehre, *J. Am. Chem. Soc.*, **104**, 1161 (1982).
2. R. Otto and H. Ostrop, *Ann. Chem.*, **141**, 370 (1867).
3. R. Otto and von Gruber, *Ann. Chem.*, **145**, 19 (1868).
4. S. Oae, K. Shinhama and Y. H. Kim, *Tetrahedron Lett.*, 3307 (1979).
5. S. Oae, Y. H. Kim, D. Fukushima and T. Takata, *Pure Appl. Chem.*, **49**, 153 (1977).
6. L. Horner and O. H. Basedow, *Ann. Chem.*, **612**, 108 (1958).
7. B. Houel, *C. R. Acad. Sci. Paris*, **250**, 3839 (1960).
8. R. M. Adlington and A. G. M. Barrett, *J. Chem. Soc., Perkin Trans. 1*, 1076 (1980).
9. P. Cocolios, P. Fournari, R. Guilard, C. Lecomte, J. Protas and J. C. Bouble, *J. Chem. Soc., Dalton Trans.*, 2081 (1980).
10. P. Cocolios, G. Lagrange, R. Guilard, H. Oumous and C. Lecomte, *J. Chem. Soc., Dalton Trans.*, 567 (1984).
11. D. Barnard, *J. Chem. Soc.*, 4547 (1957).
12. S. Oae, T. Takata and Y. H. Kim, *Tetrahedron*, **37**, 37 (1981).
13. T. Takata, Y. H. Kim and S. Oae, *Tetrahedron Lett.*, 821 (1979).
14. S. Oae, T. Takata and Y. H. Kim, *Bull. Chem. Soc. Jpn.*, **54**, 2712 (1981).
15. A. A. Frimer, I. Rosenthal and S. Hoz, *Tetrahedron Lett.*, 4631 (1977).
16. H. Berger, *Recl. Trav. Chim. Pays-Bas*, **82**, 773 (1963).
17. J. E. Semple and M. M. Joullie, *J. Org. Chem.*, **43**, 3066 (1978).
18. R. M. Hann, *J. Am. Chem. Soc.*, **57**, 2166 (1935).
19. W. E. Truce and J. F. Lyons, *J. Am. Chem. Soc.*, **73**, 126 (1951).
20. B. J. Lindberg, *Acta Chem. Scand.*, **20**, 1843 (1966).
21. R. N. Gonzalez, Ph.D. thesis, University of Delaware, 1957.
22. W. Lisowski, *Przem. Chem.*, **12**, 697 (1956); *Chem. Abstr.*, **52**, 12815e (1958).
23. J. Kenyon, H. Phillips and V. P. Pittman, *J. Chem. Soc.*, 1077 (1935).
24. O. Jacobsen, *Chem. Ber.*, **10**, 1011 (1877).
25. R. M. Coates and J. P. Chen, *Tetrahedron Lett.*, 2705 (1969).
26. U. Jager and W. Sundermeyer, *Chem. Ber.*, **119**, 3405 (1986).
27. J. L. Kice, *Adv. Phys. Org. Chem.*, 65 (1980).
28. S. Oae and T. Takata, *Chem. Lett.*, 845 (1981).
29. F. Freeman and C. N. Angeletakis, *J. Am. Chem. Soc.*, **105**, 4039 (1983).
30. F. Freeman and C. N. Angeletakis, *J. Am. Chem. Soc.*, **103**, 6232 (1981).
31. F. Freeman, C. N. Angeletakis and T. J. Maricich, *Tetrahedron Lett.*, 1867 (1981).
32. F. Freeman and C. N. Angeletakis, *J. Org. Chem.*, **50**, 793 (1985).
33. F. Freeman, C. N. Angeletakis and T. J. Maricich, *J. Org. Chem.*, **47**, 3403 (1982).
34. S. Oae, Y. H. Kim, T. Takata and D. Fukushima, *Tetrahedron Lett.*, 1195 (1977).

35. S. Oae and T. Takata, *Tetrahedron Lett.*, **21**, 3213 (1980).
36. F. Freeman and C. N. Angeletakis, *J. Org. Chem.*, **46**, 3991 (1981).
37. S. Oae, T. Takata and Y. H. Kim, *Bull. Chem. Soc. Jpn.*, **55**, 2484 (1982).
38. M. M. Chau and J. L. Kice, *J. Am. Chem. Soc.*, **98**, 7711 (1976).
39. F. Freeman and C. N. Angeletakis, *J. Am. Chem. Soc.*, **104**, 5766 (1982).
40. L. Hough and M. I. Taha, *J. Chem. Soc.*, 3546 (1957).
41. A. K. Bhattacharya and A. G. Hortmann, *J. Org. Chem.*, **43**, 2728 (1978).
42. C. N. Angeletakis, Ph.D. thesis, University of California at Irvine, 1982.
43. F. E. Hardy, P. R. H. Speakman and P. Robson, *J. Chem. Soc.* (*C*), 2334 (1969).
44. L. Field and T. F. Parsons, *J. Org. Chem.*, **30**, 657 (1965).
45. D. Barnard and E. J. Percy, *Chem. Ind.* (*London*), 1332 (1960).
46. C. Chatgilialoglu, in *The Chemistry of Sulphones and Sulphoxides* (Eds. S. Patai, Z. Rappoport and C. J. M. Stirling), Chap. 24, Wiley, Chichester, 1988.
47. S. Blackburn and A. G. Lowther, *Biochem. J.*, **49**, 554 (1951).
48. G. Toennies and T. F. Lavine, *J. Biol. Chem.*, **100**, 463 (1933).
49. W. E. Savige, J. Eager, J. A. Maclaren and C. M. Roxburgh, *Tetrahedron Lett.*, 3289 (1964).
50. G. Medes and N. F. Floyd, *Biochem. J.*, **31**, 1330 (1939).
51. G. Medes and N. F. Floyd, *Biochem. J.*, **36**, 259 (1942).
52. H. Asakawa, K. Kamuya and S. Takai, *Takeda Kenkyusho Nempo*, **29**, 610 (1970); *Chem. Abstr.*, **74**, 125603 (1971).
53. R. Otto and von Gruber, *Ann. Chem.*, **142**, 96 (1867).
54. R. Otto, *Ann. Chem.*, **145**, 317 (1868).
55. F. Ullmann and A. Lehner, *Chem. Ber*, **38**, 732 (1905).
56. E. Pfeil and O. Velten, *Ann. Chem.*, **565**, 183 (1949).
57. D. T. Gibson, C. J. Miller and S. Smiles, *J. Chem. Soc.*, 1823 (1925).
58. R. Otto and J. Troger, *Chem. Ber.*, **24**, 479 (1891).
59. K. Fries and P. Vogt, *Ann. Chem.*, **381**, 337 (1911).
60. J. Troger and M. Meine, *J. Prakt. Chem.*, [2], **68**, 317 (1903).
61. F. E. Jenkins and A. N. Hambly, *Aust. J. Chem.*, **6**, 318 (1953).
62. I. B. Douglass and T. B. Johnson, *J. Am. Chem. Soc.*, **60**, 1486 (1938).
63. I. B. Douglass, B. S. Farah and M. G. Thomas, *J. Org. Chem.*, **26**, 1996 (1961).
64. I. B. Douglass, *J. Org. Chem.*, **30**, 633 (1965).
65. R. B. Scott, J. B. Gayle, M. S. Heller and R. E. Lutz, *J. Org. Chem.*, **20**, 1165 (1955).
66. F. Asinger, B. Fell and S. Pottkaemper, *Chem. Ber.*, **97**, 3092 (1964).
67. F. Asinger, P. Laue, B. Fell and C. Gubelt, *Chem. Ber.*, **100**, 1696 (1967).
68. C. M. Starks, U.S. Patent 3134809; *Chem. Abstr.*, **61**, 4220 (1964).
69. P. R. Carter and D. H. Hey, *J. Chem. Soc.*, 147 (1948).
70. A. Matsumoto, Jap. Patent. 177918; *Chem. Abstr.*, **45**, 7592 (1951).
71. W. J. Berry, I. L. Finar and A. B. Simonds, *J. Chem. Soc.*, 4974 (1956).
72. J. L. Kice and A. R. Puls, *J. Am. Chem. Soc.*, **99**, 3455 (1977).
73. E. U. Jonsson, C. C. Bacon and C. R. Johnson, *J. Am. Chem. Soc.*, **93**, 5306 (1971).
74. C. R. Johnson and A. Wambsgans, *J. Org. Chem.*, **44**, 2278 (1979).
75. F. Wudl, C. K. Brush and T. B. K. Lee, *J. Chem. Soc., Chem. Commun.*, 151 (1972).
76. R. R. Crenshaw and L. Field, *J. Org. Chem.*, **30**, 175 (1965).
77. E. Vinkler and F. Klivenyi, *Magy. Kem. Foly.*, **60**, 95 (1954).
78. H. Limpricht, *Ann. Chem.*, **278**, 239 (1893).
79. J. Boeseken and H. W. Oeckenburg, *Recl. Trav. Chim. Pays-Bas*, **33**, 319 (1914).
80. A. D. Bliss, W. K. Cline, C. E. Hamilton and O. J. Sweeting, *J. Org. Chem.*, **28**, 3537 (1963).
81. G. Geiseler and R. Kuschmiers, *Chem. Ber.*, **93**, 2041 (1960).
82. M. Bazlen. *Chem. Ber.*, **60**, 1479 (1927).
83. R. Childs and S. Smiles, *J. Chem. Soc.*, 2696 (1926).
84. L. Field, T. F. Parsons and R. R. Crenshaw, *J. Org. Chem.*, **29**, 918 (1964).
85. Y. H. Kim, T. Takata and S. Oae, *Tetrahedron Lett.*, 2305 (1978).
86. W. Borsche and W. Lange, *Chem. Ber.*, **38**, 2766 (1905).
87. W. Borsche and W. Lange, *Chem. Ber.*, **39**, 392 (1906).
88. W. Borsche and W. Lange, *Chem. Ber.*, **40**, 2220 (1907).
89. P. W. Clutterbuck and J. B. Cohen, *J. Chem. Soc.*, 2507 (1923).

90. W. Dirscherl and K. Otto, *Chem. Ber.*, **89**, 393 (1956).
91. S. Smiles and T. P. Hilditch, *J. Chem. Soc.*, 519 (1907).
92. T.P. Hilditch, *J. Chem. Soc.*, 1524 (1908).
93. R. Childs and S. Smiles, *J. Chem. Soc.*, 2699 (1924).
94. P. Allen, L. Karger, J. D. Haygood and J. Shrensel, *J. Org. Chem.*, **16**, 767 (1951).
95. P. Allen, *J. Org. Chem.*, **7**, 23 (1942).
96. W. Von E. Doering and F. M. Beringer, *J. Am. Chem. Soc.*, **71**, 2221 (1949).
97. G. C. Denzer, P. Allen, P. Conway and J. M. Van der Veen, *J. Org. Chem.*, **31**, 3418 (1966).
98. J. Kenyon, H. Phillips and V. P. Pittman, *J. Chem. Soc.*, 1077 (1935).
99. W. Gerrard, J. Kenyon and H. Phillips, *J. Chem. Soc.*, 153 (1937).
100. Y. H. Chiang, J. S. Luloff and E. Schipper, *J. Org. Chem.*, **34**, 2397 (1969).
101. F. Kurzer, *J. Chem. Soc.*, 549 (1953).
102. H. Seiler and H. Erlenmeyer, *Helv. Chim. Acta*, **40**, 88 (1957).
103. I. Glander and A. Golloch, *J. Fluorine Chem.*, **5**, 83 (1975).
104. H. Viertler, P. D. Nachion, J. F. Ganzerli, V. L. Pardini and P. R. Schumacher, *An. Simp. Bras. Electroquim. Electroanal. 3rd.*, **1**, 173 (1982); *Chem. Abstr.*, **98**, 4279r (1983).
105. S. Oae and K. Ikura, *Bull. Chem. Soc. Jpn.*, **40**, 1420 (1967).
106. S. Oae and K. Ikura, *Bull. Chem. Soc. Jpn*, **39**, 1306 (1966).
107. T. Numata, K. Ikura, Y. Shimano and S. Oae, *Org. Prep. Proced. Int.*, **8**, 119 (1976).
108. R. A. Gancarz and J. L. Kice, *Tetrahedron Lett.*, **21**, 1697 (1980).
109. S. D. Nogare, in *Organic Analysis, Part I* (Eds. J. Mitchell, I. M. Kolthoff, E. S. Proskauer and A. Weissberger), Interscience, New York, 1953, p. 379.
110. N. I. Goldstone and M. B. Jacobs, *Ind. Eng. Chem., Anal. Ed.*, **16**, 206 (1944).
111. S. Atkin, *Anal. Chem.*, **19**, 816 (1947).
112. B. Lindberg, *Acta Chem. Scand.*, **17**, 377 (1963).
113. B. Lindberg, *Acta Chem. Scand.*, **17**, 383 (1963).
114. J. Strating and H. J. Backer, *Recl. Trav. Chim. Pays-Bas*, **69**, 638 (1950).
115. D. C. Dittmer, R. S. Henion and N. Takashina, *J. Org. Chem.*, **34**, 1310 (1969).
116. M. Cinquini and F. Cozzi, *J. Chem. Soc., Chem. Commun.*, 723 (1977).
117. S. Oae, H. Togo, T. Numata and K. Fujimori, *Chem. Lett.*, 1193 (1980).
118. G. A. Olah, S. C. Narang, L. D. Field and G. F. Salem, *J. Org. Chem.*, **45**, 4792 (1980).
119. T. H. Chan, J. P. Montillier, W. F. Van Horn and D. N. Harpp, *J. Am. Chem. Soc.*, **92**, 7224 (1970).
120. J. F. Carson and F. F. Wong, *J. Org. Chem.*, **26**, 1467 (1961).
121. H. W. Pinnick, M. A. Reynolds, R. T. McDonald and W. D. Brewster, *J. Org. Chem.*, **45**, 930 (1980).
122. M. A. Reynolds, Ph.D. thesis, University of Georgia, 1979.
123. S. Harusawa and T. Shioiri, *Tetrahedron Lett.*, **23**, 447 (1982).
124. J. R. Irving, E. Perrone and R. J. Stoodley, *Tetrahedron Lett.*, **24**, 1429 (1983).
125. I. B. Douglass and B. S. Farah, *J. Org. Chem.*, **23**, 805 (1958).
126. P. K. Singh, L. Field and B. J. Sweetman, *J. Org. Chem.*, **53**, 2608 (1988).
127. A. Schoberl and H. Grafje, *Ann. Chem.*, **617**, 71 (1958).
128. B. Boduszek and J. L. Kice, *J. Org. Chem.*, **47**, 3199 (1982).
129. H. Zahn and H. G. Otten, *Ann. Chem.*, **653**, 139 (1962).
130. H. Rheinboldt and E. Giesbrecht, *Chem. Ber.*, **88**, 1037 (1955).
131. B. Fleszar and P. Senecki, *Electrochim. Acta*, **25**, 551 (1980).
132. Y. Asahi, *Yakugaku Zasshi*, **80**, 679 (1960); *Chem. Abstr.*, **54**, 18886g (1960).
133. B. Persson, *J. Electroanal. Chem.*, **78**, 371 (1977).
134. R. Danielsson, B-L. Johansson, B. Nygard and B. Persson, *Chem. Scr.*, **20**, 19 (1982).
135. D. H. Calam and S. G. Waley, *Biochem. J.*, **85**, 417 (1962).
136. K. Fries and G. Schurmann, *Chem. Ber.*, **47**, 1195 (1914).
137. K. Fries, *Chem. Ber.*, **45**, 2965 (1912).
138. R. Otto, *Chem. Ber.*, **15**, 122 (1882).
139. L. Gattermann, *Chem. Ber.*, **32**, 1136 (1899).
140. W. P. Winter, *Am. Chem. J.*, **31**, 572 (1904).
141. B. G. Boldyrev and L. M. Grivnak, *Zh. Org. Khim.*, **4**, 1251 (1968).
142. M. Kobayashi and A. Yamamoto, *Bull. Chem. Soc. Jpn.*, **39**, 2736 (1966).

143. D. N. Harpp and J. G. MacDonald, *Tetrahedron Lett.*, **25**, 703 (1984).
144. C. Pauly and R. Otto, *Chem. Ber.*, **10**, 2181 (1877).
145. W. Autenrieth, *Ann. Chem.*, **259**, 362 (1890).
146. H. J. Barber and S. Smiles, *J. Chem. Soc.*, 1141 (1928).
147. Z. Yoshida, H. Miyoshi and K. Kawamoto, *Chem. Abstr.*, **71**, 101463 (1969).
148. L. Horner and O. H. Basedow, *Ann. Chem.*, **612**, 108 (1958).
149. P. Allen and L. Reich, *J. Phys. Chem.*, **64**, 1928 (1960).
150. J. L. Kice and K. W. Bowers, *J. Am. Chem. Soc.*, **84**, 605 (1962).
151. J. L. Kice, G. Guaraldi and C. G. Venier, *J. Org. Chem.*, **31**, 3561 (1966).
152. J. L. Kice, D. C. Hampton and A. Fitzgerald, *J. Org. Chem.*, **30**, 882 (1965).
153. J. L. Kice, B. R. Toth, D. C. Hampton and J. F. Barbour, *J. Org. Chem.*, **31**, 848 (1966).
154. C. T. Ratcliffe and J. Shreeve, *J. Am. Chem. Soc.*, **90**, 5403 (1968).
155. M. Kobayashi, H. Minato and Y. Ogi, *Bull. Chem. Soc. Jpn.*, **45**, 1224 (1972).
156. D. Barnard, *J. Chem. Soc.*, 4675 (1957).
157. P. Koch, E. Ciuffar and A. Fava, *J. Am. Chem. Soc.*, **92**, 5971 (1970).
158. J. L. Kice and J. P. Cleveland, *J. Am. Chem. Soc.*, **95**, 109 (1973).
159. J. L. Kice, C. G. Venier, G. B. Large and L. Heasley, *J. Am. Chem. Soc.*, **91**, 2028 (1969).
160. I. B. Douglass and D. A. Koop, *J. Org. Chem.*, **29**, 951 (1964).
161. H. J. Backer and H. Kloosterziel, *Recl. Trav. Chim. Pays-Bas*, **73**, 129 (1954).
162. C. R. H. I. deJonge, F. P. B. van der Maeden, M. E. F. Biemond, W. J. B. Huysmans and W. J. Mijs, *Polym. Sci., Polym. Symp.*, **57**, 197 (1976).
163. R. M. Topping and N. Kharasch, *J. Org. Chem.*, **27**, 4353 (1962).

The Chemistry of Sulphinic Acids, Esters and their Derivatives
Edited by S. Patai


CHAPTER **15**

# Synthesis and uses of isotopically labelled sulfinic acid derivatives

SHIGERU OAE

*Department of Chemistry, Okayama University of Science, Ridai-cho 1-1, Okayama, 700 Japan*

and

HIDEO TOGO

*Department of Chemistry, Faculty of Science, Chiba University, Yayoi-cho, Chiba City, 260 Japan*



## I. INTRODUCTION

Sulfinic acids and their derivatives are usually intermediates in the oxidation of sulfur compounds, for example that of thiols to sulfonic acids or in the reduction of sulfonic acids to thiols. Alkali sulfinates are also obtained by treatment of organometallic compounds with $SO_2$[1], and various sulphinic acid derivatives can be often isolated as intermediates in similar reactions. However, in general, sulfinic acids and their derivatives are less stable than the corresponding sulfonic acids and sulfonates except for sulfinate salts. Therefore, the chemistry of sulfinic acid derivatives in general has been studied less than that of the sulfonates, and the chemistry of isotopically labelled sulfinic acids and their derivatives received even less attention. In this chapter, we will summarize the limited amount of information available about isotopically labelled sulfinic acids and their derivatives.

## II. PREPARATION OF ISOTOPICALLY LABELLED SULFINIC ACID DERIVATIVES

### A. Sulfinic Acids and Sulfinate Salts

One of the simple methods of preparing $^{18}O$-labelled sulfinic acids is to treat the corresponding sulfinyl chloride with a small amount of $H_2{}^{18}O$ under cooling in an ice bath. The reaction starts immediately, and colorless crystals of the sulfinic acid deposit[2,3]. The sodium salt is obtained by addition of an equivalent amount of aqueous sodium hydroxide solution to the solution containing the sulfinic acid with vigorous stirring and cooling. After recrystallization of the precipitated crystals, the sodium sulfinate–$^{18}O$ dihydrate is obtained.

$$p\text{-}CH_3C_6H_4SOCl + H_2{}^{18}O(^{18}O{:}1.5\,\text{atom}\,\%) \longrightarrow p\text{-}CH_3C_6H_4S^{18}O_2H$$

$$\xrightarrow{\text{NaOH}} p\text{-}CH_3C_6H_4S^{18}O_2Na$$

It was confirmed in a separate experiment that no exchange of oxygen atoms occurs between the sulfinic acid and water either in neutral or in basic media. The $^{18}O$ content of the anhydrous salt was 0.70 atom %. In a second method to obtain $^{18}O$-enriched sulfinic acid, *p*-bromobenzenesulfinic acid was heated in $^{18}O$-enriched water ($^{18}O$: 1.52 atom %) at 90 °C for two hours. The product thus obtained contained 1.232 atom % excess oxygen—$^{18}O$[4]. Probably, this reaction serves as evidence for the acid-catalyzed formation of sulfinylsulfone as an intermediate.

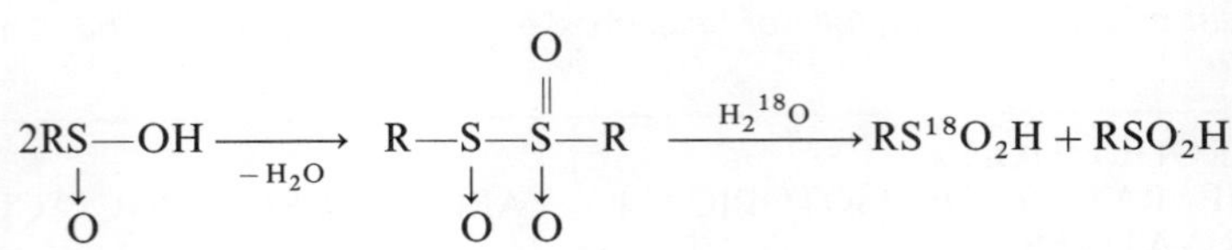

If the reaction proceeds as shown above, the sulfinic acid would be enriched to that of the whole amount of $^{18}O$ in the reaction medium. The best method to obtaine $^{18}O$-enriched sulfinic acid was reported by us[5], as in the following equations:

$$2ArS^{18}O_2Cl + 2Zn \longrightarrow (ArS^{18}O_2)_2Zn + ZnCl_2$$

$$(ArS^{18}O_2)_2Zn + Na_2CO_3 \longrightarrow ArS^{18}O_2Na \xrightarrow{\text{HCl}} ArS^{18}O_2H$$

The $^{18}O$-labelled arenesulfonyl chloride was obtained by treating the corresponding arenethiol with chlorine gas in $^{18}O$-enriched water ($^{18}O$: 1.60 atom %) under cooling. The arenesulfonyl chloride thus obtained in a good yield can be reduced by treatment with zinc[6] to give the sodium sulfinate dihydrate, which is in turn dehydrated for five hours at 120–130 °C.

The $^{18}O$-labelled arenesulfinic acids can be prepared by careful neutralization of the $^{18}O$-labelled sodium arenesulfinate with hydrochloric acid. The results of $^{18}O$ analysis are shown in Table 1. In these experiments, the $^{18}O$ analysis has been carried out by a modification of Rittenberg's method[7]. Namely, the $^{18}O$-labelled sulfinates are pyrolyzed in the presence of mercuric chloride and mercuric cyanide at 400 °C for four hours and the evolved the $CO_2$ gas, after being passed through $Pb(OAc)_2$-coated glass wool, is subjected to the mass-spectrometric analysis. From the mass peak heights of 44 and 46, the content of $^{18}O$ can be calculated[5].

The oxidation of the thiol or the disulfide by alkaline autoxidation in $H_2{}^{18}O$ gives the corresponding sulfinate-$^{18}O$ which is one of the oxidized products[8].

TABLE 1. $^{18}O$ analyses

| | ($^{18}O$ atom %) | | |
|---|---|---|---|
| Ar | $ArSO_2Cl$ | $ArSO_2Na$ | $ArSO_2H$ |
| $p$-$CH_3C_6H_4$— | 1.59 | 1.50 | 1.52 |
| $p$-$BrC_6H_4$— | 1.43 | 1.39 | 1.40 |

TABLE 2. Oxidation products ($\mu$mole)

| Experimental conditions | $O_2$ uptake | Hypotaurine | Thiotaurine | Cysteamine | Cystamine |
|---|---|---|---|---|---|
| Complete system | 48.0 | 38.0 | 8.5 | 0.0 | 3.5 |
| Enzyme boiled | 12.9 | 0.0 | 0.0 | 0.0 | 50.0 |
| Enzyme omitted | 14.0 | 0.0 | 0.0 | 8.8 | 41.0 |
| Sulfide omitted | 10.7 | 0.0 | 0.0 | 0.0 | 50.0 |

Cysteamine-$^{35}S$: 50$\mu$mole

$^{18}O$-labelled sulfoxides can be readily obtained by treating the corresponding sulfides with some amine–bromine complexes in $^{18}O$-enriched $H_2{}^{18}O$[9]. Pyrolysis of aliphatic sulfoxides bearing a $\beta$-hydrogen leads to the formation of a sulfenic acid, which can be further converted to either the thiolsulfinate or the sulfinate. This process can be utilized for preparing $^{18}O$-labelled sulfinate derivatives[10].

The enzymatic oxidation (using an enzyme isolated from horse kidney) of cysteamine $H_2NCH_2CH_2SH$ to hypotaurine $H_2NCH_2CH_2SO_2H$ and to thiotaurine $H_2NCH_2CH_2SO_2SH$[11] was reported. This reaction occurs only in the presence of elemental sulfur or the sulfide ($Na_2S$), which plays the role of catalyst, and the reaction also requires oxygen. The results in the oxidation of cysteamine-$^{35}S$ are summarized in Table 2.

$$RSH + O_2 \longrightarrow RSO_2H$$

$$nH_2S + (n-1)O \longrightarrow H_2S_n + (n-1)H_2O$$

$$RSO_2H + H_2S_n \longrightarrow RSO_2SH + H_2S_{n-1}$$

## B. Sulfinyl Halides

$^{18}O$-labelled sulfinyl chlorides have been obtained by the reaction of the corresponding sodium $^{18}O$-sulfinate with thionyl chloride[12]. Another method involves treatment of the disulfide or the thiol with chlorine gas in acetic anhydride enriched with $^{18}O$[13].

$$(PhCH_2S)_2 \xrightarrow[Cl_2]{Ac_2O(^{18}O:0.854\,\text{atom}\,\%)} PhCH_2S(^{18}O)Cl \quad 81\%$$

The second method to prepare $^{18}O$-labelled sulfinyl chlorides is the reaction of non-labelled sulfinyl chloride and $^{18}O$-labelled water ($^{18}O$: 1.5 atom %)[14]. To a dry ether solution of benzenesulfinyl chloride was added dropwise $^{18}O$-labelled water under cooling. The reaction was considerably exothermic, with evolution of HCl gas. To the benzenesulfinic acid obtained after evaporation of ether and of excess water under reduced pressure, dissolved again in ether, excess thionyl chloride was added. A vigorous reaction took place evolving gaseous $SO_2$ and HCl. The residual oil after evaporation of ether and of excess thionyl chloride was purified by distillation.

Phenynl[$^2H_2$]methanesulfinyl chloride was obtained as follows. Methyl benzoate was reduced by lithium aluminum deuteride in ether to give [$\alpha,\alpha$-$^2H_2$]benzyl alcohol, which was then treated with thiourea in hydrobromic acid (47%) solution under reflux, affording [$\alpha,\alpha$-$^2H_2$]toluene-$\alpha$-thiol. After oxidation of the thiol to the disulfide by iodine in an alkaline solution, the disulfide was treated further with acetic anhydride and an equivalent amount of chlorine to give phenyl[$^2H_2$]methanesulfinyl chloride[13].

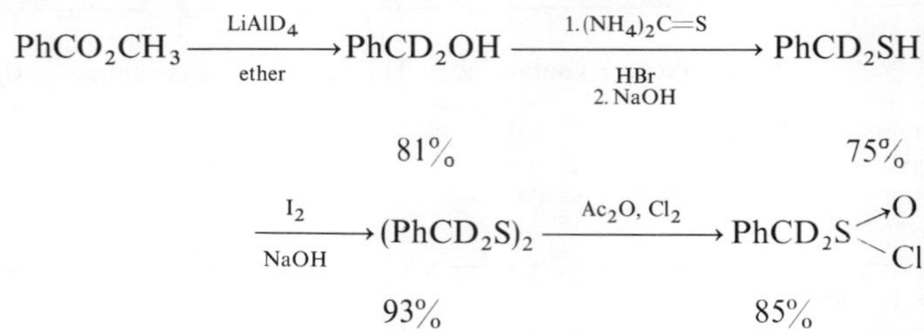

Phenyl[$^{13}C$]methanesulfinyl chloride can also be obtained by a similar procedure starting from [$\alpha$-$^{13}C$]benzyl alcohol, which is obtained by the reaction of phenylmagnesium bromide with $^{13}CO_2$ and the subsequent reduction of the benzoic[$\alpha$-$^{13}C$] acid so formed, by lithium aluminum hydride[13].

### C. Sulfinate Esters

Ethyl *p*-toluenesulfinate[sulfinyl-$^{18}O$] was obtained by treating sodium *p*-toluenesulfinate-$^{18}O$ with ethyl chlorosulfite in acetonitrile under heating[2]. The sulfinate obtained was found to contain about two-thirds of the heavy oxygen atoms originally incorporated in the starting sodium sulfinate.

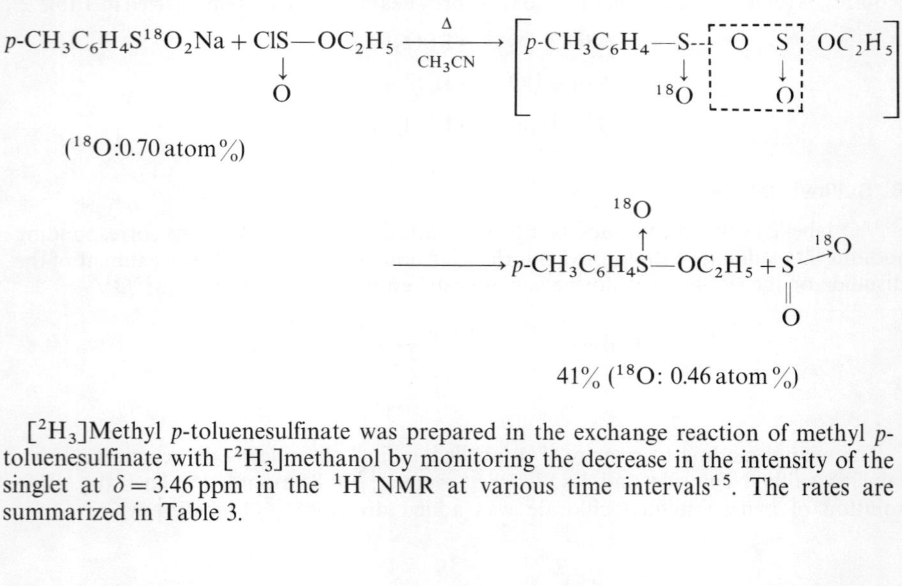

[$^2H_3$]Methyl *p*-toluenesulfinate was prepared in the exchange reaction of methyl *p*-toluenesulfinate with [$^2H_3$]methanol by monitoring the decrease in the intensity of the singlet at $\delta = 3.46$ ppm in the $^1H$ NMR at various time intervals[15]. The rates are summarized in Table 3.

TABLE 3. Rates of exchange of methanol-$d_3$ and -$d_4$ with methyl *p*-toluenesulfinate at 62 °C

| Solvent | $AcO^-/AcOH$ buffer ratio | $[AcO^-]$ (M) | $\times 10^6\ (s^{-1})$ |
|---|---|---|---|
| $CD_3OH$ | 2:1 | 0.210 | 2.82 |
| | | 0.158 | 2.38 |
| | | 0.140 | 2.26 |
| | | 0.105 | 1.95 |
| | | 0.070 | 1.74 |
| $CD_3OD$ | 2:1 | 0.210 | 1.99 |
| | | 0.158 | 1.70 |
| | | 0.140 | 1.55 |
| | | 0.105 | 1.32 |
| | | 0.070 | 1.18 |

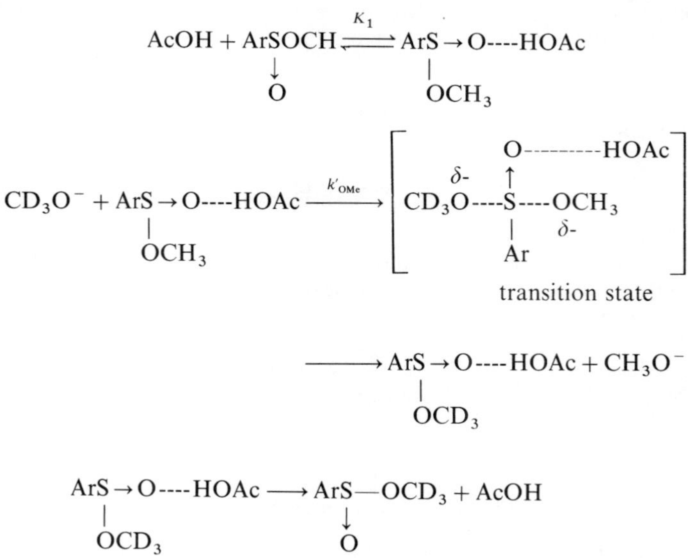

$Ar = p\text{-}CH_3C_6H_4\text{—}$

The rate-determining step of the acetate-catalyzed process is simply the attack of $CD_3O^-$, not on the sulfinate ester itself but rather on the presumably more reactive hydrogen-bonded complex of the ester with a molecule of acetic acid as shown in the above scheme. Since the process shown in Table 3 involves a general base-catalyzed mechanism, proton transfer of methanol is part of the rate determining step, i.e.:

$$AcO^- + CD_3OH(D) \rightleftharpoons AcOH(D) + CD_3O^-$$

and hence the rates in $CD_3OH$ are consistently higher than those in $CD_3OD$.

Optically active $^{18}O$-labelled (−) menthyl (−)*p*-toluenesulfinate can be prepared by the reaction of the $^{18}O$-labelled *p*-toluenesulfinyl chloride with (−) menthol in the presence of pyridine. The $^{18}O$-labelled *p*-toluenesulfinyl chloride was obtained from $^{18}O$-labelled *p*-toluenesulfinic acid as mentioned above[12].

$$p\text{-TolS}\begin{smallmatrix}\nearrow {}^{18}\text{O} \\ \searrow \text{Cl}\end{smallmatrix} \xrightarrow[\text{Py}]{(-)\text{ menthol}} p\text{-TolS}\begin{smallmatrix}\nearrow {}^{18}\text{O} \\ \searrow \text{O-menthyl}\end{smallmatrix} \longrightarrow \text{(recrystallization) Diastereomers}$$

$$\longrightarrow {}^{18}\text{O } (-) \text{ menthyl } (-)p\text{-toluenesulfinate}$$

$$\text{mp } 101\text{–}103\,^{\circ}\text{C}, [\alpha]_D - 201.5^{\circ}\ ({}^{18}\text{O}: 0.463 \text{ atom } \%)$$

## III. USES OF ISOTOPICALLY LABELLED SULFINIC ACID DERIVATIVES

### A. Sulfinic Acids

It is well known that sulfinic acids easily disproportionate to give the corresponding sulfonic acid, thiolsulfonate and $H_2O$. The mechanism and the solvent effect of this reaction have been studied with $H_2{}^{18}O$ in the following way. An ampoule containing *p*-bromobenzenesulfinic acid in water ($^{18}O$: 1.52 atom %) was heated at 90 °C for 12 hours[4]. Then the crystals precipitated were collected to obtain the corresponding thiolsulfonate, and the mother liquor gave the $^{18}O$-enriched sulfinic acid as a precipitate after evaporation, while the sulfonic acid was obtained as the crystalline S-benzyl isothiuronium salt.

$$p\text{-XC}_6\text{H}_4\text{SO}_2\text{H} \xrightarrow[\text{H}_2{}^{18}\text{O}]{} p\text{-XC}_6\text{H}_4\text{S}^{18}\text{O}_2\text{H} + p\text{-XC}_6\text{H}_4\text{S}^{18}\text{O}_2\text{SC}_6\text{H}_4\text{X-}p\, +$$

$$p\text{-XC}_6\text{H}_4\text{S}^{18}\text{O}_3\text{H} + \text{H}_2{}^{18}\text{O}$$

$$\text{X} = \text{H, Br, NO}_2$$

The results are summarized in Table 4.

The fact that both the sulfinic acids and their reaction products (the thiolsulfonate and the sulfonic acid) contain approximately the same amount of $^{18}O$ indicates that the rate of oxygen exchange is considerably faster than that of disproportionation.

$$\text{RSO}_2\text{H} + \text{RSO}_2\text{H} \underset{\text{H}_2\text{O}}{\overset{-\text{H}_2\text{O}}{\rightleftharpoons}} \text{RSO}_2\text{SOR} \underset{-\text{H}_2\text{O}}{\overset{\text{H}_2{}^{18}\text{O}}{\rightleftharpoons}} \text{RS}^{18}\text{O}_2\text{H} + \text{RSO}_2\text{H}$$

$$\text{RSO}_2\text{SOR} \xrightarrow{\text{slow}} \text{disproportionation products}$$

Thus, the above reaction scheme which was proposed by Kice and coworkers[16] seems to be quite plausible for the mechanism of the incorporation of oxygen-18 into the products. The rates of exchange are faster in solutions with greater sulfinic acid concentrations.

TABLE 4. Incorporation of $^{18}O$ into sulfinic acids and their products formed on heating in $H_2{}^{18}O$

| | Excess oxygen-18 (atom %) | | |
|---|---|---|---|
| | $p$-$BrC_6H_4SO_2H$ | $p$-$CH_3C_6H_4SO_2H$ | $p$-$O_2NC_6H_4SO_2H$ |
| $ArSO_2H$ | 1.38 | 1.02 | 1.04 |
| $ArSO_3H$ | 1.29 | 0.95 | 1.15 |
| $ArSO_2SAr$ | 1.31 | 1.04 | 1.14 |

TABLE 5. Rates of $^{18}O$ increase in arenesulfinic acids in 30% aqueous dioxene

| Sulfinic acid | | $p$-$CH_3C_6H_4SO_2H$ | $p$-$BrC_6H_4SO_2H$ | $C_6H_5SO_2H$ |
|---|---|---|---|---|
| $[ArSO_2H]$ (mol liter$^{-1}$) | | 1.32 | 1.13 | 1.30 |
| $H_2{}^{18}O$ (atom %) | | 1.52 | 1.52 | 1.52 |
| $10^7 \times k$ ($M^{-1}S^{-1}$) | at 20.0 °C | 1.37 | 2.49 | 1.42 |
| | at 40.0 °C | 12.9 | 17.4 | 8.28 |
| $E_a$(kcal mol$^{-1}$) | | 21.5 | 17.4 | 15.6 |
| $\Delta S^{\ddagger}$ (e.u.) | | −21.9 | −31.3 | −29.4 |

The rates ($k$) of increase of oxygen-18 in the sulfinic acids recovered have been measured, as shown in Table 5.

The reaction of sulfinic acids with diazomethane is well known and gives a mixture of the sulfinate ester and the sulfone. The reaction was found to be of first order in both the sulfinic acid and diphenyldiazomethane in the following reaction[17]:

$$RSO_2H + Ph_2CN_2 \longrightarrow R{-}S(\rightarrow O){-}OCHPh_2 + R{-}S(\rightarrow O)(=O){-}CHPh_2$$

The rates of the reactions and the ratios of the sulfinate ester vs. sulfone were found to vary considerably in different solvents (Table 6). The rates of the reaction are quite large in benzene or dichloromethane and much smaller in dioxane, alcohol and DMSO, as shown in Table 7.

The rates of reaction in non-polar solvents are much greater than those in polar solvents. Infrared spectroscopy shows that the sulfinic acid exists as a dimer in non-polar solvents; it seems (Table 8) that such dimeric sulfinic acids react with $Ph_2CN_2$, and protonate the latter easily as in the following equation:

TABLE 6. Molar ratios of the products in various solvents

| Solvent | $p$-TolS($\rightarrow$O)OCHPh$_2$ | $p$-TolS($\rightarrow$O)(=O)CHPh$_2$ | Ether | Total yield (%) |
|---|---|---|---|---|
| $CH_2Cl_2$ | 0 | 100 | — | 80 |
| Benzene | 20 | 80 | — | 96 |
| $CH_3CN$ | 81 | 19 | — | 100 |
| Ethanol | 60 | 14 | 26 | 100 |
| Dioxane | 83 | 17 | — | 100 |
| DMSO | 100 | 0 | — | 98 |

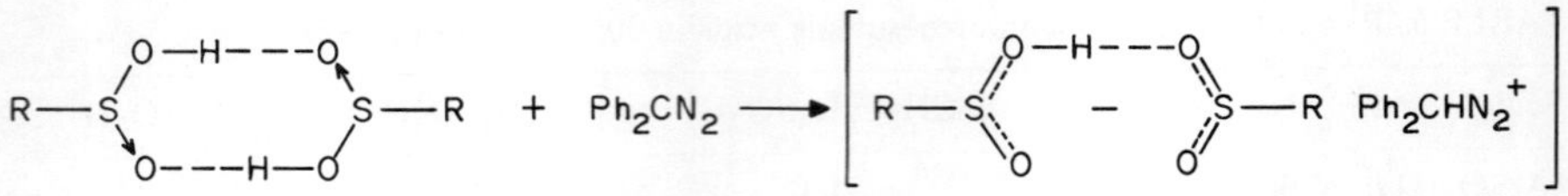


Apparently, protonation of $Ph_2CN_2$ is accelerated in non-polar solvents. Table 9 reveals that the average kinetic isotope effect, i.e. $k_H/k_D$, was about 3.0. These data indicate that the protonation of $Ph_2CN_2$ is the rate-determining step in the reaction between the sulfinic acid and the diazomethane.

### B. Sulfinyl Halides

Benzyl phenylmethanethio[$^{18}$O]sulfinate can be prepared by treating toluene-α-thiol with phenylmethane[$^{18}$O]sulfinyl chloride[13]. Other $^{18}$O-labelled unsymmetrical thiolsulfinates can be prepared by the same procedure. For example, $^{18}$O-labelled methyl benzenethiolsulfinate was prepared by treating methanethiol with $^{18}$O-labelled benzenesulfinyl chloride, which was prepared by treating diphenyl disulfide with $Cl_2$ gas in $^{18}$O-labelled acetic anhydride[3].

TABLE 7. Rate constants of the reaction between *p*-toluenesulfinic acid and diphenyldiazomethane in various solvents

| Solvent | Temperature (°C) | $k$ (liter mol$^{-1}$ s$^{-1}$) | $\Delta H^\ddagger$ (kcal mol$^{-1}$) | $\Delta S^\ddagger$ (e.u.) |
|---|---|---|---|---|
| $CH_2Cl_2$ | 20.0 | 300 | | |
| Benzene | 20.0 | 22 | | |
| $CH_3CN$ | 20.0 | 4.9 | | |
| Ethanol | 20.0 | 1.6 | | |
| Dioxane | 34.5 | 0.419 | | |
| | 30.0 | 0.321 | 13.0 | −18.1 |
| | 24.8 | 0.190 | | |
| | 19.5 | 0.314 | | |
| DMSO | 35.0 | 0.103 | | |
| | 30.0 | 0.0708 | 13.0 | −23.6 |
| | 27.0 | 0.0570 | | |
| | 19.0 | 0.0328 | | |

TABLE 8. Rates of reaction between $Ph_2CN_2$ and *p*-$TolSO_2D$ (or *p*-$TolSO_2H$) in dioxane containing 2 vol% $D_2O$ (or $H_2O$)

| Acid | Temperature (°C) | $k$ (liter mol$^{-1}$ s$^{-1}$) | $\Delta H^\ddagger$ (kcal mol$^{-1}$) | $\Delta S^\ddagger$ (e.u.) |
|---|---|---|---|---|
| | 29.5 | 0.148 | | |
| *p*-$TolSO_2D$ | 19.8 | 0.0666 | 12.8 | −18.5 |
| | 15.0 | 0.0478 | | |
| | 29.5 | 0.435 | | |
| *p*-$TolSO_2H$ | 19.5 | 0.207 | 14.0 | −16.8 |
| | 15.0 | 0.142 | | |

TABLE 9. Kinetic isotope effects in the reaction between $Ph_2CN_2$ and *p*-$TolSO_2H$

| Reaction temperature (°C) | 29.5 | 19.8 | 15.0 |
|---|---|---|---|
| $k_H/k_D$ | 2.9 | 3.1 | 3.3 |

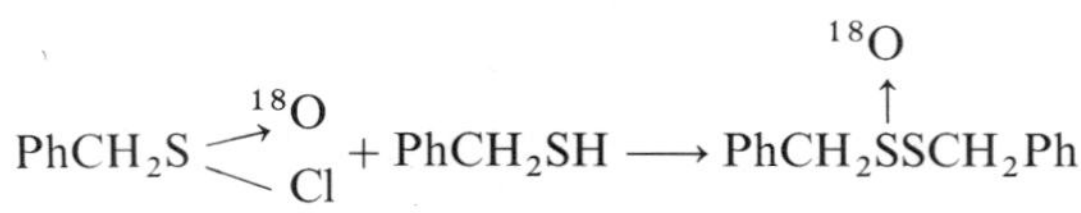

($^{18}O$: 0.0854 atom %) ($^{18}O$: 0.843 atom %)

Oxidation of disulfides is considered to proceed via various stable intermediates as shown in the accompanying equation. When the disulfides are unsymmetrical, oxidation with nucleophilic oxidants, such as superoxide anion, $O_2^-$ [18], peroxide ion[19] or $ClO^-$

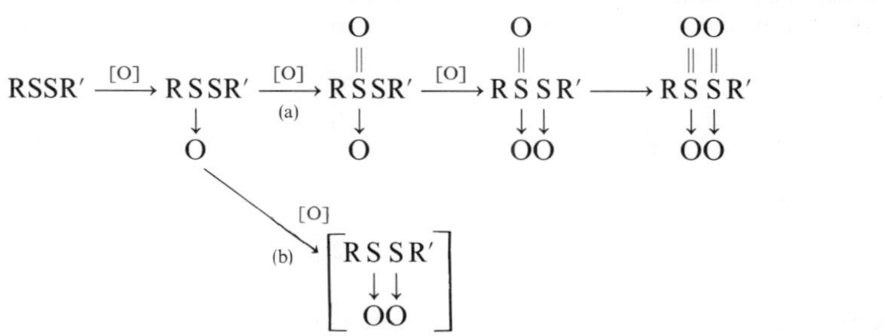

ion[20], attacks the sulfinyl sulfur atom following path a, while oxidation with electrophilic oxidants, such as peroxyacids or hydrogen peroxides, proceeds via path b, usually affording two regioisomeric oxidation products as illustrated by the equations below. The ratio of one isomer over the other depends on the electron availability of the sulfenyl sulfur atom in the electrophilic oxidation. Many examples of this nature have been known[21].

(Electrophilic)

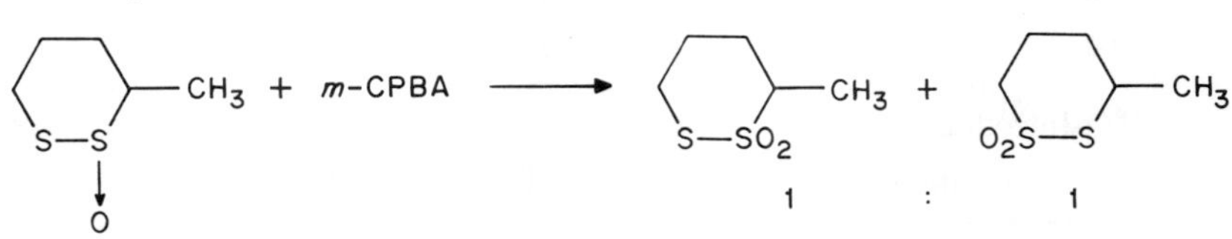

(Nucleophilic)

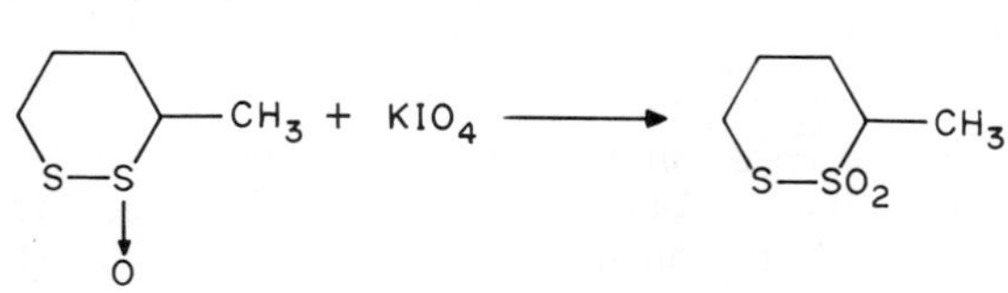

In the electrophilic reaction, only the divalent sulfenyl sulfur atom is attacked by *m*-CPBA to form incipiently the α,α′-disulfoxide, which eventually rearranges to afford the two isomeric products of the cyclic thiolsulfonate. Both mechanisms on oxidation and oxygenation have been investigated rather extensively by us[22,23].

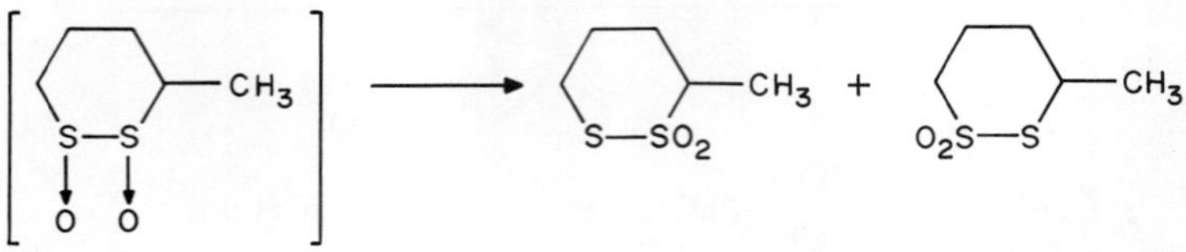

One interesting problem is whether or not it is possible to isolate or at least to detect the α,α′-disulfoxides in the process of oxidation of the thiosulfinate to the thiolsulfonate. Up to date, all attempts to observe the α,α′-disulfoxide have failed[24].

However, it was found that in the oxidation of a non-cyclic thiolsulfinate (**1**), one of the products formed predominantly was a thiolsulfonate, (**2**), usually more than 30% yield, in which the phenylsulfinyl oxygen is completely transferred into the methylsulfonyl group in the thiolsulfonate, as shown below[25]. This could happen only when the initial

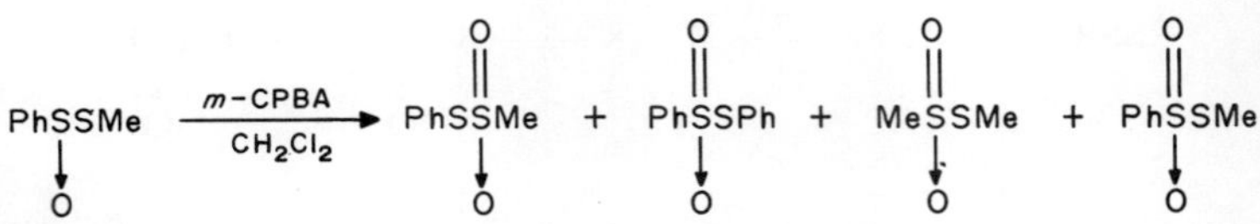

(**1**) (**2**) (**3**) (**4**) (**5**)

electrophilic oxidation takes place on the sulfenyl sulfur atom to form incipiently the α,α′-disulfoxide. In order to confirm this postulate, many $^{18}O$ tracer experiments have been carried out. Among those, only the crucial one is described in the following equation (where ● symbolizes 100% $^{18}O$ and ◒ a mixture of $^{16}O$ and $^{18}O$):

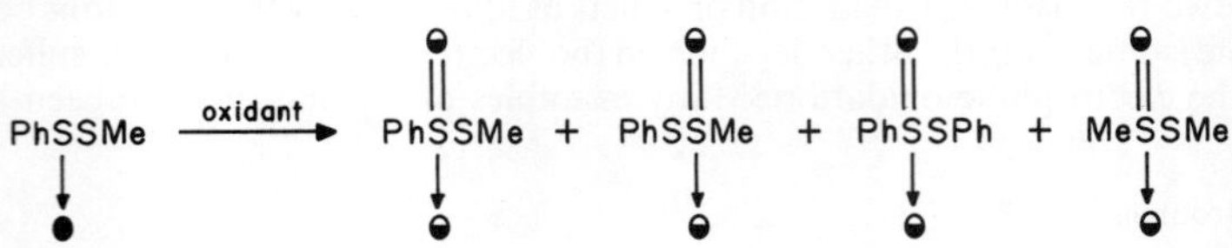

(**6**) (**7**) (**8**) (**9**) (**10**)

$^{18}O$-Introduction*

| Oxidant | 7 | 8 | 9 | 10 |
|---|---|---|---|---|
| $H_2O_2$/AcOH | 102% | 58–70% | 108–124% | 68% |
| *m*-CPBA/$CH_2Cl_2$ | — | 102–109% | 110–137% | — |

*Results for **7** and **10** were the same in many runs, while **8** and **9** varied as shown.

Indeed, when the oxidation was carried out in aprotic non-aqueous media, e.g. $CH_2Cl_2$, the resulting thiolsulfonate (**8**) was found to retain the $^{18}O$ of the original thiolsulfinate completely, clearly revealing the initial formation of the α,α-disulfoxide, as shown below. However, when the oxidation was carried out in aqueous protic media,

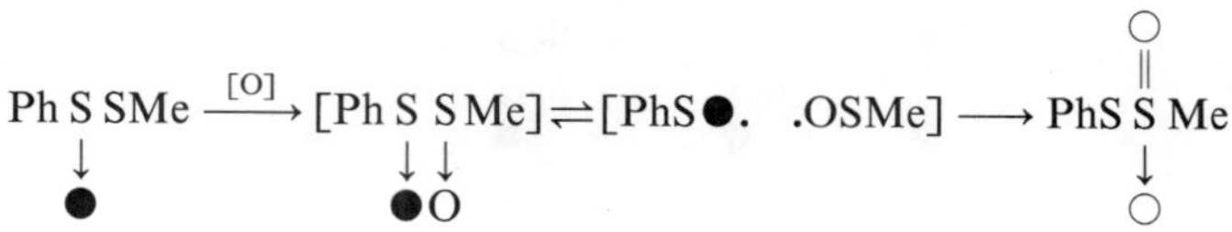

**(8)**

the extent of $^{18}O$ incorporation into the resulting thiolsulfonate was roughly two-thirds; this is obviously due to the partial or nearly complete hydrolysis of the intermediary $\alpha,\alpha'$-disulfoxide to form the two sulfenic acids, which would regenerate the thiolsulfinate upon self-condensation, thus losing a portion of $^{18}O$ in the resulting thiolsulfinate.

After all these experiments, our accumulated NMR data were used cleverly by Freeman and coworkers to lead to the NMR observation of what seems likely to be the $\alpha,\alpha'$-disulfoxide at low temperature[26].

The alkaline hydrolysis of thiolsulfinate has been studied rather extensively[27]. However, as to the initial attacking site for $OH^-$ ion, there have been some controversies[28].

Using unsymmetrical thiolsulfinates, we have found that the initial attack of the hydroxide ion takes place only on the sulfinyl sulfur atoms. Thus the overall reaction is shown in the following equation[27]:

$$R\overset{\overset{O}{\uparrow}}{S}SR' \xrightarrow{OH^-} R'SSR' + RSSR' + RSO_2^-$$

The lack of detection of $R'SO_2^-$ supports strongly the exclusive attack of hydroxide ion on the sulfinyl sulfur, and this important result was confirmed further by a tracer study using $^{18}O$-labelled thiolsulfinates. The result revealed clearly that the $^{18}O$ label of the starting thiosulfinate was retained almost completely in the sulfinate ($RSO_2Na$) formed. These observations suggest the following mechanistic scheme to be the most plausible one:

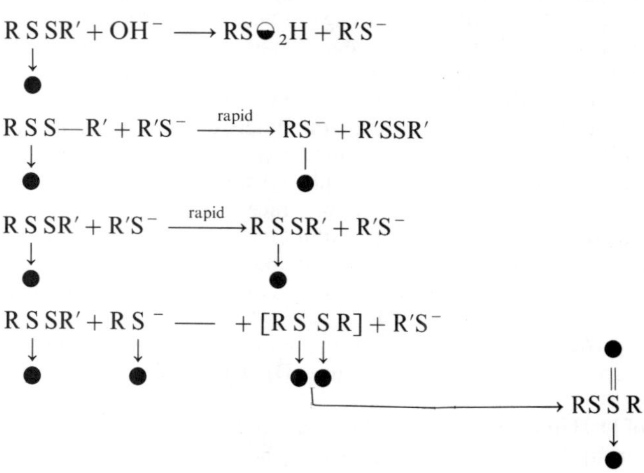

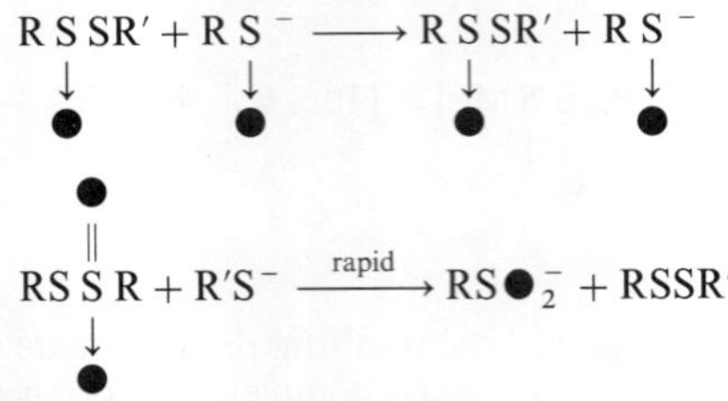

A $^{13}C$-tracer experiment using the $^{13}C$-labelled compound **11** which was prepared from $Ba^{13}CO_3$ enriched with 90% $^{13}C$ via the reaction of $PhCH_2SOCl$ and $Ph^{13}CH_2SH$, has been carried out by us.

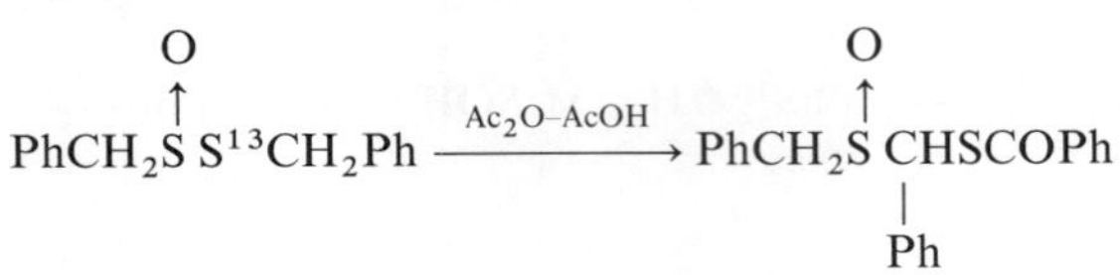

(**11**) (**12**)

This thiosulfinate **11** was treated with $Ac_2O$–AcOH for 1–2 h. After the reaction, the rearranged sulfoxide **12** and recovered **11** were separated, and then the position and amount of $^{13}C$ determined by $^{13}C$ and $^{1}H$ NMR. These results revealed that the amount of $^{13}C$ in the sulfoxide **12** and in the recovered **11** decreased to 76% and 62%, respectively, at 60% conversion of the reaction. However, when the reaction was stopped after one hour, which corresponds to 40% conversion, the amount of $^{13}C$ found in sulfoxide **12** was 96% while that in the recovered **11** was 82%[29].

Furthermore, $^{13}C$ was found solely in the sulfenyl side (i.e. at the methine position) of sulfoxide **12** and in the original position of the recovered **11**. While, based on deuterium tracer experiments, the $^{2}H$ contents of both compounds **11** and **12** were found to have decreased in the decrease in the $^{13}C$ content, however, the trend of the decrease for the two isotopes is different. Also, $^{18}O$-labelled **11** was prepared and treated under similar reaction conditions as used for the other two isotopically labelled compounds. According to the results, the contents of $^{18}O$ of both the sulfoxide **12** and the recovered **11** did not change within experimental error. These tracer experiments and the product analysis suggest that the initial step is an $E_i$ process, probably a pyrolysis of **11**, affording α-toluenesulfenic acid and thiobenzaldehyde, probably in the cage of acetic anhydride.

The sulfenic acid and thiobenzaldehyde once formed then react mainly to afford the rearranged sulfoxide **12** (path a in Scheme 1 below) or may return to the original thiolsulfinate (path b). Meanwhile, some of the sulfenic acid and thiobenzaldehyde may escape from the cage and disproportionate or trimerize to give the starting material **11** or the trithiane (**13**). The initial step is undoubtedly an equilibrium, because the recovered **11** was found to have taken up 20% D in its sulfenyl side at 26% conversion of the reaction when the reaction was carried out in the presence of $D_2O$.

Finally, when the reaction was carried out under similar conditions but in the presence of excess methyl acrylate, methyl-3-(1-phenylmethanesulfinyl)propionate, which was formed by the reaction of the sulfenic acid and methyl acrylate, was obtained in 83% yield. These observations indicate Scheme 1.

In order to prove the reaction mechanism, α,α-dideuterium-labelled thiolsulfinate (**A′**) was prepared and treated with $Ac_2O$ to give (**D**), which has a deuterium in the methine position[30].

The ratio of D:H at the methine position was 60:40. Thus, the presence of the proton in the methine group is very likely influenced by the equilibrium between **B** and **C**.

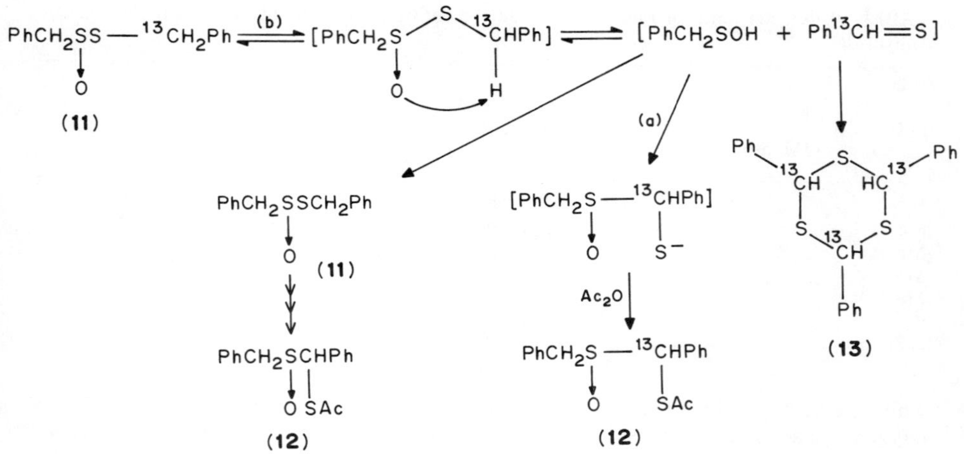


SCHEME 1

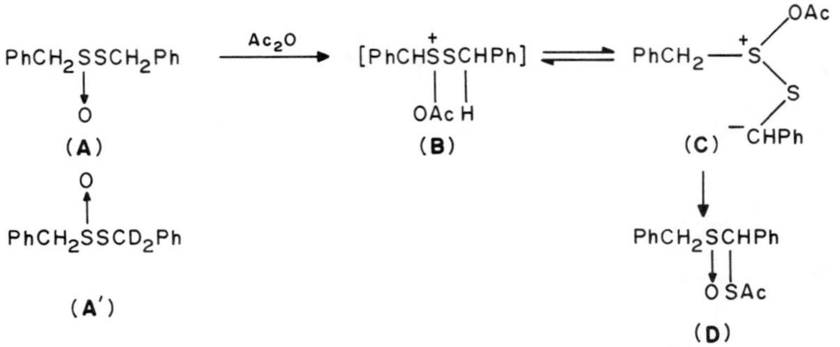


$^{35}$S-labelled symmetrical thiolsulfinates were prepared by Barnard[31] and the oxidation of Ph$^{35}$SO–SPh and PhSO–$^{35}$SPh with hydrogen peroxide in acetic acid was studied to give in good yields (>80%) the thiolsulfonates with only 66% of the activity retained in the original positions. This experiment also suggests the incipient formation of the $\alpha,\alpha'$-disulfoxide.

### C. Sulfinate Esters

The mechanism of the $^{18}$O exchange of $^{18}$(−)menthyl (−)arenesulfinate with trichloroacetic anhydride was studied by the kinetic observations of both the $^{18}$O exchange and the racemization[12]. When the sulfinate was treated with $(CCl_3CO)_2O$ in benzene solution at room temperature, the sulfinate ester was recovered quantitatively, but was found to have lost its optical activity. The kinetic data in Table 10 indicate that the rate of racemization ($k_{rac}$) of (−)menthyl (−)*p*-toluenesulfinate was about twice that of oxygen exchange ($k_{ex}$). This means that the reaction involves a Walden inversion. The energy and the entropy of activation for the racemization at 25.5 °C were found to be 14.5 kcal mol$^{-1}$ and −26.8 e.u., respectively.

The negative $\rho$ value (−1.53) indicates that the acylation is the rate-determining step,

TABLE 10. Kinetic data on racemization of (−)menthyl (−)arenesulfinate with trichloroacetic anhydride

| X in $p$-$XC_6H_4\overset{*}{S}(\rightarrow O)OMenthyl$ | Solvent | Temperature | $k_2 \times 10^4$ ($M^{-1}s^{-1}$) |
|---|---|---|---|
| H | Benzene | 25.5 | 2.98 |
| Me | Benzene | 25.5 | 5.25 |
| Me | Benzene | 35.7 | 12.1 |
| Me | Benzene | 45.1 | 24.0 |
| Me | THF | 25.5 | 2.27 |
| Cl | Benzene | 25.5 | 1.32 |

while the $k_{ex}/k_{rac}$ value of roughly 0.5 indicates that the energy barrier for the $S_N2$-like oxygen exchange process must be quite similar to that of the initial acylation. The overall process of the reaction can be illustrated as shown in Scheme 2. The rate of racemization was found to be of first order with respect to both ester and $(CCl_3CO)_2O$.

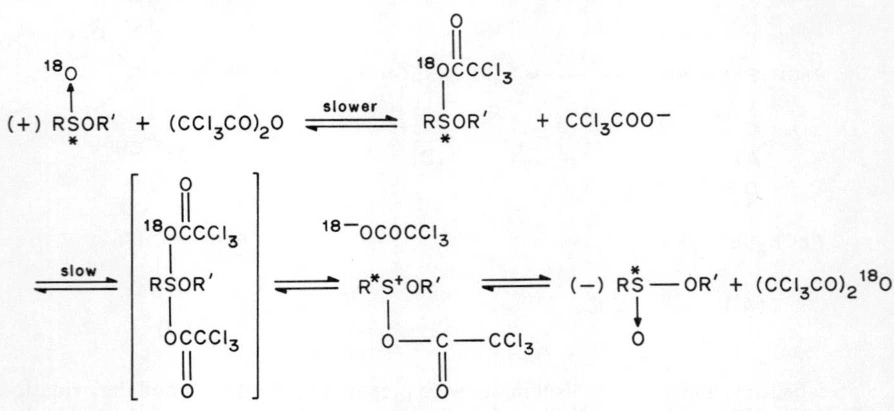


SCHEME 2

TABLE 11. Hydrolysis of diphenylmethyl $p$-toluenesulfinate in dioxane: $H_2{}^{18}O$; 60:40 v/v; 0.2 M acid or base catalyst[a]

| Isotopic abundance | HClO$_4$ | | | | | |
|---|---|---|---|---|---|---|
| Isotopic abundance | HClO$_4$ | | HBr | | NaOH | |
| $H_2O$ | | 0.92 | 0.905 | 0.70 | 0.936 | 0.705 |
| $Ph_2CHOH$ | 0.797 | 0.73 | 0.222 | 0.15 | 0.00 | 0.00 |
| alkyl–oxygen fission (%) | 85 | 80 | 25 | 21 | 0 | 0 |

[a]Isotopic abundances of oxygen are given in atom % excess above normal.

TABLE 12. Hydrolysis of methyl *p*-toluenesulfinate in dioxane: $H_2{}^{18}O$; 40:60 v/v[a]

| | $HClO_4$ | | NaOH | |
|---|---|---|---|---|
| Isotopic abundance $H_2O$ | 0.72 | 0.98 | 0.72 | 0.98 |
| MeOH | 0.00 | 0.00 | 0.00 | 0.00 |

[a]Isotopic abundances are given in atom % excess above normal.

The mechanism of the ester hydrolysis was also studied with $^{18}O$ tracer experiments. The hydrolysis of diphenylmethyl *p*-toluenesulfinate under various conditions is summarized in Table 11. The data indicate that the S—O bond is broken during the alkaline hydrolysis, while the hydrolysis catalyzed by perchloric acid leads to the predominant alkyl–O fission, whereas that catalyzed by hydrogen bromide results largely in S—O bond fission[32].

The hydrolysis of methyl *p*-toluenesulfinate is very slow in neutral aqueous dioxane, but is acid-catalyzed and is also very rapid in alkaline solution[33].

The data in Table 12 clearly reveal that the rapid second-order reaction between the ester and hydroxide ion takes place on the sulfur atom on the sulfinate ester, while in acidic conditions the reaction proceeds via the A-2 mechanism, as follows.

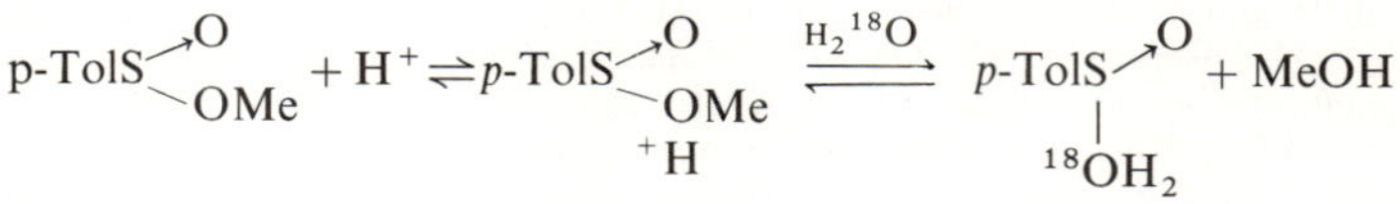


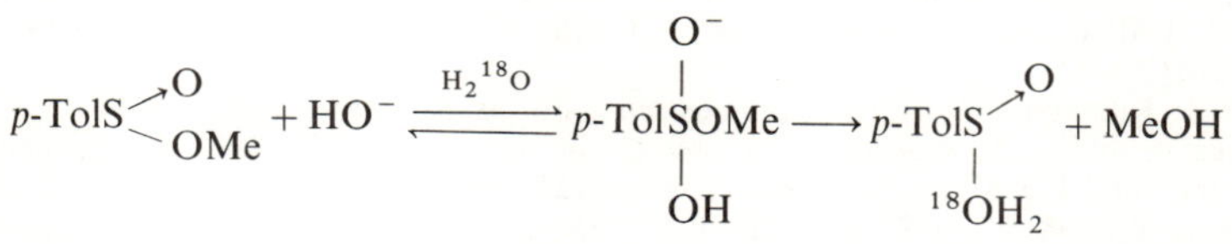


In both reaction conditions, the reactions proceed via S—OMe bond cleavage.

## IV. CONCLUSION

As mentioned in the introduction, up to date rather few studies have been performed on isotopically labelled sulfinic acid derivatives. However, with the present more facile availability of $^{13}C$ and $^{17}O$ NMR spectroscopies, together with the rapid growth of the organic chemistry of sulfur, research involving isotopically labelled sulfinic acids and sulfinates will without doubt develop more expensively in the future.

## V. REFERENCES

1. S. Oae and H. Togo, *Bull. Chem. Soc. Jpn.*, **56**, 3818 (1983).
2. M. Kobayashi, M. Terao and A. Yamamoto, *Bull. Chem. Soc. Jpn.*, **39**, 802 (1966).
3. S. Oae, Y. H. Kim, T. Takata and D. Fukushima, *Tetrahedron Lett.*, 1195 (1977).
4. M. Kobayashi, H. Minato and Y. Ogi, *Bull. Chem. Soc. Jpn.*, **45**, 1224 (1972).
5. N. Kunieda, K. Sakai and S. Oae, *Bull. Chem. Soc. Jpn.*, **41**, 3015 (1968).
6. F. C. Whitmore and F. H. Hamilton, *Org. Synth. Coll.*, **1**, 492 (1956).
7. D. Rittenberg and L. Ponticorvo, *Int. J. Appl. Radiat. Isot.*, **1**, 208 (1956).

8. S. Oae, Y. H. Kim, D. Fukushima and T. Takata, *Pure Appl. Chem.*, **49**, 153 (1977).
9. S. Oae, Y. Ohnishi, S. Kozuka and W. Tagaki, *Bull. Chem. Soc. Jpn.*, **39**, 364 (1964); T. Numata, O. Itoh, T. Yoshimura and S. Oae, *Bull. Chem. Soc. Jpn.*, **56**, 257 (1983).
10. T. Takata, Y. H. Kim and S. Oae, *Tetrahedron Lett.*, 821 (1979).
11. D. Cavallini, R. Scandurra and C. D. Marco, *J. Biol. Chem.*, **238**, 2999 (1963); D. Cavallini, C. D. Marco, R. Scandurra, S. Dupreé and M. T. Graziani, *J. Biol. Chem.*, **241**, 3189 (1966); D. Cavallini and R. Scandurra, *Biophys. Res. Commun.*, **24**, 185 (1966).
12. J. Drabowicz and S. Oae, *Tetrahedron*, **34**, 63 (1978); O. Itoh, T. Numata, T. Yoshimura and S. Oae, *Bull. Chem. Soc. Jpn.*, **56**, 266 (1983).
13. N. Furukawa, T. Morishita, T. Akasaka and S. Oae, *J. Chem. Soc., Perkin Trans. 2*, 437 (1980).
14. S. Oae, T. Takata and Y. H. Kim, *Bull. Chem. Soc. Jpn.*, **55**, 2484 (1982).
15. J. L. Kice and C. A. Walters, *J. Am. Chem. Soc.*, **94**, 590 (1972).
16. J. L. Kice and K. W. Bowers, *J. Am. Chem. Soc.*, **84**, 605 (1962); J. L. Kice and G. Guaraldi, *J. Am. Chem. Soc.*, **88**, 5242 (1966).
17. M. Kobayashi, H. Minato and H. Fukuda, *Bull. Chem. Soc. Jpn.*, **46**, 1266 (1973).
18. S. Oae, T. Takata and Y. H. Kim, *Bull. Chem. Soc. Jpn.*, **54**, 2712 (1981); S. Oae, T. Takata and Y. H. Kim, *Tetrahedron*, **37**, 37 (1981); T. Takata, Y. H. Kim and S. Oae, *Tetrahedron Lett.*, 821 (1979).
19. T. Takata, Y. H. Kim and S. Oae, *Bull. Chem. Soc. Jpn.*, **54**, 1443 (1981); S. Oae and T. Takata, *Tetrahedron Lett.*, **21**, 3213 (1980).
20. N. Furukawa, K. Akutagawa, T. Yoshimura and S. Oae, *Tetrahedron Lett.*, **22**, 3989 (1981).
21. S. Oae, T. Takata and Y. H. Kim, *Bull. Chem. Soc. Jpn.*, **55**, 2488 (1982); G. Leandri and A. Tundo, *Ann. Chim. (Rome)*, **44**, 74 (1964); W. Walter and P. M. Hell, *Justus Leibigs Ann. Chem.*, **727**, 35 (1969).
22. T. Takata and S. Oae, *Chemistry (Japan)*, **34**, 756, 891, 961 (1979); S. Oae, *Kagaku no Ryoiki (Japan)*, **34**, 344, 445 (1980); S. Oae, *The Chemistry of the Organic Sulfur Compounds—Reaction Mechanisms*, Chap. 5, Kagaku-Dojin, Kyoto, 1982.
23. S. Oae, Mechanisms of oxygenations of organic sulfur compounds, in *The Role of Oxygen in Chemistry and Biochemistry* (Eds. W. Ando and Y. Moro-oka), Elsevier, Amsterdam, 1988.
24. M. M. Chau and J. L. Kice, *J. Am. Chem. Soc.*, **98**, 7711 (1976); D. Barnard, *J. Chem. Soc.*, 4637 (1957); U. Marangelli, G. Modena and P. E. Todesco, *Gazz. Chim. Ital.*, **90**, 681 (1960); G. Modena and P. E. Todesco, *Ric. Sci.*, **30**, 1788 (1960); F. Freeman and C. N. Angeletakis, *J. Am. Chem. Soc.*, **103**, 6232 (1981).
25. S. Oae, T. Takata and Y. H. Kim, *Bull. Chem. Soc. Jpn.*, **55**, 2484 (1982).
26. F. Freeman and C. N. Angeletakis, *J. Am. Chem. Soc.*, **103**, 6232 (1982); F. Freeman, C. N. Angeletakis and T. J. Maricich, *Tetrahedron Lett.*, **22**, 1867 (1981); M. M. Chau and J. L. Kice, *J. Am. Chem. Soc.*, **98**, 7711 (1976); S. Oae, T. H. Kim, T. Takata and D. Fukushima, *Tetrahedron Lett.*, 1195 (1977); S. Oae and T. Takata, *Tetrahedron*, **36**, 3213 (1980).
27. S. Oae, T. Takata and Y. H. Kim, *Tetrahedron Lett.*, 4219 (1977).
28. J. L. Kice and T. E. Rogers, *J. Am. Chem. Soc.*, **96**, 8009 (1974).
29. N. Furukawa, T. Morishita, T. Akasaka and S. Oae, *Tetrahedron Lett.*, 1567 (1978).
30. N. Furukawa, T. Morishita, T. Akasaka and S. Oae, *Tetrahedron Lett.*, 1653 (1977); N. Furukawa, T. Morishita, T. Akasaka, and S. Oae, *J. Chem. Soc., Perkin Trans. 2*, 432 (1980).
31. D. Barnard, *J. Chem. Soc.*, 4675 (1957); D. Barnard and E. J. Percy, *Chem. Ind.*, 1332 (1960).
32. C. A. Bunton and B. N. Hendy, *J. Chem. Soc.*, 627 (1963).
33. C. A. Bunton and B. N. Hendy, *J. Chem. Soc.*, 2562 (1962).

The Chemistry of Sulphinic Acids, Esters and their Derivatives
Edited by S. Patai


CHAPTER **16**

# Thermochemistry and thermolysis of sulphinic acid derivatives

BOGDAN BUJNICKI, MARIAN MIKOŁAJCZYK and JAN OMELAŃCZUK

*Centre of Molecular and Macromolecular Studies, Polish Academy of Sciences, Department of Organic Sulphur Compounds, Sienkiewicza 112, 90-363 Łódź, Poland*



## I. INTRODUCTION

Sulphur and its inorganic and organic compounds are widely distributed in Nature. They are found in deep interstellar space and, very importantly for today's civilization, in the atmosphere, particularly under conditions of pollution. Organic sulphur compounds occur in living systems, in plants and in petroleum deposits and coal.

The problem of pollution stimulated recently much research directed towards removal of sulphur-containing compounds from oil and coal as well as from their combustion products. Thermochemistry of organic sulphur compounds plays an important role in these studies because there is a close relationship between the thermochemical parameters such as $\Delta H_f^0$, $\Delta S^0$ and $C_p^0$ and between mechanisms and kinetics of elementary reactions. The latter subject has been thoroughly treated by Benson in his monograph[1].

Although there are many reviews[2-5] and monographs[6-8] where thermochemical properties of various types of organic and inorganic sulphur compounds are discussed and collected, we were surprised to find that there is practically no information on the thermochemical data of sulphinic acid derivatives. This may be due to the fact that these compounds are not typical sulphur-containing air contaminants and are less important in modern technology. In this context, it should be noted that our literature search was mainly based on *Chemical Abstracts, Journal of Chemical Thermodynamics* and *Thermochimica Acta*. Unfortunately, some specialist periodicals such as *Bulletin of Thermodynamics and Thermochemistry* were inaccessible to us.

This chapter consists of two parts. For the reasons mentioned above, the first part devoted to thermochemistry of sulphinic acid derivatives is very short and contains a discussion on the estimation of thermochemical properties of sulphinyl derivatives. In the second part, the thermal reactions of sulphinic acid derivatives are presented.

## II. ESTIMATION OF THERMOCHEMICAL DATA BY GROUP ADDITIVITY

Group additivity is used to estimate thermochemical data of organic and inorganic compounds. This simple method originally developed by Benson and coworkers[9,10] assumes that thermochemical properties of molecules can be expressed as a sum of contributions of the individual groups that comprise the molecule. According to Benson, a group is defined as a polyvalent atom with ligancy $\geqslant 2$ in a molecule together with all its ligands. For example, the methyl group in dimethyl sulphoxide is a group where the carbon atom is connected to three hydrogens and the sulphinyl sulphur atom and is described as follows: C—$(H_3)$ (SO). The simple molecules such as $H_2O$, $CH_3Cl$ and $CH_4$ that contain only one such atom are irreducible entities and cannot be treated by group additivity.

Recently, the research team at the University of Sussex[6] modified the group additivity method and adapted it to computer systems. This new model allows one not only to calculate and store thermochemical data by computer, but also takes into account many steric and conjugative effects operating in a molecule. The model devised assumes that the standard enthalpy of formation of the ideal gaseous state is equal to the sum of contributions from substructural components within the molecule. The substructures are denoted as 'components' and their contributions to the standard enthalpy of formation as 'component enthalpies'. A component is defined as a group plus the groups to which it is formally bonded. The notation of a component consists of the code for the central group (denoted the 'principal group') followed in parentheses by the groups to which it is bonded (denoted as the 'attached groups'). The groups and their codes are given in Table 1. In Table 2 some examples of groups and components in a few molecular structures of sulphinic acid derivatives are presented.

According to the model under discussion, the standard enthalpy of formation, $\Delta H_f^0$, is given by the equation:

$$\Delta H_f^0 = \sum h\{i(j\cdots)\} \quad (1)$$

where $h\{i(j\cdots)\}$ is the enthalpy of a component, $i(j\cdots)$ in the structure; $i$ and $j$ are groups from Table 1 and dots $\cdots$ represent groups which may or may not be present depending on the valency of group $i$.

The enthalpy values of components containing the sulphinyl moiety are listed in Table 3.

However, the data so far available (see Table 3) do not allow one to calculate heats of formation of sulphinic acid derivatives owing to the lack of the basic data on enthalpy of components such as $h\{$SO(O2 1)$\}$, [SO—(O)$(CH_3)$]; $h\{$SO(= 2 2)$\}$, [SO—(O) $(CH_2)$]; $h\{$SO(1 N3)$\}$, [SO—$(CH_3)$ $(NR_2)$]; $h\{$SO(1 Cl)$\}$, [SO—(Cl)$CH_3$)]; $h\{$O2(SO 1)$\}$, [O—(SO)$(CH_3)$] and so on. In this situation it is desirable to measure experimentally heats of formation of sulphinate esters, amides or chlorides as representatives of these classes of compounds.

TABLE 1. Groups and group codes

| Group | Code | Group | Code |
|---|---|---|---|
| —$CH_3$ | 1 | —NH— | N2 |
| —$CH_2$— | 2 | —N— | N3 |
| —CH— | 3 | —NC | NC |
| —C— | 4 | —NO | NO |
| =$CH_2$ | 5 | —$NO_2$ | Nt |
| =CH— | 6 | —OH | O1 |
| =C | 7 | —O— | O2 |
| ≡C—H | 8 | —SH | S1 |
| ≡C— | 9 | —S— | S2 |
| =C= | C | —SO | SO |
| —CN | CN | —$SO_2$ | Sp |
| —CHO | K1 | —F | F |
| CO | K2 | —Cl | Cl |
| —$NH_2$ | N1 | —I | I |

TABLE 2. Examples of groups and components in some molecular structures of sulphinic acid derivatives

| | |
|---|---|
| Structure | $C_2H_5S(O)OC_2H_5$ |
| Group codes | 1—2—SO—O2—2—1 |
| Components | 1(SO)SO(O2 1)O2(SO 2)2(O2 1)1(2) |
| Structure | $(CH_3)_2CHS(O)N(C_2H_5)_2$ |
| Group codes | 1—3—SO—N3—2—1<br>   \|        \|<br>   1       2—1 |
| Components | 2 × 1(3) SO(3 N3) N3(SO 2 2) 2 × 2(N3 1) 2 × 1(2) |
| Structure | n-$C_3H_7S(O)Cl$ |
| Group codes | 1—2—2—SO—Cl |
| Components | 1(2)2(1 2)2(SO 2)SO(2 Cl)Cl(SO) |
| Structure | $(CH_3)_3CS(O)SC(CH_3)_3$ |
| Group codes | 1                1<br>\|                \|<br>1—4—SO—S2—4—1<br>\|                \|<br>1                1 |
| Components | 3 × 1(4) 4(SO 1 1 1)SO(4 S2)S2(SO 4)4(S2 1 1 1)3 × 1(4) |

TABLE 3. Component enthalpy values (group values) for $\Delta H_f^0$, of sulphinyl derivatives

| | | Enthalpy of component $\Delta H_f^0$ (kJ mol$^{-1}$) | |
|---|---|---|---|
| Component[6] | Group[5] | Reference 5 | Reference 6 |
| 1(SO) | C—(SO) $(H_3)$ | | −41.9(±0.2) |
| 2(SO 1) | C—(SO) $(H_2)$ $(CH_3)$ | | −24.1(±0.2) |
| 2(SO 2) | C—(SO) $(H_2)$ $(CH_2)$ | | −28.0(±0.2) |
| 3(SO 1 1) | C—(SO) $(C)_2$(H) | [−21.3][a] | |
| 4(SO 1 1 1) | C—(SO) $(C)_3$ | −9.25 | |
| 2(SO 6) | C—(Cd) (SO) $(H_2)$ | −27.56 | |
| | $C_B$—(SO) | 15.48 | |
| SO(1 1) | SO—$(CH_3)_2$ | | −67.4(±0.5) |
| SO(2 1) | SO—$(CH_3)$ $(CH_2)$ | | −70.6(±0.5) |
| SO(2 2) | SO—$(CH_2)$ $(CH_2)$ | | −73.8(±0.5) |
| | SO—(C) $(C_B)$ | [−72.0] | |
| | SO$(C_B)_2$ | −66.95 | |
| SO(O2 O2) | SO—$(O)_2$ | [−213.0] | |
| O1(SO) | O—(SO) (H) | −158.6 | |
| O2(SO 1) | O—(C) (SO) | −92.6 | |

[a][ ] contain estimated values taken from Reference 5.

## III. ESTIMATION OF THERMOCHEMICAL DATA FROM BOND DISSOCIATION ENERGY

Bond dissociation energy (or bond strength) also belongs to the thermochemical properties of organic compounds. Since the simple radicals undergo recombination practically without activation energy, the bond dissociation energy of the molecule A − B is equal to the activation energy and may be determined from kinetic data[3].

On the other hand, the bond dissociation energy (or bond strength) of A − B is usually defined as the enthalpy change of the reaction shown below and is expressed by the following equation:

$$A - B \longrightarrow A^{\cdot} + B^{\cdot} \tag{2}$$

$$D_{(A+B)} = \Delta H^0_{(2)} = \Delta H_f^0(A) + \Delta H_f^0(B) - \Delta H_f^0(A - B) \tag{3}$$

This equation allows one to calculate the bond dissociation energy if the heats of formation of both radicals and the compound AB are known[3]. Alternatively, one can calculate the heat of formation, $\Delta H_f^0(A - B)$, if one knows the value $D_{(A-B)}$. The latter approach for determining the heat of formation of organic compounds gives good results in many cases[5].

The present authors applied this method to predict the heat of formation of some simple thiosulphinic acid esters using the known $\Delta H_f^0$ values for disulphides[6], atomic oxygen[8] and the calculated bond dissociation energy for PhS(O)SPh taken as the standard[2]. We made the reasonable assumption that alkyl or aryl substituents do not affect the bond dissociation energy of sulphinyl compounds[5] (Table 4).

## IV. THERMOLYSIS OF SULPHINIC ACID DERIVATIVES

Thermal rearrangements and reactions of sulphinic acid derivatives are well known and have found many interesting synthetic applications. However, it should be noted that

TABLE 4. Calculated heats of formation of thiosulphinates

| Compound | $\Delta H^0$( RSSR) (kJ mol$^{-1}$)[a] | $\Delta H^{0\ f}$(RS(O)SR) (kJ mol$^{-1}$) | Uncertainty[b] (kJ mol$^{-1}$) |
|---|---|---|---|
| MeSS(O)Me | −24.2(1.0) | −125.5 | (±9.4) |
| EtSS(O)Et | −74.2(1.0) | −176.0 | (±9.5) |
| n-PrSS(O)Pr-n | −117.3(1.1) | −218.6 | (±9.5) |
| n-BuSS(O)Bu-n | −158.4(2.6) | −259.7 | (±11.0) |
| *i*-BuSS(O)Bu-*i* | −170.9(2.2) | −272.2 | (±10.6) |
| *t*-BuSS(O)Bu-*t* | −202.0(2.3) | −303.3 | (±10.9) |
| PhSS(O)Ph | 243.5(4.1) | 142.2[c] | (±8.4) |

[a] $\Delta H_f^0$(RSSR) from Reference 6.
[b] Estimated as sum of the uncertainties of the components.
[c] Value from Reference 2.

major attention was paid to the thermal rearrangements of sulphinates to sulphones. This topic has been discussed exhaustively by Braverman[11] and also by Drabowicz, Kiełbasiński and Mikołajczyk[12] in a chapter in this volume, in which emphasis was devoted to synthetic applications of the sulphinate-to-sulphone rearrangement. Therefore, in this part of our review, thermal reactions of sulphinates, especially acyclic ones, will be discussed only in a cursory manner to avoid repetition. On the other hand, more detailed descriptions of the thermal reactions of sulphinamides, thiosulphinates and sulphinyl sulphones will be given.

### A. Sulphinates

The rearrangement of sulphinic acid esters to sulphones (equation 4), first observed by Hinsberg[13] in 1917, represents one of the most widely studied reactions in sulphur chemistry[11,12]. This reaction occurs with a broad variety of sulphinates (aliphatic, aromatic) at temperatures which are strongly dependent on the sulphinate structure and on the solvent used. Allylic sulphinates undergo thermal rearrangement to allylic sulphones (equation 5), while propargylic sulphinates rearrange to allenic sulphones (equation 6).

The extensive mechanistic studies of these reactions revealed that simple sulphinates are isomerized to sulphones in general by the ion pair mechanism, while allylic and allenic sulphones are formed from their sulphinate precursors mainly via a concerted intramolecular mechanism.

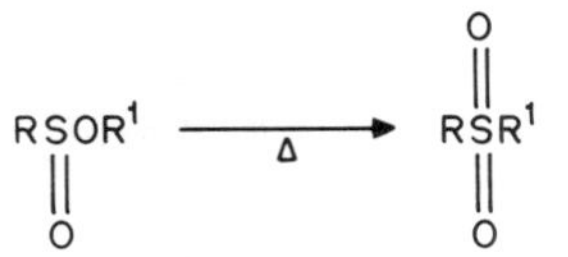

(4)

(5)

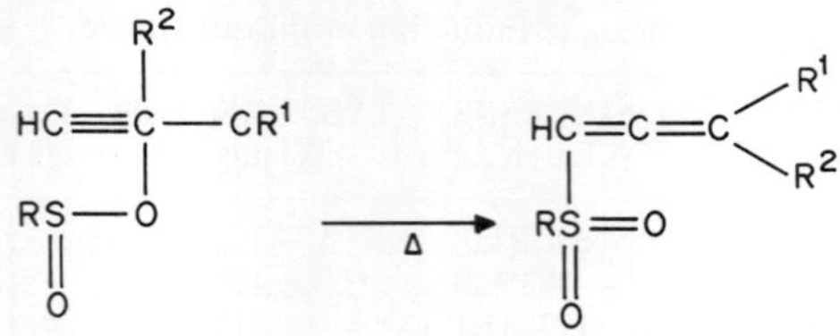


(6)

Sometimes the sulphone formation does not occur or is accompanied by other products. For example, *O*-(4-methoxy-2, 6-diphenyl)phenyl 2-phenylethanesulphinate (**1**) gives, on heating in 1,2-dichlorobenzene at 150 °C, a mixture of *S*-2-phenylethyl 2-phenylethane-thiosulphonate (**2**), *O*-(4-methoxy-2,6-diphenyl)phenyl 2-phenylethanesulphonate (**3**) and 4-methoxy-2, 6-diphenylphenol (**4**)[14,15]; see equation 7. The outcome of such a reaction is

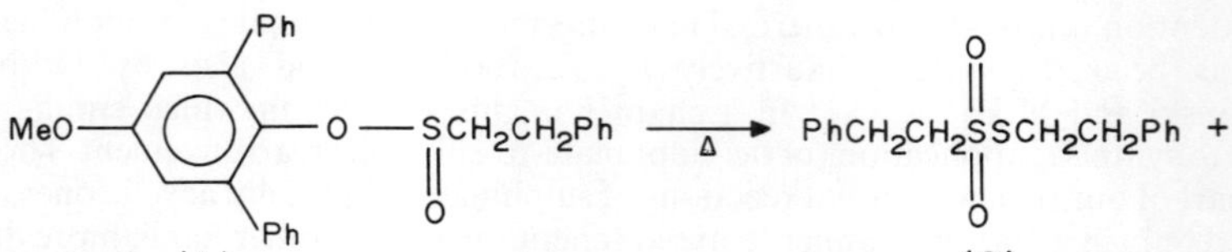


(**1**) (**2**)

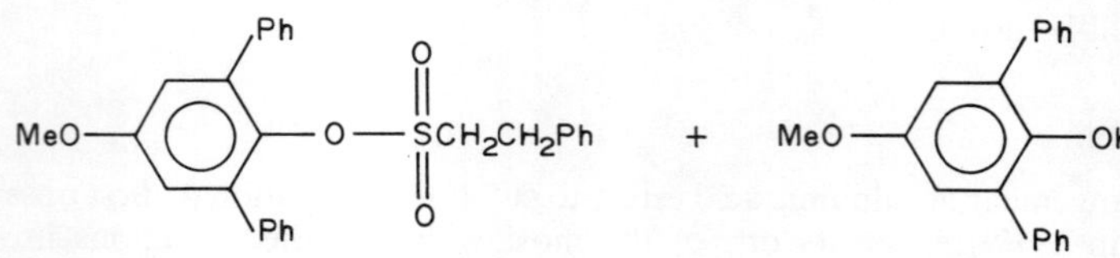


(7)

(**3**) (**4**)

best explained by assuming homolytic dissociation of the sulphur–oxygen bond in **1** leading to the phenoxy radical **5** and sulphinyl radical **6**. Dimerization of the latter results in the formation of vic-disulphoxide **7** and/or *O*-sulphenyl sulphinate **8**, which rearrange to thiosulphonate **2**. Sulphonate **3** arises from the interaction of thiosulphonate **2** and the phenoxy radical **5**.

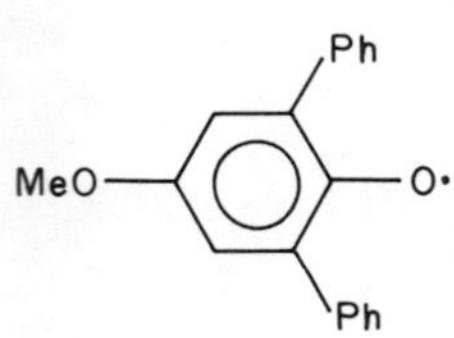


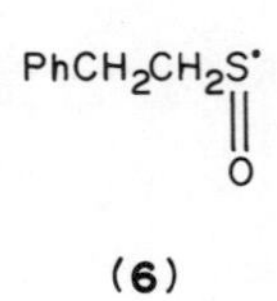


(**5**) (**6**)

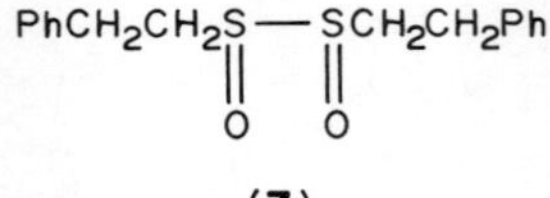


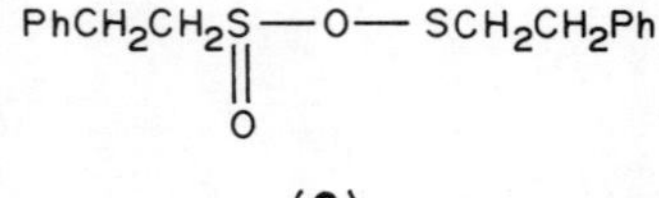


(**7**) (**8**)

Cyclic sulphinic acid esters (sultines), although less available than acyclic analogues, have also become the subject of extensive studies. Durst and coworkers investigated the photochemical[16] and thermal[17–19] reactions of monofunctionalized $\gamma$-sultines **9**. It was observed that photolytic reaction occurs only with sultines bearing a $\gamma$-phenyl substituent, affording phenylcyclopropanes **10** and sulphur dioxide as major products. Thermolysis of **9** gives rise to alkenes **11** and **12** in addition to cyclopropanes **10** (equation 8). The latter process was assumed by Durst's group to proceed via a diradical intermediate formed by consecutive cleavage of the C—O and C—S bonds. It is interesting to note that the photochemical and thermolytic breakdown of $\gamma$-lactones, which are carbon counterparts of $\gamma$-sultines, affords also **10, 11** and **12** as the reaction products[20].

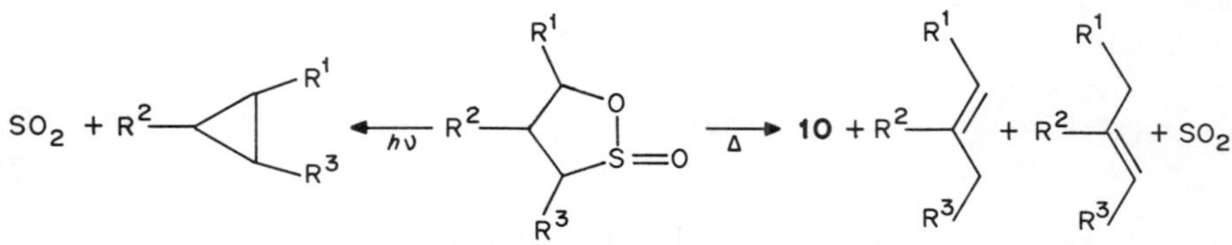

(**10**) (**9**) (**11**) (**12**)

(**a**) $R^1 = Ph$, $R^2 = R^3 = H$
(**b**) $R^1 = Ph$, $R^2 = R^3 = Me$
(**c**) $R^2 = Me$, $R^1 = R^3 = H$ (8)
(**d**) $R^2 = Ph$, $R^1 = R^3 = H$
(**e**) $R^1 = Me$, $R^2 = R^3 = H$

Benzofused sultines **13b** when heated at 750 °C gave fluorene **14** as the only product, via a diradical intermediate**15**[21] (equation 9). However, thermolysis of **13a** at 700 °C gave only the starting sultine, while at higher temperatures a multitude of unidentified products was observed.

A Dutch group[22] investigated the flash vacuum thermolyses (FVT) of the diastereoisomeric N-protected $\beta$-amino $\gamma$-sultines **16a** and **16b**. The reaction (equation 10) carried out at 700 °C gave a mixture of the N-allyl amide **17**, isomeric enamides **18a** and **18b** and amide **19**. In this case no cyclopropane derivatives have been found to be formed.

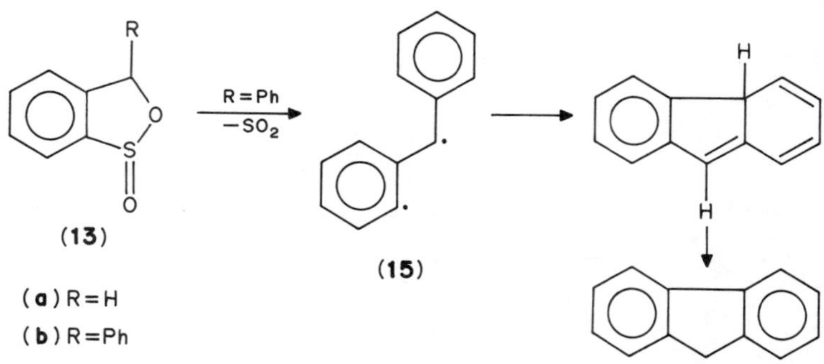

(**13**) (**15**) (**14**) (9)

(**a**) R=H
(**b**) R=Ph

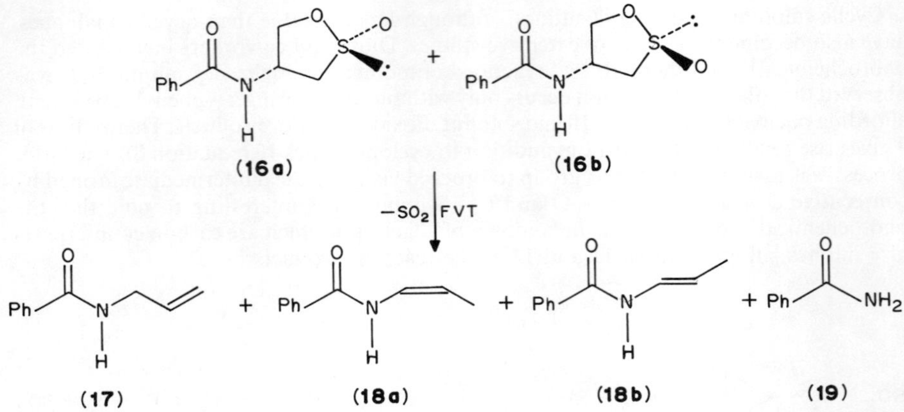


(10)

The experiments using the deuterium-labelled sultine **16a–d$_2$** (see Scheme 1) led the authors to postulate heterolytic fission of the C—S (path b) and C—O (path c) bonds in the substrate facilitated by participation of the neighbouring amide nitrogen and amide oxygen, respectively. The formation of **17** via an intermediate biradical **20** has been ruled out.

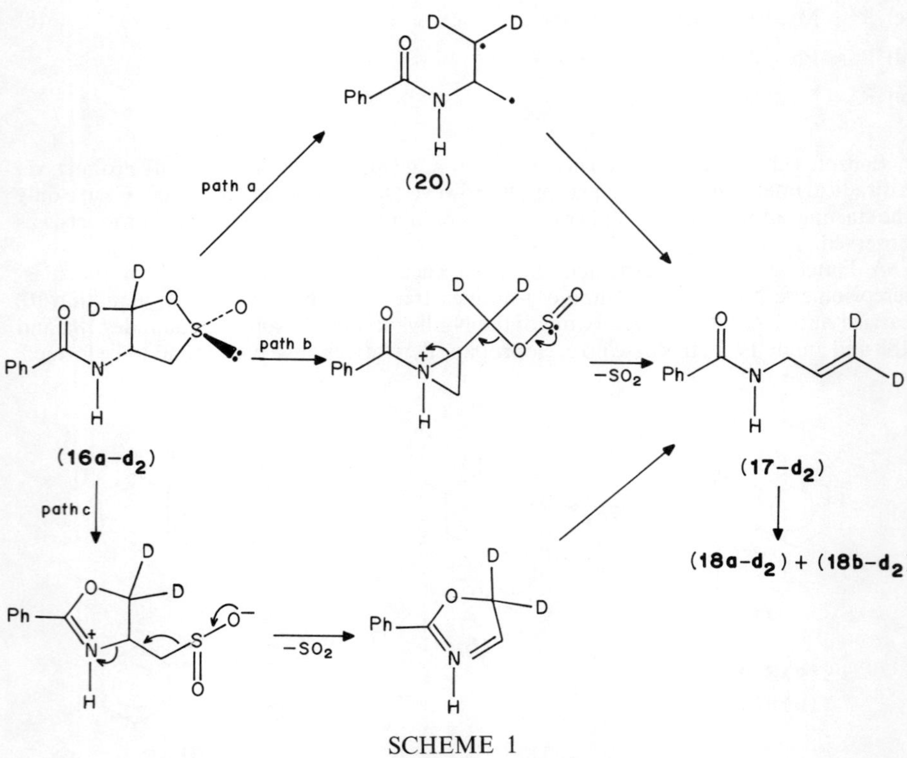


SCHEME 1

In contrast to γ-sultines, δ-sultines lose sulphur dioxide under much milder conditions. For example, sultine **21** when refluxed in benzene undergoes concerted loss of sulphur dioxide affording diene **22** as a main product[23] (equation 11).

Thermal rearrangement of sultines **23** to sulphones **24** has been shown by Durst[23] to proceed via a retro-Diels–Alder reaction. Thus, when the parent ester **23a** was heated in refluxing benzene, a clean isomerization to 1,3-dihydrobenzo[*c*]thiopene-2, 2-dioxide **24a** was observed. This reaction represents a cycloreversion of **23a** to the *o*-quinodimethane **25a** and $SO_2$ followed by a typical $SO_2$ + 1, 3-diene cycloaddition (see Scheme 2).

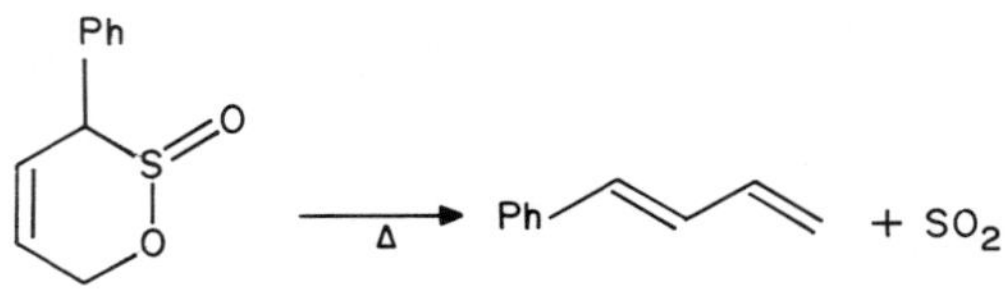


**(21)** **(22)** (11)

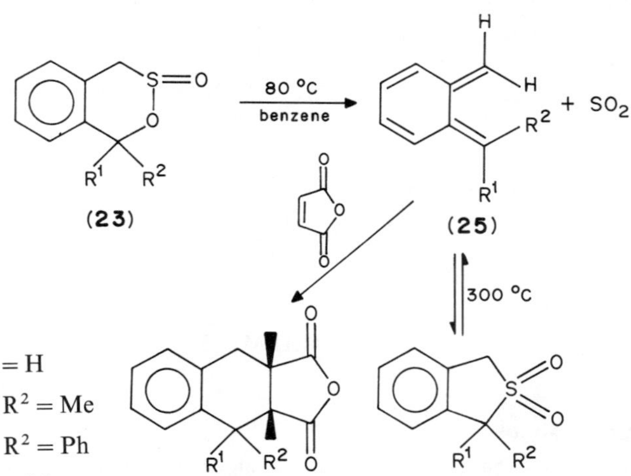


(**a**) $R^1 = R^2 = H$

(**b**) $R^1 = H, R^2 = Me$

(**c**) $R^1 = H, R^2 = Ph$

(**d**) $R^1 = R^2 = Me$

SCHEME 2

The transiently formed *o*-quinodimethanes **25** may be trapped by a very reactive dienophile, such as maleic anhydride, to give tetrahydronaphthalene derivatives **26**. The results discussed above illustrate the ease with which sulphur dioxide is lost from **23** in the retro-Diels–Alder reaction compared to the chelotropic extrusion of $SO_2$ from the isomeric sulphone **24**. The latter process requires heating in refluxing diethyl phthalate at *ca* 300 °C.

In contrast to γ- and δ-sultines, β-sultines generated according to the method of Durst[19] eliminate sulphur dioxide very readily, in the majority of cases within a few minutes at room temperature (equation 12).

This fact was utilized by Nokami and coworkers[24] who developed a highly efficient synthesis of olefins exemplified by the synthesis of **27**; see Scheme 3.

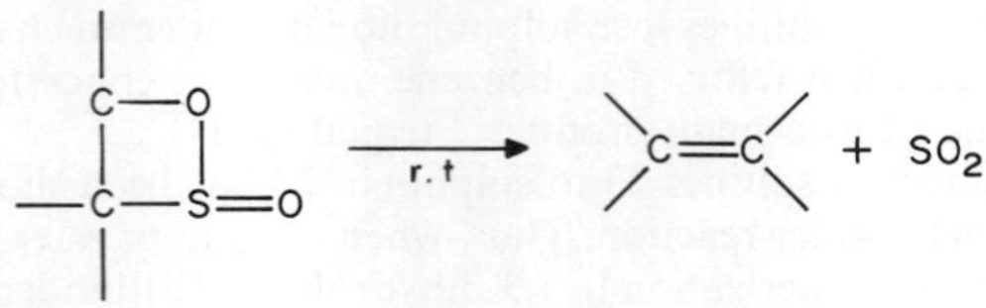


(12)

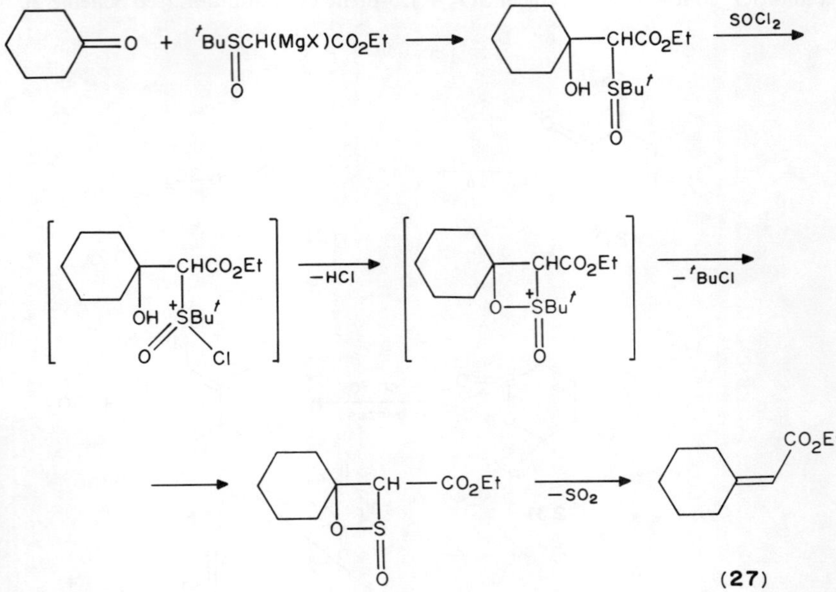


**(27)**

SCHEME 3

## B. Sulphinyl Oximes

Sulphinyl oximes **28** are a very unstable class of sulphinic acid derivatives that are formed by condensation of sulphinyl chlorides with oximes in the presence of tertiary amines (equation 13).

$$\text{>C=NOH} + \text{ClS(=O)R} \xrightarrow{R_3N} \text{>C=NOS(=O)R} \qquad (13)$$

**(28)**

Hessing and coworkers[25] have reported that *O*-alkylsulphinyl-*N*-benzoyl-*N*-phenylhydroxylamine **29** rearranges during its preparation at −70 °C to the corresponding sulphonamide **30** together with *o*- and *p*-alkylsulphonyl derivatives **31** and **32** (equation 14).

Very recently, Hudson and his coworkers[26] have shown that methylsulphinyl oximes **33** give, on warming from 0 °C to room temperature, the corresponding sulphonylimines **34** contaminated with nitriles **35** (10–20%) and products derived from the decomposition of

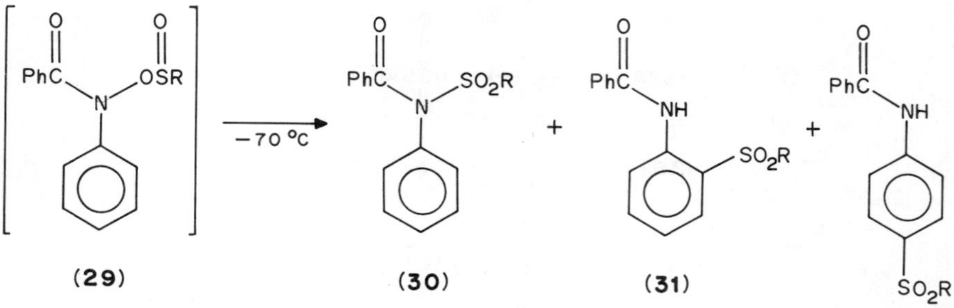


(29) (30) (31) (32)

(14)

methanesulphinic acid (equation 15). Kinetic measurements revealed that the enthalpy of activation for **33c** is 21.6 kcal mol$^{-1}$ and for **33d**, 21.3 kcal mol$^{-1}$. These values are close to that (22.4 kcal mol$^{-1}$) found for the sulphinyl oxime derived from benzophenone[27]. Positive entropies of activation (5.7 cal mol$^{-1}$ K$^{-1}$ for **33c** and 4.0 cal mol$^{-1}$ K$^{-1}$ for **33d**) strongly support the conclusion that homolytic dissociation of the N—O bond is the major pathway in this rearrangement.

$$R^1R^2C{=}NOS({=}O)Me \xrightarrow{0\rightarrow 25\,^{\circ}C} R^1R^2C{=}NS(O)_2Me + R^2C{\equiv}N + \left[MeS(O)OR^1\right] \qquad (15)$$

(33) (34) (35)

(**a**) R$^1$ = H, R$^2$ = Ph

(**b**) R$^1$ = D, R$^2$ = Ph

(**c**) R$^1$ = H, R$^2$ = *p*-Tol

(**d**) R$^1$ = D, R$^2$ = *p*-Tol

(**e**) R$^1$ = H, R$^2$ = *p*-$NO_2C_6H_4$

### C. Thiosulphinates

Aryl arenethiosulphinates **36** undergo an easy disproportionation on gentle heating to give the corresponding thiosulphonates **37** and disulphides **38**. It is now generally accepted[28–30] that the primary stage of this reaction involves the homolytic cleavage of the sulphur–sulphur bond in **36** leading to the formation of the sulphinyl and sulphenyl radicals as shown in equation 16.

On the basis of kinetic data Fava and coworkers[31] determined the energy of the S—S bond in aryl arenethiosulphinates **36** as 34.5 kcal mol$^{-1}$. In methyl methanethiosulphinate **39** the S—S bond energy was found[32] to be 46 kcal mol$^{-1}$ and is about 29 kcal mol$^{-1}$ smaller than that of the disulphide S—S linkage. A possible explanation of the weakness of the thiosulphinate S—S bond as compared, for example, with the thiosulphonate S—S bond may lie in the notable stability of sulphinyl radicals[33].

Thermal decomposition of alkyl alkanethiosulphinates is, in general, more complex.

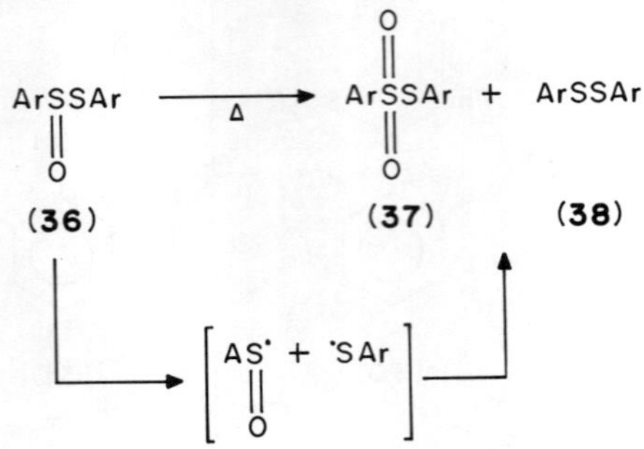

(16)

Whereas the thiosulphinate **39** affords on heating without solvents the expected disproportionation products, i.e. methyl methanethiosulphonate **40** and disulphide **41**, its thermolysis in benzene solution results in the formation of a number of additional products shown in equation 17[34].

A careful mechanistic study of Block and his group[32,34] on the thermal behaviour of alkyl thiosulphinates revealed two possible pathways for cycloelimination as a primary process (equation 18). The first route (a) results in the formation of an alkanesulphenic acid

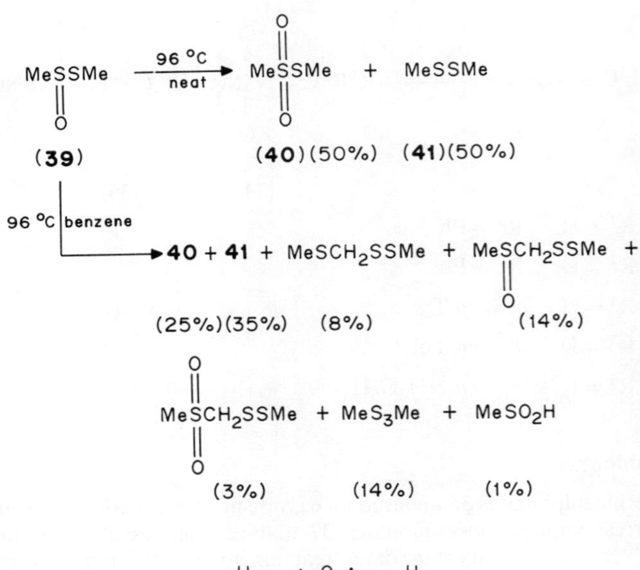

(17)

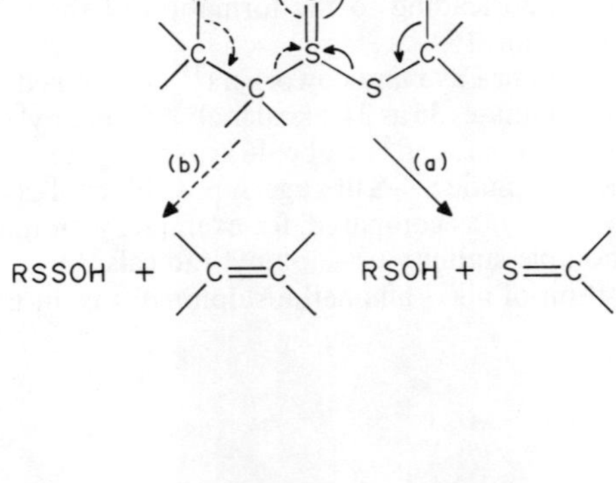

(18)

TABLE 5. Thermal stability of alkyl alkanethiosulphinates

| Structure | $t_{1/2}$ at 96 °C (min) |
|---|---|
| MeS(O)SMe | 7 |
| MeS(O)SEt | 11 |
| MeS(O)SPr$^i$ | 32 |
| EtS(O)SMe | 40 |
| n-$C_{12}H_{25}$S(O)S$C_{12}H_{25}$ | 52 |
| $^i$PrS(O)SPr$^i$ | 66 |
| $^t$BuS(O)SBu$^t$ | 148 |
| MeS(O)SBu$^t$ | ~$10^3$ |
| AdS(O)SAd$^a$ | $10^5$ |

$^a$Ad denotes adamantyl.

and a thione. The second one (b) gives an alkanethiosulphoxylic acid and an olefin. Subsequent reactions of both acids are responsible for a multitude of products formed.

Relative thermal stabilities of neat alkyl alkanethiosulphinates determined by Block and O'Connor[34] are collected in Table 5, which shows that the steric hindrance at the sulphinyl or sulphenyl sulphur retards decomposition.

Finally, it should be noted that the thermolysis of thiosulphinates was utilized in synthetic studies. For instance, Chou and coworkers[35] have succeeded in the preparation of a novel thioxo $\beta$-lactam **42** by pyrolysis of the thiosulphinate **43** (equation 19).

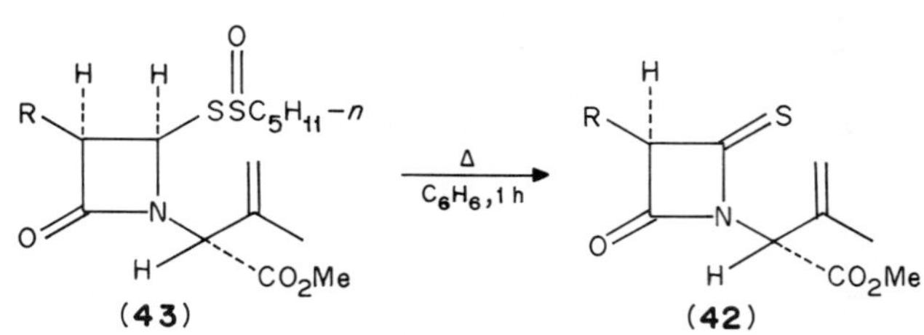


(19)

R = Me, p-$NO_2C_6H_4CH_2$

## D. Sulphinyl Sulphones

Sulphinyl sulphones **44**, which are structural isomers of sulphinic acid anhydrides **45**, decompose rapidly at 50 °C. Kice and Pawlowski[36,37] showed on the basis of kinetic data that the unimolecular decomposition of **44** involves a facile homolysis of the S—S bond to give the $ArSO_2^{\bullet}$ and $ArSO^{\bullet}$ radicals (equation 20). The enthalpy of activation of the radical scission was calculated to be 27.6 kcal mol$^{-1}$. The consecutive reactions of these radicals depend on the reaction conditions. However, sulphenyl sulphone **46** is postulated as being a reactive recombination product, especially in the absence of good radical traps. Some other processes that may occur are given in equations 21–23.

## E. Sulphinamides

Thermolysis of sulphinamides is interesting not only from the mechanistic but also from the synthetic point of view. Trost[38] has described a good method for the preparation of

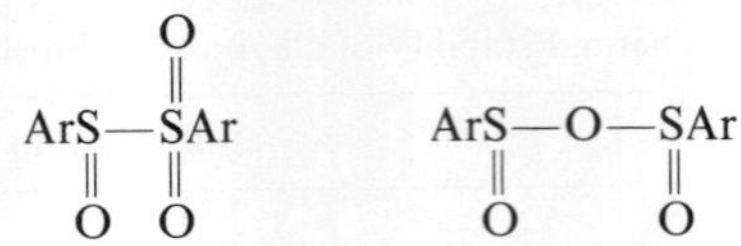

(44) (45)

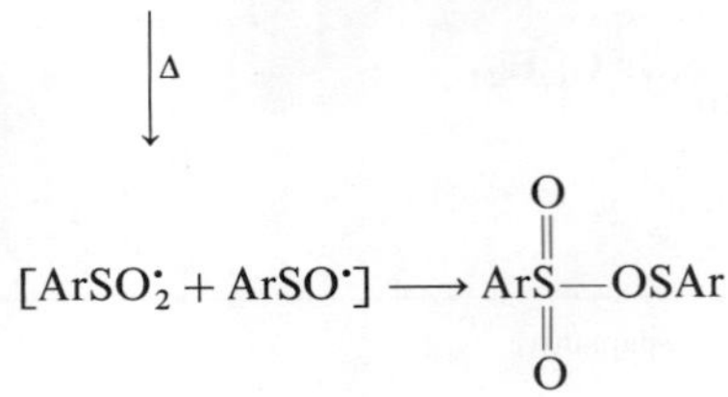

(46) (20)

$$46 + ArSO_2H \longrightarrow ArSO_3H + ArSO_2SAr \quad (21)$$

$$2ArSO^{\bullet} \longrightarrow [ArS(O)OSAr] \longrightarrow ArSO_2SAr \quad (22)$$

$$ArSO_2^{\bullet} + R{-}H \longrightarrow ArSO_2H \quad (23)$$

imines **48** via regioselective thermal dehydrosulphenylation of the easily available sulphinamides **47** (equation 24).

The reaction shown in equation 24 required the use of xylene as a solvent and proceeded efficiently at temperatures between 80 and 140 °C during 8 to 48 h. The yields were in the range from 66% (for **49d**) to 89% (for **49a**).

$\beta$-Hydroxy sulphinamides **50** undergo smooth thermolysis at 80 to 110 °C to form olefins **51** along with sulphur dioxide and the appropriate amine **52**[39] (equation 25).

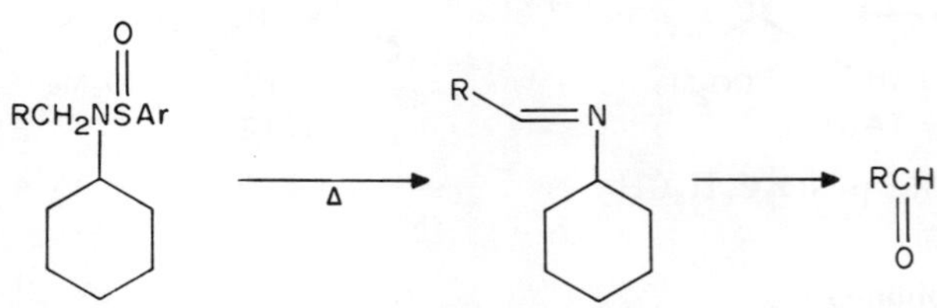


(47) (48) (49) (24)

Ar = Ph, $p$-$NO_2$–$C_6H_4$

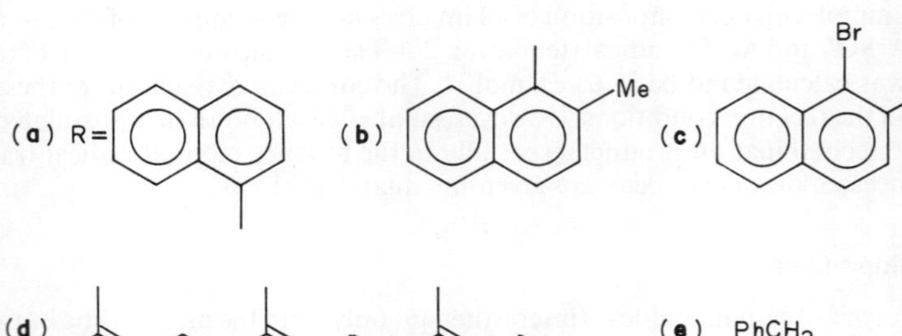

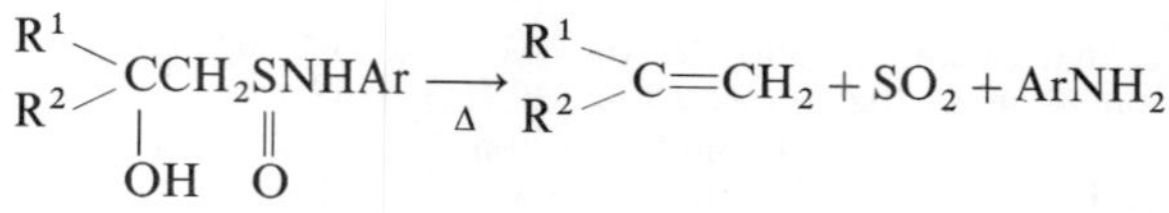

(50) (51) (52) (25)

$R^1$ and $R^2$ = H, Ph
Ar = Ph, *p*-Tol

Thermolysis of *N*-alkylidene sulphinamides **53** was examined by Davis and coworkers[40]. They found that heating **53** in benzene for 15 to 36 h affords disulphide **54**, thiosulphonate **55** and nitriles **56** as decomposition products (equation 26). The latter were isolated in 71–85% yield.

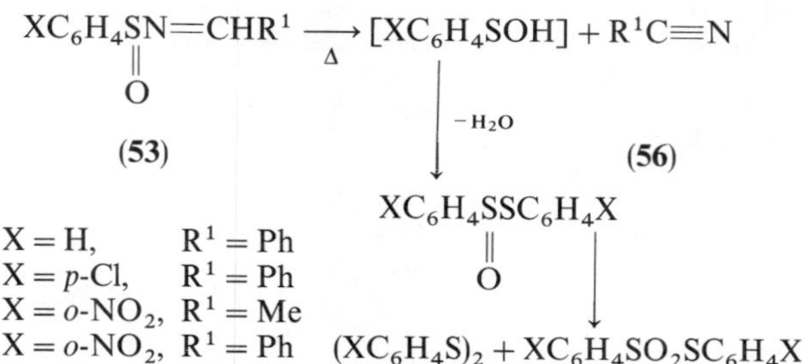

(53) (56)

(**a**) X = H, $R^1$ = Ph
(**b**) X = *p*-Cl, $R^1$ = Ph
(**c**) X = *o*-$NO_2$, $R^1$ = Me
(**d**) X = *o*-$NO_2$, $R^1$ = Ph
(**e**) X = H, $R^1$ = Me

(54) (55) (26)

## V. REFERENCES


1. S. W. Benson, in *Thermochemical Kinetic Methods for the Estimation of Thermochemical Data and Rate Parameters*, 2nd edn., Wiley, New York, 1976.
2. S. W. Benson, *Chem. Rev.*, **78**, 23 (1978).
3. J. H. Kerr, *Chem. Rev.*, **66**, 465 (1966).
4. R. Shaw, in *The Chemistry of the Thiol Group* (Ed. S. Patai), Wiley, London, 1974, p. 154.
5. J. T. Herron, in *The Chemistry of Sulphones and Sulphoxides* (Eds. S. Patai, Z. Rappoport and C. J. M. Stirling), Wiley, Chichester, 1988, p. 95.
6. J. B. Pedley, R. D. Naylor and S. P. Kirby, *Thermochemical Data of Organic Compounds*, Chapman and Hall, New York, 1986.
7. J. D. Cox and P. Pilcher, *Thermochemistry of Organic and Organometallic Compounds*, Academic Press, New York, 1970.
8. G. Pilcher, in *Thermochemistry and Thermodynamics*, Vol. 10 (Ed. M. H. Skinner), Butterworth, 1972, p. 64.
9. S. W. Benson, F. R. Cruickshank, D. M. Golden, G. R. Haugen, H. E. O'Neal, A. S. Rodgers, R. Shaw, and R. Walsh, *Chem. Rev.*, **69**, 279 (1969).
10. S. W. Benson and J. H. Buss, *J. Chem. Phys.*, **29**, 546 (1958).
11. S. Braverman, in *The Chemistry of Sulphones and Sulphoxides* (Eds. S. Patai, Z. Rappoport and C. J. M. Stirling), Wiley, Chichester, 1988, p. 665.
12. J. Drabowicz, P. Kiełbasiński and M. Mikołajczyk, in this volume.
13. O. Hinsberg, *Chem. Ber.*, **50**, 468 (1917).
14. F. Freeman, *Chem. Rev.*, **84**, 117 (1984).
15. C. R. H. I. de Jonge, F. P. B. von der Meaden, M. E. F. Biemond, W. J. B. Huysmans and W. Mijs, *J. Polym. Sci., Polym. Symp.*, **57**, 197 (1976).
16. T. Durst, J. C. Huang, N. K. Sharma, and D. J. H. Smith, *Can. J. Chem.*, **56**, 512 (1978).
17. T. Durst, J. D. Finlay and D. J. H. Smith, *J. Chem. Soc., Perkin Trans. 1*, 950 (1979).
18. T. Durst and B. P. Gimbarzewsky, *J. Chem. Soc., Chem. Commun.*, 724 (1975).

19. F. Jung, N. K. Sharma, and T. Durst, *J. Am. Chem. Soc.*, **95**, 3420 (1973).
20. R. S. Givens and W. F. Oettle, *J. Org. Chem.*, **37**, 4325 (1972).
21. T. Durst, J. D. Finlay and D. J. H. Smith, *J. Chem. Soc., Perkin Trans. 1*, 950 (1979).
22. R. M. Liskamp, H. J. Blom, J. F. Niyard, and H. C. J. Ottenheijm, *J. Org. Chem.*, **48**, 2733 (1983).
23. F. Jung, M. Molin, R. V. D. Elzen, and T. Durst, *J. Am. Chem. Soc.*, **96**, 935 (1974).
24. J. Nokami, N. Kunieda, and M. Kinoshita, *Tetrahedron Lett.*, 2179 (1975).
25. A. Hessing, W. Kleine-Hofman, and W. Mullers, *Chem. Ber.*, **113**, 152 (1980).
26. M. R. Banks, C. Brown, R. F. Hudson, and K. A. F. Record, *J. Chem. Soc., Perkin Trans. 2*, 1501 (1986).
27. C. Brown, R. F. Hudson, and K. A. F. Record, *J. Chem. Soc., Perkin Trans. 2*, 822 (1978).
28. E. Block, *Reactions of Organosulfur Compounds, Academic Press*, New York, 1978, p. 176.
29. J. L. Kice, in *Free Radicals* (Ed. J. K. Kochi), Vol. II, Wiley, New York, 1973, p. 711.
30. J. L. Kice, *Adv. Phys. Org. Chem.*, **17**, 65 (1980).
31. P. Koch, E. Ciuffarin, and A. Fava, *J. Am. Chem. Soc.*, **92**, 5971 (1970).
32. E. Block and J. O'Connor, *J. Am. Chem. Soc.*, **96**, 3921 (1974).
33. C. Chatgilialoglu, in *The Chemistry of Sulphones and Sulphoxides* (Eds. S. Patai, Z. Rappoport and C. J. M. Stirling), Wiley, Chichester, 1988, p. 1081.
34. E. Block and J. O'Connor, *J. Am. Chem. Soc.*, **96**, 3929 (1974).
35. T. S. Chou, G. A. Koppel, D. E. Dorman and J. W. Paschal, *J. Am. Chem. Soc.*, **98**, 7864 (1976).
36. J. L. Kice and N. E. Pawlowski, *J. Org. Chem.*, **28**, 1162 (1963).
37. J. L. Kice and N. E. Pawlowski, *J. Am. Chem. Soc.*, **86**, 4898 (1964).
38. B. M. Trost and G. Liu, *J. Org. Chem.*, **46**, 4617 (1981).
39. E. J. Corey and T. Durst, *J. Am. Chem. Soc.*, **90**, 5548 (1968).
40. F. A. Davis, A. J. Friedman, and E. W. Kluger, *J. Am. Chem. Soc.*, **96**, 5000 (1974).

The Chemistry of Sulphinic Acids, Esters and their Derivatives
Edited by S. Patai


CHAPTER **17**

# Electronic effects of SOOH and related groups

JOHN SHORTER

*Department of Chemistry, The University, Hull, HU6 7RX, UK*



## I. INTRODUCTION

Previous articles by the present contributor in *The Chemistry of Functional Groups* series have dealt with the electronic effects of the sulphonio group[1] and of the sulphinyl and sulphonyl groups[2]. In those cases there was copious information in the literature on which to draw. The present case in quite different: only a very few papers provide information relevant to the topic. For SOOH itself there appears to be nothing; there is a small amount of information for the related groups SOF, SOCl, SOOMe, $SONMe_2$ and $SO_2^-$ as substituents on a benzene ring.

The reason for the paucity of information probably lies in the reactive nature of these moieties. Many of the usual methods for studying substituent effects, whether on chemical reactivity or on spectroscopic properties, either cannot be applied at all or would be liable to encounter experimental difficulties. For instance, studies of directive effects in electrophilic aromatic substitution are clearly excluded; the electrophilic reagent would oxidize the substituent. A ring-substituted derivative of phenylsulphinic acid, $XC_6H_4SOOH$, is always made from the already ring-substituted $XC_6H_5$, never from $C_6H_5SOOH$. The highly acidic nature of the SOOH group would pose its own problems, which is presumably why information is only available about the effect of $SO_2^-$ as a substituent. Thus SOOH and related groups have usually been unattractive to workers studying substituent effects, linear free-energy relationships, etc., except in connection with a systematic investigation of the behaviour of sulphur substituents in the various oxidation states of sulphur.

The previous articles have discussed in detail such topics as the nature of sulphur bonding (in particular the questionable role of d orbitals), the Hammett equation and its extensions, substituent effects in aromatic systems (sigma values from studies of chemical reactivity in the ionization of benzoic acids, phenols, etc., and in electrophilic and nucleophilic substitution; sigma values from spectroscopic studies, notably $^{19}F$ NMR and infrared), substituent effects in aliphatic systems, the stabilization of carbanionic centres and the *ortho* effect[1,2]. The approach used in the present chapter will be quite different. We shall draw salient information from the previous articles where relevant and refer the reader to those articles for greater detail. In particular, the previous discussions of the behaviour of methylsulphinyl, phenylsulphinyl and trifluoromethylsulphinyl groups (with some reference to the corresponding sulphonyl groups) will be taken as the basis for approaching the behaviour of the groups SOY, where Y = F, Cl, OMe, $NMe_2$ or $O^-$. The most important part of the information about the electronic effects of these substituents is derived from $^{19}F$ NMR measurements, rather than studies of chemical reactivity, so the logical and usual order of discussion will be inverted. We shall first present and discuss the inductive and resonance parameters for these substituents. Then we shall deal with the small amount of information available from other physical and chemical studies, and with the possibility of estimating ordinary Hammett-type $\sigma$ values by an appropriate summation of inductive and resonance components.*

## II. SULPHINYL AND SULPHONYL GROUPS

### A. Sulphur Bonding

The previous discussion of sulphur bonding in these groups approached the electronic structure of the methylsulphinyl group by considering the generation of dimethyl sulphoxide from dimethyl sulphide[3]. The formation of the two single bonds by sulphur in the latter was envisaged as involving the overlap of singly occupied $3sp^3$ hybridized orbitals on sulphur with singly occupied $2sp^3$ hybridized orbitals on carbon. Two doubly occupied, localized molecular orbitals of the $\sigma$-type are thereby formed. Two unshared pairs of electrons in the valence shell of sulphur are left in the remaining $3sp^3$ orbitals. In accord with this picture, the bond angle $\angle$CSC in dimethyl sulphide is interpreted as essentially tetrahedral (109°28′; see **1**), with the contraction to the observed value of 105° being explained by postulating that the repulsion between the unshared pairs of electrons is greater than between the shared pairs. Thus the angle between the former is opened out and the angle between the latter is contracted.

Viewed from the standpoint of molecular orbital theory, as it has developed during the last decade or so, such a simple picture of the sulphur bonding in dimethyl sulphide is somewhat naïve. (As Kutzelnigg has written[4], 'The chemical bond is a highly complex

*Throughout this chapter substituent constants for benzene derivatives as originally defined by Hammett are set as '$\sigma$', while substituent constants in general are represented by 'sigma'.

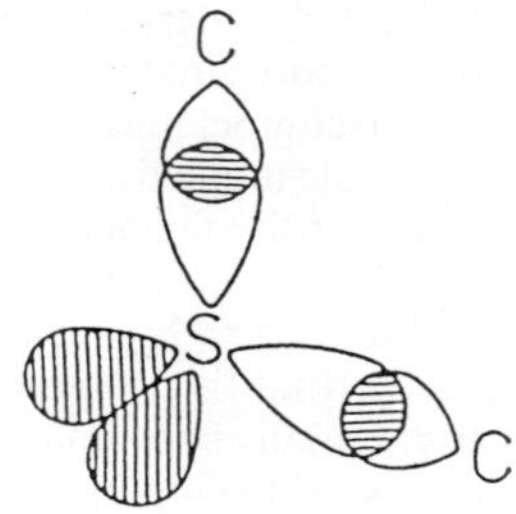


(1)

phenomenon which eludes all attempts at a simple description'.) However, this simple picture serves to introduce the subject and will act as a basis for discussing the bonding in dimethyl sulphoxide.

The formation of dimethyl sulphoxide can be pictured initially as involving a $3sp^3$ unshared pair orbital on sulphur and an empty $2sp^3$ orbital on oxygen. The bond between sulphur and oxygen is then a coordinate bond and the structure is appropriately written as **2** or **3**, with formal unit charges on sulphur and oxygen. At this point, however, the possible

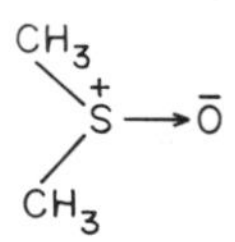

(2)

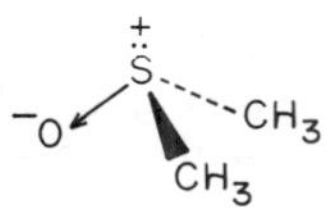

(3)

contribution of a 3d orbital on sulphur must be considered. One of the two electrons in an unshared pair $3sp^3$ orbital of dimethyl sulphide may be pictured as transferred to an appropriate 3d orbital e.g. $3d_{xy}$. The oxygen atom is considered to be in a $2sp^2$ hybridized state, with two unpaired electrons, one in one of the $2sp^2$ orbitals and the other in the unhybridized $2p_y$ orbital. A $\sigma$ bond is now formed by the end-on overlap of the singly occupied $3sp^3$ orbital of sulphur with the singly occupied $2sp^2$ orbital of oxygen, while a $\pi$(pd) bond is formed by the sideways overlap of the $3d_{xy}$ orbital of sulphur with the unhybridized $2p_y$ orbital of oxygen (see **4**). Considered in this way the structure of dimethyl sulphoxide involves a double bond and a valence shell of ten electrons for sulphur (see **5** and **6**).

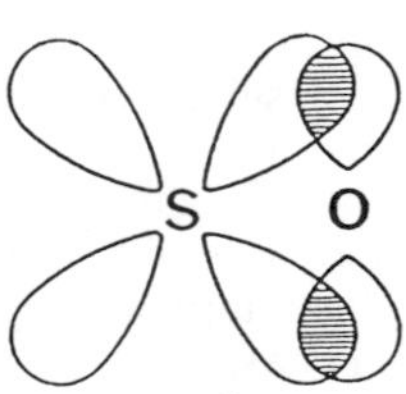


(4)

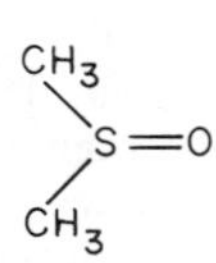

(5)

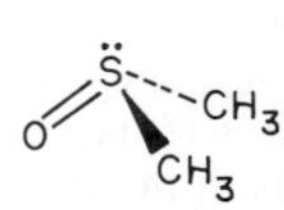

(6)

In the previous article[3] the formation of a further bond to oxygen, as in dimethyl sulphone, was then pictured in an analogous way to give tetrahedral structures involving various combinations of coordinate and double sulphur–oxygen bonds, with one structure involving a valence shell of eight electrons, two of ten electrons, and one of twelve electrons for sulphur. The nature of the sulphur–oxygen bond in sulphoxides and sulphones was then discussed in detail. It was concluded that all the evidence from bond lengths, dipole moments, bond energies, infrared spectra, molecular refraction, parachor and ultraviolet spectra may be satisfactorily interpreted in terms of a sulphur–oxygen bond which is largely, if not entirely, a coordinate bond. Some of the evidence for a coordinate bond is compelling; there is no compelling evidence for a sulphur–oxygen bond that is essentially a double bond. The dipole moment evidence, however, requires that the formal unit charges associated with the coordinate bond are partially neutralized by a shift of the bonding electrons away from oxygen towards the sulphur. The recent highly sophisticated discussion of chemical bonding in higher main group elements (to which reference has already been made[4]) agrees that the sulphur–oxygen bonds in sulphoxides and sulphones should be regarded as essentially coordinate rather than double bonds. The necessity of supposing that the valence shell of sulphur can be expanded to ten or twelve electrons by the participation of 3d orbitals in the bonding may thus be avoided for dialkyl sulphoxides and sulphones. However, this is only a temporary respite in relation both to the behaviour of a wide range of sulphoxides and sulphones and to a broader consideration of the chemistry of sulphur.

Many so-called hypervalent molecules formed by sulphur and its neighbours in the second row of the Periodic Table have traditionally been supposed to 'require' a bonding role of d orbitals, with 'octet expansion'. The best known example is, of course, sulphur hexafluoride $SF_6$. More immediately relevant to a consideration of electronic effects of substituents is the possible bonding role of 3d orbitals and octet expansion for molecules in which sulphinyl or sulphonyl groups are attached to unsaturated systems or carbanionic centres. For example, in the case of PhSOMe or $PhSO_2Me$, and more particularly for some of their ring-substituted derivatives, there is much experimental evidence of various kinds that the benzene ring is conjugated with the bond linking it to the sulphur atom. Thus canonical structures of the types shown in **7a** and **b** are usually regarded as

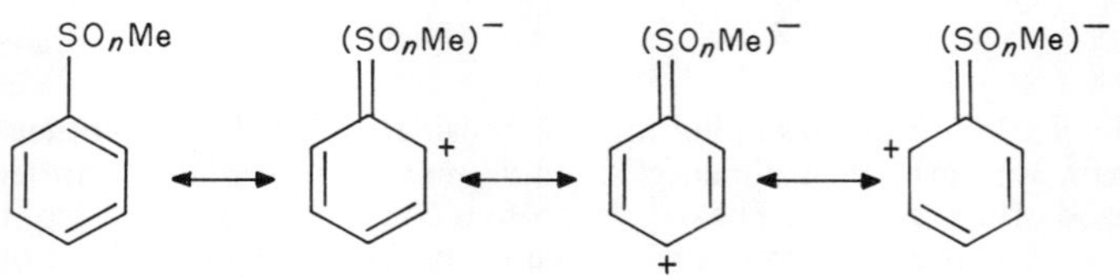


(**7**)(**a**) $n = 1$
(**b**) $n = 2$

contributing to the resonance hybrids and the groups SOMe and $SO_2Me$ may be classified as $+R$ substituents in their electronic effects on the benzene ring[5]. (For the sulphoxide there is a further complication connected with the lone pair of electrons on the sulphur, which will be dealt with later, i.e. potential behaviour of sulphinyl groups as $-R$ substituents, but we will not consider that at this stage.)

However, for many years there has been a school of thought among quantum chemists which maintains that it is not necessary to invoke a bonding role for the 3d orbitals of sulphur in some situations where this has been the traditional approach or, to put this more strongly, that it is not at all correct to do so[6,7]. (The same comment applies to the supposed bonding role of d orbitals in analogous situations for various other elements[8].) Such views did not seem to make much impact on organosulphur chemistry in general for a long time, but within the last dozen years or so there has been a move towards taking them more seriously[6–8]. This may be due partly to the accumulation of experimental evidence that d-orbital bonding does not occur in certain examples for which it would be conventionally invoked and partly to considerations of molecular orbital theory, in particular to the results of *ab initio* calculations. The previous article on the electronic effects of the sulphinyl and sulphonyl groups discussed these matters in considerable detail[9], because it was necessary to have a policy for dealing, in the rest of the chapter, with the effects of the groups in question which were traditionally ascribed to $\pi$(pd) bonding between carbon and sulphur. It does not seem possible to decide between the various alternative explanations that have been offered. The only sensible policy appears to be to use some fairly neutral descriptive term to cover the phenomena for which $\pi$(pd) bonding for long provided the conventional explanation. All parties seem to agree that these phenomena are connected with a special build-up of electron density in the vicinity of sulphur, whatever may be the precise mechanism by which this occurs. (Polarization and polarizability effects, often linked in some way with the d orbitals, are frequently invoked by those who have been led to reject a *bonding* role of sulphur d orbitals.) To that extent it seems not unsuitable to refer to these phenomena as involving 'octet expansion' of sulphur, under the *caveat* that this is done as a convenient shorthand, without prejudice to the question of the precise mechanism whereby the 'octet expansion' occurs. In appropriate historical context 'd-orbital conjugation' and '$\pi$(pd) bonding' remain suitable terms to use. Also SOMe, $SO_2Me$, etc. are still conveniently referred to as $+R$ substituents in their electronic effects.

## B. Electronic Effects of Sulphinyl and Sulphonyl Groups

### 1. Reactivity studies

The considerable dipole moment of the sulphur–oxygen bond in the sense $S^{\delta+}–O^{\delta-}$ (about 3.0*D*) would be expected to result in sulphinyl and sulphonyl groups acting as strongly electron-attracting substituents, with a similarity to $CH_3CO$, CN, $NO_2$, etc. Chemical evidence for such behaviour has been known for many years[10]. The evidence concerns in particular the acidifying influence of such groups, i.e. the incipient formation of a hydrogen ion under the electron-attracting influence of the substituent. This effect is particularly marked for sulphonyl groups[10]. The acidifying influence of sulphinyl groups, although weaker than that of sulphonyl groups, has also long been recognized. Thus dibutyl sulphoxide in alkaline $D_2O$ slowly exchanges hydrogen for deuterium[11]. The promotion of nucleophilic aromatic substitution by sulphinyl groups, analogous to the well-known effect exercised by nitro groups, has been known for half a century: Hammick and Williams showed in 1938 that *p*-iodophenyl phenyl sulphoxide was hydrolyzed by alkali under conditions in which the *meta* isomer was not affected[12].

The quantitative study of the electronic effects of sulphinyl and sulphonyl groups, as for all substituents, is much concerned with the Hammett equation and its extensions. The previous article contained a summary of the salient features of the Hammett equation and cognate linear free-energy relationships, as well as an extensive bibliography as a guide to further reading[13]. For the present chapter it will be assumed that the reader has some acquaintance with these matters, although from time to time some background material will be introduced and a brief general bibliography is provided[14–18].

Most of the information relating to sulphinyl groups is, in fact, for the methylsulphinyl group. Studies of the behaviour of SOMe with respect to the Hammett equation began in the nineteen-fifties, with the work of Price and Hydock (1952)[19] and of Bordwell and Boutan (1957)[20]. This work is discussed in detail in the former article[21]. At an early stage various problems in the assessment of the electronic effects of the *meta*- and *para*-methylsulphinyl group arose and certain aspects of these have not been altogether resolved to this day. These may be due partly to the difficulty of preparing sulphoxides completely free from traces of sulphones[20,21]. The most reliable early work seems to be that of Bordwell and Boutan[20], who measured the $pK_a$ values of substituted benzoic acids in 50% aqueous ethanol and thereby determined $\sigma_m$ and $\sigma_p$ of SOMe to be 0.51 and 0.48, respectively. (The sigma values discussed in the present section are summarized in

TABLE 1. Sigma values of sulphinyl and sulphonyl groups based on reactivity studies

| Substituent | Authors | Year | Ref. | Method | $\sigma_m$ | $\sigma_m^-$ | $\sigma_p$ | $\sigma_p^-$ |
|---|---|---|---|---|---|---|---|---|
| SOMe | Bordwell and Boutan | 1957 | 20 | p$K_a$, benzoic acids, 50% v/v EtOH–$H_2O$, 25 °C | 0.51 | — | 0.48 | — |
| | Price and Hydock | 1952 | 19 | $k$, ethyl benzoates + $OH^-$, 56% v/v $Me_2CO$–$H_2O$, 25 °C | 0.52 | — | 0.54 | — |
| | Bordwell and Boutan | 1957 | 20 | p$K_a$, phenols, $H_2O$, 25 °C | — | 0.53 | — | 0.73 |
| | Yukawa and coworkers | 1972 | 22 | $k$, ethyl benzoates + $OH^-$, 85% v/v EtOH–$H_2O$, ? °C | — | — | 0.564 | — |
| | | | | $k$, substituted-benzyl benzoates + $OH^-$, 70% v/v $Me_2CO$–$H_2O$, 25 °C | — | — | 0.573[a] | — |
| $SO_2Me$ | Bordwell and Cooper | 1952 | 23 | p$K_a$, benzoic acids, 50% v/v EtOH–$H_2O$, 25 °C | 0.65 | — | 0.72 | — |
| | Price and Hydock | 1952 | 19 | $k$, ethyl benzoates + $OH^-$, 56% v/v $Me_2CO$–$H_2O$, 25 °C | 0.65 | — | 0.76 | — |
| | Bordwell and Cooper | 1952 | 23 | p$K_a$, phenols, $H_2O$, 25 °C | — | 0.70 | — | 0.98 |
| | | | | p$K_a$, anilinium ions, $H_2O$, 25 °C | — | 0.69 | — | 1.13 |
| SOPh | Szmant and Suld | 1956 | 24 | p$K_a$, benzoic acids, 48% v/v EtOH–$H_2O$, 25 °C | — | — | 0.465 | — |
| | Meyers | 1963 | 25 | p$K_a$, phenols, 48% v/v EtOH–$H_2O$, 25 °C | — | 0.52 | — | 0.71 |
| $SO_2Ph$ | Szmant and Suld | 1956 | 24 | p$K_a$, benzoic acids, 48% v/v EtOH–$H_2O$, 25 °C | — | — | 0.70 | — |
| | Meyers | 1963 | 25 | p$K_a$, phenols, 48% v/v EtOH–$H_2O$, 25 °C | — | 0.62 | — | 0.90 |
| $SOCF_3$ | Yagupol'skii and coworkers | 1974 | 26 | p$K_a$, benzoic acids | 0.63 | — | 0.69 | — |
| | | | | p$K_a$, anilinium ions | — | 0.76 | — | 1.05 |
| $SO_2CF_3$ | Sheppard | 1963 | 27 | p$K_a$, benzoic acids | 0.79 | — | 0.93 | — |
| | | | | p$K_a$, anilinium ions | — | 1.00 | — | 1.65 |

[a] $\sigma^0$ value.

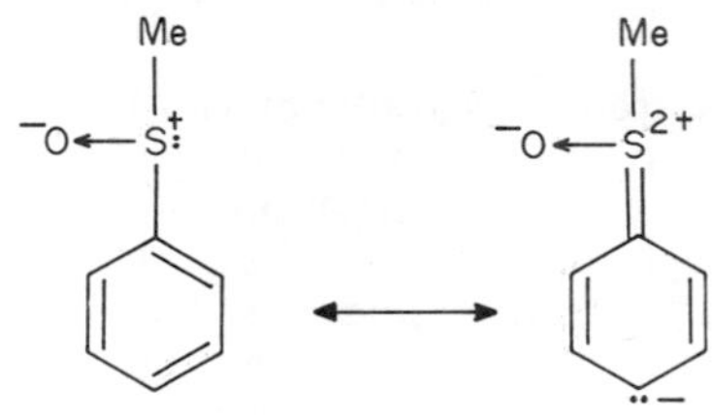

(8) (9)

Table 1.) Thus the order is $\sigma_m > \sigma_p$ (just), which is reminiscent of $+I$, $-R$ substituents such as the halogens, i.e. there is thus slight evidence for $\pi$(pp) conjugation (see **8** and **9**), the kind of electronic effect which is very prominent with the SMe group, for which the values of $\sigma_m$ and $\sigma_p$ are 0.15 and 0.00, respectively. Clearly, however, there is no very great effect of $\pi$(pp) conjugation for SOMe; presumably the appearance of a notional double positive charge on S would operate against it. Further, it needs to be said that various other determinations of $\sigma_m$ and $\sigma_p$ for SOMe have found $\sigma_p > \sigma_m$. This was the case for the values obtained by Price and Hydock[19] via studies of the rate of saponification of substituted ethyl benzoates in 56% aqueous acetone, $\sigma_m$ and $\sigma_p$ being 0.52 and 0.54, respectively. In much more recent years a high value of $\sigma_p$ at 0.564 for SOMe was also obtained in studies of the saponification of substituted ethyl benzoates by Yukawa, Tsuno and Sawada[22]. There were parallel studies on the saponification of substituted benzyl benzoates, for the determining of $\sigma_p^0$ values, i.e. 'normal' substituent constants free from any effect of $-R$ cross-conjugation. For SOMe $\sigma_p^0$ was found to be 0.573, very close to these authors' ('high') value for $\sigma_p$. Thus it seems somewhat doubtful whether there is any clear evidence for the operation of a $-R$ effect of SOMe in the reactions most often used for the determination of ordinary Hammett $\sigma$ values.

Bordwell and Boutan[20], however, observed a marked exaltation of $\sigma$ value in the effect of *p*-SOMe on the ionization of phenol: $\sigma_p^-$ is 0.73, whereas $\sigma_m^-$ at 0.53 differs little from $\sigma_m$. The authors naturally attributed the exaltation for *p*-SOMe to valence shell expansion of sulphur to ten electrons, a $+R$ effect, involving $\pi$(pd) conjugation.

For comparison with SOMe we mention that Bordwell and Cooper[23] determined $\sigma_m$ and $\sigma_p$ for $SO_2Me$ (from the ionization of substituted benzoic acids) as 0.65 and 0.72, respectively, and that Price and Hydock[19] determined closely similar values (from the saponification of substituted ethyl benzoates). The enhanced electron-attracting effect of $SO_2Me$ compared with SOMe is attributed in the first place to the inductive effect of two S–O dipoles compared with one, and secondly to enhanced $+R$ character [conventionally $\pi$(pd) conjugation]. The latter is manifested even more clearly in $\sigma_m^-$ and $\sigma_p^-$ values determined as 0.70 and 0.98, respectively, from phenol ionization and 0.69 and 1.13, respectively, from anilinium ion dissociation.

The behaviour of SOPh is fairly similar to that of SOMe, although there have been fewer studies thereof. Thus the $\sigma_p$ value of SOPh, determined from ionization of substituted benzoic acids[24], is 0.465. No exactly corresponding value for $\sigma_m$ appears to have been determined, but from the ionization of substituted phenols values of $\sigma_m^-$ and $\sigma_p^-$ were found to be 0.52 and 0.71, respectively[25]. As expected, $SO_2Ph$ is more strongly electron-attracting than SOPh, with values of $\sigma_p$, $\sigma_m^-$ and $\sigma_p^-$ equal to 0.70, 0.62 and 0.90, respectively[24,25].

The $SOCF_3$ group, as might be expected, is considerably more strongly electron-attracting than SOMe, with $\sigma_m$ and $\sigma_p$ values (benzoic acid ionization) of 0.63 and 0.69, respectively[26], i.e. $\sigma_m < \sigma_p$, a sign of much more definite $+R$ character, especially shown in the effect on anilinium ion dissociation with a $\sigma_p^-$ value of 1.05. $SO_2CF_3$ is even more strongly electron-attracting, with $\sigma_m$, $\sigma_p$, $\sigma_m^-$ and $\sigma_p^-$ values of 0.79, 0.93, 1.00 and 1.65, respectively[27,28].

The study of electrophilic aromatic substitution clearly offers the possibility of more definite evidence for $-R$ behaviour of sulphinyl groups in $\pi$(pp) conjugation[29]. In fact, many years ago it was found that the sulphinyl group was *para*-directing in the nitration or bromination of aromatic sulphoxides. In recent years the effect of sulphinyl groups on electrophilic substitution had been much studied by Marziano and colleagues[30,31]. The kinetics of nitration of diphenyl sulphoxide in strong sulphuric acid are domplex and are explained in terms of competitive nitration of two species $Ph_2SO$ and $[Ph_2SOH]^+$, the former favouring *para* substitution and the latter *meta* substitution. The results for the nitration of methyl phenyl sulphoxide are broadly similar, but PhSOMe is less reactive than $Ph_2SO$ by a factor of about ten. Molecular halogenations of methyl phenyl sulphoxide and of diphenyl sulphoxide show a great preponderance of *para* isomer in the product[32]. For chlorination in nitromethane at 25 °C there is a strong activating effect of SOPh and an effective $\sigma^+$ value of $-0.19$ is indicated.

Thus it seems clear that, in the absence of interactions with the reaction medium, sulphinyl groups tend to behave as $-R$ substituents and activate electrophilic substitution.

### 2. *Separation of inductive and resonance effects; substituent constants from spectroscopic studies*

The development of $\sigma_I$ and $\sigma_R$-type scales of substituent constants has not, of course, been a consequence solely of spectroscopic studies of organic compounds. Its origins lie in the analysis of chemically-based Hammett constants and in studies of the reactivity of aliphatic and alicyclic systems, but at an early stage the relationship of inductive and resonance parameters to spectroscopic quantities of various types acquired considerable impotance. We will begin with chemical aspects. (For a full account of all these matters, the reader is referred to the earlier article[33]; for background see also the general bibliography[15–18].

*a. Inductive and resonance constants from reactivity studies.* Taft's earliest values of $\sigma_I$ for substituents X were calculated from $\sigma^*$ values of $CH_2X$ through the relation $\sigma_I(X) = 0.45\sigma^*(CH_2X)$[34,35]. They included values for SOMe and $SO_2Me$ of 0.52 and 0.59, respectively; cf. 0.58 for CN and 0.63 for $NO_2$. [The principal sigma values for the sulphur-containing groups mentioned in this section (II.B.2) are summarized in Table 2.] These values for SOMe and $SO_2Me$ received satisfactory but rather limited testing in Taft and Lewis's examination of the general applicability of a fixed scale of inductive effects in the reactivities of *meta-* and *para*-substituted derivatives of benzene[34]. The corresponding paper on resonance effects[36] showed that no fixed scale of these was applicable, and ranges of $\sigma_R^{para}$ and $\sigma_R^{meta}$ were tabulated. It was, of course, the variability of resonance effects which ultimately led Taft and his associates to define four scales for resonance effects: $\sigma^0$, $\sigma_R$(BA), $\sigma_R^+$ and $\sigma_R^-$, each of 'limited generality' for a particular class of processes[3]. In this development the relationship of $\sigma_I$ and $\sigma_R$-type constants to spectroscopic quantities was of considerable importance but the crystallization of these ideas in the 1973 article[37] was still very largely chemically based. As far as the SOMe group is concerned, Ehrenson and coworkers[37] gave values of 0.50 and 0.00 for $\sigma_I$ and $\sigma_R^0$ [also $\sigma_R$(BA)], respectively. The $\sigma_I$ value for SOMe is thus slightly different from that given by Taft and Lewis[34]. The zero value for the resonance parameter presumably means that any tendency to octet expansion [conventionally $\pi$(pd) conjugation] is essentially cancelled by the $\pi$(pp) conjugation of the sulphur lone pair. Further, a zero value was also given for $\sigma_R^+$, i.e. no enhancement of $\pi$(pp) conjugation of SOMe was considered to occur in connection with

TABLE 2. Inductive and resonance constants of sulphinyl and sulphonyl groups

| Substituent | Authors | Year | Ref. | Method | $\sigma_I$ | $\sigma_R^0$ | $\sigma_R$(BA) | $\sigma_R^{\pm}$ |
|---|---|---|---|---|---|---|---|---|
| SOMe | Taft and Lewis | 1958 | 34 | Chemical reactivity | 0.52 | — | — | — |
| | Ehrenson and coworkers | 1973 | 37 | Chemical reactivity | 0.50 | 0.00 | 0.00 | 0.17[a] |
| | Exner | 1966 | 38 | Chemical reactivity | — | — | −0.17[b] | 0.17[c] |
| | Charton | 1981 | 39 | Chemical reactivity | — | — | 0.00[b] | −0.10[d] |
| | Taft and coworkers | 1963 | 40 | $^{19}$F NMR | 0.49[e] | — | — | — |
| | Sheppard and Taft | 1972 | 42 | $^{19}$F NMR | — | 0.00[f,g] | — | — |
| | Katritzky and coworkers | 1974 | 46 | Infrared | — | −0.07 | — | — |
| | Marriott and Topsom | 1984 | 48 | Theoretical | 0.37[h] | — | — | — |
| | Marriott and Topsom | 1985 | 49 | Theoretical | — | −0.03 | — | — |
| $SO_2Me$ | Taft and Lewis | 1958 | 34 | Chemical reactivity | 0.59 | — | — | — |
| | Ehrenson and coworkers | 1973 | 37 | Chemical reactivity | 0.59 | 0.12 | 0.12 | 0.29[i] |
| | Taft and coworkers | 1963 | 40 | $^{19}$F NMR | 0.55[e] | — | — | — |
| | Sheppard and Taft | 1972 | 42 | $^{19}$F NMR | — | 0.16[f,g] | — | — |
| | Katritzky and coworkers | 1974 | 46 | Infrared | — | 0.06 | — | — |
| | Marriott and Topsom | 1984 | 48 | Theoretical | 0.60[h] | — | — | — |
| | Marriott and Topsom | 1985 | 49 | Theoretical | — | 0.05 | — | — |
| SOPh | Charton | 1981 | 39 | Chemical reactivity | 0.51 | — | −0.07[b] | — |
| | Kaplan and Martin | 1973 | 43 | $^{19}$F NMR | 0.51 | −0.01[g] | — | — |
| $SO_2Ph$ | Charton | 1981 | 39 | Chemical reactivity | 0.56 | — | 0.12[b] | — |
| | Kaplan and Martin | 1973 | 43 | $^{19}$F NMR | 0.52 | 0.14[g] | — | — |
| | Katritzky and coworkers | 1974 | 46 | Infrared | — | 0.06[j] | — | — |
| $SOCF_3$ | Sheppard and Taft | 1972 | 42 | $^{19}$F NMR | 0.68[k] | 0.13[g,k] | — | — |
| | Ehrenson and coworkers | 1973 | 37 | Chemical reactivity | 0.64 | 0.08 | 0.08 | 0.08[d] |
| $SO_2CF_3$ | Sheppard and Taft | 1972 | 42 | $^{19}$F NMR | 0.78[f] | 0.31[f,g] | — | — |
| | Ehrenson and coworkers | 1973 | 37 | See footnote[l,m] | 0.84 | 0.24 | (0.24) | 0.41[m] |

[a]Value of $\sigma_R^-$ from phenol ionization. $\sigma_R^+ = 0.00$.
[b]Denoted $\sigma_R$; for significance see main text.
[c]Value of $\sigma_R^-$.
[d]Value of $\sigma_R^+$.
[e]In 'normal' solvents; see main text.
[f]In $CCl_4$.
[g]$\bar{\sigma}_R$ value; see main text.
[h]Value of $\sigma_F$; see main text.
[i]Value of $\sigma_R^-$ from phenol ionization. The value from anilinium ion dissociation was 0.38. $\sigma_R^+ = 0.12$.
[j]Sign not definitely established; see main text.
[k]In $CCl_3F$.
[l]Values regarded as 'secondary' substituent parameters: $\sigma_I$ value from $^{19}$F NMR, $\sigma_R$(BA) only a 'suggested' value.
[m]Value of $\sigma_R^-$ from phenol ionization. The value from anilinium ion dissociation was 0.57.

electrophilic reactivities. A $\sigma_R^-$ values of 0.17 was, however, based on phenol ionization.

It should be mentioned that Exner's procedure for separating inductive and resonance effects[38] leads to a $\sigma_R$ value [essentially equivalent to $\sigma_R$(BA) of Ehrenson and coworkers[37]] of $-0.17$ for SOMe, indicating $\pi$(pp) conjugation, cf. SMe $-0.24$. His value of 0.17 for $\sigma_R^-$ agrees with that of Ehrenson and coworkers[37].

The separation of inductive and resonance effects as carried out by Charton[39] is essentially chemically based: $\sigma_I$ values are derived from p$K_a$ values of aliphatic and alicyclic carboxylic acids and $\sigma_R$ values are obtained by subtracting $\sigma_I$ values from the corresponding $\sigma_p$ values (from the ionization of 4-substituted benzoic acids). Charton does not give a $\sigma_I$ value for SOMe, although its $\sigma_R$ value is given as 0.00, but for SOPh the $\sigma_I$ and $\sigma_R$ values are 0.51 and $-0.07$, respectively, thus giving support to appreciable $-R$ character. Further, Charton tabulated a distinctive $\sigma_R^+$ value of $-0.10$ for SOMe, in accord with the *para*-directing character of this group, cf. Ehrenson and coworkers[37].

*b. Sigma values from $^{19}$F NMR.* This subject has been associated with the development of $\sigma_I$ and $\sigma_R$-type scales almost from the start, but the first paper in which sulphinyl and sulphonyl groups played a part appears to have been one by Taft and coworkers in 1963[40]. The main object of this paper was to study the effect of solvent on the inductive order by $^{19}$F NMR measurements on a large number of *meta*-substituted fluorobenzenes in a great variety of solvents. The relationship between the NMR shielding parameter and $\sigma_I$ was established by means of selected systems such as equation 1:

$$\int_{\mathrm{H}}^{m\text{-X}} = -7.10\sigma_I + 0.60 \tag{1}$$

(The left-hand side is the $^{19}$F NMR shielding parameter for $m-$X relative to H as substituent.)

For SOMe amd $SO_2$Me the values of $\sigma_I$ as determined through chemical reactivities are quoted as 0.52 and 0.60, respectively. These provide a point of reference for consideration of the values determined through $^{19}$F NMR studies. The values for these substituents as determined in 'normal' solvents are given as 0.49 and 0.55, respectively. The term 'normal' appears to embrace a wide variety of solvents of the non-hydrogen-bonding, or not markedly hydrogen-bonding, type. For hydrogen-bonding solvents the $\sigma_I$ values are increased, the values of $0.62 \pm 0.03$ for SOMe and of $0.62 \pm 0.04$ for $SO_2$Me being quoted as relating to 'weakly protonic' solvents. Not too much quantitative significance should be attached to these values, but they indicate that hydrogen bonding of the solvent to the substituent enhances inductive electron withdrawal. This is confirmed by a value of 1.00 obtained for SOMe when trifluoroacetic acid was used as solvent.

In the related paper on $^{19}$F NMR screening parameters of *para*-substituted fluorobenzenes[41] in relation to resonance effects, a few measurements for SOMe and $SO_2$Me were recorded but no use was made of them for calculation of $\sigma_R$-type parameters. However, some years later Sheppard and Taft[42] used these data (carbon tetrachloride solution) to calculate $\bar{\sigma}_R$ values through equation 2.

$$\int_{m\text{-X}}^{p\text{-X}} = -29.5\bar{\sigma}_R \tag{2}$$

(The left-hand side is the $^{19}$F shielding parameter for *p*-X relative to *m*-X, and $\bar{\sigma}_R$ is the effective $\sigma_R$-type parameter. For $-R$ substituents the $\sigma_R$ values thereby obtained are considered to be $\sigma_R^0$ values, but $\sigma_R$ values for $+R$ substituents are slightly enhanced by the cross conjugation of the $-R$ F substituent with the $+R$ X group.) Sheppard and Taft[42] were undertaking a systematic study of the behaviour of sulphur substituents involving the various oxidation states of sulphur, and considerable use of this paper will be made later in this chapter. The $\bar{\sigma}_R$ values for SOMe and $SO_2Me$ are 0.00 and 0.16, respectively; cf. 0.00 and 0.12 suggested by Ehrenson, Brownlee and Taft[37] for $\sigma_R^0$. $^{19}$F-based values of $\sigma_I$ and $\bar{\sigma}_R$ were also given for $SOCF_3$ as 0.68 and 0.13, respectively (cf. 0.64 for $\sigma_I$ and 0.08 for $\sigma_R^0$[37]) and for $SO_2CF_3$ as 0.78 and 0.31 (cf. 0.84 for $\sigma_I$ and 0.24 for $\sigma_R^0$[37]). The solvent for $SOCF_3$ was $CCl_3F$ (infinite dilution) and for $SO_2CF_3$ was carbon tetrachloride.

Kaplan and Martin[43] determined $\sigma_I$ and $\bar{\sigma}_R$ for SOPh and $SO_2Ph$ by $^{19}$F NMR measurements. The $\sigma_I$ values showed almost no difference at 0.51 and 0.52, respectively, while the $\bar{\sigma}_R$ values were $-0.01$ and 0.14, respectively.

In passing we mention that there have been studies of the effects of SOMe as a substituent on $^1$H and $^{13}$C NMR (see earlier article[33]).

*c. The contribution of infrared spectroscopy.* The correlation of infrared frequencies or intensities with substituent constants has been practised for many years. In its more refined forms it is usual to employ $\sigma_I$ or $\sigma_R$-type constants either together in the so-called dual substituent-parameter (DSP) equation[37] or individually in special linear-regression equations which hold for particular infrared magnitudes. In this connection the work of Katritzky, Topsom and their colleagues is of particular importance. Early work by these authors established a relationship between $A_{mono}^{1/2}$, the square root of the integrated absorbance of the $\nu_{16}$ ring bands in monosubstituted benzenes and $\sigma_R^0$ for the substituents. The equation was usually written as follows[44]:

$$A_{mono} = 17{,}600(\sigma_R^0)^2 + 100 \tag{3}$$

Once the equation was well founded it became a tool for establishing a scale of resonance parameters based uniformly on infrared intensities, and particularly for measuring $\sigma_R^0$ values of substituents which had not been obtained in other ways. The above equation 3 could not give the *sign* of $\sigma_R^0$ for any given substituent, since $\sqrt{(\sigma_R^0)^2}$ can be given a positive or a negative sign. The sign has to be decided on other grounds. For instance, it was later found[45] that the integrated intensities of the $\nu_{16}$ vibration for *para*-disubstituted benzenes are correlated by equation 4:

$$A_{para} = 11{,}800\ (\sigma_R^0 1 - \sigma_R^0 2)^2 + 170 \tag{4}$$

provided the two substituents are not in donor–acceptor interaction. Suitable application of this equation enables the sign of a new $\sigma_R^0$ value to be determined.

As far as sulphinyl and sulphonyl groups are concerned, one of the papers[44] recorded $\pm\sigma_R^0$ values for SOPh, $SO_2Ph$ and $SO_2Me$ as 0.065, 0.064 and 0.069, respectively, indicating little dependence on state of oxidation or nature of hydrocarbon moiety attached to sulphur. The finding of a significant resonance effect for SOPh contrasts with other evidence regarding sulphinyl groups and the question of $\pi$(pp) versus $\pi$(pd) conjugation remained unanswered in the uncertainty as to the sign of $\sigma_R^0$. Later work resolved the question of signs[46]. It was shown that whereas $SO_2Me$ is a $+R$ group with a value of $\sigma_R^0$ equal to $+0.06$, SOMe is a $-R$ group, a net resonance donor, with a value of $\sigma_R^0$ equal to $-0.07$. However, studies of SOMe when placed *para* to a strong donor group found that it became a marked resonance acceptor, like $SO_2Me$ in the same situation. The evidence from infrared intensities seems to stand largely alone in indicating $-R$ behaviour for SOMe, but we may recall that Exner's procedure for calculating $\sigma_R$ values finds $-0.17$ for this substituent[38] (see Section II.B.2.a).

*d. Recent experimental and theoretical studies.* These matters were dealt with in some detail in the earlier article[33] and will only be summarized here in so far as they illuminate the behaviour of SOMe. The ion-cyclotron-resonance (ICR) equilibrium constant method applied to *meta*- and *para*-substituted phenols[47] has found gas-phase sigma values $\sigma_m$(g) and $\sigma_p^-$(g) for SOMe of 0.39 and 0.57, respectively, to compare with 0.52 and 0.73 for corresponding sigma values in aqueous solution. The enhancements in aqueous solution relative to the gas phase are discussed in terms of 'solvation-assisted resonance effects'. In an attempt at a separation of field/inductive and resonance effects for the gas-phase acidities of the phenols, there is reference to a $\sigma_R^0$ value of $+0.07$ for SOMe as an unpublished result of Adcock, Bromilow and Taft (cf. 0.00 from Ehrenson and coworkers[37] and $-0.07$ from Katritzky, Topsom and colleagues[46]).

Marriott and Topsom[48,49] have recently developed theoretical scales of substituent field and resonance parameters. The former correspond to the traditional 'inductive' parameters, but these authors are firm believers in the field model of the so-called inductive effect and use the symbol $\sigma_F$. The theoretical substituent field effect scale[48] is based on *ab initio* molecular orbital calculations. Various regression equations are established which become the basis for theoretical $\sigma_F$ values for about 50 substituents. These include SOMe and $SO_2Me$ at 0.37 and 0.60, respectively, which are said to agree well with 'inherent best values in the literature' of 0.36 and 0.58. However, it should be noted that $\sigma_I$ for SOMe is given as 0.50 by Ehrenson and co-workers[37].

The theoretical substituent resonance effect scale[49] is also based on *ab initio* calculations. A suitable regression equation is again established, which becomes the basis for theoretical $\sigma_R^0$ values of more than 40 substituents, including SOMe and $SO_2Me$ at $-0.03$ and 0.05, respectively. The latter agrees well with the infrared-based value of 0.06 and the former supports the occurrence of a $-R$ effect, as in the infrared value of $-0.07$[46]; cf. the $\sigma_R^0$ value of 0.00 given by Ehrenson, Brownlee, and Taft[37].

## III. ELECTRONIC EFFECTS OF GROUPS RELATED TO SOOH

### A. Introduction

In the discussion of the electronic effects of groups SOY, where Y = F, Cl, OMe or $NMe_2$, it will be assumed that the sulphur–oxygen bond is essentially a coordinate bond, as in SOMe (see Section II.A), with a formal unit positive charge on sulphur and negative charge on oxygen, as in **10**. In practice, some transfer of negative charge from oxygen to

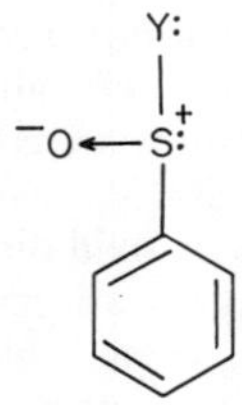

**(10)**

sulphur will no doubt occur through unequal sharing of the bonding electrons. S2p electron binding energies measured by the ESCA technique for nitrobenzenes substituted by sulphur groups in various oxidation states indicate that the positive charge carried by sulphur is rather similar in SOMe and SOOMe[50].

When such an SOY group is attached to a benzene ring, there is thus the possibility of $\pi$(pp) conjugation as with SOMe, i.e. $-R$ behaviour; see **11**. There is also the possibility of $+R$ behaviour, conventionally regarded as involving $\pi$(pd) conjugation; see **12**. There is, however, also the possibility that 'octet expansion' of sulphur in SOY may involve the use of unshared pair electrons on Y, in what would conventionally be regarded as $\pi$(pd) back-bonding, thus giving another canonical structure, **13**. Whether or not this is really a

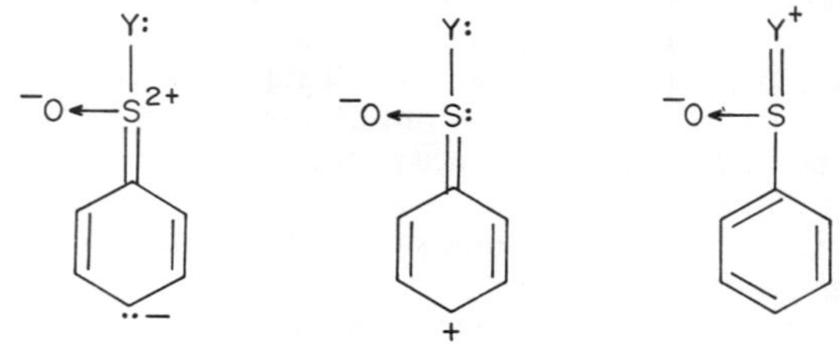

(11) (12) (13)

satisfactory formulation, a tendency for negative charge to be transferred from Y to S by polarization of the unshared pair electrons would certainly be expected. (Similarly, polarization of the unshared pair electrons on O could also contribute to the partial neutralization of the formal charges on O and S associated with the coordinate bond.)

For SOY with Y = O⁻ (the sulphinate moiety) we must suppose that the negative charge is distributed equally over the two oxygen atoms, so that the structure of $SO_2^-$ may be written as **14**.

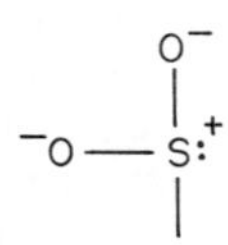

(14)

### B. Substituent Constants from ¹⁹F NMR

The paper of Sheppard and Taft, to which reference has already been made, involved the use of $^{19}F$ NMR for substituted fluorobenzenes to study the electronic effects of substituents containing sulphur in its various oxidation states[42]. Table 3 shows the $^{19}F$

TABLE 3.* Inductive and resonance constants for SOY groups from $^{19}F$ NMR of substituted fluorobenzenes[42]

| Substituent X | $\int_{H}^{m-X}$ [a] | $\int_{m-X}^{p-X}$ [a] | $\sigma_I$ [b] | $\bar{\sigma}_R$ [c] |
|---|---|---|---|---|
| SOF | −4.06 | −5.00 | 0.66 | 0.17 |
| SOCl | −4.24 | −4.22 | 0.68 | 0.14 |
| SOOMe | −2.56 | −2.66 | 0.45 | 0.09 |
| $SONMe_2$ | −1.53 | −0.87 | 0.30 | 0.03 |
| SOMe | −2.90 | −0.10 | 0.49 | 0.00 |
| $SOCF_3$ | −4.22 | −3.76 | 0.68 | 0.13 |

[a]For definition see main text. Measurements in $CCl_3F$ solution, except for X = SOMe, which was in $CCl_4$ (a 'normal' solvent[40]).
[b]Calculated from equation 1.
[c]Calculated from equation 2.

*Note:* *In connection with Table 3 it must be emphasized that the NMR magnitudes recorded are *shielding* parameters. Under the influence of the (overall) electron-attracting substituents with which we are concerned, their values are thus all negative. The negative signs in equations 1 and 2 are also a consequence of this. Nowadays it would be more usual to express the NMR magnitudes as *deshielding* parameters, i.e. the values corresponding to those in Table 3 would have positive signs and the coefficients in the equations corresponding to 1 and 2 would be positive. (See Section III.B and Table 4.)

shielding parameters obtained for SOY, with Y = F, Cl, OMe, or $NMe_2$, the groups of immediate concern to this chapter, along with the values for SOMe and $SOCF_3$ for comparison and to provide a basis for discussion. Table 3 also contains the corresponding values of $\sigma_I$ and $\bar{\sigma}_R$, calculated by means of the equations already given as 1 and 2 in Section II.B.2.b. The values of the resonance parameter obtained from $^{19}F$ NMR studies are often represented as $\sigma_R^0$ values, but for $+R$ substituents (and all of the above must be regarded as potentially of this nature) some slight enhancement by cross conjugation of F with the substituent is liable to occur.

Replacing the Me of SOMe by a more electronegative substituent would be expected to make both $\sigma_I$ and $\bar{\sigma}_R$ more positive, just as the formation of a further bond to O in $SO_2Me$ does. The $\sigma_I$ and $\bar{\sigma}_R$ values for SOY with Y=F, Cl or $CF_3$ are in accord with this expectation. However, replacing Me of SOMe by OMe or $NMe_2$ actually decreases $\sigma_I$ and the increase in $\bar{\sigma}_R$ is much less marked than for Y=F, Cl or $CF_3$. This suggests that the polarization of the unshared pairs of electrons on OMe or $NMe_2$, whether involving back-bonding or not, is having some influence in reducing the electron-attracting inductive effect and the $+R$ resonance effect of the sulphur group.

Sheppard and Taft[42] looked for a quantitative relationship between the $\sigma_I$ and $\bar{\sigma}_R$ values of the SOY groups on the one hand and the corresponding inductive and resonance parameters of Y on the other. Such a relationship was best examined through the attempted correlations of the $^{19}F$ shielding parameters, as the experimental data, with $\sigma_I$ and $\sigma_R$-type values of Y in the dual substituent-parameter equation[37]. Correlations of this general nature have been repeated for this chapter in a slightly different way, by using the extended Hammett equation, which differs from the DSP equation in permitting an intercept term (see e.g. Charton[51]). For the present purpose we will write the extended Hammett equation as equation 5:

$$S_m \quad \text{or} \quad S_p = \alpha\sigma_I + \beta\sigma_R + h \tag{5}$$

where

$$S_m = -\int_{H}^{m-X} \quad \text{and} \quad S_p = -\int_{H}^{p-X}$$

i.e. $S_m$ and $S_p$ are the negatives of the $^{19}F$ NMR shielding parameters for X = SOY relative to X = H (cf. equations 1 and 2 and Table 3. The change of sign brings the NMR magnitudes into accord with modern practice; see the note in Table 3). The regression coefficients $\alpha$ and $\beta$ give the sensitivity of $S_m$ or $S_p$ to inductive and resonance effects, respectively, while the intercept term $h$ corresponds to the value of $S_m$ or $S_p$ for the unavailable substituent SOH. The data used for the multiple regressions by a least-squares procedure are in Table 4, the values for $\sigma_I$ and $\sigma_R$-type constants of Y being from Reference 37. The details of the resulting regression equations are in Table 5, including appropriate criteria of goodness of fit. The best fits (most simply indicated by values of the multiple correlation coefficient) were obtained with the use of $\sigma_R^+$ as the resonance parameter for both the *meta* and the *para* series. For convenience of discussion the regression equations are set out as equations 6 and 7:

$$S_m = 2.652\ \sigma_I + 1.052\ \sigma_R^+ + 3.179 \tag{6}$$

$$(\pm 0.565) \quad (\pm 0.198)$$

$$n = 6 \quad R = 0.983 \quad s = 0.262 \quad \psi = 0.259$$

$$S_p = 11.346\ \sigma_I + 0.924\ \sigma_R^+ + 3.432 \tag{7}$$

$$(\pm 1.193) \quad (\pm 0.417)$$

$$n = 6 \quad R = 0.989 \quad s = 0.554 \quad \psi = 0.207$$

TABLE 4. Data for correlations of $^{19}$F NMR parameters of substituted fluorobenzenes[42] with inductive and resonance parameters of Y in substituent SOY

| Substituent Y in SOY | $S_m$ [a] | $S_p$ [a] | $\sigma_I$ [b] | $\sigma_R^0$ [b] | $\sigma_R$(BA) [b] | $\sigma_R^+$ [b] | $\sigma_R^-$ [b,c] |
|---|---|---|---|---|---|---|---|
| F | 4.06 | 9.06 | 0.50 | −0.34 | −0.45 | −0.57 | −0.45 |
| Cl | 4.24 | 8.64 | 0.46 | −0.23 | −0.23 | −0.36 | −0.23 |
| OMe | 2.56 | 5.22 | 0.27 | −0.45 | −0.61 | −1.02 | −0.45 |
| $NMe_2$ | 1.53 | 2.40 | 0.06 | −0.52 | −0.83 | −1.75 | −0.34 |
| Me | 2.90 | 3.00 | −0.11 | −0.11 | −0.11 | −0.25 | −0.11 |
| $CF_3$ | 4.22 | 7.98 | 0.45 | 0.08 | 0.08 | 0.08 | 0.17 |

[a]For definition see main text.
[b]From Ehrenson, Brownlee, and Taft[37].
[c]From dissociation of anilinium ions.

TABLE 5. Multiple regressions of $^{19}$F NMR parameters $S_m$ or $S_p$ on $\sigma_I$ and $\sigma_R$-type constants of Y in substituent SOY

| NMR parameter | $\sigma_R$-type constant | $\alpha$ [a] | $\beta$ [b] | $h$ [c] | $s_\alpha$ [d] | $s_\beta$ [e] | $R$ [f] | $s$ [g] | $\psi$ [h] |
|---|---|---|---|---|---|---|---|---|---|
| $S_m$ | $\sigma_R^0$ | 3.359 | 2.529 | 2.962 | 0.927 | 0.948 | 0.947 | 0.461 | 0.456 |
| $S_m$ | $\sigma_R$(BA) | 3.168 | 1.852 | 3.018 | 0.703 | 0.478 | 0.970 | 0.346 | 0.342 |
| $S_m$ | $\sigma_R^+$ | 2.652 | 1.052 | 3.179 | 0.565 | 0.198 | 0.983 | 0.262 | 0.259 |
| $S_m$ | $\sigma_R^-$ | 3.860 | 1.692 | 2.556 | 1.311 | 1.257 | 0.884 | 0.669 | 0.661 |
| $S_p$ | $\sigma_R^0$ | 12.050 | 1.841 | 3.118 | 1.475 | 1.509 | 0.981 | 0.734 | 0.275 |
| $S_p$ | $\sigma_R$(BA) | 11.856 | 1.487 | 3.224 | 1.321 | 0.898 | 0.985 | 0.649 | 0.243 |
| $S_p$ | $\sigma_R^+$ | 11.346 | 0.924 | 3.432 | 1.193 | 0.417 | 0.989 | 0.553 | 0.207 |
| $S_p$ | $\sigma_R^-$ | 12.426 | 0.820 | 2.722 | 1.689 | 1.621 | 0.974 | 0.862 | 0.323 |

[a]Coefficient of $\sigma_I$ in equation 5.
[b]Coefficient of $\sigma_R$ in equation 5.
[c]Intercept term in equation 5.
[d]Standard error of $\alpha$.
[e]Standard error of $\beta$.
[f]Multiple correlation coefficient.
[g]Standard error of the estimate.
[h]Exner's $\psi$ statistic of goodness of fit.

Because there are only six experimental points and thus three degrees of freedom in each regression, the high values of the multiple correlation coefficient $R$ give a misleading impression that the correlations are good. In fact the values of Exner's $\psi$ statistic[52,53], which corrects for the degrees of freedom, show that the correlations are only fair to poor, but they are acceptable as having some physical meaning. The $F$ statistics for equations 6 and 7 are 43.16 and 68.4, respectively, indicating overall significance at just over the 99% level. The $t$ statistics for the regression coefficients (essentially the ratio of each regression coefficient to its standard error) indicate significance at the 98% level or above, except for the coefficient of $\sigma_R^+$ in equation 7, which is significant at about the 90% level. The correlation coefficient giving the collinearity of $\sigma_I$ with $\sigma_R^+$ is 0.418. (See Reference 14, Chapter 1 for an elementary discussion of the statistics of multiple regression.)

The equations apparently indicate at first sight that the main role of Y in moderating the electron-attracting influence of SOY is exerted through its inductive effect, whether that influence is felt by $^{19}$F at the *meta* or the *para* position. The resonance effect of Y, reducing the overall electron-attracting effect of SOY when $\sigma_R^+$ is negative, as it is for five of the six substituents, is apparently of secondary importance. However, the proper interpretation of the regression coefficients requires a preliminary weighting thereof by the standard deviation of the corresponding explanatory variable, i.e. $\sigma_I$ or $\sigma_R^+$ [53]. When this is done the relative contributions of the inductive and resonance effects of Y to the total substituent effect are 46.9 and 53.1 percent, respectively, for the *meta* series and 81.1 and 18.9 percent, respectively, for the *para* series. Therefore the resonance effect of Y is actually slightly more important than its inductive effect in governing the behaviour of SOY in the *meta* series, while the inductive effect is not quite so overwhelmingly important in the *para* series as it appears at first sight. Thus the influence of the resonance effect of Y is relatively more important when the $+R$ effect of SOY is less important (*meta* series), and vice versa (*para* series). This seems quite reasonable if the Y and SOY resonance effects are regarded as in competition in producing octet expansion of sulphur, whatever the mechanism [$\pi$(pd) conjugation or otherwise] may be.

The occurrence of the best fits with $\sigma_R^+$ of Y rather than $\sigma_R$(BA), $\sigma_R^0$ or $\sigma_R^-$ is presumably a consequence of the influence of Y involving the polarization of electron density by a considerably positive sulphur atom, for which the polarization of a benzene ring by an electrophile (the type of process on which the $\sigma^+$ scale is based) may provide an approximate model. [It should be noted that the fit with $\sigma_R$(BA) is not much inferior, particularly for the *para* series; see Table 5.]

## C. Other Substituent Constants

### 1. Estimated sigma values

For SOY with Y = F, Cl, OMe or $NMe_2$, Exner based $\sigma_m^0$ and $\sigma_p^0$ values on the $^{19}$F measured values of $\sigma_I$ and $\bar{\sigma}_R$ [54]. (We have already commented that the $\bar{\sigma}_R$ values for $+R$ substituents can only be regarded as approximating to $\sigma_R^0$, since there will be slight enhancement from cross-conjugation of F with the substituent.) The general equations on which such estimates are based are given by Exner as 8 and 9:

$$\sigma_p = a\sigma_I + b\sigma_R \tag{8}$$

$$\sigma_m = c\sigma_I + d\sigma_R \tag{9}$$

where *a*, *b*, *c* and *d* are appropriate coefficients. The values of $\sigma_m^0$ and $\sigma_p^0$ are given in Table 6. The values of *a*, *b*, *c* and *d* used by Exner are not stated explicitly, but it appears that for Y = F, Cl, or OMe, *a*, *c* and *d* were taken as unity and *b* was taken as 0.5. Such values have long been taken to apply to the analysis of $\sigma^0$ values into inductive and resonance components[34,36]. The basis for the estimated $\sigma^0$ values of $SONMe_2$, with $\sigma_m^0 < \sigma_I$ and $\sigma_p^0 < \sigma_I$, is unclear. The situation implies that the value of $\sigma_R^0$ as determined by $^{19}$F NMR is negative, i.e. $SONMe_2$ is a $-R$ group. However, Sheppard and Taft's work[42] gives no support to this view and Exner elsewhere in his compilation of substituent constants[54] quotes $\sigma_R^0$ for $SONMe_2$ as $+0.03$.

TABLE 6. Estimated $\sigma^0$ values for SOY[54]

| Y | $\sigma_m^0$ | $\sigma_p^0$ |
|---|---|---|
| F | 0.74 | 0.83 |
| Cl | 0.75 | 0.82 |
| OMe | 0.50 | 0.54 |
| $NMe_2$ | 0.29 | 0.27 |

*2. Substituent constants from polarography*

A systematic study of the electronic effects of groups containing sulphur in its various oxidation states was undertaken by Lindberg[50]. Various physical measurements were involved, but the most important was the determination of the polarographic half-wave potentials of sulphur substituted nitrobenzenes[55]. The application of the Hammett equation was examined by employing a series of substituents such that the polarographic behaviour of the substrates could be established as comparable, i.e. the process under study was the reduction of the nitro group. At pH values of 5.0, 7.2 and 9.3 the $\rho$ values were found to be 143, 150 and 173, respectively, for 15, 15 and 18 substituents and with correlation coefficients of the Hammett plots of 0.985, 0.984 and 0.989, respectively. It appeared that for $+R$ substituents in the *para* position to the nitro group, $\sigma_p^-$ values were required for good conformity to the Hammett equation. New $\sigma_m$ or $\sigma_p^-$ values were based on the Hammett plots.

Of interest for the present chapter are values of 0.68, 0.64 and 0.65 for *m*-SOOMe at the three pH values referred to above, with a mean value of $0.66 \pm 0.02$, while $\sigma_p^-$ was given as 0.84, 0.90 and 0.93, with a mean value of $0.89 \pm 0.05$. The experimental value of $\sigma_m$ is thus considerably greater than the estimated value of 0.50 for $\sigma_m^0$ (Table 6). There is, of course, supposed to be almost no difference in the scales of $\sigma_m$ (i.e. benzoic acid-based) and $\sigma_m^0$. While it might be argued that the value of 0.66 from *m*-SOOMe should be regarded as a $\sigma_m^-$ value, an enhancement of 0.16 as between $\sigma_m$ and $\sigma_m^-$ for this substituent is not reasonable. Thus there is a definite anomaly, which might be connected with the difference in solvent: the $^{19}$F NMR measurements on which the estimated value of $\sigma_m^0$ was based were made with solutions in $CCl_3F$, while the polarographic value for $\sigma_m$ (or $\sigma_m^-$) involved experiments in aqueous ethanol. The enhanced value in the latter case could perhaps be due to hydrogen-bonding of water/ethanol to the oxygen atoms of SOOMe, thereby increasing the electron-attracting influence of the substituent. As far as the $\sigma_p^-$ value of 0.89 is concerned, an enhancement of 0.23 compared with $\sigma_m^-$ is not unreasonable; cf. $\sigma_m^-$ and $\sigma_p^-$ values of 0.53 and 0.73, respectively, for SOMe (Table 1). Note, however, that $\sigma_m^-$ and $\sigma_m$ values for SOMe are almost the same (Table 1). However, in tabulating $\sigma_p^-$ equal to 0.89 for SOOMe, Exner[54] places it in parentheses to indicate that he regards the value as in some way unreliable. The basis for this opinion probably lies in Lindberg's own discussion, in so far as *p*-SOMe shows appreciable deviation from the Hammett plots, and it may therefore be rather naïve to take values of $\sigma_p^-$ for the closely related group SOOMe off the Hammett plots, thereby assuming strict conformity to the Hammett equation on the part of this group.

It should be mentioned in passing that Lindberg[55] tabulates an apparent sigma value for *o*-SOOMe as $0.97 \pm 0.05$. The paper contains but little information on the effect of *ortho* substituents on the polarographic magnitude in question and therefore it is difficult to assess the significance of the above value. The *ortho* effect usually involves both electronic influences of substituents and various kinds of steric effect[56]. It sometimes happens, however, that steric effects are not particularly important and that the electronic effects of the substituents are fairly similar as between *para* and *ortho* positions. Such may be the case here, the value of $\sigma_p^-$ being 0.89 and the apparent *ortho* sigma value being not too different at 0.97. Lindberg also gives an estimate of $\sigma_p$ for SOOMe as 0.75, on the basis of Exner's relation for $+R$ substituents[38], $\sigma_p = 1.14\sigma_m$. As might be expected from the discussion above, this value is very much higher than the $\sigma_p^0$ value of 0.54 based on $^{19}$F NMR.

Lindberg has also included[50] *ortho*-, *meta*- and *para*-SOOMe in a study of the effect of substituents on the asymmetric and symmetric NO stretching vibrations of nitrobenzenes and in a study of S2p electron binding energies determined by ESCA for sulphur-substituted nitrobenzenes. Various correlations were presented, including Hammett treatments for the effect of substituents on the NO stretching vibrations. Unfortunately none of the correlations is sufficiently precise to permit the determination of new sigma values with any claim to reasonable precision. Thus all that can be said is that the SOOMe group appears as undoubtedly electron-attracting, not too different from SOMe in this respect and rather more weakly electron-attracting than $SO_2Me$.

Lindberg also included the unipole $SO_2^-$ in his studies, as discussed below in Section III.C.3.

### 3. *The behaviour of* $SO_2^-$

While sigma values of various types have often been measured and tabulated[54] for unipolar substituents (particularly for $NMe_3^+$ and $CO_2^-$), the whole question of the behaviour of these substituents with respect to the Hammett equation and cognate linear free-energy relationships is a complicated matter. This was discussed in detail in the earlier article on the electronic effects of the sulphonio group[1]. The substituent constants of unipolar substituents are particularly sensitive to variation of solvent and to changes in the ionic strength of the medium. The mixing of unipolar with dipolar substituents in Hammett and similar correlations is highly unwise. The formal sigma constants of unipolar substituents are mainly of interest as a general indication of their behaviour and as a more specific indication of behaviour under specified conditions.

The unipolar substituent $SO_2^-$ has attracted some attention. Thus it was included by Lindberg[55] in his polarographic studies of sulphur containing substituted nitrobenzenes. At pH values of 5.0, 7.2 and 9.3, values of $\sigma_p^-$ for $SO_2^-$ were found as 0.09, 0.05 and 0.10, respectively, giving a mean value of $0.08 \pm 0.03$. In related work Lindberg[57] determined $\sigma_m$ and $\sigma_p$ for $SO_2^-$ by potentiometric titration of the carboxyl-substituted phenylsulphinic acids in water. The apparent $\sigma$ values at an ionic strength of 0.1 were 0.31 and 0.37, respectively, but on correction to zero ionic strength the values were $-0.02$ and $-0.05$, respectively. Negatively charged substituents always tend to be more electron-releasing than closely related neutral substituents, e.g. comparable $\sigma_m$ values for $CO_2^-$ and for COOMe are $-0.01$ and 0.37, respectively. The natural tendency of the unit negative charge to be shared between the substituent and the benzene ring renders such behaviour understandable. The relative values of $\sigma_m$ and $\sigma_p$ quoted above might perhaps be taken as a slight indication of a $-R$ effect for *p*-$SO_2^-$. However, neither the effect of the negative charge nor any $-R$ effect are sufficient to outweigh the $+R$ aspect of the behaviour of $SO_2^-$ in processes which call for enhanced electron attraction by the substituent, the value of $\sigma_p^-$ as mentioned above being appreciably positive. Further, in earlier kinetic and infrared spectroscopic work by Lindberg[58], which involved $SO_2^-$ as an invariant substituent in the presence of a variable substituent, no enhancement of substituent constant for *p*-$NO_2$ or other $+R$ group was observed. Thus in that work no indication of a $-R$ effect of $SO_2^-$ was obtained (see further, below).

Lindberg[55] determined a mean sigma value for *o*-$SO_2^-$ of $-0.36 \pm 0.1$ over the three pH values studied. The considerably negative value presumably indicates the dominance of the field effect of the negative charge on the polarographic reduction of the $NO_2$ in the *ortho* position. However, the study of the effect of substituents on the NO stretching vibrations referred to above[50] indicated that the electronic effects of $SO_2^-$ in any of the three positions relative to $NO_2$ are little different from the electronic effect of H.

The $SO_2^-$ group also featured in the infrared studies of Katritzky and colleagues[44], in which the intensities of the $\nu_{16}$ bands for mono-substituted benzenes were related to the substituent $\sigma_R^0$ values (see Section II.B.2.c). For $SO_2^-$ (with $Na^+$ as counter-ion) $\pm\sigma_R^0$ is tabulated as 0.00. This presumably means that the negative charge on $SO_2^-$ suppresses $+R$ character without significantly encouraging $-R$ interaction with the benzene ring. The finding seems odd in view of the $-R$ character of SOMe as indicated by infrared studies[46], $\sigma_R^0$ being $-0.07$.

### 4. A recent study involving $SONMe_2$

Häkkinen and Ruostesuo[59] have recently studied the effect of several sulphur-containing substituents on the $^{13}C$ NMR shifts of the carbon atoms in the various positions of benzene relative to the substituent. The measurements were made with chloroform as solvent and included a study of $SONMe_2$. The results may be interpreted by reference to the extensive studies of the effect of substituents on the $^{13}C$ shifts in substituted benzenes dissolved in chloroform, which have been made by Bromilow and collaborators[60,61], who made use of the DSP equation in treating their results. For the shifts at the 3- or 5- and the 4-carbon atoms of a monosubstituted benzene, expressed relative to the shift for a carbon atom in benzene itself, and denoted $S_m$ and $S_p$ respectively, equations 10 and 11 were found to hold (see References 60 and 61, respectively):

$$S_m = 1.6\sigma_I - 1.4\sigma_R^0 \tag{10}$$

$$S_p = 4.6\sigma_I + 21.5\sigma_R^0 \tag{11}$$

By solving equations 10 and 11 as a pair of simultaneous equations, values of $\sigma_I$ and $\sigma_R^0$ may be calculated for any substituent for which the values of $S_m$ and $S_p$ have been determined but have not been used in establishing the equations. Häkkinen and Ruostesuo[59] found $S_m$ and $S_p$ for $SONMe_2$ to be 0.4 and 2.4, respectively. The corresponding values of $\sigma_I$ and $\sigma_R^0$ calculated from equations 10 and 11 are 0.30 and 0.05, in good agreement with the values of $\sigma_1$ and $\bar{\sigma}_R$ based on $^{19}F$ NMR of 0.30 and 0.03, respectively (Table 3).

## IV. ACKNOWLEDGEMENTS

I thank Professor Dr Otto Exner (Prague) for helpful correspondence and Dr David F. Ewing (Hull) for helpful discussion.

## V. REFERENCES AND NOTES

1. J. Shorter, in *The Chemistry of the Sulphonium Group* (Eds C. J. M. Stirling and S. Patai), Chap. 9, Wiley, Chichester, 1981.
2. J. Shorter, in *The Chemistry of Sulphones and Sulphoxides* (Eds S. Patai, Z. Rappoport and C. J. M. Stirling), Chap. 10, Wiley, Chichester, 1988.
3. Reference 2, p. 484.
4. W. Kutzelnigg, *Angew. Chem., Int. Ed. Engl.*, **23**, 272 (1984).
5. The symbol and sign conventions used for substituent effects are those most frequently used by writers on correlation analysis in organic chemistry (linear free-energy relationships, etc.). *I* (inductive) or *R* (resonance) effects which withdraw electrons from the ring are regarded as positive. See J. Shorter, *Correlation Analysis of Organic Reactivity*, Research Studies Press, Wiley, Chichester, 1982, pp. 229–230 for a more detailed consideration of symbol and sign conventions.
6. H. A. Bent, in *The Organic Chemistry of Sulfur* (Ed. S. Oae), Chap. 1, Plenum Press, New York, 1976.
7. E. Block, *Reactions of Organosulfur Compounds*, Academic Press, New York, 1978. See especially pp. 18–21 and 44–50.
8. H. Kwart and K. King, *d-Orbitals in the Chemistry of Silicon, Phosphorus, and Sulfur*, Springer-Verlag, Berlin, 1977.
9. Reference 2, p. 489.
10. Reference 2, p. 493.
11. W. von E. Doering and A. K. Hoffmann, *J. Am. Chem. Soc.*, **77**, 521 (1955).
12. D. L. Hammick and R. B. Williams, *J. Chem. Soc.*, 211 (1938).
13. Reference 2, p. 494.
14. J. Shorter, *Correlation Analysis of Organic Reactivity*, Research Studies Press, Wiley, Chichester, 1982.

15. N. B. Chapman and J. Shorter (Eds), *Advances in Linear Free Energy Relationships*, Plenum Press, London, 1972.
16. N. B. Chapman and J. Shorter (Eds), *Correlation Analysis in Chemistry: Recent Advances*, Plenum Press, New York, 1978.
17. O. Exner, *Correlation Analysis of Chemical Data*, Plenum Press, New York and SNTL, Prague, 1988.
18. L. P. Hammett, *Physical Organic Chemistry*, 2nd edn., McGraw-Hill, New York, 1970.
19. C. C. Price and J. J. Hydock, *J. Am. Chem. Soc.*, **74**, 1943 (1952).
20. F. G. Bordwell and P. J. Boutan, *J. Am. Chem. Soc.*, **79**, 717 (1957).
21. Reference 2, p. 503.
22. Y. Yukawa, Y. Tsuno, and M. Sawada, *Bull. Chem. Soc. Jpn.*, **45**, 1198 (1972).
23. F. G. Bordwell and G. D. Cooper, *J. Am. Chem. Soc.*, **74**, 1058 (1952).
24. H. H. Szmant and G. Suld, *J. Am. Chem. Soc.*, **78**, 3400 (1956).
25. C. Y. Meyers, *Gazz. Chim. Ital.*, **93**, 1206 (1963).
26. L. M. Yagupol'skii, A. Ya. Il'chenko., and N. B. Kondratenko, *Usp. Khim.*, **43**, 64 (1974); *Russ. Chem. Rev.*, **43**, 32 (1974).
27. W. A. Sheppard, *J. Am. Chem. Soc.*, **85**, 1314 (1963).
28. For thorough discussion of the various sigma values obtained by reactivity studies for all the groups mentioned so far in this section (and other related groups) see Reference 2, pp. 498–509.
29. Reference 2, p. 532.
30. N. C. Marziano, E. Maccarone, and R. C. Passerini, *J. Chem. Soc.* (*B*), 745 (1971).
31. N. C. Marziano, E. Maccarone, G. M. Cimino, and R. C. Passerini, *J. Org. Chem.*, **39**, 1098 (1974).
32. A. C. Boicelli, R. Danieli, A. Mangini, A. Ricci, and G. Pirazzini, *J. Chem. Soc., Perkin Trans. 2*, 1343 (1974).
33. Reference 2, pp. 509–517.
34. R. W. Taft and I. C. Lewis, *J. Am. Chem. Soc.*, **80**, 2436 (1958).
35. R. W. Taft, *J. Am. Chem. Soc.*, **79**, 1045 (1957).
36. R. W. Taft and I. C. Lewis, *J. Am. Chem. Soc.*, **81**, 5343 (1959).
37. S. Ehrenson, R. T. C. Brownlee, and R. W. Taft, *Prog. Phys. Org. Chem.*, **10**, 1 (1973).
38. O. Exner, *Collect. Czech. Chem. Commun.*, **31**, 65 (1966).
39. M. Charton, *Prog. Phys. Org. Chem.*, **13**, 119 (1981).
40. R. W. Taft, E. Price, I. R. Fox, I. C. Lewis, K. K. Andersen, and G. T. Davis, *J. Am. Chem. Soc.*, **85**, 709 (1963).
41. R. W. Taft, E. Price, I. R. Fox, I. C. Lewis, K. K. Andersen, and G. T. Davis, *J. Am. Chem. Soc.*, **85**, 3146 (1963).
42. W. A. Sheppard and R. W. Taft, *J. Am. Chem. Soc.*, **94**, 1919 (1972).
43. L. J. Kaplan and J. C. Martin, *J. Am. Chem. Soc.*, **95**, 793 (1973).
44. R. T. C. Brownlee, R. E. J. Hutchinson, A. R. Katritzky, T. T. Tidwell, and R. D. Topsom, *J. Am. Chem. Soc.*, **90**, 1757 (1968).
45. P. J. Q. English, A. R. Katritzky, T. T. Tidwell, and R. D. Topsom, *J. Am. Chem. Soc.*, **90**, 1767 (1968).
46. N. C. Cutress, T. B. Grindley, A. R. Katritzky, M. Shome, and R. D. Topsom, *J. Chem. Soc., Perkin Trans. 2*, 268 (1974).
47. M. Fujio, R. T. McIver, Jr., and R. W. Taft, *J. Am. Chem. Soc.*, **103**, 4017 (1981).
48. S. Marriott and R. D. Topsom, *J. Am. Chem. Soc.*, **106**, 7 (1984).
49. S. Marriott and R. D. Topsom, *J. Chem. Soc., Perkin Trans. 2*, 1045 (1985).
50. B. J. Lindberg and B. Schröder, *Acta Chem. Scand.*, **24**, 3089 (1970).
51. M. Charton, in Reference 16, Chap. 5.
52. O. Exner, *Collect. Czech. Chem. Commun.*, **31**, 3222 (1966).
53. See also Reference 14, Chap. 7.
54. O. Exner, in Reference 16, Chap. 10.
55. B. J. Lindberg, *Arkiv Kemi*, **32**, 317 (1970).
56. Reference 14, Chap. 4.
57. B. J. Lindberg, *Acta Chem. Scand.*, **24**, 2852 (1970).
58. B. J. Lindberg, *Acta Chem. Scand.*, **17**, 393 (1963); **20**, 1843 (1966); **21**, 2215 (1967).
59. A.-M. Häkkinen and P. Ruostesuo, *Magn. Reson. Chem.*, **23**, 424 (1985).
60. J. Bromilow, R. T. C. Brownlee, D. J. Craik, and M. Sadek, *Magn. Reson. Chem.*, **24**, 862 (1986).
61. J. Bromilow, R. T. C. Brownlee, D. J. Craik, M. Sadek, and R. W. Taft, *J. Org. Chem.*, **45**, 2429 (1980).

The Chemistry of Sulphinic Acids, Esters and their Derivatives
Edited by S. Patai


CHAPTER **18**

# Thiosulphinic acids and esters

TOSHIKAZU TAKATA and TAKESHI ENDO

*Research Laboratory of Resources Utilization, Tokyo Institute of Technology, Nagatsuta-cho Midori-ku, Yokohama 227, Japan*

## I. INTRODUCTION

Thiosulphinic acid (**1**) is a hypothetical organosulphur compound with a thiol structure which could be obtained by replacing an oxygen atom of a sulphinic acid by sulphur. Although recently the first synthesis and characterization of salts of some thiosulphinic acids have been accomplished[1], there has been no publication as yet on the isolation of free thiosulphinic acid[2,3].

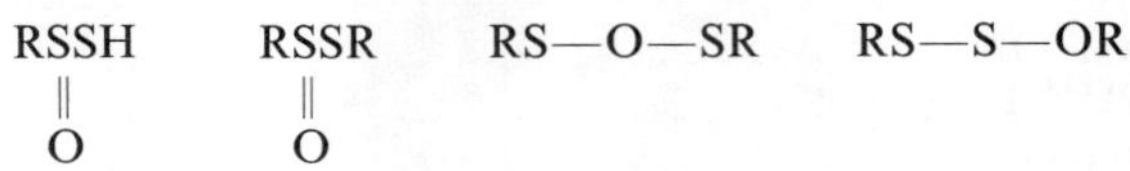

**(1)** **(2)** **(3)** **(4)**

The chemistry of thiosulphinates (sometimes called 'thio*l*sulphinates'), i.e. thiosulphinic S-esters (**2**), has only a short history starting from the isolation of *allicin* (**5**) and successive preparation of several alkyl thiosulphinates, by Cavallito and coworkers[4–6].

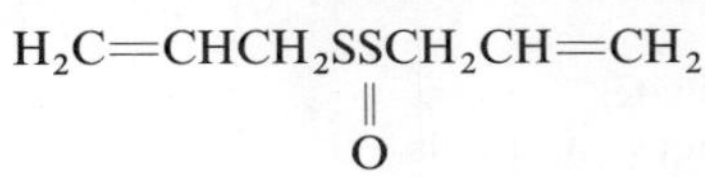

allicin (**5**)

The real structure of thiosulphinates was determined to be **2** but not **3**, **4**, or a mixture of disulphide and thiosulphonate. The interests of early papers were concerned with biological activities of **2**[7,8], antitumor[9–12], antiviral[5,13] and antifugal[6,14] activities[15] as well as antioxidant activity[16–20]. Since **2** is unstable and reactive enough to lead to complex product mixtures, the study of its chemistry was always associated with some difficulties. More recent detailed investigations have been carried out with the progress of new analytical methods and instruments, making the chemistry of thiosulphinates more clear.

In this chapter mainly more recent advances on the chemistry of thiosulphinates are described.

## II. THIOSULPHINIC ACIDS

Thiosulphinic acid may have some isomeric forms (**6**, **7**, **8**) among which **6** is believed to be the most stable and likely structure.

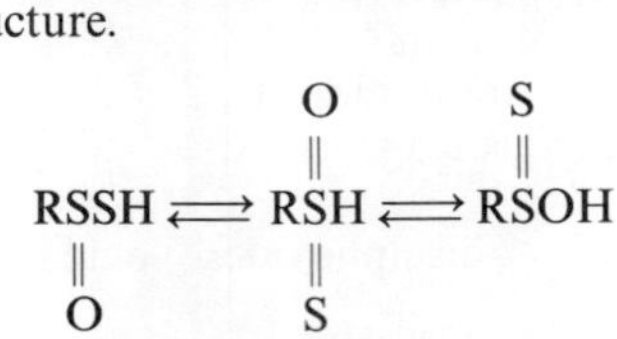

**(6)** **(7)** **(8)**

Thiosulphinic acid can be regarded as a chiral organosulphur compound[21,22], unlike sulphinic acid. No free acid is known so far[2,3] although some salts have been described quite recently[1].

### A. Synthesis and Structure

Mikołajczyk and his coworkers have synthesized and characterized some relatively stable salts of thiosulphinic acids (**9**)[1]. The stability of these salts may be attributed to steric protection by bulky *tert*-butyl (**9a**, *t*-Bu), adamantyl (**9b**, Ad) and triptycenyl (**9c**, Tr) groups bound to the central sulphinyl sulphur atom.

$$\underset{\substack{\|\\ \mathrm{O}}}{\mathrm{RSCl}} \xrightarrow[\substack{Et_2O \text{ or } CH_2Cl_2\\ -70\,^\circ C}]{H_2S/R'_3N} \underset{\substack{\|\\ \mathrm{O}\\ (\mathbf{10})}}{\mathrm{RSS^-H\overset{+}{N}R'_3}} \xrightarrow{Na_2CO_3} \underset{\substack{\|\\ \mathrm{O}\\ (\mathbf{11})}}{\mathrm{RSS^-Na^+}}$$

$$\xrightarrow[\substack{X = H_3\overset{+}{N}-C{=}NH\\ |\\ SCH_2Ph}]{X^+Cl^-} \underset{\substack{\|\\ \mathrm{O}\\ (\mathbf{9})}}{\mathrm{RSS^-X^+}} \qquad (1)$$

(**a**) R = *t*-Bu

(**b**) R = Ad (mp 135 ~ 140 °C)

(**c**) R = 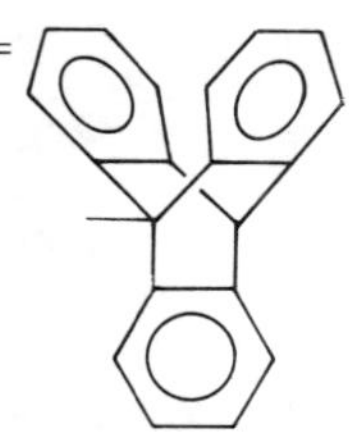 (Tr) [mp 223 ~ 225 °C (**10c**)]

Starting from the corresponding sulphinyl chloride, the corresponding ammonium salts (**10**) were prepared and isolated in 75–85% yields (equation 1). These salts were converted to the corresponding S-benzylisothiuronium salts (**9**) for better characterization, and $^{1}H$ and $^{13}C$ NMR analyses supported the proposed structures. The strongest evidence of this salt structure was given by the X-ray structure determination of **9b** (Figure 1). The authors attribute the rather longer S—O bond (1.536 Å) and shorter S—S bond (2.025 Å) to delocalization of the negative charge in the anion by resonance (equation 2).

$$\underset{\substack{\|\\ \mathrm{O}}}{\mathrm{RSS^-X^+}} \longleftrightarrow \underset{\substack{|\\ \mathrm{O^-}}}{\mathrm{RS{=}SX^+}} \rightleftarrows \mathrm{R-\overset{\substack{\mathrm{O}\\ \|}}{\underset{\substack{\|\\ \mathrm{S}}}{S}}{}^-X^+} \qquad (2)$$

Another synthetic method has also been developed by Mikołajczyk and coworkers[1]. The reaction of the stable triptycene sulphenic acid (**12c**)[23] with elemental sulphur in the

$$\underset{\substack{\|\\ \mathrm{O}\\ (\mathbf{14c})}}{\mathrm{TrSS^-Na^+}} \xleftarrow[78\%]{S_8/MeONa} \underset{(\mathbf{12c})}{\mathrm{TrSOH}} \xrightarrow[88\%]{S_8/Et_3N} \underset{\substack{\|\\ \mathrm{O}\\ (\mathbf{10c})\\ R' = Et}}{\mathrm{TrSS^-H\overset{+}{N}Et_3}} \qquad (3)$$

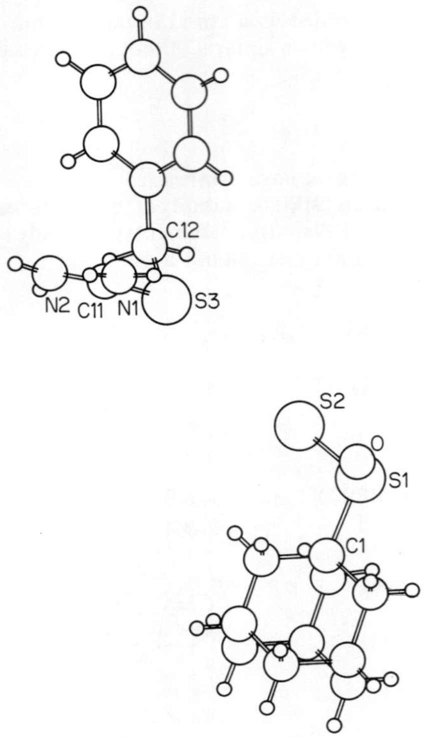


FIGURE 1. X-Ray crystal structure of **9b**

presence of bases afforded the corresponding salts of triptycenethiosulphinic acid (**10c**, **14c**) (equation 3).

### B. Reactions[1]

Several reactions of the triethylammonium salts **10** were investigated. Treatment with hydrogen chloride yielded the corresponding symmetrical thiosulphinate (**13**) (equation 4). The proposed unstable intermediates are the free thiosulphinic acid (**1**) and sulphenic acid (**12**) presumably via desulphurization. In the case of R = Tr, no self-condensation of **12** occurred and **12c** was isolated. Reaction of **10a**–**b** with triphenylphosphine as a desulphurization agent similarly gave **13a**–**b**, but the reaction mechanism was not described (equation 4). **10a**–**c** reacted with thionyl chloride to give the sulphinyl derivatives **14a**–**b**, which showed a strong S=O absorption band at $1059\,cm^{-1}$ attributable to their thiolo structure. The reactivity of the salts is characteristic of an exclusive nucleophilic attack involving the thiolate sulphur among the three possible nucleophilic sites (two sulphur and one oxygen atoms) in the reaction with electrophiles. This is unlike sulphinic acid in which the central sulphur atom is more nucleophilic than the two oxygen atoms (see the chapter by Okuyama in this volume).

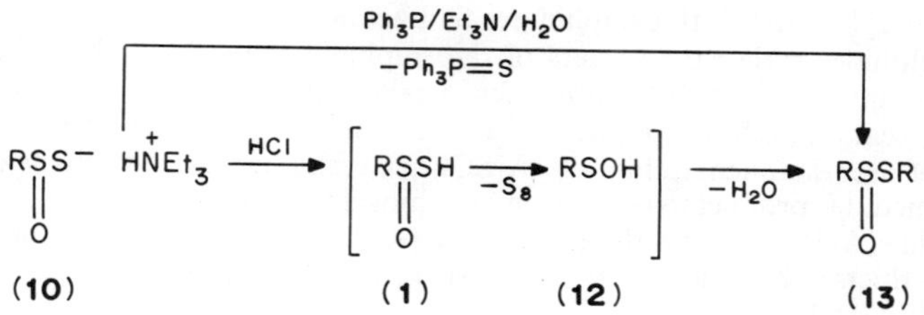

(4)

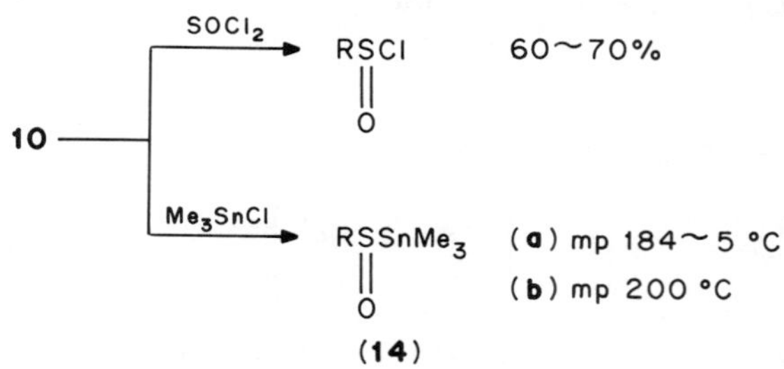

(5)

Thus, by the pioneering work of Mikołajczyk and his coworkers the chemistry of the salts of thiosulphinic acids has been clarified. The chemistry of free thiosulphinic acids will also be investigated.

## III. THIOSULPHINIC S-ESTERS (THIOSULPHINATES)

### A. Structure and Spectroscopic Characteristics

Thiosulphinic S-ester can be regarded as one of the thio derivatives of sulphinic ester whereas it may also be considered as disulphide monooxide. As described later, direct oxidation of an organic disulphide usually yields the corresponding thiosulphinic S-ester, or thiosulphinate. Since no report had appeared on free thiosulphinic acids and their salts

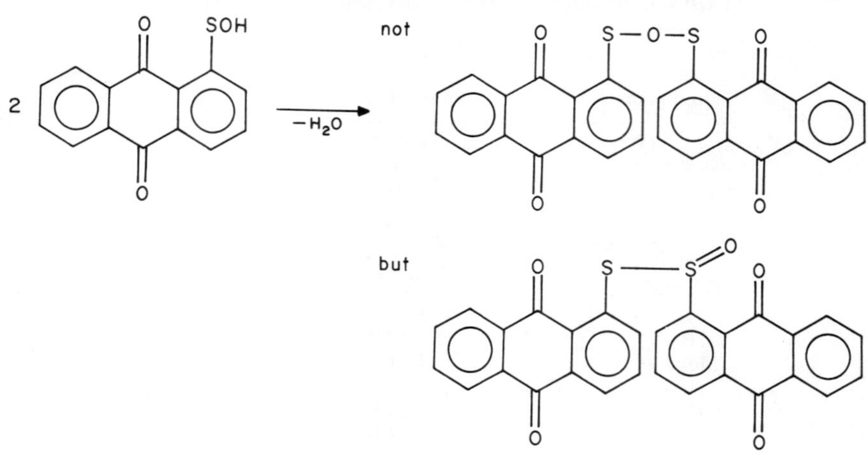

(6)

until Mikołajczyk's study[1], thiosulphinates had actually been treated as monooxides of organic disulphides rather than esters of thiosulphinic acids. There have been several fragmentary essay articles[24–31] but only a few detailed reviews[32,33] on the thiosulphinates.

Over fifty years ago, before Cavallito[4] isolated allicin (**5**), it had been reported that hydrolysis of arenesulphenyl chloride (ArSCl) gave sulphenic acid anhydride (**3**)[34–36]. Others assumed this product to be the isomeric thiosulphinate (**2**) or an equimolar mixture of a disulphide (ArSSAr) and a thiosulphonate ($ArSO_2SAr$)[37], which are the disproportionation products of **2** (equation 6)[46]. The structure of **2** was later proved[38–45] by IR[40,41] and NMR[42,43] studies.

Several naturally occurring thiosulphinates are known. Besides allicin (**5**), a few cyclic derivatives such as *β*-lipoic acid were reported[7,8,47].

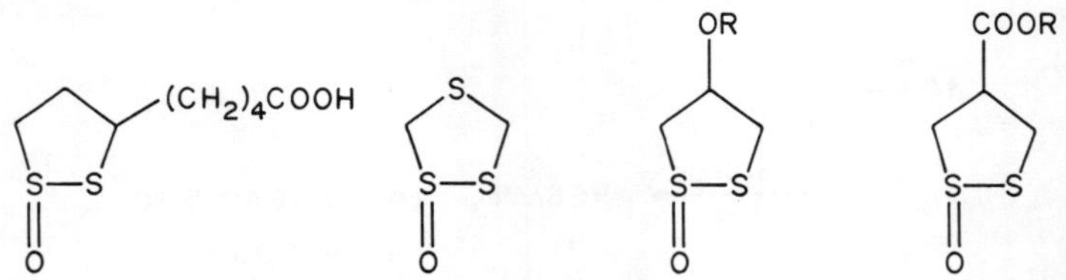


Refs. 47a, 47d     Ref. 47e     (R = H, Ac) Ref. 8     (R = H, Me) Refs. 7, 47f

### 1. *IR spectra*

In the IR spectra of thiosulphinates there appears a strong absorption band by $\nu_{S=O}$ around 1100 cm$^{-1}$. This band is located between those of sulphoxides (1055 cm$^{-1}$) and sulphinic esters (1130 cm$^{-1}$). This IR absorption may be accounted for by inductive effects rather than by a simple resonance effect. In fact, Ghersethi and Modena[41] reported that substitution of either or both methyl groups by a phenyl group in dimethyl thiosulphinate causes a blue shift of $\nu_{S=O}$ (Table 1), and this is the case also for *p*-substituted phenyl

TABLE 1. IR absorption of $\nu_{S=O}$ of selected thiosulphinates

| Thiosulphinate | | $\nu_{S=O}$ in $CCl_4$ | $\nu_{S=O}$ in $CHCl_3$ |
|---|---|---|---|
| MeS—SOMe | | 1196.5 | 1075 |
| PhS—SOMe | | 1100.5 | 1075 |
| MeS—SOPh | | 1104.0 | 1088 |
| 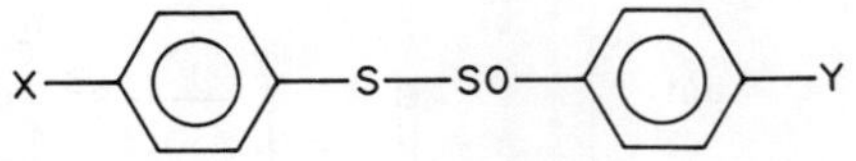 | | | |
| X = H | Y = H | 1107.5 | 1091 |
| H | Me | 1104.0 | 1092 |
| Me | H | 1106.0 | 1090 |
| H | OMe | 1098.5 | 1090 |
| H | Br | 1111.0 | 1093 |
| $NO_2$ | H | 1110.0 | 1104 |
| H | $NO_2$ | 1114.0 | 1098 |

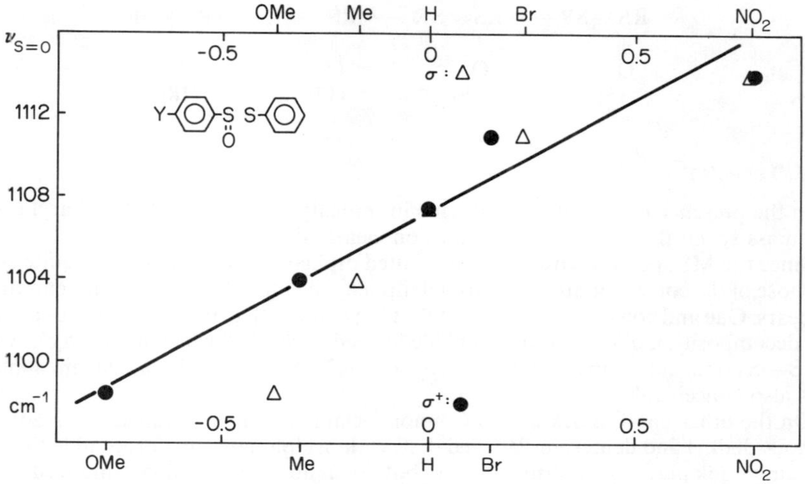


FIGURE 2. Substituent effect in IR absorption of $\nu_{S=O}$ of *S*-phenyl *p*-substituted benzenethiosulphinate

benzenethiosulphinates. As shown in Figure 2, $\nu_{S=O}$ correlates with Brown's $\sigma^+$ values much better than with Hammett's $\sigma$ values (Figure 2). $\nu_{S=O}$ values $CHCl_3$ are always smaller by 5–21 $cm^{-1}$ than those in $CCl_4$[46b].

## 2. UV spectra[48,49]

According to a detailed study by Backer and Kloosterziel[48], aliphatic thiosulphinates have two maximum absorptions around 210 nm (ε 2500) and 260 nm (ε 2050) while aromatic ones show them around 226 nm (ε 16500) and 294 nm (ε 6400) in hexane solution (Table 2). Since the absorption which appeared in the longer-wavelength region is not found in sulphoxides, it is probably based on the —SO—S— group. A small red shift (10–14 nm) is observed in alcohol solution. The resonance structures **15**–**18** were considered for the explanation of the UV spectra. When IR data are considered, the contributions of the forms **17** and **18** may be the dominant ones among the possible forms **15**–**18**[32].

TABLE 2. UV absorption of selected thiosulphinates

| Thiosulphinate | Maximum absorption λ(nm) (ε) in hexane | | in ethanol | |
|---|---|---|---|---|
| MeS—SOMe | 215 (2400) | 260 (2045) | 215 (1610) | 248 (2090) |
| EtS—SOEt | 215 (2700) | 262 (2055) | 215 (1750) | 248 (2055) |
| *i*-PrS—SOPr-*i* | 215 (2900) | 261 (2080) | 215 (2075) | 248 (2090) |
| PhS—SOPh | 226 (16500) | 294 (6400) | 224 (16100) | 284 (7400) |
| *p*-TolS—SOTol-*p* | 225 (19800) | 294 (8170) | 232 (17300) | 290 (10000) |

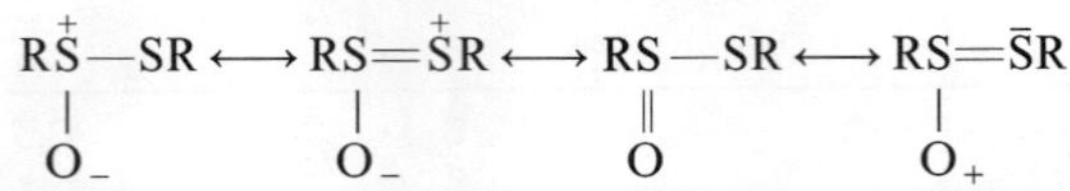

(15) (16) (17) (18)

### 3. MS spectra

In the presence of various substituents in thiosulphinates (i.e. alkyl and aryl groups) the mass spectral pattern of fragmentation clearly differs.

Since the MS spectral pattern of substituted diphenyl thiosulphinates is quite similar to those of the corresponding substituted diphenyl disulphides and since no parent peak appears, Oae and coworkers concluded that the main path of fragmentation consisted of the decomposition of the diaryl disulphide formed *in situ* and which involved cleavage of the S—S bond and elimination of the oxygen atom[50]. A minor path via thiabenzonium ion was also conceivable.

On the other hand, Block and O'Connor[51] claimed in their detailed MS study using various dialkyl and deuterium-labelled dialkyl thiosulphinates that disulphide formation certainly took place by a thermal process but not as an electron-impact-induced process. Furthermore, in the mass spectrum of ethyl ethanethiosulphinate (**19**), a parent peak ($m/z$ 138) was clearly observed and the fragmentation consisted of C—S and S—S bond cleavage via $E_i$ and homolytic processes (Scheme 1).

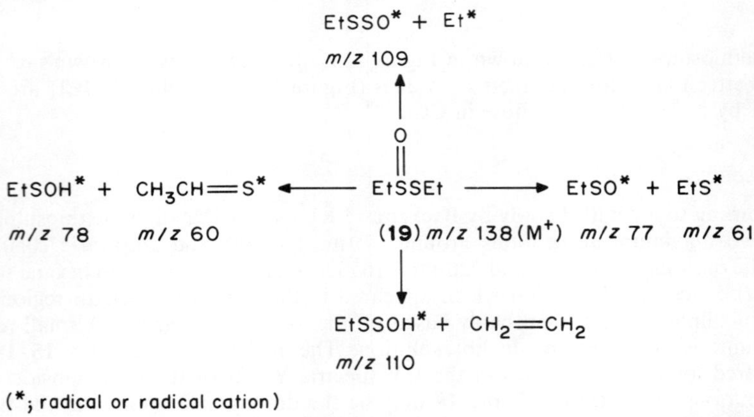


SCHEME 1

## B. Formation of Thiosulphinates

### 1. From sulphinyl chlorides and thiols

The most general and versatile method of preparing thiosulphinate esters consists of the reaction of sulphinyl chlorides and thiols in the presence of base such as a tertiaryamine[48] (equation 7). Using a metal thiolate instead of the thiol (e.g. $RS^-Na^+$) or without the amine no thiosulphinate was obtained[14,52]. This is the only method of providing unsymmetrical thiosulphinates, and there is no limitation on R and R′ groups in this method. This method is also useful for symmetrical thiosulphinates, since formation of by-products is strongly suppressed unlike in all other procedures described subsequently.

$$\underset{\underset{\displaystyle O}{\|}}{RSCl} + R'SH \xrightarrow[CCl_4,\ <0\,^{\circ}C]{\text{base}} \underset{\underset{\displaystyle O}{\|}}{RSSR'} \quad \sim 95\% \qquad (7)$$

R, R′ = alkyl, aryl, aralkyl

It should be noted that the addition of thiol into a mixture of sulphinyl chloride and a tertiaryamine leads to successful formation of RS(O)SR′, but not the inverted sequence of addition, since thiosulphinate is sensitive toward nucleophiles like thiolate anion or even thiol, both of which can easily react to give a disulfide (RSSR′)[6,52] (see Section III.D.6.a below).

An analogous reaction with sulphinic acid and triphenylphosphine in the presence of N-chlorosuccinimide was reported (equation 8)[53,54].

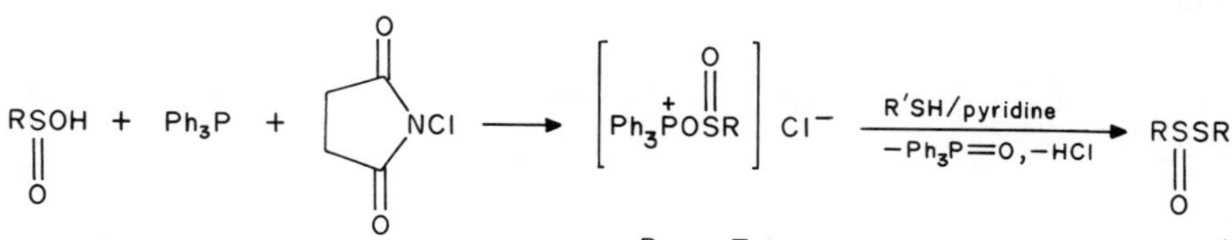


R = *p*-Tol (8)

R′= Bu(80%), $Ph_3C$(71%), Ph(35%)

Recently, it has been shown that the reaction of benzenesulphinyl azide with thiols yields benzenethiosulphinates (equation 9)[55].

$$\underset{\underset{\displaystyle O}{\|}}{PhSN_3} + RSH \xrightarrow[-20\,^{\circ}C]{} \underset{\underset{\displaystyle O}{\|}}{PhSSR} + HN_3 \qquad (9)$$

(R = Ph, $PhCH_2$, Pr, *i*-Pr, *t*-Bu, $CH_2Bu$-*t*, $CH_2CH_2OH$)

### 2. *By oxidation of disulphides*

*a. Peroxy acid oxidation.* The oxidation of disulphides[24,30] is generally used for the preparation of symmetrical thiosulphinates, since it is not selective in the oxidation of one of the two different sulphur atoms of unsymmetrical disulphides. In general it is difficult to obtain aryl arenethiosulphinates in high yields by direct oxidation of the corresponding diaryl disulphides, and therefore, synthesis by reaction of sulphinyl chlorides with thiols is the recommended procedure. On the other hand, alkyl alkanethiosulphinates are easily obtainable from dialkyl disulphides (equation 10), and can be purified by distillation (~63% yield)[6], since they are thermally relatively stable.

$$RSSR \xrightarrow[CHCl_3,\ \text{r.t.}]{\text{perbenzoic acid}} \underset{\underset{\displaystyle O}{\|}}{RSSR} \quad 20 \sim 63\% \qquad (10)$$

(R = Me, Et, Pr, *i*-Pr, Bu, Amyl)

Peracid oxidation of ethyl *t*-butyl disulphide had been reported to give selectively S-*t*-butyl ethanethiosulphinate[6,14], but the product was later shown to be a mixture of two possible thiosulphinates[51]. However, selective oxidation of an unsymmetrical disulfide is possible when the difference between the two substituents is sufficiently large (equation 11)[49,56,57], and thus methyl phenyl disulphide (**20**) gave phenyl methanethiosulphinate in more than 60% yield. The difference is probably due to electronic effects, because the selectivity is known to be influenced by the electron density on the sulphur atom, and Table 3 shows that the electron-donating ability rather than the steric effect of substituent[58,59] is the dominant factor. While usually the more electron-rich sulphur atom is preferentially oxidized, a very bulky substituent at sulphur is sometimes able to change the direction of the oxidation. The results are well consistent with a kinetic study on the oxidation which follows overall third-order kinetics, first order in disulphide and second order in peroxy acid[58]. Thus, the oxidation involves an initial nucleophilic attack of a lone pair of electrons of the disulphide at the peroxy acid oxygen. A veriety of alkyl, aryl, alicyclic and heterocyclic thiosulphinates were synthesized by this oxidation method[9,57,60–64].

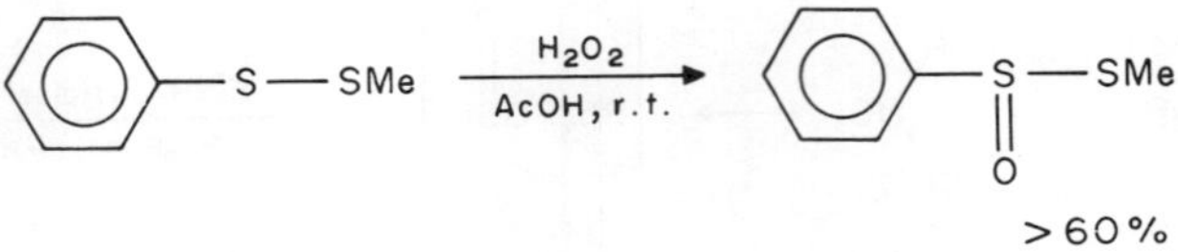


(11)

(**20**) (**21**)

Since excessive oxidation and acid-catalyzed side-reactions (yielding thiosulphonates, sulphinic and sulphonic acids)[57,64,65] often occur during the course of the oxidation of disulphide (equation 12)[65] and are unavoidable, the isolation and purification process is accompanied by special difficulties.

$$\text{PrSSPr} \xrightarrow[\text{AcOH}]{H_2O_2} \text{Pr}\overset{\overset{\text{O}}{\|}}{\text{S}}\text{SPr} + \text{Pr}\overset{\overset{\text{O}}{\|}}{\underset{\underset{\text{O}}{\|}}{\text{S}}}\text{SPr} + \text{PrSO}_2\text{H} + \text{PrSO}_3\text{H} \qquad (12)$$

However, this oxidation procedure is an excellent method to obtain cyclic thiosulphinates[49,66,67] which, with one exception[14], can only be prepared by the oxidation of the corresponding cyclic disulphides (equation 13)[66,67]. The reason may be the unusual stability or low reactivity of the cyclic thiosulphinates. In the oxidation of 3-methyl-1, 2-dithian (**22b**) Isenberg and Herbrandson could determine neither the product ratio nor the nature of isomers (diastereomers or regioisomers) by NMR[66]. Oae and Takata later estimated the product ratio (**23b**:**24b**) to be 50:50 from $^{13}$C NMR and finally the isomers were isolated by chromatography, and were shown to be regioisomers and not

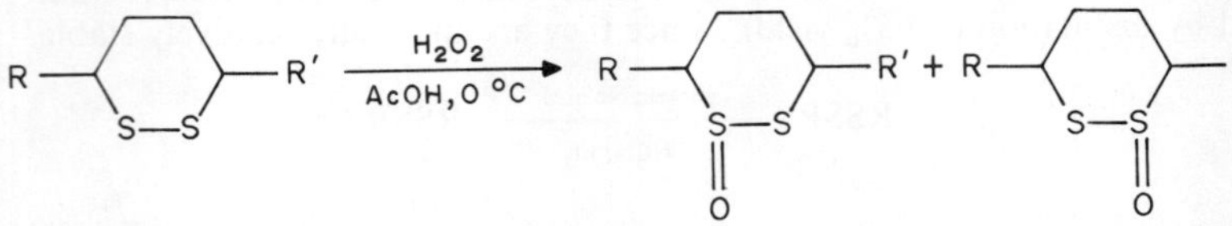


(13)

(**22**) (**23**) (**24**)

(**a**) R = R′ = H, (**b**) R = H, R′ = Me, (**c**) R = R′ = Me

TABLE 3. Oxidation products of disulphides

| Disulphide | Oxidant[a] | Product |
|---|---|---|
| 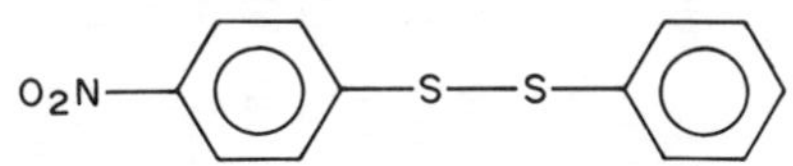 | AcOOH | 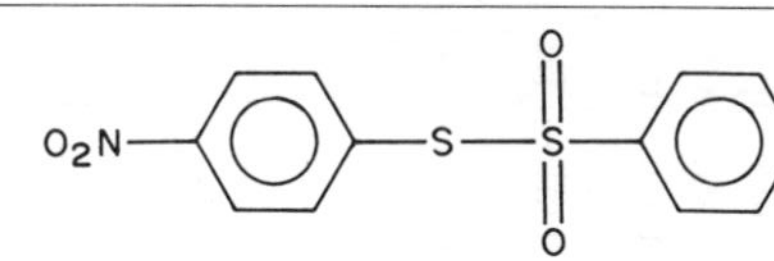 |
| 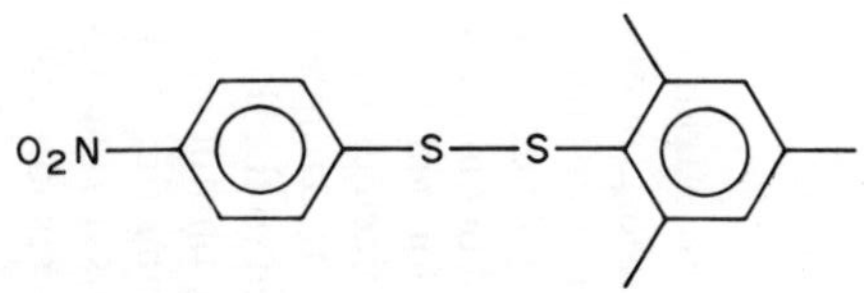 | AcOOH | 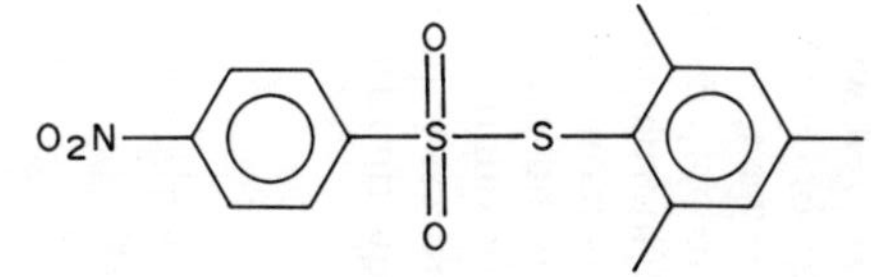 |
| MeS—SBu-*t* | AcOOH or MCBPA/$CHCl_3$ | MeS(=O)—SBu-*t*  MeS—S(=O)Bu-*t* |
| | | 1:2 |
| MeS—SBu-*t* | $h\nu/O_2$/MB/MeOH (0.1 M) | 2:1 |
| MeS—SBu-*t* | (0.25 M) | 5:1 |
| EtS—SBu-*t* | AcOOH or MCPBA | 1:1.74 |

[a]MB denotes methylene blue; MCPBA denotes *m*-chloroperbenzoic acid

diastereomers[67]. Oxidation of a diastereomeric mixture of 3,6-dimethyl-1,2-dithian (**22c**) yielded products consisting of a mixture of at least three stereoisomers[66].

*b. Photooxidation.* Disulphides are oxidized with singlet oxygen ($^1\Delta_gO_2$) to their monooxides sometimes along with corresponding *S,S*-dioxides[68–75]. $^1O_2$ is generated also by photosensitization of molecular oxygen, or by triphenylphosphite ozonide, and converts disulphides to thiosulphinates accompanied by small amounts of thiosulphonates (equations 14 and 15)[71]. Tetraphenylporphinato zinc-sensitized photooxidation of dithiolane resulted in the formation of its monoxide along with its polymer[68,69]. It is very interesting that not only dimethyl and diethyl disulphides but also di-*t*-butyl disulphide (**25**) undergoes smoothly photooxidation to give 75% yield of the corresponding thiosulphinate (**26**)[73], although the process requires considerably drastic conditions in order to obtain the monooxide **26** (equation 16)[9].

$$RSSR \xrightarrow{^{1}O_2} R\underset{\underset{O}{\|}}{S}SR + R\overset{\overset{O}{\|}}{\underset{\underset{O}{\|}}{S}}SR \qquad (14)$$

(R = Me, Et, *i*-Pr, *t*-Bu)

$h\nu/MB/O_2/MeOH/r.t$ ~75% ~13%

$(PhO)_3PO_3/CH_2Cl_2/-30 \sim -50\,^{\circ}C$ 48–100% ~5%

MB = Methylene blue

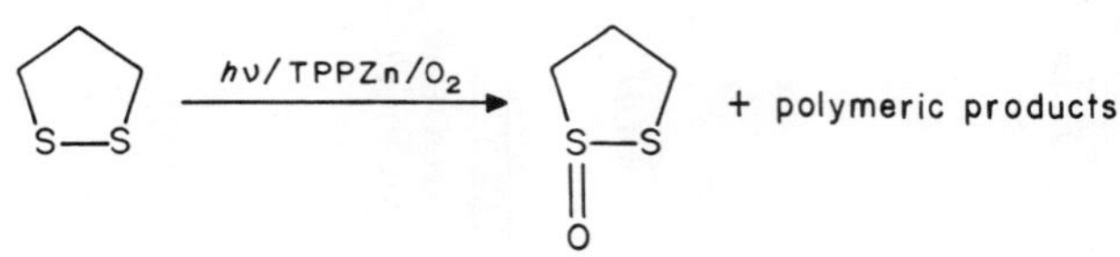


(15)

TPPZn = Tetraphenylporphinato zinc(II)

$$t\text{-}BuSSBu\text{-}t \xrightarrow[AcOH,\ 60\,^{\circ}C]{10\,eq.\ H_2O_2} t\text{-}Bu\underset{\underset{O}{\|}}{S}SBu\text{-}t \qquad (16)$$

(**25**) (**26**)

Cystine and its derivatives were also photooxidized but the oxidation was considerably suppressed, possibly by the free amino groups quenching singlet oxygen[76].

The mechanism of the photooxidation of disulphides resembles that of sulphides as reported by Foote and coworkers[77] and by Ando and Takata[78]. In the first stage, about 0.5 mol of molecular oxygen was absorbed per mol of disulphide present. As indicated in Scheme 2, the peroxy intermediate **28** formed initially with $^1O_2$ would react with another molecule of **27** to give two molecules of the thiosulphinate **19**. Additional evidence supporting this mechanism is that diphenyl disulphide, which is inactive toward photooxidation, could be oxidized to the thiosulphinate **30** along with **19**, when an equimolar mixture of **27** and **29** was photooxidized. This result also supports the absence of any S—S bond cleavage during the oxidation.

Photooxidations of lipoic acid[70] and its derivatives[68,69,74] were studied. Lipoic acid was converted to *β*-lipoic acid, i.e. the corresponding thiosulphinate[70]. The products of

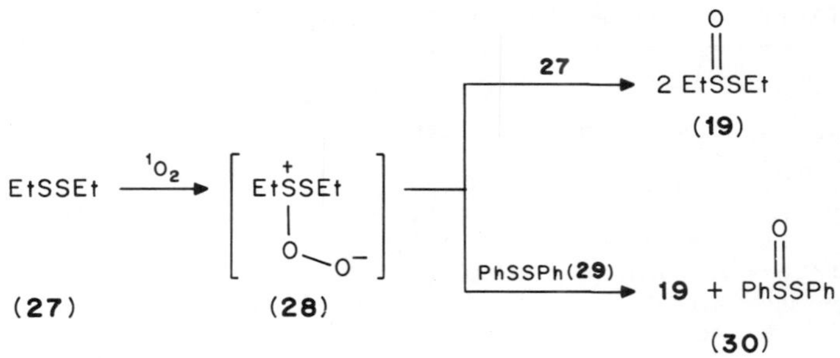


SCHEME 2

the methylene blue sensitized photooxidation consisted of four isomeric monooxides (**32**) and two regioisomeric dioxides (**33**) (Scheme 3)[74]. The product ratios were compared with those obtained by the oxidation with various oxidation systems (Table 4). In the photooxidation, more thiosulphonates were produced in aprotic solvents such as $CH_2Cl_2$. This is explained by the initially formed peroxy intermediate **34**, which is controlled profoundly by solvent polarity in accordance with the results obtained in sulphide photooxidation[77,78]. Since the thiosulphinates fromed by photooxidation are no longer active toward $^1O_2$, the formation of **33** may be attributed to intramolecular rearrangement of the intermediate **34**[77,78].

Block and O'Connor compared the regioselectivity of peroxy acid oxidation with that of photooxidation of methyl and ethyl *t*-butyl disulphides (**35**) and examined the ratio of regioisomeric thiosulphinates (**37**, **38**)[51]. In the peroxy acid oxidation the more electron-rich sulphur attached to *t*-butyl group is predominantly oxidized (Table 3) while the ratios in photooxidation appeared to be mainly influenced by steric hindrance. Probably, initial reaction of the disulphide with the small singlet oxygen molecule takes place with little regioselectivity, but in the subsequent intermolecular reaction of the peroxy intermediates

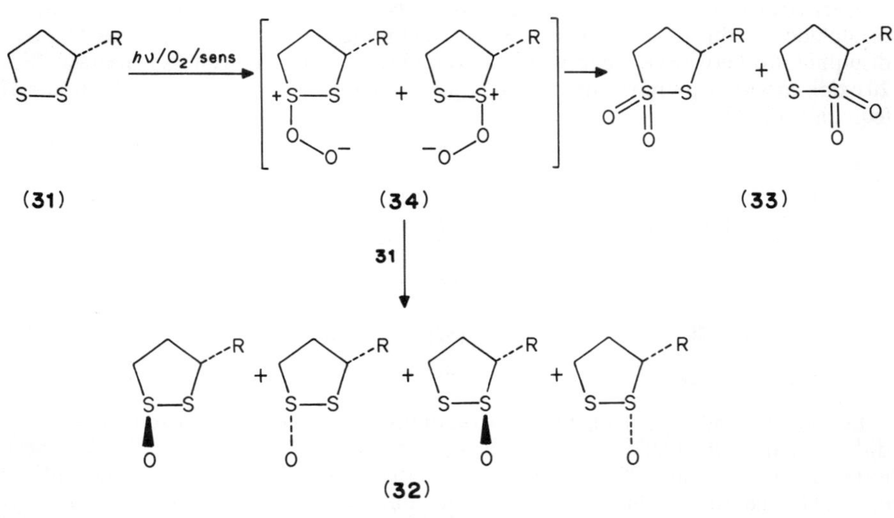


SCHEME 3

TABLE 4. Photooxidation of lipoic acid derivative **31**

| Run | Oxidation system | Solvent | Yield of product (%) Monooxides **32** | Yield of product (%) Dioxides **33** |
|---|---|---|---|---|
| 1[a] | $h\nu/O_2$/sens | $CHCl_3$ | 64 | 26 |
| 2[a] | $h\nu/O_2$/sens | MeOH | 75 | 15 |
| 3 | $(NH_4)_2S_2O_8$ | 90% EtOH | 21 | trace |
| 4 | *t*-BuOOH | MeOH | 69 | trace |
| 5[a] | AcOOH | $Et_2O$ | 42 | — |
| 6 | AcOOH | MeOH | 54 | trace |
| 7 | $(PhO)_3PO_3$ | $CH_2Cl_2$ | 26 | trace |

[a]All four monooxides and two dioxides were obtained.

**36** and **36′** with a second disulphide molecule **35** the less-hindered sulphur atom should be attacked preferentially, so that more **37** was produced than **38** (equation 17).

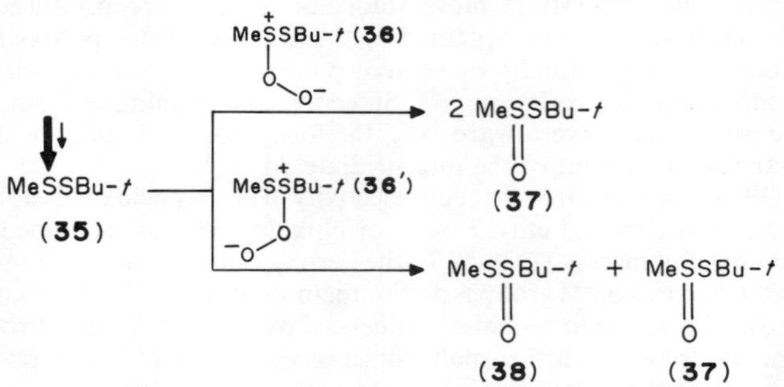


(17)

*c. Miscellaneous oxidations.* Inorganic oxidants such as sodium metaperiodate convert disulphides to thiosulphinates. However, usually these reactions are limited to cyclic disulphides, otherwise cleavage of the S—S bond takes place. For example, the disulphide **20** oxidized with two equivalents of $NaIO_4$ gave a mixture of thiosulphonates **39**, **40** and **41** (equation 18)[24,67].

$$\underset{(20)}{\text{PhSSMe}} \xrightarrow[\text{dioxane–H}_2\text{O, r.t}]{\text{2 eq. NaIO}_4} \underset{(39)\ 27\%}{\text{PhS(=O)}_2\text{SPh}} + \underset{(40)\ 25\%}{\text{MeS(=O)}_2\text{SMe}} + \underset{(41)\ 31\%}{\text{PhS(=O)}_2\text{SMe}} \tag{18}$$

Periodate and permanganate oxidations afforded monooxides of 1,8-dithiaacenaphthene (**42**)[82] and of 4,4-bis(hydroxymethyl)-1,2-dithiolane (**44**)[80], respectively (equations 19 and 20). In the latter case, the corresponding 1,1-dioxide could be obtained in neutral conditions. **42** was also oxidized to **43** with Fenton's system ($TiCl_3$–$H_2O_2$) which generates hydroxyl radicals[83].

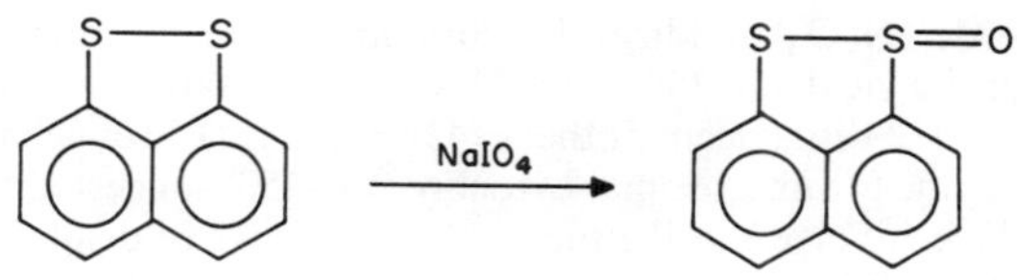


**(42)** **(43)** (19)

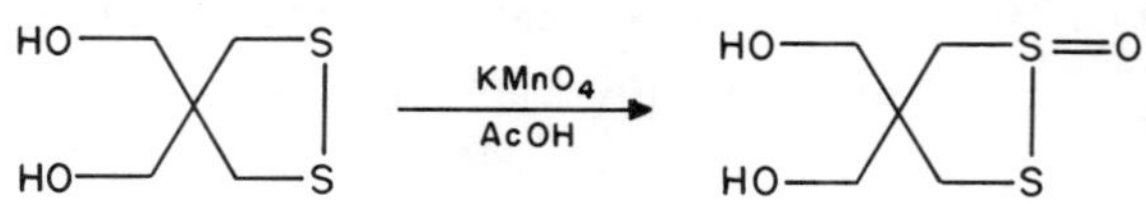


**(44)** **(45)** (20)

Periodate oxidation of 1, 2-dithian derivatives were reported by Field and Khim, and the reaction was strongly affected by the solvent system (equation 21)[84].

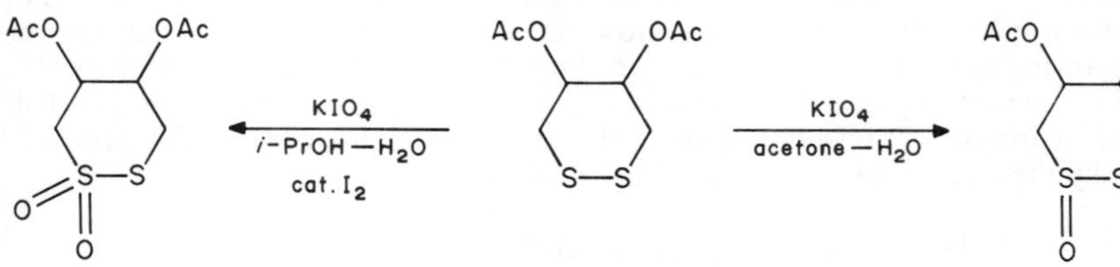


(21)

The oxidation of 1, 2-ditiolane and its derivatives with ammonium persulphate[69,79–81] in ethanol–water proceeds via initial one-electron transfer from the disulphide to the oxidant (equation 22), and follows second-order kinetics.

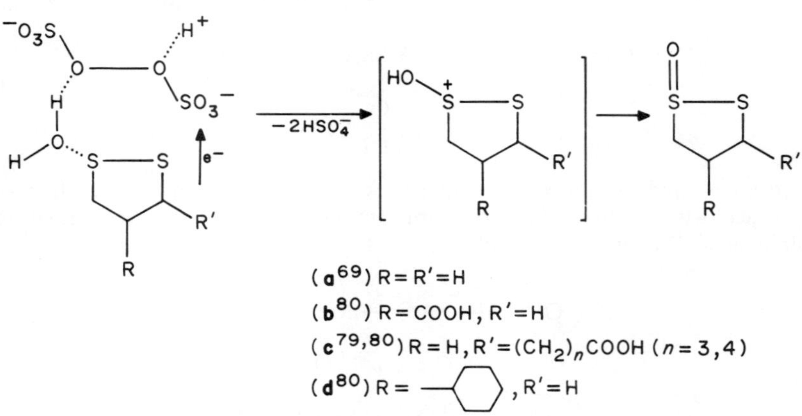


(22)

Although persulphate oxidation can be performed in an aqueous solvent, Lindberg and Bergson concluded that for the easy isolation of the thiosulphinate, hydrogen peroxide is the most suitable oxidant, and other oxidants often form undesirable by-products which make the separation difficult[80]. Dithiolanes were more easily oxidized than any other class of saturated disulphides[69,81].

Dinitrogen tetroxide ($N_2O_4$) oxidizes 1,2-dithian **22a** to its monooxide in carbon tetrachloride though the yield was low (17%)[85], while nitrosation of aromatic ring took place in the reaction of 1,2-dithiaacenaphthene (**42**)[83]. Enzymatic oxygenation of **22a** with cytochrom P-450 enzyme (or microsomal cytochrom P-450) suggested that formation of the thiosulphinate **23** is involved in the metabolic pathways of disulphides[86].

A mild oxidation procedure with 2-arenesulphonyl-3-arylazyridines has been applied to the oxidation of disulphides to thiosulphinates[87].

As a possible preparative method for thiosulphinates, reactions of salts of thiosulfinic acids[1] (see equation 1) with electrophiles such as alkyl halides may be an effective synthetic procedure (equation 23), although no report has appeared at the present time.

$$\mathrm{RSS^-X^+ + R'Y \xrightarrow[-XY]{} RS(=O)SR'} \tag{23}$$

*3. By reaction of sulphenic acids and derivatives*

Hydrolysis of aromatic sulphenyl chlorides (**46**) gives diaryl thiosulphinates (**47**) (equation 24)[38,88,89]. Benzenesulphenyl chloride was hydrolyzed to the corresponding thiosulphinate (Ar = Ph) quantitatively. However, the labile thiosulphinate may decompose, e.g. owing to the presence of hydrogen chloride formed during the hydrolysis. The reaction is believed to proceed via the unstable sulphenic acid. (see also Section III.B.4). From the hydrolysis of aryl sulphenamide (**48**), the thiosulphinate (**47**) is also produced possibly by the intermediate sulphenic acid[90,91] (equation 25).

$$\underset{(\mathbf{46})}{\mathrm{ArSCl}} + \mathrm{H_2O} \xrightarrow[-\mathrm{HCl}]{} [\mathrm{ArSOH}] \xrightarrow[-\mathrm{H_2O}]{} \underset{\substack{(\mathbf{47})\\ 80\%\ (\mathrm{Ar}=p\text{-Tol})}}{\tfrac{1}{2}\mathrm{ArS(=O)SAR}} \tag{24}$$

$$\underset{(\mathbf{48})}{\mathrm{2ArSNC_5H_{10}}} \xrightarrow{\mathrm{H_2O}} \underset{(\mathbf{47})}{\mathrm{ArS(=O)SAr}} + \mathrm{2HNC_5H_{10}} \tag{25}$$

Dehydrative condensation of two molecules of sulphenic acids, or reaction of a sulphenic acid with a sulphenyl chloride, are conveivable pathways for the formation of thiosulphinate[36,38] (equations 26 and 27).

$$\mathrm{ArS{-}OH + H{-}S(=O)Ar \longrightarrow ArS(=O)SAr + H_2O} \tag{26}$$

$$\mathrm{ArS{-}Cl + H{-}S(=O)Ar \longrightarrow ArS(=O)SAr + HCl} \tag{27}$$

Thus, sulphenic acids[92,93] play an important role in the chemistry of thiosulphinates, and have been suggested to be formed in various reactions by Kharasch[46] and Bruice[94]. Although most sulphenic acids are too unstable to be isolated, several stable ones have been reported, such as anthraquinone 1-sulphenic acid[95], 1,4- and 1,5-disulphenic acids[96,97] and others[98–106], some of which are reported to give thiosulphinates.

*4. By miscellaneous methods*

Thiosulphinates can be formed by decomposition of sulphoxides having $\beta$-protons[107–109]. Shelton and Davis reported that di-*t*-butyl sulphoxide afforded the thiosulphinate **52** with elimination of isobutene and water at 75 °C (equations 28 and 29)[107]. They obtained spectroscopic evidence for the intermediate *t*-butanesulphenic acid by NMR and for its tautomeric structures by IR. From the first-order rate constant, the Ei process involving cyclic transition state was postulated for this decomposition (equation 30). *t*-Butanesulphenic acid could be isolated as adducts with $\alpha,\beta$-unsaturated ketones and acetylenes.

$$t\text{-Bu}\overset{\overset{\displaystyle O}{\|}}{S}\text{Bu-}t \xrightarrow{\Delta(75\,^{\circ}C)} [t\text{-BuSOH}] + CH_2{=}C(CH_3)_2 \qquad (28)$$

**(49)** **(50)** **(51)**

$$2\ \mathbf{50} \xrightarrow[-H_2O]{} t\text{-Bu}\overset{\overset{\displaystyle O}{\|}}{S}\text{SBu-}t \qquad (29)$$

**(52)**

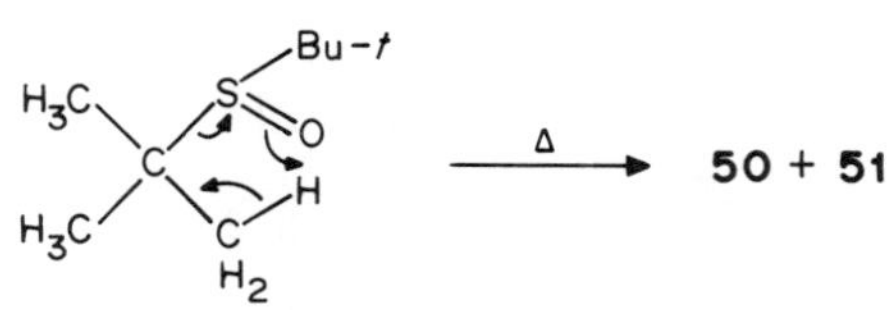


(30)

**(49)**

Stoll[110–112] studied the formation of allicin (**5**) by enzymatic decomposition of alline [(+)S-allyl-L-cysteine sulfoxide, **53**) by allinase leading to 3-propenesulphenic acid (**54**) as intermediate through a similar cyclic transition state (Scheme 4).

Dodson and coworkers[14] obtained cyclic thiosulphinates (**56**) by the reaction of butadienes with disulphur monooxide (equation 31), presumably by a Diels–Alder-type cycloaddition, but the yield was only 4%.

Kondo, Negishi and Ojima found that sulphuric acid-catalyzed ring-opening of ethylene episulphoxide (**57**) in methanol afforded a thiosulphinate, **58** (equation 32)[113]. In the acetic-acid-catalyzed reaction, in turn, a mixture of disulphide and thiosulphonate was formed owing to the disproportionation of **58** formed initially. They studied the stereochemistry of this reaction using 2-butene episulphoxide and stilbene episulphoxide, and concluded that the nucleophile was introduced stereospecifically with inversion of configuration as formulated in equation 33.

Perfluoromethyl thiosulphinate (**59**) has been prepared from trifluoromethanesulphenyl chloride and mercury difluoride according to equation 34[114]. It is unusually stable and shows no decomposition even under conditions where common thiosulphinates undergo rapid disproportionation.

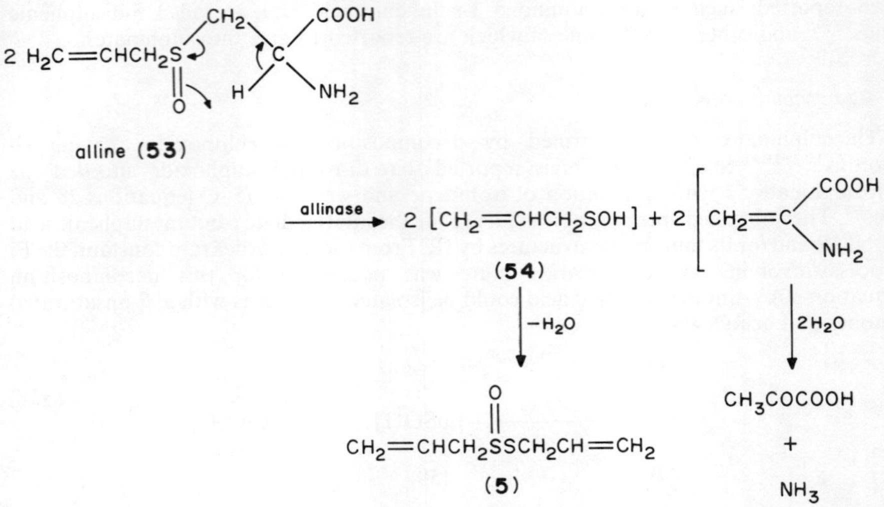


SCHEME 4

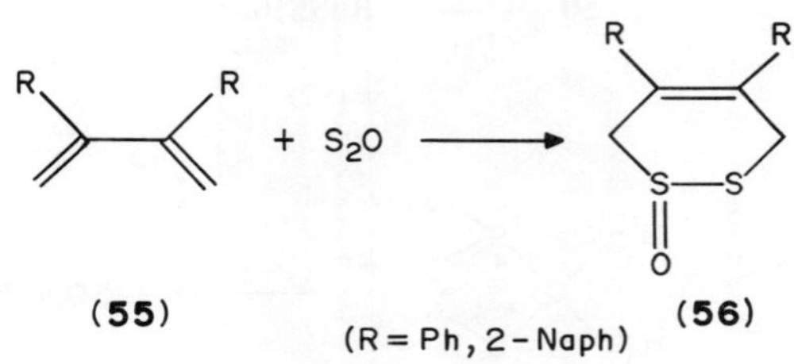


(31)

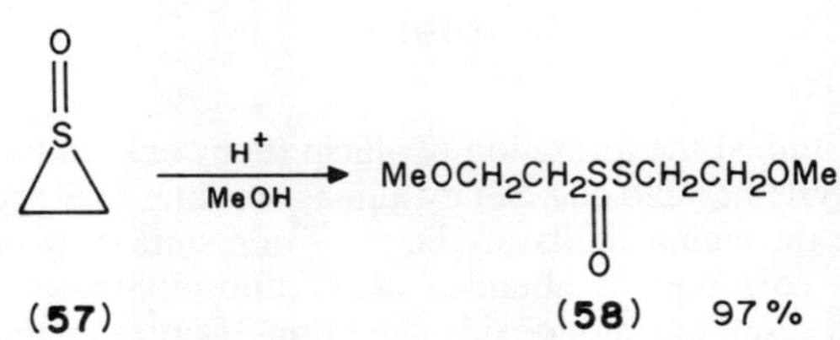


(32)

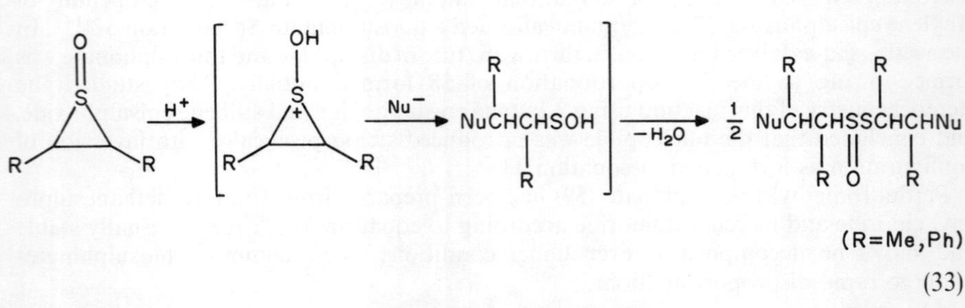


(33)

$$2CF_3SCl + HgF_2 \xrightarrow[-HgCl_2]{} CF_3SF_2SCF_3 \xrightarrow[-50\,^{\circ}C,\ -2HF]{H_2O} CF_3\overset{O}{\overset{\|}{S}}SCF_3 \qquad (34)$$

**(59)**

*5. Synthesis of optically active and oxygen-18 labelled thiosulphinates*

*a. Optically active thiosulphinates.* Optically active oxygen-containing organosulphur compounds are very convenient for study[21,22]. Thiosulphinates have a tricoordinated sulphur atom and may be optically active. Thus, Savige and collaborators separated the first optically active thiosulphinate, i.e. diastereomeric cystine S-monooxides, which had a chiral pyramidal configuration[61].

$$Cy{-}\overset{O}{\overset{\|}{S}}{-}S{-}Cy \qquad Cy = (L){-}CH_2\underset{NH_2}{\underset{|}{C}}HCOOH$$

$$[\alpha]_D^{20} + 62^{\circ},\ [\alpha]_D^{20} + 14^{\circ}$$

A few optically active thiosulphinates have been synthesized. Although the optical stability of thiosulphinates is generally low[115], 3-phenyl-4-benzoyl-[*d*]-1, 2-dithiolane-1-oxide is reported by Wudl and Gruber to be configurationally stable up to 116 °C without pyramidal inversion leading to racemization[116].

Oxidation of diphenyl disulphide with (+)-peroxycamphoric acid afforded optically active phenyl benzenethiosulphinate (**61**) ($[\alpha]_{436}$ + 8.5° to + 14.0°) (equation 35)[117–120]. However, no asymmetric induction took place with didodecyl and dibenzyl disulphides[119].

$$\underset{(\mathbf{60})}{PhS{-}SPh} \xrightarrow{\text{peroxycamphoric acid}} \underset{(\mathbf{61})}{Ph\overset{*}{\underset{\underset{O}{\|}}{S}}{-}SPh} \qquad O.P. \sim 5\% \qquad (35)$$

Thiosulphinates containing *t*-butyl groups have enhanced chemical and optical stabilities, and were obtained in optically active forms, by partial optical resolution of racemic mixtures via cyclodextrin inclusion complexes[120,121].

$$p\text{-}Tol\overset{*}{\underset{\underset{O}{\|}}{S}}{-}SBu\text{-}t \qquad t\text{-}Bu\overset{*}{\underset{\underset{O}{\|}}{S}}{-}SMe \qquad t\text{-}Bu\overset{*}{\underset{\underset{O}{\|}}{S}}{-}SBu\text{-}t$$

They were also prepared by the method described in equation 9, but carried out in the presence of a chiral amine. The method provides a small preponderance of one enantiomer[122]. The highest optical purity (*ca* 10%) was attained in the case of *t*-butyl toluenethiosulphinate (**62**) (equation 36), but the product underwent rapid racemization.

$$p\text{-}Tol\underset{\underset{O}{\|}}{S}Cl + t\text{-}BuSH \xrightarrow{N\overset{*}{R}_3} p\text{-}Tol\overset{*}{\underset{\underset{O}{\|}}{S}}{-}SBu\text{-}t \qquad (36)$$

**(62)**

A steroidal thiosulphinate has been obtained and separated into diastereomers by Kishi and coworkers[123].

An optically active amidothiosulphinate, **64**, was synthesized by asymmetric induction in the reaction of thionyl dimethylamide and *t*-butanethiol in the presence of a chiral isothiocyanate (**63**) (equation 37)[124].

$$O{=}S(NMe_2)_2 + t\text{-BuSH} \xrightarrow{(+)-\text{PhCH(Me)NCS, } \mathbf{63}} Me_2N{-}\overset{*}{\underset{\underset{O}{\|}}{S}}{-}SBu\text{-}t \qquad (37)$$

**(64)**

*b.* $^{18}O$ *labelled thiosulphinates.* Oae and his coworkers[129a,178] and Kice and Cleaveland[139] synthesized a few $^{18}O$ labelled thiosulphinates. In most preparative methods (equations 9 and 39), the key compound is an $^{18}O$ labelled sulphinyl chloride. Oae and coworkers[128,178] obtained the latter from a non-labelled sulphinyl chloride with labelled water, followed by reaction with thionyl chloride (equation 38). $^{18}O$-enriched sulphinyl chloride is also obtained by the reaction of a disulphide with chlorine in the presence of labelled acetic anhydride (equation 40), and Kice also converted sulphenyltrichloride to labelled sulphinyl chloride by the reaction with labelled acetic acid (equation 41). No $^{18}O$ exchange of the labelled thiosulphinates under acidic conditions (e.g. in acetic acid) was reported to take place[128,178].

$$\underset{\underset{O}{\|}}{RSCl} + H_2\bullet \longrightarrow \underset{\underset{◒}{\|}}{RS}◒H \xrightarrow{SOCl_2} \underset{\underset{◒}{\|}}{RSCl} \qquad (38)$$

$$\underset{\underset{◒}{\|}}{RSCl} + R'SH \xrightarrow{\text{base}} \underset{\underset{◒}{\|}}{RS}{-}SR' \qquad (39)$$

$$PhCH_2S{-}SCH_2Ph + 2(CH_3C\bullet)_2\bullet \xrightarrow{3Cl_2} 2PhCH_2\underset{\underset{\bullet}{\|}}{S}Cl \qquad (40)$$

$$PhSCl_3 + CH_3C\bullet_2H \longrightarrow Ph\underset{\underset{\bullet}{\|}}{S}Cl \qquad (41)$$

$\bullet = {}^{18}O$

### C. NMR Characteristics of Thiosulphinates

Thiosulphinates having the structure like **65** often have magnetically non-equivalent protons ($H^A$ and $B^B$) since they have an asymmetric centre at the sulphinyl sulphur, like sulfinates, sulfoxides and some other tricoordinated sulphur compounds[125–127]. Murray and his coworkers gave a detailed NMR analysis of the diastereotopic protons of diethyl (**65**, R = Et, R′ = Me) and diisopropyl thiosulphinates using an NMR shift reagent[126].

Takata, Kim and Oae obtained similar results with alkyl arenethiosulphinates[127]. In addition to the magnetic non-equivalence, the NMR chemical shift of protons and carbons

TABLE 5. $^{1}H$ NMR chemical shifts (α-methyl and methylene protons) of disulphides, thiosulphinates and thiosulphonates in $CDCl_3$ at 27 °C, δ (ppm)[a]

$$(R)\,C_{\delta}—C_{\gamma}—C_{\beta}—\underset{\underset{O_x}{\|}}{S}—S—C_{\alpha'}—C_{\beta'}—C_{\gamma'}—C_{\delta'}\,(R') \qquad (x = 0, 1, 2)$$

| Substituent R | R′ | Disulphide α, α′ | Thiosulphinate α, | Thiosulphinate α′ | Thiosulphonate α, | Thiosulphonate α′ |
|---|---|---|---|---|---|---|
| Ph | Me | 2.39 | 2.90 | 2.53 | 3.12 | 2.48 |
| Ph | Et | 2.71 | 3.10 | {3.13[b] 3.16 | 3.16 | 3.00 |
| Ph | Pr | 2.81 | 3.09 | {3.12[b] 3.09 | 3.14 | 2.97 |
| Ph | Bu | (2.65) | 3.11 | 3.14 | 3.18 | 2.99 |
| *p*-Tol | Me | — | 2.37 | 2.53 (2.30) | 2.40 | (2.21) |

[a]Values in parentheses were obtained in $CCl_4$.
[b]Magnetically non-equivalent protons.

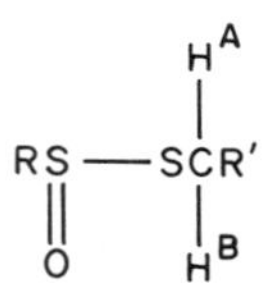


**(65)**

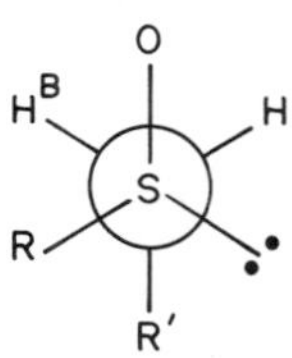


adjacent to sulphur atoms was studied in comparison with those of disulphides and thiosulphonates, using a variety of cyclic and acyclic disulphides, thiosulphinates and thiosulphonates[127,128] (Table 5). The chemical shift of α-protons reasonably shifted to low field according to the electron-withdrawing nature of neighbouring sulphur groups (—S—, —SO—, —$SO_2$—) (Table 5)[127]. A similar shift is seen in the series sulphide, sulphoxide and sulphone. However, the chemical shift of the α′-proton of the thiosulphinate always appeared at lower field than that of thiosulphonate. The relation between β- and β′-protons is approximately the same.

In the $^{13}C$ NMR spectra of these three types of compounds[127–129a], the results were different. The α′-carbon chemical shift was in the order: thiosulphonate, disulfide, thiosulphinate, although the α-carbon chemical shift showed reasonable dependence on the oxidation state of the neighbouring sulphur functions. This is also the case for the β- and β′-carbons of these compounds (Figures 3 and 4). This relation holds independently of the structure of the compounds, i.e. whether they are cyclic or acyclic; examples of cyclic derivatives are shown in Figures 3 and 4.

The phenomenon is explained by induced polarization of the C—H bond by the S=O group, via a five-membered interaction between the oxygen atom of S=O and the proton of $CH_2$ attached to the sulphenyl sulphur of the thiosulphinate. However, this type of chemical shift change was not observed in the series of oxygen analogs, i.e. sulphinates and sulphonates[127]. Therefore, a special effect of the —SO—S— moiety, namely the contribution of resonance structures (e.g. **16** or **18**), may be taken in account.

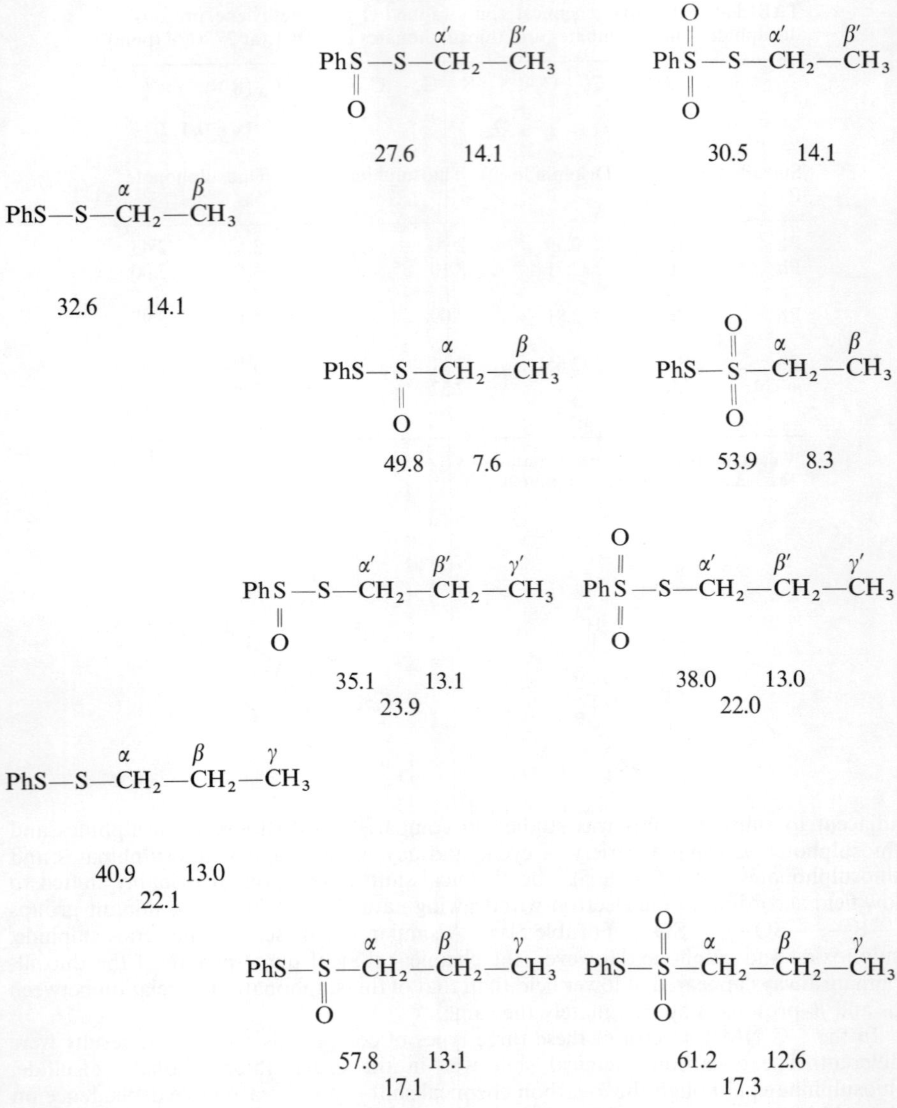

FIGURE 3. $^{13}C$ NMR chemical shifts of linear thiosulphinates ($CDCl_3$, $\delta$, TMS) at 27 °C

Furthermore, Takata and collaborators[128] measured coupling constants ($J_{C-H}$ values) of the series of ring compounds shown in Figure 5 in order to examine the acidity of the α-protons. The unusually large coupling constants of carbon-4 of the thiosulphinates and thiosulphonates suggested the contribution of the resonance structures **16** and **66**.

Through these NMR studies, Takata and coworkers concluded that cyclic unsymmetrical thiosulphinates do not show any stereoisomerism around the sulphur atom and that the oxygen of the —SO—S— group is always axially oriented[128].

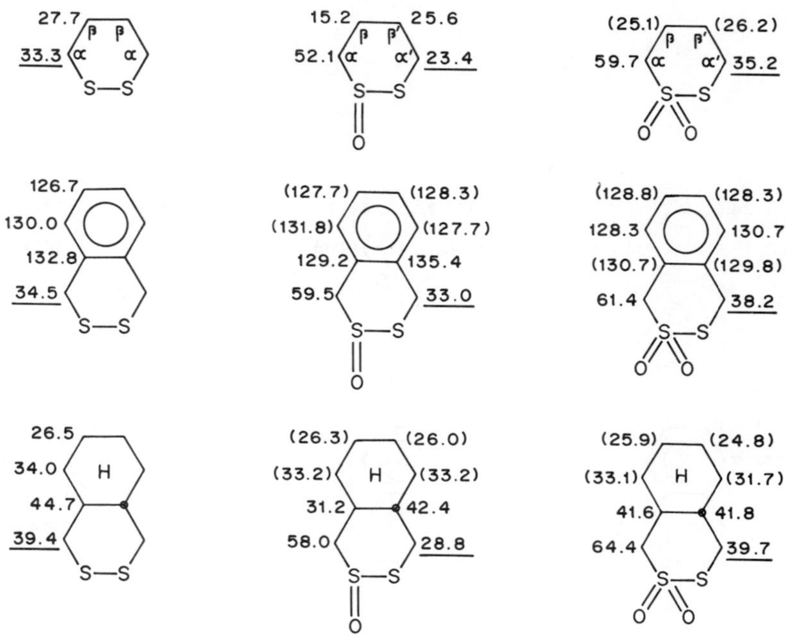


FIGURE 4. $^{13}C$ NMR chemical shifts of cyclic thiosulphinates ($CDCl_3$, $\delta$, TMS) at 27 °C

Freeman and his coworkers obtained NMR spectra of several series of organosulphur compounds including thiosulphinates mainly having neopentyl, phenyl and benzyl substituents[129b]. They have also pointed out the similar magnetic non-equivalent protons of neopentyl and benzyl derivatives. Furthermore, special deshielding electron-withdrawing and shielding effects of the thiosulphinate bond rather than the thiosulphonate bond at the $\alpha'$-position were observed in $^{1}H$ and $^{13}C$ NMR, as discussed above.

## D. Reactions of Thiosulphinates

### 1. Stability and disproportionation

As stated already, thiosulphinates are rather unstable compounds. Their S—S bond energy is unusually weak (36 kcal mol$^{-1}$ for diphenyl derivative and 46 kcal mol$^{-1}$ for dimethyl derivative) which is comparable to dialkyl peroxides and *ca* 20 kcal mol$^{-1}$ smaller than corresponding thiosulphonate, as shown in Table 6[31,51,131–133]. Therefore, the introduction of an oxygen atom into the disulphide bond leads to a bond energy decrease by 20–30 kcal mol$^{-1}$. Block and O'Connor investigated thermal stability of several alkyl thiosulphinates by measuring their half-life time (Table 7)[130]. Inspection of the data in Table 7 clearly reveals that the stability of thiosulphinates having bulky groups is enhanced as the bulk of the groups increases. This seems to indicate steric protection in terms of bulkiness against attack at the S—S bond. The same authors also studied the pyrolytic behaviour of dialkyl thiosulphinates and found the formation of intermediate alkanesulphenic and the up to then unknown alkanethiosulphoxylic acids which could be

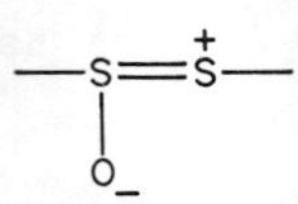


**(16)**

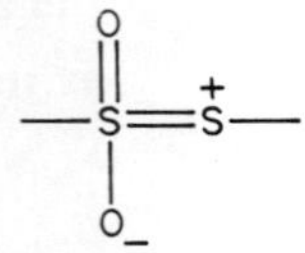


**(66)**

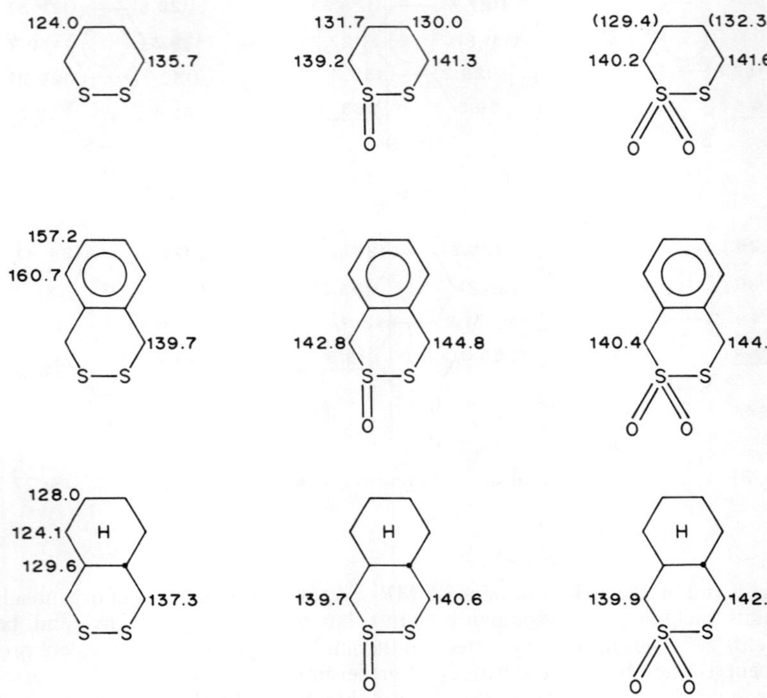


FIGURE 5. Coupling constants ($J_{C—H}$) ($CDCl_3$, Hz)

TABLE 6. Bond energy of some organosulphur compounds having S—S linkage

| Compound | Bond energy (kcal $mol^{-1}$) | References |
|---|---|---|
| HSSH | 72 | 131 |
| MeSSMe | 74, 75 | 51 |
| EtSSEt | 72 | 131 |
| PhSSPh | 55 | 132 |
| MeSOSMe | 46 | 51 |
| PhSOSPh | 36 | 133 |
| $MeSO_2SMe$ | 68 | 132 |
| HOOH | 48 | 131 |
| EtOOEt | 32 | 131 |
| *p*-TolSOSTol-*p*[a] | 34 | 31 |

[a]Activation energy ($\Delta H^{\ddagger}$ kcal $mol^{-1}$), $\Delta S^{\ddagger} = 12.0$ e.u.

TABLE 7. Relative thermal stability of alkyl alkane thiosulphinates (neat, at 96 °C)

| Thiosulphinate | Half-life time (min) |
|---|---|
| MeSO—SMe | 7 |
| MeSO—SEt | 11 |
| MeSO—SPr-*i* | 32 |
| EtSO—SMe | 40 |
| $C_{12}H_{25}$—SO—$SC_{12}H_{25}$ | 52 |
| *i*-PrSO—SPr-*i* | 66 |
| *t*-BuSO—SBu-*t* | 148 |

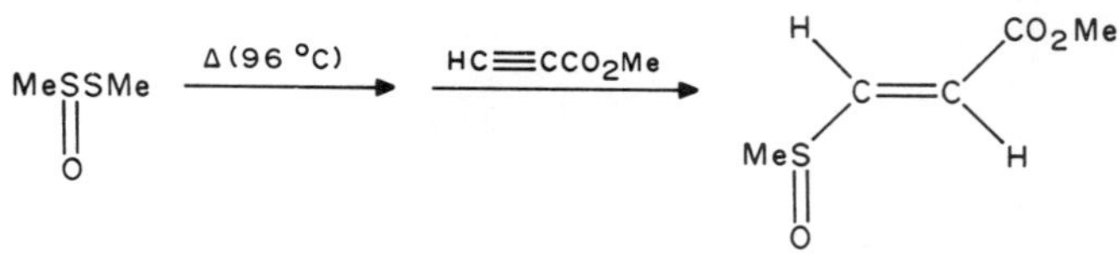

(42)

**(67)** 65%

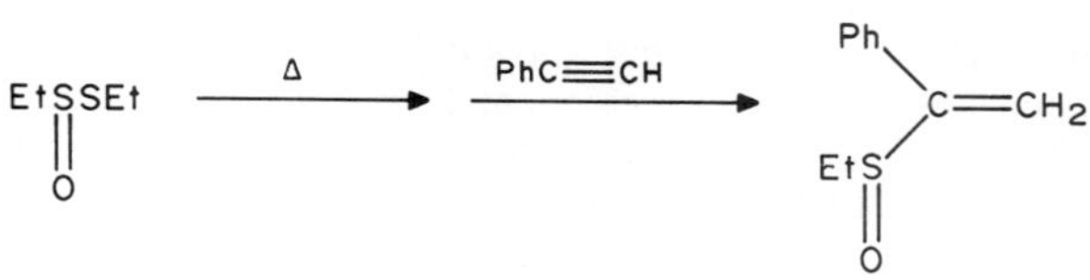

(43)

**(68)** 91%

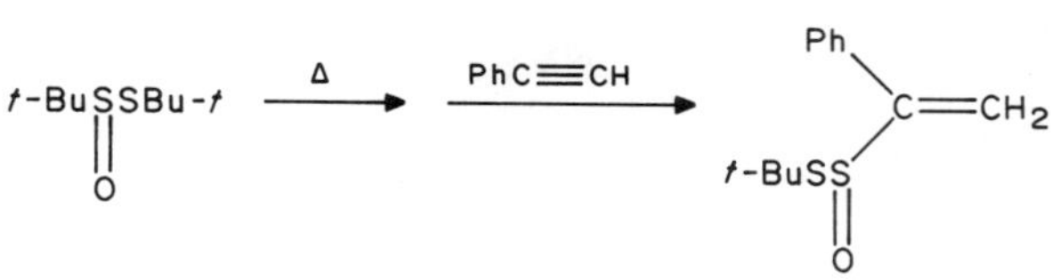

(44)

**(69)** 30%

trapped with acetylenes, giving $\alpha,\beta$-unsaturated sulphoxides (**67**, **68** and **69**) (equations 42–44)[130].

Sulphenic acids formed by $\beta$-elimination are known to add to olefins as reported with a stable penicillin sulphenic acid[106]. The mechanism of the addition is viewed as a reversible sigmatropic rearrangement. In the absence of a trapping agent, symmetrical thiosulphinates were formed in the pyrolysis of unsymmetrical ones, especially in the case of *t*-butyl substituted thiosulphinate (equation 45). When *t*-butanesulphenic acid reacts with the thioformaldehyde which is formed as a by-product of the decomposition of **70**, *t*-BuSOH as strong nucleophile gives mercaptosulfoxide **71** which further reacts with **70** to eventually afford sulphinyl disulfides (e.g. **72**, **73**) (equations 46 and 47). The products seem to be obtained via a Pummerer-type rearrangement[130]. Photochemical decomposition of alkyl thiosulphinates by a radical process induced by UV irradiation was also investigated by Block and O'Connor[130].

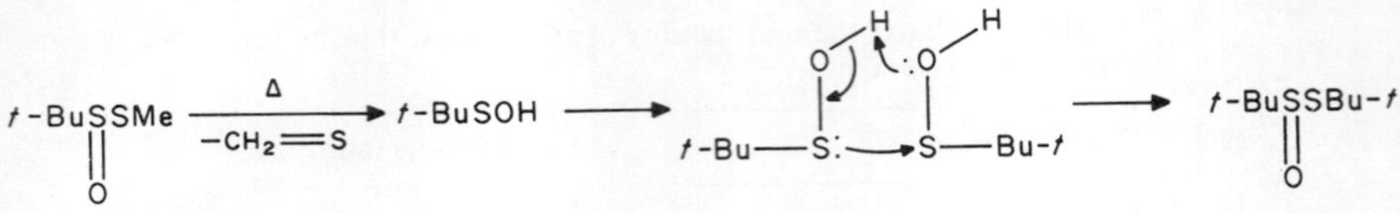

(70) (52)

(45)

$$t\text{-BuSOH} + \text{CH}_2\text{=S} \longrightarrow \underset{\textbf{(71)}}{t\text{-BuS(=O)CH}_2\text{SH}} \xrightarrow{\textbf{70}} \underset{\textbf{(72)}}{t\text{-BuS(=O)CH}_2\text{SSMe}} \quad (46)$$

$$2\,\text{MeS(=O)SMe} \xrightarrow[\text{C}_2\text{H}_6\text{–H}_2\text{O}]{\text{reflux}} 2\,\underset{\textbf{(73)}\ 84\%}{\text{MeS(=O)CH}_2\text{SSMe}} \quad (47)$$

Disproportionation of thiosulphinates to mixtures of disulphides and thiosulphonates[48,134–137] often occurs both in solution and in the solid state as well[48]. Sometimes further oxidation products such as sulphinic and sulphonic acids[130,134] are also formed (equation 48). Kice and coworkers[137] carried out an ESR study of the disproportionation of thiosulphinates. A homolytic chain mechanism including initial S—S bond cleavage is postulated, as illustrated in equations 49 and 50[130,135,136].

$$\text{MeS(=O)SEt} \xrightarrow[\text{25 °C, 4 days; conv. 34\%}]{\text{neat}} \begin{cases}\text{MeSO}_2\text{SEt (14\%)}\\ \text{MeSO}_2\text{SMe (3\%)}\end{cases} + \begin{cases}\text{MeSSMe (trace)}\\ \text{MeSSEt (9\%)}\\ \text{EtSSEt (3\%)}\end{cases} + \text{MeSOSMe (5\%)} \quad (48)$$

$$\text{ArS(=O)SAr} \longrightarrow \text{ArS(=O)}\cdot + \text{ArS}\cdot \quad (49)$$

(ArS(=O)· → $\text{ArSO}_2\text{SAr}$; ArS· → ArSSAr)

$$\text{ArS(=O)SAr} + \text{ArS}\cdot \longrightarrow \text{ArS(=O)}\cdot + \text{ArSSAr} \quad (50)$$

Regarding the optical instability of thiosulphinates, racemization of optically active and oxygen-18 labelled phenyl benzenethiosulphinate was studied[118,119,138,139]. Acid and nucleophile-catalysed $^{18}$O exchange rates were considerably slower than that of racemization. A small amount of nucleophile, such as halide ion or sulphide, dramatically accelerated the racemization[118,138,139].

*2. Hydrolysis*

Savige[30,61] reported the hydrolysis of cystine monooxide at pH 5–7 to give cystine and cysteinesulphinic acid (equation 51). Hydroxide ion mainly attacks at sulphinyl sulphur, but attack at sulphenyl sulphur may also be operative. Similar results were reported by Tsukamoto and his coworkers[140] for thiamine disulphide monooxide.

$$3\mathrm{CyS(O)SCy} + \mathrm{H_2O} \longrightarrow 2\mathrm{CySSCy} + 2\mathrm{CySO_2H} \qquad (51)$$

Thiosulphinates are readily hydrolysed under alkaline conditions. In this case, hydroxide ion was suggested to attack initially at sulphinyl sulphur[141,142]. Oae, Yoshikawa and Tagaki studied by NMR the alkaline hydrolysis of some aromatic thiosulphinates and concluded from kinetic results and product analysis that the sulphinyl sulphur is the only one attacked[142,143] (equation 52). However, Kice and Rogers claimed in a kinetic study that the initial attack of hydroxide ion occurs with almost the same rates at both suphenyl and sulphinyl sulphur[144]. To resolve this controversy, Oae, Takata and Kim carried out a reinvestigation of the alkaline hydrolysis of thiosulphinates[145,146], including a detailed product analysis and $^{18}O$ tracer experiments[145]. They found that the products contained a sulphinate derived exclusively from the sulphinyl moiety and found also both symmetrical and unsymmetrical disulphides. $^{18}O$ incorporation into the sulphinate, which was isolated as corresponding methyl sulphone by the reaction with methyl iodide, was always over 50%. These results were best explained by the following total scheme of the hydrolysis (equations 53–56). These results point to selective attack of the 'hard' $^{-}$OH on the 'hard' sulphinyl sulphur atom but not on the 'soft' sulphenyl sulphur.

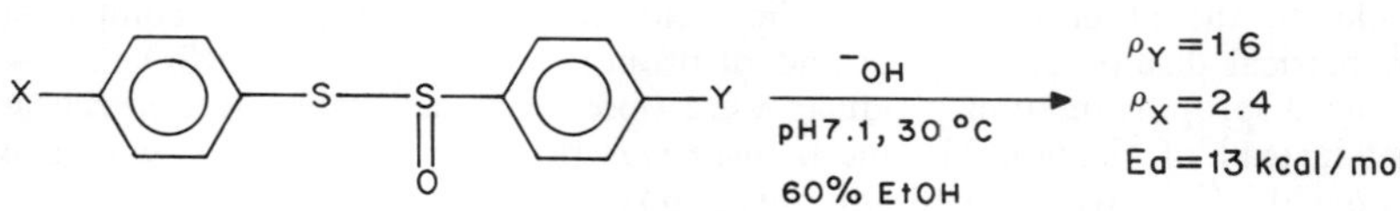


(52)

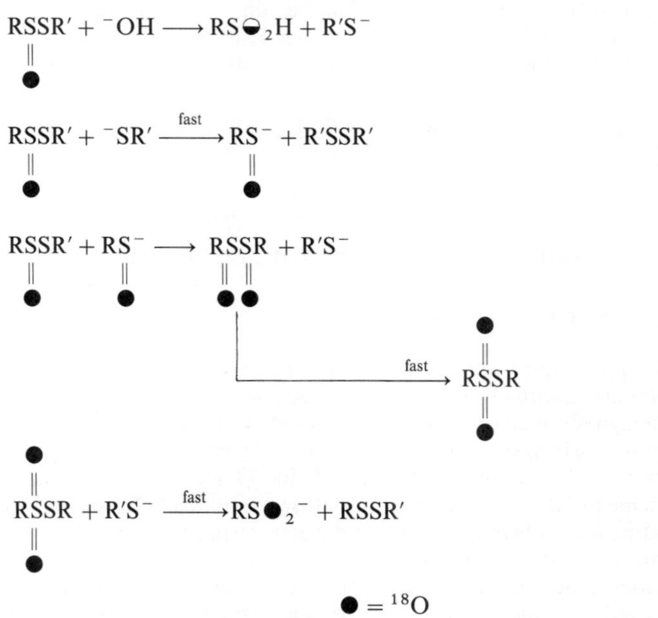


(53)

(54)

(55)

(56)

Somewhat different results were reported by Oae and collaborators in the hydrolysis of the cyclic thiosulphinate **43** which gave an equimolar mixture of the corresponding disulfide **42** and thiosulphonate **74** (equation 57) apparently without S—S bond cleavage. The reaction followed second-order kinetics, first order each in thiosulphinate and hydroxide ion[82].

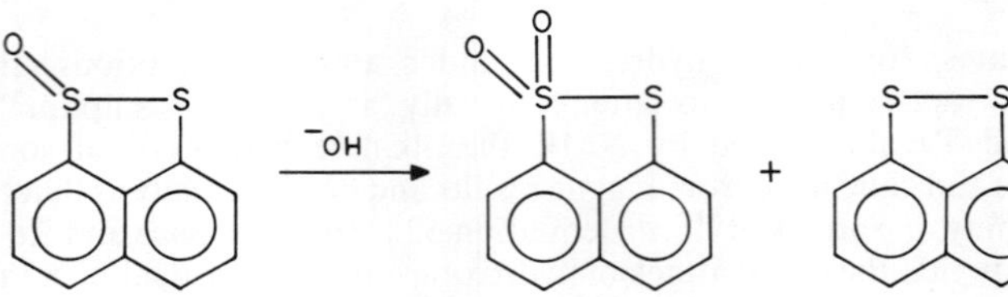


(43) (74) (42)

(57)

### 3. Alcoholysis

Convenient transformation of unstable thiosulphinates to stable sulphinates (**75**) was performed by Takata and Oae[146] (equation 58). This is a useful method for determining the structure of thiosulphinates. The replacement of the sulphenyl group by an alkoxy group is catalysed by iodine, bromine or hydrogen chloride but not by sulfuric or perchloric acid. The yields of **75** and R′SSR′ were enhanced by addition of hydrogen peroxide. In the absence of $H_2O_2$, the yield of **75** was 60–80%, accompanied by unsymmetrical disulphide (~25%) and disproportionation product $RSO_2SR$ (~8%). Only the disproportionation products were observed when using acetonitrile as the solvent instead of alcohol. The mechanism was therefore assumed to involve initial reaction with the catalyst to give sulphinyl and sulphenyl moieties by S—S bond fission, which are in turn trapped by the solvent alcohol to yield sulphinate **75** and thiol. Thiol R′SH is immediately oxidized by $H_2O_2$ to R′SSR′ or, in the absence of $H_2O_2$, reacts with thiosulphinate to lead to disproportionation products (Scheme 5).

$$\underset{\underset{\text{O}}{\|}}{\text{RS}}\text{SR}' \xrightarrow[\text{R}''\text{OH, 20 °C}]{1\sim 2\%\text{ e. cat. (XY)/H}_2\text{O}_2} \underset{\underset{\text{O}}{\|}}{\text{RS}}\text{OR}'' + \text{R}'\text{SSR}' \qquad (58)$$

**(75)** >90%

(R,R′ = *p*-Tol, Ph, Et; R″ = Me, Et, *i*-Pr; XY = $I_2$, HCl etc.)

### 4. Reaction with nucleophiles

*a. With Grignard reagents.* Vinkler and coworkers studied the reactions of a few diaryl thiosulphinates with Grignard reagents such as benzyl and propyl magnesium halides[147]. Benzylmagnesium chloride attacks nucleophilically at the sulphinyl sulphur to give the corresponding benzyl aryl sulphoxides **76**, while propylmagnesium bromide attacks at the sulphenyl sulphur to form aryl propyl sulfide **77** and arenesulphinylmagnesium bromide **78** (Scheme 6). **78** was converted to aryl benzyl sulphoxide **76** and aryl arenethiosulphinate by treatment with benzyl chloride and water, respectively (Scheme 7). The cause of these reactions is not clear at present time.

In order to determine enantiomeric excess of optically active aryl thiosulphinates (**79**) synthesized by optically active peroxy acid oxidation, reaction of **79** with benzylmag-

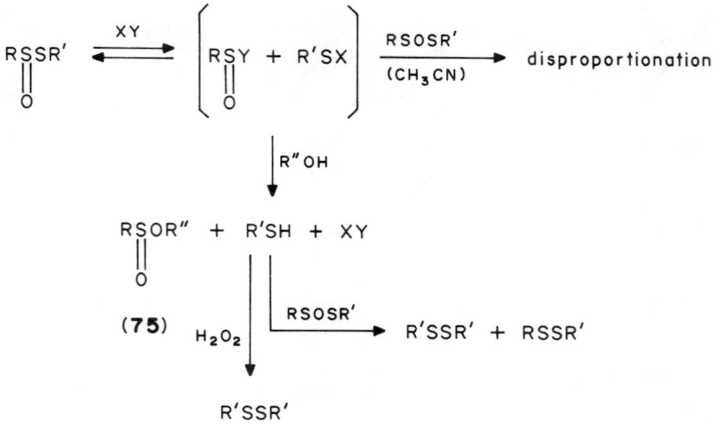


SCHEME 5

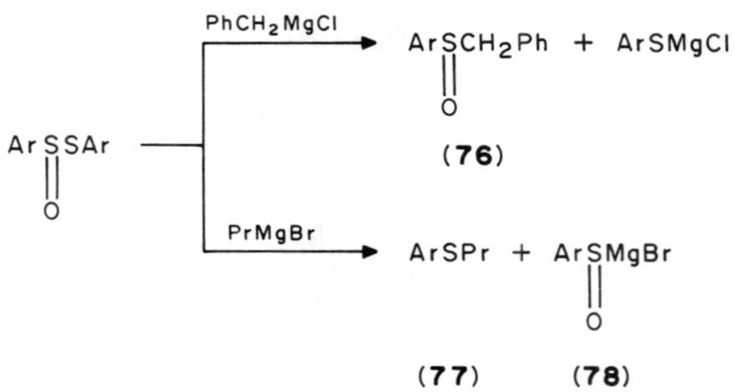


SCHEME 6

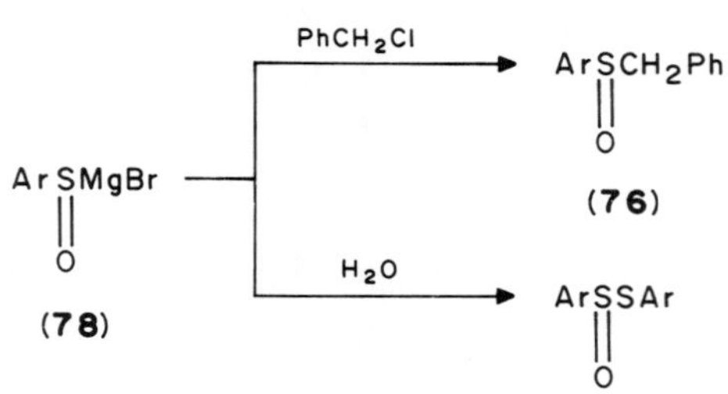


SCHEME 7

nesium chloride was carried out[117–119]. The reaction proceeded stereoselectively and with complete inversion of configuration to give the optically active sulphoxide **80**, which was identical to that derived from the similar reaction of the optically active sulphinate **81** (equation 59).

Mikołajczyk and Drabówicz used also diethylaminomagnesium bromide and N-bromosuccinimide in the reaction with the optically active unsymmetrical thiosulphinate **82** (Scheme 8)[122]. Stereochemistry of the reaction was inversion as expected.

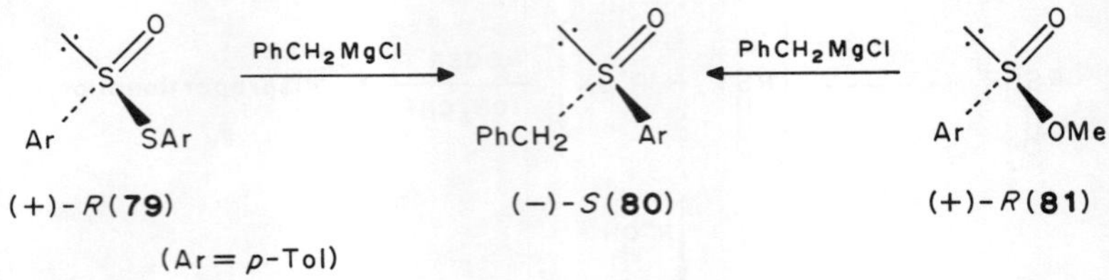


(59)

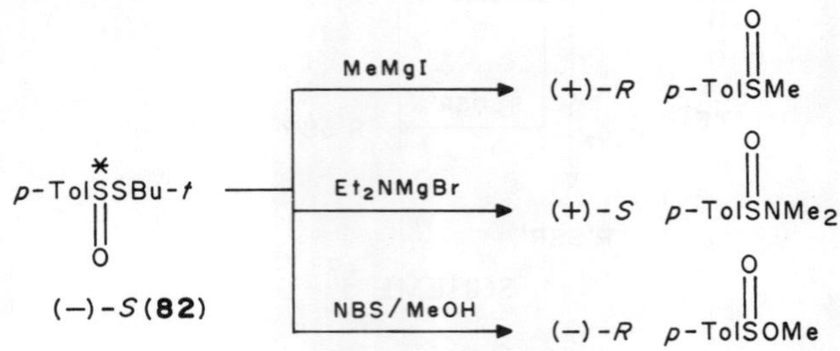


SCHEME 8

*b. With superoxide anion radical.* A detailed investigation has been undertaken by Takata and collaborators[148–151] on the reactions of a variety of organosulphur compounds including thiosulphinates, with superoxide ion ($O_2^{\overline{\cdot}}$). Aryl arenethiosulphinates reacted very rapidly at −40 to 0 °C with $KO_2$ in the presence of 18-crown-6 in pyridine or in acetonitrile alone to afford the corresponding potassium arenesulphinate and sulphonate along with the symmetrical disulphide (Scheme 9).

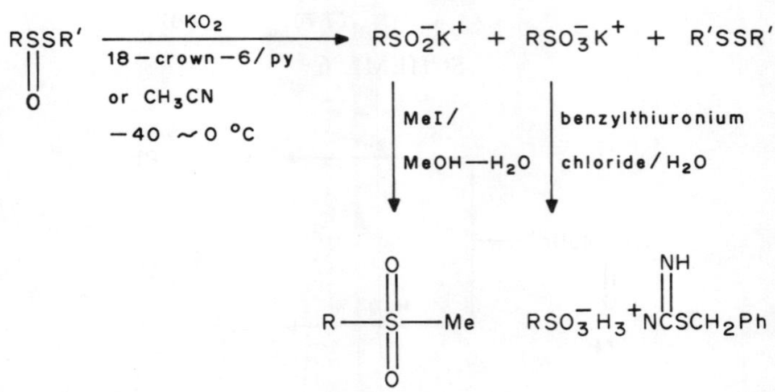


SCHEME 9

At higher (room) temperature, disulphides also react with $O_2^{\overline{\cdot}}$ to give the same products[150]. Product analysis and mechanistic investigation suggested initial nucleophilic attack by $O_2^{\overline{\cdot}}$ at the sulphinyl sulphur. This is also understood by the HSAB concept as described in the hydrolysis in the preceding section. The authors proposed the intervention of a new oxidizing species, peroxysulphinate **83**, which was proved to exist by trapping reactions with sulfoxides (equation 60), as in the cases of disulphides and other

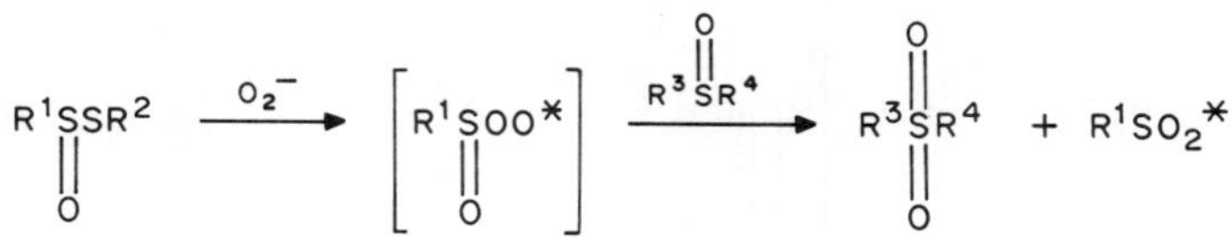


(60)

(83)

* denotes radical or anion

organosulphur compounds[150,151]. The peroxy species was assumed to be a nucleophilic oxidant because it did not oxidize sulphide. Thianthrene monooxide was converted with it to 9,9-dioxide but not 9,10-dioxide[150]. The following peroxy species were also proposed in the reactions of thiol, sodium thiolate, disulphide, thiosulphonate, sulphinyl chloride, sulphonyl chloride and sodium sulphinate with $O_2^-$ [150,151] and sometimes with molecular oxygen[24]. In some cases, the oxidation of phosphines and $\alpha,\beta$-unsaturated olefins to phosphine oxide (~37%) and epoxide (~85%), respectively, with these peroxysulphur compounds was observed.

$$RSOO^* \qquad RS(=O)OO^* \qquad RS(=O)_2O^*$$

*c. With miscellaneous reagents.* Secondary amines react with benzyl phenylmethanethiosulphonate ('dibenzyl thiosulphinate') to give a mixture of dibenzyl di- and trisulphides, thiobenzimides and benzaldehyde (equation 61)[152]. Enamines are sulphenylated by aryl benzenethiosulphinates[153].

$$PhCH_2S(=O)SCH_2Ph \xrightarrow{HNR_2} PhCH_2S_xCH_2Ph + PhC(=S)NR_2 + PhCHO \qquad (61)$$

$$x = 2,3$$

Anstad studied the formation of thiocyanates by treatment of thiosulphinates with cyanide ion[154].

### 5. Oxidation

*a. Formation of α-disulphoxides—oxidation with electrophilic reagents.* As described in Section III.B.2a, electrophilic oxidants such as peroxy acids oxidize the more electron-rich sulphenyl sulphur atom but not the sulphinyl sulphur atom of thiosulphinate. Hence the oxidation product may be the hypothetical α-disulphoxide **84**[155], but the actually isolated product was the thiosulphonate (equation 62). No α-disulphoxide has ever been isolated or trapped, although it has long been postulated as intermediate in the oxidation of disulphides and especially thiosulphinates and in some reactions of compounds with sulphinyl moieties[30,56,57,155–181]. Recently it has been proved to exist by spectroscopic detection[176,180,181].

Many attempts to prepare this elusive intermediate from cystine[156–158], alkyl or aryl thiosulphinates[164,169,171–181] and sulphinyl chlorides[161,167,168] have been unsuccessful.

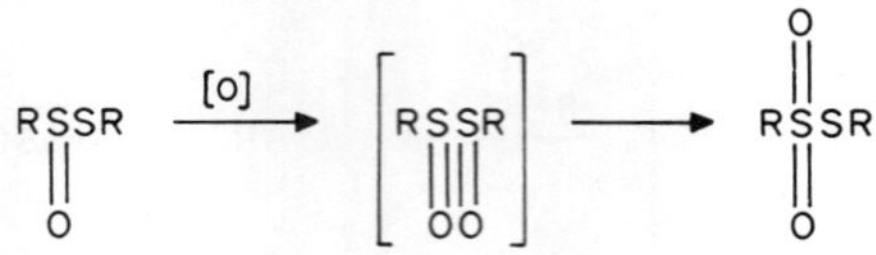

(62)

α-disulphoxide

**(84)**

Barnard[161] tried to isolate an α-disulphoxide by the reaction of benzenesulphinyl chloride with zinc, but the product was phenyl benzenethiosulphonate. A few groups[161,163,164] also noted that the final products in the peroxy acid oxidation of thiosulphinates were thiosulphonates. Modena and coworkers[163,164] concluded from kinetic studies and substituent effects in the oxidation that α-disulphoxide once formed underwent rapid isomerization to thiosulphonate. Barnard and Percy[169] suggested that fast homolytic cleavage of the S—S bond of α-disulphoxides gives a sulphinyl radical, which in turn yields with thiosulphinate the thiosulphonate. Thus, it is generally believed that α-disulphoxides are formed, but are quite unstable and collapse immediately to thiosulphonates.

In order to study the formation of α-disulphoxides Chau and Kice[171] utilized a low-temperature (−20 °C) $^{19}$F NMR technique. During the oxidation of *p*-fluorophenyl *p*-fluorobenzenethiosulphinate **85** and *p*-fluorophenyl benzenethiosulphinate **87** at −20 °C, they could find no signal to be assigned to the α-disulphoxide (equations 63 and 64). However, the product analysis suggested that at least 73% of the oxidation proceeded via a pathway involving α-disulphoxide as an intermediate. From the results, the $\Delta H^{\ddagger}$ of the decomposition of α-disulphoxide was estimated to be less than 20 kcal mol$^{-1}$ with a half-life time of less than 60 s at this temperature. They proposed as the mechanism decomposition of the α-disulphoxide yielding two sulphinyl radicals, followed by recombination to sulphenyl sulphinate which, in turn, rearranged to the thiosulphonate.

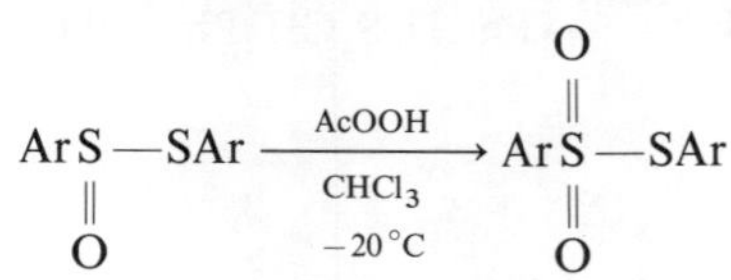

(63)

Ar = *p*-$FC_6H_4$

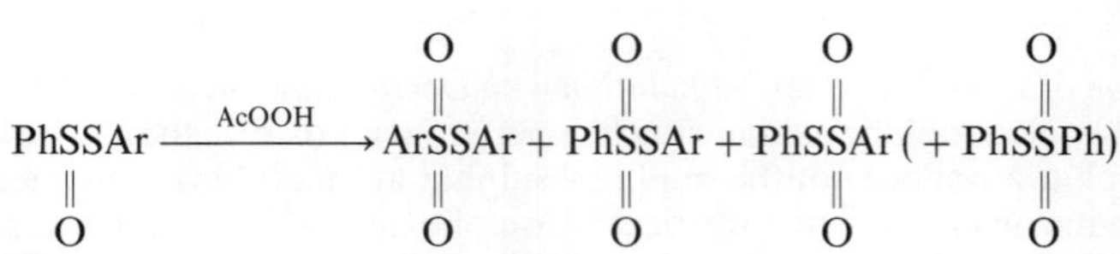

(64)

**(87)** **(88)** **(89)** **(90)**

Ar = *p*-$FC_6H_4$

Oae and coworkers[172,178] reported more concrete evidence by detailed product analysis, $^1$H NMR study and $^{18}$O tracer experiments, for the oxidation of unsymmetrical

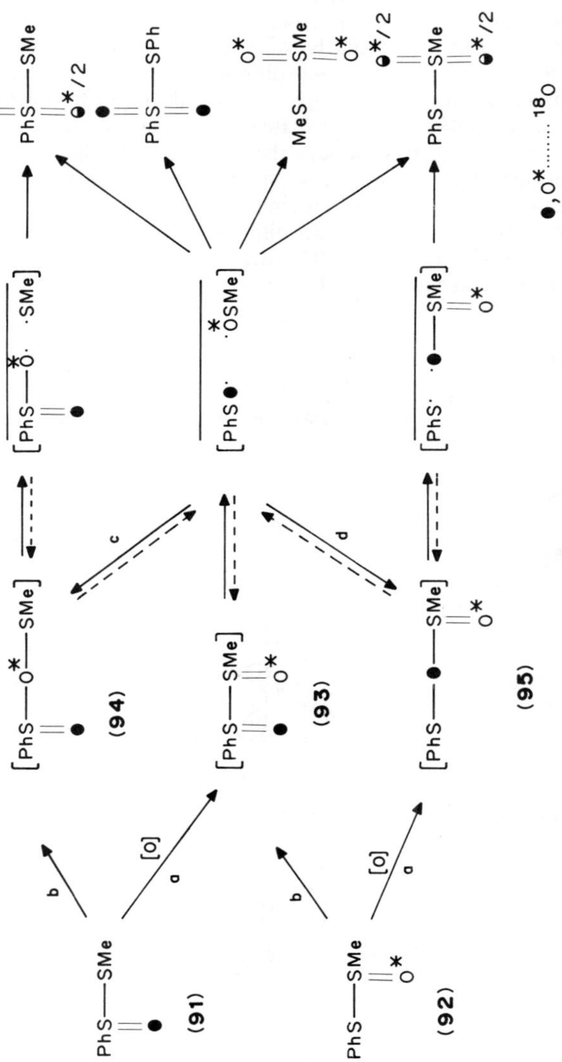

SCHEME 10. Oxidation of **91** and **92**

disulphides and thiosulphinates. They also detected no peak corresponding to any α-disulphoxide in $^{1}H$ NMR studies. In the oxidation of the thiosulphinates **91** or **92** with peroxy acids, all four possible symmetrical and unsymmetrical thiosulphonates were obtained along with some further oxidation products[177] such as sulphinic and sulphonic acids (equation 65). The results of $^{18}O$ tracer experiments using both $^{18}O$ labelled **91** and **92** (equation 65) suggested the mechanism shown in Scheme 10 involving the α-disulphoxide **93** and sulphenyl sulphinates **94** and **95** as intermediates in accordance with the mechanism proposed by Chau and Kice[171]. The initial oxidation of the sulphenyl sulphur of **91** or **92** gives an unstable α-disulphoxide **93** which probably collapses by homolytic S—S bond cleavage to two sulphinyl radicals. Head-to-tail recombination of the radicals generates both **94** and **95**, which are transformed by radical or other processes to the four stable thiosulphonates. Besides the path via α-disulphoxide (path a), direct conversion of thiosulphinate to sulphenyl sulphinate by oxidation (path b) is also conceivable (Scheme 11). Preference of either path might depend on the nature of the thiosulphinate, e.g. **91** bearing a more nucleophilic sulphenyl sulphur probably favors the intermediate α-disulphoxide (**93**), while **92** with a less nucleophilic sulphur undergoes preferentially the direct conversion to **94**.

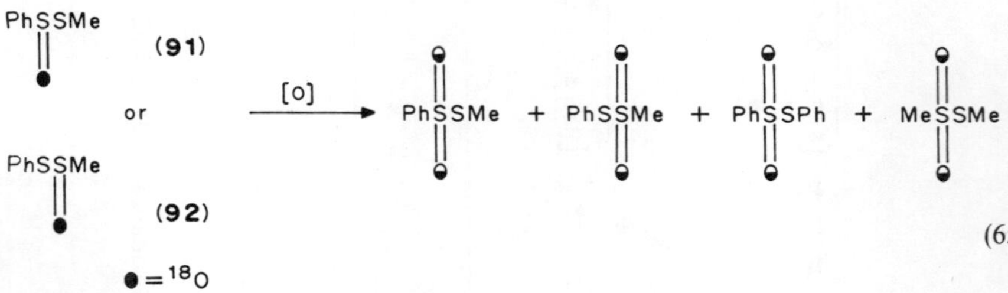


(65)

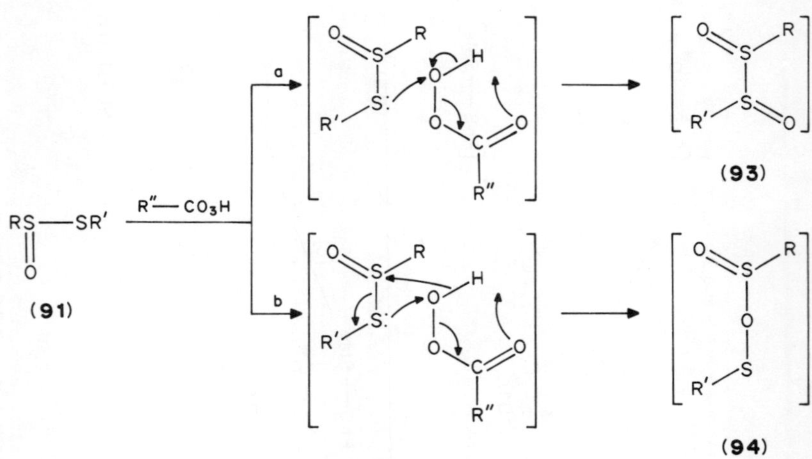


SCHEME 11

Freeman and collaborators attempted the detection of α-disulphoxide and sulphenyl sulphinate using $^{1}H$ and $^{13}C$ NMR at low temperature, and succeeded in confirming the formation of these two transient species[174–176,179–181] in the oxidations of dialkyl thiosulphinates with *m*-chloroperbenzoic acid (MCPBA) in chloroform. Although the initial attempt to detect it in the peroxy acid oxidation of phenyl phenylmethanethiosulphinate (**96**) was unsuccessful, the diastereomeric α-disulphoxides could be observed at −40 °C with several alkyl thiosulphinates (**97**–**103**; Figure 6). Sulphinic anhydride (−40 °C)[182] and sulphines (on warming to −20 °C) are also observed in most cases. The initial oxidation at the sulphinyl sulphur leading directly to thiosulphonate was ruled out from the fact that no thiosulphonate was detected at −40 °C. Figure 7 lists $^{13}C$ NMR chemical shifts of a few α-disulphoxides which should be compared with those of thiosulphinates (Figure 6). Table 8 shows the results of the MCPBA oxidation of methyl methanethiosulphinate (**97**) at −40 °C to −20 °C as an example (equation 66 in Table 8). $^{1}H$ and $^{13}C$ NMR data shown in Table 9 were obtained for each compound related to the oxidation of **97**. After 15 min reaction at −40 °C the NMR spectrum was very simple, indicating the formation of only two diastereomeric α-disulphoxides, while by warming to 0 °C these signals disappeared and instead those of thiosulphonate and sulphinic acid

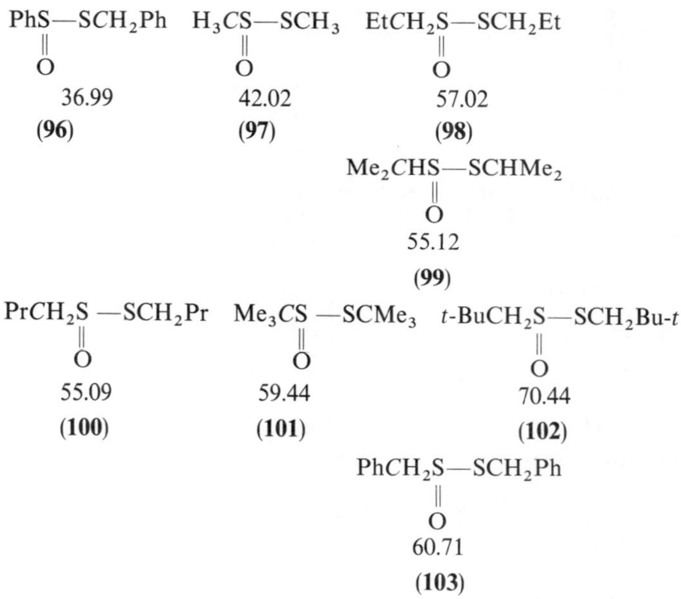


FIGURE 6. $^{13}C$ NMR chemical shifts of thiosulphinates (δ, TMS)

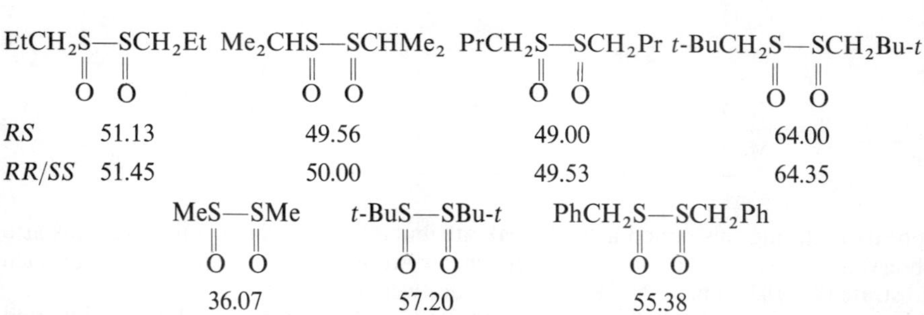


FIGURE 7. $^{13}C$ NMR chemical shifts of α-disulphoxides (δ, TMS)

TABLE 8. Oxidation of **97** with MCPBA studied by NMR

$$\underset{\parallel \atop O}{MeS}\text{—}SMe \xrightarrow[-40\,^{\circ}C]{[O]} \underset{\parallel \atop O}{MeS}\text{—}\underset{\parallel \atop O}{SMe} \xrightarrow[-20\,^{\circ}C]{} (MeSO)_2O \text{ etc.} \qquad (66)$$

| Temp (°C) | Time (min) | MeS(=O)SMe (**97**) | Product and yield (%) | | | | |
|---|---|---|---|---|---|---|---|
| | | | MeS(=O)S(=O)Me *RS* | MeS(=O)S(=O)Me *RR/SS* | $MeSSMe(O_2)$ | $(MeS)_2O$ (S=O) | MeS(=O)OH |
| −40 | 15 | 56 | 25 | 19 | — | — | — |
| −40 | 19 | 59 | 20 | 21 | — | — | — |
| −20 | 101 | 76 | — | 10 | 7 | 2 | 4 |

TABLE 9. $^1$H and $^{13}$C NMR chemical shifts of products of the oxidation of **97** with MCPBA, $\delta$ (ppm)

| Compounds | at −40 °C $^1$H | at −40 °C $^{13}$C | at −20 °C $^{13}$C |
|---|---|---|---|
| MeS(=O)—SMe (SMe) | 2.75 | 15.22 | 15.00 |
| MeS(=O)—SMe (S(O)Me) | 3.08 | 42.02 | 42.18 |
| MeS(=O)—S(=O)Me *RS* | 2.86 | 36.07 | |
| MeS(=O)—S(=O)Me *RR/SS* | 3.04 | 36.17 | 36.23 |
| MeS(=O)(=O)—SMe | | | 18.57 |
| | | | 48.63 |
| MeS(=O)—O—S(=O)Me | | | 46.47 |
| $MeSO_2H$ | | | 44.90 |

appeared. During this period a small peak attributable to sulphinic anhydride was also observed. Therefore, the main reaction scheme would be as in equation 67. In each substrate (**97**–**103**) similar NMR characteristics were observed.

Freeman and Angeletakis[181] proposed various reaction pathways which are initiated by the oxidation of sulphenyl sulphur to α-disulphoxide. Formation of sulphine, sulphenic

$$\underset{\textbf{(97)}}{CH_3S(=O)\text{—}SCH_3} \xrightarrow[-40\,^{\circ}C]{[O]} CH_3S(=O)\text{—}S(=O)CH_3 \xrightarrow[-20\,^{\circ}C]{} CH_3S(=O)\text{—}O\text{—}SCH_3$$

$$\longrightarrow CH_3S(=O)_2\text{—}SCH_3 + CH_3SO_2H + (CH_3SO)_2O \qquad (67)$$

acid and sulphinic acid are explained by cyclo-elimination of α-disulphoxide, sulphenyl sulphinate or sulphinic anhydride (equation 68), and hydrolysis of α-disulphoxide or sulphinic anhydride. Increase of thiosulphinate on warming from −40 °C to −20 °C is undoubtedly due to the condensation of the sulphenic acid formed. Thiosulphonate can be also produced by the reaction of thiosulphinate with a transient sulphinic acid (equation 69).

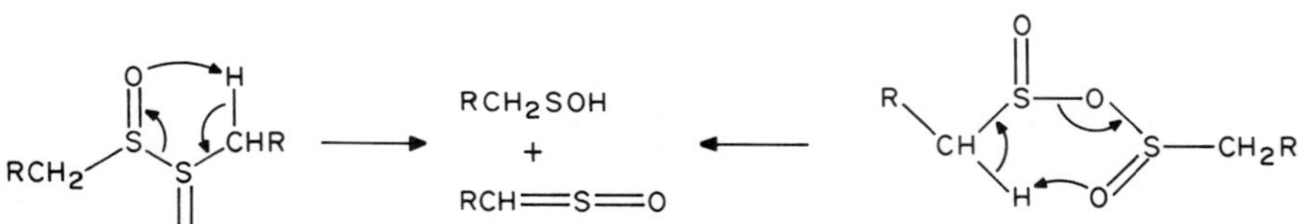


(68)

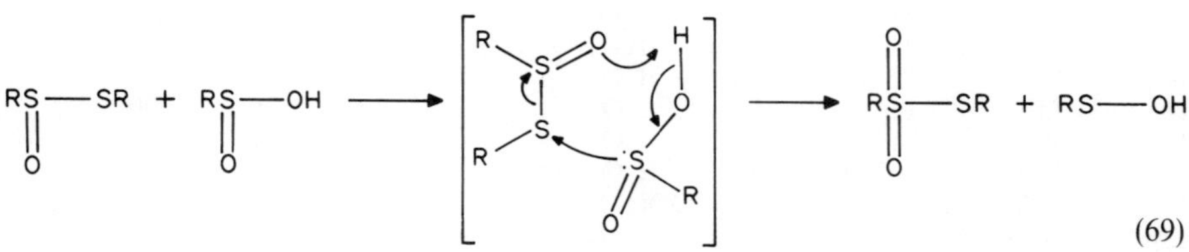


(69)

From the theoretical aspect, Freeman and his coworkers examined the structures of hydrogen persulphide (HSSH) and its monooxide (HS(O)SH), dioxide ($HS(O)_2SH$) and (HS(O)S(O)H), and tetroxide ($HS(O)_2S(O)_2H$) derivatives by *ab initio* molecular orbital calculations at HF/3–21G* and 6–31G* levels[179]. These theoretical calculations supported the mechanism proposed for the rearrangement of α-disulphoxides via sulphinyl radicals to thiosulphonates. The calculations also suggested that α-disulphoxide is sufficiently stable to be observed and/or isolated at low temperatures, in good agreement with the above results. S—S Bond lengths and angles of these species are also discussed.

In view of the above-mentioned studies α-disulphoxide has been recognized as a reactive intermediate which can be observed. It is hoped that its reactivity with some nucleophiles and electrophiles will be further studied.

*b. Selective oxidations of thiosulphinates.* In contrast to the very complex peroxy acid oxidations Takata, Kim and Oae[183,184] found that sodium (or potassium) metaperiodate in aqueous solvent oxidizes thiosulphinates to the corresponding thiosulphonates without any S—S bond fission, under mild conditions and in quantitative yields (equation 70). This is the first selective oxidation which is synthetically very useful (Table 10). The oxidation was accelerated by addition of catalytic amounts of inorganic and organic acids or halogens. In the absence of a catalyst, the oxidation suddenly started after an unspecified induction period, but again giving the thiosulphonate quantitatively. In the presence of e.g. hydrochloric acid, the reaction was about ten times faster than in the absence of the catalyst. Acetic acid could be employed as catalyst and solvent as well.

$$\underset{\underset{\displaystyle O}{\|}}{RSSR'} \xrightarrow[\text{aqueous solvent, r.t.}]{NaIO_4} \underset{\underset{\displaystyle O}{\|}}{\overset{\overset{\displaystyle O}{\|}}{RSSR'}} \quad \sim 100\% \qquad (70)$$

TABLE 10. Selective oxidation of unsymmetrical thiosulphinates with $NaIO_4$ at room temperature

| Entry | Thiosulphinate (RSO—SR′) R | R′ | Solvent | Catalyst | Time (h) | Yield[a] (%) |
|---|---|---|---|---|---|---|
| 1 | Ph | Ph | dioxane–$H_2O$ | none | 26.0 | 100 |
| 2 | Ph | *p*-Tol | $CH_3CN$–$H_2O$ | conc. HCl | 1.0 | 100 |
| 3 | *p*-Tol | Ph | $CH_3CN$–$H_2O$ | $I_2$ | 0.5 | 100 |
| 4 | *p*-Tol | Ph | $CH_3CN$–$H_2O$ | $Br_2$ | 0.5 | 95 |
| 5 | *p*-Tol | Ph | $CH_3CN$–$H_2O$ | $H_2SO_4$ | 1.0 | 100 |
| 6 | *p*-Tol | Ph | $CH_3CN$–$H_2O$ | $HClO_4$ | 1.0 | 100 |
| 7 | Ph | Me | dioxane–$H_2O$ | $CF_3COOH$ | 2.0 | 100 |
| 8 | Ph | Me | $CH_3CN$–$H_2O$ | HCOOH | 6.0 | 90 |
| 9 | Ph | Me | $CH_3COOH$–$H_2O$ | none | 0.5 | 95 |
| 10 | Me | Ph | dioxane–$H_2O$ | none | 8.0 | 98[b] |
| 11 | Et | Ph | dioxane–$H_2O$ | conc. HCl | 1.0 | 90[b] |
| 12 | Ph | *i*-Pr | dioxane–$H_2O$ | conc. HCl | 1.0 | 85[b] |
| 13 | Me | *c*-$C_6H_{11}$ | dioxane–$H_2O$ | conc. HCl | 0.5 | 90 |
| 14 | *p*-Tol | Me | $CD_3COOD$–$D_2O$ | none | 0.5 | 100[c] |
| 15 | *p*-$ClC_6H_4$ | Me | $CD_3COOD$–$D_2O$ | none | 0.5 | 100[c] |

[a]NMR yield.
[b]Isolated yield.
[c]Reaction in NMR sample tube.

When the reaction was carried out in aqueous alcohol, sulphinates were produced together with the thiosulphonates (equation 71). The yield of the sulphinates depended on the alcohol used as solvent, and when the bulkiness of the alcohol increased, the sulphinate yield decreased as shown in equation 71. $NaIO_3$, $SeO_2$, $KMnO_4$ and $NaClO_3$ were also tested as selective oxidants in these reactions[185], and usually showed the same selectivity, but the activity was rather low except for $NaIO_3$.

On the other hand, oxidation of an unsymmetrical disulphide or a mixture of two symmetrical disulphides with two equivalents of $NaIO_4$ for each equivalent of disulphide under the same conditions gave a mixture of symmetrical and unsymmetrical thiosulphonates without selectivity (equations 18 and 72). If an unsymmetrical thiosulphinate is directly formed from an unsymmetrical disulphide and then further oxidized selectively to the corresponding thiosulphonate with $NaIO_4$, symmetrical thiosulphonates should not be obtained[184]. Therefore, the oxidation without selectivity should either involve S—S

$$\underset{\underset{\displaystyle O}{\|}}{\text{PhS}}\text{SMe} \xrightarrow[\text{ROH–H}_2\text{O, r.t.}]{\text{NaIO}_4} \underset{\underset{\displaystyle O}{\|}}{\overset{\overset{\displaystyle O}{\|}}{\text{PhS}}}\text{SMe} + \underset{\underset{\displaystyle O}{\|}}{\text{PhS}}\text{OR} \ (+\text{MeSSMe}) \qquad (71)$$

| | | |
|---|---|---|
| R = Et | 42% (44) | 39% (47) |
| R = *i*-Pr | 67 | 27 |
| *i*-Bu | 82 | 7 |

( ): in the presence of HCl

$$\text{PhSSPh} + \text{MeSSMe} \xrightarrow[\text{CD}_3\text{CO}_2\text{D–D}_2\text{O, r.t. (NMR)}]{\text{4eq. NaIO}_4} \underset{\textbf{(41)}}{\text{PhS}(\text{O}_2)\text{SMe}} + \underset{\textbf{(40)}}{\text{MeS}(\text{O}_2)\text{SMe}} (+ \underset{\textbf{(39)}}{\text{PhS}(\text{O}_2)\text{SPh}}) \qquad (72)$$

bond fission in the first oxidation step of the disulphide or reaction of the disulphide with the thiosulphinate produced. This may be consistent with the unique nature of periodate, although it is also used as useful selective oxidant for sulphide to sulphoxide.

This selectivity was confirmed more clearly in the oxidation of the cyclic unsymmetrical thiosulphinates 3-methyl-1,2-dithian 1- or 2-monooxide, which were separated by chromatography by Oae and Takata[185], and converted to the corresponding thiosulphonates by $NaIO_4$ (equation 73). In contrast hydrogen peroxide or MCPBA oxidized **105** to a 1:1 mixture of thiosulphonates **106** and **107** (equation 74).

Oae and Takata therefore suggested that $NaIO_4$ oxidation is a typical 'nucleophilic oxidation' which is clearly distinguished from 'electrophilic oxidation' performed by peroxy acids[185]. The former process was speculated to involve a sulfurane intermediate (**108**) as in equation 75.

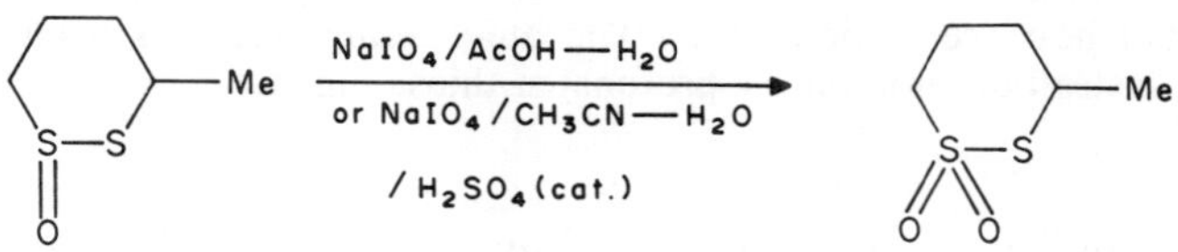


**(105)** **(106)** 100% (73)

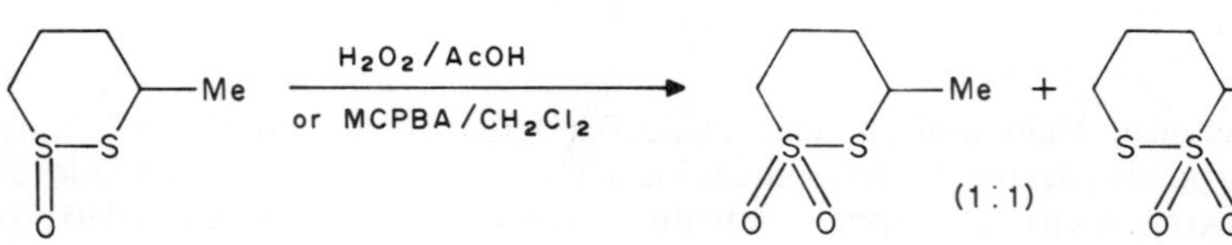


**(105)** **(106)** 100% **(107)** (74)

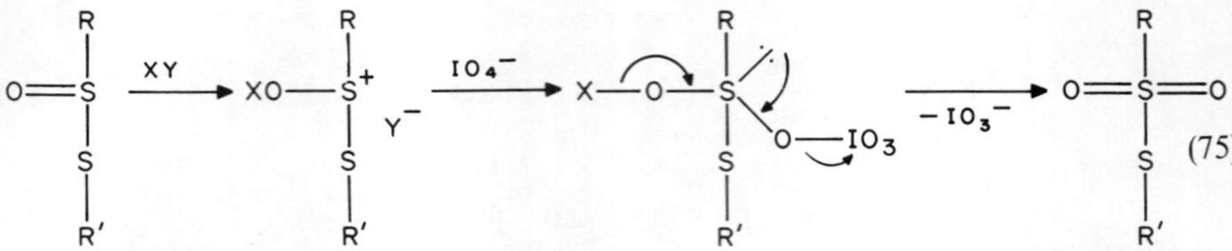


(108) (75)

c. *Miscellaneous oxidations.* Inorganic (by $N_2O_4$)[186] and biological (*in situ* by cytochrome P-450)[86] oxidations were reported.

## 6. Reduction

a. *With thiols.* Thiols react easily with thiosulphinates to give disulphides. This reaction is used for synthesis of unsymmetrical disulphides (equation 76)[6,52,187]. The mechanism is considered as shown in equation 77. Schöberl and Graefje pointed out the requirement of excess thiol since disproportionation of the produced sulphenic acid to sulphinic acid and thiol occurs[188]. It is possible if this disproportionation is competitively fast with $k_2$.

$$\underset{\substack{\| \\ \mathrm{O}}}{\mathrm{RSSR}} + 2\mathrm{R'SH} \longrightarrow 2\mathrm{RSSR'} + \mathrm{H_2O} \qquad (76)$$

$$\underset{\substack{\| \\ \mathrm{O}}}{\mathrm{RSSR}} \xrightarrow[k_1]{\mathrm{R'\ddot{S}H}} \mathrm{RSOH} + \mathrm{RSSR'} \quad (\mathrm{RSOH} \xrightarrow[k_2]{\mathrm{R'SH}} \mathrm{RSSR'}) \qquad (77)$$

The reaction with thiol takes place under either acidic or basic conditions. Kice studied the reaction in detail and concluded that the reaction follows second-order kinetics, first order in each of thiol and thiosulphinate in acidic media, via attack of thiol on the protonated thiosulphinate[189] (Scheme 12). In the sulphide-catalysed reaction with thiol[190], the kinetics are second order, first order in each sulphide and thiosulphinate, but independent of the concentration of thiol. The rate-determining step is therefore the attack of sulphide (instead of thiol) on the protonated thiosulphinate.

$$\underset{\substack{\| \\ \mathrm{O}}}{\mathrm{PhSSPh}} + \mathrm{H^+} \rightleftharpoons \underset{\substack{| \\ \mathrm{OH}}}{\overset{+}{\mathrm{PhSSPh}}} \xrightarrow{\mathrm{RSH}} \underset{\substack{| \\ \mathrm{H}}}{\overset{+}{\mathrm{RSSPh}}} + \mathrm{PhSOH} \xrightarrow{\mathrm{RSH}} \mathrm{PhSSR} \quad (\overset{+}{\mathrm{RS(H)SPh}} \xrightarrow{-\mathrm{H^+}} \mathrm{PhSSR})$$

SCHEME 12

The reaction of thiamine disulphide S-oxide (BS(O)SB) with thiols (RSH) in 80% alcohol was investigated in detail[191]. The products included the unsymmetrical disulphide (BSSR) and also two symmetrical disulphides (equation 78) and the reaction is believed to proceed via a complex mechanism. The unsymmetrical disulphide is probably formed according to equation 76, while formation of symmetrical disulphides is explained by attack of thiolate ion at the oxygen atom of the protonated thiosulphinate (to give BSSB) followed by reaction of the thiol with sulphenic acid formed (to give RSSR) (equation 79).

$$\underset{\underset{\displaystyle O}{\|}}{BS}\text{—}SB + 2RSH \longrightarrow BSSR + BSSB + RSSR \qquad (78)$$

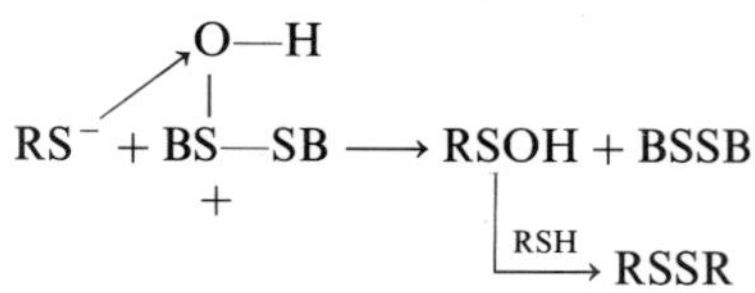


(79)

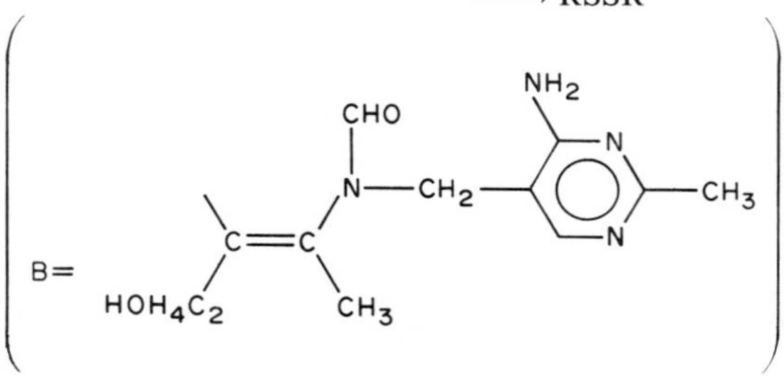


Thiosulphinate (**109**) has three reactive positions attacked by nucleophiles[32], i.e. sulphenyl sulphur, sulphinyl sulphur and sulphinyl oxygen. The mechanism is supported by the general concept that the 'soft' nucleophile thiolate ion attacks at the 'soft' sulphenyl sulphur, as Kice and Large reported[138], but attack of thiolate ion at the sulphinyl sulphur is not likely to occur. The amount of symmetrical disulphides (BSSB, RSSR) increases with decrease of acidity of thiols in good accordance with the order obtained in the oxidation of thiols with sulphoxides[192,193].

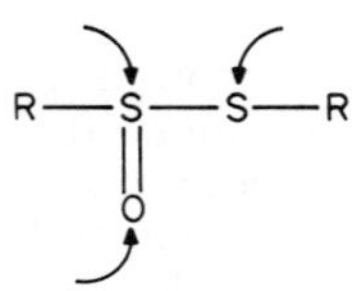


**(109)**

*b. With miscellaneous reagents.* Thiosulphinate was directly reduced with hydrogen iodide[30,194], $Na_2SO_3$[194] and triphenylphosphine (equation 80)[195,196] to afford disulphide quantitatively. More drastic conditions are needed with $Ph_3As$ or $Ph_3Sb$.

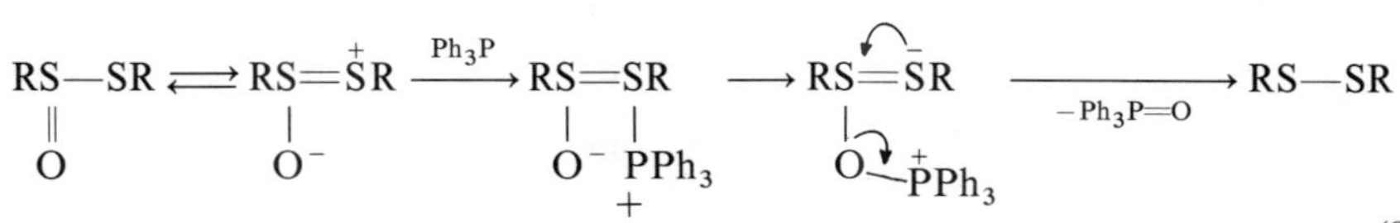


(80)

## 7. Reaction with electrophiles

*a. With acetic anhydride.* Pummerer-type rearrangements like those with sulphoxides[197] have been reported with thiosulphinates[129,198–203]. Fukui and Saito found that the reaction of α-lipoic acid monooxide (**110**) with acetic anhydride in acetonitrile gives the normal Pummerer-type product (**111**) but only in 6% yield (equation 81). *t*-Butyl methanethiosulphinate, however, did not react with acetic anhydride[202]. Kondo and Negishi reported that methoxyethyl methoxyethanethiosulphinate (**112**, X = OMe) yielded a Pummerer-like product (**113**) by heating in methanol (equation 82)[203], but the reported mechanism was quite different from the Pummerer rearrangement[197].

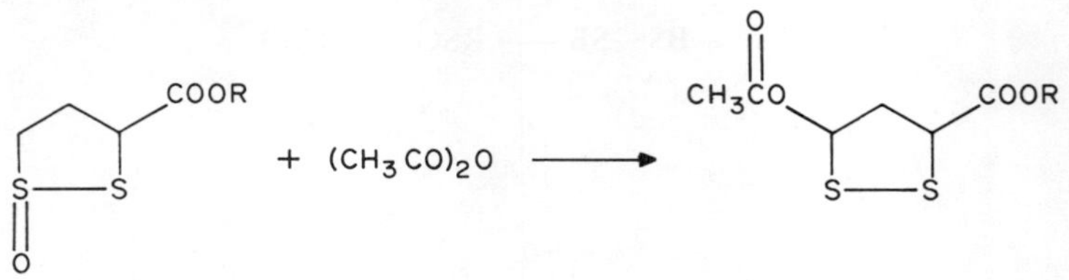


(81)

**(110)** **(111)**

$$\underset{\substack{\| \\ O}}{XCH_2CH_2SSCH_2CH_2X} \xrightarrow[\text{MeOH}]{\Delta} XCH_2\underset{\substack{| \\ OMe}}{CH}SSCH_2CH_2X \qquad (82)$$

(**112**) (**113**)

(X = H, OMe, $OCOCH_3$) 90–97%

Similar types of reactions were presented by Oae and coworkers[129,199–201]. Treatment of several thiosulphinates with acetic anhydride under the Pummerer rearrangement conditions gave new rearrangement products, α-acetylthiosulphoxides (**115**), along with a small amount of a trithian derivative (**116**) (equation 83). The reaction mechanism was determined by their detailed tracer experiments using $^{2}H$, $^{13}C$ and $^{18}O$ labelled benzyl phenylmethanethiosulphinate (**117**) (equation 84). The $^{2}H$ and $^{13}C$ contents of the main product (**118**) and recovered **117** decreased considerably, while their $^{18}O$ content remained unchanged. This meant absence of acetylation at the oxygen atom of the S=O group of **117**. The $^{2}H$ content of $PhCD_2S(O)SCH_2Ph$ (**117**-$d_2$) recovered during the reaction was not lost. These results suggested a mechanism[129] which consisted of an initial $E_i$ reaction[51,130] leading to both phenylmethanesulphenic acid (**119**) and thiobenzaldehyde (**120**) followed by their recombination to α-mercaptosulphoxide (**121**). **120** was trimerized to **122** whereas **121** was trapped with acetic anhydride to give the main product (**118**) (Scheme 13).

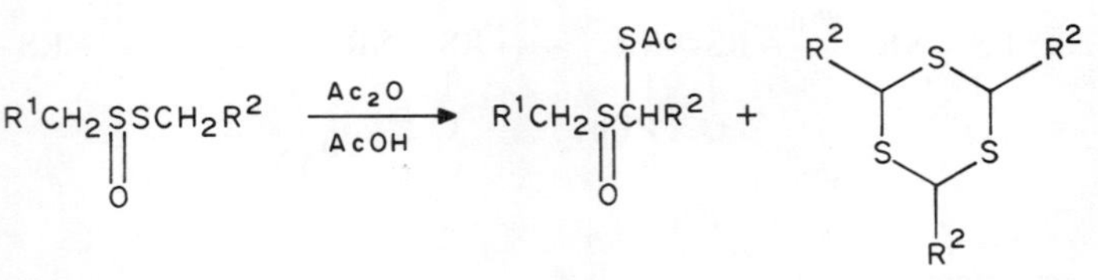


(83)

**(114)** **(115)** **(116)**

75% ($R^1$,$R^2$ = *p*-Tol)

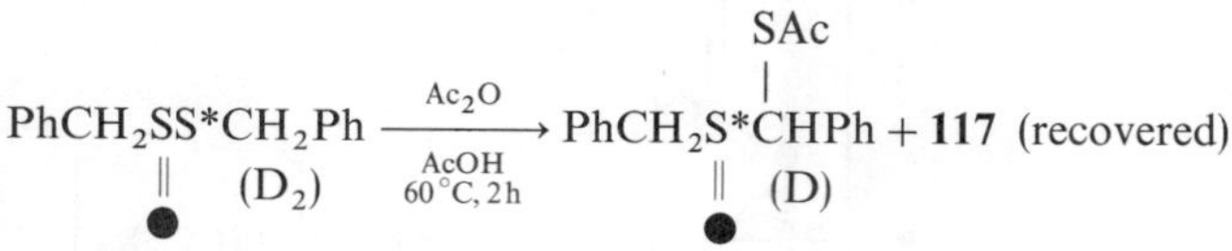


(117) (118) (84)

| Isotopic content (%) | **118** | **117** |
|---|---|---|
| $^{2}H$ (D) | 75 | 88 |
| $^{13}C$ (*C) | 76–78 | 60–62 |
| $^{18}O$ (●) | 100 | 100 |

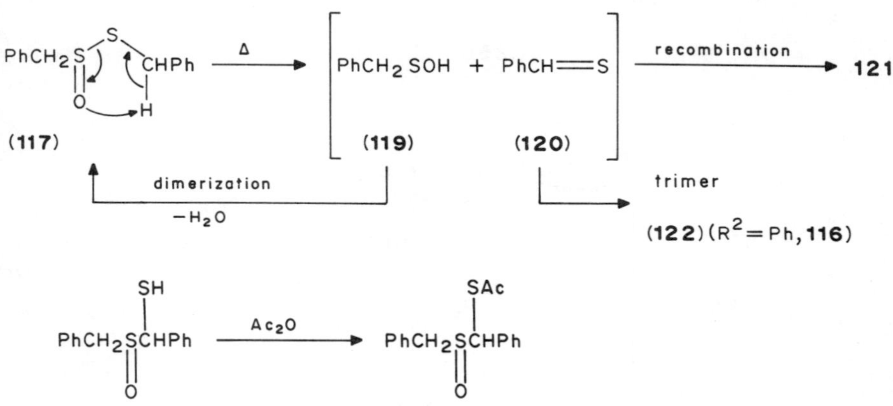


SCHEME 13

*b. With trihaloacetic anhydrides.* When stronger electrophiles such as trifluoroacetic anhydride were used, an acylation-initiated reaction was confirmed by Oae and collaborators[201,204,205]. Thiosulphinates reacted with trichloro- and trifluoroacetic anhydrides at $-10\,^{\circ}C$ in carbon tetrachloride to give an equimolar mixture of the corresponding disulphide and sulphinyl trihaloacetate (**125**), which were in equilibrium with **124**. In this system **125** was stable in solution and clearly detected by NMR (equation 85)[201,204]. The intermediate **124** could be trapped with olefins or with anisole to afford the corresponding adducts, i.e. trihaloacetyl sulphide (**126**) and *p*-methylthioanisole (**127**), respectively (equation 86). This means that **124** acts as a good methylthio cation source. The structure of **125** was suggested by IR and NMR, and also by comparison with authentic samples prepared by the reaction of silver carboxylates and sulphinyl chlorides. Since the addition of sulphenyl trihalocarboxylates **124** to olefins proceeded regio- and stereospecifically (*trans* addition), many adducts were synthesized and the synthetic utility was also discussed (equation 87)[205].

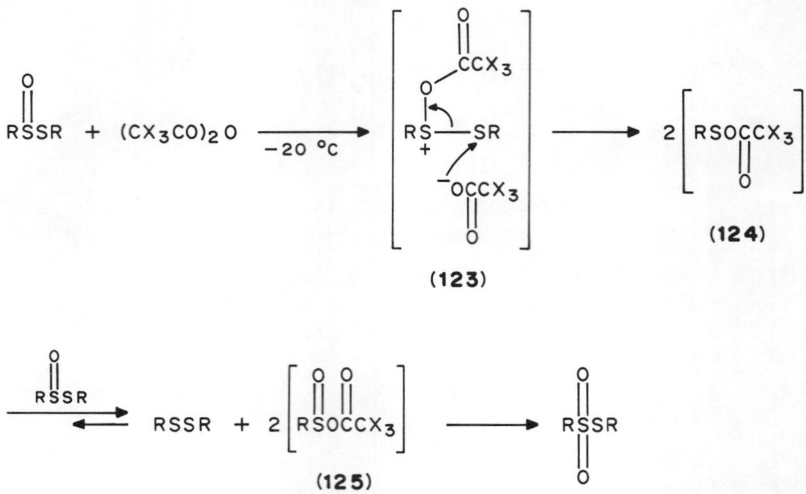

(123) (124) (125)

(85)

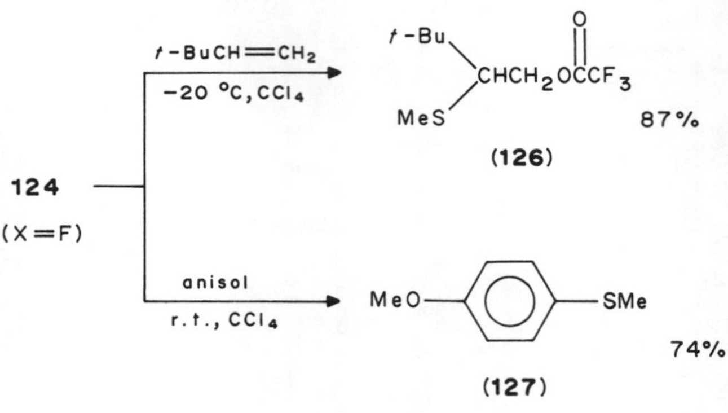

(126) 87% (127) 74%

(86)

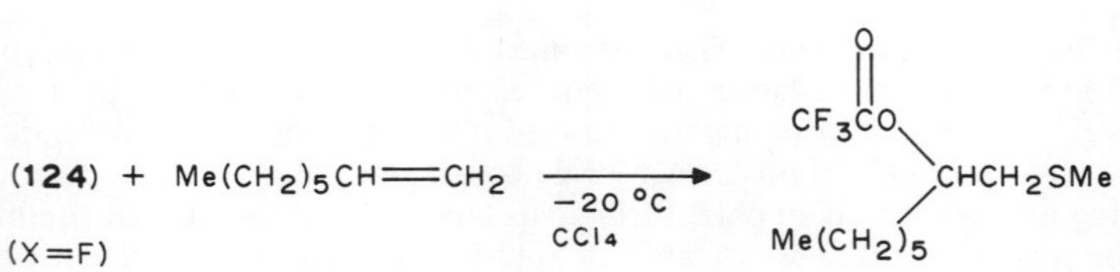

91% (87)

*8. Miscellaneous reactions*

Some other reactions of thiosulphinates with 2,4-dinitrofluoro[206] and chlorobenzenes[207], N-ethylmaleimide[30], sulphinic acids[25,208], hydrogen sulfide[30], trichloromethanesulphenyl chloride[209], maleinimide in the presence of water[30] and a sigmatropic rearrangement of diisobutenyl thiosulphinate[43] were also reported.

## IV. REFERENCES


1. M. Mikołajczyk, P. Łyzwa, J. Drabówicz, M. Wieczorek and G. Bujacz, *Angew. Chem. Int. Ed. Engl.*, **28**, 97 (1989).
2. E. Krauthausen, *Methoden der Organischen Chemie*, Vol. XI, 5th edn. (Ed. D. Klamann), Georg Thieme Verlag, Stuttgart, 1985, pp. 614–618.
3. K. K. Anderson, *Comprehensive Organic Chemistry*, Vol. 3 (Ed. D. N. Jones), Pergamon Press, Oxford, 1979, pp. 317–329.
4. C. J. Cavallito and J. H. Bailey, *J. Am. Chem. Soc.*, **66**, 1950 (1944).
5. C. J. Cavallito, J. S. Buck and C. M. Suter, *J. Am. Chem. Soc.*, **66**, 1952 (1944).
6. L. D. Small, J. H. Bailey and C. J. Cavallito, *J. Am. Chem. Soc.*, **69**, 1710 (1947).
7. A. Kato and M. Numata, *Tetrahedron Lett.*, 203 (1972).
8. H. Yanagawa, T. Kato and Y. Kitahara, *Tetrahedron Lett.*, 1173 (1973).
9. T. Kametani, K. Fukumoto and S. Umezawa, *Yakugaku Kenkyuu*, **31**, 60, 132 (1959); **33**, 125 (1960).
10. A. S. Weisberger and J. Pensky, *Cancer Res.*, **18**, 10 (1958).
11. A. S. Weisberger and J. Pensky, *Science*, **126**, 1112 (1957).
12. A. F. Hirsch, C. Piantadori and J. L. Irvin, *J. Med. Chem.*, **8**, 10 (1965).
13. E. D. Wills, *Biochem. J.*, **63**, 514 (1956).
14. R. M. Dodson, V. Srinivasan, K. S. Sharma and R. F. Sauers, *J. Org. Chem.*, **37**, 2367 (1972).
15. N. Kharasch and A. S. Arora, *Phosphorus Sulfur*, **2**, 1 (1976).
16. D. Barnard, L. Bateman, E. R. Cole and J. I. Cunneen, *Chem. Ind.* (*London*), 918 (1958).
17. D. Barnard, L. Bateman, M. E. Cain, J. Colclough and J. I. Cunneen, *J. Chem. Soc.*, 5339 (1961).
18. L. Bateman, M. Cain, J. Colclough and J. I. Cunneen, *J. Chem. Soc.*, 3570 (1962).
19. A. Rahman and A. Williams, *J. Chem. Soc.* (*B*), 1391 (1970).
20. J. I. Cunneen and D. F. Lee, *J. Appl. Polym. Sci.*, **8**, 699 (1964).
21. M. Mikołajczyk and J. Drabówicz, *Top. Stereochem.*, **13**, 333 (1982).
22. A. Nudelman, *Int. J. Sulfur Chem., B*, **7**, 241 (1972); *Phosphorus Sulfur*, **2**, 51 (1976); **9**, 1 (1980).
23. N. Nakamura, *J. Am. Chem. Soc.*, **105**, 7172 (1983).
24. S. Oae and T. Takata, *Kagaku* (*Chemistry Japan*), **34**, 756, 891, 961 (1979).
25. J. L. Kice, *Acc. Chem. Res.*, **1**, 58 (1968).
26. A. Schöberl and A. Wegner, *Methoden der Organischen Chemie*, **B9** (Ed. E. Müller), George Thieme Verlag, Stuttgart, 1955, pp. 691–693.
27. Reference 2, pp. 651–654.
28. G. C. Barnett, *Org. Compd. Sulphur, Selenium, Tellurium*, **1**, 106 (1970); **2**, 97 (1973); **3**, 82 (1975); **4**, 75 (1977); **5**, 67 (1979); **6**, 76 (1981).
29. S. Oae, *Organic Sulfur Chem.* (*Japanese*), Chap. 5, Kagaku Dojin, Kyoto, 1982, p. 131.
30. W. E. Savige and J. A. Maclaren, *The Chemistry of Organic Sulfur Compounds*, Vol. 2 (Eds. N. Kharasch and C. Y. Meyers), Chap. 15, Pergamon Press, London, p. 367.
31. J. L. Kice, *Sulphur in Organic and Inorganic Chemistry*, Vol. 1 (Ed. A. Senning), Chap. 6, Marcel Dekker, New York, 1971, p. 153.
32. S. Oae and G. Tsukamoto, *Kagaku* (*Chemistry Japan*), **26**, 172 (1971).
33. N. Isenberg and M. Grdinic, *Int. J. Sulfur. Chem.*, **8**, 307 (1973).
34. T. Zinke and F. Farn, *Ann. Chem.*, **391**, 55 (1912).
35. (a) T. Zinke and S. Lenhardt, *Ann. Chem.*, **400**, 1 (1913).
    (b) N. Kharasch, W. King and T. C. Bruice, *J. Am. Chem. Soc.*, **77**, 931 (1955).
36. T. Zinke and K. Eismayer, *Chem. Ber.*, **51**, 751 (1981).
37. O. Hinsberg, *Chem. Ber.*, **41**, 2838 (1908); **42**, 1278 (1909).
38. E. Vinkler and F. Klivenyi, *Acta Chim. Acad. Sci. Hung.*, **11**, 15 (1957); *Chem. Abstr.*, **52**, 6242 (1958).
39. E. Vinkler and F. Klivenyi, *Magy. Kem. Foly.*, **62**, 48 (1956); *Chem. Abstr.*, **53**, 273i (1959).
40. B. G. Boldyrev, L. P. Slesarchuk and T. A. Trovimova, *Khim. Seraorg. Soedin., Soderzh. Neftyakh Nefteprod.*, **8**, 108 (1968); *Chem. Abstr.*, **72**, 54934n (1970).
41. G. Ghersetti and G. Modena, *Spectrochim. Acta*, **19**, 1809 (1963).
42. P. Allen, Jr., J. Berner and E. R. Malinowski, *Chem. Ind.* (*London*), 1164 (1961); 208 (1963).
43. J. E. Baldwin, G. Koefle and S. E. Chun Choi, *J. Am. Chem. Soc.*, **93**, 2810 (1971).
44. S. Oae and S. Kawamura, *Bull. Chem. Soc. Jpn.*, **35**, 1156 (1962).
45. A. Stoll and E. Seebeck, *Adv. Enzym.*, **11**, 377 (1951).

46. (a) N. Kharasch, S. J. Petempa and H. L. Wehrweister, *Chem. Rev.*, **39**, 269 (1946).
(b) D. Barnard, J. M. Fabin and H. P. Koch, *J. Chem. Soc.*, 2442 (1949).
47. (a) E. L. Patterson, J. A. Brockman, Jr. and F. P. Day, *J. Am. Chem. Soc.*, **73**, 5919 (1951).
(b) L. J. Reed, I. C. Gunsalus, G. H. F. Schnakerberg, Q. F. Soper, H. E. Boaz, S. F. Kernan and T. L. Parke, *J. Am. Chem. Soc.*, **75**, 1267 (1953).
(c) L. J. Reed and I. C. Gunsalus, *U.S. Patent*, 3,049,549 (1962); *Chem. Abstr.*, **59**, 11498a (1963).
(d) L. J. Reed, *The Chemistry of Organic Sulfur Compounds*, Vol. 1 (Ed. N. Kharasch), Chap. 36, Pergamon Press, 1961, pp. 443–452.
(e) S. J. Wratten and D. J. Faulkner, *J. Org. Chem.*, **41**, 2465 (1976).
(f) A. Kato and T. Okutani, *Tetrahedron Lett.*, 2959 (1972).
48. H. J. Backer and H. Kloosterziel, *Recl. Trav. Chim. Pays-Bas.*, **73**, 129 (1954).
49. S. Oae and T. Takata, *Kagaku* (*Chemistry Japan*), **34**, 891 (1979).
50. S. Kozuka, H. Takahashi and S. Oae, *Bull. Chem. Soc. Jpn.*, **43**, 129 (1970).
51. E. Block and J. O'Connor, *J. Am. Chem. Soc.*, **96**, 3921 (1974).
52. L. D. Small, J. H. Bailey and C. J. Cavallito, *J. Am. Chem. Soc.*, **71**, 3565 (1949).
53. Y. Noguchi, M. Isoda, K. Kuroki and M. Furukawa, *Chem. Pharm. Bull.*, **30**, 1646 (1982).
54. M. Furukawa, T. Ohkawara, Y. Noguchi, M. Isoda and T. Hitoshi, *Synthesis*, 937 (1980).
55. T. J. Maricichi and C. N. Angeletakis, *J. Org. Chem.*, **49**, 1931 (1984).
56. S. Oae, T. Takata and Y. H. Kim, *Bull. Chem. Soc. Jpn.*, **55**, 2484 (1982).
57. A. K. Bhattacharya and A. G. Hortman, *J. Org. Chem.*, **43**, 2728 (1978).
58. G. Leandri and A. Tundo, *Ann. Chim.* (*Rome*), **44**, 74 (1954).
59. W. Walter and P. M. Hell, *Justus Liebigs Ann. Chem.*, **727**, 35, 50 (1969).
60. G. Jacini and F. Laurica, *Gazz. Chim. Ital.*, **80**, 762 (1950).
61. W. E. Savige, J. Eager, J. A. MacLaren and C. M. Roxburgh, *Tetrahedron Lett.*, 3289 (1964).
62. A. Schöberl, H. Tausent and H. Graéfje, *Angew. Chem.*, **68**, 213 (1956).
63. P. Allen, Jr. and J. W. Brook, *J. Org. Chem.*, **27**, 1019 (1962).
64. D. N. Harpp and A. Granata, *Synthesis*, 782 (1978).
65. H. Nogami, J. Hasegawa and K. Aoki, *Chem. Pharm. Bull.*, **19**, 2471 (1971).
66. N. Isenberg and H. F. Herbrandson, *Int. J. Sulfur Chem, A*, **1**, 179 (1971).
67. S. Oae and T. Takata, *Tetrahedron Lett.*, **21**, 3213 (1980).
68. J. A. Barltrop, P. M. Hayes and M. Calvin, *J. Am. Chem. Soc.*, **76**, 4348 (1954).
69. M. Calvin, H. Grizaback and R. C. Fuller, *J. Am. Chem. Soc.*, **77**, 2659 (1955).
70. K. Fujii, S. Shimizu and S. Fukui, *Vitamins*, **34**, 357 (1966).
71. R. W. Murray, R. D. Smetana and E. Block, *Tetrahedron Lett.*, 299 (1971).
72. R. W. Murray and S. L. Jindal, *Photochem. Photobiol.*, **16**, 147 (1972).
73. R. W. Murray and S. L. Jindal, *J. Org. Chem.*, **37**, 3516 (1972).
74. F. E. Stary, S. L. Jindal and R. W. Murray, *J. Org. Chem.*, **40**, 58 (1975).
75. R. W. Murray and M. L. Kaplan, *J. Am. Chem. Soc.*, **90**, 537, 4161 (1968); **91**, 5358 (1969).
76. L. Weil, *Arch. Biochem. Biophys.*, **110**, 57 (1965).
77. J.-J. Liang, C.-L. Gu, M. L. Kacher and C. S. Foote, *J. Am. Chem. Soc.*, **105**, 4717 (1983).
78. W. Ando and T. Takata, *Singlet $O_2$*, Vol. III (Ed. A. A. Frimer), Chap. 1, CRC Press, Florida, 1985, p. 1.
79. G. Bergson, *Acta Chem. Scand.*, **15**, 1611 (1961).
80. B. Lindberg and G. Bergson, *Ark. Kémi.*, **23**, 319 (1965).
81. C. Frisell and G. Bergson, *Ark. Kémi*, **25**, 263 (1966).
82. S. Tamagaki, H. Hirota and S. Oae, *Bull. Chem. Soc. Jpn.*, **46**, 1247 (1973).
83. S. Oae, T. Nabeshima and T. Takata, *Heterocycles*, **18**, 41 (1982).
84. L. Field and Y. H. Khim, *J. Org. Chem.*, **37**, 2710 (1970).
85. Ref. 24, p. 961.
86. D. Fukushima, Y. H. Kim, T. Iyanagi and S. Oae, *J. Biochem.*, **83**, 1019 (1978).
87. F. Davis, R. Jenkins and S. G. Yoklorich, *Tetrahedron Lett.*, 5171 (1978).
88. E. Vinkler, F. Klivenyi and J. Szabo, *Acta Chim. Acad. Sci. Hung.*, **15**, 385 (1958).
89. L. DiNunno, G. Modena and G. Scorrano, *Ric. Sci.*, **36**, 825 (1966).
90. H. J. Backer, *Recl. Trav. Chim. Pays-Bas.*, **70**, 95, 99 (1951).
91. H. J. Backer, *Recl. Trav. Chim. Pays-Bas.*, **71**, 418 (1952).
92. W. S. Allison, *Acc. Chem. Res.*, **9**, 293 (1976).
93. D. R. Hogg, *Comprehensive Organic Chemistry*, Vol. 3 (Ed. D. N. Jones), Pergamon Press, Oxford, 1979, p. 261.

94. T. C. Bruice and A. B. Sayigh, *J. Am. Chem. Soc.*, **81**, 3416 (1959).
95. K. Fries, *Chem. Ber.*, **45**, 965 (1912).
96. T. C. Bruice and R. T. Markiw, *J. Am. Chem. Soc.*, **79**, 3150 (1957).
97. W. Jenny, *Helv. Chim. Acta*, **41**, 317 (1958).
98. M. Ikehara, M. Kaneko, Y. Osigo and T. Morii, *Proc. Third Int. Congr. Heterocyclic Chem.*, Sendai, Japan, 1971, p. 98.
99. S. P. Kulkolja and S. R. Lammert, *J. Am. Chem. Soc.*, **94**, 7169 (1972).
100. T. S. Chou, J. R. Burgtorf, A. L. Ellis, S. R. Lammert and S. P. Kukolja, *J. Am. Chem. Soc.*, **96**, 1609 (1974).
101. A. Heckel and W. Pfleiderer, *Tetrahedron Lett.*, **24**, 5047 (1983).
102. T. C. Bruice and A. B. Sayigh, *J. Am. Chem. Soc.*, **81**, 3416 (1959).
103. R. T. Abraham, L. M. Benson and J. Jardine, *J. Med. Chem.*, **26**, 1523 (1983).
104. W. Walter and K. D. Bode, *Justus Liebigs Ann. Chem.*, **698**, 122 (1966).
105. W. Walter and K. Wohlers, *Justus Liebigs Ann. Chem.*, **752**, 115 (1971).
106. D. H. R. Barton, D. G. T. Greig, G. Lucente, P. G. Sammes, M. V. Taylor, G. H. Hewitt, B. E. Looker, A. Mowatt, C. A. Robson and W. G. E. Underwood, *J. Chem. Soc., Perkin Trans. 1*, 1187 (1973).
107. J. R. Shelton and K. E. Davis, *J. Am. Chem. Soc.*, **89**, 718 (1967).
108. T. Colclough and J. I. Cunneen, *Chem. Ind. (London)*, 626 (1960).
109. T. J. Maricichi and C. K. Harrington, *J. Am. Chem. Soc.*, **94**, 5115 (1972).
110. A. Stoll and E. Seebeck, *Experientia*, **3**, 114 (1947).
111. A. Stoll and E. Seebeck, *Helv. Chim. Acta*, **31**, 189 (1948).
112. A. Stoll and E. Seebeck, *Helv. Chim. Acta*, **32**, 197 (1949).
113. K. Kondo, A. Negishi and I. Ojima, *J. Am. Chem. Soc.*, **94**, 5786 (1972).
114. W. Gombler, *Angew. Chem., Int. Ed. Engl.*, **16**, 723 (1977).
115. J. D. Morrison and H. S. Mosher, *Asymmetric Organic Reactions*, Prentice-Hall, Englewood Cliffs, New Jersey, 1971, pp. 349–351.
116. F. Wudl and R. Gruber, *Tetrahedron Lett.*, 2133 (1969).
117. L. Sagramora, P. Koch, A. Garbesi and A. Fava, *J. Chem. Soc., Chem. Commun.*, 985 (1967).
118. J. L. Kice and G. B. Large, *Tetrahedron Lett.*, 3537 (1965).
119. W. E. Savige and A. Fava, *Chem. Commun.*, 417 (1965).
120. M. Mikołajczyk and J. Drabówicz, *J. Am. Chem. Soc.*, **100**, 2510 (1978).
121. M. Mikołajczyk, J. Drabówicz and F. Cramer, *J. Chem. Soc. (D)*, 317 (1971).
122. M. Mikołajczyk and J. Drabówicz, *J. Chem. Soc., Chem. Commun.*, 220 (1976).
123. M. Kishi, S. Ishihara and T. Komeno, *Tetrahedron*, **30**, 2135 (1974).
124. M. Mikołajczyk and J. Drabówicz, *J. Chem. Soc., Chem. Commun.*, 775 (1974).
125. W. H. Pirkle, S. D. Beere and R. L. Muntz, *J. Am. Chem. Soc.*, **91**, 4575 (1969).
126. L. E. Legler, S. L. Jindal and R. W. Murray, *Tetrahedron Lett.*, 3907 (1972).
127. T. Takata, Y. H. Kim and S. Oae, *Tetrahedron Lett.*, 4303 (1978).
128. T. Takata, K. Iida and S. Oae, *Heterocycles*, **15**, 847 (1981).
129. (a) N. Furukawa, T. Morishita, T. Akasaka and S. Oae, *J. Chem. Soc., Perkin Trans. 2*, 432 (1980).
(b) F. Freeman, C. N. Angeletakis and T. J. Maricichi, *Org. Magn. Reson.*, **17**, 53 (1981).
130. E. Block and J. O'Connor, *J. Am. Chem. Soc.*, **96**, 3929 (1974).
131. T. L. Cottrell, *The Strength of Chemical Bonds*, Academic Press, New York, 1954.
132. E. Block, *Reactions of Organosulfur Compds*, Academic Press, New York, 1978, p. 302.
133. P. Koch, E. Ciuffarin and A. Fava, *J. Am. Chem. Soc.*, **92**, 5971 (1970).
134. D. Barnard, *J. Chem. Soc.*, 4675 (1957).
135. L. Senatore, E. Ciuffarin and A. Fava, *J. Am. Chem. Soc.*, **92**, 3035 (1970).
136. P. Koch, E. Ciuffarin and A. Fava, *Int. J. Sulfur Chem., C*, **6**, 167 (1971).
137. J. L. Kice, C. G. Venier, G. B. Large and L. Heasley, *J. Am. Chem. Soc.*, **91**, 2028 (1969).
138. J. L. Kice and G. B. Large, *J. Am. Chem. Soc.*, **90**, 4069 (1968).
139. J. L. Kice and J. P. Cleaveland, *J. Am. Chem. Soc.*, **95**, 104, 109 (1973).
140. I. Utsumi, T. Watanabe, K. Harada, K. Kohno and G. Tsukamoto, *J. Vitaminol. (Kyoto)*, **13**, 26 (1967).
141. I. Utsumi, T. Watanabe, K. Harada and G. Tsukamoto, *Chem. Pharm. Bull.*, **15**, 1485 (1967).
142. S. Oae, Y. Yoshikawa and W. Tagaki, *Bull. Chem. Soc. Jpn.*, **42**, 2899 (1969).

143. S. Oae, K. Nomura, Y. Yoshikawa and W. Tagaki, *Bull. Chem. Soc. Jpn.*, **42**, 2903 (1969).
144. J. L. Kice and T. E. Rogers, *J. Am. Chem. Soc.*, **96**, 8009, 8015 (1974).
145. S. Oae, T. Takata and Y. H. Kim, *Tetrahedron Lett.*, 4219 (1977).
146. T. Takata and S. Oae, *Bull. Chem. Soc. Jpn.*, **55**, 3937 (1982).
147. (a) E. Vinkler, F. Klivenyi and E. Klivenyi, *Acta Chim. Acad. Sci. Hung.*, **16**, 247 (1958); *Chem. Abstr.*, **53**, 8041a (1959).
(b) E. Vinkler, F. Klivenyi and J. Pentye, *Acta Chim.* (*Budapest*), **65**, 333 (1970); *Chem. Abstr.*, 3383r (1971).
148. T. Takata, Y. H. Kim and S. Oae, *Tetrahedron Lett.*, 821 (1979).
149. S. Oae, T. Takata and Y. H. Kim, *Tetrahedron*, **37**, 37 (1981).
150. S. Oae, T. Takata and Y. H. Kim, *Bull. Chem. Soc. Jpn.*, **54**, 2712 (1981).
151. S. Oae and T. Takata, *Tetrahedron Lett.*, **21**, 3689 (1980).
152. M. Furukawa, S. Tsuji, Y. Kojima and S. Hayashi, *Chem. Pharm. Bull.*, **21**, 2391 (1973).
153. M. Furukawa, S. Tsuji, Y. Kojima and S. Hayashi, *Chem. Pharm. Bull.*, **21**, 1965 (1973).
154. T. Anstad, *Acta Chem. Scand., Ser. A*, **29**, 241 (1975).
155. S. Oae, *Kagaku* (*Chemistry Japan*), **33**, 240 (1978).
156. T. F. Lavine, G. Toennies and E. D. Wagner, *J. Am. Chem. Soc.*, **56**, 242 (1934).
157. G. Toennies and T. F. Lavine, *J. Biol. Chem.*, **113**, 571 (1936).
158. T. F. Lavine, *J. Biol. Chem.*, **113**, 583 (1936).
159. G. Medes and N. Floyd, *Biochem. J.*, **31**, 1330 (1939).
160. R. Consden and A. H. Gordon, *Biochem. J.*, **46**, 8 (1950).
161. D. Barnard, *J. Chem. Soc.*, 4673 (1957).
162. R. Emiliozzi and L. Pichat, *Bull. Soc. Chim. Fr.*, 1887 (1959).
163. U. Marangeli, G. Modena and P. E. Todesco, *Gazz. Chim. Ital.*, **90**, 681 (1960).
164. G. Modena and P. E. Todesco, *Ric. Sci.*, **30**, 1788 (1960).
165. G. E. Utzinger, *Experientia*, **17**, 374 (1961).
166. R. Ratz and O. Sweeting, *J. Org. Chem.*, **28**, 1612 (1963).
167. F. Wudl, D. A. Lightner and D. J. Cram, *J. Am. Chem. Soc.*, **89**, 4099 (1967).
168. R. V. Norton, G. M. Beverly and I. B. Douglass, *J. Org. Chem.*, **32**, 3645 (1967).
169. D. Barnard and E. J. Percy, *Chem. Ind.* (*London*), 1332 (1969).
170. A. Padwa and R. Gruber, *J. Org. Chem.*, **35**, 1781 (1970).
171. M. M. Chau and J. L. Kice, *J. Am. Chem. Soc.*, **98**, 7711 (1976).
172. S. Oae, Y. H. Kim, T. Takata and D. Fukushima, *Tetrahedron Lett.*, 1195 (1977).
173. B. C. Gilbert, B. Gill and M. J. Ramsden, *Chem. Ind.* (*London*), 283 (1979).
174. F. Freeman, C. N. Angeletakis and T. J. Maricichi, *Tetrahedron Lett.*, **22**, 1867 (1981).
175. F. Freeman and C. N. Angeletakis, *J. Org. Chem.*, **46**, 3991 (1981).
176. F. Freeman and C. N. Angeletakis, *J. Am. Chem. Soc.*, **103**, 6232 (1981).
177. H. Noguchi, J. Hasegawa and K. Aoki, *Chem. Pharm. Bull.*, **19**, 2471 (1971).
178. S. Oae, T. Takata and Y. H. Kim, *Bull. Chem. Soc. Jpn.*, **55**, 2484 (1982).
179. F. Freeman, C. N. Angeletakis, W. J. Pietro and W. J. Hehre, *J. Am. Chem. Soc.*, **104**, 1161 (1982).
180. F. Freeman and C. N. Angeletakis, *J. Am. Chem. Soc.*, **104**, 5766 (1982).
181. F. Freeman and C. N. Angeletakis, *J. Am. Chem. Soc.*, **105**, 4039 (1983).
182. J. L. Kice and K. Ikura, *J. Am. Chem. Soc.*, **90**, 7378 (1968).
183. Y. H. Kim, T. Takata and S. Oae, *Tetrahedron Lett.*, 2305 (1978).
184. T. Takata, Y. H. Kim and S. Oae, *Bull. Chem. Soc. Jpn.*, **54**, 1443 (1981).
185. S. Oae and T. Takata, *Tetrahedron Lett.*, **21**, 3213 (1980).
186. S. Oae, D. Fukushima and Y. H. Kim, *Chem. Lett.*, 297 (1978).
187. A. Schöberl, H. Tausent and H. Graefje, *Angew. Chem.*, **68**, 213 (1956).
188. A. Schöberl and H. Graefje, *Ann. Chem.*, **617**, 71 (1958).
189. J. L. Kice and G. B. Large, *J. Org. Chem.*, **33**, 1940 (1968).
190. J.-L. Ju, J. L. Kice and C. G. Venier, *J. Org. Chem.*, **44**, 610, 1918 (1979).
191. G. Tsukamoto, T. Watanabe and T. Utsumi, *Bull. Chem. Soc. Jpn.*, **42**, 2566 (1969).
192. T. J. Wallace, *J. Am. Chem. Soc.*, **86**, 2018 (1964).
193. M. S. Kharasch, W. Nundnberg and G. J. Mantell, *J. Org. Chem.*, **16**, 524 (1951).
194. H. Bretschneider and W. Klotzer, *Monatsch. Chem.*, **81**, 589 (1950).
195. H. Kawasaki, *J. Pharm. Sci. Jpn.* (*Yakugaku Zasshi*), **73**, 712 (1953).
196. J. F. Carson and F. F. Wong, *J. Org. Chem.*, **26**, 1467 (1961).

197. S. Oae and T. Numata, *Isotopes in Organic Chemistry*, Vol. 5 (Eds. E. Buncel and E. C. Lee), Chap. 2, Elsevier, Amsterdam, 1980, pp. 45–145.
198. I. Saito and S. Fukui, *J. Vitaminol.* (*Kyoto*), **12**, 244 (1966).
199. N. Furukawa, T. Morishita, T. Akasaka and S. Oae, *Tetrahedron Lett.*, 1653 (1977).
200. N. Furukawa, T. Morishita, T. Akasaka and S. Oae, *Tetrahedron Lett.*, 1567 (1978).
201. N. Furukawa, T. Morishita, T. Akasaka and S. Oae, *Tetrahedron Lett.*, 3973 (1979).
202. E. Block, *J. Org. Chem.*, **39**, 734 (1974).
203. K. Kondo and A. Negishi, *Chem. Lett.*, 1525 (1974).
204. T. Morishita, N. Furukawa and S. Oae, *Tetrahedron*, **37**, 2539 (1981).
205. T. Morishita, N. Furukawa and S. Oae, *Tetrahedron*, **37**, 3115 (1981).
206. J. F. Carson and F. F. Wong, *J. Org. Chem.*, **26**, 3028 (1961).
207. E. Vinkler and F. Klivenyi, *Acta Chim. Acad Sci. Hung.*, **57**, 91 (1968).
208. J. L. Kice, C. G. Venier and L. Heasley, *J. Am. Chem. Soc.*, **89**, 3557 (1967).
209. J. Pintyc, G. Stajer and E. Vinkler, *Acta Chim.* (*Budapest*), **95**, 307 (1977).

The Chemistry of Sulphinic Acids, Esters and their Derivatives
Edited by S. Patai


CHAPTER **19**

# Sulphinyl chlorides and sulphinic anhydrides

J. G. TILLETT

*Department of Chemistry and Biological Chemistry, University of Essex, Colchester, CO4 3SQ, UK*



## I. SULPHINYL CHLORIDES

### A. Synthesis

The syntheses of a large number of sulphinyl chlorides have been reported[1–3]. The lower molecular weight aliphatic analogues and benzenesulphinyl chloride are liquids; substituted arenesulphinyl chlorides are generally solids. Although various sulphinyl chlorides have been distilled as part of the purification procedures, this should only be carried out with care at low pressure, particularly for arenesulphinyl chlorides for which explosions have been reported[4].

Sulphinyl chlorides are particularly sensitive to moisture. Thus, in the presence of a limited amount of water (mole ratio 3:1) methanesulphinyl chloride decomposes to form methanesulphonyl chloride **1** and methylmethanesulphonate **2** according to the stoichiometry shown in equation 1[5]. In the presence of larger amounts of water, initially considerable quantities of methanesulphinic acid, $CH_3SO_2H$ (**3**), are formed although the final products are mainly **1** and **2** with only a small quantity of **3**.

$$3CH_3SOCl + H_2O \longrightarrow \underset{(1)}{CH_3SO_2Cl} + \underset{(2)}{CH_3SO_2SCH_3} + 2HCl \quad (1)$$

Alkanesulphinyl chlorides cannot be safely stored at room temperature in sealed containers[6]. Disproportionation occurs according to equation 2.

$$2CH_3SOCl \longrightarrow CH_3SO_2Cl + CH_3SCl \tag{2}$$

Subsequent decomposition of the sulphenyl halide can form a variety of decomposition products including hydrogen chloride, so that high pressure many develop[7].

The earliest reported methods for the preparation of sulphinyl chlorides involved the reaction of thionyl chloride with the free sulphinic acid[8–11] or its sodium salt[12]. These methods suffered from a number of disadvantages. They were often difficult to reproduce; in many cases the sulphinic acids required were not commercially available and the product was often difficult to extract from the reactants[13,14].

A greatly improved set of synthetic methods for sulphinyl chlorides was developed by Douglass and his coworkers. The first of these evolved from a study of the solvolysis of organosulphur trichlorides. These readily undergo reaction with any hydroxylic solvent such as water, an alcohol or a carboxylic acid to give sulphinyl chlorides in good yield (*ca* 90% for R = Me, Et) (equation 3)[15]. Handling of the very reactive trichloride and the use of the large volumes of solvent necessary in this method are avoided by the formation of the trichloride *in situ* by chlorination of exactly one mole of a disulphide in two moles of a carboxylic acid as solvent (equation 4)[16]. On addition of chlorine, the disulphide is initially converted to the characteristically reddish orange sulphenyl chloride, RSCl, which is subsequently transformed as shown in equation 5. Yields for simple alkanesulphinyl halides and for benzenesulphinyl chlorides were *ca* 90%. Water or alcohols cannot be used as solvents for this method because a variety of different side-products are formed[17].

$$RSCl_3 + R'OH \longrightarrow RS{\stackrel{\displaystyle O}{\diagup\!\!\diagup}\atop\diagdown Cl} + HCl + R'Cl \tag{3}$$

$$RSSR + 2CH_3CO_2H + 3Cl_2 \longrightarrow 2RS{\stackrel{\displaystyle O}{\diagup\!\!\diagup}\atop\diagdown Cl} + 2CH_3COCl + 2HCl \tag{4}$$

$$RSSR \xrightarrow{Cl_2} RSCl \xrightarrow{Cl_2} RSCl_3 \xrightarrow{CH_3CO_2H} RS{\stackrel{\displaystyle O}{\diagup\!\!\diagup}\atop\diagdown Cl} + CH_3COCl \tag{5}$$

Both the previous methods involve the production of large volumes of hydrogen chloride. To minimize handling problem, Douglass and Norton changed the solvent to acetic anhydride so that no gaseous products are formed (equation 6)[4]. With the chlorination reactions, whatever the solvent, it was found to be important to use strictly stoichiometric quantities of reagents, otherwise contamination of the product by side-reactions readily occurs. A further modification has been suggested in which sulphinyl chlorides can be prepared in > 80% yield by chlorination of thiolesters in acetic anhydride (equation 7)[17]. This method proved to be particularly effective for the synthesis of benzenesulphinyl chloride (> 98% yield).

$$RSSR + 2(CH_3CO)_2O + 3Cl_2 \longrightarrow 2RS{\stackrel{\displaystyle O}{\diagup\!\!\diagup}\atop\diagdown Cl} + 4CH_3COCl \tag{6}$$

$$CH_3\overset{\displaystyle O}{\overset{\|}{C}}SCH_3 \xrightarrow{Cl_2,\ Ac_2O} CH_3S{\stackrel{\displaystyle O}{\diagup\!\!\diagup}\atop\diagdown Cl} \tag{7}$$

None of the chlorination methods described above, however, proved successful for the synthesis of *t*-butylsulphinyl chloride. Chlorination of both *t*-butyl disulphide and *t*-butyl thiolacetate gives *t*-butyl chloride as the major product. *t*-Butylsulphinyl chloride can, however, be prepared either by the reaction of thionyl chloride with 2-methylpropane-2-sulphinic acid[18] or by chlorination of the corresponding thiolsulphinate via the route shown in equation 8[19].

$$\mathrm{Bu^tS{-}SBu^t} \xrightarrow[0\,^\circ\mathrm{C}]{30\%\ \mathrm{H_2O_2}} \mathrm{Bu^tS(=O){-}SBu^t} \xrightarrow[10\,^\circ\mathrm{C}]{\mathrm{Cl_2/CHCl_3}} \mathrm{Bu^tS(=O)Cl} \qquad (8)$$

More recently, Hermann and his group have shown that chlorination of disulphides can be effected at low temperatures with the more conveniently handled sulphuryl chloride and leads to the formation of sulphinyl chlorides in almost quantitative yields (equation 9)[20]. It is interesting to note that this method does work for the synthesis of *t*-butylsulphinyl chloride which is formed in 86% yield. The method provides a cleaner general route to sulphinyl chlorides, providing that no acid sensitive groups are present in the disulphide. In a modification of this method, Hermann has shown that sulphinyl chlorides can also be prepared in essentially quantitative yield by low temperature (−40 °C) chlorination of thiols by sulphinyl chloride (equation 10)[21]. Netscher and Prinzbach showed that the sterically hindered 2,2,2-trifluoro-1,1-diphenylethanesulphinyl chloride, **4**, can also be prepared in high yield from the corresponding thiol (equation 11)[22].

$$\mathrm{RSSR + 3SO_2Cl_2 + 2CH_2CO_2H \longrightarrow 2RS(=O)Cl + 2CH_3COCl} \qquad (9)$$

R = Me, $Pr^i$, $Bu^t$, Ph, $PhCH_2$

$$\mathrm{RSH + 2SO_2Cl_2 + CH_3CO_2H \xrightarrow{-40\,^\circ C} RS(=O)Cl + CH_3COCl + SO_2 + HCl} \qquad (10)$$

R = Et, $Pr^i$, Ph, $PhCH_2$, 4-Tol

$$\mathrm{Ph_2C(CF_3)SH} \xrightarrow[30\,^\circ\mathrm{C}]{\mathrm{SO_2Cl_2}} \mathrm{Ph_2C(CF_3)SCl} \xrightarrow[0\,^\circ\mathrm{C}\rightarrow\mathrm{r.t.}]{\mathrm{CF_3CO_2H}} \mathrm{Ph_2C(CF_3)S(=O)Cl} \qquad (11)$$

(**4**)

## B. Chiral Properties

Like other sulphinyl systems, sulphinyl chlorides **5** have a chiral sulphur centre arising from the tetrahedral orientation of groups around sulphur. Such chirality was first demonstrated by King and Beatson who showed that the NMR spectrum of the sulphinyl chloride **6**, obtained from 1-chloroethanesulphinic acid, indicated the presence of two diastereomers arising from chiral centres at both carbon and sulphur[23]. Magnetic non-equivalence of protons or methyl groups in ethyl and isopropyl sulphinyl chlorides containing only a sulphur chiral centre has also been observed (**7**, **8**)[24].

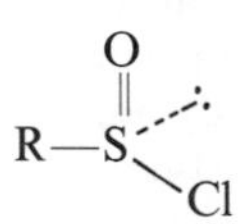


(**5**)

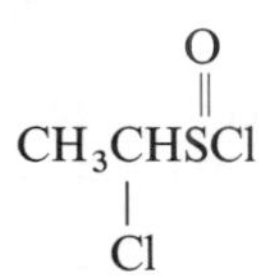


(**6**)

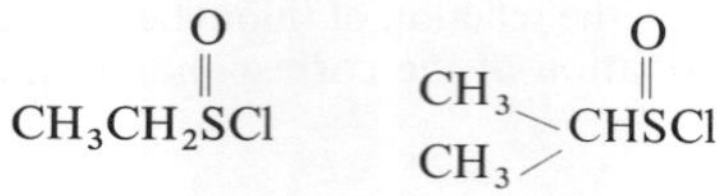


(7) (8)

In a detailed study Pizey and his coworkers showed that the NMR spectra of a series of $\beta$-ketosulphinyl chlorides are both solvent and temperature dependent[25]. Magnetic non-equivalence of $\alpha$ *gem*-dimethyl groups was, in general, observed. The chiral effect of the chlorosulphinyl group in **9** on the protons $H_A$ and $H_B$ could be detected in benzene ($\Delta_{H_B-H_A} \sim 4$ Hz) but not in carbon tetrachloride. The temperature-dependent $^1$H and $^{13}$C NMR spectra of the sulphinyl chlorides **7** and **8** were also found to be consistent with a chiral sulphur centre[26]. Treatment of the penicillin **10** ($R^1$—phthalimido, $R^2$ = Me) with *N*-chlorosuccinimide gave an almost quantitative mixture of the two diastereomers of the sulphinyl chloride **11**[27,28].

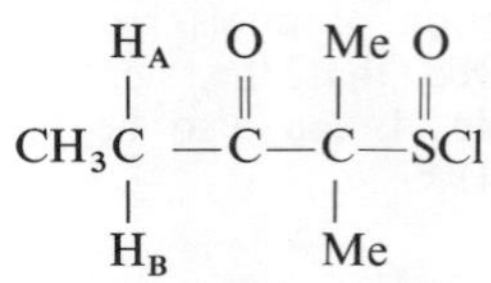


(9)

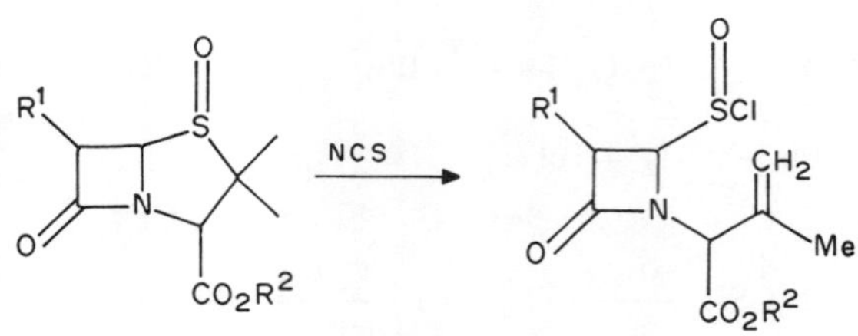


(10) (11)

### C. Reaction with Alcohols and Thiols

The reaction of sulphinyl chlorides with alcohols or phenols in the presence of bases such as pyridine or potassium carbonate provides the general method for the synthesis of sulphinate esters[1–3,29,30]. It has been clearly demonstrated by deuterium labelling experiments that methanolysis of sulphinyl chlorides in the presence of a base (pyridine or triethylamine) follows an $S_N2(S)$ reaction and does not proceed via a sulphine intermediate (equation 12)[31].

The preparation of chiral sulphinates has assumed increasing importance because of their use in the synthesis of other chiral sulphur compounds, e.g. sulphoxides, used in asymmetric reactions and stereochemical correlations.

Phillips was the first to prepare a mixture of the diastereomeric menthyl *p*-toluenesulphinates **12** and was able to isolate what is now known to be (−)-menthyl(−)-(*S*)-*p*-toluenesulphinate in a pure state (equation 13)[32]. This is a solid and can be separated from the liquid lower melting diastereomer (−)-menthyl (+)-(*R*)-*p*-toluenesulphinate.

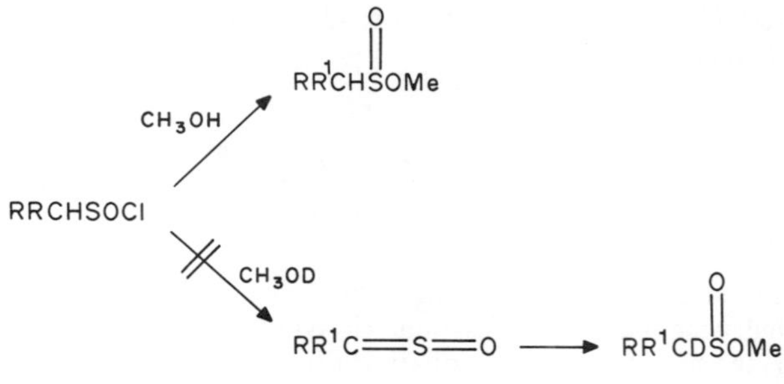 (12)

(R=R[1]=H; R=H,R[1]=Et)

$$p\text{-Tol}\overset{\overset{\displaystyle O}{\|}}{S}\text{Cl} + (-)\text{MenOH} \xrightarrow{\text{py}} p\text{-Tol}\overset{\overset{\displaystyle O}{\|}}{S}\text{OMen} \qquad (13)$$

(Men = Menthyl) **(12)** $[\alpha]_D^{25} - 210\,°C$

Mislow and his group have used asymmetric synthesis of menthyl sulphinates to establish the absolute configuration of sulphinate esters and sulfoxides[33,34]. Reaction of *p*-toluenesulphinyl chloride with a variety of optically active secondary alcohols gives a mixture of diastereomers which can be converted by a Grignard reaction to optically active methyl *p*-tolyl sulphoxide in high yield (equation 14)[35].

$$p\text{-Tol}\overset{\overset{\displaystyle O}{\|}}{S}\text{Cl} \xrightarrow[-78\,°C,\ \text{py}]{R^1OH} p\text{-Tol}\overset{\overset{\displaystyle O}{\|}}{S}\text{OR}^1 \xrightarrow{R^2MgBr} R^2\overset{\overset{\displaystyle O}{\|}}{S}\text{Tol-}p \qquad (14)$$

Because of the importance of (−)-(*S*)-**12**, considerable effort has been expended to improve its synthesis. Mislow and his group showed that the (−)-(*S*)-**12** diastereomer formed from *p*-toluenesulphinyl chloride is the minor product and the composition of the products is kinetically controlled[33,34]. Addition of hydrogen chloride gas has been used to epimerize the liquid (+)-(*R*)-diastereomer to (−)-(*S*)-**12**[36].

Whilst chiral diaryl and alkyl aryl sulphoxides can be synthesized relatively easily as shown in equation 14, it has not been possible until quite recently to utilize this method for the synthesis of dialkyl sulphoxides because the required sulphinates (e.g. menthyl methanesulphinate) are oils and not easily purified to produce a single epimer[37]. Solladie and his group recently devised experimental conditions under which diastereomeric sulphinate esters in acidic media readily undergo epimerization in favour of the less soluble isomer[38]. By this method (−)-menthyl (−)-(*S*)-*p*-toluenesulphinate was obtained from *p*-toluenesulphinic acid in 90% yield (equation 15). Andersen and his group have overcome the difficulties associated with the formation of menthyl sulphinates by using instead the

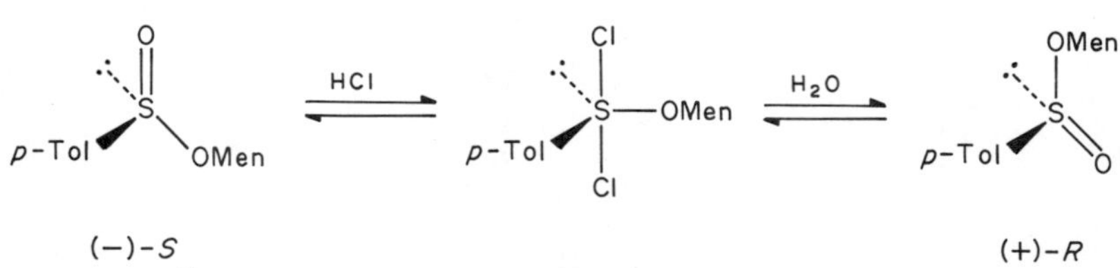 (15)

crystalline cholesteryl methanesulphinate to produce dialkyl sulphoxides of high enantiomeric purity (equation 16)[39].

$$\text{CholOH} \xrightarrow[\text{Et}_3\text{N, ether}]{\text{MeSOCl, 0–5 °C}} \text{CH}_3\overset{\text{O}}{\overset{\|}{\text{S}}}\text{OChol} \xrightarrow{\text{RMgX}} \text{R}\overset{\text{O}}{\overset{\|}{\text{S}}}\text{CH}_3 \quad (16)$$

(Chol = cholesteryl)

The reaction of sulphinyl chlorides with achiral alcohols in the presence of optically active amines has been used for the asymmetric synthesis of chiral sulphinates with the sulphur atom as the sole chiral centre (equation 17)[40]. The chiral-inducing amine is easily removed as the hydrochloride. The extent of asymmetric induction is comparable to that observed in the reaction of sulphinyl chlorides with achiral alcohols. Thus the reaction of *p*-toluenesulphinyl chloride with methanol in the presence of (−)-*N*,*N*-dimethylmenthylamine gave the methyl ester with optical purity 20.6%.

Harpp and his group have described the use of the trimethylsilyl group in the synthesis of sulphinate esters (equation 18)[41]. Thus benzenesulphinyl chloride reacts with neat menthoxytrimethylsilane to give the crude diastereomeric mixture of **13** in 91% yield (equation 19). Two different mechanistic pathways were considered for this reaction involving either nucleophilic attack of the ether oxygen at sulphinyl sulphur[42] or a four-centre transition state (equation 20)[43].

$$\text{R}^1\overset{\text{O}}{\overset{\|}{\text{S}}}\text{Cl} + \text{R}^2\text{OH} \xrightarrow[-70\,°\text{C, ether}]{\text{Me}_2{}^*\text{NR}^3} \text{R}^1\overset{\text{O}}{\overset{\|}{\text{S}}}\text{OR}^2 \quad (17)$$

$$\text{R}\overset{\text{O}}{\overset{\|}{\text{S}}}\text{Cl} + \text{R}^1\text{OSiMe}_3 \longrightarrow \text{R}\overset{\text{O}}{\overset{\|}{\text{S}}}\text{OR}^1 + \text{Me}_3\text{SiCl} \quad (18)$$

$$\text{Ph}\overset{\text{O}}{\overset{\|}{\text{S}}}\text{Cl} + \text{Me}_3\text{SiOMen} \longrightarrow \text{Ph}\overset{\text{O}}{\overset{\|}{\text{S}}}\text{OMen} + \text{Me}_3\text{SiCl} \quad (19)$$

**(13)**

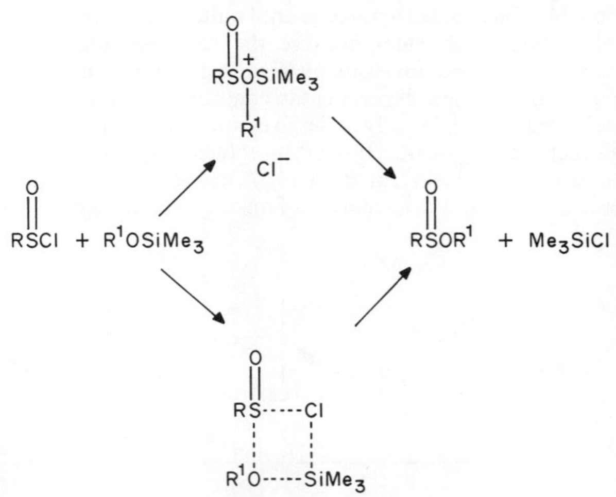


(20)

The relatively small solvent effects observed on reaction rates were thought to be consistent with a four-centre non-ionic transition state. More recently, however, a more extensive kinetic study of the reaction of methanesulphinyl chloride with *p*-substituted aryloxytrimethylsilanes has been reported[43,44]. The negative $\rho$ value observed ($\rho = -1.44$), the Arrhenius parameters and the solvent effects (covering a wider range of dielectric constant than the earlier study) were considered to be more consistent with an ionic mechanism, although the exact details of this mechanism remain to be established.

Early attempts to synthesize thiolsulphinates from the reaction of alkanesulphinyl halides and thiols were unsuccessful[45,46]. *t*-Butyl 2-methylpropane-2-thiolsulphinate **14** and a variety of unsymmetrical dialkyl thiolsulphinates have been subsequently prepared by modification of the conditions (equation 21)[18,47]. Alkyl or arylthioesters of aromatic sulphinic acids may be readily synthesized from arylsulphinyl chlorides and the corresponding thiols[48].

$$t\text{-Bu}\overset{\text{O}}{\overset{\|}{\text{S}}}\text{Cl} + \text{Bu}^t\text{SH} \longrightarrow t\text{-Bu}\overset{\text{O}}{\overset{\|}{\text{S}}}\text{SBu-}t \qquad (21)$$

Mikolajczyk and Drabowicz have shown that optically active thiolsulphinate *S*-esters can be prepared by the asymmetric condensation of sulphinyl chlorides with thiols in the presence of optically active tertiary amines (equation 22)[49].

$$\text{R}^1\overset{\text{O}}{\overset{\|}{\text{S}}}\text{Cl} + \text{R}^2\text{SH} \xrightarrow[-70\,^{\circ}\text{C, ether}]{\text{Me}_2\text{NR}} \text{R}^1\overset{\text{O}}{\overset{\|}{\text{S}}}\text{SR}^2 \qquad (22)$$

(R = amphetamine or α-fenchylamine; $R^1 = Bu^t$ or Ar)

Thiolsulphinates containing the *t*-BuS group were found to be optically stable at room temperature for several weeks, whilst those with other alkyl substituents (e.g. $R^2 = Et, Pr^i$) racemized within hours.

### D. Reaction with Nitrogen Nucleophiles/Bases

The reaction of sulphinyl chlorides with amines forms the principal method for the synthesis of both aliphatic and aromatic sulphinamides[1,2,14,46,50]. This method has also been used for the synthesis of *N,N*-dialkylalkanesulphinamides (equation 23)[51,52]. *N*-aryl-*N'*-arylsulphinylureas have been prepared by the analogous reaction (equation 24)[53]. The diastereomeric sulphinamides **15** and **16** have been prepared by the reaction of the appropriate racemic sulphinyl chloride and chiral amines as shown in equations 25 and 26[54,55].

$$\text{R}\overset{\text{O}}{\overset{\|}{\text{S}}}\text{Cl} + \text{R}^1\text{R}^2\text{NH} \longrightarrow \text{R}\overset{\text{O}}{\overset{\|}{\text{S}}}\text{NR}^1\text{R}^2 \qquad (23)$$

$$p\text{-Tol}\overset{\text{O}}{\overset{\|}{\text{S}}}\text{Cl} + \text{PhNHCONH}_2 \longrightarrow p\text{-TolSONHCONHPh} \qquad (24)$$

$$\text{Ph}\overset{\text{O}}{\overset{\|}{\text{S}}}\text{Cl} + \text{CH}_3\text{NHCH}_2\overset{*}{\text{C}}\text{H(CH}_3\text{)Ph} \longrightarrow \text{Ph}\overset{\text{O}}{\overset{\|}{\text{S}}}{}^{*}\text{N(CH}_3\text{)CH}_2\overset{*}{\text{C}}\text{H(CH}_3\text{)Ph} \qquad (25)$$

(**15**)

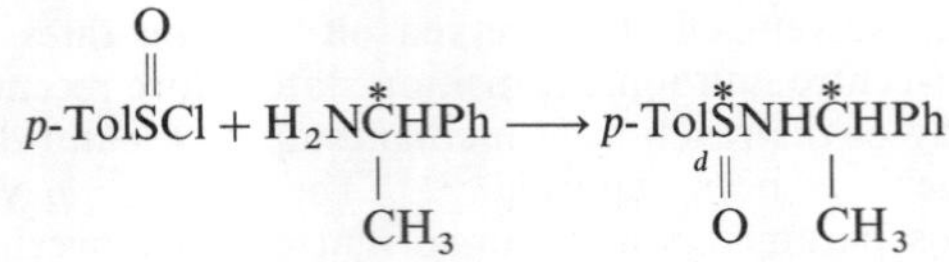
(26)

**(16)**

Whalen and Jones were the first to show that the reaction of hydroxylamine with a sulphinyl chloride does not form the expected *N*-sulphinyl derivative but the corresponding sulphonamide (equation 27)[13]. This reaction in fact provides a convenient synthesis of sulphonamides. Thus primary, secondary and tertiary tert-alkanesulphonamides can be obtained from *t*-butylsulphinyl chloride and the corresponding hydroxylamine (equation 28)[56].

$$\mathrm{PhSOCl + NH_2OH \not\longrightarrow PhSONHOH + HCl \qquad \searrow PhSO_2NH_2} \qquad (27)$$

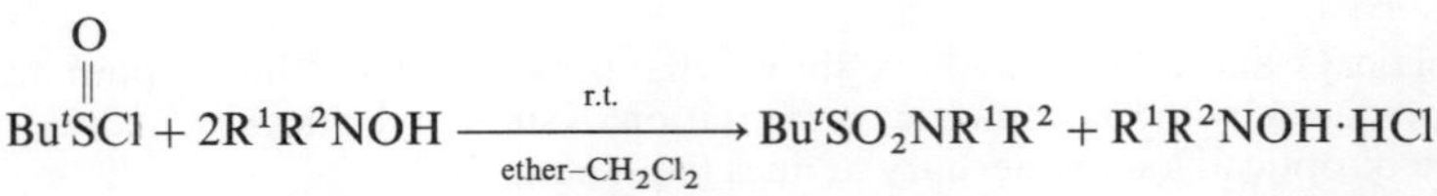
(28)

Several groups of workers have now established that the reaction of hydroxylamine and related compounds with sulphinyl chloride occurs via an *O*-sulphinylated intermediate which subsequently undergoes decomposition via a radical mechanism. Engberts and his group showed that the intermediate **17** formed in the reaction of *t*-butylsulphinyl chloride with ethyl *N*-hydroxycarbonate and leading to the formation of *N*-*t*-butylsulphonylcarbonate could be detected by NMR spectroscopy (equation 29)[57]. The reaction of *t*-butylsulphinyl chloride with *N*-hydroxyureas is considered to involve a similar intermediate **18** (equation 30)[58]. The large $^1$H-CIDNP effects observed during the

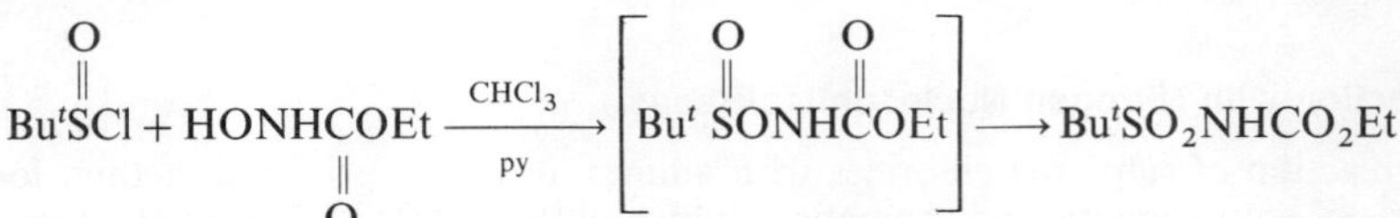
(29)

**(17)**

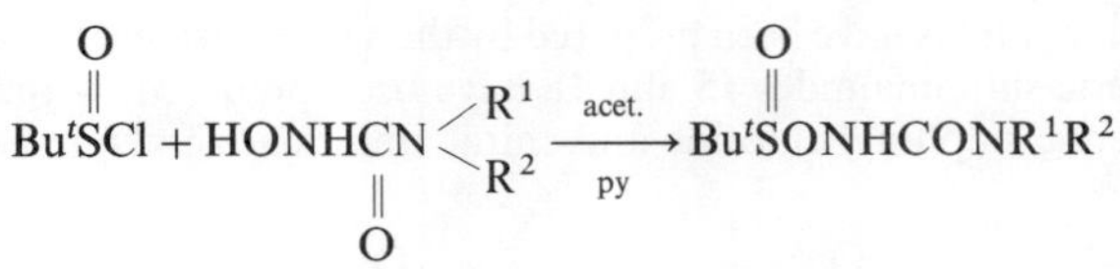
(30)

(e.g. $R^1 = H$; $R^2 = Me$) **(18)**

reaction are consistent with homolytic breakdown of **18** into carbamoylamino and *t*-butylsulphonyl radicals. Typically, products corresponding to both in-cage recombination (*N*-*t*-butylsulphonylureas) and escape products (alkyl ureas and sulphonic acids) were observed (equation 31). Hudson and his group proposed a similar mechanism for the reaction of sulphinyl chlorides with aldoximes and ketoximes (equation 32)[59–61]. The sulphinyl oximes **19** could be prepared in the presence of triethylamine in ether at − 20 °C and characterized by $^1$H and $^{13}$C NMR spectroscopy. They are reported to decompose explosively at room temperature but are quite stable in solution or as solids below − 30 °C. Above 0 °C they rearrange to the *N*-sulphonylimines.

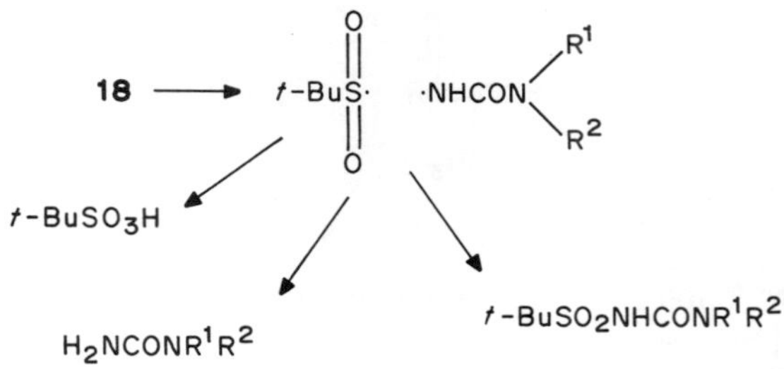


(31)

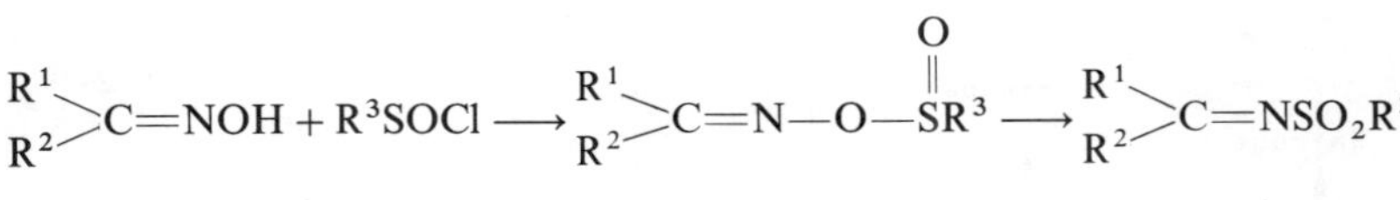

(19)

(32)

The *O*-sulphinylated intermediates **20** in the reaction of *N,N*-dialkylhydroxylamines with methyl and phenyl sulphinyl chloride have also been isolated and characterized spectroscopically (equation 33)[62,63].

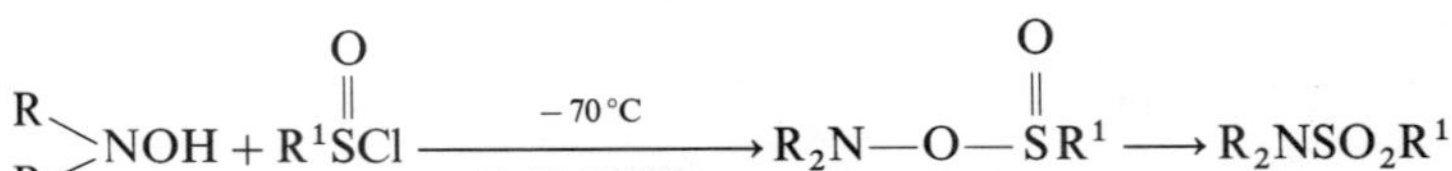

(20)

(33)

(R = Me, Et or $PhCH_2$; $R^1$ = Me or Ph)

*N*-Phenylhydroxamic acids have also been shown to react with sulphinyl chlorides to give *N*-acylsulphonamides via a radical pair mechanism[64]. More recently *O*-sulphinylated hydroxamic acids **21** were obtained as solids from *N*-methylbenzohydroxamic acids and sulphinyl chlorides (equation 34)[65].

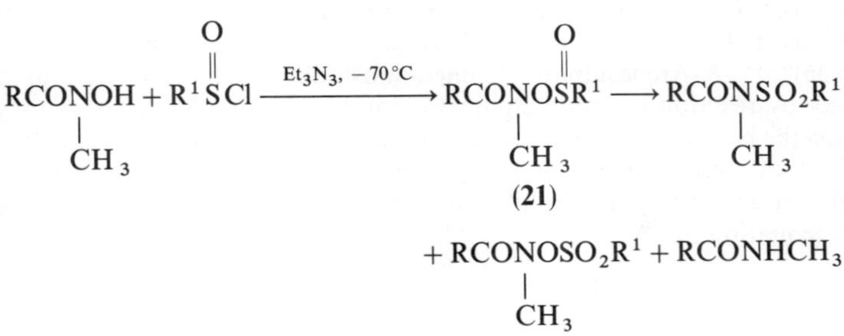

(34)

The intermediate decomposes at room temperature to form the isomeric *N*-acyl-*N*-methylsulphonamide and *N*-methyl-*O*-sulphonylhydroxamic acid. The reaction products are explained by the simultaneous low-temperature homolysis of two bonds in a molecule (Scheme 1)[65]. Initial nucleophilic attack at sulphinyl sulphur to form the *O*-sulphinylated intermediate **22** has also been proposed for the reaction of *t*-butylsulphinyl chloride and *N*-hydroxymethane and *N*-hydroxybenzenesulphonamides and their *N*-methyl substituted derivatives (equation 35)[66].

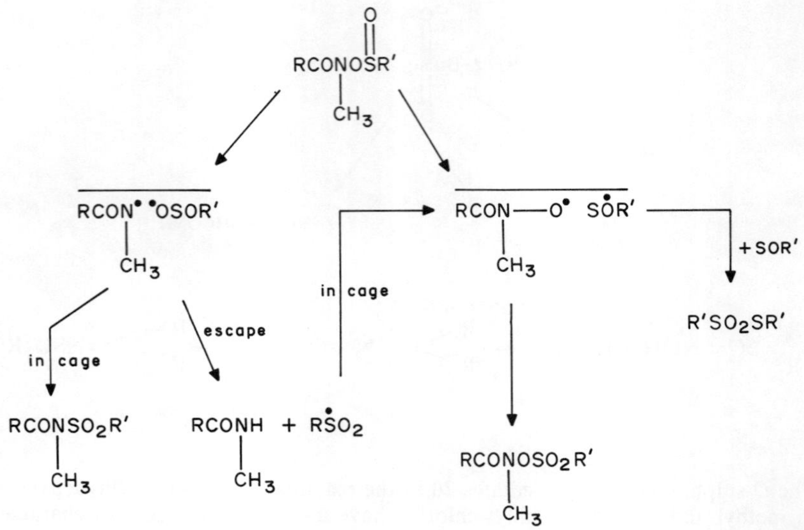


SCHEME 1

$$t\text{-BuS(O)Cl} + \text{R}^1\text{SO}_2\text{N(OH)R}^2 \xrightarrow[\text{acetone}]{\text{py, 20}\,^\circ\text{C}} t\text{-BuS(O)ON(R}^2\text{)SO}_2\text{R}^1\ (\mathbf{22}) \longrightarrow t\text{-BuSO}_2\text{NR}^2\text{SO}_2\text{R}^1 + \text{other products} \quad (35)$$

The reaction of sulphinyl chloride with positive *N*-halogen compounds (or their salts) has been used for the synthesis of arene and alkanesulphonimidoyl chlorides (equation 36)[67,68]. *N*-Arenesulphonylareneiminosulphonyl chlorides were prepared in a similar way by reaction of arenesulphinyl chloride with dry *N*-arenesulphonyl chloramides[69]. On the basis of a $^{36}$Cl-tracer study of the reaction of *p*-toluenesulphinyl chloride–$^{36}$Cl with chloramine T, Oae and his group proposed a mechanism involving initial nucleophilic attack at sulphinyl sulphur by the chloramine T anion followed by chlorine migration (equation 37)[70].

$$\text{RS(O)Cl} + \text{R}^1\text{NCl}_2 \longrightarrow \text{RS(O)(=NR}^1\text{)Cl} + \text{Cl}_2 \quad (36)$$

$$\text{ArS(O)Cl} + {}^-\text{N(Cl)SO}_2\text{Ar}^1 \longrightarrow \text{ArS(O)N(Cl)SO}_2\text{Ar}^1 + \text{NaCl} \longrightarrow \text{ArS(O)(=NSO}_2\text{Ar}^1\text{)Cl} \quad (37)$$

The related *N*-arylsulphinyldimethylsulphoximides have been prepared from lithium or sodium dimethylsulphoximide (equation 38)[71].

The formation of diphenyldiazomethane from the reaction of benzophenone hydrazone with methanesulphinyl chloride in the presence of two equivalents of base is considered to arise from base-catalyzed α-elimination from the methylsulphinylhydrazone intermediate **23** (Scheme 2)[72]. When the reaction was carried out at $-60\,^{\circ}$C, the sulphinyl hydrazone **23** could be isolated. NMR data on such compounds possessing a hydrogen atom α to the sulphinyl group show that **23** exists predominantly as the iminosulphinic acid tautomer **24**[73]. Elimination of the methylsulphenate ion from the anion of **24** leads to the formation of diphenyldiazomethane. The reaction pathway proposed is further supported by the independent generation of the butyl analogue of **25** (equation 39) from which the same product is rapidly formed[72].

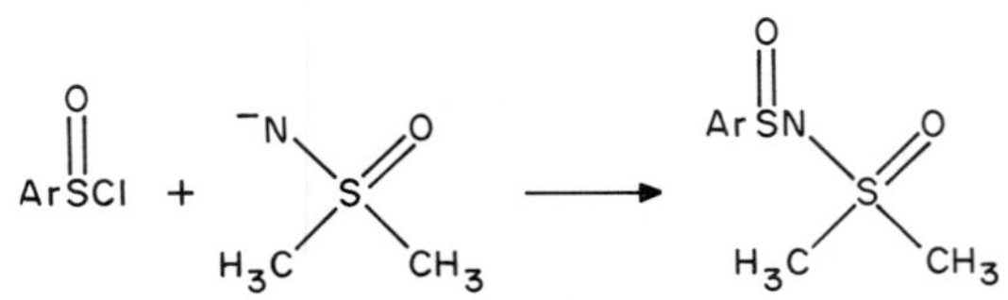


(38)

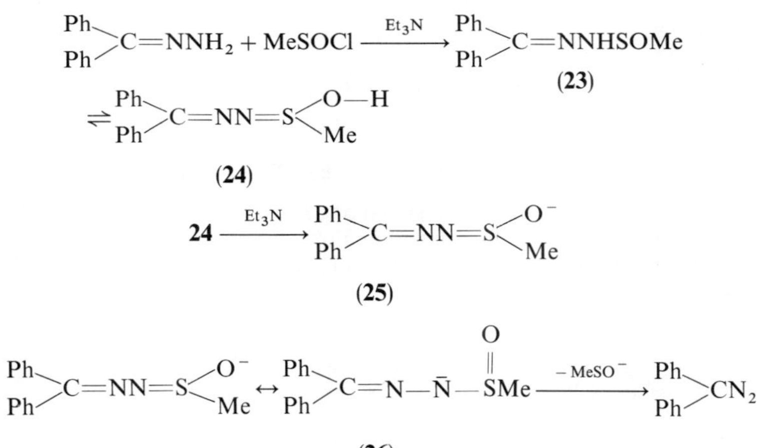


SCHEME 2

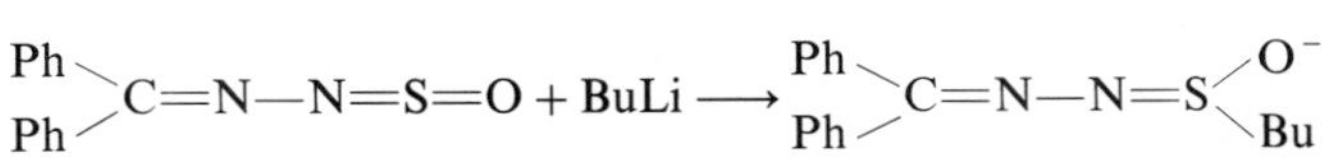


(39)

The formation of phenyl diazomethyl sulphoxide from benzenesulphinyl chloride and diazomethane has also been reported (equation 40)[74] and developed into a general method for the synthesis of α-chlorosulphoxides[75]. Maricich and his group were the first to isolate arylsulphinyl azides from the reaction of azide ions and arylsulphinyl chlorides (equation 41)[71].

$$\mathrm{Ph\overset{O}{\overset{\|}{S}}Cl + CH_3N_2 \longrightarrow Ph\overset{O}{\overset{\|}{S}}CH_2Cl + Ph\overset{O}{\overset{\|}{S}}CHN_2} \qquad (40)$$

$$\mathrm{ArS(O)Cl} + \mathrm{NaN_3} \xrightarrow{-20\,^\circ\mathrm{C}} \mathrm{ArS(O)N_3} + \mathrm{NaCl} \qquad (41)$$

$$(\mathrm{Ar} = \mathrm{Ph},\ p\text{-}\mathrm{ClC_6H_4},\ p\text{-}\mathrm{NO_2C_6H_4})$$

Benzenesulphinyl chloride when heated sequentially with triphenylphosphine and sodium azide gives the same products produced from the reaction of the phosphine with benzenesulphinyl azide (equation 42)[71,72]. A mechanism involving a sulphenyl chloride intermediate has been suggested (equation 43)[71]. The second step could involve formation of a sulphenyl azide intermediate as suggested by equation 42.

$$\mathrm{PhS(O)Cl} \xrightarrow[2.\ \mathrm{NaN_3}]{1.\ \mathrm{Ph_3P}} \mathrm{PhSSPh} + \mathrm{Ph_3PO}$$

$$\mathrm{PhS(O)N_3} \xrightarrow{\mathrm{Ph_3P}} \mathrm{PhSSPh} + \mathrm{Ph_3PO} \qquad (42)$$

$$\mathrm{PhS(O)Cl} + \mathrm{Ph_3P} \longrightarrow \mathrm{PhSCl} + \mathrm{Ph_3PO}$$

$$\mathrm{PhSCl} + \mathrm{NaN_3} \longrightarrow \mathrm{PhSSPh} + \mathrm{NaCl} \qquad (43)$$

Sheppard and Diekmann were the first to report the synthesis of a sulphine (**27**) from the reaction of triethylamine with 9-fluorenesulphinyl chloride (equation 44)[76].

Almost simultaneously a Dutch group reported the isolation of the thioaldehyde *S*-oxide **28** (equation 45)[77].

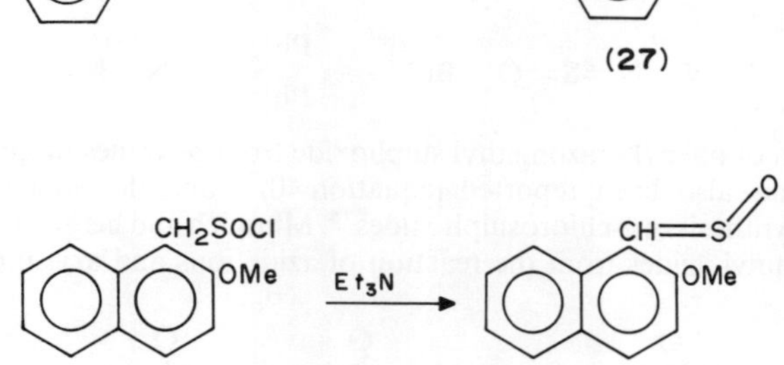


**(27)** (44)




**(28)** (45)

Whilst aliphatic sulphines are generally too unstable to be isolated at room temperature, Sheppard and Diekmann[76] reported the formation in solution of the dimethylsulphine **29** (equation 46) which was detected by its NMR spectrum and by trapping with chlorine to form 2-chloro-2-propanesulphinylchloride. The parent sulphine $CH_2{=}S{=}O$ was eventually generated as a short-lived species ($t_{1/2} \sim 30$–$60\,\text{min}$) by Block and his coworkers using flash vacuum pyrolysis of methanesulphinyl chloride (equation 47)[78]. Interestingly, pyrolyses of ethanesulphinyl chloride and 2-propanesulphinyl chloride form the corresponding alkenes possibly via a Cope elimination (equation 48)[78].

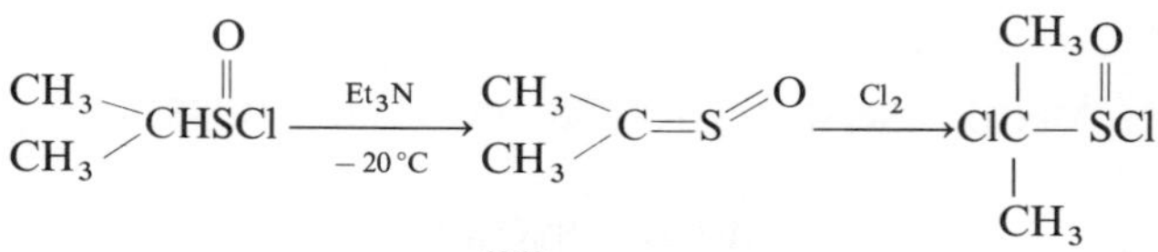

(46)

(**29**)

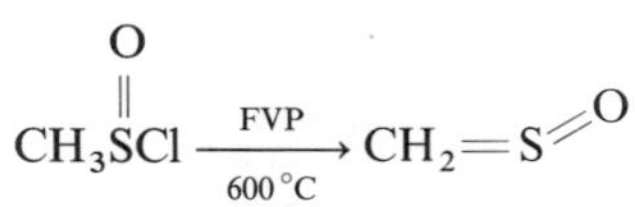

(47)

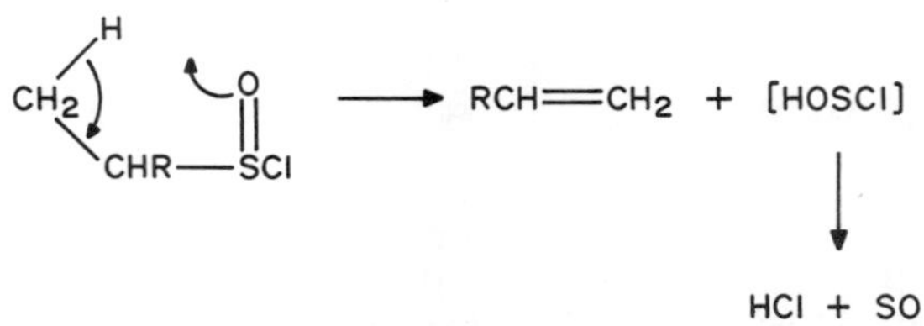

(48)

Whilst the reaction of equimolar quantities of sulphinyl chlorides and base leads to the formation of sulphines, reaction of two equivalents of sulphinyl chloride with one of base leads cleanly and in good yield to α-chloroalkyl alkanethiolsulphonates (equation 49)[79].

$$2MeCH_2SOCl + Et_3N \longrightarrow MeCHClSSO_2CH_2Me + Et_3\overset{+}{N}HCl^- \qquad (49)$$

Block and Bazzi[79] suggested that initial nucleophilic attack of the rapidly formed sulphine occurred on unreacted sulphinyl chloride to produce an intermediate adduct which could rearrange to the product in two different ways (equation 50). Similarly, Freeman and Keindl observed that sulphinyl chlorides possessing an α-hydrogen react with dimethylformamide (DMF) under nitrogen to give both the α-chlorothiolsulphonate as the major product (observed by Block and Bazzi) and the corresponding thiolsulphonate as the minor product (equation 51)[80]. Thiolsulphonate formation was considered to occur via radical decomposition of one of the intermediates **30** involved in α-chlorothiolsulphonate formation. In the case of methanesulphinyl chloride the corresponding intermediate **32** decomposes according to Scheme 3. Support for a free radical mechanism comes from the observation that methanesulphinyl chloride with DMF in the presence of a radical inhibitor gives only the corresponding α-chlorothiolsulphonate.

An example of β-chloroalkanethiolsulphonate formation is to be found in the reaction of ethanesulphinyl chloride with thiirane *S*-oxide (equation 52)[81].

Earlier Senning had demonstrated that dimethyl sulphoxide reacts exothermically with 2,2,2-trichloroethanesulphinyl chloride to form the corresponding sulphonyl chloride (equation 53)[82].

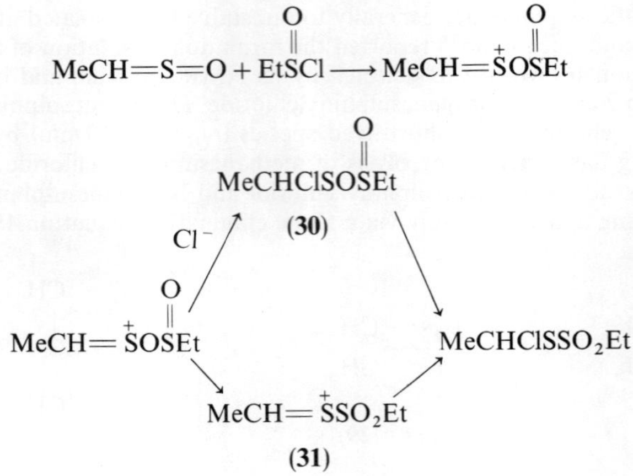


(50)

$$\mathrm{RCH_2\overset{O}{\overset{\|}{S}}Cl + DMF \longrightarrow RCH_2SSO_2CH_2R + \underset{\displaystyle Cl}{\underset{|}{RCH}}SSO_2CH_2R} \qquad (51)$$

(R = H, Me, Pr)

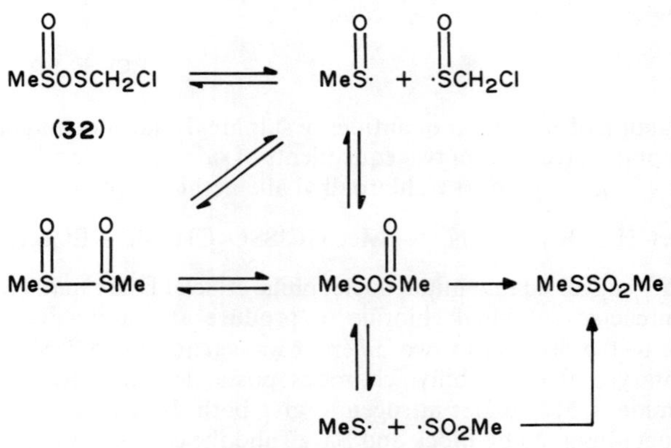


SCHEME 3

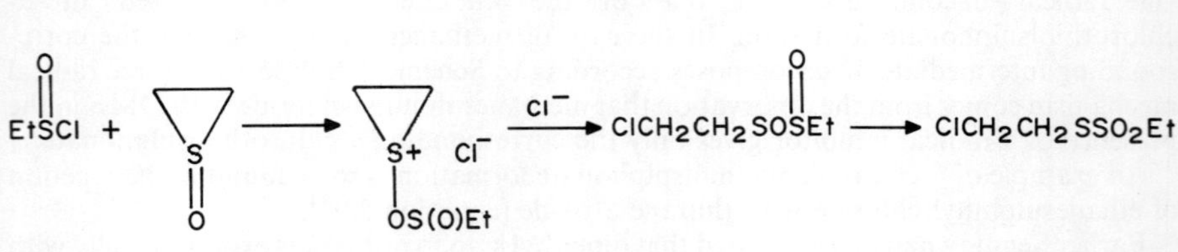


(33)

(52)

$$Cl_3CCH_2S(O)Cl + MeS(O)Me \xrightarrow{-HCl} Cl_3CCH_2SO_2Cl \qquad (53)$$

More recently, Oae and his group have shown that simple dialkyl and diaryl sulphoxides react smoothly at room temperature with methanesulphinyl chloride to form the sulphide and methanesulphonyl chloride (equation 54)[83]. Reduction of the sulphinyl chloride was assumed to occur via either a covalent intermediate or the sulphonium ion **34** formed by nucleophilic displacement of chloride ion from the sulphinyl chloride. Rearrangements of **34** to **35** could lead to the observed products. Sulphinic esters can be obtained in high yields by the reaction of sulphinyl chlorides with chlorosulphites in the presence of hexamethyldisiloxane and assisted by the addition of small quantities of dimethyl sulphoxide (equation 55)[84]. Whilst the mechanism of this reaction has not been clearly established, both **36** and **37** may be involved as intermediates.

$$RS(O)R^1 + MeS(O)Cl \longrightarrow RSR^1 + MeSO_2Cl \qquad (54)$$

$$R\overset{+}{S}(OS(O)Me)R^1\ Cl^- \quad \textbf{(34)} \qquad R\overset{+}{S}(Cl)R^1\ \ MeSO_2^- \quad \textbf{(35)}$$

$$RS(O)Cl + R^1OS(O)Cl + (Me_3Si)_2O \xrightarrow{DMSO} RS(O)OR^1 + 2Me_3SiCl + SO_2 \qquad (55)$$

$$RS(O)O\overset{+}{S}R_2{}^2Cl^- \quad \textbf{(36)} \qquad R^1OS(O)CO(SiMe_3)_2Cl^- \quad \textbf{(37)}$$

The reaction of two moles of a sulphinyl chloride with one mole of the active methylene compound **38** (e.g. $R^1 = R^2 = COCH_3$) in the presence of base yields the unsymmetric dithioacetal dioxides as the major products (equation 56)[85]. The authors suggest that the initially formed sulphoxide reacts with excess sulphinyl chloride to form an adduct which can disproportionate to **39** (equation 57).

$$\underset{\textbf{(38)}}{R^1R^2CH_2} \xrightarrow[0\,°C,\ Et_3N]{RSOCl} R^1R^2CHS(O)R \xrightarrow{RSOCl} \underset{\textbf{(39)}}{R^1R^2C(SR)(SO_2R)} \qquad (56)$$

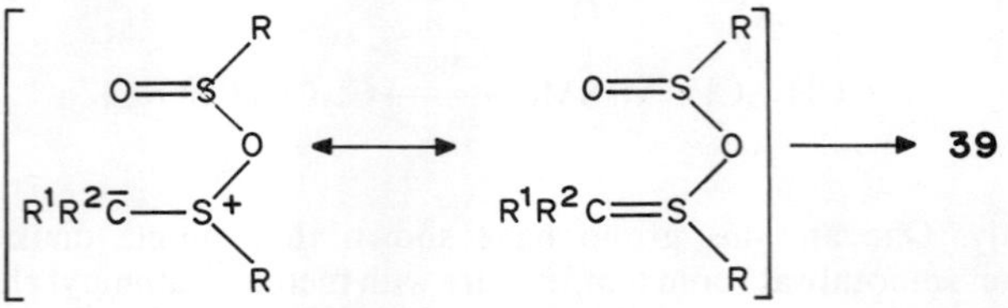

(57)

**(40)**

### E. Reaction with Metals

Both aromatic and aliphatic sulphinyl chlorides react with activated zinc powder to form thiolsulphonates (equation 58)[86–89]. Barnard assumed that reaction occurred via a *vic*-disulphoxide which subsequently rearranges[87]. More recently, Freeman and his group have used $^1H$ and $^{13}C$ NMR to confirm the existence of these elusive species[90,91]. A detailed study of the reaction of methanesulphinyl chloride with zinc in anhydrous ether identified a large number of products (equation 59)[92]. The reaction probably involves both sulphinyl radicals and sulphenate anions (equation 60). Either of these species could lead to the formation of *vic*-disulphoxides and sulphenyl sulphinates both of which are thought to be intermediates (equation 61). Either the *vic*-disulphoxide or the sulphenyl sulphinate could react with traces of water to form the sulphinic acid (equation 62). This in turn could react with sulphinyl chloride to form the sulphinyl sulphone (equation 63); the other products could be accounted for by similar reactions.

$$R\overset{O}{\overset{\|}{S}}Cl + Zn \longrightarrow RSSO_2R \qquad (58)$$

$$CH_3\overset{O}{\overset{\|}{S}}Cl \xrightarrow{Zn} CH_3\underset{\underset{O}{\|}}{\overset{\overset{O}{\|}}{S}}SCH_3 + CH_3\overset{O}{\overset{\|}{S}}OH \text{ or } (CH_3SO_2)_2Zn$$

$$+ CH_3SO_2Cl + (CH_3)_2S + CH_3\overset{O}{\overset{\|}{S}}SCH_3 + CH_3\overset{O}{\overset{\|}{S}}{-}\underset{\underset{O}{\|}}{\overset{\overset{O}{\|}}{S}}CH_3 \qquad (59)$$

$$CH_3\overset{O}{\overset{\|}{S}}Cl + Zn \longrightarrow CH_3\overset{O}{\overset{\|}{S}}\cdot + [ZnCl]$$

$$CH_3\overset{O}{\overset{\|}{S}}Cl + Zn \longrightarrow CH_3\overset{O}{\overset{\|}{S}}{}^{-} + ZnCl^+ \qquad (60)$$

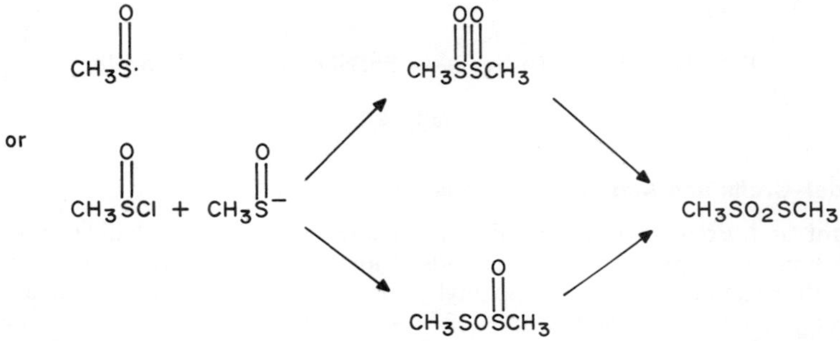


(61)

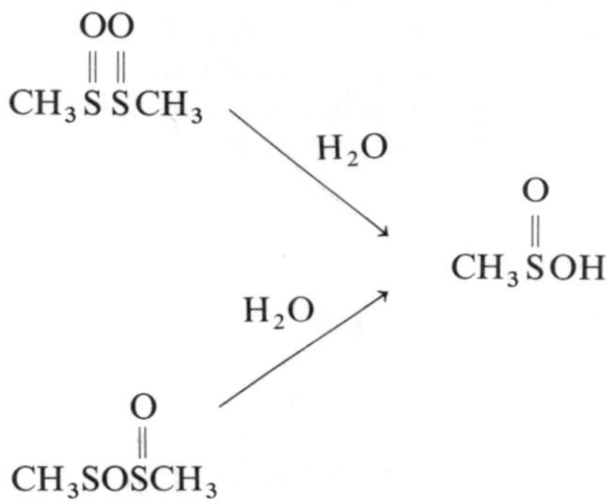


(62)

$$\mathrm{CH_3\overset{O}{\overset{\|}{S}}Cl + CH_3\overset{O}{\overset{\|}{S}}OH \longrightarrow CH_3\overset{O}{\overset{\|}{S}}\underset{\underset{O}{\|}}{\overset{O}{\overset{\|}{S}}}CH_3} \tag{63}$$

Sulphinyl chlorides also react with other metals such as copper[93,94] and silver[95,96] in organic solvents to form thiolsulphonates. Harpp has recently reported that butanesulphinyl chloride reacts with tributyltin lithium to form the thiolsulphonate in good yield (equation 64)[97]. The $^{13}$C NMR spectrum of the reaction solution showed the presence of the *vic*-disulphoxide **41** and the sulphenyl sulphinate **42**. The likelihood that *vic*-disulphoxides undergo rearrangement via formation of sulphinyl radicals is supported by *ab initio* calculations[98].

$$\mathrm{2\text{-}Bu\overset{O}{\overset{\|}{S}}Cl + Bu_3SnLi \xrightarrow{-60^\circ} BuSSO_2Bu + Bu_3SnCl + LiCl} \tag{64}$$

$$\mathrm{Bu\overset{O}{\overset{\|}{S}}\overset{O}{\overset{\|}{S}}Bu} \qquad \qquad \mathrm{Bu\overset{O}{\overset{\|}{S}}OSBu}$$

**(41)** **(42)**

Reaction of benzenesulphinyl chloride with the persulphide complex **43** gives diphenyl disulphide in 68% yield[97]. It seems likely that, as in the case of other metals, the thiolsulphonate is initially formed and that this itself is reduced to the disulphide (equation 65).

$$\mathrm{Ph\overset{O}{\overset{\|}{S}}Cl + (NH_4)_2[Mo_2S_{12}] \longrightarrow PhSSO_2Ph \xrightarrow{\mathbf{43}} PhSSPh} \qquad (65)$$

**(43)**

### F. Friedel–Crafts and Addition Reactions

Courtot and Frenkiel were the first to report that sulphinyl halides react with hydrocarbons in the presence of Lewis acids to form sulphoxides (equation 66)[99]. More recently, the reaction of *p*-toluenesulphinyl chloride with anisole was examined in more detail (equation 67)[100]. Reaction at −15 °C in carbon disulphide in the presence of aluminium trichloride, antimony pentachloride or stannic chloride produced the sulphoxide in good yield after hydrolytic decomposition of the complex formed. On the other hand, when zinc chloride, iron powder or boron trifluoride etherate were used, the products were the corresponding sulphide and thiolsulphonate. Sulphide formation is considered to occur by reduction of the sulphoxide by excess sulphinyl chloride (equation 68).

$$o\text{-TolSOCl} + p\text{-TolH} \longrightarrow o\text{-Tol}\overset{O}{\overset{\|}{S}}\text{Tol-}p \qquad (66)$$

$$p\text{-Tol}\overset{O}{\overset{\|}{S}}\text{Cl} + \text{AnH} \longrightarrow p\text{-Tol}\overset{O}{\overset{\|}{S}}\text{An-}p \quad \text{or} \quad p\text{-TolSAn-}p + p\text{-TolSO}_2\text{Cl} \qquad (67)$$

$$p\text{-Tol}\overset{O}{\overset{\|}{S}}\text{An-}p + p\text{-Tol}\overset{O}{\overset{\|}{S}}\text{Cl} \longrightarrow p\text{-TolSAn-}p + p\text{-TolSO}_2\text{Cl} \qquad (68)$$

Competition experiments on the aluminium chloride catalysed arylsulphinylation of benzene and polymethylbenzenes in nitromethane indicated high positional selectivity, although the exact meaning of a positive $\rho$ value (+0.25) for this aromatic substitution reaction is unclear[101].

A related reaction is the zinc chloride catalysed addition of sulphinyl chlorides to styrene to form the corresponding $\beta$-chloroalkyl sulphoxides (equation 69)[102]. A facile ene reaction of alkenes with sulphinyl chlorides to form a six-membered cyclic allylic sulphoxide has also been reported (equation 70)[28].

$$p\text{-Tol}\overset{O}{\overset{\|}{S}}\text{Cl} + \text{PhCH{=}CH}_2 \xrightarrow[\text{ether}]{\text{ZnCl}_2} p\text{-Tol}\underset{\underset{\text{Cl}}{|}}{\text{C}}\text{HCH}_2\text{SOPh} \qquad (69)$$

The more general reaction of arenesulphinyl chlorides with alkenes catalysed by ethylaluminium dichloride also gives allylic sulphoxides (equation 71)[103]. The choice of

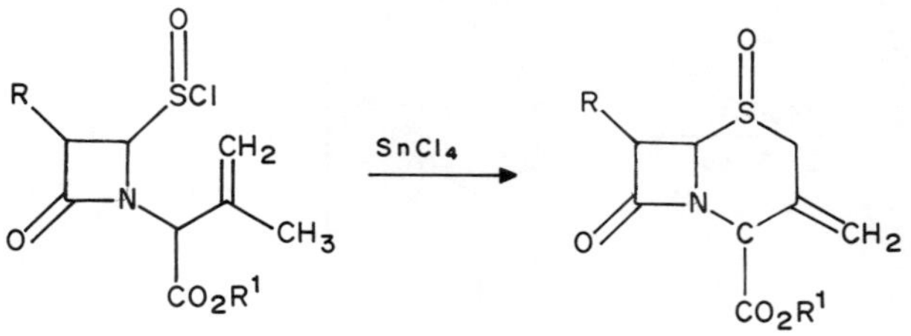


(70)

(R=phthalimido; $R^1$=Me)

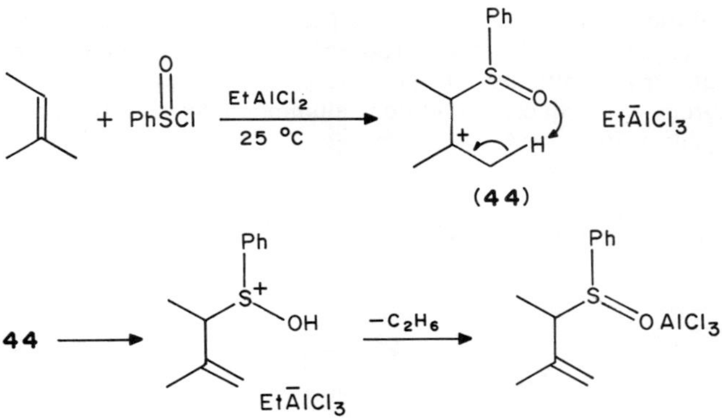


(71)

conditions for this reaction is critical. Ethylaluminium dichloride behaves both as a Lewis acid and an acid scavenger, removing the hydrogen chloride produced.

Moiseenkov and his group have shown that the addition of phenylsulphinyl chloride to 1,3-dienes in the presence of zinc chloride produces **45** as the major product (equation 72)[104]. When zinc chloride is replaced by silver borofluoride, **45** is formed as the sole product in 75% yield probably via ene addition through the six-membered transition state **47**. High pressure forces [4 + 2] cycloaddition, which is considered to proceed via the cyclic sulphurane **48** (equation 73).

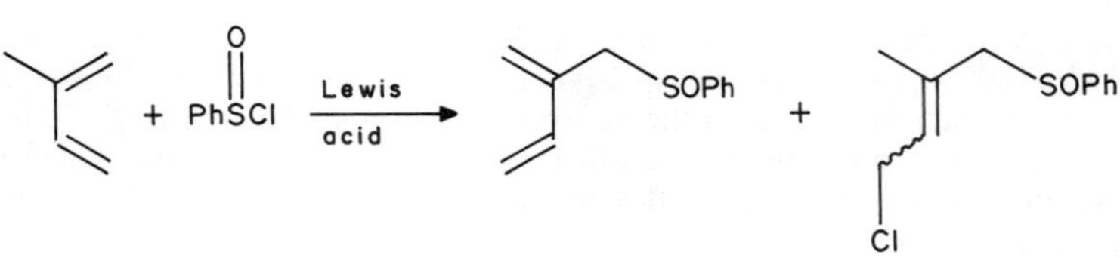


(72)

**(45)** **(46)** *E/Z*

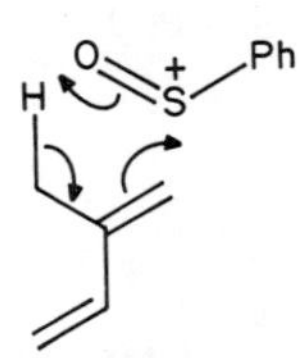

**(47)**

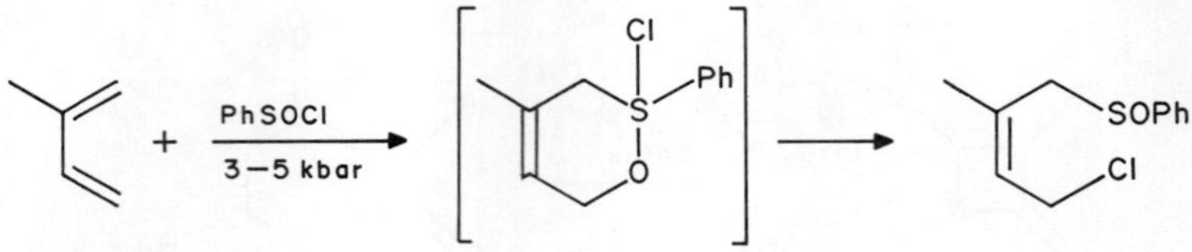


(48)

(73)

## G. Miscellaneous Reactions

Sulphinyl chlorides react with carbanions such as Grignard reagents (equation 74)[105] and cyanide ion (equation 75)[106]. It is interesting that the corresponding sulphinyl cyanide cannot be isolated from *n*-butane- or *t*-butane-sulphinyl chloride. In the presence of thiols, sulphinyl chlorides are reduced by lithium aluminium hydride to the corresponding disulphide (equation 76)[107].

$$p\text{-Tol}\overset{\overset{\displaystyle O}{\|}}{S}\text{Cl} + \text{PhMgBr} \longrightarrow p\text{-Tol}\overset{\overset{\displaystyle O}{\|}}{S}\text{Ph} + p\text{-TolSPh} \qquad (74)$$

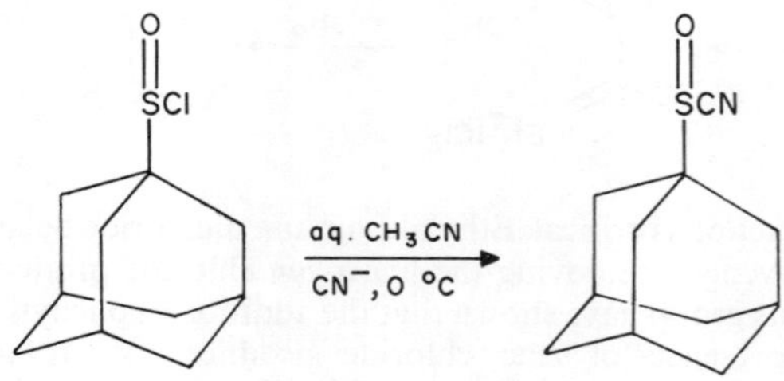


(75)

$$R\overset{\overset{\displaystyle O}{\|}}{S}Cl \xrightarrow{\text{LAH}} [RSCl] \xrightarrow{+\text{RSH}} RSSR \qquad (76)$$

The reaction of triphenylphosphine with sulphinyl chlorides has been referred to previously[71,72]. Trialkyl phosphites react with *p*-chlorophenylsulphinyl chloride to form a sulphinyl phosphite ester probably via the phosphonium ion **49** (equation 77)[108]. It has been suggested that sulphinyl chlorides are formed as intermediates in the reaction of triethyl phosphite and sulphonyl chlorides in the presence of an alcohol to form a sulphinate ester (equation 78)[109].

$$p\text{-ClC}_6\text{H}_4\overset{\overset{\displaystyle O}{\|}}{S}\text{Cl} \xrightarrow{\text{P(OMe)}_3} \left[p\text{-ClC}_6\text{H}_4\overset{\overset{\displaystyle O}{\|}}{S}-\overset{+}{P}(\text{OMe})_3\right]\text{Cl}^- \longrightarrow p\text{-ClC}_6\text{H}_4\overset{\overset{\displaystyle O}{\|}}{S}\text{PO(OMe)}_2$$

(49)

(77)

$$R^1SO_2Cl \xrightarrow[\text{CH}_2\text{Cl}_2]{(\text{MeO})_3\text{P, Et}_3\text{N}} [R^1SOCl] \xrightarrow{\text{ROH}} R^1\overset{\overset{\displaystyle O}{\|}}{S}OR \qquad (78)$$

*p*-Nitrobenzenesulphinyl chloride has been reported to react with pyridine *N*-oxide at room temperature to form colourless crystals thought to be the salt **50**[110]. This in turn appears to cleave homolytically to form the relatively stable *p*-nitrobenzenesulphonyl radical from which a variety of products are generated (equation 79). The reaction with pyridine (and α-picoline) *N*-oxide closely resembles the corresponding reaction with *p*-nitrobenzenesulphenyl chloride[111] but is in distinct contrast to the reaction of amine oxides with tosyl chloride which appears to react via heterolytic N—O cleavage of the initially formed salt[112].

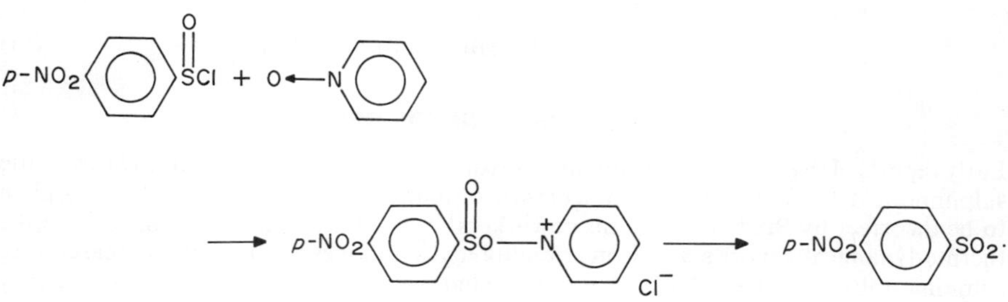

**(50)**

(79)

Bleeker and Engberts showed that *t*-butylsulphinyl chloride reacts smoothly with *t*-butyl hydroperoxide at 0 °C in ether in the presence of pyridine to form a variety of products (Scheme 4)[113]. A general mechanism involving initial nucleophilic attack by the hydroperoxide at sulphinyl sulphur to form a peroxysulphinate (**51**), which subsequently undergoes homolysis via a radical cage process, is supported by the CIDNP effects observed during the reaction. This mechanism closely resembles that proposed for the reaction of sulphinyl chlorides with substituted hydroxylamines (see Section I.D). Under similar conditions, reaction of *t*-butylsulphinyl chloride with *t*-butyl hydrosulphide produces *t*-butylsulphinyl *t*-butyl disulphide **52** in 90% yield (equation 80). Clearly **52** is much more stable than **51**. At high temperature a thiolsulphonate and tetrasulphide are formed from homolytic decomposition of **52** for which the driving force is the generation of the stable *t*-BuSS· radical (equation 81).

$$t\text{-Bu}\overset{\overset{\text{O}}{\|}}{\text{S}}\text{Cl} + t\text{-BuOOH} \longrightarrow \left[t\text{-Bu}\overset{\overset{\text{O}}{\|}}{\text{S}}\text{OOBu-}t\right]$$

**(51)**

$$\mathbf{51} \longrightarrow t\text{-Bu}\underset{\underset{\text{O}}{\|}}{\overset{\overset{\text{O}}{\|}}{\text{S}}}\cdot + \cdot\text{Bu-}t \longrightarrow t\text{-Bu}\underset{\underset{\text{O}}{\|}}{\overset{\overset{\text{O}}{\|}}{\text{S}}}\text{O}\underset{\underset{\text{O}}{\|}}{\overset{\overset{\text{O}}{\|}}{\text{S}}}\text{OBu-}t + t\text{-BuS}\underset{\underset{\text{O}}{\|}}{\overset{\overset{\text{O}}{\|}}{\text{S}}}\text{Bu-}t$$

$$+\ t\text{-BuSO}_2\text{Cl} + t\text{-BuOH} + \text{Me}_2\text{CO}$$

SCHEME 4

$$t\text{-BuS(=O)Cl} + t\text{-BuSSH} \xrightarrow[\text{py}]{\text{CHCl}_3,\ 0\text{–}5\,^{\circ}\text{C}} t\text{-BuS(=O)SSBu-}t \quad \textbf{(52)} \tag{80}$$

$$t\text{-BuS(=O)SSBu-}t \longrightarrow t\text{-BuS(=O)}\cdot + t\text{-BuSS}\cdot \;\to\; t\text{-BuSO}_2\text{SBu-}t \quad t\text{-BuSSSSBu-}t \tag{81}$$

## II. SULPHINIC ANHYDRIDES

Early reports of the synthesis of sulphinic anhydrides **53**, either by dehydration of aromatic sulphinic acids[114] or from the reaction of their sodium salts with phosgene[115], were shown to be incorrect by Bredereck and his coworkers[116,117]. The products obtained were the thermodynamically more stable isomeric sulphinyl sulphones **54**. In only a few cases have sulphinic anhydrides been detected as intermediates and in only three instances have they been isolated.

$$\text{RS(=O)—O—S(=O)R} \quad \textbf{(53)} \qquad \text{RSO}_2\text{S(=O)R} \quad \textbf{(54)}$$

Kice and Ikura were able to isolate 2-methyl-2-propanesulphinic anhydride **55**, m.p. 45–46 °C in 50% yield (equation 82)[118]. In contrast to the hydrolysis of sulphinyl sulphones (equation 83) which is not sensitive to acid catalysis, the hydrolysis of **55** exhibits both spontaneous and acid-catalysed hydrolysis. The kinetic solvent isotope effect for the latter reaction is consistent with attack of water on the conjugate acid of **55** in the slow step (equation 84).

$$\text{Bu}^t\text{SO}_2\text{Ag} + \text{Bu}^t\text{S(=O)Cl} \xrightarrow[\text{ether}]{-5\text{ to }-10\,^{\circ}\text{C}} \text{Bu}^t\text{S(=O)—O—S(=O)Bu}^t + \text{AgCl} \quad \textbf{(55)} \tag{82}$$

$$\text{RS(=O)—S(=O)}_2\text{R} + \text{H}_2\text{O} \longrightarrow 2\text{RSO}_2\text{H} \tag{83}$$

$$\text{H}_2\ddot{\text{O}} \curvearrowright \text{RS(=O)—O—}\overset{+}{\text{S}}\text{(OH)R} \xrightarrow{\text{slow}} \text{RSO}_2\text{H}_2^{\ +} + \text{RSO}_2\text{H} \tag{84}$$

The cyclic 1,2-ethanedisulphinic anhydride **56** has been synthesized by controlled hydrolysis of ethanebisdisulphinyl chloride (equation 85)[119].

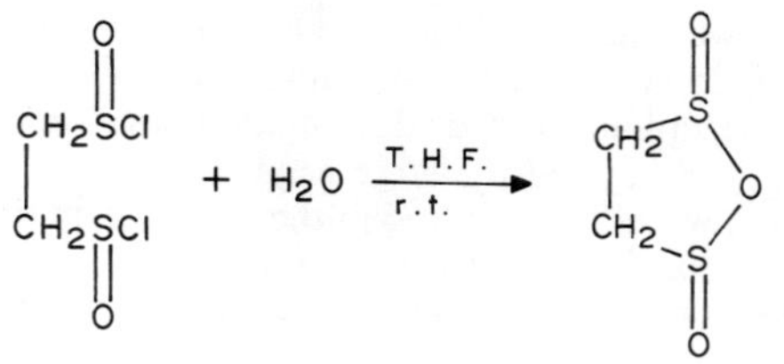


(85)

(56)

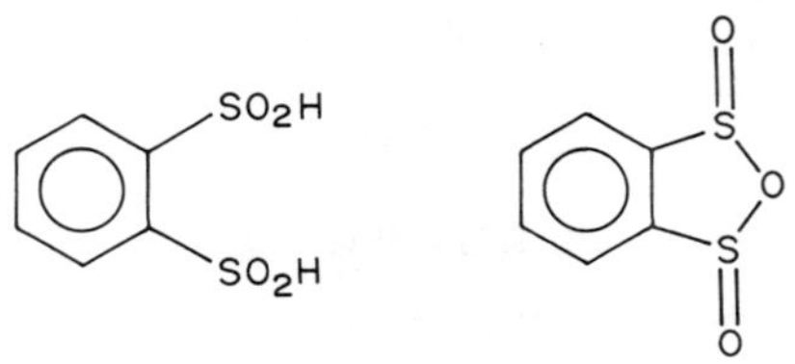


(86)

(57) (58)

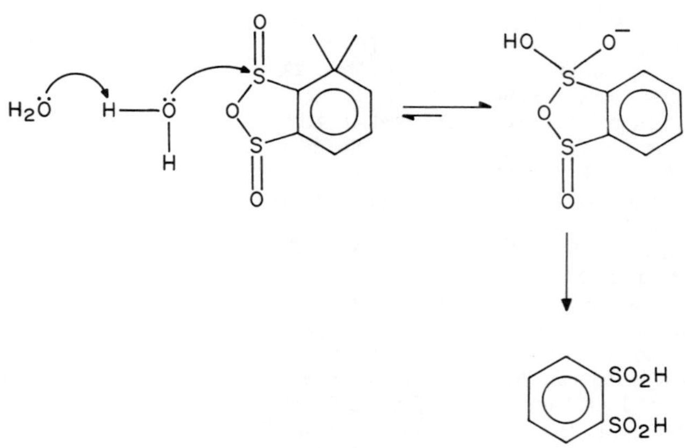


(87)

$$\mathrm{R\overset{O}{\overset{\|}{S}}{-}SR \longrightarrow R\overset{O}{\overset{\|}{S}}{-}\overset{O}{\overset{\|}{S}}R \longrightarrow RSO_2SR} \qquad (88)$$

(59)

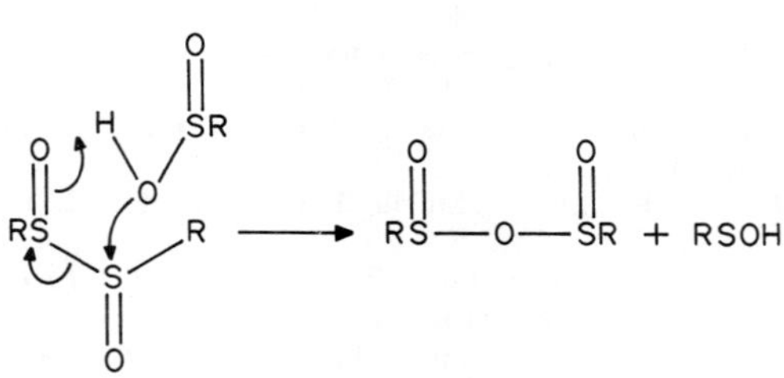


(89)

Kice and Liao[120] showed that the compound originally thought[121] to be benzene-1,2-disulphinic acid **57** was in fact the corresponding anhydride **58** and as such is the first example of an aromatic sulphinic anhydride (equation 86). The large value of the solvent isotope effect ($k_{H_2O}/k_{D_2O} = 2.3$) and the large negative value of $\Delta S^{\ddagger}$ ($-49.7$ eu) suggests that several molecules of water are involved in the slow step in the hydrolysis of **58** to **57**. Either a classic 'proton-bridge' mechanism (equation 87) or one involving a cyclic transition state with several molecules of water is suggested as a possible mechanism[120].

The peroxidation of *S*-aryl and *S*-alkyl arenethiolsulphinates to the corresponding thiolsulphonates is thought to occur by a variety of mechanisms via a *vic* disulphoxide (equation 88)[122,123]. Freeman and his group[122,123] showed that sulphinic anhydride intermediates could be detected in the low temperature oxidation ($-40$ °C) of **59** (R = Me, $Pr^i$, $Bu^n$ and $Bu^t$). These could arise by oxidation of sulphenyl sulphinate intermediates or by reaction of any sulphinic acids formed with disulphoxides (equation 89).

## III. REFERENCES


1. A. M. Quaedlieg, in *Methoden der Organischen Chemie* (*Houben-Weyl*), Vol. IX, Georg Thieme Verlag, Stuttgart, 1955, p. 295.
2. B. F. Muth, in *Methoden der Organischen Chemie* (*Houben-Weyl*), Vol. IX, Georg Thieme Verlag, Stuttgart, 1955, p. 295.
3. C. J. M. Stirling, *Int. J. Sulfur Chem.* (*B*), **6**, 277 (1971).
4. Cf. I. B. Douglass and R. V. Norton, *J. Org. Chem.*, **33**, 2104 (1968).
5. R. V. Norton, G. M. Beverley and I. B. Douglass, *J. Org. Chem.*, **32**, 3645 (1967).
6. I. B. Douglass and D. A. Koop, *J. Org. Chem.*, **29**, 951 (1964).
7. E. Schneider, *Chem. Ber.*, **84**, 911 (1951).
8. T. P. Hilditch and S. Smiles, *Chem. Ber.*, **41**, 4115 (1908); *J. Chem. Soc.*, 2585 (1910).
9. J. von Braun and W. Kaiser, *Chem. Ber.*, **56**, 553 (1923).
10. J. von Braun and K. Weissbach, *Chem. Ber.*, **63**, 2856 (1930).
11. A. Meuwsan and H. Gebhardt, *Chem. Ber.*, **69**, 937 (1936).
12. F. Kurzer, *J. Chem. Soc.*, 549 (1953).
13. H. F. Whalen and L. W. Jones, *J. Chem. Soc.*, 1353 (1925).
14. L. C. Raiford and S. E. Haslet, *J. Am. Chem. Soc.*, **57**, 2172 (1935).
15. I. B. Douglass and D. R. Poole, *J. Org. Chem.*, **22**, 536 (1957).
16. I. B. Douglass, B. S. Farrah and E. G. Thomas, *J. Org. Chem.*, **26**, 1996 (1961).
17. M. L. Kee and I. B. Douglass, *Org. Prep. Proced. Int.*, **2**, 235 (1970).
18. D. Barnard, L. Bateman, M. E. Cain, T. Colclough, and J. I. Cunneen, *J. Chem. Soc.*, 5339 (1961).
19. E. Block and J. O'Connor, *J. Am. Chem. Soc.*, **96**, 3922 (1974).
20. J.-H. Youn and R. Herrmann, *Tetrahedron Lett.*, **27**, 1493 (1986).
21. J.-H. Youn and R. Herrmann, *Synthesis*, 72 (1987).
22. T. Netscher and H. Prinzbach, *Synthesis*, 683 (1987).
23. J. F. and R. P. Beatson, *J. Chem. Soc., Chem. Commun.*, 663 (1970).
24. G. Canalini, G. Maccagnani, and F. Taddei, *Tetrahedron Lett.*, 3035 (1971).
25. R. P. Gupta, J. S. Pizey, and K. Symeonides, *Tetrahedron*, **32**, 1971 (1976).
26. Von D. Rinne and A. Blaschette, *Z. Anorg. Allg. Chem.*, **638**, 237 (1977).
27. S. Kukolja and S. R. Lammert, *Angew. Chem., Int. Ed. Engl.*, **12**, 67 (1973).
28. S. Kukolja, S. R. Lammert, M. R. B. Gilesner, and A. I. Ellis, *J. Am. Chem. Soc.*, **98**, 5041 (1976).
29. J. von Braun and K. Weissbach, *Chem. Ber.*, **63**, 2840 (1930).
30. I. B. Douglass, *J. Org. Chem.*, **30**, 633 (1965).
31. A. Heering, M. Jaspers, and I. Schwermann, *Chem. Ber.*, **112**, 2903 (1979).
32. H. Phillips, *J. Chem. Soc.*, 2552 (1925).
33. K. Mislow, M. M. Green, P. Laur, J. T. Melillo, T. Simmons and A. L. Ternay, *J. Am. Chem. Soc.*, **87**, 1958 (1965).
34. M. Axelrod, P. Bickert, J. Jacobus, M. Green, and K. Mislow, *J. Am. Chem. Soc.*, **90**, 4835 (1968).
35. M. M. Green, M. Axelrod, and K. Mislow, *J. Am. Chem. Soc.*, **88**, 861 (1986).
36. H. F. Herbrandson and R. T. Dickerson, *J. Am. Chem. Soc.*, **81**, 4102 (1959).

37. Cf. R. E. Estep and D. F. Tavares, *Int. J. Sulfur Chem.*, **8**, 279 (1973).
38. C. Mioskowski and G. Solladie, *Tetrahedron*, **36**, 227 (1980).
39. K. K. Anderson, B. Bujnicki, J. Drabowicz, M. Mikolajczyk, and J. B. O'Brien, *J. Org. Chem.*, **49**, 4070 (1984).
40. M. Mikolajczyk and J. Drabowicz, *J. Chem. Soc., Chem. Commun.*, 547 (1974).
41. D. N. Harpp, B. T. Frielander, C. Larsen, K. Stelion, and A. Stockton, *J. Org. Chem.*, **43**, 3481 (1978).
42. Cf. J. G. Tillett, *Chem. Rev.*, **76**, 747 (1976).
48. S. Kozaka and T. Higashino, *Tetrahedron Lett.*, **21**, 2067 (1980).
44. S. Kozuka and T. Higashino, *Bull. Chem. Soc. Jpn.*, **54**, 1183 (1981).
45. La V. D. Small, J. H. Bailey, and C. J. Cavallito, *J. Am. Chem. Soc.*, **69**, 1710 (1947).
46. I. B. Douglass and B. S. Fareh, *J. Org. Chem.*, **23**, 805 (1958).
47. E. Block and J. O'Connor, *J. Am. Chem. Soc.*, **96**, 3921 (1974).
48. H. J. Backer and H. Kloosterziel, *Recl. Trav. Chim. Pays-Bas*, **73**, 129 (1954).
49. M. Mikolajczyk and J. Drabowicz, *J. Chem. Soc., Chem. Commun.*, 220 (1976).
50. H. Gilman and H. L. Morris, *J. Am. Chem. Soc.*, **48**, 2399 (1926).
51. Y. H. Chiang, J. S. Luloff, and E. Schipper, *J. Org. Chem.*, **34**, 2397 (1969).
52. R. M. Moriarty, *J. Org. Chem.*, **30**, 600 (1965).
53. F. Kurzer, *J. Chem. Soc.*, 549 (1953).
54. J. Jacobus and K. Mislow, *J. Chem. Soc., Chem. Commun.*, 253 (1968).
55. D. J. Cram and A. Nudelman, *J. Am. Chem. Soc.*, **90**, 3869 (1968).
56. K. Hovius and J. B. F. N. Engberts, *Tetrahedron Lett.*, 181 (1972).
57. W. J. Bourma and J. B. F. N. Engberts, *J. Org. Chem.*, **41**, 143 (1976).
58. I. P. Bleeker and J. B. F. N. Engberts, *Recl. Trav. Chim. Pays-Bas*, **98**, 120 (1979).
59. R. F. Hudson and K. A. F. Record, *J. Chem. Soc., Chem. Commun.*, 831 (1976).
60. C. Brown, R. F. Hudson, and K. A. F. Record, *J. Chem. Soc., Chem. Commun.*, 540 (1977).
61. C. Brown, R. F. Hudson, and K. A. F. Record, *J. Chem. Soc., Perkin Trans. 2*, 822 (1978).
62. M. R. Banks and R. F. Hudson, *J. Chem. Soc., Chem. Commun.*, 799 (1985).
63. M. R. Banks and R. F. Hudson, *J. Chem. Soc., Perkin Trans. 2*, 151 (1986).
64. A. Heesing, W. K. Homann, and W. Müllers, *Chem. Ber.*, **113**, 152 (1980).
65. M. R. Banks and R. F. Hudson, *J. Chem. Soc., Perkin Trans. 2*, 1211 (1986).
66. I. P. Bleeker and J. B. F. N. Engberts, *J. Org. Chem.*, **46**, 1012 (1981).
67. E. Z. Levchenko and A. V. Kirsanov, *Zh. Obshch. Khim.*, **30**, 1553 (1960); *J. Gen. Chem. USSR*, **30**, 1562 (1960).
68. C. R. Johnson and E. U. Jonsson, *J. Am. Chem. Soc.*, **92**, 3815 (1970).
69. E. S. Levchenko, N. Ya Derkash and A. V. Kirsanov, *Zh. Obshch, Khim.*, **30**, 1971 (1960).
70. S. Oae, M. Nakai, N. Furukawa, and R. Kiritani, *Bull. Chem. Soc. Jpn.*, **45**, 1268 (1972).
71. T. J. Maricich and V. L. Hoffman, *J. Am. Chem. Soc.*, **96**, 7770 (1974).
72. J. G. Shelnut, S. Mataka, and J.-P. Anselme, *J. Chem. Soc., Chem. Commun.*, 114 (1975).
73. S. Mataka and J.-P. Anselme, *J. Chem. Soc., Chem. Commun.*, 554 (1974).
74. C. G. Venier, H. J. Barager III, and M. A. Ward, *J. Am. Chem. Soc.*, **97**, 3238 (1975).
75. C. G. Venier, H.-H. Hsieh, and H. J. Barager III, *J. Org. Chem.*, **38**, 17 (1973).
76. W. A. Sheppard and J. Diekmann, *J. Am. Chem. Soc.*, **86**, 1893 (1964).
77. J. Strating, L. Thijs, and B. Zwanenburg, *Recl. Trav. Chim. Pays-Bas*, **83**, 631 (1964).
78. E. Block, R. E. Penn, R. J. Olsen, and P. F. Sherwin, *J. Am. Chem. Soc.*, **98**, 1264 (1976).
79. E. Block and A. A. Bazzi, *Tetrahedron Lett.*, **23**, 4569 (1982).
80. F. Freeman and M. C. Keindl, *J. Chem. Soc., Chem. Commun.*, 138 (1984).
81. K. Kando, A. Negishi, and G. Tsuchihashi, *Tetrahedron Lett.*, 2743 (1969).
82. A. Senning, *J. Chem. Soc., Chem. Commun.*, 64 (1967).
83. T. Numata, K. Ikura, Y. Shimano, and S. Oae, *Org. Prep. Proced. Int.*, **8**, 119 (1976).
84. J. Drabowicz, *Chem. Lett.*, 1753 (1981).
85. K. Schank and S. Bueglar, *Sulphur Letters*, **1**, 63 (1982).
86. R. Schiller and P. Otto, *Chem. Ber.*, **9**, 1584 (1876).
87. D. Barnard, *J. Chem. Soc.*, 4673 (1957).
88. F. Freeman and C. N. Angeletakis, *Tetrahedron Lett.*, **23**, 491 (1982).
89. F. Freeman and M. C. Keindl, *Synthesis*, 913 (1983).
90. F. Freeman and C. N. Angeletakis, *J. Am. Chem. Soc.*, **104**, 5766 (1982).
91. F. Freeman and C. N. Angeletakis, *J. Am. Chem. Soc.*, **103**, 6232 (1981).
92. F. Freeman, C. N. Angeletakis, and M. C. Keindl, *J. Org. Chem.*, **49**, 454 (1984).
93. C. M. C. da Silva Correa, and W. A. Waters, *J. Chem. Soc. (C)*, 1874 (1968).

94. G. Leandri, *Ann. Chim.* (*Rome*), **44**, 330 (1954).
95. B. N. Trivedi, *J. Indian Chem. Soc.*, **33**, 359 (1956).
96. G. Leandri and A. Tunelo, *Ann. Chim.* (*Rome*), **47**, 575 (1957).
97. D. N. Harpp, in *Perspectives in the Organic Chemistry of Sulphur* (Eds. B. Zwanberg and A. J. H. Klunder), Invited Lectures of the Twelfth International Symposium in the Organic Chemistry of Sulphur, Nijmegen, 1986.
98. F. Freeman, C. N. Angeletakis, W. J. Pietro, and W. J. Hehre, *J. Am. Chem. Soc.*, **104**, 1164 (1982).
99. C. Courtot and J. Frenkiel, *C.R. Acad. Sci.*, **109**, 557 (1934).
100. T. Fujisarva, M. Kakutani, and N. Kobayashi, *Bull. Chem. Soc. Jpn.*, **46**, 3615 (1973).
101. G. A. Olah and J. Nishimura, *J. Org. Chem.*, **39**, 1203 (1974).
102. G. Glaros and S. Sullivan, *Synth. Commun.*, **6**, 495 (1976).
103. B. Snider, *J. Org. Chem.*, **46**, 3155 (1981).
104. A. M. Moiseenkov, V. V. Veselovsky, Z. G. Makarova, V. M. Zhulin, and W. A. Smit, *Tetrahedron Lett.*, **25**, 5929 (1984).
105. H. Burton and W. A. Davy, *J. Chem. Soc.*, 528 (1948).
106. A. Boerma-Markerink, J. C. Jagt, H. Meyer, J. Wildeman, and A. M. van Leusan, *Synth. Commun.*, **5**, 147 (1975).
107. J. Strating and H. J. Backer, *Recl. Trav. Chim. Pays-Bas*, **69**, 638 (1950).
108. G. Schrader and W. Lorenz, U. S. Patent, 2,910,500 (1959); *Chem. Abstr.*, **54**, 3318 (1960).
109. J. M. Klunder and K. B. Sharpless, *J. Org. Chem.*, **52**, 2598 (1987).
110. S. Oae and K. Ikura, *Bull. Chem. Soc. Jpn.*, **39**, 1306 (1968).
111. S. Oae and K. Ikura, *Bull. Chem. Soc. Jpn.*, **38**, 58 (1965).
112. S. Oae, T. Kitao and Y. Kitaoka, *Tetrahedron*, **19**, 827 (1963).
113. I. P. Bleeker and J. B. F. N. Engberts, *Recl. Trav. Chim. Pays-Bas*, **100**, 459 (1981).
114. E. Knoevenagel and L. Polack, *Chem. Ber.*, **41**, 3323 (1908).
115. R. Otto, *Chem. Ber.*, **20**, 3337 (1987).
116. H. Bredereck, A. Wagner, E. H. Beck, and R. J. Klein, *Chem. Ber.*, **93**, 2736 (1960).
117. H. Bredereck, A. Wagner, E. H. Beck, H. Berlinger, and K.-G. Kottenhaln, *Angew. Chem.*, **70**, 268 (1958).
118. J. L. Kice and K. Ikura, *J. Am. Chem. Soc.*, **90**, 7378 (1968).
119. W. H. Mueller and M. B. Dines, *J. Chem. Soc., Chem. Commun.*, 1205 (1969).
120. J. L. Kice and S.-T. Liao, *J. Org. Chem.*, **46**, 2691 (1981).
121. J. B. Hendrickson, S. Okano, and R. K. Bloom, *J. Org. Chem.*, **34**, 3434 (1969).
122. F. Freeman, C. N. Angeletakis, and T. J. Maricich, *J. Org. Chem.*, **47**, 3403 (1982).
123. F. Freeman and C. N. Angeletakis, *J. Am. Chem. Soc.*, **105**, 4039 (1983).

The Chemistry of Sulphinic Acids, Esters and their Derivatives
Edited by S. Patai


CHAPTER **20**

# Sulphinamides

J. G. TILLETT

*Department of Chemistry and Biological Chemistry, University of Essex, Colchester, CO4 3SQ, UK*



## I. SYNTHESIS

### A. Formation from Sulphinyl Chlorides

Von Braun and his coworkers[1,2] showed that sulphinamides could be readily prepared from the reaction of sulphinyl chlorides with amines and this reaction provides the most direct method for the synthesis of both alkane- and arenesulphinamides (equation 1)[3–5].

$$\mathrm{RS(=O)Cl} + \mathrm{R^1R^2NH} \longrightarrow \underset{(\mathbf{1})}{\mathrm{RS(=O)NR^1R^2}} + \mathrm{HCl} \qquad (1)$$

Some typical examples are shown in Table 1.

Sulphinyl chlorides have been used in the synthesis of a variety of chiral sulphinamides. Thus Jacobus and Mislow showed that racemic benzenesulphinyl chloride reacts with (+)-(*S*)-deoxyephedrine (*N*-Methyl-1-phenyl-2-propylamine), **2**, to form a mixture of the diastereomeric sulphinamides **3**, the ratio of diastereomers depending significantly on the temperature used (equation 2)[7]. Cram and Nudelman prepared the diastereomeric sulphinamide **4** from the lithium salt of α-methylbenzyl amine (equation 3)[8]. Chiral

TABLE 1. Preparation of sulphinamides **1** from sulphinyl chlorides

| R | R[1] | R[2] | m.p. (°C) | Yield (%) | Reference |
|---|---|---|---|---|---|
| Ph | H | p-$ClC_6H_4$ | 155.5 | 78 | 3 |
| Ph | H | p-Tol | 100–101 | 56 | 3 |
| Ph | H | $PhCH_2$ | 100–104 | 26 | 3 |
| Ph | H | p-An | 131 | 89 | 3 |
| *p*-Tol | H | m-An | 87–88 | 84 | 4 |
| me | me | m-An | oil | 100 | 4 |
| Me | H | Ph | 87 | 71 | 5 |
| Me | Me | Me | 38/1.2[a] | 16 | 6 |
| Pr | Me | Me | 55/1.1[a] | 20 | 6 |

[a]b.p./m.m.

hydroxysulphinamides have been synthesized from L-ephedrine via formation of the chiral oxathiazolidene 2-oxides **5**, **6**. The diastereomers can be interconverted to one form **5** which can be hydrolysed to **7** (equation 4)[9,10].

$$\underset{(\mathbf{2})}{PhS(=O)Cl + CH_3NHCH_2\overset{*}{C}H(CH_3)Ph} \longrightarrow \underset{(\mathbf{3})}{Ph\overset{*}{S}(=O)N(CH_3)CH_2\overset{*}{C}H(CH_3)Ph} \tag{2}$$

$$\underset{(\pm)}{p\text{-}TolS(=O)Cl} + Ph\overset{*}{C}H(CH_3)NHLi \longrightarrow \underset{(\mathbf{4})}{p\text{-}TolS(=O)NH\overset{*}{C}H(CH_3)Ph} \tag{3}$$

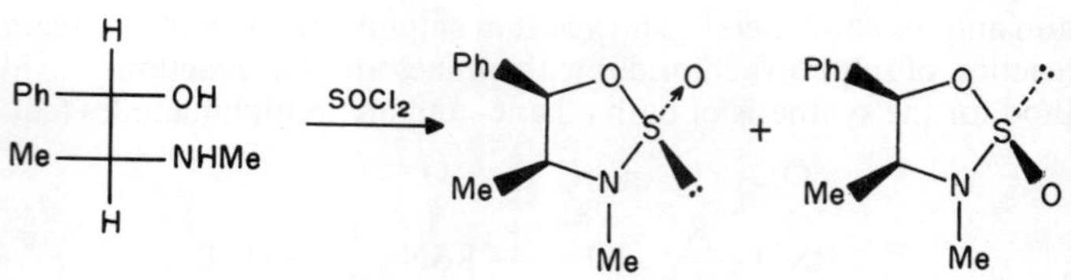


**(5)** **(6)**

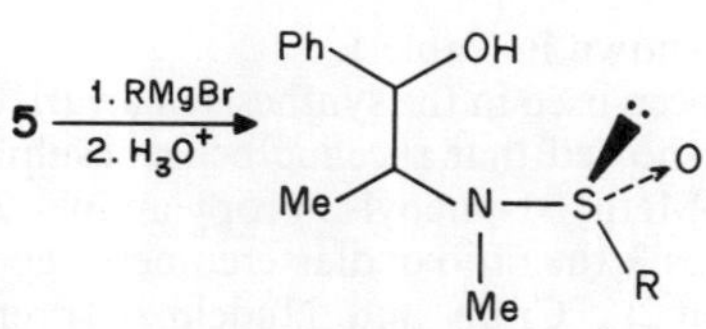


**(7)**

(4)

Arenesulphinyl chlorides react with arylureas in pyridine at moderate temperatures to form $N$-aryl-$N^1$-arylsulphinyl- and sulphenyl-ureas (equation 5)[11].

$$\mathrm{ArS(O)Cl + Ar^1NHCONH_2 \longrightarrow ArS(O)NHCONHAr^1 + ArSNHCONHAr^1} \qquad (5)$$

The synthesis of a number of $N$-alkoxyarenesulphinamides from the appropriate sulphinyl chlorides and the corresponding alkoxyamines has been reported (equation 6)[12,13].

$$\mathrm{RS(O)Cl + R^1NHOR^2 \longrightarrow RS(O)N(R^1)OR^2} \qquad (6)$$

$R = Ph, R^1 = H, R^2 = PhCH_2$; 45%

$R = p\text{-}NO_2C_6H_4, R^1 = H, R^2 = Me$; 28%

This method has also been used to synthesize several $N$-alkoxy-alkanesulphinamides[13,14] including $N$-methyl-$N$-methoxy-$t$-butanesulphinamide **8** in 68% yield (equation 7)[14]. A variety of $\beta$-ketosulphinamides, **9**, has also been prepared from the corresponding sulphinyl chlorides[15].

$$\mathrm{Bu^tS(O)Cl + CH_3NHOCH_3 \longrightarrow Bu^tS(O)N(CH_3)OCH_3} \qquad (7)$$

(**8**)

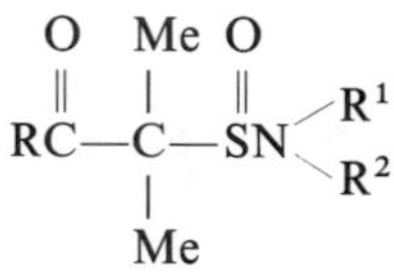

(**9**)

$R = Me, R^1 = H, R^2 = Ph$

$R = Me, R^1 = H, R^2 = Et$

$R = Me, R^1 = Et, R^2 = Et$

### B. Formation from *N*-Sulphinylamines

Sonn and Schmidt were the first to report that sulphinamides can be prepared from the reaction of a Grignard reagent with $N$-sulphinylamines[16]. This method has been extended to a variety of alkane- and arenesulphinamides[17,18].

$$\mathrm{RN{=}S{=}O \xrightarrow[2.\ H_2O]{1.\ R^1MgX} R^1S(O)NHR}$$

$R = R^1 = Ph$

$R = C_6H_{11}, R^1 = Ph$

$R = R^1 = Bu$

$R = Ph, R^1 = C_6H_{11}$

Gilman and Morris proposed a mechanism involving initial addition of the Grignard reagent across the S=O bond and rearrangement of the sulphenic acid formed by hydrolytic decomposition of this adduct (equation 8)[17]. Support for this mechanism (rather than for addition across the N=S bond) comes from the observation that phenylmagnesium bromide reacts readily with sterically-hindered sulphinylamines such as *N*-sulphinylmesidine and *N*-sulphinyl-*t*-butylamine[19].

$$\mathrm{PhN{=}S{=}O + PhMgBr \longrightarrow \underset{(\mathbf{10})}{PhN{=}S(Ph)OMgBr} \longrightarrow \left[PhN{=}S(Ph)OH\right] \longrightarrow PhS(O)NHPh} \quad (8)$$

*N*-Sulphinylamines have also been used for the synthesis of *N*-haloalkylsulphinamides (equation 9)[20] and of *N*-aryl-1-alkenylsulphinamides (equation 10)[21]. Allenylcopper(I) species add to *N*-sulphinylamines to form 2-alkynylsulphinamides (equation 11)[22]. Acetylenic sulphinamides have also been prepared via the corresponding acetylenic Grignard reagent (equation 12)[23].

$$\mathrm{Cl(CH_2)_nN{=}S{=}O \xrightarrow[2.\ NH_2Cl]{1.\ PhMgBr} PhS(O)NH(CH_2)_nCl} \quad (9)$$

$(n = 2,3)$

$$\mathrm{PhN{=}S{=}O \xrightarrow[2.\ H_2O,\ KHSO_4]{1.\ CH_2{=}CHMgBr} CH_2{=}CHS(O)NHPh} \quad (10)$$

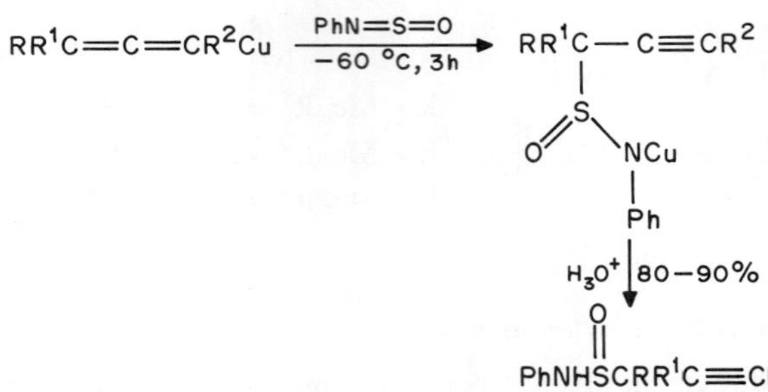


(11)

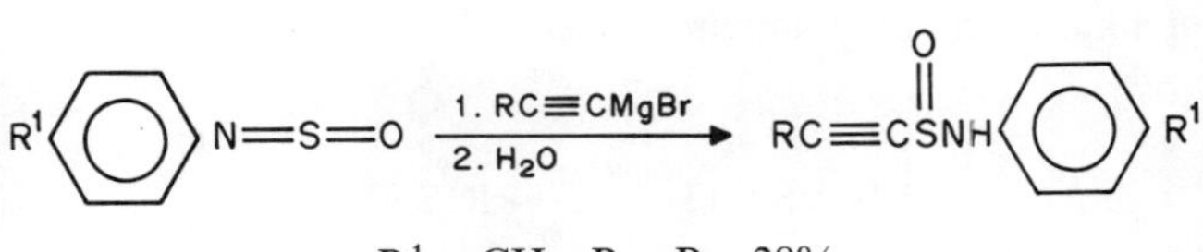


(12)

$R^1 = CH_3$, R = Bu; 28%

$R^1 = CH_3$, R = Ph; 20%

$R^1 = OMe$, R = 4-$ClC_6H_4$; 21%

### C. Formation from Sulphinylphthalimides

Harpp and his coworkers have developed the use of *N*-alkyl- and *N*-arenesulphinylphthalimides as sulphinyl transfer agents[24,25]. These compounds are conveniently made by *m*-chloroperbenzoic acid (MCPBA) oxidation of the corresponding thiophthalimides (equation 13). Sulphinylphthalimides are formed in high yield by this reaction (e.g. R = Me, 90%; R = *t*-Bu, 100%; R = Ph, 89%). Subsequent reaction of the sulphinylphthalimides with primary or secondary amines in an inert solvent gives high yields of the corresponding sulphinamide (equation 14). The generally high yields obtained make this a superior synthetic method to those previously described.

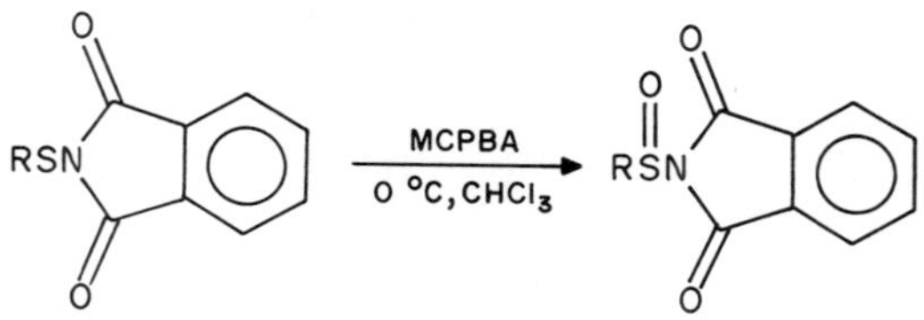


(13)

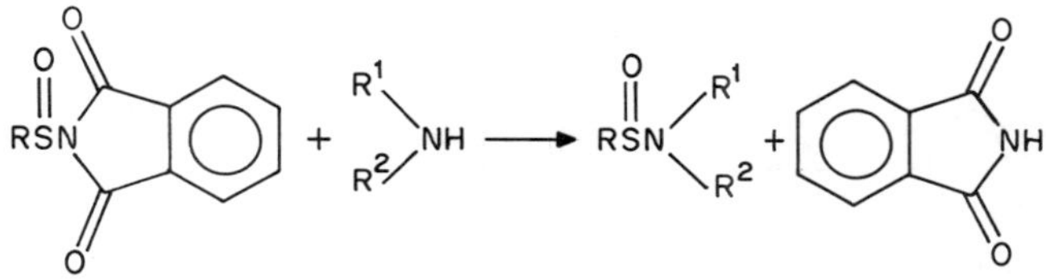


(14)

R = Me, $R^1 = R^2$ = Et; 80%
R = $Pr^i$, $R^1$ = Me, $R^2$ = Bu; 80%
R = Me, $R^1$ = H, $R^2$ = Ph; 88%
R = Ph, $R^1$ = H, $R^2 = C_6H_{11}$; 89%

## D. Formation from Sulphinic Acids

Furukawa and his group have developed several synthetic methods which use the free sulphinic acid. The first of these involves reaction with an amine in the presence of dicyclohexylcarbodiimide (DCC) as a dehydrating agent (equation 15)[26].

$$\mathrm{RS(O)OH} + \mathrm{R^1R^2NH} \xrightarrow[\text{dioxane}]{\text{DCC, rt}} \mathrm{RS(O)NR^1R^2} \quad (15)$$

R = *p*-Tol, $R^1$ = H, $R^2$ = Ph; 56%
R = $C_{12}H_{25}$, $R^1$ = H, $R^2$ = Ph; 65%

Drabowicz and Pacholczyk have suggested that the first step in this reaction involves the formation of an unstable *O*-sulphinylisourea[27]. If a chiral diimide such as *N,N'*-di-α-phenylethylcarbodiimide is used, the *O*-sulphinylurea formed should consist of a diastereomeric mixture. This intermediate **11** can then react either directly with the amine or with another molecule of sulphinic acid to form a chiral sulphinic anhydride, which can also react with the amine; in either case a chiral sulphinamide results (equation 16)[27].

This reaction therefore provides a method for the direct conversion of prochiral sulphinic acids into chiral sulphinamides, albeit in modest enantiomeric excess.

The second method introduced by Furukawa and his group involves the reaction of a sulphinic acid with the appropriate amine in the presence of 2-chloro-1-methylpyridinium iodide as a coupling reagent[26]. The *O*-sulphinylated intermediate **12** is considered to react with any amine present (equation 17).

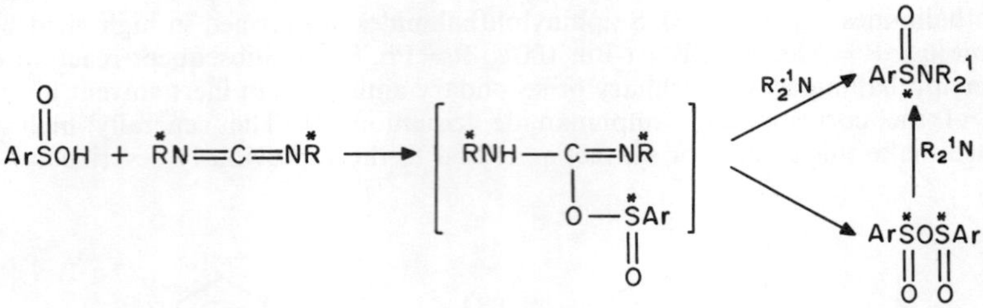

**(11)**

(16)

Ar = Ph, $R^1$ = Et; 66%; e.e. 3.0%
Ar = *p*-Tol, $R^1$ = Pr; 40%; e.e. 2.8%

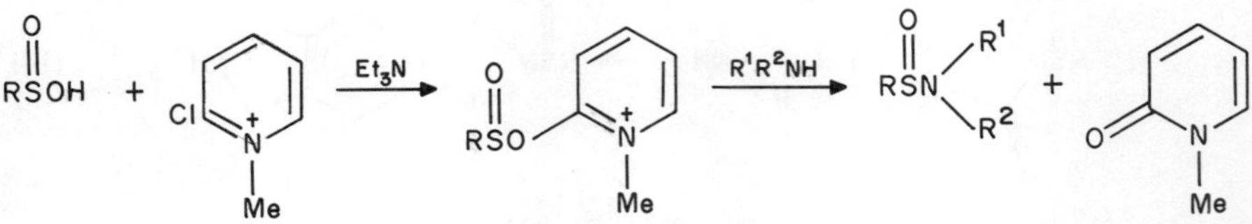

**(12)**

(17)

R = *p*-Tol, $R^1$ = H, $R^2 = C_6H_5CH_2$; 49%
R = $C_{12}H_{25}$, $R^1$ = H, $R^2 = C_6H_5CH_2$; 39%

Furukawa and his group have also investigated the 'one-pot' synthesis of sulphinamides from the reaction of *p*-toluenesulphinic acid and amines in the presence of a variety of activating agents such as phenyl phosphorodichloride, diphenylphosphorochloridate, triphenylphosphine-*N*-chlorosuccinimide and 3-(phthalimidoxy) 1,2-benzoisothiazole 1,1-dioxide[28]. Thus when *p*-toluenesulphinic acid was treated with an equivalent quantity of an amine and diphenylphosphorochloridate in the presence of excess pyridine, yields of sulphinamides in the range 0–36% were obtained. The reaction is assumed to proceed via formation of a pyridinium phosphate intermediate **13** (equation 18).

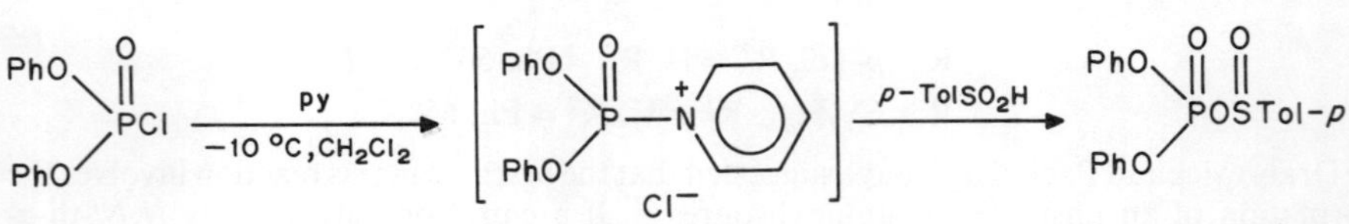

**(13)**

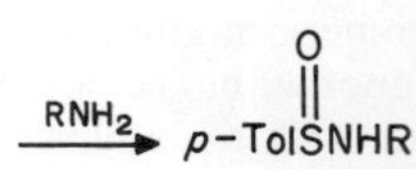

(18)

## E. Formation from Sulphinates

Cram and Nudelman were the first to show that (−)-(*S*)-menthyl *p*-toluenesulphinate **14** is converted by phenyl lithium with high stereospecificity to give the sulphinamide of opposite configuration (equation 19)[8]. (+)-*N*-α-naphthalenesulphinamide was prepared in a similar way in 34% yield from (−)-menthyl α-naphthalenesulphinate and α-naphthylamine in the presence of butyllithium[29].

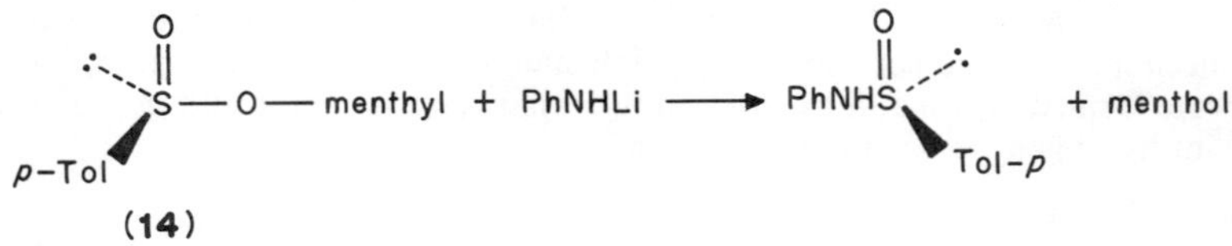

(19)

Montanari and his coworkers showed that (−)-(*S*)-menthyl *p*-toluenesulphinate reacts with dialkylaminobromomagnesium to form the corresponding sulphinamides with predominant inversion (equation 20)[30]. Chiral *N*-*p*-toluenesulphinylmorpholine has also been synthesized from (−)-(*S*)-14 and morpholinemagnesium bromide (equation 21)[31]. A series of optically active *N*-alkylidenesulphinamides of high optical purity were obtained from the reaction of imino-Grignard reagents with chiral menthyl *p*-toluenesulphinate (equation 22)[32]. Subsequent reduction by lithium aluminium hydride produces a diastereomeric mixture of sulphinamides in which substantial asymmetric induction occurs at the amine carbon atom (equation 23)[33].

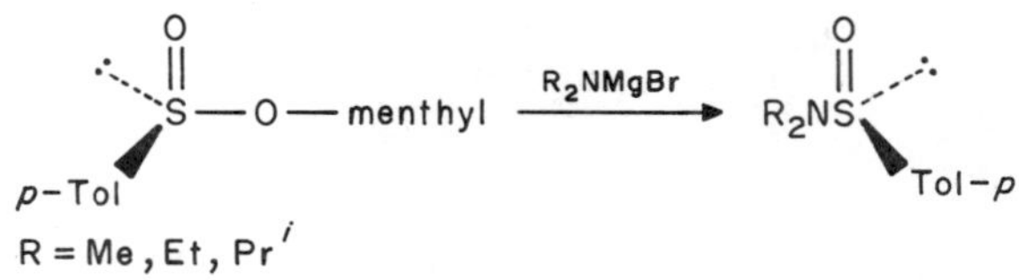

(20)

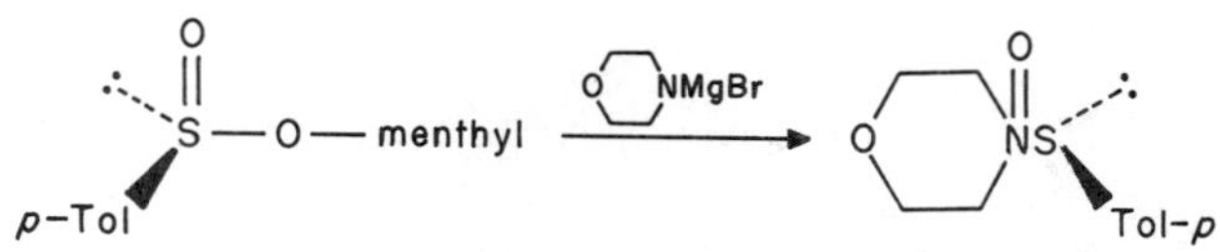

(21)

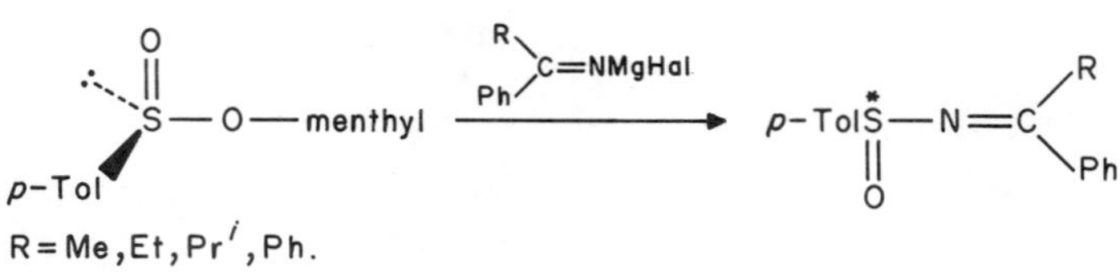

(22)

$$p\text{-Tol}\overset{*}{\underset{\|\atop \text{O}}{\text{S}}}\text{—N}=\text{C}\begin{smallmatrix}\text{R}\\ \text{Ph}\end{smallmatrix} \xrightarrow{\text{LiAlH}_4} p\text{-Tol}\overset{*}{\underset{\|\atop \text{O}}{\text{S}}}\text{NH—}\overset{*}{\underset{\text{Ph}}{\overset{\text{R}}{\text{CH}}}} \qquad (23)$$

R = Me, Et, α-Naph

## F. Oxidation of Sulphenamides

Fava and his coworkers were the first to synthesize an optically active sulphinamide. Oxidation of *p*-toluenesulphenylpiperidine with (+)-monopercamphoric acid produced the corresponding sulphinamide with low stereoselectivity (equation 24)[34]. The oxidation of the corresponding sulphenamides has also been used as a synthetic route to *N*-alkylidenesulphinamides (equation 25)[35]. Haake and his group have described a 'one-pot' synthesis of *N*,*N*-dialkylsulphinamides which utilizes oxidation with *N*-chlorosuccinimide of the corresponding sulphenamide[36]. The intermediate sulphonium salts **15** are formed *in situ* and hydrolyzed to sulphinamides by the addition of aqueous potassium hydrogen carbonate (equation 26).

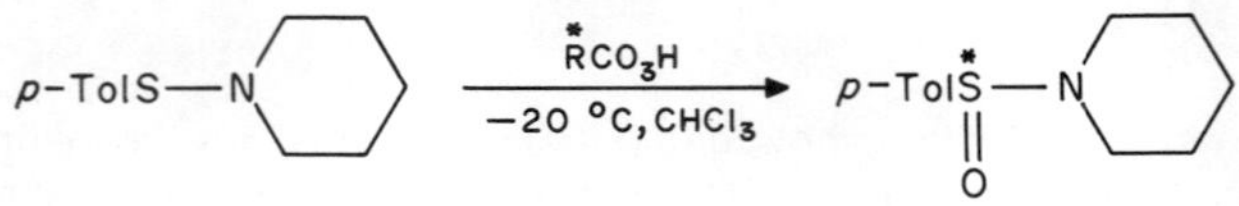


(24)

$$XC_6H_4SN{=}C\begin{smallmatrix}R\\R^1\end{smallmatrix} \xrightarrow{\text{MCPBA}} XC_6H_4\overset{\overset{O}{\|}}{S}N{=}C\begin{smallmatrix}R\\R^1\end{smallmatrix} \qquad (25)$$

$X = H, R = R^1 = CH_3$

$X = H, R = H, R^1 = Ph$

$X = 4\text{-Cl}, R = H, R^1 = Ph$

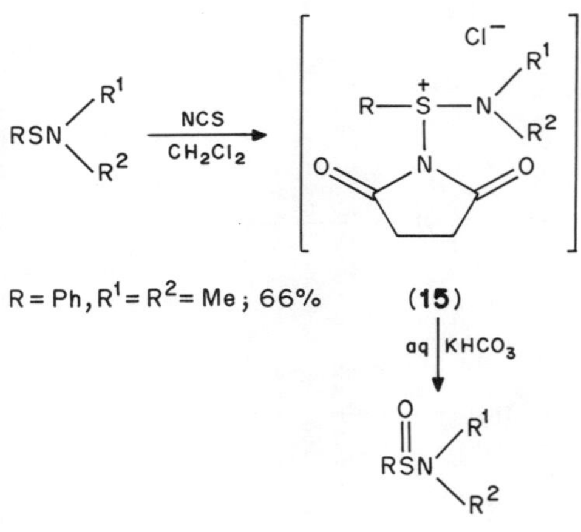


(26)

## G. Miscellaneous Methods

Johnson and his coworkers have shown that sulphinamides are readily formed from sulphoxonium salts (equation 27)[37]. Optically active *N*,*N*-dimethylsulphinamides can be generated from chiral oxosulphonium salts via the ylide (equation 28)[38]. In a similar way, reduction of (+)-(*S*)-*N*,*S*-dimethyl-*S*-phenylsulphoximine with aluminium analgam results in cleavage of the sulphur–alkyl bond to give (+)-(*S*)-*N*-methylbenzenesulphinamide (equation 29)[39]. Demethylation of sulphoximides with tosyl chloride produces a low yield of the corresponding chiral *N*-tosylsulphinamide (equation 30)[40]. Oxidation of an alkaline solution of benzyl mercaptan with chloramine produces benzylsulphinamide in 70% yield (equation 31)[41]. A 'one-pot' synthesis of arenesulphinamides from 4-nitrophenyl-substituted phenyl sulphoxides with elemental sulphur has also been described in which yields varied from 38–65% (equation 32)[42].

$$\mathrm{PhS(=O)(N(CH_3)_2)^{+}CH_3BF_2^{-}} \xrightarrow{\mathrm{CH_3ONa}} \mathrm{PhSO_2CH_3} + \mathrm{PhS(=O)N(CH_3)_2} + \mathrm{PhSO_2CH_3} \qquad (27)$$

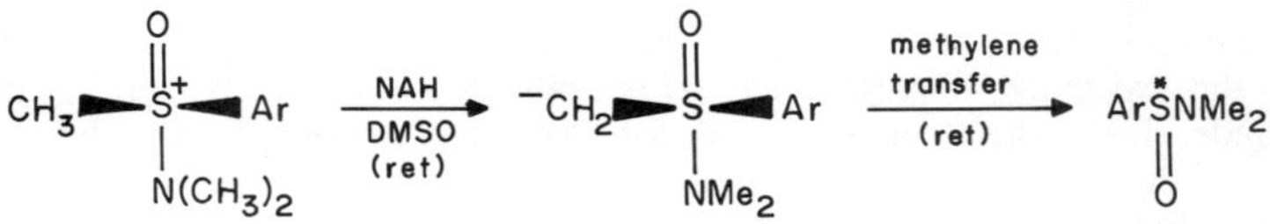


(28)

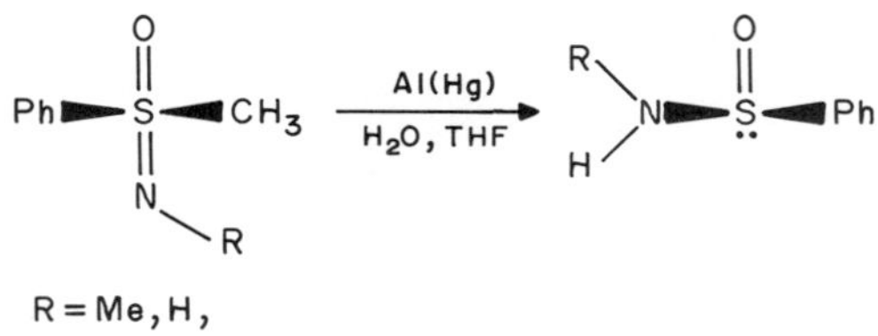


(29)

$$p\text{-Tol}\overset{*}{\mathrm{S}}(\mathrm{=NCH_3})(\mathrm{=O})\mathrm{CH_3} \xrightarrow[\text{(ret)}]{\text{TsCl}} p\text{-Tol}\overset{*}{\mathrm{S}}(\mathrm{=O})\mathrm{N(CH_3)Ts} \qquad (30)$$

$$\mathrm{PhCH_2SNa} + \mathrm{ClNH_2} \longrightarrow \mathrm{PhCH_2S(=O)NH_2} \qquad (31)$$

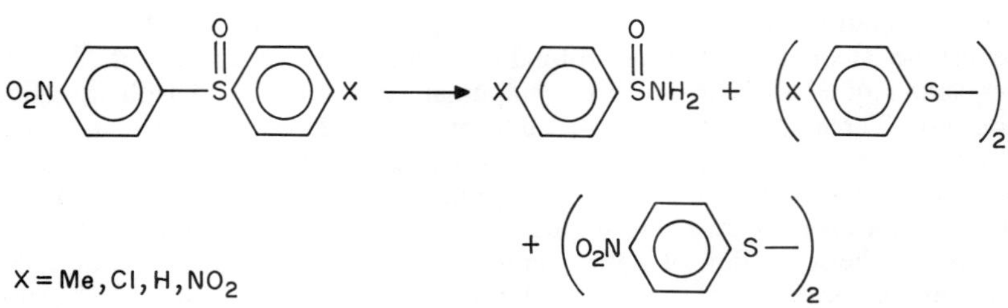


(32)

## II. STEREOCHEMISTRY

The chirality of the sulphur atom in sulphinamides has been confirmed by $^1$H NMR data. Geminal protons adjacent to the sulphinamido group are magnetically non-equivalent. Thus the methylene protons of *N*, *N*-diethylmethanesulphinamide **16** give rise to a 16-line spectrum and the methyl protons of the isopropyl groups in **17** appear as a quartet[43]. Comparison of the relative shielding effect of nuclei in sulphinamides with those in sulphonamides has been made from studies of $^{13}C$, $^{15}N$ and $^{17}O$ NMR chemical shifts[44].

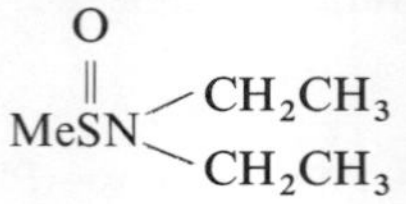

**(16)**

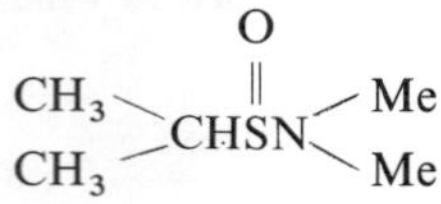

**(17)**

The barrier to internal rotational rotation about the N–S bond in sulphinamides has also been investigated by $^{1}H$-NMR studies[45,46]. Such a barrier is assumed to arise from the double-bond character of the N–S bond originating from $p_\pi$–$d_\pi$ delocalization implying the existence of resonance structures **18**–**20** for *N*,*N*-dimethylmethanesulphinamide. Both the *N*-methyl and *S*-methyl groups of **18** and the *N*-methyl group of *N*,*N*-dimethyl-*p*-toluenesulphinamide appeared as singlets at $-60\,^{\circ}C$. This suggests that because of the multiple degeneracy of sulphur 3d orbitals, $p_\pi$–$d_\pi$ overlap does not have the strict conformational requirements of $p_\pi$–$p_\pi$ overlap and that essentially free rotation may exist with almost continuous overlap.

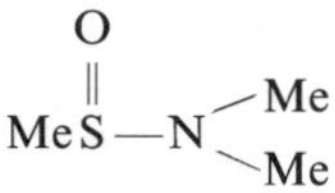

**(18)**

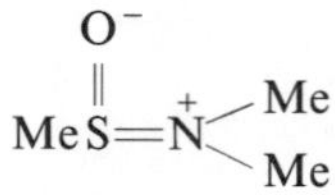

**(19)**

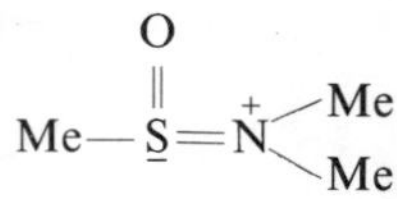

**(20)**

Mikolajczyk and Drabowicz and their coworkers have investigated the stereochemical course of the reaction of chiral sulphinamides with both alcohols and thiols. The acid-catalysed alcoholysis of *N*, *N*-diethyl *p*-toluenesulphinamide **21** was found to occur with full or predominant inversion of configuration (equation 33)[47]. The reduced stereospecificity observed for secondary and tertiary alcohols, for which nucleophilic attack at sulphur was slowed down by steric hindrance, was attributed to partial racemization of **21** under the acidic reaction conditions. These experiments were repeated with the optically stable *N*,*N*-diisopropyl system **22** (equation 34)[48]. Whilst reaction with primary alcohols was again observed to proceed mainly with inversion (R = Me, 69% inv.; R = Et, 54%; R = $Pr^n$, 58%), with secondary alcohols predominant retention was observed (e.g. R = $Pr^i$, 59% ret.). The stereochemical and kinetic features of these reactions were rationalized by an addition–elimination mechanism in which the sulphurane intermediate **23**a is formed by attack of the alcohol on the *N*-protonated sulphinamide (Scheme 1)[48–50]. Direct decomposition of **23**a will produce the sulphinate ester (−)-(*S*)-**24** with inversion of configuration. Three consecutive Berry pseudorotations of **23**, however, lead to formation of the sulphurane **23**d, which decomposes to the sulphinate with overall retention of configuration. This is the first example of retention at sulphur which does not involve formation of a four-membered ring sulphurane. An alternative reaction sequence which would explain the formation of both enantiomers of **24** is shown in equation 35 and involves parallel formation of the two sulphurane intermediates **23**a and **23**c[50]. Another intriguing feature of the trifluoroacetic acid-catalysed alcoholysis of sulphinamides is the effect of inorganic salts on the inversion-to-retention ratio. The addition of silver perchlorate greatly increases the percentage of inversion product (R = Me, 100% inv.; R = Et, 91%) and in the presence of secondary alcohols the reaction switches from predominantly retention to inversion (R = $Pr^i$, 82% inv.)[48]. The effect of the added silver salt on the stereochemistry of the reaction was attributed to complex formation between the silver ion and the sulphurane **23**a in which silver coordinates with sulphur (**25**). This is expected to both assist direct S–N bond-fission and to increase the energy of pseudorotation of **25** compared to that the **23**a. Both cations and anions have an important influence on the stereochemical course of this reaction. Thus whilst the addition of $COCl_2$, $NiC_2O_4$ and $Ag_2SO_4$ causes predominant retention, the addition of $CO(NO_3)_3$, $Ni(NO_3)_3$ and $AgNO_3$ favours inversion at sulphur[49,50]. The exact cause of these specific effects remains to be explained.

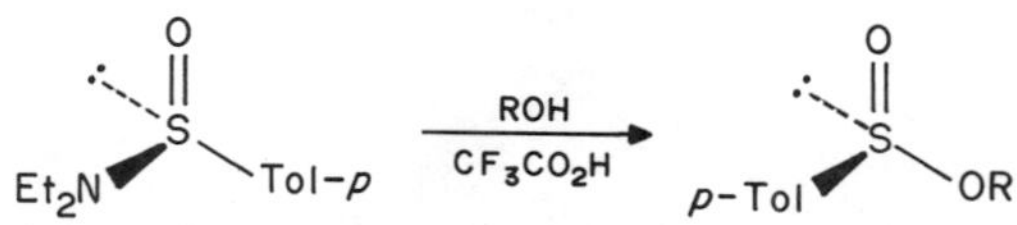


**(21)**
R = Me, Et, Pr$^i$, Bu$^t$

(33)

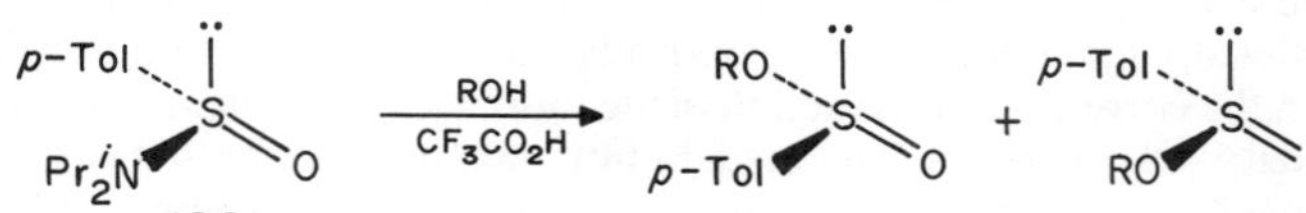


**(22)**

(34)

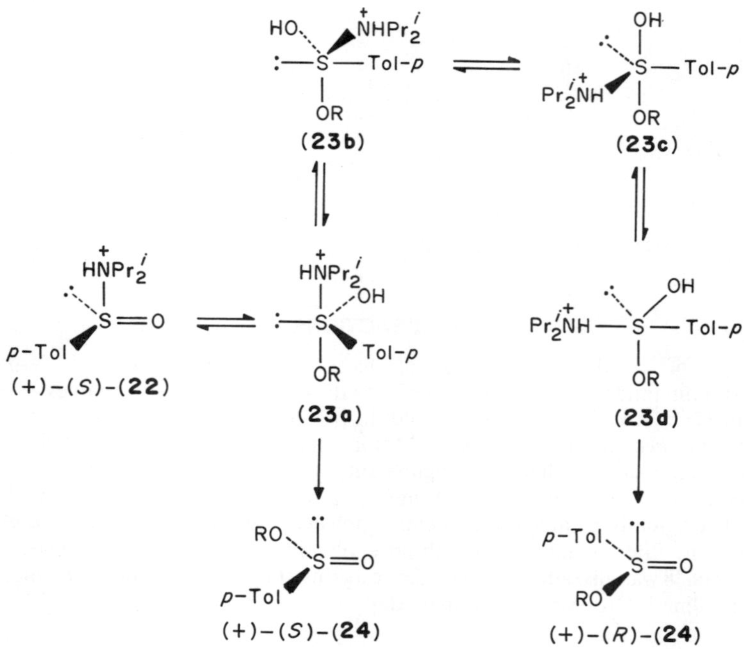


SCHEME 1

$$(+)\text{-}(S)\text{-}\mathbf{22} \xrightarrow{\text{ROH}} \begin{cases} \mathbf{23a} \longrightarrow (-)\text{-}(S)\text{-}\mathbf{24} \\ \mathbf{23c} \rightleftharpoons \mathbf{23d} \longrightarrow (+)\text{-}(R)\text{-}\mathbf{24} \end{cases} \qquad (35)$$

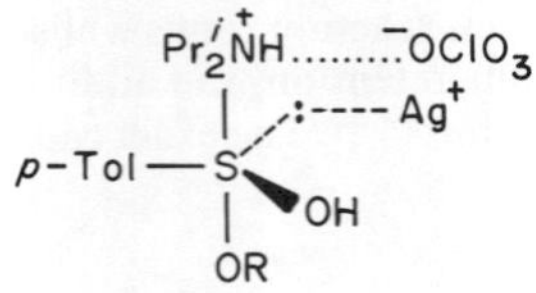

**(25)**

The stereoselective synthesis of optically active thiosulphinates was achieved by the acid-catalysed reaction of chiral sulphinamides with thiols (equation 36)[51]. The chiral sulphinates were obtained in high chemical yield with predominant inversion (typically 30–80%), the stereospecificity depending on the nature of both the thiol and sulphinamide used. The thiosulphinates formed were optically stable under the conditions used and, by analogy with the corresponding reaction of alcohols with sulphinamides, the variation in stereospecificity was attributed to an addition–elimination mechanism in which the initially formed sulphurane intermediate either decomposes directly to the product of inversion or via three pseudorotations to the product with retention of configuration.

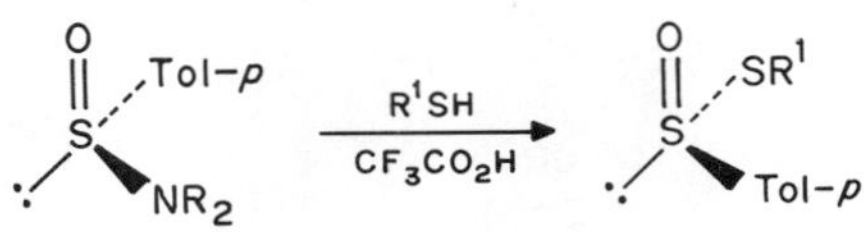

(36)

(+)-(*S*)-(**26**) (−)-(*S*)-(**27**)

R = Et, Pr$^i$

R$^1$ = Pr$^n$, Pr$^i$, Bu$^t$

## III. REACTIONS

The rates of alkaline hydrolysis of *meta*- and *para*-substituted *N*-menthylbenzenesulphinamides correlate well with Hammett $\sigma$ values ($\rho = 1.3$) (equation 37)[52]. Alkaline hydrolysis could, in principle, proceed via either an $S_N2$-type displacement mechanism or via an addition–elimination mechanism. Andersen and Biasotti[52] were unable to detect any significant $^{18}O$ incorporation into the sulphinamide recovered from partial hydrolysis of *N*-mesityl-*p*-toluenesulphinamide (**28**) in 20 atom% $H_2{}^{18}O$. As originally pointed out by Bender, however, this does not rule out the existence of a covalent intermediate[53]. The absence of any exalted substituent effect for the hydrolysis of **28** was also adduced as evidence against an addition–elimination mechanism for the alkaline hydrolysis of sulphinamides.

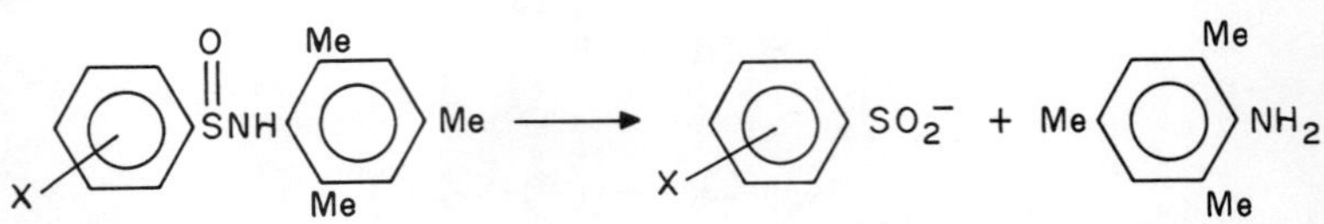

(37)

Tillett and Asefi showed that the acid-catalysed hydrolyses of some *N*-arylarenesulphinamides in hydrochloric or hydrobromic acids proceed concurrently via an acid-catalysed (A-2) mechanism and a hydrogen ion-dependent nucleophile-catalysed reaction (Scheme 2)[54].

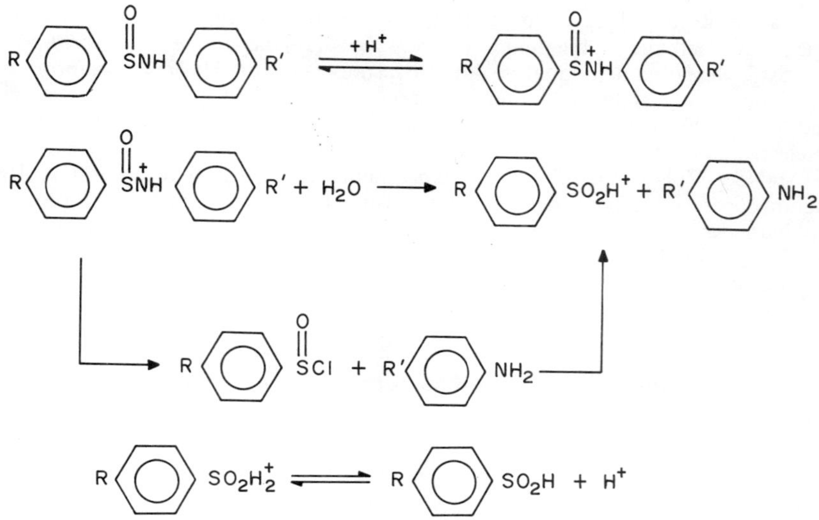

SCHEME 2

The overall effect of substituents on the rate of hydrolysis in perchloric acid is small ($\rho = -0.44$), as expected for an A-2 process. The role of halide ions is to provide an additional acid-catalysed reaction pathway by converting the sulphinamide into the more reactive sulphinyl halide. The rate-determining steps could proceed as shown by a synchronous mechanism or alternatively via a trigonal bipyramidal intermediate.

As the temperature is increased at which the hydrolysis of *p*-tolyltoluene-*p*-sulphinamide is carried out, a nucleophile-catalysed spontaneous reaction is also observed. Similar behaviour has been reported for the halide-catalysed hydrolysis of arylsulphinyl sulphones[55].

Several groups have demonstrated that the reactions of chiral sulphinamides with organolithium compounds proceed stereospecifically with inversion, e.g. the reaction of methyl lithium with *p*-toluenesulphinamide (equation 38)[30] and *N*-phenyl-*p*-toluenesulphinamide (equation 39). The latter reaction has a key role in one of Cram's triligostatic stereochemical cycles[56]. Jacobus and Mislow also used this method to obtain chiral methyl phenyl sulphoxides from the diastereomers **3** (see Section I)[7].

(38)

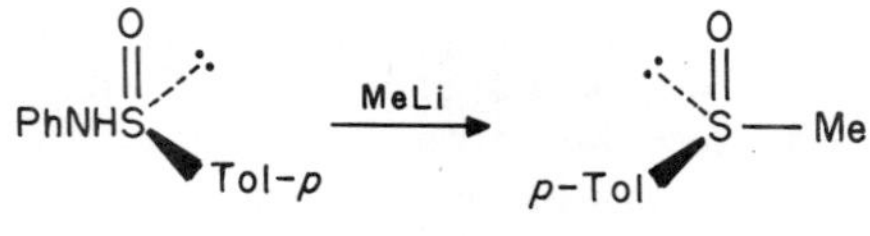

(39)

An attempt to induce endocyclic initiated rearrangement in *N*-methyl-*N*-aryl-*p*-toluene- or methanesulphinamides was unsuccessful; the aniline and corresponding sulphoxide were obtained (equation 40)[3]. In the presence of dry HCl in chloroform, certain sulphinamides were, however, found to undergo rearrangement (equation 41). A necessary condition for this to occur was the presence of an additional *ortho–para* directing group (for electrophilic aromatic substitution) in the aniline ring and this group must be *meta* to the MeNS(O)Tol group. Crossover experiments confirmed that no intramolecular reaction occurs. A mechanism involving initial attack of chloride ion on the *N*-protonated sulphinamide to form *p*-toluenesulphinyl chloride (and an *N*-methylaniline) was suggested, since sulphinyl chlorides can act as electrophilic sulphinylating agents in the presence of aluminium trichloride (see the previous chapter).

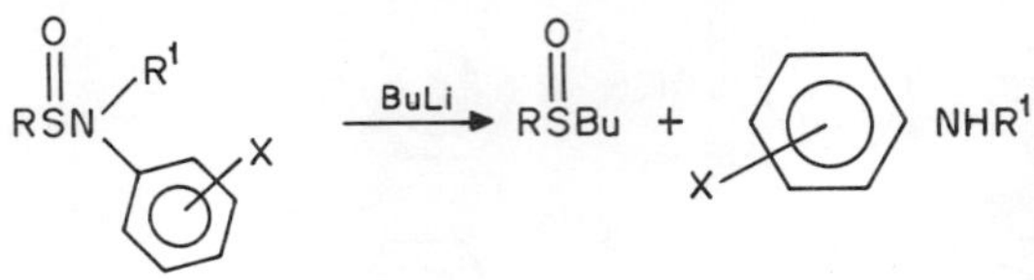


(40)

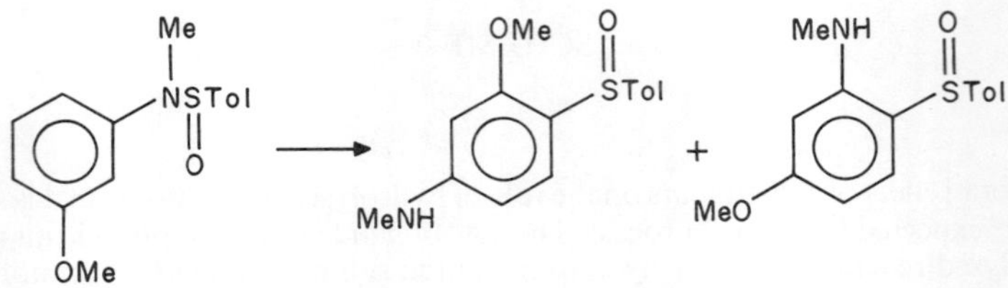


(41)

The reaction of α-lithio sulphinamide derivatives with aldehydes or ketones forms β-hydroxysulphinamides[57,58]. Thermal decomposition of these adducts forms a convenient synthetic route to alkenes (equation 42). A stereospecific *cis*-elimination pathway has been proposed for the elimination mechanism (equation 43).

$$\mathrm{MeS(=O)NHTol}\text{-}p \xrightarrow{\mathrm{2BuLi}} \mathrm{CH_2(Li)S(=O)N(Li)Tol}\text{-}p$$

$$\xrightarrow{\mathrm{RR^1CO}} \mathrm{RR^1C(O^-Li^+)CH_2S(=\overset{+}{O}Li)N(Li)Tol}\text{-}p \xrightarrow{\mathrm{H_2O}} \underset{\mathbf{(29)}}{\mathrm{RR^1C(OH)CH_2SONHTol}\text{-}p} \qquad (42)$$

$$\mathbf{29} \longrightarrow \mathrm{RR^1C{=}CH_2 + SO_2 + \textit{p}\text{-}TolNH_2}$$

Cram and Booms showed that the racemization of sulphinamides **30** and **31** was unlike that of other sulphinyl compounds and proceeded via a radical chain mechanism characterized by varying induction periods and inhibited by di-*t*-butyl nitroxide[29]. That S–N bond-fission occurs in racemization was demonstrated by cross-breeding experiments in which racemization of a mixture of equal concentrations of **30** and **31** produced the cross products **32** and **33**. The chain carrier was found to be ArN· rather than ArSO· and racemization involves radical substitution on sulphur probably via the symmetric transition state **34** (equation 44)[29].

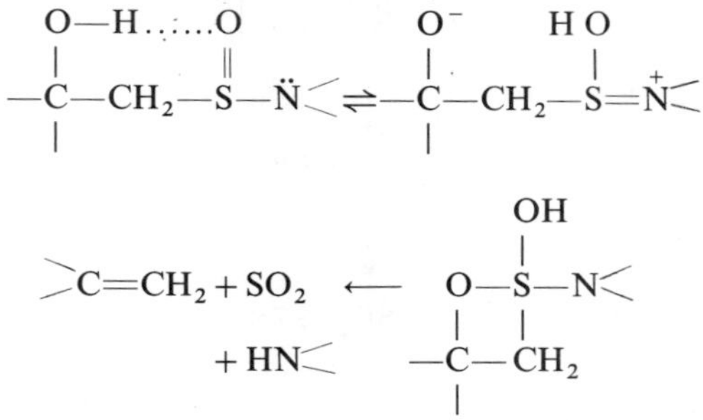

(43)

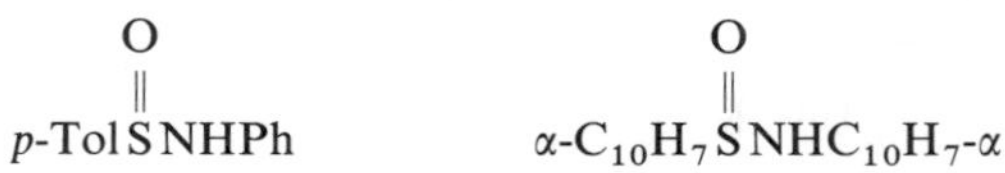

**(30)** **(31)**

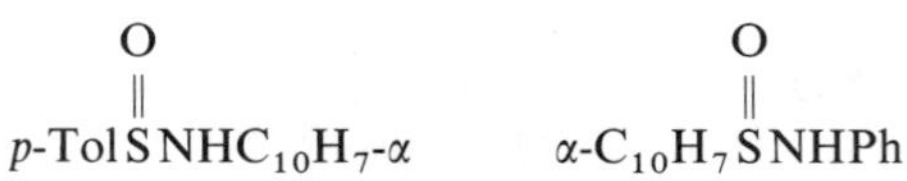

**(32)** **(33)**

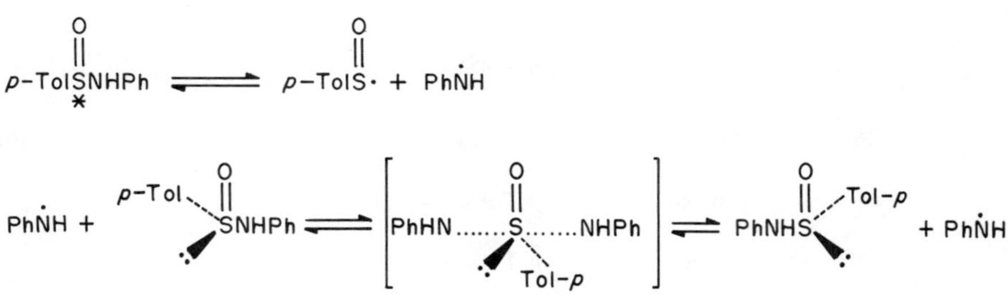

(+)-(*S*)-(**30**) (**34**) (−)-(*R*)-(**30**)

(44)

Sulphinamides are much more sensitive to light than the corresponding sulphinate esters. In aprotic solvents, *p*-toluenesulphinamides readily undergo homolysis of the S—N bond and a variety of products are formed resulting mainly from recombination and disproportionation of sulphinyl radicals formed (equation 45)[59]. It is interesting to note that photolysis of *N*-phenyl-*p*-toluenesulphinamide in methanol leads to formation of methyl *p*-toluenesulphinate in 30–40% yield[59].

The products of aprotic diazotization of *p*-toluenesulphinamide by isopentyl nitrite in different aromatic hydrocarbons ($Ar^1H$) have been rationalized by an ionic mechanism (equation 46)[60].

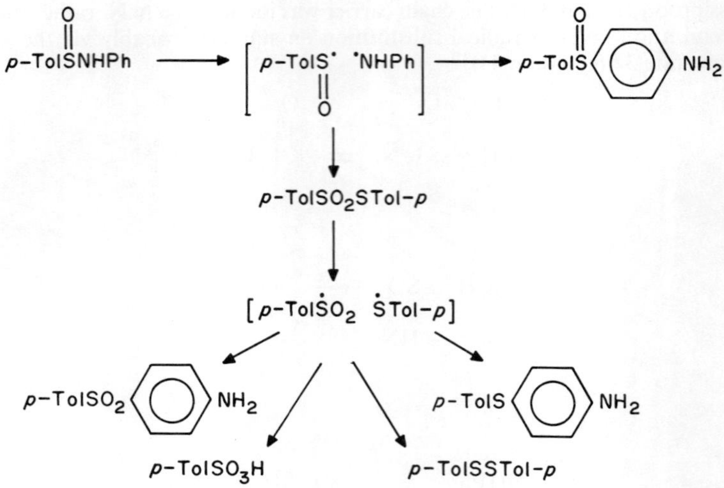

(45)

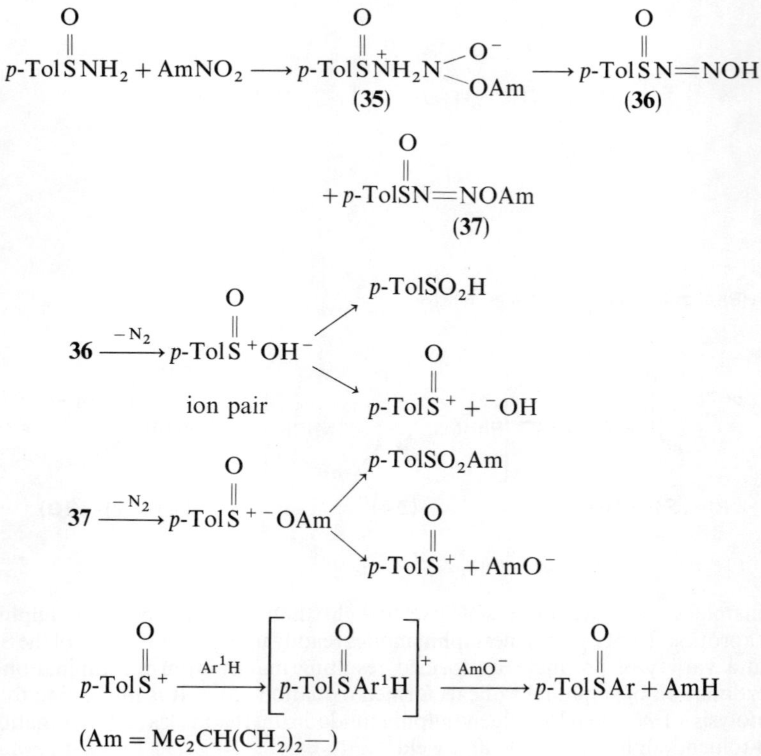

(46)

*N*-unsubstituted alkoxysulphinamides are a novel type of alkylating agent which involve the migration of an alkoxy group from nitrogen to sulphur and initial formation of an *O*-alkylsulphonimidate intermediate (equation 47)[12]. Maricich and his group[12] proposed that formation of **39** occurs by a dissociative rearrangement process to allow the migrating alkoxy group to exchange with the alcohol solvent (equation 48). In contrast, *N*-alkoxy-*N*-alkylbenzenesulphinamides decompose on heating in toluene via homolytic cleavage of the S–N bond (equation 49)[12].

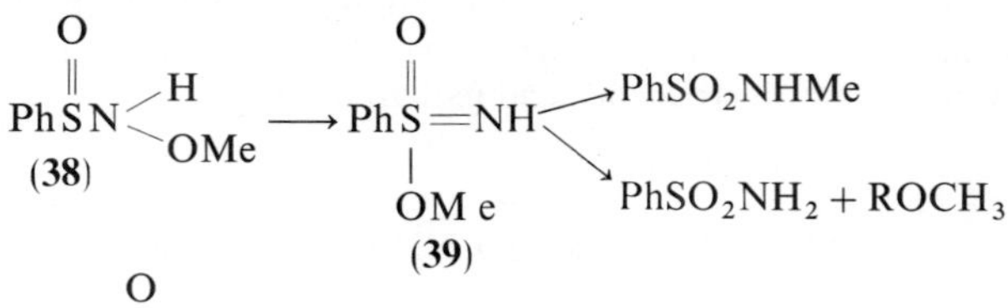

(47)

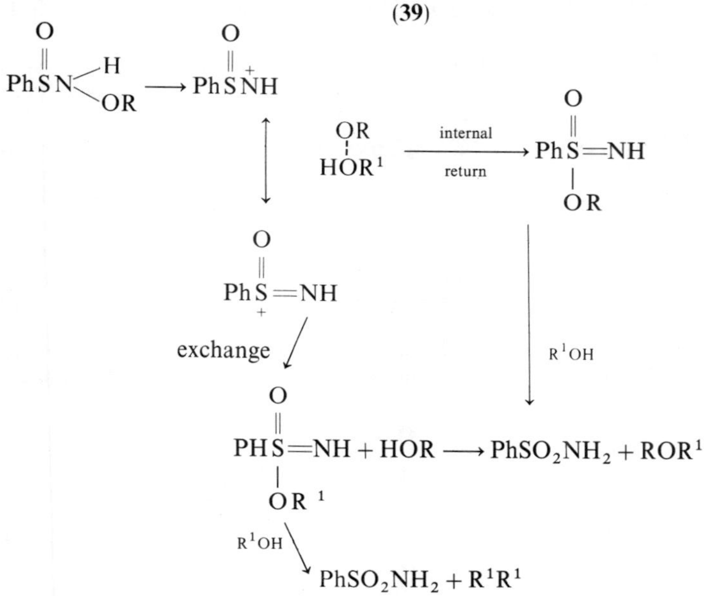

(48)

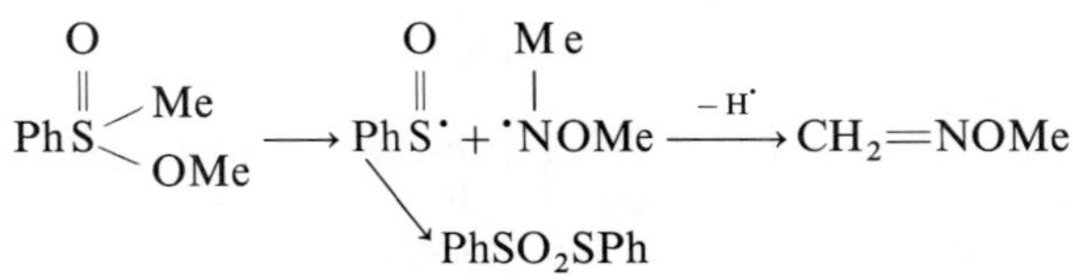

(49)

The chlorination of sulphinamides has been developed as a synthetic route to sulphinimidoyl chlorides (equation 50)[61–63]. Other chlorinating agents which have been used for this purpose include *N*-chlorotriazole[63,64] and *t*-butyl hypochlorite[65].

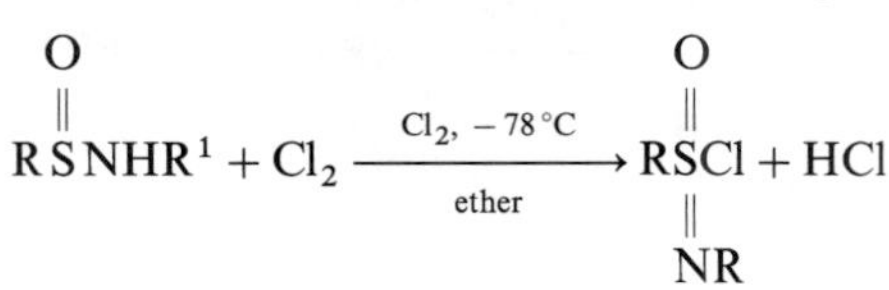

(50)

R = Me, $R^1$ = *p*-$TolSO_2$, 89%

R = Ph, $R^1$ = H, 69%

R = $PhCH_2$, $R^1$ = *p*-$ClC_6H_4$—, 52%.

The decomposition of methanesulphinamides at high temperatures was attributed to oxidation, although the products were not isolated (equation 51)[5]. Chiang and his coworkers also found that a variety of sulphinamides were readily oxidized by $KMnO_4$ but the expected sulphonamides could not be isolated[6]. Kurzer, however, was able to isolate sulphonyl ureas in good yield from the oxidation of sulphinyl ureas (equation 52)[11]. Although generally rather unstable, certain acetylenic sulphinamides have been successfully oxidized with *m*-chloroperbenzoic acid (equation 53)[23].

$$\mathrm{Me\overset{O}{\overset{\|}{S}}NHR} \longrightarrow \mathrm{MeSO_2NHR + Me\overset{}{S}NHR} \tag{51}$$

$R = PhNH$, $p$-TolNH

$$\mathrm{R\overset{O}{\overset{\|}{S}}NHCONHR^1} \xrightarrow[\text{r.t., 93\%}]{KMnO_4} \mathrm{RSO_2NHCONHR} \tag{52}$$

$R = p$-Tol, $R^1 = Ph$

$$\mathrm{BuC{\equiv}C\overset{O}{\overset{\|}{S}}NHTol\text{-}\mathit{p}} \xrightarrow[72\%]{CPA} \mathrm{BuC{\equiv}CSO_2NHTol\text{-}\mathit{p}} \tag{53}$$

Decomposition of $\beta$-ketosulphinamides with water or ethanol leads to cleavage of the carbon–sulfur bond (unlike the normal acid-catalysed decomposition of sulphinamides) (equation 54)[15].

$$\mathrm{Me\overset{O}{\overset{\|}{C}}{-}\underset{Me}{\underset{|}{\overset{Me}{\overset{|}{C}}}}{-}\overset{O}{\overset{\|}{S}}NHPh} \begin{cases} \xrightarrow{H_2O} \mathrm{Me\overset{O}{\overset{\|}{C}}CHMe_2 + PhNH_2} \\ \xrightarrow{EtOH} \mathrm{Me\overset{O}{\overset{\|}{C}}CHMe_2 + PhNH_2 + (EtO)_2SO} \end{cases} \tag{54}$$

The 2-alkenylsulphinamides **40** undergo a retro-ene cleavage in aprotic solvents to give allenes and *N*-sulphinylaniline (equation 55)[21]. Base-catalysed cyclization of **40** leads to formation of the 2, 5-dihydroisothiazole *S*-oxides, **41** (equation 56)[22]. An alternative route to **41** is shown in equation **57**[22].

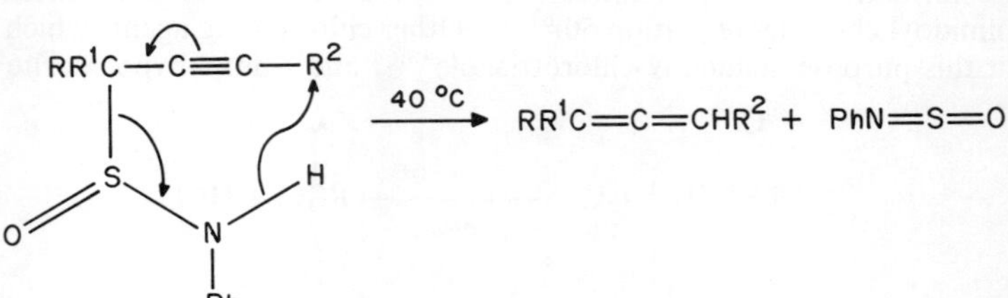


(55)

**(40)**

$R = R^1 = Me$, $R^2 = H$

$R = R^2$, $R^1 = Bu$

$R = R^1 = Me$, $R^2 = Ph$

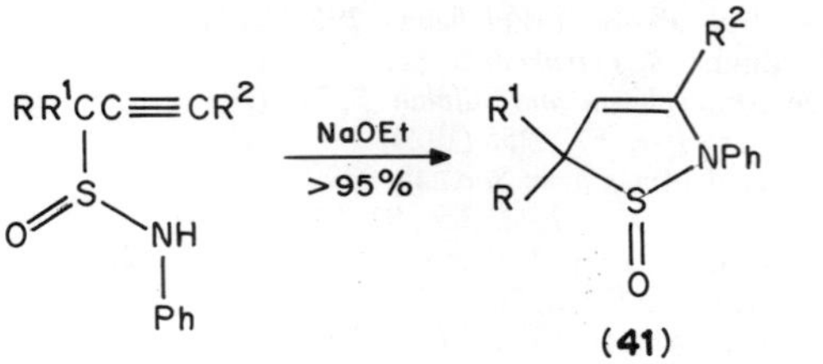


(56)

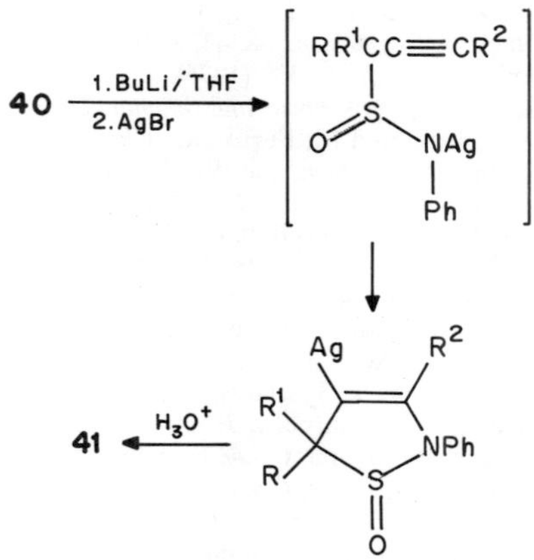


(57)

Thermal decomposition of *N*-aryl-1-alkenylsulphinamides in benzene or toluene leads to formation of the corresponding indoles possibly via a [3.3] sigmatropic rearrangement (equation 58)[21].

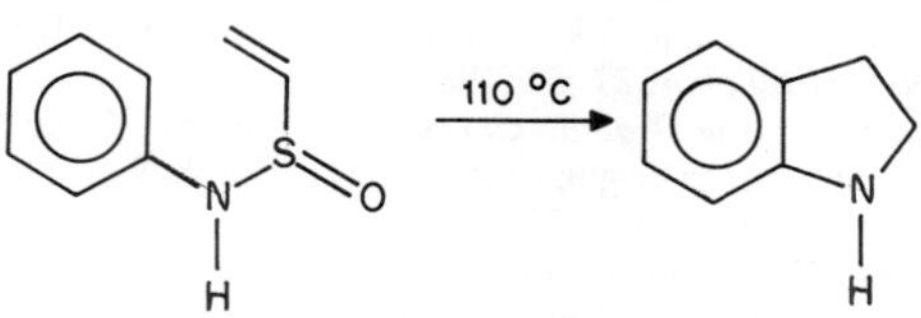


(58)

## IV. REFERENCES


1. J. von Braun and W. Kaiser, *Chem. Ber.*, **56**, 549 (1923).
2. J. von Braun and K. Weissbach, *Chem. Ber.*, **63**, 2836 (1930).
3. L. C. Raiford and S. E. Haslet, *J. Am. Chem. Soc.*, **57**, 2172 (1935).
4. K. K. Andersen and O. Malver, *J. Org. Chem.*, **48**, 4803 (1983).
5. I. B. Douglass and B. S. Farah, *J. Org. Chem.*, **23**, 805 (1958).
6. Y. H. Chiang, J. S. Luloff and E. Schipper, *J. Org. Chem.*, **34**, 2397 (1969).
7. J. Jacobus and K. Mislow, *J. Chem. Soc., Chem. Commun.*, 253 (1968).
8. A. Nudelman and D. J. Cram, *J. Am. Chem. Soc.*, **90**, 3869 (1968).
9. F. Wudl and T. B. K. Lee, *J. Chem. Soc., Chem. Commun.*, 61 (1972).
10. F. Wudl and T. B. K. Lee, *J. Am. Chem. Soc.*, **95**, 6349 (1973).
11. F. Kurzer, *J. Chem. Soc.*, 549 (1953).
12. T. M. Maricich, R. A. Jourdenais and T. A. Ulbright, *J. Am. Chem. Soc.*, **95**, 5831 (1973).

13. G. Zimmer and W. Ritter, *Arch. Pharm. (Weinheim)*, **296**, 681 (1963).
14. K. Hovius and J. B. F. N. Engberts, *Tetrahedron Lett.*, 181 (1972).
15. R. P. Gupta and J. S. Pizey, *Phosphorus and Sulphur*, **7**, 325 (1979).
16. A. Sonn and E. Schmidt, *Chem. Ber.*, **57**, 1355 (1924).
17. H. Gilman and H. L. Morris, *J. Am. Chem. Soc.*, **48**, 2399 (1926).
18. D. Klamann, C. Sars and M. Zelenka, *Chem. Ber.*, **92**, 1910 (1959).
19. W. T. Smith, P. A. Thio and M. Grasley, *J. Org. Chem.*, **27**, 692 (1962).
20. W. T. Smith and M. Grasley, *Chim. Ther.*, 266 (1968).
21. J.-B. Baudin and S. A. Julia, *Tetrahedron Lett.*, **27**, 837 (1986).
22. K. Ruitenberg and P. Vermeer, *J. Organomet. Chem.*, **256**, 175 (1983).
23. H. A. Selling and H. J. Mak, *Synth. Commun.*, 129 (1976).
24. D. N. Harpp and T. G. Back, *Tetrahedron Lett.*, 5313 (1972).
25. D. N. Harpp and T. G. Back, *J. Org. Chem.*, **38**, 4328 (1973).
26. M. Furukawa and T. Okawara, *Synthesis*, 339 (1976).
27. J. Drabowicz and M. Pacholczyk, *Phosphorus and Sulphur*, **29**, 257 (1987).
28. Y. Noguichi, M. Isoda, K. Kuroki and M. Furukawa, *Chem. Pharm. Bull.*, **30**, 1646 (1982).
29. R. E. Booms and D. J. Cram, *J. Am. Chem. Soc.*, **94**, 5438 (1972).
30. S. Colonna, R. Giovini and F. Montanari, *J. Chem. Soc., Chem. Commun.*, 865 (1968).
31. K. Okuma, H. Minato and M. Kobayashi, *Bull. Chem. Soc. Jpn.*, **53**, 435 (1980).
32. M. Cinquinni and F. Cozzi, *J. Chem. Soc., Chem. Commun.*, 502 (1977).
33. M. Cinquinni and F. Cozzi, *J. Chem. Soc., Chem. Commun.*, 723 (1977).
34. L. Sagramora, P. Koch, A. Garbesi and A. Fava, *J. Chem. Soc., Chem. Commun.*, 985 (1967).
35. F. A. Davis, A. J. Friedman and E. W. Kluger, *J. Am. Chem. Soc.*, **96**, 5000 (1974).
36. M. Haake, H. Gebbing and H. Benack, *Synthesis*, 97 (1979).
37. C. R. Johnson, M. Haake and C. W. Schroeck, *J. Am. Chem. Soc.*, **92**, 6594 (1970).
38. C. R. Johnson and C. W. Schroeck, *J. Am. Chem. Soc.*, **95**, 7418 (1973).
39. C. W. Schroeck and C. R. Johnson, *J. Am. Chem. Soc.*, **93**, 5305 (1971).
40. T. R. Williams, R. E. Booms and D. J. Cram, *J. Am. Chem. Soc.*, **93**, 7338 (1971).
41. H. Seiler and H. Erlenmeyer, *Helv. Chim. Acta*, **40**, 88 (1957).
42. R. Saito, S. Chiba, Y. Takikawa, S. Takizawara and M. Sarto, *Chem. Lett.*, 535 (1983).
43. R. M. Moriarty, *J. Org. Chem.*, **30**, 600 (1965).
44. P. Ruostesuo, A.-M. Hakkinen and T. Mattila, *Magn. Res. in Chem.*, **25**, 189 (1987).
45. R. M. Moriarty, *J. Org. Chem.*, **28**, 1296 (1963).
46. R. M. Moriarty, *Tetrahedron Lett.*, 509 (1964).
47. M. Mikolajczyk, J. Drabowicz and B. Bujnicki, *J. Chem. Soc., Chem. Commun.*, 568 (1976).
48. M. Mikolajczyk, J. Drabowicz and B. Bujnicki, *Tetrahedron Lett.*, **26**, 5699 (1985).
49. M. Mikolajczyk, *Phosphorus and Sulphur*, **27**, 31 (1986).
50. M. Mikolajczyk, in *Perspectives in the Organic Chemistry of Sulphur*, Invited Lectures of the Twelfth International Symposium on the Organic Chemistry of Sulphur, Nijmegen, 1986 (Eds. B. Zwannenburg and A. J. H. Klunder), 1986, p. 23.
51. J. Drabowicz and M. M. Kolayczyk, *Tetrahedron Lett.*, **26**, 5703 (1985).
52. J. B. Biasotti and K. K. Andersen, *J. Am. Chem. Soc.*, **93**, 1178 (1971).
53. M. L. Bender, *J. Am. Chem. Soc.*, **73**, 1626 (1951).
54. H. Asefi and J. G. Tillett, *J. Chem. Soc., Perkin Trans. 2*, 1579 (1979).
55. J. L. Kice and G. Guaraldi, *J. Am. Chem. Soc.*, **89**, 4113 (1967).
56. T. R. Williams, A. Nudelman, R. E. Booms and D. J. Cram, *J. Am. Chem. Soc.*, **94**, 4684 (1972).
57. E. J. Corey and T. Durst, *J. Am. Chem. Soc.*, **88**, 5656 (1966).
58. E. J. Corey and T. Durst, *J. Am. Chem. Soc.*, **90**, 5548, 5553 (1968).
59. H. Tsudo, H. Minato and M. Kobayashi, *Chem. Lett.*, 149 (1976).
60. G. D. Luca, G. Renzi, V. Bartocci and C. Panattoni, *Chem. Ind.*, 1054 (1975).
61. C. R. Johnson and E. U. Jonsson, *J. Am. Chem. Soc.*, **92**, 3815 (1970).
62. E. U. Jonsson, C. C. Bacon and C. R. Johnson, *J. Am. Chem. Soc.*, **93**, 5306 (1971).
63. C. R. Johnson, E. U. Jonsson and C. C. Bacon, *J. Org. Chem.*, **44**, 2055 (1979).
64. F. Wudl, C. K. Brush and T. B. K. Lee, *J. Chem. Soc., Chem. Commun.*, 151 (1972).
65. C. R. Johnson and A. Wambsgans, *J. Org. Chem.*, **44**, 2278 (1979).

The Chemistry of Sulphinic Acids, Esters and their Derivatives
Edited by S. Patai


CHAPTER **21**

# Mechanism of nucleophilic displacement reactions of sulfinic acid derivatives

TADASHI OKUYAMA

*Faculty of Engineering Science, Osaka University, Toyonaka, Osaka 560, Japan*



## I. INTRODUCTION

Transformations of various sulfinic acid derivatives take place through nucleophilic substitution. The present chapter deals with those mechanistic aspects of the reaction which are common to the whole class of compounds. The nucleophilic displacement reaction of sulfinic acid derivatives (equation 1) formally resembles that of carboxylic acid derivatives (equation 2), where Y/Nu = OR′, SR′, $NR'_2$, halogens and $SO_2R'$. The addition–elimination (A–E) mechanism involving a tetrahedral intermediate is well established as a general pathway of the latter reaction. However, the intermediacy of a hypervalent tetracoordinate sulfur species (sulfurane) formed by the addition of a nucleophile to the sulfinyl sulfur atom has not been demonstrated conclusively for the former reaction.

$$\mathrm{R}\overset{\mathrm{O}}{\overset{\|}{\mathrm{S}}}\mathrm{Y} + \mathrm{NuH} \rightleftharpoons \mathrm{R}\underset{\mathrm{Nu}}{\underset{|}{\overset{\mathrm{OH}}{\overset{|}{\mathrm{S}}}}}\mathrm{Y} \rightleftharpoons \mathrm{R}\overset{\mathrm{O}}{\overset{\|}{\mathrm{S}}}\mathrm{Nu} + \mathrm{YH} \quad (1)$$

$$\mathrm{R}\overset{\mathrm{O}}{\overset{\|}{\mathrm{C}}}\mathrm{Y} + \mathrm{NuH} \rightleftharpoons \mathrm{R}\underset{\mathrm{Nu}}{\underset{|}{\overset{\mathrm{OH}}{\overset{|}{\mathrm{C}}}}}\mathrm{Y} \rightleftharpoons \mathrm{R}\overset{\mathrm{O}}{\overset{\|}{\mathrm{C}}}\mathrm{Nu} + \mathrm{YH} \quad (2)$$

A contrasting difference between sulfinic and carboxylic acid derivatives is in the stereochemistry at the central atom. The sulfinyl sulfur has a stable pyramidal arrangement of the ligands while the carbonyl carbon is planar. As a consequence the sulfinyl derivatives are chiral and the stereochemical course of the reaction is closely associated with its mechanism. This is in turn closely related to the stereochemical nature of the sulfurane intermediate, if it does intervene.

Most of the nucleophilic substitutions of sulfinicacid derivatives occur by predominant inversion. This stereochemical course can be accounted for by a sulfurane intermediate in which an incoming nucleophile and an outgoing leaving group occupy the apical positions. An alternative pathway involving the inversion of configuration may be a one-step displacement similar to the $S_N2$ reaction at saturated carbon, i.e. bond formation and bond breaking are occurring synchronously in the rate-determining step, and the structure of the transition state is similar to that of the sulfurane (equation 3).

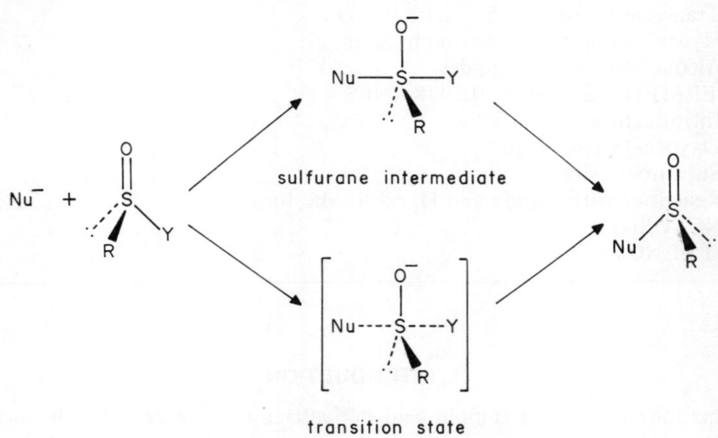


(3)

The important problems to be solved are: whether the reaction is concerted ($S_N2$-like) without any intermediate or whether it is stepwise (A–E mechanism) with sulfurane as a discrete intermediate, and how the nature of the intermediate (transition state) affects the stereochemical course of the reaction. Discussion will be focused on these problems in this chapter. Recent reviews are concerned with general features of the nucleophilic substitution at sulfur[1, 2] and with the stereochemical aspects of the reaction[3–5].

## II. STEREOCHEMISTRY OF SULFURANE INTERMEDIATES

### A. Hypervalent Bonding

Let us first examine the stereochemical nature of a potential sulfurane intermediate in order to understand the relationship between the stereochemistry of the reaction and its mechanism. The tetracoordinate sulfur intermediate, sulfurane (**1**), has an electronic structure involving a formal expansion of the valence shell octet of the central sulfur atom and is called a hypervalent species. Although this class of compounds is not usually stable, a number of stable derivatives have recently been isolated[6]. The stable form of the structure is established to be a pseudotrigonal bipyramid (Ψ-TBP) with a pair of unshared electrons in an equatorial position.

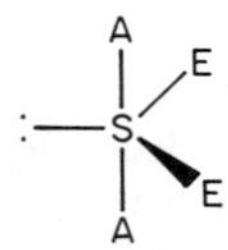


(**1**)

Two linear apical bonds A—S—A are modelled by a three-center four-electron bond, which is termed hypervalent bonding[7]. This approximate molecular orbital model shows that the electron-rich delocalized sigma bonds are analogous to the delocalized $\pi$ bonds in the allyl anion as shown in Figure 1. That is, the first two of the four electrons of the hypervalent bonds occupy the bonding molecular orbital while the second two occupy the nonbonding orbital which has no contribution from the central atom. Although the symmetry of this nonbonding molecular orbital is compatible with a contribution from a 3d ($d_{z^2}$) orbital of the sulfur, this contribution must be very small because of a large energy gap between the 3d orbitals and the p orbitals of the apical ligands. Theoretical studies in

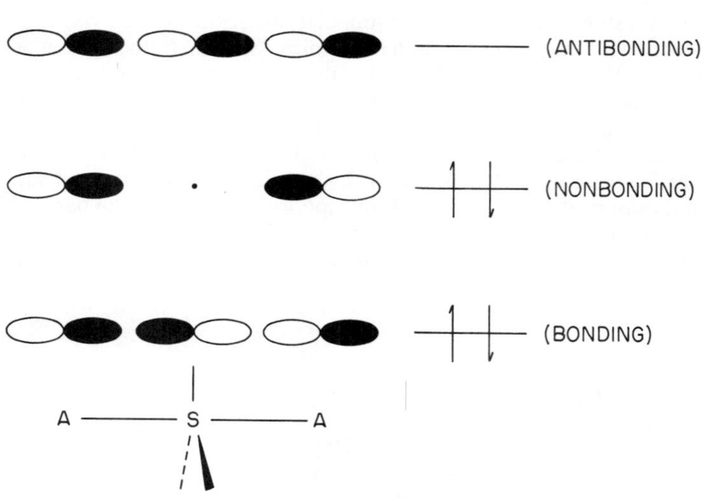


FIGURE 1. Molecular orbital model of hypervalent bonding in sulfurane

fact show that a qualitative picture of the hypervalent bonding is well described by a delocalized three-center four-electron $\sigma$ bond without considering the contribution from sulfur 3d functions[8,9]. The electron distribution in the nonbonding orbital predicts relatively negative charge on the apical ligands and positive charge on the central sulfur atom. This situation may also be visualized by a qualitative valence bond description involving no-bond resonance structures.

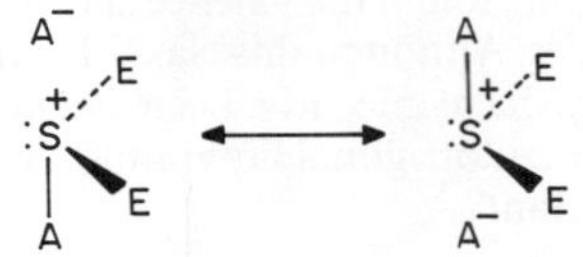

(1a) (1b)

These models rationalize that more electronegative ligands prefer apical positions (apicophilicity)[10]. Both nucleophiles and leaving groups involved in nucleophilic substitutions are generally electronegative and tend to occupy an apical position. Furthermore, the apical bonds are long and weak since these two delocalized bonds contain only two electrons in the bonding molecular orbital and the bond order is expected to be low. This consideration predicts that the nucleophile generally enters from one apical position and the leaving group departs from the other apical position, resulting in inversion of configuration.

### B. Pseudorotation

However, the apical and equatorial ligands can interchange with each other to result in isomerization. The nondissociative permutational isomerization is considered to take place through pseudorotation[11]. The pseudorotation occurs by pairwise exchange of two equatorial and two apical ligands via a square pyramidal transition state (Figure 2). A closely related mechanism called the 'turnstile' rotation has been proposed by Ugi and coworkers[12], but this was shown to be a higher-energy process[13]. Far infrared spectral data of $SF_4$ in fact showed that the permutational isomerization occurs via a $C_{4v}$ transition state[14] in accord with the pseudorotation mechanism. Such an isomerization occurs often quite readily. The $^{19}$F NMR studies show that the barrier to the interchange of apical and equatorial fluorines in $SF_4$ is 11–12 kcal mol$^{-1}$ [15–17] in accord with the result obtained by IR spectra (10.2 kcal mol$^{-1}$)[14]. The tetraoxyspirosulfuranes **2**[18] and **3**[19] were found to undergo pseudorotation with barriers of about 7.5 and 9 kcal mol$^{-1}$, respectively. However, the barrier to pseudorotation of this class of Ψ-TBP sulfur species is usually higher than that of pentacoordinate TBP phosphorus species (e.g. the barrier for $PF_5$ is less than 5 kcal mol$^{-1}$ [20]) and is not always low enough to ensure the rapid interchange of apical and equatorial ligands of sulfurane intermediates of nucleophilic substitutions, if any.

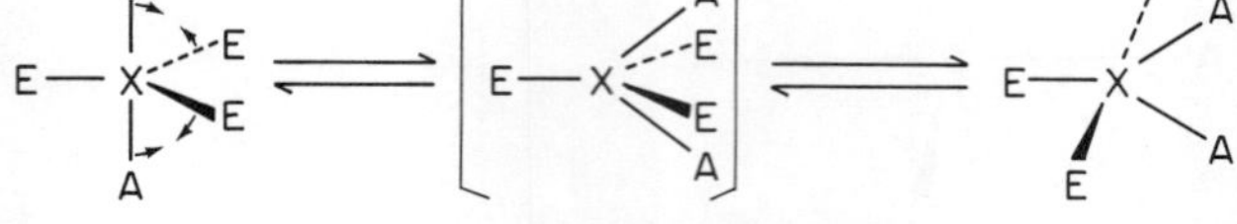

FIGURE 2. Pseudorotation of a trigonal bipyramidal compound

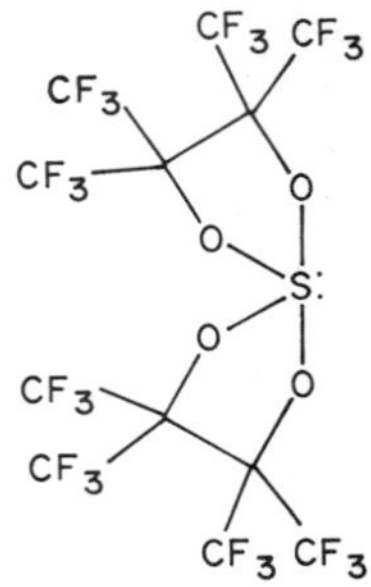

**(2)**

**(3)**

### C. Stereochemical Courses

Nucleophilic displacement reactions involving a trigonal bipyramidal intermediate in principle take place in three ways. When both an incoming and an outgoing group react in the apical positions, inversion of configuration of the central atom results. From the above considerations, this would be the most probable stereochemical course of the reaction. Since this steric arrangement is also followed by $S_N2$-like reactions where the bond-making and bond-breaking are synchronous, it is important to distinguish the stepwise reactions from synchronous ones.

The inversion of configuration may also occur when both nucleophilic attack and leaving-group departure take place at the equatorial sites. Such a special case is noted by Cram and Day[21], where the equatorial–equatorial arrangement of entering and leaving groups may be preferred by formation of a six-membered ring system. On the other hand, the reactions at the apical and equatorial positions result in retention of configuration. Such an example was first presented by Oae and coworkers[22], in which a four-membered ring system involving entering and outgoing atoms was thought to favor the apical–equatorial arrangement. For some other reactions proceeding with retention of configuration, four-membered cyclic structures with apical–equatorial arrangement were also postulated[23,24]. These situations, however, become complicated when intramolecular ligand exchange (pseudorotation) occurs rapidly before the decomposition of the Ψ-TBP intermediate.

## III. STEREOCHEMISTRY OF NUCLEOPHILIC SUBSTITUTION

### A. Transesterification of Sulfinate Esters

Nucleophilic substitutions at chiral sulfinyl derivatives generally proceed with inversion of configuration. The first reported example is the thermal transesterification of (−) ethyl *p*-toluenesulfinate with butanol to give (+) butyl *p*-toluenesulfinate[25] (equation 4). The reaction involves inversion but the stereospecificity was quite low. The same reaction was reexamined later by Mikolajczyk and coworkers[26] and the product they obtained under the same conditions was always completely racemic. However, they established more rigorously that the methanol exchange reaction of methyl *p*-toluenesulfinate occurs stereospecifically with inversion of configuration under kinetic conditions. Using an optically active sulfinate labelled with carbon-14, the rates of both racemization and isotopic methoxy–methoxy exchange in methanol were measured in the presence of trifluoroacetic acid as an acid catalyst (equation 5). It was found within experimental error that the rate of racemization is twice as large as that of loss of radioactivity of the sulfinate[26]. This means that every methoxy exchange must occur with inversion of configuration at the sulfinyl group.

$$(-)\ \mathrm{Tol}\overset{\overset{\displaystyle O}{\|}}{\mathrm{S}}\mathrm{OEt} + \mathrm{BuOH} \longrightarrow (+)\ \mathrm{Tol}\overset{\overset{\displaystyle O}{\|}}{\mathrm{S}}\mathrm{OBu} + \mathrm{EtOH} \qquad (4)$$

$$(+)\ \mathrm{Tol}\overset{\overset{\displaystyle O}{\|}}{\mathrm{S}}\mathrm{O}^{14}\mathrm{CH_3} + \mathrm{CH_3OH} \xrightarrow{\mathrm{H^+}} (\pm)\ \mathrm{Tol}\overset{\overset{\displaystyle O}{\|}}{\mathrm{S}}\mathrm{OCH_3} + {}^{14}\mathrm{CH_3OH} \qquad (5)$$

However, the base-catalyzed transesterification was found to be nonstereospecific[27]. Diastereoisomerically pure (−) menthyl (−) arenesulfinates were converted into racemic ethyl sulfinates in ethanol in the presence of sodium ethoxide (equation 6).

$$(-)\ \mathrm{Ar}\overset{\overset{\displaystyle O}{\|}}{\mathrm{S}}\mathrm{OMenthyl} + \mathrm{EtOH} \xrightarrow{\mathrm{EtONa}} (\pm)\ \mathrm{Ar}\overset{\overset{\displaystyle O}{\|}}{\mathrm{S}}\mathrm{OEt} + \mathrm{MenthylOH} \qquad (6)$$

Ar = Ph or Tol

The *N*-bromosuccinimide-catalyzed alcoholysis of a thiolsulfinate was noted to occur with predominant inversion[3], but the similar NBS-catalyzed alcohol exchange of sulfinates was found to proceed with complete racemization[28].

The reactions of chiral sulfinate esters with organometallic compounds to form sulfoxides also belong to those nucleophilic displacement reactions which occur with inversion of configuration. The Andersen synthesis of optically active sulfoxides[29] and the closely related reactions of thiolsulfinates[30] and sulfinamides[31] with Grignard and organolithium reagents are known all to proceed with inversion of configuration and high stereospecificity.

### B. Hydrolysis of Alkoxysulfonium Salts

A closely related reaction which proceeds with complete inversion of configuration is the alkaline hydrolysis of the alkoxysulfonium salt **5**, obtained by O-alkylation of the

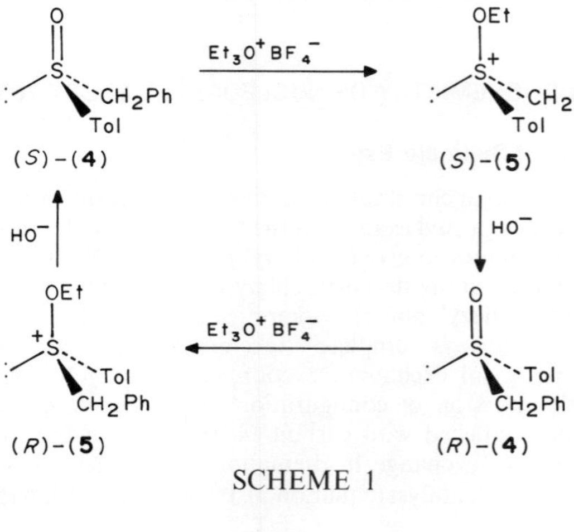


SCHEME 1

sulfoxide **4**[32] (Scheme 1). The O-alkylated derivative of sulfinamide **6** was also found to undergo alkaline hydrolysis (equation 7) mostly with inversion (> 91%)[33].

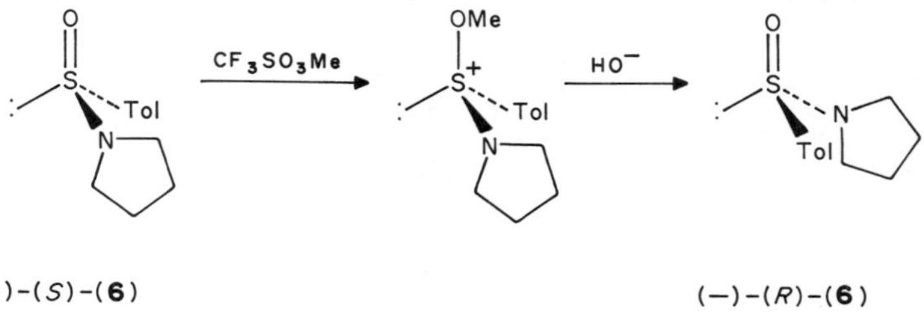


(7)

Alkaline hydrolysis of dialkoxysulfonium salts may proceed via sulfurane intermediates which have the same structure as that in transesterification of sulfinate esters. Hence knowledge of the former reaction will be very informative as to the latter. The stereochemistry of hydrolysis of some sulfonium salts was examined by Mikolajczyk and coworkers[34]. A series of chiral alkoxymethoxyisopropylsulfonium triflates **7**, obtained *in situ* by methylation of the alkyl sulfinates **8** with methyl triflate, were used as substrates, and it was found that the hydrolysis of the sulfonium salts **7** gives two possible sulfinates **8** and **9**, which have respectively a configuration opposite to the starting sulfinates **8** (equation 8). The displacement of both alkoxy groups at sulfur of **7** with predominant inversion can be accommodated by the simultaneous formation of two different sulfurane intermediates **10** and **11** (or transition states of these arrangements) which undergo decomposition before pseudorotation. Rapid pseudorotation of the intermediate may result in lesser stereoselectivity.

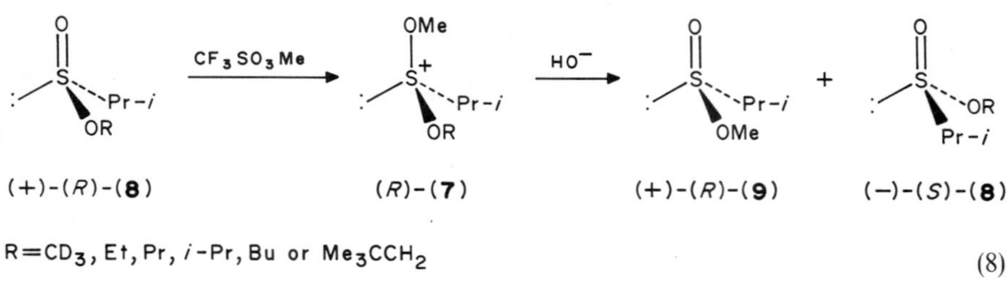


(8)

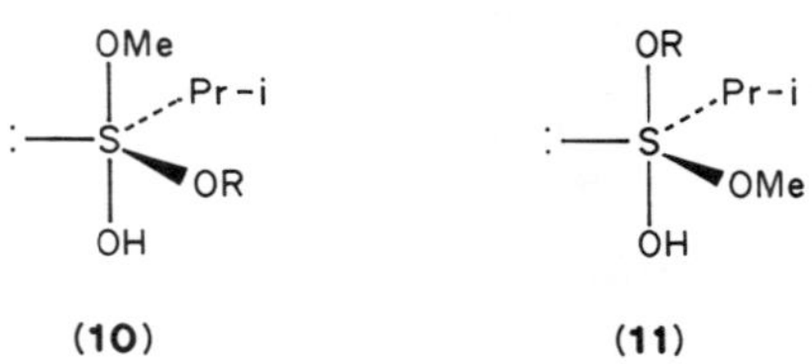


### C. Alcoholysis of Sulfinamides

Acid-catalyzed alcoholysis of *N, N*-diethyl *p*-toluenesulfinamide (**12a**) was found to take place with complete or predominant inversion of configuration[35]. The decreased stereoselectivity observed for secondary and tertiary alcohols was considered to be due to a partial racemization of the substrate **12a** under acidic reaction conditions (equation 9). However, examination of the alcoholysis of the *N, N*-diisopropyl sulfinamide **12b** (equation 10), which is optically stable under the reaction conditions, showed that the stereoselectivity of the reaction is not so good and is largely dependent on the structure of the alcohols (from 69% inversion with methanol to 74% retention with cyclohexanol)[36].

$$\underset{(\mathbf{12a})}{\text{TolS(=O)NEt}_2} + \text{ROH} \xrightarrow{\text{CF}_3\text{CO}_2\text{H}} \text{TolS(=O)OR} + \text{Et}_2\text{NH} \qquad (9)$$

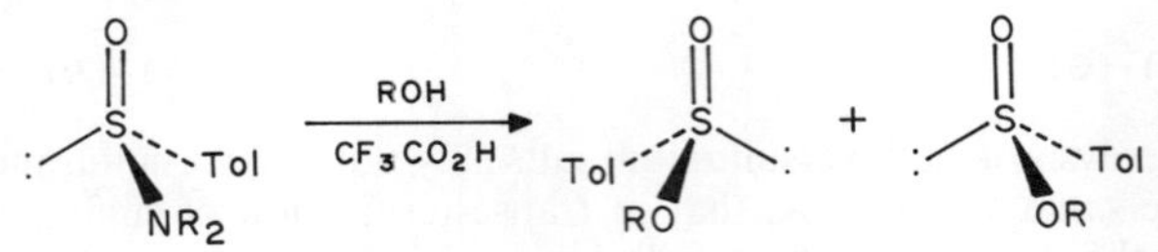


(10)

(+)-(*S*)-(**12b**) (−)-(*S*)-(**13**) (+)-(*R*)-(**13**)

R= *i*-Pr

The steric course of this reaction was also greatly influenced by added inorganic salts. Among various salts examined, silver perchlorate very much enhanced the formation of the inversion product (e.g. 100% inversion with methanol and 65.5% inversion with cyclohexanol). These diverse stereochemical results may best be rationalized by assuming intermediate formation of a sulfurane which undergoes pseudorotation during the reaction (Scheme 2). Various configurations of the sulfurane intermediate in Scheme 2 can

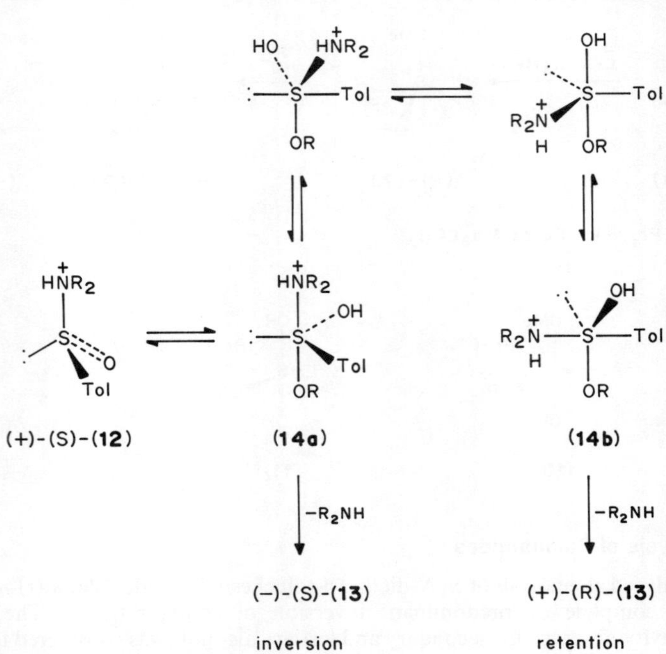


SCHEME 2

be interchanged by the Berry pseudorotation[11] and the structure with more electronegative ligands in the apical positions would be more stable. The sulfuranes of structures **14a** and **14b** could be formed in parallel rather than by permutational isomerization via pseudorotations. Although the results may best be accommodated by Scheme 2 with the sulfurane intermediate which undergoes rapid pseudorotation during the reaction, one-step reactions involving parallel reaction pathways for the inversion (**15a**) and retention (**15b**) mechanisms cannot be completely ruled out.

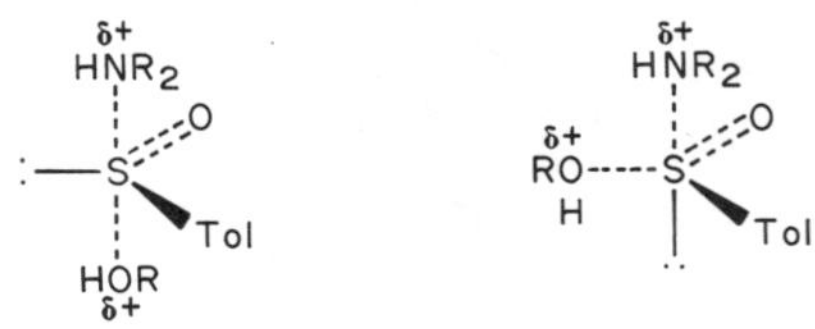


**(15a)** **(15b)**

Reactions of sulfinamides with thiols to give thiolsulfinates (equation 11) also proceed with predominant inversion, the stereoselectivity decreasing in the order R′ = Pr > *i*-Pr > *t*-Bu, from more than 80% to about 30%[37]. Both the starting sulfinamides **12** and products **16** were ascertained to be optically stable under the reaction conditions. The stereochemical results may be again accommodated by the addition–elimination mechanism involving a sulfurane intermediate similar to that outlined in Scheme 2. Variable stereoselectivities observed may be accounted for by variable ease of pseudorotation of the intermediate.

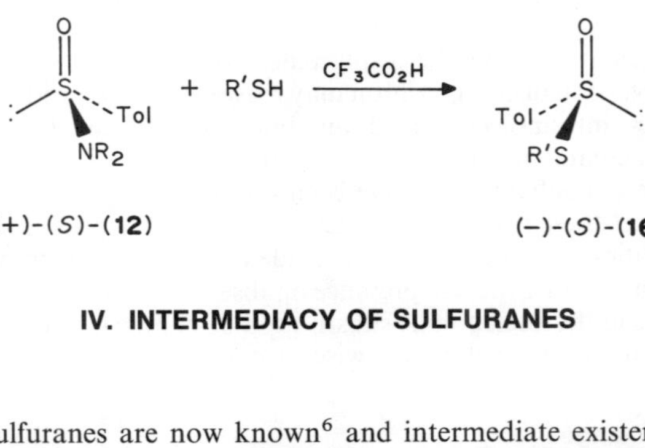


(11)

(+)-(*S*)-(**12**) (−)-(*S*)-(**16**)

## IV. INTERMEDIACY OF SULFURANES

### A. Introduction

Many stable sulfuranes are now known[6] and intermediate existence of this type of species can be expected for nucleophilic substitution reactions at sulfur. One of the most pertinent models for the sulfurane intermediate in nucleophilic substitution of sulfinic acid derivatives may be a sulfuranide oxide. The sulfuranide oxide **17** was recently isolated as an ammonium salt and the Ψ-TBP structure was demonstrated by X-ray analysis[38]. Dynamic $^{19}F$ NMR spectroscopic observations show that this hypervalent species is in equilibrium with the ring-opened sulfinate **18** (equation 12) in solution. The $pK_a$ of the conjugate acid of **18** was also determined titrimetrically ($pK_a = 5.0$). The equilibrium (equation 12) is a degenerate intramolecular nucleophilic substitution (transesterification) of a sulfinate ester **18**, and the intermediate bicyclic hypervalent species **17** was found to be more stable than the open-chain sulfinate **18**. The sulfinate alcohol **19**, an analogue of **18** with $CF_3$ groups replaced by $CH_3$ groups, was also described[39]. The NMR spectrum of a solution of **19** showed a singlet for the aromatic protons, suggesting the formation of the sulfuranide oxide **20** (equation 13).

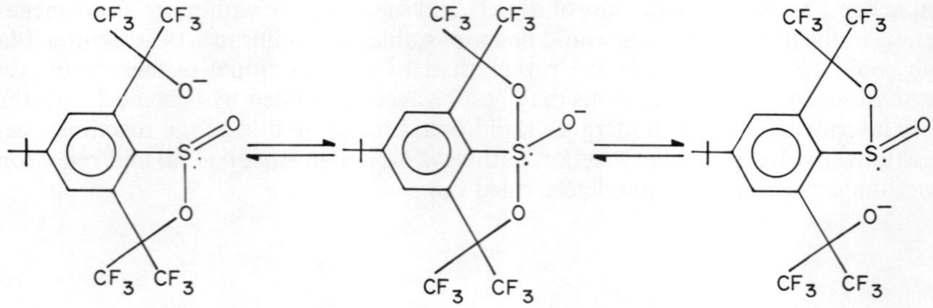


(18) (17) (18)

(12)

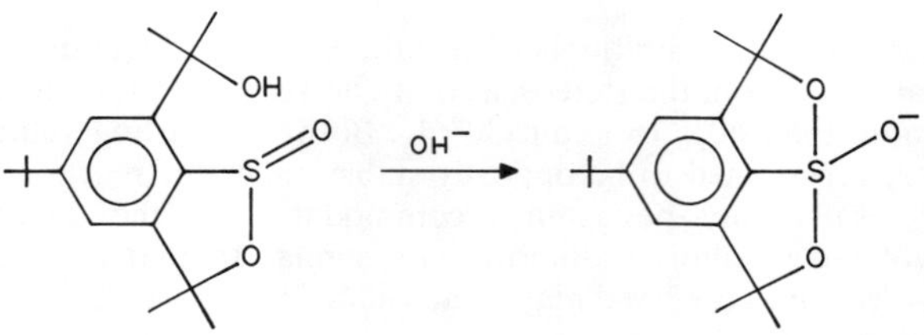


(19) (20)

(13)

Usual intermolecular nucleophilic displacement reactions of sulfinic acid derivatives proceeding with inversion of configuration may reasonably be considered to take place in a similar way to these intramolecular reactions, but there is no direct evidence for existence of such an intermediate in the intermolecular reactions. Stereochemical pathways involving retention of configuration have been interpreted by a sulfurane intermediate. Nevertheless, none of these results can be taken as conclusive proof for the intermediate: alternative possibilities are not completely excluded. In this section, we will elaborate how far we can go to demonstrate the real presence or absence of a discrete intermediate on the reaction coordinate in the nucleophilic substitution of sulfinic acid derivatives, and try to answer the question: Is the reaction stepwise or concerted?

## B. Oxygen-18 Exchange

The classic example for demonstrating the existence of a tetrahedral intermediate in the hydrolysis of carboxylate esters was presented by Bender[40]. He showed that in alkaline

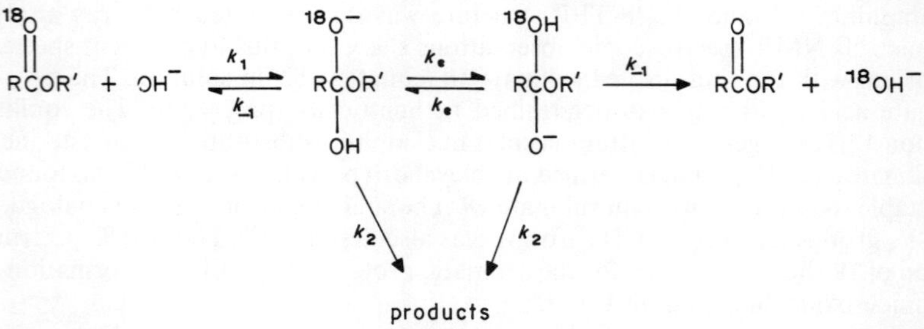


(14)

hydrolysis of a labelled ester RC($^{18}$O)OR′ the substrate ester recovered after partial hydrolysis underwent substantial loss of oxygen-18 label (equation 14). That is, exchange of oxygen-18 occurs during hydrolysis.

Similar experiments with sulfinate esters have been carried out to see if such evidence can be obtained for the presence of a sulfurane intermediate[41,42]. Two such attempts reported are concerned with alkaline hydrolysis of a five-membered cyclic sulfite[41] and a similar sulfinate[42] in $^{18}$O-enriched water.

However, in neither case was there detected any significant incorporation of oxygen-18 into the sulfinyl group of the substrate ester recovered after partial hydrolysis. These results could not demonstrate the existence of any intermediate, but do not necessarily imply that there is no intermediate and the reaction is concerted.

In order to be able to detect the $^{18}$O exchange, it is required not only that the intermediate be actually formed but also that the equilibration of oxygen in the intermediate ($k_e$) be fast as compared to the pathway for return of the intermediate ($k_{-1}$), which in turn must be no slower than the breakdown to products ($k_2$). In the sulfinate hydrolysis there is reason to believe that the equilibration of oxygen in the sulfurane intermediate might be slower than the return to the substrate[43]. In the preferred conformation of the Ψ-TBP intermediate the two more electronegative groups occupy the apical positions and a stable structure should be **21**. Proton transfer from the apical —OH to the equatorial —O$^-$ results in an energetically unfavorable form **22** of the intermediate with the apical —O$^-$, and the equilibration of oxygen could well be slower than the breakdown of the intermediate by loss of either $^{18}$OH$^-$ or R′O$^-$. Hence the $^{18}$O exchange would not be observed even though the intermediate was being formed. Thus the failure to detect oxygen-18 incorporation into the unreacted ester does not rule out the mechanism involving the intermediate. Neither was any $^{18}$O incorporation detected in the alkaline hydrolysis of *N*-mesityl-*p*-toluenesulfinamide in $^{18}$O-enriched water[44].

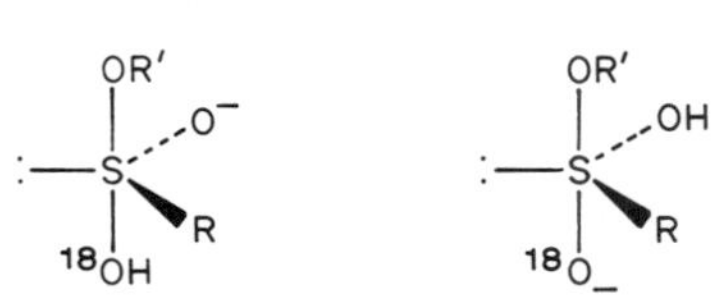

**(21)** **(22)**

Several kinetic criteria have been used as evidence for the existence of tetrahedral intermediates in nucleophilic displacement reactions of carboxylic acid derivatives[45]. These are concerned with a change in the rate-determining step with changing reaction conditions. If there is an intermediate on a reaction pathway, the overall reaction necessarily consists of a sequence of at least two steps, one of which is rate determining (with the highest transition state of the individual steps). The transition-state energy for each step may be affected differently by a change in reaction conditions like pH and concentrations of catalysts (general acid and base). A systematic structural change of the substrate can also be a probe to detect a change in the rate-determining step. Therefore, we may observe a break in a pH–rate profile, dependence on buffer concentrations, and/or the substituent effect correlation owing to a change in rate-determining step. Various such observations were reported for carboxylic acid derivatives[45]. However, such investigations have never been undertaken successfully for nucleophilic substitutions at sulfur atom. We have recently found a break in a pH–rate profile for the acid-catalyzed hydrolysis of ethyl benzenesulfenate (**23**) to suggest the existence of a hypervalent intermediate involving tricoordinate sulfur atom (equation 15)[46]. Kinetic investigations along this line are still awaited for sulfinic acid derivatives.

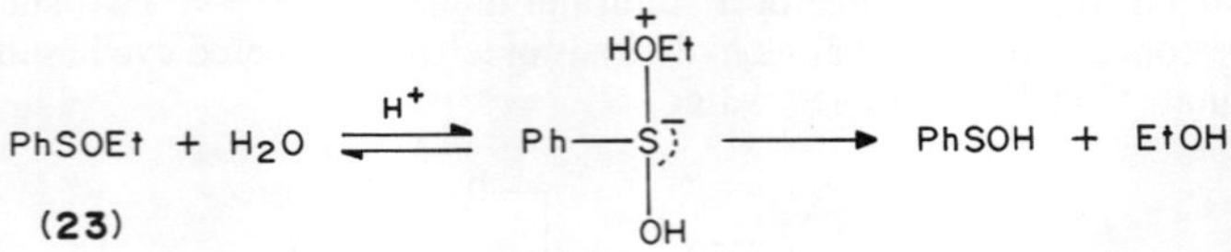


(**23**) (15)

### C. Substituent Effects

Substituent effects observed for alkaline hydrolyses of sulfinate esters[47] and sulfinamides[44] are not definitive in differentiating the two possible mechanisms with and without a sulfurane intermediate. Alkaline hydrolysis of ethyl arenesulfinates (**24**) in 40% aqueous ethanol gave the Hammett $\rho$ value of $+1.60$ at 20 °C (equation 16), while the $\rho$ value for alkaline hydrolysis of *N*-mesitylarenesulfinamides (**25**) in 95% ethanol was $+1.3$ at 50 °C (equation 17). The *p*-nitro group did not show any exalted rate enhancement in the latter reaction, which should be expected if a significant resonance stabilization was exerted by this group in the transition state. The absence of the expected *p*-nitro substituent effect was taken to argue against the existence of the sulfurane intermediate together with the observed absence of oxygen-18 exchange during the hydrolysis[44]. The magnitude of the $\rho$ value ($+1.3$) was smaller than that for a similar hydrolysis of substituted benzoate esters ($\rho = +2.51$)[48], and this was also considered to suggest a concerted mechanism.

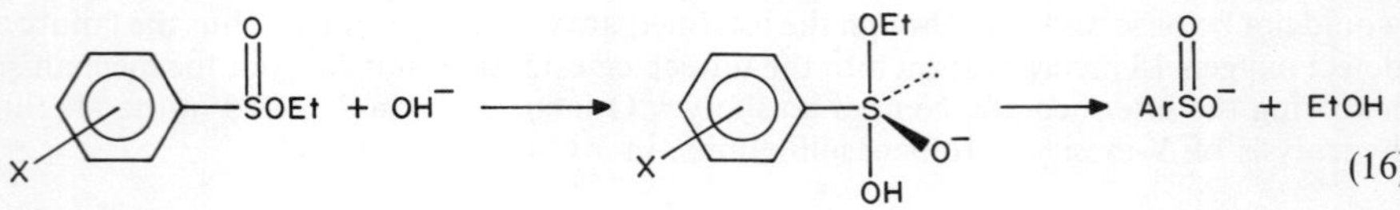


(16)

(**24**)

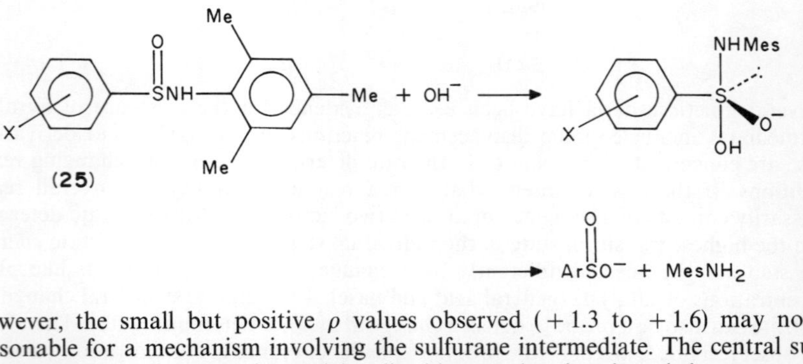


(**25**) (17)

However, the small but positive $\rho$ values observed ($+1.3$ to $+1.6$) may not be unreasonable for a mechanism involving the sulfurane intermediate. The central sulfur atom of the intermediate constitutes delocalized hypervalent bonds and electrons tend to reside on the apical ligands, while the central carbon of a tetrahedral intermediate of carboxylate hydrolysis is an sp$^3$ carbon simply bound to a negative oxygen. The $\rho$ value for sulfinate hydrolysis may well be smaller than that for carboxylate hydrolysis. Substituent effects for a concerted reaction similar to the $S_N2$ reaction would be still smaller. The $S_N2$ reactions of substituted benzyl halides where bond-making and bond-breaking are synchronous show either no correlation with the Hammett equation or very small positive values of $\rho$ (+0.5 to +0.8)[49,50].

Acid-catalyzed hydrolyses of both sulfinates (ArSOOEt)[47] and sulfinamides (ArSONHTol)[51] showed very small negative $\rho$ values (−0.54 to −0.44). These reactions are composites of pre-equilibrium protonation and nucleophilic reaction, and the negative $\rho$ values reflect the protonation step.

Hydrolysis of sulfinate esters usually proceeds by S—O bond cleavage as expected for a nucleophilic reaction at the sulfur atom[52]. However, when the alkyl group can produce a stable carbocation, then C—O cleavage becomes the main course of the reaction. This is an $S_N1$ reaction at saturated carbon with sulfinate anion as a leaving group. Such examples include benzhydryl[53] and cumyl[54] sulfinates.

### D. Reactions with Halide and Hypochlorite Ions

Acid-catalyzed hydrolysis of sulfinic acid derivatives was found to be accelerated by added halide ions[51,55–58]. The nucleophilic catalysis by halide ions can be formulated by intermediate formation of sulfinyl halides formed either directly or via a sulfurane intermediate. Interesting to note here is that acid-independent halide catalysis was also observed in some cases[51,58]. Hydrolysis of sulfinamides undergo such acid-independent halide catalysis[51]. If such a catalysis occurred by a one-step concerted reaction, the departure of an amide anion should have to occur simultaneously with the attack of halide ion at the sulfur in the rate-determining step. Formation of a highly basic amide anion is unlikely and an alternative and more plausible pathway would be that involving rate-determining formation of a sulfurane intermediate followed by a rapid departure of amine by assistance of an acid catalyst (equation 18).

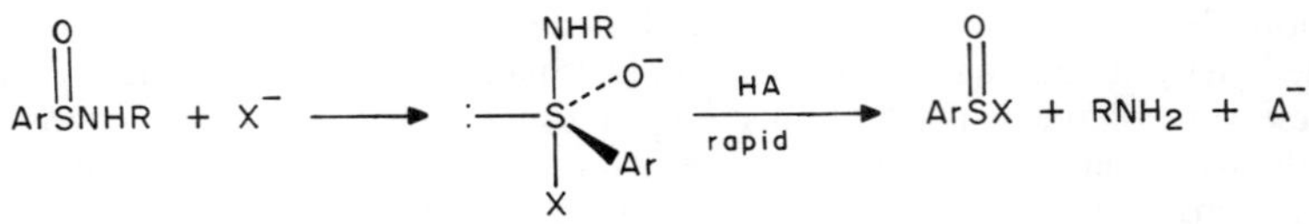


(18)

In the reaction of hypochlorite ion with phenyl benzenesulfinyl sulfone (**26**), the major products were found to be formed from S—S bond cleavage while 10% of phenyl α-disulfone (**28**) was formed by O—Cl cleavage (equation 19)[59]. This was explained by the competitive breakdown of a common intermediate formed by the initial attack of $OCl^-$ on

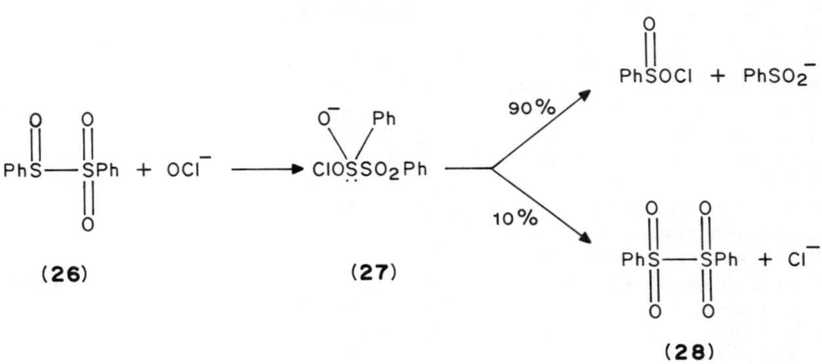


(19)

the sulfinyl sulfur. The two sets of products must be formed via two transition states of similar energy, but these two transition states are not necessarily preceded by a common intermediate. The direct reaction of the sulfinyl sulfone **26** and hypochlorite ion by nucleophilic attack on the hypochlorite oxygen, via the transition state like **29**, which would lead to the α-disulfone **28**, was suggested to be less plausible[59], but is not completely ruled out[38].

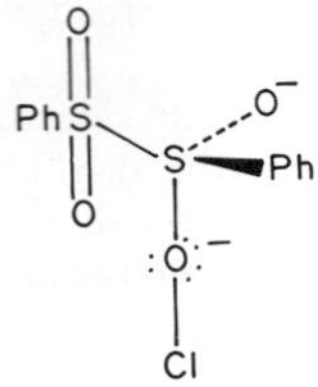

**(29)**

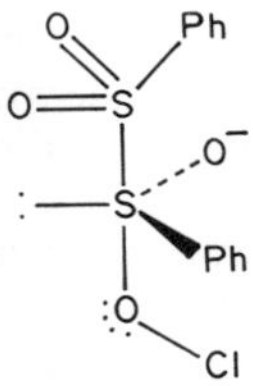

**(27)**

## V. CONCLUSION

Stereochemical investigations on nucleophilic substitutions of sulfinate esters and sulfinamides have provided a variety of results which strongly suggest the existence of the sulfurane intermediate. However, in a strict sense, they can only be taken as support, but not as conclusive proof, for the intermediacy of a sulfurane.

There is no question that the hypervalent sulfurane intermediate can exist in nucleophilic displacement reactions of sulfinic acid derivatives since various stable compounds of this structure have been isolated[6]. The isolated sulfuranide oxide **17** was characterized by X-ray analysis and was spectroscopically demonstrated to be the intermediate of intramolecular transesterification of a sulfinate ester in solution (equation 12)[38]. However, a question still remains as to whether acyclic sulfurane intermediates generally have a long enough lifetime to make them kinetically significant and deserving of the name 'intermediate' in usual intermolecular reactions. Kinetic methods may offer the best means of resolving this question, and should be used in investigations on nucleophilic displacement reactions of sulfinic acid derivatives.

## VI. REFERENCES


1. J. G. Tillet, *Chem. Rev.*, **76**, 747 (1976).
2. J. L. Kice, *Adv. Phys. Org. Chem.*, **17**, 115 (1980).
3. M. Mikolajczyk and J. Drabowicz, *Top. Stereochem.*, **13**, 333 (1982).
4. M. Mikolajczyk, *Phosphorus Sulfur*, **27**, 31 (1986).
5. M. Mikolajczyk, in *Perspectives in the Organic Chemistry of Sulfur* (Eds. B. Zwanenburg and A. J. H. Klunder), Elsevier, Amsterdam, 1987, pp. 23–40.
6. R. A. Hayes and J. C. Martin, in *Organic Sulfur Chemistry* (Eds. E. Bernardi, I. G. Csizmadia and A. Mangini), Elsevier, Amsterdam, 1985, pp. 408–483.
7. J. I. Musher, *Angew. Chem., Int. Ed. Engl.*, **8**, 54 (1969).
8. P. J. Hay, *J. Am. Chem. Soc.*, **99**, 1003 (1977).
9. H. Oberhammer and J. E. Boggs, *J. Mol. Struct.*, **56**, 107 (1979).
10. E. L. Muetterties and R. A. Schunn, *Quart. Rev. Chem. Soc.*, **20**, 245 (1966).
11. R. S. Berry, *J. Chem. Phys.*, **32**, 933 (1960).
12. I. Ugi, D. Marquarding, H. Klusacek, P. Gillespie and F. Ramirez, *Acc. Chem. Res.*, **4**, 288 (1971).
13. J. A. Altmann, K. Yates, and I. G. Csizmadia, *J. Am. Chem. Soc.*, **98**, 1450 (1976).
14. I. W. Levin and W. C. Harris, *J. Chem. Phys.*, **55**, 3048 (1971).
15. F. Seel and W. Gombler, *J. Fluorine Chem.*, **4**, 327 (1974).

16. W. G. Klemperer, J. K. Krieger, M. D. McCreary, E. L. Muetterties, D. D. Traficante and G. M. Whitesides, *J. Am. Chem. Soc.*, **97**, 7023 (1975).
17. C. A. Spring and N. S. True, *J. Am. Chem. Soc.*, **105**, 7231 (1983).
18. G. W. Astrologes and J. C. Martin, *J. Am. Chem. Soc.*, **98**, 2895 (1976).
19. B. A. Belkind, D. B. Denney, D. Z. Denney, Y. E. Hsu and G. E. Wilson, Jr., *J. Am. Chem. Soc.*, **100**, 6327 (1978).
20. (a) F. Ramirez and I. Ugi, *Bull. Soc. Chim. Fr.*, 453 (1974).
(b) See for another example, R. Tang and K. Mislow, *J. Am. Chem. Soc.*, **91**, 5644 (1969).
21. J. Day and D. J. Cram, *J. Am. Chem. Soc.*, **87**, 4398 (1965).
22. S. Oae, M. Yokoyama, M. Kise and N. Furukawa, *Tetrahedron Lett.*, 4131 (1968).
23. M. Mikolajczyk and J. Drabowicz, *J. Chem. Soc., Chem. Commun.*, 775 (1974).
24. J. Drabowicz, B. Bujnick and M. Mikolajczyk, *J. Org. Chem.*, **46**, 2788 (1981).
25. H. Phillips, *J. Chem. Soc.*, **127**, 2552 (1925).
26. M. Mikolajczyk, J. Drabowicz and H. Slebocka-Tilk, *J. Am. Chem. Soc.*, **101**, 1302 (1979).
27. H. F. Herbrandson and R. T. Dickerson, *J. Am. Chem. Soc.*, **81**, 4102 (1959).
28. J. Drabowicz, *Phosphorus Sulfur*, **31**, 123 (1987).
29. K. K. Andersen, *Tetrahedron Lett.*, 93 (1962); K. K. Andersen, W. Gaffield, N. E. Papanikolaou, J. W. Folly and R. I. Perkins, *J. Am. Chem. Soc.*, **86**, 5637 (1964); K. Mislow, M. M. Green, P. Lauer, J. T. Melillo, T. Simmons and A. L., Ternay, Jr., *J. Am. Chem. Soc.*, **87**, 1958 (1965).
30. L. Sagramora, P. Koch, A. Garbesi and A. Fava, *J. Chem. Soc., Chem. Commun.*, 985 (1967); M. Mikolajczyk and J. Drabowicz, *J. Chem. Soc., Chem. Commun.*, 220 (1976).
31. J. Jacobus and K. Mislow, *J. Chem. Soc., Chem. Commun.*, 253 (1968); S. Colonna, R. Giovini and F. Montanari, *J. Chem. Soc., Chem. Commun.*, 865 (1968); A. Nudelman and D. J. Cram, *J. Am. Chem. Soc.*, **90**, 3869 (1968).
32. C. R. Johnson and D. McCants, Jr., *J. Am. Chem. Soc.*, **87**, 5404 (1965).
33. M. Mikolajczyk, B. Bujnicki and J. Drabowicz, *Bull. Acad. Pol. Sci., Ser. Chem.*, **25**, 267 (1977).
34. M. Mikolajczyk, J. Drabowicz and B. Bujnicki, unpublished results cited in Reference 5.
35. M. Mikolajczyk, J. Drabowicz and B. Bujnicki, *J. Chem. Soc., Chem. Commun.*, 568 (1976).
36. M. Mikolajczyk, J. Drabowicz and B. Bujnicki, *Tetrahedron Lett.*, **26**, 5699 (1985).
37. J. Drabowicz and M. Mikolajczyk, *Tetrahedron Lett.*, **26**, 5703 (1985).
38. C. W. Perkins, S. R. Wilson and J. C. Martin, *J. Am. Chem. Soc.*, **107**, 3209 (1985).
39. P. H. W. Lau and J. C. Martin, *J. Am. Chem. Soc.*, **100**, 7077 (1978).
40. M. L. Bender, *J. Am. Chem. Soc.*, **73**, 1626 (1951).
41. C. A. Bunton, P. D. B. de la Mare, P. M. Greasely, D. R. Llewellyn, N. H. Pratt and J. G. Tillet, *J. Chem. Soc.*, 4751 (1958).
42. A. A. Najam and J. G. Tillet, *J. Chem. Soc., Perkin Trans. 2*, 858 (1975).
43. J. L. Kice and C. A. Walters, *J. Am. Chem. Soc.*, **94**, 590 (1972).
44. J. B. Biasotti and K. K. Andersen, *J. Am. Chem. Soc.*, **93**, 1178 (1971).
45. See, e.g., W. P. Jencks, *Catalysis in Chemistry and Enzymology*, McGraw-Hill, New York, 1969, pp. 463–490.
46. T. Okuyama, T. Nakamura and T. Fueno, presented in 13th International Symposium on the Organic Chemistry of Sulfur, 7–12 August, 1988, Odense, Denmark, Abstract, p. 52.
47. M. Kobayashi, R. Nishi and H. Minato, *Bull. Chem. Soc. Jpn.*, **47**, 888 (1974).
48. K. Kindler, *Ann. Chem.*, **450**, 1 (1926).
49. R. F. Hudson and G. Klopman, *J. Chem. Soc.*, 1062 (1962).
50. R. Fuchs and D. M. Carlton, *J. Am. Chem. Soc.*, **85**, 104 (1963).
51. H. Asefi and J. G. Tillet, *J. Chem. Soc., Perkin Trans. 2*, 1579 (1979).
52. C. A. Bunton and B. N. Hendy, *J. Chem. Soc.*, 2562 (1962).
53. C. A. Bunton and B. N. Hendy, *J. Chem. Soc.*, 627 (1963).
54. X. Creary, *J. Org. Chem.*, **50**, 5080 (1985).
55. C. A. Bunton, P. B. D. de la Mare and J. G. Tillet, *J. Chem. Soc.*, 4754 (1958).
56. C. A. Bunton, P. B. D. de la Mare and J. G. Tillet, *J. Chem. Soc.*, 1766 (1959).
57. C. A. Bunton and B. N. Hendy, *J. Chem. Soc.*, 2567 (1962).
58. J. L. Kice and G. Guaraldi, *J. Am. Chem. Soc.*, **89**, 4113 (1967).
59. J. L. Kice and A. R. Puls, *J. Am. Chem. Soc.*, **99**, 3455 (1977).

The Chemistry of Sulphinic Acids, Esters and their Derivatives
Edited by S. Patai


CHAPTER **22**

# Sulfinate ions as nucleophiles

TADASHI OKUYAMA

*Faculty of Engineering Science, Osaka University, Toyonaka, Osaka 560, Japan*



## I. INTRODUCTION

Sulfinic acid shows duality in its reactivity; that is, it can react either as a nucleophile or as an electrophile. Rapid equilibrium formation of sulfinyl sulfone from two molecules of sulfinic acid[1] provides an example of the dual reactivity, one molecule reacting as a nucleophile and the other as an electrophile (equation 1). Nonetheless, sulfinic acids act more generally as nucleophiles.

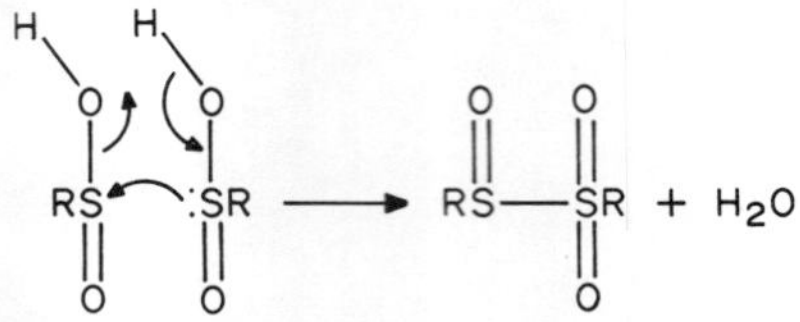

(1)

Sulfinate ions are good, ambident nucleophiles. They can react with various electrophiles either at the sulfur atom, to give sulfonyl derivatives, or at the oxygen end, to lead to sulfinate esters. Although the negative charge seems to be mostly on the oxygen atom, the sulfur has been considered as the main nucleophilic center (equation 2). However, the oxygen atom can also be the nucleophilic center toward a certain class of electrophiles. Most of the latter examples have been found in the last two decades. This duality of nucleophilicity is accommodated by the hard–soft acid–base (HSAB) concept proposed by Pearson[2]. The oxygen and sulfur atoms of sulfinate are respectively considered to be hard and soft nucleophilic centers. Hard electrophiles may react at the oxygen end while soft ones may attack at the central sulfur.

$$\mathrm{R\overset{O}{\overset{\|}{\ddot{S}}}O^-} \leftrightarrow \mathrm{R\overset{O^-}{\overset{|}{\ddot{S}}}{=}O} \leftrightarrow \mathrm{R\overset{O^-}{\overset{|}{\underset{+}{\ddot{S}}}}{-}O^-} \leftrightarrow \mathrm{R\overset{O}{\overset{\|}{\ddot{S}}}{=}O} \xrightarrow{E^+} \mathrm{R\overset{O}{\overset{\|}{\underset{\underset{O}{\|}}{S}}}E} + \mathrm{R\overset{O}{\overset{\|}{S}}OE} \qquad (2)$$

Reactions of sulfinate ions as nucleophiles are described in several review articles[1,3–5] as well as in other chapters of this volume. In the present chapter, we will summarize various nucleophilic reactions of sulfinate ions according to types of electrophiles, and special attention will be focused on the ambident nature of these ions. Quantitative evaluation of the nucleophilicity of the sulfinate ion will be considered in the final section.

## II. DISPLACEMENT AT SATURATED CARBON

### A. Alkylating Agents—Effects of Leaving Groups

Reactions of primary and secondary alkyl halides with sulfinate salts have long been used as general methods for the synthesis of sulfones; these reactions proceed predominantly, if not exclusively, through S-alkylation of sulfinates[6] (equation 3).

$$\mathrm{RSO_2^-} + \mathrm{R'X} \longrightarrow \mathrm{R\overset{O}{\overset{\|}{\underset{\underset{O}{\|}}{S}}}R'} + \mathrm{X^-} \qquad (3)$$

However, ethyl chloroformate was found in 1885 to give ethyl sulfinate by O-alkylation with concomitant decarboxylation[7] (equation 4). In spite of this early work[7], the possibility of O-alkylation had been neglected for a long time, and S-alkylation was considered to be usually the sole reaction of sulfinate ions until the mid-1960s, when the O-alkylation was clearly demonstrated in several other examples and the ambident nucleophilicity was rationalized by the hard–soft acid–base (HSAB) theory[2].

$$\mathrm{RSO_2^-} + \mathrm{Cl\overset{O}{\overset{\|}{C}}OEt} \longrightarrow \mathrm{R\overset{O}{\overset{\|}{S}}OEt} + \mathrm{CO_2} + \mathrm{Cl^-} \qquad (4)$$

Kobayashi[8] first showed a definitive example when he found that alkylation of arenesulfinate ions with triethyloxonium fluoroborate resulted exclusively in formation of ethyl sulfinates (equation 5). This work was initiated because the author deduced from the results of the reaction of sulfinates with acyl chlorides[9] that a highly reactive alkylating agent might attack the sulfinate ion at the oxygen rather than at the sulfur atom.

$$\mathrm{ArSO_2^- + Et_3O^+ \longrightarrow Ar\overset{\overset{\displaystyle O}{\|}}{S}OEt + Et_2O} \qquad (5)$$

A similar O-alkylation has also been noted to occur during the reaction of a thiolsulfinate with triethyl phosphite[10]. Intermediate formation of an ion pair of a sulfinate anion and an alkoxyphosphonium ion, leading to a sulfinate ester product, was suggested for this reaction (equation 6).

$$\mathrm{RSO_2SR' + (EtO)_3P \longrightarrow RSO_2^- + (EtO)_3P^+SR'}$$

$$\mathrm{\longrightarrow R\overset{\overset{\displaystyle O}{\|}}{S}OEt + (EtO)_2\overset{\overset{\displaystyle O}{\|}}{P}SR' + R\underset{\underset{\displaystyle O}{\|}}{\overset{\overset{\displaystyle O}{\|}}{S}}Et} \qquad (6)$$

Meek and Fowler[11] found also formation of a sulfinate ester in the reaction between 1, 2-bis-(*p*-toluenesulfonyl)ethene and trimethyl phosphite. The reaction can be formulated as in equation 7, involving intermediate formation of a sulfinate–alkoxyphosphonium pair. It occurred to these authors that the ambident reactivity of the sulfinate ions should be accommodated by the HSAB concept[2]. Although a soft alkylating agent may alkylate sulfinate ions at the softer sulfur electrophilic center, a hard reagent may react at the harder oxygen atom.

$$\mathrm{TolSO_2CH{=}CHSO_2Tol + (MeO)_3P \longrightarrow TolSO_2\bar{C}H{-}CH\Big\langle{}^{SO_2Tol}_{\overset{+}{P}(OMe)_3}}$$

$$\mathrm{\longrightarrow TolSO_2CH{=}CHP^+(OMe)_3 + TolSO_2^-}$$

$$\mathrm{\longrightarrow Tol\overset{\overset{\displaystyle O}{\|}}{S}OMe + TolSO_2CH{=}CHP(OMe)_2 + \underset{(trace)}{Tol\underset{\underset{\displaystyle O}{\|}}{\overset{\overset{\displaystyle O}{\|}}{S}}Me}} \qquad (7)$$

They examined alkylations of *p*-toluenesulfinate ion with various alkylating agents of varying soft–hard character[11]. The products were generally mixtures of a sulfinate ester and a sulfone. The observed product ratios (or O/S selectivities) are summarized in Table 1 together with those obtained by Kobayashi and Toriyabe[12]. Alkyl halides, such as methyl iodide and benzyl bromide, mostly or exclusively alkylate the sulfinate at the sulfur atom to give sulfones. Allyl chloride and bromide were also found to give solely the sulfone[12]. These results are in accord with earlier observations. However, other harder alkylating agents in fact give increasing fractions of the esters by O-alkylation. The fraction of the ester product (O selectivity) increases in the order:

$$\mathrm{RX < MeS^+Ph_2 < MeS^+(O)Ph_2 < MeOSO_2Y < (MeO)_3P^+R < Et_3O^+,}$$
$$\mathrm{CH_2N_2(MeN_2^+)}$$

(X = halogen) (Y = Tol, OMe, F)

TABLE 1. Alkylation of *p*-toluenesulfinate ion by various alkylating agents

| Substrate | Alkylating agent | Solvent | Temp. (°C) | Ester (%) | Sulfone (%) | Ref. |
|---|---|---|---|---|---|---|
| $TolSO_2K$ | $PhCH_2Br$ | MeCN | r.t. | 0 | 100 | 12 |
| $TolSO_2Na$ | MeI | DMF | 25 | 7 | 93 | 11 |
| $TolSO_2Na$ | MeI | MeOH | 63 | 2 | 98 | 11 |
| $TolSO_2K$ | MeI | MeCN | r.t. | 0 | 100 | 12 |
| $TolSO_2Na$ | $MeOSO_2Tol$ | DMF | 25 | 77 | 23 | 11 |
| $TolSO_2Na$ | $MeOSO_2Tol$ | MeOH | 63 | 54 | 46 | 11 |
| $TolSO_2Na$ | $(MeO)_2SO_2$ | DMF | 25 | 88 | 12 | 11 |
| $TolSO_2Na$ | $(MeO)_2SO_2$ | MeOH | 63 | 69 | 31 | 11 |
| $TolSO_2K$ | $(MeO)_2SO_2$ | $CH_2Cl_2$ | r.t. | 50 | 50 | 12 |
| $TolSO_2K$ | $MeOSO_2F$ | DMF | r.t. | 77 | 23 | 12 |
| $TolSO_2K$ | $MeOSO_2F$ | $CH_2Cl_2$ | r.t. | 40 | 60 | 12 |
| $TolSO_2K$ | $MeS^+Ph_2ClO_4^-$ | $CH_2Cl_2$ | r.t. | 44 | 56 | 12 |
| $TolSO_2K$ | $MeS^+(O)Ph_2ClO_4^-$ | $CH_2Cl_2$ | r.t. | 56 | 44 | 12 |
| $TolSO_2K$ | $MeS^+(O)Ph_2ClO_4^-$ | DMF | r.t. | 24 | 76 | 12 |
| $TolSO_2^-$ | $(MeO)_3P^+CH{=}CHTs$ | none | 25 | 95 | 5 | 11 |
| $TolSO_2Na$ | $Et_3O^+BF_4^-$ | $CH_2Cl_2$ | r.t. | 100 | 0 | 8 |
| $TolSO_2H$ | $CH_2N_2$ | $Et_2O$–MeOH(10:1) | 25 | 100 | 0 | 11 |

This order of reactivity seems to conform to the order of increasing hardness according to the HSAB concept[2]. It was previously reported that the reaction of diazomethane with the sulfinic acid gives only the methyl sulfinate[13]. This reaction may be assumed to occur between the methyldiazonium ion and the sulfinate (equation 8). Methyl sulfonate derivatives $MeOSO_2Y$ seem to show similar reactivity irrespective of Y groups as considerably hard reagents. However, the results are strongly dependent on the solvent used. The counter cation, sodium or potassium, apparently does not have much influence.

$$RSO_2H + CH_2N_2 \longrightarrow RSO_2^- + CH_3{-}N_2^+ \longrightarrow R\overset{\overset{\displaystyle O}{\|}}{S}OCH_3 + N_2 \qquad (8)$$

Mikolajczyk and coworkers[14] have recently investigated the alkylation of benzenesulfinic acid with various *O*-alkyl-*N*, *N'*-dicyclohexylisoureas. The reactions give predominantly O-alkylation products as summarized in Table 2. The reactions were examined in THF as well as in some other solvents, but solvent effects on the product ratio are not straightforward (equation 9). The reaction is believed to proceed through a preequilibrium protonation, and the alkylation of the sulfinate ion with the protonated *O*-alkylisourea, which is considered to be a hard electrophile, may account for the predominant formation of sulfinate esters (equation 10). Another factor which is considered to be responsible for the preferential O-alkylation is the steric effect exerted by the large electrophile. The terminal oxygen atom may be less susceptible to such steric effects than the sulfur center of the sulfinate ion.

$$PhSO_2H + H_{11}C_6N{=}\overset{\overset{\displaystyle OR}{|}}{C}{-}NHC_6H_{11} \longrightarrow Ph\overset{\overset{\displaystyle O}{\|}}{S}OR + Ph\underset{\underset{\displaystyle O}{\|}}{\overset{\overset{\displaystyle O}{\|}}{S}}R + H_{11}C_6NH\overset{\overset{\displaystyle O}{\|}}{C}NHC_6H_{11} \qquad (9)$$

TABLE 2. Product ratio in the alkylation of benzenesulfinic acid with *O*-alkylisourea (equation 9) in THF[14]

| R | Ester (%) | Sulfone (%) |
|---|---|---|
| Me | 75 | 25 |
| Et | 90 | 10 |
| $PhCH_2$ | 61 | 39 |
| *i*-Pr | 100 | (trace) |
| *sec*-Bu | 100 | 0 |
| 2-Hexyl | 100 | 0 |

$$PhSO_2H + R'N{=}\overset{\displaystyle OR}{\overset{|}{C}}NHR' \rightleftharpoons PhSO_2^{\,-} + R'NH\underset{+}{\overset{\displaystyle OR}{\overset{|}{C}}}NHR' \longrightarrow \text{products} \qquad (10)$$

The second step of this reaction is taken as a nucleophilic substitution at the carbon which can in principle proceed via the $S_N2$ or $S_N1$ mechanism. They examined the reaction with an optically active *O*-2-hexylisourea and found that the reaction takes place essentially through inversion (99%). Reactions with *O*-alkylisoureas bearing optically active substituents at the nitrogen atoms were also carried out. The obtained sulfinate esters were optically active (with a chiral sulfur) although the enantiomeric excess was less than 10%.

In addition to the above-mentioned alkylating agents, epoxides[15], β-propiolactone[16] and Mannich bases ($Me_2NCH_2CH_2COR$)[17] were reported to give sulfones. Various addition reactions of sulfinates to unsaturated bonds were found also to lead to sulfones, as will be discussed in the following section. Some of these results must, however, be taken carefully in terms of the ambident nucleophilicity of sulfinate ions. In addition to the possible rearrangement of some sulfinate esters[18–20], alkyl sulfinates in general undergo hydrolysis quite rapidly in acidic and alkaline aqueous solutions[21,22]. The sulfinate ester product could thus easily be lost during the usual workup procedure. Some of the reported results, especially early ones, might be affected by this possibility unless care was taken.

The 'organic syntheses' method for preparation of methyl *p*-tolyl sulfone by the reaction of sodium *p*-toluenesulfinate with dimethyl sulfate[23] must involve this problem. The reaction medium used is aqueous bicarbonate solution in which the methyl sulfinate formed should hydrolyze very rapidly and the sulfinate ion be regenerated for further alkylation, while the sulfone product is stable in this medium. Field and Clark[23b] in fact mention that the reactions of sodium arenesulfinates with methyl sulfate in organic solvents gave lower yields of the desired sulfone.

### B. Medium Effects

Data given in Table 1 show that reaction media influence considerably the O/S selectivity in alkylation of the sulfinate. The effects of solvents on the alkylation of potassium *p*-toluenesulfinate as well as those of added crown ether were examined in more detail by Kobayashi and Toriyabe[12]. The results are given in Table 3. The data with methyl fluorosulfonate (in parentheses) clearly show that a polar aprotic solvent increases the fraction of O-alkylation. The reaction in dichloromethane gives 40% of the ester (very probably less in benzene or carbon tetrachloride), while the reaction in HMPA leads

TABLE 3. Effects of solvents and crown ether on alkylation of potassium *p*-toluenesulfinate at room temperature[a]

| Alkylating agent | Solvent | Ester (%) | Sulfone (%) |
|---|---|---|---|
| $PhCH_2Br$ | $C_6H_6$ | 0 | 100 |
| | MeCN | 0(0) | 100(100) |
| $(MeO)_2SO_2$ | $CH_2Cl_2$ | 58(50) | 42(50) |
| $MeOSO_2F$ | $C_6H_6$ | 48 | 52 |
| | $CCl_4$ | 52 | 48 |
| | $CH_2Cl_2$ | 70(40) | 30(60) |
| | diglyme | 79(76) | 21(24) |
| | DMF | 82(77) | 18(23) |
| | HMPA | 94(100) | 6(0) |
| $MeS^+Ph_2ClO_4^-$ | $CH_2Cl_2$ | 40(44) | 60(56) |
| $MeS^+(O)Ph_2ClO_4^-$ | $CH_2Cl_2$ | 25(56) | 75(44) |

[a]Product distributions in the presence of 80–200% of the substrate concentration of 18-crown-6 are given and values in its absence are in parentheses. Data are taken from Reference 12.

essentially to the ester alone. The added crown ether (18-crown-6) in general tends to increase the ester fraction, but the effects are small. The reactions using methylsulfonium and oxosulfonium salts suffer opposite effects of solvent polarity ($DMF/CH_2Cl_2$ in Table 1) and of added crown ether. Alkylation of the adamantane-1-sulfinate ion was also examined and the results are similar to those obtained with *p*-toluenesulfinate[12].

The solvent effects may be accommodated by the nature of ion pairs of sulfinate anions and alkali metal cations. In nonpolar solvents, an alkali metal cation which is a hard acid in the sense of HSAB may bind mostly to the oxygen atoms, and therefore the electrophile would attack the anion preferentially at the sulfur atom. In polar aprotic solvents, the sulfinate anion may be present more in a free form and able to accept the attack at the oxygen end. The crown ether may have the same effect by separating the alkali metal cation from the sulfinate anion by complexation. Cryptands, which bind the cation more effectively, have somewhat greater effects[12]. The effects of polar aprotic solvents and added crown ethers which decrease the fraction of *O*-alkylation by sulfonium salts cannot clearly be explained. The alkylating properties of these positively charged reagents may be influenced greatly by the medium effects.

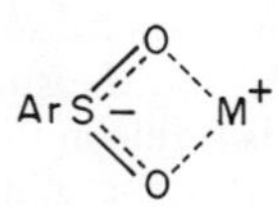


Data in Table 1 show that methanol has a distinct tendency to decrease the O-alkylation as compared with *N*,*N*-dimethylformamide (although the dielectric constants are similar; MeOH, 32.6 and DMF, 36.7 D). This may be attributed to the hydrogen bonding of methanol with the oxygen atoms of the sulfinate anion, thus making them less available for alkylation. Possible hydrolysis of the ester product, by water present as impurity in the solvent methanol, was also suggested to be responsible for the lower yield of the ester[11]. However, the yield and the fractional distribution of the products given in the paper[11] indicate obviously that this is not the sole reason. Certain sulfinate esters are known to rearrange easily to the sulfones[18–20], but methyl *p*-toluenesulfinate was confirmed to be stable under the reaction conditions[11]. In 50% aqueous dioxane, 2-nitro-

4-trifluoromethylbenzenesulfinate was found to undergo S-alkylation with methyl iodide but preferentially O-alkylation with methyl fluorosulfonate[24].

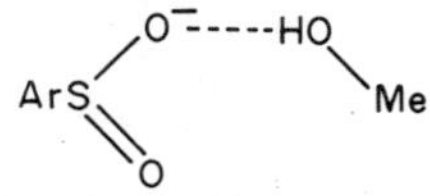


Effects of phase-transfer agents on alkylation of sulfinate ions have been extensively studied in recent years in order to improve the method for the synthesis of sulfones[25–30].

### C. Effects of Counter Ions

Effects of counter cations of sulfinate salts were not observed with $Na^+$ and $K^+$. Silver *p*-toluenesulfinate was found to behave similarly to alkali metal salts in the reaction with methyl iodide in DMF (7% of ester formation compared with 9% of ester formation by the sodium salt)[11]. Kondratenko and coworkers[31] examined in more detail the reactions of some silver sulfinates with benzyl iodides in acetonitrile in comparison with those of the potassium salts. All the reactions with potassium salts gave sulfones as sole products in acetonitrile (equation 11). However, the same reactions with silver salts (equation 12)

$$RSO_2K + XC_6H_4CH_2I \xrightarrow{MeCN} XC_6H_4CH_2\overset{\overset{\displaystyle O}{\|}}{\underset{\underset{\displaystyle O}{\|}}{S}}R \qquad (11)$$

$(R = Me, Ph, CF_3;\ X = H, p\text{-}NO_2)$

$$RSO_2Ag + ArCH_2I \longrightarrow ArCH_2\overset{\overset{\displaystyle O}{\|}}{\underset{\underset{\displaystyle O}{\|}}{S}}R + ArCH_2\overset{\overset{\displaystyle O}{\|}}{S}OR \qquad (12)$$

resulted in formation of a considerable amount of the sulfinate ester as summarized in Table 4. Silver ion clearly enhances the O-alkylation. This may be accommodated by greater interaction of the solft cation $Ag^+$ with the softer sulfur center of the sulfinate to inhibit the S-alkylation. The four-membered cyclic transition state shown below involving

TABLE 4. Reactions of benzyl iodides with silver salts of sulfinic acids in acetonitrile[a]

| | $PhCH_2I$ | | $p\text{-}NO_2C_6H_4CH_2I$ | |
|---|---|---|---|---|
| Sulfinates | Ester (%) | Sulfone (%) | Ester (%) | Sulfone (%) |
| $MeSO_2Ag$ | 24<br>(24) | 76<br>(76) | 0<br>(0) | 80<br>(100) |
| $PhSO_2Ag$ | 36<br>(36) | 63<br>(64) | 14<br>(17) | 80<br>(83) |
| $CF_3SO_2Ag$ | 56.6<br>(58) | 41.3<br>(42) | 50<br>(56) | 40<br>(44) |

[a]Data are taken from Reference 31. Values given are percent yields and those in parentheses show calculated percent fractions.

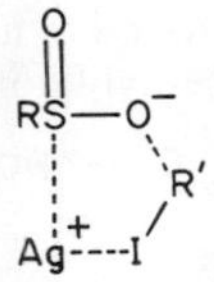


coodination of $Ag^+$ on the oxygen atom was proposed by Kondratenko and coworkers[31], but the above mentioned S-coordination mechanism seems to be more reasonable.

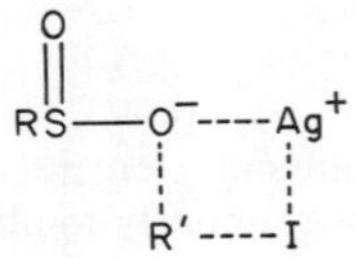


### D. Structural Effects

The structures of both the silver sulfinates and the benzyl iodides influence the product ratio in the above reaction. An electron-withdrawing group in the sulfinate seems to increase the fraction of the ester, while the *p*-nitro substituent in benzyl iodide tends to decrease the O-alkylation. The reasons for these structural effects are not obvious, but one possible explanation may be that the $S_N1$ character of the reaction will enhance the tendency of O-alkylation. The positive charge on the potential carbocation may enhance its hardness to facilitate the O-attack. Electron withdrawal in the nucleophile which may diminish its nucleophilicity and electron donation in the alkyl halide which may stabilize the potential carbocation would both make the reaction more $S_N1$-like and thus favor the O-alkylation. A similar tendency was found in the alkylation with *O*-alkylisoureas, where secondary alkyl groups lead more easily to O-alkylation than primary alkyl groups (Table 2)[14]. However, the reaction involving optically active secondary alkyl derivative was found to undergo inversion ($S_N2$).

Structural effects of alkyl halides on the O/S selectivity in the reaction with sulfinate ions have been discussed in terms of $S_N1$–$S_N2$ character of the displacement reaction by Schank[32–35]. He examined reactions of arenesulfinate salts with α-haloethers. The product ester/sulfone ratios for reactions of sodium arenesulfinates with some halomethoxymethanes in $CFCl_3$ at 0 °C were determined by NMR spectroscopy[34]. The ratios obtained under the same conditions were variable between 55/45 and 65/35 because of instability of the ester. In the presence of a small amount of sodium hydride, the results became reproducible and the maximum fraction of the ester obtained in the reaction of *p*-chlorobenzenesulfinate with bromomethoxymethane was 86% (equation 13). Owing to the instability of the ester, only the sulfone was initially obtained[32] and only later was

$$p\text{-}ClC_6H_4SO_2Na + BrCH_2OMe \xrightarrow{CFCl_3,\ 0\,^\circ C} \underset{14\%}{p\text{-}ClC_6H_4\overset{O}{\underset{O}{\overset{\|}{\underset{\|}{S}}}}CH_2OMe} + \underset{86\%}{p\text{-}ClC_6H_4\overset{O}{\overset{\|}{S}}OCH_2OMe} \quad (13)$$

formation of the ester detected by the IR spectra of the product mixtures[33]. Such instability of the sulfinate ester was also noted by Mulder and coworkers[36]. Methoxymethyl arenesulfinates can readily rearrange to the sulfone in the presence of acids and undergo easily hydrolysis by the moisture present[21].

Substituents on the sulfinate (*p*-Me, H and *p*-Cl) have little influence on the O/S selectivity, but a change of the halogen atom in the haloether affects the selectivity, increasing the O-alkylation in the order: Cl < Br < I. When a class of alkyl halides can produce a stable methoxy carbocation, the reaction may have a high $S_N1$ character to favor the O-alkylation. A better leaving group further enhances the $S_N1$ character of the reaction[33].

Substitution at the α-position of the haloether by the electron-withdrawing acetyl group increases the S-alkylation in accord with the decreasing $S_N1$ character[35]. The sulfinate ester products **2** are again unstable and lead to thiolsulfonates **3** under the reaction conditions (equation 14). Final yields of the sulfone and the thiolsulfonate were determined under various conditions[37]. The sulfinate structure (from *p*-MeO to *p*-$NO_2$) and metal ions ($Li^+$, $Na^+$ and $K^+$) had some effect on the product ratio. The nature of the solvents used and the reactant concentrations affected considerably the product yields. However, all these effects are not straightforward. Some concentration effects seem to have arisen from solubility problems, and the reactions are partly heterogeneous. Effects of leaving halogen atoms on the ethers are opposite to those observed with the simple α-haloethers. In the case of the α-acetyl haloethers, yields of the sulfone increase always in the order: Cl < Br < I as shown in Table 5[37].

$$\mathrm{ArSO_2M + XCH(OMe)COMe \longrightarrow \underset{\textbf{(1)}}{ArS(=O)(=O)CH(OMe)COMe} + \underset{\textbf{(2)}}{[ArS(=O)OCH(OMe)COMe]}}$$

$$\textbf{(2)} \longrightarrow \underset{\textbf{(3)}}{\mathrm{ArSO_2SAr}} \qquad (14)$$

This tendency was accommodated by assuming participation of the carbonyl oxygen atom as observed in the related displacement reactions[38]. The α-acetyl chloroether may undergo substitution as shown in equation 15. However, the nucleophile cannot attack the carbonyl carbon of the acetyl bromo- and iodoethers because of the much greater atomic

TABLE 5. Product yields (%) in the reactions of sodium salts of *p*-substituted benzenesulfinic acids with α-acetyl haloethers in acetone (equation 14)[a]

| | X = Cl | | X = Br | | X = I | |
|---|---|---|---|---|---|---|
| *p*-substituent | **1** | **3** | **1** | **3** | **1** | **3** |
| $No_2$ | 9 | 23.5 | 29.3 | 22.5 | 47 | 6.9 |
| Cl | 1.3 | 29.3 | 9 | 43.5 | 53 | 7.4 |
| Me | — | 33 | 22 | 11 | 49 | 10 |
| OMe | 1.3 | 34.4 | 9 | 36.5 | 34 | 15 |

[a]Data are taken from Reference 37.

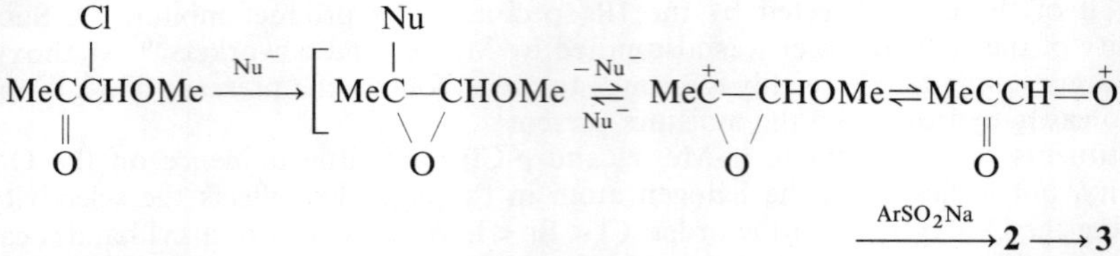

(15)

volumes of Br and I, and the halogen atom may partially interact with the carbonyl carbon. These factors inhibit participation of the carbonyl oxygen and favor $S_N2$-type displacement of the bromo- and iodoethers.

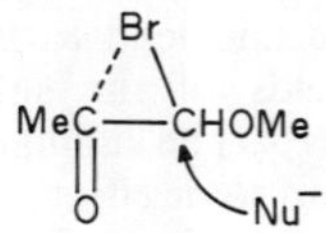

As a whole, alkyl derivatives which undergo favorable $S_N1$ displacement seem to react with sulfinate ions preferentially at the oxygen atom. However, the potential alkyl sulfinates, which can provide a stable carbocation by ionization, may very readily rearrange to the more stable sulfone. Even in cases when the kinetic product, the sulfinate ester, is formed, this is difficult to isolate and the thermodynamic product, the sulfone, may result. Reactions of isolated carbocation salts with sodium *p*-toluenesulfinate gave only the corresponding sulfones[39] (equation 16). Fava and coworkers[40] could determine the rate ratio of O- and S-alkylation in ion-pair return in their investigation on the racemization and rearrangement of benzhydryl *p*-toluenesulfinate in acetic acid ($k_O/k_S = 0.8$; equation 17).

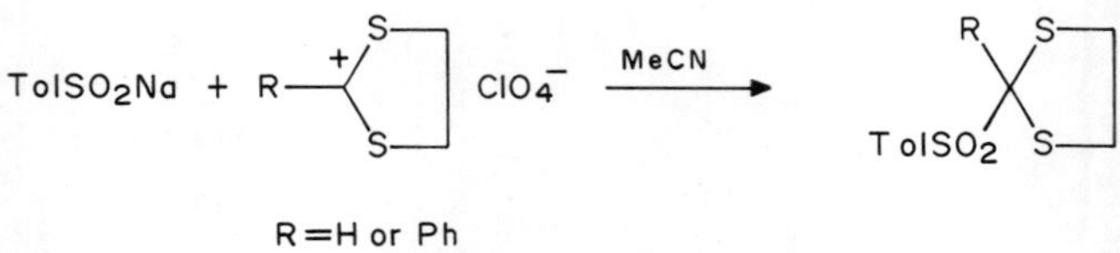

(16)

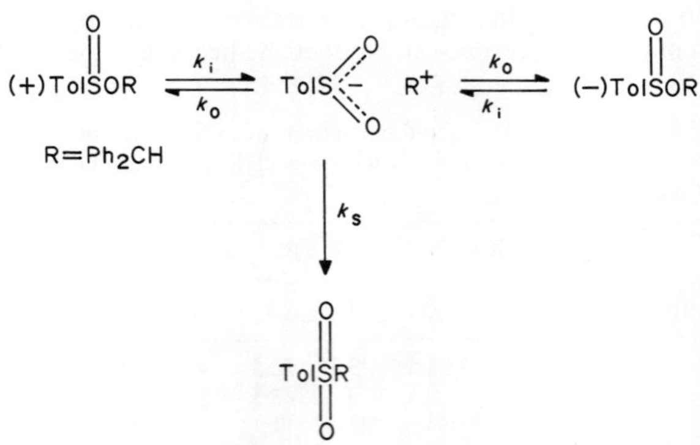

(17)

Acid-catalyzed reaction of an alcohol with a nucleophile proceeds typically by an $S_N1$ mechanism. However, the reaction of an alcohol with a sulfinic acid under acidic conditions cannot afford the sulfinate ester, since sulfinic acid readily undergoes disproportionation to yield thiolsulfonate and sulfonic acid under the reaction conditions.

Tertiary alkyl halides do not alkylate alkali metal sulfinates but undergo an elimination reaction leading to alkenes[41]. This result must be due to the basic reaction conditions employed. By contrast, nucleophilic attack by the sulfinate on tertiary carbocations has been observed in $S_N1$ reactions, and formation of sulfones in the presence of *p*-toluenesulfinate has been exploited as diagnostic for carbocation formation from peroxides[42], alcohols[42] and esters[43]. The isolation of sulfones in these reactions may be ascribed to the instability of *tert*-alkyl sulfinates.

Reactions of sulfinate ion with haloforms yield exclusively dihalo sulfones in the presence of aqueous alkali[44]. Dihalocarbenes must be initially formed in this reaction and react as true electrophiles with the sulfinate (equation 18). The exclusive formation of sulfones by this route conforms to the HSAB principle.

$$RSO_2^- + CHX_3 \xrightarrow{OH^-} RSO_2^- + {:}CX_2 \longrightarrow RS(=O)_2CHX_2 \qquad (X = Cl, Br) \tag{18}$$

Schank and Schmitt[45] examined reactions of various acetals and sulfinic acids in the presence of boron trifluoride etherate. In the presence of acid, acetal provides an alkoxy carbocation and reaction of this ambident cation with sulfinic acid (or sulfinate ion) resulted in formation of the methyl sulfinate ester (with generation of the aldehyde) and the $\alpha$-alkoxysulfone (equation 19). The expected $\alpha$-methoxyalkyl sulfinate was not obtained probably because of its instability. Although the total yield of products was poor and methyl sulfinate was the main product in the absence of $BF_3$, addition of $BF_3$ etherate increased markedly the yield of the sulfone. In some examples, the $\alpha$-alkoxy sulfone was exclusively obtained in 91% yield.

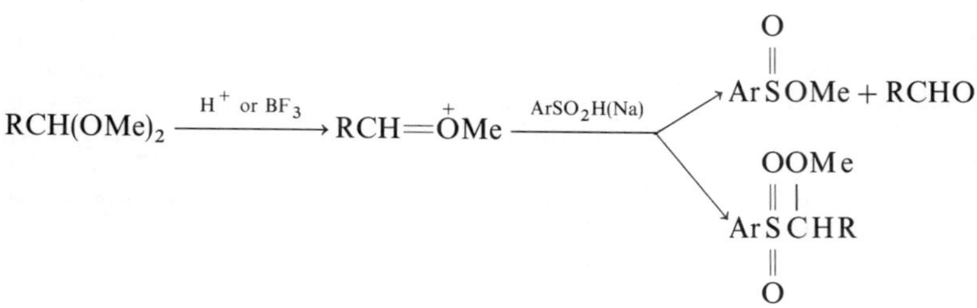

(19)

## III. REACTIONS AT UNSATURATED CARBON

### A. Addition to Carbon–Carbon Unsaturated Bonds

Michael-type additions of sulfinate ions to olefins having a variety of electron-withdrawing groups are reported to give $\beta$-substituted sulfones as isolated products[3,4,46–48] (equation 20). The reactions were usually carried out in aqueous or alcoholic solutions. Failure in observing the formation of sulfinate ester products may be due in part to the hydrogen-bonding solvation of the sulfinate ions and/or the instability of the potential ester products[21].

$$RSO_2H + CH_2{=}CHX \longrightarrow RSO_2CH_2CH_2X \tag{20}$$

Addition of benzenesulfinic acid to *N*-phenylmaleimide gives a sulfone (equation 21). Substituent effects on the rate constant for this reaction were examined[49,50]. Substitution in the N-phenyl group of maleimide resulted in a U-shaped Hammett correlation[49], while that in benzenesulfinic acid gave a negative $\rho$ value as expected for a nucleophilic reaction[50].

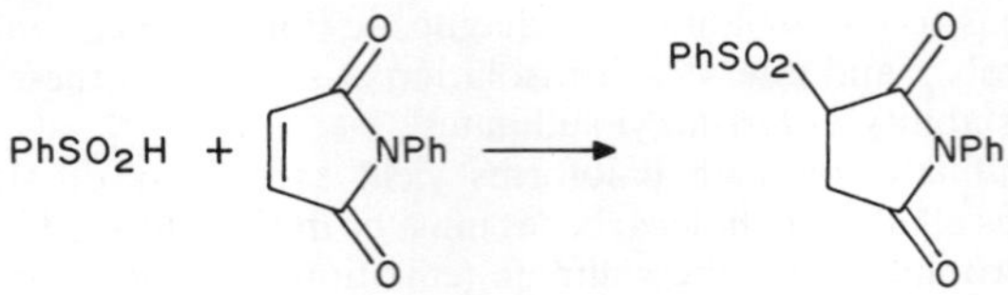


(21)

Reaction of sulfinate ion with *p*-benzoquinone can also be formulated as a Michael-type reaction which is followed by enolization to give 2, 5-dihydroxyphenyl sulfones[51,52] (equation 22). The mechanism of this reaction was established by Ogata and coworkers[52]. The rate-determining step changes from the nucleophilic addition of sulfinate ion (pH $< 3.1$) to the deprotonation of the intermediate adduct ($4.0 < \text{pH} < 5.7$) with increasing pH of the reaction medium.

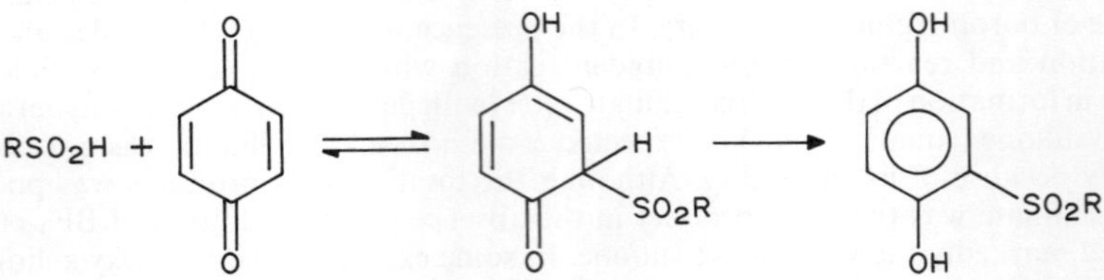


(22)

Analogous reactions with quinone diimines have also been reported[53–55] (equation 23). In the addition of arenesulfinates to *N*, *N*-dialkylquinone diimines[55], products are formed from additions to both carbon and nitrogen of the diimine (equation 24). Both reactions occur at the sulfinate sulfur. The product ratio changes markedly with pH. The reason for this change is that the initial addition to carbon is reversible but not the one to nitrogen. At lower pH, the ring-substituted product (attack at carbon) is obtained in significant yield, since the initial adduct is protonated to facilitate a loss of a ring proton leading to the stable ring-substituted product.

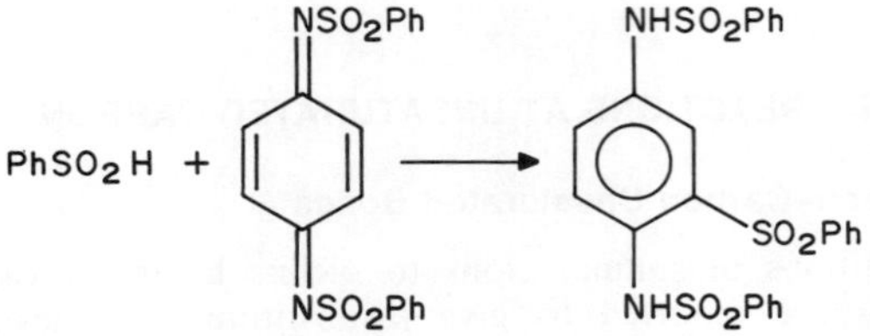


(23)

Addition of sulfinic acids or sulfinate ions to acetylenes with an electron-withdrawing group also occurs readily to yield unsaturated sulfones[56]. In all these addition reactions one notes that only the sulfone products are obtained. None of the studies shows any indication of the reaction occurring at the sulfinate oxygen atom, although careful examinations appear to be lacking.

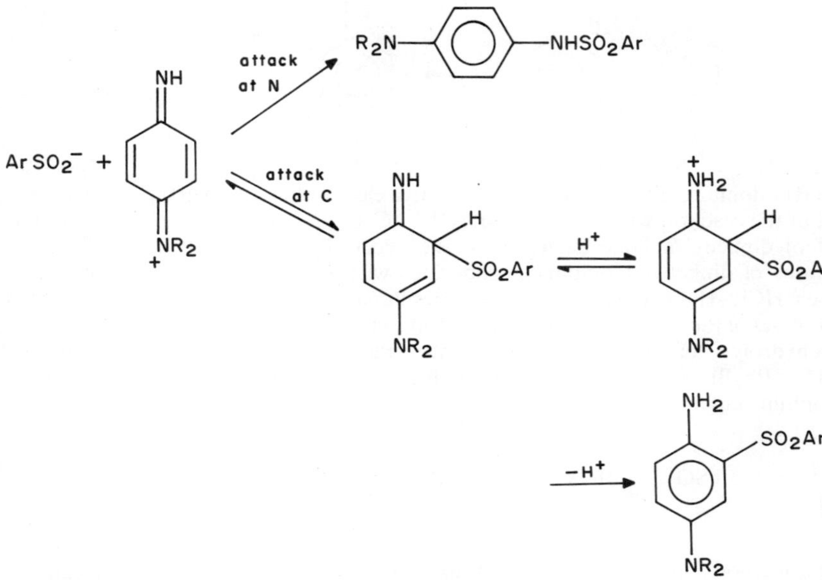


(24)

**B. Vinyl Substitution**

Nucleophilic vinyl substitution takes place in the reaction of sulfinate ions with haloalkenes carrying electron-withdrawing group(s)[57]. β-Halovinyl ketones[58] and β-chloroacrylonitrile[59] are typical substrates. The reaction proceeds through the addition–elimination mechanism[57] and the initial step closely resembles that of the Michael addition (equation 25).

$$\mathrm{ClCH{=}CHCN} + \mathrm{ArSO_2}^- \longrightarrow \mathrm{ArSO_2}\overset{\overset{\displaystyle Cl}{|}}{\underset{\underset{\displaystyle H}{|}}{C}}\!-\!\bar{C}\!\begin{smallmatrix}H\\CN\end{smallmatrix} \xrightarrow{-Cl^-} \mathrm{ArSO_2CH{=}CHCN} \qquad (25)$$

Although reaction of β-chloroacrylonitrile with sodium p-toluenesulfinate was found to give β-(p-toluenesulfonyl)acrylonitrile in high yield[59], similar reactions of α, β- and β, β-dichloroacrylonitriles resulted in C—C bond cleavage to give acetonitrile derivatives owing to instability of the disulfonylacrylonitriles under the reaction conditions[60]. Similar results were also obtained in the reaction of dichlorovinyl ketones[58]. However, introduction of an alkyl or aryl group in the α position of β, β-dichloroacrylonitrile considerably reduced the reactivity and resulted in isolation of only the normal monosubstitution product[61].

**C. Aromatic Substitution**

Sulfinate ion is a fairly weak nucleophile in aromatic substitutions[62] and only a few examples of such reactions are known. The reverse Smiles rearrangement is facilitated because of the intramolecular nature of the reaction[63] (equation 26).

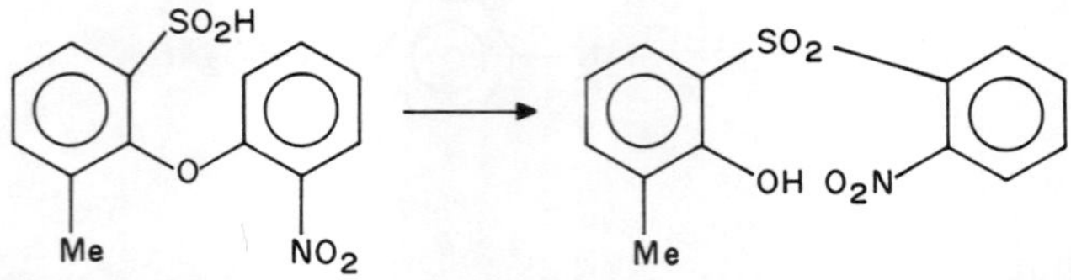


(26)

Diayliodonium salts, which are very reactive electrophiles, give diaryl sulfones in high yield in the reaction with arenesulfinates[64,65]. Grushin and coworkers[65] found that the yield of diphenyl sulfone decreases in the order: $Ph_2I^+ > Ph_2Br^+ > Ph_2Cl^+$, in the reactions of diphenylhalonium fluoroborates with sodium benzenesulfinate in a two-phase $CHCl_3$–$H_2O$ system. Phenyl benzenesulfinate as well as phenol was also obtained in the cases of the bromonium and chloronium salts (equation 27). Phenol must originate from hydrolysis of the sulfinate ester. That is, the O-phenylation increases in the order: $Ph_2I^+ < Ph_2Br^+ < Ph_2Cl^+$, in accord with the hardness of the phenyl carbon linked to the onium center.

$$PhSO_2^- + Ph_2X^+ \longrightarrow Ph\overset{\overset{\displaystyle O}{\|}}{\underset{\underset{\displaystyle O}{\|}}{S}}Ph + Ph\overset{\overset{\displaystyle O}{\|}}{S}OPh(+PhOH) + PhX \qquad (27)$$

In a method for the preparation of alkyl sulfinates, sulfinic acids were treated with alcohols in the presence of 1-methyl-2-chloropyridinium iodide and triethylamine[66]. The initial step of this synthesis is postulated to be the displacement of chloride by the sulfinate to give an intermediate ester (equation 28). The pyridinium ion reacts at the oxygen atom of the sulfinate ion.

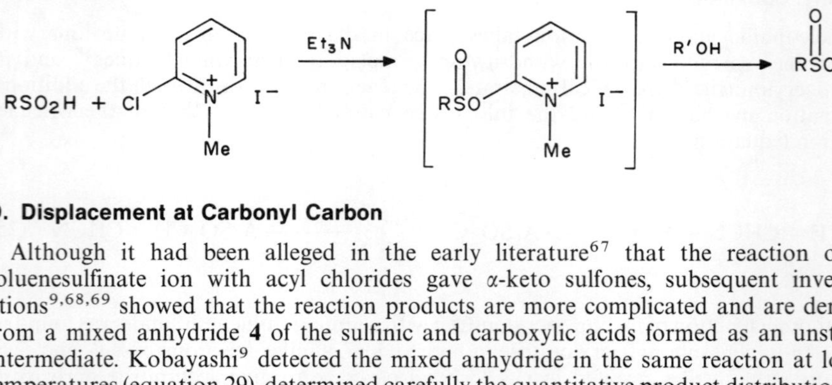


(28)

## D. Displacement at Carbonyl Carbon

Although it had been alleged in the early literature[67] that the reaction of *p*-toluenesulfinate ion with acyl chlorides gave α-keto sulfones, subsequent investigations[9,68,69] showed that the reaction products are more complicated and are derived from a mixed anhydride **4** of the sulfinic and carboxylic acids formed as an unstable intermediate. Kobayashi[9] detected the mixed anhydride in the same reaction at lower temperatures (equation 29), determined carefully the quantitative product distributions at higher temperatures, and formulated a reaction sequence which rationalizes the stoichiometry of the overall reaction. The initial reaction is acylation of the sulfinate ion at the oxygen atom (O-acylation). A similar mixed anhydride could be isolated from the reaction in dioxane solution in the presence of pyridine[70]. These results (equations 30 and 31) are concordant with the HSAB concept since the carbonyl carbon may be considered to be a hard acid.

$$ArSO_2Na + RCOCl \longrightarrow \underset{\textbf{(4)}}{Ar\overset{\overset{\displaystyle O}{\|}}{S}O\overset{\overset{\displaystyle O}{\|}}{C}R} + NaCl \qquad (29)$$

$$\mathbf{4} \longrightarrow \mathrm{Ar}\overset{\mathrm{O}}{\overset{\|}{\mathrm{S}}}\mathrm{O}\overset{\mathrm{O}}{\overset{\|}{\mathrm{S}}}\mathrm{Ar} + (\mathrm{RCO})_2\mathrm{O} \qquad (30)$$

**(5)**

$$\mathbf{5} \longrightarrow \mathrm{Ar}\underset{\underset{\mathrm{O}}{\|}}{\overset{\overset{\mathrm{O}}{\|}}{\mathrm{S}}}-\overset{\overset{\mathrm{O}}{\|}}{\mathrm{S}}\mathrm{Ar} \longrightarrow \mathrm{Ar}\underset{\underset{\mathrm{O}}{\|}}{\overset{\overset{\mathrm{O}}{\|}}{\mathrm{S}}}\mathrm{OSAr} \xrightarrow{\mathrm{ArSO_2Na}} \mathrm{Ar}\underset{\underset{\mathrm{O}}{\|}}{\overset{\overset{\mathrm{O}}{\|}}{\mathrm{S}}}\mathrm{SAr} + \mathrm{ArSO_3Na} \qquad (31)$$

Although acyl chlorides react at the oxygen atom of sulfinate ions, thioacyl chlorides seem to lead to S-acylation[71,72] (equation 32).

$$\mathrm{RSO_2}^- + \mathrm{Y}\overset{\overset{\mathrm{S}}{\|}}{\mathrm{C}}\mathrm{Cl} \longrightarrow \mathrm{R}\underset{\underset{\mathrm{O}}{\|}}{\overset{\overset{\mathrm{O}}{\|}}{\mathrm{S}}}-\overset{\overset{\mathrm{S}}{\|}}{\mathrm{C}}\mathrm{Y} + \mathrm{Cl}^- \qquad (32)$$

$(\mathrm{Y} = \mathrm{Me_2N}, \mathrm{MeS} \text{ or } \mathrm{PhS})$

The reaction of chlorosulfines **6** with *p*-toluenesulfinate ion leads to products in which the C=S=O function is replaced by a $CH_2$ group[73] (equation 33). The first step of this reductive substitution is a nucleophilic displacement of chloride by the sulfinate (as a sulfur nucleophile). The intermediate sulfonylsulfines subsequently undergo attack by the sulfinate at the sulfur atom to yield sulfinylsulfones, which upon hydrolysis provide the apparent reduction products.

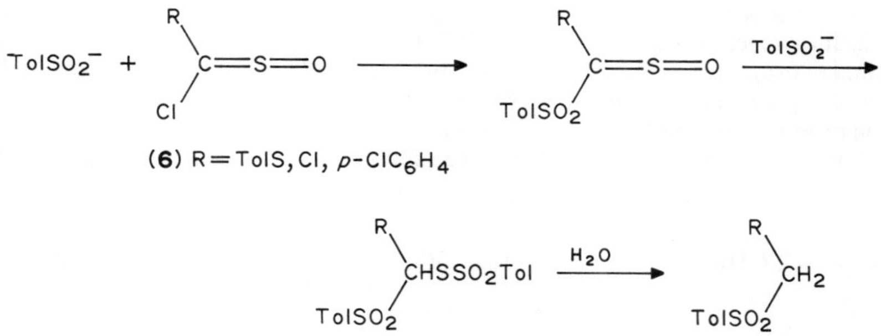

(33)

Reactions of ethyl chloroformate with sulfinate ions were found in early work[7] to give mostly the O-ethylation products. This reaction was reinvestigated by Kobayashi and Terao[74]. Reactions were carried out in various alcohols (primary and secondary), and the alkoxyl group in the product alkyl sulfinates was found to originate from the alcohol used as solvent but not from the chloroformate. The intermediate formation of a mixed anhydride **7** of sulfinic acid and monoethyl carbonate is postulated (equation 34). The alcoholysis of the anhydride **7** would give the alkyl sulfinates (equation 35). The initial reaction is again O-acylation.

$$\mathrm{ArSO_2Na} + \mathrm{Cl}\overset{\overset{\mathrm{O}}{\|}}{\mathrm{C}}\mathrm{OEt} \longrightarrow \mathrm{Ar}\overset{\overset{\mathrm{O}}{\|}}{\mathrm{S}}\mathrm{O}\overset{\overset{\mathrm{O}}{\|}}{\mathrm{C}}\mathrm{OEt} + \mathrm{NaCl} \qquad (34)$$

**(7)**

$$\mathbf{7} + \text{ROH} \longrightarrow \text{Ar}\overset{\text{O}}{\overset{\|}{\text{S}}}\text{OR} + \text{CO}_2 + \text{EtOH} \tag{35}$$

In summary, acylation occurs primarily at the oxygen but thioacylation at the sulfur atom of sulfinates.

## E. Addition to Carbonyl Groups

α-Hydroxy sulfones have been obtained by the addition of sulfinic acid to aldehydes[75–76]. The reactions were carried out either in aqueous solution (formaldehyde) or in ether solution. Although the mechanism of this addition is not clear, a possible reaction sequence may involve a protonated aldehyde and sulfinate ion (equation 36). In this reaction, O-hydroxyalkylation may also occur due to the hardness of the carbon atom of the protonated carbonyl group. However, this possible reaction must be reversible (equation 37), and the hemiacetal-type adduct cannot be isolated owing to its instability.

$$\text{RSO}_2\text{H} + \text{R}'\overset{\text{O}}{\overset{\|}{\text{C}}}\text{H} \rightleftharpoons \text{RSO}_2^- + \text{R}'\overset{^+\text{OH}}{\overset{\|}{\text{C}}}\text{H} \longrightarrow \text{R}\underset{\underset{\text{O}}{\|}}{\overset{\text{O}}{\overset{\|}{\text{S}}}}\overset{\text{OH}}{\overset{|}{\text{C}}}\text{HR}' \tag{36}$$

$$\text{RSO}_2^- + \text{R}'\overset{^+\text{OH}}{\overset{\|}{\text{C}}}\text{H} \rightleftharpoons \text{R}\overset{\text{O}}{\overset{\|}{\text{S}}}\text{O}\underset{\underset{\text{R}}{|}}{\overset{\text{OH}}{\overset{|}{\text{C}}}}\text{H} \tag{37}$$

In the presence of amines, aldehydes give α-amino sulfones by a Mannich-type condensation reaction with sulfinic acids[77,78]. This reaction is similar to the hydroxyalkylation and may involve an iminium ion as an intermediate (equation 38). The potential, and presumably preferred, reaction at the oxygen atom of sulfinates must be again reversible, and the product could not be isolated. Amines examined in this reaction include primary and secondary amines, ureas, carboxamides, sulfonamides and carbamates[78].

$$\text{RSO}_2\text{H} + \text{R}'\text{CHO} + \text{HNXY} \rightleftharpoons \text{RSO}_2^- + \text{R}'\overset{\text{X}\overset{+}{\text{N}}\text{Y}}{\overset{\|}{\text{C}}}\text{H} + \text{H}_2\text{O} \longrightarrow \text{R}\underset{\underset{\text{O}}{\|}}{\overset{\text{O}}{\overset{\|}{\text{S}}}}\overset{\text{NXY}}{\overset{|}{\text{C}}}\text{HR}' + \text{H}_2\text{O} \tag{38}$$

Kobayashi and coworkers[79] found that the reaction of arenesulfinic acids and alcohols leading to alkyl arenesulfinates can be accomplished in the presence of dicyclohexylcarbodiimide (DCC). Formation of an adduct between the sulfinic acid and DCC or the sulfinic anhydride as an intermediate was considered to be the first step of this reaction (equation 39). This reaction closely resembles that with *O*-alkylisourea (equation 9)[14].

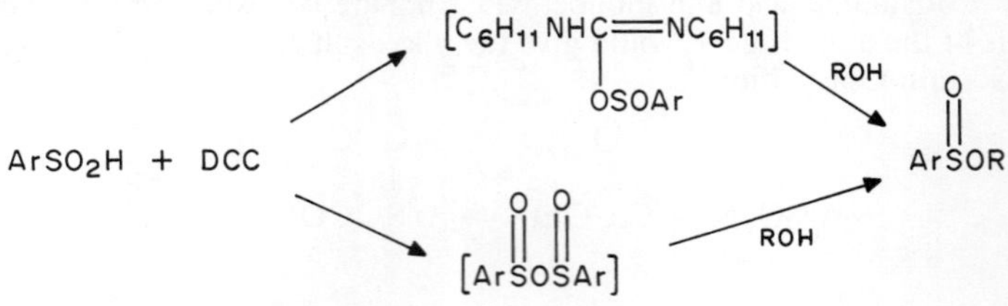

(39)

However, the possibility of the initial formation of the *O*-alkylisourea from the reaction of alcohol with DCC may be excluded for this reaction, since the authors found that $^{18}O$ of the labelled sulfinic acid was transferred to the urea product[79] (equation 40).

$$ArS^{18}O_2H + EtOH + DCC \longrightarrow Ar\overset{^{18}O}{\overset{\|}{S}}OEt + C_6H_{11}NH\overset{^{18}O}{\overset{\|}{C}}NHC_6H_{11} \qquad (40)$$

## IV. REACTIONS AT HETEROATOMS

### A. Reactions at Sulfur

#### 1. Substitution at sulfenyl sulfur

Sulfinic acids react rapidly with sulfenyl halides to give thiolsulfonates in high yields[80] (equation 41). One of a few stable sulfenic acids, anthraquinone-2-sulfenic acid, reacts directly with sulfinic acids to yield thiosulfonates[81]. Selenenyl halides similarly give selenosulfonates on reaction with alkali sulfinates[82].

$$RSO_2H + R'SCl \longrightarrow R\underset{\underset{O}{\|}}{\overset{\overset{O}{\|}}{S}}SR' \qquad (41)$$

Thiosulfonates are also subject to nucleophilic attack and undergo exchange of the sulfonyl group with sulfinate ions[83] (equation 42). Thiolsulfinates undergo a similar nucleophilic reaction. Phenyl benzenethiolsulfinate reacts rapidly with arenesulfinic acids in acid solution in the presence of alkyl sulfides to yield phenyl arenethiolsulfonates **8**[84] (equation 43). The kinetics of the reaction show that the rate-determining step of the reaction is a nucleophilic attack by the alkyl sulfide on the protonated thiolsulfinate to give benzenesulfenic acid and an intermediate dialkylphenylthiosulfonium ion **9**, which then reacts rapidly with the sulfinic acid to yield the thiolsulfonate **8**. The reactions involved are summarized in equations 44–48.

$$ArSO_2^- + Ar'SSO_2Ar' \longrightarrow ArSO_2SAr' + Ar'SO_2^- \qquad (42)$$

$$2ArSO_2H + PhS\overset{\overset{O}{\|}}{S}Ph \xrightarrow[R_2S]{H^+} 2PhS\underset{\underset{O}{\|}}{\overset{\overset{O}{\|}}{S}}Ar + H_2O \qquad (43)$$

**(8)**

$$PhS\overset{\overset{O}{\|}}{S}Ph + H^+ \rightleftharpoons PhS\underset{+}{\overset{\overset{OH}{|}}{S}}Ph \qquad (44)$$

$$Ph\underset{+}{\overset{\overset{OH}{|}}{S}}SPh + R_2S \longrightarrow PhSOH + PhS\underset{+}{S}R_2 \qquad (45)$$

**(9)**

$$9 + ArSO_2H \xrightarrow{\text{rapid}} \mathbf{8} + R_2S \tag{46}$$

$$PhSOH + R_2S + H^+ \rightleftharpoons \mathbf{9} + H_2O \tag{47}$$

$$PhSOH + ArSO_2H \longrightarrow \mathbf{8} + H_2O \tag{48}$$

*2. Substitution at sulfinyl sulfur*

Like sulfenyl halides, sulfinyl chlorides react with alkali sulfinates to yield the corresponding sulfinyl sulfones **10** instead of the mixed anhydride **11**[85] (equation 49). The products isolated are solely the sulfinyl sulfones **10** with a sulfur–sulfur bond. However, whether this results because the reaction at the sulfinate sulfur is kinetically preferred, or rather comes about because the anhydride **11** resulting from the initial attack on oxygen is readily converted to the thermodynamically more stable isomer **10**, is another problem to be solved. Although the bivalent sulfur involved in sulfenyl derivatives is a reasonably soft electrophile able to react at the sulfinate sulfur, a sulfur atom of higher valency would be harder and the reaction at the oxygen end might be possible. A possible reaction leading to **10** from **11** may be described by equation 50, but this possibility does not seem to have been definitely established. In this case, the exchanged product **10′** should also be formed if R′ ≠ R.

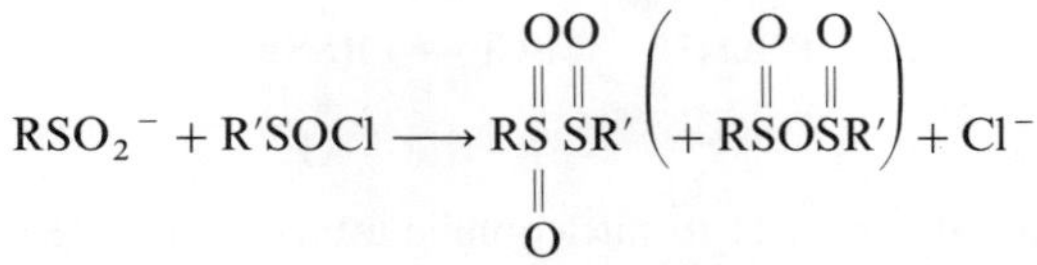
**(10)** **(11)**

(49)

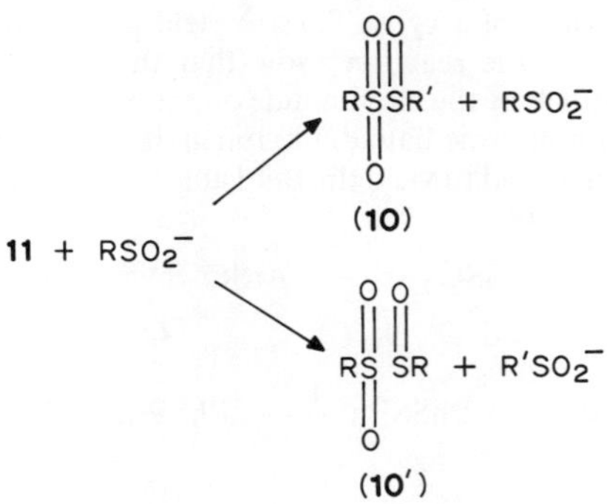

(50)

Sulfinic acids easily dimerize to form sulfinyl sulfones[86]. This reaction must involve a nucleophilic attack by a sulfinic acid (or sulfinate) on the sulfur atom of another molecule of sulfinic acid, but the reaction mechanism has not been investigated. However, the kinetics of the reverse reaction, namely the hydrolysis of sulfinyl sulfone, was examined in detail[87]. Considering microscopic reversibility, the transition state for the reaction may be something like the structure given in equation 51 or 52.

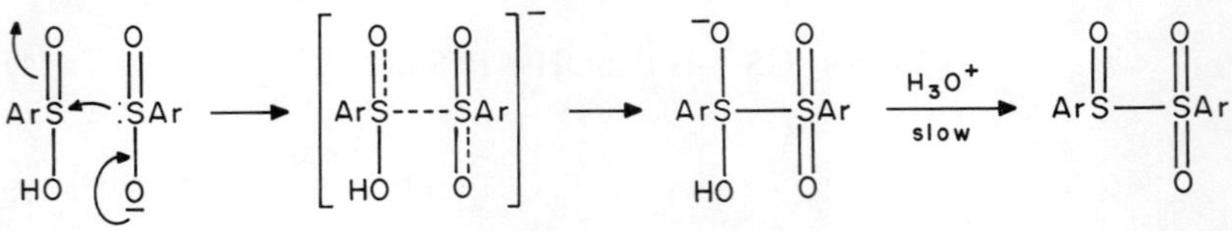

(51)

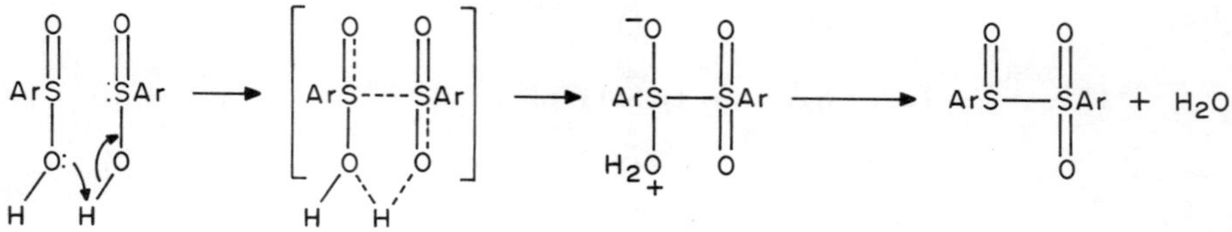

(52)

### 3. *Addition to sulfines*

The addition of sulfinate ions to sulfines occurs at the sulfur atom of the sulfine to form sulfinyl sulfones[73] (equation 53).

$$ArSO_2^- + \rangle C{=}S{=}O + H^+ \longrightarrow -\overset{|}{C}H\overset{O}{\overset{\|}{S}}-\underset{\underset{O}{\|}}{\overset{O}{\overset{\|}{S}}}Ar \qquad (53)$$

### 4. *Substitution at sulfonyl sulfur*

The reaction of sulfinate salts with sulfonyl chlorides has been reported to give α disulfones in aqueous solution[67], but the yield of the product was quite low[88] (equation 54). In a more recent report[89], the reaction of *p*-toluenesulfonyl chloride with sodium *p*-toluenesulfinate in acetonitrile was found to proceed according to the stoichiometry given in equation 55. This process was formulated as involving a sequence in which the initial reaction of sulfonyl chloride with sulfinate mostly occurs at the sulfinate oxygen (equation 56). The ratio of the reactions occurring at the oxygen end and at the sulfur atom of the sulfinate was estimated to be about 95/5. Formation of the isolated product may result from the reaction of the intermediate anhydride **12** with the sulfinate (equation 57).

$$RSO_2^- + R'SO_2Cl \longrightarrow R\underset{\underset{O}{\|}}{\overset{O}{\overset{\|}{S}}}-\underset{\underset{O}{\|}}{\overset{O}{\overset{\|}{S}}}R' \qquad (54)$$

$$3TolSO_2Na + TolSO_2Cl \xrightarrow{MeCN} Tol\underset{\underset{O}{\|}}{\overset{O}{\overset{\|}{S}}}STol + 2TolSO_3Na + NaCl \qquad (55)$$

$$TolSO_2Na + TolSO_2Cl \longrightarrow \underset{(\mathbf{12})\ 95\%}{Tol\underset{\underset{O}{\|}}{\overset{O}{\overset{\|}{S}}}O\overset{O}{\overset{\|}{S}}Tol} + \underset{5\%}{Tol\underset{\underset{O}{\|}}{\overset{O}{\overset{\|}{S}}}-\underset{\underset{O}{\|}}{\overset{O}{\overset{\|}{S}}}Tol} \qquad (56)$$

$$\mathbf{12} \xrightarrow{\text{TolSO}_2^-} \text{TolS(=O)}_2\text{—S(=O)Tol} \xrightarrow{\Delta} \text{TolS(=O)OS(=O)}_2\text{Tol} \xrightarrow{\text{TolSO}_2^-} \text{TolS(=O)}_2\text{STol} \qquad (57)$$

In summary, the sulfinate ion attacks at sulfenyl sulfur exclusively as an S nucleophile while it reacts at sulfonyl sulfur mostly as an O nucleophile. This tendency is in accord with the hardness of higher-valent sulfur.

## B. Displacement at Oxygen

Oxidations of sulfinic acids with hydrogen peroxide[90] and hypochlorite[91] to sulfonic acids involve the rate-determining nucleophilic attack of sulfinate ion (as an S-nucleophile) on the oxygen atom. In the oxidation with hydrogen peroxide, the reactive oxidizing agent is neutral $H_2O_2$, rather than $HO_2^-$, in the pH region 2–9. Displacement of hydroxide by sulfinate takes place as shown in equation 58. The effects of ring substituents in arenesulfinic acids show a modest negative $\rho$ value ($-0.5$), consistent with reaction 58.

$$\text{ArSO}_2^- + \text{HO—OH} \longrightarrow \text{ArSO}_3\text{H} + \text{OH}^- \xrightarrow{\text{rapid}} \text{ArSO}_3^- + \text{H}_2\text{O} \qquad (58)$$

In the hypochlorite oxidation, both the neutral (HOCl) and anionic species ($^-$OCl) can serve as reactive oxidizing agents, with the anion being about 300 times more reactive than HOCl[91]. The slower oxidation by HOCl involves the rate-determining nucleophilic attack of the sulfinate on HOCl in a similar manner to the oxidation by $H_2O_2$ (equation 59). The rate is several orders of magnitude larger than that with $H_2O_2$, reflecting the greater leaving ability of $Cl^-$ (than $OH^-$). By contrast, the fast oxidation by the anion $^-$OCl involves a nucleophilic attack by $^-$OCl on the sulfur atom of the sulfinate to form a sulfurane intermediate **13**, which decomposes rapidly to arenesulfonate and chloride ions (equation 60).

$$\text{ArSO}_2^- + \text{HO—Cl} \longrightarrow \text{ArSO}_3\text{H} + \text{Cl}^- \longrightarrow \text{ArSO}_3^- + \text{H}^+ + \text{Cl}^- \qquad (59)$$

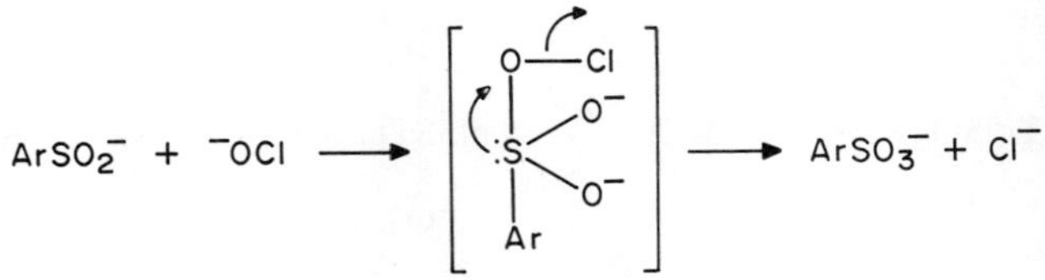


(60)

**(13)**

## C. Reactions at Nitrogen

Additions of sulfinate ions to nitroso and azo compounds occur at the nitrogen atom. Treatment of a sulfinic acid with nitrous acid yields bis(alkanesulfonyl)hydroxylamine[92] (equation 61a). Arenesulfinate ions add to aromatic nitroso compounds at pH0–3 to give *N*-hydroxysulfonamides[93]. The reaction can be reversed at higher pH(> 8) probably through the anion $ArSO_2N(O^-)Ar'$ (equation 61b).

$$2RSO_2H + HONO \longrightarrow (RSO_2)_2NOH + H_2O \qquad (61a)$$

$$ArSO_2^- + H^+ + Ar'N{=}O \longrightarrow ArSO_2N(OH)Ar' \qquad (61b)$$

Addition of arene sulfinic acids to 1-nitroso-2-naphthol was found to give the sulfone **15** probably from the initial addition product **14** (equation 62). 2-Nitroso-1-naphthol also gave a similar product.

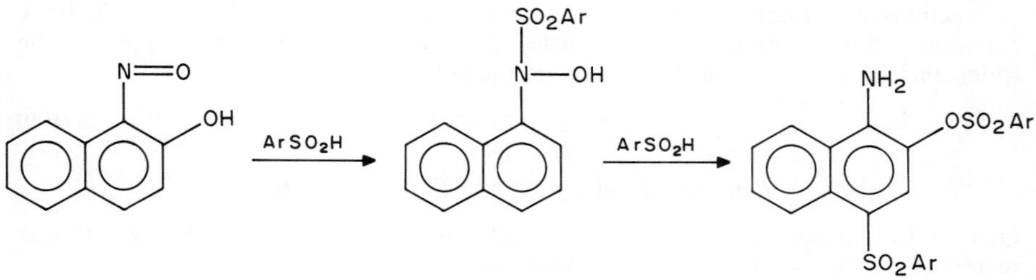


**(14)** **(15)** (62)

Arenesulfinic acids add to azobenzene to give hydrazo derivatives at room temperature, while in refluxing ethanol the products isolated were 4-arenesulfonyl derivatives of azobenzene[94,95]. The migration from the initial addition product was postulated to lead to the final product. However, the reaction involved seem to be more complicated[4] (equation 63).

$$ArSO_2H + PhN{=}NPh \longrightarrow PhNHN(SO_2Ar)Ph \xrightarrow{-2H} PhN{=}N{-}C_6H_4{-}SO_2A \qquad (63)$$

A similar reaction is also reported with phenazine[96]. The authors suggest that the initial addition product rearranges to the ring-substituted product, with accompanying loss of hydrogen atoms (equation 64).

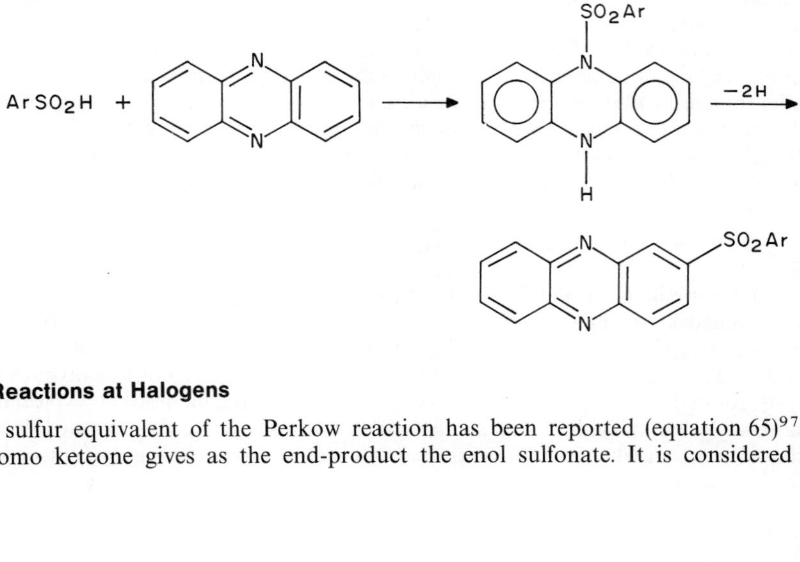


(64)

### D. Reactions at Halogens

A sulfur equivalent of the Perkow reaction has been reported (equation 65)[97]. An α-bromo keteone gives as the end-product the enol sulfonate. It is considered that the reaction is initiated by attack of the sulfinate ion on the bromine atom, followed by O-sulfonation by the resultant sulfonyl bromide.

$$PhSO_2^- + ArC(=O){-}CBr(COOR)_2 \longrightarrow PhSO_2Br + ArC(OH){=}C(COOR)_2 \longrightarrow ArC(OSO_2Ph){=}C(COOR)_2 \qquad (65)$$

Reactions of sulfinate ions with halogens to give sulfonyl halides are known[3,80]. In these reactions, sulfinate ions react as sulfur nucleophiles. Soft halogen electrophiles react at the softer nucleophilic center of sulfinate (equation 66).

$$RSO_2^- + X_2 \longrightarrow ArSO_2X + X^- \qquad (66)$$

## V. NUCLEOPHILICITY OF SULFINATE IONS

Quantitative nucleophilic reactivities of sulfinate ions in displacement or addition reactions can only be evaluate from limited results of kinetic investigations.

Lindberg[98] measured rate constants for the reaction of a series of arenesulfinate ions with sodium bromoacetate and bromoacetamide in aqueous solution at 60 °C. The Hammett $\rho$ values are negative and small ($\rho = -0.712$ and $-0.914$ for bromoacetate and bromoacetamide, respectively). The second-order rate constant obtained for bromoacetate ($2.4 \times 10^{-3}\,M^{-1}\,s^{-1}$) is close to that estimated for the reaction with hydroxide ion ($k_{OH} \sim 2 \times 10^{-3}\,M^{-1}\,s^{-1}$). The basis for this estimation is the $k_{OH}$ value determined for chloroacetate at 56 °C ($3.44 \times 10^{-5}\,M^{-1}\,s^{-1}$)[99] and the relative leaving ability of Br/Cl in $S_N2$ reactions which is considered to be about 50[100]. This estimation shows that the nucleophilicity of the benzenesulfinate ion is similar to that of hydroxide ion in $S_N2$ reactions in aqueous solution (equation 67).

$$BrCH_2COO^- + Nu^- \xrightarrow{H_2O} NuCH_2COO^- + Br^- \qquad (67)$$

Ogata and coworkers[101] measured rates for the addition of arenesulfinate ions to acrylonitrile in aqueous solution at 50 °C (equation 68), the Hammett $\rho$ value being $-1.15$. The rate constant for benzenesulfinate is compared with those for other related nucleophiles in Table 6. Data in this Table show that the nucleophilic reactivity of the sulfinate is of the same order of magnitude as those of hydroxide and an amine but much smaller than those of thiolate and sulfite ions. It was also noted that logarithms of the rate coefficients are linearly correlated with the $pK_a$ of the conjugate acids of sulfur nucleophiles with a slope of 0.59. Thiols[104] and amines[105] with secondary and tertiary alkyl groups exhibited considerable steric effects in this reaction.

$$CH_2{=}CHCN + Nu^- \xrightarrow{H_2O} NuCH_2CH_2CN \qquad (68)$$

The nucleophilic reaction of benzenesulfinate with arenediazonium ions (equation 69) was examined in methanol solution[107]. The results are summarized in terms of $N_+$ $[= \log(k_{Nu}/k_{H_2O})]$[108,109], some typical examples of which are listed in Table 7. We have recently measured rates of nucleophilic reactions of 2-phenyl-1,3-dithiolanylium ion in 50% aqueous ethanol at 25 °C by means of the flash-photolytic method[110]. Typical results are also included in Table 7. The sulfinate is much less reactive than methoxide ion in methanol solution while it is 10 times as reactive as hydroxide ion in aqueous ethanol.

TABLE 6. Second-order rate constants for addition of nucleophiles to acrylonitrile in aqueous solution

| Nucleophile | $pK_a$[a] | $10^3k_2$ ($M^{-1}s^{-1}$) 30 °C | 50 °C | Rel. rate | Ref. |
|---|---|---|---|---|---|
| $PhSO_2^-$ | 1.84, 2.16 | 0.0741 | 0.585 | 1.0 | 101 |
| $S_2O_3^{2-}$ | 1.72 | | 0.160 | 0.27 | 102 |
| $SO_3^{2-}$ | 7.21 | 220 | | $3.0 \times 10^3$ | 103 |
| $^-O_2CCH_2S^-$ | 10.68 | 4260 | | $5.7 \times 10^4$ | 104 |
| $^-O_2CCH_2NH_2$ | 9.6 | 0.534 | | 7.2 | 105 |
| $OH^-$ | 15.7 | | 1.93 | 3.3 | 106 |

[a]$pK_a$ for the conjugate acid, taken from *Handbook of Biochemistry* (Ed. H. A. Sober), CRC Press, Cleveland, OH, 1968.

TABLE 7. Nucleophilicity of some nucleophiles in reactions with cationic species

| Nucleophile | $N_+(H_2O)$[a] | $N_+(MeOH)$[a] | $\log(k_{Nu}/k_{H_2O})$[b] |
|---|---|---|---|
| $PhSO_2^-$ | | 3.67 | 6.2[c] |
| $CN^-$ | 4.12 | 5.94 | 5.9 |
| $MeO^-$ | 7.28 | 7.51 | |
| $N_3^-$ | 7.54 | 8.78 | |
| $PhS^-$ | 9.10 | 10.41 | 7.7[d] |
| $HOCH_2CH_2S^-$ | 8.87 | | |
| $HO^-$ | 4.75 | | 5.1 |

[a]Data are taken from Reference 109.
[b]Obtained in 50% (v/v) aqueous ethanol at 25 °C[110] (see equation 70).
[c]Value for *p*-toluenesulfinate.
[d]Value for *p*-chlorobenzenethiolate.

Similarly, the reactivity of the sulfinate is comparable to that of cyanide ion in aqueous ethanol solution although the former is $10^2$ times smaller than the latter in methanol. Effects of solvent may be relatively small upon the nucleophilicity of a complex large ion like sulfinate as compared with those on that of anionic small nucleophiles like $RO^-$ and $CN^-$. In conclusion, the nucleophilicity of sulfinate is comparable to that of $OH^-$ in aqueous solution but smaller than that of alkoxide in alcohol. Thiolate ion is much more reactive than sulfinate.

$$ArN_2^+ + Nu^- \xrightarrow{MeOH} ArN{=}NNu \quad (69)$$

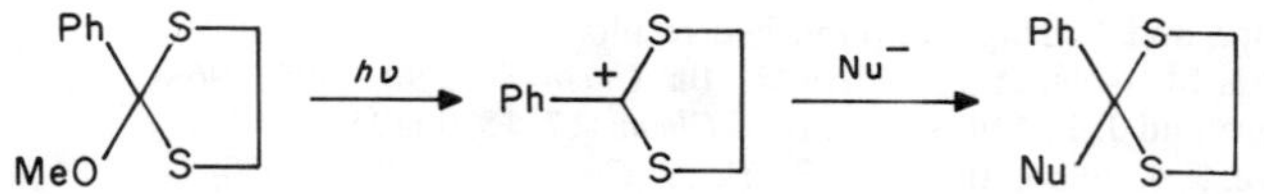


(70)

## VI. REFERENCES

1. J. L. Kice, *Adv. Phys. Org. Chem.*, **17**, 65 (1980).
2. R. G. Pearson and J. Songstad, *J. Am. Chem. Soc.*, **89**, 1827 (1967); *J. Org. Chem.*, **32**, 2899 (1967).

3. C. J. M. Stirling, *Int. J. Sulfur Chem., B*, **6**, 277 (1971).
4. S. Oae and N. Kunieda, in *Organic Chemistry of Sulfur* (Ed. S. Oae), Plenum, New York, 1977, pp. 625–637.
5. K. K. Andersen, in *Comprehensive Organic Chemistry*, Vol. 3 (Eds. D. Barton and D. N. Jones), Pergamon, Oxford, 1979, p. 320.
6. R. Otto, *Chem. Ber.*, **13**, 1150 (1880); F. Ullmann and G. Pasdermadjian, *Chem. Ber.*, **34**, 1150 (1901).
7. R. Otto and A. Rossing, *Chem. Ber.*, **18**, 2493 (1885); *J. Prakt. Chem.*, [2] **47**, 152 (1893); R. Otto, *Chem. Ber.*, **26**, 308 (1893).
8. M. Kobayashi, *Bull. Chem. Soc. Jpn.*, **39**, 1296 (1966).
9. M. Kobayashi, *Bull. Chem. Soc. Jpn.*, **39**, 967 (1966).
10. J. Michalski, T. Modro, and J. Wieczorkowski, *J. Chem. Soc.*, 1665 (1960).
11. J. S. Meek and J. S. Fowler, *J. Org. Chem.*, **33**, 3422 (1968).
12. M. Kobayashi and K. Toriyabe, *Sulfur Lett.*, **3**, 117 (1985).
13. F. Arndt and A. Scholz, *Ann. Chem.*, **510**, 62 (1934).
14. P. Kiełbasiński, R. Zurawiński, J. Drabowicz, and M. Mikołajczyk, *Tetrahedron*, in press. The author thanks Professor Mikołajczyk for the manuscript.
15. (a) C. C. J. Culvenor, W. Davies, and N. S. Heath, *J. Am. Chem. Soc.*, **71**, 278 (1949).
    (b) I. W. J. Still and F. J. Ablenas, *Synth. Commun.*, **12**, 1103 (1982).
16. T. L. Gresham, J. E. Jensen, F. W. Shaver, M. R. Frederick, F. T. Fiedorek, R. A. Bankert, J. T. Gregory, and W. L. Beears, *J. Am. Chem. Soc.*, **74**, 1323 (1952).
17. P. Messinger and H. Greve, *Synthesis*, 259 (1977); *Arch. Pharmaz.*, **311**, 280 (1978).
18. A. H. Wragg, J. S. McFadyen, and T. S. Stevens, *J. Chem. Soc.*, 3603 (1958).
19. D. Darwish and R. McLaren, *Tetrahedron Lett.*, 1231 (1962).
20. E. Ciuffarin, M. Isola, and A. Fava, *J. Am. Chem. Soc.*, **90**, 3594 (1968).
21. C. A. Bunton and B. N. Hendy, *J. Chem. Soc.*, 2562 (1962); 627 (1963).
22. M. Kobayashi, R. Nishi, and H. Minato, *Bull. Chem. Soc. Jpn.*, **47**, 888 (1974).
23. (a) L. Field and R. D. Clark, in *Organic Syntheses*, Coll. Vol. IV (Ed. N. Rabjohn), Wiley, New York, 1963, p. 674.
    (b) L. Field and R. D. Clark, *J. Org. Chem.*, **22**, 1129 (1957).
24. D. R. Hogg and A. Robertson, *J. Chem. Soc., Perkin Trans. 1*, 1125 (1979).
25. G. E. Vennstra and B. Zwanenburg, *Synthesis*, 519 (1975).
26. F. Manescalchi, M. Orena, and D. Savoia, *Synthesis*, 445 (1979).
27. K. Sukata, *Bull. Chem. Soc. Jpn.*, **57**, 613 (1984).
28. J. K. Crandall and C. Pradat, *J. Org. Chem.*, **50**, 1327 (1985).
29. J. V. Weber, M. Schneider, D. Paquer, and P. Faller, *Sulfur Lett.*, **3**, 45 (1985).
30. G. Bram, A. Loupy, M. C. Roux-Schmitt, J. Sansoulet, T. Strazalko, and J. Seyden-Penne, *Synthesis*, 56 (1987).
31. N. V. Kondratenko, V. P. Sambur, and L. M. Yagupol'skii, *Zh. Org. Khim.*, **7**, 2382 (1971).
32. K. Schank, *Ann. Chem.*, **702**, 75 (1967).
33. K. Schank, *Ann. Chem.*, **714**, 117 (1968).
34. K. Schank and H.-G. Schmitt, *Chem. Ber.*, **107**, 3026 (1974).
35. K. Schank, *Ann. Chem.*, **716**, 87 (1968).
36. R. J. Mulder, A. M. van Leusen, and J. Strating, *Tetrahedron Lett.*, 3061 (1967).
37. K. Schank and A. Weber, *Chem. Ber.*, **105**, 2188 (1972).
38. W. A. Cowdrey, E. D. Hughes, and C. K. Ingold, *J. Chem. Soc.*, 1208 (1937); J. F. Lane and H. W. Heine, *J. Am. Chem. Soc.*, **73**, 1348 (1951); K. R. Henery-Logan and T. L. Fridinger, *Chem. Commun.*, 130 (1968).
39. T. Okuyama and N. Haga, unpublished results.
40. E. Ciuffarin, M. Isola, and A. Fava, *J. Am. Chem. Soc.*, **90**, 3594 (1968).
41. W. E. Truce and J. P. Milionis, *J. Org. Chem.*, **17**, 1529 (1952).
42. M. P. Balfe, J. Kenyon, and C. E. Searle, *J. Chem. Soc.*, 3309 (1950).
43. L. Kenyon, M. P. Balfe, E. A. W. Downer, A. A. Evans, J. R. Poplett, C. E. Searle, and A. L. Tarnoky, *J. Chem. Soc.*, 797 (1946).
44. W. Middelbos, J. Strating, and B. Zwanenburg, *Tetrahedron Lett.*, 351 (1971).
45. K. Schank and H.-G. Schmitt, *Chem. Ber.*, **110**, 3235 (1977).
46. O. Achmatowicz, E. Maruszewska-Wieczorkowska, and J. Michalski, *Rocz. Chem.*, **29**, 1029 (1955); O. Achmatowicz and J. Michalski, *Rocz. Chem.*, **30**, 243 (1956).

47. V. Mikhailova and A. I. Filippova, *Zh. Org. Khim.*, **1**, 1621 (1965). V. Mikhailova, N. Borisova, and D. Stankevich, *Zh. Org. Khim.*, **2**, 1437 (1966). M. V. Kalnins and B. Miller, *Chem. Ind.* (London), 555 (1966).
48. A. R. Katritzky and O. Rubio, *J. Org. Chem.*, **48**, 4017 (1983).
49. I. Matsuda, K. Akiyama, T. Toyoshima, S. Kato, and M. Mizuta, *Bull. Chem. Soc. Jpn.*, **48**, 3675 (1975).
50. I. Matsuda, K. Akiyama, H. Furuta, and M. Mizuta, *Bull. Chem. Soc. Jpn.*, **57**, 219 (1984).
51. W. B. Price and S. Smiles, *J. Chem. Soc.*, 3154 (1928).
52. Y. Ogata, Y. Sawaki, and M. Isono, *Tetrahedron*, **25**, 2715 (1969); **26**, 731 (1970).
53. R. Adams and W. Moje, *J. Am. Chem. Soc.*, **74**, 5560 (1952).
54. R. Adams and T. E. Young, *J. Am. Chem. Soc.*, **75**, 3235 (1953).
55. K. T. Finley, R. S. Kaiser, R. L. Reeves, and G. Werimint, *J. Org. Chem.*, **34**, 2083 (1969).
56. C. J. M. Stirling, *J. Chem. Soc.*, 5856 (1964).
57. Z. Rappoport, *Adv. Phys. Org. Chem.*, **7**, 1 (1969).
58. A. Nesmeyanov, O. Reutov, and A. Gudkov, *Izv. Akad. Nauk SSSR, Otd. Khim. Nauk*, 260 (1961); *Chem. Abstr.*, **55**, 23375 (1961).
59. F. Scotti and F. J. Frazza, *J. Org. Chem.*, **29**, 1800 (1964).
60. B. Miller and M. V. Kalnins, *Tetrahedron*, **23**, 1145 (1967).
61. J. J. Sepiol, J. A. Sepiol, and R. L. Soulen, *J. Org. Chem.*, **49**, 1125 (1984).
62. J. F. Bunnett and R. E. Zahler, *Chem. Rev.*, **49**, 273 (1951).
63. R. R. Coats and D. T. Gibson, *J. Chem. Soc.*, 442 (1940).
64. F. M. Beringer, A. Brierley, M. Drexler, E. M. Gindler, and C. C. Lumpkin, *J. Am. Chem. Soc.*, **75**, 2708 (1953).
65. V.V. Grushin, M. M. Kantor, T. P. Tolstaya, and T. M. Shcherbina, *Izv. Akad. Nauk SSSR, Ser. Khim.*, 2332 (1984).
66. M. Furukawa, T. Okawara, Y. Noguchi, and M. Nishikawa, *Synthesis*, 441 (1978).
68. H. T. Hookway, *J. Am. Chem. Soc.*, **71**, 3240 (1949).
69. L. Panizzi and R. A. Nicolaus, *Gazz. Chim. Ital.*, **80**, 431 (1950).
70. H. Böhme and K.-H. Meyer-Dulheuer, *Ann. Chem.*, **688**, 78 (1965).
71. A. Senning, O. N. Sørensen, and Ch. Jacobsen, *Angew. Chem., Int. Ed. Engl.*, **7**, 734 (1968).
72. N. H. Nilsson, C. Jacobasen, and A. Senning, *J. Chem. Soc.* (*D*), 314 (1971).
73. G. E. Veenstra and B. Zwanenburg, *Recl. Trav. Chim. Pays-Bas*, **95**, 28 (1976).
74. M. Kobayashi and M. Terao, *Bull. Chem. Soc. Jpn.*, **39**, 1292 (1966).
75. E. v. Meyer, *J. Prakt. Chem.*, **63**, 167 (1901).
76. H. Bredereck and E. Bäder, *Chem. Ber.*, **87**, 129 (1954).
77. E. Bäder and H. D. Hermann, *Chem. Ber.*, **88**, 41 (1955).
78. H. Meijer, T. J. Strating, and J. B. F. N. Engbert, *Recl. Trav. Chim. Pays-Bas*, **92**, 72 (1973) and references cited therein.
79. Y. Miyaji, H. Minato, and M. Kobayashi, *Bull. Chem. Soc. Jpn.*, **44**, 862 (1971).
80. C. J. M. Stirling, *J. Chem. Soc.*, 3597 (1957).
81. E. Vinkler and F. Klivenyi, *Magy. Kem. Foly.*, **61**, 103 (1955).
82. O. Foss, *J. Am. Chem. Soc.*, **69**, 2236 (1947).
83. J. D. Loudon and A. Livingston, *J. Chem. Soc.*, 896 (1935).
84. J. L. Kice, C. G. Venier, and L. Heasley, *J. Am. Chem. Soc.*, **89**, 3357(1967).
85. H. Bredereck, A. Wagner, H. Beck, and R. J. Klein, *Chem. Ber.*, **93**, 2736 (1960).
86. J. L. Kice, G. Guaraldi, and C. G. Venier, *J. Org. Chem.*, **31**, 3561 (1966); J. L. Kice and G. Guaraldi, *J. Org. Chem.*, **31**, 3568 (1966).
87. J. L. Kice and G. Guaraldi, *J. Am. Chem. Soc.*, **89**, 4113 (1967).
88. Y. Urushibara and G. Koga, *Nippon Kagaku Zasshi*, **81**, 1615 (1960).
89. F. P. Corson and R. G. Pews, *J. Org. Chem.*, **36**, 1654 (1971).
90. B. J. Lindberg, *Acta Chem. Scand.*, **20**, 1843 (1966).
91. J. L. Kice and A. R. Puls, *J. Am. Chem. Soc.*, **99**, 3455 (1977).
92. C. S. Marvel and R. S. Johnson, *J. Org. Chem.*, **13**, 822 (1948).
93. A. Darchen and C. Moinet, *J. Chem. Soc., Chem. Commun.*, 820 (1976).
94. W. Bradley and J. D. Hannon, *J. Chem. Soc.*, 2713 (1962).
95. W. Bradley and J. D. Hannon, *Chem. Ind.* (London), 540 (1959).
96. W. Bradley and J. D. Hannon, *J. Chem. Soc.*, 4438 (1962).
97. I. Fleming and C. R. Owen, *J. Chem. Soc., Chem. Commun.*, 1402 (1970).

98. B. Lindberg, *Acta Chem. Scand.*, **17**, 393 (1963).
99. E.-H. M. Diefallah, *Can. J. Chem.*, **54**, 1687 (1976).
100. A. Streitwiesser, *Chem. Rev.*, **56**, 571 (1956).
101. Y. Ogata, Y. Sawaki, and M. Isono, *Tetrahedron*, **26**, 3045 (1970).
102. R. Kerber and J. Sarnik, *Tetrahedron Lett.*, 3007 (1966).
103. M. Morton and H. Landfield, *J. Am. Chem. Soc.*, **74**, 3523 (1952).
104. M. Friedman, J. F. Cavins, and J. S. Wall, *J. Am. Chem. Soc.*, **87**, 3672 (1965).
105. M. Friedman and J. S. Wall, *J. Am. Chem. Soc.*, **86**, 3735 (1964).
106. M. Wronski and J. Bogdanski, *Zeszyl Nauk. Uniw. Lodg.*, **14**, 153 (1963); cited in Reference 101.
107. C. D. Ritchie and P. O. I. Virtanen, *J. Am. Chem. Soc.*, **94**, 4966 (1972).
108. C. D. Ritchie, *J. Am. Chem. Soc.*, **97**, 1170 (1975).
109. C. D. Ritchie, *Can. J. Chem.*, **64**, 2239 (1986).
110. T. Okuyama, N. Haga, S. Takane, and T. Fueno, presented in the Ist Int. Conf. Heteroatom Chem., Kobe, July 17–24, 1987.

The Chemistry of Sulphinic Acids, Esters and their Derivatives
Edited by S. Patai


CHAPTER **23**

# Biological activity of sulfinic acid derivatives

ASHER KALIR

*Department of Physiology and Pharmacology, Sackler Faculty of Medicine, Tel Aviv University, Tel Aviv 69978, Israel*

and

HENRY H. KALIR

*Department of Neurobiology, Mount Sinai School of Medicine, One Gustave L. Levy Place, New York, NY 10029, USA*



## ABBREVIATIONS

| | | | |
|---|---|---|---|
| Asp | aspartic acid | HCSA | homocysteinesulfinic acid |
| CNS | central nervous system | HHT | homohypotaurine |
| CSA | cysteinesulfinic acid | HT | hypotaurine |

| | | | |
|---|---|---|---|
| Cys | cysteine | HTC | hypotaurocyamine |
| FAS | formamidinosulfinic acid | NMDA | N-methyl-D-aspartic acid |
| GABA | $\gamma$-aminobutyric acid | TA | taurine |
| Glu | glutamic acid | | |

## I. INTRODUCTION

Sulfinic acid derivatives are found in all living systems. The most important compounds in this series are aminoalkylsulfinic acids and particularly 3-sulfino-L-alanine (**2**), better known as cysteinesulfinic acid (CSA), and 2-aminoethylsulfinic acid or hypotaurine (HT) (**3**). These derivatives are closely linked with taurine (TA) (**12**) and cysteine (Cys) (**1**), and both possess diverse physiological activities that were and are intensively investigated. Much relevant material can be found in the excellent review of Jacobsen and Smith[1] and in several books dealing with these and related subjects[2-5]. The first part of this review describes the properties of the above compounds and of their congeners.

Many synthetic derivatives were tested for their activities as plant growth regulators[6,7], injection stabilizers[8,9], radioprotectors[10] and cytotoxic drugs[11]. These agents are described in the second part of this chapter.

## II. L-CYSTEINESULFINIC ACID (3-SULFINO-L-ALANINE, CSA)

The acid was first reported by Lavin in 1936[12]. As the main metabolite of Cys[13,14] it is widespread in bacteria, plants and mammalian tissues. CSA attracted considerable attention because of its neuroexcitatory action[15] similar to that of aspartic acid (Asp) (**7**). The preparations of sulfinic acids[16] and of L-($^{35}$S)CSA[17] have been described.

### A. Biosynthesis and Metabolism

CSA is formed from Cys by the action of Cys dioxygenase[1,13,18]. In rats the prefrontal cortex contains the highest concentration of CSA[19]. It is decarboxylated to HT by CSA decarboxylase[20-22]. The activity of this enzyme is considerably lower in plasma in the female than in the male rat[1]. Its distribution in liver and brain of rat, dog, cat, rhesus monkey and man has been summarized by Rassin[23]. Two different forms of this decarboxylase have been isolated from rat brain by Legay and colleagues, and it is suggested that only one form plays a role in the biosynthesis of taurine (TA) **12**[24]. Recently Weinstein and Griffith reported the resolution of this decarboxylase from male rat liver, brain and kidney into five distinct species. Their activities have not been determined. D-CSA is not decarboxylated by this enzyme but acts as inhibitor[25]. Antibodies to CSA decarboxylase were prepared, and served to locate this enzyme in nerve endings by radioimmunoassay[26]. CSA is transaminated to the unstable $\beta$-sulfinyl pyruvate **5** *in vitro* and *in vivo*[20,27], and desulfinated to alanine **6** by bacterial aspartate $\beta$-decarboxylases[14]. Oxidation of CSA to cysteic acid **4** followed by its decarboxylation yields TA[1,14,28]. It occurs, for instance, in chicken embryo and in the mollusc *Rangia cuneata*[1] (Scheme 1).

The metabolic rates are different for various species. About 85% of CSA is converted to HT in mice[29], less in humans[30] and very little in cats, that develop blindness when deprived of TA in food[31,32]. Isolation and determination of CSA and of its metabolites in biological samples has been published[33-36].

### B. Biochemistry and Physiology

CSA is a structural analog of the neuroexcitatory L-Asp and is a substrate for the same enzymes. It is also a substrate for tryptophanase[38], and glutamic acid (Glu) **8**

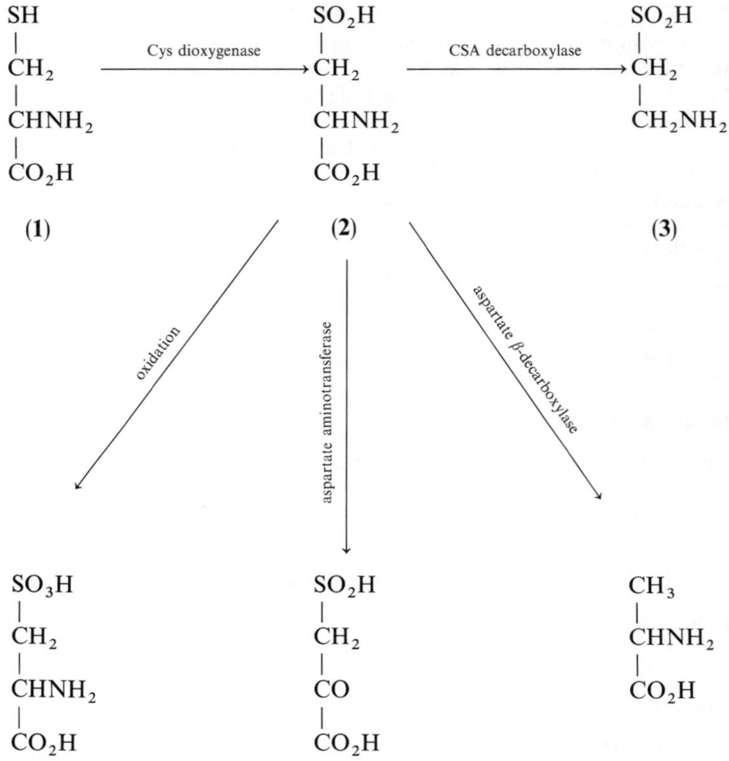


SCHEME 1. Biotransformations of CSA

decarboxylase from *E. Coli*[39]. The p$K_a$ of the —SOOH group is 1.50 and of —COOH, 2.38[37]. For Asp the values are 2.10 and 3.86[40].

$$HOOCCH_2CH(NH_2)COOH$$
(7)

$$HOOC(CH_2)_2CH(NH_2)COOH \qquad HOOCCH_2CH(NHMe)COOH$$
(8) (9)

CSA, Asp and Glu act similarly in the central nervous system (CNS)[41]. This effect was reported in a number of papers. CSA accelerates GABA liberation from brain hippocampus. $Zn^{2+}$ (0.1 mM) or insulin (10 $\mu$M) depresses this effect[42]. Agonist efficacy is greater than that of L-Glu when determined by chick retinal excitotoxicity and $Na^+$ efflux from rat brain slices and it has been characterized as a broad spectrum agonist at excitatory amino acid receptors with a potency in the functional assay greater than that of L-Glu[43]. The neuroexcitatory effect of amino acids and aminoalkylsulfinic acids on mammalian neurons was studied by Curtis and Watkins who found them very active[15].

CSA inhibits the uptake of labeled D-Asp by $P_2$ rat synaptosomes[44], of Glu by rat striatal homogenates[45], and of labeled L-Asp and L-Glu binding to membranes prepared from frozen human cerebellar cortex[46]. It activates N-methyl-D-aspartic acid (NMDA) **9** channels in mouse central neurons in culture[47]. Study of responses evoked by L-CSA and L-Asp on the membrane potential of cat caudate neurons provide evidence that the compounds interact with both NMDA and non-NMDA excitatory amino acid receptors[48]. L-CSA increases $Ca^{2+}$ permeability of plasma membrane of synaptosomes from rat brain. The effect seems to be mediated by highly specific receptors[49].

L-CSA given subcutaneously to 250 Webster Swiss albino mice (12 mmol $kg^{-1}$) produces retinal and hypothalamus lesions equal to those of L-Glu and L-Asp[50]. Wu and Dowling suggested that L-Asp is likely to act as a cone photoreceptor transmitter in the carp (*Cyprinus carpio*) retina and that CSA matches all of the action of L-Asp on the horizontal cells[51].

CSA causes EEG seizures and convulsions after intracerebroventricular injection (100 $\mu$g) into mice. The convulsions are inhibited by TA[52]. Stimulation of the formation of cyclic AMP in brain slices was also reported[53–56]. A possibility that CSA acts as an excitatory neurotransmitter has been suggested[42,56,57].

### C. Additional Effects of CSA

CSA supplemented to diet produced a weight gain in mice[58]. It has been mentioned as a component of lotions and ointments for prevention and treatment of disturbed keratinization of skin[59].

## III. 2-AMINOETHYLSULFINIC ACID (HYPOTAURINE, HT)

The compound was reported for the first time by Chatagner and Bergeret in 1951[20,60]. Cavallini and coworkers prepared it in pure state and defined its mp as 175–177 °C[61].

### A. Biosynthesis and Metabolism

The formation of HT from its precursors was discussed by Eldjarn and collaborators[62] and Jacobsen and Smith[1]. As mentioned above, CSA is decarboxylated to HT by CSA decarboxylase[20–22]. A second pathway involves conversion of Cys to pantetheine, hydrolysis to cysteamine **10**[63] and its oxidation to HT[64] by the action of cysteamine dioxygenase[65]. It has been found that in most animal tissues TA (and of course HT) is produced preferentially from Cys bound to phosphopantothenate rather than from the free amino acid (through CSA) when both forms are present at equal concentrations[66]. Liver homogenate converted 50% of **10** to HT and TA during 4 h incubation[67].

An additional biosynthetic possibility is the oxidation of cystine **11** to cystine disulfoxide **13** (or the isomeric thiosulfonate[68]), decarboxylation to the corresponding cystamine derivative **14** and conversion to HT[1,61]. This transformation has been observed after intravenous injection of cystamine into mice and rats[69].

HT is oxidized to TA, probably with the help of HT oxidase[13]. Recently Fellman presented evidence that HT is first oxidized by a hydroxyl radical to bis-aminoethyl-$\alpha$-disulfone (**15**). The hydroxyl radical is generated by a liver microsomal NADPH oxidase. **15** has been prepared from HT in the presence of chemically or enzymatically generated radicals. It has been found in male sexual tissue which contains HT and TA both in high concentrations[28]. A minor part of HT is converted to thiotaurine **16** or hypotaurocyanamine **17**. HT has been found to undergo transamination to the unstable $\alpha$-sulfinylacetaldehyde **18**[29,70]. Fellman and Roth determined the HT aminotransferase activity in particulate fraction of rat tissues. The highest value of the enzyme has been found in brain followed by liver and testes. Pyridoxal-5′-phosphate acts as a coenzyme for this aminotransferase[71].

The transport and metabolism of TA and HT in brain was reviewed by Kontro[72].

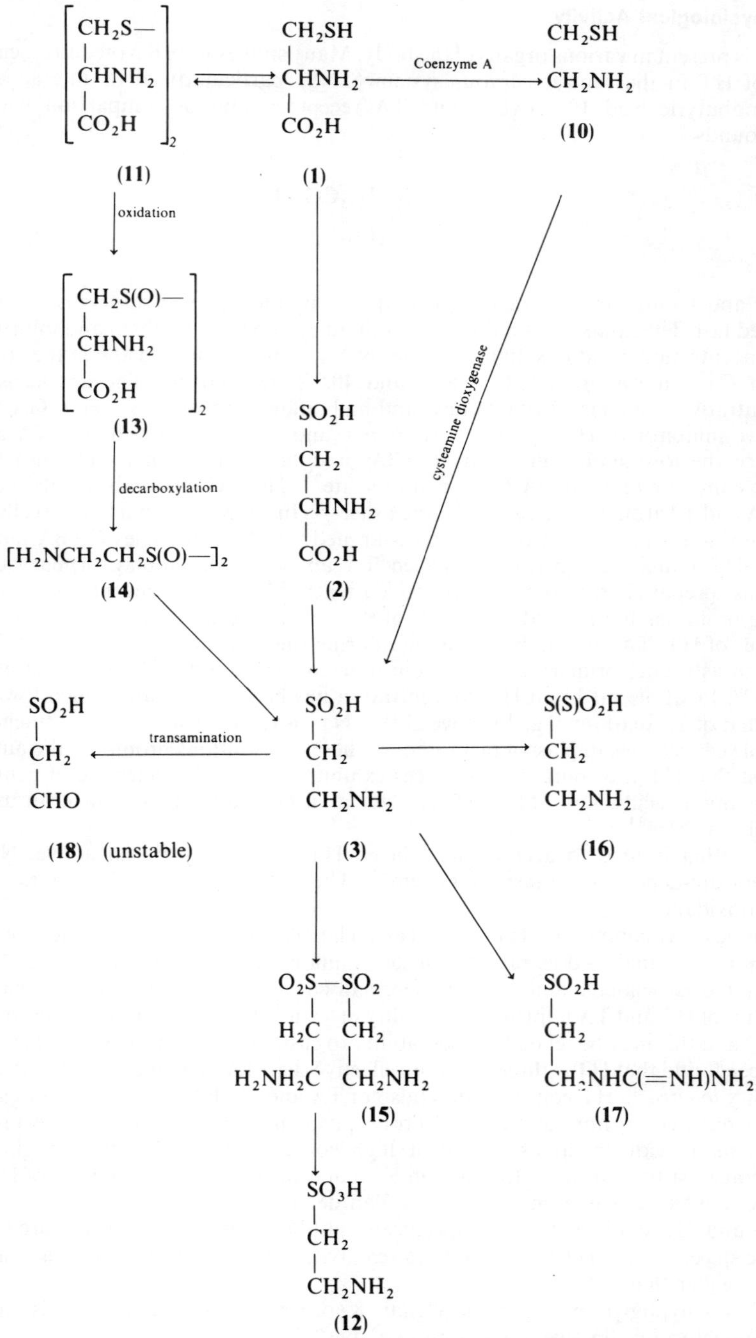


SCHEME 2. Biosyntheses and biotransformations of hypotaurine (HT)

### B. Physiological Activity

HT is present in various organs of the body. Many studies were devoted to elucidate the role of HT in the central nervous system (CNS), particularly of its interaction with $\gamma$-aminobutyric acid **19** (GABA) and TA receptors, and of comparison with these compounds.

$$H_2N(CH_2)_3CO_2H$$

**(19)**

Oja and Kontro investigated the HT uptake by mouse brain slices and found it to proceed fast. The uptake was reduced by sodium cyanide and ouabain and influenced by the concentration of cations. In the absence of $Na^+$ it was abolished and in the absence of $K^+$ or $Ca^{2+}$ it was inhibited by 63% and 40.0%, respectively. The uptake is highly concentrative and consists of two low- and high-affinity transport systems. GABA is an effective inhibitor of HT uptake[73]. Malminen and Kontro reported that TA and HT displace the low- and high-affinity GABA binding in rat brain membranes through possible interaction with GABA recognition site[74]. These compounds facilitate efflux of GABA and TA from mouse cerebral cortex slices[75]. In neuroblastoma C 1300 cells GABA uptake is almost abolished by HT[76]. It is argued that HT resembles GABA more than TA[77]. HT inhibits competitively labeled TA uptake in developing primary cultured neurons, prepared from mouse cerebral cortex[78]. Maximum concentration of HT in mouse brain has been found at the age of three weeks, and in serum at one week. The uptakes of HT, TA and GABA were high during the first three weeks of life[79]. The HT value in astroglial primary cultures from different brain regions was highest after two weeks[80]. In whole rat brain HT concentration has been found to be much lower (0.07) than that of TA (6.61 $\mu$mol $g^{-1}$ wet weight)[81]. $Na^+$ is required when HT is attached to its possible carrier sites in plasma membranes. These observations prompted the authors to suggest that HT transport in brain slices exhibits features characteristic of neurotransmitter amino acids, and HT itself may act as a false inhibitory neurotransmitter or modulator[73,82–84].

Rat retina is able to accumulate labeled HT, apparently by an active, $Na^+$ and temperature-dependent transport system[85]. The authors suppose that it may act as an antioxidant.

Among other compounds HT is found at a relatively high level in the epididymal plasma of various mammals—dog, rabbit, hamster, stallion, rhesus monkey and others[86]. It has been found associated mainly with the spermatozoa and less in the seminal plasma[87]. The presence of HT and TA improves the quality of fertilization of bovine follicular oocytes *in vitro*[88] and this may be related to their ability to sustain sperm mobility and fertility[89]. It has been found that HT is threefold more effective than TA in maintaining hamster sperm mobility *in vitro*[90]. HT concentration (also of TA and GABA) decreases after castration and is restored to normal by testosterone propionate[91]. HT and TA are present in mammalian oviductal fluids, and their high concentration (0.5–2 mM) might protect sperm against the harmful effect of high $K^+$ concentrations[92]. The activity of HT in the reproductive tract has been reviewed by Van der Horst[93].

TA and HT could induce nonspecifically antibody production in cultured DOA/2 mouse spleen cells[94]. HT tested for cross-reactivity with TA antiserum interacts about 15 times weaker than TA[95].

HT is a hypoglycemic agent in Wistar–Kyoto rats and prevents the rise in serum immunoactive insulin levels. TA is more active[96].

HT and TA induce hyperthermia when injected to rats[97].

### C. Additional Effects of HT and Derivatives

HT is a component of cosmetics with skin-whitening effect[98].

HT and $PtCl_2$ yield a Pt complex, dichloro(2-aminoethylsulfonyl)Pt (**20**), which has a specific cytostatic–cytotoxic activity against adriamycin-resistant cancer cells[11].

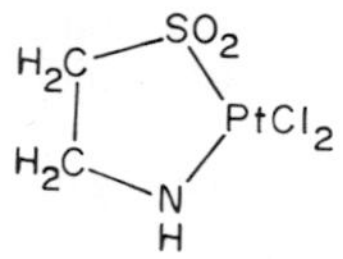

(**20**)

## IV. HYPOTAUROCYAMINE, HTC

Hypotaurocyamine or 2-guanidoethylsulfinic acid **17** is found in some of the invertebrate phyla, e.g. Phascolosoma[99] and in cridaria[100]. It is synthesized in invertebrates by transamidination between arginine and HT and appears to be formed in the viscera[101]. Syntheses of HTC have been reported[99,102,103], mp 183–184 °C. HTC can be oxidized chemically to taurocyamine (2-guanidoethylsulfonic acid) **21**.

$$HN{=}C(NH_2)NHCH_2CH_2SO_3H$$

(**21**)

## V. HOMOLOGS OF CSA AND HT

### A. Homocysteinesulfinic Acid

Homocysteinesulfinic acid or 2-amino-4-sulfinobutanoic acid **22** (HCSA) is similar in many aspects to its lower homolog CSA and is a close analog of Asp **7**. The value of $pK_a$ of the —SOOH group is 1.66 and of —COOH, 2.6[17,37].

$$HO_2S(CH_2)_2CH(NH_2)COOH \qquad NH_2(CH_2)_3SO_2H$$

(**22**) (**23**)

HCSA is decarboxylated to homohypotaurine **23** (HHT) by rat brain homogenate (kinetic studies indicate that the reaction is carried out by L-Glu decarboxylase)[104]. The L-isomer is decarboxylated by preparation of *Clostridium welchii* or *E. coli*[105], and transaminated to 2-oxo-4-sulfinobutanoic acid **24** during incubation with a keto acid (e.g. pyruvic acid) and rat liver homogenate or L-aspartate 2-oxoglutarate aminotransferase[106]. HCSA is an endogenous substance and is released from various rat brain regions (together with other sulfur-containing amino acids) in a $Ca^{2+}$-dependent manner. The highest concentration of HCSA is found in striatum[107].

The neuroexcitatory action in CNS has been tested on isolated spinal cord of frog. The activity of D-HCSA is almost equal to that of NMDA and estimated as very strong, while DL-HCSA is less active[15,41]. A possible role in CNS transmission has been suggested[107].

HCSA was detected in the urine of patients suffering from homocystinuria[17,108], probably as a result of deficiency of cystathionine $\beta$-synthase activity[109].

The preparation of L- and DL-HCSA has been described in several papers[13,110–112].

### B. Homohypotaurine

Homohypotaurine or 3-aminopropropanesulfinic acid **23** (HHT) is mentioned as a product of HCSA decarboxylation by rat-brain homogenate. It is suggested that the reaction is carried out by L-Glu decarboxylase[104]. HT and homotaurine **25**, but not HT and TA, are transaminated by cell-free extracts of *Pseudomonas fluorescens* in the presence of α-ketoglutarate[113]. HHT was investigated (together with other related compounds) for its cross-reactivity with GABA receptor binding and found to be fairly active. The measurements were carried out by GABA radioreceptor assay with receptors isolated from the brain of male rats[114].

Synthesis of HHT from homocystamine (3-aminopropyl disulfide) **26** and $H_2O_2$ has been published[115].

$$HO_2SCH_2CH_2COCO_2H \quad H_2N(CH_2)_3SO_3H \quad [H_2N(CH_2)_3S{-}]_2$$

**(24)** **(25)** **(26)**

## VI. SYNTHETIC SULFINATES OF PHARMACOLOGICAL INTEREST

### A. Methanesulfinic Acid

Only one unsubstituted derivative of methanesulfinic acid **27** (R = H) of biological origin is mentioned in the literature. Ethyl methanesulfinate (**27**, R = Et) has been found in volatile compounds obtained from Japanese radish, processed by fermentation with rice brain[116].

Aromatic esters of **27** possess insecticidal properties. 3,4-Dichlorophenyl methylsulfinate **27** (R = 3,4-$ClC_6H_3$—) controls the corn rootworm (*Diabrotica virgifera*) larvae in soil[117].

$$CH_3SO_2R \quad HOCH_2SO_2H \quad ClCH_2SO_2H \quad Cl_2CHSO_2H$$

**(27)** **(28)** **(29)** **(30)**

Hydroxymethanesulfinic acid **28** and its sodium salt (rongalite) are used as stabilizers of aqueous solutions of various drugs such as oxytetracycline[9,118], adrenaline[8] and sodium salicylate[119]. The Zn salt is a component of an antidandruff hair preparation[120]. The acid ($10\,mg\,kg^{-1}$, intraperitoneally) has been found to reduce the severity of symptoms of experimental allergic encephalomyelitis, induced by brain/spinal cord antigen[121].

Esters and Zn salts of chloromethyl- **29** and dichloromethanesulfinic acid **30** possess acaricidal activity[122].

Sodium 2-isonicotinoylhydrazinomethanesulfinate **31** has been patented as a low-toxicity agent against leprosy and tuberculosis[123]. 6-Chrysenylaminomethanesulfinic acid **32** reduced tumor growth when given orally or intraperitoneally to rats with transplanted rhabdomyosarcoma BA 112[124].

Formamidinosulfinic acid **33** (FAS) has been mentioned by Kennedy and colleagues as

**(31)**

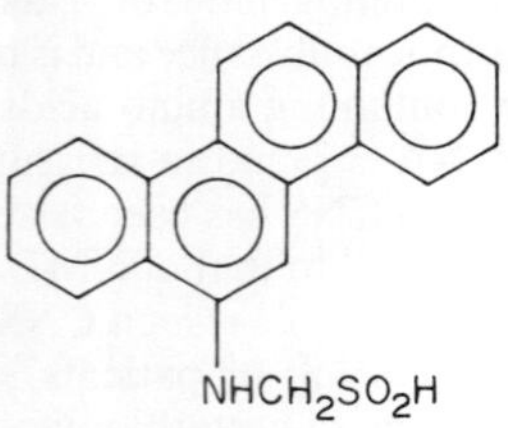


**(32)**

$H_2NC({=}NH)SO_2H$

**(33)**

an activator of aconitase from beef heart mitochondria to 55–75% of its maximum activity during 0.5 h. It proceeds on reduction of its Fe—S cluster[125]. *Cyanobacterium synechococcus* 6301 is able to use FAS as a source of sulfur for its growth demands[126]. The Ca salt has been patented as a neoplasm inhibitor, effective intraperitoneally and orally for inhibition of adenocarcinoma CA-755 and sarcomas 180 and HS-1 in mice[127].

### B. Butanesulfinic Acid

Derivatives of butanesulfinic acid **34**, such as sodium salt of 4,4′-dithiobisbutanesulfinic acid **35** and of the corresponding trisulfide **36** are potent antiradiation agents. Compound **36** conferred 87% protection on white mice (30 days survival after irradiation) after 75 mg kg$^{-1}$ intraperitoneally and 100% after 300 mg kg$^{-1}$ orally[128].

$CH_3(CH_2)_3SO_2H$ **(34)** $[-S(CH_2)_4SO_2Na]_2$ **(35)** $S[S(CH_2)_4SO_2Na]_2$ **(36)**

### C. Aromatic Sulfinic Acids

Sodium salts of benzene **37** (R,R′,R″ = H) and 4-toluenesulfinic (R = Me, R′,R″ = H) acids are mentioned as components of dentin adhesives for dental repair[129–131]. Esters of substituted benzenesulfinic acid, such as 2-nitro-5-aryloxy **37** (R′ = ArO—, R = H, R″ = 2-$NO_2$)[132], and derivatives of 4-bromobenzenesulfinamide **38**[133] were tested as

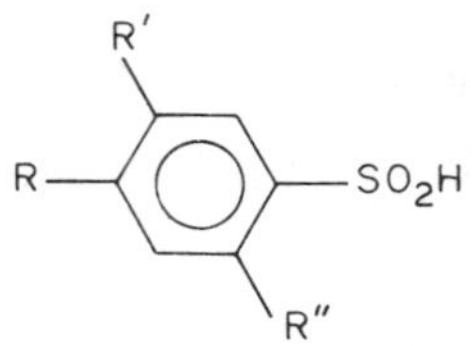


**(37)**

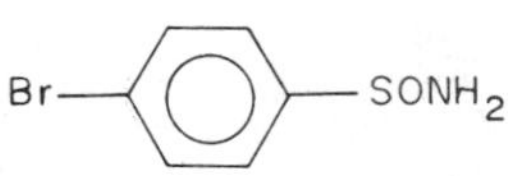


**(38)**

herbicides. Esters of 2-amino-4-methylbenzenesulfinic acid **37** (R = M, R′ = H, R″ = $NH_2$) were found active against gram-positive bacteria[134], and 4-chloro (R = Cl, R′, R″ = H) and 3-nitro (R′ = $NO_2$, R,R″ = H) derivatives are patented as bactericides[135].

2,3,4,5-Tetrachloro- **39** and 2,3,5,6-tetrachlorobenzenesulfinic acids **40** were detected in faeces of squirrel monkey as metabolites of 1,2,3,4- and 1,2,3,5-tetrachlorobenzene, respectively[136].

2,3,4,5-$Cl_4C_6HSO_2H$ **(39)** 2,3,5,6-$Cl_4C_6HSO_2H$ **(40)**

Sodium 2,2′-[trithiobis(methylene)]bis(benzenemethanesulfinate) **41** and 2′,2′-trithiobis(2-biphenylsulfinate) **42** are antiradiation agents. 100 mg kg$^{-1}$ of **41** or 20 mg kg$^{-1}$ of **42** intraperitoneally gave about 80% protection against irradiation of white mice[128,137]. The activity of these and related compounds has been summarized[10].

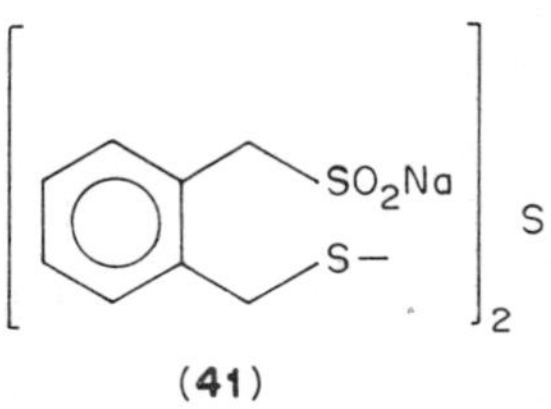


**(41)**

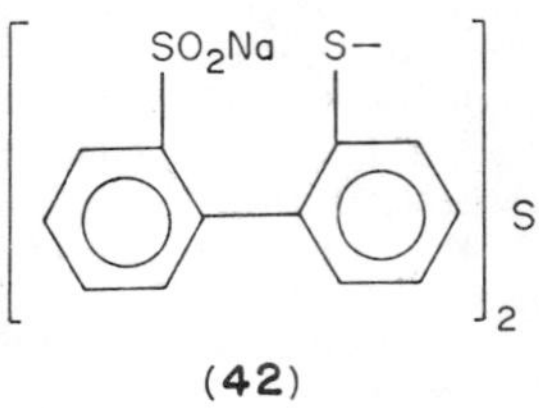


**(42)**

## VII. REFERENCES

1. J. G. Jacobsen and L. H. Smith, *Physiol. Rev.*, **48**, 424 (1968).
2. R. J. Huxtable and H. Pasantes-Morales (Eds.), *Taurine in Nutrition and Neurology*, Plenum Press, New York, 1982; *Adv. Exp. Med. Biol.*, **139** (1982).
3. S. S. Oja, L. Ahtee, P. Kontro, and M. K. Paasonen (Eds.), *Taurine: Biological Actions and Clinical Pespectives*, Alan R. Liss, New York, 1985.
4. W. B. Jacoby and O. W. Griffith (Eds.), *Methods in Enzymology*, Vol. 143, Academic Press, New York, 1987.
5. R. J. Huxtable, F. Franconi, and A. Giotti (Eds.), *The Biology of Taurine: Methods and Mechanisms*), Plenum Press, New York, 1987; *Adv. Exp. Med. Biol.*, **217** (1987).
6. M. J. Brown, U. S. Patent 4,359,334; *Chem. Abstr.*, **98**, 67,123 (1983).
7. M. J. Brown, U. S. Patent 4,427,439; *Chem. Abstr.*, **100**, 116,482 (1984).
8. H. Wollmann and G. Raether, *Pharmazie*, **38**, 37 (1983).
9. W. Hacke and H. Herman, U. S. Patent 4,386,083; *Chem. Abstr.*, **99**, 76,789 (1983).
10. G. T. Bowman, J. J. Clement, D. E. Davidson, V. Eswarakrishnan, L. Field, J. M. Holi, H. A. Musallan, R. O. Pick, R. Ravichandran, and P. K. Srivastava, *Chem.-Biol. Interact.*, **57**, 161 (1986).
11. T. Sugawara, T. Iida, A. Kohno, A. Miyashida, and T. Mitamura, *Nippon Kagaku Kaishi*, 719 (1987).
12. T. F. Lavine, *J. Biol. Chem.*, **113**, 583 (1936).
13. O. W. Griffith, Reference 4, p. 366.
14. K. Soda, Reference 4, p. 453.
15. D. R. Curtis and J. C. Watkins, *J. Physiol.*, **166**, 1 (1963).
16. O. W. Griffith, Reference 4, p. 274.
17. O. W. Griffith and C. L. Weinstein, Reference 4, p. 270.
18. K. Yamaguchi and Y. Hosokawa, Reference 4, p. 395.
19. I. C. Kilpatrick and L. S. Mozley, *Neurol. Neurobiol.*, **24**, 181 (1987).
20. F. Chatagner and B. Bergeret, *Compt. Rend.*, **232**, 448 (1951).
21. C. Fromageot, in *Methods of Enzymology*, Vol. II (Eds. S. P. Colovick and N. O. Kaplan), Academic Press, New York, 1955, p. 324.
22. C. L. Weinstein and O. W. Griffith, Reference 4, p. 404.
23. D. K. Rassin, *Adv. Exp. Med. Biol.*, **139**, 257 (1982).
24. F. Legay, D. Lecestre, and M. Tappaz, *J. Neurochem.*, **48**, 340 (1987).
25. C. L. Weinstein and O. W. Griffith, *J. Biol. Chem.*, **262**, 7254 (1987).
26. V. Chan-Palay, C. T. Lin, S. L. Palay, M. Yamamoto, and J. Y. Wu, *Proc. Natl. Acad. Sci. U.S.A.*, **79**, 2695 (1982).
27. T. P. Singer and A. B. Kearney, *Biochem. Biophys. Acta*, **14**, 570 (1954).
28. J. H. Fellman and E. S. Roth, *Adv. Exp. Med. Biol.*, **217**, 39 (1987).
29. O. W. Griffith, *J. Biol. Chem.*, **258**, 1591 (1983).
30. J. G. Jacobsen, L. L. Thomas, and L. H. Smith, *Biochim. Biophys. Acta*, **85**, 103 (1964).
31. K. Knopf, J. A. Sturman, M. Armstrong, and K. C. Hayes, *J. Nutr.*, **108**, 773 (1978).
32. J. De La Rosa and M. H. Stipanuk, *Comp. Biochem. Physiol. B*, **B81**, 565 (1985).
33. L. L. Hirschberger, J. De La Rosa, and M. H. Stipanuk, *J. Chromatogr.*, **343**, 303 (1985).
34. M. H. Stipanuk, L. L. Hirschberger, and J. De La Rosa, Reference 4, p. 155.
35. F. J. Leinweber and K. J. Monty, Reference 4, p. 160.
36. K. Kuriyama and Y. Tanaka, Reference 4, p. 164.
37. F. Palmieri, I. Stipani, and V. Iacolazzi, *Biochim. Biophys. Acta*, **555**, 531 (1979).
38. T. Watanabe and E. E. Snell, *J. Biochem.* (*Tokyo*), **82**, 733 (1977).
39. B. S. Sukhareva and L. G. Malikova, *Med. Biol.* (*Moscow*), **11**, 394 (1977).
40. *CRC Handbook of Chemistry and Physics*, 64th Edn., Chemical Rubber Publishing Co., Boca Raton, Florida, 1983/84, p. C 719.
41. K. N. Mewett, D. J. Oakes, H. J. Olverman, D. A. S. Smith, and J. C. Watkins, *Adv. Biochem. Psychopharmacol.*, **37**, 163 (1983).
42. H. Iwata, Y. Koyama, and A. Baba, *Ganryu Aminosan*, **8**, 289 (1985).
43. L. M. Pullan, J. W. Olney, M. T. Price, R. P. Compton, W. F. Hood, J. Michel, and J. B. Monahan, *J. Neurochem.*, **49**, 1301 (1987).
44. G. Takayaki, *NATO Adv. Study Inst. Ser. A*, **357** (1977); *Chem. Abstr.*, **91**, 121, 143 (1979).

45. S. R. Vincent and E. M. McGeer, *J. Pharm. Pharmacol.*, **31**, 703 (1979).
46. A. Cross, W. Skan, and P. Slater, *J. Neurochem.*, **47**, 1463 (1986).
47. P. Ascher, P. Bregestovski, and L. Novak, *J. Physiol.* (*London*), **399**, 207 (1988).
48. W. A. Turski, P. L. Herrling, and K. Q. Do, *Brain Res.*, **414**, 330 (1987).
49. A. Pastuszko and D. E. Wilson, *Neurosci. Lett.*, **82**, 71 (1987).
50. J. W. Olney, D. L. Ho, and V. Rhee, *Exp. Brain Res.*, **14**, 61 (1971).
51. S. M. Wu and J. E. Dowling, *Proc. Natl. Acad. Sci. U.S.A.*, **75**, 5205 (1978).
52. H. Iwata, A. Baba, and S. Yamagami, Adv. Pharmacol. Res. Pract., Proc. 3rd Congr. Hung. Pharmacol. Soc., 417 (1979); *Chem. Abstr.*, **94**, 189,200 (1981).
53. J. A. Ferendelli, M. G. Chang, and D. A. Kinscherf, *J. Neurochem.*, **22**, 535 (1974).
54. H. Shimizu and Y. Yamamura, *J. Neurochem.*, **28**, 383 (1977).
55. A. Baba, E. Lee, T. Tatsuno, and H. Iwata, *J. Neurochem.*, **38**, 1280 (1982).
56. H. Iwata and A. Baba, *Adv. Exp. Med. Biol.*, **139**, 211 (1982).
57. R. J. Huxtable, *Adv. Exp. Med. Biol.*, **139**, 207 (1982).
58. M. Friedman and M. R. Gumbman, *J. Nutr.*, **114**, 2301 (1984).
59. E. J. Van Scott and J. R. Yu, U.S. Patent 4,224,339; *Chem. Abstr.*, **108**, 198, 746 (1988).
60. B. Bergeret and F. Chatagner, *Biochem. Biophys. Acta*, **13**, 313 (1954).
61. D. Cavallini, B. Mondovi, and C. De Marco, *J. Biol. Chem.*, **216**, 577 (1955).
62. L. Eldjarn, A. Pihl, and A. Sverdrup, *J. Biol. Chem.*, **223**, 353 (1956).
63. G. D. Novelli, F. J. Schmetz, and N. O. Kaplan, *J. Biol. Chem.*, **206**, 533 (1954).
64. M. W. Duffel, D. J. Logan, and D. M. Ziegler, Reference 4, p. 149.
65. R. B. Richerson and D. M. Ziegler, Reference 4, p. 410.
66. R. Scandurra, L. Politi, S. Dupre, M. Moriggi, D. Barra, and D. Cavallini, *Bull. Mol. Biol. Med.*, **2**, 172 (1977).
67. G. Federici, G. Ricci, L. Santoro, A. Antonucci, and D. Cavallini, in *Natural Sulfur Compounds* (Eds. D. Cavallini, G. E. Gaull and V. Zappia), Plenum, New York, 1980, p. 187.
68. T. Ubuka, S. Yuasa, M. Kinuta, and R. Agaki, *Physiol. Chem. Phys.*, **12**, 3 (1980).
69. L. G. Tarnopolskaya, *Farmakol. Toksikol.* (*Moscow*), **41**, 93 (1978); *Chem. Abstr.*, **88**, 83, 354 (1978).
70. J. H. Fellman, Reference 4, p. 183.
71. J. H. Fellman and E. S. Roth, *Adv. Exp. Med. Biol.*, **139**, 99 (1982).
72. P. Kontro, *Acta Univ. Ouluensis, Ser. A*, **88**, 1 (1980).
73. S. S. Oja and P. Kontro, *Adv. Exp. Med. Biol.*, **139**, 115 (1982).
74. O. Malminen and P. Kontro, *Neurochem. Res.*, **11**, 85 (1986).
75. A. N. Clements and T. E. May, *J. Exp. Biol.*, **61**, 421 (1974).
76. I. Holopainen, P. Kontro, H. J. Frey, and S. S. Oja, *J. Neurosci. Res.*, **10**, 83 (1983).
77. E. R. Korpi, P. Kontro, K. Nieminen, K. M. Marnela, and S. S. Oja, *Life Sci.*, **29**, 811 (1981).
78. M. Kishi, S. Ohkuma, M. Kimori, and K. Kuriyama, *Biochem. Biophys. Acta*, **939**, 605 (1988).
79. P. Kontro, K. M. Marnela, and S. S. Oja, *Acta Physiol. Scand., Suppl.*, **537**, 71 (1984).
80. A. Lehmann and E. Hansson, *Neurochem. Res.*, **12**, 797 (1897).
81. T. L. Perry and S. Hansen, *J. Neurochem.*, **21**, 1009 (1973).
82. P. Kontro and S. S. Oja, *Neurochem. Res.*, **8**, 1377 (1983).
83. L. Holopainen, P. Kontro, and S. S. Oja, *Neurochem. Int.*, **6**, 217 (1984).
84. P. Kontro and S. S. Oja, *J. Neurochem.*, **37**, 297 (1981).
85. H. Pasantes-Morales, J. Moran, and J. H. Fellman, *J. Neurosci. Res.*, **15**, 101 (1986).
86. R. Jones, *Comp. Biochem. Physiol. B*, **61B**, 365 (1978).
87. C. J. G. Van der Horst and H. J. G. Grooten, *Biochim. Biophys. Acta*, **117**, 495 (1966).
88. G. D. Ball, M. L. Liebfried, R. W. Lenz, R. L. Ax, B. D. Bavister, and N. L. First, *Biol. Reprod.*, **28**, 717 (1983).
89. R. J. Mrsny and S. Meisel, *Life Sci.*, **36**, 271 (1985).
90. R. B. L. Gwatkin, *Gamete Res.*, **7**, 347 (1983).
91. C. D. Kochakian, *Am. J. Physiol.*, **228**, 1231 (1975).
92. R. J. Mrsny, L. Waxman, and S. Meisel, *J. Exp. Zool.*, **210**, 123 (1979).
93. C. J. G. Van der Horst, in *Natural Sulfur Compounds* (Ed. D. Cavallini), Plenum, New York, 1980, p. 225.
94. S. Ishizaka, K. Kitagami, M. Yoshikawa, H. Ito, I. Sugawara, and T. Tsuji, *Ganryu Aminosan*, **6**, 291 (1983).
95. S. Madsen, O. P. Ottersen, and J. Storm-Mathisen, *Adv. Exp. Med. Biol.*, **217**, 275 (1987).

96. E. C. Kulakowski and J. Maturo, *Biochem. Pharmacol.*, **33**, 2835 (1984).
97. R. W. Kerwin and C. J. Pycock, *J. Pharm. Pharmacol.*, **31**, 466 (1979).
98. T. Taki and T. Komukai, Jap. Patent 6,169,707; *Chem. Abstr.*, **105**, 48,377 (1986).
99. Y. Robin, N. van Thoai, L. A. Pradel, and J. Roche, *Biochim. Biophys. Acta*, **63**, 481 (1962).
100. Y. Robin and Y. Guillon, *Oceanis*, **5**, 575 (1980).
101. N. van Thoai, S. Zappacosta, and Y. Robin, *Compt. Biochem. Physiol.*, **10**, 209 (1963).
102. B. W. Shapiro and E. A. Dickens, *Radiat. Res.*, **13**, 857 (1960).
103. G. Desvages and N. van Thoai, *C.R. Acad. Sci. Paris, Ser. C*, **267**, 1868 (1968).
104. B. Jolles-Bergeret and M. H. De Vaucher, *J. Neurochem.*, **20**, 1797 (1973).
105. B. Jolles-Bergeret and M. Charton, *Biochimie*, **53**, 553 (1971).
106. B. Jolles-Bergeret and M. Marty-Lopez, *Compt. Rend., Ser. D*, **262**, 930 (1966).
107. K. Q. Do, M. M. Henberger, P. Streit, and M. Cuenod, *J. Neurochem.*, **46**, 779 (1986).
108. S. Ohmori, H. Kodama, T. Ikegami, S. Mizuhara, T. Oura, G. Isshiki, and I. Uemura, *Physiol. Chem. Phys.*, **4**, 286 (1972).
109. S. H. Mudd and H. L. Levy, in *The Metabolic Basis of Inherited Disease* (Eds. J. B. Stanbury, J. B. Wyngaarden, D. S. Frederickson, J. L. Goldstein, and M. S. Brown), McGraw-Hill, New York, 1983, p. 522.
110. J. C. Watkins, *J. Med. Pharm. Chem.*, **5**, 1187 (1962).
111. B. Jolles-Bergeret, *Bull. Soc. Chim. Biol.*, **48**, 1265 (1966).
112. P. Luchi and C. De Marco, *Anal. Biochem.*, **45**, 236 (1971).
113. D. Gomez de Garcia and B. Jolles-Bergeret., *Biochim. Biophys. Acta*, **49**, 315 (1973).
114. S. Tsukamoto, N. Ogawa, S. Mizuno, and A. Mori, *Neurosciences* (*Japan*), **8**, 58 (1982).
115. C. De Marco and A. Rinaldi, *Anal. Biochem.*, **17**, 265 (1973).
116. A. Kjaer, J. O. Oegaard, Y. Maeda, Y. Ozawa, and Y. Uda, *Agric. Biol. Chem.*, **42**, 1989 (1978).
117. R. D. Partos, U.S. Patent 3,764,698; *Chem. Abstr.*, **80**, 92,009 (1974).
118. W. W. Armstrong and S. A. Desai, German Patent 2,659,152; *Chem. Abstr.*, **87**, 172,890 (1977).
119. R. Lukoserviciene and S. A. Minina, *Khim.-Farm. Zh.*, **11**, 106 (1977).
120. P. Flemming, K. Giede, and Greb, German Patent 2,634,677; *Chem. Abstr.*, **88**, 177214 (1978).
121. D. Dobrescu, A. Cristea, and E. E. Mariko, *Farmacia* (*Bucharest*), **34**, 99 (1986).
122. T. Wakamori, T. Kitagaku, and H. Ito, Jap. Patent 74,35,415; *Chem. Abstr.*, **82**, 150,542 (1975).
123. A. German, T. Ba Loc and Y. Peruvin, French Patent M1120; *Chem. Abstr.*, **58**, 1320g (1963).
124. G. Franchi, L. Moretti, and S. Garattini, *Eur. J. Cancer*, **6**, 441 (1970).
125. M. C. Kennedy, M. H. Emptage, J. L. Dreyer, and H. Beinert, *J. Biol. Chem.*, **258**, 11098 (1983).
126. A. Schmidt, I. Erdle, and H. P. Koest, *Z. Naturforsch.*, **37C**, 870 (1982).
127. K. V. Rao, U.S. Patent 3,051,626; *Chem. Abstr.*, **15**, 15257d (1962).
128. L. Field and Y. H. Yong, *J. Med. Chem.*, **15**, 312 (1972).
129. N. Nakabayashi and S. Honda, Jpn. Patent 62,161,709; *Chem. Abstr.*, **108**, 44,069 (1988).
130. S. A. Taylor, Eur. Pat. Appl. EP 247,726; *Chem. Abstr.*, **108**, 188,074 (1988).
131. T. Nakahara and K. Kusumoto, Jpn. Patent 61,155,403; *Chem. Abstr.*, **105**, 232, 488 (1986).
132. A. Parg, H. Ziegler, and G. Hamprecht, German Patent 3,434,315; *Chem. Abstr.*, **105**, 6310 (1986).
133. J. A. Friedman, *J. Agric. Food Chem.*, **31**, 127 (1983).
134. L. Ferenczy, J. Zsolt, E. Vinkler, and F. Klivenyi, *Acta Biol. Acad. Sci. Hung.*, **12**, 121 (1961).
135. G. A. Tolstikov, V. M. Dzhemilev, R. K. Andreson, R. V. Kunakova, B. A. Andreson, L. A. Propadushchaya, and R. L. Gaisin, USSR Patent 1,099,966; *Chem. Abstr.*, **101**, 177,518 (1984).
136. H. Schwartz, I. Chu, D. C. Villeneuve, and F. M. Benoit, *J. Toxicol. Environ. Health*, **22**, 341 (1987).
137. P. K. Srivastava, L. Field, and M. M. Grenan, *J. Med. Chem.*, **18**, 798 (1975).

# Author index

This author index is designed to enable the reader to locate an author's name and work with the aid of the reference numbers appearing in the text. The page numbers are printed in normal type in ascending numerical order, followed by the reference numbers in parentheses. The numbers in *italics* refer to the pages on which the references are actually listed.

# Subject index